Subsets of the Real Numbers

Natural numbers: $\{1, 2, 3, 4, 5, 6, \ldots\}$
Whole numbers: $\{0, 1, 2, 3, 4, 5, 6, \ldots\}$
Integers: $\{\ldots -3, -2, -1, 0, 1, 2, 3, \ldots\}$
Rational: $\left\{\dfrac{a}{b}: a, b \text{ are integers and } b \neq 0\right\}$
Irrational: The set of real numbers that are not rational.

CONSTANTS: $\pi \approx 3.14159,\ e \approx 2.71828$

Operations with Fractions

Addition: $\dfrac{a}{b} + \dfrac{c}{b} = \dfrac{a + c}{b}$ for $b \neq 0$

Subtraction: $\dfrac{a}{b} - \dfrac{c}{b} = \dfrac{a - c}{b}$ for $b \neq 0$

Reducing fractions: $\dfrac{ac}{bc} = \dfrac{a}{b}$ for $b \neq 0$ and $c \neq 0$

Multiplication: $\dfrac{a}{b} \cdot \dfrac{c}{d} = \dfrac{a \cdot c}{b \cdot d}$ for $b \neq 0$ and $d \neq 0$

Division: $\dfrac{a}{b} \div \dfrac{c}{d} = \dfrac{a}{b} \cdot \dfrac{d}{c} = \dfrac{a \cdot d}{b \cdot c}$ for $b \neq 0$, $c \neq 0$, and $d \neq 0$

Properties of the Real Numbers

Commutative properties:

of addition $\qquad a + b = b + a$

of multiplication $\qquad ab = ba$

Associative properties:

of addition $\qquad (a + b) + c = a + (b + c)$

of multiplication $\qquad (ab)(c) = (a)(bc)$

Distributive properties of multiplication over addition:

$a(b + c) = ab + ac$

$(b + c)a = ba + ca$

Identities:

additive $\qquad a + 0 = 0 + a = a$

multiplicative $\qquad a \cdot 1 = 1 \cdot a = a$

Inverses:

additive $\qquad a + (-a) = (-a) + a = 0$

multiplicative $\qquad a \cdot \dfrac{1}{a} = \dfrac{1}{a} \cdot a = 1$ for $a \neq 0$

Order of Operations

Step 1. Start with the expression within the innermost pair of grouping symbols.
Step 2. Perform all exponentiations.
Step 3. Perform all multiplications and divisions as they appear from left to right.
Step 4. Perform all additions and subtractions as they appear from left to right.

D0151131

Interval Notation

Inequality Notation	Verbal Meaning	Graph	Interval Notation
$x > a$	x is greater than a		(a, ∞)
$x \geq a$	x is greater than or equal to a		$[a, \infty)$
$x < a$	x is less than a		$(-\infty, a)$
$x \leq a$	x is less than or equal to a		$(-\infty, a]$
$a < x < b$	x is greater than a and less than b		(a, b)
$a < x \leq b$	x is greater than a and less than or equal to b		$(a, b]$
$a \leq x < b$	x is greater than or equal to a and less than b		$[a, b)$
$a \leq x \leq b$	x is greater than or equal to a and less than or equal to b		$[a, b]$
$-\infty < x < \infty$	x is any real number		$(-\infty, \infty)$

Beginning and Intermediate Algebra

THE LANGUAGE AND SYMBOLISM OF MATHEMATICS

THIRD EDITION

James W. Hall

Parkland College

Brian A. Mercer

Parkland College

Mc Graw Hill

Connect
Learn
Succeed™

BEGINNING AND INTERMEDIATE ALGEBRA: THE LANGUAGE AND SYMBOLISM
OF MATHEMATICS, THIRD EDITION

This book is printed on acid-free paper.

1 2 3 4 5 6 7 8 9 0 DOW/DOW 1 0 9 8 7 6 5 4 3 2 1 0

ISBN 978–0–07–338424–5
MHID 0–07–338424–0

ISBN 978–0–07–729688–9 (Instructor's Edition)
MHID 0–07–729688–5

Vice President, Editor-in-Chief: *Marty Lange*
Vice President, EDP: *Kimberly Meriwether David*
Director of Development: *Kristine Tibbetts*
Editorial Director: *Stewart K. Mattson*
Executive Editor: *David Millage*
Developmental Editor: *Adam Fischer*
Marketing Manager: *Victoria Anderson*
Senior Project Manager: *April R. Southwood*
Lead Production Supervisor: *Sandy Ludovissy*
Lead Media Project Manager: *Judi David*
Designer: *Tara McDermott*
Cover/Interior Designer: *Ellen Pettergell*
(USE) Cover Image: © *Gavin Hellier/Getty Images*
Senior Photo Research Coordinator: *John C. Leland*
Compositor: *Aptara, Inc.*
Typeface: *10.5/12 Times*
Printer: *R. R. Donnelley*

Library of Congress Cataloging-in-Publication Data

Hall, James W.
  Beginning and intermediate algebra : the language and symbolism of mathematics / James W. Hall, Brian A.
Mercer. –3rd ed.
      p. cm.
  Includes index.
  ISBN 978–0–07–338424–5 — ISBN 0–07–338424–0 (hard copy : alk. paper) 1. Algebra–Textbooks.
I. Mercer, Brian A.   II. Title.
  QA152.3.H28   2011
  512.9–dc22

                                                             2009023208

www.mhhe.com

Letter from the Authors

Dear Colleagues,

We believe strongly in the importance of developmental mathematics. Beginning and intermediate algebra contain key mathematical concepts and can be an important component of the critical thinking skills acquired by college students. Learning how to learn is one of the most important skills that a college student can master. One way this skill is emphasized in *Beginning and Intermediate Algebra: The Language and Symbolism of Mathematics* is by using multiple perspectives. Examining concepts algebraically, graphically, numerically, and verbally not only promotes the understanding of the mathematical concepts but also develops a way of thinking about mathematics.

Mathematics, like many other subject areas, has its own vocabulary and symbolism. Students who master the language of mathematics will recognize the use of mathematics in their chosen profession and will develop an appreciation for mathematics in everyday life. This textbook has a unique emphasis on using language and symbolism exercises and multiple perspectives to master the topics from both beginning algebra and intermediate algebra courses.

Why Did We Write This Book?

The short answer is that we wrote this book to address the needs of our students at Parkland College. Specifically, our goal was to produce a book that:

- Presents exposition, definitions, examples, and exercises using multiple perspectives. Concepts are examined algebraically, graphically, numerically, and verbally.
- Provides examples with solutions in exactly the same format that we want students to use. Sidebar explanations can be used by students as needed.
- Answers student questions about why we are covering material and how it can be used.
- Presents solid, carefully-constructed exercise sets, and realistic applications.
- Provides a geometry review and integrates geometry into the exercise sets.
- Provides guidance not only in how to use technology but also guidance in using technology appropriately.
- Includes calculator-specific keystrokes in the book right where it is needed rather than in a separate supplement. This material makes it easier for students, faculty, and departments to use technology to examine concepts from multiple perspectives.

The third edition of this book contains many tweaks and refinements from the second edition. All of these changes are based on numerous student surveys and suggestions from our colleagues at Parkland, at AMATYC, and from all of our reviewers. We appreciate your many insightful suggestions. You have helped shape this book and you continue to help us refine the way we teach beginning and intermediate algebra.

Our sincere thanks for using our textbook! We welcome your comments!

James W. Hall
Parkland College
wesley1946@gmail.com

Brian A. Mercer
Parkland College
bmercer@parkland.edu

About the Authors

My wife and I enjoy traveling and seeing the wonders of the world, both natural and man-made.

JAMES W. HALL

- B.S. and M.A. in mathematics from Eastern Illinois University and Ed.D. from Oklahoma State University
- 35 years teaching college mathematics with 31 years in the community college system
- Chair of the Mathematics Department at Parkland College in Champaign, Illinois, for 7 years
- Author of 19 mathematics books in developmental education
- Member of AMATYC (American Mathematical Association for Two-Year Colleges) for 34 years, Midwest Regional Vice President 1987–1989, chair of the editorial review committee 1991–1995, and writing team chair for Chapter 6 on Curriculum and Program Development of *Beyond Crossroads*.
- President of IMACC (Illinois Mathematics Association of Community Colleges) 1995–1996

My wife, Nikki, and I stay busy with our two small children!

BRIAN A. MERCER

- B.S. in mathematics from Eastern Illinois University and M.S. in mathematics from Southern Illinois University
- 11 years teaching community college mathematics
- Author of four mathematics books in developmental education
- Member of AMATYC and NADE (National Association for Developmental Mathematics)
- Board member of IMACC 2002–2005

Preface

"The Universe is a grand book which cannot be read until one first learns to comprehend the language and become familiar with the characters it which it is composed. It is written in the language of mathematics."

—*Galileo Galilei*

Beginning and Intermediate Algebra: The Language and Symbolism of Mathematics emphasizes what great mathematicians, like Galileo Galilei, have identified for generations—mathematics is everywhere! Authors James Hall and Brian Mercer believe active student involvement remains the key to learning algebra. In this edition of *Beginning and Intermediate Algebra: The Language and Symbolism of Mathematics*, topics from both beginning algebra and intermediate algebra courses are well integrated to help students become fluent in algebra—the language and symbolism of mathematics.

Hall-Mercer Approach

Unique Emphasis on the "Rule of Four" and Multiple Perspectives

"This is a very well written text with an integration of the verbal, symbolic, numerical, and graphical approach. The examples, illustrations, and applications in the exercises are well thought out and highly motivating."

—*Mark Sigfrids, Kalamazoo Valley Community College*

The "rule of four" is a phrase that describes how topics are examined: algebraically, graphically, numerically, and verbally. These multiple perspectives have been integrated throughout the textbook in definitions, exposition, examples, and exercises to better facilitate student learning, making it less likely for students to memorize steps and more likely to retain the material they understand while applying mathematics outside the classroom. (See the AMATYC Standard for Intellectual Development: Linking Multiple Perspectives.)

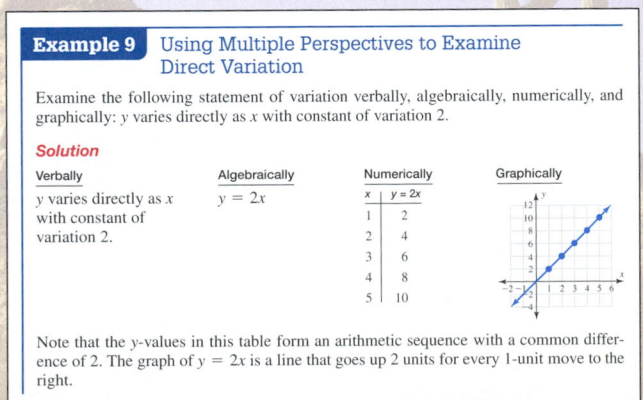

Tables and Graphs

"The use of the Rule of Four enhances this chapter as do the graphing calculator screen captures."

—Joanna K. Pruden, Pennsylvania College of Technology

To further emphasize the use of the "rule of four," we have incorporated tables of numerical values and graphs to illustrate ideas and explore concepts. When the main point of the exercise is to create a graph or a table, we usually expect students to demonstrate their knowledge using pencil and paper. In other problems, calculators or spreadsheets can be used to create tables and graphs while requiring students to interpret the information. Technology Perspectives make use of commonly used technologies such as calculators and spreadsheets to further illustrate how mathematics is implemented in the workplace.

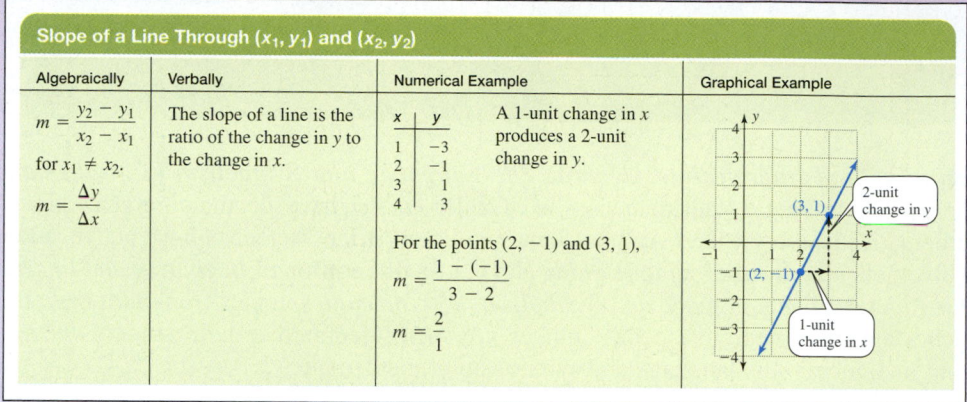

Appropriate Use of Technology

"It is a well written text that integrates technology with the algebraic solution of problems. The readability is excellent, and the order of the topics is quite good."

—Lois Clemens, Angelina College

One of the Standards for Pedagogy stated in AMATYC's *Beyond Crossroads* is: "Mathematics faculty will model the use of appropriate technology in the teaching of mathematics so that students can benefit from the opportunities technology presents as a medium of instruction." Given this emphasis, one of the goals of this textbook is to give guidance through examples, exercises, and suggestions on what usage of technology is appropriate for optimal student learning. For example, using a calculator or a spreadsheet to evaluate $\sqrt{25}$ is not appropriate, whereas approximating $\sqrt{26}$ to the nearest hundredth is appropriate. Similarly, we expect students to be able to quickly sketch the graph of $y = x^2$ or to use technology to graph $y = x^3 + 2x^2 - x - 2$. Students completing this book should be able to work almost all of the exercises in this book using only pencil and paper. On the other hand, selected examples, the Technology Perspectives, and Applying Technology Exercises can be used to explore concepts, enrich topics, show the power of technology, and better prepare students for the workplace. Either way, we expect students to be able to create and interpret graphs and tables and to verbally relate this interpretation to the corresponding algebraic expression.

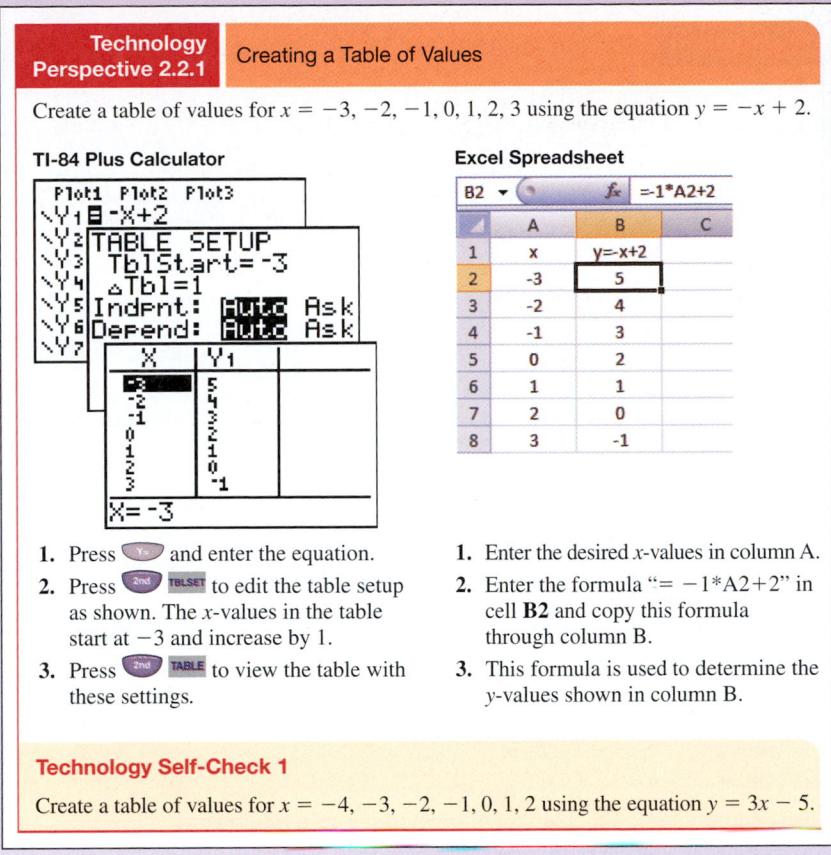

Technology Perspective 2.2.1 — Creating a Table of Values

Create a table of values for $x = -3, -2, -1, 0, 1, 2, 3$ using the equation $y = -x + 2$.

TI-84 Plus Calculator

1. Press [Y=] and enter the equation.
2. Press [2nd] [TBLSET] to edit the table setup as shown. The x-values in the table start at -3 and increase by 1.
3. Press [2nd] [TABLE] to view the table with these settings.

Excel Spreadsheet

1. Enter the desired x-values in column A.
2. Enter the formula "$= -1*A2+2$" in cell **B2** and copy this formula through column B.
3. This formula is used to determine the y-values shown in column B.

Technology Self-Check 1

Create a table of values for $x = -4, -3, -2, -1, 0, 1, 2$ using the equation $y = 3x - 5$.

Function Notation in Beginning Algebra/Function Concept in Intermediate Algebra

The Beginning Algebra portion of the textbook introduces function notation in Chapter 2 and gives the student many opportunities to become familiar with this notation. This helps students to gradually build a solid foundation with the input-output concept.

Classroom data show that beginning algebra students can successfully evaluate $f(2)$ for $f(x) = 5x - 4$. The same data show less success with using the formal definition of a function early in a beginning algebra course. The material from the Beginning Algebra portion of the textbook concentrates primarily on linear functions, with nonlinear material reserved for the Intermediate Algebra portion of the textbook, giving students a solid foundation before they encounter the formal definition of a function in Chapter 8. It also allows each institution considerable flexibility to either increase or decrease the emphasis on functions in the first seven chapters. (See AMATYC Standard for Content: Function.)

Active Student Involvement

"I really like the presentation, with the questions, "What are we doing?" and "Why are we doing it?" I also like how the author anticipates student questions and asks them, "What if ....?"
—Elizabeth Fochs. Lake Superior College

A key factor in the success of each student is time spent on task and the level of attention during that time period. This textbook and supplements are organized to help instructors achieve active student involvement.

- A **Self-Check** is paired with each example in the textbook to encourage reading and listening with pencil in hand. Data show improved student learning of material when students try a problem of their own directly after an example.

Example 4 Solving an Absolute Value Equation

Solve $|2x - 3| = 31$.

Solution

Left 31 Units	or	Right 31 Units	
$2x - 3 = -31$	or	$2x - 3 = +31$	To obtain all solutions for this equation, we must consider both of these equations.
$2x = -28$		$2x = 34$	
$x = -14$		$x = 17$	

Answer: $x = -14$ or $x = 17$ Do these values check?

Self-Check 4

Solve $|4x - 6| = 10$.

- The **Example** format provides students with a clear model they can use to work through exercises. The sidebar explanations can be used by the students as needed. This format keeps the text from being verbose and lets students with different ability levels use the examples in different ways.
- **Group Exercises** and **Group Projects** are available to promote mathematical discussion among students.

Chapter 2 Group Project

Risks and Choices: Shopping for Interest Rates on a Home Loan

The formula for determining the monthly payment for a loan is

$$P = \frac{AR\left(1 + \dfrac{R}{12}\right)^{12n}}{12\left(1 + \dfrac{R}{12}\right)^{12n} - 12}$$

where A represents the amount of the loan, R represents the interest rate, n represents the number of years, and P represents the monthly payment required to pay off the loan in n years.

1. Use this formula and a spreadsheet or a calculator to complete the following tables.

Consider a $100,000 loan at several different interest rates.

Interest Rate (%)	Loan Amount ($)	Number of Years	Monthly Payment ($)	Total of All Payments ($)
5.50	100,000	30		
6.00	100,000	30		

- The **Lecture Guides** provide the structure for excellent class notes and strongly promote active participation in class.
- The **Language and Symbolism of Mathematics Exercises** can be used to encourage students to communicate using the language of mathematics.

2.2 Using the Language and Symbolism of Mathematics

1. The notation $f(x)$ is called _____ notation.

2. The notation $f(x) = 8x - 2$ is read "_____ of _____ equals eight x minus two."

3. In the notation $f(x) = 8x - 2$, the input variable is represented by _____ and $f(x)$ represents the _____ variable.

4. In the notation $f(5) = 9$, the input value is _____ and the output value is _____.

5. The graph of $f(x) = 8x - 2$ is a straight _____.

6. The function $f(x) = mx + b$ is called a _____ function.

7. Creating an equation or a function to describe an application is called mathematical _____.

Mathematical Modeling and Word Equations

The residual value of mathematics—mathematics that students can use years after taking a course—is not a collection of tricks or memorized steps but an understanding that allows students to see mathematics as useful in improving their daily lives. Most people encounter mathematics through words, either orally or in writing, not through equations. Word equations help students to bridge the gap between the statement of a word problem and the formation of an algebraic equation that models the problem. Students must model real problems in a course if we expect them to use mathematics on their own. To that end, the book presents many realistic examples and exercises involving data. (See the Index of Applications as well as the AMATYC Standards for Intellectual Development: Modeling and for Content: Symbolism and Algebra.)

Systems of Equations

Word problems that involve two unknowns are solved in Chapter 3 using two variables rather than one variable. This approach has been well received by the students, who often have more trouble identifying two unknowns using one variable than using a separate variable for each unknown. This approach also received favorable feedback from the users of the first two editions. Later in the textbook, we examine alternate approaches that build on creating functional models and the relating of one variable to another.

Integration of Geometry

Section 1.8 reviews terminology and concepts from geometry. Examples and exercises using this material are then integrated into the examples and exercises throughout the rest of the textbook. These exercises include problems involving perimeter, area, volume, and the measurement of angles.

Textbook Organization

Definitions

Many definitions are given algebraically, graphically, numerically, and verbally.

Examples and Self-Checks

The examples, uninterrupted by dialogue, provide a clear model that we expect students to follow. The sidebar explanations can be used as little or as much as students need. A self-check is paired with each example.

Technology Perspectives and Self-Checks

A self-check is paired with each Technology Perspective.

Design of Exercises

The exercises for each section are preceded by Language and Symbolism of Mathematics exercises and by five Quick Review exercises. Following the section exercises, there are five Cumulative Review exercises. Section exercises are organized by the section objectives; this makes it easy to tailor assignments to meet specific goals. Exercises as well as examples combine multiple perspectives so students work with problems algebraically, graphically, numerically, and verbally. Exercise levels range from skill and drill to discovery questions. Categories of problems include Estimate Then Calculate, Simplify versus Solve, Expand versus Factor, and paired exercises such as Simplify **a.** $(5 + 7)^2$ and **b.** $5^2 + 7^2$.

Group Discussion Questions and Group Projects

Each exercise set has Group Discussion Questions, and there is a Group Project at the end of each of the first eleven chapters. Some of these exercises build bridges to past or future material. Some of them are more challenging for the students. Many of these problems seek to engage the students in communicating with mathematics orally and in writing and to involve the students in interactive and collaborative learning. (See AMATYC Standards for Pedagogy: Active Learning.)

Key Concepts, Chapter Review, and Mastery Tests

"I feel that this is a text that demonstrates clear learning objectives throughout the book. Concepts learned are reinforced by incorporating self-checks, chapter key concepts, cumulative reviews, group projects, and chapter mastery test. There is no lack of exercises."
—*Sheeny Behmard, Chemeketa Community College*

Each chapter ends with these features. The Key Concepts outline the main points covered in that chapter. The Chapter Review contains a selection of exercises designed to help the students review the material from the chapter and to gauge their readiness for an exam. The Chapter Review is longer than an hour exam, and the order of the questions may not parallel the order of each topic within the chapter. The Mastery Test is more limited in its purpose. Each question matches an objective stated at the beginning of one of the sections in the chapter. The students can use this Mastery Test in a diagnostic way to determine which sections and objectives they have mastered and which may still need more work.

Reviews

The Cumulative Review of Chapters 1–5 after Chapter 5 can be used as a final exam review by Beginning Algebra students. The Preparing for Intermediate Algebra review is designed to emphasize key concepts needed by students starting the book in Chapter 6. Some instructors may use this material to create a brief review before starting in Chapter 6. The Cumulative Review of Chapters 6–11 after Chapter 11 can be used as a final exam review by Intermediate Algebra students.

Mathematical Notes

"I like to see both ancient and modern history notes. It really allows students to know that math is continually progressing."
—*Lois Clemens, Angelina College*

Mathematical Notes are placed throughout the book to give the students some sense of historical perspective and to help connect mathematics to other disciplines. These short vignettes give the origin of some of the symbols and terms that we now use and provide brief glimpses into the lives of some of the men and women of mathematics. (See AMATYC Standards for Pedagogy: Making Connections.)

> **A Mathematical Note**
>
> π is defined as the ratio of the circumference of a circle to its diameter. $\pi \approx 3.14159265$ ($\approx$ means "approximately equal to"). However, the decimal form of π does not terminate or repeat. In 1897, House Bill #246 was introduced in the Indiana legislature to make 3.2 the value of π in that state. Fortunately, better judgment prevailed, and this foolish bill was defeated.

What's New in the Third Edition?

- Sections are more closely organized around the objectives for the section.
- The exercises for each section are more closely organized around the objectives for the section.

- Each exercise set is preceded by five Quick Review exercises and followed by five Cumulative Review exercises.
- The organization, font size, and graphs have been changed to improve the appearance and readability of the book.
- The book has a clearer focus on helping students model the appropriate use of technology.
- Technology Perspectives include calculator and spreadsheet commands where appropriate.
- Section 1.3 expands and combines the review of positive fractions into one location.
- Section 1.8 includes a review of geometry.
- The presentation on factoring in Chapter 6 has been revised in response to user input. The strong emphasis on the role of the distributive property and the connection between multiplying polynomials and factoring polynomials has been continued.
- Properties of square roots are now located in Section 7.1, so students are prepared to give answers to quadratic equations in simplified radical form or as decimal approximations.
- Chapter 8 focuses on the concept of a function, the properties of graphs, and the characteristics of families of functions.
- Chapter 8 contains additional review of key topics from linear and absolute value functions, including solving linear equations and inequalities, writing equations of lines, and solving absolute value equations and inequalities.

Supplements for the Instructor

Instructor's Edition

The Instructor's Edition (IE) contains answers to problems and exercises in the text, including answers to all Language and Symbolism of Mathematics vocabulary questions, all end-of-section exercises, all end-of-chapter review exercises, and all end-of-chapter mastery tests.

Instructor's Solutions Manual

The Instructor's Solutions Manual, prepared by Mark Smith of College of Lake County, provides comprehensive **worked-out solutions** to exercises in the text.

Lecture Guide

This supplement by Brian Mercer and James Hall, with the assistance of Kelly Bails of Parkland College, provides instructors with the framework of day-by-day class activities for each section in the book. Each lecture guide can help instructors make more efficient use of class time and can help keep students focused on active learning. Students who use the lecture guides have the framework of well-organized notes that can be completed with the instructor in class.

Instructor's Testing and Resource Online

These supplements provide a wealth of resources for the instructor, including a computerized test bank utilizing Brownstone Diploma® algorithm-based testing software to create customized exams quickly. This user-friendly program enables instructors to search for questions by topic, format, or difficulty level; to edit existing questions or to add new ones; and to scramble questions and answer keys for multiple versions of a single test. Hundreds of text-specific, open-ended, and multiple-choice questions are included in the question bank. Sample chapter tests are also provided. CD available upon request.

Video Lectures on Digital Video Disk

In the videos, the authors, James Hall and Brian Mercer, work through selected problems from the textbook, following the solution methodology employed in the text. The video series is available on DVD or online as an assignable element of MathZone (see next). The

DVDs are closed-captioned for the hearing impaired, subtitled in Spanish, and meet the Americans with Disabilities Act Standards for Accessible Design. Instructors may use them as resources in a learning center, for online courses, and/or to provide additional help to students who require extra practice.

MathZone www.mathzone.com

McGraw-Hill's **MathZone** is a complete online tutorial and course management system for mathematics and statistics, designed for greater ease of use than any other system available. Available with selected McGraw-Hill textbooks, the system enables instructors to **create and share courses and assignments** with colleagues and adjuncts with only a few clicks of the mouse. All assignments, questions, e-professors, online tutoring, and video lectures are directly tied to **text-specific** materials.

MathZone courses are customized to your textbook, but you can edit questions and algorithms, import your own content, and **create** announcements and due dates for assignments.

MathZone has **automatic grading** and reporting of easy-to-assign, algorithmically generated homework, quizzing, and testing. All student activity within **MathZone** is automatically recorded and available to you through a **fully integrated grade book** that can be downloaded to Excel.

MathZone offers:

- **Practice exercises** based on the textbook and generated in an unlimited number for as much practice as needed to master any topic you study.
- **Videos** of classroom instructors giving lectures and showing you how to solve exercises from the textbook.
- **e-Professors** to take you through animated, step-by-step instructions (delivered via on-screen text and synchronized audio) for solving problems in the book, allowing you to digest each step at your own pace.
- **NetTutor,** which offers live, personalized tutoring via the Internet.

New ALEKS Instructor Module

The new ALEKS Instructor Module features enhanced functionality and a streamlined interface based on research with ALEKS instructors and homework management instructors. Paired with powerful assignment-driven features, textbook integration, and extensive content flexibility, the new ALEKS Instructor Module simplifies administrative tasks and makes ALEKS more powerful than ever. Features include:

- **Gradebook**   Instructors can seamlessly track student scores on automatically graded assignments. They can also easily adjust the weighting and grading scale of each assignment.
- **Course Calendar**   Instructors can schedule assignments and reminders for students.
- **Automatically Graded Assignments**   Instructors can easily assign homework, quizzes, tests, and assessments to all or select students. Deadline extensions can also be created for select students.
- **Set-Up Wizards**   Instructors can use wizards to easily set up assignments, course content, textbook integration, etc.
- **Message Center**   Instructors can use the redesigned Message Center to send, receive, and archive messages; input tools are available to convey mathematical expressions via e-mail.

MathZone Reporting

Visual Reporting

The new dashboard-like reports will provide the progress snapshot instructors are looking for to help them make informed decisions about their students.

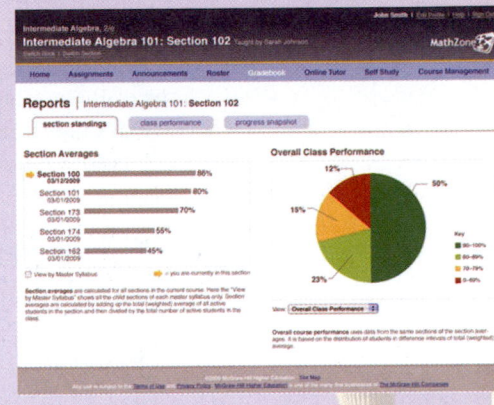

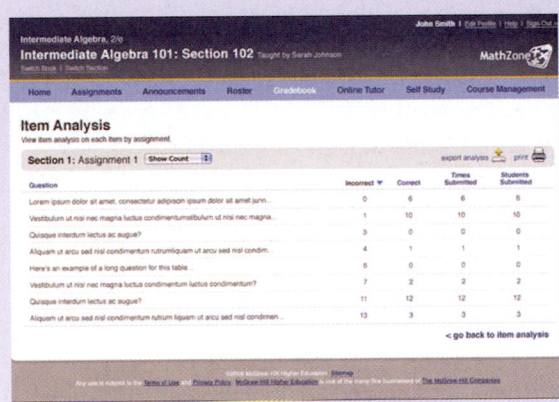

Item Analysis

Instructors can view detailed statistics on student performance at a learning objective level to understand what students have mastered and where they need additional help.

Managing Assignments for Individual Students

Instructors have greater control over creating individualized assignment parameters for individual students, special populations and groups of students, and for managing specific or ad hoc course events.

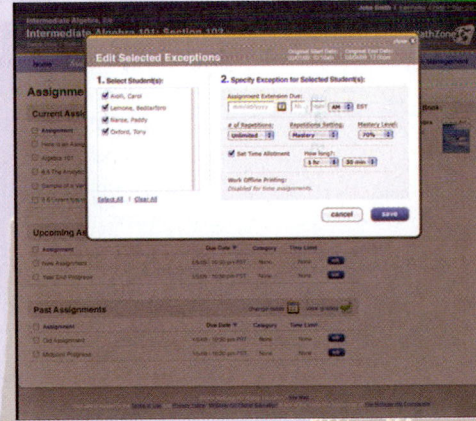

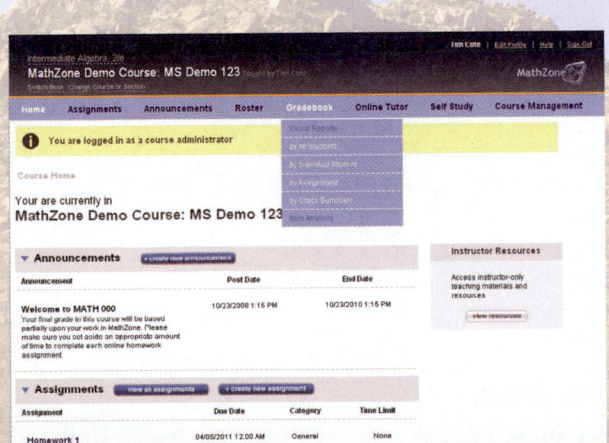

New User Interface

Designed by You! Instructors and students will experience a modern, more intuitive layout. Items used most commonly are easily accessible through the menu bar such as assignments, visual reports, and course management options.

Supplements for the Student

Student's Solutions Manual

The Student's Solutions Manual, by Mark Smith of College of Lake County, provides comprehensive, **worked-out solutions** to odd-numbered exercises. The steps shown in the solutions match the style of solved examples in the textbook.

Lecture Guide

This supplement by Brian Mercer and James Hall, with the assistance of Kelly Bails of Parkland College, provides instructors with the framework of day-by-day class activities for each section in the book. Each lecture guide can help instructors make more efficient use of class time and can help keep students focused on active learning. Students who use the lecture guides have the framework of well-organized notes that can be completed with the instructor in class.

MathZone www.mathzone.com

McGraw-Hill's MathZone is a powerful Web-based tutorial for homework, quizzing, testing, and multimedia instruction. Also available in CD-ROM format, MathZone offers:

- **Practice exercises** based on the text and generated in an unlimited quantity for as much practice as needed to master any objective
- **Video** clips of classroom instructors showing how to solve exercises from the text, step by step
- **e-Professor** animations that take the student through step-by-step instructions, delivered on-screen and narrated by a teacher on audio, for solving exercises from the textbook; the user controls the pace of the explanations and can review as needed
- **NetTutor,** which offers personalized instruction by live tutors familiar with the textbook's objectives and problem-solving methods

Every assignment, exercise, video lecture, and e-Professor is derived from the textbook.

ALEKS Prep for Developmental Mathematics

ALEKS Prep for Beginning Algebra and Prep for Intermediate Algebra focus on prerequisite and introductory material for Beginning Algebra and Intermediate Algebra. These prep products can be used during the first 3 weeks of a course to prepare students for future success in the course and to increase retention and pass rates. Backed by two decades of National Science Foundation funded research, ALEKS interacts with students much like a human tutor, with the ability to precisely assess a student's preparedness and provide instruction on the topics the student is most likely to learn.

ALEKS Prep Course Products Feature:

- Artificial intelligence targets gaps in individual students' knowledge
- Assessment and learning directed toward individual students' needs
- Open response environment with realistic input tools
- Unlimited online access—PC & Mac compatible

Free trial at www.aleks.com/free_trial/instructor

NetTutor

Available through MathZone, NetTutor is a revolutionary system that enables students to interact with a live tutor over the Web. NetTutor's Web-based, graphical chat capabilities enable students and tutors to use mathematical notation and even to draw graphs as they work through a problem together. Students can also submit questions and receive answers, browse previously answered questions, and view previous sessions. Tutors are familiar with the textbook's objectives and problem-solving styles.

Video Lectures on Digital Video Disk (DVD)

The video series is based on exercises from the textbook. The authors, James Hall and Brian Mercer, work through selected problems, following the solution methodology employed in the text. The video series is available on DVD or online as part of MathZone. The DVDs are closed-captioned for the hearing impaired, subtitled in Spanish, and meet the Americans with Disabilities Act Standards for Accessible Design.

Experience Student Success!

ALEKS®

ALEKS is a unique online math tool that uses adaptive questioning and artificial intelligence to correctly place, prepare, and remediate students . . . all in one product! Institutional case studies have shown that **ALEKS has improved pass rates by over 20% versus traditional online homework, and by over 30% compared to using a text alone**.

By offering each student an individualized learning path, ALEKS directs students to work on the math topics that they are ready to learn. Also, to help students keep pace in their course, instructors can correlate ALEKS to their textbook or syllabus in seconds.

To learn more about how ALEKS can be used to boost student performance, please visit www.aleks.com/highered/math or contact your McGraw-Hill representative.

ALEKS Pie

Each student is given her or his own individualized learning path.

Easy Graphing Utility!

Students can answer graphing problems with ease!

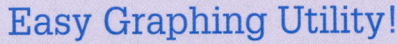

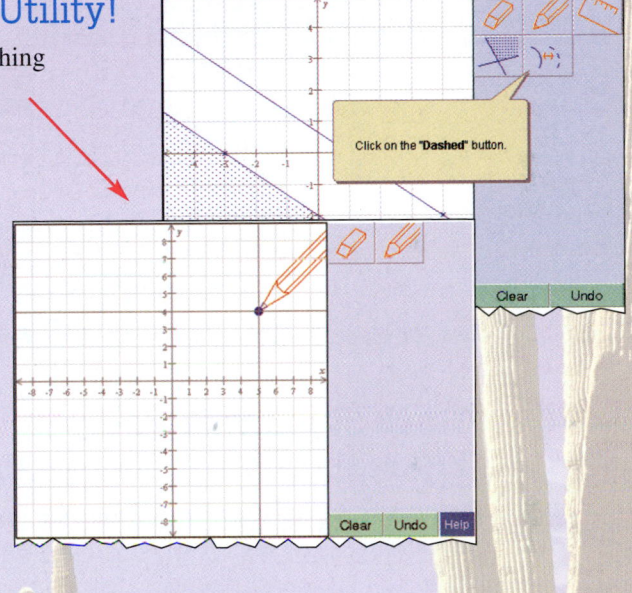

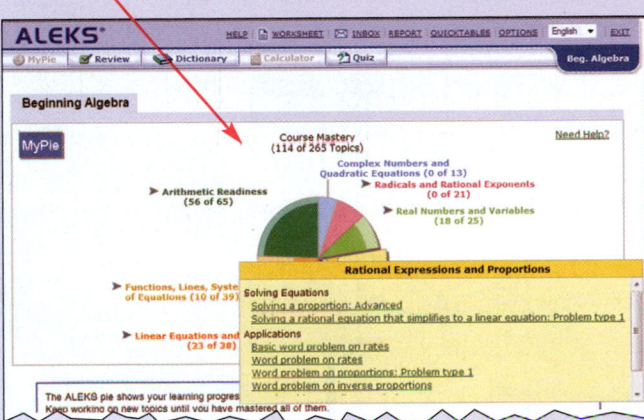

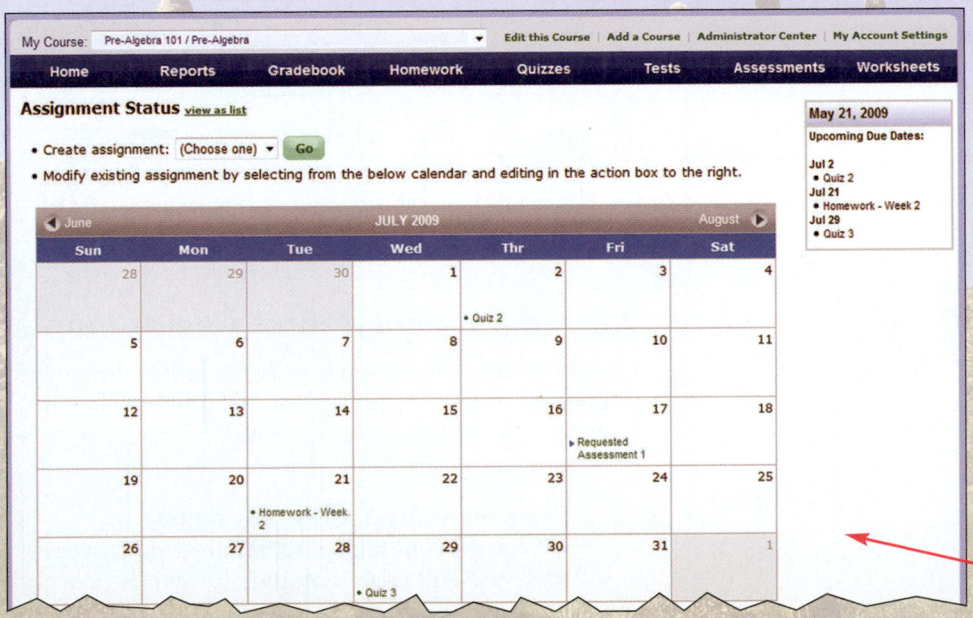

Course Calendar

Instructors can schedule assignments and reminders for students.

New ALEKS Instructor Module

Enhanced Functionality and Streamlined Interface Help to Save Instructor Time

ALEKS®

The new ALEKS Instructor Module features enhanced functionality and a streamlined interface based on research with ALEKS instructors and homework management instructors. Paired with powerful assignment-driven features, textbook integration, and extensive content flexibility, the new ALEKS Instructor Module simplifies administrative tasks and makes ALEKS more powerful than ever.

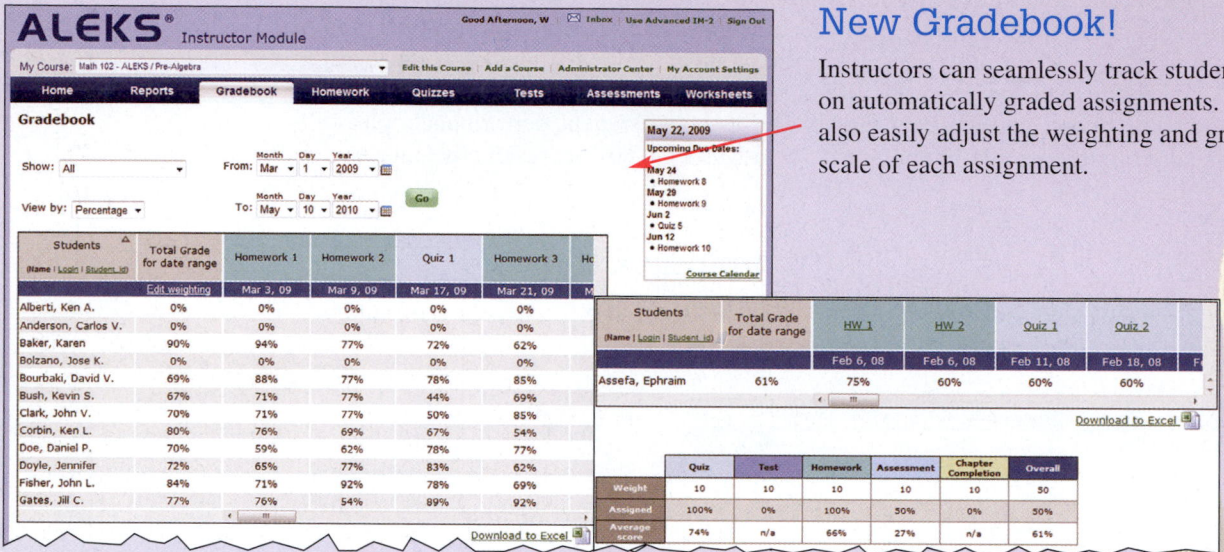

New Gradebook!

Instructors can seamlessly track student scores on automatically graded assignments. They can also easily adjust the weighting and grading scale of each assignment.

Gradebook view for all students   **Gradebook view for an individual student**

Track Student Progress Through Detailed Reporting

Instructors can track student progress through automated reports and robust reporting features.

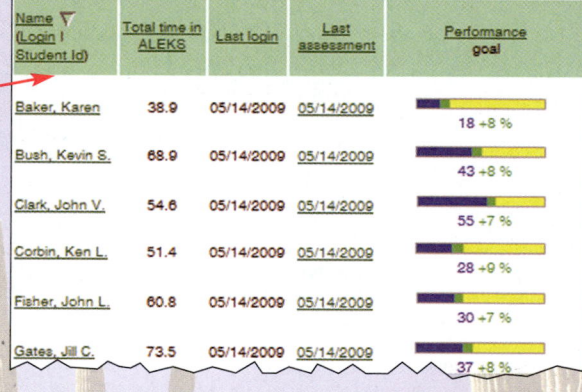

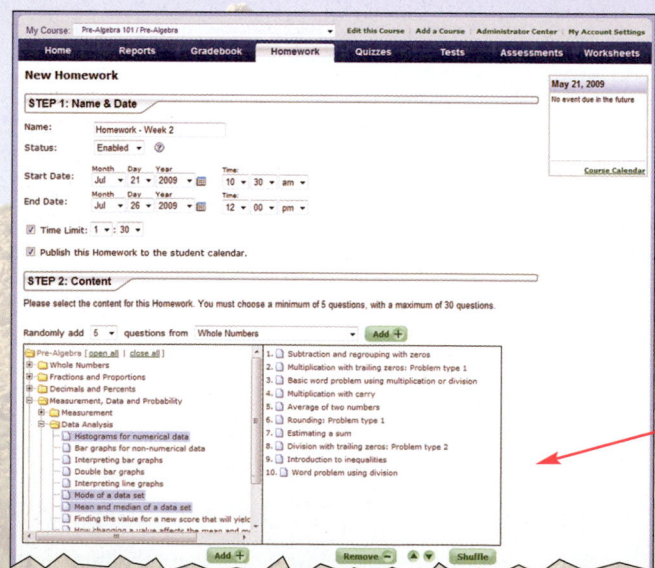

Select topics for each assignment

Automatically Graded Assignments

Instructors can easily assign homework, quizzes, tests, and assessments to all or select students. Deadline extensions can also be created for select students.

Learn more about ALEKS by visiting www.aleks.com/highered/math or contact your McGraw-Hill representative.

360° Development Process

McGraw-Hill's 360° Development Process is an ongoing, never-ending, market-oriented approach to building accurate and innovative print and digital products. It is dedicated to continual large-scale and incremental improvement driven by multiple customer feedback loops and checkpoints. This is initiated during the early planning stages of our new products, intensifies during the development and production stages, and then begins again upon publication, in anticipation of the next edition.

A key principle in the development of any mathematics text is its ability to adapt to teaching specifications in a universal way. The only way to do so is by contacting those universal voices—and learning from their suggestions. We are confident that our book has the most current content the industry has to offer, thus pushing our desire for accuracy to the highest standard possible. In order to accomplish this, we have moved through an arduous road to production. Extensive and open-minded advice is critical in the production of a superior text.

Listening to You...

This textbook has been reviewed by over 300 teachers across the country. Our textbook is a commitment to your students, providing a clear explanation, concise writing style, step-by-step learning tools, and the best exercises and applications in developmental mathematics. How do we know? You told us so!

Teachers *just like you* are saying great things about the Hall-Mercer developmental mathematics series:

"This text does a nice job of utilizing the AMATYC standards, including technology explanations and using the 'rule of four'."

—*Helen Smith, South Mountain Community College*

"I like more and more the use of the — — student perspective questions leading into topics. I think it potentially breaks up a passive-reading mindset students often lock themselves into and may get their brain 'thinking' as they read. The brain processes questions differently than straight narrative and it is the commutation of little things like this that can make an overall large improvement."

—*Ken Anderson, Chemeketa Community College*

Acknowledgments and Reviewers

The development of this textbook series would never have been possible without the creative ideas and feedback offered by many reviewers. We are especially thankful to the following instructors for their careful review of the manuscript.

Board of Advisors

Ken Anderson, *Chemeketa Community College*
Kelly Bails, *Parkland College*
Douglas Darbro, *Shawnee State University*
Cynthia Gubitose, *Southern Connecticut State University*
Jack Rotman, *Lansing Community College*

Reviewers

Sheeny Behmard, *Chemeketa Community College*
Eric Bennett, *Michigan State University*
Becky Blackwell, *Pellissippi State Technical Community College*
Ashley Boone, *Pellissippi State Technical Community College*
Susan Bradley, *Angelina College*
Shawna Bynum, *Napa Valley College*
Susan Caldiero, *Cosumnes River College*

Nancy Casten, *Front Range Community College–Fort Collins*
Lois Clemens, *Angelina College*
Brandi Cooke, *Santa Fe Community College*
Aharon Dagan, *Santa Fe Community College*
Dr. Emmett C. Dennis, *Southern Connecticut State University*
Jeffrey Dyess, *Bishop State Community College*
David Ellingson, *Napa Valley College*
Barbara Fisher, *Pennsylvania College of Technology*
Elizabeth Fochs, *Lake Superior College*
Thomas Fox, *Cleveland State Community College*
Lauren Rhodes Gordon, *Pennsylvania College of Technology*
Catherine Griffin, *Lansing Community College*
Christina Gundlach, *Eastern Connecticut State University*
Jeri Hamilton, *Yavapai College–Prescott*
Natasha Haydel, *Lone Star College CyFair*
Andrea Hoagland, *Lansing Community College*
Irene Hollman, *Southwestern College*
Byron Hunter, *College of Lake County*
Karen Inkelis, *Las Positas College*
Joanna Kendall, *Lone Star College CyFair*
Weiping Li, *Walsh University*
Nicole Lloyd, *Lansing Community College*
Diane L. Martling, *Harper College*
David Matthews, *Minnesota West Community and Technical College*
Julie Mays, *Angelina College*
Toni McCall, *Angelina College*
Ashley McHale, *Las Positas College*
Ellen Musen, *Brookdale Community College*
Nicole Newman, *Kalamazoo Valley Community College*
Samuel Ofori, *Cleveland State Community College*
Luc Patry, *University of Arkansas Pine Bluff*
Jody Peeling, *Santa Fe Community College*
Nancy Pevey, *Pellissippi State Technical Community College*
Joanna K. Pruden, *Pennsylvania College of Technology*
Jeffery M. Rabish, *Pennsylvania Highlands Community College*
George Reed, *Angelina College*
Pamelyn Reed, *Lone Star College CyFair*
Peter Rosnick, *Greenfield Community College*
John J. Salak, *Tallahassee Community College*
Kristina Sampson, *Lone Star College CyFair*
Mark Sigfrids, *Kalamazoo Valley Community College*
Helen Smith, *South Mountain Community College*
Karen Smydra, *Lansing Community College*
Malissa Trent, *Northeast State*
Virginia VanDeusen, *Lansing Community College*
Shannon Vaughn, *University of Arkansas*
Ellen Vilas, *York Technical College*
Dr. Kim Ward, *Eastern Connecticut State University*
Danae Watson, *University of Arkansas*
Michelle Watts, *Lone Star College Tomball*
Zbigniew Wdowiak, *Lake Superior College*
Michelle Whitmer, *Lansing Community College*
Jackie Wing, *Angelina College*
Kristine Woods, *Las Positas College*
Landra Young, *Lone Star College CyFair*

Brief Contents

Contents

11 Exponential and Logarithmic Functions   843

12 A Preview of College Algebra (Available online at www.mhhe.com/hallmercer)

Index of Applications

Operations with Real Numbers and a Review of Geometry

Chapter Outline

Universal Product Code, UPC

Universal product code (UPC) labels are on most of the products that we buy at stores. These codes were originally created to speed up the checkout at grocery stores and to help track inventory. The success of this system was so persuasive that it has been adopted by almost all manufacturers. The design of the machinery to produce and read these bar codes involved a significant amount of mathematics.

There is also some mathematics involved each time one of these bar codes is read by a scanner. Note that UPC codes, like the one shown here for a 12 pack of *7 UP*, have both a machine-readable bar code and a 12-digit numeric code for individuals to read.

The group project at the end of this chapter will use a five-step process to calculate the check digit for the UPC code 0 78000 01180 7 seen in the photo—a check that is done every time a UPC code is scanned. This process uses only the arithmetic skills reviewed in this chapter; however, the speedy application of this process by scanning devices saves businesses millions of dollars every year. Exercises 57 and 58 in Section 1.4 will also involve calculating the check digit for UPC codes.

Section 1.1	Preparing for an Algebra Class

The Universe is a grand book which cannot be read until one first learns to comprehend the language and become familiar with the characters in which it is composed. It is written in the language of mathematics.   —GALILEO GALILEI ("THE FATHER OF MODERN SCIENCE")

Objective:

1. Understand the class syllabus and textbook features.

Can I Be Successful If I Have Math Anxiety?

YES! Almost everyone has anxiety related to some activity. For many it is public speaking, flying, appearing on live television, or performing mathematics. Several famous actresses and actors suffer from intense anxiety before every single performance. Their anxiety is never eliminated, yet they learn to deal with it quite successfully and to great reward. If you have any math anxiety, you can still perform well and be successful in mathematics. Even if you encounter some difficulty in algebra, you control the most important factors in your success. Your positive attitude, determination, and work ethic will be the primary factors in this success. We encourage you to consult with your instructor who is a key source of information and assistance.

1. Understand the Class Syllabus and Textbook Features

With All the Materials Available to Me, Where Do I Start?

Start with your syllabus and materials provided by your instructor. These materials may contain information on office hours or other supplemental help that is available on your campus. Be sure to consult with your instructor if you have any questions about placement, tutoring services, or supplements to this textbook.

The success tips and study hints given below are based on many years of experience and research used by the authors. We are sure that some of these hints will be familiar to you, and we also expect that some of them will be new things for you to try. We urge you to consider all these tips and use them in this course and in other courses.

How Can I Dramatically Increase My Odds for Success in This Course?

Set a specific academic goal. Students with clearly defined academic goals outperform students with only vague goals by a wide margin. As you set your goals, remember that your education can never be taken from you. Your ability to know how to learn will be with you in all future economic circumstances.

Students with specific goals often overcome the conflicts of work and other time commitments to reach their goals. Goals provide a direction and a motivation for the work and time necessary to meet the challenge presented by a mathematics class. If you have not already established some academic goals for the term, it would help you to set some now. Clearly establish a realistic goal, even if it is tentative or short term, and routinely remind yourself of this goal to help filter out distractions. Check with your faculty advisor to be sure that you select courses that will help you meet your goals and that you understand the requirements.

What Is the Most Important Factor in My Course Success?

Several studies have shown that the time you spend on task is the most important factor in your success in a mathematics class. The more quality time you spend, the more you will gain. Two places to use your time are attending class daily and doing homework daily. The importance of these two elements to your learning cannot be overstated. To do both consistently requires time management. Perhaps the most frequent mistake made by beginning college students is underestimating the time required to study and learn.

It is possible to shortchange yourself in sleep, study, or any area for a short time. However, frequently shortchanging any of these areas can have serious consequences. As a college student, it is critical that you budget enough time for studying in addition to attending classes. A good rule of thumb is to allow about 2 hours of study for each hour of class. Therefore if you are enrolled in 16 credit hours, you should plan to study another 32 hours per week. For academics, this represents a total of 48 hours a week or 8 hours per day for 6 of the days of the week. This leaves only 16 hours of each of these 6 days for family, sleep, work, and recreation. To decide the amount of time to allow for each activity, you must set priorities.

How important is an education to you? If it is one of your top priorities, then you must allow adequate time to study. You may need to decrease the time spent on other commitments to gain sufficient time to study. If you cannot do this, then maybe you should consider taking fewer courses this term. Keep in mind that your education is of long-term benefit. What you invest in it now will determine how qualified and competitive you will be in the marketplace. If you have any concerns about your schedule, it is advisable to consult with your college advisor or counselor for assistance.

Why Should I Show My Work? Isn't the Answer All That Is Important?

No, actually your work is very important. Teachers want to ensure that students have mastered the concepts and procedures for solving all problems of a given type, and therefore they will often examine the steps and logic of a student's work to be sure that the problems have been worked correctly. Showing your work is also good professional development for many jobs. For example, engineers must document their work so that their designs can be examined by other engineers for compliance with safety standards, government contracts, and insurance requirements.

It will help you both in class and in your future career if you organize your work so that others can understand it easily. A paper that is well organized gives the impression that its writer understands the material. A messy or disorganized paper gives the reader many negative impressions about the writer, whether they are true or not. A well-organized paper makes it easier for you to proof your own work. The following samples illustrate some pointers for organizing mathematics problems. The problem to be worked is:

$$\text{Simplify } \frac{-3 - \sqrt{3^2 - 4(-2)(2)}}{2(-2)}.$$

Here is a sample of a well-organized problem:

$$\frac{-3 - \sqrt{3^2 - 4(-2)(2)}}{2(-2)} = \frac{-3 - \sqrt{9 + 16}}{-4}$$

The problem is recopied so that the starting point is clear.

$$= \frac{-3 - \sqrt{25}}{-4}$$

The equals signs are aligned to help the reader see the connection between the steps.

$$= \frac{-3 - 5}{-4}$$

Each expression following an equals symbol is equal to the previous expression and thus to the original expression.

$$= \frac{-8}{-4}$$

The location of the last step makes the answer clear.

$$= 2$$

In contrast, here is a sample of a poorly organized problem:

$$-3 - \sqrt{3^2 - 4(-2)(2)} = 9 + 16 = 25 = \sqrt{25} = 5$$
$$2(-2) = -4$$

$$-3 - 5 = \frac{-8}{-4} = 2$$

The problem is not recopied so the relation of the work to the question is not clear. The equals signs are not aligned and are used incorrectly. Note that

$$25 \neq \sqrt{25} \quad \text{and} \quad -3 - 5 \neq \frac{-8}{-4}$$

The expressions written are just pieces of the original problem. Confusion is likely for the reader regarding the relationship of these pieces.

The answer to this problem is not clear with this organization.

You should check all your homework to verify that your practice is correct. Any problems with wrong answers should be circled so that you can ask questions about them in class. It is important to review and then review some more. As you review, look over the problems you have circled, and be sure that you have mastered any troublesome points. By keeping track of the sources of your errors, you can be self-correcting on both homework and tests. If you have extra time on a test, double-check your work, especially for any of the errors that you are prone to make.

Many teachers recommend a three-ring binder for your homework, quizzes, tests, and other papers. Bring this binder, your textbook, and your calculator to each class.

How Can I Get the Most Out of Lectures?

Start by attending every class to avoid a gap in your understanding. To maximize your gain from lectures, we suggest that you *preview*, *participate*, and *review*. Students who preview the material before the lecture have a big advantage over students who wait until after the lecture to examine the material. If you have read the material, you can focus your attention on any points you find troublesome. You can also concentrate on listening for concepts and the connections between different ideas rather than worrying about details. Participate by listening actively rather than passively. So that your mind does not wander, try to predict ahead of time the point your instructor is going to make, and keep pace with the lecture. Watch for your instructor's nonverbal signals, which will clarify meanings or intent. Ask questions; your instructor will appreciate your interest, and other students are likely to have the same questions you do. Review; all review is beneficial, especially an early review of your lecture notes. Before you start your homework, make sure that all the examples covered in class make sense. Clarify the organization of your notes, fill in any steps that you may have omitted, and be sure to highlight any hints that your instructor may have given regarding possible test questions.

Will a Study Group Benefit Me?

For most students, the answer is yes. Research projects have shown that studying and doing some of your homework with two or three of your classmates can improve your grade. Organizing your questions or your responses to questions well enough to verbalize them to others can help you clarify your understanding of the material. Exchange phone numbers and e-mail addresses with members of your study group, and consider meeting them to discuss the major purposes of an assignment or the best way to approach certain problems. You can even practice testing each other prior to an exam. Having other points of view to consider is itself part of the educational process; few people in business, education, or industry work alone. Learning in a group situation will help prepare you for working in a group environment.

How Can I Best Use a Calculator or a Spreadsheet?

Integrate calculators and spreadsheets into your course work as you examine new mathematical concepts. This will help you become better prepared for your career. Do not try to substitute calculator usage for understanding the material, but do use a calculator or a spreadsheet to expedite your work and to experiment and consider alternatives.

Some general suggestions on calculator usage follow. For specific instructions, consult this textbook or your calculator's operating manual. If you run into problems, ask your instructor for help.

1. Practice predicting the answer by using your estimation skills.

2. Use your calculator to work a problem with simple values whose result is known before you undertake the calculation of similar problems whose results are unknown.

3. During calculations, use as many digits of accuracy as your calculator will allow. Since most calculators store more digits than they display, you can increase accuracy by observing the following guidelines.

   a. Leave intermediate values in the calculator rather than copying down the display digits and then reentering these values. Learn to use the memory and parentheses keys.

   b. Enter e and π by using the special keys on your calculator. Such entries will typically be accurate to two or three more digits than are shown on the display.

How Can I Best Prepare for a Mathematics Test?

You will benefit most if the majority of your test preparation occurs well before the test. However, near test time you may find it helpful to take some practice tests to gauge your level of preparation. You can use the Key Concepts and the Mastery Test at the end of each chapter for this purpose. There are also sample quizzes available on the website for this text. If you work with a study group, you can gain by testing each other. One way to overcome test anxiety is to test yourself frequently so that a test is no longer a big deal to you.

Another strategy for testing yourself is as follows: Each day, copy a few representative problems from your homework assignment on the front of 3×5 cards. Put the solution and/or a page reference on the back of each card. Review this stack of cards for a few minutes after each assignment. You don't have to actually rework the problems. Just shuffle the cards, and then mentally ask yourself what steps you would follow if this problem were given on a test. You will be practicing the discrimination skills that are required on tests and that are easily overlooked on the day-to-day homework. Many students have used this strategy to raise their grade by 10% to 20%.

How Can I Monitor My Own Progress?

After each assignment, review your progress and note your gains, as well as problems that may still require further practice. Take pride in what you do understand, and ask yourself how this material is similar to (or different from) your new material. Try to understand ideas and build connections between different ideas—avoid memorizing without understanding. Maintain a positive attitude; it really does help. Many students are harsher critics of themselves than anyone else is. No one understands every topic immediately; basketball, cooking, golf, mathematics, and piano all take practice. Work steadily, and don't expect to master all the information presented the first time through it. By working in groups you can help one another, and you will observe how others must also practice to gain their understanding.

Almost all the mathematics problems that you will encounter outside the classroom will be stated in words. Thus to profit from your algebra, you must be able to work with word problems. Numerous suggestions accompany the word problems worked in the textbook. Many of these suggestions are based on research articles describing successful study techniques.

What Textbook Features and Supplements Should I Use?

Some textbook features should be used for each assignment. For example, it is wise to use the Self-Checks, the Language and Symbolism Exercises, and the Quick Review Exercises in each section. Some supplements such as ALEKS and the textbook website provide abundant supplemental help if you ever need it. Realize that no one understands each new topic immediately. If one topic should cause you particular concern, start by going over the examples and key points that are boxed in the text. Then you may need to use some of the textbook supplements. Also feel free to go to your instructor with well-organized questions.

Instructors enjoy students who work hard and are usually willing to provide these students with additional help during office hours. Instructors may also have recommendations regarding useful text supplements. The main point is to obtain this help as soon as you need it.

How Do I Prepare for a Final Exam?

To maximize the benefit of your review for the final, you should spread this review over several days. It should definitely not consist of one long marathon session. Start by spending a couple of hours going over your first hour exam with particular emphasis on correcting any careless errors that you may have made. On the next day do the same thing with the second hour exam. By the end of the week you will have gone over all your hour exams. Then review the key concepts at the end of each chapter, and practice on some of the exercises in the cumulative reviews in the text. It would be wise to go through your notes and look over the troublesome problems you circled earlier. However, it is generally unwise to try to cram in new material that could leave you confused and cause you to miss reviewing concepts that you have mastered. Just review what you already know, and don't miss any problems that you can work. The final exam will then give an accurate measure of your knowledge.

How Will This Algebra Class Help Me Prepare for My Future?

One of the reasons algebra is required of so many students is that it does more than prepare you for advanced mathematical topics. It also helps you see how mathematics can be applied to real-life problems, and it aids you in improving your reasoning and critical-thinking skills.

A significant part of a college education is learning how to learn. Many of the job titles that are common today did not exist 25 years ago, and many people must change jobs at least once during their careers. Although you may have a specific job in mind now, you must educate yourself as broadly as possible rather than focus yourself narrowly.

As you prepare for your next term, try to keep in mind your long-term goal of improving yourself through education. Review all the other study hints given in this book, and reflect on the learning strategies that you have developed during this term. Then try to apply these strategies right from the start of your next term.

What Should I Do with My College Textbooks?

As you advance in your education, it is important to begin to build your own professional library of reference books. In addition to a good dictionary, thesaurus, and college handbook, you should include basic textbooks that you have studied in detail. You are already familiar with their style, language, and format, so when you refer to these books in later sequential courses, it will be like working with an old friend. Any financial gain that you may receive by selling these books now may cost you much time and effort later if you have to adjust to the new style and format of another book. This *Beginning and Intermediate Algebra* book has been designed with several goals in mind. One important goal is to prepare you for subsequent mathematics courses and to serve as a resource for you in these courses. We hope that you find this book a valuable reference and addition to your professional library.

1.1	**Exercises**

Objective 1 Understand the Class Syllabus and Textbook Features

1. What is the name of your instructor? _____.

2. Your instructor's office is in _____.

3. Your instructor's office hours are _____.

4. Your instructor's phone number is _____.

5. Your instructor's e-mail address is _____.

6. Is there a mathematics tutoring lab on your campus? _____ If so, it is located in _____.

7. I realize that to succeed in this course I will need to work about _____ hours per day outside of class.

8. The date and time for the final exam (if known) are _____. Mark this date on your calendar. Also record the dates of the hour exams if these are known.

9. The calculator used in this class is the _____.

10. In addition to the textbook, calculator, and notebook, the supplies recommended by your instructor are _____.

11. The textbook has _____-Check Exercises by the examples in the text for you to check your own understanding as you read through the text.

12. The *Student's Solutions Manual* for this book contains the worked-out solutions to the _____-numbered problems. (See the Preface.)

13. The _____ *Guide* for this book has the key language and symbolism, definitions, and examples structured in the framework of a class lecture. (See the Preface.)

14. The video series for this book has a separate video for each _____. (See the Preface.)

15. The _____ website provides multiple resources, including video lectures, linked to this text. (See the Preface.)

16. After rereading the tips on time management, fill out your schedule for the term including your class time, your study time, and work or other time commitments.

	Sun	Mon	Tue	Wed	Thu	Fri	Sat
7:00 A.M.							
8:00 A.M.							
9:00 A.M.							
10:00 A.M.							
11:00 A.M.							
12:00 P.M.							
1:00 P.M.							
2:00 P.M.							
3:00 P.M.							
4:00 P.M.							
5:00 P.M.							
6:00 P.M.							
7:00 P.M.							
8:00 P.M.							
9:00 P.M.							

17. ALEKS is an artificial intelligence-based system that can interact with you much as a human _____ would. (See the Preface.)

18. One hint given under the heading "How Can I Best Prepare for a Mathematics Test?" suggests that you copy a few representative problems on a _____ for each assignment. One advantage of this format is that the questions can be shuffled and you can review and give yourself a quick practice test each day.

19. When you form your study group, record the names, phone numbers, and e-mail addresses of your study group here.

Name	Phone Number	E-mail Address

Group discussion questions

20. Have each person in your group briefly share his or her greatest anxiety. Discuss how you think an actress who has an anxiety about going on stage deals with her anxiety. Discuss how you think a teacher who has an anxiety about getting in front of a group deals with his anxiety. Discuss how you think a student who has an anxiety about mathematics tests can deal with this anxiety.

21. A quote attributed to Henry Ford states, "If you think you can, you can. If you think you can't, you can't. In either case you are right." Discuss what you think is meant by this quote.

22. Two very common comments by students are "I knew this material, my mind just went blank" and "I can do all the problems, I just don't test well." The purpose of this group exercise is to remind you of test-taking strategies that you may have used successfully in the past, to share new test-taking strategies, and to improve your test scores.

a. In your group, discuss and list two kinds of test-taking strategies.

  i. Strategies to use before the test. (For example: Do the chapter review a couple of days before the test.)

  ii. Strategies to use during the test. (For example: Never change an answer unless you are very confident that you have a better answer.)

b. In class, share two strategies to use before the test and two strategies to use during the test that your group likes best. Perhaps the teacher can record the contributions from each group and give each student a copy of all these strategies.

c. Decide on at least one strategy that you will use before the test and one strategy that you will use during the test, and then use these strategies.

d. After the exam have each member of your group share her or his perceptions of the results that each experienced while using these strategies. What worked and what did not work? What will you do differently for the next exam?

| **Section 1.2** | **The Real Number Line** |

Objectives:

1. Identify additive inverses.
2. Evaluate absolute value expressions.
3. Use inequality symbols and interval notation.
4. Mentally estimate square roots and use a calculator to approximate square roots.
5. Identify natural numbers, whole numbers, integers, rational numbers, and irrational numbers.

What Are the Origin and the Meaning of the Word *Algebra*?

The word *algebra* comes from the book *Hisab al-jabr*, written by a Persian mathematician, Al-Khowarizmi, in A.D. 830; translations of this text on solving equations became widely known in Europe as *al-jabr*. Algebra, however, is concerned with more than solving equations; it is a generalization of arithmetic. Algebra is used to explore, explain, and to model life's questions from a quantitative viewpoint. We will examine mathematical modeling throughout this book.

One of the primary uses of algebra is to manipulate mathematical expressions into a more desirable form. Three of the reasons that we rewrite an algebraic expression follow.

1. To put the expression in simplest form.
2. To simplify an equation in order to solve the equation.
3. To place a function in a form that reveals more information about the graph of the function.

This book will also focus on the relationship between algebra and geometry, including various graphs. Graphs are used widely in newspapers, magazines, and mathematics texts. Graphs can help us to understand data, to quickly note patterns and trends, or to visualize abstract concepts. Common graphs are histograms, bar graphs, pie charts, line graphs, the real number line, and the Cartesian plane.

To interpret or create graphs, we need to understand the terminology and concepts reviewed in this section. The number line is one of the most used graphs in mathematics. It provides a model of the real numbers and is used when students are first introduced to a concept. In arithmetic we use only constants such as -13, 0, $\frac{2}{3}$, $\sqrt{2}$, and π, which have a fixed value. Algebra uses not only constants but also variables such as a, b, P, W, x, y, and z. A **variable** is a letter that can be used to represent different numbers. Variables also are used to identify cells in a spreadsheet. The constants and variables that we use most often will represent real numbers.

What Is a Real Number?

Every number associated with a point on the real number line shown here is called a **real number.** The number of real numbers is infinite.

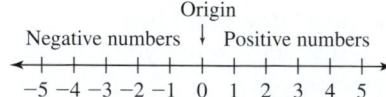

The real number line

Origin

Negative numbers ↓ Positive numbers

−5 −4 −3 −2 −1 0 1 2 3 4 5

The numbers to the right of 0 are called the **positive numbers,** and those to the left of 0 are called the **negative numbers.** The number 0 is neither positive nor negative. The set of numbers $\{1, 2, 3, 4, \ldots\}$ associated with the points marked off to the right of the origin is called the set of **natural numbers.** The three dots inside this set notation indicate that these numbers continue without end and thus the set is infinite. The set $\{0, 1, 2, 3, 4, \ldots\}$ is called the set of **whole numbers.**

A Mathematical Note

π is defined as the ratio of the circumference of a circle to its diameter. $\pi \approx 3.14159265$ ($\approx$ means "approximately equal to"). However, the decimal form of π does not terminate or repeat. In 1897, House Bill #246 was introduced in the Indiana legislature to make 3.2 the value of π in that state. Fortunately, better judgment prevailed, and this foolish bill was defeated.

1. Identify Additive Inverses

What Is the Relationship Between the Positive and Negative Real Numbers?

Real numbers the same distance from the origin, but on opposite sides of the origin, are called **opposites** of each other, or **additive inverses.** For example, -2 (negative two) and 2 are additive inverses of each other.

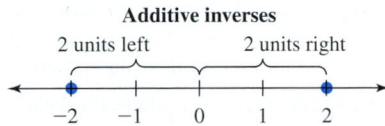

Additive inverses

The sum of a number and its additive inverse is zero. For example, $-2 + 2 = 0$ and $5 + (-5) = 0$. The additive inverse of 0 is 0. Zero is called the **additive identity** because 0 is the only real number with the property that $a + 0 = a$ and $0 + a = a$ for every real number a.

Definitions and the solution of problems will be presented by using multiple representations throughout this book. The definition of additive inverses is given algebraically, verbally, numerically, and graphically. Comparing these representations can help you increase your understanding of mathematics.

Opposites or Additive Inverses

Algebraically	Verbally	Numerical Example	Graphical Example
1. If a is a real number, the *opposite* of a is $-a$.	Except for zero, the additive inverse of a real number is formed by changing the sign of the number.	-5 is the opposite of 5 5 is the opposite of -5 0 is the opposite of 0	Opposites -5 0 5
2. $a + (-a) = 0$ $-a + a = 0$	The sum of a real number and its additive inverse is zero.	$5 + (-5) = 0$ $-5 + 5 = 0$ $0 + 0 = 0$	

The set $\{\ldots, -4, -3, -2, -1, 0, 1, 2, 3, 4, \ldots\}$ is called the set of **integers.** The set of integers contains all the whole numbers and their opposites, or additive inverses. As illustrated in the figure, the integers are spaced 1 unit apart on the number line.

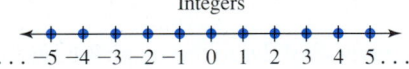

Integers
$\ldots -5 -4 -3 -2 -1 \ 0 \ 1 \ 2 \ 3 \ 4 \ 5 \ldots$

The opposite of any positive number is a negative number; for example, the opposite of 4 is -4. The opposite of any negative number is a positive number; for example, the opposite of -4 is 4. The opposite of x is $-x$. If x is -4, then $-x$ is 4. Thus $-(-4) = 4$ and, in general, $-(-x) = x$.

Double Negative Rule

Algebraically	Verbally	Numerical Example
For any real number a, $-(-a) = a$.	The opposite of the additive inverse of a is a.	$-(-5) = 5$

Example 1 Writing Additive Inverses

Write the additive inverse of each of the following real numbers.

Solution

Number	Additive Inverse	
(a) 7	-7	Except for zero, the additive inverse is formed by changing the sign of the number.
(b) -3	$-(-3) = 3$	The opposite of negative three is three.
(c) 0	0	Zero is its own additive inverse.
(d) $\dfrac{4}{17}$	$-\dfrac{4}{17}$	
(e) π	$-\pi$	The opposite of pi is negative pi.
(f) $3x$	$-3x$	The opposite of $3x$ is represented by $-3x$.

Self-Check 1

Write the additive inverse of -5, $\dfrac{7}{3}$, and x.

Graphing calculators have a special key $(-)$ to designate the opposite of a number. This key is distinct from the $-$ key, which is used to subtract one number from another. However, in spreadsheets the same minus symbol is used to indicate the opposite of a number and to indicate the subtraction of one number from another.

Graphing calculators and spreadsheets also have special keys or functions to represent some of the frequently used constants such as π. In Technology Perspective 1.2.1, the calculator screen and the spreadsheet show an excellent approximation of -2π, but neither approximation is the exact value of -2π. Note the difference between the calculator entry and the spreadsheet entry. The spreadsheet requires the multiplication symbol to be entered between -2 and pi but the calculator does not. Technology Perspectives will occur throughout the text and will integrate mathematical concepts with the appropriate use of technology. We will also point out some of the limitations of technology and how it can vary from calculators to spreadsheets.

Technology Perspective 1.2.1 Evaluating Additive Inverses

Approximate the additive inverse of 2π to the nearest thousandth.

TI-84 Plus Calculator

```
-2π
        -6.283185307
```

Excel Spreadsheet

A1		f_x	=-2*PI()
	A	B	C
1	-6.28319		
2			

1. Enter the expression by pressing the pi key as shown here:

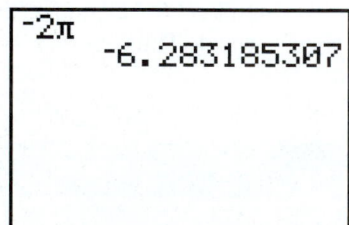

1. Enter "=−2 * PI( )" in cell **A1.**
2. The formula bar at the top of the spreadsheet displays this entry with the asterisk denoting multiplication.

2. Press ⬭ to evaluate this expression.

Answer: $-2\pi \approx -6.283$.

3. This entry produces a decimal approximation for -2π that is shown in cell **A1.**

Technology Self-Check 1

Approximate the additive inverse of 5π to the nearest hundredth.

Example 2 examines the sum of a number and its additive inverse and the sum of a number and 0.

| **Example 2** | Adding Additive Inverses and the Additive Identity |

Evaluate each of the following sums.

(a) $7 + (-7)$ **(b)** $-7 + 7$ **(c)** $-\pi + \pi$

(d) $3x + (-3x)$ **(e)** $19 + 0$

Solution

(a) $7 + (-7) = 0$ The sum of a number and its additive inverse is 0.

(b) $-7 + 7 = 0$

(c) $-\pi + \pi = 0$

(d) $3x + (-3x) = 0$

(e) $19 + 0 = 19$ Zero is the additive identity; $a + 0 = a$ for all real numbers a.

Self-Check 2

a. Evaluate $-45 + 0$.

b. Evaluate $-45 + 45$.

---

2. Evaluate Absolute Value Expressions

What Is the Absolute Value of a Number?

Additive inverses are the same distance from the origin, so these numbers are said to have the same magnitude, or absolute value. The **absolute value** of a real number x, denoted by $|x|$, is the distance on the number line between 0 and x. Distance is never negative, so $|x|$ is never negative.

Absolute Value

Algebraically	Verbally	Numerical Example	Graphical Example
$|x| = \begin{cases} x & \text{if } x \text{ is nonnegative} \\ -x & \text{if } x \text{ is negative} \end{cases}$	The absolute value of x is the distance between 0 and x on the number line.	$\begin{aligned} |2| &= 2 \\ |-2| &= 2 \end{aligned}$	2 units left 2 units right $\overset{}{\underset{-2\ \ -1\ \ \ 0\ \ \ 1\ \ \ 2}{\longleftrightarrow}}$

We use this definition of absolute value in Example 3 to evaluate four absolute value expressions.

Example 3   Evaluating Absolute Value Expressions

Evaluate each of the following absolute value expressions.

(a) $|26|$ **(b)** $|-4.29|$ **(c)** $-|5.8|$ **(d)** $|0|$

Solution

(a) $|26| = 26$ $|x| = x$ if x is nonnegative. 26 is 26 units to the right of 0 on the number line.

(b) $|-4.29| = 4.29$ $|x|$ is the opposite of x if x is negative. -4.29 is 4.29 units to the left of 0 on the number line.

(c) $-|5.8| = -5.8$ The additive inverse symbol is outside the absolute value symbols.

(d) $|0| = 0$ 0 is 0 units from 0 on the number line.

Self-Check 3

a. Evaluate $|-45|$.

b. Evaluate $|45|$.

The algebraic notation for the absolute value of x is $|x|$. Calculators and computers have their own notation for representing absolute value. The absolute value of x is represented by **abs( )** on a TI-84 Plus calculator. Note that the first option under **NUM** within the **MATH** menu is **abs(**. This option includes the left parenthesis. The expression within the absolute value is completed by entering the right parenthesis key. In an Excel spreadsheet, the absolute value of x is represented by **ABS( )**.

| Technology Perspective 1.2.2 | Evaluating Absolute Value Expressions |

Evaluate $|-4.29|$.

TI-84 Plus Calculator

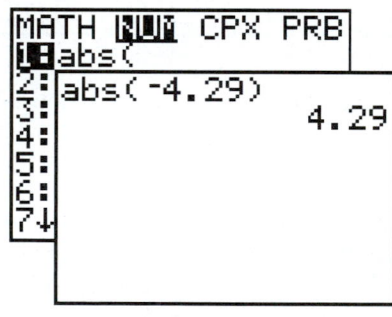

Excel Spreadsheet

1. Press [MATH] [▸] [ENTER] to access the absolute value function.
2. Enter -4.29.
3. Press [ )] [ENTER] to complete and evaluate this expression.

Answer: $|-4.29| = 4.29$.

1. Enter "=ABS (-4.29)" in cell **A1.**
2. The formula bar at the top of the spreadsheet displays this entry.
3. This entry produces the value of $|-4.29|$ that is shown in cell **A1.**

Technology Self-Check 2

Approximate $|-9.7\pi|$ to the nearest hundredth.

3. Use Inequality Symbols and Interval Notation

Can I Use the Number Line to Visualize Inequalities?

Yes, the number line also provides an excellent model for examining the inequalities *less than* and *greater than*. Because these inequalities describe the order of numbers on the number line, they are also known as the **order relations.** Table 1.2.1 illustrates all possible orders for arranging two numbers x and y on the number line.

Table 1.2.1   Equality and Inequality Symbols

Algebraic Notation	Verbal Meaning	Graphical Relationship on the Number Line
$x = y$	x equals y	x and y are the same point.
$x \approx y$	x is approximately equal to y	x and y are "close" but are not the same point.
$x \neq y$	x is not equal to y	x and y are different points.
$x < y$	x is less than y	Point x is to the left of point y.
$x \leq y$	x is less than or equal to y	Point x is on or to the left of point y.
$x > y$	x is greater than y	Point x is to the right of point y.
$x \geq y$	x is greater than or equal to y	Point x is on or to the right of point y.

The statements $x < y$ and $y > x$ are equivalent because they both specify that point x is to the left of point y.

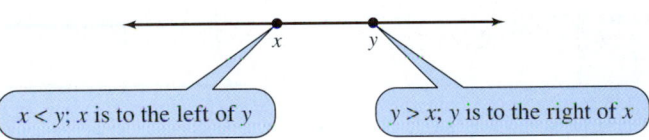

$x < y$; x is to the left of y $y > x$; y is to the right of x

---

Example 4   Showing Order Relationships on the Number Line

Plot each pair of numbers on a number line, and determine the order relationship between the numbers.

Solution

(a) -4 and -2 $-4 < -2$ -4 is to the left of -2; thus -4 is less than -2.

(b) π and 3 $\pi > 3$ $\pi \approx 3.14159$; thus π is to the right of 3 and is greater than 3.

(c) 0 and -3 $0 > -3$ 0 is to the right of -3.

Self-Check 4

Determine whether each statement is true (T) or false (F).

a. $9 \geq 9$ **b.** $9 > 9$ **c.** $-7 \leq -3$

d. $-5 \geq 0$ **e.** $4\frac{1}{3} > 4\frac{1}{2}$ **f.** $\frac{1}{3} > 0.3$

A Mathematical Note

The symbol ∞ was used to represent infinity by John Wallis in *Arithmetica Infinitorum* in 1655. The Romans had commonly used this symbol to represent 1,000. Likewise, we now use the word *myriad* to mean any large number, although the Greeks used it to mean 10,000.

The real number line extends infinitely to both the left and the right. The **infinity symbol** ∞ is not a specific real number; rather it signifies that the values continue through extremely large values without any end or bound. The symbol $-\infty$ indicates values unbounded to the left, and $+\infty$ (or just ∞) indicates values unbounded to the right. The interval notation $(3, \infty)$ is a compact notation to represent all real numbers greater than 3.

Inequality Notation	Verbal Meaning	Graph	Interval Notation
$x > 3$	x is greater than 3		$(3, \infty)$
$x \leq 5$	x is less than or equal to 5		$(-\infty, 5]$

Note in the interval notation defined in the following table that a parenthesis indicates that a value is not included in the interval and a bracket indicates that a value is included in the interval. An infinity symbol at either end of an interval is enclosed with parentheses since this denotes an unbounded interval that continues without any ending point.

Interval Notation

Inequality Notation	Verbal Meaning	Graph	Interval Notation
$x > a$	x is greater than a		(a, ∞)
$x \geq a$	x is greater than or equal to a		$[a, \infty)$
$x < a$	x is less than a		$(-\infty, a)$
$x \leq a$	x is less than or equal to a		$(-\infty, a]$
$a < x < b$	x is greater than a and less than b		(a, b)
$a < x \leq b$	x is greater than a and less than or equal to b		$(a, b]$
$a \leq x < b$	x is greater than or equal to a and less than b		$[a, b)$
$a \leq x \leq b$	x is greater than or equal to a and less than or equal to b		$[a, b]$
$-\infty < x < \infty$	x is any real number		$(-\infty, \infty)$

The interval $(-\infty, \infty)$, which represents the set of all real numbers, can also be represented by $\mathbb{R}$. In the interval $[a, b]$, a is called the left endpoint and b is called the right endpoint. The smaller value is always the left endpoint, and the larger value is always the right endpoint. For example, $[2, 5]$ is proper notation while $[5, 2]$ is improper notation because the larger value is listed first. The interval $(-1, 3)$ contains all points between -1 and 3.

Example 5	Writing Multiple Representations for Intervals

Express each interval in inequality notation, in verbal form, and graphically.

Solution

Interval Notation	Inequality Notation	Verbally	Graphically
(a) $[2, \infty)$	$x \geq 2$	x is greater than or equal to 2	
(b) $(-\infty, \pi)$	$x < \pi$	x is less than π	
(c) $(-2, 1]$	$-2 < x \leq 1$	x is greater than -2 and less than or equal to 1	
(d) $[-3, 2]$	$-3 \leq x \leq 2$	x is greater than or equal to -3 and less than or equal to 2	

Self-Check 5

Express each of these intervals by using interval notation.

a. All real numbers greater than or equal to -2

b. $x < 4$ **c.** $-5 < x \leq 3$ **d.**

4. Mentally Estimate Square Roots and Use a Calculator to Approximate Square Roots

Although calculators are used frequently to approximate square roots, it is important to be able to make rough mental estimates of square roots. Before we evaluate some square roots, let's examine the notation that is used.

What Does $\sqrt{x}$ Represent?

For $x \geq 0$, the **principal square root of x,** denoted by $\sqrt{x}$, is a nonnegative real number r so that $r^2 = x$. For example, $\sqrt{4} = 2$ since $2^2 = 4$. We use $-\sqrt{4}$ to represent -2, a negative number that is also a square root of 4. The key point is that $\sqrt{x}$ denotes only one of the square roots of x, and $-\sqrt{x}$ denotes the other.

   To estimate $\sqrt{5}$, note that $\sqrt{4} < \sqrt{5} < \sqrt{9}$. Thus $2 < \sqrt{5} < 3$. A calculator approximation yields $\sqrt{5} \approx 2.2361$, a value consistent with our mental estimation.

   The algebraic notation for the principal square root of x is $\sqrt{x}$. The bar over x denotes that the square root applies to the whole expression under the radical. The TI-84 Plus calculator denotes this by $\sqrt{(x)}$. The left parenthesis is included when the square root button is pressed. Then the expression under the square root is entered and completed by pressing the right parenthesis key. Note that the square root ⎷ is the secondary feature of the x² key.

   Excel uses **SQRT(x)** to denote the square root of x. Again parentheses are used to designate that the square root function applies to the expression x.

Technology Perspective 1.2.3 Approximating Square Roots

Approximate $\sqrt{47}$ to the nearest thousandth.

TI-84 Plus Calculator

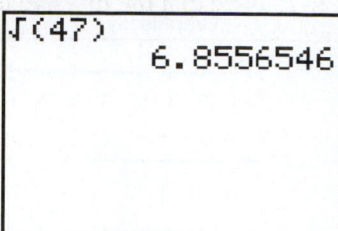

Excel Spreadsheet

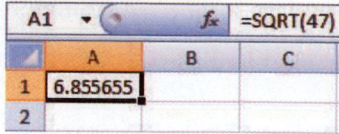

1. Press [2nd] [√] to access the square root function.
2. Enter 47.
3. Press [)] [ENTER] to complete and evaluate this expression.

1. Enter "=SQRT(47)" in cell **A1.**
2. The formula bar at the top of the spreadsheet displays this entry.
3. This entry produces a decimal approximation of $\sqrt{47}$ that is shown in cell **A1.**

Answer: $\sqrt{47} \approx 6.856$.

Technology Self-Check 3

Approximate $\sqrt{35.89}$ to the nearest tenth.

Example 6   Estimate Then Calculate

Estimate $\sqrt{15.4}$ to the nearest integer, and then use a calculator to approximate value to the nearest thousandth. Is the calculator value a reasonable answer?

Solution

Estimated Value

$$15.4 \approx 16$$
$$\sqrt{15.4} \approx \sqrt{16} = 4$$

The calculator approximation seems reasonable because $\sqrt{15.4} \approx 4$.

Answer: $\sqrt{15.4} \approx 3.924$

Calculator Approximation

√(15.4)
    3.924283374

Self-Check 6

a. Mentally estimate $\sqrt{99.7}$ to the nearest integer.
b. Use a calculator to approximate $\sqrt{99.7}$ to the nearest hundredth.

If you pick up a mathematics or science textbook from 1970 or earlier, you will likely find several tables in the back of the book. One of the most common tables is a square root table because these values were needed frequently. Advances in technology have made the inclusion of these tables unnecessary as we can quickly use a calculator or computer to determine square roots. In Section 2.2, we will illustrate how to use technology to create these tables.

5. Identify Natural Numbers, Whole Numbers, Integers, Rational Numbers, and Irrational Numbers

What Is the Distinction Between the Rational and Irrational Numbers?

The real numbers are classified as either rational or irrational. A real number that can be written as a ratio of two integers is called a **rational number.** (Notice that the first five letters of the word *rational* spell *ratio.*) The other real numbers, which cannot be written as the ratio of two integers, are called **irrational numbers.**

Rational and Irrational Numbers			
Algebraically	Numerically	Numerical Examples	Verbal Examples
Rational: A real number x is rational if $x = \dfrac{a}{b}$ for integers a and b, with $b \neq 0$.	In decimal form, a rational number is either a terminating decimal or an infinite repeating decimal.	$\dfrac{1}{2} = 0.5$ $\dfrac{1}{3} = 0.333\ldots = 0.\overline{3}$ $\dfrac{5}{33} = 0.151515\ldots = 0.\overline{15}$	$\dfrac{1}{2}$ in decimal form is a terminating decimal. $\dfrac{1}{3}$ in decimal form is a repeating decimal. $\dfrac{5}{33}$ in decimal form is a repeating decimal.
Irrational: A real number x is irrational if it cannot be written as $x = \dfrac{a}{b}$ for integers a and b.	In decimal form, an irrational number is an infinite nonrepeating decimal.	$\sqrt{2} \approx 1.414214$ $\pi \approx 3.141593$ $0.1010010001\ldots$	$\sqrt{2}$ cannot be written as a rational fraction—it is an infinite nonrepeating decimal. π cannot be written as a rational fraction—it is an infinite nonrepeating decimal. This irrational number does exhibit a pattern, but it does not terminate and it does not repeat.

All integers are also rational numbers, since every integer can be written as the ratio of itself to 1. For example, -3, 0, and 7 can be written as $\dfrac{-3}{1}, \dfrac{0}{1}$, and $\dfrac{7}{1}$. Another form that can be interpreted as a ratio is percent notation. For example, 17%, read as "17 percent," means 17 parts out of 100; $17\% = \dfrac{17}{100}$. Other examples of rational numbers are $\dfrac{1}{7}, \dfrac{-2}{13}$, and $\dfrac{479}{-22}$. The numbers $4\dfrac{2}{5}, 0.17$, and $0.333\ldots$ are also rational since each of these numbers can be written as a ratio of two integers: $4\dfrac{2}{5} = \dfrac{22}{5}, 0.17 = \dfrac{17}{100}$, and $0.333\ldots = \dfrac{1}{3}$. The repeating decimal form for $\dfrac{1}{3}$ can be written as either $0.\overline{3}$ or as $0.333\ldots$. Numbers such as π, $\sqrt{2}, \sqrt{3}$, and $-\sqrt{5}$ are irrational since they cannot be written as the ratio of two integers.

A number is not automatically irrational just because it has a square root symbol. For example, $\sqrt{4}$ is another notation for the rational number 2. All numbers with rational numbers as their square roots are called **perfect squares**. The integers 1, 4, 9, 16, and 25 are perfect squares since they have as their square roots 1, 2, 3, 4, and 5, respectively. The number $\frac{4}{81}$ is also a perfect square since $\sqrt{\frac{4}{81}} = \frac{2}{9}$. However, $\sqrt{2}$ is irrational, since it can be shown that 2 is not a perfect square of any rational number.

Example 7   Identifying Rational and Irrational Numbers

Classify each of the following real numbers as either rational or irrational.

Solution

(a) $1\frac{3}{5}$ Rational $1\frac{3}{5}$ can be written as the fraction $\frac{8}{5}$.

(b) 1.35   Rational   1.35 is a terminating decimal.

(c) 0.666…   Rational   All repeating decimals are rational numbers. This number can also be written as $\frac{2}{3}$.

(d) 0.101101110…   Irrational Although there is a pattern (the number of 1s is increasing between the 0s), this is not a repeating decimal.

(e) $\sqrt{\dfrac{36}{49}}$ Rational $\sqrt{\frac{36}{49}} = \frac{6}{7}$ since $\left(\frac{6}{7}\right)^2 = \frac{36}{49}$.

(f) $\sqrt{0}$ Rational $\sqrt{0} = 0$ since $0^2 = 0$.

(g) $\sqrt{5}$ Irrational 5 is not the square of any rational number.

(h) $\sqrt{0.25}$ Rational $\sqrt{0.25} = 0.5$ since $0.5^2 = 0.25$.

Self-Check 7

a. Classify $\sqrt{9}$ as either a rational number or an irrational number.

b. Classify $\sqrt{10}$ as either a rational number or an irrational number.

The relationship of the important subsets of the real numbers that have been defined in this section is summarized by the accompanying figure and side comments.

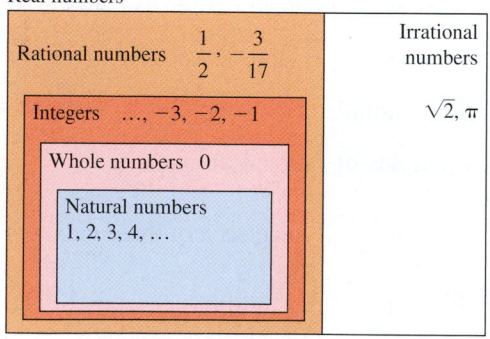

Note that this figure shows that:

The natural numbers consist of 1, 2, 3, . . .

The whole numbers consist of 0, 1, 2, 3, . . .

The integers consist of . . ., −3, −2, −1, 0, 1, 2, 3, . . .

All natural numbers are whole numbers.

All whole numbers are integers.

All integers are rational numbers.

No real number can be both rational and irrational.

Example 8 Classifying Real Numbers

List each real number from the set $\left\{-5, -3.2, -\sqrt{3}, 0, \frac{2}{3}, 2, \pi, 4\frac{1}{3}, 6\right\}$ that is a(n):

Solution

(a) Natural number	2, 6	The natural numbers consist of 1, 2, 3, . . .
(b) Whole number	0, 2, 6	The whole numbers consist of 0, 1, 2, . . .
(c) Integer	$-5, 0, 2, 6$	The integers consist of . . . , $-2, -1$, 0, 1, 2, . . .
(d) Rational number	$-5, -3.2, 0, \frac{2}{3}, 2, 4\frac{1}{3}, 6$	The rational numbers consist of the integers, fractions, and terminating or repeating decimals.
(e) Irrational number	$-\sqrt{3}, \pi$	Irrational numbers consist of the real numbers that are not rational.

Self-Check 8

List each real number from the set $\left\{-4, -\sqrt{2}, -1, -\frac{3}{7}, 0, 4, 5\frac{3}{8}, 7\right\}$ that is a(n):

a. Natural number   **b.** Whole number   **c.** Integer

d. Rational number   **e.** Irrational number

Self-Check Answers

1. $5, -\frac{7}{3}, -x$

2. **a.** -45 **b.** 0

3. **a.** 45   **b.** 45

4. **a.** T   **b.** F   **c.** T   **d.** F   **e.** F   **f.** T

5. **a.** $[-2, \infty)$ **b.** $(-\infty, 4)$ **c.** $(-5, 3]$ **d.** $[-\pi, 1)$

6. **a.** 10   **b.** 9.98

7. **a.** rational   **b.** irrational

8. **a.** 4, 7   **b.** 0, 4, 7   **c.** $-4, -1, 0, 4, 7$
   d. $-4, -1, -\frac{3}{7}, 0, 4, 5\frac{3}{8}, 7$ **e.** $-\sqrt{2}$

Technology Self-Check Answers

1. -15.71 2. 30.47   3. 6.0

1.2	Using the Language and Symbolism of Mathematics

1. A _____ is a letter that can be used to represent different numbers.

2. The notation $-a$ is read "the additive inverse of a" or "_____ of a."

3. The additive identity is _____.

4. The additive inverse of $-a$ is _____.

5. If one number is seven units to the right of the origin, its additive inverse is _____ units to the _____ of the origin.

6. The notation $|x|$ is read "the _____ value of x." $|x|$ represents the _____ between 0 and x on the number line.

7. The notation $x \geq y$ is read "x is _____ than or _____ to y."

8. The notation $x < y$ is read "x is _____ than y." On the number line, x is to the _____ of y.

9. The notation $x \neq y$ is read "x is not _____ to y."

10. The notation $x \approx y$ is read "x is _____ _____ to y."

11. The notation $\sqrt{x}$ is read "the principal _____ _____ of x."

12. A real number that is a terminating decimal is a _____ number.

13. A real number that is a repeating decimal is a _____ number.

14. A real number that is an infinite nonrepeating decimal is an _____ number.

15. The symbol π is the symbol for _____.

16. The symbol ∞ is the symbol for _____.

17. In interval notation, a parenthesis indicates that a value **is/is not** included in the interval. (Select the correct choice.)

18. In interval notation, a bracket indicates that a value **is/is not** included in the interval. (Select the correct choice.)

19. The set of all real numbers can be represented by the interval notation $(-\infty, \infty)$ or by _____.

20. The notation $[2, -3]$ is incorrect notation because the larger value is listed first. The correct notation for $-3 \le x \le 2$ is _____.

21. The interval notation to represent $a < x \le b$ is _____.

1.2 | Quick Review

1. The digit in the hundreds place of 9,876.54321 is _____.

2. The digit in the tens place of 9,876.54321 is _____.

3. The digit in the tenths place of 9,876.54321 is _____.

4. The digit in the hundredths place of 9,876.54321 is _____.

5. Round 9,876.54321 to the nearest hundred.

1.2 | Exercises

Objective 1 Identify Additive Inverses

In Exercises 1 and 2, write the additive inverse of each number.

1. a. 9   **b.** -13 **c.** $-\dfrac{3}{5}$

    d. 7.23   **e.** $-\sqrt{13}$ **f.** 0

2. a. 8   **b.** -30 **c.** $\dfrac{5}{7}$

    d. -8.94 **e.** $-\pi$ **f.** 0.03

3. Simplify each expression to either 7 or -7.
    a. $-(-7)$ **b.** $-[-(-7)]$ **c.** $0 + 7$

4. Simplify each expression to either 9 or -9.
    a. $-(-9)$ **b.** $-[-(-9)]$ **c.** $-9 + 0$

In Exercises 5–8, simplify each expression.

5. a. $17 + (-17)$ **b.** $-17 + 17$
    c. $-\pi + \pi$ **d.** $x + (-x)$

6. a. $2.3 + (-2.3)$ **b.** $-2.3 + 2.3$
    c. $e + (-e)$ **d.** $-y + y$

7. a. $-6x + 6x$ **b.** $6x + 0$
    c. $6x + (-6x)$ **d.** $-6x + 0$

8. a. $2y + (-2y)$ **b.** $0 + 2y$
    c. $-2y + 2y$ **d.** $0 + (-2y)$

Objective 2 Evaluate Absolute Value Expressions

In Exercises 9 and 10, evaluate each expression.

9. a. $|29|$ **b.** $|-29|$
    c. $-|29|$ **d.** $-|-29|$

10. a. $|37|$ **b.** $|-37|$
    c. $-|37|$ **d.** $-|-37|$

Objective 3 Use Inequality Symbols and Interval Notation

In Exercises 11–16, insert $<$, $=$, or $>$ in the blank to make each statement true.

11. a. 83____38 **b.** -83____-38

12. a. -5.4____-8.5 **b.** 5.4____8.5

13. a. 0____$\dfrac{1}{3}$ **b.** 0____$-\dfrac{1}{3}$

14. a. $\dfrac{1}{2}$____0 **b.** $-\dfrac{1}{2}$____0

15. a. $|-5|$____$-|5|$ **b.** $|-5|$____$-|-5|$

16. a. $-|7|$____$-|-7|$ **b.** $-|-7|$____$|7|$

Multiple Representations

In Exercises 17–24, use the given information to complete each row as illustrated by the example in the first row.

	Inequality Symbols	Verbally	Graphically	Interval Notation
Example:	$x > 2$	x is greater than 2.		$(2, \infty)$
17.	$x > 1$			
18.	$1 \le x \le 5$			
19.		x is greater than or equal to -3.		
20.		x is greater than or equal to 0 and less than 2.		
21.				
22.				
23.				$(-\infty, 4]$
24.				$(-3, 5)$

Objective 4 Mentally Estimate Square Roots and Use a Calculator to Approximate Square Roots

In Exercises 25 and 26, mentally evaluate each expression.

25. a. $\sqrt{25}$ **b.** $-\sqrt{25}$
    c. $\sqrt{0.25}$ **d.** $\sqrt{2{,}500}$

26. a. $\sqrt{36}$ **b.** $-\sqrt{36}$
    c. $\sqrt{0.36}$ **d.** $\sqrt{3{,}600}$

	Integer Estimate	Inequality	Approximation
Example: $\sqrt{5}$	2	$2 < \sqrt{5}$	2.236
31. $\sqrt{50.6}$			
32. $\sqrt{8.92}$			
33. $\sqrt{0.976}$			
34. $\sqrt{1.23}$			

Estimation Skills

In Exercises 27–30, mentally estimate the value of each square root and then select the choice that is closest to your estimate.

27. $\sqrt{17}$ **A.** 3.88   **B.** 4.12
          C. 4.92   **D.** 9.24

28. $\sqrt{24}$ **A.** 4.90   **B.** 5.10
          C. 12.01   **D.** 12.99

29. $\sqrt{62.41}$ **A.** 6.7   **B.** 7.1
          C. 7.9   **D.** 8.9

30. $\sqrt{123.21}$ **A.** 61.6   **B.** 58.4
          C. 12.3   **D.** 11.1

Estimate then Calculate

In Exercises 31–34, complete the following table by:

a. estimating each square root to the nearest integer;

b. determining whether this integer estimate is less than or greater than the actual value;

c. using a calculator or a spreadsheet to approximate each expression to the nearest thousandth. (*Hint:* See Technology Perspective 1.2.3 on Approximating Square Roots.)

Objective 5 Identify Natural Numbers, Whole Numbers, Integers, Rational Numbers, and Irrational Numbers

In Exercises 35–38, list each real number from the given set that is a(n):

a. Natural number   **b.** Whole number
c. Integer   **d.** Rational number
e. Irrational number

35. $\left\{ -11, -4.8, -\sqrt{9}, 0, 1\frac{3}{5}, \sqrt{5}, 15 \right\}$

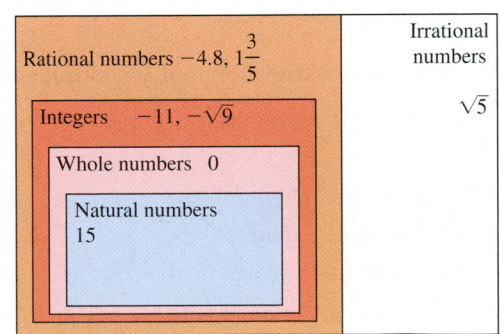

36. $\left\{-9.3, -5, 0, \sqrt{\dfrac{25}{36}}, 1, 2, \sqrt{7}, 5\dfrac{1}{9}\right\}$

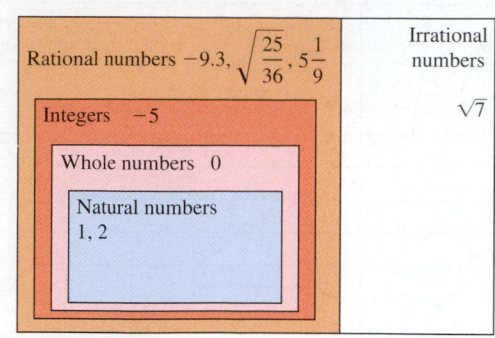

37. $\left\{-\sqrt{5}, -\sqrt{4}, -\sqrt{\dfrac{9}{25}}, \sqrt{0}, \sqrt{1}, \sqrt{6}, \sqrt{16}\right\}$

38. $\left\{-\sqrt{9}, -\sqrt{8}, -\sqrt{\dfrac{4}{9}}, \sqrt{0}, \sqrt{2}, \sqrt{4}, \sqrt{36}\right\}$

In Exercises 39–46, list every set to which the given real number belongs. The choices are the natural numbers, whole numbers, integers, rational numbers, and irrational numbers.

39. -18 **40.** $-\dfrac{1}{8}$ **41.** 81 **42.** $\sqrt{81}$

43. $\sqrt{8}$ **44.** 8 **45.** $5\dfrac{3}{7}$ **46.** 0

47. Plot these numbers on a real number line:

$$-5, -3.5, 0, 1\dfrac{3}{4}, 4$$

48. Plot these numbers on a real number line:

$$-4, -2, -\dfrac{1}{2}, 0, 1, 2\dfrac{1}{2}, 5$$

In Exercises 49–52, identify each real number as either rational or irrational.

49. a. $\sqrt{9}$ **b.** $\sqrt{10}$
50. a. $\sqrt{20}$ **b.** $\sqrt{25}$
51. a. $0.171717\ldots$ **b.** $0.171771777\ldots$
52. a. $0.232232223\ldots$ **b.** $0.133133133\ldots$

Multiple Representations and Concept Development

In Exercises 53–60, write each verbal statement in algebraic form.

53. Twenty percent is equal to one-fifth.

54. Seventy-five percent is equal to 0.75.

55. The absolute value of x is equal to y.

56. The opposite of x is less than or equal to negative two.

57. The square root of x is equal to y.

58. The principal square root of 16 equals 4.

59. π is greater than 3.14 and less than 3.15.

60. The square root of 2 is greater than one and less than two.

61. A variety of notations can be used to represent the same real number. For example, 0.25, $\dfrac{1}{4}$, and 25% all represent the same real number.
 a. Plot 0.25 on this number line:

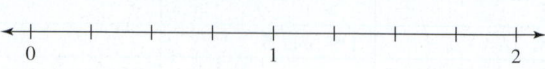

 b. Shade $\dfrac{1}{4}$ of this region:

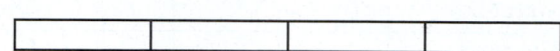

 c. Shade 25% of this region:

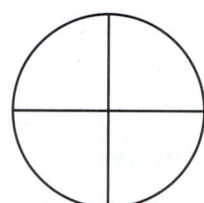

62. Complete the table to represent each of these rational numbers in fractional form, in decimal form, and as a percent.

	Fraction	Decimal	Percent
Example:	$\dfrac{1}{2}$	0.50	50%
a.	$\dfrac{1}{4}$		
b.		0.10	
c.			5%

63. What value of x makes the statement $x < 7$ false but makes the statement $x \le 7$ true?

64. What value of x makes the statement $x > 9$ false but makes the statement $x \ge 9$ true?

65. What value of x makes both $x \le 3$ and $x \ge 3$ true statements?

66. What value of x makes both $x \le -5$ and $x \ge -5$ true statements?

67. a. What whole number is not a natural number?
 b. List two integers that are not natural numbers.
 c. List two negative rational numbers that are not integers.
 d. List two rational numbers between 1 and 2.

68. a. What number is its own additive inverse?
 b. List two positive rational numbers that are not integers.
 c. List two real numbers that are not rational numbers.
 d. List two rational numbers between 0 and 1.

Applying Technology

In Exercises 69 and 70, use a calculator or a spreadsheet to complete each table. If the square root is not an integer, round this value to the nearest hundredth.

69.

x	$\sqrt{x}$
2	
3	
4	
5	

70.

x	$\sqrt{x}$
50	
75	
100	
125	

Group discussion questions

Concept Development

71. a. Give an example of a number x so that $\sqrt{x} < x$.
  b. Give an example of a number x so that $\sqrt{x} = x$.
  c. Give an example of a number x so that $\sqrt{x} > x$.
  d. Compare your results with others in your group for parts **a, b,** and **c,** and discuss any general statements that can be made for each of these three cases.

72. a. Use the accompanying figure and appropriate shading to demonstrate to a classmate that $\dfrac{1}{2} = \dfrac{3}{6}$.
  b. Use the figure and appropriate shading to demonstrate to a classmate that $\dfrac{1}{2} + \dfrac{1}{3} = \dfrac{5}{6}$.

Communicating Mathematically

73. There are two square roots of 19. Write an algebraic representation for each of these square roots.

74. a. Using 3.14 as an approximation for π, one student reports the area of this circle as 12.56 cm². Is this student's answer a rational number or an irrational number? Give a verbal justification for your answer.

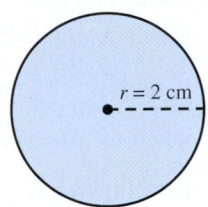

  b. Another student reports the area of this circle as 4π cm². Is this student's answer a rational number or an irrational number? Give a verbal justification for your answer.

75. a. Describe in your own words how to evaluate the absolute value of a real number.
  b. Is there any real number whose absolute value is not a positive number?

76. Write a paragraph describing the relationship of the natural numbers, whole numbers, integers, rationals, irrationals, and real numbers.

Section 1.3   Operations with Positive Fractions

Objectives:

1. Reduce fractions to lowest terms.
2. Multiply and divide fractions.
3. Add and subtract fractions with the same denominator.
4. Add and subtract fractions with different denominators.
5. Perform operations with mixed numbers.

The real numbers used in everyday applications can appear as either decimals, fractions, or percents. We use decimals to a much greater extent than we did 50 years ago. For example, the NASDAQ stock index switched to decimal pricing of stocks in April 2001. However, in many applications, fractions are still preferred.

Why Are Fractions Preferred over Decimals in Some Applications?

When we want to represent a part of a whole, a fraction can do this exactly. On the other hand, some fractions can only be approximated by decimals with a finite number of decimal places.

In this figure, the fraction $\dfrac{1}{3}$ indicates that 1 out of 3 equal parts is shaded. This is a

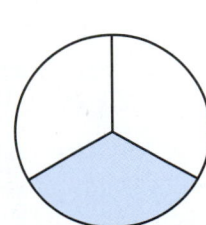

number that cannot be represented exactly by 0.33 or 0.333 or by any finite number of decimal digits. A background in fractions is also important before working with rational expressions in Chapter 9.

Numerator

$\dfrac{1}{3}$

Denominator

In general, a quotient of two numbers written in the form $\dfrac{a}{b}$ with $b \neq 0$ is called a **fraction.** The top number, a, is called the **numerator,** and the bottom number, b, is called the **denominator.** In the fraction $\dfrac{1}{3}$ the numerator is 1 and the denominator is 3.

1. Reduce Fractions to Lowest Terms

Why Do We Reduce Fractions to Lowest Terms?

The quick answer to this question is for simplicity and consistency. It is easier to represent $\dfrac{2}{4}, \dfrac{3}{6}, \dfrac{4}{8}$, and $\dfrac{222}{444}$ by the equivalent fraction $\dfrac{1}{2}$. It is also easier to communicate with each other when we use a consistent representation for numbers.

The following figures illustrate that $\dfrac{2}{4}, \dfrac{3}{6}$, and $\dfrac{4}{8}$ are each equal to $\dfrac{1}{2}$. In each case, $\dfrac{1}{2}$ of the figure is shaded.

$\frac{1}{2}$ of the figure is shaded

$\frac{2}{4}$ of the figure is shaded

$\frac{3}{6}$ of the figure is shaded

$\frac{4}{8}$ of the figure is shaded

Each of the fractions $\dfrac{2}{4}, \dfrac{3}{6}$, and $\dfrac{4}{8}$ can be reduced to lowest terms as the fraction $\dfrac{1}{2}$. A number that is an exact divisor of both the numerator and the denominator of a fraction is called a **common factor.** Every fraction can be reduced to lowest terms by dividing both the numerator and denominator by their common factors. A positive fraction is in **lowest terms** if the numerator and the denominator are positive and have no common factor greater than 1.

Example 1   Reducing Fractions to Lowest Terms

Reduce each fraction to lowest terms.

(a) $\dfrac{12}{15}$ (b) $\dfrac{12}{23}$

Solution

(a) $\dfrac{12}{15} = \dfrac{\overset{4}{\cancel{12}}}{\underset{5}{\cancel{15}}}$ Divide both the numerator and the denominator by 3. Since 4 and 5 have no common factor greater than 1, $\dfrac{4}{5}$ is in lowest terms.

$\quad = \dfrac{4}{5}$

(b) $\dfrac{12}{23}$ is in lowest terms   Since 12 and 23 have no common factor greater than 1, $\dfrac{12}{23}$ is in lowest terms.

Self-Check 1

Reduce each fraction to lowest terms.

a. $\dfrac{6}{9}$ **b.** $\dfrac{12}{30}$

Can I Just "Cancel" by Drawing a Line Through Numbers in the Numerator and the Denominator?

We suggest you avoid the term *cancel*. As an alternative, remind yourself that to reduce a fraction you must *divide* the whole numerator and denominator by the same number. "Canceling" CANNOT be used with sums or differences. We do not use the term *canceling* in this textbook because it can easily be misinterpreted. Instead, we describe the justification for each step that we use and then we may draw lines through the numerator and denominator to illustrate this step.

If you have trouble recognizing a common factor of the numerator and the denominator of a fraction, then you may wish to start by writing the prime factors of the numerator and the denominator. Recall that a **prime number** is a natural number that has exactly two positive factors, 1 and the number itself. Examples of prime numbers include 2, 3, 5, 7, 11, 13, and 17.

Example 2   Reducing a Fraction to Lowest Terms

Reduce $\dfrac{60}{84}$ to lowest terms.

Solution

$$\frac{60}{84} = \frac{\overset{1}{\cancel{2}} \cdot \overset{1}{\cancel{2}} \cdot \overset{1}{\cancel{3}} \cdot 5}{\underset{1}{\cancel{2}} \cdot \underset{1}{\cancel{2}} \cdot \underset{1}{\cancel{3}} \cdot 7}$$

The factorization of each number into prime factors can be done in stages if you wish. For example, $60 = 2 \cdot 30 = 2 \cdot 2 \cdot 15 = 2 \cdot 2 \cdot 3 \cdot 5$ and $84 = 2 \cdot 42 = 2 \cdot 2 \cdot 21 = 2 \cdot 2 \cdot 3 \cdot 7$.

$$= \frac{5}{7}$$

Self-Check 2

Reduce $\dfrac{30}{75}$ to lowest terms.

Note that, in the steps used to reduce the fraction in Example 2, $\dfrac{2}{2} = 1$ and $\dfrac{3}{3} = 1$. In general, the fraction $\dfrac{a}{a} = 1$ for $a \neq 0$. To convert fractions to a common denominator we will frequently use a form like $\dfrac{2}{2}$ or $\dfrac{3}{3}$ to represent 1.

Although it is important to understand how to reduce fractions (this skill is also used heavily when we examine rational expressions), there are situations for which it is appropriate to use a calculator or a spreadsheet to do this. Most calculator and spreadsheets have a limited range of values for the numerator and denominator and will not be able to simplify fractions with values outside this range. In Technology Perspective 1.3.1, we illustrate how to reduce the fraction given in Example 2.

For a TI-84 Plus calculator use the **Frac** option in the **MATH** menu to reduce a fraction to lowest terms. In an Excel spreadsheet, a similar result can be obtained by using the **Fraction** option in the **Number** menu to format a cell.

Technology Perspective 1.3.1	Reducing a Fraction to Lowest Terms

Reduce $\dfrac{60}{84}$ to lowest terms.

TI-84 Plus Calculator   **Excel Spreadsheet**

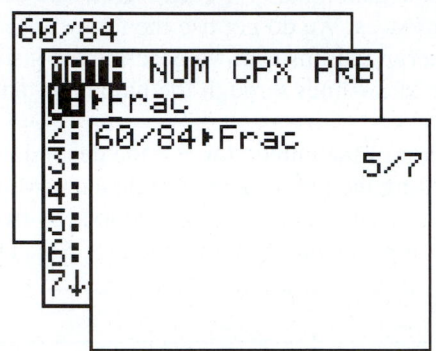

1. Enter the fraction using the division symbol for the fraction bar:

    [6] [0] [÷] [8] [4]

2. Press [MATH] [ENTER] to access the fraction option.

3. Press [ENTER] to execute this option.

1. Format cell **B1** using the **Fraction** option in the **Number** menu.

2. To reduce $\dfrac{60}{84}$ enter "=60/84" in cell **B1**. The fraction formatting will cause the result to be displayed as a reduced fraction.

3. Entering "60/84" in cell **A1** without the "=" will cause this fraction to display as text and not as a formula to be calculated.

Answer: $\dfrac{60}{84} = \dfrac{5}{7}$

Technology Self-Check 1

Reduce each fraction to lowest terms.

a. $\dfrac{34}{153}$ b. $\dfrac{360}{315}$ c. $\dfrac{678}{904}$

---

2. Multiply and Divide Fractions

How Do I Multiply and Divide Fractions?

The procedure is described in the following box. Before presenting this box, we shall examine some of the terminology used. Dividing a number by 3 is equivalent to multiplying the number by $\dfrac{1}{3}$. For example, $24 \div 3 = 8$ and $(24)\left(\dfrac{1}{3}\right) = 8$. The number $\dfrac{1}{3}$ is called the **reciprocal** or the **multiplicative inverse** of 3. In general, the multiplicative inverse of $\dfrac{a}{b}$ is $\dfrac{b}{a}$ for $a \neq 0$ and $b \neq 0$.

Multiplication and Division of Fractions

Verbally	Algebraically	Numerical Example
To multiply two fractions, multiply the numerators and multiply the denominators.	$\dfrac{a}{b} \cdot \dfrac{c}{d} = \dfrac{ac}{bd}$ for $b \neq 0$ and $d \neq 0$.	$\dfrac{2}{7} \cdot \dfrac{3}{5} = \dfrac{2 \cdot 3}{7 \cdot 5} = \dfrac{6}{35}$
To divide two fractions, multiply the first fraction by the reciprocal of the second fraction.	$\dfrac{a}{b} \div \dfrac{c}{d} = \dfrac{a}{b} \cdot \dfrac{d}{c} = \dfrac{ad}{bc}$ for $b \neq 0, c \neq 0,$ and $d \neq 0$.	$\dfrac{3}{4} \div \dfrac{4}{5} = \dfrac{3}{4} \cdot \dfrac{5}{4} = \dfrac{15}{16}$

The two fractions that are multiplied in the box have no common factors for the numerator and the denominator, so the product could not be reduced either before or after the product was formed. If the numerator and the denominator have common factors greater than 1, then it is easier to reduce before multiplying the numerators and the denominators, as illustrated in Example 3.

Example 3   Multiplying Fractions

Determine the following products, expressing each product in lowest terms:

(a) $2 \cdot \dfrac{5}{7}$ (b) $\dfrac{8}{21} \cdot \dfrac{7}{13} \cdot \dfrac{5}{12}$

Solution

(a) $2 \cdot \dfrac{5}{7} = \dfrac{2}{1} \cdot \dfrac{5}{7}$

$= \dfrac{2 \cdot 5}{1 \cdot 7}$

$= \dfrac{10}{7}$

Multiply the numerators and multiply the denominators.

Since 10 and 7 have no common factor greater than 1, $\dfrac{10}{7}$ is in lowest terms. It is okay to have the numerator larger than the denominator in a reduced fraction.

(b) $\dfrac{8}{21} \cdot \dfrac{7}{13} \cdot \dfrac{5}{12} = \dfrac{\overset{2}{\cancel{8}}}{\underset{3}{\cancel{21}}} \cdot \dfrac{\overset{1}{\cancel{7}}}{13} \cdot \dfrac{5}{\underset{3}{\cancel{12}}}$

$= \dfrac{2 \cdot 1 \cdot 5}{3 \cdot 13 \cdot 3}$

$= \dfrac{10}{117}$

Divide 8 and 12 by 4, and divide 7 and 21 by 7. Then multiply the factors in the numerators and multiply the factors in the denominators.

Self-Check 3

Determine the following products, expressing each product in lowest terms.

a. $18 \cdot \dfrac{12}{30}$ b. $\dfrac{8}{21} \cdot \dfrac{49}{13} \cdot \dfrac{26}{24}$

In the proper context, the word *of* often indicates the operation of multiplication. In Example 4, the phrase "three-fourths of the freshmen" is interpreted to mean $\dfrac{3}{4}$ times the number of freshmen.

Example 4 Using Multiplication to Calculate the Number of Parking Permits

The Office of Safety and Security estimates that three-fourths of the freshmen enrolling at Clayton University will apply for an automobile parking permit. If 560 freshmen enroll, estimate how many parking permits will be required.

PARKING PERMIT
403

Solution

$$\frac{3}{4} \cdot 560 = \frac{3}{\overset{1}{\cancel{4}}} \cdot \frac{\overset{140}{\cancel{560}}}{1}$$ $\dfrac{3}{4}$ of the 560 freshmen indicates the product $\dfrac{3}{4} \cdot 560$. Divide both 4 and 560 by 4. Then multiply the factors in the numerators and multiply the factors in the denominators.

$$= \frac{420}{1}$$

$$= 420$$

Answer: Approximately 420 permits will be required by the freshmen.

Self-Check 4

A recipe for a drink calls for $\dfrac{5}{8}$ of a quart of tomato juice. A quart contains 32 ounces. How many ounces of tomato juice are required for this recipe?

The key phrase to remember when you divide fractions is "invert the divisor and multiply." This is illustrated in Example 5. When we invert or form the reciprocal of $\dfrac{3}{4}$ we obtain its multiplicative inverse, $\dfrac{4}{3}$.

Example 5 Dividing Fractions

Determine the following quotients, expressing each quotient in lowest terms:

(a) $\dfrac{2}{3} \div \dfrac{3}{4}$ **(b)** $\dfrac{15}{28} \div \dfrac{9}{14}$

Solution

(a) $\dfrac{2}{3} \div \dfrac{3}{4} = \dfrac{2}{3} \cdot \dfrac{4}{3}$ Invert the divisor $\dfrac{3}{4}$ and then multiply by $\dfrac{4}{3}$.

$$= \frac{2 \cdot 4}{3 \cdot 3}$$

$$= \frac{8}{9}$$

(b) $\dfrac{15}{28} \div \dfrac{9}{14} = \dfrac{\overset{5}{\cancel{15}}}{\underset{2}{\cancel{28}}} \cdot \dfrac{\overset{1}{\cancel{14}}}{\underset{3}{\cancel{9}}}$ Invert the divisor $\dfrac{9}{14}$ and then multiply by $\dfrac{14}{9}$. Divide both 15 and 9 by 3.

Then divide both 14 and 28 by 14.

$= \dfrac{5 \cdot 1}{2 \cdot 3}$

$= \dfrac{5}{6}$

Self-Check 5

Determine the following quotients, expressing each quotient in lowest terms:

a. $\dfrac{5}{6} \div \dfrac{2}{3}$ **b.** $\dfrac{15}{14} \div \dfrac{35}{16}$

One concept involving division is the splitting of an item into parts. If a word problem asks how many parts result when a whole is split into equal parts, then this problem may be indicating that division is the appropriate operation. In Example 6, a bolt of material is split into many separate parts to form a number of uniforms.

Example 6   Using Division to Calculate the Number of Uniforms

A bolt of material contains 100 yards of material, which will be used to make uniforms for the children in a school play. How many uniforms can be made from the bolt if each uniform requires $\dfrac{5}{8}$ yard?

Solution

$100 \div \dfrac{5}{8} = \dfrac{\overset{20}{\cancel{100}}}{1} \cdot \dfrac{8}{\underset{1}{\cancel{5}}}$ The bolt of material is being divided into many pieces, each requiring $\dfrac{5}{8}$ yard.

To divide by $\dfrac{5}{8}$, invert and then multiply by $\dfrac{8}{5}$. Then divide both the

$= \dfrac{20 \cdot 8}{1}$ numerator and the denominator by 5.

$= 160$

Answer: The bolt of material can be used to make 160 uniforms.

Self-Check 6

If a patient is to consume 18 ounces of liquid by taking $\dfrac{2}{3}$ ounce at a time, how many times will the patient take the liquid?

---

3. Add and Subtract Fractions with the Same Denominator

How Do I Add and Subtract Fractions with the Same Denominator?

A fraction can be interpreted as counting a number of parts out of a whole. If we add two fractions with the same denominator, this can be interpreted as adding the number of parts from the first fraction to the number of parts from the second fraction. This idea is illustrated in the following figure, where each whole has been divided into seven equal parts.

Adding the 2 parts shaded in the first figure to the 4 parts shaded in the second figure, we have a total of 6 of the 7 parts shaded. Thus $\dfrac{2}{7} + \dfrac{4}{7} = \dfrac{6}{7}$.

$\dfrac{2}{7}$ of the figure is shaded.

$\dfrac{4}{7}$ of the figure is shaded.

A total of $\dfrac{6}{7}$ of the figure is shaded.

The procedure for adding and subtracting fractions is summarized in the box for your reference. Remember that the result should be expressed in lowest terms.

Addition and Subtraction of Fractions

Verbally	Algebraically	Numerical Example
To add fractions with the same denominator, add the numerators and use the common denominator.	$\dfrac{a}{b} + \dfrac{c}{b} = \dfrac{a+c}{b}$ for $b \neq 0$	$\dfrac{1}{5} + \dfrac{3}{5} = \dfrac{1+3}{5} = \dfrac{4}{5}$
To subtract fractions with the same denominator, subtract the numerators and use the common denominator.	$\dfrac{a}{b} - \dfrac{c}{b} = \dfrac{a-c}{b}$ for $b \neq 0$	$\dfrac{5}{7} - \dfrac{3}{7} = \dfrac{5-3}{7} = \dfrac{2}{7}$

Example 7   Adding and Subtracting Fractions with the Same Denominator

Perform the indicated operations and reduce each result to lowest terms:

(a) $\dfrac{7}{21} + \dfrac{5}{21}$ (b) $\dfrac{7}{21} - \dfrac{5}{21}$

Solution

(a) $\dfrac{7}{21} + \dfrac{5}{21} = \dfrac{7+5}{21}$ First add the numerators and use the common denominator of 21.

$= \dfrac{\overset{4}{\cancel{12}}}{\underset{7}{\cancel{21}}}$ Then reduce the fraction by dividing both the numerator and the denominator by 3.

$= \dfrac{4}{7}$

(b) $\dfrac{7}{21} - \dfrac{5}{21} = \dfrac{7-5}{21}$ First subtract the numerators and use the common denominator of 21.

$= \dfrac{2}{21}$ This fraction is in lowest terms because 2 and 21 do not have any common factors greater than 1.

Self-Check 7

Perform the indicated operations and reduce each result to lowest terms:

a. $\dfrac{9}{16} + \dfrac{5}{16}$ b. $\dfrac{9}{16} - \dfrac{5}{16}$

4. Add and Subtract Fractions with Different Denominators

How Do I Add and Subtract Fractions with Different Denominators?

The first step is to express the fractions in terms of a common denominator.

We now illustrate the underlying logic of this with two figures of the same size, each separated into 20 equal parts (20 is the common denominator for $\frac{1}{4}$ and $\frac{2}{5}$). We represent both $\frac{1}{4}$ and $\frac{2}{5}$ by shading a specific number of parts of each figure. Then we can add the number of parts shaded for one fraction to the number of parts shaded for the other fraction. The first figure has 1 of 4 rows shaded, and the second figure has 2 of 5 columns shaded. Counting the smaller squares in each figure, we note that $\frac{1}{4} = \frac{5}{20}$ and $\frac{2}{5} = \frac{8}{20}$.

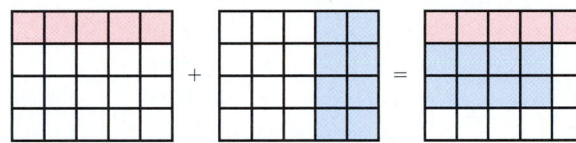

Thus $\frac{1}{4} + \frac{2}{5} = \frac{5}{20} + \frac{8}{20} = \frac{13}{20}$.

How Do I Express Two Fractions in Terms of a Common Denominator?

We can rewrite fractions with the same denominator by multiplying a given fraction by 1 in the form $\frac{a}{a}$ for $a \neq 0$. This does not change the value of the fraction, but it does express the fraction in a form with the denominator that we want. This procedure is illustrated in Example 8.

Example 8   Rewriting Fractions with a Common Denominator

Rewrite **(a)** $\frac{1}{4}$ and **(b)** $\frac{2}{5}$ so that they share a common denominator of 20.

Solution

(a) $\frac{1}{4} = \frac{1}{4} \cdot \frac{5}{5}$ Multiplying by 1 does not change the value of the fraction. However, multiplying by 1 in the form $\frac{5}{5}$ does produce an equivalent fraction with a denominator of 20.

$= \frac{5}{20}$ (Note that $20 \div 4 = 5$.)

(b) $\frac{2}{5} = \frac{2}{5} \cdot \frac{4}{4}$ Multiplying by 1 does not change the value of the fraction. However, multiplying by 1 in the form $\frac{4}{4}$ does produce an equivalent fraction with a denominator of 20.

$= \frac{8}{20}$ (Note that $20 \div 5 = 4$.)

Self-Check 8

Rewrite each of these fractions so they share a common denominator of 16.

a. $\frac{3}{4}$ **b.** $\frac{5}{8}$

To add or subtract fractions, it is generally most convenient to express the fractions in terms of their least common denominator. The **least common denominator (LCD)** for a set of fractions is the smallest number that is a multiple of each of the given denominators. Although you will likely be able to determine the LCD for many fractions by inspection, we illustrate a procedure involving prime factorizations in Example 9.

Example 9   Subtracting Fractions with Different Denominators

Subtract $\dfrac{4}{15}$ from $\dfrac{5}{6}$.

Solution

$$\frac{5}{6} - \frac{4}{15} = \frac{5}{6} \cdot \frac{5}{5} - \frac{4}{15} \cdot \frac{2}{2}$$

$$= \frac{25}{30} - \frac{8}{30}$$

$$= \frac{25 - 8}{30}$$

$$= \frac{17}{30}$$

Although you may be able to determine the LCD of 30 by inspection, the prime factorization procedure is shown here for your reference.

$$6 = 2 \cdot 3$$
$$15 = 3 \cdot 5$$
$$\textbf{LCD} = 2 \cdot 3 \cdot 5 = 30$$

Convert each fraction to an equivalent fraction with a denominator of 30 and then subtract.

Self-Check 9

Subtract $\dfrac{7}{10}$ from $\dfrac{5}{6}$.

In Example 9, we could have used many different common denominators for $\dfrac{4}{15}$ and $\dfrac{5}{6}$, including the common denominator of $6 \cdot 15 = 90$. The disadvantage of using a common denominator other than the LCD is that the work will involve larger denominators and an extra step will be needed at the end when the difference must be reduced to lowest terms.

The procedure for adding two fractions can be extended to adding several fractions. Example 10 illustrates the addition of three fractions.

Example 10   Adding Fractions with Different Denominators

Simplify $\dfrac{1}{4} + \dfrac{3}{40} + \dfrac{7}{50}$.

Solution

$$\frac{1}{4} + \frac{3}{40} + \frac{7}{50} = \frac{1}{4} \cdot \frac{50}{50} + \frac{3}{40} \cdot \frac{5}{5} + \frac{7}{50} \cdot \frac{4}{4}$$

$$= \frac{50 + 15 + 28}{200}$$

$$= \frac{93}{200}$$

First determine the LCD.

$$4 = 2 \cdot 2$$
$$40 = 2 \cdot 2 \cdot 2 \cdot 5$$
$$50 = 2 \cdot 5 \cdot 5$$
$$\textbf{LCD} = 2 \cdot 2 \cdot 2 \cdot 5 \cdot 5 = 200$$

Convert each fraction to an equivalent fraction with a denominator of 200 and then add.

Self-Check 10

Simplify $\dfrac{1}{6} + \dfrac{4}{9} + \dfrac{1}{4}$.

Should I Avoid Improper Fractions?

No! A fraction such as $\dfrac{1}{2}$ or $\dfrac{7}{10}$ that has a numerator less than the denominator is called a **proper fraction.** A fraction such as $\dfrac{2}{2}$ or $\dfrac{13}{9}$ that has a numerator equal to or greater than the denominator is called an **improper fraction.** There is nothing wrong with an improper fraction; this is just a term to designate that the fraction represents a number greater than or equal to 1 whole unit. Usually we rewrite $\dfrac{2}{2}$ as 1, but we often leave $\dfrac{13}{9}$ as an improper fraction rather than rewriting it as the mixed number $1\dfrac{4}{9}$. One reason for this is that it is easier to multiply or divide mixed numbers if they are first written as improper fractions. A **mixed number** such as $1\dfrac{4}{9}$ represents the sum of a whole number and a proper fraction. For example, $1\dfrac{4}{9} = 1 + \dfrac{4}{9}$. To enter a mixed number like $1\dfrac{4}{9}$ into a calculator or a spreadsheet, enter $1 + \dfrac{4}{9}$.

5. Perform Operations with Mixed Numbers

In Example 11, we subtract two mixed numbers and review the step of borrowing that is sometimes necessary.

| **Example 11** | Using Subtraction to Determine Fuel Usage |

A Cessna 340 private airplane left Sugarland, Texas, with $50\dfrac{1}{2}$ gallons of fuel. When it returned later, the gauge showed that $31\dfrac{7}{8}$ gallons remained in the tank. How many gallons of fuel were used?

Solution

$$50\dfrac{1}{2} - 31\dfrac{7}{8} = 50\dfrac{4}{8} - 31\dfrac{7}{8}$$

$$= 49\dfrac{12}{8} - 31\dfrac{7}{8}$$

$$= 18\dfrac{5}{8}$$

To find the amount of fuel used, we subtract the fuel remaining from the original amount. The common denominator for the two fractions is 8. Since $\dfrac{7}{8}$ is greater than $\dfrac{4}{8}$, borrow 1 from 50 in the form of $\dfrac{8}{8}$. Use this to rewrite $50\dfrac{4}{8}$ as $49\dfrac{12}{8}$. Note that $49 - 31 = 18$ and $\dfrac{12}{8} - \dfrac{7}{8} = \dfrac{5}{8}$.

Answer: The airplane used $18\dfrac{5}{8}$ gallons of fuel.

Self-Check 11

A plumber needs two lengths of three-quarter-inch copper pipe. The length of the first piece is $5\dfrac{1}{4}$ ft and the length of the second piece is $11\dfrac{1}{2}$ ft. What is the total length of the two pieces?

Example 12 shows how to multiply two mixed numbers using pencil and paper and then how to use a calculator to check the answer. Note that when we use pencil and paper, we first convert the mixed numbers to improper fractions. This is not necessary when we use a calculator. Note however, the importance of the parentheses when the two numbers are entered into the calculator.

Example 12 Multiplying Mixed Numbers

Multiply $1\frac{9}{16}$ by $1\frac{7}{25}$ using pencil and paper, and then check the result using a calculator.

Solution

$$\left(1\frac{9}{16}\right)\left(1\frac{7}{25}\right) = \left(1+\frac{9}{16}\right)\left(1+\frac{7}{25}\right)$$

$$= \left(\frac{16}{16} + \frac{9}{16}\right)\left(\frac{25}{25} + \frac{7}{25}\right)$$

$$= \left(\frac{\overset{1}{\cancel{25}}}{\underset{1}{\cancel{16}}}\right)\left(\frac{\overset{2}{\cancel{32}}}{\underset{1}{\cancel{25}}}\right)$$

$$= 2$$

Calculator Check

```
(1+9/16)(1+7/25)
                2
```

Answer: $\left(1\frac{9}{16}\right)\left(1\frac{7}{25}\right) = 2$

Self-Check 12

Multiply $1\frac{4}{11}$ by $7\frac{1}{3}$.

Self-Check Answers

1. **a.** $\frac{2}{3}$ **b.** $\frac{2}{5}$

2. $\frac{2}{5}$

3. **a.** $\frac{36}{5}$ **b.** $\frac{14}{9}$

4. The recipe requires 20 ounces of tomato juice.

5. **a.** $\frac{5}{4}$ **b.** $\frac{24}{49}$

6. The patient will take the liquid 27 times.

7. **a.** $\frac{7}{8}$ **b.** $\frac{1}{4}$

8. **a.** $\frac{12}{16}$ **b.** $\frac{10}{16}$

9. $\frac{2}{15}$

10. $\frac{31}{36}$

11. The total length of the two pieces of pipe is $16\frac{3}{4}$ feet.

12. 10

Technology Self-Check Answers

1. **a.** $\frac{2}{9}$ **b.** $\frac{8}{7}$ **c.** $\frac{3}{4}$

1.3 Using the Language and Symbolism of Mathematics

1. In the fraction $\frac{a}{b}$ with $b \neq 0$, the numerator is _____.

2. In the fraction $\frac{a}{b}$ with $b \neq 0$, the denominator is _____.

3. A fraction can be interpreted as counting a number of _____ out of a whole. For example, $\frac{2}{5}$ can be interpreted as 2 of 5 equal _____.

4. A number that is an exact divisor of both the numerator and the denominator of a fraction is called a common _____.

5. A positive fraction is in lowest terms if the numerator and the denominator are positive and have no common factor greater than _____.

6. A _____ number is a natural number such as 2, 3, 5, 7, 11, 13, and 17 that has exactly two factors, 1 and the number itself.

7. The number $\frac{3}{7}$ is the multiplicative inverse or the _____ of $\frac{7}{3}$.

8. To divide $\frac{4}{15}$ by $\frac{6}{11}$, we multiply $\frac{4}{15}$ by _____.

9. To add or subtract two fractions, the first step is to express the fractions in terms of a common _____.

10. LCD stands for _____ common denominator.

11. A fraction that has a numerator less than the denominator is called a _____ fraction.

12. A fraction that has a numerator equal to or greater than the denominator is called an _____ fraction.

13. A _____ number represents the sum of a whole number and a proper fraction.

14. The fraction $\frac{7}{7}$ can also be written as the whole number _____.

15. If you multiply a number by _____, you do not change the value of the number.

1.3 Quick Review

1. The next prime number after 17 is _____.

The prime factorization of 15 is $3 \cdot 5$. Write the prime factorization of each of these natural numbers.

2. 35   3. 36   4. 45   5. 88

1.3 Exercises

Objective 1 Reduce Fractions to Lowest Terms

In Exercises 1–6, write a fraction in lowest terms to represent the shaded portion of each figure.

1.

2.

3.

4.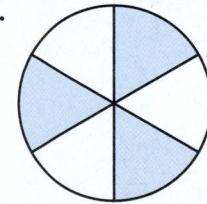

5.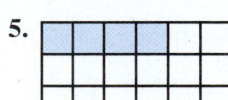

6.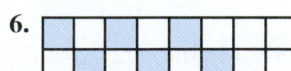

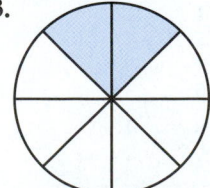

In Exercises 7–12, reduce each fraction to lowest terms.

7. $\dfrac{10}{15}$ **8.** $\dfrac{14}{22}$

9. $\dfrac{35}{7}$ **10.** $\dfrac{21}{7}$

11. $\dfrac{11}{99}$ **12.** $\dfrac{99}{121}$

Objective 2 Multiply and Divide Fractions

In Exercises 13–16, write the multiplicative inverse (reciprocal) of each number.

13. a. $\dfrac{5}{11}$ **b.** $\dfrac{11}{5}$

14. a. $\dfrac{9}{2}$ **b.** $\dfrac{2}{9}$

15. a. 6 **b.** $\dfrac{1}{6}$

16. a. $\dfrac{1}{9}$ **b.** 9

In Exercises 17–40, perform the indicated multiplications and divisions and write the answer in lowest terms.

17. $\dfrac{3}{8} \cdot \dfrac{5}{7}$ **18.** $\dfrac{4}{7} \cdot \dfrac{6}{11}$

19. $\dfrac{3}{8} \cdot \dfrac{4}{7}$ **20.** $\dfrac{14}{35} \cdot \dfrac{5}{7}$

21. $\dfrac{18}{15} \cdot \dfrac{4}{27}$ **22.** $\dfrac{10}{49} \cdot \dfrac{14}{25}$

23. $6 \cdot \dfrac{3}{2}$ **24.** $20 \cdot \dfrac{4}{5}$

25. $\dfrac{2}{5} \cdot \dfrac{2}{5} \cdot \dfrac{2}{5}$ **26.** $\dfrac{3}{10} \cdot \dfrac{3}{10} \cdot \dfrac{3}{10}$

27. $\dfrac{2}{15} \cdot \dfrac{5}{12} \cdot \dfrac{36}{11}$ **28.** $\dfrac{5}{8} \cdot \dfrac{14}{15} \cdot \dfrac{10}{7}$

29. $\dfrac{3}{8} \div \dfrac{5}{7}$ **30.** $\dfrac{2}{3} \div \dfrac{4}{5}$

31. $\dfrac{3}{8} \div 6$ **32.** $15 \div \dfrac{5}{3}$

33. $\dfrac{16}{27} \div \dfrac{18}{24}$ **34.** $\dfrac{16}{18} \div \dfrac{24}{22}$

35. Determine $\dfrac{3}{5}$ of 360.   **36.** Determine $\dfrac{5}{8}$ of 400.

37. Divide 42 by $\dfrac{3}{7}$.   **38.** Divide 144 by $\dfrac{2}{3}$.

39. Multiply $\dfrac{10}{21}$ by 35.   **40.** Multiply $\dfrac{15}{27}$ by 25.

Objective 3 Add and Subtract Fractions with the Same Denominator

In Exercises 41–50, perform the indicated additions and subtractions and express the result in lowest terms.

41. $\dfrac{2}{7} + \dfrac{3}{7}$ **42.** $\dfrac{4}{13} + \dfrac{6}{13}$

43. $\dfrac{5}{8} - \dfrac{1}{8}$ **44.** $\dfrac{13}{15} - \dfrac{8}{15}$

45. $\dfrac{21}{48} + \dfrac{19}{48}$ **46.** $\dfrac{8}{30} + \dfrac{25}{30}$

47. $\dfrac{91}{126} - \dfrac{27}{126}$ **48.** $\dfrac{25}{30} - \dfrac{9}{30}$

49. $\dfrac{5}{36} + \dfrac{11}{36} + \dfrac{17}{36}$ **50.** $\dfrac{7}{48} + \dfrac{13}{48} + \dfrac{25}{48}$

Objective 4 Add and Subtract Fractions with Different Denominators

In Exercises 51–56, write each fraction as an equivalent fraction with the given denominator.

51. $\dfrac{5}{11} = \dfrac{?}{33}$ **52.** $\dfrac{7}{8} = \dfrac{?}{24}$

53. $\dfrac{4}{15} = \dfrac{?}{60}$ **54.** $\dfrac{11}{21} = \dfrac{?}{147}$

55. $\dfrac{9}{26} = \dfrac{?}{130}$ **56.** $\dfrac{41}{45} = \dfrac{?}{135}$

In Exercises 57–66, perform the indicated additions and subtractions and express the result in lowest terms.

57. $\dfrac{1}{4} + \dfrac{3}{8}$ **58.** $\dfrac{1}{3} + \dfrac{1}{27}$

59. $\dfrac{5}{12} + \dfrac{7}{20}$ **60.** $\dfrac{7}{10} + \dfrac{13}{15}$

61. $\dfrac{5}{7} - \dfrac{3}{14}$ **62.** $\dfrac{3}{5} - \dfrac{4}{15}$

63. $\dfrac{13}{18} - \dfrac{3}{14}$ **64.** $\dfrac{5}{26} - \dfrac{4}{39}$

65. $\dfrac{5}{6} + \dfrac{3}{10} + \dfrac{7}{15}$ **66.** $\dfrac{5}{8} + \dfrac{5}{12} + \dfrac{5}{18}$

Objective 5 Perform Operations with Mixed Numbers

In Exercises 67–76, perform the indicated operations and express the result as a mixed number in lowest terms.

67. $4\dfrac{1}{2} + 2\dfrac{1}{3}$ **68.** $4\dfrac{1}{2} - 2\dfrac{1}{3}$

69. $7\dfrac{2}{3} - 5\dfrac{3}{4}$ **70.** $4\dfrac{5}{8} + 7\dfrac{3}{4}$

71. $5\dfrac{1}{2} + 2\dfrac{2}{3} + 7\dfrac{5}{6}$ **72.** $6\dfrac{3}{4} + 8\dfrac{2}{3} + 5\dfrac{5}{6}$

73. $7\frac{1}{2} \times 1\frac{2}{3}$

74. $7\frac{1}{2} \div 1\frac{2}{3}$

75. $1\frac{7}{8} \div 1\frac{5}{7}$

76. $1\frac{7}{8} \times 1\frac{5}{7}$

Connecting Concepts to Applications

77. Student Absences An elementary school had 360 of 960 students absent because of a respiratory illness. Write in reduced form the fraction that gives the portion of students who were absent.

78. Batting Record An outfielder had 68 hits in his first 272 official at bats. Write in reduced form the fraction that gives the portion of his at bats that resulted in hits.

79. Pounds of Muscle Muscles normally account for approximately $\frac{2}{5}$ of a person's body weight. Approximately how many pounds of muscle does a 155-lb person have?

80. Height of a Stack of Books A bookstore stacked biology lab manuals in stacks of 24 each. Each lab manual is $\frac{3}{4}$ in thick. What is the height of each stack?

81. Fertilizer Usage A woman has 12 cups of rose fertilizer. The fertilizer box recommends $\frac{1}{4}$ cup for each rose bush. How many rose bushes can she fertilize with the 12 cups?

82. Slices of Bacon A butcher is slicing a slab of bacon that is 6 inches thick into slices that are each $\frac{3}{16}$ in thick. How many slices will this slab yield?

83. Total Sunflower Seeds A health food store has three partially filled barrels of sunflower seeds. One barrel is $\frac{3}{8}$ full, another $\frac{1}{3}$ full, and the third is $\frac{1}{4}$ full. If all these seeds are placed on one barrel, what fractional part of the barrel would be filled?

84. Monthly Withholdings Approximately $\frac{1}{5}$ of an employee's monthly check is withheld for federal taxes and $\frac{3}{40}$ for Social Security. Another $\frac{1}{20}$ is withheld for state taxes and $\frac{3}{40}$ for a retirement fund. What fractional part of the salary is withheld?

Where your salary goes

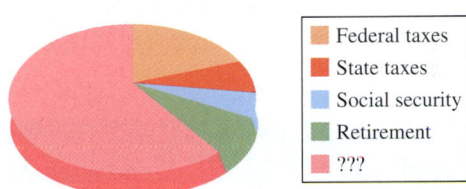

■	Federal taxes
■	State taxes
■	Social security
■	Retirement
■	???

85. Portion of a Whole The shaded portion of the circle represents $\frac{9}{40}$ of the circle. What fraction represents the portion of the circle that is not shaded?

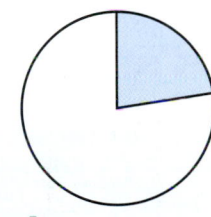

86. Sugar in a Recipe A recipe calls for $\frac{5}{8}$ cup of sugar. If the cook has already put $\frac{1}{3}$ cup of sugar in the mix, how much more sugar is needed?

87. Pieces of Pipe A welder has a piece of pipe $19\frac{1}{4}$ ft long that needs to be cut into pieces each of length $1\frac{3}{4}$ ft. How many pieces can be cut from this pipe? (Assume no waste.)

Group discussion questions

88. Fraction Concepts

    a. Equivalent Fractions Use the figure and appropriate shading to demonstrate to a classmate that $\dfrac{4}{6} = \dfrac{2}{3}$.

    b. Adding Fractions with Different Denominators Use the figure and appropriate shading to demonstrate to a classmate that $\dfrac{2}{3} + \dfrac{1}{5} = \dfrac{13}{15}$.

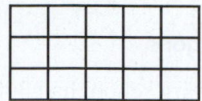

89. Challenge Question Simplify

$$\frac{11}{12} \cdot \frac{12}{13} \cdot \frac{13}{14} \cdot \frac{14}{15} \cdot \frac{15}{16} \cdot \frac{16}{17} \cdot \frac{17}{18} \cdot \frac{18}{19} \cdot \frac{19}{20} \cdot \frac{20}{21} \cdot \frac{21}{22}.$$

1.3   Cumulative Review

1. The additive inverse of $-\dfrac{5}{8}$ is _____.

2. List all integers between –5 and 5.

3. List all natural numbers between –5 and 5.

4. Express the inequality $-2 < x \le 3$ using interval notation.

5. Determine the only natural number n that makes the inequality $n < \sqrt{7} < n + 1$ a true statement.

Section 1.4   Addition and Subtraction of Real Numbers

Objectives:

    1. Add positive and negative real numbers.

    2. Use the commutative and associative properties of addition.

    3. Subtract positive and negative real numbers.

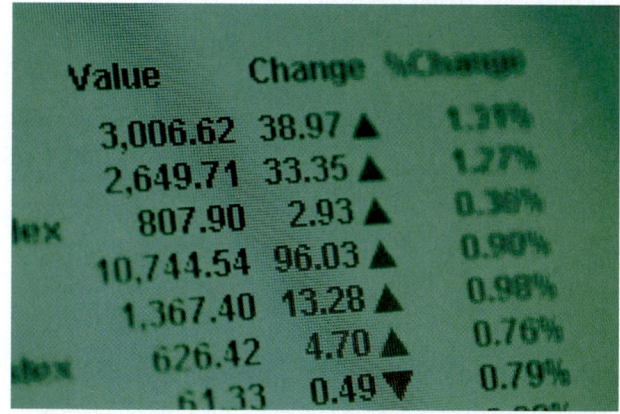

Positive and negative numbers both occur in many real-life applications: the price changes of a stock, the profit or loss of a business, the temperature on a cold day, the elevation of an object, the net yardage on a football play, and the directed distance as shown on a number line.

1. Add Positive and Negative Real Numbers

In a business situation, a positive number represents a profit and a negative number represents a loss. We will now examine combining profits and losses using addition. First, recall the terminology for the addition $5 + 8 = 13$; the numbers 5 and 8 are called **terms** or **addends,** and the result 13 is called the **sum.**

How Do I Add Positive and Negative Real Numbers?

Table 1.4.1 illustrates all possible cases of adding positive and negative real numbers. The word *total* indicates that the values from both branches have been added. Note that profits and losses tend to offset each other. If the profit is greater than the loss, then the net result is a profit. If the loss is greater than the profit, then the net result is a loss.

A Mathematical Note

The symbols + and −, used to indicate addition and subtraction, were introduced by Johannes Widman, a Bohemian mathematician, in the 15th century. The words *plus* and *minus* are from the Latin for more and less, respectively.

Table 1.4.1 Branch Results for H&M Productions

	Quarter 1	Quarter 2	Quarter 3	Quarter 4
Arizona Branch	+\$35,000	+\$35,000	−\$35,000	−\$35,000
Illinois Branch	+\$10,000	−\$10,000	+\$10,000	−\$10,000
Total	+\$45,000	+\$25,000	−\$25,000	−\$45,000

In Table 1.4.1, the sum for the first quarter is +\$45,000. This is usually just written as \$45,000 and is understood to represent a positive forty-five thousand dollars. Note that we use the "+" symbol for two distinct meanings. One meaning indicates that a number is positive, as in +3. A second meaning is to indicate the addition of two numbers, as in 3 + 5.

Is There a Way for Me to Visualize the Addition of Positive and Negative Numbers?

Yes, the addition of real numbers can be represented by combining directed distances on the number line. View each term as a trip in a specific direction and addition as the combining of these individual trips. This is illustrated in the following box, which gives the rules for adding signed numbers.

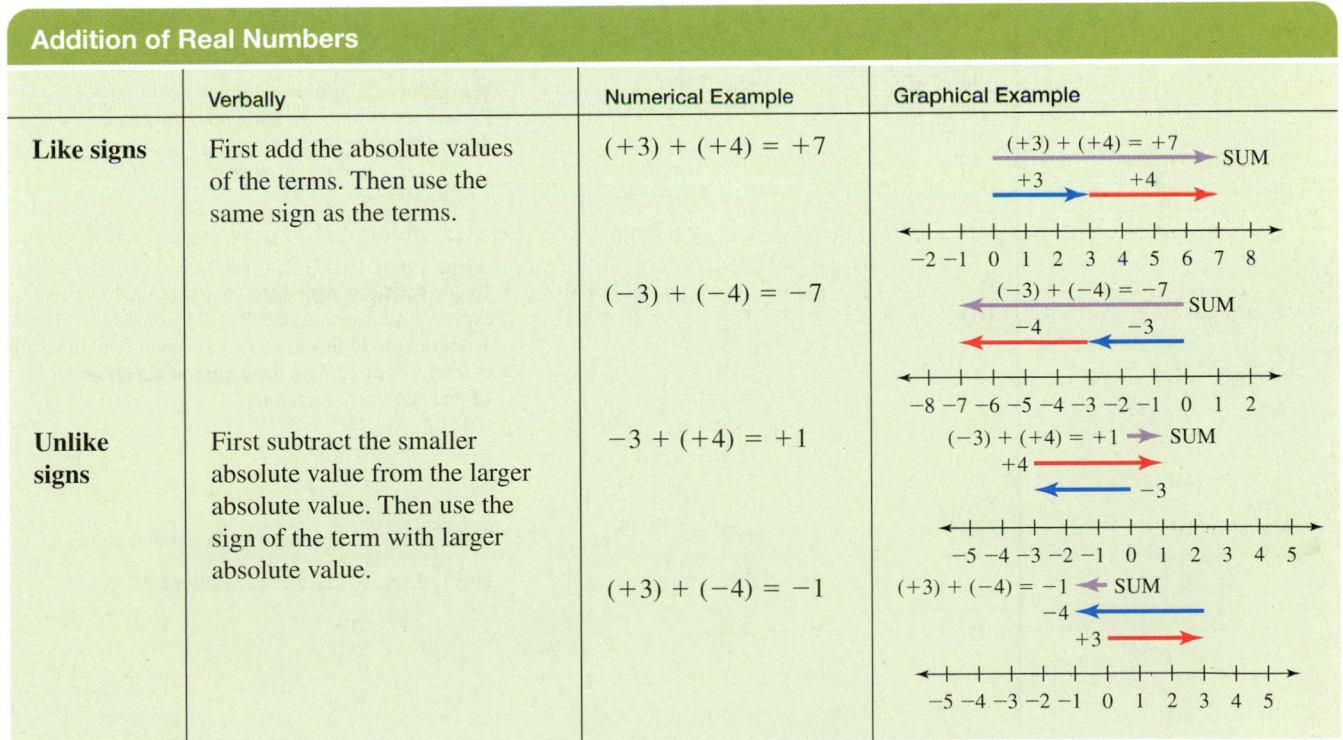

Addition of Real Numbers			
	Verbally	Numerical Example	Graphical Example
Like signs	First add the absolute values of the terms. Then use the same sign as the terms.	$(+3) + (+4) = +7$ $(-3) + (-4) = -7$	
Unlike signs	First subtract the smaller absolute value from the larger absolute value. Then use the sign of the term with larger absolute value.	$-3 + (+4) = +1$ $(+3) + (-4) = -1$	

Example 1 illustrates the addition of terms with both like signs and unlike signs.

Example 1 | Adding Terms with Like and Unlike Signs

Calculate the following sums:

(a) $17 + 23$ (b) $-17 + (-23)$ (c) $-17 + 23$ (d) $17 + (-23)$

Solution

(a) $17 + 23 = 40$ Both terms and the sum are positive.

(b) $-17 + (-23) = -40$ Both terms and the sum are negative.

(c) $-17 + 23 = 6$ The terms have unlike signs. The sum is positive like the term with the larger absolute value, 23.

(d) $17 + (-23) = -6$ The terms have unlike signs. The sum is negative like the term with the larger absolute value, -23.

Self-Check 1

Calculate each sum.

a. $22 + 55$ b. $-22 + (-55)$ c. $-22 + 55$ d. $22 + (-55)$

Example 2 | Adding Fractions with Like and Unlike Denominators

Calculate the following sums:

(a) $-\dfrac{5}{12} + \dfrac{7}{12}$ (b) $-\dfrac{1}{4} + \left(-\dfrac{1}{3}\right)$

Solution

(a) $-\dfrac{5}{12} + \dfrac{7}{12} = \dfrac{-5 + 7}{12}$ To add fractions with the same denominator, add the numerators and use the common denominator. Reduce the sum by dividing both the numerator and the denominator by 2.

$$= \dfrac{2}{12}$$

$$= \dfrac{1}{6}$$

(b) $-\dfrac{1}{4} + \left(-\dfrac{1}{3}\right) = -\dfrac{1}{4} \cdot \dfrac{3}{3} + \left(-\dfrac{1}{3}\right) \cdot \dfrac{4}{4}$ To add fractions with different denominators, first express each fraction in terms of a common denominator. In this case, express each term in terms of the LCD of 12. Use the common denominator of 12 and add the numerators.

$$= -\dfrac{3}{12} + \left(-\dfrac{4}{12}\right)$$

$$= \dfrac{-3 + (-4)}{12}$$

$$= \dfrac{-7}{12}$$ Both $\dfrac{-7}{12}$ and $-\dfrac{7}{12}$ represent the same number. This is discussed further in Section 1.5.

$$= -\dfrac{7}{12}$$

Self-Check 2

Calculate each sum.

a. $-\dfrac{5}{13} + \left(\dfrac{-7}{13}\right)$ b. $-\dfrac{2}{5} + \dfrac{3}{4}$

The ability to mentally estimate answers is an important skill that can be used in a variety of real-world applications. This is particularly true when we are working with decimals. It is very easy to make an error on a keystroke when entering a decimal into a calculator or a spreadsheet. Example 3 illustrates how we can add two decimals and how we can mentally estimate sums in order to check our pencil-and-paper calculations or calculator and spreadsheet results for keystroke errors.

Example 3   Estimate Then Calculate

Mentally estimate the value of each sum by first rounding each term to the nearest integer. Then calculate the exact sum by aligning the decimal points and adding the terms.

(a) $-24.937 + (-15.054)$ **(b)** $61.150 + (-9.823)$

Solution

Estimated Value   Exact Value

(a) $-24.937 \approx -25$ -25 -24.937
        $-15.054 \approx -15$ $(+)$ -15 $(+)$ -15.054
                                        $\overline{-40}$ $\overline{-39.991}$

(b) $61.150 \approx 61$ 61 61.150
        $-9.823 \approx -10$ $(+)$ -10 $(+)$ -9.823
                                        $\overline{51}$ $\overline{51.327}$

Both actual results are very close to the estimated values so these results seem reasonable.

Self-Check 3

a. Mentally estimate $20.05 + (-79.96)$ by first rounding each term to the nearest integer.

b. Calculate the exact sum $20.05 + (-79.96)$.

How Do I Add the Terms in An Expression with Grouping Symbols?

When grouping symbols are used, the implied order of operations is to perform the operations within the innermost grouping symbols first. The most commonly used grouping symbols in textbooks are parentheses "( )" and brackets "[ ]". We will examine order of operations in more detail in Section 1.6. Use only parentheses when entering expressions into a calculator or a spreadsheet because brackets have a special meaning to a TI-84 Plus calculator and are considered invalid characters in an expression by Excel.

Example 4   Simplifying Expressions with Grouping Symbols

Determine the following sums:

(a) $-7 + [-8 + (+5)]$ **(b)** $[-9 + (-7 + 3)] + (-11)$

Solution

(a) $-7 + [-8 + (+5)] = -7 + (-3)$ The indicated order of operations is to first add the
                    $= -10$ terms inside the brackets.

(b) $[-9 + (-7 + 3)] + (-11)$ Add the terms inside the inner parentheses first; then
    $= [-9 + (-4)] + (-11)$ add the terms inside the brackets.
    $= -13 + (-11)$
    $= -24$

Self-Check 4

Calculate the sum $[-9 + (-5 + 8)] + [-3 + (-4 + 11)]$.

2. Use the Commutative and Associative Properties of Addition

Does the Order That I Add Terms Affect the Sum?

No; for example, $9 + 5 = 5 + 9$. When we purchase two items at a store, the total price will be the same no matter which item is charged first. (Can you imagine the confusion that would result if the order did affect the total price?) We say that addition is **commutative** to describe the fact that the order of the terms does not change the sum. That is, $a + b = b + a$.

When we purchase several items at a store, we may cluster them together in different ways. For example, all canned goods and all fresh produce might be totaled separately. Again if we subtotaled these groups separately, we are confident that the total price of all items will not change. We say that addition is **associative** to describe the fact that terms can be regrouped without changing the sum. For example, $(3 + 4) + 5$ and $3 + (4 + 5)$ both equal 12. In general, the associative property of addition says $(a + b) + c = a + (b + c)$.

The properties of addition, subtraction, multiplication, and division are presented as needed. These properties are summarized in the Key Concepts for Chapter 1 after we have reviewed all of these operations.

Properties of Addition

	Algebraically	Verbally	Numerical Example
Commutative property of addition	$a + b = b + a$	The sum of two terms in either order is the same.	$4 + 5 = 5 + 4$
Associative property of addition	$(a + b) + c = a + (b + c)$	Terms can be regrouped without changing the sum.	$(3 + 4) + 5 = 3 + (4 + 5)$

Together the commutative and associative properties allow us to add terms in any order we desire. People who are very skilled at mental arithmetic often take advantage of this to add terms producing multiples of 10 or other convenient values. For expressions with several terms, it is often convenient to first add all terms with like signs.

Example 5   Using the Properties of Addition

Calculate these sums:

(a) $47 + 19 + 53 + 31 + 7$ **(b)** $3 + (-4) + 5 + (-6) + 7$

Solution

a. $47 + 19 + 53 + 31 + 7$

$= (47 + 53) + (19 + 31) + 7$

$= 100 + 50 + 7$

$= 157$

For convenience, we use the commutative and associative properties to regroup the terms in order to pair terms that produce multiples of 10. With practice you may be able to do this entire computation mentally. For most problems, writing some of these steps is usually a good idea.

b. $3 + (-4) + 5 + (-6) + 7$

$= (3 + 5 + 7) + [(-4) + (-6)]$

$= 15 + (-10)$

$= 5$

For convenience, we first regroup terms with like signs together.

Self-Check 5

Calculate each sum.

a. $-11 + [-9 + 5]$ **b.** $[-2 + (-8 + 25)] + (-16)$

3. Subtract Positive and Negative Real Numbers

How Do I Subtract Positive and Negative Real Numbers?

Table 1.4.2 illustrates all possible cases of subtracting positive and negative real numbers. The word *change* indicates the operation of subtraction. The **change** from *a* to *b* is determined by the subtraction $b - a$. The result of this subtraction is called their **difference.** The change from the 8 A.M. temperature to the noon temperature is:

$$\text{Noon temperature} - 8 \text{ A.M. temperature} = \text{Change in temperature}$$

Table 1.4.2 Temperatures at Harper College

Temperatures	Monday	Tuesday	Wednesday	Thursday
Noon	$+5°$	$+5°$	$-5°$	$-5°$
8 A.M.	$+2°$	$-2°$	$-2°$	$+2°$
Change	$+3°$	$+7°$	$-3°$	$-7°$

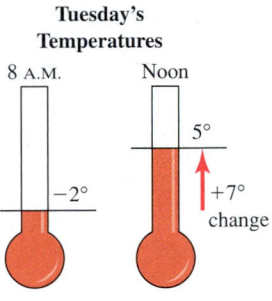

Tuesday's Temperatures

A definition of subtraction that covers all four cases illustrated in Table 1.4.2 is given in the following box. We say that subtraction is the inverse of addition and define subtraction as the addition of an additive inverse.

Definition of Subtraction

Algebraically	Verbally	Numerical Example
For any real numbers x and y, $x - y = x + (-y)$.	To subtract y from x, add the opposite of y to x.	$5 - 2 = 5 + (-2)$ $= 3$

When you use a calculator or a spreadsheet to perform a subtraction, there is no need to rewrite the problem as an addition problem in order to obtain the correct result. However, for pencil-and-paper calculations or mental computations, you may find it wise either to write or to think of the subtraction in terms of addition, as illustrated in Example 6. Remember that you can check the solution to each subtraction problem by using addition. For example, $5 - (-2) = 7$ checks because $7 + (-2) = 5$.

Example 6   Subtracting Two Real Numbers

Determine the following differences.

(a) $9 - 13$ **(b)** $17 - (-5)$ **(c)** $-8 - 4$ **(d)** $-6 - (-11)$

Solution

(a) $9 - 13 = 9 + (-13)$
$\qquad\quad = -4$

Change the operation to addition, and change 13 to its additive inverse, -13. Do not change the sign of 9.

(b) $17 - (-5) = 17 + (5)$
$\qquad\qquad\quad = 22$

Change the operation to addition, and change -5 to its additive inverse, 5. Do not change the sign of 17.

(c) $-8 - 4 = -8 + (-4)$
$\qquad\qquad = -12$

Change the operation to addition, and change 4 to its additive inverse, -4. Do not change the sign of -8.

(d) $-6 - (-11) = -6 + (11)$
$\qquad\qquad\qquad = 5$

Change the operation to addition, and change -11 to its additive inverse, 11. Do not change the sign of -6.

Self-Check 6

a. Determine the difference $13 - (-7)$.

b. Check this result using addition.

For subtraction problems involving negative terms, it is wise to write the step showing the conversion to an addition problem until your proficiency level is very high. After your proficiency level is very high, it is natural to want to use shortcuts. Many problems can be done mentally, as illustrated in Example 7.

Example 7 Subtracting Two Real Numbers

Mentally calculate these differences.

(a) $-6 - 5$ **(b)** $-8 - (-3)$ **(c)** $11 - (-5)$

Solution

(a) $-6 - 5 = -11$ Think "-6 *plus* -5."

(b) $-8 - (-3) = -5$ Think "-8 *plus* $+3$."

(c) $11 - (-5) = 16$ Think "11 *plus* $+5$."

Self-Check 7

Determine the following differences.

a. $-7 - 8$ **b.** $-7 - (-8)$ **c.** $11 - (-17)$

How Do I Know Which Operation to Use When I Read a Word Problem?

There are many words used in the English language that imply either addition or subtraction. Many of these words are antonyms since addition and subtraction are opposites. For example, we sometimes use the word *increase* to indicate addition and the word *decrease* to indicate subtraction. In Example 8, the word *change* implies that subtraction is the appropriate operation.

Table 1.4.3 Phrases Often Used To Indicate Addition and Subtraction*

	Key Phrase	Verbal Example	Algebraic Example
Addition	Plus	"12 plus 8"	$12 + 8$
	Total	"The total of \$25 and \$40"	$\$25 + \$40$
	Sum	"The sum of x and y"	$x + y$
	Increased by	"An interest rate r is increased by 0.5%"	$r + 0.005$
	More than	"7 more than x"	$x + 7$
Subtraction	Minus	"x minus y"	$x - y$
	Subtracted from	"x is subtracted from y"	$y - x$
	Difference from	"The difference from \$8 to \$12"	$\$12 - \$8$
	Decreased by	"An interest rate r is decreased by 0.5%"	$r - 0.005$
	Change from	"The change from 70° to 96°"	$96° - 70°$
	Less than	"7 less than x"	$x - 7$

*The context of each phrase must be considered. For example, "x is more than 7" does not indicate addition but the inequality $x > 7$.

Example 8   Subtracting Decimals

Determine the change in elevation from −4.05 meters to an elevation of 7.7 meters.

Solution

$$7.7 \text{ m} - (-4.05 \text{ m}) = 7.7 \text{ m} + 4.05 \text{ m}$$

Change the operation to addition, and change −4.05 to its additive inverse, 4.05. Do not change the sign of 7.7.

$$= 11.75 \text{ m}$$

The change in elevation is an increase of 11.75 meters.

Self-Check 8

Determine the change in elevation from 2.34 meters to an elevation of −13.78 meters.

Example 9   Subtracting Two Fractions

Determine the difference from $-\dfrac{4}{15}$ to $\dfrac{7}{10}$.

Solution

$$\frac{7}{10} - \left(-\frac{4}{15}\right) = \frac{7}{10} + \left(+\frac{4}{15}\right)$$

Change the operation to addition, and change $-\dfrac{4}{15}$ to its additive inverse, $+\dfrac{4}{15}$. Do not change the sign of $\dfrac{7}{10}$.

$$= \frac{7}{10} \cdot \frac{3}{3} + \frac{4}{15} \cdot \frac{2}{2}$$

To add the fractions with different denominators, first express each fraction in terms of a common denominator. In this case, express each fraction in terms of the least common denominator of 30 and add the numerators.

$$= \frac{21}{30} + \frac{8}{30}$$

$$= \frac{21 + 8}{30}$$

$$= \frac{29}{30}$$

Self-Check 9

Determine the difference from $-\dfrac{5}{12}$ to $-\dfrac{4}{15}$.

The operations of addition and subtraction are also often indicated when one needs to extract specific information from given tables and graphs. Newspapers and magazines often use bar graphs to display both positive and negative results. In Example 10 we use the information given in a bar graph to determine the change in a company's profits.

Example 10 Determining Changes from a Bar Graph

This bar graph compares Ford Motor Company's 2006 and 2007 second quarter profit results. All profits are given in millions of dollars. Use this information to determine the change in second quarter profits from 2006 to 2007.

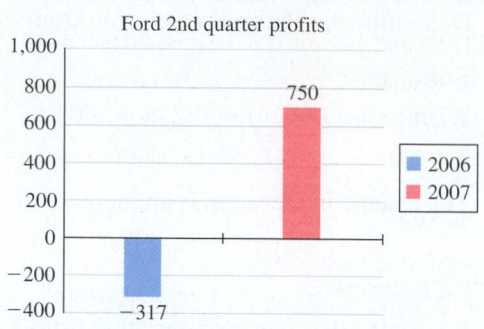

Ford 2nd quarter profits

Solution

$750 - (-317) = 1,067$ The change from -317 to 750 is determined by subtraction.

Answer: The profits for Ford from the second quarter of 2006 to the second quarter of 2007 increased by $1,067 million dollars or $1,067,000,000.

Self-Check 10

In the second quarter of 2008, Ford Motor Company lost $8,667 million dollars. What was the change from the second quarter profits of $750 million dollars in 2007 to the second quarter profits in 2008?

Self-Check Answers

1. **a.** 77   **b.** -77
   c. 33   **d.** -33
2. **a.** $-\dfrac{12}{13}$ **b.** $\dfrac{7}{20}$
3. **a.** -60 **b.** -59.91
4. **a.** -2
5. **a.** -15 **b.** -1
6. **a.** 20   **b.** $20 + (-7) = 13$
7. **a.** -15 **b.** 1   **c.** 28
8. The change in elevation is a decrease of 16.12 meters.
9. $\dfrac{3}{20}$
10. The profits decreased by $9,417,000,000.

1.4 Using the Language and Symbolism of Mathematics

1. In the addition $27 + (-13) = 14$, 27 and -13 are called addends or _____ and 14 is called the _____.

2. The property that says that $a + b = b + a$ for all real numbers a and b is called the _____ property of addition.

3. The property that says that $(a + b) + c = a + (b + c)$ for all real numbers a, b, and c is called the _____ property of addition.

4. If x and y have the same sign, then the sum $x + y$ will have the _____ sign as both terms.

5. If x and y have unlike signs, then the sum $x + y$ will have the same sign as the term with the _____ absolute value.

6. Together the _____ and _____ properties of addition allow us to add terms in any order and to obtain the same sum.

7. In the subtraction $11 - 4 = 7$, the result 7 is called the _____ of 11 minus 4.

8. To subtract y from x, add the additive inverse or the _____ of y to x.

9. The phrase "x increased by 5" is represented algebraically by _____.

10. The phrase "x decreased by 5" is represented algebraically by _____.

11. The phrase "the difference of x from 5" is represented algebraically by _____.

12. The phrase "17 less than w" is represented algebraically by _____.

13. The change from x_1 to x_2 is represented algebraically by _____.

14. The change from y_1 to y_2 is represented algebraically by _____.

15. The phrase "the total of x, y, and z" is represented algebraically by _____.

1.4   Quick Review

1. Round 19.85 to the nearest integer.

2. Round 6.23 to the nearest integer.

3. Which has the greater absolute value, -15.12 or 12.15?

4. Which has the greater absolute value, -5.73 or -7.35?

5. Add $\dfrac{2}{5} + \dfrac{2}{7}$.

1.4   Exercises

Objective 1 Add Positive and Negative Real Numbers

In Exercises 1–20, calculate each sum using only pencil and paper.

1. **a.** $8 + (+5)$ **b.** $-8 + (+5)$
   c. $8 + (-5)$ **d.** $-8 + (-5)$

2. **a.** $9 + (+11)$ **b.** $-9 + (+11)$
   c. $9 + (-11)$ **d.** $-9 + (-11)$

3. **a.** $-45 + 45$ **b.** $45 + (-45)$

4. **a.** $97 + (-97)$ **b.** $-97 + 97$

5. $-8 + (-7) + (-6)$ 6. $-3 + (-9) + (-2)$

7. $-17 + [(-11) + 9]$ 8. $[(-12) + (-9)] + 8$

9. $-4.8 + 0 + 6.09$ 10. $-5.4 + 0 + 8.6$

11. $\dfrac{7}{11} + \left(-\dfrac{3}{11}\right)$ 12. $-\dfrac{11}{17} + \dfrac{5}{17}$

13. $-\dfrac{1}{2} + \left(-\dfrac{1}{3}\right) + \left(-\dfrac{1}{4}\right)$ 14. $-\dfrac{2}{5} + \left(-\dfrac{2}{4}\right) + \dfrac{2}{3}$

15. $-3\dfrac{2}{5} + 7\dfrac{3}{4}$ 16. $-4\dfrac{2}{3} + 3\dfrac{3}{5}$

17. $[11 + (-17)] + [(-8) + (-4)]$

18. $[-19 + 15] + [17 + (-30)]$

19. $[-5 + (-6) + (-7)] + [8 + 4 + 3]$

20. $[11 + 12 + 13] + [(-10) + (-15) + (-16)]$

Objective 2 Use the Commutative and Associative Properties of Addition

21. The property that says $(5 + 7) + 9 = (7 + 5) + 9$ is the _____ property of addition.

22. The property that says $(5 + 7) + 9 = 5 + (7 + 9)$ is the _____ property of addition.

23. Use the associative property of addition to rewrite $11 + (12 + 13)$.

24. Use the commutative property of addition to rewrite $17 + 13$.

Objective 3 Subtract Positive and Negative Real Numbers

In Exercises 25–40, calculate the value of each expression using only pencil and paper.

25. **a.** $7 - 13$ **b.** $7 - (-13)$
    c. $-7 - 13$ **d.** $-7 - (-13)$

26. **a.** $-15 - (-8)$ **b.** $15 - 8$
    c. $-15 - 8$ **d.** $15 - (-8)$

27. **a.** $9 - 9$ **b.** $-9 - (-9)$

28. **a.** $8 - 8$ **b.** $-8 - (-8)$

29. $-18 - 21$ 30. $15 - 17$

31. $-18 - (-21)$ 32. $-15 - 17$

33. $-\dfrac{2}{3} - \dfrac{5}{3}$ 34. $-\dfrac{3}{7} - \dfrac{2}{7}$

35. $\dfrac{3}{5} - \left(-\dfrac{5}{6}\right)$ 36. $-\dfrac{4}{7} - \left(-\dfrac{1}{2}\right)$

37. $-14.8 - (-21.9)$ 38. $-240.8 - 500.9$

39. $-3\dfrac{2}{5} - 7\dfrac{3}{4}$ 40. $-4\dfrac{2}{3} - 3\dfrac{3}{5}$

Connecting Concepts to Applications

41. **Business Profit or Loss** Complete the last row of the table using a positive number to represent a total profit for the two stores and a negative number to represent a total that is a loss.

Quarterly Profit/Loss Statement for Emmett's Barbeque

	Quarter 1	Quarter 2	Quarter 3	Quarter 4
Springfield Store	+$7,000	+$7,000	-$7,000	-$7,000
Washington Store	+$5,000	-$5,000	+$5,000	-$5,000
Total				

42. Tuition Bill The table gives tuition comparisons for Austin Community College and the University of Texas at Austin. How much can a student save by taking 12 semester hours at ACC rather than at UTA?

Semester Hours	Tuition at ACC	Tuition at UTA
6	$192	$504
12	$384	$1,008
18	$576	$1,512

Thermometer readings

43. A thermometer at Parkland College in Champaign, Illinois, registers −12 degrees Fahrenheit on one February morning. The temperature rises 21 degrees that afternoon. What does the thermometer then register?

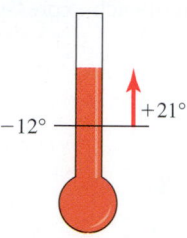

44. A thermometer registers −32 degrees Fahrenheit. The temperature rises 30 degrees. What does the thermometer then register?

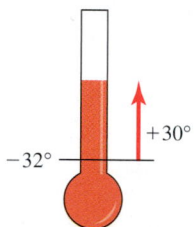

In Exercises 45 and 46, determine the change in Fahrenheit temperature from 8 A.M. to noon.

45. 8 A.M.   Noon   **46.** 8 A.M.   Noon

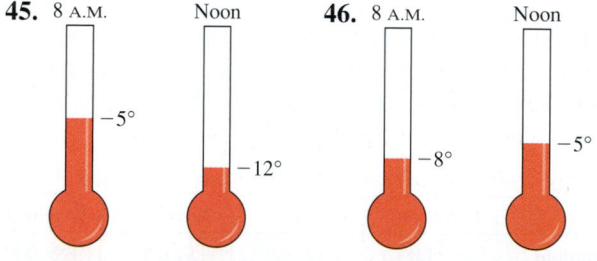

In Exercises 47 and 48, find each temperature change.

47. a. Find the change in temperature from 2° to 8°.
  b. Find the change in temperature from 8° to 2°.
  c. Find the change in temperature from −2° to 8°.
  d. Find the change in temperature from −2° to −8°.

48. a. Find the change in temperature from 7° to 3°.
  b. Find the change in temperature from 3° to 7°.
  c. Find the change in temperature from −7° to 3°.
  d. Find the change in temperature from −7° to −3°.

iPod sales

49. Use this bar graph to determine the total sales of iPods for 2005.

50. Use this bar graph to determine the total sales of iPods for 2007.

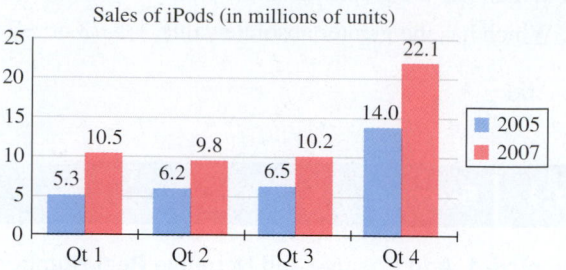

Net income

51. Use this bar graph to determine the change in net income for Ford Motor Company from 2005 to 2006.

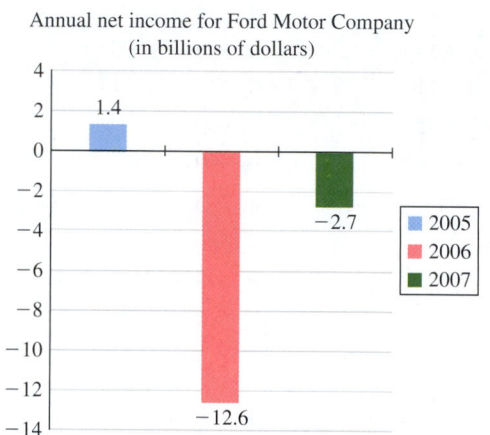

52. Use this bar graph to determine the change in net income for Ford Motor Company from 2006 to 2007.

Tuition bill

The table gives the 2009–2010 tuition bill for an Indiana Purdue Fort Wayne University business administration student.

	Hours	Tuition
Fall	12	$2,251.80
Spring	14	$2,627.10
Summer	6	$1,125.90

53. Calculate the total number of semester hours taken by this student in 2009–2010.

54. Calculate the total tuition billed to this student for 2009–2010.

55. Football Yardage A football team lost 7 yards on the first play, gained 10 yards on the second play, and gained $2\frac{1}{2}$ yards on the third play. What was the net yardage for the three plays?

56. Submarine Depth A submarine was submerged to a depth of 85 feet. It ascended 40 feet, descended 25 feet, and then descended another 50 feet. What was its new depth?

UPC bar code

57. The universal product code (UPC) is shown here for a box of cereal.
  a. Starting from left to right with the first eleven digits of the code 0 30000 06430 6, add all of the digits in the odd-numbered positions.
  b. Multiply the sum in part **a** by 3.
  c. Starting from left to right with the first eleven digits of 0 30000 06430 6, add all of the digits in the even-numbered positions.
  d. Add the results from parts **b** and **c**.
  e. What number must be added to the result from part **d** to produce the next multiple of 10?
  f. Does the number in part **e** equal the check digit of 6 shown on the UPC code?

58. Repeat Exercise 57 for the UPC code 0 27541 00760 2 from a bottle of drinking water.

Connecting Algebra to Geometry

The perimeter of a geometric figure is the distance around the figure. For a triangle, the perimeter is the sum of the lengths of all sides. In Exercises 59 and 60 calculate the perimeter of each triangle. All units are in centimeters.

59.

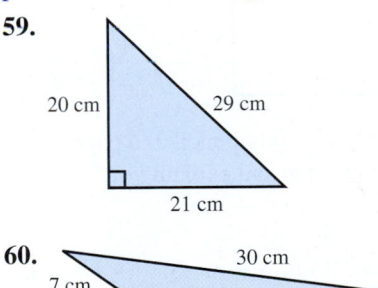

20 cm   29 cm
21 cm

60.   30 cm
7 cm
24 cm

To order the crown molding trim for a basement recreation room, a contractor must calculate the total length of trim needed. To do this, the contractor calculates the perimeter of the room. In Exercises 61 and 62, calculate the perimeter of each room. (*Hint:* The perimeter can be calculated using the given measurements even though the measurements of some sides are not given.)

61.

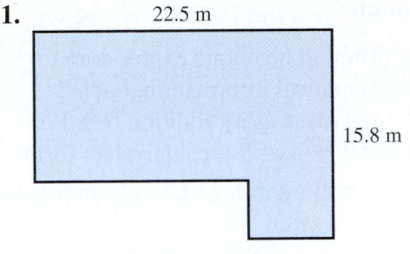

22.5 m
15.8 m

62.

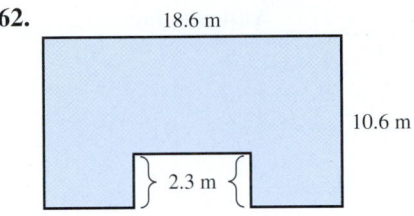

18.6 m
10.6 m
2.3 m

Estimation Skills

In Exercises 63–66, mentally estimate each sum or difference, and then select the most appropriate answer.

63. $-35.764 + 66.056$
  A. -100 **B.** $+100$
  C. -30 **D.** 30

64. $\pi - 3.15$
  A. Less than -1 **B.** Between -1 and 0
  C. Between 0 and 1   **D.** Greater than 1

65. $\dfrac{1}{3} - 0.3333$
  A. Less than -1 **B.** Between -1 and 0
  C. Between 0 and 1   **D.** Greater than 1

66. $\dfrac{1}{2} - \dfrac{1}{8}$
  A. Less than -1 **B.** Between -1 and 0
  C. Between 0 and 1   **D.** Greater than 1

In Exercises 67–70, first mentally determine the correct sign of the sum and then calculate the sum.

Problem	Sign of Sum	Sum
67. $-47.6 + 53.4$		
68. $47.6 + (-53.4)$		
69. $-\dfrac{5}{4} + \left(-\dfrac{3}{8}\right)$		
70. $\dfrac{5}{4} + \left(-\dfrac{3}{8}\right)$		

Multiple Representations

There are often many different algebraic expressions to represent the same mathematical information. Each subtraction fact can be rewritten as an addition fact. For example, $7 - 4 = 3$ and $3 + 4 = 7$ are equivalent forms.

In Exercises 71–74, use addition to rewrite each subtraction in an equivalent form.

Subtraction fact	Addition fact

71. $a - b = c$

72. $11 - 15 = -4$

73. $-17 - (-25) = 8$

74. $\dfrac{1}{2} - \dfrac{3}{16} = \dfrac{5}{16}$

In Exercises 75–80, write each verbal statement in algebraic form.

75. a. a plus b equals c.   **b.** a minus b equals c.

76. a. x decreased by eight equals z.
 b. x increased by eight equals z.

77. a. The change from y to z equals x.
 b. The difference of a from b equals c.

78. a. The sum of a and b equals the sum of b and a.
 b. The sum of a real number x and its additive inverse is zero.

79. a. m is 11 more than n.   **b.** m is more than n.

80. If n represents an integer, write an expression for:
 a. the next integer   **b.** the previous integer

Group discussion questions

81. Multiple Representations Represent the addition numerically, verbally, or graphically as indicated. (*Hint:* See the following example.)

Numerically Verbally Graphically

$4 + 2 = 6$ The sum of four and two is six.

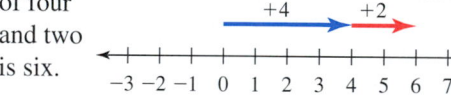

a. Give the numerical and verbal representations for the addition illustrated here.

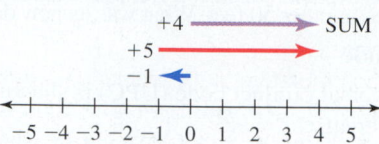

b. Represent the addition $-3 + 5 = 2$ both verbally and graphically.

c. Represent the addition "The sum of two and three is five" both numerically and graphically.

Communicating Mathematically

82. a. Explain the difference between the meanings of the symbol $+$ in the expressions $-7 - (+4)$ and $-7 + (-4)$.
 b. Explain the difference between the meanings of the symbol $-$ in the expressions $8 - 3$ and $8 + (-3)$.
 c. Explain the difference between the ⊖ key and the ⊖ (-) key on a calculator.

83. Without consulting this textbook or other sources, make a list of words or phrases that are often used to indicate the operation of:
 a. addition   **b.** subtraction

84. Error Analysis If you use a TI-84 Plus calculator to evaluate the expression $-7 + (-8) + 5$ as shown here, you will obtain an error message. What is the source of the syntax error?

85. Challenge Question Using only pencil and paper, determine the following sums and explain your reasoning for part **c.**
 a. $1 + (-2) + 3 + (-4)$
 b. $1 + (-2) + 3 + (-4) + 5 + (-6)$
 c. $1 + (-2) + 3 + (-4) + \cdots + 99 + (-100)$

1.4   Cumulative Review

1. Evaluate $|-17| + |17|$.

2. Evaluate $\sqrt{16} - \sqrt{25}$.

3. Write the additive inverse of the first three natural numbers and then find the sum of these additive inverses.

4. Reduce $-\dfrac{35}{49}$ to lowest terms.

5. Use interval notation to represent the real numbers shown in the graph.

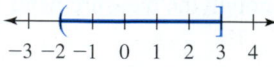

Section 1.5 | Multiplication and Division of Real Numbers

Objectives:

1. Use the commutative and associative properties of multiplication.
2. Multiply positive and negative real numbers.
3. Divide positive and negative real numbers.
4. Express ratios in lowest terms.

What Are the Terminology and Notation for a Multiplication Problem?

In the multiplication $5 \times 8 = 40$, both 5 and 8 are called **factors** of the **product** 40. Using the correct mathematical vocabulary gives you an advantage throughout your course work. Remember that factors are multiplied and terms are added.

The times sign "$\times$" is used frequently in arithmetic to indicate the product of constants. However, in algebra this symbol is rarely used. This is due, in part, to the possible confusion with the variable x. The multiplication symbol used most frequently with variables is the dot, as in $x \cdot y$. In many cases, the product of x times y is indicated by just xy with no symbol between the factors x and y. We also use parentheses to separate the factors as shown in the following box. The symbol "$*$" is used more on calculator displays and in spreadsheets than in written mathematics. Graphing calculators allow the multiplication between the factors to be implied, as in xy, $(x)(y)$, $x(y)$, or $(x)y$. However, spreadsheets require the multiplication symbol "$*$" to be used when indicating multiplication. Failure to follow this convention will result in an error.

Notations for the Product of the Factors x and y
xy $x \cdot y$ $(x)(y)$ $x(y)$ $(x)y$ $x*y$

> **A Mathematical Note**
>
> The symbol $\times$ for multiplication was introduced by the English mathematician William Oughtred (1574–1660).

1. Use the Commutative and Associative Properties of Multiplication

Can Multiplication Be Described As Repeated Addition?

Yes, multiplication by whole numbers is often introduced this way. For example, the multiplication 3×4 can be interpreted as the repeated addition $4 + 4 + 4$. Likewise, 4×3 can be interpreted as the repeated addition $3 + 3 + 3 + 3$. Thus both $3 \times 4 = 12$ and $4 \times 3 = 12$. In general, $a \cdot b = b \cdot a$. We describe this by saying that **multiplication is commutative. Multiplication is also associative,** that is, $(ab)c = a(bc)$. These two properties are summarized in the following box.

Properties of Multiplication	Algebraically	Verbally	Numerical Example
Commutative property of multiplication	$ab = ba$	The product of two factors in either order is the same.	$4 \cdot 5 = 5 \cdot 4$
Associative property of multiplication	$(ab)(c) = (a)(bc)$	Factors can be regrouped without changing the product.	$(4 \cdot 5)(6) = (4)(5 \cdot 6)$

2. Multiply Positive and Negative Real Numbers

How Do I Multiply Positive and Negative Real Numbers?

The rules for multiplying signed numbers depend on the sign of each factor. We can develop the rules for doing this by considering the following four cases. Also recall that the product of 0 and any other factor is 0.

Case 1:   *A Positive Factor Times a Positive Factor*
We are already familiar with this case. The product of two positive factors is positive. For example, $3 \cdot 4 = 12$.

Case 2:   *A Positive Factor Times a Negative Factor*
Because multiplication can be interpreted as repeated addition, $3(-4)$ can be interpreted as $(-4) + (-4) + (-4) = -12$. Thus the product of a positive factor and a negative factor is negative.

Case 3:   *A Negative Factor Times a Positive Factor*
Because multiplication is commutative, $(-4)(3)$ equals $3(-4) = -12$. Thus the product of a negative factor and a positive factor is negative.

Case 4:   *A Negative Factor Times a Negative Factor*
Examine the pattern illustrated by the following products:

$$3(-3) = -9 \quad \text{As the factor to the left decreases}$$
$$2(-3) = -6 \quad \text{by 1, the product to the right}$$
$$1(-3) = -3 \quad \text{increases by 3.}$$
$$0(-3) = 0$$
$$-1(-3) = ?$$

To continue this pattern, $(-1)(-3) = 3$. This suggests that the product of two negative factors is positive. For example, $(-3)(-4) = 12$.

These four cases can be summarized by examining the product of like signs and unlike signs as shown in the following box.

Multiplication of Two Real Numbers		
	Verbally	Numerical Example
Like signs:	Multiply the absolute values of the two factors and use a positive sign for the product.	$(+3)(+6) = +18$ $(-3)(-6) = +18$
Unlike signs:	Multiply the absolute values of the two factors and use a negative sign for the product.	$(+3)(-6) = -18$ $(-3)(+6) = -18$
Zero factor:	The product of 0 and any other factor is 0.	$(3)(0) = 0$ $(0)(-6) = 0$

Example 1 illustrates the multiplication of two real numbers, including the multiplication of fractions.

Example 1   Multiplying Two Real Numbers

Determine the following products:

(a) $5(-9)$ **(b)** $-6(7)$ **(c)** $\left(\dfrac{-2}{3}\right)\left(\dfrac{-4}{5}\right)$ **(d)** $(-17)(0)$

Solution

(a) $5(-9) = -45$ The product of two factors with unlike signs is negative.

(b) $-6(7) = -42$

(c) $\left(\dfrac{-2}{3}\right)\left(\dfrac{-4}{5}\right) = \dfrac{(-2)(-4)}{3 \cdot 5}$ To multiply two fractions, multiply the numerators and multiply the denominators. The product of two factors with like signs is positive.

$\qquad\qquad\quad = \dfrac{8}{15}$

(d) $(-17)(0) = 0$ The product of 0 and any other factor is 0.

Self-Check 1

Determine the following products.

a. $-11(7)$ **b.** $-11(-7)$ **c.** $\left(\dfrac{4}{7}\right)\left(\dfrac{-5}{3}\right)$

What Is the Most Efficient Way to Multiply Several Negative Factors?

Many students find it easiest to determine the sign of the product first and then multiply the absolute values of the factors. The product of several negative factors is either positive or negative, as illustrated in Table 1.5.1 and summarized in the accompanying box. When a product is positive, it is customary not to write the plus symbol. For example, in the table $+2$ is usually written as just 2.

Table 1.5.1   Sign Pattern for Negative Factors

Number of Negative Factors	Product	Sign of Product
2 (even)	$(-1)(-2) = +2$	Positive
3 (odd)	$(-1)(-2)(-3) = -6$	Negative
4 (even)	$(-1)(-2)(-3)(-4) = +24$	Positive
5 (odd)	$(-1)(-2)(-3)(-4)(-5) = -120$	Negative
6 (even)	$(-1)(-2)(-3)(-4)(-5)(-6) = +720$	Positive

Product of Negative Factors

Even number: The product is positive if the number of negative factors is even.

Odd number: The product is negative if the number of negative factors is odd.

Thus to multiply several signed factors, determine the sign of the product first and then multiply the absolute values of the factors. The sign pattern of negative factors is reexamined when exponents are covered.

Example 2　Determining Products of Negative Factors

Determine the following products:

(a) $(-2)(3)(-5)(-6)$　　　**(b)** $(-17)(33)(0)(-45)$　　　**(c)** $(-8)(-5)(2)(-2)(-10)$

Solution

(a) $(-2)(3)(-5)(-6) = -180$　　The product is negative because the number of negative factors is odd (three).

(b) $(-17)(33)(0)(-45) = 0$　　Since one factor is 0, the product is 0. Zero is neither negative nor positive.

(c) $(-8)(-5)(2)(-2)(-10) = 1,600$　　The product is positive because the number of negative factors is even (four). The product is written as 1,600, which represents $+1,600$.

Self-Check 2

Determine the following products.

a. $-10(-10)(-10)$　　　**b.** $-10(10)(-10)(-10)$　　　**c.** $-10(0)(-10)(-10)$

The following illustration shows the effects of a factor of 1 and a factor of -1:

Factor of 1

$1(+5) = +5$
$1(-5) = -5$

Factor of -1

$-1(+5) = -5$
$-1(-5) = +5$

One is called the **multiplicative identity** because 1 is the only real number with the property that $1 \cdot a = a$ and $a \cdot 1 = a$ for every real number a. Also note that the product of -1 and a number results in the opposite of that number.

A Factor of 1 or −1

Algebraically	Verbally	Numerical Example
For any real number a: $1 \cdot a = a$	The product of one and any real number is that same real number.	$1 \cdot 3 = 3$ and $1 \cdot (-3) = -3$
$-1 \cdot a = -a$	The product of negative one and any real number is the opposite of that real number.	$-1 \cdot 3 = -3$ and $-1 \cdot (-3) = 3$

Example 3　Multiplying by 0, 1, and −1

Multiply each of these algebraic expressions by 0, 1, and -1:

(a) π　　　**(b)** x　　　**(c)** $\dfrac{4}{11}$　　　**(d)** -6.58

Solution

Factor of 0	Factor of 1	Factor of −1
(a) $0 \cdot \pi = 0$	$1 \cdot \pi = \pi$	$-1 \cdot \pi = -\pi$
(b) $0 \cdot x = 0$	$1 \cdot x = x$	$-1 \cdot x = -x$
(c) $0\left(\dfrac{4}{11}\right) = 0$	$1\left(\dfrac{4}{11}\right) = \dfrac{4}{11}$	$-1\left(\dfrac{4}{11}\right) = -\dfrac{4}{11}$
(d) $0(-6.58) = 0$	$1(-6.58) = -6.58$	$-1(-6.58) = 6.58$

Self-Check 3

Multiply $-\dfrac{2}{13}$ by:

a. 0 **b.** 1 **c.** -1

What Is a Multiplicative Inverse and What Is It Used For?

The **reciprocal** or **multiplicative inverse** of any nonzero real number a can be represented by $\dfrac{1}{a}$. The product of a number and its multiplicative inverse is the multiplicative identity 1.

For example, $3\left(\dfrac{1}{3}\right) = 1$ and $-\dfrac{3}{5}\left(-\dfrac{5}{3}\right) = 1$. Reciprocals are used to perform many divisions, especially problems involving fractions. We also extensively use multiplicative inverses and the multiplicative identity 1 as tools to solve linear equations.

Reciprocals or Multiplicative Inverses

Algebraically	Verbally	Numerical Examples
For any real number $a \neq 0$: $a \cdot \dfrac{1}{a} = 1$	For any real number a other than zero, the product of the number a and its multiplicative inverse $\dfrac{1}{a}$ is 1.	$-4 \cdot \left(-\dfrac{1}{4}\right) = 1$, and $\dfrac{4}{3} \cdot \dfrac{3}{4} = 1$
$\dfrac{1}{0}$ is undefined	Zero has no multiplicative inverse.	$\dfrac{0}{1} = 0$ but $\dfrac{1}{0}$ is undefined

The number 0 has no multiplicative inverse. That is, $\dfrac{1}{0}$ is undefined. For a number a to have a multiplicative inverse, it must be true that $a \cdot \dfrac{1}{a} = 1$. This is impossible for $a = 0$ because the product of 0 and any real number is always 0.

Example 4   Using Multiplicative Inverses

Multiply each number by its multiplicative inverse.

(a) $\dfrac{2}{7}$ **(b)** $-\dfrac{1}{5}$ **(c)** y (for $y \neq 0$)   **(d)** 0

Solution

(a) $\left(\dfrac{2}{7}\right)\left(\dfrac{7}{2}\right) = 1$ The reciprocal or multiplicative inverse of $\dfrac{2}{7}$ is $\dfrac{7}{2}$.

(b) $\left(-\dfrac{1}{5}\right)(-5) = 1$ The reciprocal of $-\dfrac{1}{5}$ is $-\dfrac{5}{1}$, which is normally written as just -5.

(c) $y\left(\dfrac{1}{y}\right) = 1$ The reciprocal of y or $\dfrac{y}{1}$ is $\dfrac{1}{y}$.

(d) 0 has no multiplicative inverse.

Self-Check 4

Multiply each number by its multiplicative inverse.

a. -8 **b.** $\dfrac{3}{8}$

It is easy to make keystroke errors when we enter values into a calculator or a spreadsheet. You often can spot serious keystroke errors by performing a quick mental estimate of the answer. Example 5 illustrates how we can mentally estimate a product. Note that multiplication is implied by the parentheses between the factors in this example.

Example 5   Estimate Then Calculate

Estimate the product $-25.32(-3.89)$ and then use a calculator to perform this multiplication.

Solution

Estimated Value

$$
\begin{array}{r}
-25.32 \approx -25 \\
-3.89 \approx -4
\end{array}
\times
\begin{array}{r}
-25 \\
(-4) \\
\hline
+100
\end{array}
$$

Exact Value

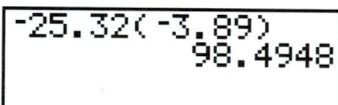

```
-25.32(-3.89)
            98.4948
```

Answer: 98.4948 seems reasonable based on the estimated value of 100.

Self-Check 5

Estimate the product $8.97(-10.02)$ and then calculate the exact product.

What Are the Terminology and Notation for a Division Problem?

For the division $16 \div 8 = 2$, 16 is called the **dividend,** 8 is called the **divisor,** and 2 is called the **quotient.** The division symbol "$\div$" is used frequently in arithmetic to indicate the division of constants. In algebra, we also frequently use the fraction form $\dfrac{x}{y}$ to indicate the division $x \div y$. Other notations used for division are given in the box.

A Mathematical Note

The symbol $\div$ for division is an imitation of fractional division; the dots indicate the numerator and the denominator of a fraction. This symbol was invented by the English mathematician John Pell (1611–1685).

Notations for the Quotient of *x* Divided by *y* for $y \neq 0$

$$x \div y \qquad \frac{x}{y} \qquad x/y \qquad x : y$$

The form x/y is seldom used because it is easy to misinterpret. This becomes very apparent when we enter expressions like $\dfrac{x}{y + z}$ into a calculator or a spreadsheet. We will consider expressions like $\dfrac{x}{y + z}$ again in Section 1.6 when the order of operations is covered. The form $x : y$ is rarely used other than to denote the ratio of two constants.

What Is the Relationship Between Division and Multiplication?

Dividing by 2 produces the same result as multiplying by $\dfrac{1}{2}$. For example, if two people agree to share a \$50 restaurant bill equally, we can divide the bill by 2; $\$50 \div 2 = \$25$. Equivalently, we could say that each person gets one-half of the bill; $\dfrac{1}{2}(\$50) = \$25$. Either way, the bill for each person is \$25. The division $50 \div 2 = 25$ checks because $2 \cdot 25 = 50$. We say that division is the inverse of multiplication, and we can describe division as multiplication by the multiplicative inverse.

Relationship Between Division and Multiplication

Algebraically	Verbally	Numerical Example
For any real numbers x and y with $y \neq 0$, $$x \div y = x\left(\frac{1}{y}\right) = \frac{x}{y}$$	Dividing two real numbers is the same as multiplying the first number by the multiplicative inverse of the second number.	$$6 \div 2 = 6\left(\frac{1}{2}\right) = 3$$

---

3. Divide Positive and Negative Real Numbers

How Do I Divide Positive and Negative Real Numbers?

Since division can be described in terms of multiplication, the rules for dividing signed numbers are derived from the rules for multiplying signed numbers.

Division of Two Real Numbers

Like signs: Divide the absolute values of the two numbers and use a positive sign for the quotient.

Unlike signs: Divide the absolute values of the two numbers and use a negative sign for the quotient.

Zero dividend: $\dfrac{0}{x} = 0$ for $x \neq 0$.

Zero divisor: $\dfrac{x}{0}$ is undefined for every real number x.

Example 6 illustrates the division of two real numbers, including the division of fractions.

Example 6   Dividing Real Numbers

Determine the following quotients:

(a) $21 \div (-7)$ **(b)** $-36 \div (+4)$ **(c)** $-45 \div (-9)$ **(d)** $\dfrac{-3}{5} \div \dfrac{12}{7}$

Solution

(a) $21 \div (-7) = -3$ The quotient of two numbers with unlike signs is negative.

(b) $-36 \div (+4) = -9$

(c) $-45 \div (-9) = 5$ The quotient of two numbers with like signs is positive.

(d) $\dfrac{-3}{5} \div \dfrac{12}{7} = \dfrac{\overset{-1}{\cancel{-3}}}{5} \cdot \dfrac{7}{\underset{4}{\cancel{12}}}$ To divide by $\dfrac{12}{7}$, multiply by its multiplicative inverse, $\dfrac{7}{12}$.

$= \dfrac{-1 \cdot 7}{5 \cdot 4}$

$= \dfrac{-7}{20}$

$= -\dfrac{7}{20}$

Self-Check 6

Determine the following quotients.

a. $-9 \div (-3)$ **b.** $-9 \div \left(\dfrac{1}{3}\right)$ **c.** $9 \div \left(-\dfrac{3}{2}\right)$

Since $1 \div (-2) = -\dfrac{1}{2}$ and $(-1) \div 2 = -\dfrac{1}{2}$, we can write $\dfrac{1}{-2} = \dfrac{-1}{2} = -\dfrac{1}{2}$. The relationship of the sign of the numerator, the sign of the denominator, and the sign of a fraction is given in the following box. This relationship is given not just for rational numbers such as $\dfrac{3}{5}$, but for all algebraic fractions $\dfrac{a}{b}$ where a and b are algebraic expressions representing real numbers and $b \neq 0$.

Three Signs of a Fraction

Algebraically	Verbally	Numerical Example
For all real numbers a and b with $b \neq 0$, $$\frac{-a}{b} = \frac{a}{-b} = -\frac{-a}{-b} = -\frac{a}{b} \text{ and}$$ $$\frac{-a}{-b} = -\frac{-a}{b} = -\frac{a}{-b} = \frac{a}{b}$$	Each fraction has three signs associated with it. Any two of these signs can be changed and the value of the fraction will stay the same.	$$\frac{-2}{3} = \frac{2}{-3} = -\frac{-2}{-3} = -\frac{2}{3}$$ and $$\frac{-2}{-3} = -\frac{-2}{3} = -\frac{2}{-3} = \frac{2}{3}$$

Why Do Mathematicians Make Such a Big Deal about Division by 0?

To define division by 0 would create inconsistencies in our algebraic system. It is important that the basic operations of addition, subtraction, multiplication, and division have unique answers. Engineers and others making important calculations must be able to trust their results. Thus we must have division by 0 as an undefined value. Both calculators and spreadsheets will produce error messages if you try to divide by 0. The logic for having division by 0 remain undefined is examined in the following paragraph.

The division $50 \div 2 = 25$ checks because $2 \cdot 25 = 50$. Likewise if $3 \div 0 = a$, then $0 \cdot a = 3$ must check. However, this is impossible because $0 \cdot a = 0$ for all values of a. Thus this division by 0 is undefined. Next, we will examine $0 \div 0 = b$, which will check if $0 \cdot b = 0$. However, this is true for all real numbers b. We say that $\frac{0}{0}$ is **indeterminate** since there is no reason to select or determine one value of b as preferable to any other value of b. Thus, this division by 0 is also undefined.

Example 7   Dividing Expressions Involving Zero

Determine the following quotients.

(a) $\dfrac{10}{5}$ (b) $\dfrac{0}{5}$ (c) $\dfrac{5}{0}$ (d) $\dfrac{0}{0}$

Solution

(a) $\dfrac{10}{5} = 2$ *Check:* $2 \cdot 5 = 10$.

(b) $\dfrac{0}{5} = 0$ *Check:* $0 \cdot 5 = 0$.

(c) $\dfrac{5}{0}$ is undefined   Division by 0 is undefined. There is no value to multiply times 0 that will produce 5.

(d) $\dfrac{0}{0}$ is undefined   Division by 0 is undefined. All values multiplied times 0 will yield 0, thus there is no unique quotient.

Self-Check 7

Determine the following quotients.

a. $\dfrac{0}{-10}$ b. $\dfrac{-10}{0}$

How Do I Know Which Operation to Use When I Read a Word Problem?

There are many words used in the English language that imply either multiplication or division. Some of these words are given in Table 1.5.2.

Table 1.5.2 Phrases Used To Indicate Multiplication and Division

	Key Phrase	Verbal Example	Algebraic Example
Multiplication	Times	"x times y"	xy
	Product	"The product of 5 and 7"	$5 \cdot 7$
	Multiplied by	"The rate r is multiplied by the time t"	rt
	Percent of	"Twenty percent of x"	$0.20x$
	Twice	"Twice y"	$2y$
	Double	"Double the price P"	$2P$
	Triple	"Triple the coupon value V"	$3V$
Division	Divided by	"x divided by y"	$\dfrac{x}{y}$
	Quotient	"The quotient of 5 and 3"	$5 \div 3$
	Ratio	"The ratio of x to 2"	$x : 2$ or $\dfrac{x}{2}$

In addition to the words and phrases listed in Table 1.5.2 to indicate multiplication and division, many applications are worked by using multiplication or division. Example 8 calculates the area of a rectangle by multiplying its width times its length.

Example 8   Calculating the Area of a Rectangle

Calculate the area of this rectangle in square ft (ft²).

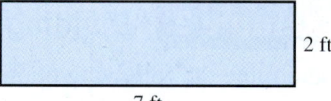

2 ft

7 ft

Solution

$(2 \text{ ft})(7 \text{ ft}) = 14 \text{ ft}^2$ Multiply the width times the length.

Self-Check 8

Determine the area of a rectangle that measures 4.5 ft by 8.2 ft.

---

4. Express Ratios in Lowest Terms

How Are Ratios Expressed Algebraically?

A Mathematical Note

The colon has been used by the French to indicate division. The use of the colon in the ratio $a : b$ also indicates $a \div b$.

The ratio of a to b is the quotient $a \div b$. This ratio is sometimes written as $a : b$ and read as "the ratio of a to b." Since $a \div b = \dfrac{a}{b}$, ratios are frequently written as fractions in reduced form. Sometimes both the numerator and the denominator of a ratio can be expressed in terms of the same unit. In this case, the reduced form of the ratio is a number free of units, as illustrated in Example 9.

Example 9   Determining the Ratio of Two Measurements

The flagpole shown is 8 meters tall and casts a 2-meter shadow. Determine the ratio of the height of the flagpole to the length of its shadow.

Solution

$$\frac{\boxed{\text{Height of flagpole}}}{\boxed{\text{Length of shadow}}} = \frac{8 \text{ m}}{2 \text{ m}}$$

Divide both the numerator and denominator of this ratio by 2 m.

$$= \frac{4}{1}$$

The reduced form of the ratio is a number free of units.

$$= 4$$

The flagpole is taller than the shadow by a factor of 4.

Answer: The ratio of the height of the flagpole to the length of its shadow is $\frac{4}{1} = 4 : 1$.

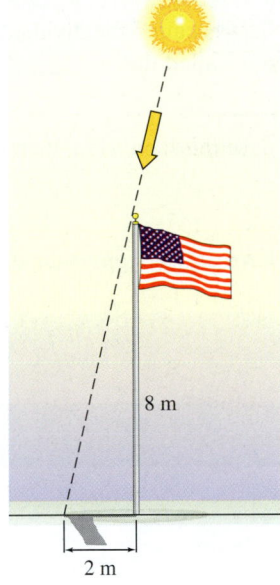

8 m

2 m

Self-Check 9

The front gear of a bicycle has 52 teeth, and the rear gear has 16 teeth. What is the ratio of the teeth on the front gear to those on the rear gear?

Self-Check Answers

1. **a.** -77 **b.** 77 **c.** $-\dfrac{20}{21}$

2. **a.** $-1{,}000$ **b.** $-10{,}000$ **c.** 0

3. **a.** 0 **b.** $-\dfrac{2}{13}$ **c.** $\dfrac{2}{13}$

4. **a.** $(-8)\left(-\dfrac{1}{8}\right) = 1$ **b.** $\left(\dfrac{3}{8}\right)\left(\dfrac{8}{3}\right) = 1$

5. $8.97(-10.02) \approx 9(-10) = -90,\ 8.97(-10.02) = -89.8794$
6. **a.** 3 **b.** -27 **c.** -6
7. **a.** 0 **b.** undefined
8. 36.9 ft^2
9. $13 : 4$

1.5   Using the Language and Symbolism of Mathematics

1. In the multiplication $9 \cdot 11 = 99$, 9 and 11 are called _____ and 99 is called the _____.

2. The property that says that $ab = ba$ for all real numbers is called the _____ property of multiplication.

3. The property that says that $(ab)(c) = (a)(bc)$ for all real numbers is called the _____ property of multiplication.

4. If either x or y is zero, the product $xy =$ _____.

5. Factors are _____ and terms are added.

6. If x and y have the same sign, then the sign of the product xy is _____.

7. If x and y have unlike signs, then the sign of the product xy is _____.

8. If an even number of negative factors are multiplied, the product will be _____.

9. If an odd number of negative factors are multiplied, the product will be _____.

10. If $12 * 5$ is shown on a calculator display, then the result will be _____.

11. The real numbers $-\dfrac{4}{7}$ and $-\dfrac{7}{4}$ are reciprocals or multiplicative _____ of each other.

12. The product of a number and its reciprocal is _____ .

13. The only real number that does not have a reciprocal is _____ .

14. The multiplicative identity is _____ .

15. In the division $24 \div 3 = 8$, 24 is called the dividend, 3 is called the divisor, and 8 is called the _____ .

16. Division by zero is _____ .

17. If x and y are nonzero and have the same sign, then the quotient $\frac{x}{y}$ is _____ .

18. If x and y are nonzero and have unlike signs, then the quotient $\frac{x}{y}$ is _____ .

19. The notation $a : b$ is read the _____ of a to b.

20. To divide $\frac{5}{6}$ by $\frac{2}{3}$, we can multiply $\frac{5}{6}$ by $\frac{3}{2}$, the multiplicative inverse or the _____ of $\frac{2}{3}$.

21. The result of the addition "a plus b" is called their _____ .

22. The result of the subtraction "a minus b" is called their _____ .

23. The result of the multiplication "a times b" is called their _____ .

24. The result of the division "a divided by b" is called their _____ .

1.5 Quick Review

1. Reduce $\frac{8}{20}$ to lowest terms.

2. Reduce $\frac{75}{60}$ to lowest terms.

In Exercises 3–5, perform the indicated operations.

3. $\frac{2}{5} \cdot \frac{2}{7}$ **4.** $\frac{2}{5} \div \frac{3}{4}$ **5.** $\frac{6}{35} \div \frac{10}{21}$

1.5 Exercises

Objective 1 Use the Commutative and Associative Properties of Multiplication

1. $6 \times 7 = 7 \times 6$ is an example of the _____ property of multiplication.

2. $6 \cdot (7 \cdot 8) = (6 \cdot 7) \cdot 8$ is an example of the _____ property of multiplication.

In Exercises 3–6, match each expression with the appropriate property.

3. a. $5 \cdot (7 \cdot 9) = 5 \cdot (9 \cdot 7)$ **A.** Associative property
   b. $5 \cdot (7 \cdot 9) = (5 \cdot 7) \cdot 9$ of multiplication
                                         B. Commutative property
                                            of multiplication

4. a. $(3 \cdot 8) \cdot 9 = 3 \cdot (8 \cdot 9)$ **A.** Associative property
   b. $(3 \cdot 8) \cdot 9 = (8 \cdot 3) \cdot 9$ of multiplication
                                         B. Commutative property
                                            of multiplication

5. a. $wx + yz = yz + wx$ **A.** Commutative property
   b. $wx + yz = wx + zy$ of addition
                              B. Commutative property
                                 of multiplication

6. a. $w + (x + yz) = (w + x) + yz$ **A.** Associative
   b. $w + x(yz) = w + (xy)(z)$ property of
                                              addition
                                         B. Associative
                                              property of
                                              multiplication

Objective 2 Multiply Positive and Negative Real Numbers

In Exercises 7–16, calculate each product using only pencil and paper.

7. a. $7(-11)$ **b.** $-7(11)$
   c. $(-7)(-11)$ **d.** $-(-7)(-11)$

8. a. $-5(12)$ **b.** $-(5)(12)$
   c. $(5)(-12)$ **d.** $(-5)(-12)$

9. a. $2(-3)(10)$ **b.** $2(-3)(-10)$
   c. $(-2)(-3)(-10)$ **d.** $-(-2)(-3)(0)$

10. a. $2(5)(-20)$ **b.** $2(-5)(-20)$
   c. $(-2)(-5)(-20)$ **d.** $-(-2)(-5)(0)$

11. a. $-0.1(1,234)$ **b.** $-100(1,234)$
   c. $-0.001(1,234)$ **d.** $-1,000(1,234)$

12. a. $-0.01(-45.96)$ **b.** $-10(-45.96)$
   c. $-100(45.96)$ **d.** $-0.001(45.96)$

13. a. $\dfrac{1}{2} \cdot \dfrac{3}{5}$ **b.** $-\dfrac{1}{2} \cdot \dfrac{3}{5}$

14. a. $\dfrac{1}{3} \cdot \dfrac{4}{5}$ **b.** $-\dfrac{1}{3} \cdot \dfrac{4}{5}$

15. a. $\left(\dfrac{1}{5}\right)(5)$ **b.** $\left(\dfrac{1}{5}\right)(-5)$

16. a. $\left(-\dfrac{3}{2}\right)\left(-\dfrac{2}{3}\right)$ **b.** $\left(-\dfrac{2}{3}\right)\left(\dfrac{3}{2}\right)$

Objective 3 Divide Positive and Negative Real Numbers

In Exercises 17–26, calculate each quotient using only pencil and paper.

17. a. $48 \div (-6)$ **b.** $-48 \div (-6)$
   c. $-48 \div 6$ **d.** $0 \div 6$

18. a. $56 \div (-8)$ **b.** $-56 \div (-8)$
   c. $-56 \div 8$ **d.** $0 \div (-8)$

19. a. $-48 \div \left(\dfrac{1}{2}\right)$ **b.** $-48 \div 2$

   c. $-48 \div \left(-\dfrac{1}{2}\right)$ **d.** $-48 \div (-2)$

20. a. $48 \div \left(-\dfrac{1}{4}\right)$ **b.** $48 \div (-4)$

   c. $-48 \div \left(-\dfrac{1}{4}\right)$ **d.** $-48 \div (-4)$

21. a. $123 \div (0.1)$ **b.** $123 \div (-0.01)$
   c. $123 \div (-1,000)$ **d.** $-123 \div (-10)$

22. a. $-123 \div (0.01)$ **b.** $-123 \div (100)$
   c. $123 \div (-0.001)$ **d.** $123 \div (-10,000)$

23. a. $\dfrac{2}{3} \div \left(\dfrac{1}{6}\right)$ **b.** $-\dfrac{2}{3} \div \dfrac{1}{6}$

   c. $-\dfrac{2}{3} \div (-6)$ **d.** $\dfrac{2}{3} \div (-1)$

24. a. $\dfrac{4}{5} \div 20$ **b.** $-\dfrac{4}{5} \div \dfrac{1}{20}$

   c. $-\dfrac{4}{5} \div (-20)$ **d.** $\dfrac{1}{20} \div \left(-\dfrac{4}{5}\right)$

25. a. $0 \div (-7)$ **b.** $-7 \div 0$
   c. $7 \div 0$ **d.** $0 \div 0$

26. a. $0 \div (-11)$ **b.** $-11 \div 0$
   c. $11 \div 0$ **d.** $-0.01 \div 0$

Objective 4 Express Ratios in Lowest Terms

In Exercises 27–30, write each ratio in lowest terms.

27. $8 : 20$ **28.** $77 : 33$

29. $24 : 36$ **30.** $75 : 60$

Connecting Concepts to Applications

31. Diamonds in a Deck of Cards Thirteen of 52 cards in a deck of cards are diamonds. What is the ratio of diamonds to all cards in the deck?

32. Defective Computer Chips An inspection of 1,056 experimental computer chips found 132 defective. What is the ratio of defective chips to all the chips inspected?

33. Tree Shadow A tree 24 meters tall casts an 8-meter shadow. Determine the ratio of the height of the tree to the length of its shadow.

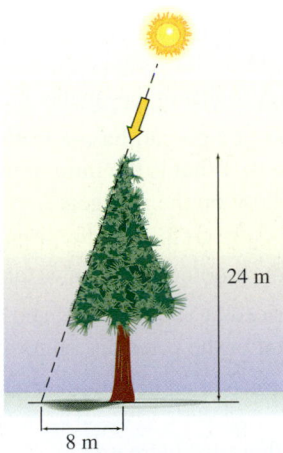

24 m

8 m

34. Shadow of a Man A young man 2 meters tall casts a 6-meter shadow. Determine the ratio of his height to the length of his shadow.

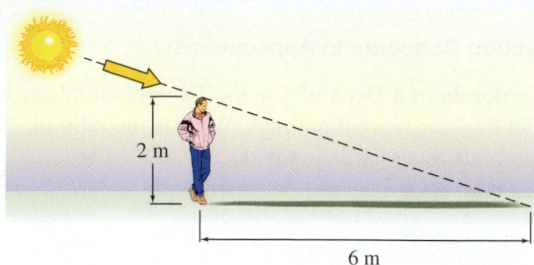

35. Banded Ducks A wildlife study involved banding ducks and then studying those birds when they were recaptured. Of the 85 captured ducks, 17 had already been banded.
  a. What is the ratio of banded ducks to the total captured?
  b. What is the ratio of banded ducks to those not banded?

36. Banded Rabbits A wildlife study involved banding rabbits' ears and then studying those banded rabbits when they were recaptured. Of the 117 captured rabbits, 26 had already been banded.
  a. What is the ratio of banded rabbits to the total captured?
  b. What is the ratio of banded rabbits to those not banded?

37. Pulley Ratio The radius of one pulley is 24 cm and the radius of a second pulley is 15 cm. Determine the ratio of the radius of the larger pulley to that of the smaller pulley.

38. Gear Ratio The front gear of a go-cart has 48 teeth and the rear gear has 16 teeth. What is the ratio of the teeth on the front gear to those on the rear gear?

39. Sales Tax The sales tax on a $180 item is 8% of the sales amount. Determine the amount of the sales tax.

40. Price Discount A shirt priced at $80 is discounted 30%. Determine the amount of the price discount.

41. Percent of Increase The price of an item was increased from $200 to $250.
  a. What is the amount of the price increase?
  b. What is the percent of increase—that is, what percent of the original price is the increase?

42. Percent of Decrease The price of an item was decreased from $250 to $200.
  a. What is the amount of the price decrease?
  b. What is the percent of decrease—that is, what percent of the original price is the decrease?

43. Basketball Ticket Payments The spreadsheet shows a partially completed order form for college basketball tickets. Complete the last column of this spreadsheet by calculating the cost for each game and then determine the total cost of the tickets for all four games.

	A	B	C	D
1	Game	Number of Tickets	Price per Ticket ($)	Cost ($)
2	1	4	$16	
3	2	2	$16	
4	3	3	$25	
5	4	6	$25	

44. Length of a Board A board is cut into two pieces so that the length of the shorter piece is x cm. Write an expression for the length of the longer piece if the longer piece is:
  a. 26 cm longer than the shorter piece
  b. 4 times as long as the shorter piece

Average score

The **mean** of a set of numerical scores is an average calculated by dividing the sum of scores by the number of scores. For example, the mean of test scores of 84 and 76 is
$$\frac{84 + 76}{2} = \frac{160}{2} = 80.$$

45. A student made five grades of 75, 88, 94, 78, and 91 on 100-point tests. Find the mean of these test scores.

46. A student scored 86, 78, 82, and 80 on four algebra exams. What is the student's average for these four exams?

Distance traveled

The distance an object travels can be calculated by finding the product of its rate and the time it travels. In Exercises 47 and 48, compute the distance each object travels.

47. A plane flies 420 miles per hour for 3 hours.

48. An ant crawls 2 meters per minute for 6.5 minutes.

Rate of travel

The average rate at which an object travels can be calculated by dividing the distance it travels by the time it takes to travel that distance. A car that drives 300 miles in 5 hours travels at an average rate of $\dfrac{300 \text{ miles}}{5 \text{ hours}} = 60 \text{ mi/h}.$

In Exercises 49 and 50, calculate the rate at which each object travels.

49. A plane flies 870 miles in 3 hours.

50. A hiker walks 40 miles in 2 days.

Rate of work

The rate at which a task is completed depends on the time required to complete the task. If a painter can paint 1 room in 3 hours, then the rate of work is $\frac{1}{3}$ room per hour. In Exercises 51 and 52, divide the amount of work by the time worked to compute the rate of work for each situation.

51. An assembly line produces 2,400 lightbulbs in 8 hours.

52. A bricklayer lays 500 bricks in 10 hours.

Connecting Algebra to Geometry

Area of a rectangle

The area of a rectangle can be calculated by multiplying its width times its length. In Exercises 53 and 54, determine the area of each rectangle in cm^2.

53.

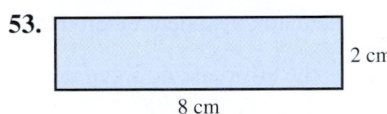

8 cm, 2 cm

54.

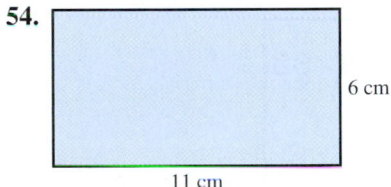

11 cm, 6 cm

Volume of a box

The volume of a rectangular box can be calculated by multiplying the length times the width times the height.

55. The mini-refrigerator shown here has interior dimensions of 14 in by 15 in by 16 in. Determine the interior volume of this refrigerator in cubic inches (in^3).

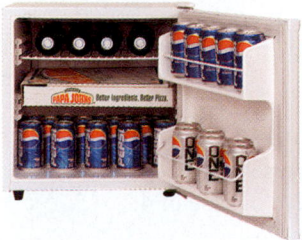

56. A box that contains 10 reams of printer paper has dimensions of 9 inches by 11.5 inches by 18 inches.

Determine the interior volume of this box in cubic inches (in^3).

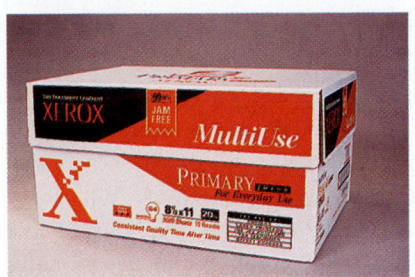

Estimation Skills

In Exercises 57–60, mentally estimate the value of each expression and then select the most appropriate answer.

57. $(-9.9)(15.12)$
A. -149.688 B. -249.688 C. 0
D. 149.688   E. 249.688

58. $(93.45)(187.8)(0)(-35.3)$
A. $-620,000$ B. $-260,000$ C. 0
D. 260,000   E. 620,000

59. $-28.5 \div 11.4$
A. -17.5 B. 17.5   C. -2.5
D. 2.5   E. -5.2

60. $45.1 \div (-0.11)$
A. -610 B. 610   C. -410
D. 410   E. 31

Multiple Representations

There are often many different algebraic expressions to represent the same mathematical information. For example, $4 \cdot 5$ and $5 + 5 + 5 + 5$ are equivalent forms. In Exercises 61–64, complete the table so that each expression is written both as a multiplication and as an addition.

Addition	Multiplication
61. $x + x + x + x$	
62. $8 + 8 + 8 + 8 + 8 + 8$	
63.	$5(-4)$
64.	$3w$

In Exercises 65–70, write each verbal statement in algebraic form.

65. The product of a and b equals the product of b and a.

66. The product of a real number x and its multiplicative inverse is 1.

67. The product of three and x is less than nine.

68. The quotient of any nonzero real number x divided by itself is 1.

69. The quotient of a and b is greater than or equal to c.

70. The quotient of 0 divided by x is 0 if $x \neq 0$.

71. If n represents an integer, write an expression for twice this integer.

72. The initial value of an investment is P. Write an expression for the new value of this investment if this investment triples in value.

73. Preparation for Factoring Trinomials When we factor trinomials in Section 6.2, it is very useful to be able to mentally add two numbers and also to multiply those numbers. For example, the sum of 3 and 4 is 7 and the product of 3 and 4 is 12. Complete this table by first computing the sum of x and y and then computing the product of x and y.

x	y	$x + y$	xy
5	8		
-5	8		
5	-8		
-5	-8		

Group discussion questions

Communicating mathematically

74. The number 1 is its own multiplicative inverse. The only other number that is its own multiplicative inverse is _____.

75. Discuss why the "*" symbol is used to denote multiplication on computers rather than the symbols "×" and "·", which were already common conventions for multiplication.

76. Which operations are commutative?
   a. Is addition commutative—that is, does $a + b = b + a$? If the answer is no, give a supporting example.
   b. Is subtraction commutative—that is, does $a - b = b - a$? If the answer is no, give a supporting example.
   c. Is multiplication commutative—that is, does $ab = ba$? If the answer is no, give a supporting example.
   d. Is division commutative—that is, does $a \div b = b \div a$? If the answer is no, give a supporting example.

Challenge Questions

77. Multiply $(-10)(-9)(-8) \cdot \cdots \cdot (8)(9)(10)$. (The notation $\cdots$ means that the pattern continues and includes the factors that are not typed.)

78. a. Find two numbers with a product of 24 and a sum of 11. Describe the method that you used to find these numbers.
   b. Find two numbers with a product of 24 and a sum of -11. Describe the method that you used to find these numbers.

79. Error Analysis In trying to evaluate the expression $\dfrac{2}{5} \div \dfrac{3}{8}$ on a TI-84 Plus calculator, a student entered the expression shown. What is the error and how can it be corrected?

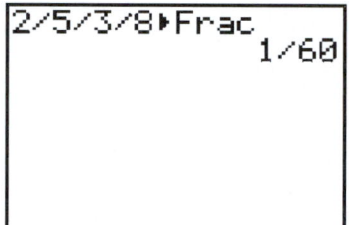

---

1. a. Write an algebraic equation describing the commutative property of addition.
   b. Write a sentence describing the commutative property of addition.

2. Simplify $4\dfrac{2}{5} + 3\dfrac{2}{7}$.

3. Which of the numbers $0, 1, 2, \sqrt{9}, |-4|$ is *not* a natural number?

4. Which of these notations is *not* acceptable interval notation?
   A. $(2, 3)$
   B. $(3, 2]$
   C. $[2, 3)$
   D. $[2, 3]$

5. a. Write the additive inverse of 4.
   b. Write the multiplicative inverse of 4.

Section 1.6 | Natural Number Exponents and Order of Operations

Objectives:

1. Use natural number exponents.
2. Use the standard order of operations.
3. Use the distributive property of multiplication over addition.
4. Simplify expressions by combining like terms.

1. Use Natural Number Exponents

What Is the Advantage of Using Exponential Notation?

Exponential notation is a shortcut symbolism used to denote repeated multiplication by the same factor. The expression 5^2 means $5 \cdot 5$ and is read as "five squared." In the expression 5^2, 5 is called the **base** and 2 is called the **exponent.** The expression b^n indicates that b is used as a factor n times, where the exponent n can be 1, 2, 3, 4, or any other natural number. Note how much simpler it is to write the exponential form 4^6 than it is to write the expanded form $4 \cdot 4 \cdot 4 \cdot 4 \cdot 4 \cdot 4$. Taking this another step, think how much more convenient is to write x^{88} than it would be to write the corresponding expanded form.

The compact use of exponential notation to indicate repeated multiplication is defined in the box. In Chapter 5 we will also develop the rules that serve as tremendous shortcuts for simplifying exponential expressions.

Exponential Notation

Algebraically	Verbally	Numerical Example
For any natural number n, $b^n = \underbrace{b \cdot b \cdot \, \cdots \, \cdot b}_{n \text{ factors of } b}$ with base b and exponent n.	For any natural number n, b^n is the product of b used as a factor n times. The expression b^n is read as "b to the nth power."	$7^4 = 7 \cdot 7 \cdot 7 \cdot 7$ $(-3)^2 = (-3)(-3)$

Example 1   Evaluating Exponential Expressions

Evaluate each of the following exponential expressions.

(a) 3^4 **(b)** 2^5 **(c)** 1^6 **(d)** $\left(\dfrac{2}{5}\right)^3$ **(e)** 0.12^2

Solution

(a) $3^4 = 3 \cdot 3 \cdot 3 \cdot 3$
$\qquad = 81$

The base is 3. The exponent of 4 indicates that 3 is used as a factor 4 times. This is read as "3 to the fourth power equals 81."

(b) $2^5 = 2 \cdot 2 \cdot 2 \cdot 2 \cdot 2$
$\qquad = 32$

"2 to the fifth power equals 32." The expanded form of 2^5 is $2 \cdot 2 \cdot 2 \cdot 2 \cdot 2$.

(c) $1^6 = 1 \cdot 1 \cdot 1 \cdot 1 \cdot 1 \cdot 1$
$\qquad = 1$

"1 to the sixth power equals 1."

(d) $\left(\dfrac{2}{5}\right)^3 = \dfrac{2}{5} \cdot \dfrac{2}{5} \cdot \dfrac{2}{5}$
$\qquad = \dfrac{8}{125}$

The base of $\dfrac{2}{5}$ is used as a factor 3 times. This is read, "Two-fifths cubed equals eight one-hundred-twenty-fifths."

(e) $0.12^2 = (0.12) \cdot (0.12)$ "Twelve hundredths squared equals one hundred forty-four
$\quad\quad = 0.0144$ ten-thousandths."

Self-Check 1

a. Evaluate 2^3.   **b.** Evaluate 3^2.

Since an exponent counts the number of times the base is used as a factor, the expression x is understood to mean the same as x^1. For example, xy^2 has one factor of x and two factors of y. An exponent is understood to refer to only the constant or variable that immediately precedes it. This can lead to misconceptions when expressions such as $(-3)^2$ and -3^2 are considered.

Caution: Note in Example 2 that $(-3)^2$ and -3^2 are not the same.

Example 2   Evaluating Exponential Expressions

Evaluate each of the following exponential expressions.

(a) $(-3)^2$ **(b)** -3^2 **(c)** $(-2)^4$ **(d)** -2^4 **(e)** $(-1)^{117}$

Solution

(a) $(-3)^2 = (-3)(-3)$ The base is -3. The product is positive since the
$\quad\quad\quad = 9$ number of negative factors is even (2).

(b) $-3^2 = -(3)(3)$ The base is 3. Square 3, and then form its
$\quad\quad\quad = -9$ additive inverse.

(c) $(-2)^4 = (-2)(-2)(-2)(-2)$ The base is -2. The product is positive since the
$\quad\quad\quad = 16$ number of negative factors is even (4).

(d) $-2^4 = -(2 \cdot 2 \cdot 2 \cdot 2)$ The base is 2. Raise 2 to the fourth power, and
$\quad\quad\quad = -16$ then form its additive inverse. Contrast this to part
$\quad\quad\quad\quad\quad\quad\quad\quad\quad\quad\quad$ (c) where the base is -2.

(e) $(-1)^{117} = -1$ The base of -1 is used as a factor an odd number
$\quad\quad\quad\quad\quad\quad\quad$ of times (117), so the result is negative.

Self-Check 2

Evaluate each of the following exponential expressions.

a. -3^4 **b.** $(-3)^4$ **c.** -1^{22} **d.** $(-1)^{22}$

As we noted in the Caution before Example 2, the expressions -3^2 and $(-3)^2$ do not have the same value. However, the expressions -2^3 and $(-2)^3$, which represent different expanded forms, do have the same value: $-2^3 = -(2 \cdot 2 \cdot 2) = -8$ and $(-2)^3 = (-2)(-2)(-2) = -8$. The correct algebraic interpretation in each case is to always apply an exponent to only the constant or variable that immediately precedes it.*

*See the warning in Technology Perspective 1.6.1.

What Terminology Is Used in the Real World to Indicate the Use of Exponents?

Some of the words and phrases used in word problems and applications to indicate exponentiation are given in Table 1.6.1.

Table 1.6.1   Phrases Used To Indicate Exponentiation

Key Phrase	Verbal Example	Algebraic Example
To a power	"3 to the 6th power"	3^6
Raised to	"y raised to the 5th power"	y^5
Squared	"4 squared"	4^2
Cubed	"x cubed"	x^3

A Mathematical Note

x^2, which can be read as "x to the second power," is usually read as "x squared." This is the area of a square with sides of length x.

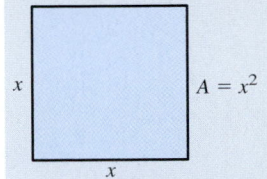

x^3, which can be read as "x to the third power," is usually read as "x cubed." This is the volume of a cube with sides of length x.

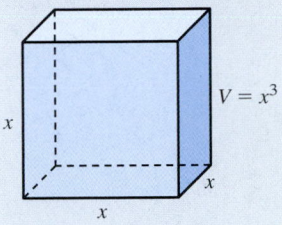

Example 3   Using Multiple Perspectives to Examine Exponential Expressions

Write each verbal expression in both exponential form and expanded form.

Solution

Verbal Expression	Exponential Form	Expanded Form
(a) x to the 7th power	x^7	$x \cdot x \cdot x \cdot x \cdot x \cdot x \cdot x$
(b) the square of the sum of x and y	$(x + y)^2$	$(x + y)(x + y)$
(c) x cubed times y	$x^3 y$	$x \cdot x \cdot x \cdot y$
(d) -9 raised to the 5th power	$(-9)^5$	$(-9)(-9)(-9)(-9)(-9)$

Self-Check 3

a. Write $x \cdot x \cdot x \cdot x \cdot x \cdot x$ in exponential form.

b. Write $(2w)(2w)(2w)$ in exponential form.

c. Write the square of x plus the cube of y in exponential form.

d. Write $x^2 y^3$ as a verbal expression.

How Can I Use a Calculator or a Spreadsheet to Evaluate Exponential Expressions?

Graphing calculators and spreadsheets have special keys or functions for squaring a value or raising an expression to other powers. On a calculator, the squaring key is very useful because this is the most common power that you will need. For other powers we recommend using the caret key ⌃. To square a value, press the x^2 button after entering the expression to be squared. For exponents other than 2, use the ⌃ followed by the desired exponent. The expression x^4 can be read as "x raised to the 4th power" and can be represented by x^4.

In an Excel spreadsheet you also can use the caret (^) key to enter exponents. When you use the caret key, we highly recommend that you use parentheses to specify exactly what you want. This is especially true when the expression involves additive inverses. Technology Perspective 1.6.1 illustrates evaluating both $(-3)^2$ and -2^4.

Evaluate $(-3)^2$ and -2^4.

TI-84 Plus Calculator

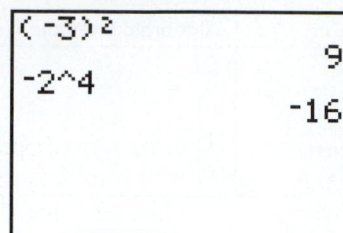

Excel Spreadsheet

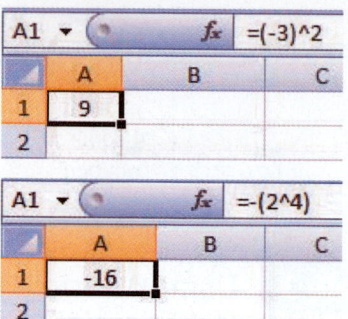

1. Enter the base of -3 within parentheses, then press to evaluate $(-3)^2$.

2. Enter -2 and press to evaluate -2^4. The base is 2 and the exponent is 4.

1. Enter "=(-3)^2" into cell **A1** of the first spreadsheet to evaluate $(-3)^2$. The parentheses enclose the base of -3 and the caret is used to enter the exponent of 2.

2. Enter "=$-(2$^4)" into cell **A1** of the second spreadsheet to evaluate -2^4. The parentheses are used to ensure the base is 2 and the caret is used to enter the exponent of 4.

Warning: To correctly enter -2^4 using Excel, you must enter "=$-(2$^4)". Excel will interpret "=-2^4" as $(-2)^4$.

Technology Self-Check 1

Calculate 4.2^5.

There are two points that we would like to emphasize regarding using any technology to evaluate mathematical expressions.

Point 1: Work a very simple problem whose answer is known before you undertake a similar important calculation.

Point 2: The importance of the order of operations that mathematicians have established is clarity. So if you really want to be clear about a result, use parentheses to force the expression to be evaluated in the order that you want.

2. Use the Standard Order of Operations

Why Do Many Mathematical Expressions Contain Parentheses and Brackets?

Symbols of grouping are used in mathematical expressions to separate signs, to enclose an expression to be treated as one quantity, and to indicate the order of operations within an expression. The common grouping symbols are the fraction bar ——, which separates the numerator from the denominator; parentheses ( ); brackets [ ]; and braces { }. For problems involving more than one operation, mathematicians have developed the following order of operations so that each expression will have a unique and agreed-upon interpretation.

Standard Order of Operations

Step 1   Start with the expression within the innermost pair of grouping symbols.

Step 2   Perform all exponentiations.

Step 3   Perform all multiplications and divisions as they appear from left to right.

Step 4   Perform all additions and subtractions as they appear from left to right.

Because calculators or spreadsheets can have subtle variations from this order of operations, you should use some test values to verify that any expression that you enter is following this standard order of operations. If not, then insert parentheses to force the expression to be evaluated in the order that you want.

Example 4   Using the Standard Order of Operations

Evaluate each of the following expressions, using the standard order of operations.

(a) $3 + 4 \cdot 7$ (b) $(3 + 4)(7)$ (c) $5 \cdot 8 - 12 \div 3 + 1$

(d) $24 \div 6 \cdot 2$ (e) $\dfrac{12 - 3(2)}{(12 - 3)(2)}$

Solution

(a) $3 + 4 \cdot 7 = 3 + 28$ Multiplication has priority over addition.

$\qquad\qquad = 31$

(b) $(3 + 4)(7) = 7(7)$ First simplify the expression within the

$\qquad\qquad = 49$ parentheses. Note the distinction between

$\qquad\qquad$ part (a) and part (b) of this example.

(c) $5 \cdot 8 - 12 \div 3 + 1 = 40 - 12 \div 3 + 1$ Multiplication and division are performed

$\qquad\qquad\qquad = 40 - 4 + 1$ as they appear from left to right. Then

$\qquad\qquad\qquad = 36 + 1$ addition and subtraction are performed as

$\qquad\qquad\qquad = 37$ they appear from left to right.

(d) $24 \div 6 \cdot 2 = 4 \cdot 2$ Multiplication and division are performed

$\qquad\qquad = 8$ as they appear from left to right.

(e) $\dfrac{12 - 3(2)}{(12 - 3)(2)} = \dfrac{12 - 6}{9(2)}$ Simplify the numerator and the denominator

$\qquad\qquad\qquad$ separately, as the fraction bar indicates this

$\qquad\qquad\qquad$ grouping.

$\qquad\quad = \dfrac{6}{18}$ Then reduce the fraction to lowest terms.

$\qquad\quad = \dfrac{1}{3}$

Self-Check 4

Evaluate each expression.

a. $18 - 4 \cdot 3$ b. $(18 - 4)(3)$

c. $\dfrac{20 - 45 \div 3}{8 + 6 \cdot 2}$ d. $5 \cdot 6 - 8 \div 2$

When we use the term *quantity* in a verbal statement of an algebraic expression, we are indicating that parentheses group the sum or difference that follows. This is significant because the expressions shown below have two distinct meanings.

Verbal Expression	Algebraic Expression
x plus *y* to the fourth power	$x + y^4$
The quantity *x* plus *y* to the fourth power	$(x + y)^4$

As Example 5 illustrates, order-of-operations errors can occur subtly in expressions involving exponents or square roots. It is important to be clear whether the exponents apply to individual factors or to a whole quantity.

Example 5   Using the Standard Order of Operations

Evaluate each of the following expressions.

(a) $5^2 + 12^2$ (b) $(5 + 12)^2$

(c) $\sqrt{25} + \sqrt{144}$ (d) $\sqrt{25 + 144}$

Solution

(a) $5^2 + 12^2 = 25 + 144$ Exponentiation has priority over addition.
$ = 169$

(b) $(5 + 12)^2 = 17^2$ First simplify within the parentheses, then square. Notice that
$ = 289$ this is a different answer than in part (a) of this example.

(c) $\sqrt{25} + \sqrt{144} = 5 + 12$ First evaluate each separate square root and then add these
$\phantom{\sqrt{25} + \sqrt{144}} = 17$ two terms.

(d) $\sqrt{25 + 144} = \sqrt{169}$ First add the terms underneath the square root symbol and then
$\phantom{\sqrt{25 + 144}} = 13$ take the square root of the result. Notice that this is a different answer than in part (c) of this example.

Self-Check 5

Evaluate each of the following expressions.

a. $2^2 + 3^2$ **b.** $(2 + 3)^2$ **c.** $\sqrt{25 - 9}$ **d.** $\sqrt{25} - \sqrt{9}$

How Do I Simplify Expressions That Contain Multiple Levels of Grouping Symbols?

Whenever more than one set of grouping symbols appears in an expression, you should start from the innermost pair of symbols and work toward the outermost pair of symbols. In Example 6 we first simplify within the parentheses and then within the brackets.

Example 6   Evaluating Expressions with Grouping Symbols

Evaluate $20 + 3[(7 + 8) - (3 + 6)]$.

Solution

$20 + 3[(7 + 8) - (3 + 6)]$ Simplify first within the parentheses and then within
$= 20 + 3[15 - 9]$ the brackets.
$= 20 + 3[6]$ Multiplication has priority over addition.
$= 20 + 18$
$= 38$

Self-Check 6

Evaluate $[14 - 5(13 - 11)][6 + 2(14 - 3)]$.

How Do I Enter Fractions or Expressions That Contain Multiple Levels of Grouping Symbols into a Calculator or Spreadsheet?

Very carefully! Written mathematics uses different levels (as in fractions) or different size parentheses and brackets. However, expressions entered into calculators, spreadsheets, and computer programs must be entered as one horizontal line. Brackets are not allowed, so nested parentheses must be used. In this case, you must be careful to keep track of pairs of left and right parentheses. These expressions are evaluated by starting within the inner-most grouping pair of parentheses. When you are translating a fraction to this horizontal one-line format, remember that the fraction bar is also a grouping symbol that separates all the terms in the numerator from all the terms in the denominator. Thus it is wise to include a set of parentheses around the entire numerator and the entire denominator.

Example 7   Entering Algebraic Expressions into a Calculator or Spreadsheet

Write each algebraic expression in the horizontal one-line format used by calculators or spreadsheets.

Solution

Algebraic Expression	Horizontal One-Line Format
(a) $20 + 3[(7 + 8) - (3 + 6)]$	$20 + 3((7 + 8) - (3 + 6))$
(b) $\dfrac{12 - 3(2)}{(12 - 3)(2)}$	$(12 - 3(2))/((12 - 3)(2))$

Self-Check 7

Write $\dfrac{a(b + c)}{x - y}$ in the horizontal one-line format used by calculators or spreadsheets.

---

3. Use the Distributive Property of Multiplication over Addition

Can I Rewrite Expressions That Have Parentheses So That Parentheses Are Not Needed?

Sometimes you can. Many expressions involving both multiplication and addition can be rewritten without parentheses if you use the distributive property of multiplication over addition. Compare the following expressions.

$$5(2 + 6) = 5(8) \qquad 5(2) + 5(6) = 10 + 30$$
$$= 40 \qquad\qquad\qquad = 40$$

Thus $5(2 + 6) = 5(2) + 5(6)$. In general, the **distributive property** of multiplication over addition says $a(b + c) = ab + ac$.

A Mathematical Note

The French mathematician Servois (ca. 1814) introduced the terms *commutative* and *distributive*. The term *associative* is attributed to the Irish mathematician William R. Hamilton (1805–1865).

Distributive Property of Multiplication Over Addition		
Algebraically	**Verbally**	**Numerical Examples**
For all real numbers a, b, and c, $a(b + c) = ab + ac$ and $(b + c)a = ba + ca$	Multiplication distributes over addition.	$5(2 + 6) = 5 \cdot 2 + 5 \cdot 6$ $5 \cdot 8 = 10 + 30$ $40 = 40$ $(2 + 6)5 = 2 \cdot 5 + 6 \cdot 5$ $8 \cdot 5 = 10 + 30$ $40 = 40$

Consider both $5(6 + 7)$ and $5(x + 7)$. We can use the distributive property to rewrite $5(6 + 7)$ as $5(6) + 5(7)$ and $5(x + 7)$ as $5x + 35$. Using the distributive property is not crucial with constants such as $5(6 + 7)$, but it is necessary to expand $5(x + 7)$ as $5x + 35$. We use the distributive property extensively both to expand expressions such as $5(x + 7)$ to $5x + 35$ and to factor expressions such as $5x + 35$ to $5(x + 7)$.

Example 8   Using the Distributive Property to Expand an Expression

Use the distributive property to expand each expression.

(a) $9(x + 4)$ **(b)** $11(2y - 3)$ **(c)** $-(7x - 3y)$

Solution

(a) $9(x + 4) = 9 \cdot x + 9 \cdot 4$ Distribute the factor of 9 to both the x term and the
$\qquad\qquad = 9x + 36$ 4 term.

(b) $11(2y - 3) = 11(2y) - 11(3)$ Multiplication distributes over both addition and
$\qquad\qquad = 22y - 33$ subtraction. We can first think of this expression as
$\qquad\qquad\qquad\qquad\qquad\qquad\qquad\qquad\qquad$ $11[2y + (-3)]$.

(c) $-(7x - 3y) = -1(7x - 3y)$ The factor of -1 is understood. Distribute the factor
$\qquad\qquad = (-1)(7x) - (-1)(3y)$ of -1 to both terms.
$\qquad\qquad = -7x + 3y$

Self-Check 8

Use the distributive property to expand each expression.

a. $5(2x - 4y)$ **b.** $-(2x - 3y + 4z)$

4. Simplify Expressions by Combining Like Terms

Can I Use the Distributive Property to Simplify Expressions Containing Variables?

Yes, the distributive property plays a key role in adding like terms and simplifying algebraic expressions. **Like terms** have exactly the same variable factors. Using the distributive property, we can rewrite $5x + 7x$ as $(5 + 7)x$ or $12x$. The constant factor in a term is called the **numerical coefficient** of the term. For example, the numerical coefficient of $5x$ is 5 and the numerical coefficient of $7x$ is 7. The coefficient of x is understood to be 1; that is, $1x = x$. The process of adding like terms is sometimes called **collecting like terms** or **combining like terms.** This is illustrated in Example 9.

Example 9   Simplifying an Expression by Combining Like Terms

Use the distributive property to combine the like terms.

(a) $13x + 8x$ **(b)** $13x - 8x$

(c) $2a + (3b + 4a)$ **(d)** $(9x - 2) - (7x - 5)$

Solution

(a) $13x + 8x = (13 + 8)x$ Usually we skip writing the middle step and perform this addition
$\qquad\qquad = 21x$ mentally. It is the distributive property that justifies what we are
$\qquad\qquad\qquad\qquad\qquad\qquad$ doing.

(b) $13x - 8x = (13 - 8)x$
$\qquad\qquad = 5x$

(c) $2a + (3b + 4a) = 2a + (4a + 3b)$

$\qquad\qquad\qquad = (2a + 4a) + 3b$

$\qquad\qquad\qquad = (2 + 4)a + 3b$

$\qquad\qquad\qquad = 6a + 3b$

Reorder by using the commutative property of addition. Then regroup by using the associative property of addition. Then add like terms, using the distributive property. Again, it is common to perform this addition mentally.

(d) $(9x - 2) - (7x - 5) = 9x - 2 - 7x + 5$

$\qquad\qquad\qquad\qquad = 9x + (-2) + (-7x) + 5$

$\qquad\qquad\qquad\qquad = 9x + (-7x) + (-2) + 5$

$\qquad\qquad\qquad\qquad = (9 + (-7))x + (-2 + 5)$

$\qquad\qquad\qquad\qquad = 2x + 3$

Distribute to remove grouping symbols and rewrite each subtraction in terms of addition. Use the commutative and associative properties of addition to reorder and regroup the terms. Finally, combine like terms. Again, it is common to perform most of these steps mentally.

Self-Check 9

Use the distributive property to add like terms.

a. $-8y + 17y$ **b.** $5x + (2x - 3)$

Self-Check Answers

1. a. 8   **b.** 9   **c.** $x^2 + y^3$ **5. a.** 13   **b.** 25   **8. a.** $10x - 20y$

2. a. -81 **b.** 81   **d.** x squared times y cubed   **c.** 4   **d.** 2   **b.** $-2x + 3y - 4z$

    c. -1 **d.** 1   **4. a.** 6   **b.** 42   **6.** 112   **9. a.** $9y$ **b.** $7x - 3$

3. a. x^6 **b.** $(2w)^3$ **c.** $\dfrac{1}{4}$ **d.** 26   **7.** $(a(b + c))/(x - y)$

Technology Self-Check Answer

1. 1,306.91232

1.6   Using the Language and Symbolism of Mathematics

1. In the expression w^4, the base is w and the exponent is _____.

2. _____ is read "x to the 7th power."

3. y^5 is read "y to the _____ power."

4. a^8 means that a is used as a _____ 8 times.

5. In the expression $(-x)^2$, the base is _____.

6. In the expression $-x^2$, the base is _____.

7. In the expression x, the exponent on the variable x is understood to be _____.

8. In the expression $2 - 5 \cdot 7$, the first operation to perform is _____.

9. In the expression $(2 - 5) \cdot 7$, the first operation to perform is _____.

10. In the expression $2 + 5^2 \cdot 3$, the first operation to perform is _____.

11. In the expression $3x^5$, the base is _____ and the exponent is 5.

12. In the expression $(3x)^5$, the base is _____ and the exponent is 5.

13. In the expression $7v$, 7 is the _____ of v.

14. In the expression w^3, the coefficient is understood to be _____.

15. Considering the commutative, associative, and distributive properties, the property that involves two operations is the _____ property of multiplication over addition.

16. The _____ property is used to add like terms.

17. The process of adding like terms is sometimes called collecting like terms or _____ like terms.

1.6 Quick Review

In Exercises 1–5, simplify each expression.

1. $3 \cdot 3 \cdot 3 \cdot 3$　　**2.** $-1 \cdot 3 \cdot 3 \cdot 3 \cdot 3$　　**3.** $(-4)(-4)$　　**4.** $-1 \cdot 4 \cdot 4$　　**5.** $\sqrt{64}$

1.6 Exercises

Objective 1 Use Natural Number Exponents

In Exercises 1 and 2, write each expression in exponential form.

1. a. $5 \cdot 5 \cdot 5 \cdot 5$　　　**b.** $(-4)(-4)(-4)$
　c. $y \cdot y \cdot y \cdot y \cdot y$　　**d.** $(3z)(3z)(3z)(3z)(3z)(3z)$
2. a. $7 \cdot 7 \cdot 7$　　　　　**b.** $(-6)(-6)(-6)(-6)$
　c. $w \cdot w \cdot w \cdot w \cdot w \cdot w$　**d.** $(-5z)(-5z)$

In Exercises 3 and 4, write each exponential expression in expanded form.

3. a. 4^3　　　　　　　　**b.** $(-3)^4$
　c. $(-x)^3$　　　　　　**d.** $(2y)^3$
4. a. x^3　　　　　　　　**b.** $(3y)^3$
　c. 3^4　　　　　　　　**d.** $(-4)^3$

In the exponential expression x^9, the base is x and the exponent is 9. For each expression in Exercises 5–8, determine the base and the exponent, and write the expression in expanded form.

Exponential Expression	Base	Exponent	Expanded Form
5. a. 3^4			
b. 4^3			
6. a. w^3			
b. $-w^3$			
7. a. $-3x^2$			
b. $(-3x)^2$			
8. a. $2w^3$			
b. $(2w)^3$			

In Exercises 9–16, mentally evaluate each expression.

9. a. $(-3)^2$　　　　　　**b.** -3^2
10. a. $(-5)^2$　　　　　**b.** -5^2
11. a. 0^4　　　　　　　**b.** $(-1)^4$
12. a. 0^{10}　　　　　　**b.** $(-1)^{10}$
13. a. $(0.1)^3$　　　　　**b.** $(-10)^3$
14. a. -0.1^4　　　　　**b.** $(-10)^4$
15. a. $\left(\dfrac{1}{2}\right)^3$　　　　**b.** $\left(-\dfrac{1}{2}\right)^3$
16. a. $\left(\dfrac{1}{2}\right)^5$　　　　**b.** $\left(-\dfrac{1}{2}\right)^5$

Objective 2 Use the Standard Order of Operations

In Exercises 17–52, calculate each expression using only pencil and paper.

17. a. $5 + 2 \cdot 8$　　　　　　**b.** $(5 + 2) \cdot 8$
18. a. $11 + 3 \cdot 9$　　　　　**b.** $(11 + 3) \cdot 9$
19. a. $17 - 3 \cdot 5$　　　　　**b.** $(17 - 3)(5)$
20. a. $41 - 6 \cdot 4$　　　　　**b.** $(41 - 6)(4)$
21. a. $(6 + 4)^2$　　　　　　**b.** $6^2 + 4^2$
22. a. $(6 - 4)^2$　　　　　　**b.** $6^2 - 4^2$
23. a. $25 \div 5 \cdot 5$　　　　　**b.** $25 \div (5 \cdot 5)$
24. a. $25 \div 5 \div 5$　　　　　**b.** $25 \div (5 \div 5)$
25. a. $19 - 19 \cdot 2$　　　　　**b.** $(19 - 19) \cdot 2$
26. a. $10 - 10 \cdot 3$　　　　　**b.** $(10 - 10) \cdot 3$
27. a. $4 - 7^2$　　　　　　　**b.** $(4 - 7)^2$
28. a. $1 - 9^2$　　　　　　　**b.** $(1 - 9)^2$
29. a. $(3 + 4)^2$　　　　　　**b.** $3^2 + 4^2$
30. a. $(5 + 7)^2$　　　　　　**b.** $5^2 + 7^2$
31. a. $3 \cdot 5 - 6 \cdot 7$　　　　**b.** $3 \cdot (5 - 6) \cdot 7$
32. a. $4 \cdot 9 - 8 \cdot 11$　　　**b.** $4 \cdot (9 - 8) \cdot 11$
33. a. $15 - 3^2 - 6$　　　　**b.** $15 - (3^2 - 6)$
34. a. $(15 - 3)^2 - 6$　　　**b.** $(15 - 3 - 6)^2$
35. a. $5^3 - 2^3$　　　　　　**b.** $(5 - 2)^3$
36. a. $10^3 - 5^3$　　　　　**b.** $(10 - 5)^3$
37. a. $36 - 24 \div 3 + 5$　　**b.** $36 - (24 \div 3 + 5)$
38. a. $24 \div 8 \cdot 2 - 4$　　　**b.** $24 \div 8 \cdot (2 - 4)$
39. a. $\dfrac{15 - 3(4)}{(15 - 3)(4)}$　　　　**b.** $\dfrac{15 \div 3(4)}{(15 - 3)(4)}$
40. a. $\dfrac{18 + 2(5)}{(18 + 2)(5)}$　　　**b.** $\dfrac{18 + 2^5}{(18 + 2)(5)}$
41. a. $\sqrt{9} + \sqrt{16}$　　　　**b.** $\sqrt{9 + 16}$
42. a. $\sqrt{169} - \sqrt{144}$　　**b.** $\sqrt{169 - 144}$
43. a. $\sqrt{169} - \sqrt{144} - \sqrt{16}$
　b. $\sqrt{169 - 144 - 16}$
44. a. $\sqrt{81} - \sqrt{64} - \sqrt{16}$
　b. $\sqrt{81 - 64 - 16}$

45. a. $|-23| + |17|$

 b. $|-23 + 17|$

46. a. $|36| + |-22|$

 b. $|36 - 22|$

47. $12 + 2[8 - 3(7 - 5)]$

48. $15 + 5[19 - 4(11 - 9)]$

49. $113 + 5[8 + 2(13 - 3^2) - 7]$

50. $217 + 6[11 - 3(50 - 7^2)]$

51. $\dfrac{5}{16} + \dfrac{1}{2} \cdot \dfrac{3}{8}$

52. $\dfrac{6}{5} - \dfrac{1}{5}\left(\dfrac{7}{8} + \dfrac{3}{8}\right)$

Objective 3 Use the Distributive Property of Multiplication over Addition

In Exercises 53–56, use the distributive property to expand each expression.

53. a. $7(x + 5)$ **b.** $-7(x + 5)$

54. a. $3(y + 9)$ **b.** $-3(y + 9)$

55. a. $-1(2x - 3y)$ **b.** $-(2x - 3y)$

56. a. $-1(9x - 4y)$ **b.** $-(9x - 4y)$

Objective 4 Simplify Expressions by Combining Like Terms

In Exercises 57–60, use the distributive property to combine like terms.

57. a. $2x + 5x$ **b.** $2x - 5x$

58. a. $3y - 8y$ **b.** $3y + 8y$

59. a. $(2a + 3b) + (4a - 5b)$

 b. $(2a + 3b) - (4a - 5b)$

60. a. $(3v - 4w) + (5v - 6w)$

 b. $(3v - 4w) - (5v - 6w)$

Multiple Representations

In Exercises 61 and 62, write each expression in terms of multiplication.

61. a. z^4 **b.** $z + z + z + z$

62. a. w^5 **b.** $w + w + w + w + w$

In Exercises 63–65, write each verbal expression as an algebraic expression.

Verbal Expression	Algebraic Expression

63. a. 5 times the quantity x plus $3y$

 b. $5x$ plus $3y$

64. a. x cubed minus y cubed

 b. x minus y the quantity cubed

65. a. The quantity $2xy$ raised to the 5th power

 b. $2x$ times y to the 5th power

In Exercises 66–68, write each algebraic expression as a verbal expression.

Algebraic Expression	Verbal Expression

66. a. $7x - y$

 b. $7(x - y)$

67. a. $x^2 + y^2$

 b. $(x + y)^2$

68. a. $3xy^6$

 b. $(3xy)^6$

Applying Technology

In Exercises 69–72, write each expression in the horizontal one-line format used by calculators. (*Hint:* See Example 7.)

69. $2[5 - (8 - 3)]$

70. $12 - 5[(11 - 4) - (9 - 7)]$

71. $\dfrac{5 - 7}{11 - 7}$

72. $\dfrac{15 - 3(7 - 1)}{-11 - (4 + 3)}$

In Exercises 73–76, each expression is given in the horizontal one-line format used by calculators. Rewrite each expression in the standard algebraic format.

73. $(x + 5y\text{^}3)/(x\text{^}2 + 9y)$

74. $w + x/y - z$

75. $(w + x)/(y - z)$

76. $2(x\text{^}2 - (y + z)\text{^}2)((x\text{^}2 - y\text{^}2) + z)$

Estimation Skills

In Exercises 77 and 78, mentally estimate the value of each expression and then select the most appropriate answer.

77. $(6.07 + 3.98)^2$

   A. 20.0025   **B.** 52.6853   **C.** 72.6853
   D. 101.0025   **E.** 221.0025

78. $(6.07)^2 + (3.98)^2$

   A. 20.0025   **B.** 52.6853   **C.** 72.6853
   D. 101.0025   **E.** 221.0025

Preparation for Solving Linear Equations

In Exercises 79–82, use the distributive property to simplify each expression. We use these skills to solve equations in Exercises 2.5.

79. $7(x - 1) - 4(2x + 3)$

80. $2(x + 1) - 3(4 - x)$

81. $5(2x - 1) + 3(x - 3)$

82. $-4(x - 6) + 2(13 - 3x)$

Group discussion questions

83. Error Analysis A student described 5^4 by saying, "This means four multiplications are performed with the factor 5." What is the error is this statement? Make a correct statement describing 5^4.

84. Error Analysis In trying to evaluate the expression $\dfrac{16}{2 + 8}$ on a TI-84 Plus calculator, a student enters the expression as shown. What is the error and how can it be corrected?

```
16/2+8
            16
```

85. Order of Operations Match each expression with the value of this expression.

a. $(2 + 3)(4)$		**A.** 9
b. $2 + 3(4)$		**B.** 12
c. $2 + 3 + 4$		**C.** 14
d. $2^3 + 4$		**D.** 20
e. $(2 + 3)^4$		**E.** 83
f. $2 + 3^4$		**F.** 625

1.6   Cumulative Review

1. Reduce $\dfrac{30}{75}$ to lowest terms.

2. Simplify $3\dfrac{2}{7} - 1\dfrac{2}{5}$.

3. Which of the operations of addition, division, multiplication, and subtraction are *not* commutative? Give examples to support your choices.

4. Correctly place either the $<$ or the $>$ symbol between each pair of expressions.
   a. 2^3 ___ 3^2 **b.** 2^5 ___ 5^2

5. Which of the symbols $=$, $<$, $\leq$, $>$, $\geq$, $\neq$, and $\approx$ can correctly be placed in the blank in the expression that follows? $\dfrac{2}{3}$ ___ 0.666

Section 1.7   Using Variables and Formulas

Objectives:

1. Evaluate an algebraic expression for specific values of the variables.
2. Use algebraic formulas.
3. Use subscript notation.
4. Check a possible solution of an equation.

1. Evaluate an Algebraic Expression for Specific Values of the Variables

One main distinction between arithmetic and algebra is that arithmetic uses constants while algebra uses both constants and variables. **Algebraic expressions** can be as simple as a single constant or a variable or can be more involved when we combine constants and variables with operations. Examples of algebraic expressions are:

$$-\frac{7}{12}, \pi, x, x + y, 2x + y - 11, x^2 - 3xy - y^2, |x| + |y|, \sqrt{x + y}$$

How Do I Evaluate Algebraic Expressions for Given Values of the Variables?

To **evaluate an algebraic expression** for given values of the variables means to replace each variable by the specific value given for that variable and then to simplify this expression. When substituting a constant for a variable, it is wise to use parentheses to avoid a careless error in sign or an incorrect order of operations.

Example 1   Evaluating an Algebraic Expression

Evaluate $3x - y + 4$ for $x = 7$ and $y = -5$.

Solution

$$3x - y + 4 = 3(7) - (-5) + 4$$

Position parentheses in place of the variables and then substitute 7 for x and -5 for y.

$$= 21 + 5 + 4$$
$$= 26 + 4$$
$$= 30$$

Self-Check 1

Evaluate $-7x - (2y + 5)$ for $x = 3$ and $y = -2$.

Example 2   Evaluating an Algebraic Expression

Evaluate $\dfrac{2x - yz}{3xy - z}$ for $x = 2$, $y = -5$, and $z = 6$.

Solution

$$\frac{2x - yz}{3xy - z} = \frac{2(2) - (-5)(6)}{3(2)(-5) - (6)}$$

Position parentheses in place of the variables and then substitute 2 for x, -5 for y, and 6 for z.

$$= \frac{4 + 30}{-30 - 6}$$

Simplify both the numerator and the denominator, and reduce the fraction to lowest terms.

$$= \frac{34}{-36}$$
$$= -\frac{17}{18}$$

Self-Check 2

Evaluate $\dfrac{x^2 - 3x - 5}{(x - 4)(2x + 1)}$ for $x = -1$.

2. Use Algebraic Formulas

An **equation** states that two expressions are equal. A **formula** is an equation that states a relationship between different quantities. Formulas play an important role in business and the sciences. For example, the formula $A = l \cdot w$ relates the area of a rectangle, A, to its length, l, and its width, w. Each variable used in a formula has a selected meaning and represents every possible value to which it applies.

Why Do Formulas Often Use Letters Other Than *x* and *y*?

It is helpful to use representative letters for variables, such as V for volume, P for perimeter, r for radius, and t for time. A list of common formulas, including many formulas from geometry, is given on the inside back cover of this book.

Example 3 | Calculating the Interest on an Investment

Use the formula $I = PRT$ to calculate the interest on an investment of $5,000 at a rate of 8% per year for 1 year.

Solution

$$I = PRT$$
$$I = (5,000)(0.08)(1)$$
$$I = 400$$

Substitute into the interest formula $5,000 for the principal P, 0.08 for the interest rate R of 8%, and 1 for the time T for 1 year.

Answer: The interest on this investment for 1 year is $400.

Self-Check 3

Use the formula $I = PRT$ to determine the interest on an investment of $13,500 at 8.5% for 1 year.

Many formulas involve squares or cubes, especially those involving geometric shapes. Example 4 illustrates the formula for the volume of a right circular cylinder.

Example 4 | Calculating the Volume of a Cylinder

Use the formula $V = \pi r^2 h$ to calculate the exact volume of this cylinder. Then approximate this value to the nearest 10 cm³.

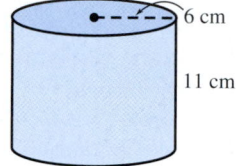
6 cm

11 cm

Solution

$$V = \pi r^2 h$$
$$V = \pi (6 \text{ cm})^2 (11 \text{ cm})$$
$$= \pi (36 \text{ cm}^2)(11 \text{ cm})$$
$$= 396\pi \text{ cm}^3$$
$$\approx 1,240 \text{ cm}^3$$

Substitute 6 cm for the radius r and 11 cm for the height h.

Then simplify this expression and use a calculator to make the approximation.

Self-Check 4

Use the formula $S = 4\pi r^2$ to calculate the exact surface area of a sphere with a radius of 5 cm. Then approximate this value to the nearest cm².

Can I Use Variables with a Graphing Calculator or in a Spreadsheet?

Yes, graphing calculators and spreadsheets allow us to input both constants and variables. Each cell in a spreadsheet is a variable that can store different values. The TI-84 Plus calculator also has variables that can store different real numbers at different times. To store values under any variable on a TI-84 Plus calculator, use the STO▸ and ALPHA keys, followed by any letter you need. The variable x is used so frequently that the X,T,θ,n key is provided to enter this variable with a single keystroke. Once the values have been stored under each variable, you may enter the expression to be evaluated.

Technology Perspective 1.7.1	Storing Values Under Variable Names

Evaluate $x^2 + 2x - 5$ for $x = 3$.

TI-84 Plus Calculator

```
3→X
                           3
X²+2X−5
                          10
```

Excel Spreadsheet

B1	▼	f_x	=A1^2+2*A1-5

	A	B	C
1	3	10	
2			

1. Press ③ (STO▸) (X,T,θ,n) (ENTER) to store the value of 3 for x.

2. Entering (X,T,θ,n) (x²) (+) ② (X,T,θ,n)

   (−) ⑤ (ENTER) will then evaluate this expression for $x = 3$.

1. Enter 3 into cell **A1**.

2. Entering "=A1^2+2*A1−5" into cell **B1** will evaluate this expression for the value of 3 stored in cell **A1**.

Technology Self-Check 1

Evaluate $6.2x^2 - 5.8x + 12.4$ for $x = 1.5$.

3. Use Subscript Notation

What Are Subscripts and Where Will I Use Them?

A small number or letter written next to and below a variable is called a **subscript**. For example, the subscripted variable x_1 is read "x sub 1" and a_n is read "a sub n." One use of subscripts is to distinguish among several possible cases or variables. For example, (x_1, y_1) and (x_2, y_2) can be used to refer to two distinct points both with x and y coordinates. In Example 5, we use subscripts in the formula for the slope of a line through two points. This formula is important to our study of lines in Chapter 3.

Example 5	Calculating the Slope of a Line

A line passes through the points $(x_1, y_1) = (2, 1)$ and $(x_2, y_2) = (6, 3)$. Use the formula $m = \dfrac{y_2 - y_1}{x_2 - x_1}$ to calculate the value of m, the slope of this line.

Solution

$m = \dfrac{y_2 - y_1}{x_2 - x_1}$ Substitute the appropriate values for (x_1, y_1) and (x_2, y_2) in the formula. Then simplify this fraction.

$\quad = \dfrac{3 - 1}{6 - 2}$

$\quad = \dfrac{2}{4}$

$\quad = \dfrac{1}{2}$

The slope is $\dfrac{1}{2}$.

Self-Check 5

A line passes through the points $(x_1, y_1) = (-3, 2)$ and $(x_2, y_2) = (5, -4)$. Use the formula $m = \dfrac{y_2 - y_1}{x_2 - x_1}$ to calculate the value of m, the slope of this line.

Another common use for subscript notation is to denote the terms of a sequence. A **sequence** is an ordered set of numbers with a first term, a second term, a third term, etc. The terms a_1, a_2, and a_n represent the first, second, and nth terms in the sequence. If a sequence follows a predictable pattern, then we may be able to describe this pattern with an equation for a_n. A typical application for a sequence is illustrated in Example 6.

Example 6 | Total of Monthly Automobile Payments

An automobile lease requires an initial payment of $1,200 followed by monthly payments of $300 for 2 years. The equation $a_n = 300n + 1,200$ gives the total of all payments made on this lease after n months. Use this equation to calculate the total paid after each of the first 3 months.

Solution

$a_n = 300n + 1,200$ Substitute 1, 2, and 3 for n.

$a_1 = 300(1) + 1,200 = 1,500$

$a_2 = 300(2) + 1,200 = 1,800$ Note that this sequence has a constant change of

$a_3 = 300(3) + 1,200 = 2,100$ 300 from term to term.

The totals after each of the first 3 months are $1,500, $1,800, and $2,100.

Self-Check 6

Use the equation $a_n = 4n - 3$ to calculate the first three terms of this sequence.

Another place that we use subscript-like notation is to denote the cells in spreadsheets. For example, A1 and A2 denote the first and second cells in column A. This is very similar to the notation mathematicians use for sequences. A TI-84 Plus calculator also uses Y1 and Y2 to denote two different functions that can be defined using the Y= key.

4. Check a Possible Solution of an Equation

A Mathematical Note

The first recorded use of $=$ as the equals symbol was by Robert Recorde, an English author, in 1557. His justification for this notation was "No two things coulde be more equalle than two straight lines."

What Does It Mean for a Value of a Variable to Be a Solution of an Equation?

An equation that contains a variable may or may not be a true statement when a value is substituted for the variable. A value of a variable that makes the equation true is called a **solution** of the equation, and this solution is said to **satisfy the equation.** The collection of all solutions is called the **solution set.**

For example, 3 is a solution of $x + 2 = 5$ because $3 + 2 = 5$; 3 satisfies the equation. To substitute a value into the equation to determine whether it satisfies the equation is referred to as a **check** of the value or as **checking a possible solution.**

In this section we will not solve equations. We will only practice checking possible solutions and establishing what it means to be a solution of an equation. We will use the notation $\overset{?}{=}$ to show that we are checking to determine whether the two sides of the equation are equal.

Example 7   Checking Possible Solutions of an Equation

Substitute 2 and 3 for x to determine whether either is a solution of $x + 5 = x + x + 2$.

Solution

Check $x = 2$:

$$x + 5 = x + x + 2$$
$$2 + 5 \overset{?}{=} 2 + 2 + 2 \qquad \text{Evaluate each side of the equation for } x = 2.$$
$$7 \overset{?}{=} 4 + 2$$
$$7 \overset{?}{=} 6 \quad \text{does not check} \qquad \text{Because the statement } 7 = 6 \text{ is false, 2 does not satisfy the equation.}$$

Answer: 2 is not a solution.

Check $x = 3$:

$$x + 5 = x + x + 2$$
$$3 + 5 \overset{?}{=} 3 + 3 + 2 \qquad \text{Evaluate each side of the equation for } x = 3.$$
$$8 \overset{?}{=} 6 + 2$$
$$8 \overset{?}{=} 8 \quad \text{checks} \qquad \text{Because the statement } 8 = 8 \text{ is true, 3 satisfies the equation.}$$

Answer: 3 is a solution.

Self-Check 7

Substitute -4 and 4 for x to determine whether either is a solution of $3(x - 1) + 2 = 2x + 3$.

In Chapter 6 we will solve equations that contain variables with an exponent of 2; such equations are called quadratic equations. Many quadratic equations have two distinct solutions. We will examine an instance of this in Example 8.

Example 8   Checking Possible Solutions of a Quadratic Equation

Substitute –6 and 5 for x to determine whether either is a solution of the quadratic equation $x^2 + 6x = 5x + 30$.

Solution

Check $x = -6$:

$$x^2 + 6x = 5x + 30$$
$$(-6)^2 + 6(-6) \overset{?}{=} 5(-6) + 30 \qquad \text{Evaluate both sides of the equation for } x = -6.$$
$$36 - 36 \overset{?}{=} -30 + 30 \qquad \text{Because the statement } 0 = 0 \text{ is true, } -6 \text{ satisfies the equation.}$$
$$0 \overset{?}{=} 0 \quad \text{checks}$$

Check $x = 5$:

$$x^2 + 6x = 5x + 30$$
$$(5)^2 + 6(5) \overset{?}{=} 5(5) + 30 \qquad \text{Evaluate both sides of the equation for } x = 5.$$
$$25 + 30 \overset{?}{=} 25 + 30 \qquad \text{Because the statement } 55 = 55 \text{ is true, 5 satisfies the equation.}$$
$$55 \overset{?}{=} 55 \quad \text{checks}$$

Both -6 and 5 are solutions of $x^2 + 6x = 5x + 30$.

Self-Check 8

Substitute -3, 2, and 6 for x to determine which two values satisfy the quadratic equation $x^2 - 12 = 3(x + 2)$.

You may find it useful in the workplace to use either a calculator or a spreadsheet to evaluate both sides of an equation. Then you can compare these values to determine whether the two sides of the equation are equal.

Self-Check Answers

1. -22
2. $-\dfrac{1}{5}$
3. $\$1,147.50$
4. $100\pi \text{ cm}^2 \approx 314 \text{ cm}^2$
5. $m = -\dfrac{3}{4}$
6. $1, 5, 9$
7. 4 is a solution
8. -3 and 6 are solutions

Technology Self-Check Answer

1. 17.65

1.7   Using the Language and Symbolism of Mathematics

1. To _____ an algebraic expression for given values of the variables means to replace each variable by the specific value given for that variable and then to simplify this expression.

2. An _____ is a statement that two expressions are equal.

3. A formula is an _____ that states the relationship between different quantities.

4. A value of a variable that makes an equation a true statement is called a _____ of the equation.

5. A solution of an equation is said to _____ the equation.

6. To substitute a value into an equation to determine whether it makes the equation true is referred to as _____ a possible solution of the equation.

7. The collection of all solutions of an equation is called the _____ set.

8. In formulas, we often use representative letters for each _____, such as V for volume and w for width.

9. A small number or letter written next to and below a variable is called a _____.

10. The algebraic notation for the verbal expression "x sub n" is _____.

11. An ordered set of numbers with a first term, a second term, a third term, and so on is called a _____.

1.7   Quick Review

Convert each percent to a decimal.

1. 23%
2. 1.4%
3. 0.04%

Convert each decimal to a percent.

4. 0.125
5. 1.4

Objective 1 Evaluate an Algebraic Expression for Specific Values of the Variables

In Exercises 1–4, evaluate $x^2 - 5x + 3$ for each value of x.

1. $x = 6$　　　　　　**2.** $x = -6$

3. $x = -10$　　　　　**4.** $x = 10$

In Exercises 5–8, evaluate each expression for $x = 3$ and $y = -5$.

5. $2x - 7y$　　　　　**6.** $2(x - 7y)$

7. $x^2 - y^2$　　　　　**8.** $(x - y)^2$

In Exercises 9–12, evaluate each expression for $w = 7$, $x = -3$, and $y = -6$.

9. $w + x + y$　　　　**10.** $-(w + x + y)$

11. $-w + (-x) + y$　　**12.** $w + (-x) + (-y)$

In Exercises 13–16, evaluate each expression for $x = 9$ and $y = 16$.

13. $\sqrt{x} + \sqrt{y}$　　　　**14.** $\sqrt{x + y}$

15. $\sqrt{25 - x}$　　　　**16.** $\sqrt{25} - \sqrt{x}$

In Exercises 17–20, evaluate each expression for $x = -4$ and $y = -6$.

17. $\dfrac{x + 3y}{x - 2y}$　　　　**18.** $\dfrac{5x - 4y}{4x + 5y}$

19. $\dfrac{x^2 - xy - y^2}{x^2 + 3xy + 2y^2}$　　**20.** $\dfrac{(x + y)^2}{x^2 + y^2}$

Objective 2 Use Algebraic Formulas

Simple Interest

In Exercises 21–24, use a calculator and the formula $I = PRT$ to calculate the interest for each investment.

21. An investment of $4,000 at a rate of 7% for 1 year

22. An investment of $4,000 at a rate of 7.25% for 1 year

23. An investment of $7,000 at a rate of 8.5% for 1 year

24. An investment of $7,000 at a rate of 8% for 1 year

Distance Traveled

The distance, D, an object travels can be calculated by finding the product of its rate, R, and the time, T, it travels. In Exercises 25–28, use the formula $D = RT$ to compute the distance each object travels.

25. A plane flies 450 miles per hour for 4 hours.

26. An explorer walks 3 days and averages 15 miles per day.

27. An armadillo crawls 5 meters per minute for 4.4 minutes.

28. A spacecraft flies 6,200 feet per second for 120 seconds.

Active Ingredient

The amount A of active ingredient in a solution can be calculated using the formula $A = RB$. This means that the percent of concentration R of the active ingredient is multiplied by the volume B of the solution. In Exercises 29–32, find the amount of active ingredient in each solution.

29. 500 milliliters of a 15% juice solution

30. 200 gallons of a 3.8% insecticide solution

31. 3 liters of a 22.5% hydrochloric acid solution

32. 1.5 gallons of a 70% antifreeze solution

Objective 3 Use Subscript Notation

In Exercises 33–36, use the formula $m = \dfrac{y_2 - y_1}{x_2 - x_1}$ to calculate the value of m, the slope of the line through each pair of points.

33. (3, 8) and (6, 17)　　**34.** (4, 1) and (2, 6)

35. (−3, 1) and (4, −5)　　**36.** (−2, −1) and (−4, −2)

In Exercises 37 and 38, calculate the first three terms of each sequence.

37. $a_n = 2n - 4$　　　**38.** $a_n = 10n - 7$

39. Automobile Lease An automobile lease requires an initial payment of $1,000 followed by monthly payments of $400 for 2 years. The equation $a_n = 400n + 1,000$ gives the total of all payments made on this lease after n months. Use this equation to calculate the total paid after each of the first 3 months.

40. Apartment Lease An apartment lease requires an initial deposit of $800 followed by monthly payments of $900. The equation $a_n = 900n + 800$ gives the total of all payments made on this lease after n months. Use this equation to calculate the total paid after each of the first 3 months.

Objective 4 Check a Possible Solution of an Equation

In Exercises 41–46, substitute both $x = 4$ and $x = 5$ into each equation to check whether either is a solution of the equation.

41. $x + 7 = 12$

42. $-x - 7 = -11$

43. $3(x - 1) = 2(x + 1) - 1$

44. $2(x + 3) = 3(x + 2) - 5$

45. $\dfrac{2x + 1}{x - 1} = 3$

46. $\dfrac{3(x - 1)}{x + 3} = \dfrac{3}{2}$

In Exercises 47–52, substitute both $x = -2$ and $x = 8$ into each equation to check whether either is a solution of the equation.

47. $2x + 1 = 3(x + 1)$ **48.** $4x - 3 = 3x + 5$

49. $(x + 2)(x - 8) = 0$ **50.** $x^2 - 6x - 16 = 0$

51. $\dfrac{2x - 7}{x - 5} = 3$ **52.** $\dfrac{3x + 6}{11x - 5} = 0$

In Exercises 53–56, substitute both $x = \dfrac{1}{2}$ and $x = \dfrac{3}{5}$ into each equation to check whether either is a solution of the equation.

53. $4x + 3 = 2(x + 2)$

54. $10(x - 1) = 5(3x - 2) - 3$

55. $10x^2 - 11x + 3 = 0$

56. $(2x - 1)(5x - 3) = 0$

Connecting Algebra to Geometry

57. Circumference of a Circle Substitute $r = 6.7$ into the formula $C = 2\pi r$ to determine the exact circumference of this circle in cm. Then approximate this length to the nearest tenth of a cm.

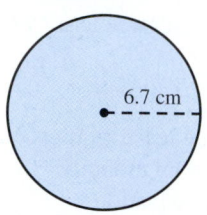
6.7 cm

58. Volume of a Cube Substitute $s = 9$ into the formula $V = s^3$ to calculate the volume in cm^3 of this cube.

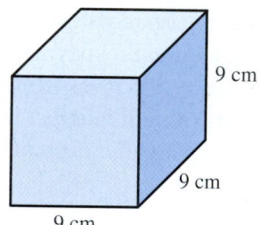
9 cm
9 cm
9 cm

Estimate then Calculate

The formula for determining the area of a rectangle is $A = LW$. Use the given width and length in Exercises 59 and 60 to mentally estimate the area of each rectangle. Then calculate the exact area.

59.

6.1 cm
8.8 cm

60.
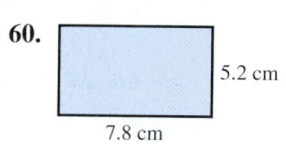
5.2 cm
7.8 cm

Applying Technology

In Exercises 61–64, use a calculator or a spreadsheet to evaluate each expression for $x = 12.45$.

61. $(2.5x - 3.3)(3.4x + 7.1)$

62. $(4.4x - 5.5)(3.3x + 6.6)$

63. $400x^2 - 70x - 30$

64. $500x^2 + 200x - 750$

Review and Concept Development

65. Percent of Decrease The price of an item was decreased from $400 to $300.
   a. What is the amount of the price decrease?
   b. What is the percent of decrease—that is, what percent of the original price is the decrease?

66. Percent of Increase The price of an item was increased from $300 to $390.
   a. What is the amount of the price increase?
   b. What is the percent of increase—that is, what percent of the original price is the increase?

Group discussion questions

67. Error Analysis A student produced the following work while trying to evaluate the expression $x^2 - 5x + 8$ for $x = -2$. What is the error and what would you suggest the student do to avoid similar errors in the future?

$$(-2)^2 - 5 - 2 + 8 = 4 - 7 + 8$$
$$= -3 + 8$$
$$= 5$$

68. Pattern Recognition For each of these tables, first use the pattern shown to complete the next two values in the table. Then describe verbally how you would obtain one more entry. Finally, write an equation for a_n, the nth term of the sequence.

a.

n	a_n
1	2
2	4
3	6
4	8
5	
6	

b.

n	a_n
1	3
2	5
3	7
4	9
5	
6	

c.

n	a_n
1	7
2	12
3	17
4	22
5	
6	

69. Risks and Choices Being a good citizen involves making personal decisions based on facts. Consider the following situation, which involves alcohol consumption. Three workers each had a drink after work. One drank a 12-ounce bottle of beer that was 5% alcohol; the second had a 4-ounce glass of wine that was 15% alcohol; and the third had a 1.5-ounce shot that was 40% alcohol. Which worker consumed the most alcohol? (*Hint:* See Exercises 29–32.)

70. Risks and Choices A web search for the "Karvonen formula" will produce information about the target heart rate for a person who is exercising. In this formula, a is the age of the person, r is the resting heart rate, and T is the target heart rate for fitness level training. Use the formula $T = r + 0.65(220 - a - r)$ to compute the fitness zone target heart rate for 3 or 4 people in your group. You can determine your heart rate by placing your index finger on the right side of your neck between the middle of your collarbone and your jaw line. Count the heartbeats from the carotid artery for 60 seconds.

1.7	**Cumulative Review**

1. Write 1.25 as:
  a. an improper fraction
  b. a mixed number
  c. a percent

2. Which is farther to the right on the real number line, 5.63 or $|-6.35|$?

3. Simplify $\dfrac{12}{35} \div \dfrac{18}{14}$.

4. Simplify $5 - 4(7 - 2)$.

5. Simplify $(5^2 - 4^2)^2$.

Section 1.8	**Geometry Review**

Objectives:

1. Classify and name common polygons.
2. Calculate the perimeter of common geometric figures.
3. Calculate the area of common geometric regions.
4. Calculate the volume of a rectangular solid.
5. Classify and name common angles.

What Is Geometry?

The word geometry literally means measurement of the earth (from the Greek words *geo* meaning earth and *metro* meaning measure). Geometry is the mathematics of shape and space. It's about the properties of objects: their length, area, volume, angles, and orientation. Geometry surrounds us in the design of buildings, automobiles, airplanes, books, computers, and dinnerware. Nature also abounds in geometric structures that are as small as molecular crystals or as large as spheres like the earth, moon, and sun. In fact, it is difficult to find objects in our universe that do not have a geometrical structure.

   The purpose of this section is to review several vocabulary items from geometry and to refresh some basic concepts. We will examine length, which is a one-dimensional measurement; area, which is a two-dimensional measurement; and volume, which is a three-dimensional measurement. Area is a measure of the amount of surface covered, and volume is a measure of the amount of space enclosed.

1. Classify and Name Common Polygons

A **polygon** is a closed geometric figure formed by straight line segments. Some common polygons are given here.

Sample measurements:

Length: _____ 1 cm

Area: 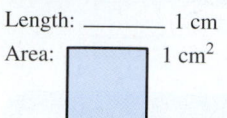 1 cm²

Volume: 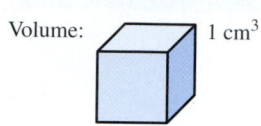 1 cm³

Triangle:	Quadrilateral:	Trapezoid:	Parallelogram:	Rhombus:
three sides	four sides	quadrilateral with two opposite sides parallel	quadrilateral with opposite sides parallel	parallelogram with four equal sides

Rectangle:	Square:	Pentagon:	Hexagon:	Octagon:
parallelogram with adjacent sides perpendicular	rectangle with four equal sides	five sides	six sides	eight sides

2. Calculate the Perimeter of Common Geometric Figures

How Do I Calculate the Perimeter of a Geometric Figure?

The **perimeter** of a geometric figure is the distance around the figure. If you were to loop a string snuggly around a figure, the length of this string would be the perimeter of the figure. The perimeter of a polygon can be calculated by adding the lengths of the sides. Some common units of length are: inches (in), feet (ft), miles (mi), centimeters (cm), meters (m), and kilometers (km).

Example 1 Calculating the Perimeter of an Isosceles Trapezoid

An isosceles trapezoid has two equal sides. Using P to represent the perimeter of this trapezoid, calculate P.

Isosceles Trapezoid

3 m

5 m   5 m

7 m

Solution

$P = (7\text{ m}) + (5\text{ m}) + (3\text{ m}) + (5\text{ m})$ Add the lengths of the 4 sides.

$= 20\text{ m}$

The perimeter is 20 m.

Self-Check 1

Calculate the perimeter of this rectangle.

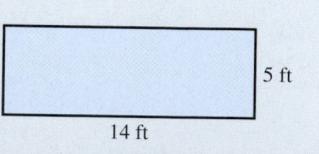

5 ft

14 ft

A **regular polygon** is a polygon whose sides are all the same length and whose angles are all the same. Regular polygons are often pleasing to the eye because they exhibit visual symmetry. In Example 2 we will calculate the perimeter of a regular hexagon.

Example 2   Calculating the Perimeter of a Regular Hexagon

A regular hexagon has 6 equal sides. Using P to represent the perimeter of this hexagon, calculate P.

Regular Hexagon

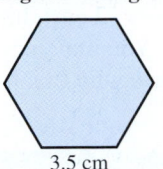

3.5 cm

Solution

$P = 6(3.5 \text{ cm})$ Each of the 6 equal sides is of length 3.5 cm.

$\quad = 21 \text{ cm}$

The perimeter is 21 cm.

Self-Check 2

Calculate the perimeter of this rhombus.

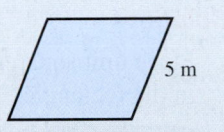

5 m

The perimeter of a circle is also called its **circumference** and can be calculated by multiplying the radius by 2π. We can express this by writing the formula $C = 2\pi r$, where C represents the circumference and r is the length of a radius.

Circle

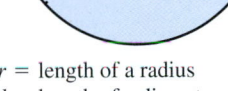

r = length of a radius
d = length of a diameter

Example 3   Calculating the Circumference of a Circle

Determine the exact circumference of this circle. Then use a calculator to approximate this circumference to the nearest hundredth of a cm.

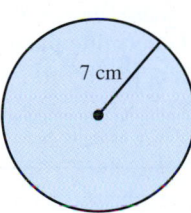

7 cm

Solution

$C = 2\pi r$ To calculate the circumference,
$\quad = 2\pi(7 \text{ cm})$ multiply the radius by 2π.
$\quad = 14\pi \text{ cm}$

The circumference is exactly 14π cm.

$C = 2\pi(7 \text{ cm})$ Then use a calculator with a π key to approximate the circumference.
$\quad \approx 43.98 \text{ cm}$ (*Hint:* The use of the pi key is illustrated in Technology Perspective 1.2.1.)

The circumference is approximately 43.98 cm.

Self-Check 3

Determine the exact circumference of this circle. Then approximate this length to the nearest tenth of a cm.

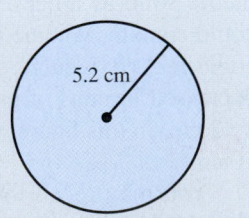

5.2 cm

3. Calculate the Area of Common Geometric Regions

What Does the Area of a Geometric Region Measure?

Rectangle: $A = 12$ cm^2

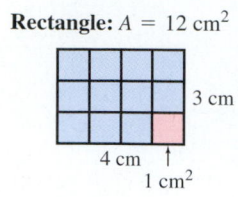

The **area** of a region is a measure of the surface enclosed by the region. Area is usually expressed in terms of unit squares such as square inches (in^2), square feet (ft^2), square miles (mi^2), square centimeters (cm^2), square meters (m^2) or square kilometers (km^2). The rectangle shown measures 3 cm by 4 cm and contains 12 squares, each 1 cm on a side. That is, the area is 12 cm^2. In this figure you can count each of the 12 unit squares.

You may recall the formula $A = LW$ for calculating the area of a rectangle. Example 4 shows why this formula works.

Example 4   Determining the Area of a Rectangle

Determine the area of the rectangle with a width of 2 cm and a length of 5 cm.

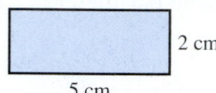

Solution

Place unit squares (1 cm on a side) in the rectangle.

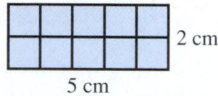

With a rectangle of width 2 cm and length 5 cm, we can position 2 rows of squares 1 cm on a side. Each row will contain 5 of these squares.

We can count 10 of these unit squares. Thus the area is 10 cm^2.

Also note that using the formula $A = LW$ gives $A = (5 \text{ cm})(2 \text{ cm}) = 10 \text{ cm}^2$.

Self-Check 4

Determine the area of the rectangle with a width of 4 ft and a length of 5 ft.

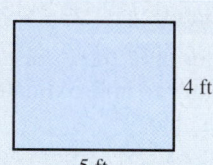

The rectangle shown in Example 4 with 2 rows and 5 columns has $2 \times 5 = 10$ unit squares. In general, a rectangle with L rows and W columns has $L \cdot W$ unit squares. Thus the area of a rectangle is $A = LW$. This formula and other area formulas are located on the inside jackets of this book.

4. Calculate the Volume of a Rectangular Solid

What Does the Volume of a Geometric Solid Measure?

Rectangular solid:
$V = 10$ cm^3

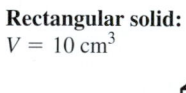

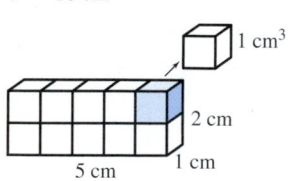

The **volume** of a solid is a measure of the space enclosed by the solid. If you think of a geometric solid as a bucket of some shape, its volume would measure the amount of liquid it could contain. Volume is usually expressed in terms of unit cubes such as cubic inches (in^3), cubic feet (ft^3), cubic centimeters (cm^3), or cubic meters (m^3). Liquid volume can also be expressed in units of gallons or liters. The rectangular solid shown measures 5 cm by 1 cm by 2 cm. In this figure you can count each of the 10 unit cubes. This means that the volume is 10 cm^3.

You may recall that the formula for calculating the volume of a rectangular solid is $V = LWH$. Example 5 shows why this formula works.

Example 5   Determining the Volume of a Rectangular Solid

Determine the volume of a rectangular solid measuring 2 cm by 3 cm by 5 cm.

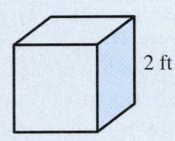

Solution

Place unit cubes (1 cm on a side) on both the first and second layers of the solid.

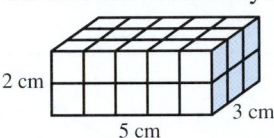

We can count 15 of these unit cubes on the first level and another 15 on the second level for a total volume of 30 cm³.

The first layer has a width of 3 cm and a length of 5 cm. We can position 3 rows of cubes 1 cm on a side. Each row will contain 5 of these cubes. Counting these cubes, we can determine that the first layer has 15 unit cubes. The second level likewise has 15 unit cubes.

Also note that using the formula $V = LWH$ gives $V = (5 \text{ cm})(3 \text{ cm})(2 \text{ cm}) = 30 \text{ cm}^3$.

Self-Check 5

Determine the volume of a cube that is 2 ft on each side.

The rectangular solid shown in Example 5 has 2 layers with each layer containing 3 rows and 5 columns of unit cubes. Thus this solid has $2 \times 3 \times 5$ unit cubes. In general, a rectangular solid with dimensions L, W, and H has $L \cdot W \cdot H$ unit cubes. Thus the volume of a rectangular solid is $V = LWH$. Formulas for the volumes of other solids are located on the inside jackets of this book.

5. Classify and Name Common Angles

How Do I Measure an Angle?

An angle cannot be measured with a ruler because the measure of an angle is not determined by measuring one of its sides but by measuring the opening between its sides. One common unit of measure for an angle is the **degree** (°). A symbol for an angle is ∠. If the measure of angle XYZ is 40 degrees, we can represent this by $m\angle XYZ = 40°$. This **protractor** shows that $m\angle XYZ = 40°$.

We classify angles according to their measures. Two angles are **equal** (or **congruent**) if they have the same measure. Note that the symbol "□" is used at the vertex of an angle to indicate a right angle.

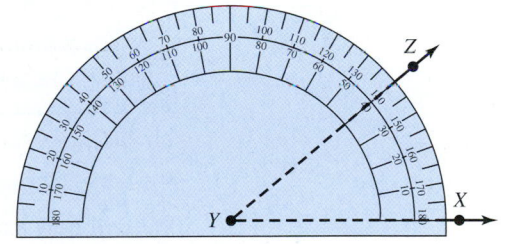

Acute Angle:	Right Angle:	Obtuse Angle:	Straight Angle:
between 0° and 90°	90°	between 90° and 180°	180°

The relationship of two angles to each other is very important in the components of many machines, buildings, and bridges. In the box we define complementary and supplementary angles.

Complementary and Supplementary Angles

Verbally	Algebraically	Graphically
Complementary: Two angles are complementary if the sum of their measures is 90°.	$m\angle a + m\angle b = 90°$	
Supplementary: Two angles are supplementary if the sum of their measures is 180°.	$m\angle a + m\angle b = 180°$	

Example 6 Determining the Measure of a Complementary Angle

A hinged waterproof hatchway on a submarine can hinge 90° between a closed position and an open position on a side wall. Thus the angle between the door and hatchway is complementary to the angle between the door and side wall. Represent this by saying $\angle A$ and $\angle B$ are complementary. If $m\angle A = 35°$, determine $m\angle B$.

Solution

$$m\angle B = 90° - m\angle A$$
$$= 90° - 35°$$
$$= 55°$$

If angles A and B are complementary,

$m\angle A + m\angle B = 90°$ or
$m\angle B = 90° - m\angle A$.

Self-Check 6

A metal fabrication shop must assemble a part so that $\angle A$ and $\angle B$ are supplementary. If $m\angle A = 47°$, determine $m\angle B$.

How Do I Classify Angles Formed by Intersecting Lines?

We will now examine some of the terminology for angles formed by intersecting lines. In the figure, lines L_1 and L_2 intersect to form angles a, b, c, and d.

Adjacent and Vertical Angles

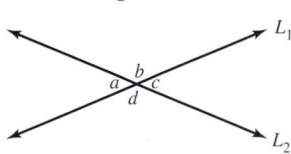

Adjacent angles are distinct angles that share a common vertex and one common side. $\angle a$ is adjacent to $\angle b$ and to $\angle d$. Likewise, $\angle b$ is adjacent to $\angle a$ and to $\angle c$.
Vertical angles are nonadjacent angles that share a common vertex and are opposite of each other. $\angle a$ and $\angle c$ are vertical angles. Likewise, $\angle b$ and $\angle d$ are vertical angles. In this figure, note that the adjacent angles are supplementary angles because they form a straight angle. Also, the measures of vertical angles are equal.

Example 7 Determining the Measure of Adjacent and Vertical Angles

Given that $m\angle a = 43°$, determine $m\angle b$, $m\angle c$, and $m\angle d$.

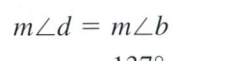

Solution

$m\angle b = 180° - m\angle a$ $\angle a$ and $\angle b$ form a straight angle
$\qquad = 180° - 43°$ and are therefore supplementary.
$\qquad = 137°$ That is, $m\angle a + m\angle b = 180°$ or $m\angle b = 180° - m\angle a$.

$m\angle c = m\angle a$
$\qquad = 43°$ $m\angle c = m\angle a$ because they are vertical angles.

$m\angle d = m\angle b$
$\qquad = 137°$ $m\angle d = m\angle b$ because they are vertical angles.

Self-Check 7

Given that $m\angle b = 128°$ in the figure in Example 7, determine $m\angle a$, $m\angle c$, and $m\angle d$.

How Do I Classify Angles Formed by One Line Intersecting Parallel Lines?

In the figure, line L_3 intersects the parallel lines L_1 and L_2. Line L_3 is called a **transversal.** Two angles that have the same relative positions are called **corresponding angles.** For example, $\angle a$ and $\angle e$ are corresponding angles. The measures of corresponding angles are equal. **Alternate interior angles** lie on opposite sides of the transversal and between the parallel lines. For example, $\angle d$ and $\angle f$ are alternate interior angles. The measures of alternate interior angles are equal. **Alternate exterior angles** lie on opposite sides of the transversal and outside the parallel lines. For example, $\angle a$ and $\angle g$ are alternate exterior angles. The measures of alternate exterior angles are equal.

Corresponding Angles, Alternate Interior Angles, Alternate Exterior Angles

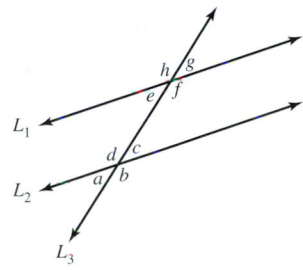

Example 8 Determining the Measure of a Corresponding Angle

A power line crosses the eastbound lane of an interstate highway so that $m\angle a = 40°$. The westbound lane of the interstate is parallel to the eastbound lane. Determine $m\angle e$.

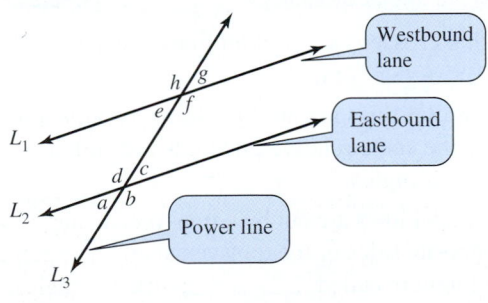

Solution

$m\angle e = m\angle a$ $\angle a$ and $\angle e$ are corresponding angles.
$\qquad = 40°$ Therefore their measures are equal.

Self-Check 8

Determine $m\angle c$ in Example 8.

The purpose of this section has been to review terminology and some basic concepts of length, area, volume, and angle relationships. We will use these concepts throughout this book. In particular, we will use the relationship of similar geometric figures in Section 2.7 when we examine applications of proportions.

Self-Check Answers

1. 38 ft
2. 20 m
3. 10.4π cm ≈ 32.7 cm
4. 20 ft^2
5. 8 ft^3

6. 133°
7. $m\angle a = 52°$,
   $m\angle c = 52°$,
   $m\angle d = 128°$
8. $m\angle c = 40°$

1.8   Using the Language and Symbolism of Mathematics

1. The word *geometry* literally means _____ of the earth.

2. A _____ is a closed geometric figure formed by straight line segments.

3. A _____ is a polygon with three sides.

4. A _____ is a polygon with four sides.

5. A _____ is a polygon with five sides.

6. A _____ is a polygon with six sides.

7. An _____ is a polygon with eight sides.

8. A _____ is a polygon with four equal sides.

9. A _____ is a quadrilateral with opposite sides parallel.

10. A _____ is a parallelogram with adjacent sides perpendicular.

11. A rhombus that is also a rectangle is a _____.

12. A _____ polygon is a polygon whose sides are all the same length and whose angles are all the same.

13. The _____ of a geometric figure is the distance around the figure.

14. The perimeter of a circle is also called its _____.

15. An approximation of π to the nearest hundredth is _____.

16. The _____ of a region is a measure of the surface enclosed by the region.

17. The _____ of a solid is a measure of the space enclosed by the solid.

18. A protractor is used to measure _____.

19. Two angles are _____ or congruent if they have the same measure.

20. An angle whose measure is between 0° and 90° is called an _____ angle.

21. An angle whose measure is 90° is called a _____ angle.

22. An angle whose measure is between 90° and 180° is called an _____ angle.

23. An angle whose measure is 180° is called a _____ angle.

24. Two angles are _____ if the sum of their measures is 90°.

25. Two angles are _____ if the sum of their measures is 180°.

26. _____ angles are distinct angles that share a common vertex and one common side.

27. If two lines intersect, they form four angles. The nonadjacent angles are called _____ angles.

28. A line that crosses two parallel lines is called a _____.

29. If two parallel lines are cut by a transversal, angles that have the same relative positions are called _____ angles.

30. If two parallel lines are cut by a transversal, angles that lie on opposite sides of the transversal and between the parallel lines are called _____ interior angles.

31. If two parallel lines are cut by a transversal, angles that lie on opposite sides of the transversal and outside the parallel lines are called alternate _____ angles.

1.8 | Quick Review

1. The irrational number π can be defined as the ratio of the circumference of a circle to its _____.

2. In decimal form, π is an infinite _____ decimal.

3. Given $L = 15$ and $W = 4$, use the formula $A = LW$ to evaluate A.

4. Given $s = 5$, use the formula $V = s^3$ to evaluate V.

5. Given $d = 11$, use the formula $C = \pi d$ to approximate C to the nearest hundredth.

1.8 | Exercises

Objective 1 Classify and Name Common Polygons

In Exercises 1–10, match each description with the most appropriate figure.

1. A three-sided polygon
2. A four-sided polygon
3. A five-sided polygon
4. A six-sided polygon
5. An eight-sided polygon
6. A regular quadrilateral
7. A quadrilateral with two opposite sides parallel
8. A quadrilateral with opposite sides parallel
9. A parallelogram with four equal sides
10. A parallelogram with adjacent sides perpendicular

   A. Hexagon
   B. Octagon
   C. Parallelogram
   D. Pentagon
   E. Quadrilateral
   F. Rectangle
   G. Rhombus
   H. Square
   I. Trapezoid
   J. Triangle

Objective 2 Calculate the Perimeter of Common Geometric Figures

In Exercises 11–16, calculate the perimeter of each figure.

11. **Perimeter of a Regular Pentagon**
Each side of the regular pentagon (five equal sides) shown is 9.4 cm long. Determine its perimeter.

9.4 cm

12. **Perimeter of a Rhombus** Each side of the rhombus (four equal sides) shown is 6.85 cm long. Determine its perimeter.

6.85 cm

13. **Perimeter of a Rectangle** Determine the perimeter of this rectangle.

3.9 cm

11.2 cm

14. **Perimeter of an Isosceles Triangle** An isosceles triangle has two equal sides. Determine the perimeter of this isosceles triangle.

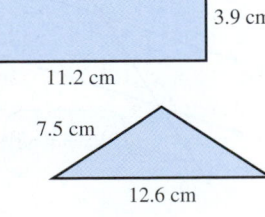

7.5 cm

12.6 cm

15. **Circumference of a Circle**
a. Determine the exact circumference of this circle.
b. Approximate this length to the nearest tenth of a cm.

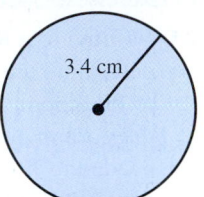

3.4 cm

16. **Circumference of a Circle**
a. Determine the exact circumference of this circle.
b. Approximate this length to the nearest tenth of a cm.

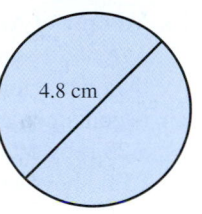

4.8 cm

Objective 3 Calculate the Area of Common Geometric Regions

In Exercises 17–22, calculate the area of each region.

17. **Area of a Soccer Field** A soccer field in Spain measures 75 m by 110 m. Determine the area of this rectangular field.

18. **Area of a Carpet** A rectangular bedroom measures 12 ft by 14 ft. Determine the number of square feet of carpet needed to cover the floor of this room.

19. **Area of the Base of a Cylindrical Tank** A cylindrical storage tank is installed with a special epoxy coating on its circular base. The radius of this tank is 3 m. Approximate to the nearest tenth of a square meter the area of the base that will be covered by this epoxy.

20. **Area of a Manhole Cover** The radius of a standard sanitary manhole cover in Columbus, Ohio, is 11.125 inches. Approximate to the nearest square inch the area of the top of this circular manhole cover.

21. **Area of a Semicircle** A semicircle is one-half of a circle. If the radius of a semicircle is 6 m, determine the exact area within this semicircle.

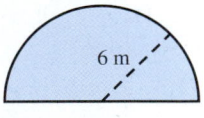

6 m

22. Area of a Region
Twenty-five percent of a
circular mat is painted red
and the rest is painted
blue. If the radius of this
circle is 7 m, determine to
the nearest tenth of a
square meter the area
painted red.

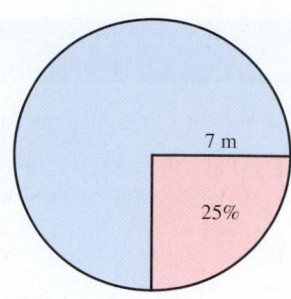

7 m

25%

Objective 4 Calculate the Volume of a Rectangular Solid

In Exercises 23–26, calculate the volume of each solid.

23. Volume of a Rectangular Swimming Pool To
determine the number of gallons of water needed
to fill a pool, one can first calculate the volume of
the pool in cubic feet. How many cubic feet are
needed to fill a 20 ft by 40 ft rectangular pool to a
depth of 8 ft?

24. Volume of Air in a Storage Room To determine the
size of air conditioner necessary to cool a storage room,
the contractor will first calculate the volume of air that
is contained in the empty room. What is the volume of
a 25 ft by 60 ft rectangular room that has a flat roof
15 ft above the floor?

25. Volume of a Spherical Water Tank The interior
radius of a spherical water tank is 15 ft.

   a. Use the formula $V = \dfrac{4}{3}\pi r^3$ to approximate the
   interior volume to the nearest ft^3.

   b. Each cubic foot contains approximately 7.48 gallons
   of water. Approximate to the nearest hundred gallons
   the capacity of this tank.

26. Volume of a Cylindrical Water Tank The radius of
this cylindrical tank is 8 ft and its height is 120 ft.
   a. Use the formula $V = \pi r^2 h$ to approximate the
   interior volume to the nearest ft^3.
   b. Each cubic foot contains approximately 7.48 gallons
   of water. Approximate to the nearest hundred gallons
   the capacity of this tank.

Objective 5 Classify and Name Common Angles

In Exercises 27–36, match each description with the most
appropriate choice. For Exercises 33–36, assume that L_1 is
parallel to L_2 in the figure.

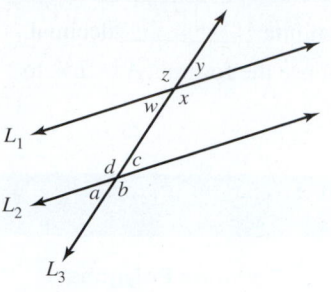

27. An angle that measures
between 0° and 90°
28. An angle that measures
between 90° and 180°
29. An angle that
measures 180° is a _____.
30. An angle that
measures 90° is a _____.
31. Two angles are
supplementary if the sum
of their measures is _____.
32. Two angles are
complementary if the sum
of their measures is _____.
33. Angles w and y are _____.
34. Angles w and x are _____.
35. Angles b and z are _____.
36. Angles c and w are _____.

A. 90°
B. 180°
C. Acute angle
D. Adjacent angles
E. Alternate exterior
   angles
F. Alternate interior
   angles
G. Obtuse angle
H. Right angle
I. Straight angle
J. Vertical angles

In Exercises 37–42, assume that $m\angle a = 36°$ in the figure
for Exercises 27–36. Determine the measure of each of the
following angles.

37. $\angle b$ **38.** $\angle c$ **39.** $\angle d$
40. $\angle x$ **41.** $\angle y$ **42.** $\angle z$

Buried Pipe

A water pipe is buried underneath an interstate highway so that
it crosses the eastbound lane with $m\angle a = 40°$. The westbound lane of the interstate is parallel to the eastbound lane.

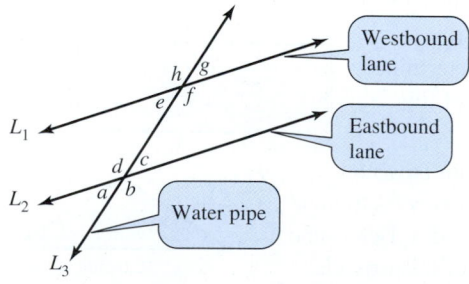

43. Determine $m\angle d$.   **44.** Determine $m\angle g$.

45. Supplementary Angles A metal fabrication shop must
assemble a part so that $\angle A$ and $\angle B$ are supplementary.
If $m\angle A = 21°$, determine $m\angle B$.

46. Complementary Angles A cabinet maker must assemble a part on a drawer so that $\angle A$ and $\angle B$ are complementary. If $m\angle A = 18°$, determine $m\angle B$.

Review and Concept Development

In Exercises 47–50, match each description with the most appropriate term.

47. A measurement of area
48. A measurement of length
49. A measurement of volume
50. A measurement of cost

A. Cent
B. Centimeter
C. Cubic foot
D. Square foot

Group discussion questions

51. Baseball Diamond
  a. Is the shape of a baseball diamond a polygon?
  b. If so, is it a quadrilateral?
  c. If so, is it a rhombus?
  d. If so, is it a rectangle?
  e. If so, is it a square?

52. Shapes of Traffic Signs Working with your group, list the shapes of several traffic signs and the meanings of the signs that have these shapes.

53. Inscribed Polygons A square, a regular pentagon, and a regular hexagon are each inscribed in a circle of radius 12 cm. Without doing any calculations, determine which polygon has:
  a. the least area
  b. the greatest area
  c. Justify your answers.

54. Area of a Triangle Use this figure and triangle *BCD* to write a justification that the area of a triangle is $A = \dfrac{1}{2}bh$. Assume that you know the formula for the area of a rectangle is $A = LW$.

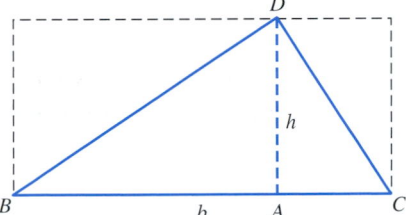

1.8   Cumulative Review

1. Match each equation with the property that is illustrated.
  a. $a + 2(xy + z) = a + 2xy + 2z$
  b. $a + 2(xy + z) = a + 2(yx + z)$
  c. $a + 2(xy + z) = a + 2(z + xy)$
  d. $a + (2x + z) = (a + 2x) + z$

  A. Associative property of addition
  B. Commutative property of addition
  C. Commutative property of multiplication
  D. Distributive property of multiplication over addition

Evaluate each of these expressions for $x = -3$ and $y = 8$.

2. $(x + y)^2$ **3.** $x^2 + y^2$ **4.** xy^2 **5.** x^2y^2

Chapter 1   Key Concepts

1. Preparing for an Algebra Class:
  • Set a specific educational goal for this academic term.
  • Become familiar with your class syllabus and textbook features.
  • Spend quality time in class and on homework outside of class.

  • Organize the steps in your homework in preparation for exams and for using mathematics outside the classroom.
  • Form a study group.
  • Use the textbook hints to prepare for tests.

2. Real Numbers: The real numbers consist of the rational numbers and the irrational numbers.

Rational Numbers:
- Real numbers that can be written as a ratio of two integers
- Real numbers that can be written as either terminating decimals or repeating decimals

Irrational Numbers:
- Real numbers, such as $\sqrt{2}$ or π, that cannot be written as a ratio of two integers
- Real numbers that can be written decimally only as infinite nonrepeating decimals

Subsets of the Rational Numbers:
- **Natural numbers:** 1, 2, 3, 4, 5, 6, ...
- **Whole numbers:** 0, 1, 2, 3, 4, 5, 6, ...
- **Integers:** ..., −3, −2, −1, 0, 1, 2, 3, ...

3. Properties of Addition and Multiplication

Additive identity: Zero is called the additive identity because 0 is the only real number with the property that $a + 0 = a$ and $0 + a = a$ for every real number a.

Multiplicative identity: One is called the multiplicative identity because 1 is the only real number with the property that $1 \cdot a = a$ and $a \cdot 1 = a$ for every real number a.

Additive inverses: Real numbers that are the same distance from the origin but on opposite sides of the origin are called opposites of each other or additive inverses. If a is a real number, the opposite of a is $-a$. The opposite of $-a$ is a. The numbers -2 and 2 are opposites of each other. The sum of a number and its opposite is 0. Both $2 + (-2) = 0$ and $-2 + 2 = 0$.

Multiplicative inverse: The reciprocal or multiplicative inverse of any nonzero real number a can be represented by $\dfrac{1}{a}$. The product of a number and its multiplicative inverse is positive 1. Both $\left(\dfrac{3}{5}\right)\left(\dfrac{5}{3}\right) = 1$ and $\left(\dfrac{5}{3}\right)\left(\dfrac{3}{5}\right) = 1$.

Commutative properties: Properties dealing with order of operations:
- $a + b = b + a$ Addition is commutative.
- $ab = ba$ Multiplication is commutative.

Associative properties: Properties dealing with grouping:
- $(a + b) + c = a + (b + c)$ Addition is associative.
- $(ab)(c) = a(bc)$ Multiplication is associative.

Distributive property of multiplication over addition: Property describing the relationship between multiplication and addition:
- $a(b + c) = ab + ac$ Multiplication distributes over addition.
- $(b + c)a = ba + ca$

4. Operations with Fractions

Addition: $\dfrac{a}{b} + \dfrac{c}{b} = \dfrac{a + c}{b}$ for $b \neq 0$

Subtraction: $\dfrac{a}{b} - \dfrac{c}{b} = \dfrac{a - c}{b}$ for $b \neq 0$

Multiplication: $\dfrac{a}{b} \cdot \dfrac{c}{d} = \dfrac{a \cdot c}{b \cdot d}$ for $b \neq 0$ and $d \neq 0$

Division:
$\dfrac{a}{b} \div \dfrac{c}{d} = \dfrac{a}{b} \cdot \dfrac{d}{c} = \dfrac{a \cdot d}{b \cdot c}$ for $b \neq 0,\, c \neq 0,$ and $d \neq 0$

Reducing fractions: $\dfrac{ac}{bc} = \dfrac{a}{b}$ for $b \neq 0$ and $c \neq 0$

5. Ratio: The ratio a to b can be denoted by either $a : b$ or $\dfrac{a}{b}$ for $b \neq 0$.

6. Addition of Real Numbers

Like signs: Add the absolute values of the numbers, and use the common sign of these terms.

Unlike signs: Find the difference of the absolute values of the numbers, and use the sign of the term that has the larger absolute value.

7. Subtraction of Real Numbers

To subtract y from x, add the opposite of y to x, using the rule for adding real numbers.

8. Multiplication of Real Numbers

Like signs: Multiply the absolute values of the two factors and use a positive sign for the product.

Unlike signs: Multiply the absolute values of the two factors and use a negative sign for the product.

Zero factor: The product of 0 and any other factor is 0.

Negative factors: The product is positive if the number of negative factors is even. The product is negative if the number of negative factors is odd.

9. Division of Real Numbers

Like signs: Divide the absolute values of the two numbers and use a positive sign for the quotient.

Unlike signs: Divide the absolute values of the two numbers and use a negative sign for the quotient.

Zero dividend: $\dfrac{0}{x} = 0$ for $x \neq 0$.

Zero divisor: $\dfrac{x}{0}$ is undefined for every real number x.

10. Exponential Notation: For any natural number n,

$$b^n = \underbrace{b \cdot b \cdot \cdots \cdot b}_{n \text{ factors of } b} \text{ with base } b \text{ and exponent } n.$$

11. Standard Order of Operations

Step 1 Start with the expression within the innermost pair of grouping symbols.

Step 2 Perform all exponentiations.

Step 3 Perform all multiplications and divisions as they appear from left to right.

Step 4 Perform all additions and subtractions as they appear from left to right.

12. Inequality Symbols and Interval Notation

Inequality Symbols	Meaning	Interval Notation
$x > a$	x is greater than a	(a, ∞)
$x \geq a$	x is greater than or equal to a	$[a, \infty)$
$x < a$	x is less than a	$(-\infty, a)$
$x \leq a$	x is less than or equal to a	$(-\infty, a]$
$a < x < b$	x is greater than a and less than b	(a, b)
$a < x \leq b$	x is greater than a and less than or equal to b	$(a, b]$
$a \leq x < b$	x is greater than or equal to a and less than b	$[a, b)$
$a \leq x \leq b$	x is greater than or equal to a and less than or equal to b	$[a, b]$

13. Absolute Value: The absolute value of a real number is the distance between this number and the origin on the real number line.

$$|x| = \begin{cases} x & \text{if } x \text{ is nonnegative} \\ -x & \text{if } x \text{ is negative} \end{cases}$$

14. Square Roots: If x is a positive real number, then $\sqrt{x}$ denotes a positive number r such that $r^2 = x$. For example, $\sqrt{9} = +3$ since $3^2 = 9$.

15. Adding Like Terms
- **Like terms:** Like terms have exactly the same variable factors.
- **Coefficient:** The constant factor for a term is called the numerical coefficient.
- **Using the distributive property:** To add like terms, use the distributive property to add the coefficients. For example, we can rewrite $5x + 7x$ as $(5 + 7)x$ or $12x$.

16. Evaluating an Algebraic Expression: To evaluate an algebraic expression for given values of the variables means to replace each variable by the specific value given for that variable and then simplify this expression. When a constant is substituted for a variable, it is wise to use parentheses to avoid careless errors.

17. Formula: A formula is an equation that states the relationship between different quantities. It is helpful to use representative letters for variables, such as V for volume or I for interest.

18. Subscript: A small number or letter written next to and below a variable is called a subscript. The subscripted variable x_1 is read "x sub 1" and a_n is read "a sub n." The notation used to identify cells like A1 in a spreadsheet is very similar to subscript notation.

19. Sequence: A sequence is an ordered set of numbers with a first term, a second term, a third term, etc. The terms a_1, a_2, and a_n represent the first, second, and nth terms in the sequence. Similarly A1, A2, and A3 represent the first three cells in column A of a spreadsheet.

20. Solution of an Equation
- **Solution:** A value of a variable that makes an equation true is called a solution of the equation, and this solution is said to satisfy the equation. The collection of all solutions is called the solution set.
- **Check:** To substitute a value into the equation to determine whether it satisfies the equation is referred to as a check of the value or as checking a possible solution.

21. Geometry Review
- **Polygon:** A polygon is a closed geometric figure formed by straight line segments.
- **Perimeter:** The perimeter of a geometric figure is the distance around the figure.
- **Area:** The area of a region is a measure of the surface enclosed by the region. Area is usually expressed in terms of unit squares such as cm^2 or ft^2.
- **Volume:** The volume of a solid is a measure of the space enclosed by the solid. Volume is usually expressed in terms of unit cubes such as cm^3 or ft^3.
- **Angles:** One common unit of measure for an angle is the degree (°). A protractor can be used to measure an angle. Acute angles measure less than 90°, a right angle measures 90°, an obtuse angle measures between 90° and 180°, and a straight angle measures 180°. Two angles are complementary if the sum of their measures is 90°. Two angles are supplementary if the sum of their measures is 180°.

Chapter 1 | Review Exercises

Operations with Real Numbers

In Exercises 1–24, perform the indicated operations.

1. a. $-16 + 4$ **b.** $-16 + (-4)$
   c. $-16 - 4$ **d.** $-16 - (-4)$

2. a. $-16(4)$ **b.** $-16(-4)$
   c. $-16 \div 4$ **d.** $-16 \div (-4)$

3. a. $-7 + 0$ **b.** $-7(0)$
   c. $\dfrac{0}{-7}$ **d.** $\dfrac{-7}{0}$

4. a. $24 + (-6)$ **b.** $24 - (-6)$
   c. $24(-6)$ **d.** $24 \div (-6)$

5. a. $9 + 0.01$ **b.** $9 - 0.01$
   c. $9(0.01)$ **d.** $9 \div 0.01$

6. a. $-4.5 + 1{,}000$ **b.** $-4.5 - 1{,}000$
   c. $-4.5(1{,}000)$ **d.** $-4.5 \div 1{,}000$

7. a. $\dfrac{2}{3} + \dfrac{3}{4}$ **b.** $\dfrac{2}{3} - \dfrac{3}{4}$
   c. $\left(\dfrac{2}{3}\right)\left(\dfrac{3}{4}\right)$ **d.** $\left(\dfrac{2}{3}\right) \div \left(\dfrac{3}{4}\right)$

8. a. $-\dfrac{14}{15} + \dfrac{21}{25}$ **b.** $-\dfrac{14}{15} - \dfrac{21}{25}$
   c. $-\dfrac{14}{15} \cdot \dfrac{21}{25}$ **d.** $-\dfrac{14}{15} \div \dfrac{21}{25}$

9. a. 0^{37} **b.** 0^{38}
   c. $(-1)^{37}$ **d.** $(-1)^{38}$

10. a. $-7 - 8 + 9$ **b.** $-7 - (8 + 9)$

11. a. $-6 + 8 - 11 - 15$ **b.** $(-6 + 8) - (11 - 15)$

12. a. $(15 - 6)(9 - 11)$ **b.** $15 - 6(9 - 11)$

13. a. $-36 \div 4 \cdot 3$ **b.** $-36 \div (4 \cdot 3)$

14. a. $(-10)^2$ **b.** -10^2

15. a. $(3 + 4)^2$ **b.** $3^2 + 4^2$

16. a. 5^2 **b.** 2^5

17. a. $|3 - 11|$ **b.** $|3| - |11|$

18. a. $\sqrt{64} + \sqrt{36}$ **b.** $\sqrt{64 + 36}$

19. a. $(-1)(-2)(-3)(-4)$ **b.** $(-1)(-2)(-3)(-4)(-5)$

20. $-48 + 12 \div 6 + 3 - 1$

21. $14 - 2[11 - 5(13 - 10)]$

22. $19 + 3[(40 - 7) - (4 + 3^2)]$

23. $\dfrac{-3 + 7}{8 - 10}$

24. $\dfrac{14 - 7(-5)}{-3(9) - 2(-10)}$

In Exercises 25 and 26, write a fraction to represent the shaded portion of each figure and then reduce this fraction to lowest terms.

25. 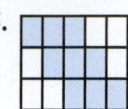   **26.**

Using the Language and Symbolism of Mathematics

In Exercises 27–31, match each algebraic equation with the property it illustrates.

27. $(xy)z = x(yz)$
28. $(x + y) + z = x + (y + z)$
29. $w(xy + z) = w(yx + z)$
30. $w(xy + z) = wxy + wz$
31. $w(xy + z) = w(z + xy)$

A. Associative property of addition
B. Associative property of multiplication
C. Commutative property of addition
D. Commutative property of multiplication
E. Distributive property of multiplication over addition

In Exercises 32–35, match each verbal description with the number that is the most appropriate choice.

32. The additive identity   **A.** -0.2
33. The multiplicative identity   **B.** 0
34. The multiplicative inverse of 0.2   **C.** 1
35. The additive inverse of 0.2   **D.** 5

In Exercises 36–39, match each verbal description with the number that is the most appropriate choice.

36. A natural number   **A.** 0
37. An integer that is not a natural number   **B.** $\dfrac{3}{4}$
38. An irrational number
39. A rational number that is not an integer   **C.** $\sqrt{4}$
40. List each real number from   **D.** $\sqrt{5}$
$$\left\{ -8.1, -7, 0, \sqrt{9}, \pi, \frac{15}{2}, 15 \right\} \text{ that is a(n):}$$

  a. Natural number   **b.** Whole number
  c. Integer   **d.** Rational number
  e. Irrational number

Multiple Representations

In Exercises 41–49, write each verbal statement in algebraic form.

41. The opposite of x equals eleven.

42. The absolute value of x is less than or equal to seven.

43. The square root of twenty-six is greater than five.

44. The sum of x and five is equal to four.

45. The product of negative three and y is negative one.

46. The quotient of x and three is 12.

47. The ratio of x to y equals three-fourths.

48. Five times the quantity three x minus four equals thirteen.

49. The change from x_1 to x_2.

In Exercises 50–53, write each algebraic expression in verbal form.

50. $x^2 - y^2$ **51.** $(x - y)^2$

52. $7(x - 5)$ **53.** $7x - 5$

54. Match each expression in the first column with an alternate representation for this expression from the second column.

a. $4 \cdot 3$	**A.** $0.333\ldots$
b. 4^3	**B.** 0.75
c. 3^4	**C.** 50%
d. -3^4	**D.** $3 + 4 = 7$
e. $(-3)^4$	**E.** $3 \cdot 5 = 15$
f. $15 \div 5 = 3$	**F.** $4 \cdot 4 \cdot 4$
g. $7 - 4 = 3$	**G.** $3 + 3 + 3 + 3$
h. $\dfrac{3}{4}$	**H.** $(-3)(-3)(-3)(-3)$
	I. $-3 \cdot 3 \cdot 3 \cdot 3$
i. $\dfrac{1}{3}$	**J.** $3 \cdot 3 \cdot 3 \cdot 3$
j. $\dfrac{1}{2}$	

In Exercises 55 and 56, write each expression in the horizontal one-line format used by calculators.

55. $\dfrac{17 + 7^2}{20 - 9}$

56. $13 - 4[(2 - 8) - 3(7 - 2)]$

In Exercises 57–60, use the given information to complete each row as illustrated by the example in the first row.

	Inequality Symbols	Verbally	Graphically	Interval Notation
Example:	$x > 2$	x is greater than 2.	⟵(⟶ 2	$(2, \infty)$
57.	$x < 3$			
58.		x is greater than or equal to -1		
59.			⟵[ ] ⟶ 4 10	
60.				$(-2, 3]$

Estimate Before You Calculate

In Exercises 61–67, do not calculate the result of each expression. Instead only mentally determine what the sign of the result will be.

Problem	Sign of the result
61. $-312 + (-221)$	
62. $-41 - (-72)$	
63. $(53)(-37)(-21)$	
64. $(-123.3)(0)(-893.8) + 1$	
65. $(-3)^2 \cdot 5 \cdot (-7)$	
66. $\sqrt{196} - \sqrt{361}$	
67. $(-7.2)^{12}$	

In Exercises 68–73, mentally estimate the value of each expression and then select the most appropriate answer.

68. $-5.96 + 2.01$
  A. -11.98 **B.** -7.97
  C. -3.95 **D.** -2.97

69. $-5.96 - 2.01$
  A. -11.98 **B.** -7.97
  C. -3.95 **D.** -2.97

70. $-5.96(2.01)$
  A. -11.98 **B.** -7.97
  C. -3.95 **D.** -2.97

71. $-5.96 \div 2.01$
  A. -11.98 **B.** -7.97
  C. -3.95 **D.** -2.97

72. $\dfrac{1}{2} - \dfrac{1}{8}$
  A. Less than -1 **B.** Between -1 and 0
  C. Between 0 and 1   **D.** Greater than 1

73. $0.6666 - \dfrac{2}{3}$
  A. Less than -1 **B.** Between -1 and 0
  C. Between 0 and 1   **D.** Greater than 1

Algebraic Expressions Containing Variables

In Exercises 74–77, simplify each expression by using the distributive property to add like terms.

74. $2x - 5y + 7x$

75. $5a - (3a - b)$

76. $5(3a + 2) - 4(2a - 3)$

77. $8(2x - 1) - 3(3x - 5)$

In Exercises 78 and 79, evaluate each expression for $x = -8$ and $y = 7$.

78. $x^2 - 2xy - y^2$

79. $\dfrac{(x - y)(x + y)}{2x - 3y}$

80. Use the formula $m = \dfrac{y_2 - y_1}{x_2 - x_1}$ to calculate the value of m, which gives the slope of a line through the points $(5, -4)$ and $(2, 2)$.

81. Calculate the first three terms of the sequence with $a_n = 10n - 3$.

In Exercises 82–85, substitute both $x = -7$ and $x = 4$ into each equation to determine whether either is a solution of the equation.

82. $5x - 17 = x - 1$

83. $4x + 29 = x + 8$

84. $(x + 7)(x - 4) = 0$

85. $\dfrac{4x - 1}{2x - 3} = 3$

Connecting Concepts to Applications

86. Total Cost Determine the total cost of three items whose prices are $45, $35, and $58.

87. Temperature Change Determine the change in temperature from $-6°$ to $5°$.

88. Price Discount A store advertised a 1-day sale with every item 40% off the marked price. If an item is marked at $65, how much is the 1-day discount on this item?

89. Percent of Increase The price of an item was increased from $600 to $630.
   a. What is the amount of the price increase?
   b. What is the percent of increase—that is, what percent of the original price is the increase?

90. Pulley Ratio The radius of one pulley is 32 cm and the radius of a second pulley is 20 cm. Determine the ratio of the radius of the larger pulley to that of the smaller pulley.

91. Active Ingredient Calculate the amount of the active ingredient in each mixture.
   a. 400 liters of a 10% insecticide mixture
   b. 600 gallons of a 5% ethanol mixture

92. Rate of Work Use the formula $R = \dfrac{W}{T}$ to compute the rate of work for each situation.
   a. An assembly line produces 1,600 golf balls in 8 hours.
   b. A pipe can fill one brewery vat in 4 hours.

93. Perimeter and Area
   a. Determine the perimeter of the volleyball court shown here.
   b. Determine the area of the volleyball court shown here.

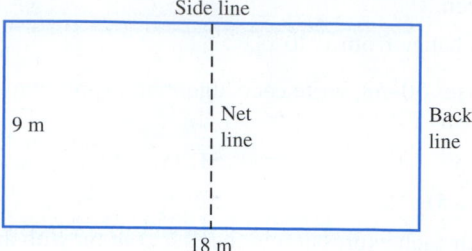

94. Surface Area and Volume The crate in the photo has a length of 6 ft, a width of 5 ft, and a height of 4.5 ft.

   a. Use the formula $V = LWH$ to determine the volume of this crate.
   b. Use the formula $S = 2LW + 2HL + 2HW$ to determine the surface area of this crate.

95. Circumference and Area The top of a circular patio table has a radius of 95 cm.
   a. What is the circumference of the top of this table?
   b. What is the area of the top of this table?

96. Complementary Angles Two of the angles formed on the flap of an airplane control surface are complementary angles. One of these angles measures $63°$. What is the measure of the other angle?

Chapter 1 | Mastery Test

Objective 1.1.1 Understand the Class Syllabus and Textbook Features

1. a. The name of your instructor is _____.
   b. Your instructor's office is in _____.
   c. Your instructor's e-mail address is _____.
   d. The back of the book contains the answers to the _____-numbered exercises.

Objective 1.2.1 Identify Additive Inverses

2. Write the additive inverse of each real number.
   a. -2
   b. 1.5
   c. $\dfrac{1}{5}$
   d. 0

Objective 1.2.2 Evaluate Absolute Value Expressions

3. Evaluate each of these absolute value expressions.
   a. $|23|$
   b. $|-23|$
   c. $|23 - 23|$
   d. $|23| + |-23|$

Objective 1.2.3 Use Inequality Symbols and Interval Notation

4. Express each of these intervals using interval notation.
   a.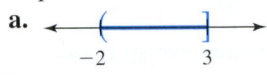

   $-2 \qquad 3$

   b.

   -1

   c. $x < 4$
   d. $5 \leq x < 9$

Objective 1.2.4 Mentally Estimate Square Roots and Use a Calculator to Approximate Square Roots

5. Mentally estimate to the nearest integer the value of each of the following square roots. Then use a calculator or spreadsheet to approximate each value to the nearest hundredth.
   a. $\sqrt{26}$
   b. $\sqrt{99}$
   c. $\sqrt{4 + 4}$
   d. $\sqrt{4} + \sqrt{4}$

Objective 1.2.5 Identify Natural Numbers, Whole Numbers, Integers, Rational Numbers, and Irrational Numbers

6. Match each number with the most appropriate description.
   a. -5
   b. $-\dfrac{2}{3}$
   c. 0
   d. $\sqrt{9}$
   e. $\sqrt{11}$

   A. A whole number that is not a natural number
   B. An integer that is not a whole number
   C. A rational number that is not an integer
   D. A positive number that is an irrational number
   E. A natural number

Objective 1.3.1 Reduce Fractions to Lowest Terms

7. Write a fraction in lowest terms that represents the shaded portion of each figure.
   a.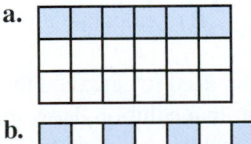

   b.

8. Reduce each of these fractions to lowest terms.
   a. $\dfrac{66}{99}$
   b. $\dfrac{75}{120}$

Objective 1.3.2 Multiply and Divide Fractions

9. Perform the indicated multiplications and divisions and write the answer in lowest terms.
   a. $\dfrac{1}{3} \cdot \dfrac{1}{5}$
   b. $\dfrac{3}{7} \cdot \dfrac{7}{9}$
   c. $\dfrac{2}{5} \div \dfrac{14}{15}$
   d. $\dfrac{12}{35} \div \dfrac{14}{45}$

Objective 1.3.3 Add and Subtract Fractions with the Same Denominator

10. Perform the indicated additions and subtractions and write the answer in lowest terms.
    a. $\dfrac{3}{8} + \dfrac{1}{8}$
    b. $\dfrac{5}{6} - \dfrac{1}{6}$
    c. $\dfrac{1}{42} + \dfrac{13}{42}$
    d. $\dfrac{13}{42} - \dfrac{1}{42}$

Objective 1.3.4 Add and Subtract Fractions with Different Denominators

11. Perform the indicated additions and subtractions and write the answer in lowest terms.
    a. $\dfrac{2}{5} + \dfrac{3}{8}$
    b. $\dfrac{5}{6} - \dfrac{3}{5}$
    c. $\dfrac{2}{7} + \dfrac{5}{6}$
    d. $\dfrac{7}{15} - \dfrac{3}{35}$

Objective 1.3.5 Perform Operations with Mixed Numbers

12. Perform each operation and write the answer in lowest terms.
    a. $11\dfrac{1}{8} + 5\dfrac{3}{8}$
    b. $11\dfrac{1}{8} - 5\dfrac{3}{8}$
    c. $\left(2\dfrac{1}{7}\right)\left(4\dfrac{2}{3}\right)$
    d. $\left(2\dfrac{1}{7}\right) \div \left(1\dfrac{2}{3}\right)$

Objective 1.4.1 Add Positive and Negative Real Numbers

13. Calculate each sum.
    a. $17 + (-11)$
    b. $-17 + (-11)$
    c. $-\dfrac{3}{8} + \dfrac{1}{8}$
    d. $57.3 + (-57.3)$

Objective 1.4.2 Use the Commutative and Associative Properties of Addition

14. a. The property that says that $5(6 + 7) = 5(7 + 6)$ is the _____ property of addition.
    b. The property that says that $5 + (6 + 7) = (5 + 6) + 7$ is the _____ property of addition.
    c. Use the associative property of addition to rewrite $(x + y) + 5$.
    d. Use the commutative property of addition to rewrite $5(x + y)$.

Objective 1.4.3 Subtract Positive and Negative Real Numbers

15. Calculate each difference.
    a. $15 - 8$
    b. $-15 - 8$
    c. $\dfrac{1}{2} - \left(-\dfrac{1}{3}\right)$
    d. $-7.35 - (-3.75)$

Objective 1.5.1 Use the Commutative and Associative Properties of Multiplication

16. a. The property that says that $5 \cdot (6 + 7) = (6 + 7) \cdot 5$ is the _____ property of multiplication.
    b. The property that says that $5 \cdot (6 \cdot 7) = (5 \cdot 6) \cdot 7$ is the _____ property of multiplication.
    c. Use the associative property of multiplication to rewrite $(5x)(y)$.
    d. Use the commutative property of multiplication to rewrite $x(a + b)$.

Objective 1.5.2 Multiply Positive and Negative Real Numbers

17. Calculate each product.
    a. $5(-6)$
    b. $-5(-6)$
    c. $-5(0)$
    d. $\left(-\dfrac{6}{35}\right)\left(-\dfrac{14}{15}\right)$

Objective 1.5.3 Divide Positive and Negative Real Numbers

18. Calculate each quotient.
    a. $-12 \div 4$
    b. $-12 \div (-4)$
    c. $0 \div 4$
    d. $-4{,}578.91 \div (-0.001)$

Objective 1.5.4 Express Ratios in Lowest Terms

19. Write each ratio in lowest terms.
    a. $16{:}12$
    b. **Aces in a Deck of Cards** Four of the 52 cards in a deck of cards are aces. What is the ratio of aces to all cards in the deck?
    c. **Gear Ratio** The front gear of a bicycle has 52 teeth and the rear gear has 24 teeth. What is the ratio of the teeth on the front gear to those on the rear gear?
    d. **Defective Keyboard Trays** The computer support services at Parkland College reported that 21 of 105 keyboard trays in an open computer lab had to be repaired one year. What is the ratio of trays that were repaired to all of the trays in the lab?

Objective 1.6.1 Use Natural Number Exponents

20. Calculate the value of each exponential expression.
    a. 2^5
    b. $(-5)^2$
    c. -5^2
    d. $(-1)^{219}$

Objective 1.6.2 Use the Standard Order of Operations

21. Calculate the value of each expression.
    a. $15 - 2(8 - 5)$
    b. $4 \cdot 7 - 2 \cdot 3 + 5$
    c. $15 - 63 \div 3^2 + 11$
    d. $\dfrac{14 - 3 \cdot 6}{(14 - 3) \cdot 6}$
    e. $(3 + 5)^2$
    f. $3^2 + 5^2$
    g. $-8 - 3[5 - 4(6 - 9)]$
    h. $\sqrt{25 - 16} - (\sqrt{25} - \sqrt{16})$

Objective 1.6.3 Use the Distributive Property of Multiplication over Addition

22. Use the distributive property to expand each expression.
    a. $3(5x + 9)$
    b. $-4(8x - 5)$
    c. $\dfrac{1}{2}(4x - 6y)$
    d. $-(11x - 13y)$

Objective 1.6.4 Simplify Expressions by Combining Like Terms

23. Use the distributive property to simplify each expression by combining like terms.
    a. $20x - 11x + 3x$
    b. $20x - 11(x + 3)$
    c. $(-5x - 9) - (4x - 3)$
    d. $3(x - 1) - 2(x + 2)$

Objective 1.7.1 Evaluate an Algebraic Expression for Specific Values of the Variables

24. Evaluate each expression for $x = -1$, $y = -2$, and $z = -3$.
    a. $x + y + z$
    b. $-x + yz$
    c. $-2x - 3(2y - z - 1)$
    d. $\dfrac{2x - y}{2x + z}$

Objective 1.7.2 Use Algebraic Formulas

25.

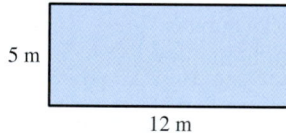

5 m

12 m

    a. Use the formula $A = LW$ to calculate the area of this rectangle.
    b. Use the formula $P = 2L + 2W$ to calculate the perimeter of this rectangle.
    c. The formula $A = RB$ can be used to calculate the amount of active ingredient in a solution. How much insecticide is in a full 500-gallon tank that contains a 1.5% insecticide mixture?
    d. The formula $A = P(1 + r)^2$ can be used to calculate the amount that will result if a principal is left to earn yearly interest compounded for 2 years. Determine the value of a \$1,200 investment earning simple yearly interest of 6% compounded for 2 years.

Objective 1.7.3 Use Subscript Notation

26. A line passes through the points $(x_1, y_1) = (-1, 7)$ and $(x_2, y_2) = (3, -7)$. Use the formula $m = \dfrac{y_2 - y_1}{x_2 - x_1}$ to calculate the value of m, the slope of this line.

27. Calculate the first three terms of the sequence with $a_n = -2n + 11$.

Objective 1.7.4 Check a Possible Solution of an Equation

28. Substitute $x = -5$ into each equation to determine whether it is a solution of the equation.

a. $3x + 2 = 2x - 3$ **b.** $2(x + 1) = x + 4$

c. $(x - 3)(x + 5) = 0$ **d.** $\dfrac{2x + 13}{1 - x} = \dfrac{1}{2}$

Objective 1.8.1 Classify and Name Common Polygons

29. a. A polygon with four sides is called a _____.
b. A parallelogram with adjacent sides perpendicular is called a _____.
c. A polygon with six sides is called a _____.
d. A regular pentagon has _____ sides that are all equal in length.

Objective 1.8.2 Calculate the Perimeter of Common Geometric Figures

30.

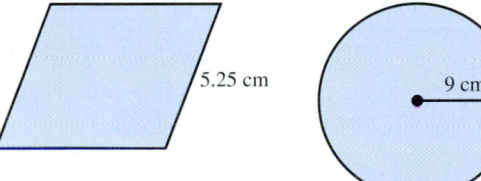

5.25 cm   9 cm

a. Rhombus Calculate the perimeter of a rhombus that has one side of length 5.25 cm.
b. Circle Calculate the exact circumference of a circle with a radius of 9 cm. Then approximate this length to the nearest tenth of a cm.

Objective 1.8.3 Calculate the Area of Common Geometric Regions

31.

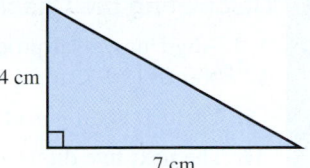

4 cm

7 cm

a. Triangle Use the formula $A = \dfrac{1}{2}bh$ to calculate the area of this triangle.
b. Circle Calculate the exact area of a circle with a radius of 9 cm. Then approximate this area to the nearest tenth of a cm^2.

Objective 1.8.4 Calculate the Volume of a Rectangular Solid

32. Volume of a Freezing Compartment The interior of the freezing compartment of a refrigerator measures 25 in wide by 20 in deep by 15 in tall. What is the storage capacity of this compartment in cubic inches?

Objective 1.8.5 Classify and Name Common Angles

33. a. A 90° angle is called a _____ angle.
b. A straight angle measures _____ degrees.
c. If the sum of the measures of two angles is 90°, then these angles are _____.
d. _____ angles are distinct angles that share a common vertex and one common side.
e. If two parallel lines are cut by a transversal, then two angles that lie on opposite sides of the transversal and outside the parallel lines are called _____ exterior angles.

SU-7212 7PHC.00107D

0   78000 01180   7

Chapter 1 | Group Project

Calculating the Check Digit for a 12-Digit UPC Number

A 12-digit universal product code (UPC) has three parts. The UPC code, 0 78000 01180 7, for a 12 pack of 7 UP is shown here.

I. The first six digits of this code—0 78000—identify the manufacturer.

II. The next five digits—01180—is the item number, identifying a specific product from this manufacturer.

III. The 12th digit—7—is a **check digit.** Each time an item is scanned, calculations are performed on the first 11 digits of the UPC and the result is compared against the check digit. If the results are different, then the scanner will not accept the results of this scan.

The calculation of the check digit is a five-step process as detailed in the box.

Calculation of a UPC Check Digit

Step 1. Starting from left to right with the first 11 digits from the code, add all of the digits in the odd-numbered positions.

Step 2. Multiply the result from Step 1 by 3.

Step 3. Starting from left to right with the first 11 digits from the code, add all of the digits in the even-numbered positions.

Step 4. Add the results of Steps 2 and 3.

Step 5. The check digit is the digit that must be added to the result from Step 4 to produce the next multiple of 10.

(a) Calculate the check digit for 0 78000 01180 7. Show each step of your work. Does the check digit that you calculated agree with the one shown?

(b) Have each person in your group bring a UPC label for the group to use. Write the UPC number for each item and then calculate the check digit for this item. Does the check digit that you calculated agree with the one shown on the label?

Linear Equations and Patterns

CHAPTER 2

Chapter Outline

Lake Mead

Lake Mead Could Run Dry by 2020!

The white ring you see in the photo is where the water level used to be in Lake Mead. Held back by the Hoover Dam between Arizona and Nevada, this lake is currently less than half full. Some scientists claim that there is a good chance this lake could run dry by 2020.

This chapter will examine graphing points and solving linear equations. These skills are needed to describe data and to make predictions.

Water Level at Hoover Dam

Year	Depth (ft)
1999	142.89
2000	144.26
2001	127.27
2002	107.94
2003	83.33
2004	70.39
2005	67.40
2006	69.46
2007	59.55
2008	46.46

Exercises 31–34 in Section 2.1 use line graphs like that shown for the water level at Hoover Dam.

Section 2.1 | The Rectangular Coordinate System

"A mathematical equation is a brush stroke used to paint one of the wonders of nature. We should look at it as being as beautiful as art or literature or music." —WALTER ISAACSON

Objectives:

1. Plot ordered pairs on a rectangular coordinate system.
2. Draw a scatter diagram of a set of data points.
3. Identify data whose graph forms a linear pattern.
4. Interpret a line graph.

The primary focus of this section is on plotting data points and creating and interpreting graphs.

What Is the Advantage of Graphing Data Points?

Graphs of data points often make it easier to notice patterns or to spot trends. The wind-chill data from the accompanying newspaper article illustrate this and provide a context for the topics examined in this section. Each point in this table pairs a Fahrenheit temperature with the wind chill in a 10-mi/h wind. The graph visually displays these points. The dashed line through the points is not part of the graph but is inserted to emphasize the linear pattern formed by the points.

Fahrenheit Temperature	Wind Chill for 10-mi/h Winds
−5	−21
0	−15
5	−9
10	−3
15	3
20	9

"Low Temperatures and Winds Create Dangerous Conditions for Tomorrow"

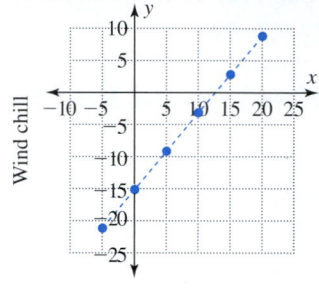

What Type of Graph Is Used for the Wind-Chill Data Given in the Previous Table?

This graph was shown on a **rectangular coordinate system,** also known as the **Cartesian coordinate system,** or the **coordinate plane.** On the rectangular coordinate system, the horizontal number line is called the **x-axis,** the vertical number line is called the **y-axis,** and the point where they cross is called the **origin.** These axes divide the plane into four **quadrants,** labeled counterclockwise as I, II, III, and IV, as illustrated in Figure 2.1.1. The points on the axes are not considered to be in any of the quadrants.

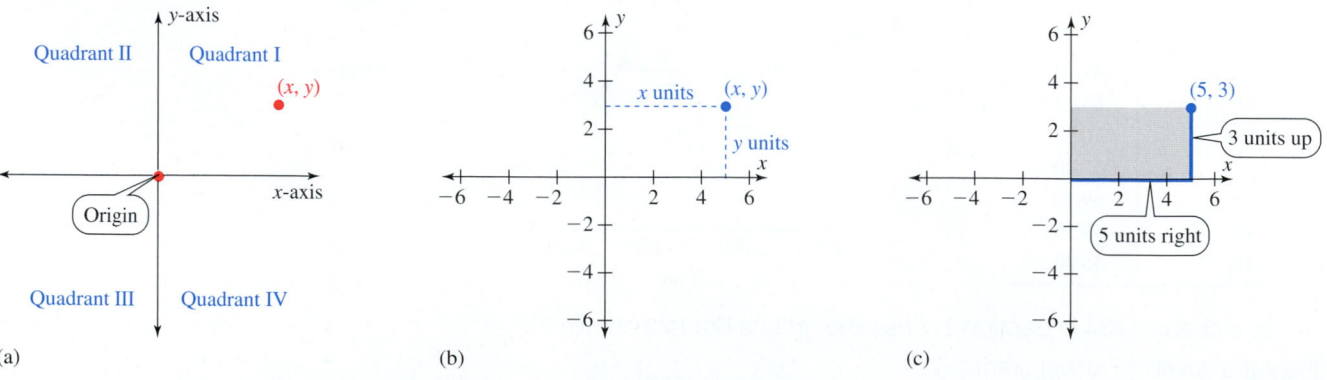

(a) (b) (c)

Figure 2.1.1 Cartesian coordinate system.

1. Plot Ordered Pairs on a Rectangular Coordinate System

On the *x*-axis, points to the right of the origin are positive and those to the left of the origin are negative. On the *y*-axis, points above the origin are positive and those below are negative. Any point in the plane can be uniquely identified by specifying its horizontal and vertical location with respect to the origin. We identify a particular point by giving an ordered pair (x, y) with **coordinates** x and y. The first coordinate is called the ***x*-coordinate,** and the second coordinate is called the ***y*-coordinate.** The ordered pair $(0, 0)$ identifies the origin.

The point in Figure 2.1.1(c) is identified by the ordered pair $(5, 3)$ in quadrant I. The *x*-coordinate is 5 and the *y*-coordinate is 3. The identification of this point can be associated with the horizontal and vertical segments of the shaded rectangle—hence the name *rectangular coordinate system.*

How Do I Tell the Difference Between an Ordered Pair and an Interval?

Parentheses are used to indicate both ordered pairs like the point $(-3, 4)$ in Example 1 and intervals like the interval $(-3, 4)$:

Each meaning is generally clear because of the context in which it occurs.

> **A Mathematical Note**
>
> René Descartes (1596–1650), born to a noble French family, was known for his studies in anatomy, astronomy, chemistry, physics, and philosophy as well as mathematics. Prior to Descartes, algebra was concerned with numbers and calculations, and geometry was concerned with figures and shapes. Descartes merged the power of these two areas into analytic geometry—his most famous discovery.

Example 1   Plotting Points on a Rectangular Coordinate System

Plot $(-3, 4)$ and $(2, -5)$ on a rectangular coordinate system.

Solution

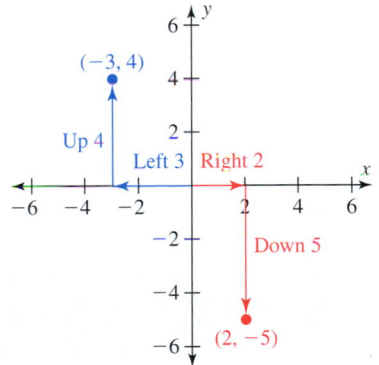

To plot $(-3, 4)$, start at the origin and move 3 units left and then 4 units up.

To plot $(2, -5)$, start at the origin and move 2 units right and then 5 units down.

Self-Check 1

Plot $(-2, 0)$ and $(3, -4)$ on a rectangular coordinate system.

Example 2   Identifying Coordinates of Points

Identify the coordinates of each of the points *A* through *E* in the following figure, and give the quadrant in which each point is located.

Solution

A: (5, 1); quadrant I
B: (−2, 5); quadrant II
C: (−4, −2); quadrant III
D: (3, −6); quadrant IV
E: (4, 0) is on the *x*-axis and thus is
   not in any quadrant.

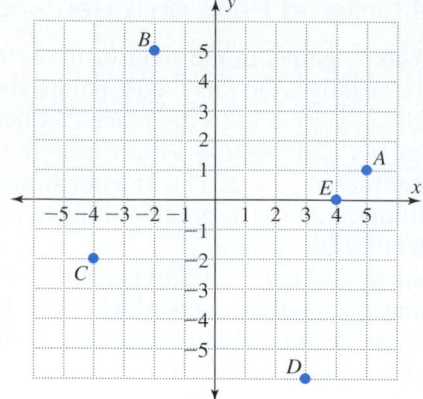

Self-Check 2

Give the quadrant for $(1, -3)$, $(-1, -3)$, $(-1, 3)$, and $(1, 3)$.

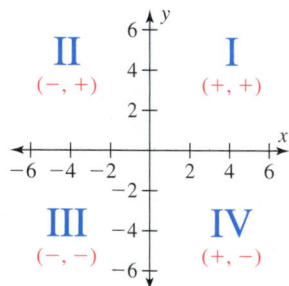

Figure 2.1.2 Quadrant sign pattern.

Is There a Pattern to the Points on the Rectangular Coordinate System?

Yes, all the points within the same quadrant have the same sign pattern. For example, in quadrant I both coordinates are positive, which we can denote by $(+, +)$. The sign pattern for each quadrant is shown in Figure 2.1.2. Knowing the sign pattern in each quadrant is useful in trigonometry. These patterns can also be helpful in analyzing scatter diagrams in statistics.

2. Draw a Scatter Diagram of a Set of Data Points

What Is a Scatter Diagram?

A **scatter diagram** for a set of data points is simply a graph of these points. It allows us to examine the data for some type of visual pattern. For example, if the data points all lie near a line, then the pattern exhibited is called a **linear relationship.**

Example 3   Drawing a Scatter Diagram

Draw a scatter diagram for the data points given in the following table.

Solution

x	y
−15	−12
−10	−4
−5	−1
0	6
5	8
9	15
15	18

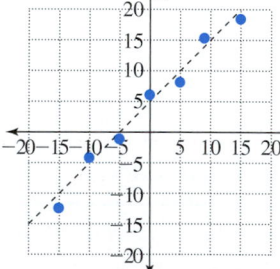

The table format is often used to give ordered pairs. The first ordered pair from this table is $(-15, -12)$.

The dashed line is not part of the graph. It is placed on the graph to show that the data points all lie near this line. Therefore, the relationship between *x* and *y* is approximately a linear relationship.

Self-Check 3

Draw a scatter diagram for the data points given in the table.

x	y
−2	5
−1	2
0	−1
1	−4
2	−7

Can I Adjust the Scale on Each Axis?

Yes, the size of the coordinates is used to determine the scale on each axis. If the coordinates differ significantly in size, then a different scale can be used on each axis. This practice is common in statistics, where the *x*- and *y*-variables often represent quantities measured in different units. This is illustrated in Example 4, which also shows that we sometimes start the input with a gap between the origin and the first value.

Example 4   Drawing a Scatter Diagram

Draw a scatter diagram for the data in these financial reports from Harley-Davidson.

Solution

Year x	Net Income y, Millions of $
2003	761
2004	890
2005	960
2006	1,043
2007	930

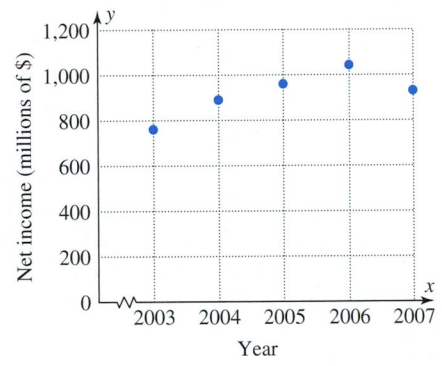

Business results often defy perfect patterns and have a nasty habit of changing patterns just when this information is used to make investment decisions. Nonetheless, the pattern given here by this very limited set of data shows a business that is growing and doing well.

Self-Check 4

Draw a scatter diagram for the financial data in the given table.

Year x	Net Income y, Millions of $
2002	507
2004	465
2006	402
2008	347
2010	299

All the data that we have graphed in the previous examples have both an *x*- and a *y*-coordinate. Sometimes the data that we are examining is simply a list of values.

Can I Graph a List of Values on a Rectangular Coordinate System?

Yes, If the data are collected in a particular order or over a period of time, then the data items form a sequence and we can form paired values to plot points. We can use the sequence position as the first coordinate and the value in this position as the second coordinate. For example, if $a_1 = 3$ then we can graph the ordered pair (1, 3). This is illustrated in Example 5.

Example 5   Graphing a Sequence of Data Values

Consider the sequence 3, 5, 7, 9, 11.

(a) Represent this sequence by using subscript notation.

(b) Represent this sequence by using ordered pair notation.

(c) Form a table of values for this sequence.

(d) Draw a scatter diagram for this sequence.

Solution

(a) $a_1 = 3$, $a_2 = 5$, $a_3 = 7$, $a_4 = 9$, $a_5 = 11$ The first term is 3 and can be represented by $a_1 = 3$ or by the ordered pair (1, 3).

(b) (1, 3), (2, 5), (3, 7), (4, 9), (5, 11)   Continuing in this manner, the fifth term is 11 and can be represented by $a_5 = 11$ or by the ordered pair (5, 11).

(c)

n	1	2	3	4	5
a_n	3	5	7	9	11

(d)

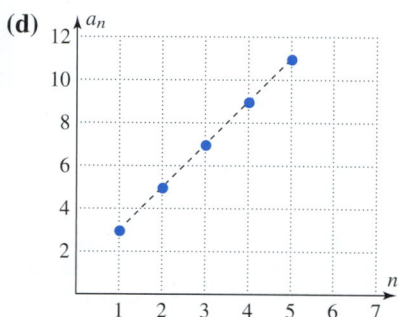

Then plot these ordered pairs and note that the points all lie on the straight line that is superimposed on the graph.

Self-Check 5

a. Represent the sequence –5, –2, 1, 4, 7 by using subscript notation.

b. Represent this sequence by using ordered pair notation.

c. Form a table of values for this sequence.

d. Draw a scatter diagram for this sequence.

You will see tables of values in every newspaper from the business section to the sports page. Some of these tables will be arranged vertically like the one shown in Example 4 and others will be arranged horizontally like the one shown in Example 5. Mathematically, both are completely acceptable.

3. Identify Data Whose Graph Forms a Linear Pattern

The graph of the sequence 3, 5, 7, 9, 11 in Example 5 consists of individual disconnected points that all lie on the same line. Linear patterns are common in many fields of study and it is very useful to examine this pattern thoroughly. We will start this examination now and then spend several sections working with linear equations.

Can I Determine Whether a Sequence Forms a Linear Pattern by Examining the Numerical Values in the Sequence?

Yes, for the points to lie on a line, there must be a constant change from term to term. In the sequence 3, 5, 7, 9, 11, the change from one term to the next is +2. Mathematicians call a sequence with a constant change from one term to the next term an **arithmetic sequence** and the constant change the **common difference**. The common difference is often denoted by d.

Example 6   Identifying a Linear Pattern

Determine whether each sequence is an arithmetic sequence and whether its graph forms a linear pattern.

(a) $-4, -1, 2, 5, 8$ **(b)** $2, 4, 7, 8, 6$

Solution

(a) Numerical Difference Between Terms

$$-1 - (-4) = 3$$
$$2 - (-1) = 3$$
$$5 - 2 = 3$$
$$8 - 5 = 3$$
$$d = 3$$

Verbally

This is an arithmetic sequence because there is a common difference of 3. The points establish a linear pattern with the points rising 3 units from one term to the next.

Graphically

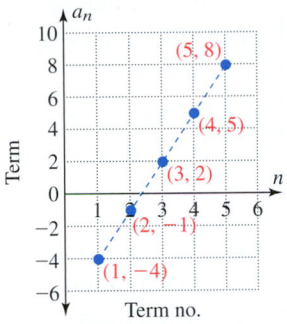

(b) Numerical Difference Between Terms

$$4 - 2 = 2$$
$$7 - 4 = 3$$
$$2 \neq 3$$

Verbally

This is not an arithmetic sequence because the change from term to term is not constant. The points do not form a linear pattern. The height change between consecutive terms is not constant.

Graphically

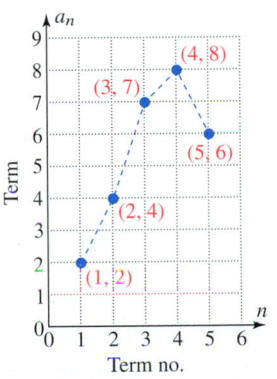

Self-Check 6

Determine whether each sequence is an arithmetic sequence and whether its graph forms a linear pattern.

a. $1, 0, -1, -2, -3$

b. $1, 0, -1, 0, 1$

---

4. Interpret a Line Graph

What Is a Line Graph?

A **line graph** connects individual data points with line segments and is very useful to show trends for data changing over time. The graph shown in Example 6(b) is a typical appearance of a line graph. The information available in a graph may also be available in a table of values, but many people understand and relate to mathematics better when they can visualize the material. Graphs show not only the points from a table but also their relationship to each other. Example 7 illustrates how to use a line graph to determine a maximum value and to determine trends in data.

Example 7 Interpreting a Line Graph

The line graph shown gives the distance of a taxi from a dispatching office at a given time. The time x is given in minutes from the start of the shift of one driver, and the distance y is given in miles away from the dispatching office. Answer each of these questions by examining this graph.

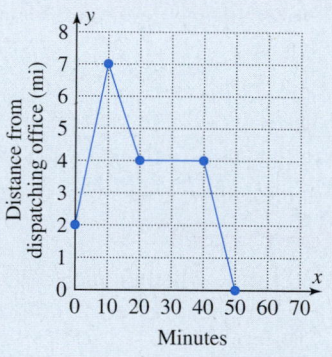

(a) At a time of 0 min, how far is the taxi from the dispatching office?

(b) When did the taxi arrive at the dispatching office?

(c) What was the maximum distance the taxi was from the office?

(d) The taxi was parked for a period of time. During what time was the taxi parked?

Solution

(a) The taxi is 4 mi from the office at a time of 0 min.

The point (0, 4) indicates that when $x = 0$, $y = 4$.

(b) The taxi arrived at the office after 60 min.

The point (60, 0) indicates that after 60 min the distance from the office is 0 mi.

(c) The maximum distance the taxi was from the office was 6 mi.

The highest point on the graph is the point (40, 6). Thus the maximum y-value is $y = 6$.

(d) The taxi was parked from 10 to 20 min after the start of the driver's shift.

From $x = 10$ to $x = 20$ the y-value is a constant 3 mi. This corresponds to the taxi being parked and not changing its distance.

Self-Check 7

The line graph shown gives the distance of a taxi from a dispatching office at a given time. The time x is given in minutes from the start of the shift of one driver, and the distance y is given in miles away from the dispatching office. Answer each of these questions by examining this graph.

a. At a time of 0 minutes, how far is the taxi from the dispatching office?

b. When did the taxi arrive at the dispatching office?

c. What was the maximum distance the taxi was from the office?

d. The taxi was parked for a period of time. During what time was the taxi parked?

Self-Check Answers

1.

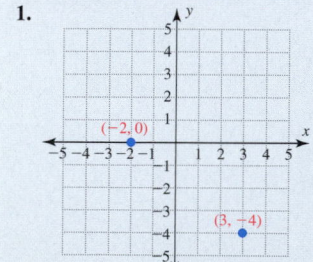

2. (1, −3); quadrant IV
(−1, −3); quadrant III
(−1, 3); quadrant II
(1, 3); quadrant I

3.

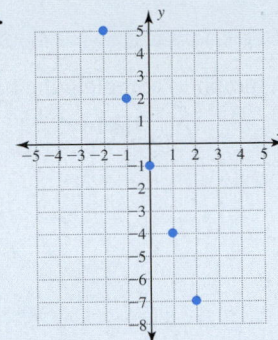

4.

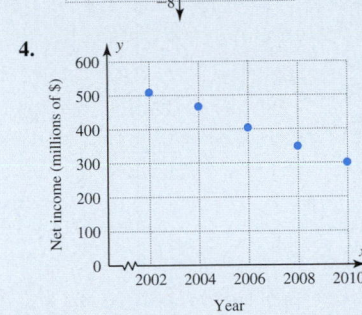

5. a. $a_1 = -5$, $a_2 = -2$, $a_3 = 1$, $a_4 = 4$, $a_5 = 7$
b. $(1, -5), (2, -2), (3, 1), (4, 4), (5, 7)$

c.

n	a_n
1	−5
2	−2
3	1
4	4
5	7

d.

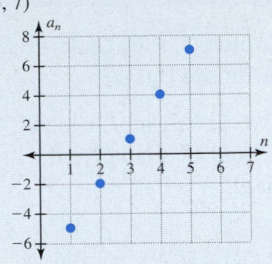

6. a. $1, 0, -1, -2, -3$ is an arithmetic sequence with $d = -1$. The points will all lie on the same line.
b. $1, 0, -1, 0, 1$ is not an arithmetic sequence. The points will not all lie on the same line.

7. a. The taxi is 2 mi from the office at a time of 0 min.
b. The taxi arrives at the office after 50 min.
c. The maximum distance the taxi was from the office was 7 mi.
d. The taxi was parked from 20 to 40 min after the start of the driver's shift.

2.1 Using the Language and Symbolism of Mathematics

1. The rectangular coordinate system is also called the _____ coordinate system.

2. On the rectangular coordinate system, the horizontal axis is called the _____-axis, and the vertical axis is called the _____-axis.

3. The point where the horizontal and vertical axes cross is called the _____ and has coordinates_____.

4. All points with an x-coordinate of 0 are on the _____-axis.

5. All points with a y-coordinate of 0 are on the _____-axis.

6. The four quadrants are numbered I, II, III, and IV in a **clockwise/counterclockwise** direction. (Select the correct choice.)

7. In quadrant _____ both coordinates are positive.

8. In quadrant _____ both coordinates are negative.

9. A _____ diagram for a set of data points is a graph of those points.

10. A sequence with a constant change from term to term is called an _____ sequence.

11. In an arithmetic sequence, d represents the common _____.

12. The graph of an arithmetic sequence consists of distinct disconnected points that form a _____ pattern.

2.1 Quick Review

Plot each number on a real number line.

1. 4   **2.** −4

Graph each interval on a real number line.

3. $(-2, 3]$ **4.** $(-1, 4)$

5. Use interval notation to represent the interval with a graph of 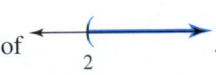 .

Objective 1 Plot Ordered Pairs on a Rectangular Coordinate System

In Exercises 1 and 2, identify the coordinates of points A, B, C, and D. Also give the quadrant in which each point is located.

1.   **2.**

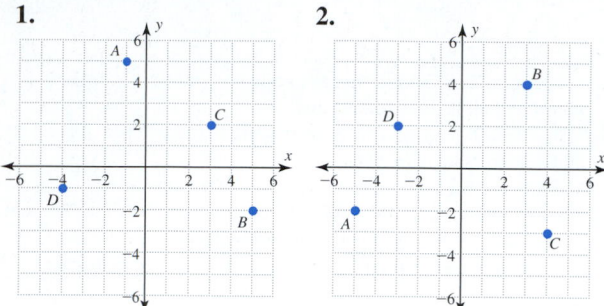

In Exercises 3 and 4, plot and label the points whose coordinates are given. Plot these points on the same coordinate system.

3. $A(5, 4)$ $B(-4, -5)$ $C(3, -4)$
     $D(-6, 1)$ $E(4, 0)$ $F(0, -5)$

4. $A(2, 6)$ $B(-3, -6)$ $C(4, -2)$
     $D(-5, 3)$ $E(-3, 0)$ $F(0, 6)$

In Exercises 5 and 6, use the sign pattern of the coordinates to determine the quadrant in which each point is located. Do not plot these points.

5. a. $(-1.3, 3.7)$ **b.** $(-5.8, -9.3)$
   c. $(6.7, -2.1)$ **d.** $(0.4, 0.1)$

6. a. $(-193, -217)$ **b.** $\left(\dfrac{2}{3}, -\dfrac{3}{5}\right)$
   c. $(801, 913)$ **d.** $(-\pi, \sqrt{2})$

In Exercises 7 and 8, determine the axis on which each point lies. Do not plot these points.

7. a. $(0, 8)$ **b.** $(-8, 0)$
   c. $(-\sqrt{3}, 0)$ **d.** $(0, \sqrt{2})$

8. a. $\left(\dfrac{1}{8}, 0\right)$ **b.** $(0, 81)$
   c. $(0, -\pi)$ **d.** $(-\sqrt{5}, 0)$

Objective 2 Draw a Scatter Diagram of a Set of Data Points

In Exercises 9 and 10, draw a scatter diagram for each set of data.

9.

x	-2	-1	0	1	2	3
y	-10	-9	-4	1	6	8

10.

x	-2	-1	0	1	2	3
y	8	6	4	4	2	-2

In Exercises 11 and 12, use the graph to complete the table.

11. Graph

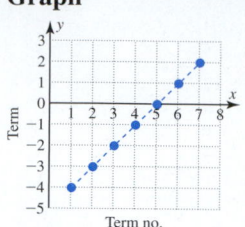

Table

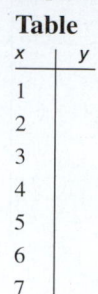

x	y
1	
2	
3	
4	
5	
6	
7	

12. Graph

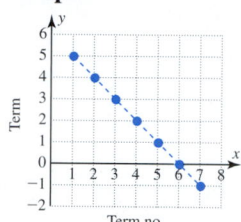

Table

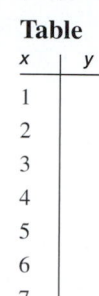

x	y
1	
2	
3	
4	
5	
6	
7	

Objective 3 Identify Data Whose Graph Forms a Linear Pattern

In Exercises 13 and 14,
a. represent the sequence by using subscript notation and
b. form a table of x-y values for this sequence.

13. 5, 13, 19, 8, 6   **14.** 2, 9, 15, 24, 4

In Exercises 15–18,
a. represent the sequence using ordered pairs and
b. graph the sequence.

15. 8, 5, 2, -1, -4   **16.** -5, -2, 1, 4, 7
17. -2, 4, 1, 1, 3   **18.** 3, 5, 2, 2, 4

In Exercises 19 and 20, use the graph of each arithmetic sequence to write the first five terms of each sequence.

19.   **20.**

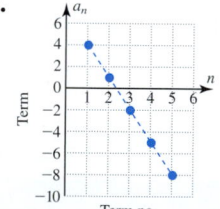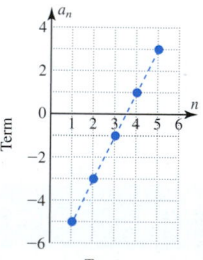

In Exercises 21–26, determine whether the sequence is an arithmetic sequence and whether its graph forms a linear pattern.

21. $-3, 1, 5, 9, \ldots$ **22.** $-11, -6, -1, 4, 9, \ldots$
23. 40, 30, 20, 10, 0   **24.** 20, 17, 14, 11, 8, 5
25. 3, 5, 8, 12, 17   **26.** 2, 4, 8, 16, 32, 64

In Exercises 27–30, determine whether the sequence shown in each graph is an arithmetic sequence. For those that are arithmetic sequences, find the common difference d.

27.
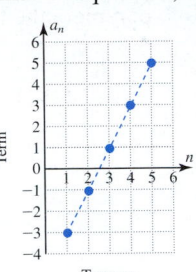

28.
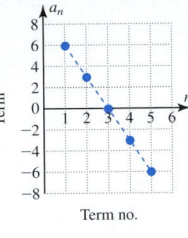

29.
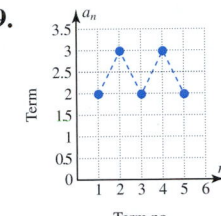

30.
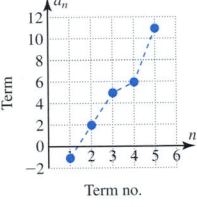

Objective 4 Interpret a Line Graph

31. Depth of Water in a Reservoir The depth of water in a city reservoir was carefully recorded for a 6-month period. The results are displayed in the given graph. The time x is given in months, and the depth y is given in feet. Use this graph to answer the following questions.

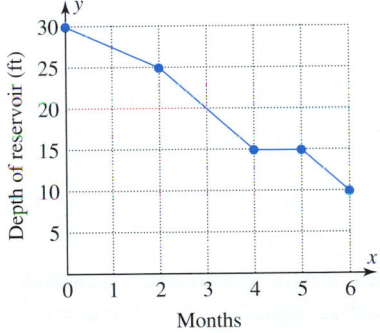

a. At the start of the recording time, a time of 0 months, what was the depth of the water in the reservoir?

b. At the end of 5 months, what was the depth of the water in the reservoir?

c. When did the depth of the water reach 25 ft?

d. What was the minimum depth of the water during this 6-month period?

32. Weight of a Patient The weight of a patient was recorded for an 8-week period by a physical therapist. The results are displayed in the given graph. The time x is given in weeks, and the weight y is given in pounds. Use this graph to answer the following questions.

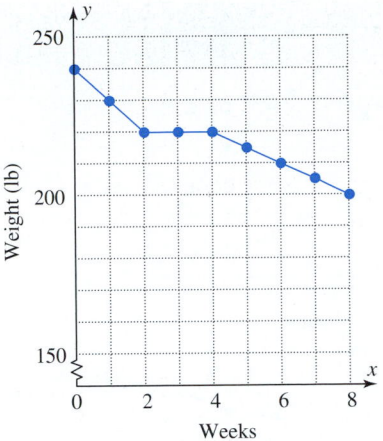

a. At the start of the recording time, a time of 0 weeks, what was the weight of the patient?

b. At the end of 1 week, what was the weight of the patient?

c. When did the weight of the patient reach 210 lb?

d. The patient maintained a steady weight of 220 lb for a period of time. During what time period did the patient weigh 220 lb?

33. Price of a Stock Each graph corresponds to the price of a stock over a 6-month period. The time x is given in months, and the price y is given in dollars. Match each of these descriptions with the corresponding graph.

a. The price is increasing for the entire 6 months.

b. The price is decreasing for the entire 6 months.

c. The price is constant for the entire 6 months.

d. The price has a maximum after 3 months.

A.
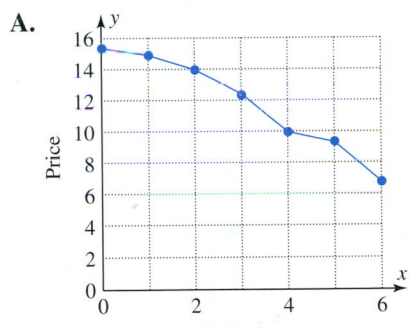

B.
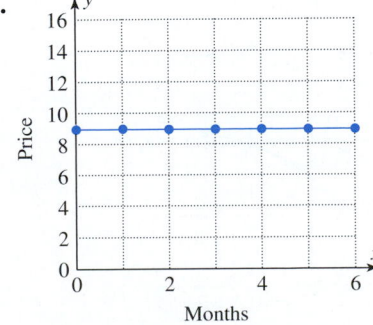

C.

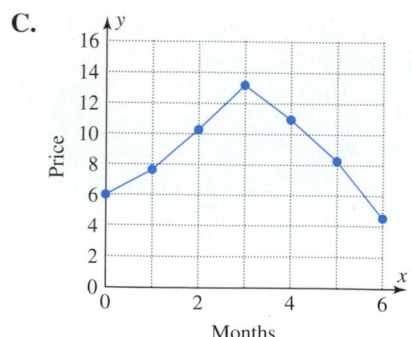

Months

D.

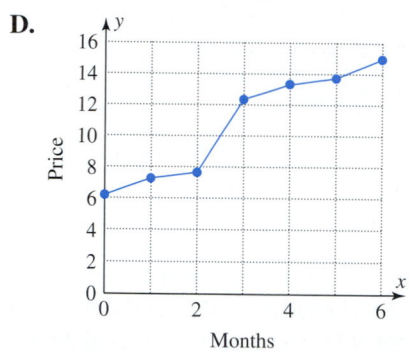

Months

C.

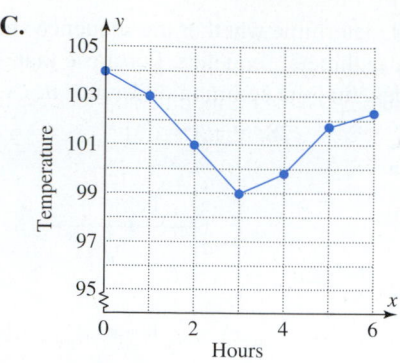

Hours

D.

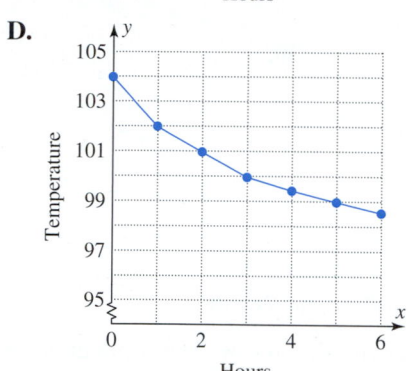

Hours

34. Temperature of a Child Each graph corresponds to the temperature of a sick child over a 6-hour period. The time x is given in hours, and the temperature y is given in degrees Fahrenheit. Match each of these descriptions with the corresponding graph.

a. The temperature is increasing for the entire 6 hours.
b. The temperature is decreasing for the entire 6 hours.
c. The temperature is constant for the entire 6 hours.
d. The temperature reaches a minimum after 3 hours and then increases again.

A.

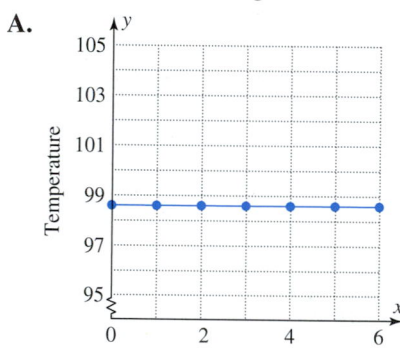

Hours

B.

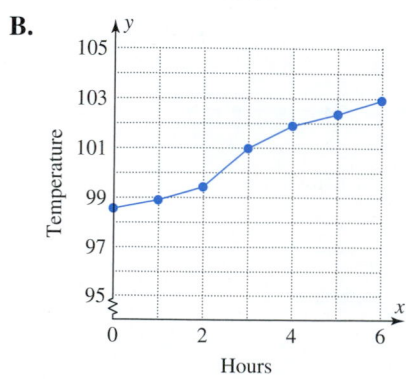

Hours

Review and Concept Development

In Exercises 35–38, complete each inequality by filling in each blank with either $<$ or $>$.

35. If a point (x, y) is in quadrant I, then x _____ 0 and y _____ 0.

36. If a point (x, y) is in quadrant II, then x _____ 0 and y _____ 0.

37. If a point (x, y) is in quadrant III, then x _____ 0 and y _____ 0.

38. If a point (x, y) is in quadrant IV, then x _____ 0 and y _____ 0.

39. If $xy > 0$, then the point (x, y) is in either quadrant _____ or quadrant _____.

40. If $xy < 0$, then the point (x, y) is in either quadrant _____ or quadrant _____.

41. Consider an arithmetic sequence that has a positive common difference. Does the line through the graph of this sequence go up or down as the points move to the right?

42. Consider an arithmetic sequence that has a negative common difference. Does the line through the graph of this sequence go up or down as the points move to the right?

43. Are the even integers an arithmetic sequence? If so, what is the common difference d?

44. Are the odd integers an arithmetic sequence? If so, what is the common difference d?

Automobile lease

In Exercises 45 and 46, the equation $a_n = 450n + 2,400$ gives the total payments in dollars after n months on the lease for a luxury sports utility vehicle. Calculate and interpret each value of a_n. (*Hint:* See Example 6 in Section 1.7.)

45. a. a_1 **b.** a_{18} **c.** a_{24}

46. a. a_0 **b.** a_6 **c.** a_{12}

Connecting Algebra to Geometry

47. Determine the area and perimeter of a rectangle whose corners are located at $(0, 0)$, $(5, 0)$, $(5, 4)$, and $(0, 4)$.

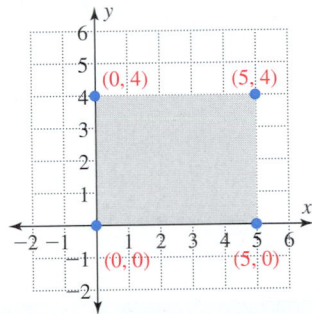

48. Determine the area and perimeter of a rectangle whose corners are located at $(0, -3)$, $(2, -3)$, $(2, 3)$, and $(0, 3)$.

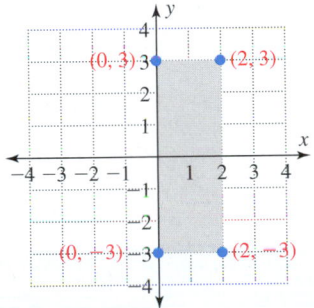

Connecting Concepts to Applications

Water Bills In Exercises 49–52, the water bills of a homeowner are recorded under a variety of situations over a 6-month period. The time x is given in months, and the water bill y is given in dollars. Sketch a separate graph for each of the following verbal descriptions.

49. The homeowner has limited water usage and each month has the same minimum usage charge.

50. The homeowner has had family members leave the house and has gradually reduced her water bill each month.

51. The homeowner has increased her water bill each month as she has watered the yard during an extended dry period.

52. The homeowner has increased her water bill for 3 months as she has watered the yard during a dry period, and then a series of rains allowed the homeowner to gradually reduce the water usage to the minimal level by the end of the sixth month.

Blood Pressure In Exercises 53–56, the diastolic blood pressure of a man is recorded under a variety of situations over a 5-minute period. The time x is given in minutes, and the blood pressure y is given in millimeters (mm) of mercury. Assuming the man starts the period with a diastolic blood pressure of 90 (representing 90 mm of mercury), sketch a separate graph for each of the following verbal descriptions.

53. The man is inactive, and the blood pressure remains steady.

54. The man has just completed a physical activity, and as he rests, his blood pressure steadily drops.

55. The man has just initiated an exercise on a treadmill, and his blood pressure is steadily increasing.

56. The man is watching a movie, and at first a calm scene helps to lower his blood pressure, but then an intense scene causes his blood pressure to rise.

Group discussion questions

57. Risks and Choices Suppose an employer offered you a $30,000 starting salary for your first year of employment with raises for the first 5 years determined by either option A or option B.
  Option A: A $1,500 raise for each of the next 5 years.
  Option B: A 5% raise for each of the next 5 years.
  a. Can you determine without any calculations which option will produce a graph with a linear pattern? Explain your reasoning.
  b. Write the sequence of salaries for each option.
  c. Use the changes from term to term in each sequence to explain which option you would prefer.

58. Writing Mathematically A basketball player scored 5, 10, and 15 points in his first three games. If you graph his scores for the first eight games, do you think the graph will form a linear pattern? Explain your reasoning.

59. Writing Mathematically Write a story to accompany the given graph, and label each axis to match your story. Then use this story to ask four questions that the members of your group can answer by using this graph. (*Hint:* See Example 7.)

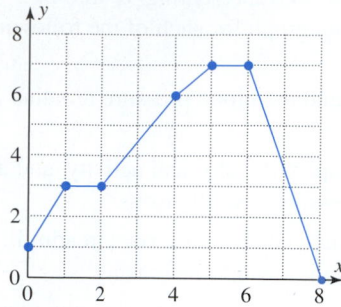

	A	B
1	Time (h)	Cost ($)
2	1	$125.00
3	2	$175.00
4		$225.00
5	4	$275.00
6	5	

60. Applying Technology The given graph was created from an Excel spreadsheet that displays the cost of operating a paint sprayer for different time periods. Use the graph to complete the missing entries in this spreadsheet.

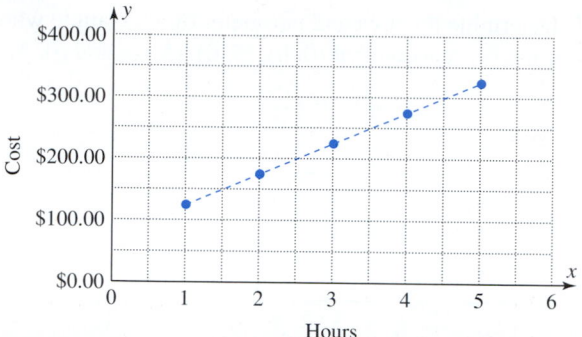

2.1 Cumulative Review

Use the distributive property to combine like terms.

1. $5x + 7x$

2. $12v - 7v + 2v$

3. $5a + 3 - 2(2a - 4)$

4. Use the figure to determine the measure of angle A.

5. Use the figure to determine the measure of angle B.

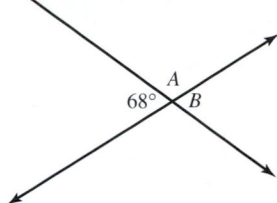

Section 2.2   Function Notation and Linear Functions

Objectives:

1. Use function notation.
2. Use a linear equation to form a table of values and to graph a linear equation.
3. Write a linear function to model an application.

1. Use Function Notation

The primary purpose of this section is to introduce function notation and to illustrate how linear functions can be used to model applications. This introduction will enable us to use function notation throughout this text. By the time you reach Chapter 8, you will be very familiar with this notation and the concept of input-output pairs. Then we will examine the function concept more formally with more emphasis on domains, ranges, and families of functions.

A table of values and a graph of a few points both give valuable representations for the relationship between a set of *x-y* pairs. For many *x-y* pairs, it may be inconvenient or impossible to put all these points in a table. Thus it is useful to be able to express the relationship between *x* and *y* by using an equation. One notation for writing an equation relating *x-y* pairs is called function notation. The notation $f(x)$ is referred to as **function notation.**

A Mathematical Note

The history of mathematics has many noted women mathematicians. One of the earliest women about whom anything is written is Hypatia (ca. 360–415). She studied in Athens and then taught geometry and astronomy at Alexandria. She also wrote on algebra. History records that she was killed in a religious riot.

How Do I Read Function Notation?

The notation $f(x)$ is read **"f of x"** or **"f(x) is the output value for an input value of x."** *Caution:* $f(x)$ does not mean f times x in this context. Function notation is used to give an equation that describes a unique output that we can calculate for each input of *x*. To evaluate

$f(x)$ for a specific value of x, replace x on both sides of the equation by this specific value, as illustrated in Example 1.

Example 1   Using Function Notation to Evaluate a Function

Evaluate each expression for $f(x) = 3x + 5$.

(a) $f(0)$ **(b)** $f(4)$ **(c)** $f(-4)$

Solution

(a) $f(x) = 3x + 5$ Substitute each input value of x into the equation
$$f(0) = 3(0) + 5$$ $f(x) = 3x + 5$.
$$f(0) = 0 + 5$$

Answer: $f(0) = 5$ For an input of 0, the output is 5.

(b) $f(x) = 3x + 5$
$$f(4) = 3(4) + 5$$ Substitute 4 for x in the equation.
$$f(4) = 12 + 5$$

Answer: $f(4) = 17$ For an input of 4, the output is 17.

(c) $f(x) = 3x + 5$
$$f(-4) = 3(-4) + 5$$ Substitute -4 for x in the equation.
$$f(-4) = -12 + 5$$

Answer: $f(-4) = -7$ For an input of -4, the output is -7.

Self-Check 1

Evaluate each expression for $f(x) = 4x + 11$.

a. $f(0)$ **b.** $f(10)$ **c.** $f(-10)$

---

2. Use a Linear Equation to Form a Table of Values and to Graph a Linear Equation

How Can I Graph a Function?

Start by plotting points. Although we will develop more sophisticated methods for graphing different families of functions, every graph in this book can be sketched if you plot enough points. In Example 2, we will arbitrarily select some input values of x and use the function $f(x) = 2x + 1$ to calculate the corresponding output values. The graph of $f(x) = 2x + 1$ is the continuous line that is drawn through all of these points.

Example 2   Graphing a Function

Use the function $f(x) = 2x + 1$ to complete a table for x-values of -2, -1, 0, 1, and 2. Then use these input-output pairs to graph the line through these points.

Solution

Algebraically

$$f(x) = 2x + 1$$
$$f(-2) = 2(-2) + 1 = -3$$
$$f(-1) = 2(-1) + 1 = -1$$
$$f(0) = 2(0) + 1 = 1$$
$$f(1) = 2(1) + 1 = 3$$
$$f(2) = 2(2) + 1 = 5$$

Numerically

x	$f(x)$
-2	-3
-1	-1
0	1
1	3
2	5

Graphically

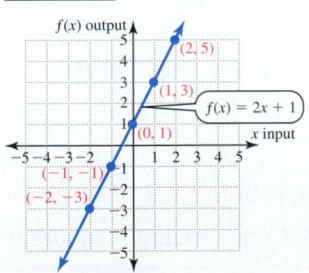

Verbally

The function $f(x) = 2x + 1$ defines a continuous line (no gaps) that contains the points in this table.

Self-Check 2

a. Use the function $f(x) = -x + 3$ to complete this table:

x	-2	-1	0	1	2
$f(x)$					

b. Graph this function.

The equation in Example 2, $f(x) = 2x + 1$, is an excellent example of a function of the form $f(x) = mx + b$. A function of this form is called a **linear function** because its graph is a straight line. We will examine linear functions in detail in Section 3.2. The linear function $f(x) = mx + b$ can also be represented by $y = mx + b$. Using the subscript notation introduced in Section 1.7, graphing calculators can denote several functions using Y_1, Y_2, Y_3, and so on. You should become comfortable using either y or $f(x)$ to represent an output value.

How Can I Use a Calculator or a Spreadsheet to Examine a Function?

Once the equation that defines a function is entered into a graphing calculator or a spreadsheet, there are many ways the function can be examined. One way is to create a table of values. Technology perspective 2.2.1 illustrates how to produce a table of values for the equation $y = -x + 2$. To create a table of values on a TI-84 plus calculator, we must enter the function using the [Y=] key and then access the **TblSet** and **Table** features. The [TBLSET] feature is the secondary function of the [WINDOW] key, and the [TABLE] feature is the secondary function of the [GRAPH] key. Spreadsheets are designed to use tables and can create them rather easily.

Technology Perspective 2.2.1	Creating a Table of Values

Create a table of values for $x = -3, -2, -1, 0, 1, 2, 3$ using the equation $y = -x + 2$.

TI-84 Plus Calculator

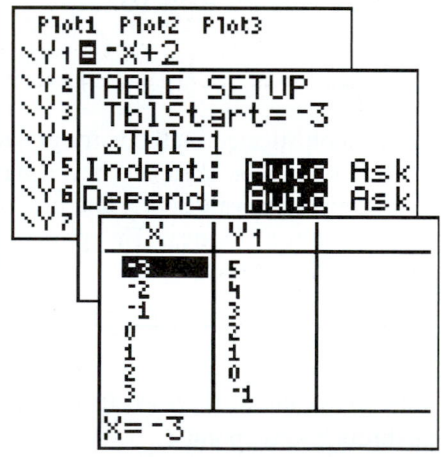

Excel Spreadsheet

B2 ▾	f_x	=-1*A2+2	
	A	B	C
1	x	y=-x+2	
2	-3	5	
3	-2	4	
4	-1	3	
5	0	2	
6	1	1	
7	2	0	
8	3	-1	

1. Press [Y=] and enter the equation.

2. Press [2nd] [TBLSET] to edit the table setup as shown. The x-values in the table start at -3 and increase by 1.

3. Press [2nd] [TABLE] to view the table with these settings.

1. Enter the desired x-values in column A.

2. Enter the formula "$= -1*A2+2$" in cell **B2** and copy this formula through column B.

3. This formula is used to determine the y-values shown in column B.

Technology Self-Check 1

Create a table of values for $x = -4, -3, -2, -1, 0, 1, 2$ using the equation $y = 3x - 5$.

Another way that we can use a graphing calculator to examine a function is to graph the function. Technology Perspective 2.2.2 illustrates how to produce a graph for the equation $y = -x + 2$. The notation below the calculator graph indicates the minimum and maximum x- and y-values used as well as the scale for each axis. The notation $[-10, 10, 1]$ by $[-10, 10, 1]$ means the x-values and the y-values extend from -10 to 10 with each mark on each axis representing 1 unit. On a TI-84 Plus calculator, this viewing rectangle is referred to as the **Standard Window** and it can be produced by pressing ⬭ZOOM ⬭6 . We will examine the use of the ⬭WINDOW key to manually change the window settings in Section 2.3.

When creating a graph for a report in a class or for a project at work, there are many graphing applications from which to choose. Excel has the ability to create graphs of many types. There are also many computer programs and web-based applications with graphing capabilities. For purposes of examining algebraic concepts, we will limit our focus to graphing calculators. Keep in mind that most of what we illustrate can be done with other forms of technology. The important thing is to understand the concepts relating to each graph.

Technology Perspective 2.2.2	Creating a Graph

Create a graph using the equation $y = -x + 2$.

TI-84 Plus keystrokes

1. Press ⬭Y= and enter the equation.
2. Press ⬭ZOOM ⬭6 to create the graph with the standard window of $[-10, 10, 1]$ by $[-10, 10, 1]$.

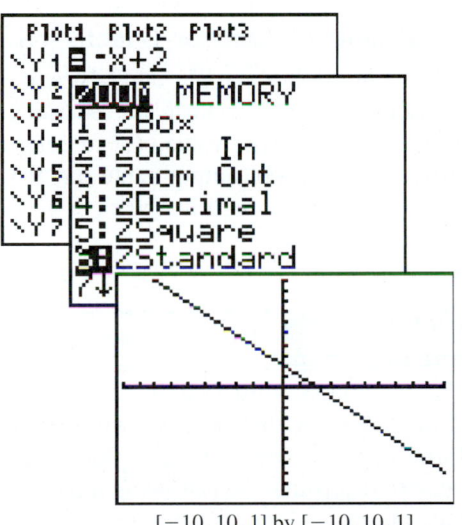

$[-10, 10, 1]$ by $[-10, 10, 1]$

Technology Self-Check 2

Create a graph using the equation $y = 3x - 5$.

Example 3   Using Multiple Perspectives to Examine a Linear Function

Use a graphing calculator to examine the linear function $f(x) = 2x - 3$ numerically and graphically. Then verbally describe the relationship given by this algebraic equation. Use the x-values of $-3, -2, -1, 0, 1, 2,$ and 3 to form the table of values.

Solution

Algebraically	Numerically	Graphically

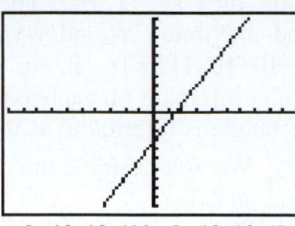

$[-10, 10, 1]$ by $[-10, 10, 1]$

Verbally

The function defines a table of points that all lie on a straight line. Note from the table and the graph that the y-values increase by 2 for every 1-unit increase in x.

Self-Check 3

Use a graphing calculator and $y = 1.75x - 2.25$ to:

a. Form a table of values for the input values of x of $-1, 0, 1, 2, 3, 4,$ and 5.

b. Graph the line defined by this equation, using the window $[-10, 10, 1]$ by $[-10, 10, 1]$.

Can I Evaluate a Function Directly from a Graph or a Table?

Yes, we can use both graphs and tables to identify input and output values. There are two types of questions that we can ask regarding these input-output pairs. The first type of question uses a given input value of x to determine the output value $f(x)$. The second type of question uses a given output value $f(x)$ to determine input value(s) of x that produce this output. This is illustrated in Example 4. Again we can use either $f(x)$ or y to identify the output value.

Example 4 Evaluating a Function from a Graph and a Table

The accompanying graph and table both define the same function, one whose algebraic form is $f(x) = mx + b$. Use this graph and table to determine the following input and output values.

(a) $f(-2) = \underline{?}$

(b) $f(4) = \underline{?}$

(c) $f(x) = -1; x = \underline{?}$

(d) $f(x) = 2; x = \underline{?}$

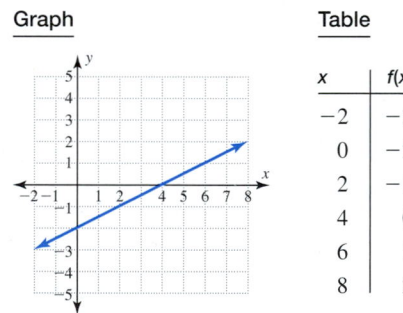

Graph	Table	
	x	$f(x)$
	-2	-3
	0	-2
	2	-1
	4	0
	6	1
	8	2

Solution

(a) $f(-2) = -3$ — Both the graph and the table contain the point $(-2, -3)$. The function pairs the input value of -2 with the output value of -3.

(b) $f(4) = 0$ — Both the graph and the table contain the point $(4, 0)$. The function pairs the input value of 4 with the output value of 0.

(c) $f(x) = -1; x = 2$ — Both the graph and the table contain the point $(2, -1)$. The function pairs the input value of 2 with the output value of -1.

(d) $f(x) = 2; x = 8$ — Both the graph and the table contain the point $(8, 2)$. The function pairs the input value of 8 with the output value of 2.

Self-Check 4

The given graph and table both define the same function, one whose algebraic form is $f(x) = mx + b$. Use this graph and table to determine the following input and output values.

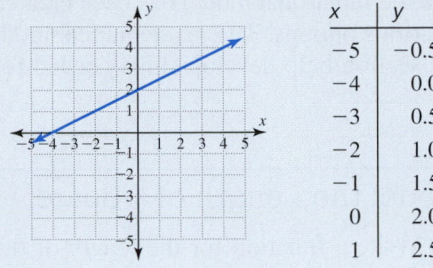

x	y
-5	-0.5
-4	0.0
-3	0.5
-2	1.0
-1	1.5
0	2.0
1	2.5

a. $f(0) = \underline{?}$ **b.** $f(1) = \underline{?}$ **c.** $f(x) = 0;\ x = \underline{?}$ **d.** $f(x) = 1;\ x = \underline{?}$

Does a Linear Function Produce a Continuous Graph or Just a Set of Disconnected Points?

A linear function can have any real number as its input, and its graph is a continuous line with no breaks or gaps. This is quite different from the graph of a sequence of distinct data values that we produced in Section 2.1. In Example 5, we compare a graph of disconnected points to the graph of a continuous line.

Example 5 Comparing a Disconnected Set of Points to a Continuous Line

(a) Graph the sequence of data values –1, 1, 3, 5, 7.

(b) For the function $f(x) = 2x - 3$, plot the points with x-coordinates of 1, 2, 3, 4, and 5. Then sketch the line through these points.

Solution

(a) Disconnected Points   **(b)** Continuous Line

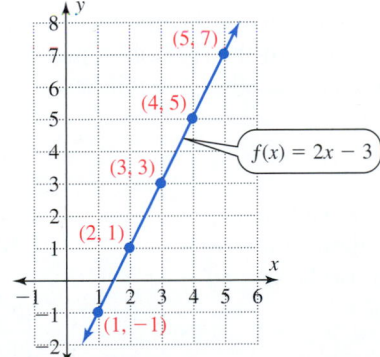

The data values in part (a) correspond to the five points $(1, -1)$, $(2, 1)$, $(3, 3)$, $(4, 5)$, and $(5, 7)$. Plot these five distinct points.

For the line in part (b), plot selected points and then draw the continuous line through these points.

Self-Check 5

a. Graph the sequence of data values 4, 1, –2, –5, –8.

b. Graph the linear function $f(x) = -3x + 7$.

3. Write a Linear Function to Model an Application

Functions are used frequently to describe the relationship between two variables. The function concept is useful because we can observe the changes to the output value of the function as we change the input value. Creating an equation or a function to describe an application is called **mathematical modeling.** Once the functional model has been created in Example 6, we can create a table and examine various options. The mathematical models presented in the examples and exercises in this section help develop skills needed for word problems that occur later in this book.

Example 6   Using a Function to Model the Length of a Board

A board 12 ft long has a piece x ft long cut off. Write a function for the length of the remaining piece in terms of x, and examine this function numerically and verbally.

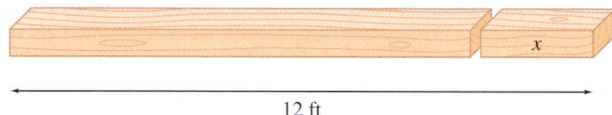

12 ft

Solution

Algebraically

$f(x) = 12 - x$ If x ft is cut off, then $12 - x$ ft remains.

Numerically

x	$f(x) = 12 - x$
0	12
2	10
4	8
6	6
8	4
10	2
12	0

Select values from 0 to 12 for x and use $f(x) = 12 - x$ to calculate the output values.

Verbally

From the table we observe that if the first piece is 2 ft, then the remaining piece is 10 ft. If the first piece is 4 ft, then the remaining piece is 8 ft. The input-output pairs in the table can be used to examine other possibilities.

Self-Check 6

a. The two angles shown here are complementary; their sum is 90°. Write a function for the second angle in terms of x.

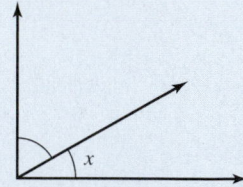

b. Use this function to complete this table.

x	$f(x)$
10	
20	
30	
40	
50	
60	
70	

As functions increase in complexity or as tables increase in size, it becomes more important to apply technology to assist us. Spreadsheets are especially designed to assist in this task. In Technology Perspective 2.2.3, we will illustrate some capabilities using the function from Example 6.

To create this table, we need to make adjustments to the table setup. Note that the *x*-values for this table start at 0 and increase by 2. To create this table with a TI-84 Plus calculator, we must adjust the settings in the **Table Setup** menu. The value next to **TblStart** represents the first *x*-value in the table and the value next to **ΔTbl** represents the change between consecutive *x*-values. In an Excel spreadsheet, the *x*-values can be created by typing them one at a time or by copying and pasting to continue an existing pattern. The *y*-values in the calculator table are created based on the equation typed on the **Y=** screen. The *y*-values in a spreadsheet can be created by entering a formula in one cell at a time or by copying and pasting an existing pattern.

Technology Perspective 2.2.3	Changing the Table Settings

Use the equation $f(x) = 12 - x$ from Example 6 to create a table of values for $x = 0, 2, 4, 6, 8, 10, 12$.

TI-84 Plus Calculator

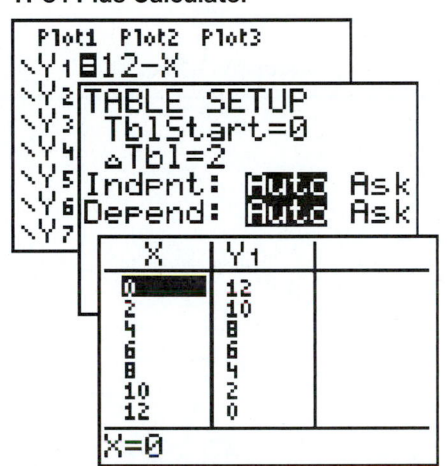

Excel Spreadsheet

	A	B	C	D
1	x	y=12-x		
2	0	12		
3	2	10		
4	4	8		
5	6	6		
6	8	4		
7	10	2		
8	12	0		

B2 · fx =12-A2

1. Press **Y=** and enter the equation.
2. Press **2nd** **TBLSET** to edit the table setup as shown. The *x*-values in the table from Example 6 start at 0 and increase by 2.
3. Press **2nd** **TABLE** to view the table with these settings.

1. Enter the desired *x*-values in column A.
2. Enter the formula "= 12−A2" in cell **B2** and copy this formula through column B.
3. This formula is used to determine the *y*-values shown in column B.

Technology Self-Check 3

Use the equation $y = 5x - 20$ to create a table of values for $x = -5, 0, 5, 10, 15, 20, 25$.

Example 7   Carpet Prices

A carpet wholesaler is increasing its prices by 7% on its top five styles of carpeting. It has the price per square yard posted on a table on its website.

(a) Write a function for the new price of a carpet with an original cost of *x* dollars per square yard.

(b) Complete the table for the new price of each item whose original cost is given.

x($)	f(x)($)
35.95	
42.90	
44.89	
48.99	
51.69	

Solution

(a) Word Equation

New price = Original price + 7% of Original price

$$f(x) = x + 0.07x$$
$$f(x) = 1.07x$$

As problems become more difficult, it is useful to start by stating the equation in words. Then translate the word equation into an algebraic equation.

(b)

	A	B
1	x	f(x) = 1.07x
2	$ 35.95	$ 38.47
3	$ 42.90	$ 45.90
4	$ 44.89	$ 48.03
5	$ 48.99	$ 52.42
6	$ 51.69	$ 55.31

We could produce this table using pencil and paper, a graphing calculator, or a spreadsheet. In the workplace, the most likely choice would be a spreadsheet.

Self-Check 7

Rework Example 7 assuming the wholesaler is increasing prices by 4.5%.

Self-Check Answers

1. **a.** $f(0) = 11$ **b.** $f(10) = 51$ **c.** $f(-10) = -29$

2. **a.**

x	-2	-1	0	1	2
f(x)	5	4	3	2	1

b.

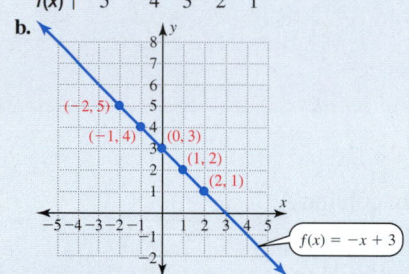

3. **a.**

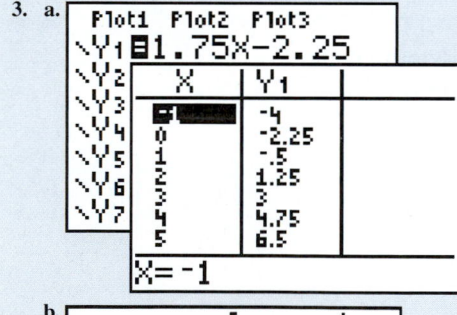

b.

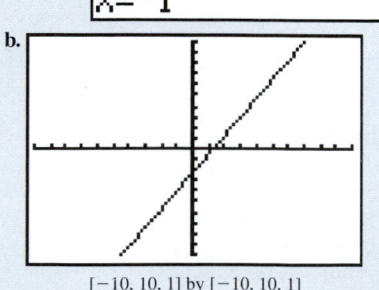

$[-10, 10, 1]$ by $[-10, 10, 1]$

4. **a.** 2 **b.** 2.5 **c.** −4 **d.** −2

5.

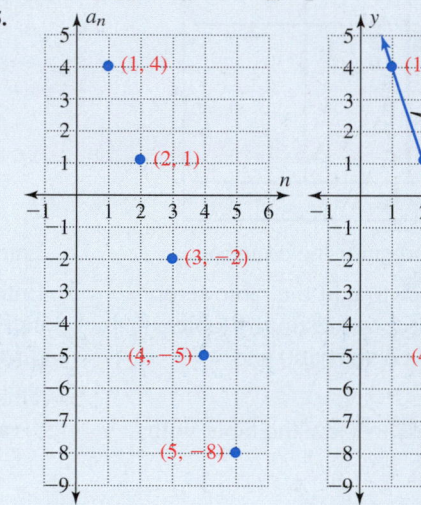

6. **a.** $f(x) = 90 - x$

b.

x	f(x)
10	80
20	70
30	60
40	50
50	40
60	30
70	20

7. **a.** $f(x) = 1.045x$

b.

x($)	f(x)($)
35.95	37.57
42.90	44.83
44.89	46.91
48.99	51.19
51.69	54.02

Technology Self-Check Answers

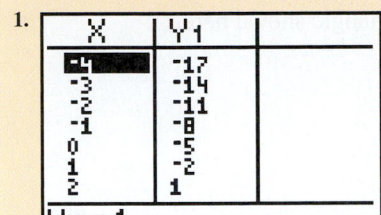

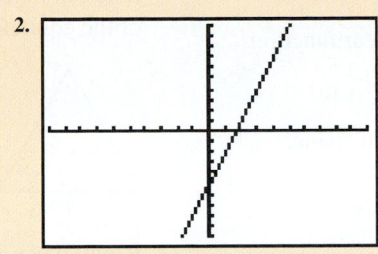

$[-10, 10, 1]$ by $[-10, 10, 1]$

2.2 | Using the Language and Symbolism of Mathematics

1. The notation $f(x)$ is called _____ notation.

2. The notation $f(x) = 8x - 2$ is read "_____ of _____ equals eight x minus two."

3. In the notation $f(x) = 8x - 2$, the input variable is represented by _____ and $f(x)$ represents the _____ variable.

4. In the notation $f(5) = 9$, the input value is _____ and the output value is _____.

5. The graph of $f(x) = 8x - 2$ is a straight _____.

6. The function $f(x) = mx + b$ is called a _____ function.

7. Creating an equation or a function to describe an application is called mathematical _____.

2.2 | Quick Review

Evaluate each expression using only pencil and paper.

1. $-\dfrac{1}{2}(-6) + 3$

2. $-\dfrac{1}{2}(4) + 3$

Evaluate each expression for $x = -2$.

3. $4 + 2(3x - 8)$

4. $(4x - 7) - 2(3x - 8)$

Write the following expression in the one-line horizontal format used by calculators.

5. $\dfrac{x + 4}{2}$

2.2 | Exercises

Objective 1 Use Function Notation

In Exercises 1–4, use the linear function $f(x) = 3x + 7$ to evaluate each expression.

1. $f(0)$
2. $f(1)$
3. $f(-1)$
4. $f(5)$

In Exercises 5–8, use the linear function $f(x) = 5x - 6$ to evaluate each expression.

5. $f(2)$
6. $f(-2)$
7. $f(-10)$
8. $f(10)$

In Exercises 9–12, use the linear function $f(x) = -6x + 4$ to evaluate each expression.

9. $f(-4)$
10. $f(4)$
11. $f\left(\dfrac{1}{6}\right)$
12. $f\left(-\dfrac{1}{6}\right)$

In Exercises 13–16, use the linear function $f(x) = -8x + 5$ to evaluate each expression.

13. $f\left(\dfrac{1}{4}\right)$
14. $f\left(-\dfrac{1}{2}\right)$
15. $f\left(-\dfrac{1}{8}\right)$
16. $f(0)$

Objective 2 Use a Linear Equation to Form a Table of Values and to Graph a Linear Equation

17. Use the linear function $f(x) = -2x + 4$ to complete this table.

x	f(x)
-2	
-1	
0	
1	
2	

18. Use the linear function $f(x) = \dfrac{x-3}{2}$ to complete this table.

x	f(x)
-2	
-1	
0	
1	
2	

19. Use the linear function $f(x) = \dfrac{x+3}{2}$ to complete this table and to graph this function.

x	f(x)
0	
1	
2	
3	

20. Use the linear function $f(x) = -\dfrac{1}{2}x + 1$ to complete this table and to graph this function.

x	f(x)
-2	
-1	
0	
1	
2	

21. For the function $f(x) = x + 2$, plot the points with x-coordinates of –4, 0, and 2. Then sketch the line through these three points.

22. For the function $f(x) = -x + 1$, plot the points with x-coordinates of –4, 0, and 2. Then sketch the line through these three points.

Objective 3 Write a Linear Function to Model an Application

23. Length of a Board A board 18 ft long has three pieces each x ft long cut off. Write a function for the length of the remaining piece in terms of x.

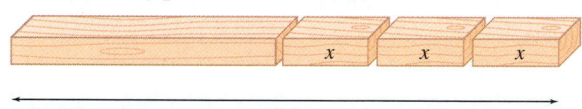

18 ft

24. Supplementary Angles The two angles shown in the figure are supplementary; their sum is 180°. Write a function for the second angle in terms of x.

25. Perimeter of an Isosceles Triangle Write a function in terms of x for the perimeter in centimeters of the isosceles triangle shown here.

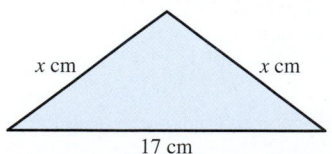

26. Perimeter of an Equilateral Triangle Write a function in terms of x for the perimeter in centimeters of the equilateral triangle shown here.

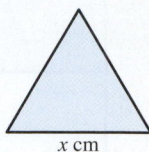

x cm

27. Perimeter and Area of a Rectangle Write a function in terms of x for

a. The perimeter in centimeters of the rectangle shown in the figure.

b. The area in square centimeters of the rectangle shown in the figure.

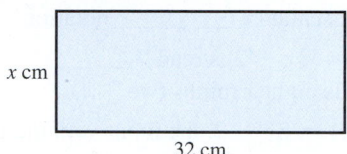

x cm

32 cm

28. Perimeter and Area of a Rhombus Write a function in terms of x for

a. The perimeter in cm of the rhombus shown here.

b. The area of the rhombus shown here.

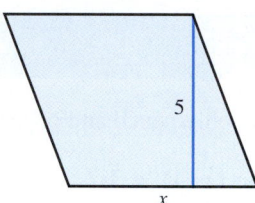

29. Rate and Distance of a Boat A boat has a speed of x mi/h in still water. Write a function in terms of x for

a. The rate of this boat traveling downstream in a river with a current of 5 mi/h.

b. The rate of this boat traveling upstream in a river with a current of 5 mi/h.

c. The distance the boat travels in 2 hours going downstream in a river with a current of 5 mi/h.

30. Rate and Distance of an Airplane An airplane has a speed of x mi/h in calm skies. Write a function in terms of x for

a. The rate of this airplane traveling in the same direction as a 30-mi/h wind.

b. The rate of this airplane traveling in the opposite direction of a 30-mi/h wind.

c. The distance the airplane travels in 2 hours going in the same direction as a 30-mi/h wind.

Concept Development

31. Use the given graph to complete this table.

x	f(x)
-2	
-1	
0	
1	
2	

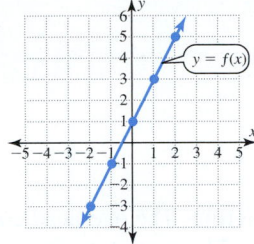

32. Use the given graph to complete this table.

x	f(x)
-2	
-1	
0	
1	
2	

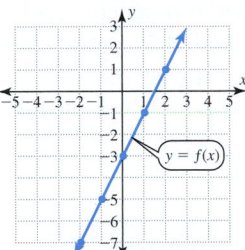

33. Use the given graph to complete this table.

x	f(x)
-5	
-3	
0	
2	

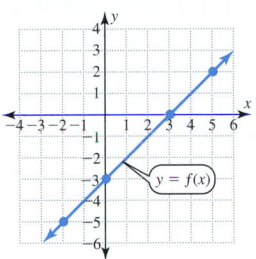

34. Use the given graph to complete this table.

x	f(x)
-3	
-1	
0	
3	

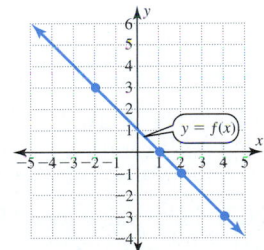

35. Use the given table of values to determine the missing input and output values.

x	f(x)
0	-10
1	-8
6	2
8	6

a. $f(1) = $ _____
b. $f(6) = $ _____
c. $f(x) = 6; x = $ _____
d. $f(x) = -10; x = $ _____

36. Use the given table of values to determine the missing input and output values.

x	f(x)
0	2
2	3
6	5
8	6

a. $f(0) = $ _____
b. $f(8) = $ _____
c. $f(x) = 3; x = $ _____
d. $f(x) = 5; x = $ _____

37. Use the given graph to determine the missing input and output values.

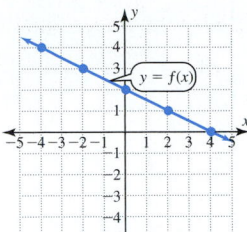

a. $f(-2) = $ _____
b. $f(2) = $ _____
c. $f(x) = 2; x = $ _____
d. $f(x) = 0; x = $ _____

38. Use the given graph to determine the missing input and output values.

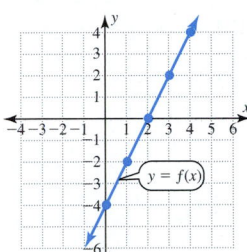

a. $f(0) = $ _____
b. $f(2) = $ _____
c. $f(x) = 0; x = $ _____
d. $f(x) = -2; x = $ _____

39. a. Graph the sequence of data values 4, 3, 2, 1, 0.
b. For the linear function $f(x) = -x + 5$, plot the points with x-coordinates of 1, 2, 3, 4, and 5, and then sketch the line through these points.
c. Compare the graphs in parts **a** and **b**.

40. a. Graph the sequence of data values -3, -2, -1, 0, 1.
b. For the linear function $f(x) = x - 4$, plot the points with x-coordinates of 1, 2, 3, 4, and 5, and then sketch the line through these points.
c. Compare the graphs in parts **a** and **b**.

Applying Technology

In Exercises 41 and 42, use a calculator and the given linear equation to:
a. complete the table and
b. graph the equation on the window [-10, 10, 1] by [-10, 10, 1]. (*Hint:* See Technology Perspectives 2.2.1 and 2.2.2.)

41. $y = -\dfrac{x}{2}$

x	-6	-4	-2	0	2	4	6
y							

42. $y = \dfrac{x}{3}$

x	-9	-6	-3	0	3	6	9
y							

In Exercises 43–46, use the given function and a calculator or spreadsheet to complete each table.

43. $y = 2.7x + 4.5$

a.

x	y
-2	
-1	
0	
1	
2	

b.

x	y
	0
	10
	20
	30
	40

44. $y = -6.8x + 3.7$

a.

x	y
0	
1	
2	
3	
4	

b.

x	y
-5	
0	
5	
10	
15	

45. $y = \dfrac{1}{4}x - 2$

a.

x	y
-8	
-4	
0	
4	
8	

b.

x	y
	0
	20
	40
	60
	80

46. $y = -\dfrac{1}{2}x + 4$

a.

x	y
-4	
-2	
0	
2	
4	

b.

x	y
-40	
-20	
0	
20	
40	

Connecting Concepts to Applications

47. Price Discount A coupon for a restaurant entitled the user to a 20% discount on any entree.
  a. Write a function for the amount of discount on an entree priced at x dollars.
  b. Evaluate and interpret $f(25)$.

48. Gratuity A restaurant automatically adds an 18% gratuity to the food and beverage total on all bills.
  a. Write a function for the gratuity added to a food and beverage total of x dollars.
  b. Evaluate and interpret $f(40)$.

49. Price Discount A discount store had a holiday special with a 15% discount on the price of all sporting goods.
  a. Write a function for the new price of an item with an original price of x dollars.
  b. Complete the table for the new price of each item whose original price is given.

x	f(x)
20	17
44	
60	
68	
90	

50. Price Increase A wholesaler is increasing the price of all items 2% to cover increased costs.
  a. Write a function for the new price on an item with an original price of x dollars.

  b. Complete the table for the new price of each item whose original price is given.

x	f(x)
20	20.40
35	
44	
80	
90	

51. Cost of Production The overhead cost for a company is $500 per day. The cost of producing each item is $15. The total cost of production is the sum of the overhead cost and the cost of producing each item.
  a. Write a function that gives the total cost of producing x units per day.
  b. Use the function to complete this table.

x	0	100	200	300	400	500
f(x)						

  c. Evaluate and interpret $f(250)$.

52. Length Remaining on a Roll of Wire A roll of wire 100 ft long has five pieces each of length x cut from it.
  a. Write a function that models the length of wire remaining on the roll.
  b. Use the function to complete this table.

x	f(x)
0	
5	
10	
15	
20	
25	
30	

  c. Evaluate and interpret $f(20)$.
  d. Determine and interpret the value of x for which $f(x) = 50$.
  e. Is 25 a practical input value for x? Explain.

53. Angles in an Isosceles Triangle An isosceles triangle has two equal angles each with x degrees. The sum of all the angles in a triangle is 180°.

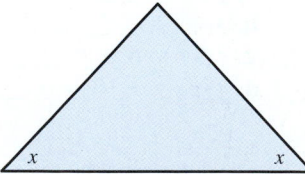

  a. Write a function that models the number of degrees in the third angle.

b. Use the function to complete this table.

x	f(x)
10	
20	
30	
40	
50	
60	
70	

c. Evaluate and interpret $f(40)$.

d. Determine and interpret the value of x for which $f(x) = 40$.

e. Is 100 a practical input value for x? Explain.

Group discussion questions

54. Challenge Question A carpenter uses x pieces from a 12-foot board. If each of these x pieces is 2 feet long, then the function $f(x) = 12 - 2x$ models the length of what remains from the board.

a. What is the smallest practical value of x for this application?

b. What is the largest practical value of x for this application?

c. List all practical values of x.

55. Discovery Question Using the function $y = \dfrac{x}{2}$:

a. Complete this table.

x	−6	−4	−2	0	2	4	6
y							

b. Graph this function using a window of $[-10, 10, 1]$ by $[-10, 10, 1]$.

c. For each increase of 2 units in x in the table, what is the change in y?

d. For each 2-unit movement to the right on the graph, what is the change in the y-coordinate?

56. Discovery Question Use a graphing calculator to graph $f(x) = \dfrac{x}{2}, f(x) = \dfrac{x}{2} + 1, f(x) = \dfrac{x}{2} + 3$, and $f(x) = \dfrac{x}{2} - 3$. What relationship do you observe among these graphs?

2.2	**Cumulative Review**

1. Use the formula $I = PRT$ to determine the interest on $5,000 at 7% for 1 year.

2. Use the formula $D = RT$ to determine the distance traveled at 60 mi/h for 3.5 hours.

3. Evaluate $2^4 - 4^2$.

4. Evaluate $2^5 - 5^2$.

5. Evaluate $(2 - 5)^2$.

Section 2.3	Graphs of Linear Equations in Two Variables

Objectives:

1. Check possible solutions of a linear equation.
2. Determine the intercepts from a graph.
3. Determine the point where two lines intersect.

The two lease options given here are summaries of two advertisements by two different automobile agencies, both trying to lease the same automobile.

Lease option A: A $2,000 initial down payment followed by 36 monthly payments of $250.

Lease option B: A $2,500 initial down payment followed by 36 monthly payments of $200.

In this section, we will examine a method for comparing these two options. In Example 8, we will examine which of these two options is the more costly. To prepare for this we will examine linear equations in greater detail.

1. Check Possible Solutions of a Linear Equation

What Is a Solution of a Linear Equation?

A **solution** of a linear equation of the form $y = mx + b$ is an ordered pair (x, y) that makes the equation a true statement. A solution (x, y) is said to **satisfy** the equation. Example 1 illustrates the fact that an ordered pair (x, y) satisfies an equation if and only if that point lies on the graph of that equation. *Key point:* A solution of $y = mx + b$ is an ordered pair—not just an x-value or a y-value.

Example 1   Checking Ordered Pairs in an Equation

Determine algebraically whether **(a)** $(0, -3)$ or **(b)** $(2, 3)$ is on the graph of $y = 2x - 3$.

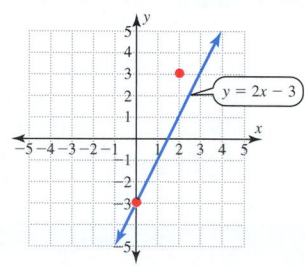

Solution

(a) $y = 2x - 3$ To check $(0, -3)$, substitute
   $-3 \stackrel{?}{=} 2(0) - 3$ -3 for y and 0 for x and then
   $-3 \stackrel{?}{=} 0 - 3$ simplify.
   $-3 \stackrel{?}{=} -3$ checks
   $(0, -3)$ is on the graph   Examine the graph of $y = 2x - 3$ to
   of $y = 2x - 3$.   note that $(0, -3)$ is a point on this graph.

(b) $y = 2x - 3$ To check $(2, 3)$, substitute 3 for y and 2 for x and
   $3 \stackrel{?}{=} 2(2) - 3$ then simplify.
   $-3 \stackrel{?}{=} 4 - 3$
   $-3 \stackrel{?}{=} 1$ *is false*
   $(2, 3)$ is not on the graph   Examine the graph of $y = 2x - 3$ to note that
   of $y = 2x - 3$. $(2, 3)$ is not a point on this graph.

Self-Check 1

Determine whether **a.** $(2, -2)$ or **b.** $(1, 5)$ is on the graph of $y = -3x + 4$.

In the linear equation $y = mx + b$, x represents the input variable and y represents the output variable. Linear equations are also frequently given using function notation in the form $f(x) = mx + b$. In this notation, x represents the input variable and $f(x)$ represents the output variable. Both y and $f(x)$ are used to denote the output value.

A linear equation in two variables has an infinite set of ordered pairs that are solutions of the equation. Graphically all the solutions lie on a line, and all points not on the line are not solutions. In Example 2, we check selected points on the line defined by $f(x) = -2x + 4$ to confirm that they satisfy the equation.

Example 2 Checking the Coordinates of a Point on a Line

Determine whether the points A, B, and C satisfy
the equation $f(x) = -2x + 4$.

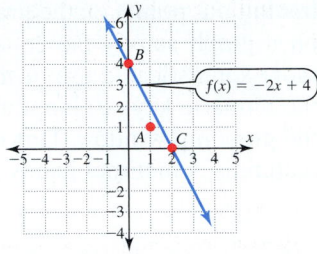

Solution

(a) Point A has coordinates $(1, 1)$.

$f(x) = -2x + 4$

$1 \overset{?}{=} -2(1) + 4$

$1 \overset{?}{=} -2 + 4$

$1 \overset{?}{=} 2$ is false

$(1, 1)$ does not satisfy the equation.

Substitute 1 for the input value of x and 1 for the output value of $f(x)$.

This point is not on the line and does not satisfy the equation.

(b) Point B has coordinates $(0, 4)$.

$f(x) = -2x + 4$

$4 \overset{?}{=} -2(0) + 4$

$4 \overset{?}{=} 0 + 4$

$4 \overset{?}{=} 4$ is true

$(0, 4)$ satisfies the equation.

Substitute 0 for the input value of x and 4 for the output value of $f(x)$.

This point is on the line and does satisfy the equation. The point $(0, 4)$ lies on the y-axis and is called the y-intercept of the graph.

(c) Point C has coordinates $(2, 0)$.

$f(x) = -2x + 4$

$0 \overset{?}{=} -2(2) + 4$

$0 \overset{?}{=} -4 + 4$

$0 \overset{?}{=} 0$ is true

$(2, 0)$ satisfies the equation.

Substitute 2 for the input value of x and 0 for the output value of $f(x)$.

This point is on the line and does satisfy the equation. The point $(2, 0)$ lies on the x-axis and is called the x-intercept of the graph.

Self-Check 2

Determine whether the points A and B satisfy the equation $y = \dfrac{1}{3}x - 1$.

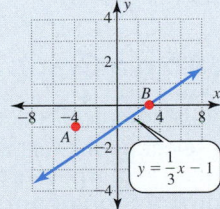

If you have created a graph with a calculator, then it is likely that the calculator has some powerful features for examining the graph. One feature allows us to quickly examine many ordered pairs that lie on the graph. Technology Perspective 2.3.1 illustrates how to do this using the **TRACE** feature on the TI-84 Plus calculator. The **TRACE** feature is not available in most spreadsheets.

Once a graph has been created, pressing the [TRACE] key allows us to use the left and right arrow keys to move along the graph while displaying the x- and y-coordinates. By tracing out points on the graph, we can dynamically illustrate the connection that points on a graph satisfy the equation for the graph. Technology Perspective 2.3.1 makes a more sophisticated use of the **TRACE** feature by pressing the [TRACE] key and then using the numeric keys to input an x-value between the minimum and maximum x-values on the graphing window. This enables us to evaluate a function at a particular x-value. This feature is sometimes handy for checking given values in an equation.

| **Technology Perspective 2.3.1** | **Using a TRACE Feature** |

Use the equation $y = 2x - 3$ from Example 1 and the TRACE feature on the TI-84 Plus calculator to determine whether the points $(0, -3)$ and $(2, 3)$ are on the graph of $y = 2x - 3$.

TI-84 Plus Keystrokes

1. Press [Y=] and enter the equation.

2. Press [ZOOM] [6] to display the graph in the standard viewing window.

3. Press [TRACE] to access the **TRACE** feature.

   Press [0] [ENTER] to move the cursor to the location on the graph where $x = 0$. We see that $(0, -3)$ is a point on the graph.

4. Press [2] [ENTER] to move the cursor to the location on the graph where $x = 2$. When $x = 2$, $y = 1$. Thus, $(2, 3)$ is not a point on the graph.

TI-84 Plus Calculator

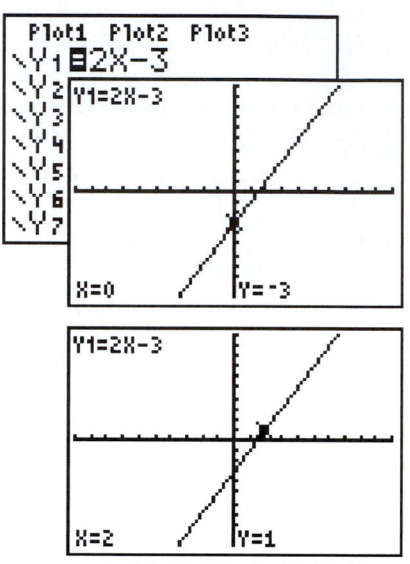

$[-10, 10, 1]$ by $[-10, 10, 1]$

Technology Self-Check 1

Determine whether the points $(4, 6)$ and $(-1, -8)$ are on the graph of $y = 3x - 5$.

---

2. Determine the Intercepts from a Graph

The line defined by a linear equation is completely determined by any two points on the line. To graph the line, we will often select points that are easy to calculate.

Do Any Points on a Line Have Special Significance?

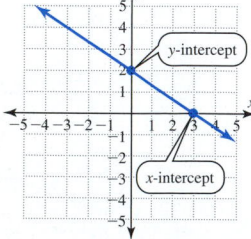

Yes, points where a graph intersects, or crosses, the axes are called the **intercepts.** The **x-intercept** is often denoted by $(a, 0)$ and the **y-intercept** by $(0, b)$. The intercepts are points with both x- and y-coordinates. These points frequently have special significance in applications, as illustrated in Example 3.

| **Example 3** | Interpreting the Intercepts of a Graph: Overhead Costs and Break-Even Values |

In the graph to the right, the input value of x is the number of units a machine produces. The output y is the profit generated by the sale of these units when they are produced. Determine the intercepts and interpret the meaning of these points.

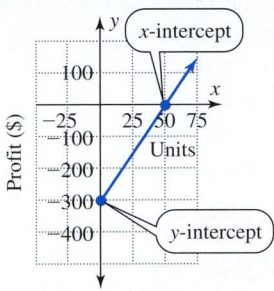

Solution

The y-intercept is $(0, -300)$. If 0 units are produced, then \$300 will be lost. (Often there are **overhead costs** of rent, electricity, etc., even if no units are produced.)

The x-intercept is $(50, 0)$. If 50 units are produced, then \$0 will be made. (This is the **break-even value** for this machine—the number of units the machine must make before a profit can be realized.)

Self-Check 3

Using the same assumptions as those given in Example 3, determine the intercepts of the graph and interpret the meaning of these intercepts.

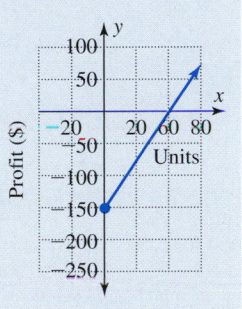

If the intercepts are listed in a table of values, then they can be determined by inspection. Generally we will not rely on tables to determine the intercepts because only carefully selected tables will display both intercepts.

| **Example 4** | Determining Intercepts from a Table |

Use this table to determine the intercepts of the graph of $f(x) = 4.2x - 8.4$.

x	$f(x) = 4.2x - 8.4$
-4	-25.2
-2	-16.8
0	-8.4
2	0.0
4	8.4

Solution

x-intercept: $(2, 0)$ The x-intercept has a y-coordinate of 0.

y-intercept: $(0, -8.4)$ The y-intercept has an x-coordinate of 0.

Self-Check 4

Use this table to determine the intercepts of the graph of $f(x) = 4.4x + 6.6$.

x	$f(x) = 4.4x - 6.6$
-2.0	-2.2
-1.5	0.0
-1.0	2.2
-0.5	4.4
0.0	6.6

The graph of a linear function $f(x) = mx + b$ can be obtained by using any two points that lie on the line. One particularly easy point to determine is the y-intercept, which can be found by evaluating $f(0)$. Another point may be obtained by substituting any other value for x. It is also wise to plot a third point as a double-check on our work. Since all points should be on the line, a third point not on the line is a sure indication of a computation error or an error in plotting the points.

Example 5   Graphing a Linear Equation

Use the y-intercept and another point to graph the line defined by $f(x) = \dfrac{1}{2}x - 2$. Plot a third point as a check on your work.

Solution

y-intercept

$f(x) = \dfrac{1}{2}x - 2$

$f(0) = \dfrac{1}{2}(0) - 2$ To find the y-intercept, substitute 0 for x.

$f(0) = 0 - 2$

$f(0) = -2$

$(0, -2)$ is the y-intercept.

Second Point

$f(x) = \dfrac{1}{2}x - 2$

$f(4) = \dfrac{1}{2}(4) - 2$ To find a second point, find y when x is 4.

$f(4) = 2 - 2$

$f(4) = 0$

Note that $(4, 0)$ is the x-intercept.

Third Point

$f(x) = \dfrac{1}{2}x - 2$

$f(6) = \dfrac{1}{2}(6) - 2$ To find a third point to double-check our work, find y when x is 6.

$f(6) = 3 - 2$

$f(6) = 1$

$(6, 1)$ is also on the line.

Table

x	y
0	-2
4	0
6	1

Graph

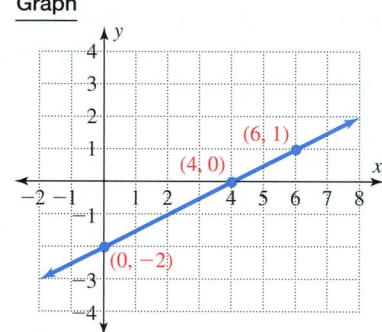

Plot these points and sketch the line through the points.

Self-Check 5

Use the y-intercept and another point to graph the line defined by $f(x) = -\dfrac{2}{3}x + 4$.

Plot a third point as a check on your work.

---

3. Determine the Point Where Two Lines Intersect

Sometimes it is useful to examine two graphs simultaneously or to compare two different options. When we refer to two or more equations at the same time, we refer to this as a **system of equations. A solution of a system of linear equations in two variables** is an ordered pair that satisfies each equation in the system.

What Is the Significance of the Point Where Two Lines Cross?

If there is a unique solution to a system of linear equations, it is represented graphically by their point of intersection, the point that is on both graphs.

Solution of a System of Two Linear Equations

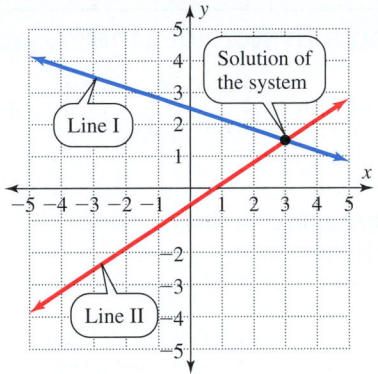

To solve a system of equations graphically, graph each equation and then estimate the coordinates of the point of intersection. Because serious errors of estimation can occur, it is wise to check an estimated solution by substituting it into each equation of the system.

Example 6   Solving a System of Linear Equations Graphically

The graphs of $y = x - 1$ and $y = \dfrac{x}{3} + 1$ are shown here.

Determine their point of intersection by inspecting the graph. Then check this point in both equations.

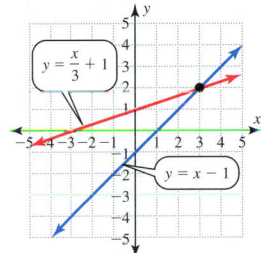

Solution

An estimate of the solution is (3, 2).

Check:

Frist Equation	Second Equation

$y = x - 1$ $\qquad y = \dfrac{x}{3} + 1$

$2 \overset{?}{=} 3 - 1$

$2 \overset{?}{=} 2$ checks. $\qquad 2 \overset{?}{=} \dfrac{3}{3} + 1$ Substitute (3, 2) into both equations to check this point.

$\qquad\qquad\qquad\qquad 2 \overset{?}{=} 1 + 1$

$\qquad\qquad\qquad\qquad 2 \overset{?}{=} 2$ checks.

Answer: (3, 2) is a solution of this system of linear equations because it checks in both equations.   As you can verify by checking any other point, this is the only point that checks in both equations.

Self-Check 6

The graphs of $y = \dfrac{1}{2}x + 5$ and $y = -x - 1$ are given.
Determine their point of intersection by inspecting both graphs. Does this point check in both equations?

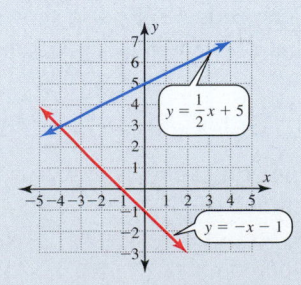

In Example 6, we were given the graphs of two linear equations. If instead we are given a system of two linear equations, we can find the solution by first graphing these equations. In Example 7, we start by forming a table of points so we can graph the lines.

Example 7   Solving a System of Linear Equations Graphically

Solve the system of linear equations $\left\{\begin{array}{l} y = 2x + 5 \\ y = -2x - 3 \end{array}\right\}$ graphically.

Solution

First graph both lines on the same coordinate system.

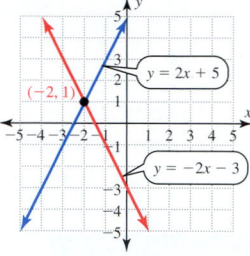

$y = 2x + 5$

x	y
0	5
$-\dfrac{5}{2}$	0
-1	3

$y = -2x - 3$

x	y
0	-3
$-\dfrac{3}{2}$	0
-1	-1

Set $x = 0$ to find the y-intercept. Then find the coordinates of two other points. Note that we also selected the x-intercept. A third point was calculated by letting $x = -1$.

Then estimate the coordinates of the point of intersection. The point of intersection appears to be $(-2, 1)$.

Check:

First Equation

$y = 2x + 5$

$1 \overset{?}{=} 2(-2) + 5$

$1 \overset{?}{=} -4 + 5$

$1 \overset{?}{=} 1$ checks.

Second Equation

$y = -2x - 3$

$1 \overset{?}{=} -2(-2) - 3$

$1 \overset{?}{=} 4 - 3$

$1 \overset{?}{=} 1$ checks.

Substitute $(-2, 1)$ into both equations to check this point.

Answer: $(-2, 1)$ is a solution of this system of linear equations because it checks in both equations.

Self-Check 7

Solve the system of linear equations $\left\{\begin{array}{l} y = 3x - 5 \\ y = -x + 3 \end{array}\right\}$ graphically.

Technology Perspective 2.3.2 shows how we can approximate a point of intersection of two lines by using the **Intersect** feature on a TI-84 Plus calculator. An important consideration is that this approximation is sometimes exactly correct and other times it has a small error. If you must have an exact answer, you may need to use the algebraic methods that we will cover later. Sometimes it is necessary to select a viewing window other than the standard viewing window so that we can display key features of the graph. Technology Perspective 2.3.2 also shows how we can select an arbitrary viewing window. The window that is selected is $[-5, 5, 1]$ by $[-5, 5, 1]$.

Technology Perspective 2.3.2 | Approximating a Point of Intersection

Solve the system of linear equations $\begin{Bmatrix} y = 2x + 5 \\ y = -2x - 3 \end{Bmatrix}$ from Example 7 by using the **Intersect** feature.

TI-84 Plus Keystrokes

1. First press  and enter the equations.
2. To change the graphing window, press `WINDOW` and edit the window settings as shown.
3. Press `GRAPH` to view the graphs. (Make sure you can see the point of intersection in the window you have selected.)
4. To calculate the point of intersection, select the **Intersect** feature (option 5 in the `CALC` menu). To do this,

   press `2nd` `TRACE` `5` `ENTER` `ENTER` `ENTER`.

Answer: $(-2, 1)$

TI-84 Plus Calculator

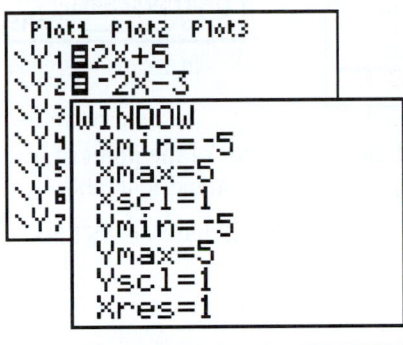

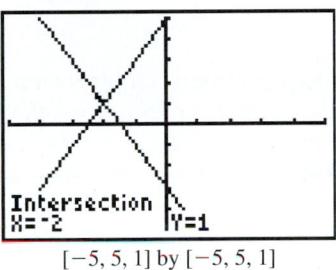

$[-5, 5, 1]$ by $[-5, 5, 1]$

Technology Self-Check 2

Use the **Intersect** feature on a graphing calculator to solve the system of linear equations $\begin{Bmatrix} y = 3x + 7 \\ y = -x + 3 \end{Bmatrix}$.

How Can I Use Systems of Equations to Compare the Two Lease Options Given at the Beginning of This Section?

Example 8 will illustrate how to do this. We will assume there are no factors other than the given costs that will affect the selection between these two options.

Example 8 | Determining Equivalent Costs

Examine these lease options numerically and graphically, and determine whether the costs will ever be the same.

> **Lease option A:** A $2,000 initial down payment followed by 36 monthly payments of $250.
> **Lease option B:** A $2,500 initial down payment followed by 36 monthly payments of $200.

Solution

We will let y represent the total cost incurred by the xth month.

Verbally | Total cost | = | Total of monthly payments | + | Initial payment |

Algebraically

Option A:	y	$=$	$250x$	$+$	2,000
Option B:	y	$=$	$200x$	$+$	2,500

The total of the monthly payments is calculated by multiplying the monthly payment by the number of months this payment has been made.

Numerically

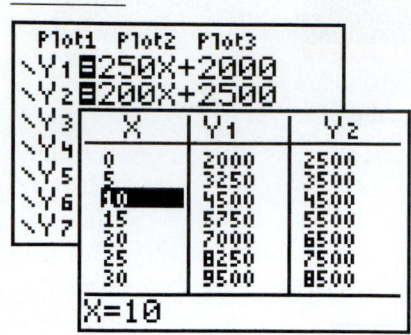

Graphically

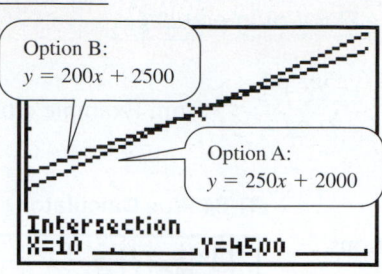

Option B:
$y = 200x + 2500$

Option A:
$y = 250x + 2000$

Intersection
X=10   Y=4500

[0, 20, 5] by [0, 7000, 1000]

A graphical approximation of the coordinates of the point of intersection is (10, 4,500); this value is supported by the numerical table.

Answer: After 10 months, both options will have cost $4,500.

Note from both the table and the graph that option A starts with the lower initial cost, and the output values (the cost) are lower until the tenth month. After the tenth month, the higher monthly cost for option A will produce output values (the cost) that are higher than the output values for option B.

Self-Check 8

Rework Example 8 assuming lease option A requires an initial down payment of $1,000 followed by 36 monthly payments of $300 and the conditions for lease option B remain the same.

Self-Check Answers

1. **a.** (2, −2) is on the graph.
   b. (1, 5) is not on the graph.
2. **a.** A: (−4, −1) does not satisfy the equation.
   b. B: (3, 0) satisfies the equation.
3. The y-intercept is (0, −150). If 0 units are produced, then $150 will be lost. The x-intercept is (60, 0). If 60 units are produced, then $0 will be made.
4. x-intercept: (−1.5, 0); y-intercept: (0, 6.6)

5.

$f(x) = -\dfrac{2}{3}x + 4$

6. The point of intersection is (−4, 3). This point checks in both equations.
7. (2, 1)
8. After 15 months, both options will have cost $5,500.

Technology Self-Check Answers

1. (4, 6) is not on the graph of $y = 3x − 5$, but (−1, −8) is on the graph of $y = 3x − 5$.

2. The solution is (−1, 4).

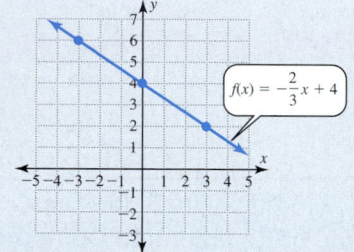

Plot1 Plot2 Plot3
\Y1=3X+7
\Y2=-X+3
\Y3=
\Y4=
\Y5=
\Y6=
\Y7=

Intersection
X=-1   Y=4

[−10, 10, 1] by [−10, 10, 1]

2.3 Using the Language and Symbolism of Mathematics

1. The point $(a, 0)$ on the graph of a line is called the _____ .

2. The point $(0, b)$ on the graph of a line is called the _____ .

3. A solution of a linear equation $y = mx + b$ is an ordered pair (x, y) that makes the equation a _____ statement.

4. A linear equation of the form $y = mx + b$ has an _____ number of solutions.

5. An ordered pair that satisfies each equation in a system of linear equations is called a _____ of the system of linear equations.

6. A unique solution to a system of two linear equations is represented graphically by their point of _____ .

7. If a point is the x-intercept of a graph, then the _____ -coordinate is zero.

8. If a point is the y-intercept of a graph, then the _____ -coordinate is zero.

9. If a point satisfies $f(x) = mx + b$ and $x = 0$, then this point is the _____ -intercept of the graph of this linear function.

10. If a point satisfies $f(x) = mx + b$ and $f(x) = 0$, then this point is the _____ -intercept of the graph of this linear function.

11. Even when no units are produced there are costs for rent, electricity, and so on. These costs are called _____ costs.

12. The number of units a machine must make before a profit can be realized is called the _____ value for this machine.

2.3 Quick Review

Give the coordinates of each point.

1. Point A

2. Point B

3. Point C

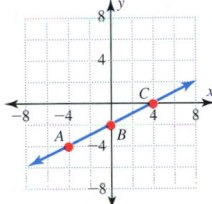

Use the given table of values to determine the missing input and output values.

4. $f(17) =$ ____

5. $f(x) = 11; x =$ ____

x	$f(x)$
1	17
5	11
11	3
17	2

2.3 Exercises

Objective 1 Check Possible Solutions of a Linear Equation

In Exercises 1–8, determine whether each ordered pair is a solution of $y = 6x - 1$.

1. $(1, 5)$ 2. $(2, 4)$ 3. $\left(\dfrac{1}{6}, 6\right)$

4. $(0, -1)$ 5. $(2, 11)$ 6. $(1, -1)$

7. $\left(\dfrac{1}{6}, 0\right)$ 8. $(0, 3)$

In Exercises 9–16, determine whether each point is on the graph $f(x) = -\dfrac{x}{2} + 5$.

9. $(0, 6)$ 10. $(10, 0)$ 11. $(-6, -8)$

12. $(4, 3)$ 13. $(0, 5)$ 14. $(6, 0)$

15. $(12, -1)$ 16. $(-8, -9)$

In Exercises 17–20, determine whether the points A, B, C, and D on the graph are solutions of the given equation.

17. $y = \dfrac{x}{4} + 1$

18. $y = -2x + 10$

19. $y = 3$

20. $x = -4$

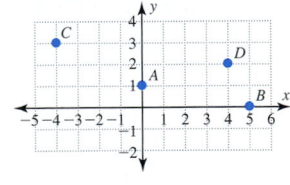

Objective 2 Determine the Intercepts from a Graph

In Exercises 21 and 22, identify the x- and y-intercepts on the given graph.

21.
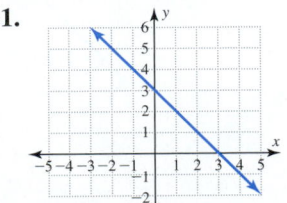

22.
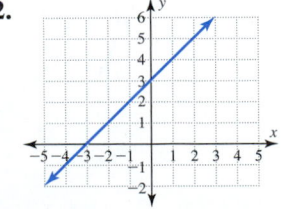

In Exercises 23 and 24, determine the intercepts of each line and interpret the meaning of these points. In each graph, the input value of x is the number of units of production by one assembly line at a factory, and the output value y is the profit in dollars generated by the sale of these units when they are produced. (*Hint:* See Example 3.)

23.

Profit in Dollars

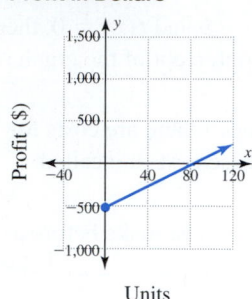

Units

24.

Profit in Dollars

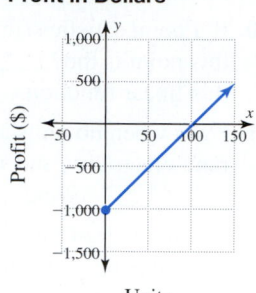

Units

In Exercises 25 and 26, use the table for the linear equation $y = mx + b$ to identify the x- and y-intercepts of the graph of the line through these points.

25.

x	y
-4	1
-3	0
-2	-1
-1	-2
0	-3
1	-4

26.

x	y
-2	-4
-1	-3
0	-2
1	-1
2	0
3	1

Objective 3 Determine the Point Where Two Lines Intersect

In Exercises 27–32, determine the point of intersection of the two lines.

27.

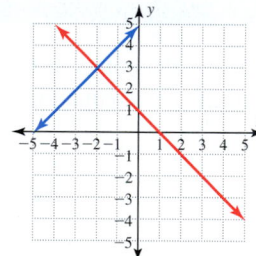

28.

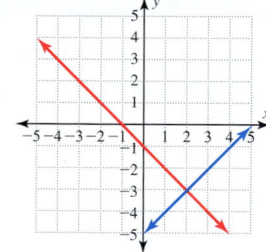

29.

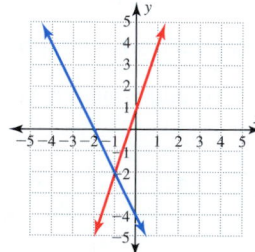

30.

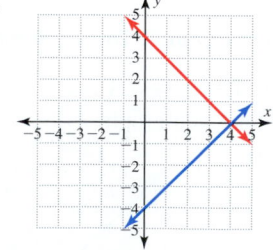

31.

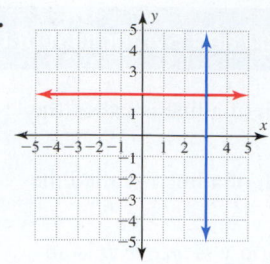

32.

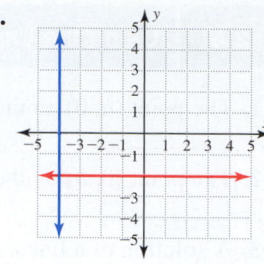

Using and Interpreting Graphs

33. Graph the line with x-intercept $(3, 0)$ and y-intercept $(0, 2)$.

34. Graph the line with x-intercept $(-2, 0)$ and y-intercept $(0, -3)$.

35. Graph the line if every point on the line has an x-coordinate of -4.

36. Graph the line if every point on the line has an x-coordinate of 3.

37. Graph the line if every point on the line has a y-coordinate of 4.

38. Graph the line if every point on the line has a y-coordinate of -3.

In Exercises 39–46, plot the y-intercept of each line and one other point to graph the line. Select a third point to double-check your work.

39. $y = x + 2$

40. $y = x - 3$

41. $y = \dfrac{x}{3} - 1$

42. $y = \dfrac{x}{4} + 1$

43. $f(x) = -x + 4$

44. $f(x) = x - 2$

45. $f(x) = 2x + 3$

46. $f(x) = -\dfrac{x}{2} + 1$

In Exercises 47–50, solve each system of linear equations by graphing each equation on the same coordinate system and determining the point of intersection. Check the coordinates of this point in both of the linear equations.

47. $y = 2x - 5$
$y = -2x - 1$

48. $y = x - 5$
$y = -x + 3$

49. $y = 2x - 1$
$y = \dfrac{x}{2} + 2$

50. $y = -x + 5$
$y = \dfrac{x}{3} + 1$

Applying Technology

In Exercises 51 and 52, use a calculator to complete the table of values and then identify the x- and y-intercepts of the graph of the line through these points.

51. $y = 3x - 9$

52. $y = -5x + 10$

x	y
-3	
-2	
-1	
0	
1	
2	
3	

In Exercises 53 and 54, use a calculator with a **Trace** feature to graph the linear function on the window $[-10, 10, 1]$ by $[-10, 10, 1]$, and then identify the x- and y-intercepts of the line. (*Hint:* See Technology Perspective 2.3.1.)

53. $f(x) = 1.2x - 6$ **54.** $f(x) = -1.4x + 7$

In Exercises 55 and 56, use a calculator with an **Intersect** feature to graph each system of equations and to determine their point of intersecion. (*Hint:* See Technology Perspective 2.3.2.)

55. $\begin{cases} y = 0.5x \\ y = 1.5x - 2 \end{cases}$ **56.** $\begin{cases} y = -\dfrac{2}{3}x \\ y = \dfrac{2}{3}x + 4 \end{cases}$

Connecting Concepts to Applications

Car payments

In Exercises 57 and 58, use a calculator or a spreadsheet and the given equation to prepare a table of the total payments for each of months 1 through 6. In the equation, x represents the number of months, and $f(x)$ represents the total payment in dollars at the end of x months.

57. $f(x) = 333.33x + 457.50$

58. $f(x) = 287.67x + 393.80$

Equivalent costs

In Exercises 59–62, determine when both options will have the same cost; also determine what this cost will be.

59. The input variable x represents the number of units produced, and the output variable $f(x)$ represents the cost of this production.
 Option A: $f(x) = x + 5$
 Option B: $f(x) = 2x + 3$

60. The input variable x represents the number of units produced, and the output variable $f(x)$ represents the cost of this production.
 Option A: $f(x) = 1.2x + 3.4$
 Option B: $f(x) = 0.8x + 5.0$

61. The input variable x represents the number of months an apartment has been rented. The output variable $f(x)$ represents the rental cost by the end of the xth month including an initial nonrefundable deposit. (*Hint:* Use a calculator window of $[0, 5, 1]$ by $[0, 1500, 500]$.)
 Option A: $f(x) = 250x + 50$
 Option B: $f(x) = 225x + 100$

62. The input variable x represents the number of months an apartment has been rented. The output variable $f(x)$ represents the rental cost by the end of the xth month including an initial nonrefundable deposit. (*Hint:* Use a calculator window of $[0, 4, 1]$ by $[0, 1000, 200]$.)
 Option A: $f(x) = 400x + 100$
 Option B: $f(x) = 450x$

Group discussion questions

63. Discovery Question Graph $y = x + 1$, $y = x + 5$, and $y = x + 9$ all on the same coordinate system. Compare these graphs. What is the same? What is different? What can you predict about the graph from the form $y = mx + b$?

64. Discovery Question Graph $y = x + 1$, $y = \dfrac{x}{2} + 1$, and $y = 2x + 1$ all on the same coordinate system. Compare these graphs. What is the same? What is different? What can you predict about the graph from the form $y = mx + b$?

65. Discovery Question Graph $y = -2$, $y = 1$, and $y = 3$ all on the same coordinate system. Compare these graphs. What is the same? What is different? What can you predict about the graph from the form $y = mx + b$?

| **2.3** | **Cumulative Review** |

Evaluate each expression using only pencil and paper.

1. $5 - 2(6 - 2^2)$

2. $5 - 2(6 - 2)^2$

3. $(1 - 2)^3$

4. $1^3 - 2^3$

5. $\dfrac{3 - (4 - 10)}{6 - 3(4)}$

| **Section 2.4** | **Solving Linear Equations in One Variable Using the Addition-Subtraction Principle** |

Objectives:

1. Solve linear equations in one variable using the addition-subtraction principle.
2. Use graphs and tables to solve a linear equation in one variable.
3. Identify a linear equation as a conditional equation, an identity, or a contradiction.

One purpose of Sections 2.1 through 2.3 was to develop the concepts and skills to examine linear equations by using both graphs and tables. We will now use these skills to solve linear equations containing only one variable. We will concentrate on linear equations in one variable for the rest of this chapter and return to linear equations in two variables in Chapter 3.

1. Solve Linear Equations in One Variable Using the Addition-Subtraction Principle

In this section we will concentrate on developing algebraic skills for solving linear equations. We have already noted that $y = mx + b$, an equation with two variables, has a graph that is a straight line. Thus, we refer to $y = mx + b$ as a linear equation. Note that the exponent on both x and y is understood to be 1; we refer to this by saying that the equation is **first degree in both x and y.** If each variable in an equation is first degree, then we extend the terminology and call the equation a **linear equation.**

How Can I Identify a Linear Equation in One Variable?

It is one that can be written in the form $Ax = B$, where x is first degree.

Linear Equation in One Variable		
Algebraically	**Verbally**	**Algebraic Example**
A **linear equation in x** is an equation that can be written in the form $Ax = B$, where A and B are real constants and $A \neq 0$.	A linear equation in one variable is first degree in this variable.	$2x = 24$

Example 1 Identifying Linear Equations

Determine which of the following choices are linear equations in one variable.

Solution

(a) $3x - 1 = 23$ — Linear equation in one variable — The exponent of the variable x is understood to be 1. This is a first-degree equation.

(b) $x^2 = 24$ — Not a linear equation — The exponent on the variable x is 2.

(c) $3(y - 5) = 4y + 7$ — Linear equation in one variable — The only variable is y, and the exponent on y is 1 on both sides of the equation. This equation can be simplified to the form $y = -22$.

(d) $7x - 2 = 5y + 1$ — Not a linear equation in one variable — This is a linear equation, but it contains two variables, x and y.

(e) $5x - 3 + 4(x - 1)$ — Not a linear equation — This is an algebraic expression, but it is not an equation since it has no symbol of equality.

Self-Check 1

Determine which of the following choices are linear equations in one variable.

a. $w + 6 = 4$ **b.** $2(w + 6) - 4$ **c.** $5(3x - 1) = 2(7x + 5)$ **d.** $(x - 1)^2 = 4$

What Is the Basic Algebraic Strategy Used to Solve Linear Equations?

To solve an equation whose solution is not obvious, we form simpler equivalent equations until we obtain an equation whose solution is obvious. **Equivalent equations** have the same solution set.

Think of a Balance Scale When You Solve Equations

When thinking of an equation, you may find it helpful to use the concept of a balance scale that has the left side in balance with the right side. We must always perform the same operation on both sides to preserve this balance.

The general strategy for solving a linear equation, regardless of its complexity, is to isolate the variable whose value is to be determined on one side of the equation and to place all other terms on the other side. One of the principles used to accomplish this is given in the following box. Another principle, the Multiplication-Division Principle of Equality, is covered in Section 2.5. Many students benefit by practicing on the material in this section before using the multiplication-division principle.

Addition-Subtraction Principle of Equality

Verbally	Algebraically	Numerical Example
If the same number is added to or subtracted from both sides of an equation, the result is an equivalent equation.	If a, b, and c are real numbers, then $a = b$ is equivalent to $a + c = b + c$ and to $a - c = b - c$.	$x + 2 = 5$ is equivalent to: $x + 2 - 2 = 5 - 2$ $x = 3$

Example 2 Solving a Linear Equation with One Variable Term

Solve $x - 3 = 8$.

Solution

$$x - 3 = 8$$
$$x - 3 + 3 = 8 + 3$$
$$x = 11$$

Use the addition-subtraction principle of equality to add 3 to both sides of the equation in order to isolate the variable term on the left side and the constant terms on the right side.

Check: $x - 3 = 8$
$$11 - 3 \overset{?}{=} 8$$
$$8 \overset{?}{=} 8 \text{ checks}$$

Answer: $x = 11$

Self-Check 2

Solve $w + 6 = 4$.

To remove a constant or variable from one side of an equation, we add its additive inverse to both sides of the equation. This is illustrated by both Examples 2 and 3.

Example 3 contains constants and variables on both sides of the equation. We will use the addition-subtraction principle of equality to isolate the variables on the left side of the equation and the constant terms on the right side of the equation.

Example 3   Solving a Linear Equation with Variable and Constant Terms on Both Sides

Solve $6v - 7 = 5v - 3$.

Solution

$$6v - 7 = 5v - 3$$
$$6v - 7 - 5v = 5v - 3 - 5v \qquad \text{Subtract } 5v \text{ from both sides of the equation to isolate the}$$
$$v - 7 = -3 \qquad\qquad\qquad \text{variable term on the left side of the equation.}$$
$$v - 7 + 7 = -3 + 7 \qquad \text{Add 7 to both sides of the equation to isolate the constant}$$
$$v = 4 \qquad\qquad\qquad\quad \text{term on the right side of the equation.}$$

Check: $6v - 7 = 5v - 3$
$$6(4) - 7 \overset{?}{=} 5(4) - 3$$
$$24 - 7 \overset{?}{=} 20 - 3$$
$$17 \overset{?}{=} 17 \quad \text{checks}$$

Answer: $v = 4$

Self-Check 3

Solve $8a + 9 = 7a + 15$.

Example 4 illustrates an equation that contains grouping symbols on both sides of the equation. The strategy is first to simplify the left and right sides of the equation and then to isolate the variable terms on one side of the equation and the constant terms on the other side. To solve a linear equation, try to make each step produce a simpler equation than in the previous step.

Example 4   Solving a Linear Equation Containing Parentheses

Solve $4(2b - 3) = 7(b - 2)$.

Solution

$$4(2b - 3) = 7(b - 2) \qquad \text{Use the distributive property } a(b + c) = ab + ac \text{ to}$$
$$4(2b) - 4(3) = 7b - 7(2) \qquad \text{remove the parentheses, and then simplify both sides of}$$
$$8b - 12 = 7b - 14 \qquad\qquad \text{the equation.}$$
$$8b - 12 - 7b = 7b - 14 - 7b \qquad \text{Subtract } 7b \text{ from both sides of the equation.}$$
$$b - 12 = -14$$
$$b - 12 + 12 = -14 + 12 \qquad \text{Add 12 to both sides of the equation.}$$
$$b = -2$$

Answer: $b = -2$ $\qquad\qquad\qquad$ Does this answer check?

Self-Check 4

Solve $5(3x - 1) = 2(7x + 5)$.

2. Use Graphs and Tables to Solve a Linear Equation in One Variable

In Example 5, we use multiple representations to solve the linear equation $2x - 5 = x - 2$. We examine this equation algebraically, graphically, numerically, and verbally. To examine this equation graphically and numerically, we make use of Technology Perspective 2.3.2, where we determined the point of intersection of two lines. The main purpose for examining linear equations in two variables in Section 2.3 was to gain a visual perspective for solving linear equations in one variable. To create this visual perspective in Example 5, we create two equations, $Y_1 = 2x - 5$ and $Y_2 = x - 2$, and look for their point of intersection. On the graph, the x-value of this point of intersection is the solution of $2x - 5 = x - 2$. In the table, the solution of $2x - 5 = x - 2$ is the x-value at which Y_1 is equal to Y_2.

Example 5 — Using Multiple Representations to Solve a Linear Equation

Solve $2x - 5 = x - 2$.

Solution

Algebraic Solution

$$2x - 5 = x - 2$$
$$2x - 5 - x = x - 2 - x$$
$$x - 5 = -2$$
$$x - 5 + 5 = -2 + 5$$
$$x = 3$$

Graphical Solution

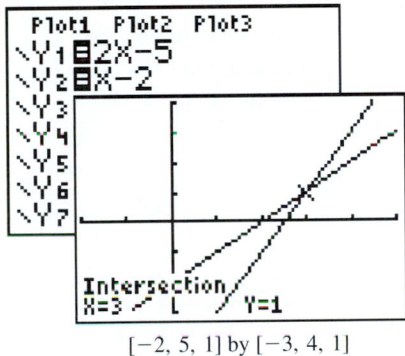

$[-2, 5, 1]$ by $[-3, 4, 1]$

Numerical Check

X	Y1	Y2
-2	-9	-4
-1	-7	-3
0	-5	-2
1	-3	-1
2	-1	0
3	**1**	**1**
4	3	2

X=3

Verbal Description

The value of x that satisfies the equation is $x = 3$. From the graph, the x-coordinate of the point of intersection is also $x = 3$. This is confirmed by the table, which shows that $Y_1 = Y_2$ for $x = 3$.

Self-Check 5

Use a graph to solve $4.7x - 1.5 = 3.7x + 1.5$ and a table to check this solution.

---

3. Identify a Linear Equation as a Conditional Equation, an Identity, or a Contradiction

Do All Linear Equations Have Exactly One Solution?

The linear equations we have examined thus far have all had exactly one solution. These equations are examples of conditional equations. A **conditional equation** is true for some values of the variable and false for other values.

To complete our discussion, we must consider two other types of equations. An equation that is true for all values of the variable is called an **identity.** An equation that is false for all values of the variable is called a **contradiction.** These two types of equations are considered in Examples 6 and 7, respectively.

Conditional Equation, Identity, and Contradiction

Verbally	Algebraic Example	Graphical Example	Numerical Example
Conditional Equation: A conditional equation is true for some values of the variable and false for other values.	$2x = x + 3$ *Solution:* $x = 3$ The only value of x that checks is $x = 3$.	 The lines intersect at an x-value of 3.	(table below)

x	$y = 2x$	$y = x + 3$
-1	-2	2
0	0	3
1	2	4
2	4	5
3	6	6
4	8	7

The table values of $2x$ and $x + 3$ are equal for $x = 3$.

Verbally	Algebraic Example	Graphical Example
Identity: An identity is an equation that is true for all values of the variable.	$2x = x + x$ *Solution:* All real numbers. All real numbers will check. $x + x$ is always $2x$.	 You see only one line because the lines coincide for all values of x.

x	$y = 2x$	$y = x + x$
-1	-2	-2
0	0	0
1	2	2
2	4	4
3	6	6
4	8	8

The table values of $2x$ and $x + x$ are equal for all values of x.

Verbally	Algebraic Example	Graphical Example
Contradiction: A contradiction is an equation that is false for all values of the variable.	$x = x + 3$ *Solution:* No solution. No real numbers will check because no real number is 3 greater than its own value.	 These lines have no points in common.

x	$y = x$	$y = x + 3$
-1	-1	2
0	0	3
1	1	4
2	2	5
3	3	6
4	4	7

The table values of x and $x + 3$ will never be equal for any value of x.

Example 6 Solving an Identity

Solve $4x + 1 - x = 2x + 1 + x$.

Solution

$$4x + 1 - x = 2x + 1 + x$$
First simplify both sides of the equation.

$$3x + 1 = 3x + 1$$

$$3x + 1 - 3x = 3x + 1 - 3x$$
Then subtract $3x$ from both sides of the equation.

$$1 = 1 \text{ is a true statement.}$$
The last equation is an identity, so the original equation is also an identity.

Answer: Every real number is a solution since the equation is an identity.

Test a couple of values to verify that all real numbers will check.

Self-Check 6

Solve $3x + 2 = 5x + 2 - 2x$.

Example 7   Solving a Contradiction

Solve $4v - 2 = 4v + 1$.

Solution

$$4v - 2 = 4v + 1$$

$$4v - 2 - 4v = 4v + 1 - 4v$$ Subtract $4v$ from both sides of the equation.

$$-2 = 1 \text{ is a false statement.}$$ This last equation is a contradiction, so the original equation is also a contradiction.

Answer: Because the equation is a contradiction, there is no solution.   No matter what values you may test, none of them will check.

Self-Check 7

Solve $3x + 4 = 2 + 3x + 1$.

In Examples 6 and 7, all variables were eliminated in the process of solving the equation. If all the variables are eliminated and an identity results, then every real number is a solution of the equation. If all the variables are eliminated and a contradiction results, then the equation has no solution.

Example 8 illustrates how to estimate the solution of a linear equation.

Example 8   Estimate Then Calculate

Estimate the solution of $5x + 529.995 = 4x - 1.001$ to the nearest integer and then calculate the exact solution.

Solution

Estimated Solution

$$5x + 530 = 4x - 1$$

$$5x - 4x + 530 = 4x - 4x - 1$$

$$x + 530 = -1$$

$$x + 530 - 530 = -1 - 530$$

$$x = -531$$

Exact Solution

$$5x + 529.995 = 4x - 1.001$$

$$5x - 4x + 529.995 = 4x - 4x - 1.001$$

$$x + 529.995 = -1.001$$

$$x + 529.995 - 529.995 = -1.001 - 529.995$$

$$x = -530.996$$

Self-Check 8

Estimate the solution of $x + 27.888 = 57.904$ to the nearest integer and then calculate the exact solution.

In Example 9, we solve a verbally stated problem by first rephrasing the problem into a word equation and then translating this word equation into algebraic form.

Example 9 Solving a Verbally Stated Equation

If twice a number is decreased by seven, the result is the same as four more than the number. Find this number.

Solution

Twice a number decreased by 7   is   4 more than the number

$$2n \quad - \quad 7 \qquad = \qquad n + 4$$

$2n - 7 - n = n + 4 - n$	Let n represent the number and translate the **word equation** into an algebraic equation. Subtract n from both sides of the equation.
$n - 7 = 4$	
$n - 7 + 7 = 4 + 7$	Add 7 to both sides of the equation.
$n = 11$	Does this value check?

Answer: The number is 11.

Self-Check 9

If three times a number is increased by 5, the result is 4 less than twice the number. Find this number.

Self-Check Answers

1. **a.** Linear equation in one variable
   b. Not a linear equation
   c. Linear equation in one variable
   d. Not a linear equation
2. $w = -2$
3. $a = 6$
4. $x = 15$
5. $x = 3$

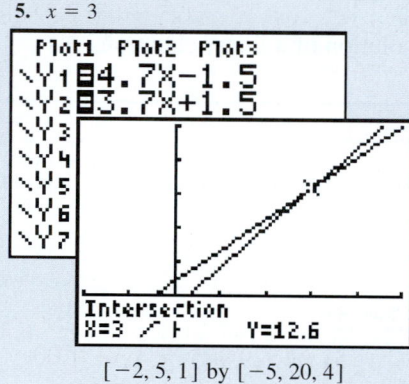

$[-2, 5, 1]$ by $[-5, 20, 4]$

6. Every real number is a solution.
7. No solution
8. $x \approx 30; x = 30.016$
9. The number is -9.

2.4 Using the Language and Symbolism of Mathematics

1. An equation is first degree in x if the exponent on x is _____.

2. An equation that is first degree in each variable is called a _____ equation.

3. The equation $y = mx + b$ is a _____ equation in two variables.

4. The equation $Ax = B$ is a _____ equation in one variable.

5. Equations with the same solution are called _____ equations.

6. By the _____-subtraction principle of equality, $a = b$ is equivalent to $a + c = b + c$.

7. By the addition-subtraction principle of equality, $a = b$ is equivalent to $a - c =$ _____.

8. A _____ equation is true for some values of the variable and false for other values.

9. An equation that is true for all values of the variable is called an _____.

10. An equation that is false for all values of the variable is called a _____.

2.4 | Quick Review

1. Check both $x = 1$ and $x = -4$ to determine whether either is a solution of $4(3x + 2) = 11x + 4$.

2. For the expression $3x^2$, what is the exponent on x?

3. For the expression $3x$, what is the exponent on x?

4. For the expression $3x$, what is the coefficient of x?

5. Translate the verbal statement into algebraic form. "Five less than three times x equals thirteen."

2.4 | Exercises

Objective 1 Solve Linear Equations in One Variable Using the Addition-Subtraction Principle

In Exercises 1 and 2, determine which of the following are linear equations in one variable.

1. **A.** $5x + 7$ **B.** $5x + 7 = 4x$
   C. $y = 5x + 7$ **D.** $x^2 = 25$

2. **A.** $x^2 = 4$ **B.** $4x + 5$
   C. $y = 4x + 5$ **D.** $4x + 5 = 3x$

In Exercises 3–12, solve each equation, and then check your solution.

3. $x - 11 = 13$ 4. $x + 11 = 13$

5. $v + 6 = 2$ 6. $v - 6 = -9$

7. $2y = y - 1$ 8. $3y = 2y + 4$

9. $3z + 7 = 2z - 4$ 10. $5z - 1 = 4z + 3$

11. $10y + 17 = 9y + 21$ 12. $8y + 17 = 7y - 21$

In Exercises 13–40, solve each equation.

13. $6x + 2 = 5x + 18$ 14. $7x - 4 = 6x + 28$

15. $9y - 11 = 8y + 52$ 16. $14y + 21 = 13y - 6$

17. $5v + 7 = 4v + 9$ 18. $7v - 8 = 6v + 11$

19. $12y + 2 = 11y + 2$ 20. $24y - 8 = 23y - 8$

21. $5m - 7 + 3m = 19 + 7m$

22. $9m - 7 = 13m - 11 - 5m$

23. $12n - 103 = 9n - 4 + 2n$

24. $17n - 18 - 7n = 9n + 18$

25. $3t + 3t + 3t = 4t + 4t$ 26. $5t + 5t + 5t = 7t + 7t$

27. $10v - 8v = 9v - 8v$ 28. $12v - 10v = 14v - 13v$

29. $3x - 4 + 5x = 6x + 3 + x$

30. $2x + 7 + 9x = 4x - 3 + 6x$

31. $3(v - 5) = 2(v + 5)$ 32. $5(v + 2) = 4(v - 6)$

33. $5(w - 2) = 4(w + 3)$ 34. $9(w + 4) = 8(w - 1)$

35. $8(m - 2) = 7(m - 3)$ 36. $7(m + 5) = 6(m + 3)$

37. $4(n - 6) = 3(n - 5)$ 38. $8(n - 2) = 7(n - 4)$

39. $3(5y + 2) = 2(7y - 3)$ 40. $7(3y + 2) = 5(4y - 3)$

Objective 2 Use Graphs and Tables to Solve a Linear Equation in One Variable

In Exercises 41–44, use the given graph to solve each equation. Then check your solution.

41. $1.5x - 2 = 0.5x - 1$

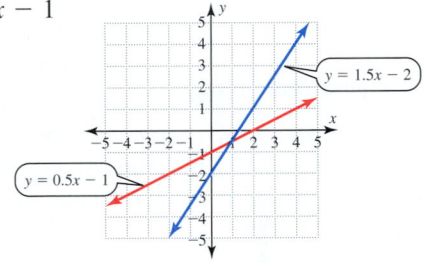

42. $0.5x - 1 = 1 - 0.5x$

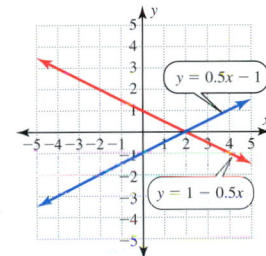

43. $1.5x - 2 = 1$

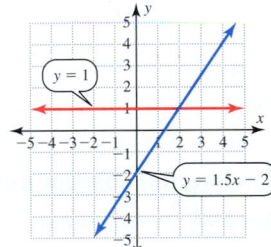

44. $0.5x - 1 = -1$

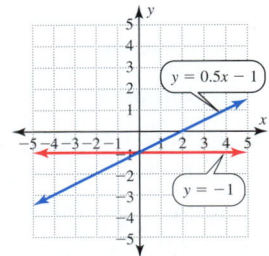

In Exercises 45–48, use the given table to solve each equation.

45. $0.5x - 1 = 1 - 0.5x$

x	$0.5x - 1$	$1 - 0.5x$
-1	-1.5	1.5
0	-1.0	1.0
1	-0.5	0.5
2	0.0	0.0
3	0.5	-0.5

46. $1.5x - 2 = 0.5x - 1$

x	$1.5x - 2$	$0.5x - 1$
-1	-3.5	-1.5
0	-2.0	-1.0
1	-0.5	-0.5
2	1.0	0.0
3	2.5	0.5

47. $1.3x + 1.7 = 0.3x + 2.7$

x	$1.3x + 1.7$	$0.3x + 2.7$
-1	0.4	2.4
0	1.7	2.7
1	3.0	3.0
2	4.3	3.3
3	5.6	3.6

48. $0.8x - 1.4 = 0.6 - 0.2x$

x	$0.8x - 1.4$	$0.6 - 0.2x$
-1	-2.2	0.8
0	-1.4	0.6
1	-0.6	0.4
2	0.2	0.2
3	1.0	0.0

Applying Technology

In Exercises 49–54, use a calculator or a spreadsheet to solve each equation by letting Y_1 represent the left side of the equation and Y_2 represent the right side of the equation.

49. $2x - 2 = x - 3$

50. $2x + 4 = x + 1$

51. $1.8x - 4.6 = 0.8x - 2.6$

52. $2.5x - 2.3 = 1.5x + 0.7$

53. $\dfrac{1}{3}x - \dfrac{2}{3} = \dfrac{7}{3} - \dfrac{2}{3}x$

54. $\dfrac{2}{5}x - \dfrac{4}{7} = \dfrac{3}{7} - \dfrac{3}{5}x$

Objective 3 Identify a Linear Equation as a Conditional Equation, an Identity, or a Contradiction

In Exercises 55–58, each equation is a conditional equation, an identity, or a contradiction. Identify the type of equation and solve it.

55. a. $2x = x$ **b.** $x = x + 2$ **c.** $x + 2 = x + 2$

56. a. $2x + 1 = x$ **b.** $2x = x + x$ **c.** $x + 3 = 3 + x$

57. a. $v = v + 4$ **b.** $4 + v = v + 4$ **c.** $4v = 3v + 4$

58. a. $x + x = 2x + 1$ **b.** $2(x + 1) = 2x + 2$
c. $x + x = x + 2$

Review and concept development

Simplify Versus Solve

In Exercises 59–64, simplify the expression in the first column by adding like terms, and solve the equation in the second column.

Simplify	**Solve**
59. a. $5x + 1 + 4x - 6$	**b.** $5x + 1 = 4x - 6$
60. a. $8x + 5 + (7x + 2)$	**b.** $8x + 5 = 7x + 2$
61. a. $6x - 4 - (5x + 3)$	**b.** $6x - 4 = 5x + 3$
62. a. $12x - 5 - (11x + 1)$	**b.** $12x - 5 = 11x + 1$
63. a. $3.4x - 1.7 + 2.4x + 2.3$	**b.** $3.4x - 1.7 = 2.4x + 2.3$
64. a. $2.6x - 1.9 - (1.6x + 4.1)$	**b.** $2.6x - 1.9 = 1.6x + 4.1$

Estimate Then Calculate

In Exercises 65–68, mentally estimate the solution of each equation to the nearest integer, and then calculate the exact solution.

Problem	Mental Estimate	Exact Solution
65. $x - 0.918 = 0.987$		
66. $x + 39.783 = 70.098$		
67. $2x + 354.916 = x + 855.193$		
68. $5x - 1.393 = 4x + 5.416$		

Multiple Representations

In Exercises 69–74, first write an algebraic equation for each verbal statement, using the variable m to represent the number, and then solve for m.

69. Seven more than three times a number equals eight less than twice the number.

70. If three is subtracted from five times a number, the result equals eight more than four times the number.

71. Twelve minus nine times a number is the same as two minus ten times the number.

72. Seven minus six times a number is the same as twelve minus seven times the number.

73. Twice the quantity three times a number minus nine is the same as five times the sum of the number and thirteen.

74. Four times the sum of a number and eleven is equal to three times the difference of the number and five.

Connecting Algebra to Geometry

75. Perimeter of a Triangle The perimeter of the triangle shown in the figure is 28 cm. Find the value of a.

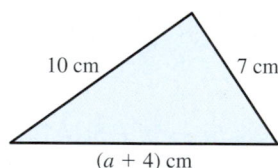

10 cm   7 cm
$(a + 4)$ cm

76. Perimeter of a Triangle The perimeter of the triangle shown in the figure is 31 cm. Find the value of a.

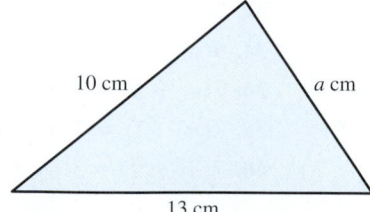

10 cm a cm
13 cm

77. Perimeter of a Basketball Court The perimeter of the basketball court shown in the figure is $(x + 35)$ ft. Find the value of x.

50 ft

94 ft

78. Perimeter of a Soccer Field The perimeter of the soccer field shown in the figure is $(x + 40)$ yd. Find the value of x.

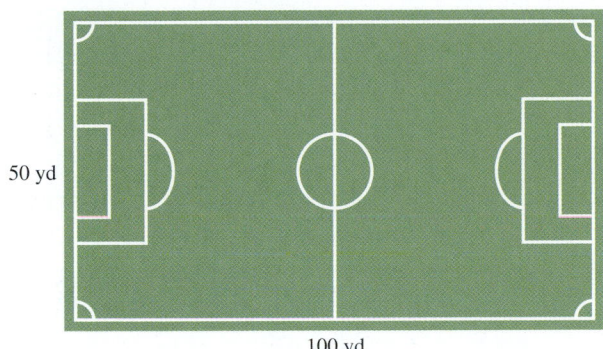

50 yd

100 yd

Group discussion questions

79. Writing Mathematically Write a paragraph using the analogy of a balance scale to describe the addition-subtraction principle.

80. Challenge Question
   a. Complete the equation $8x + 7 = 7x + \underline{?}$ so that the solution is $x = 4$.
   b. Complete the equation $12x - 3 = 11x + \underline{?}$ so that the solution is $x = 8$.

81. Challenge Question
   a. Complete the equation $3x + 5 = 2x + \underline{?}$ so that the solution is $x = 7$.
   b. Complete the equation $3x + 5 = 2x + \underline{?}$ so that the equation is a contradiction.
   c. Complete the equation $3x + 5 = 2x + \underline{?}$ so that the equation is an identity.

82. Discovery Question Let Y_1 represent the left side of $x^2 = 5x - 6$ and Y_2 represent the right side. Then use a graphing calculator to explore this equation by using the same strategy used for linear equations. Record your observations and explanation of these observations.

2.4 Cumulative Review

1. Given $f(x) = 25x - 100$, evaluate $f(10)$.

In Exercises 2 and 3, use the given table to complete each question.

x	$f(x)$
-1	-3.5
0	-2.0
1	-0.5
2	1.0
3	2.5

2. $f(1) = $ _____

3. $f(x) = 1; x = $ _____

In Exercises 4 and 5, use the given graph to complete each question.

4. $f(-1) = $ _____

5. $f(x) = 0; x = $ _____

| Section 2.5 | Solving Linear Equations in One Variable Using the Multiplication-Division Principle |

Objective:

1. Solve linear equations in one variable using the multiplication-division principle.

1. Solve Linear Equations in One Variable Using the Multiplication-Division Principle

This section expands our ability to solve linear equations in one variable. We solve equations of the form $Ax = B$ by obtaining a coefficient of 1 for x. To obtain the coefficient of 1, we will use the multiplication-division principle of equality.

Is It Okay to Perform Any Operation on Both Sides of an Equation?

No, multiplying or dividing both sides of an equation by zero is an exception. This is especially tricky if both sides of an equation are multiplied by a variable. If this variable is zero, this can cause problems such as extraneous values. We will examine this situation later in the book.

Multiplication-Division Principle of Equality

Verbally	Algebraically	Numerical Example
If both sides of an equation are multiplied or divided by the same nonzero number, the result is an equivalent equation.	If a, b, and c are real numbers and $c \neq 0$, then $a = b$ is equivalent to $ac = bc$ and to $\dfrac{a}{c} = \dfrac{b}{c}$.	$\dfrac{x}{2} = 5$ is equivalent to $2\left(\dfrac{x}{2}\right) = 2(5)$; and $3x = 12$ is equivalent to $\dfrac{3x}{3} = \dfrac{12}{3}$.

To solve $Ax = B$ with $A \neq 0$, we will either divide both sides of the equation by A or equivalently multiply by $\dfrac{1}{A}$, the multiplicative inverse of A. If A is a rational fraction, then we usually multiply by the reciprocal of A. Otherwise, we usually use division.

Example 1   Solving a Linear Equation with an Integer Coefficient

Solve $7y = 91$ and check the solution.

Solution

$$7y = 91$$

$$\frac{7y}{7} = \frac{91}{7} \qquad \text{Divide both sides of the equation by 7, the coefficient of } y.$$

$$y = 13 \qquad \text{This produces a coefficient on } y \text{ of 1. Note that } 1 \cdot y \text{ is usually written as just } y.$$

Check: $7y = 91$

$$7(13) \overset{?}{=} 91$$

$$91 \overset{?}{=} 91 \text{ checks.}$$

Answer: $y = 13$

Self-Check 1

Solve $6v = -42$.

How Do I Know When a Linear Equation Is Solved?

A linear equation is not solved until the coefficient of the variable is 1. If the coefficient is -1, the equation is not yet solved.

Example 2 Solving a Linear Equation with a Coefficient of -1

Solve $-t = 17$.

Solution

$$-t = 17 \qquad \text{The coefficient of } t \text{ is } -1.$$

$$-t(-1) = 17(-1) \qquad \text{Multiply both sides of the equation by } -1.$$
$$\text{(The multiplication inverse of } -1 \text{ is } -1.)$$

$$t = -17 \qquad \text{Does this value check?}$$

Self-Check 2

Solve $-v = 4$.

The ability to estimate answers is an important skill not only in recognizing incorrect answers from keystroke errors on a calculator, but also in making routine consumer decisions. Families make many choices from the dimensions of bedspreads and carpet to room additions that can be facilitated by good estimation skills. Example 3 illustrates how to estimate the solution of a linear equation.

Example 3 Estimate Then Calculate

Mentally estimate the solution of $-7.15m = 42.042$ to the nearest integer, and then calculate the exact solution.

Solution

Estimate by considering the equation $-7m = 42$.

Estimated Solution	Exact Solution

$$-7m = 42$$

$$\frac{-7m}{-7} = \frac{42}{-7}$$

$$m = -6$$

```
42.042/(-7.15)
              -5.88
```

Answer: The solution $m = -5.88$ seems reasonable based upon the estimated solution of $m \approx -6$.

Self-Check 3

Mentally estimate the solution of $6.2x = 18.786$ to the nearest integer, and then calculate the exact solution.

Linear equations often contain variables and constants on both sides of the equation. To solve such linear equations, first isolate the variable terms on one side of the equation by using the addition-subtraction principle. Once the equation has been simplified to the form $Ax = B$, use the multiplication-division principle to solve for the variable. This is illustrated in Example 4.

We also use Technology Perspective 1.7.1, Storing Values Under Variable Names, to check the solution.

Example 4   Solving a Linear Equation with Variables on Both Sides

Solve $5v - 4 = 7v + 10$, and check the solution on a calculator.

Solution

$$5v - 4 = 7v + 10$$

$$5v - 4 - 7v = 7v + 10 - 7v$$
Subtract $7v$ from both sides of the equation to isolate the variables on the left side.

$$-2v - 4 = 10$$

$$-2v - 4 + 4 = 10 + 4$$
Add 4 to both sides of the equation to isolate the constants on the right side.

$$-2v = 14$$

$$\frac{-2v}{-2} = \frac{14}{-2}$$
Divide both sides of the equation by -2.

$$v = -7$$

Check:

```
-7→X
           -7
5X-4
          -39
7X+10
          -39
```

Using Technology Perspective 1.7.1, store the value to be checked under the variable x. Then calculate the value of both sides of the equation.

If we use -7 for the value of the variable, both sides of the equation have the same value, so this solution checks.

Answer: $v = -7$

Self-Check 4

Solve $7x + 2 = 9x + 10$.

The strategy used in the previous examples is described in the following box. We will then use this strategy to solve equations containing parentheses and fractions. A good rule of thumb to use when solving a linear equation is to try to produce simpler expressions with each step.

Strategy for Solving Linear Equations

Step 1. Simplify each side of the equation.
   a. If the equation contains fractions, simplify by multiplying both sides of the equation by the least common denominator (LCD) of all the fractions.
   b. If the equation contains grouping symbols, simplify by using the distributive property to remove the grouping symbols and then combine like terms.

Step 2. Using the addition-subtraction principle of equality, isolate the variable terms on one side of the equation and the constant terms on the other side.

Step 3. Using the multiplication-division principle of equality, solve the equation produced in step 2.

Example 5 Solving a Linear Equation Containing Parentheses

Solve $4y + 3(y - 2) = 2(y + 4) - (2y - 7)$.

Solution

$$4y + 3(y - 2) = 2(y + 4) - (2y - 7)$$
$$4y + 3y - 6 = 2y + 8 - 2y + 7$$
Remove parentheses by using the distributive property, and then combine like terms.

$$7y - 6 = 15$$

$$7y - 6 + 6 = 15 + 6$$
Isolate the constant terms on the right side of the equation by adding 6 to both sides.

$$7y = 21$$

$$\frac{7y}{7} = \frac{21}{7}$$
Solve for y by dividing both sides of the equation by 7.

Answer: $y = 3$ Does this value check?

Self-Check 5

Solve $3(x + 2) - (3x - 5) = 5x - 2(x + 2)$.

If a linear equation contains fractions, then we can simplify it by converting it to an equivalent equation that does not involve fractions. To do this, multiply both sides of the equation by the LCD of all the terms. In Example 6, the LCD is 12.

Example 6 Solving a Linear Equation Containing Fractions

Solve $\dfrac{z}{6} + 2 = \dfrac{z}{4}$.

Solution

$$\frac{z}{6} + 2 = \frac{z}{4}$$
$$
\begin{aligned}
6 &= 2 \cdot 3 \\
4 &= 2 \cdot 2 \\
\text{LCD} &= 2 \cdot 2 \cdot 3 = 12
\end{aligned}
$$

$$12\left(\frac{z}{6} + 2\right) = 12\left(\frac{z}{4}\right)$$
Multiply both sides of the equation by the LCD, 12.

$$12\left(\frac{z}{6}\right) + 12(2) = 12\left(\frac{z}{4}\right)$$
Use the distributive property to remove the parentheses, and then simplify.

$$2z + 24 = 3z$$

$$2z + 24 - 2z = 3z - 2z$$
Subtract $2z$ from both sides of the equation.

$$24 = z \text{ or}$$

Answer: $z = 24$ Does this solution check?

Self-Check 6

Solve $\dfrac{y}{3} - \dfrac{y}{12} = -\dfrac{1}{2}$.

Observe in Examples 6 and 7 the execution of our basic strategy—at each step of the solution process we use operations that will produce simpler expressions.

Example 7   Solving a Linear Equation Containing Fractions

Solve $\dfrac{3v - 3}{6} = \dfrac{4v + 1}{15} + 2$ algebraically and graphically.

Solution

Algebraic Solution

$$\frac{3v - 3}{6} = \frac{4v + 1}{15} + 2$$

$$30\left(\frac{3v - 3}{6}\right) = 30\left(\frac{4v + 1}{15} + 2\right)$$

$$\frac{30}{6}(3v - 3) = \frac{30}{15}(4v + 1) + 30(2)$$

$$5(3v - 3) = 2(4v + 1) + 30(2)$$

$$15v - 15 = 8v + 2 + 60$$

$$15v - 15 = 8v + 62$$

$$15v - 15 - 8v = 8v + 62 - 8v$$

$$7v - 15 = 62$$

$$7v - 15 + 15 = 62 + 15$$

$$7v = 77$$

$$\frac{7v}{7} = \frac{77}{7}$$

$$v = 11$$

$6 = 2 \cdot 3$
$15 = 3 \cdot 5$
LCD $= 2 \cdot 3 \cdot 5 = 30$

Multiply both sides of the equation by the LCD, 30.

Use the distributive property to remove parentheses and then combine like terms.

Subtract $8v$ from both sides of the equation.

Then add 15 to both sides of the equation.

Divide both sides of the equation by 7.

Graphical Solution

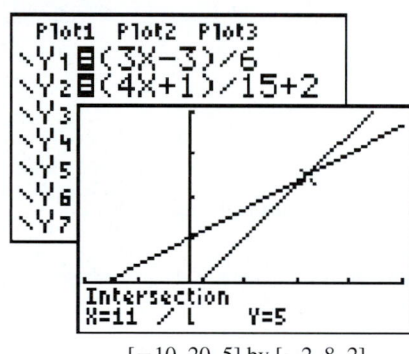

$[-10, 20, 5]$ by $[-2, 8, 2]$

Note that we use the variable x in place of v for the calculator.

Knowing the algebraic solution makes it easier to select a suitable window for the graph.

The Y_1 value equals the Y_2 value for $x = 11$.

The graph confirms the solution is $v = 11$.

Answer: $v = 11$

Self-Check 7

Solve $1 - \dfrac{4x + 2}{5} = \dfrac{3 - 7x}{2}$.

Self-Check Answers

1. $v = -7$
2. $v = -4$

3. Estimate: $x \approx 3$
   Exact: $x = 3.03$

4. $x = -4$
5. $x = 5$

6. $y = -2$

7. $x = \dfrac{1}{3}$

2.5 | Using the Language and Symbolism of Mathematics

1. Equations with the same solution are called _____ equations.

2. By the _____-division principle, $a = b$ is equivalent to $ac = bc$ for $c \neq 0$.

3. By the multiplication-division principle of equality, $a = b$ is equivalent to $\dfrac{a}{c} =$ _____ for $c \neq 0$.

4. The equation $-x = 9$ is equivalent to $x =$ _____.

5. A _____ equation is true for some values of the variable and false for other values.

6. The statement "You can always multiply both sides of an equation by the same number to produce an

equivalent equation" is false because we do not obtain equivalent equations if we multiply both sides of the equation by _____.

7. The _____ property is often used to help us remove parentheses from expressions in an equation.

8. To clear an equation of fractions, we can multiply both sides of the equation by the _____ _____ _____ of all the terms in the equation.

9. The multiplicative inverse of $\dfrac{1}{4}$ is _____.

10. The multiplicative inverse of $-\dfrac{1}{4}$ is _____.

2.5 | Quick Review

1. What is the LCD for the fractions $\dfrac{5}{12}$ and $\dfrac{4}{15}$?

Simplify each expression by removing grouping symbols and combining like terms.

2. $5x - 2 - (5x + 2)$

3. $5x - 2(5x + 2)$

4. $12\left(5 - \dfrac{3 - x}{6}\right)$

5. Write the formula for the perimeter of a rectangle.

2.5 | Exercises

Objective 1 Solve Linear Equations in One Variable Using the Multiplication-Division Principle

In Exercises 1–6, solve each equation, and then check your solution.

1. a. $7x = 42$ b. $x + 7 = 42$

2. a. $-5y = 45$ b. $y - 5 = 45$

3. a. $-v = 8$ b. $v - 1 = 8$

4. a. $-m = -6$ b. $m - 1 = -6$

5. a. $\dfrac{3}{4}t = -36$ b. $\dfrac{4}{3}t = -36$

6. a. $-\dfrac{2}{3}m = 48$ b. $-\dfrac{3}{2}m = 48$

In Exercises 7–40, solve each equation.

7. $3(2t - 1) = 7t + 1$

8. $-3(4t - 1) = -(t - 14)$

9. $4(2 - 3x) = 3 - 13x$

10. $-2(5x + 4) = -(x - 10)$

11. $\dfrac{y}{3} + \dfrac{2}{3} = \dfrac{y}{2} + \dfrac{3}{2}$ 12. $\dfrac{v}{4} - \dfrac{2}{5} = \dfrac{v}{10} + \dfrac{1}{2}$

13. $2 - 6(y + 1) = 4(2 - 3y) + 6$

14. $13 + 3(2v - 5) = 1 - 2(3 - 6v)$

15. $24 = -8z$ 16. $-15 = 25z$

17. $-1 = 9m$ 18. $1 = -7m$

19. $-47w = 0$ 20. $0 = 31w$

21. $\dfrac{-5v}{3} = 25$ 22. $-\dfrac{2v}{11} = 66$

23. $8x - 1 = 13x - 1$ 24. $-17x + 5 = 17x + 5$

25. $4(3y - 5) = 5(4y + 4)$ 26. $7(2y - 3) = 3(6y + 5)$

27. $0.12a = 13.2$ 28. $0.07a = -3.5$

29. $2.3x + 29.3 = 1.2(4 - x)$

30. $5.7x + 35.7 = 1.2(x - 4)$

31. $6 - 3(2v + 2) = 4(8 - v)$

32. $2(1 - v) + 9 = 3(2v + 1)$

33. $4(x + 1) + 3(x + 2) = 9(x + 1) - 5$

34. $11(x + 3) + 4(2x - 1) = 5(3x - 2) + 13$

35. $\dfrac{w + 1}{3} = \dfrac{w - 5}{5}$ 36. $\dfrac{24w - 67}{60} = \dfrac{3w - 8}{12}$

37. $4 - \dfrac{x+2}{3} = \dfrac{x}{2}$ **38.** $\dfrac{2x}{7} = 1 - \dfrac{2x+1}{3}$

39. $7(x-1) - 4(2x+3) = 2(x+1) - 3(4-x)$

40. $5(2x-1) + 3(x-3) = -4(x-6) + 2(13-3x)$

Review and Concept Development

In Exercises 41 and 42, each equation is a conditional equation, an identity, or a contradiction. Identify the type of each equation and then solve it.

41. a. $3(x+1) = 3x$ **b.** $3(x+1) = 3x + 3$
    c. $3(x+2) = 2(x+3)$

42. a. $4(x+5) = 4x$ **b.** $4(x+5) = 4x + 20$
    c. $4(x+5) = 5(x+4)$

Simplify Versus Solve

In Exercises 43–46, simplify the expression in the first column, and solve the equation in the second column.

Simplify	Solve

43. a. $3(2x-4) - 5(x-2)$ **b.** $3(2x-4) = 5(x-2)$

44. a. $3(2x-4) + 5(x-2)$ **b.** $3(2x-4) = -5(x-2)$

45. a. $1.5(4x-6) + 2.5(6x-4)$ **b.** $1.5(4x-6) = -2.5(6x-4)$

46. a. $1.5(4x-6) - 2.5(6x-4)$ **b.** $1.5(4x-6) = 2.5(6x-4)$

Connecting Concepts to Applications

Children are often prescribed the same drugs used for adults. Two commonly used formulas for adjusting the dosage to account for the age of the child are Cowling's formula and Young's formula.

47. Cowling's formula for a 10-year-old child is $y = 0.8x$, where x is the adult dosage and y is the child dosage. What is the adult dosage if the child dosage of a medication is 5 mg?

48. Young's formula for a 12-year-old child is $y = 0.5x$, where x is the adult dosage and y is the child dosage. What is the adult dosage if the child dosage of a medication is 4 mg?

Estimation Skills

49. Perimeter of a Square $(P = 4s)$
The perimeter of a square is 99.837 cm. Mentally estimate to the nearest centimeter the length of each side.

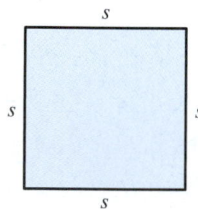

50. Perimeter of an Equilateral Triangle $(P = 3s)$ The perimeter of an equilateral triangle is 99.378 m. Mentally estimate to the nearest meter the length of each side.

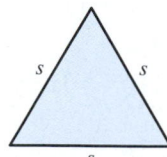

51. Hotel Room Tax $(T = 0.10C)$ A hotel tax on a bill is 10% of the room charges. Mentally estimate the pretax room charges for a room with a tax charge of $7.58.

52. Gratuity $(G = 0.20B)$ A customer added a gratuity of $13 to the restaurant charge. If the gratuity was approximately 20% of the restaurant bill, mentally estimate the bill before the gratuity was added.

Estimate Then Calculate

In Exercises 53–56, mentally estimate the solution of each equation to the nearest integer, and then calculate the exact solution.

Problem	Mental Estimate	Exact Solution
53. $-2.1x = 8.82$		
54. $4.9x = -14.945$		
55. $-0.49x = -2.009$		
56. $-0.24x = 2.01$		

In Exercises 57 and 58, Y_1 represents the left side of the equation and Y_2 represents the right side. Use the table to solve the equation.

57. $5.8x - 1.71 = 2.7x + 2.94$

X	Y_1	Y_2
0.0	-1.71	2.94
0.5	1.19	4.29
1.0	4.09	5.64
1.5	6.99	6.99
2.0	9.89	8.34
2.5	12.79	9.69
3.0	15.69	11.04

58. $7.2x + 4.2 = 4.7x + 0.95$

X	Y_1	Y_2
-1.5	-6.60	-6.10
-1.4	-5.88	-5.63
-1.3	-5.16	-5.16
-1.2	-4.44	-4.69
-1.1	-3.72	-4.22
-1.0	-3.00	-3.75
-0.9	-2.28	-3.28

Applying Technology

In Exercises 59 and 60, let Y_1 equal the left side of the equation and Y_2 equal the right side. Use a graphing calculator to graph these two equations and to solve the given equation for x.

59. $2.4(4x - 1) = 1.8(2x + 1)$

60. $2.5(6x + 1) = 12x + 10.9$

Multiple Representations

In Exercises 61–68, write an algebraic equation for each verbal statement, and then solve for the variable.

61. Five times the sum of x and two is the same as seven times the quantity x minus three.

62. Three times the sum of x and six equals eleven times the quantity x minus two.

63. Twice the quantity three v minus two equals four times the quantity v plus nine.

64. Six times the quantity seven minus v is the same as five times the quantity two v plus nine.

65. Twice the difference of three m minus five is four less than three times the sum of m and six.

66. Five times the sum of two m plus eleven is three more than two times the quantity m minus nine.

67. One-third of the quantity two x plus five is the same as the quantity four x plus two.

68. One-half of the quantity three x plus seven equals the quantity six x plus two.

Connecting Algebra to Geometry

69. Perimeter of a Wrestling Mat The perimeter of each square wrestling mat shown in the figure is 168 ft. The length of one side of a mat is $(4x + 18)$ ft. Find the value of x.

70. Perimeter of a Yield Sign The perimeter of the yield sign shown in the figure is 60 in. Assuming this sign is approximately an equilateral triangle, find the value of x.

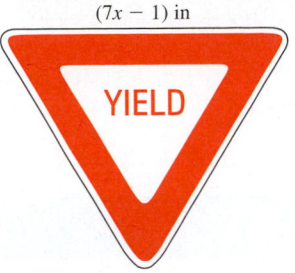

$(7x - 1)$ in

71. Perimeter of a Rectangle The perimeter of the rectangle shown in the figure is 17 ft. Find the value of a.

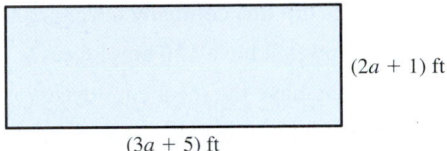

$(2a + 1)$ ft

$(3a + 5)$ ft

72. Perimeter of a Parallelogram The perimeter of the parallelogram shown in the figure is 18 cm. Find the value of a.

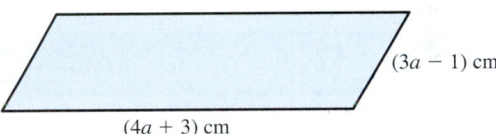

$(3a - 1)$ cm

$(4a + 3)$ cm

73. Area of a Triangle The area of the triangle shown in the figure is 51 cm² (square centimeters). Find the value of x. $\left(A = \dfrac{1}{2}bh \right)$

6 cm

$(5x + 7)$ cm

74. Area of a Rectangle The area of the rectangle shown in the figure is 102 cm². Find the value of x. $(A = lw)$

6 cm

$(4x - 3)$ cm

Group discussion questions

75. Writing Mathematically Solve $5(x - 3) + 4 = 4(2x + 1) - 9$, and then write an explanation of each step of the solution. Also explain why you performed the steps in the order in which you have them listed rather than in some other order.

76. Challenge Question

   a. Complete the equation $\dfrac{4}{5}x + 11 = \underline{?}$ so that the solution is $x = 31$.

   b. Complete the equation $5(x - 3) + 1 = 2(x + 4) + \underline{?}$ so that the solution is $x = 3$.

77. Error Analysis Examine the following argument and explain the error in the reasoning. If $x = 0$, then $2x = 0$; thus $x = 2x$. Dividing both sides of this equation by x, we conclude that $1 = 2$.

2.5	Cumulative Review

Use this graph to answer each question.

  1. How many employees did this company have in 2004?

  2. How many employees did this company have in 2006?

  3. When did the company first have 150 employees?

  4. When did the company have the most employees?

  5. How many employees were lost from 2008 to 2009?

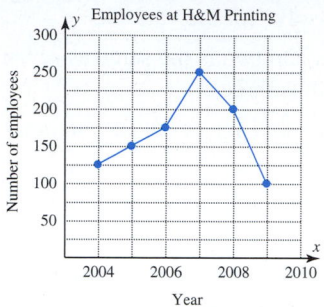

Section 2.6	Calculating Intercepts and Rearranging Formulas

Objectives:

  1. Rewrite a linear equation in the form $y = mx + b$.

  2. Calculate the x- and y-intercepts of a line.

  3. Solve an equation for a specified variable.

To enter information into calculators or computers, we often must rearrange equations so that the information we have fits the required format. We will develop this ability in this section by using the addition-subtraction and the multiplication-division principles of equality. One form of a linear equation that is particularly useful is the form $y = mx + b$.

1. Rewrite a Linear Equation in the Form $y = mx + b$

Can a First-Degree Linear Equation Always Be Written in the Form $y = mx + b$?

Yes, an equation that is first degree in both x and y can always be written in this form. We will illustrate how to do this in Examples 1 and 2. We will examine this form in greater detail in Section 3.2, where we describe the meaning of both m and b.

Example 1	Solving a Linear Equation for y

Solve the linear equation $4x + 2y = 6$ for y, and then use a graphing calculator to graph this equation.

Solution

Algebraically

$$4x + 2y = 6$$

$$4x + 2y - 4x = 6 - 4x$$

Use the addition principle of equality to subtract $4x$ from both sides of the equation to isolate the y-term on the left side of the equation.

$$2y = -4x + 6$$

$$2y = 2(-2x + 3)$$

Use the distributive property $ab + ac = a(b + c)$ to rewrite the right side of the equation.

$$\frac{2y}{2} = \frac{2(-2x + 3)}{2}$$

Then use the division principle of equality to divide both sides of the equation by 2 to solve for y.

$$y = -2x + 3$$

Graphically

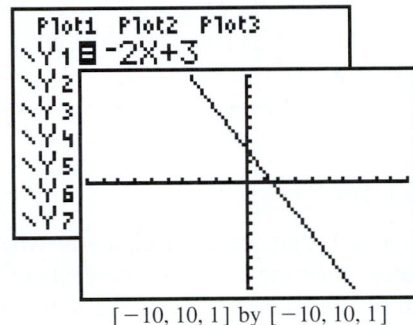

$$[-10, 10, 1] \text{ by } [-10, 10, 1]$$

Now that the linear equation is in the form $y = mx + b$, enter this equation into a graphing calculator and graph this equation.

Self-Check 1

Solve the linear equation $x - 2y = 2$ for y, and then use a graphing calculator to graph this equation.

Once a linear function is rewritten in the form $y = mx + b$, we can use a calculator or a spreadsheet to examine the function either numerically or graphically. In Example 2, we will create a table of values for the given linear function.

Example 2   Solving a Linear Equation for y

Solve the linear equation $3x - y = 3(2x - y) - 1$ for y, and then use a graphing calculator to complete a table of values for the x-values $-3, -2, -1, 0, 1, 2, 3$.

Solution

Algebraically

$$3x - y = 3(2x - y) - 1$$

First use the distributive property to remove the parentheses from the right side.

$$3x - y = 6x - 3y - 1$$

$$3x - 3x - y = 6x - 3x - 3y - 1$$

Subtract $3x$ from both sides to move all x-terms to the right side.

$$-y = 3x - 3y - 1$$

$$-y + 3y = 3x - 3y - 1 + 3y$$

Add $3y$ to both sides to isolate all y-terms on the left side.

$$2y = 3x - 1$$

$$\frac{2y}{2} = \frac{3x - 1}{2}$$

Then divide both sides of the equation by 2 to solve for y.

Answer: $y = \dfrac{3}{2}x - \dfrac{1}{2}$

Numerically

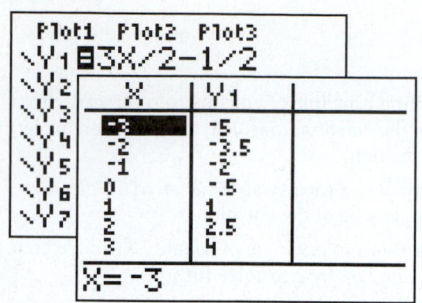

Enter the given equation into a graphing calculator to create the table shown.

Self-Check 2

Solve the linear equation $5(3y + 2x) - 4 = 11y + 7x - 12$ for y, and then use a graphing calculator to complete a table of values for the x-values $-3, -2, -1, 0, 1, 2, 3$.

---

2. Calculate the x- and y-Intercepts of a Line

The ability to solve an equation for either x or y is a skill that we use frequently when working with linear equations. In Section 2.3, we determined the x- and y-intercepts of a line by examining either a table of values or a graph of the linear function. We will now examine how to determine the exact values of these intercepts algebraically. Remember that the intercepts are points on the line and should be written as ordered pairs. In Example 3, we use the fact that the x-intercept is of the form $(a, 0)$ with a y-coordinate of zero, and the y-intercept is of the form $(0, b)$ with an x-coordinate of zero. We can then use these intercepts as a quick way to graph the line.

Example 3 Using the Intercepts to Graph a Line

Calculate the x- and y-intercepts of $3x - 2y = 6$, and use these intercepts to sketch the graph of this equation.

Solution

Calculation of the x-Intercept

$$3x - 2y = 6$$
$$3x - 2(0) = 6$$
$$3x - 0 = 6$$
$$3x = 6$$
$$x = 2$$
$(2, 0)$ is the x-intercept.

To calculate the x-intercept, set $y = 0$ and solve for x.

Simplify the left side of the equation and then divide both sides by 3.

The x-intercept is not 2 but $(2, 0)$.

Calculation of the y-Intercept

$$3x - 2y = 6$$
$$3(0) - 2y = 6$$
$$0 - 2y = 6$$
$$-2y = 6$$
$$y = -3$$
$(0, -3)$ is the y-intercept.

To calculate the y-intercept, set $x = 0$ and solve for y.

Simplify the left side of the equation and then divide both sides of the equation by -2.

The y-intercept is not -3 but $(0, -3)$.

Graph of $3x - 2y = 6$

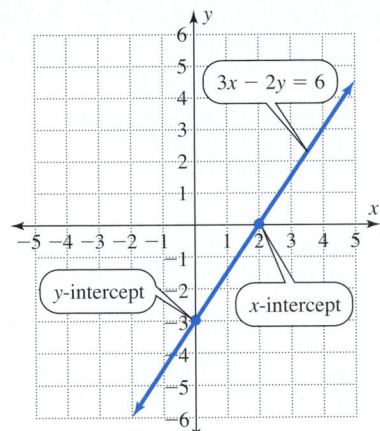

To sketch the graph of $3x - 2y = 6$, plot both the x- and y-intercepts and then draw the line through these two points.

Self-Check 3

Calculate the x- and y-intercepts of $5x - 2y = 20$, and use these intercepts to sketch the graph of this equation.

Example 3 presented the third way that we have graphed a line in this chapter.

- First, we plotted all of the points in a table and sketched the line through these points.
- Next, we strategically selected two input values and determined their corresponding output values. We sketched the line through these two points and plotted a third point as a check.
- Finally, we calculated the intercepts and sketched the line through these two points.

In Chapter 3, we will take an even more sophisticated look at graphing lines using the slope and y-intercept. Remember that the graph of a line passes through *all* the points that satisfy the equation. Find any two of them and you can graph the line.

Recall that the x- and y-intercepts are key points on the graph of an equation with important interpretations in real-world problems that are modeled by these equations. Thus it is useful to be able to calculate the coordinates of these points exactly rather than to approximate them from a graph. One example of this is the examination of the overhead cost and the break-even value of a profit function. The **fixed costs** or the **overhead costs** for a company can include insurance, rent, electricity, and other expenses that must be paid even when 0 units are produced. The **break-even value** is the number of units that must be produced and sold to pay for the overhead costs and production costs and to produce a profit of $0. Example 4 examines both the overhead cost and the break-even value for the production at a pizza parlor.

Example 4   Profit on Pizzas

A pizza parlor in a small town has daily overhead costs of $150. They cover production costs and make a profit of $2 on each pizza that they sell.

(a) Write an equation that gives the net profit for this business when they sell x pizzas.

(b) Using the equation from part (a), determine the x-intercept of the graph of this equation and interpret this point.

(c) Using the equation from part (a), determine the y-intercept of the graph of this equation and interpret this point.

Solution

Let $x =$ number of pizzas sold

$y =$ net profit made by selling these pizzas

(a) Verbally

$$\boxed{\text{Net profit}} = \boxed{\begin{array}{c}\text{Profit made by} \\ \text{selling } x \text{ pizzas}\end{array}} - \boxed{\text{Overhead costs}}$$

We first concisely state the word equation that we will use to form the algebraic equation.

Algebraically

$$y = 2x - 150$$

The profit per pizza is $2, and the overhead costs are $150.

(b) Calculation of the *x*-intercept

$$y = 2x - 150$$
$$0 = 2x - 150$$

To calculate the *x*-intercept, set $y = 0$ and solve for *x*.

$$0 + 150 = 2x - 150 + 150$$
$$150 = 2x$$
$$75 = x \quad \text{or}$$
$$x = 75$$

First add 150 to both sides of the equation, and then divide both sides of the equation by 2.

The *x*-intercept is $(75, 0)$. Producing 75 pizzas will result in a profit of $0. Thus 75 pizzas is the break-even value for this business.

The break-even value is the number of units that must be produced to pay for the overhead costs and production costs and produce a profit of $0.

(c) Calculation of the *y*-intercept

$$y = 2x - 150$$
$$y = 2(0) - 150$$
$$y = 0 - 150$$
$$y = -150$$

To calculate the *y*-intercept, set $x = 0$ and solve for *y*.

The *y*-intercept is $(0, -150)$. Producing 0 pizzas will result in a loss of $150. The given overhead costs for this business are $150.

Self-Check 4

Rework Example 4 assuming daily overhead costs of $200 and a profit of $2.50 per pizza.

3. Solve an Equation for a Specified Variable

Can I Rearrange a Formula so That It Is More Convenient to Use?

Yes, many common formulas are given in a standard form and then rearranged depending on the variable that is needed for a particular application.

When we specify the variable in an equation that we wish to solve for, that variable is called the **specified variable.** To solve for a specified variable, isolate this variable on the left side of the equation with a coefficient of 1. In Example 5, we rearrange the formula for the perimeter of a trapezoid to solve for the variable *c*.

Example 5   Solving the Formula for the Perimeter of an
Isosceles Trapezoid for One Side

Solve $P = a + b + 2c$ for c.

Solution

$$P = a + b + 2c$$

An isosceles trapezoid (as shown in the figure) has two
parallel sides and two equal nonparallel sides.

$$P - a - b = a + b + 2c - a - b$$

Subtract a and b from both sides to isolate the term
containing c on one side of the equation.

$$P - a - b = 2c$$

$$\frac{P - a - b}{2} = \frac{2c}{2}$$

Then divide both sides of the equation by 2, the
coefficient of c.

$$\frac{P - a - b}{2} = c \quad \text{or}$$

$$c = \frac{P - a - b}{2}$$

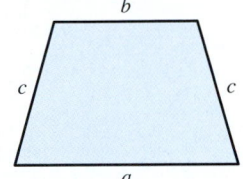

Self-Check 5

Solve $P = a + b + c + 2d$ for b.

In Example 6, there are four variables: S, n, a_1, and a_n. We will solve this equation for
a_1. We will revisit this equation in Section 12.6.

Example 6   Solving the Formula for the Sum of an Arithmetic
Sequence for the First Term

Solve $S = \dfrac{n}{2}(a_1 + a_n)$ for a_1.

Solution

$$S = \frac{n}{2}(a_1 + a_n)$$

$$\left(\frac{2}{n}\right)S = \left(\frac{2}{n}\right)\left(\frac{n}{2}\right)(a_1 + a_n)$$

Multiply both sides of the equation by $\dfrac{2}{n}$, the reciprocal of $\dfrac{n}{2}$.

$$\frac{2S}{n} = a_1 + a_n$$

$$\frac{2S}{n} - a_n = a_1 + a_n - a_n$$

To isolate a_1 on the right side, subtract a_n (read as "a sub n")
from both sides of the equation.

$$\frac{2S}{n} - a_n = a_1 \quad \text{or}$$

$$a_1 = \frac{2S}{n} - a_n$$

Self-Check 6

Solve $6ab + 3 = 3cd$ for d. Assume $c \neq 0$.

The equation $I = PRT$ allows us to directly calculate the simple interest I on a principal P at an interest rate R for a time T. In Example 7, we solve this equation for P. If we have a budgeted amount to spend on interest, then this form of the equation allows us to directly calculate the principal corresponding to that interest amount.

Example 7 Solving the Simple Interest Formula for P

Solve $I = PRT$ for P. Then prepare a table of principal amounts that can be borrowed when the monthly interest is $33. Use interest rates of 7%, 7.5%, 8%, 8.5%, 9%, 9.5%, and 10%.

Solution

$$I = PRT$$

$$\frac{I}{RT} = \frac{PRT}{RT}$$ Divide both sides of the equation by RT to isolate the specified variable P on the right side of the equation.

$$\frac{I}{RT} = P \quad \text{or}$$

$$P = \frac{I}{RT}$$ This is an alternate form of the equation that has been solved for P.

$$P = \frac{33}{R\left(\dfrac{1}{12}\right)}$$ To determine the principal that can be paid for with a monthly interest payment of $33, substitute $33 for the interest I and $\dfrac{1}{12}$ for T for a time of $\dfrac{1}{12}$ year.

$$P = \frac{33(12)}{R\left(\dfrac{1}{12}\right)(12)}$$ Then simplify by multiplying the numerator and denominator by 12.

$$P = \frac{396}{R}$$

B2		f_x	=396/A2

	A	B
1	Interest Rate	Principal
2	0.070	$5,657.14
3	0.075	$5,280.00
4	0.080	$4,950.00
5	0.085	$4,658.82
6	0.090	$4,400.00
7	0.095	$4,168.42
8	0.100	$3,960.00

Answer: The amount of principal that can be borrowed for a $33 monthly interest payment varies from $5,657.14 at 7% to $3,960.00 at 10%. The greater the interest rate, the less one can borrow on a fixed budget.

Self-Check 7

Rework Example 7 assuming the monthly interest is $45. Use interest rates of 6%, 6.5%, 7%, 7.5%, 8%, 8.5%, and 9%.

If you are trying to solve for a specified variable, you may find it useful to highlight this variable. In more complicated equations, this will help you clearly focus on the variable that needs to be isolated on the left side of the equation.

Self-Check Answers

1. $y = \dfrac{1}{2}x - 1$

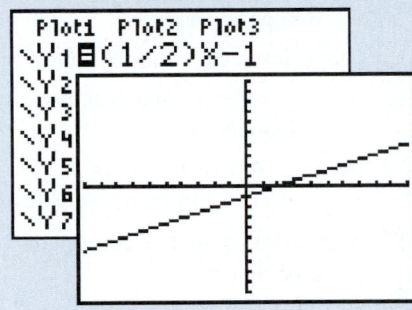

$[-10, 10, 1]$ by $[-10, 10, 1]$

2. $y = -\dfrac{3}{4}x - 2$

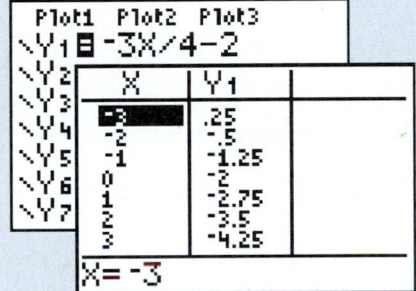

3. x-intercept: $(4, 0)$
y-intercept: $(0, -10)$

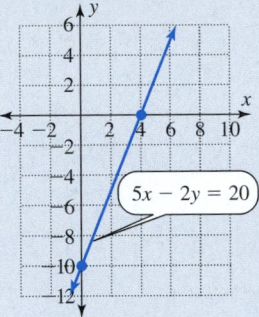

$5x - 2y = 20$

4. a. $y = 2.50x - 200$
b. The x-intercept is $(80, 0)$. The break-even value for this business is 80 pizzas.
c. The y-intercept is $(0, -200)$. The overhead costs for this business are \$200.

5. $b = P - a - c - 2d$

6. $d = \dfrac{2ab + 1}{c}$

7. $P = \dfrac{540}{R}$

Plot1 Plot2 Plot3		
\Y1⊟540/X		
\Y2=	X	Y1
\Y3=	.06	9000
\Y4=	.065	8307.7
\Y5=	.07	7714.3
\Y6=	.075	7200
\Y7=	.08	6750
	.085	6352.9
	.09	6000
X=.06		

2.6   Using the Language and Symbolism of Mathematics

1. Rewriting an equation to solve for one particular variable is called solving for a _____ variable.

2. The _____ property of multiplication over addition allows us to rewrite $2(3x + y)$ as $6x + 2y$.

3. In a profit function, the _____-_____ value produces a profit of \$0.

4. The _____ costs for a company can include insurance, rent, electricity, and other fixed expenses when 0 units are produced.

5. The point on the graph of a linear equation with a y-coordinate of zero is the _____-intercept.

6. The point on the graph of a linear equation with an x-coordinate of zero is the _____-intercept.

2.6   Quick Review

1. Given $I = PRT$, evaluate I for $P = \$5,000$, $R = 6\%$, and $T = 1$ year.

2. Is $(3, 0)$ a solution of $5x - 3y = 15$?

3. Is $(0, 5)$ a solution of $5x - 3y = 15$?

4. Solve $3x - (2x - 5) + 3 = 5x$.

5. Solve $\dfrac{x}{5} - \dfrac{3x}{2} = 13$.

2.6   Exercises

Objective 1 Rewrite a Linear Equation in the Form $y = mx + b$

In Exercises 1–14, solve each equation for y.

1. $2x + y = 7$

2. $5x + y = 8$

3. $3x - y = 2$

4. $4x - y = 3$

5. $-6x + 3y = -9$

6. $-30x - 5y = 20$

7. $\dfrac{x}{2} - \dfrac{y}{4} = -1$

8. $\dfrac{x}{3} + \dfrac{y}{12} = -\dfrac{1}{6}$

9. $-0.3x - 0.1y = 0.2$

10. $-0.4x + 0.8y = -0.2$

11. $5x - 2y = x - 3y + 4$

12. $-6x + 5y = 8x - 2y - 21$

13. $2(3x - y + 1) = 3(4x - y - 2)$

14. $4(2x + 3y - 5) = 3(2x + 5y + 4)$

In Exercises 15–18, solve each equation for y. Then use a graphing calculator to complete a table of values for the x-values $-3, -2, -1, 0, 1, 2, 3$ and to graph the equation by using the window. $[-10, 10, 1]$ by $[-10, 10, 1]$. (*Hint:* See Examples 1 and 2.)

15. $2x + 3y = 4x + 2y - 3$

16. $7x - 5y = 6(x - y) + 1$

17. $-5x + 4y = -2(2x - 3y) + 4$

18. $-4x + 7y = -5(x - y) + 3$

Objective 2 Calculate the *x*- and *y*-Intercepts of a Line

In Exercises 19–24, calculate the x- and y-intercepts of each line, and then use the intercepts to graph the line.

19. $6x - 3y = 12$

20. $4x + 8y = 24$

21. $\dfrac{x}{2} - \dfrac{y}{8} = 1$

22. $-\dfrac{x}{3} + \dfrac{y}{7} = 1$

23. $0.2x + 0.3y = 0.6$

24. $0.4x - 0.3y = 2.4$

Objective 3 Solve an Equation for a Specified Variable

In Exercises 25–48, solve each literal equation for the variable specified. The context in which you may encounter these formulas is indicated in parentheses. Assume all variables are nonzero.

25. $A = lw$ for l (area of a rectangle)

26. $A = \dfrac{1}{2}bh$ for h (area of a triangle)

27. $C = 2\pi r$ for r (circumference of a circle)

28. $V = lwh$ for w (volume of a rectangular box)

29. $V_1 T_2 = V_2 T_1$ for V_1 (Charles's law in chemistry)

30. $P_1 V_1 = P_2 V_2$ for P_2 (Boyle's law in chemistry)

31. $V = \dfrac{1}{3}\pi r^2 h$ for h (volume of a cone)

32. $I = PRT$ for T (interest formula)

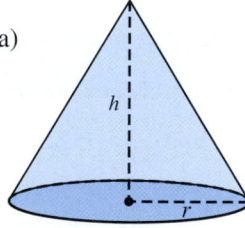

33. $P = a + b + c$ for a (perimeter of a triangle)

34. $V - E + F = 2$ for E (Euler's theorem)

35. $F = \dfrac{9}{5}C + 32$ for C (Fahrenheit and Celsius temperatures)

36. $x = \dfrac{x_1 + x_2}{2}$ for x_1 (x-coordinate of midpoint)

37. $A = \dfrac{1}{2}h(a + b)$ for b (area of trapezoid)

38. $A = \dfrac{1}{2}h(a + b)$ for a (area of trapezoid)

39. $C = \dfrac{5}{9}(F - 32)$ for F (Fahrenheit and Celsius temperatures)

40. $y = mx + b$ for x (slope-intercept form of a line)

41. $l = a + (n - 1)d$ for a (last term of an arithmetic sequence)

42. $l = a + (n - 1)d$ for $d; n \neq 1$ (last term of an arithmetic sequence)

43. $l = a + (n - 1)d$ for n (last term of an arithmetic sequence)

44. $S = 2\pi r^2 + 2\pi rh$ for h (surface area of a cylinder)

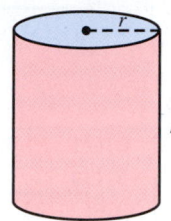

45. $y = mx + b$ for m (slope-intercept form of a line)

46. $S = \dfrac{n}{2}(a + l)$ for n (sum of an arithmetic sequence)

47. $P = 2l + 2w$ for w (perimeter of a rectangle)

48. $E = mc^2$ for m (Einstein's formula)

49. Sales Tax The sales tax T on a purchase amount P at a sales tax rate R is given by the formula $T = PR$.

  a. Solve this formula for P.

  b. Write the specific formula for P that is used for a store in a state with a sales tax rate of 8%.

  c. The accompanying table gives various sales tax amounts collected by the store. Prepare a table of purchase amounts that correspond to these sales tax amounts.

Sales Tax T, $	Purchase Amount P, $
100	
200	
300	
400	
500	
600	

50. Dimensions of a Building The area of a rectangular building is given by the formula $A = LW$.

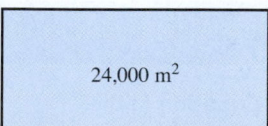

24,000 m²

  a. Solve this formula for W.

  b. Write the specific formula that is used for a rectangular building that will have an area of 24,000 m².

c. The table gives various possible lengths for this building. Determine the widths that correspond to these lengths.

Length L, m	Width W, m
100	
110	
120	
130	
140	
150	

51. Dimensions of a Building The perimeter of a building is given by the formula $P = 2W + 2L$.
a. Solve this formula for W.
b. Write the specific formula that is used for a rectangular building that will have a perimeter of 800 m.
c. The table gives various possible lengths for this building. Determine the widths that correspond to these lengths.

Length L, m	Width W, m
100	
120	
140	
160	
180	
200	

52. Isosceles Triangle An isosceles triangle has two equal sides and two equal angles. In the isosceles triangle shown, $x + 2y = 180$ (degrees).
a. Solve this formula for y.
b. The table gives the number of degrees for several possible values for angle x. Determine the corresponding values of y.

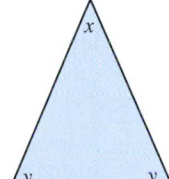

x, Degrees	y, Degrees
5	
20	
30	
40	
60	
85	

Connecting Concepts to Applications

53. Profit on Ice Cream An ice cream shop in a small town has daily overhead costs of $120. It makes $0.75 on each ice cream product that it sells.
a. Write an equation that gives the profit for this business when it sells x orders for its ice cream products.
b. Using the equation from part **a**, determine the x-intercept of the graph of this equation and interpret this point.

c. Using the equation from part **a**, determine the y-intercept of the graph of this equation and interpret this point.

54. Profit on Coffee A small drive-through coffee franchise has daily overhead costs of $75. It makes $0.50 on each cup of coffee that it sells.
a. Write an equation that gives the profit for this business when it sells x cups of coffee.
b. Using the equation from part **a**, determine the x-intercept of the graph of this equation and interpret this point.
c. Using the equation from part **a**, determine the y-intercept of the graph of this equation and interpret this point.

55. Simple Interest Formula A bank uses an annual percent rate (APR) of R and a time of $\frac{1}{12}$ year to determine the monthly interest it pays customers on their certificates of deposit. Solve $I = PRT$ for R. Then prepare a table of interest rates that will result in a monthly interest of $50 for CDs in these amounts: $10,000; $12,000; $14,000; $16,000; $18,000; $20,000; and $22,000 (*Hint:* See Example 7.)

56. Determining the Length of a Copper Wire A copper wire of a fixed radius is really a very long cylinder whose volume is given by the formula $V = \pi r^2 h$, where r is the radius of the wire and the height h of the cylinder is the length L of the wire. Solve this formula for h and then complete this table for a wire of radius 0.8 cm. For each volume of copper delivered to a wire extrusion machine, the table will give the length of wire that can be produced.

Volume of Copper V, cm³	Length of Wire L, cm
1,000	
2,000	
3,000	
4,000	
5,000	

Group discussion questions

57. Writing Mathematically Write a paragraph describing why it is handier to have the formula $F = \dfrac{9}{5}C + 32$ for some applications, whereas the transformed equation $C = \dfrac{5}{9}(F - 32)$ is handier for other applications. Then complete the following table.

Celsius	Fahrenheit	
100°		The temperature at which water boils.
94°		Is this hot?
	98.6°	Normal body temperature.
	72°	A comfortable room temperature.
0°		Water freezes.
	0°	Zero on Fahrenheit scale.
		Both temperatures are equal.

58. Challenge Question It is approximately 5 ft between the centers of the front tires on a sport utility vehicle (SUV).

a. A driver of this SUV drives around so that the inside tire traces out a circle with a radius of 40 ft. How much farther does the outside tire travel than the inside tire?

b. If the radius of the circle formed by the inside tire is 400 ft, predict without calculating how much farther the outside tire travels.

c. Check the prediction made for part **b** by calculating this distance.

d. Explain the connection between the result in parts **a** and **c** by analyzing the formula for the circumference of a circle.

59. Challenge Question A sheet metal worker had two identical rectangular sheets of metal that he used to form two cylindrical heating ducts. One of the cylindrical ducts is 50 cm long with a circumference of 80 cm. The second cylindrical duct is 80 cm long with a circumference of 50 cm. If you unroll each cylinder as illustrated in the figure, you can determine the dimensions of each original rectangular sheet of metal.

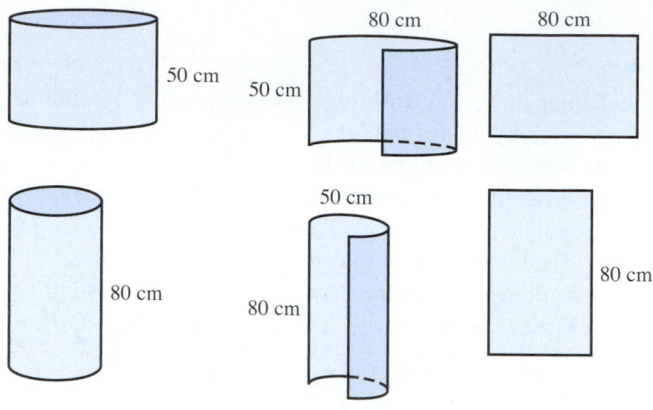

a. Calculate the volume of the cylinder that is 50 cm long. ($V = \pi r^2 h$)

b. Calculate the volume of the cylinder that is 80 cm long.

c. Are the volumes the same? If not, use the formula for volume to describe why one of these has a larger volume.

d. Determine the dimensions of each original rectangular sheet of metal.

e. Calculate the area of each sheet of metal.

f. Is the area of each sheet of metal the same? Use the formula $A = lw$ to explain your answer.

60. Applying Technology Example 7 presented a spreadsheet for principal amounts that could be borrowed with a monthly interest payment of $33 at different interest rates. Prepare a similar spreadsheet for principal amounts that can be borrowed when the monthly interest is $500. Use interest rates of 4.5%, 5%, 5.5%, 6%, 6.5%, 7%, and 7.5%.

2.6 Cumulative Review

1. An electrical cable 50 ft long has six equal pieces each of length x cut off. Write a function for the length of the remaining piece in terms of x.

2. Write the ratio 45:60 in lowest terms.

3. Simplify $3\frac{1}{3} + 4\frac{1}{5}$.

4. Simplify $\left(3\frac{1}{3}\right)\left(4\frac{1}{5}\right)$.

5. Use the distributive property of multiplication over addition to complete the equation $7x - 14y = 7(\ \ ?\ \ )$.

Section 2.7 | Proportions and Direct Variation

Objectives:

1. Solve problems involving proportions.
2. Solve problems involving direct variation.

The ability to solve real-world problems often requires us to create a mathematical model of the problem. Linear equations are used to model many problems. In this section we will use proportions to form linear equations to model some applications. A classic type of problem is to determine the height of one object by comparing it to the height of a second object. We will use proportions to accomplish this.

What Is a Proportion?

A **proportion** is an equation that states that two ratios are equal. For example, the proportion $\frac{a}{b} = \frac{c}{d}$ is read "a is to b as c is to d." The four numbers a, b, c, and d are called the **terms of the proportion:** a is the first term, b is the second term, c is the third term, and d is the fourth term. The first and fourth terms are called the **extremes,** and the second and third terms are called the **means.** The word *mean* has now been used in two contexts: as an average of a set of values and as one of the middle terms of a proportion.

> *A Mathematical Note*
>
> Emmy Noether (1882–1935) was called by Albert Einstein the greatest of all the women who were creative mathematical geniuses. Emmy completed her doctoral dissertation in algebra at the University of Erlangen in 1907. She taught at the University of Göttingen in Germany from 1915 until 1933. She then moved to the United States and taught at Bryn Mawr for 2 years until her death in 1935.

Proportion

Algebraically	Verbally	Numerical Example
$\frac{a}{b} = \frac{c}{d}$ or $a:b = c:d$	This proportion is read "a is to b as c is to d." The extremes are a and d, and the means are b and c.	$\frac{2}{5} = \frac{40}{100}$ or $2:5 = 40:100$

1. Solve Problems Involving Proportions

Since proportions are equations, we can solve them by using the same rules and procedures that apply to all equations.

Example 1 | Solving a Proportion

Solve $\dfrac{x}{3} = \dfrac{5}{4}$.

Solution

$$\frac{x}{3} = \frac{5}{4}$$

$$12\left(\frac{x}{3}\right) = 12\left(\frac{5}{4}\right) \quad \begin{array}{l} \text{LCD} = 3 \cdot 4 = 12 \\ \text{Multiply both sides of the equation by the LCD of 12.} \end{array}$$

$$4x = 3(5)$$

$$4x = 15$$

$$\frac{4x}{4} = \frac{15}{4} \quad \text{Divide both sides of the equation by 4.}$$

$$x = \frac{15}{4}$$

Answer: $x = \dfrac{15}{4}$ Does this solution check?

Self-Check 1

Solve $\dfrac{11}{6} = \dfrac{x}{3}$.

Example 2   Solving a Proportion

Solve $\dfrac{35}{6m} = \dfrac{5}{9}$.

Solution

$$\frac{35}{6m} = \frac{5}{9} \qquad \begin{array}{l} 6m = 2 \cdot 3 \cdot m \\ 9 = 3 \cdot 3 \\ \text{LCD} = 2 \cdot 3 \cdot 3 \cdot m = 18m \end{array}$$

$$18m\left(\frac{35}{6m}\right) = 18m\left(\frac{5}{9}\right) \qquad \text{Multiply both sides by the LCD of } 18m.$$

$$\frac{\overset{3}{\cancel{18m}}}{\underset{1}{\cancel{6m}}}(35) = \frac{\overset{2m}{\cancel{18m}}}{\underset{1}{\cancel{9}}}(5) \qquad \text{Then simplify both sides of the equation.}$$

$$3(35) = (2m)(5)$$

$$105 = 10m$$

$$\frac{105}{10} = \frac{10m}{10} \qquad \text{To solve for } m, \text{ divide both sides by 10.}$$

$$10.5 = m \quad \text{or}$$

Answer: $m = 10.5$ Does this value check?

Self-Check 2

Solve $\dfrac{10}{x} = \dfrac{8}{12}$.

Ratios and proportions are so useful that many employment tests, including civil service tests, have problems that test your ability to use proportions. A typical problem involves a constant ratio under two different situations; that is,

$$\boxed{\text{Ratio for first situation}} = \boxed{\text{Ratio for second situation}}.$$

This type of problem is illustrated in Example 3, which compares the ratio of distances on a first map reading to distances on a second map reading.

Example 3 Distance Using a Map

On a particular map, 2 cm represents a distance of 230 km. What distance corresponds to 6.5 cm on the map?

Solution

Let d = distance in kilometers corresponding to 6.5 cm.

Verbally

$$\frac{\text{First map reading}}{\text{First distance}} = \frac{\text{Second map reading}}{\text{Second distance}}$$

First stating a word equation is an excellent way to start a word problem.

Algebraically

$$\frac{2}{230} = \frac{6.5}{d}$$

Use the given values to translate the word equation into an algebraic model of the problem.

$$230d\left(\frac{2}{230}\right) = 230d\left(\frac{6.5}{d}\right)$$

Multiply both sides of the equation by the LCD of $230d$.

$$2d = 230(6.5)$$

$$\frac{2d}{2} = \frac{1{,}495}{2}$$

Divide both sides by 2.

$$d = 747.5$$

Answer: On this map, 6.5 cm represents 747.5 km.

Does this answer seem reasonable?

Self-Check 3

A recipe for old-fashioned vegetable soup suggests using 4 cups of water to prepare a serving for 6 people. How much water should be used to prepare a serving for 10?

A common question regarding Example 3 is to inquire whether it is acceptable to form the ratio

$$\frac{\text{First map reading}}{\text{Second map reading}} = \frac{\text{First distance}}{\text{Second distance}}.$$

This is a good question and the answer is yes. More generally, the equation $\frac{a}{b} = \frac{c}{d}$ is equivalent to $\frac{a}{c} = \frac{b}{d}$.

2. Solve Problems Involving Direct Variation

Are There Other Ways to Indicate That Quantities Are Proportional?

Yes, describing the relationship among variables is a very important part of mathematics. Some relationships are so common that it is not surprising that we have developed multiple ways to describe the relationships. The concept of direct variation, which we examine now, is closely related to arithmetic sequences, linear equations, and ratios and proportions.

Direct Variation

If x and y are real variables and k is a real constant with $k \neq 0$, then:

Verbally	Algebraically	Numerical Example	
y varies directly as x with the constant of variation k.	$y = kx$ Example: $y = 3x$	x	$y = 3x$
		1	3
		2	6
		3	9
		4	12
		5	15

Example 4 — Translating Statements of Variation

Solution

(a) Translate $C = \pi d$ into a verbal statement of variation.

C varies directly as d with the constant of variation π.

The circumference of a circle varies directly as the diameter with π being the constant of variation.

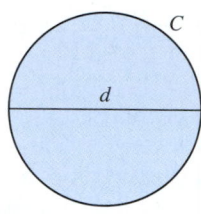

(b) As a result of Hooke's law, we know that the distance d a spring stretches varies directly as the mass m attached to the spring. Translate this statement of variation into an algebraic equation.

$d = km$

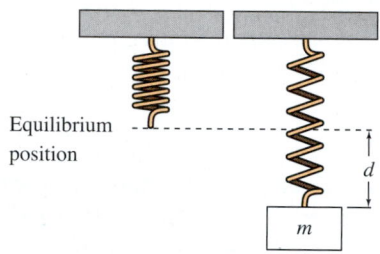

Equilibrium position

Self-Check 4

a. Translate $D = kR$ into a verbal statement of variation.

b. The height h of a plant varies directly as the number of days n after it has been planted. Translate this statement of variation into an algebraic equation.

Example 5 — Solving a Direct Variation Problem

If y varies directly as x, and y is 12 when x is 9, find y when x is 15.

Solution

$y = kx$ Write the statement of direct variation as an equation and substitute in the given values.

$12 = k(9)$

$\dfrac{12}{9} = k$

$k = \dfrac{4}{3}$ Then solve for k.

$$y = \frac{4}{3}x$$ Substitute this value of k into the variation equation and then evaluate this equation for $x = 15$.

$$y = \frac{4}{3}(15)$$

$$y = 20$$

Answer: $y = 20$ when $x = 15$.

Self-Check 5

If y varies directly as x, and y is 18 when x is 22, find y when x is 33.

Is the Constant of Variation Ever Important?

Yes, sometimes the constant of variation is so important that it becomes well known or has a special symbol assigned to it. An example of this is π, which is the ratio of the circumference of a circle to its diameter. In Example 6, we solve for the constant of variation that represents the exchange rate between two currencies.

   As we shall see in Example 7, problems involving direct variation can be solved by using proportions. However, Example 6 illustrates that using direct variation is more useful for situations where the constant of variation has a significant meaning or where we want to generate a table of values.

Example 6   Determining a Currency Exchange Rate

Traveling in New Zealand during 2009, one could exchange 800 USD (U.S. dollars) for 1,552 NZD (New Zealand dollars). The number of NZD received varies directly as the number of USD exchanged. The constant of variation is called the exchange rate.

(a) What was the exchange rate at this time?

(b) Use the exchange rate to write an equation relating the number of NZD received to the number of USD exchanged.

(c) Use this equation to create a table of values displaying exchange amounts in $50 increments starting at $50 USD.

Solution

Let $U =$ number of U.S. dollars

$N =$ number of New Zealand dollars

$k =$ constant of variation

Verbally

The number of NZD varies directly as the number of USD.	State a precise word equation that is easily translated into algebraic form.

Algebraically

$N = kU$	Translate the word equation into algebraic form by using the variables identified above.
$1{,}552 \text{ NZD} = k(800 \text{ USD})$	Substitute in the given values for N and U.
$\dfrac{1{,}552 \text{ NZD}}{800 \text{ USD}} = k$	Then solve for k, the exchange rate.
$k = 1.94$	This means we would receive 1.94 NZD for each 1 USD exchanged.

(a) The exchange rate was 1.94 NZD for 1 USD.

(b) Equation: $N = 1.94U$

(c)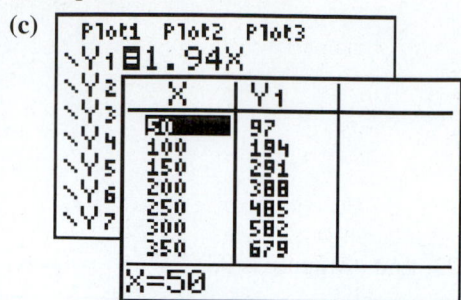

To generate a table of values, enter the equation $N = 1.94U$ as $y = 1.94x$. Then use 50 for the **TblStart** and **ΔTbl** values.

Self-Check 6

A traveler exchanged 350 USD (U.S. dollars) for 402.5 CAD (Canadian dollars). The number of CAD received varies directly as the number of USD exchanged.

a. What was the exchange rate at this time?

b. Use the exchange rate to write an equation relating the number of CAD received to the number of USD exchanged.

c. Use this equation to create a table of values displaying exchange amounts in $50 increments starting at 50 USD.

Example 7 illustrates another way to work Example 5. This example shows how to solve a problem involving direct variation without calculating the constant of variation. If $y = kx$, then $k = \dfrac{y}{x}$. When y varies directly as x, k is a constant while x and y vary. For two sets of values we can denote the two x-values by x_1 (read "x sub 1") and x_2 (read "x sub 2"). Likewise the two y-values can be denoted by y_1 and y_2. If $k = \dfrac{y_1}{x_1}$ and $k = \dfrac{y_2}{x_2}$ for two sets of values, then $\dfrac{y_1}{x_1} = \dfrac{y_2}{x_2}$. This can be described by saying that x and y are *directly proportional*.

Example 7   Solving a Problem Involving Direct Proportionality

If x and y are directly proportional, and y is 12 when x is 9, find y when x is 15.

Solution

$$\frac{y_1}{x_1} = \frac{y_2}{x_2}$$
x and y are directly proportional.

$$\frac{12}{9} = \frac{y_2}{15}$$
Substitute 12 for y_1, 9 for x_1, and 15 for x_2.

$$\frac{4}{3} = \frac{y_2}{15}$$
Reduce the left side of the equation by dividing both the numerator and the denominator by 3.

$$15\left(\frac{4}{3}\right) = 15\left(\frac{y_2}{15}\right)$$
Multiply both sides by the LCD of 15.

$$5(4) = y_2$$

$$20 = y_2 \quad \text{or}$$
Then simplify both sides of the equation.

$$y_2 = 20$$
When you compare this solution to the one given in Example 5, which do you prefer?

Check: $\dfrac{12}{9} \overset{?}{=} \dfrac{20}{15}$

$\dfrac{4}{3} \overset{?}{=} \dfrac{4}{3}$ checks.

Since $\dfrac{y_1}{x_1} = \dfrac{y_2}{x_2}$, x and y are directly proportional.

Answer: $y = 20$ when $x = 15$.

Self-Check 7

If x and y are directly proportional, and y is 16 when x is 20, find y when x is 25.

Ratios and proportions are the basis for many indirect measurements. Example 8 illustrates how to determine the height of a tree by measuring the length of a shadow. This same technique has been used to determine the heights of mountains on Mars.

Example 8   Modeling the Height of a Tree by Using Its Shadow

At a fixed time of the day, the length of a shadow and the height of the object casting the shadow are directly proportional. If a person 2 m tall casts a 5-m shadow, how tall is a fir tree that casts a 40-m shadow?

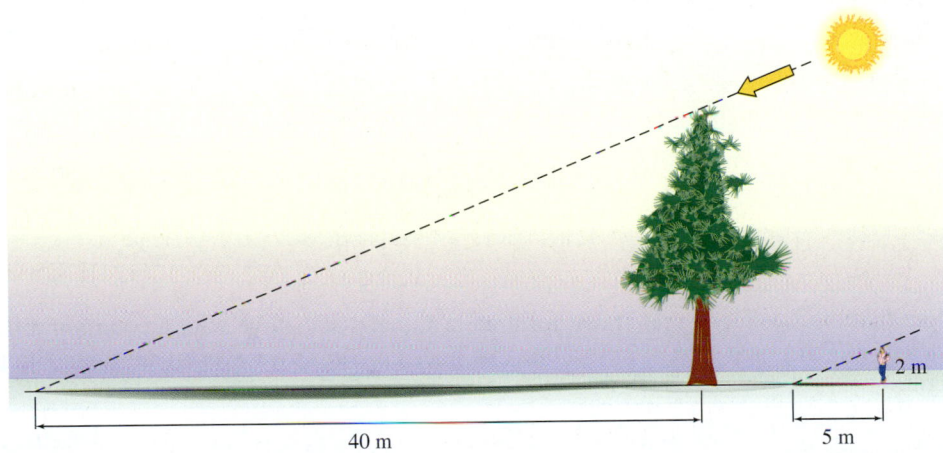

40 m   5 m   2 m

Solution

Let t = height of tree in meters.

Verbally

$$\dfrac{\text{Height of person}}{\text{Length of person's shadow}} = \dfrac{\text{Height of tree}}{\text{Length of tree's shadow}}$$

First form a word equation based on the given proportion. Also note that this proportion could be formed by using corresponding sides of similar right triangles.

Algebraically

$$\dfrac{2}{5} = \dfrac{t}{40}$$

$$40\left(\dfrac{2}{5}\right) = 40\left(\dfrac{t}{40}\right)$$

$$8(2) = t$$

$$16 = t \quad \text{or}$$

$$t = 16$$

Substitute the given values to translate the word equation into an algebraic model of the problem.

Multiply both sides of the equation by the LCD of 40.

Then simplify both sides of the equation.

Does this answer seem reasonable?

Answer: The tree is 16 m tall.

Self-Check 8

Rework Example 8, assuming that the tree casts a 35-m shadow.

When y varies directly as x, increasing magnitudes of x result in increasing magnitudes of y. Likewise, decreasing magnitudes of x result in decreasing magnitudes of y. For example, a taller tree produces a longer shadow, and a shorter tree produces a shorter shadow.

We will revisit direct variation in Section 9.6 when we examine inverse and joint variation.

Example 9 examines the concept of direct variation from multiple perspectives. Note the connection of this concept to linear equations, arithmetic sequences, and straight lines. The sequence of y-values shown in this table forms an arithmetic sequence with a common difference that is equal to the constant of variation.

Example 9	**Using Multiple Perspectives to Examine Direct Variation**

Examine the following statement of variation verbally, algebraically, numerically, and graphically: y varies directly as x with constant of variation 2.

Solution

Verbally	Algebraically	Numerically	Graphically
y varies directly as x with constant of variation 2.	$y = 2x$		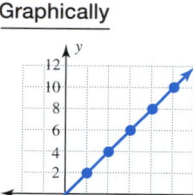

x	$y = 2x$
1	2
2	4
3	6
4	8
5	10

Note that the y-values in this table form an arithmetic sequence with a common difference of 2. The graph of $y = 2x$ is a line that goes up 2 units for every 1-unit move to the right.

Self-Check 9

Rework Example 9 by examining the statement y varies directly as x with constant of variation –3.

Self-Check Answers

1. $x = \dfrac{11}{2}$

2. $x = 15$

3. Use $6\dfrac{2}{3}$ cups of water.

4. **a.** D varies directly as R with constant of variation k.
   b. $h = kn$

5. $y = 27$ when $x = 33$.

6. **a.** 1.15 CAD for 1 USD
   b. $C = 1.15U$

c.

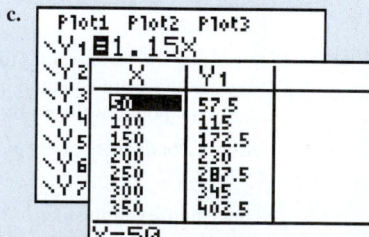

7. $y = 20$

8. The tree is 14 m tall.

9. **Verbally:** y varies directly as x with constant of variation -3.
   Algebraically: $y = -3x$
   Numerically:

x	$y = -3x$
1	-3
2	-6
3	-9
4	-12
5	-15

Graphically:

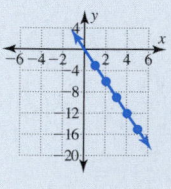

2.7   Using the Language and Symbolism of Mathematics

1. An equation that states two ratios are equal is called a _____.

2. In the proportion $\dfrac{a}{b} = \dfrac{c}{d}$, a, b, c, and d are called the _____ of the proportion.

3. The means of $\dfrac{a}{b} = \dfrac{c}{d}$ are _____ and _____.

4. The extremes of $\dfrac{a}{b} = \dfrac{c}{d}$ are _____ and _____.

5. If x and y are directly proportional, then y varies _____ as x.

6. If y varies directly as x, then the equation expressing this is a _____ equation.

7. If y varies directly as x and x increases in magnitude, then y _____ in magnitude.

8. If y varies directly as x and x decreases in magnitude, then y _____ in magnitude.

9. If y varies directly as x, then inputting consecutive natural numbers for x will produce output y-values that form an _____ sequence.

2.7   Quick Review

1. Evaluate $\dfrac{1}{2} + \dfrac{2}{3} - \dfrac{3}{4}$ using only pencil and paper.

2. Solve $\dfrac{2}{3}x = 12$.

3. Solve $0.45x = 0.54$.

4. Use the equation $y = 2.5x - 10$ to complete the table.

x	0		20	
y		0		15

5. Division by ____ is undefined.

2.7   Exercises

Objective 1 Solve Problems Involving Proportions

In Exercises 1–22, solve each proportion.

1. $\dfrac{x}{20} = \dfrac{3}{5}$

2. $\dfrac{x}{16} = \dfrac{3}{2}$

3. $\dfrac{3}{5} = \dfrac{6}{x}$

4. $\dfrac{2}{7} = \dfrac{10}{x}$

5. $\dfrac{5x}{2} = \dfrac{45}{6}$

6. $\dfrac{11x}{3} = \dfrac{154}{21}$

7. $\dfrac{9}{2y} = \dfrac{27}{66}$

8. $\dfrac{6}{3y} = \dfrac{5}{6}$

9. $\dfrac{5}{6} = \dfrac{40}{3y}$

10. $\dfrac{13}{5} = \dfrac{52}{10y}$

11. $a : 3 = 4 : 6$

12. $7z : 10 = 7 : 2$

13. $5 : 2v = 3 : 4$

14. $5 : 11 = 15 : b$

15. $\dfrac{x+1}{5} = \dfrac{8}{10}$

16. $\dfrac{x+1}{20} = \dfrac{3}{5}$

17. $\dfrac{z+2}{3} = \dfrac{z}{4}$

18. $\dfrac{t-1}{5} = \dfrac{t}{2}$

19. $\dfrac{x}{3} = \dfrac{2x-3}{5}$

20. $\dfrac{3w}{4} = \dfrac{w-5}{1}$

21. $\dfrac{2v-1}{5} = \dfrac{2v-5}{1}$

22. $\dfrac{x-1}{6} = \dfrac{2x-4}{9}$

Objective 2 Solve Problems Involving Direct Variation

In Exercises 23 and 24, write an equation for each statement of variation, using k as the constant of variation.

23. At a fixed speed, the distance d that a car travels varies directly as the time t it travels.

24. At a fixed pressure, the volume V of a gas varies directly as the absolute temperature T.

In Exercises 25 and 26, write a statement of variation for each equation, assuming k is the constant of variation.

25. $v = kw$

26. $b = ka$

In Exercises 27–32, solve each of these problems involving direct variation.

27. If v varies directly as w, and $v = 22$ when $w = 77$, find v when $w = 35$.

28. If m varies directly as n, and $m = 12$ when $n = 20$, find n when $m = 18$.

29. If p is directly proportional to g, and p is 0.375 when g is 1, find g when p is 1.5.

30. If b is directly proportional to d, and b is 0.3125 when d is 1, find b when d is 0.4.

31. If v varies directly as w, and $v = 12$ when $w = 15$, find the constant of variation k.

32. If b varies directly as a, and $b = \dfrac{2}{3}$ when $a = \dfrac{4}{5}$, find the constant of variation k.

Connecting Concepts to Applications

33. Recipe Proportions A recipe for 50 people used 3 cups of sugar. How many cups of sugar are needed for 75 people?

34. Recipe Proportions A recipe for six adults called for $\dfrac{3}{4}$ tsp of salt. If this recipe is used to cook for four people, how much salt is required?

35. Microscope Magnification Under a microscope, an insect antenna that measures 1.5 mm appears to be 7.5 mm. Find the apparent length under this magnification of an object that is 2 mm long.

36. Distance on a Map If 6 cm on a map represents 300 km, what distance does 1 cm represent?

37. Defective Lightbulbs A factory that produces lightbulbs finds that 2.5 out of every 500 bulbs are defective. In a group of 10,000 bulbs, how many would be expected to be defective?

38. Gold Ore A mining company was able to recover 3 oz of gold from 10 tons of ore. How many tons would be needed to recover 12 oz of gold?

39. Bricks for a Wall If 118 bricks are used in constructing a 2-ft section of a wall, how many bricks will be needed for a similar wall that is 24 ft long?

40. Sand in Concrete If 11 ft³ of sand is needed to make 32 ft³ of concrete, how much sand is needed to make 224 ft³ of concrete?

41. Scale Drawing An architect's sketch of one room in a new computer lab is shown here. The dimensions of the sketch are given in centimeters. The architect used a constant scale factor of 1 cm representing 2.5 m. What will be the actual dimensions of the room?

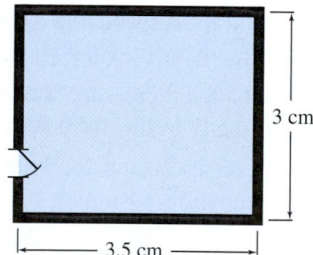

3 cm

3.5 cm

42. Scale Drawing For the computer lab sketch in Exercise 41, determine the actual width of the door if the door in the sketch is 0.4 cm wide.

43. Grass Seed A homeowner used 3 lb of grass seed to cover an area of 100 ft². How many pounds would be needed for 450 ft² if the amount of seed used and the area covered are directly proportional?

44. Quality Control A quality control inspector found 4 defective computer chips in the 250 that were tested. How many defective chips would be expected in a shipment of 10,000 chips?

45. Face Cards in Blackjack A blackjack dealer in a casino dealt 39 cards. Each of the decks from which he dealt had 12 face cards for every 52 cards. Approximately how many of these 39 cards would you expect to be face cards?

46. Maglev Train Plans for an operational scale model of a maglev (magnetically levitating) train call for a scale of 1 to 50. The two-car train will be about 28 in long and will sell to hobbyists for $2,000 to $5,000. What is the length of the maglev train on which this scale model is based?

47. Sales Commission Rate The commission earned by a salesman varies directly as his gross sales. In one month, he earned a commission of $2,700 on $18,000 in gross sales.

  a. What was the sales commission rate?

  b. Use the sales commission rate to write an equation relating the commission earned to the amount of gross sales for that month.

  c. Use this equation to complete the table of values.

	A	B
1	Gross Sales ($)	Commission ($)
2	$16,000	
3	$18,000	
4	$20,000	
5	$22,000	
6	$24,000	
7	$26,000	
8	$28,000	

48. Running Rate The distance run by a runner varies directly as the amount of time she runs. She runs 3 mi in 30 minutes.

  a. What was her running rate in miles per hour?

  b. Use the running rate to write an equation relating the distance she ran to her running time.

  c. Use this equation to complete the table of values.

	A	B
1	Time (h)	Distance (mi)
2	0.5	
3	1.0	
4	1.5	
5	2.0	
6	2.5	
7	3.0	
8	3.5	

49. Currency Conversion Rate The number of EUR (euros) received varies directly as the number of USD (U.S. dollars) exchanged. In 2009, one could exchange 500 USD for 390 EUR.

  a. What was the exchange rate at that time?

  b. Use the exchange rate to write an equation relating the number of EUR received to the number of USD exchanged.

c. Use this equation and a graphing calculator to create a table of values displaying exchange amounts in $50 increments starting at 50 USD.

50. **Currency Conversion Rate** The number of AUD (Australian dollars) received varies directly as the number of USD (U.S. dollars) exchanged. In 2009, one could exchange 700 USD for 1,085 AUD.
    a. What was the exchange rate at that time?
    b. Use the exchange rate to write an equation relating the number of AUD received to the number of USD exchanged.
    c. Use this equation and a graphing calculator to create a table of values displaying exchange amounts in $50 increments starting at 50 USD.

51. **Weight of Jet Fuel** The weight of jet fuel is approximately 6.7 lb/gal. The amount of jet fuel used at cruising altitude by an MD80 passenger jet is approximately 11.4 gal/min. The weight of fuel used varies directly as the time.
    a. Write an equation that gives W, the pounds of fuel used, in terms of t, the time in minutes.
    b. What does the constant of variation represent?

52. **Snow Cone Revenue** There are 128 one-ounce servings of snow cone juice in 1 gal. Each snow cone serving produces revenue of $1.75. The revenue generated by selling snow cones varies directly as the number of gallons of juice used in a day.
    a. Write an equation that gives D, the dollars of revenue generated, in terms of G, the gallons of snow cone juice served.
    b. What does the constant of variation represent?

53. **Width of a River** Use the dimensions shown in the figure to determine the width of the river. Triangles AB_1C_1 and AB_2C_2 are similar.

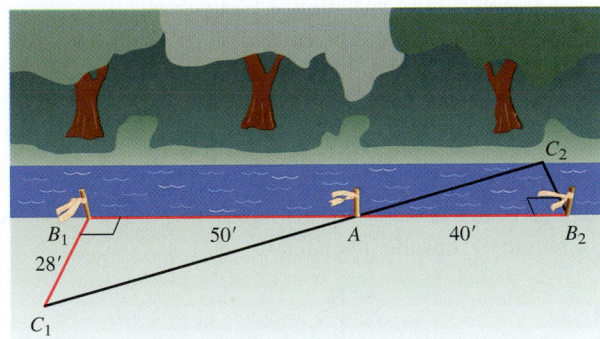

54. **Dosage Instructions** A pharmaceutical company testing a new drug finds that the ideal dosages appear to be 292.5 mg for a patient weighing 45 kg; 390 mg for a patient weighing 60 kg; and 487.5 mg for a patient weighing 75 kg. Determine the correct dosage instructions for this drug.

55. **Photograph Enlargement** A photograph is 7 cm wide and 10 cm long. What would be the width of the photograph if it were enlarged so that the length was 25 cm?

56. **Photocopier Reduction** A photocopier is to be used to reduce the printed material on a page 8.5 in wide and 11 in high so that it will occupy a page 4.25 in wide. How high will be the reduced page of material?

57. **Height of a Flagpole** The length of a shadow and the length of the object casting the shadow are directly proportional. A boy 5 ft tall casts a 6-ft shadow. How high is the flagpole beside him that casts a 21-ft shadow? (See the figure.)

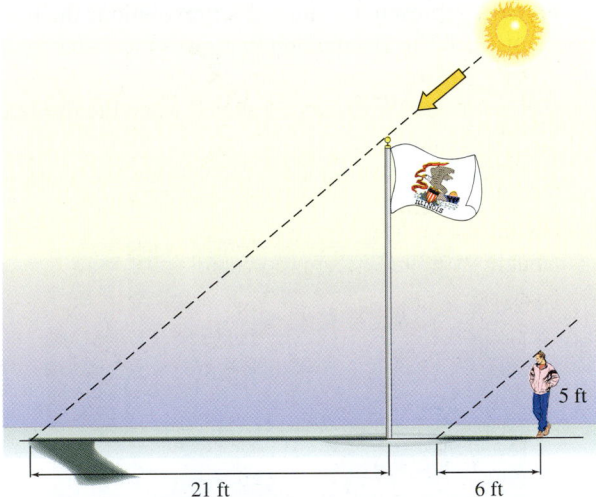

58. **Height of a Building** An architect who is designing a building is concerned with the shadow that will be created by this building during a certain time of the day. If a yardstick (3 ft) casts a 10-ft shadow, what height of building would cast a 400-ft shadow?

59. **Height of a Television Tower** The slope of a cable is the ratio of the rise to the run. A woman wished to determine the height of a television tower. The distance (or run) from the bottom of the cable to the bottom of the tower measured 40 ft. She then measured a rise of 8 ft and found a run of 6 ft. What is the height of the tower whose tip is at the top end of the cable?

60. **Scale Drawing of a House** The scale drawing of the house in the figure reveals that the roof rises 3 in over a run of 12 in. When the roof is actually built, how far will the roof rise over a run of 16 ft?

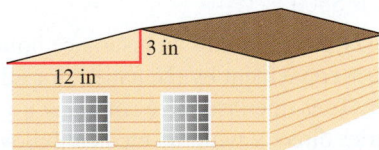

61. **Baseball ERA** A baseball pitcher's ERA (earned run average) is computed by determining the number of earned runs he allowed while recording 27 outs (one game). If a pitcher allowed two earned runs while recording only six outs in his first game, what would his ERA be?

62. Baseball ERA In his first major league game, a pitcher allowed five earned runs while recording nine outs. What was his ERA at the end of this game? (See Exercise 61.)

63. Distance Traveled by a Truck Tire The radius of a truck tire is 50 cm.
  a. To the nearest tenth of a centimeter, how far will the truck travel when the tire makes 1 revolution?
  b. To the nearest hundred thousandth of a kilometer, how far will the truck travel when the tire makes 1 revolution?
  c. Let x represent the number of revolutions the tire turns. Write an equation that gives the distance that the truck travels in kilometers.
  d. Determine the distance traveled when the tire makes 500 revolutions.
  e. Determine the number of revolutions necessary to travel 1 km.

64. Health Care Costs A patient has an insurance policy that requires a 10% copayment by the patient for all medical expenses. Thus, the patient's responsibility varies directly with the medical bill with a constant of variation of 0.10. Let x represent the dollar cost of a medical bill.
  a. Write an equation that gives the patient's responsibility for this medical bill.
  b. Determine the patient's responsibility for a $500 medical bill.
  c. Determine the medical bill if the patient's responsibility is $500.

65. Health Care Costs A patient has an insurance policy that requires a 20% copayment by the patient for all medical expenses. Thus, the insurance company's responsibility varies directly with the medical bill with a constant of variation of 0.80. Let x represent the dollar cost of a medical bill.
  a. Write an equation that gives the insurance company's responsibility for this patient.
  b. Determine the insurance company's responsibility for a $500 medical bill.

  c. Determine the medical bill if the insurance company's responsibility is $500.

In Exercises 66 and 67, y varies directly as x. Use the given table of values to determine the constant of variation.

66.

x	3	4	5	6	7
y	4.5	6	7.5	9	10.5

67.

x	3	4	5	6	7
y	2.4	3.2	4	4.8	5.6

Multiple Representations

In Exercises 68–71, represent each direct variation algebraically, numerically, verbally, or graphically as described in each problem. (*Hint:* See the following example.)

Algebraically	Numerically	Graphically	Verbally
$y = -2x$	<table><tr><td>x</td><td>y</td></tr><tr><td>1</td><td>-2</td></tr><tr><td>2</td><td>-4</td></tr><tr><td>3</td><td>-6</td></tr><tr><td>4</td><td>-8</td></tr><tr><td>5</td><td>-10</td></tr></table>	(graph of $y = -2x$)	y varies directly as x with constant of variation -2.

68. Represent the direct variation "y varies directly as x with constant of variation $\dfrac{1}{2}$" algebraically, numerically, and graphically.

69. Represent the direct variation $y = -\dfrac{x}{2}$ numerically, graphically, and verbally.

70. Represent the direct variation $y = 3x$ numerically, graphically, and verbally.

71. Represent the direct variation "y varies directly as x with constant of variation -1" algebraically, numerically, and graphically.

Connecting Algebra to Geometry

The lengths of corresponding sides of similar geometric figures are directly proportional.

72. The triangles shown in the figure are similar. Find the length of the side labeled a.

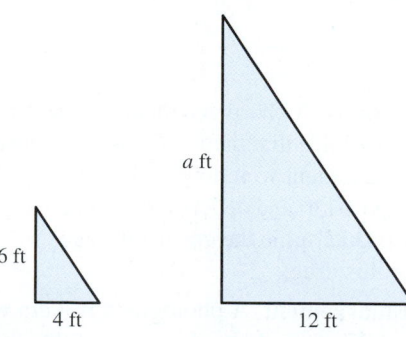

73. The two rectangles shown in the figure are similar. Find the length of the side labeled w.

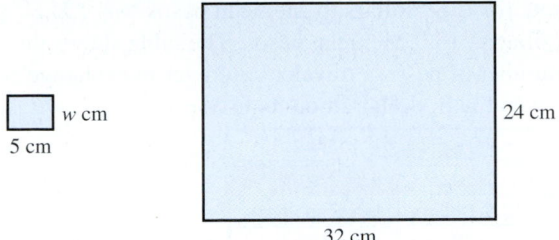

74. Use similar triangles and proportional sides to find the value of x.

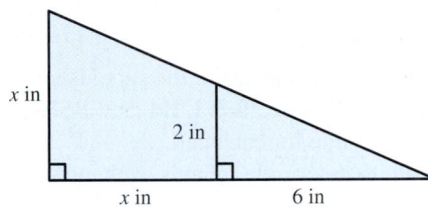

75. Use similar triangles and proportional sides to find the value of x.

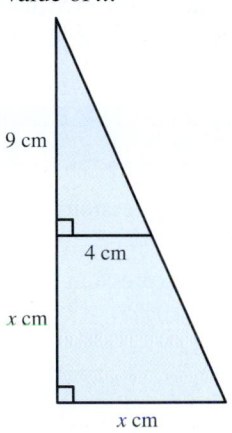

76. Similar Triangles Similar triangles have the same shape, and the corresponding angles have the same measure. The lengths of the corresponding sides are proportional. The sum of the measures of the interior angles of any triangle is 180°. Triangles ABC and $A'B'C'$ are similar.

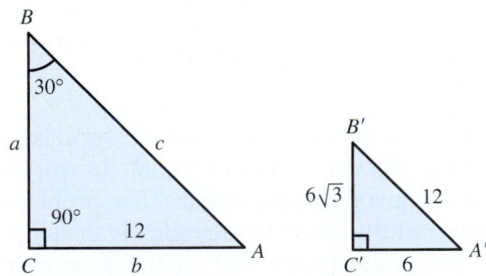

a. Determine the measure of angle A.
b. Determine the measure of angle A'.

c. Determine the measure of angle B'.
d. Determine the length of side c in triangle ABC.
e. Determine the length of side a in triangle ABC.

Group discussion questions

77. Solve Then Explain Use the proportion $\dfrac{x}{a} = \dfrac{b}{x}$ from the semicircle shown to answer the following question. An architect is planning a semicircular arch over a hallway as illustrated. The minimum height of the hallway is at the sides. The architect has set this measurement from building codes to be 8 ft. To design this arch, the architect has selected an imaginary point 4 ft from the bottom of the wall for this semicircle to pass through, if the arch were to continue.

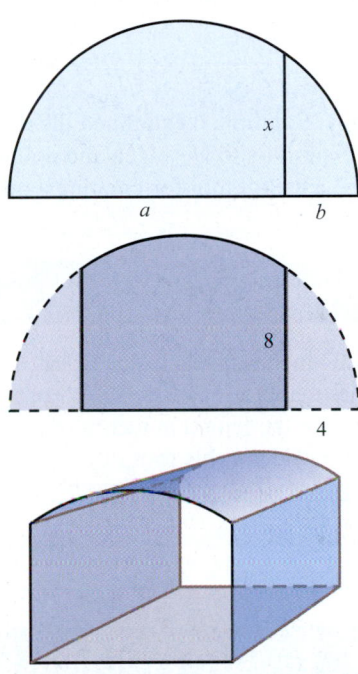

a. Determine the height of the hallway at the center.
b. Determine the width of this hallway.
c. Do these height and width values seem reasonable?
d. Prepare an explanation of the logic you used to solve this problem that you can present to other groups in your class.

78. Challenge Question Complete the equation
$$\frac{2}{m-4} = \frac{4}{2m-?}$$
so that it is a proportion for all values of $m \neq 4$.

79. Challenge Question An observer located at point *A* has a clear line of sight over level ground to points *B* and *C*. Describe a procedure using linear measurements that could be employed to determine indirectly the distance across the lake from point *B* to point *C*. (*Hint:* Use similar triangles.)

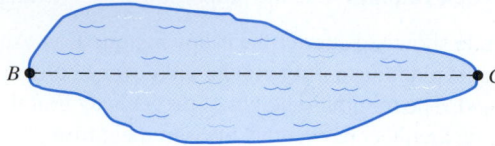

80. Applying Technology Example 6 examined the exchange rate of U.S. dollars to New Zealand dollars. A spreadsheet can be a useful tool for creating a quick reference sheet to use when making purchases in a foreign country. At one time in 2009, the exchange rate for U.S. dollars to Mexican pesos was 1 U.S. dollar = 15.0 Mexican pesos. The table shows the number of pesos a traveler could get in exchange for several U.S. dollar amounts in Mexico.

B2		f_x	=15*A2
	A	B	
1	U.S. Dollars	Mexican Pesos	
2	10	150	
3	20		
4	50		
5	100		
6	500		

a. Complete the missing entries in the spreadsheet by writing the formula for cells **B3**, **B4**, and **B5** and then giving the value for each cell.

b. Use a website to determine today's exchange rate for U.S. dollars to euros.

c. Then create a table similar to the one shown. You may want to consider creating the given table and using column *C* to display the conversions to euros.

2.7 | Cumulative Review

1. A square window is 24 inches on each side. It has a wood trim on all sides that is *x* inches wide. Write a function in terms of *x* for the length in inches of the perimeter around the outside of this trim.

2. Simplify $\sqrt{4} + \sqrt{4} + \sqrt{4} + \sqrt{4}$.

3. Simplify $\sqrt{4 + 4 + 4 + 4}$.

4. The price on an item is *P*. Write an expression for the new price if the price increased by 10%.

5. The price on an item is *P*. Write an expression for the new price if the price decreased by 10%.

Section 2.8 | More Applications of Linear Equations

Objectives:

1. Determine the restrictions on a variable in an application.

2. Use the mixture principle to create a mathematical model.

Are There Any General Principles or Strategies That I Can Use to Help Me Solve Word Problems?

YES. It is likely that almost all the mathematics you will encounter outside the classroom will be stated in words. Thus to profit from your algebra, you must be able to work with word problems. The applications in Section 2.7 involved proportions. The problems in this section use the mixture principle and are carefully selected to gradually develop and broaden your problem-solving skills. Some problems are selected for the same purpose as drill exercises on a piano or in basketball—to hone the skills needed for a real performance that will come later. One of the important goals of this book is to enable you to write equations that model real applications. An important consideration when we write an equation to model an application is the values that are practical to use for the input variable.

1. Determine the Restrictions on a Variable in an Application

What Is Meant by Restrictions on a Variable?

We use the phrase **restrictions on the variable** to refer to the values that are permissible in an application. In Example 1, we illustrate an application with the input values restricted to the whole numbers $\{0, 1, 2, 3, \ldots, 20\}$.

Example 1   Amount of Guttering Remaining on a Roll

A roofing company installs seamless guttering from a roll of aluminum 900 ft long. Working in a new subdivision, an installer cuts several 45-ft pieces of aluminum from the roll.

(a) Identify an input variable x so that we can write a function f so that $f(x)$ represents the length remaining on the roll.

(b) Write a function f so that $f(x)$ represents the length of aluminum remaining on the roll after the pieces have been cut from the roll.

(c) What restrictions should be placed on the variable x?

Solution

(a) Let $x =$ number of 45-ft pieces of aluminum cut from the roll
   $f(x) =$ length in feet remaining after x pieces are cut off

(b) Verbal Equation

   | Length remaining | = | Total length | − | Length cut off |

   Algebraic Equation

   $$f(x) = 900 - 45x$$

   The starting length is 900 ft, and 45 ft is used for each length cut off.

(c) The variable x must be restricted to the values $\{0, 1, 2, 3, \ldots, 20\}$.

   The number of pieces cut off cannot be less than 0 pieces. If 20 pieces of length 45 ft are cut from the roll, this will use all 900 ft. Thus x cannot exceed 20. The value of x must be one of the following whole numbers: $\{0, 1, 2, \ldots, 20\}$.

Self-Check 1

Rework Example 1 assuming the installer cuts several 50-ft pieces from the 900-ft roll.

Are the Restrictions on a Variable Always Stated Explicitly?

In many applications, restrictions on the variable are implied from the given problem rather than stated explicitly. Example 2 illustrates some possibilities. These restrictions are important to recognize so that we do not present impractical solutions in the workplace. Knowing the restrictions on the variable can also help us to select an appropriate viewing window on a graphing calculator. We will illustrate this in Section 3.3.

Example 2 Identifying the Restrictions on a Variable

Identify the restrictions on the variable for each of these applications.

Solution

(a) The function $L(x) = 100 - 4x$ models the length of wire remaining on a 100-ft roll after an electrician has cut off four equal pieces of length x.

The value of x is any real number in the interval $(0, 25)$.

The length x that is cut off cannot be less than 0 ft. Also this length cannot be more than 25 ft because four pieces of 25 ft will use all the available wire. Thus x can be any real number between 0 and 25.

(b) The function $C(x) = 225x + 200$ models the total rental cost of an apartment for a 1-year lease, where x is the number of months after the lease started.

The value of x can be any of these integers: $\{0, 1, 2, \ldots, 11, 12\}$.

The input values of x can be only the discrete values 0, 1, 2, 3, 4, 5, 6, 7, 8, 9, 10, 11, or 12 because the rent is paid only once a month. The value 0 is included to represent the time when the lease is signed. Because the lease is only for 1 year, the x-values stop at 12. If the lease is extended, then a new model would apply and the values could be extended beyond 12.

(c) The function $D(x) = 55x$ models the distance in miles that a car can travel in 1 day averaging 55 mi/h. The input variable x is the number of hours the car travels.

The value of x is any real number in the interval $[0, 24]$.

The car could travel up to 24 hours in the day, but x cannot assume any negative value or any value more than 24 for this specific application.

Self-Check 2

a. The function $f(x) = 10 - 2x$ models the length of board after x pieces each 2 ft long have been cut from this board. Identify the restrictions on the variable for this application.

b. The function $f(x) = 10 - 2x$ models the length of board after two pieces each x ft long have been cut from this board. Identify the restrictions on the variable for this application.

A basic strategy for solving word problems is outlined in the following box. Until you become quite proficient in solving word problems, we suggest you use the steps in this box as a checklist. Learning how to break down a seemingly complicated problem into a series of manageable steps is an important lifetime skill that extends well beyond word problems. When you start the first step, read actively—not passively. Circle key words, underline key phrases, make sketches, or jot down key facts as you read.

A Mathematical Note

The problem-solving strategy given here and used throughout this book is not new or unique. George Polya (1887–1985) is well known for teaching problem-solving techniques. His best-selling book *How to Solve It* includes four steps to problem solving:

Step 1 Understand the problem.
Step 2 Devise a plan.
Step 3 Carry out the plan.
Step 4 Check back.

Compare these steps to those given in the box.

Strategy for Solving Word Problems

Step 1. Read the problem carefully to determine what you are being asked to find.

Step 2. Select a variable to represent each unknown quantity. Specify precisely what each variable represents and note any restrictions on each variable.

Step 3. If necessary, make a sketch and translate the problem into a word equation or a system of word equations. Then translate each word equation into an algebraic equation.

Step 4. Solve the equation or system of equations, and answer the question completely in the form of a sentence.

Step 5. Check the reasonableness of your answer.

2. Use the Mixture Principle to Create a Mathematical Model

The mixture principle is a great example of how one general mathematical principle can be used to form equations for seemingly unrelated problems. One of the main goals of this book is to help you see these connections. Mathematics can be difficult if you try to memorize it bit by bit; it is much easier when you understand how to use its general principles.

Mixture Principle for Two Ingredients:
Amount in first + Amount in second = Amount in mixture

Applications of the Mixture Principle

1. Amount of product A + Amount of product B = Total amount of mixture

2. Variable cost + Fixed cost = Total cost

3. Interest on bonds + Interest on CDs = Total interest

4. Distance by first plane + Distance by second plane = Total distance

5. Antifreeze in first solution + Antifreeze in second solution = Total amount of antifreeze

---

The overhead cost for a company includes such expenses as rent, salaries, and insurance. This cost does not change with the number of items produced and is commonly referred to as the fixed cost or overhead cost. The cost involved in making each unit of a product is commonly called the variable cost because this cost varies with the number of items produced. The variable cost is determined by multiplying the cost per item by the number of items produced.

Example 3   **Fixed and Variable Costs for a Company**

For July 2010, a snow cone business had a fixed monthly cost of $120 and a variable cost of $0.30 per snow cone. How many snow cones did it sell in July 2010 if the total cost was $405 for that month?

Solution

Define the Variable

Let x = number of snow cones sold in July 2010.

Restrictions on the Variable

The variable x must take on values from the set of whole numbers.

Word Equation

| Variable cost | + | Fixed cost | = | Total cost |

Algebraic Equation

$$0.30x + 120 = 405$$
$$0.30x = 285$$
$$x = 950$$

Answer: The business sold 950 snow cones in July 2010.

Assuming the business does not sell partial snow cones, the variable x is restricted to the whole numbers $\{0, 1, 2, 3, \dots \}$. Although there is clearly some upper limit to the actual number of snow cones that can be made, this upper limit is unknown, so an upper limit was not placed on this set of values.

The variable cost for the month is $0.30x$, the cost per item times the number of items. The fixed cost for the month is $120. The total cost for the month is $405.

This answer seems reasonable. Does it check?

Self-Check 3

Rework Example 3, assuming that the total cost for the month was $450. Practice using each of the steps illustrated in Example 3. Then write your answer as a full sentence.

The distance formula $D = R \cdot T$ allows us to calculate the distance an object has traveled by multiplying its rate by the time it has traveled. In Example 4 the distance formula is used to determine the distance flown by each plane.

Example 4   Time It Takes Two Planes to Fly a Given Distance

Two passenger airplanes depart from an airport simultaneously. One flies east at 300 mi/h, and the other flies west at 450 mi/h. Determine how long it will take the planes to be 2,250 mi apart. (Fuel capacity limits the flight time to a maximum of 5 hours for each plane.)

Solution

Define the Variable

Let t = time in hours each plane has been flying.

Restrictions on the Variable

The variable t is restricted to the set of real numbers $[0, 5]$.

Word Equation

$$\boxed{\begin{array}{c}\text{Distance flown} \\ \text{by first plane}\end{array}} + \boxed{\begin{array}{c}\text{Distance flown} \\ \text{by second plane}\end{array}} = \boxed{\begin{array}{c}\text{Total distance} \\ \text{between planes}\end{array}}$$

The planes can fly from 0 to their 5-hour maximum. In many cases it is not clear exactly what the upper limit of the variable may be, but in most physical applications there is some upper limit.

Note the verbal equation is based on the mixture principle.

Algebraic Equation

$$300t + 450t = 2{,}250$$
$$750t = 2{,}250$$
$$t = 3$$

If we use $D = RT$ and a rate of 300 mi/h, the distance flown by the eastbound plane after t hours is $300t$. If we use $D = RT$ and a rate of 450 mi/h, the distance flown by the westbound plane after t hours is $450t$.

Answer: After 3 hours, the two planes will be 2,250 mi apart.

This answer seems reasonable and it checks. In 3 hours, one plane flies $3(300) = 900$ mi and the other plane flies $3(450) = 1{,}350$ mi for a total of 2,250 mi.

Self-Check 4

Rework Example 4 assuming the plane flying east travels 225 mi/h and the plane flying west travels 275mi/h. Assume the same 5-hour time limit on fuel.

It is very important to check answers to be sure they meet the restrictions on the variable. If the planes in Example 4 were far enough apart to require 5.5 hours of flight, this would exceed the fuel capacity of the planes and this would not be an acceptable answer for a nonstop flight. Another situation in which it is important to check the restrictions on the variable is given in Example 5. The answer to the algebraic equation will check in the equation, but it is not practical in this application.

Example 5   Amount of a Pesticide in a Solution

A farmer needs to apply a pesticide mixture that is 0.5% pesticide. From an earlier application, she has a 500-gallon tank that is partially filled with 350 gallons of a mixture that is 0.3% pesticide. How much of a previously mixed 0.9% pesticide solution does she need to add to the 500-gallon tank to have the desired concentration? Does the tank have this much capacity?

Solution

Define the Variable

Let x = the number of gallons of 0.9% pesticide solution to be added to the tank.

Restrictions on the Variable

The variable x could take on any real number from 0 to 150.

The smallest amount of solution the farmer could add is 0 gallons. She could add up to 150 gallons—the amount of empty space remaining in the tank.

Word Equation

| Pesticide in existing mixture | + | Pesticide in added mixture | = | Total pesticide in desired mixture |

Note the word equation is based on the mixture principle.

Algebraic Equation

$$0.003(350) + 0.009(x) = 0.005(350 + x)$$
$$1.05 + 0.009x = 1.75 + 0.005x$$
$$0.004x = 0.70$$
$$x = 175$$

The volume of pesticide in 350 gallons of 0.3% pesticide solution is 0.003(350), the product of the concentration and the volume of mixture.

Answer: The farmer would need to add 175 gallons of the 0.9% mixture to increase the mixture to 0.5% insecticide. This is not possible because there is not enough capacity in the 500-gallon tank.

The volume of pesticide added is 0.009x. Note that the new volume of the mixture would be $(350 + x)$ gallons. The volume of pesticide in $(350 + x)$ gallons of 0.5% pesticide solution is $0.005(350 + x)$.

Self-Check 5

A farmer needs to apply a pesticide mixture that is 0.4% pesticide. From an earlier application, she has a 400-gallon tank that is partially filled with 150 gallons of a mixture that is 0.2% pesticide. How much of a previously mixed 1% pesticide solution does she need to add to the 400-gallon tank to have the desired concentration?

Some mixture applications that appear much like Example 5 have an added twist. The solution that is added contains either 0% or 100% of the active ingredient. In Example 6 the solution that is added contains only water and therefore contains 0% pesticide.

Example 6   Amount of a Pesticide in a Solution

A farmer needs to apply a pesticide mixture that is 0.2% pesticide. From an earlier application, she has 200 gal of a more concentrated mixture that is 0.6% pesticide in a 600-gal tank. Rather than discard the existing mixture, she decides to dilute it with water. How much water does she need to add to the 600-gal tank to have the desired concentration? Does she have enough capacity in this tank?

Solution

Define the Variable

Let x = number of gallons of water that should be added to tank.

Restrictions on the Variable

The variable x could take on any real number from 0 to 400.

The smallest amount of water the farmer could add is 0 gal. She could add up to 400 gal of water—the amount of empty space remaining in the tank.

Word Equation

| Pesticide in existing mixture | + | Pesticide in water added | = | Total pesticide in desired mixture |

Note the word equation is based on the mixture principle.

Algebraic Equation

$$0.006(200) + 0.0(x) = 0.002(200 + x)$$
$$1.2 = 0.4 + 0.002x$$
$$0.8 = 0.002x$$
$$400 = x$$
$$x = 400$$

The volume of pesticide in 200 gal of 0.6% pesticide solution is 0.006(200), the product of the concentration and the volume of the mixture. There is 0.0% pesticide in the water added, so the volume of pesticide in x gal of water is 0.0(x). Note that the new volume of the mixture would be $(200 + x)$ gal. The volume of pesticide in $(200 + x)$ gal of 0.2% pesticide solution is $0.002(200 + x)$.

Answer: The farmer would need to add 400 gal of water to dilute the mixture to 0.2% insecticide. She has just enough capacity in the tank.

Self-Check 6

A chemist needs to decrease the concentration of a salt solution. She has 4 liters (L) of a 30% salt solution in a 6-L container. How much water should she add to the container to decrease the concentration to 15%?

The interest formula $I = PRT$ can be used to calculate the interest earned by an investment P at an annual rate R for a time of T years. In Example 7, we use this formula to determine the income from two different investments.

Example 7   Interest Earned on an Investment

A young couple had savings of $4,800. They invested $1,440 with an insurance company that produced a 2% annual income. A portion of the remaining amount was invested in a CD that paid 5%. If the total of these two investments resulted in a 4% income on the total of the two investments, how much was invested in the CD?

Solution

Define the Variable

Let x = number of dollars invested in the CD.

Restrictions on the Variable

The variable x may take on values from $0 to $3,360.

The minimum they could invest in the CD is $0. The maximum they could invest is $3,360, the amount remaining from $4,800 after the insurance investment of $1,440.

Word Equation

| Interest from insurance | + | Interest from CD investment | = | Total interest earned |

Note the word equation is based on the mixture principle.

Algebraic Equation

$$0.02(1,440) + 0.05(x) = 0.04(1,440 + x)$$
$$28.8 + 0.05x = 57.6 + 0.04x$$
$$0.01x = 28.8$$
$$x = 2,880$$

The interest earned from $1,440 invested at 2% is 0.02(1,440), the product of the interest rate and the amount invested at that rate. The interest earned from x dollars invested at 5% is 0.05(x). A combined return of 4% on the two is represented by 0.04(1,440 + x).

Answer: $2,880 was invested in the CD.

Note the value of $2,880 falls within the stated restrictions of $0 to $3,360.

Self-Check 7

A retiree receives a yearly income of $20,000 from her IRA to help fund her retirement. She has placed $40,000 of this IRA in a secure Treasury bond earning 5% yearly interest. If she earns an 8% rate of return on the rest of this IRA through an investment in bank stocks, how much of the IRA is invested in bank stocks?

As you solve problems throughout this book, try to recognize how problems share general principles of mathematics. Students who recognize general principles can profit from their mathematics more than those who memorize individual problems.

Self-Check Answers

1. **a.** Let x = the number of 50-ft pieces of aluminum cut from the roll
   $f(x)$ = length in feet remaining after x pieces are cut off
   b. $f(x) = 900 - 50x$
   c. The variable x must be restricted to the values $\{0, 1, 2, 3, \ldots, 18\}$.
2. **a.** $\{0, 1, 2, 3, 4, 5\}$
   b. $(0, 5)$

3. The business sold 1,100 snow cones in July 2010.
4. After 4.5 hours, the planes will be 2,250 mi apart.
5. The farmer would need to add 50 gallons of the 1% pesticide solution.
6. There is not enough room in the 6-L container to add 4 L of water.
7. $225,000 is invested in bank stocks.

2.8 | Using the Language and Symbolism of Mathematics

1. We use the phrase _____ *on the variable* to refer to the values of the variable that are permissible in an application.

2. The first step in the word problem strategy given in this book is to read the problem carefully to determine what you are being asked to _____.

3. The second step in the word problem strategy given in this book is to select a _____ to represent each unknown quantity.

4. The third step in the word problem strategy given in this book is to model the problem verbally with word equations and then to translate these word equations into _____ equations.

5. The fourth step in the word problem strategy given in this book is to _____ the equation or system of equations and answer the question completely in the form of a sentence.

6. The fifth step in the word problem strategy given in this book is to check your answer to make sure the answer is _____.

7. The _____ principle states that the amount obtained by combining two parts is equal to the amount obtained from the first part plus the amount obtained from the second part.

8. In the formula $D = RT$, D represents distance, R represents _____, and T represents _____.

9. In the formula $I = PRT$, I represents interest, P represents _____, R represents _____, and T represents _____.

10. The cost for a company that does not change with the number of units produced is called the _____ cost or the overhead cost.

11. The cost for a company that depends on how many units the company produces is called the _____ cost.

2.8 | Quick Review

1. Determine the distance a plane will fly if it flies at a rate of 250 mi/h for 3 hours.

2. Solve $1.50x + 3.25x = 323$.

Identify each equation as a conditional equation, a contradiction, or an identity. Where possible, give the solution of each equation.

3. $3x - 6 = 2x + 7$ 4. $3x - 6 = 3(x - 2)$ 5. $3x - 6 = 3(x - 4)$

2.8 | Exercises

Objective 1 Determine the Restrictions on a Variable in an Application

In Exercises 1–4, match each problem with the restricted values for the variable in this application.

A. $\{0, 1, 2, 3, \ldots, 30\}$ **B.** $\{0, 1, 2, 3, \ldots, 10\}$
C. $(0, 10]$ **D.** $[0, 30]$

1. The function $f(x) = 500 - 50x$ models the length of optical fiber remaining on a spool after x pieces each 50 ft long have been removed from this spool.

2. The function $f(x) = 500 - 50x$ models the length of optical fiber remaining on a spool after 50 pieces each of length x have been removed from this spool.

3. The function $f(x) = 0.05(70) + 0.08x$ models the amount of acid in a 100-gal tank that has x gal of an 8% acidic solution poured into a tank that already contains 70 gal of a 5% acidic solution.

4. The function $f(x) = 100x + 250$ models the cost in dollars to lease a highway messaging system for x days during the month of June.

In Exercises 5–8, determine the restrictions on the variable for each application.

5. **Quick Pass Account** The function $V(x) = 100 - 2x$ models the number of dollars remaining on a $100 toll road Quick Pass account. Each trip on this toll road costs $2, and x represents the number of trips on this toll road.

6. **Cost of Gasoline** The function $C(x) = 3.10x$ models the total cost of filling a 20-gal tank with x gal of gasoline costing $3.10 per gallon.

7. **Acidic Solution** The function $f(x) = 0.07(175) + 0.12x$ models the number of gallons of acid in a 400-gal tank. The 400-gal tank originally contained 175 gal of a 7% acidic solution, and then x gal of a stronger solution was added to the tank.

8. **Cost of Renting a Tent** The function $C(x) = 250x$ models the cost of renting a tent for x days in one week. The rental company will not rent the tent for partial days.

In Exercises 9 and 10, solve each problem by completing each of the given steps.

9. **Width of a Brake Pad** The brake pad on a new truck is 11.2 mm thick. The maintenance shop that services this truck estimates that about 0.2 mm will wear from the pad each month. How many months can the truck be used before the thickness of the brake pad is reduced to 4.0 mm?
   a. Select a variable to represent the unknown quantity, and identify this variable.
   b. Write a word equation that models this problem.
   c. Translate this word equation into an algebraic equation.
   d. Solve this equation.
   e. Is this solution within the restrictions on the variable, and does it seem reasonable?
   f. Write a sentence that answers the problem.

10. **Depth of a Tire Tread** The depth of the tread on a new tire on a taxi is 9.8 mm. It is estimated that the tire will wear about 0.3 mm per month. Determine the number of months until the depth of the tread on these tires is 2.6 mm.
    a. Select a variable to represent the unknown quantity and identify this variable.
    b. Write a word equation that models this problem.
    c. Translate this word equation into an algebraic equation.
    d. Solve this equation.
    e. Is this solution within the restrictions on the variable, and does it seem reasonable?
    f. Write a sentence that answers the problem.

Objective 2 Use the Mixture Principle to Create a Mathematical Model

In Exercises 11–30, solve each problem by using each step of the word problem strategy outlined in this section.

11. **Telephone Costs** A long-distance phone package charges a fixed amount of $24.95 per month plus a variable charge of $0.05 per minute. How many minutes were billed if the long-distance phone charges were $42.20 for a month of service?

12. **Taxi Costs** The cost of a taxi ride involves a fixed cost of $3.00 and a variable cost of $4.00 per mile. If a taxi fare was $25.40, how many miles did the taxi cover in this trip?

13. **Fixed and Variable Costs of Production** The manufacturer of custom sofa cushions has fixed daily costs of $150 and a variable cost of $6 per cushion. Determine how many cushions the manufacturer needs to make per day to meet the goal of an average cost of $9 per cushion. Note that the total cost can also be calculated by multiplying the average cost per cushion by the number of cushions produced.

14. **Fixed and Variable Costs of Production** The manufacturer of custom bricks places names of individuals on donor bricks sold as fund-raisers. The manufacturer has fixed daily costs of $180 and a variable cost of $1.50 per brick. Determine how many bricks the company needs to make per day to meet its goal of an average cost of $2 per brick. Note that the total cost can also be calculated by multiplying the average cost per brick by the number of bricks produced.

15. **Income Earned on an IRA Investment** A retiree needs a yearly income of $10,200 from his $150,000 IRA to help fund his retirement. He has placed $60,000 of this account in a secure Treasury bond earning 5% yearly interest. What rate of return must he earn on the rest of this investment to reach his $10,200 income goal?

16. **Income Earned on an IRA Investment** A retiree receives a yearly income of $13,000 from her IRA to help fund her retirement. She has placed $50,000 of this IRA in a secure Treasury bond earning 5% yearly interest. If she earns a 7% rate of return on the rest of this IRA through an insurance annuity, what is the principal that is invested in the insurance annuity?

17. **Income Earned on an Educational Account** A student receives a yearly income of $4,000 from an account set up by her grandparents to help fund her education. From this educational account $10,000 is invested in a secure Treasury bond earning 5.5% yearly interest. If she earns a 6% rate of return on the rest of this educational account through a corporate bond, what is the principal that is invested in the corporate bond?

18. **Income Earned on an Educational Account** A student receives a yearly income of $4,900 from a $70,000 account set up by his grandparents to help fund his education. From this educational account $20,000 is invested in a secure Treasury bond earning 4.5% yearly interest. What rate of return must he earn on the rest of this investment to produce the yearly income of $4,900?

19. **Travel Time** Two boats leave a dock at the same time. One travels downstream at a rate of 12 mi/h, and the other travels upstream 4 mi/h slower than the first boat. Determine the number of hours it will take the boats to be 64 mi apart.

20. **Flight Time** At the time a refueling request is made by a jet pilot, his plane is 175 mi from a refueling tanker. The two planes head toward each other so the jet can refuel. The jet flies 450 mi/h, and the tanker flies 200 mi/h slower. How long will it take the two to meet?

21. **Travel Time** A car and a truck simultaneously leave two towns 364 mi apart and head toward each other. The car travels at a rate of 70 mi/h, and the truck travels 10 mi/h slower. How long will it take the two to meet?

22. Exercise Time Two friends go to a park to exercise. One starts walking at a rate of 4 mi/h on the 7.5-mi path around the park, and the other starts from the same point, jogging at a rate of 6 mi/h in the opposite direction on this path. How many minutes will it be before they meet on this path?

23. Amount of a Pesticide in a Solution A crop-dusting service needs to prepare a 0.8% pesticide solution. It has an airplane with a 300-gal tank that contains 100 gal of a 0.5% solution. How much of a previously mixed 1% pesticide solution should be added to the tank to increase the concentration to the desired 0.8% level?

24. Amount of a Pesticide in a Solution A crop-dusting service needs to prepare a 0.6% pesticide solution. It has an airplane with a 200-gal tank that contains 80 gal of a 0.4% solution. How much of a previously mixed 1% pesticide solution should be added to the tank to increase the concentration to the desired 0.6% level?

25. Mixture of Gasohol A gasoline distributor has a 20,000-gal tank to store a gasohol mix. The tank already contains 12,000 gal of gasohol that contains 4% ethanol by volume. How many gallons of pure ethanol must be added to the tank to produce a mix that is 10% ethanol?

26. Mixture of Gasohol A gasoline distributor has a 15,000-gal tank to store a gasohol mix. The tank already contains 9,000 gal of gasohol that contains 6% ethanol by volume. How many gallons of pure ethanol must be added to the tank to produce a mix that is 10% ethanol?

27. Amount of a Pesticide in a Solution A crop-dusting service needs to prepare a 0.4% pesticide solution. It has an airplane with a 200-gal tank that contains 75 gal of a 1% solution. How much water should be added to the tank to decrease the concentration to the desired 0.4% level?

28. Amount of Acid in a Solution A chemist needs to create a 20% acid solution. She has 4 L of a 30% acid solution in a 20-L container. How much water should she mix with this solution to decrease the concentration to the desired 20%?

29. Amount of Sand in a Concrete Mix A company that mixes and sells concrete has a contract to deliver several loads of concrete containing an aggregate that is 20% sand by weight. The company has a pile that contains 360,000 lb of aggregate that is 15% sand. How much sand will be required to mix with the aggregate in this pile to produce a mixture that is 20% sand?

30. Amount of Lean Beef in Hamburger The meat department in a grocery store purchased 800 lb of hamburger that is 90% lean beef. How many pounds of hamburger that is 75% lean must be mixed with this 90% lean mix to produce hamburger that is 85% lean beef?

Skill and Concept Development

31. Electricity Generated by Windmills One growing community in California currently uses 2 megawatts (MW) of electricity per year. All this electricity is generated from coal. The community plans to add x MW of windmill generation capacity to its system to meet increasing demand. How many megawatts of windmill generation capacity must be built in order that 20% of the community's total electricity usage comes from windmills?

32. Electricity Generated by Windmills One small island community in the Pacific currently uses 1.8 megawatts (MW) of electricity per year. All this electricity is generated from plants fired by natural gas. The island community plans to add x MW of offshore windmill generation capacity to its system. How many megawatts of windmill generation capacity must be built in order that 25% of the island community's total electricity usage comes from windmills?

In Exercises 33–38, use the given information to solve each part of the problem. (*Hint:* Some of your proposed answers may not meet the restrictions placed on the variable.)

33. Cost of Manufacturing Ceramic Tile A manufacturer of custom ceramic tiles has the capacity to produce 8,000 tiles per day. It has a fixed daily cost of $2,000 plus a variable cost of $0.50 per tile.
   a. How many tiles were created in a batch that had a total daily cost of $5,750?
   b. The total cost of the ceramic tiles can also be calculated by multiplying the average cost per tile by the number of tiles produced. If the company needs to reach an average cost of $0.70 per tile, how many tiles must it produce per day?

34. Cost of Manufacturing Mugs A manufacturer of custom drinking mugs has the capacity to produce 1,800 mugs per day. It has a fixed daily cost of $1,000 plus a variable cost of $1.50 per mug.
   a. How many mugs were created in a batch that had a total daily cost of $2,800?
   b. The total cost of the mugs can also be calculated by multiplying the average cost per mug by the number of mugs produced. If the company needs to reach an average cost of $2.00 per mug, how many mugs must it produce per day?

35. Mixture of Gasohol A gasoline distributor has a 20,000-gal tank to store a gasohol mix. The tank already contains 19,000 gal of gasohol that contains 4% ethanol by volume.

  a. How many gallons of ethanol must be added to the tank to produce a mix that is 8% ethanol? (Round to the nearest gallon.)

  b. How many gallons of ethanol must be added to the tank to produce a mix that is 10% ethanol? (Round to the nearest gallon.)

36. Amount of a Pesticide in a Solution A crop-dusting service needs to decrease the concentration of the pesticide in a tank. It has an airplane with a 200-gal tank that contains 80 gal of a 1% solution.

  a. How much water should be added to the tank to decrease the concentration to 0.3%?

  b. How much water should be added to the tank to decrease the concentration to 0.4%?

37. Amount of a Pesticide in a Solution A crop-dusting service needs to increase the concentration of the pesticide in a tank. It has an airplane with a 350-gal tank that contains 80 gal of a 0.5% solution.

  a. How much of a previously mixed 1% pesticide solution should be added to the tank to increase the concentration to 0.9%?

  b. How much of a previously mixed 1% pesticide solution should be added to the tank to increase the concentration to 0.8%?

38. Amount of Acid in a Solution A chemist needs to decrease the concentration of an acid solution. She has 5 L of a 25% acid solution on hand. She will store the resulting solution in a 10-L container.

  a. How much water should she add to the mixture to decrease the concentration to 20%?

  b. How much water should she add to the mixture to decrease the concentration to 10%?

39. Cost of Grass Seed A landscaping company has two popular types of grass seed mixes for its area, a premium mix that sells for $1.25 per pound and a standard mix that sells for $0.95 per pound.

  a. If the store manager dumps a 20-lb bag of each mix into a large barrel, what is the total value of the grass seed in the barrel?

  b. What price per pound should the company sell this new mix for to produce the same income as selling the two mixes separately?

  c. If the store manager started with 20 lb of the premium mix in a large barrel, how many pounds of the standard mix should be added to create a mix that could be sold for $1.00 per pound?

  d. How many pounds of grass seed must the barrel hold in order to contain all the mixture created in part **c**?

  e. How many pounds of the standard mix should be put in a barrel with 20 lb of the premium mix to create a mix worth $1.05 per pound?

40. Cost of Paving Bricks A construction company used two different types of paving bricks to create a design around a water fountain. The company charges $1.50 each for the rectangular bricks and $2.00 each for the hexagonal bricks. These bricks are stored on pallets with each pallet of the rectangular bricks containing 144 bricks and each pallet of the hexagonal bricks containing 120 bricks.

  a. If a forklift operator places one pallet of each type of bricks on a truck, what is the total value of the bricks on this truck?

  b. What is the average price per brick for all the bricks on this load?

  c. If the design for the fountain calls for 2,000 hexagonal bricks, how many rectangular bricks should be added to create a mix that costs an average of $1.60 per brick?

  d. If the design for the fountain calls for 2,000 rectangular bricks, how many hexagonal bricks should be added to create a mix that costs an average of $1.60 per brick?

  e. If the design for the fountain calls for 2,000 hexagonal bricks and the budget for the cost of the bricks will allow for an expense of $9,250, how many rectangular bricks can be used?

Group discussion questions

41. Challenge Question A farmer raising watermelons for a contest produced a 200-lb watermelon that was 98% water. After the watermelon was moved, weighed, and left in the sun, some of the moisture evaporated. The result was that the watermelon was now only 95% water. What is the new weight of the watermelon?

42. Challenge Question The antifreeze mixture in a car radiator in Michigan should be kept between 40% and 60%. The cooling system of a car holds 6 L of a 40% antifreeze mixture. To raise the antifreeze concentration to 60%, how much of this mixture should be drained from the radiator and replaced with pure antifreeze?

43. Communicating Mathematically Write a word problem that can be solved by using the equation $0.06(4,000) + 0.10x = 0.07(4,000 + x)$ and that models
   a. The income earned from an investment.
   b. The amount of ethanol in a gasohol mixture.

44. Communicating Mathematically Write a word problem that can be solved by using the equation $50,000 - 8,500x = 16,000$ and that models
   a. The number of meters of transmission cable on an installation truck.
   b. The number of pounds of jet fuel on an airplane.

2.8 | Cumulative Review

1. A polygon with equal sides and equal angles is called a _____ polygon.

2. A hexagon has _____ sides.

3. A trapezoid has _____ sides.

4. If the sum of the measures of two angles is $90°$, the angles are _____.

5. If two parallel lines are cut by a transversal, then two angles that are on opposite sides of the transversal and between the parallel lines are called _____ _____ angles.

Chapter 2 | Key Concepts

1. Cartesian Coordinate System: The sign pattern for the coordinates in each quadrant of the rectangular coordinate system is shown in the figure.

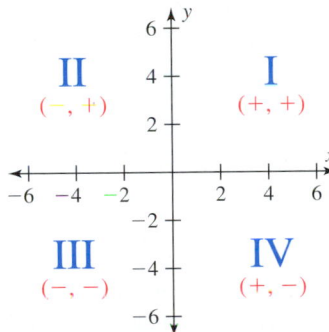

2. Scatter Diagram: A scatter diagram for a set of data points is a graph of these points that can be used to examine the data for some type of visual pattern.

3. Arithmetic Sequences and Linear Equations
   - A sequence is an ordered set of numbers.
   - An arithmetic sequence is a sequence with a constant change from term to term.
   - The graph of the terms of an arithmetic sequence consists of discrete points that lie on a straight line.

4. Function Notation: The function notation $f(x)$ is read "f of x." $f(x)$ represents a unique output value for each input value of x.

5. Solution of a Linear Equation: A solution of a linear equation of the form $y = mx + b$ is an ordered pair (x, y) that makes the equation a true statement.

6. Intercepts of a Line
   - The x-intercept, often denoted by $(a, 0)$, is the point where the line crosses the x-axis.
   - The y-intercept, often denoted by $(0, b)$, is the point where the line crosses the y-axis.

7. Graphing a Line
   - Create a table of values and graph the line through all points in the table.
   - Select two input values and determine their corresponding output values. Then graph the line through these two points. Plotting a point from a third input value can be used as a check.
   - Calculate the x- and y-intercepts, and graph the line through the two intercepts.

8. Solution of a System of Linear Equations
   - A solution of a system of linear equations is an ordered pair that satisfies each equation in the system.
   - A solution of a system of linear equations is represented graphically by a point that is on both lines in the system.
   - A solution is represented in a table of values by an x-value for which both y-values are the same.

9. Linear Equation in One Variable: A linear equation in one variable x is an equation that can be written in the form $Ax = B$, where A and B are real constants and $A \neq 0$

10. Equivalent Equations: Equivalent equations have the same solution set.

11. Addition-Subtraction Principle of Equality: If the same number is added to or subtracted from both sides of an equation, the result is an equivalent equation.
   - $a = b$ is equivalent to $a + c = b + c$.
   - $a = b$ is equivalent to $a - c = b - c$.

12. Multiplication-Division Principle of Equality: If both sides of an equation are multiplied or divided by the same nonzero number, the result is an equivalent equation.
   - $a = b$ is equivalent to $ac = bc$ for $c \neq 0$.
   - $a = b$ is equivalent to $\dfrac{a}{c} = \dfrac{b}{c}$ for $c \neq 0$.

13. Classification of Equations
- **Conditional Equation:** A conditional equation is true for some values of the variable and false for other values.
- **Contradiction:** An equation that is false for all values of the variable is a contradiction.
- **Identity:** An equation that is true for all values of the variable is an identity.

14. Strategy for Solving Linear Equations Algebraically

Step 1 Simplify each side of the equation.
- **a.** If the equation contains fractions, simplify by multiplying both sides of the equation by the least common denominator (LCD) of all the fractions.
- **b.** If the equation contains grouping symbols, simplify by using the distributive property to remove the grouping symbols and then combine like terms.

Step 2 Using the addition-subtraction principle of equality, isolate the variable terms on one side of the equation and the constant terms on the other side.

Step 3 Using the multiplication-division principle of equality, solve the equation produced in step 2.

15. Solving Linear Equations by Using Tables: Enter the left side of the equation as Y_1 and the right side of the equation as Y_2. Then select x-values and form a table displaying the values for x, Y_1, and Y_2. The x-coordinate that causes Y_1 to equal Y_2 is the solution of the original linear equation in x. (*Caution:* This value of x may not appear on the first table that you select.)

16. Solving Equations by Using Graphs: Enter the left side of the equation as Y_1 and the right side as Y_2. Then graph the equations corresponding to Y_1 and Y_2. The x-coordinate at the point of intersection of these two lines is the solution of the original linear equation in x.

17. Solving for a Specified Variable: To solve for a specified variable, isolate this variable on one side of the equation with all other variables on the other side of the equation.

18. Proportions
- A proportion is an equation that states two ratios are equal.
- In the proportion $\frac{a}{b} = \frac{c}{d}$, the terms a and d are called the *extremes* and the terms b and c are called the *means*.

19. Direct Variation
- If x and y are variables and k is a real constant with $k \neq 0$, then stating "y varies directly as x with constant of variation k" means $y = kx$.
- If y varies directly as x, then x and y are directly proportional. That is, $\frac{y_1}{x_1} = \frac{y_2}{x_2}$.
- If y varies directly as x, then the graph of the (x, y) pairs will lie on a straight line.

20. Restrictions on a Variable: The restrictions on a variable refer to the values that are permissible in an application.

21. Strategy for Solving Word Problems

Step 1 Read the problem carefully to determine what you are being asked to find.

Step 2 Select a variable to represent each unknown quantity. Specify precisely what each variable represents, and note any restrictions on each variable.

Step 3 If necessary, make a sketch and translate the problem into a word equation or a system of word equations. Then translate each word equation into an algebraic equation.

Step 4 Solve the equation or system of equations, and answer the question completely in the form of a sentence.

Step 5 Check the reasonableness of your answer.

22. Mixture Principle:

Amount in first + Amount in second = Amount in mixture

Chapter 2 | Review Exercises

The Rectangular Coordinate System

1. Identify the coordinates of the points A through D in the figure, and give the quadrant in which each point is located.

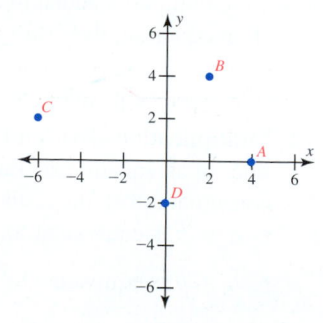

2. Draw a scatter diagram for the data points in the given table.

x	y
-3	-2
-2	0
-1	2
0	1
2	-1

3. a. Write the arithmetic sequence 1, −1, −3, −5, −7 using ordered pairs.
  b. Graph the five ordered pairs from part **a.**

In Exercises 4–7, determine whether each sequence is arithmetic and whether the graph of its points forms a linear pattern.

4. 4, 7, 10, 13, 16

5. 0, 1, 3, 6, 10

6.

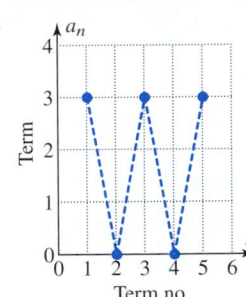

7.

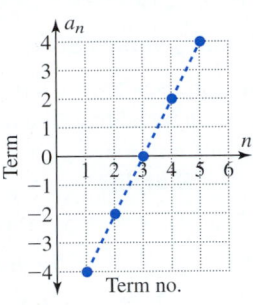

8. Water Elevation at Lake Powell The depth of water at Lake Powell is dependent on the snow runoff from the Colorado Rockies. The results from recent years are displayed in the line graph. Use this graph to answer the following questions.
  a. What was the elevation of the water in 2009?
  b. Did the water level increase or decrease from 2004 to 2006?
  c. What year from 2000 to 2009 was the water level the lowest?

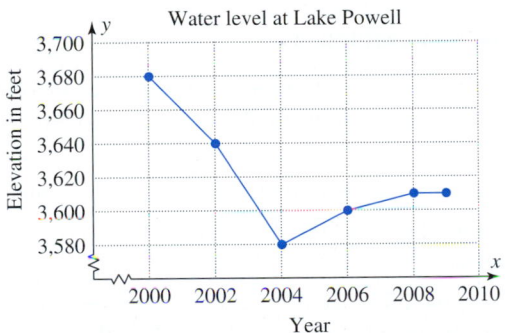

Linear Equations in Two Variables

In Exercises 9–12, determine whether either $(3, -2)$ or $(-3, 2)$ is a solution of each equation.

9. $y - 3 = 4(x + 2)$

10. $y = x + 5$

11.

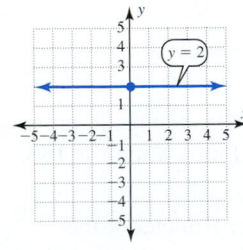

12.

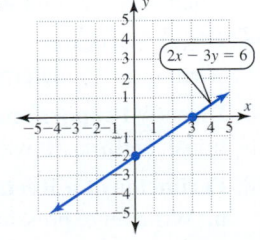

13. Determine the x- and y-intercepts of the line in Exercise 12.

14. Determine the x- and y-intercepts of $y = x + 5$ and graph this line.

15. Determine the x- and y-intercepts of $y - 3 = 4(x + 2)$ and graph this line.

In Exercises 16–19, match each exercise with the most appropriate choice.

16. $5x + 3 = 2(x + 4)$

17. $5x^2 = 3$

18. $y = 2x + 4$

19. $5x + 3 - 2(x + 4)$

A. A first-degree expression in x but not a linear equation

B. A linear equation in x

C. An equation that is not linear

D. A linear equation in two variables

In Exercises 20 and 21, determine the point of intersection of the two lines.

20.

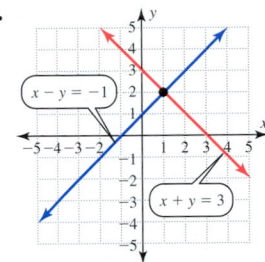

21.

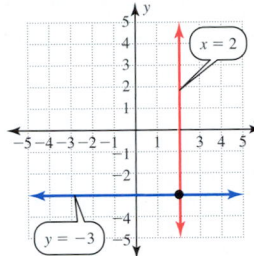

Solving Linear Equations in One Variable

In Exercises 22 and 23, check $x = 2$ and $x = 3$ as possible solutions of the given equation.

22. $3x + 5 = 4x + 2$

23. $3(x + 5) = 4(x + 3) + 1$

In Exercises 24 and 25, Y_1 represents the left side of the equation and Y_2 represents the right side of the equation. Use the table shown from a graphing calculator to solve this equation.

24. $3.6x + 1.2 = 1.8x + 5.7$ **25.** $8.5x - 55 = 10.5x - 103$

x	Y_1	Y_2
0.0	1.2	5.7
0.5	3.0	6.6
1.0	4.8	7.5
1.5	6.6	8.4
2.0	8.4	9.3
2.5	10.2	10.2
3.0	12.0	11.1

x	Y_1	Y_2
20	115.0	107.0
21	123.5	117.5
22	132.0	128.0
23	140.5	138.5
24	149.0	149.0
25	157.5	159.5
26	166.0	170.0

In Exercises 26 and 27, use the given graph to solve each equation.

26. $-\dfrac{1}{3}x + 2 = \dfrac{2}{3}x - 1$ **27.** $0.6x + 0.3 = -0.4x - 0.7$

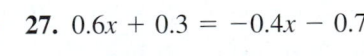

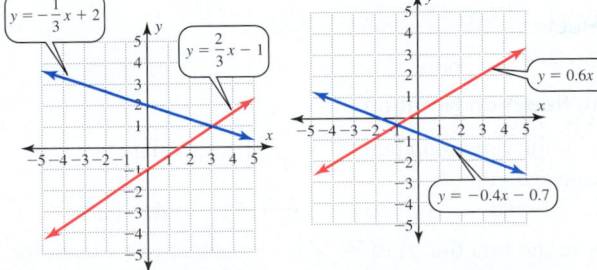

In Exercises 28–52, solve each equation.

28. $x + 2 = 17$

29. $5v + 7 = 4v + 9$

30. $7m = 6m$

31. $-x = 3$

32. $-17m = -34$

33. $\dfrac{n}{3} = -12$

34. $-\dfrac{2y}{7} = 28$

35. $0.045w = -0.18$

36. $117 = 0.01t$

37. $0 = -147b$

38. $5x - 8 = 27$

39. $-8z + 7 - 2z = 5z + 4 + z$

40. $-r - r - r - r - r = r$

41. $3(2x + 7) = x + 1$

42. $(6x - 9) - (3x + 8) = 4(x - 11)$

43. $7y - 5(2 - y) = 3(2y + 1) + 5$

44. $9(2y - 3) - 13(5y - 1) = 14(3y - 1)$

45. $0.55(w - 10) = 0.80(w + 3)$

46. $\dfrac{t}{8} - 5 = \dfrac{t}{12} - 6$

47. $\dfrac{7m + 27}{6} = \dfrac{4m + 6}{5}$

48. $\dfrac{167 - 3n}{10} = 6 - \dfrac{5n - 1}{4}$

49. $1{,}293y = 1{,}294y + 1{,}295$

50. $0.289x - 5 = 0.488 - 0.78(2 - x)$

51. $\dfrac{3v}{10} - \dfrac{5v}{6} = -\dfrac{8v}{15}$

52. $1.8a - 7.8a + 1.97a - 8.3a = 0$

Applying Technology

In Exercises 53 and 54, use a graphing calculator to solve each equation by letting Y_1 represent the left side of the equation and Y_2 represent the right side of the equation.

53. $1.5x - 1.2 = 2.7x - 3.0$

54. $8.5x + 24 = 12.5x + 33.6$

Classifying Linear Equations

In Exercises 55–58, each equation is a conditional equation, an identity, or a contradiction. Identify the type of equation, and then solve it.

55. $7(y - 3) - 4(y + 2) = 3(y - 7)$

56. $3(2v + 1) = 2(3v + 1)$

57. $4(w - 1) + (7 - 2w) = 2(w + 1) + 1$

58. $2v + 2v + 2v = 5v$

Using Function Notation

59. Use the function $f(x) = 7x - 11$ to evaluate the following.

  a. $f(-10)$ **b.** $f(0)$ **c.** $f(9)$ **d.** $f(\pi)$

60. Use the function $f(x) = 5(4x - 3) + 1$ to evaluate the following.

  a. $f(-3)$ **b.** $f(0)$ **c.** $f(6)$ **d.** $f(100)$

Solving for a Specified Variable

In Exercises 61–64, solve each equation for x.

61. $v + w = x + y$

62. $vw = xy$

63. $3x - 5y = 7z$

64. $2(x - y) + 3(2x + y) = 38$

Simplify Versus Solve

In Exercises 65–68, simplify the expression in the first column by adding like terms, and solve the equation in the second column.

Simplify	Solve
65. a. $3x + 4 + 5x + 6$	**b.** $3x + 4 = 5x + 6$
66. a. $3x + 4 - (5x + 6)$	**b.** $3x + 4 = -5x + 6$
67. a. $-2x + 3 + 4(x - 1)$	**b.** $-2x + 3 = -4(x - 1)$
68. a. $5(x + 1) - 6(x + 2)$	**b.** $5(x + 1) = 6(x + 2)$

Estimate Then Calculate

In Exercises 69 and 70, mentally estimate the solution of each equation to the nearest integer, and then calculate the exact solution.

Problem	Mental Estimate	Exact Solution
69. $0.99x - 5.1 = 4.8495$		
70. $11x - 20 = 9x + 19.9$		

Multiple Representations

In Exercises 71–74, first write an algebraic equation for each verbal statement, using the variable m to represent the number, and then solve for m.

71. Four more than five times a number is forty-nine.

72. One-third of the sum of a number and seven equals twelve.

73. Twice the quantity of two more than a number is the opposite of the quantity of three less than the number.

74. The sum of a number, one more than the number, and two more than the number is ninety-three.

Connecting Concepts to Applications

75. Direct Variation If v varies directly as w, and $v = 22$ when $w = 77$, find v when $w = 35$.

76. Direct Variation If v varies directly as w, and $v = 12$ when $w = 15$, determine the constant of variation.

77. Directly Proportional If x and y are directly proportional, and y is 20 when x is 6, find y when x is 15.

78. Perimeter of a Rectangle

  a. Write an expression for the perimeter of the rectangle shown in terms of the variable w.

  b. Determine the perimeter of this rectangle when $w = 12$.

  c. Write an equation that represents a perimeter of 48 cm for this rectangle.

d. Solve the equation in part **c** for w.

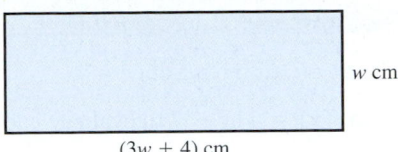

(3w + 4) cm

79. **Fixed and Variable Costs** A small business makes custom ornamental pottery. The fixed cost is $160 per week, and the variable cost is $26 per piece of pottery made. The company's sales are seasonal, so the company has a weekly budget of $1,200 for making the pottery pieces. It then stores this pottery until orders are received. Let x represent the number of pieces of pottery made per week.
   a. Write an expression for the cost of making x pieces of pottery per week.
   b. Determine the weekly cost of making 60 pieces of pottery per week.
   c. Write an equation that represents a total weekly cost of $1,200.
   d. Solve the equation in part **c**.

80. **Distance on a Map** On a particular map of the United States, 1 in represents 20 mi. If the distance on the map between Louisville, Kentucky, and Columbus, Ohio, is 10.5 in, find the distance in miles between these two cities.

81. **Similar Polygons** The two figures shown are similar. Thus, their corresponding parts are proportional. Find the lengths of sides a, b, and c.

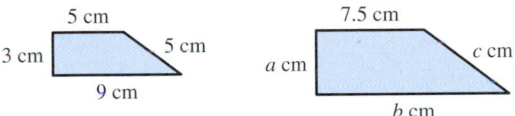

82. **Height of a Tree** If a shrub 0.64 m tall casts a 6.72-m shadow, how tall is a tree that casts a 46.2-m shadow? (See the figure.)

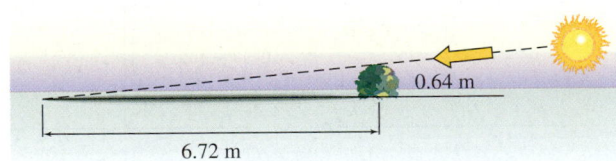

83. **Airspeed of a Plane** Two ground observers are 20 mi apart. They communicate with each other by cell phone and note that a passenger plane has passed over each of them exactly 5 minutes apart. If this plane is flying into a 50-mi/h head wind, what is the airspeed of this plane?

84. **Quality Control** A quality control inspector discovered 3 defective display panels in the 75 handheld computing devices that he tested. How many defective display panels would be expected in a shipment of 500 of these items?

In Exercises 85–88, match each problem with the appropriate restrictions on the variable for this application.
   A. $\{0, 1, 2, 3, \ldots, 365\}$ **B.** $\{0, 1, 2, 3, \ldots, 20\}$
   C. $[0, 20]$ **D.** $[0, 365]$

85. **Newsprint** The function $f(x) = 1,000 - 50x$ models the length of paper remaining on a spool of newsprint after x pieces each 50 ft long have been removed from this spool.

86. **Newsprint** The function $f(x) = 1,000 - 50x$ models the length of paper remaining on a spool of newsprint after 50 pieces each of length x have been removed from this spool.

87. **Gasohol** A 1,000-gal tank contains 635 gal of gasohol that is 8% ethanol by volume. The function $f(x) = 0.08(635) + 0.12x$ models the amount of ethanol that will be in the tank after x gal of gasohol that is 12% by volume is pumped into this tank.

88. **Sign Rental** The function $f(x) = 100x + 250$ models the cost in dollars to lease a highway messaging system for x days during the year 2010.

89. **Time Traveled** Two passenger airplanes depart from an airport simultaneously. One flies east at 400 mi/h, and the other flies west at 450 mi/h. Determine how long it will take the planes to be 2,975 mi apart. (Fuel capacity limits the flight time to a maximum of 5 hours for each plane.)

90. **Amount of Pesticide in a Solution** A crop-dusting company needs to apply a pesticide mixture that is 0.2% pesticide. From an earlier application, it already has 300 gal of a more concentrated mixture that is 0.3% pesticide. This mixture is in a 500-gal tank on the plane that it plans to use. Rather than discard the existing mixture, the company decides to dilute the existing mixture with water. How much water must be added to the 500-gal tank to have the desired concentration? Is there enough capacity in this tank?

Chapter 2 | Mastery Test

Objective 2.1.1 Plot Ordered Pairs on a Rectangular Coordinate System

1. Identify the coordinates of the points *A* through *D* in the figure and give the quadrant in which each point is located.

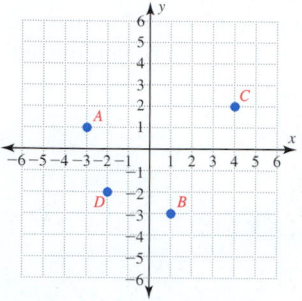

Objective 2.1.2 Draw a Scatter Diagram of a Set of Data Points

2. Draw a scatter diagram for the data points in the table.

x	y
−2	3
0	−2
1	2
3	1
4	−3

Objective 2.1.3 Identify Data Whose Graph Forms a Linear Pattern

3. Determine whether each sequence is an arithmetic sequence and whether the graph of the data forms a linear pattern.
   a. 0, 2, 4, 6, 8   b. 1, 2, 4, 8, 16
   c. 4, 4, 4, 4, 4

Objective 2.1.4 Interpret a Line Graph

4. **Blood Pressure** A patient was placed on blood pressure medication by his doctor. The blood pressure readings from five consecutive Mondays were recorded after he started his medication. Use this graph to answer the following questions.
   a. What was the highest blood pressure recorded?
   b. What was the lowest blood pressure recorded?
   c. What week was the lowest blood pressure recorded?

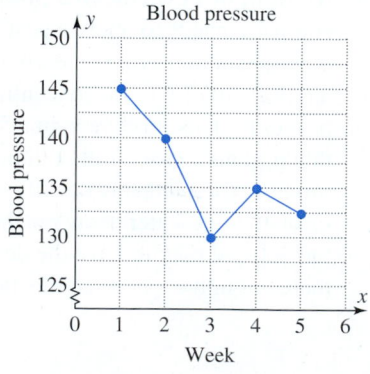

Objective 2.2.1 Use Function Notation

5. Use the function $f(x) = 11x - 7$ to evaluate each expression.
   a. $f(0)$ b. $f(-1)$ c. $f(7)$ d. $f(10)$

Objective 2.2.2 Use a Linear Equation to Form a Table of Values and to Graph a Linear Equation

6. Use each equation to complete a table with input values of 1, 2, 3, 4, and 5. Then graph these points and the line through these points.
   a. $y = x + 1$ b. $y = -x + 2$
   c. $y = 2x - 5$ d. $y = -2x + 4$

Objective 2.2.3 Write a Linear Function to Model an Application

7. **Car Loan** A new car loan requires an $800 down payment and a monthly payment of $375.
   a. Write a function that gives the total paid by the end of the *x*th month.
   b. Use this function to complete this table.

x	0	6	12	18	24	30
f(x)						

   c. Evaluate and interpret $f(25)$.

Objective 2.3.1 Check Possible Solutions of a Linear Equation

8. Determine whether $(-2, 3)$ is a solution of each equation.
   a. $y = 2x + 7$ b. $y = -x - 1$
   c. $y = 3$ d. $x = -2$

Objective 2.3.2 Determine the Intercepts from a Graph

9. Determine the *x*- and *y*-intercepts of these lines.
   a.   b.

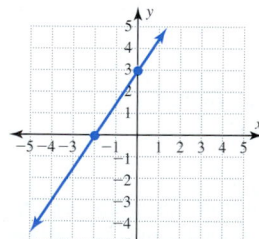

   c.   d.

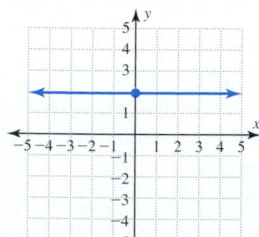

Objective 2.3.3 Determine the Point Where Two Lines Intersect

10. Determine the point of intersection of the lines in each graph, and then check this point in both equations.

a.   **b.**

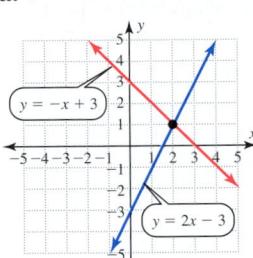

Objective 2.4.1 Solve Linear Equations in One Variable Using the Addition-Subtraction Principle

11. Solve each linear equation.
  a. $x + 7 = 11$ **b.** $2x + 7 = x + 21$
  c. $7x + 1 = 6x + 1$ **d.** $x + 2(x + 1) = 4x - 4$

Objective 2.4.2 Use Graphs and Tables to Solve a Linear Equation in One Variable

12. Let Y_1 equal the left side of each equation and Y_2 equal the right side. Then use a graphing calculator to solve each equation by examining graphs and tables for Y_1 and Y_2.

  a. $\dfrac{x - 4}{2} = \dfrac{-3x + 4}{2}$

  b. $2.3x + 1.6 = 1.7x + 0.4$

Objective 2.4.3 Identify a Linear Equation As a Conditional Equation, an Identity, or a Contradiction

13. Each of these equations is a conditional equation, an identity, or a contradiction. Identify the type of each equation and then solve it.
  a. $4(x + 2) = 4x + 8$ **b.** $4(x + 2) = 4x + 2$
  c. $4(x + 2) = 2(x + 4)$

Objective 2.5.1 Solve Linear Equations in One Variable Using the Multiplication-Division Principle

14. Solve each linear equation.
  a. $-x = 3x - 12$ **b.** $-\dfrac{5}{21} = \dfrac{3y}{7}$

  c. $3(2v + 4) = 4(v + 3) + 3$ **d.** $\dfrac{w - 1}{6} = \dfrac{w - 3}{5}$

Objective 2.6.1 Rewrite a Linear Equation in the Form $y = mx + b$

15. Solve the linear equation $2x - y = 2(3x - y - 1) + 1$ for y, and then use a graphing calculator to complete

this table and to graph the equation, using a window of $[-10, 10, 1]$ by $[-10, 10, 1]$.

X	Y1	
-3		
-2		
-1		
0		
1		
2		
3		
X= -3		

Objective 2.6.2 Calculate the *x*- and *y*-Intercepts of a Line

16. Calculate the x- and y-intercepts of the graphs of these equations, and use the intercepts to sketch the graph of the line.
  a. $y = 2x - 6$ **b.** $y = 4$
  c. $2x - 5y = 10$ **d.** $3(2x - 1) = 4(3y - 2) - 7$

Objective 2.6.3 Solve an Equation for a Specified Variable

17. Solve each formula for the variable specified.
  a. $A = \dfrac{1}{2}h(a + b)$ for a **b.** $y = mx + b$ for x

Objective 2.7.1 Solve Problems Involving Proportions

18. Solve the proportions in parts **a** and **b**.
  a. $\dfrac{m - 1}{m - 7} = \dfrac{2}{5}$ **b.** $\dfrac{2x - 1}{7} = \dfrac{x + 1}{4}$
  c. On a certain map, 2 cm corresponds to a distance of 75 km. What distance corresponds to a distance of 5 cm on the map?

Objective 2.7.2 Solve Problems Involving Direct Variation

19. a. If y varies directly as x, and y is 33 when x is 22, find y when x is 8.
  b. The number of MXP (Mexican pesos) varies directly as the number of USD (U.S. dollars) exchanged. In 2010, one could exchange 50 USD for 475 MXP. What was the constant of variation (the exchange rate) at that time?

Objective 2.8.1 Determine the Restrictions on a Variable in an Application

20. Determine the restrictions on the variable for each problem.
  a. The function $C(x) = 17.50x$ models the cost of filling a 30-gal sprayer tank with x gal of insecticide for a pest control service.
  b. The function $C(x) = 17.50x$ models the cost of renting a paint sprayer for x days during the month of June. The rental company will not rent the sprayer for partial days.

c. The function $f(x) = 0.06(200) + 0.10x$ models the number of gallons of acid in a 500-gal tank. The 500-gal tank originally contained 200 gal of a 6% acidic solution, and then x gal of a stronger solution was added to the tank.

d. The function $f(x) = 250 - 25x$ models the length of speaker wire remaining on a spool after x pieces each 25 ft long have been removed from this spool.

Objective 2.8.2 Use the Mixture Principle to Create a Mathematical Model

21. Mixture of Gasohol A gasoline distributor has a 12,000-gal tank to store a gasohol mix. The tank already contains 5,000 gal of gasohol that contains 5% ethanol by volume.

a. If 1,000 gal of pure ethanol is added to the tank, what percent of alcohol will be in the resulting mixture? (Round to the nearest tenth of a percent.)

b. How many gallons of pure ethanol must be added to the tank to produce a mix that is 8% ethanol? (Round to the nearest gallon.)

c. How many gallons of pure ethanol must be added to the tank to produce a mix that is 10% ethanol? (Round to the nearest gallon.)

Chapter 2 | **Group Project**

Risks and Choices: Shopping for Interest Rates on a Home Loan

The formula for determining the monthly payment for a loan is

$$P = \frac{AR\left(1 + \dfrac{R}{12}\right)^{12n}}{12\left(1 + \dfrac{R}{12}\right)^{12n} - 12}$$

where A represents the amount of the loan, R represents the interest rate, n represents the number of years, and P represents the monthly payment required to pay off the loan in n years.

1. Use this formula and a spreadsheet or a calculator to complete the following tables.

Consider a $100,000 loan at several different interest rates.

Interest Rate (%)	Loan Amount ($)	Number of Years	Monthly Payment ($)	Total of All Payments ($)
5.50	100,000	30		
6.00	100,000	30		
6.50	100,000	30		
7.00	100,000	30		
7.50	100,000	30		
8.00	100,000	30		

Consider a $50,000 loan at 7.00% interest for several different time periods.

Interest Rate (%)	Loan Amount ($)	Number of Years	Monthly Payment ($)	Total of All Payments ($)
7.00	50,000	10		
7.00	50,000	15		
7.00	50,000	20		
7.00	50,000	25		
7.00	50,000	30		

2. Use these facts to discuss the advantages and disadvantages for the borrower for each of these options.

Lines and Systems of Linear Equations in Two Variables

Chapter Outline

Wind Power

Fossil fuels cannot sustain our energy needs forever, and they produce pollution. Wind power doesn't produce pollutants and indirectly captures heat energy radiating from our sun. Thus this energy source won't run out or even diminish for billions of years. Wind power is one of the fastest-growing energy resources. Since 1995, global wind-generating capacity has increased by almost 500%.

Mathematics is used to design more efficient towers and wind turbines and to reduce the power loss over transmission lines. Increases of efficiency by even 1% can produce tremendous savings and make this energy source more competitive with other energy sources.

The European Union has long been a leader in the production of wind power. The accompanying graph compares the wind power capacity of the European Union with that of the rest of the world. Exercise 67 in Section 3.3 analyzes the intersection of these two line graphs.

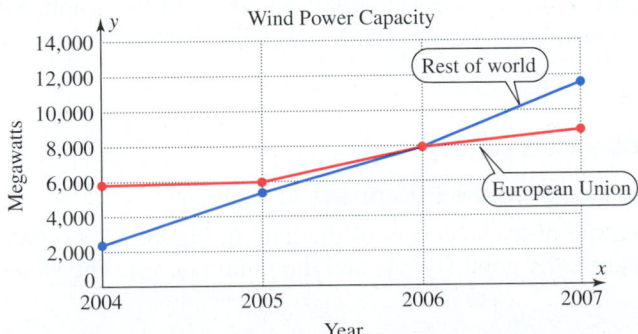

Source: European Wind Energy Association.

Section 3.1 | Slope of a Line and Applications of Slope

Humans are basically pattern recognition devices. Our eyes take in the world, but what we really see are intricate patterns of lines and curves and colors and brightness. —DAVID BLATNER

Objectives:

1. Determine the slope of a line.
2. Use slopes to determine whether two lines are parallel, perpendicular, or neither.
3. Calculate and interpret rates of change.

Highway engineers must consider safety, construction difficulty, cost, and many other factors when designing new highways. The slope or grade of a highway influences all the decisions related to these factors. In Example 7, we will examine a section of highway that cannot exceed a 6% grade. An important part of mathematics involves analyzing trends and change. This chapter will examine the slope of a line and other key skills needed to study trends and change.

Which of the Following Lines Is the Steepest or Represents the Greatest Change in Share Price Per Year?

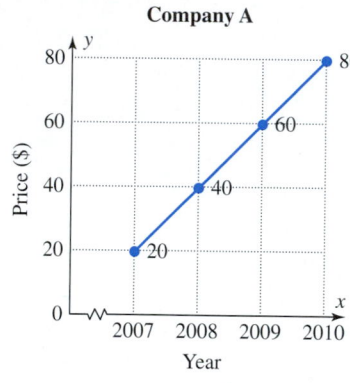

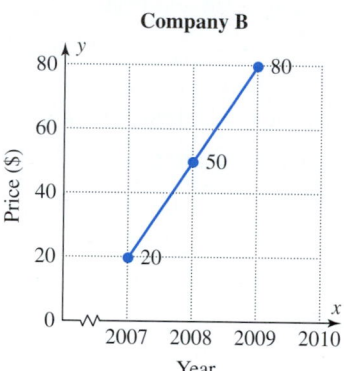

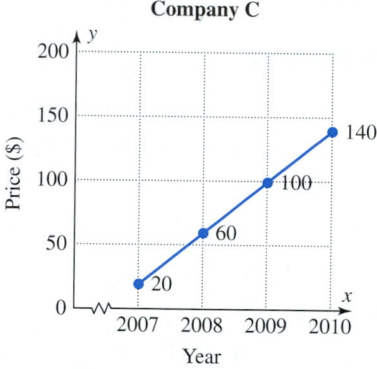

The graph for company B appears steeper than the graph for company A and accurately reflects that the share price of company B is increasing faster than that of company A ($30 per year for B versus $20 per year for A). Although the graph for company C does not appear as steep as the other two graphs, it is company C whose share price is increasing most rapidly—at $40 per year. The reason this graph does not appear as steep is that a different scale was used on the vertical axis for this graph.

This example illustrates the importance of looking beyond the picture portion of a graph and also examining the numerical information or the scale of the graph. Since our eyes can be misled when examining the steepness, we will also look at this concept algebraically.

A Mathematical Note

The origin of the use of *m* to designate slope is unknown. In his book *Mathematical Circles Revisited,* Howard Eves says that *m* might have been used because slopes were first studied with respect to mountains. Some suggest that *m* may have been derived from the French word *monter,* which means "to mount, to climb, or to slope up."

1. Determine the Slope of a Line

How Do Mathematicians Describe the Steepness of a Line?

The **slope of a line** is a measure of the steepness of the line. In Figure 3.1.1, consider the steepness of the line connecting the point (x_1, y_1) and the point (x_2, y_2). The slope of this line is defined to be the ratio of the change in y (the rise) to the change in x (the run). Slope is usually represented by the letter m.

$$m = \frac{\text{change in } y}{\text{change in } x}$$

$$m = \frac{y_2 - y_1}{x_2 - x_1}$$

The delta symbol Δ is often used to designate change. In Technology Perspective 2.2.3, the Δ**Tbl** notation was used to note the change from one input value to the next in a table setup on a TI-84 Plus calculator. Many books also use Δx, read "delta x," to denote a change in the x-variable and Δy, read "delta y," to denote a change in the y-variable.

Using this notation, we can represent the slope of line by $m = \dfrac{\Delta y}{\Delta x}$.

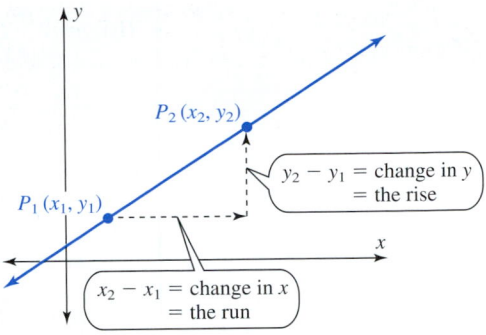

Figure 3.1.1

Slope of a Line Through (x_1, y_1) and (x_2, y_2)

Algebraically	Verbally	Numerical Example	Graphical Example
$m = \dfrac{y_2 - y_1}{x_2 - x_1}$ for $x_1 \neq x_2$. $m = \dfrac{\Delta y}{\Delta x}$	The slope of a line is the ratio of the change in y to the change in x.	A 1-unit change in x produces a 2-unit change in y. $\begin{array}{c\|c} x & y \\ \hline 1 & -3 \\ 2 & -1 \\ 3 & 1 \\ 4 & 3 \end{array}$ For the points $(2, -1)$ and $(3, 1)$, $m = \dfrac{1 - (-1)}{3 - 2}$ $m = \dfrac{2}{1}$	

A key point is that slope measures the steepness of a line and compares the rate of change of y with respect to x. Example 1 gives the slope in decimal form, thus making it easy to determine the change in y for each 1-unit change in x.

Example 1   Calculating the Slope of a Line Through Two Points

Calculate the slope of the line through the given points.

(a) $(-5, 8)$ and $(3, -2)$ **(b)** $(3, -2)$ and $(-5, 8)$

Solution

(a) $m = \dfrac{y_2 - y_1}{x_2 - x_1}$ Substitute the given points into the formula for slope with $x_1 = -5$, $y_1 = 8$, $x_2 = 3$, and $y_2 = -2$. Note the use of parentheses in the denominator to prevent an error in sign.

$m = \dfrac{-2 - 8}{3 - (-5)}$

$m = \dfrac{-10}{8}$ Express the slope as a fraction in reduced form.

$m = -\dfrac{5}{4}$ or y decreases 5 units for each 4-unit increase in x, or y decreases 1.25 units for every 1-unit increase in x.

$m = -1.25$

Answer: $m = -1.25$

(b) $m = \dfrac{y_2 - y_1}{x_2 - x_1}$ Although the order of these two points is different,
these are the same points as those in part **a**, and the
slope is also the same. Substitute $x_1 = 3$,
$y_1 = -2$, $x_2 = -5$, and $y_2 = 8$.

$m = \dfrac{8 - (-2)}{-5 - 3}$

$m = \dfrac{10}{-8}$

$m = -\dfrac{5}{4}$ or

$m = -1.25$

Answer: $m = -1.25$

Self-Check 1

Calculate the slope of the line through $(5, -9)$ and $(-4, -21)$.

The slope of a line is the same no matter which two points on the line are used to calculate the slope. As shown in Example 1, this slope is also the same no matter which point is taken first. This is true because the slope is the ratio of the change in y to the corresponding change in x. However, it is crucial that the x- and y-values be kept in the correct pairings. Do not match the x-coordinate of one point with the y-coordinate of another point.

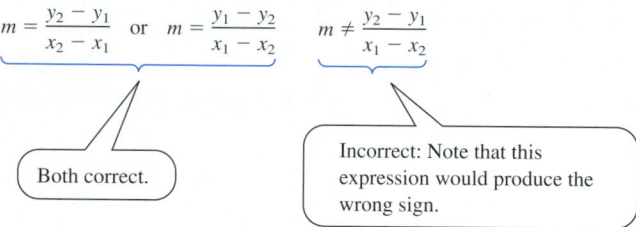

$$m = \dfrac{y_2 - y_1}{x_2 - x_1} \quad \text{or} \quad m = \dfrac{y_1 - y_2}{x_1 - x_2} \qquad m \neq \dfrac{y_2 - y_1}{x_1 - x_2}$$

Both correct.

Incorrect: Note that this
expression would produce the
wrong sign.

Example 2   Calculating the Slope of a Line Through Given Points

Calculate the slope of each of the following lines, using the points labeled on the line.

Solution

(a)

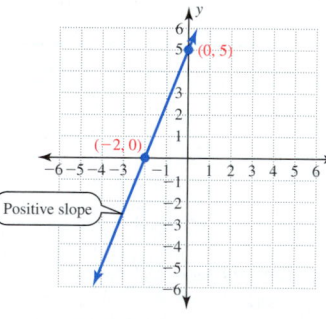

$m = \dfrac{y_2 - y_1}{x_2 - x_1}$ The line slopes upward to the right;
its slope is a positive number.

$m = \dfrac{5 - 0}{0 - (-2)}$ Use $x_1 = -2$, $y_1 = 0$, $x_2 = 0$, and
$y_2 = 5$.

$m = \dfrac{5}{2}$ or y increases 5 units for each 2-unit
increase in x, or y increases
2.5 units for each 1-unit increase
in x.

$m = 2.5$

Answer: $m = 2.5$

(b)

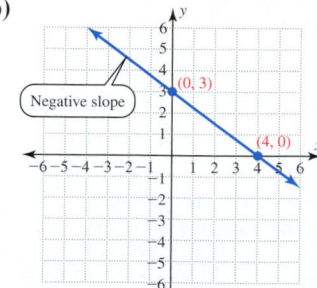

$m = \dfrac{y_2 - y_1}{x_2 - x_1}$ The line slopes downward to the
right; its slope is a negative number.

$m = \dfrac{0 - 3}{4 - 0}$ Use $x_1 = 0$, $y_1 = 3$, $x_2 = 4$, and
$y_2 = 0$.

$m = -\dfrac{3}{4}$ or y decreases 3 units for each 4-unit
increase in x, or y decreases 0.75
unit for each 1-unit increase in x.

$m = -0.75$

Answer: $m = -0.75$

Self-Check 2

Calculate the slope of the line, using the points labeled on
the graph.

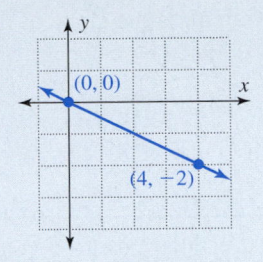

How Can I Use a Table of Values to Calculate the Slope of a Line?

Although only two points are needed to determine the slope of a line, it is sometimes easy
to determine the slope of a line by examining a table of values of points on the line. This is
illustrated in Example 3. If you are creating a table of values for a linear equation with

slope $m = \dfrac{\Delta y}{\Delta x}$, often it is convenient to let the change in x in the table be the same as Δx in

the fraction $\dfrac{\Delta y}{\Delta x}$. *Caution:* Be sure all points in the table lie on the same line before calcu-

lating the slope of the line.

| **Example 3** | Determining Slope from a Table of Values |

Calculate the slope of the line containing the points in the table.

Solution

(a)

x	y
0	−4
5	−2
10	0
15	2
20	4
25	6
30	8

$m = \dfrac{2}{5}$

For each increase of 5 units for x, y increases by
2 units. Thus $\dfrac{\Delta y}{\Delta x} = \dfrac{2}{5}$.

(b)

x	y
0	7
4	4
8	1
12	−2
16	−5
20	−8
24	−11

$m = -\dfrac{3}{4}$

For each increase of 4 units for x, y decreases by
3 units. Thus $\dfrac{\Delta y}{\Delta x} = -\dfrac{3}{4}$.

(c)

x	y
−3	0.5
−2	1.0
−1	1.5
0	2.0
1	2.5
2	3.0
3	3.5

$m = 0.5$

For each increase of 1 unit for x, y increases by
0.5 unit. Thus $\dfrac{\Delta y}{\Delta x} = \dfrac{0.5}{1}$.

What Is the Significance of a Positive or a Negative Slope?

All lines with positive slope go upward to the right since the *x*- and *y*-coordinates increase together. All lines with negative slope go downward to the right since the *y*-coordinate decreases as the *x*-coordinate increases. If we compare two lines graphed on the same coordinate system, the steeper line will have the slope with the larger absolute value.

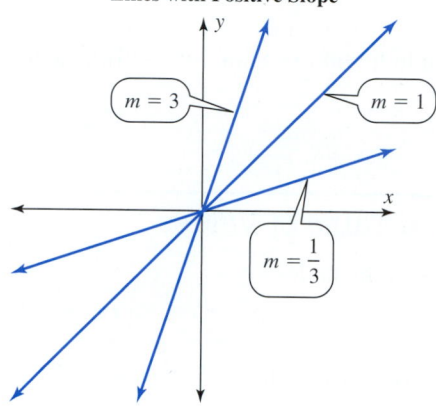

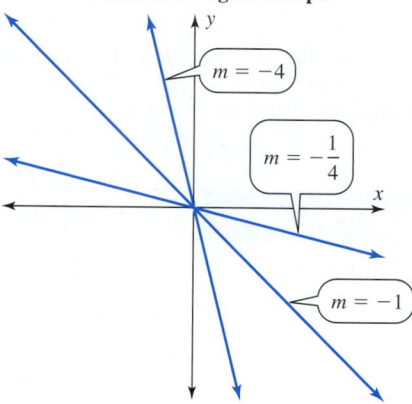

Example 4 shows that the slope of a horizontal line is 0 and the slope of a vertical line is undefined.

Example 4   Calculating the Slopes of Horizontal and Vertical Lines

Calculate the slope of each of the following lines, using the points labeled on the line.

Solution

(a)

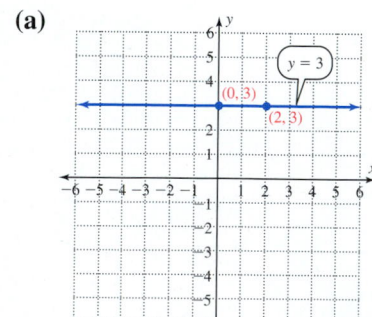

$$m = \frac{y_2 - y_1}{x_2 - x_1}$$

$$m = \frac{3 - 3}{2 - 0}$$ Use $x_1 = 0$, $y_1 = 3$, $x_2 = 2$, and $y_2 = 3$.

$$m = \frac{0}{2}$$ For any change in *x*, the change in *y* is 0.

Answer: $m = 0$ This line is horizontal, and its slope is 0.

(b)

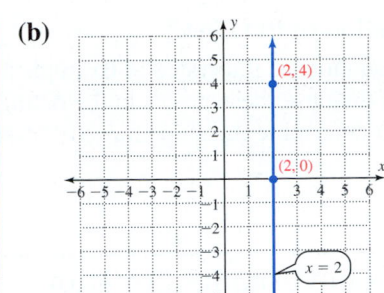

$$m = \frac{y_2 - y_1}{x_2 - x_1}$$

$$m = \frac{4 - 0}{2 - 2}$$ Use $x_1 = 2$, $y_1 = 0$, $x_2 = 2$, and $y_2 = 4$.

$$m = \frac{4}{0}$$

Answer: m is undefined.   This line is vertical, and its slope is undefined since division by 0 is undefined.

Self-Check 4

Calculate the slope of:

a. The vertical line through $(5, 1)$ and $(5, 4)$.   **b.** The horizontal line through $(5, 1)$ and $(2, 1)$.

The following table shows how we can classify lines by using information about their slopes. On the other hand, if we can see the graph of a line, then we will know whether the slope is positive, negative, zero, or undefined.

Classifying Lines by Their Slopes

Numerically	Verbally	Graphically
m is positive	The line slopes upward to the right.	
m is negative	The line slopes downward to the right.	
m is zero	The line is horizontal.	
m is undefined	The line is vertical.	

Can I Calculate the Slope of a Line Using an Equation of the Line?

Yes, the slope of a line can also be determined from its equation, as illustrated in Example 5. One approach is to determine two points on the line and then use these points in the definition of the slope. In Section 3.2 we will examine how to determine the slope directly from a linear equation.

Example 5   Calculating the Slope of a Line from Its Equation

Determine the slope of each of the following lines.

(a) $2x - 9y = 18$ (b) $y = 4$ (c) $x = -3$

Solution

(a) $2x - 9y = 18$ $2x - 9(0) = 18$

      $2(0) - 9y = 18$ $2x = 18$

           $-9y = 18$ $x = 9$

               $y = -2$

Determine the y-intercept by substituting 0 for x. Then determine the x-intercept by substituting 0 for y.

x	y
0	-2
9	0

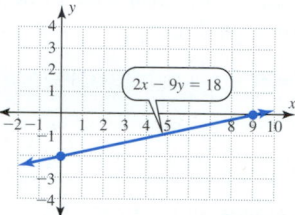

$$m = \frac{y_2 - y_1}{x_2 - x_1}$$

$$m = \frac{0 - (-2)}{9 - 0}$$

Substitute $(0, -2)$ for (x_1, y_1) and $(9, 0)$ for (x_2, y_2).

Answer: $m = \dfrac{2}{9}$

(b)

x	y
0	4
3	4

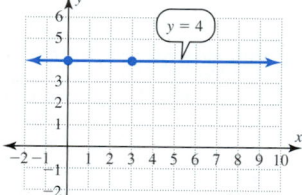

$$m = \frac{y_2 - y_1}{x_2 - x_1}$$

$$m = \frac{4 - 4}{3 - 0}$$

$$m = \frac{0}{3}$$

All points on this line have a y-coordinate of 4.

This is a horizontal line with a slope of 0. The y-coordinate does not change when the x-coordinate changes.

The equation $y = 4$ can also be written as $0x + 1y = 4$. Check the points $(0, 4)$ and $(3, 4)$ in the equation $0x + 1y = 4$.

Answer: $m = 0$

(c)

x	y
-3	0
-3	2

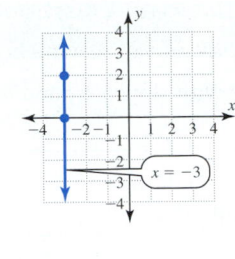

$$m = \frac{y_2 - y_1}{x_2 - x_1}$$

$$m = \frac{2 - 0}{-3 - (-3)}$$

$$m = \frac{2}{0}$$

Answer: m is undefined

All points on this line have an x-coordinate of -3.

This is a vertical line whose slope is undefined. The x-coordinate does not change.

The equation $x = -3$ can also be written as $1x + 0y = -3$. Check the points $(-3, 0)$ and $(-3, 2)$ in the equation $1x + 0y = -3$.

Self-Check 5

Calculate the slope of the lines defined by these equations.

a. $x + y = 5$ **b.** $y = 5x$ **c.** $y = 5$ **d.** $x = 5$

---

2. Use Slopes to Determine Whether Two Lines Are Parallel, Perpendicular, or Neither

Parallel lines are lines that lie in the same plane but never intersect. **Perpendicular lines** form a 90° angle when they intersect. (See Figure 3.1.2.)

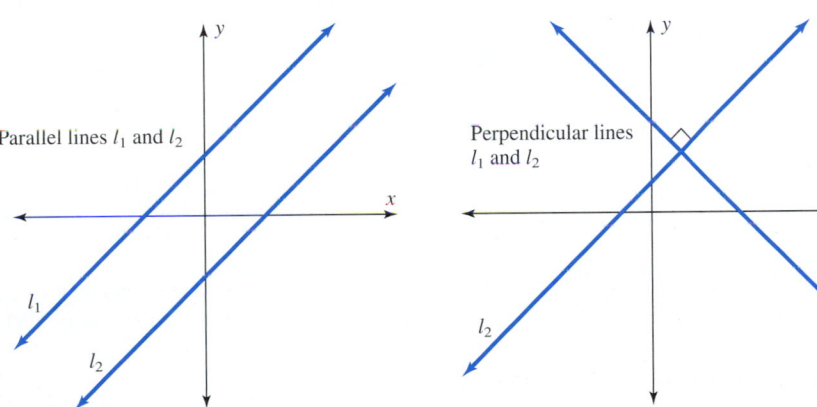

Figure 3.1.2 Parallel and perpendicular lines.

How Are the Slopes of Parallel and Perpendicular Lines Related?

Parallel lines have the same slope because they rise or fall at the same rate. Perpendicular lines have slopes whose product is -1. If one line slopes upward to the right, then a line perpendicular to it will slope downward to the right. Perpendicular lines have slopes that are opposite reciprocals. Opposite reciprocals, such as $\frac{2}{3}$ and $-\frac{3}{2}$, have a product of -1, as in $\left(\frac{2}{3}\right)\left(-\frac{3}{2}\right) = -1$. All vertical lines are parallel to one another, and all vertical lines are perpendicular to all horizontal lines.

Parallel and Perpendicular Lines

If l_1 and l_2 are distinct nonvertical* lines with slopes m_1 and m_2, respectively, then:

Algebraically	Verbally	Graphically
$m_1 = m_2$	Lines l_1 and l_2 are parallel because they have the same slope.	
$m_1 = -\dfrac{1}{m_2}$ or $m_1 m_2 = -1$	Lines l_1 and l_2 are perpendicular because their slopes are opposite reciprocals.	

*Also note that all vertical lines are parallel to one another, and all vertical lines are perpendicular to all horizontal lines.

Example 6 Determining Whether Two Lines Are Parallel or Perpendicular

Determine whether the line through the first pair of points is parallel to, perpendicular to, or neither parallel nor perpendicular to the line through the second pair of points.

Solution

(a) $(1, 3)$ and $(-1, -1)$
$(2, 1)$ and $(3, 3)$

$$m = \frac{y_2 - y_1}{x_2 - x_1} \qquad m = \frac{y_2 - y_1}{x_2 - x_1}$$

$$m = \frac{-1 - 3}{-1 - 1} \qquad m = \frac{3 - 1}{3 - 2}$$

$$m = \frac{-4}{-2} \qquad m = \frac{2}{1}$$

$$m = 2 \qquad m = 2$$

Calculate the slope of each line by substituting in the given points.

Since the slopes are equal, the lines are parallel.

Answer: The two lines are parallel.

(b) $(3, 1)$ and $(6, 3)$
$(2, -1)$ and $(4, -4)$

$$m = \frac{y_2 - y_1}{x_2 - x_1} \qquad m = \frac{y_2 - y_1}{x_2 - x_1}$$

$$m = \frac{3 - 1}{6 - 3} \qquad m = \frac{-4 - (-1)}{4 - 2}$$

$$m = \frac{2}{3} \qquad m = -\frac{3}{2}$$

Since $\left(\dfrac{2}{3}\right)\left(-\dfrac{3}{2}\right) = -1$, the two lines are perpendicular.

Answer: The two lines are perpendicular.

(c) $(1, 7)$ and $(-1, -1)$
$(1, 1)$ and $(-1, 9)$

$$m = \frac{y_2 - y_1}{x_2 - x_1} \qquad m = \frac{y_2 - y_1}{x_2 - x_1}$$

$$m = \frac{-1 - 7}{-1 - 1} \qquad m = \frac{9 - 1}{-1 - 1}$$

$$m = \frac{-8}{-2} \qquad m = \frac{8}{-2}$$

$$m = 4 \qquad m = -4$$

The slopes are not equal, and $4(-4) \neq -1$. Thus the lines are neither parallel nor perpendicular.

Answer: The lines are neither parallel nor perpendicular.

Self-Check 6

Determine whether the line through $(5, 3)$ and $(-5, -1)$ is parallel to, perpendicular to, or neither parallel nor perpendicular to the line through $(2, -3)$ and $(-2, 7)$.

3. Calculate and Interpret Rates of Change

Are There Uses for Slope Other Than for Measuring the Steepness of a Line?

Yes, a key concept is that slope represents a rate of change. Slope is used in many applications to compare how one quantity is changing with respect to another quantity. Some of these applications use the terms *rise* and *run*. Other applications involving slope may use the terms *angle of elevation* or *grade*. Example 7 involves the grade of a highway.

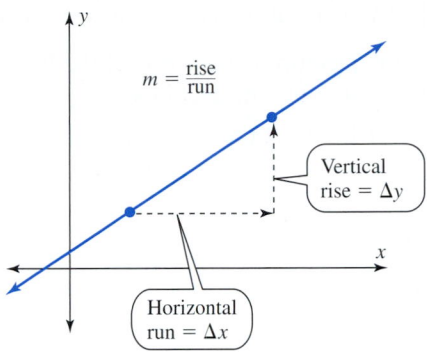

A Mathematical Note

The concept of slope as a rate of change is one of the most important concepts in algebra. The term *slope* is widely used in many disciplines. A quick search on the Internet can reveal many uses of the term *slope* from the "glide slope" of an airplane to the "bunny slope" at a ski resort.

Example 7   Modeling the Slope of a Highway

For several reasons, including highway safety, an engineer has determined that the slope of a particular section of highway should not exceed a 6% grade (slope). If this maximum grade is allowed on a section covering a horizontal distance of 2,000 m, how much change in elevation is permitted on this section? (The change of elevation can be controlled by topping hills and filling low places.)

Solution

Let y = change in elevation.

$m = \dfrac{y_2 - y_1}{x_2 - x_1}$	Sketch the problem, using a convenient placement of the origin to simplify the computations (see the figure).
$\dfrac{6}{100} = \dfrac{y - 0}{2,000 - 0}$	Use the formula for slope. Substitute in the changes for y and x, and write 6% as $\dfrac{6}{100}$.
$\dfrac{6}{100} = \dfrac{y}{2,000}$	
$2,000\left(\dfrac{6}{100}\right) = 2,000\left(\dfrac{y}{2,000}\right)$	To solve this proportion for y, multiply both sides by the LCD of 2,000.
$20(6) = y$	
$120 = y$ or	
$y = 120$	

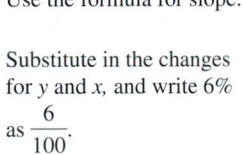
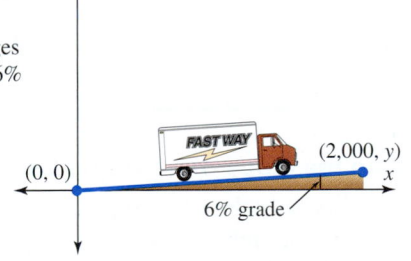

Answer: This section of road could change 120 m in elevation.

Self-Check 7

Calculate the elevation change that is permitted in Example 7 if the highway is restricted to a 5% grade.

In Example 7, the slope represents a rate of change of elevation. In Example 8, the slope represents a rate of change in the depth of water of 3 in/h.

Example 8   Rate of Change of the Depth of Water in a Pool

A pool is being filled with water from a hose flowing at a constant rate. The depth of the water is measured every 2 hours. These depths recorded in inches are shown in the table.

(a) Determine the slope of the line through these points.

(b) Interpret the meaning of the slope in this application.

Time (hr)	Depth (in)
0	20
2	26
4	32
6	38
8	44
10	50
12	56

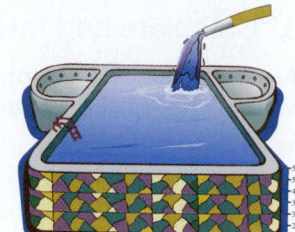

Solution

(a) $m = \dfrac{6}{2}$

$= 3$

From the table, the depth increases 6 inches every 2 hours. The slope of the line is $\dfrac{3}{1} = 3$.

(b) The depth of the water increases at a rate of 3 in/h.

Self-Check 8

A tool rental company posted the following chart that gives the cost to rent a concrete mixer.

a. Determine the slope of the line through these points.

b. Interpret the meaning of the slope in this application.

Time (hr)	Rental Cost
0.5	$18
1.0	$36
1.5	$54
2.0	$72
2.5	$90
3.0	$108

Can the Slope of a Line Be Used to Help Graph the Line?

Yes, not only does the slope of a line indicate whether the line is sloping upward or downward to the right, it also gives us information that can help us graph the line. In review, we have already graphed lines by:

- plotting two given points
- calculating two points from an equation and then plotting these points
- calculating the *x*- and *y*-intercepts from an equation and then plotting these intercepts

Now we will use one given point and the slope of a line to graph the line. Start by plotting the given point, and then determine the changes for *x* and *y* in the $\dfrac{\Delta y}{\Delta x}$ form for the slope.

Use the change Δx to move horizontally from the given point and then use the change Δy to move vertically to plot a new point. Then draw the line through the given point and the new point that you have plotted. This is illustrated in Example 9. We encourage you to practice graphing lines by this method to increase your understanding of lines and slope.

Example 9   Using the Slope and *y*-Intercept to Graph a Line

A line has a *y*-intercept of (0, 1) and a slope of $\dfrac{5}{3}$. Use this information to determine

another point on the line and to graph this line.

Solution

y-intercept: (0, 1)

Second point: $(0 + 3, 1 + 5) = (3, 6)$

Start with the *y*-intercept and use the slope of $\dfrac{5}{3}$ to determine a second point. Do this by increasing *y* by 5 units for a 3-unit increase in *x*.

Plot both (0, 1) and (3, 6), and sketch the line through these points. There is a 5-unit rise for each run of 3 units.

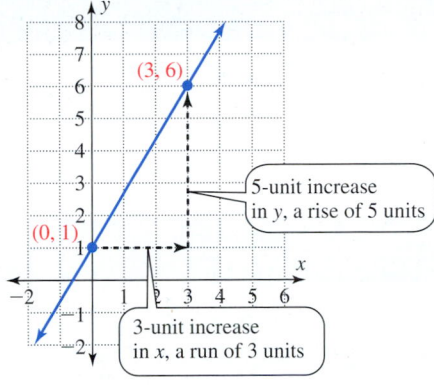

5-unit increase in *y*, a rise of 5 units

3-unit increase in *x*, a run of 3 units

Check: $m = \dfrac{y_2 - y_1}{x_2 - x_1}$

$m = \dfrac{6 - 1}{3 - 0}$

$m = \dfrac{5}{3}$ checks.

Self-Check 9

The slope of a line is $-\dfrac{2}{5}$.

a. What change in *y* will a 5-unit increase in *x* produce?

b. Graph a line with this slope and a *y*-intercept of (0, 3).

Self-Check Answers

1. $m = \dfrac{4}{3}$

2. $m = -\dfrac{1}{2}$

3. $m = \dfrac{2}{3}$

4. a. *m* is undefined   b. *m* = 0

5. a. *m* = −1   b. *m* = 5
    c. *m* = 0
    d. *m* is undefined

6. The lines are perpendicular.

7. This section of road could change 100 m in elevation.

8. a. *m* = 36
    b. The cost to rent the concrete mixer is $36 per hour.

9. a. A 2-unit decrease in *y*.

b.

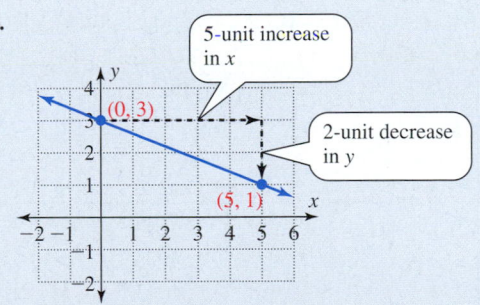

5-unit increase in *x*

2-unit decrease in *y*

3.1   **Using the Language and Symbolism of Mathematics**

1. The rise between two points on a line refers to the change in the _____ variable.

2. The run between two points on a line refers to the change in the _____ variable.

3. The letter that is used to represent the slope of a line is _____ .

4. The formula for the slope of a line through (x_1, y_1) and (x_2, y_2) is _____ .

5. The slope of a line gives the change in *y* for each _____ -unit change in *x*.

6. The slope of a horizontal line is _____ .

7. The slope of a vertical line is _____ .

8. The symbol Δx represents the _____ in x.

9. The symbol Δy represents the _____ in y.

10. Slope represents a _____ of change.

11. If the slope of a line is positive and the x-coordinate increases, then the y-coordinate _____.

12. If the slope of a line is negative and the x-coordinate increases, then the y-coordinate _____.

13. If two lines have the same slope, the lines are _____.

14. If the slopes of two lines are opposite reciprocals, the lines are _____.

15. The slope of a highway is sometimes referred to as the _____ of the highway.

3.1 Quick Review

Reduce each fraction to lowest terms.

1. $\dfrac{6}{8}$　　**2.** $\dfrac{2}{-8}$　　**3.** $\dfrac{-9}{12}$

Rewrite each fraction as a decimal.

4. $\dfrac{5}{4}$　　**5.** $-\dfrac{7}{8}$

3.1 Exercises

Objective 1 Determine the Slope of a Line

In Exercises 1–12, calculate the slope of the line through the given points.

1. a. $(4, 3)$ and $(3, 1)$　　**b.** $(3, 1)$ and $(4, 3)$

2. a. $(5, 8)$ and $(2, 2)$　　**b.** $(2, 2)$ and $(5, 8)$

3. $(2, -7)$ and $(-7, 4)$　　**4.** $(-6, 2)$ and $(2, 9)$

5. $(2, -7)$ and $(4, -7)$　　**6.** $(-6, 2)$ and $(9, 2)$

7. $(2, -7)$ and $(2, -10)$　　**8.** $(-2, -1)$ and $(-2, -10)$

9. $(1.40, 0.56)$ and $(0, 0)$　　**10.** $(0, 0)$ and $(0.51, 0.34)$

11. $(0, 0)$ and $\left(\dfrac{1}{3}, \dfrac{1}{5}\right)$　　**12.** $\left(\dfrac{1}{2}, \dfrac{1}{4}\right)$ and $(0, 0)$

In Exercises 13 and 14, calculate the slope of each line.

13. a.

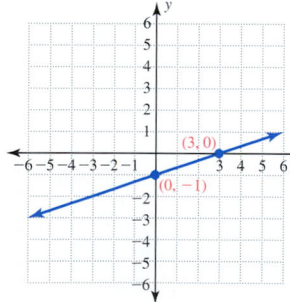

b.

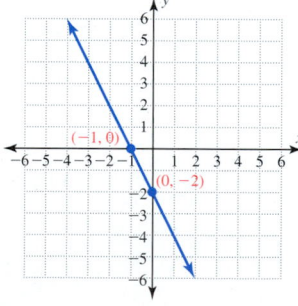

c.

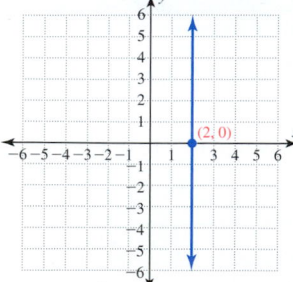

d.

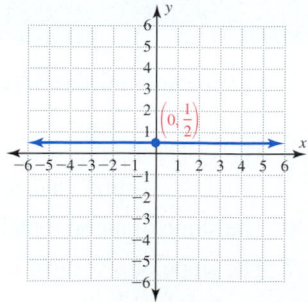

14. a.

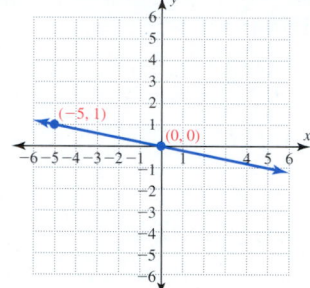

b.

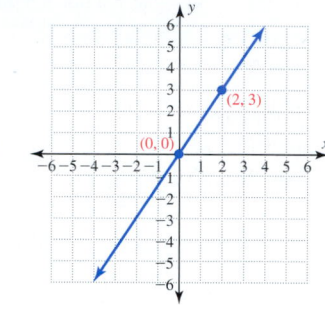

c.

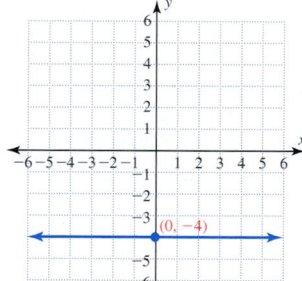

d.

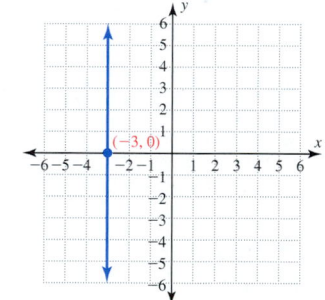

In Exercises 15 and 16, calculate the slope of the line containing the points in the table.

15.

x	y
0	1
2	4
4	7
6	10
8	13
10	16
12	19

16.

x	y
0	11
7	9
14	7
21	5
28	3
35	1
42	-1

In Exercises 17 and 18, determine two points on each line and then calculate the slope of the line.

17. $3x + 5y = 15$　　**18.** $4x + 3y = 24$

19. Slope of a Wheelchair Ramp Determine the slope of the wheelchair ramp shown in the figure.

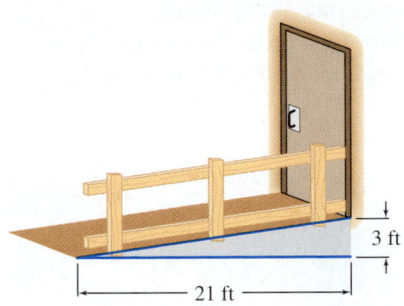

3 ft

21 ft

20. Slope of a Roof Determine the slope of the roof on the gable of the house whose cross section is sketched.

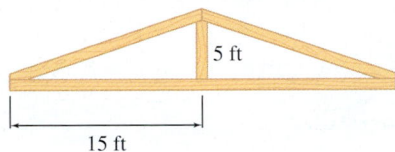

5 ft

15 ft

Objective 2 Use Slopes to Determine Whether Two Lines Are Parallel, Perpendicular, or Neither

In Exercises 21–24, m_1 is the slope of line l_1; m_2 is the slope of line l_2; and m_3 is the slope of line l_3. Line l_1 is parallel to l_2, and l_1 is perpendicular to l_3. Complete the following table.

	m_1	m_2	m_3
21.	$\dfrac{3}{7}$		
22.	$-\dfrac{3}{7}$		
23.	0		
24.	Undefined		

In Exercises 25–34, determine whether the line through the first pair of points is parallel to, perpendicular to, or neither parallel nor perpendicular to the line through the second pair of points. Assume that no line passes through both pairs of points.

25. $(-3, 5)$ and $(-6, 3)$
   $(3, -3)$ and $(6, -1)$

26. $(4, 4)$ and $(-4, -2)$
   $(3, 1)$ and $(-3, 9)$

27. $(5, 4)$ and $(10, 1)$
   $(6, 8)$ and $(3, 3)$

28. $(7, -1)$ and $(-7, -7)$
   $(7, 5)$ and $(-7, -1)$

29. $(0, 6)$ and $(8, 0)$
   $(-5, 0)$ and $(0, -7)$

30. $(0, -11)$ and $(9, 0)$
   $(-6, 0)$ and $(0, 6)$

31. $(0, 6)$ and $(4, 6)$
   $(0, 6)$ and $(0, 0)$

32. $(0, 0)$ and $(0, -3)$
   $(0, 0)$ and $(3, 0)$

33. $(5, 3)$ and $(5, -3)$
   $(7, 6)$ and $(7, -6)$

34. $(5, 3)$ and $(-5, 3)$
   $(7, 6)$ and $(-7, 6)$

Objective 3 Calculate and Interpret Rates of Change

35. Grade of a Highway For several reasons, including highway safety, an engineer has determined that the slope of a particular section of highway should not exceed a 5% grade (slope). If this maximum grade is allowed on a section covering a horizontal distance of 1,800 m, how much change in elevation is permitted on this section? (The change of elevation can be controlled by topping hills and filling in low places. See the given figure.)

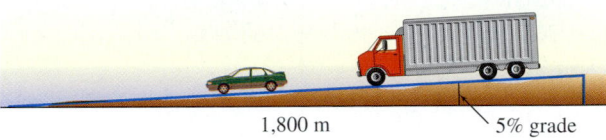

1,800 m   5% grade

36. Grade of a Highway How much change in elevation would be permitted on a section covering a horizontal distance of 2,400 m on the highway in Exercise 35?

37. A linear equation gives the production cost in dollars (y) in terms of the number of units produced (x). The slope of this line is 35. Interpret the meaning of this rate of change.

38. A linear equation gives the kilowatts (y) of electricity generated by a set of solar panels in terms of the number of hours (x) of sunlight. The slope of this line is 4.5. Interpret the meaning of this rate of change.

39. Distance Traveled The graph gives the distance an automobile has traveled after different periods of time. Determine the slope of this line, and then interpret the meaning of this rate of change.

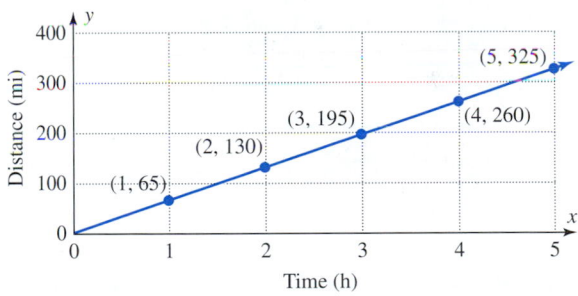

40. Cost of Shirts The graph gives the total order and shipping costs for an order of shirts from an online vendor. Determine the slope of this line, and then interpret the meaning of this rate of change.

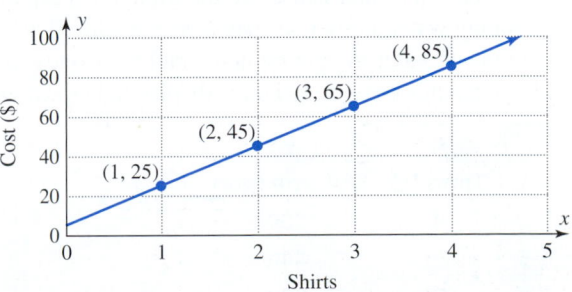

In Exercises 41–48, complete the following table involving the change in x, the change in y, and the slope of the line defined by $y = mx + b$.

	Change in x	Change in y	Slope
41.	-5	8	
42.	-7	-2	
43.	3		$\dfrac{2}{3}$
44.	3		$-\dfrac{2}{3}$
45.		2	$\dfrac{2}{3}$
46.		2	$-\dfrac{2}{3}$
47.	6		0
48.	-6		0

Connecting Concepts to Applications

49. Height of a Roof Brace The roof of the mountain cabin in the figure rises 14 ft over a run of 12 ft. Determine the height h of the brace placed 3 ft from the side of the house.

50. Height of a Tree A man who is 6 ft tall is standing so that the tip of his head is exactly in the shadow line of a tree. His shadow is 4 ft long, and the shadow of the tree is 34 ft long. How tall is the tree?

51. Rate of Flow The water tank on a firetruck holds 500 gal of water. This water is used so the firefighters can begin pumping water as soon as they arrive at a fire. The volume of water remaining in the tank x seconds after the pump has been turned on is displayed in the table.

	A	B
1	Time (s)	Volume (gal)
2	0	500
3	20	440
4	40	380
5	60	320

a. Determine the rate of change of the volume with respect to time.
b. Interpret the meaning of this value.
c. At this rate, how long do the firefighters have to connect to a hydrant before the water in the tank runs out?

52. Rate of Descent At a local flight school, the air traffic control tower began recording the altitude of an airplane flown by a student pilot x seconds after the plane began its descent. These data are displayed in the given table.

	A	B
1	Time (s)	Altitude (ft)
2	0	2800
3	15	2740
4	30	2680
5	45	2620

a. Determine the rate of change of the altitude with respect to time.
b. Interpret the meaning of this value.
c. At this rate, how long after the plane begins its descent will it land?

Review and Concept Development

53. If the slope of the line through points P and Q is $\dfrac{3}{7}$, determine the slope of the line through points Q and P.

54. If the slope of the line through points P and Q is $-\dfrac{5}{2}$, determine the slope of the line through points Q and P.

In Exercises 55 and 56, complete each table so that the points all lie on a line with the given slope.

55. $m = -\dfrac{2}{3}$

a.

x	y
0	5
3	
6	
9	
12	

b.

x	y
0	5
6	
12	
18	
24	

56. $m = \dfrac{2}{3}$

a.

x	y
0	5
3	
6	
9	
12	

b.

x	y
0	5
9	
18	
27	
36	

Applying Technology

57. Given the linear equation $y = \dfrac{5}{3}x - 2$:

    a. Use this equation and a calculator or spreadsheet to complete the table of values shown.

    b. What is the value of Δx used in this table?

    c. What is the value of Δy that is produced in this table?

    d. What is the slope of the line through these points?

x	y
0	
3	
6	
9	
12	
15	
18	

58. Given the linear equation $y = -\dfrac{2}{5}x + 1$:

    a. Use this equation and a calculator or spreadsheet to complete the table of values shown.

    b. What is the value of Δx used in this table?

    c. What is the value of Δy that is produced in this table?

    d. What is the slope of the line through these points?

x	y
0	
5	
10	
15	
20	
25	
30	

In Exercises 59–62, a line has the given y-intercept and slope. Use this information to determine another point on the line and to graph the line.

59. $(0, 2); m = \dfrac{3}{4}$ **60.** $(0, 2); m = -\dfrac{3}{4}$

61. $(0, -3); m = -2$ **62.** $(0, -3); m = 2$

In Exercises 63–68, draw a line through the point $(1, 3)$ that has the given slope.

63. $m = 0$

64. $m = 1$

65. $m = 2$

66. $m = -1$

67. $m = -2$

68. m is undefined

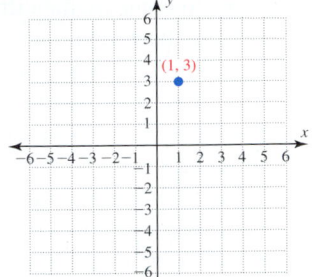

In Exercises 69 and 70, the table gives x and y values for a linear equation $y = mx + b$. Determine
    a. The x-intercept of this line
    b. The y-intercept of this line
    c. The slope of this line

69.

x	y
-3	10
-2	8
-1	6
0	4
1	2
2	0
3	-2

70.

x	y
-3	-6
-2	-3
-1	0
0	3
1	6
2	9
3	12

71. What is the slope of the x-axis?

72. What is the slope of the y-axis?

73. A line passes through quadrants I, II, and III but not quadrant IV. Is the slope of this line positive or negative?

74. A line passes through quadrants I, II, and IV but not quadrant III. Is the slope of this line positive or negative?

75. A line passes through quadrants II, III, and IV but not quadrant I. Is the slope of this line positive or negative?

76. A line passes through quadrants I, III, and IV but not quadrant II. Is the slope of this line positive or negative?

77. A line passes through quadrants I and III but not quadrants II and IV. Is the slope of this line positive or negative?

78. A line passes through quadrants II and IV but not quadrants I and III. Is the slope of this line positive or negative?

79. Variable y varies directly as x with a constant of variation 2. What is the slope of the line connecting these (x, y) data points?

80. Variable y varies directly as x with a constant of variation -3. What is the slope of the line connecting these (x, y) data points?

Group discussion questions

81. **Communicating Mathematically** Use the language of slope to compare the difficulty in pedaling for a bicycle rider in these situations. Assume the rider is moving from left to right as you present your descriptions. (Thanks to Fred Worth of Henderson State University for suggesting this problem.)
    a. A flat road
    b. A road that goes up 10 ft over 1 mi of roadway
    c. A road that goes up 10 ft over 50 ft of roadway
    d. A road that goes up 20 ft over 200 ft of roadway
    e. A road that goes down 10 ft over 1 mi of roadway

82. **Discovery Question** The equation of a line is $f(x) = mx + b$.
    a. Calculate the y-intercept of this line.
    b. Calculate the x-intercept of this line.
    c. Use the intercepts and the formula for slope to calculate the slope of this line.
    d. Use the results obtained in parts **a** to **c** to determine by inspection the slope and y-intercept of $f(x) = 5x - 3$.

83. **Communicating Mathematically** Can you calculate the slope between $(-2, -10)$ and $(-2, -10)$? Explain why or why not.

84. **Error Analysis** A student examined the graph on the calculator display shown here and concluded the line was vertical and therefore the slope was undefined. Describe the error the student has made.

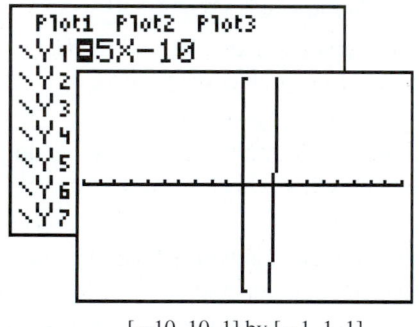

$[-10, 10, 1]$ by $[-1, 1, 1]$

85. **Error Analysis** A student examined the graphs on the calculator display shown here and concluded that the lines were parallel. Describe the error the student has made.

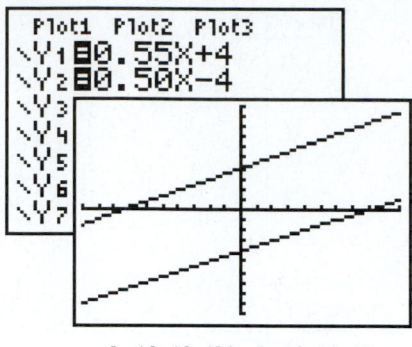

$[-10, 10, 1]$ by $[-10, 10, 1]$

86. **Communicating Mathematically** The equation $a_n = 500n + 1{,}500$ gives the total payment in dollars for a house for n months.
    a. Determine the sequence of total payments for the first 4 months.
    b. Graph this sequence.
    c. Determine the slope of the line through these points.
    d. Interpret the meaning of the rate of change in this application.
    e. Calculate a_0, and interpret the meaning of a_0.

3.1 | **Cumulative Review**

In Exercises 1–5, match each algebraic equation with the property it illustrates.

1. $(ab)c = a(bc)$
2. $(a + b) + c = a + (b + c)$
3. $a(bc + d) = a(cb + d)$
4. $a(bc + d) = abc + ad$
5. $a(bc + d) = a(d + bc)$

   A. Associative property of addition
   B. Associative property of multiplication
   C. Commutative property of addition
   D. Commutative property of multiplication
   E. Distributive property of multiplication over addition

Objectives:

1. Use the slope-intercept form to write and graph linear equations.
2. Use the point-slope form to write and graph linear equations.
3. Use the special forms of equations for horizontal and vertical lines.

We use a variety of graphs—bar graphs, pie graphs, graphs in the Cartesian plane, etc.—to examine mathematical concepts from a graphical perspective. Each type of graph has a context in which it is more informative than other graphs. Bar graphs are good for quick comparisons of data; pie charts are good for comparing parts of a whole; and line graphs are good for showing trends.

We also use a variety of algebraic forms of linear equations. Again, the context will determine which form is the most useful. The forms of linear equations that we will examine in this section are $y = mx + b$, $y - y_1 = m(x - x_1)$, $y = b$, $x = a$, and $Ax + By = C$.

1. Use the Slope-Intercept Form to Write and Graph Linear Equations

Which Form of a Linear Equation Is Best to Use for Entering an Equation into a Calculator or Spreadsheet?

One of the most useful forms of a linear equation is the form $y = mx + b$. We have already used this form to enter equations into a graphing calculator. This form is also the function form for a linear equation and can be written as $f(x) = mx + b$ to stress the relationship between the x-y input-output pairs.

The form $y = mx + b$ is called the **slope-intercept form** because this form displays the slope m and the y-intercept $(0, b)$ directly from the equation. This form can be developed by using the slope m, the y-intercept $(0, b)$, and an arbitrary point (x, y) in the formula for slope.

$$m = \frac{y_2 - y_1}{x_2 - x_1}$$ Start with the formula for slope and substitute in m for the slope and $(0, b)$ and (x, y) for the two points.

$$m = \frac{y - b}{x - 0}$$

$$m = \frac{y - b}{x}$$

$$mx = y - b$$ Then multiply both sides by the LCD x.

$$mx + b = y \quad \text{or}$$ Add b to both sides.

$$y = mx + b$$ This form of a linear equation is referred to as the slope-intercept form.

$$f(x) = mx + b$$ This is the function form for a linear function.

> **A Mathematical Note**
>
> W. W. Sawyer makes the point in his 1943 book *Mathematician's Delight* that the first mathematicians were practical men who built or made things. Some of our terminology can be traced to this source. For example, the word *straight* comes from Old English for "stretched," whereas the word *line* is the same as that for "linen thread." Thus a straight line is literally a stretched linen thread—as anyone who is planting potatoes or laying bricks knows.

Slope-Intercept Form

Algebraically	Algebraic Example	Verbal Example	Graphical Example
$y = mx + b$ is the equation of a line with slope m and y-intercept $(0, b)$.	$y = \frac{1}{2}x + 3$	This line has slope $\frac{1}{2}$ and a y-intercept of $(0, 3)$.	

Example 1   **Writing the Equation of a Line, Given Its Slope and y-Intercept**

Write in slope-intercept form the equation of a line satisfying the given conditions.

Solution

(a) $m = \dfrac{2}{7}$ and

y-intercept is $(0, 4)$

$y = mx + b$ Use the slope-intercept form.

$y = \dfrac{2}{7}x + 4$ Substitute $\dfrac{2}{7}$ for m and 4 for b.

Answer: $y = \dfrac{2}{7}x + 4$

(b) $m = -\dfrac{5}{8}$ and

y-intercept is $(0, -2)$

$y = mx + b$ Use the slope-intercept form.

$y = -\dfrac{5}{8}x + (-2)$ Substitute $-\dfrac{5}{8}$ for m and -2 for b.

Answer: $y = -\dfrac{5}{8}x - 2$

(c) $m = 0$, and
y-intercept is $(0, 3)$

$y = mx + b$ Use the slope-intercept form.

$y = 0x + 3$ Substitute 0 for m and 3 for b.

Answer: $y = 3$ This is the equation of a horizontal line with slope 0.

Self-Check 1

Write in slope-intercept form the equation of a line with slope $\dfrac{3}{4}$ and y-intercept $(0, -1)$.

One of the advantages of the slope-intercept form of a line is that we can determine by inspection two key pieces of information about the line—both the slope and the y-intercept. This is illustrated in Example 2.

Example 2   **Using the Slope-Intercept Form**

Determine the slope and y-intercept of the following lines.

(a) $f(x) = -\dfrac{5}{13}x + \dfrac{1}{3}$ **(b)** $f(x) = -6x$ **(c)** $2x + 5y = -6$

Solution

(a) $f(x) = -\dfrac{5}{13}x + \dfrac{1}{3}$ This equation is in the slope-intercept form, $f(x) = mx + b$, with $m = -\dfrac{5}{13}$ and $b = \dfrac{1}{3}$.

Answer: $m = -\dfrac{5}{13}$, and y-intercept is $\left(0, \dfrac{1}{3}\right)$

(b) $f(x) = -6x$ This equation is in the slope-intercept form, with b understood to be 0.

$f(x) = -6x + 0$

Answer: $m = -6$, and y-intercept is $(0, 0)$

(c) $2x + 5y = -6$

$$5y = -2x - 6$$

$$y = -\frac{2}{5}x - \frac{6}{5}$$

To put the equation in slope-intercept form, solve for y.

Subtract $2x$ from both sides and then divide both sides by 5.

Answer: $m = -\dfrac{2}{5}$, and y-intercept is $\left(0, -\dfrac{6}{5}\right)$

Self-Check 2

a. Determine the slope and y-intercept of $y = \dfrac{7}{11}x - 8$.

b. Determine the slope and y-intercept of $3x + 4y = 12$.

Can I Use the Slope-Intercept Form to Graph a Line?

Yes, we can sketch the graph of the line directly from the slope-intercept form without forming a table of values. In Example 3, we do this by plotting the y-intercept and then using the slope to move over horizontally and vertically. This is the skill that we introduced in Example 9 of Section 3.1.

Example 3   Graphing a Line by Using Slope-Intercept Form

Given the line defined by $f(x) = \dfrac{5}{3}x - 2$:

(a) Determine by inspection the slope of the line.

(b) Determine by inspection the y-intercept.

(c) Use this information to graph the line.

Solution

$$f(x) = \frac{5}{3}x - 2$$

(a) Slope: $\qquad\qquad m = \dfrac{5}{3}$

Use the slope-intercept form $f(x) = mx + b$ to determine the slope and y-intercept.

(b) y-intercept: $\qquad (0, -2)$

Second point: $\qquad (0 + 3, -2 + 5)$

$$= (3, 3)$$

Plot the y-intercept $(0, -2)$ and then use the slope to determine a second point. Move 3 units to the right and 5 units up to plot a second point, $(3, 3)$. Sketch the line through these two points.

(c) Graph:

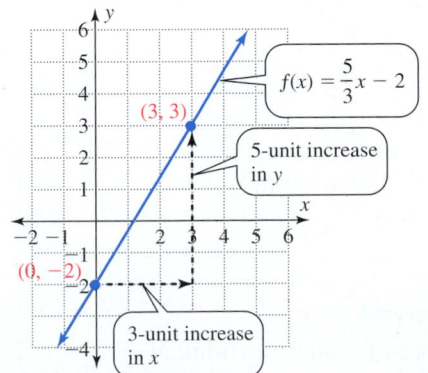

Self-Check 3

Given the line defined by $y = -\dfrac{1}{2}x + 3$:

a. Determine by inspection the slope of the line.
b. Determine by inspection the y-intercept.
c. Use this information to graph the line.

Can I Use a Table of Values Containing the y-Intercept to Write the Equation of a Line?

Yes, in Example 4 we use the slope-intercept form $y = mx + b$ and the table of values to write the equation of the line containing the points in the table. This is easily done by using inspection to determine the slope and y-intercept of this line. This is a very useful skill.

Example 4   **Using a Table of Values to Write the Equation of a Line**

Use this table of values for a linear function to do the following:

x	y
0	-2
3	3
6	8
9	13
12	18
15	23
18	28

(a) Write the y-intercept of this line.
(b) Find the slope of this line.
(c) Write the equation of this line in slope-intercept form.
(d) Use this equation and a graphing calculator to produce the original table as a check on this result.

Solution

(a) The y-intercept is $(0, -2)$.

From the first row in the table, the y-intercept is $(0, -2)$.

(b) $m = \dfrac{5}{3}$

From one row to the next in the table, x increases by 3 units and y increases by 5 units. $m = \dfrac{\Delta y}{\Delta x}$ so the slope is $\dfrac{5}{3}$.

(c) $y = \dfrac{5}{3}x - 2$

Substitute this y-intercept and the slope into the slope-intercept form $y = mx + b$.

After entering this equation into a graphing calculator, we can produce the given table.

(d)

```
Plot1 Plot2 Plot3
\Y1 =(5/3)X-2
\Y2=        X    | Y1
\Y3=      0      | -2
\Y4=      3      | 3
\Y5=      6      | 8
\Y6=      9      | 13
\Y7=      12     | 18
          15     | 23
          18     | 28
         X=0
```

Self-Check 4

Use this table of values for a linear function to do the following:

x	y
-6	6
-3	4
0	2
3	0
6	-2
9	-4
12	-6

a. Write the y-intercept of this line.
b. Find the slope of this line.
c. Write the equation of this line in slope-intercept form.
d. Use this equation and a graphing calculator to produce the original table as a check on this result.

2. Use the Point-Slope Form to Write and Graph Linear Equations

Are There Any Other Forms of Linear Equations That Are Widely Used?

Yes, the point-slope form that is developed here is often used to write the equation of a line, given either a point on the line and its slope or two points on the line. Suppose that a line passes through the point (x_1, y_1) and has slope m. To determine the equation that relates x_1 and y_1 to any other point (x, y) on this line, we substitute (x, y) for (x_2, y_2) into the formula for slope to obtain

$$m = \frac{y_2 - y_1}{x_2 - x_1}$$ Start with the formula for slope. Then substitute m and the two points (x_1, y_1) and (x, y) into this formula.

$$m = \frac{y - y_1}{x - x_1}$$

$$m(x - x_1) = y - y_1 \quad \text{or}$$ To produce the point-slope form, multiply both sides of this equation by $x - x_1$.

$$y - y_1 = m(x - x_1)$$

Point-Slope Form $y - y_1 = m(x - x_1)$

Algebraically	Algebraic Example	Graphical Example	Verbal Example
$y - y_1 = m(x - x_1)$ is the equation of a line through (x_1, y_1) with slope m.	$y - 1 = \dfrac{1}{3}(x - 1)$		This line passes through the point $(1, 1)$ with slope $\dfrac{1}{3}$.

Can I Write an Equation of a Line or Graph a Line Directly from a Given Point and Slope?

Yes, this is illustrated in Example 5 and is particularly useful when the slope is given as a fraction.

Example 5 Writing the Equation of the Line Through a Point with a Given Slope

Write the point-slope equation of the line that passes through $(1, -2)$ with slope $m = \dfrac{3}{4}$.

Then sketch the graph of this line.

Solution

Point-slope form:

$$y - y_1 = m(x - x_1)$$ Use the point-slope form.

$$y - (-2) = \frac{3}{4}(x - 1)$$ Since the line passes through the point $(1, -2)$ with slope $m = \dfrac{3}{4}$, substitute 1 for x_1, -2 for y_1, and $\dfrac{3}{4}$ for m.

$$y + 2 = \frac{3}{4}(x - 1)$$

Graph:

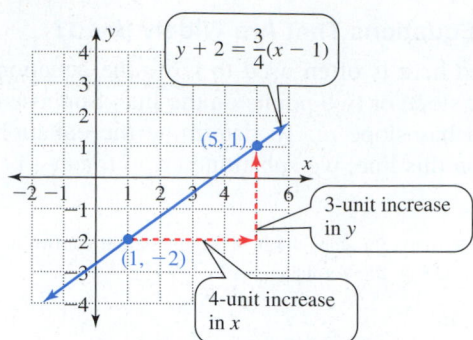

$$y + 2 = \frac{3}{4}(x - 1)$$

(5, 1)

3-unit increase in y

(1, −2)

4-unit increase in x

To graph this line, plot the point $(1, -2)$ and then use the slope to determine a second point. Move 4 units to the right and 3 units up to the point $(5, 1)$. Sketch the line through these two points.

Self-Check 5

Write the point-slope equation of the line that passes through $(-3, 5)$ with slope $m = -\dfrac{4}{3}$. Then sketch the graph of this line.

When I Am Trying to Write the Equation of a Line, Which Form Should I Use?

The answer depends on the two pieces of information that you are given. You can always use the slope-intercept form $y = mx + b$, but this may not always be the easiest method. In Example 6, we illustrate two methods for writing the equation of a line through two points.

Example 6   ### Writing the Equation of the Line Through Two Points

Write in slope-intercept form the equation of the line through $(-6, -4)$ and $(-2, 2)$.

Solution

Method I

Using the point-slope form:

$$\frac{y_2 - y_1}{x_2 - x_1} = \frac{2 - (-4)}{-2 - (-6)}$$ Step 1: Since the slope is not given, calculate m by using the two given points.

$$m = \frac{6}{4} = \frac{3}{2}$$

$$y - y_1 = m(x - x_1)$$ Step 2: Use the point-slope form.

$$y - (-4) = \frac{3}{2}(x - (-6))$$ Substitute $(-6, -4)$ for (x_1, y_1) and $\frac{3}{2}$ for m.

$$y + 4 = \frac{3}{2}(x + 6)$$

$$2(y + 4) = 3(x + 6)$$ Multiply both sides by the LCD, 2. Simplify and write the equation in the slope-intercept form, $y = mx + b$.

$$2y + 8 = 3x + 18$$

$$2y = 3x + 10$$

$$y = \frac{3}{2}x + 5$$

Method II

Using the slope-intercept form:

$$\frac{y_2 - y_1}{x_2 - x_1} = \frac{2 - (-4)}{-2 - (-6)}$$ Step 1: Since the slope is not given, calculate m by using the two given points.

$$m = \frac{6}{4} = \frac{3}{2}$$

$$y = mx + b$$ Step 2: Use the slope-intercept form to calculate b.

$$-4 = \frac{3}{2}(-6) + b$$ Substitute $(-6, -4)$ for (x_1, y_1) and $\frac{3}{2}$ for m.

$$-4 = -9 + b$$ Simplify and solve for b.

$$b = 5$$

$$y = \frac{3}{2}x + 5$$ Substitute the values for m and b into the slope-intercept form to write the equation satisfying the given conditions.

Answer: $y = \frac{3}{2}x + 5$ Can you use a graphing calculator to check that this is the correct equation?

Self-Check 6

Write in slope-intercept form the equation of the line through $(5, -3)$ and $(-2, 1)$.

In Example 6, the same equation can be obtained by substituting in the point $(-2, 2)$ instead of the point $(-6, -4)$. It is worthwhile to note that both points were used to calculate the slope of the line. Thus either point can be used to select the specific line that has this slope.

Can the Equation of a Line Be Used to Identify Parallel and Perpendicular Lines?

In Section 3.1, we noted that parallel lines have the same slope and that perpendicular lines have slopes that are opposite reciprocals. If the equations of lines are written in slope-intercept form, it is easy to identify their slopes and to determine whether the lines are parallel or perpendicular.

Example 7   Determining Whether Two Lines Are Parallel or Perpendicular

Determine whether the first line is parallel to, perpendicular to, or neither parallel nor perpendicular to the second line.

Solution

First Equation		Second Equation	
(a) $f(x) = \frac{2}{3}x + 7$	$f(x) = \frac{2}{3}x + 7$	$f(x) = -\frac{3}{2}x + 5$	Each slope can be determined by inspection since the equations are in slope-intercept form.
$f(x) = -\frac{3}{2}x + 5$	Slope: $m = \frac{2}{3}$	Slope: $m = -\frac{3}{2}$	

Answer: Since $\left(\frac{2}{3}\right)\left(-\frac{3}{2}\right) = -1$, these lines are perpendicular.   Lines are perpendicular if the product of their slopes is -1.

	First Equation	**Second Equation**	

(b) $5x - 3y = 12$

$5x = 3y + 8$

First Equation

$5x - 3y = 12$

$-3y = -5x + 12$

$$y = \frac{5}{3}x - 4$$

Slope: $m = \dfrac{5}{3}$

Second Equation

$5x = 3y + 8$

$5x - 8 = 3y$

$$\frac{5}{3}x - \frac{8}{3} = y$$

$$y = \frac{5}{3}x - \frac{8}{3}$$

Slope: $m = \dfrac{5}{3}$

Write each equation in slope-intercept form so the slope can be determined by inspection.

The lines are parallel, but they are not the same line because the y-intercepts are different.

Answer: Since the slopes are equal, the lines are parallel.

Self-Check 7

Determine whether the first line is parallel to, perpendicular to, or neither parallel nor perpendicular to the second line.

a. $f(x) = \dfrac{2}{5}x - 8$; $f(x) = \dfrac{5}{2}x + 8$ **b.** $f(x) = \dfrac{2}{5}x - 8$; $f(x) = \dfrac{2}{5}x + 8$

c. $4x - 3y = 9$; $3x + 4y = 8$

In Example 8, we write the equation of a line perpendicular to a given line. Before reading this example, recall that either two points on the line or one point on the line and the slope are required to determine the equation of a line. In this case, one point on the line is given. The word "perpendicular" is a clue about the slope.

Example 8   Writing the Equation of a Line Perpendicular to a Given Line

Write in slope-intercept form the equation of the line passing through $(-5, 6)$ and perpendicular to the line $y = -\dfrac{5}{3}x - 6$. Then use a graphing calculator to check that your equation is correct.

Solution

$$y = -\frac{5}{3}x - 6$$

From the slope-intercept form of the given line, its slope is $-\dfrac{5}{3}$.

$$\text{Slope of given line} = -\frac{5}{3}$$

Because the new line is perpendicular to the given line, its slope is $\dfrac{3}{5}$.

Slope of perpendicular line: $m = \dfrac{3}{5}$

$$y - y_1 = m(x - x_1)$$

Use the point-slope form to determine the equation of the new line. Substitute $(-5, 6)$ for (x_1, y_1) and $\dfrac{3}{5}$ for m.

$$y - 6 = \frac{3}{5}[x - (-5)]$$

$$5(y - 6) = 3(x + 5)$$

Multiply both sides by the LCD, 5.

$$5y - 30 = 3x + 15$$

$$5y = 3x + 45$$

Simplify, and write the equation in slope-intercept form, $y = mx + b$.

Answer: $y = \dfrac{3}{5}x + 9$

Graphical Check

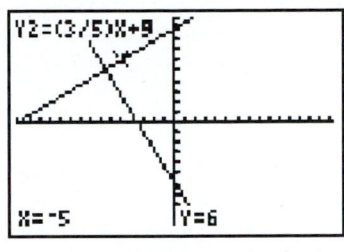

[−15.2, 15.2, 1] by [−10, 10, 1]

This graph visually displays a slope of $\dfrac{3}{5}$ and passes through the point $(-5, 6)$ and the Y_2-intercept $(0, 9)$. On a TI-84 Plus calculator, using ZOOM **ZStandard** followed by **ZSquare** creates a screen that squares the rectangular display and makes the perpendicular lines really appear perpendicular on the display.

Self-Check 8

Write the equation of the line that is parallel to $y = \dfrac{4}{7}x - 2$ and passes through $(4, 8)$.

---

3. Use the Special Forms of Equations for Horizontal and Vertical Lines

Are There Special Forms for Horizontal and Vertical Lines?

Yes, we examined horizontal and vertical lines in Section 3.1. Now we will examine their equations. Recall that all horizontal lines have a slope of zero—there is no rise between two points on a horizontal line. The equation of a horizontal line is developed next by substituting $m = 0$ and a y-intercept of $(0, b)$ into the slope-intercept form $y = mx + b$.

$y = mx + b$ Start with slope-intercept from.

$y = 0x + b$ Substitute in 0 for m.

$y = b$

The equation of a horizontal line can always be written in the form $y = b$ or $f(x) = b$. The y-coordinate of every point on a horizontal line is the same—thus y is a constant.

The equation of a vertical line has a similar form. The x-coordinate of every point on a vertical line is the same—thus x is a constant. A vertical line with an x-intercept $(a, 0)$ is defined by the equation $x = a$.

Horizontal and Vertical Lines

Algebraically	Numerical Example		Graphical Example	Verbally
$y = b$ is the equation of a horizontal line with y-intercept $(0, b)$. **Example:** $y = 3$	**x** −2 −1 0 1 2	**y** 3 3 3 3 3		This horizontal line has a y-intercept of $(0, 3)$ and a slope of 0.
$x = a$ is the equation of a vertical line with x-intercept $(a, 0)$. **Example:** $x = -2$	**x** −2 −2 −2 −2 −2	**y** −2 −1 0 1 2		This vertical line has an x-intercept of $(-2, 0)$, and its slope is undefined.

Example 9 Writing the Equations of Horizontal and Vertical Lines

Write the equation of each of these lines.

Solution

(a)

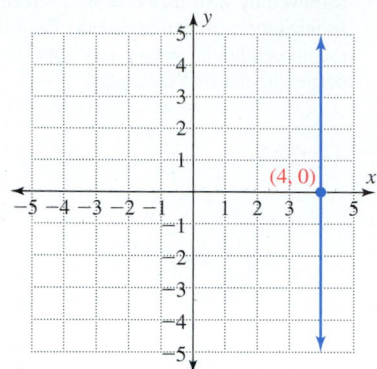

Answer: $x = 4$

This is a vertical line with an x-intercept of $(4, 0)$. Use the form $x = a$ with 4 as the value of a.

(b)

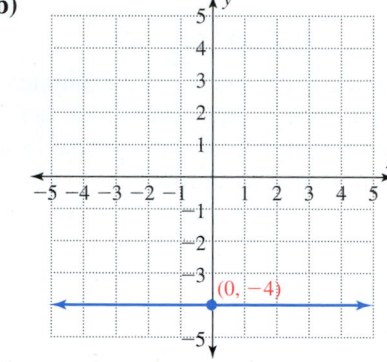

Answer: $y = -4$

This is a horizontal line with a y-intercept of $(0, -4)$. Use the form $y = b$ with -4 as the value of b.

Self-Check 9

a. Write the equation of a horizontal line through $(6, -1)$.

b. Write the equation of a vertical line through $(6, -1)$.

We now examine two other forms of linear equations—the general form and the point-slope form.

How Can I Recognize When an Equation Is a Linear Equation?

Every linear equation with variables x and y is a first-degree equation in both x and y. A first-degree equation in both x and y means that the exponent on both of these variables is 1. Thus a linear equation can be written in the **general form** $Ax + By = C$, where A, B, and C are real-number constants. The general form is used to represent linear equations in many disciplines because this form allows us to write many equations in a nice, clean form that is free of fractions. The general form is generally written so that the coefficient of x is positive. (Sometimes the form $Ax + By + C = 0$ is also referred to as general form.)

Example 10 Writing a Linear Equation in General Form

Rewrite the equation $y = \dfrac{2}{3}x + \dfrac{4}{5}$ in the general form $Ax + By = C$ with integer coefficients.

Solution

$$y = \frac{2}{3}x + \frac{4}{5}$$

$$15y = 15\left(\frac{2}{3}x + \frac{4}{5}\right) \qquad \text{Multiply both sides of the equation by 15, the LCD of all these terms.}$$

$$15y = 15\left(\frac{2}{3}x\right) + 15\left(\frac{4}{5}\right) \qquad \text{Distribute the factor of 15 and then simplify.}$$

$$15y = 5(2x) + 3(4)$$

$$15y = 10x + 12$$

$$-12 = 10x - 15y \quad \text{or} \qquad \text{Subtract } 15y \text{ from both sides of the equation, and subtract 12 from both sides of the equation.}$$

Answer: $10x - 15y = -12$

Self-Check 10

Rewrite $y = -\dfrac{1}{2}x + \dfrac{3}{7}$ in general form with integer coefficients.

Summary

A line is completely determined by two points. In fact, given two appropriate facts about the line, we can write an equation of the line. The following box gives some possible scenarios and some comments on each of these possibilities.

Strategy for Writing the Equation of a Line

Given Information	Method	Comments
Horizontal line through a given point	Use the special form $y = b$ and use the y-coordinate of the given point.	Very easy to use, but this is a very special case and will work only in limited cases.
Vertical line through a given point	Use the special form $x = a$ and use the x-coordinate of the given point.	Very easy to use, but this is a very special case and will work only in limited cases.
Slope and y-intercept	Use the form $y = mx + b$ and substitute in the given values of m and b.	This information fits the slope-intercept form perfectly.
Point and slope	**a.** Substitute the given values of m, x, and y into the form $y = mx + b$. Calculate b and then use m and b to write the equation in the form $y = mx + b$. **b.** Substitute the given values of m, x_1, and y_1 into the form $y - y_1 = m(x - x_1)$. This equation can be rewritten in another form if desired.	**a.** This method uses the function form of a linear equation. The form $y = mx + b$ is widely used in mathematics and to enter equations into calculators and spreadsheets. **b.** This information fits the point-slope form perfectly but it will require extra steps to convert this equation to the form $y = mx + b$ or to the general form.
Two points	Use the given values to calculate the slope m. Then use the method for finding the equation for a given point and slope as shown above.	Both given points are used to calculate the slope of the line. Then either point can be used with this slope to determine an equation of the line.

Self-Check Answers

1. $y = \dfrac{3}{4}x - 1$

2. **a.** $m = \dfrac{7}{11}$, y-intercept $(0, -8)$

   b. $m = -\dfrac{3}{4}$, y-intercept $(0, 3)$

3. **a.** $m = -\dfrac{1}{2}$ **b.** y-intercept $(0, 3)$

   c.

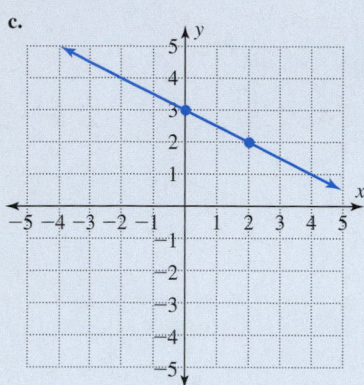

4. **a.** $(0, 2)$ **b.** $m = -\dfrac{2}{3}$ **c.** $y = -\dfrac{2}{3}x + 2$

   d.

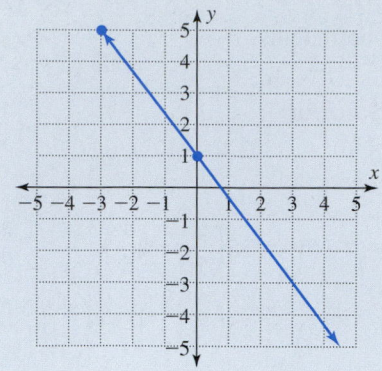

5. $y - 5 = -\dfrac{4}{3}(x + 3)$

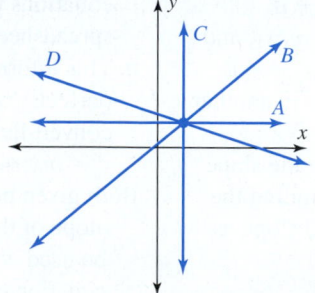

6. $y = -\dfrac{4}{7}x - \dfrac{1}{7}$

7. **a.** Neither parallel nor perpendicular
   b. Parallel
   c. Perpendicular

8. $y = \dfrac{4}{7}x + \dfrac{40}{7}$

9. **a.** $y = -1$ **b.** $x = 6$

10. $7x + 14y = 6$

3.2   Using the Language and Symbolism of Mathematics

1. The slope-intercept form of the equation of a line is _____.

2. Lines with the same slope are _____.

3. Lines whose slopes are opposite reciprocals of each other are _____.

4. The line defined by $x = 5$ is a _____ line whose x-intercept is _____.

5. The line defined by $y = -5$ is a _____ line whose y-intercept is _____.

6. The general form of the equation of a line is _____.

7. When we say that a linear equation in x and y is a first-degree equation, we mean that the exponent on both x and y is _____.

8. The equation $y - y_1 = m(x - x_1)$ is the _____-_____ form of a linear equation with (x_1, y_1) representing a _____ on the line and m representing the _____ of the line.

3.2   Quick Review

1. Even though no scale is given in this graph, the slope of each line is listed below. Match each slope with the corresponding line.
   a. undefined
   b. −0.5
   c. 0.0
   d. 1.0

2. Solve $2x + 5y + 20 = 0$ for y.

3. Determine the x- and y-intercepts of the line defined by $2x + 5y + 20 = 0$.

4. What is the least common denominator of the fractions $\dfrac{5}{6}$ and $\dfrac{2}{15}$?

5. What is the multiplicative inverse of $\dfrac{2}{15}$?

3.2 Exercises

Objective 1 Use the Slope-Intercept Form to Write and Graph Linear Equations

In Exercises 1 and 2, determine by inspection the slope and y-intercept of each line.

1. a. $f(x) = 2x + 5$ **b.** $f(x) = -\dfrac{3}{11}x - \dfrac{4}{5}$
   c. $f(x) = 6x$

2. a. $f(x) = -3x + 7$ **b.** $f(x) = \dfrac{4}{9}x - \dfrac{2}{3}$
   c. $f(x) = -6x$

In Exercises 3–8, write in the slope-intercept form $y = mx + b$ the equation of the line with the given slope and y-intercept.

3. $m = 4$; $(0, 7)$ **4.** $m = 3$; $(0, 5)$

5. $m = -\dfrac{2}{11}$; $(0, 5)$ **6.** $m = -\dfrac{5}{7}$; $(0, -2)$

7. $m = 0$; $(0, -6)$ **8.** $m = 0$; $(0, 4)$

In Exercises 9–14, use the equation of each line to:

a. Determine by inspection the slope of the line.
b. Determine by inspection the y-intercept.
c. Graph the line using pencil and paper.

9. $f(x) = \dfrac{2}{3}x - 4$ **10.** $f(x) = \dfrac{3}{4}x - 5$

11. $f(x) = -\dfrac{5}{3}x + 2$ **12.** $f(x) = -\dfrac{5}{2}x + 3$

13. $f(x) = 2$ **14.** $f(x) = -3$

In Exercises 15–18, use the graph of each line to:

a. Determine the slope of the line.
b. Determine the y-intercept.
c. Write the equation of the line in slope-intercept form $f(x) = mx + b$.

15.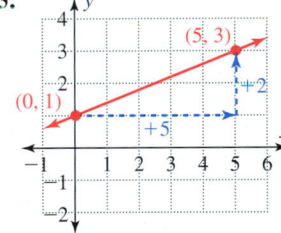

16.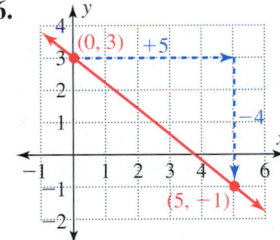

17.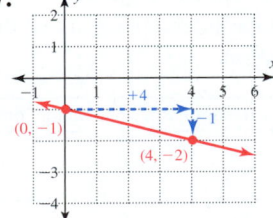

18.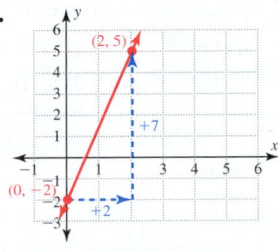

In Exercises 19–22, use the given table of values for a linear function to determine the following:

a. Value of Δx shown in the table
b. Value of Δy produced in the table
c. Slope of this line
d. The y-intercept of this line
e. Equation of this line in slope-intercept form

19.

x	y
−3	−20
−2	−13
−1	−6
0	1
1	8
2	15
3	22

20.

x	y
−3	18
−2	13
−1	8
0	3
1	−2
2	−7
3	−12

21.

x	y
−4	10
−2	9
0	8
2	7
4	6
6	5
8	4

22.

x	y
−10	−6
−5	−5
0	−4
5	−3
10	−2
15	−1
20	0

Objective 2 Use the Point-Slope Form to Write and Graph Linear Equations

In Exercises 23–26, write in point-slope form the equation of the line passing through the given point with the slope specified. Then sketch a graph of the line.

23. $(2, 3)$; $m = -4$ **24.** $(5, 2)$; $m = -2$

25. $(-1, 4)$; $m = \dfrac{3}{5}$ **26.** $(3, -2)$; $m = \dfrac{1}{2}$

In Exercises 27–30, determine by inspection the slope of the line and one point on the line. Then sketch a graph of the line.

27. $y - 4 = 2(x - 3)$ **28.** $y - 1 = -3(x - 2)$

29. $y + 5 = -\dfrac{3}{2}(x - 2)$ **30.** $y - 3 = \dfrac{1}{3}(x + 1)$

In Exercises 31–38, write in slope-intercept form the equation of the line passing through the given points.

31. $(0, 6), (-3, 4)$ **32.** $(-4, 3), (0, 5)$

33. $(-4, 2), (4, 4)$ **34.** $(2, 3), (-2, 1)$

35. $(-2, 1), (3, 2)$ **36.** $(-5, -2), (-2, 5)$

37. $(-4, -2), (-1, 3)$ **38.** $(5, 3), (8, 1)$

In Exercises 39–42, determine the slope and one point on each line; then write the equation of the line in slope-intercept form.

39.

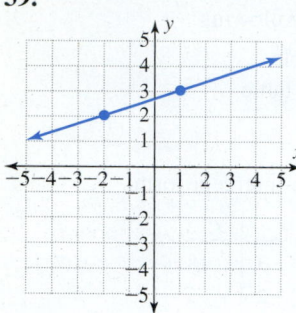

40.

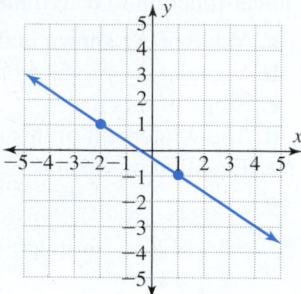

41.

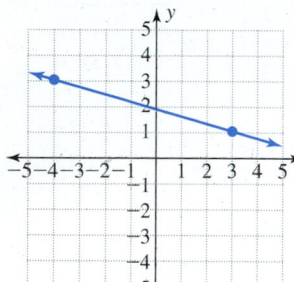

42.

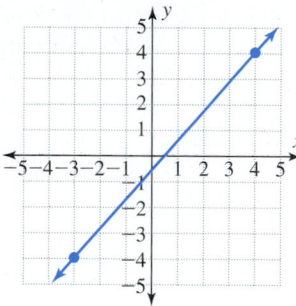

Objective 3 Use the Special Forms of Equations for Horizontal and Vertical Lines

In Exercises 43–46, write the equation of the line passing through the given points. Then sketch a graph of the line.

43. $(2, 7), (2, 4)$

44. $(-1, 6), (-1, -3)$

45. $(3, -5), (5, -5)$

46. $(-2, 1), (4, 1)$

In Exercises 47–50, write the equation of each line.

47.

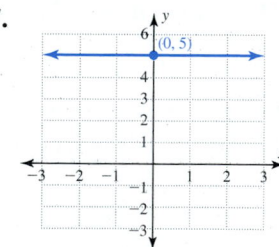

48.

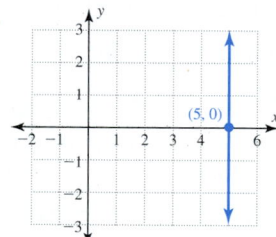

49.

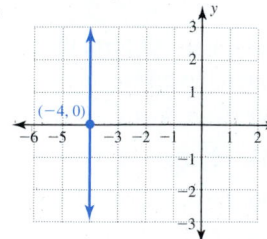

50.

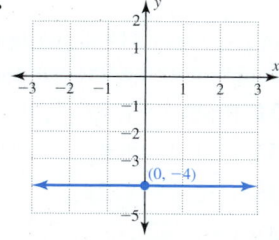

51. Write an equation of the line passing through $(4, -8)$ and parallel to the x-axis.

52. Write an equation of the line passing through $(4, -8)$ and perpendicular to the x-axis.

Review and Concept Development

In Exercises 53–60, determine whether the first line is parallel to, perpendicular to, or neither parallel nor perpendicular to the second line.

53. $y = 5x + 11$
$y = -5x + 11$

54. $y = 5x + 11$
$y = 5x - 11$

55. $y = \dfrac{2}{3}x + 7$
$y = -\dfrac{3}{2}x - 2$

56. $y = \dfrac{3}{13}x - 4$
$y = \dfrac{13}{3}x + 4$

57. $2x + 5y = 8$
$6x + 15y - 7 = 0$

58. $3x - 4y + 8 = 0$
$4x + 3y = -15$

59. $y - 8 = 4(x + 5)$
$y + 4 = 5(x - 3)$

60. $y + 6 = 2(x - 1)$
$y - 1 = 2(x + 3)$

61. Write in slope-intercept form the equation of the line passing through $(2, 3)$ and parallel to $y = \dfrac{3}{7}x - 1$.

62. Write in slope-intercept form the equation of the line passing through $(2, 3)$ and perpendicular to $y = \dfrac{3}{7}x - 1$.

63. Write in slope-intercept form the equation of the line passing through $(-5, 1)$ and perpendicular to $y = -\dfrac{4}{3}x + 7$.

64. Write in slope-intercept form the equation of the line passing through $(-5, 1)$ and parallel to $y = -\dfrac{4}{3}x + 7$.

65. The slope of a line is $m = \dfrac{3}{4}$, and its y-intercept is $(0, -2)$. Use this information to complete this table.

x	y
0	
4	
8	
12	
16	

66. The slope of a line is $m = -\dfrac{4}{3}$, and its y-intercept is $(0, 2)$. Use this information to complete this table.

x	y
0	
3	
6	
9	
12	

Comparing Equivalent Forms of Linear Equations In Exercises 67–70, complete the slope-intercept form and the general form for each linear equation.

Point-Slope Form	Slope-Intercept Form	General Form
Example:		
$y - 1 = 2(x - 3)$	$y = 2x - 5$	$2x - y = 5$
67. $y - 3 = 4(x + 1)$		
68. $y + 2 = -3(x - 1)$		
69. $y + 4 = -\dfrac{2}{3}(x - 5)$		
70. $y - 5 = \dfrac{7}{2}(x + 3)$		

Connecting Concepts to Applications

71. Distance Traveled This graph gives the distance from the dispatch center for a truck at different times. Use this graph to determine the linear equation $f(x) = mx + b$ of this line. Then interpret the meaning of m and b in this problem.

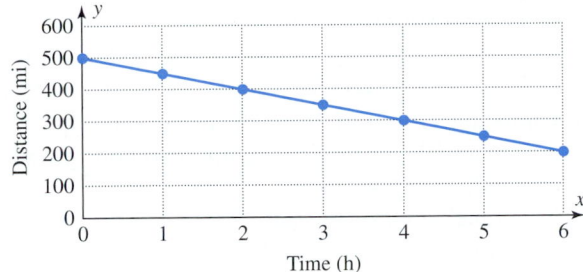

72. Cost of Telephone Minutes The graph gives the monthly cost for a cell phone based on the number of minutes used. Use this graph to determine the linear equation $f(x) = mx + b$ of this line. Then interpret the meaning of m and b in this problem.

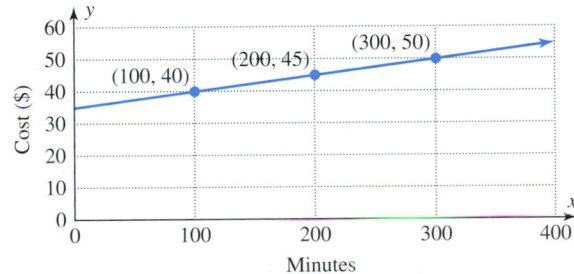

73. Jet Fuel Usage The main fuel tank on one aircraft contains 3,500 gal of jet fuel when fuel begins to be pumped from this tank. The volume of fuel remaining in the tank x minutes after the fuel pump has been turned on is displayed in the table.

	A	B
1	Time (min)	Volume (gal)
2	0	3,500
3	15	3,320
4	30	3,140
5	45	2,960

a. Determine the slope of the line containing these points.
b. Interpret the meaning of the slope from part **a**.
c. Write a function f so that $f(x)$ gives the number of gallons of fuel left after x minutes.
d. Is 300 a practical input value for x? Explain.

74. Postage Meter Usage A postage meter has had $100 of credit placed into it at a post office. The number of dollars of credit remaining in the postage meter after x letters have had postage applied by this meter is displayed in the given table.

	A	B
1	Letters	Credit ($)
2	0	$100.00
3	25	$89.00
4	50	$78.00
5	75	$67.00

a. Determine the slope of the line containing these points.
b. Interpret the meaning of the slope from part **a**.
c. Write a function f so that $f(x)$ gives the number of dollars of credit left after x letters have had postage applied.
d. Is 300 a practical input value for x? Explain.

75. Cost of a Service Call The table displays the dollar cost of a service call by Reliable Heating and Air Conditioning, based on the number of hours the service person is at the customer's location. Use this table to determine the linear equation $f(x) = mx + b$ for these data points. Then interpret the meaning of m and b in this problem.

Hours x	Cost y ($)
0	75
0.5	100
1.0	125
1.5	150
2.0	175
2.5	200

76. Salary of a Salesperson The table displays the total monthly salary of a salesperson at an automobile dealership. This salary consists of a base salary and a commission based on this person's sales for the month. Use this table to determine the linear equation $f(x) = mx + b$ for these data points. Then interpret the meaning of m and b in this problem.

Sales x ($)	Salary y ($)
0	1,200
20,000	1,400
40,000	1,600
60,000	1,800
80,000	2,000
100,000	2,200

77. Taxi Fares The table displays the dollar cost of a taxi ride based on the number of miles traveled. Write the linear equation $f(x) = mx + b$ for the line that contains these data points. Then interpret the meaning of m and b in this problem. (*Hint:* Although you would not pay for a ride of 0 mi, the y-intercept still has a meaningful interpretation.)

Miles x	Cost y ($)
1.5	4.45
4.0	6.45
7.0	8.85
12.5	13.25

78. Cost of Books The table displays the dollar cost of a shipment of math books from a publisher to a campus bookstore. Write the linear equation $f(x) = mx + b$ for the line that contains these data points. Then interpret the meaning of m and b in this problem. (*Hint:* Although you

Books x	Cost y ($)
70	4,210
140	8,410
250	15,010
400	24,010

would not pay for a shipment of zero books, the *y*-intercept still has a meaningful interpretation.)

79. Displacement of a Spring A spring is 8 cm long. The spring stretches 2 cm for each kilogram attached. Express the new length of the spring *y* in terms of the mass *x*. Write this equation in slope-intercept form. What is the significance of the slope and the *y*-intercept?

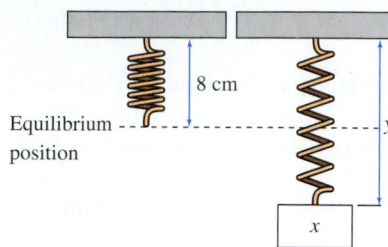

80. Fixed and Variable Production Costs The monthly cost *y* for producing toner cartridges at a factory is $17,500 plus $5 for each of the *x* cartridges produced. Write a linear equation in slope-intercept form that expresses the relationship between *x* and *y*. What is the significance of the slope and the *y*-intercept?

81. Duke Energy Dividends An investor had the annual dividend results for Duke Energy for the years 2002 and 2005. In 2002 the annual dividend per share was $1.10, and in 2005 the annual dividend per share was $1.17.
   a. Using *x* for the number of years after 2000 and *y* as the annual dividend, write a linear equation to relate the dividend to the year.
   b. Use this equation to estimate the dividend per share that the investor might receive in 2006. The actual dividend in 2006 was $1.19. Do you think this was a reasonable way to estimate this dividend?

Group discussion questions

82. Communicating Mathematically The profit *y* of a company over a period of time *x* is graphed on a rectangular coordinate system. Write a paragraph describing your interpretation when the slope is **(a)** negative, **(b)** zero, and **(c)** positive.

83. Challenge Question
   a. Determine by inspection which point is not on the same line as all the other points in this table.
   b. Explain your reasoning, using the notation $m = \dfrac{\Delta y}{\Delta x}$.

x	y
−22	−3
−11	4
0	11
11	18
22	21

84. Challenge Question Write in general form the equation of the line satisfying the conditions given.
   a. Parallel to *x*-axis through (2, 3)
   b. Parallel to *y*-axis through (2, 3)
   c. Perpendicular to *x*-axis through (−5, 8)
   d. Perpendicular to *y*-axis through (−5, 8)
   e. Parallel to $y = \dfrac{3}{7}x + 7$ through (2, −3)
   f. Parallel to $y = \dfrac{2}{9}x - 5$ through (2, −3)
   g. Perpendicular to $y - 1 = \dfrac{1}{2}(x + 3)$ through (3, 5)
   h. Perpendicular to $y + 2 = -\dfrac{1}{3}(x - 7)$ through (4, −9)

85. Challenge Question Write in general form the equation of the line satisfying the conditions given.
   a. *y*-intercept (0, −5) with $m = \dfrac{1}{5}$
   b. *y*-intercept (0, 5) with $m = -\dfrac{1}{5}$
   c. *x*-intercept (3, 0) with $m = 7$
   d. *x*-intercept (2, 0) with $m = -4$
   e. Through (−7, −2) with $m = 0$
   f. Through (−7, 5) with slope undefined

3.2 Cumulative Review

In Exercises 1–5, match each verbal description with the number that is the most appropriate choice.

1. A natural number   **A.** 0

2. A whole number that is not a natural number   **B.** 0.25

3. An irrational number   **C.** $-\sqrt{9}$

4. A rational number that is not an integer   **D.** $\sqrt{3}$

5. An integer that is not a whole number   **E.** $\sqrt{16}$

| **Section 3.3** | Solving Systems of Linear Equations in Two Variables Graphically and Numerically |

Objectives:

1. Check possible solutions of a linear system.
2. Solve a system of linear equations by using graphs and tables.
3. Identify inconsistent systems and systems of dependent linear equations.

It is often useful for a business to examine two different options or situations simultaneously and to be able to compare these options. We will illustrate this in Example 4 when we compare two different options for producing pasta. If we use equations to model both options, then we can examine both equations simultaneously. When we consider two or more equations at the same time, we refer to this as a **system of equations.** A **solution of a system of linear equations** in two variables is an ordered pair that is a solution of each equation in the system. Each solution of a linear equation in two variables is represented by a point on the line. If there is a unique solution to a system of linear equations, it is represented by their point of intersection, the point that is on both graphs.

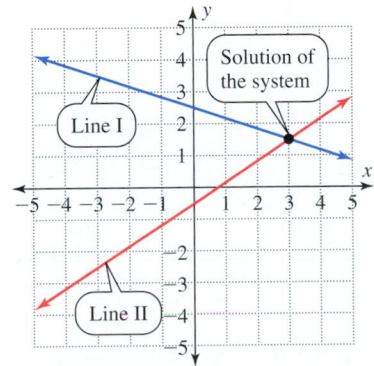

The graphical method for solving systems of linear equations, first covered in Section 2.3, is an excellent way to develop a good understanding of what the solution to a system of linear equations represents. It is easy to "see" the point of intersection of two lines, the ordered pair that satisfies the equation of both lines. Although it may be easy to approximate the coordinates of this point of intersection, you may need to use the algebraic methods covered in Sections 3.4 and 3.5 to obtain the exact values of these coordinates.

Solution of a system of two linear equations

1. Check Possible Solutions of a Linear System

In Example 1, we check possible solutions of a system of linear equations by substituting the x- and y-coordinates into each equation. A solution of a linear system will check in each equation.

Example 1   Checking Possible Solutions of a System of Linear Equations

Determine whether each ordered pair is a solution of the linear system $\begin{cases} 3x + y = 9 \\ x + 2y = 8 \end{cases}$.

(a) $(4, -3)$ **(b)** $(6, 1)$ **(c)** $(2, 3)$

Solution

(a) Check $(4, -3)$

First Equation

$$3x + y = 9$$
$$3(4) + (-3) \overset{?}{=} 9$$
$$12 - 3 \overset{?}{=} 9$$
$$9 \overset{?}{=} 9 \text{ Checks}$$

$(4, -3)$ is not a solution of the system.

Second Equation

$$x + 2y = 8$$
$$(4) + 2(-3) \overset{?}{=} 8$$
$$4 - 6 \overset{?}{=} 8$$
$$-2 \overset{?}{=} 8 \text{ Does not check}$$

Substitute the coordinates for x and y in each equation.

Since $(4, -3)$ does not check in both equations, this point is not a solution of the system.

(b) Check $(6, 1)$

First Equation

$$3x + y = 9$$
$$3(6) + (1) \overset{?}{=} 9$$
$$18 + 1 \overset{?}{=} 9$$
$$19 \overset{?}{=} 9 \text{ Does not check}$$

$(6, 1)$ is not a solution of the system.

Second Equation

$$x + 2y = 8$$
$$(6) + 2(1) \overset{?}{=} 8$$
$$6 + 2 \overset{?}{=} 8$$
$$8 \overset{?}{=} 8 \text{ Checks}$$

Substitute the coordinates for x and y in each equation.

Since $(6, 1)$ does not check in both equations, this point is not a solution of the system.

(c) Check $(2, 3)$

First Equation

$$3x + y = 9$$
$$3(2) + (3) \overset{?}{=} 9$$
$$6 + 3 \overset{?}{=} 9$$
$$9 \overset{?}{=} 9 \text{ Checks}$$

$(2, 3)$ is a solution of the system.

Second Equation

$$x + 2y = 8$$
$$(2) + 2(3) \overset{?}{=} 8$$
$$2 + 6 \overset{?}{=} 8$$
$$8 \overset{?}{=} 8 \text{ Checks}$$

Substitute the coordinates for x and y in each equation.

Since $(2, 3)$ checks in both equations, this point is a solution of the system.

Self-Check 1

Determine whether each ordered pair is a solution of the linear system $\begin{cases} 3x - 2y = -9 \\ 2x + 4y = 10 \end{cases}$.

a. $(1, 6)$ **b.** $(-1, 3)$ **c.** $(3, 1)$

2. Solve a System of Linear Equations by Using Graphs and Tables

How Can I Solve a System of Linear Equations Graphically?

To solve a system of equations graphically, graph each equation and then estimate the coordinates of the point of intersection. Because serious errors of estimation can occur, it is wise to check an estimated solution by substituting it into each equation of the system. This is illustrated in Example 2.

Example 2   Solving a System of Linear Equations Graphically

Solve the linear system $\begin{cases} 2x - 3y = 3 \\ 4x - 3y = 9 \end{cases}$ graphically.

Solution

First Equation

$2x - 3y = 3$

$-3y = -2x + 3$

$y = \dfrac{2}{3}x - 1$

Second Equation

$4x - 3y = 9$

$-3y = -4x + 9$

$y = \dfrac{4}{3}x - 3$

Solve each equation for y.

Graph:

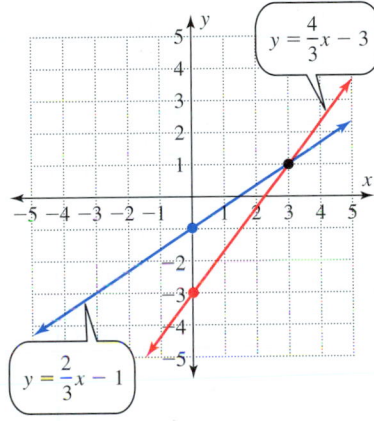

Sketch the graph of each line.

The graph yields an approximate solution of $(3, 1)$.

Check:

$2x - 3y = 3$

$2(3) - 3(1) \overset{?}{=} 3$

$6 - 3 \overset{?}{=} 3$

$3 \overset{?}{=} 3$ Checks

$4x - 3y = 9$

$4(3) - 3(1) \overset{?}{=} 9$

$12 - 3 \overset{?}{=} 9$

$9 \overset{?}{=} 9$ Checks

Check by substituting the coordinates of the approximate solution into each of the original equations.

Answer: $(3, 1)$

Self-Check 2

Solve the linear system $\begin{cases} 4x - 3y = -13 \\ x + 3y = -7 \end{cases}$ graphically.

The solution to a system of two linear equations should be written in ordered-pair notation. Writing the solution to the system in Example 2 as $(3, 1)$ emphasizes there is only one point of intersection, which has two coordinates.

If the solution for a system of linear equations is not an ordered pair of small integers, then the graphical method may require some effort to obtain the solution. We will examine a couple of techniques that you can use to facilitate graphical solutions. In Technology Perspective 1.3.1 on Reducing a Fraction to Lowest Terms, we showed how to convert a decimal to fractional form. We use this feature in Technology Perspective 3.3.1 to express both coordinates of the solution in fractional form. Note that the [QUIT] feature is the secondary function of the [MODE] key.

| Technology Perspective 3.3.1 | Determining Rational Solutions of a System of Linear Equations |

Solve the system of equations $\begin{Bmatrix} y = 2x - 3 \\ y = -5x + 6 \end{Bmatrix}$ on a TI-84 Plus calculator and convert the decimal coordinates of x and y to fractions.

TI-84 Plus Keystrokes

1. First press [Y=] and enter the equations.

2. Press [ZOOM] [6] to view the standard viewing window. Adjust the window as needed if you don't "see" the point of intersection.

3. Press [2nd] [CALC] [5] [ENTER] [ENTER] [ENTER] to calculate the point of intersection.

4. Press [2nd] [QUIT] to return to the home screen.

5. Press [X,T,θ,n] [MATH] [ENTER] [ENTER] to convert the x-coordinate to a fraction.

6. Press [ALPHA] [Y] [MATH] [ENTER] [ENTER] to convert the y-coordinate to a fraction.

TI-84 Plus Calculator

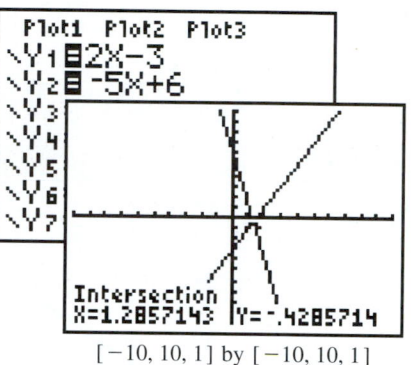

$[-10, 10, 1]$ by $[-10, 10, 1]$

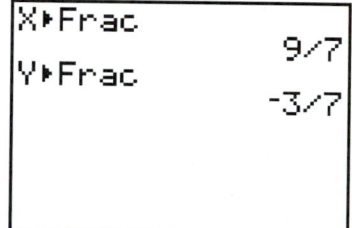

Answer: The solution of the system $\begin{Bmatrix} y = 2x - 3 \\ y = -5x + 6 \end{Bmatrix}$ is the ordered pair $\left(\dfrac{9}{7}, -\dfrac{3}{7} \right)$.

Technology Self-Check 1

Solve the system of equations $\begin{Bmatrix} 5x - y = 6 \\ 3x + 2y = 4 \end{Bmatrix}$ on a TI-84 Plus calculator and convert the decimal coordinates of x and y to fractions.

In Example 3, we solve a system of linear equations by creating a table of values. In this example, the equations are simple enough that we can create the table without assistance from technology. In most cases, it would be more practical to use a calculator or a spreadsheet to generate this table. The main purpose of this example is to point out how to recognize a solution of a system if it is displayed in a table of values.

Example 3 Solving a System of Linear Equations Numerically

A metal fabrication shop must assemble a part so that the angles shown in the figure are supplementary. To function properly, the specifications for the part require that the larger angle be twice the smaller angle. Use a table of values to determine the number of degrees in each angle.

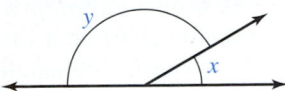

Solution

Let x = number of degrees in smaller angle   Select a variable to represent each unknown.

y = number of degrees in larger angle

Word Equations	Algebraic Equations
1. The angles are supplementary.	$x + y = 180$ or $y = 180 - x$
2. The larger angle is twice the smaller angle.	$y = 2x$

Angles are supplementary if the total of their measures is 180°.

Specifications require that the larger angle be twice the smaller angle.

Complete a table of values for these equations, and examine the table to determine the solution of the system.

x	$y_1 = 180 - x$	$y_2 = 2x$
0	180	0
10	170	20
20	160	40
30	150	60
40	140	80
50	130	100
60	120	120

It is not practical in this application for either angle to be negative. Also their total is 180°. Thus the x-variable is restricted to the values 0 to 180.

It is okay to stop generating values once the table displays the point where y_1 and y_2 have the same value.

Answer: The smaller angle is 60° and the larger angle is 120°.

Note that a smaller angle of 60° and a larger angle of 120° meet the conditions of the problem:
(1) The two angles have a sum of 180°: $60 + 120 = 180$
(2) The larger angle is twice the smaller angle: $120 = 2(60)$

Self-Check 3

Rework Example 3 with the same conditions except that the angles must be complementary (sum is 90°).

One of the disadvantages of solving a system of linear equations by using a table or a graph is that it is sometimes difficult to select the appropriate table setup or the appropriate viewing window on a calculator. Fortunately, our work in Section 2.8 concerning the restricted values on a variable in an application can assist us in selecting an appropriate table or viewing rectangle. Technology Perspective 3.3.2 explores an option that can help us create a viewing rectangle to display the point of intersection of two lines. In this perspective, we enter the restrictions from 0 to 180 for x and then use **ZoomFit** to select the y-values for the viewing window. The **ZoomFit** option is obtained by pressing [ZOOM] [0].

Technology Perspective 3.3.2 | **Using the ZoomFit Option to Select an Appropriate Viewing Window**

Restrict the x-values from 0 to 180, and use the **ZoomFit** option to set an appropriate viewing window for the system of linear equations $\begin{Bmatrix} x + y = 180 \\ y = 2x \end{Bmatrix}$ from Example 3. Then solve this system graphically.

TI-84 Plus Keystrokes

1. First press [Y=] and enter the equations.

2. Press [WINDOW] and set the **Xmin** and **Xmax** values based on the restrictions on the x-values.

3. Press [ZOOM] [0], which will automatically set the **Ymin** and **Ymax** values for the viewing window.

4. Press [2nd] [CALC] [5] [ENTER] [ENTER] [ENTER] to calculate the point of intersection.

Answer: The solution of this system of equations is (60, 120).

TI-84 Plus Calculator

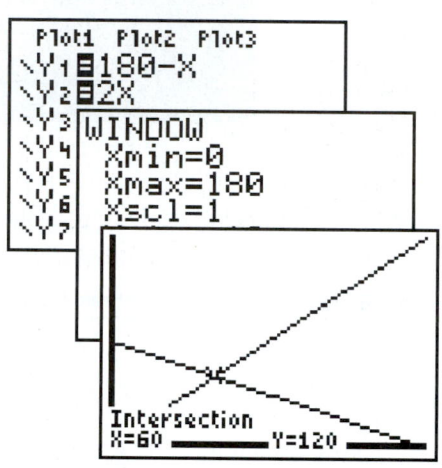

Note that [ZOOM] [0] will select the **ZoomFit** option which will set the viewing window to an appropriate range of y-values based on the x-values we choose. The accompanying window displays these y-values.

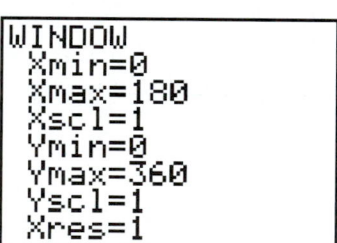

Technology Self-Check 2

Restrict the x-values from 0 to 100, and use the **ZoomFit** option to set an appropriate viewing window for the system of linear equations $\begin{Bmatrix} x + y = 180 \\ y = 2x \end{Bmatrix}$. Give the values for **Ymin** and **Ymax.**

Example 4 examines two options that a pizza business can use to prepare its pasta. The mixture principle is used to form the equation: Total cost = Variable cost + Fixed cost. Then the **ZoomFit** feature on a calculator is used to examine these two options.

Example 4 Business Options with Fixed and Variable Costs

A small family restaurant needs to be able to prepare up to 20 batches of its specialty pasta per day. There are two different machines used to make this pasta. For the first machine, there is a fixed daily cost of $46, plus a variable cost of $5 for each order of pasta produced. The second machine has a fixed daily cost of $22 and a variable cost of $8 for each order produced. Using x to represent the number of orders of pasta produced and $C(x)$ to represent the total daily cost of producing these orders, determine the number of orders for which the daily costs will be the same on the two machines.

Solution

Let x = number of orders of pasta produced
$C(x)$ = total daily cost of producing x orders of pasta

Word Equations	Algebraic Equations	
Machine 1: Total daily cost equals variable cost plus fixed cost.	$C(x) = 5x + 46$	The function notation $C(x)$ stresses that for the input x, the number of pasta orders, we can calculate the output $C(x)$, the daily cost.
Machine 2: Total daily cost equals variable cost plus fixed cost.	$C(x) = 8x + 22$	

Graphical Solution

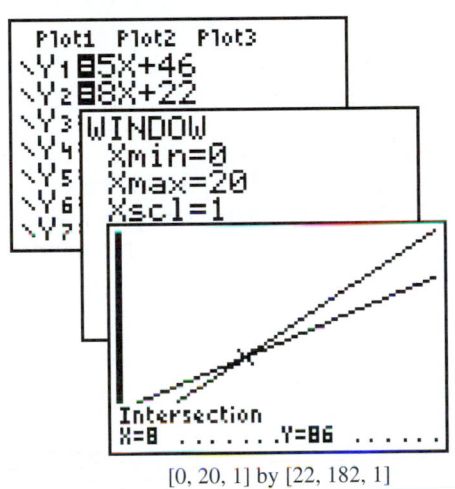

[0, 20, 1] by [22, 182, 1]

Numerical Check

X	Y₁	Y₂
3	61	46
4	66	54
5	71	62
6	76	70
7	81	78
8	86	86
9	91	94

X=8

From the problem statement, the variable x is restricted to the values from 0 to 20. Use the ⬡ key to enter these values for x, and then use the **ZoomFit** option to select the window shown here. The **Intersect** feature is used with this window to determine the point of intersection. The calculator window can be used to suggest an appropriate table of values to examine. This table confirms that the costs are both $86 when eight batches are produced.

Answer: The total daily costs are the same on the two machines when eight batches are produced.

Write the answers to word problems as full sentences. This helps reinforce that you understand the question, and it increases your ability to use mathematics in the workplace.

Self-Check 4

One machine has a fixed daily cost of $120 and a variable cost of $4 per item produced, whereas a second machine has a fixed daily cost of $80 and a variable cost of $4.50 per item produced. Using y to represent the total daily costs of these items, determine the number of items x for which the total daily costs will be the same. Assume that the number of units that can be produced per day is limited to 100 units.

3. Identify Inconsistent Systems and Systems of Dependent Linear Equations

Will Two Lines Always Have a Point of Intersection?

No. Examples 2–4 in this section have illustrated systems of linear equations with exactly one simultaneous solution. There are two other possibilities that we examine now.

Classification of Linear Systems

Two lines in a plane can be related in three ways:

1. **One solution:** The lines are distinct and intersect at a single point. This point represents the only simultaneous solution.
2. **No solution:** The lines are parallel and distinct. There is no point of intersection and no simultaneous solution.
3. **An infinite number of solutions:** The lines coincide (both equations represent the same line). There is an infinite number of points on this shared line, and each represents a solution. Only these points are solutions; the points not on the line are not solutions.

One Solution	No Solution	An Infinite Number of Solutions
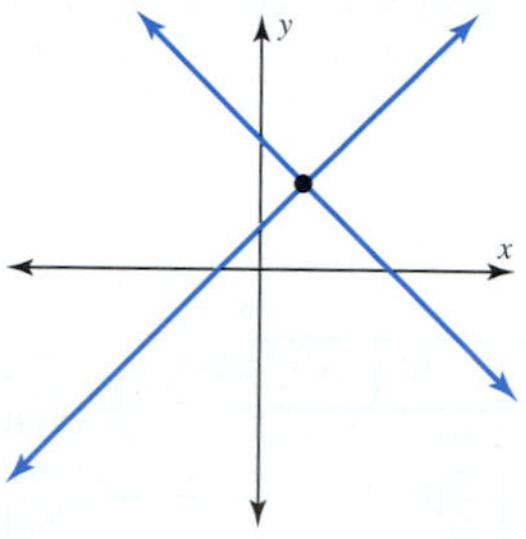	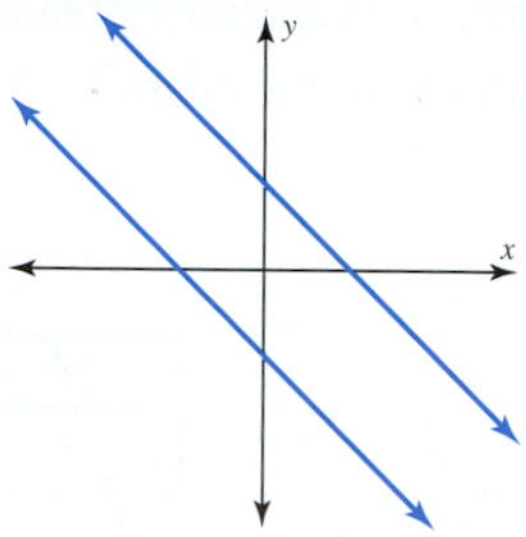	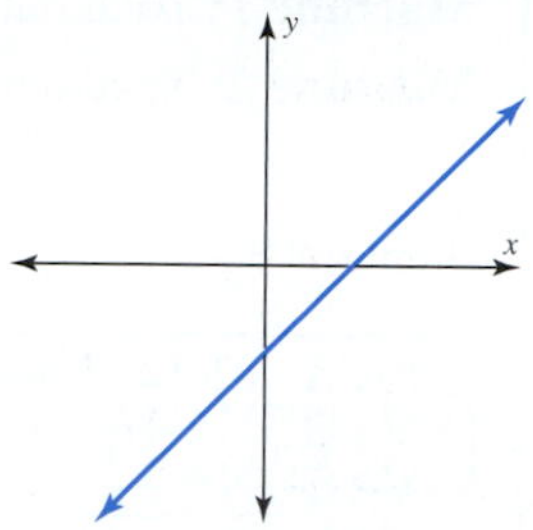

If a system of two equations in two variables has a solution, the system is called **consistent;** otherwise it is called **inconsistent.** If the equations have distinct graphs, the equations are called **independent;** if the graphs coincide, the equations are called **dependent.** Example 5 presents an inconsistent system of equations.

Example 5   Solving an Inconsistent System

Solve the linear system $\begin{Bmatrix} x + 2y = 2 \\ 2x + 4y = 8 \end{Bmatrix}$ by analyzing the equations and by examining a table and a graph.

Solution

Algebraically

First Equation

$$x + 2y = 2$$
$$2y = -x + 2$$
$$y = -\frac{1}{2}x + 1$$

The slope is $-\frac{1}{2}$ and the y-intercept is $(0, 1)$.

Second Equation

$$2x + 4y = 8$$
$$4y = -2x + 8$$
$$y = -\frac{1}{2}x + 2$$

The slope is $-\frac{1}{2}$ and the y-intercept is $(0, 2)$.

We are writing each equation in slope-intercept form both to facilitate an algebraic comparison and to prepare the equations for graphing either by hand or by using a graphing calculator.

Thus the two lines are parallel and distinct.

Numerically	Graphically	
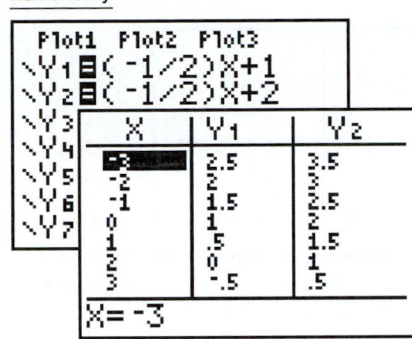	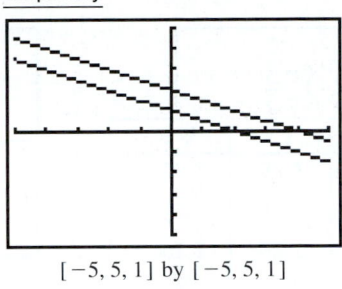 $[-5, 5, 1]$ by $[-5, 5, 1]$	These parallel lines are always the same distance apart. This is shown both by the table and by the graph. Since these parallel lines never meet, the system has no solution and is classified as inconsistent.

For the same x-value, the Y_1- and Y_2-values are always 1 unit apart.

The lines appear parallel and are always the same distance apart.

Answer: No solution; this is an inconsistent system.

Self-Check 5

Solve the linear system $\left\{ \begin{array}{r} 3x - y = 5 \\ -6x + 2y = 6 \end{array} \right\}$.

Example 6 presents the third possibility that can occur with a system of linear equations. This is a system of dependent linear equations.

Example 6 Solving a System of Dependent Equations

Solve the linear system $\left\{ \begin{array}{r} 3x - 4y = 12 \\ 6x - 8y = 24 \end{array} \right\}$ by analyzing the equations and by examining a table and a graph.

Solution

Algebraically

First Equation	Second Equation	
$3x - 4y = 12$	$6x - 8y = 24$	Rewrite each equation in slope-intercept form. Note that both equations have the same slope-intercept form, revealing that they are equations of the same line.
$-4y = -3x + 12$	$-8y = -6x + 24$	
$y = \dfrac{3}{4}x - 3$	$y = \dfrac{3}{4}x - 3$	

The slope is $\dfrac{3}{4}$ and the y-intercept is $(0, -3)$.

The slope is $\dfrac{3}{4}$ and the y-intercept is $(0, -3)$.

The two equations produce the same line.

Numerically   Graphically

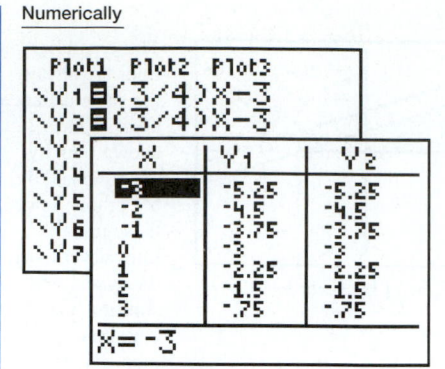

$[-5, 5, 1]$ by $[-5, 5, 1]$

For each x-value the Y_1- and   The equations produce the
Y_2-values are the same.   same line.

Answer: There is an infinite number of solutions; the equations are dependent.

Self-Check 6

Solve the linear system $\begin{Bmatrix} 2x - 6y = 4 \\ x - 3y = 2 \end{Bmatrix}$.

Although the original equations in Example 6 look different, the lines coincide. Thus every point satisfying one equation will also satisfy the other. However, only the points on this line are solutions of the linear system of equations. All other points are not solutions of the system.

It is now time to reflect on the three classifications of systems of linear equations. Compare the slope-intercept forms of the equations in the following three classifications of linear systems. Can you predict the classification by using the slope-intercept form?

Classification of Systems of Linear Equations

The linear system $\begin{Bmatrix} y = m_1 x + b_1 \\ y = m_2 x + b_2 \end{Bmatrix}$ is:

Verbally	Algebraically	Numerical Example	Graphical Example
1. A **consistent system of independent linear equations** having exactly one solution	$m_1 \neq m_2$ *Example:* $\begin{Bmatrix} y = 2x - 1 \\ y = -3x - 6 \end{Bmatrix}$	Only one x-value has matching y-values.	One distinct point of intersection $[-10, 10, 1]$ by $[-10, 10, 1]$
2. An **inconsistent system of linear equations** having no solution	$m_1 = m_2$ and $b_1 \neq b_2$ *Example:* $\begin{Bmatrix} y = \dfrac{1}{2}x + 1 \\ y = \dfrac{1}{2}x - 3 \end{Bmatrix}$	For each x-value, Y_2 is 4 units less than Y_1.	Parallel lines with no point of intersection $[-10, 10, 1]$ by $[-10, 10, 1]$

3. A **consistent system of dependent linear equations** having an infinite number of solutions	$m_1 = m_2$ and $b_1 = b_2$ *Example:* $$\begin{cases} y = \dfrac{3}{2}x + \dfrac{1}{2} \\ y = \dfrac{3}{2}x + \dfrac{1}{2} \end{cases}$$	For each x-value the Y_1- and Y_2-values are the same. 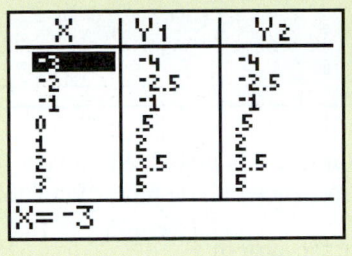	Infinite number of common points on these coincident lines 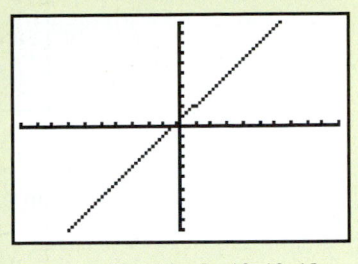 $[-10, 10, 1]$ by $[-10, 10, 1]$

Example 7 uses the slope-intercept form and the information from the previous box to determine the number of solutions of each of the following systems. Note that this work does not tell us which points satisfy both equations, only how many points will be solutions.

Example 7   Determining the Number of Solutions of a Linear System

Determine the number of solutions of each of the following systems.

(a) $y = \dfrac{2}{3}x - 5$

$\quad\; y = \dfrac{3}{4}x + 7$

(b) $3x - y = 5$

$\quad\; 2y = 6x - 10$

(c) $2x - 3y = 15$

$\quad\; 2x - 3y = -6$

Solution

First Equation

$y = \dfrac{2}{3}x - 5$

$m_1 = \dfrac{2}{3}$

Second Equation

$y = \dfrac{3}{4}x + 7$

$m_2 = \dfrac{3}{4}$

Determine the slope from the slope-intercept form $y = mx + b$.

Because the slopes are different, the lines will intersect at exactly one point.

Answer: Because $m_1 \neq m_2$, the system has exactly one solution.

First Equation

$3x - y = 5$

$\quad -y = -3x + 5$

$\quad\;\; y = 3x - 5$

$\quad m_1 = 3, b_1 = -5$

Second Equation

$2y = 6x - 10$

$\;\; y = 3x - 5$

$m_2 = 3, b_2 = -5$

First write each equation in slope-intercept form. Then use the slopes and y-intercepts to determine the number of solutions.

Answer: Because $m_1 = m_2$ and $b_1 = b_2$, the equations are dependent and the system of equations has an infinite number of solutions.

Both equations represent the same line. Only the points on this line are solutions of the system of linear equations.

First Equation

$2x - 3y = 15$

$\quad -3y = -2x + 15$

$\quad\;\;\; y = \dfrac{2}{3}x - 5$

$\quad m_1 = \dfrac{2}{3}, b_1 = -5$

Second Equation

$2x - 3y = -6$

$\quad -3y = -2x - 6$

$\quad\;\;\; y = \dfrac{2}{3}x + 2$

$\quad m_2 = \dfrac{2}{3}, b_2 = 2$

Because the slopes are equal, the lines are parallel. Because the y-intercepts are different, the lines do not coincide.

Answer: Because $m_1 = m_2$ but $b_1 \neq b_2$, the system is inconsistent and has no solution.

Self-Check 7

Determine the number of solutions to each of the following systems of linear equations.

a. $y_1 = 2x - 7$
$\quad y_2 = 2x + 7$

b. $y_1 = 2x - 7$
$\quad y_2 = 7x - 2$

c. $y_1 = 2x - 7$
$\quad y_2 = 2x - 7$

Example 8 is included to reveal some of the weaknesses of the graphical method and an overreliance on graphing calculators. Before you read through the text explanation, try to solve the system $\begin{cases} 2x - 3y = 10 \\ 3x - 5y = 0 \end{cases}$ by using a graphing calculator. Please note the solution you obtain and any difficulties you encounter. In Example 8, we will need to adjust the calculator viewing window. This was first shown in Technology Perspective 2.3.2 on Approximating a Point of Intersection.

Example 8   Using a Graphing Calculator to Solve a System of Linear Equations

Solve the linear system $\begin{cases} 2x - 3y = 10 \\ 3x - 5y = 0 \end{cases}$ by using a graphing calculator.

Solution

First Equation

$$2x - 3y = 10$$
$$-3y = -2x + 10$$
$$y = \frac{2}{3}x - \frac{10}{3}$$

Second Equation

$$3x - 5y = 0$$
$$-5y = -3x$$
$$y = \frac{3}{5}x$$

Write each equation in slope-intercept form to prepare the equations for entry into a graphing calculator.

First Viewing Window

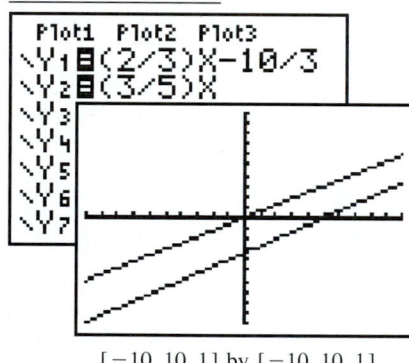

$[-10, 10, 1]$ by $[-10, 10, 1]$

Adjusted Window

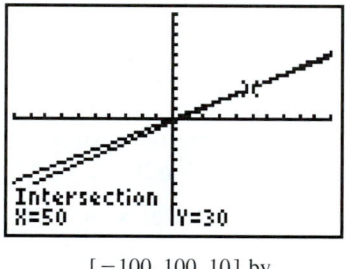

$[-100, 100, 10]$ by
$[-100, 100, 10]$

Although the lines may appear to be parallel on this window, we know the lines are not parallel because they do not have the same slope. Thus the lines must have a point of intersection that is not revealed on this window.

The new viewing rectangle contains the point of intersection, which is located by using the **Intersect** feature on a graphing calculator.

The point of intersection is (50, 30).

Check:

First Equation

$$2x - 3y = 10$$
$$2(50) - 3(30) \overset{?}{=} 10$$
$$100 - 90 \overset{?}{=} 10$$
$$10 \overset{?}{=} 10 \text{ checks.}$$

Second Equation

$$3x - 5y = 0$$
$$3(50) - 5(30) \overset{?}{=} 0$$
$$150 - 150 \overset{?}{=} 0$$
$$0 \overset{?}{=} 0 \text{ checks.}$$

Answer: (50, 30).

Self-Check 8

Solve the linear system $\left\{ \begin{array}{l} 2x + y = 10 \\ 3x + y = 0 \end{array} \right\}$ by using a graphing calculator window of $[-20, 5, 5]$ by $[-5, 40, 5]$.

Because of the difficulties in determining approximate window or table settings on a calculator, the algebraic methods for solving linear systems presented in Sections 3.4 and 3.5 may be preferred for many problems.

Self-Check Answers

1. **a.** Not a solution   **b.** A solution   **c.** Not a solution
2. $(-4, -1)$
3. The smaller angle is 30° and the larger angle is 60°.
4. The total daily costs are the same for the two machines when 80 units are produced.

5. No solution; this is an inconsistent system.
6. There is an infinite number of solutions; the equations are dependent.
7. **a.** No solution   **b.** Exactly one solution   **c.** An infinite number of solutions
8. $(-10, 30)$

Technology Self-Check Answers

1. $\left(\dfrac{16}{13}, \dfrac{2}{13} \right)$

2. **Ymin** = 0; **Ymax** = 200

3.3	Using the Language and Symbolism of Mathematics

1. An ordered pair that satisfies each equation in a system of linear equations is called a _____ of the system of linear equations.

2. Graphically, the solution to a system of two linear equations is represented by their point of _____.

3. An _____ system of two linear equations has no solution. Graphically, these lines are _____ with no point in common.

4. A consistent system of two independent linear equations has exactly _____ solution. Graphically, these lines intersect at exactly _____ point.

5. A consistent system of two dependent linear equations has an _____ number of solutions. Graphically, these lines _____.

6. Use the linear system $\left\{ \begin{array}{l} y = m_1x + b_1 \\ y = m_2x + b_2 \end{array} \right\}$ to select the correct choice that describes the relationship between the two lines given.

   a. $m_1 = m_2$ and $b_1 = b_2$ The lines intersect at **zero/one/infinitely many** point(s).

   b. $m_1 = m_2$ and $b_1 \neq b_2$ The lines **are/are not** parallel. The lines intersect at **zero/one/infinitely many** point(s).

   c. $m_1 \neq m_2$ The lines **are/are not** parallel. The lines intersect at **zero/one/infinitely many** point(s).

3.3 | Quick Review

For each exercise, consider the linear equation $y = \frac{2}{3}x - 3$.

1. Determine the slope of this line.

2. Determine the y-intercept of this line.

3. Sketch the graph of this line.

4. Use this equation to complete the table of values.

x	$y = \frac{2}{3}x - 3$
-3	
0	
3	
6	

5. Use Technology Perspective 2.3.2 on Approximating a Point of Intersection to determine the point of intersection of $y = -x + 2$ and $y = \frac{2}{3}x - 3$.

3.3 | Exercises

Objective 1 Check Possible Solutions of a Linear System

In Exercise 1–8, determine whether the given ordered pair is a solution of the system of linear equations.

1. $(-1, -2)$
$7x - 3y + 1 = 0$
$5x - 4y - 3 = 0$

2. $(-4, 2)$
$4x + 5y + 6 = 0$
$6x + 5y + 14 = 0$

3. $\left(\frac{1}{2}, \frac{1}{3}\right)$
$4x - 3y = 1$
$6x - 6y = 1$

4. $\left(\frac{1}{4}, -\frac{2}{3}\right)$
$8x + 3y = 0$
$4x - 3y = 3$

5. $(0.1, -0.2)$
$4x - 3y = 1$
$2x + y = 0$

6. $(-0.3, 0.5)$
$5x + 3y = 0$
$6x - y = -2.3$

7. $(0, 6)$
$2x - y = -6$
$3x + y = 3$

8. $(9, 0)$
$3x - 8y = -24$
$2x + 9y = 18$

Objective 2 Solve a System of Linear Equations by Using Graphs and Tables

In Exercises 9–12, determine by inspection the point of intersection of the two lines.

9.
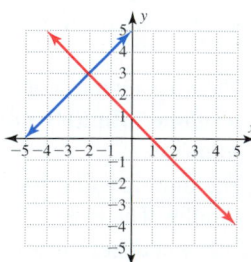

10.
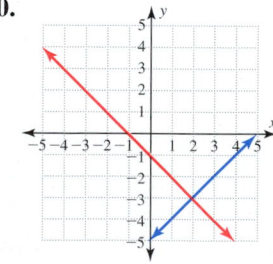

11.
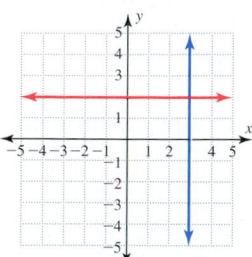

12.
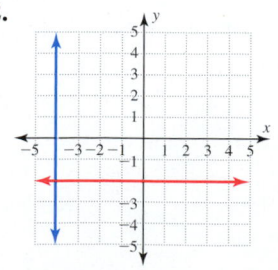

In Exercises 13–16, determine the point of intersection. Then check this ordered pair in both equations to verify that it is a solution of the system of linear equations.

13.
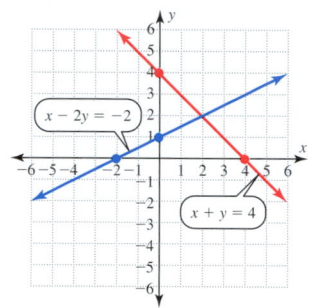

14.
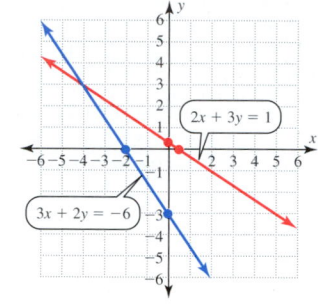

15.
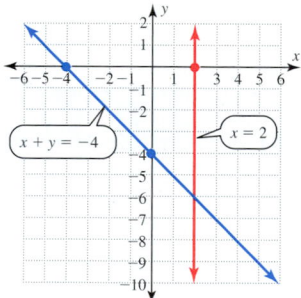

16.
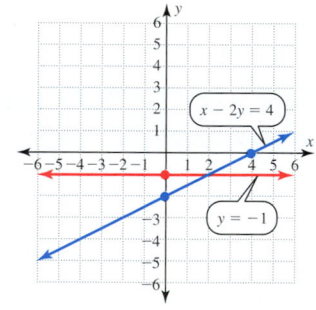

In Exercises 17–20, use the tables to solve each system of linear equations.

17. $y_1 = -\frac{7}{4}x + \frac{3}{4}$

$y_2 = -\frac{9}{5}x + 1$

x	y_1	y_2
0	0.75	1.0
1	-1.00	-0.8
2	-2.75	-2.6
3	-4.50	-4.4
4	-6.25	-6.2
5	-8.00	-8.0
6	-9.75	-9.8

18. $y_1 = 5x$
$y_2 = 4x + 20$

x	y_1	y_2
0	0	20
10	50	60
20	100	100
30	150	140
40	200	180
50	250	220
60	300	260

19. $y_1 = -\dfrac{20}{3}x + \dfrac{2}{3}$

$y_2 = -\dfrac{15}{2}x + \dfrac{1}{4}$

x	y_1	y_2
−1.0	7.33	7.75
−0.9	6.67	7.00
−0.8	6.00	6.25
−0.7	5.33	5.50
−0.6	4.67	4.75
−0.5	4.00	4.00
−0.4	3.33	3.25

20. $y_1 = x + 2$

$y_2 = \dfrac{3}{2}x - \dfrac{13}{2}$

x	y_1	y_2
15	17	16.0
16	18	17.5
17	19	19.0
18	20	20.5
19	21	22.0
20	22	23.5
21	23	25.0

In Exercises 21–26, solve each system of linear equations by graphing each line.

21. $y = -\dfrac{x}{3} + 2$

$y = -\dfrac{x}{2} + 1$

22. $y = -\dfrac{x}{2} + 2$

$y = \dfrac{x}{2}$

23. $x + 3 = 0$
$y - 2 = 0$

24. $y = 2$
$x - 5 = 0$

25. $y = -2x + 10$
$x = 6$

26. $y = 2x - 6$
$y = -4$

Objective 3 Identify Inconsistent Systems and Systems of Dependent Linear Equations

In Exercises 27–34, use the slope-intercept form of each line to determine the number of solutions of the system; then classify each system as a consistent system of independent equations, an inconsistent system, or a consistent system of dependent equations.

27. $y = \dfrac{3}{7}x - 17$

$y = -\dfrac{7}{8}x + 8$

28. $y = \dfrac{5}{9}x + 4$

$y = \dfrac{9}{5}x - 4$

29. $y = \dfrac{3}{8}x + 11$

$y = \dfrac{3}{8}x - 5$

30. $y = -\dfrac{9}{5}x + 7$

$y = -\dfrac{9}{5}x + 4$

31. $6x + 2y = 4$
$9x = 6 - 3y$

32. $8x = 12 - 20y$
$6x + 15y = 9$

33. $2x + 3y = 3$
$4x - 3y = -3$

34. $\dfrac{x}{6} + \dfrac{y}{4} = \dfrac{1}{3}$

$\dfrac{x}{3} + \dfrac{y}{2} = \dfrac{1}{4}$

In Exercises 35–38, use the table for each system of two linear equations, $y_1 = m_1 x + b_1$ and $y_2 = m_2 x + b_2$ to classify the system as a consistent system of independent equations, an inconsistent system, or a consistent system of dependent equations.

35.

x	y_1	y_2
1.0	−1.0	1.0
1.2	−.3	.7
1.4	.4	.4
1.6	1.1	.1
1.8	1.8	−.2
2.0	2.5	−.5
2.2	3.2	−.8

36.

x	y_1	y_2
1.0	−1.0	−1.0
1.2	−.3	−.3
1.4	.4	.4
1.6	1.1	1.1
1.8	1.8	1.8
2.0	2.5	2.5
2.2	3.2	3.2

37.

x	y_1	y_2
1.0	−1.0	1.0
1.2	−.3	1.7
1.4	.4	2.4
1.6	1.1	3.1
1.8	1.8	3.8
2.0	2.5	4.5
2.2	3.2	5.2

38.

x	y_1	y_2
−8	24	−24
−4	16	−16
0	8	−8
4	0	0
8	−8	8
12	−16	16
16	−24	24

Review and Concept Development

In Exercises 39–46, solve each system of linear equations by graphing each line. (*Hint:* Some systems may have no solution and some may have an infinite number of solutions.)

39. $y = 2x - 7$

$y = -\dfrac{3}{2}x + 7$

40. $y = -3x + 7$

$y = -\dfrac{2}{5}x + \dfrac{9}{5}$

41. $2x - 3y - 6 = 0$
$4x - 3y + 3 = 0$

42. $5x - 4y + 16 = 0$
$3x - 2y - 6 = 0$

43. $y = 3x$
$y = 3x - 6$

44. $y = 2x + 1$
$y = 2x - 1$

45. $x - 2y - 2 = 0$
$2x - 4y - 4 = 0$

46. $2x - 3y - 3 = 0$
$6x - 9y - 9 = 0$

Applying Technology

In Exercises 47–50, solve each system of linear equations by using a calculator to graph each equation; then determine the point of intersection. Check the coordinates of this point in both of the linear equations. (*Hint:* See Technology Perspective 2.3.2 on Approximating a Point of Intersection.)

47. $y = 2x - 5$
$y = -2x - 1$

48. $y = x - 5$
$y = -x + 3$

49. $y = 2x - 1$
$y = \dfrac{x}{2} + 2$

50. $y = -x + 5$
$y = \dfrac{x}{3} + 1$

In Exercises 51–54, solve each system of linear equations and express both coordinates of the answer in fractional form. (*Hint:* See Technology Perspective 3.3.1.)

51. $14x - 7y = -5$
$7x + 21y = 29$

52. $13x + 7y = 8$
$13x - 14y = -7$

53. $6x - 11y = 7$
$3x + 22y = -1$

54. $6x + 12y = -7$
$6x - 8y = 5$

In Exercises 55–58, solve each system of linear equations by using the given restriction on the variable x and the **ZoomFit** option and the **Intersect** feature on a calculator. (*Hint:* See Technology Perspective 3.3.2.)

55. $x + y = 1{,}000$
$2x + 3y = 2{,}700$

The variable x is restricted to the interval [0, 1,000].

56. $x + y = 8{,}000$
$0.06x + 0.03y = 435$

The variable x is restricted to the interval [0, 8,000].

57. $x + y = 1$
$8x + 11y = 10.1$

The variable x is restricted to the interval [0, 1].

58. $x + y = 1$
$4x - 5y = 2.65$

The variable x is restricted to the interval [0, 1].

In Exercises 59–62, solve each system of linear equations by using a calculator or a spreadsheet to complete the following table of values.

59. $y = 2x - 5$
$y = -2x - 1$

60. $y = 2x - 5$
$y = -2x - 5$

61. $y = 5x + 3$
$y = -5x + 3$

62. $y = 5x + 3$
$y = -5x - 7$

x	y_1	y_2
-3		
-2		
-1		
0		
1		
2		
3		

In Exercises 63 and 64, use a calculator or a spreadsheet to complete each table and to solve the given system of linear equations.

63. $4x - y = 80$
$3x + y = 130$

64. $5x + 2y = 5$
$7x - 2y = 175$

x	$y_1 = 4x - 80$	$y_2 = -3x + 130$
-10		
0		
10		
20		
30		
40		
50		

x	$y_1 = -2.5x + 2.5$	$y_2 = 3.5x - 87.5$
0		
5		
10		
15		
20		
25		
30		

Connecting Concepts to Applications

65. Rental Truck Costs The graph compares the cost of a one-day rental for a moving truck at two different rental companies. The cost is based on the number of miles driven. Give the solution to the corresponding system of equations. Then interpret the meaning of the x- and y-coordinates of this solution.

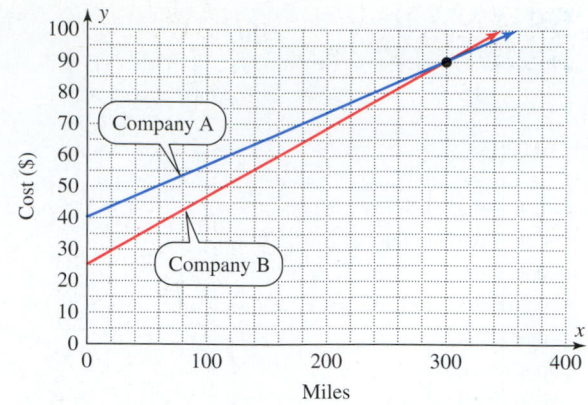

66. Digital Phone Plans The graph compares the monthly cost of two different digital phone plans based on the number of minutes of use. Give the solution to the corresponding system of equations. Then interpret the meaning of the x- and y-coordinates of this solution.

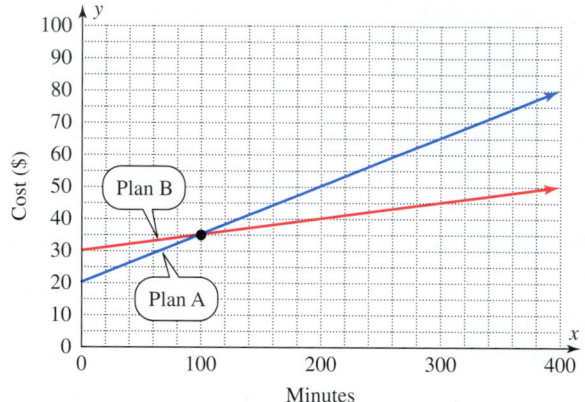

67. Wind Power Capacity The wind power capacity of the European Union and that of the rest of the world are displayed in the accompanying graph. Determine the point of intersection of these two line graphs. Then interpret the meaning of the x- and y-coordinates of this point.

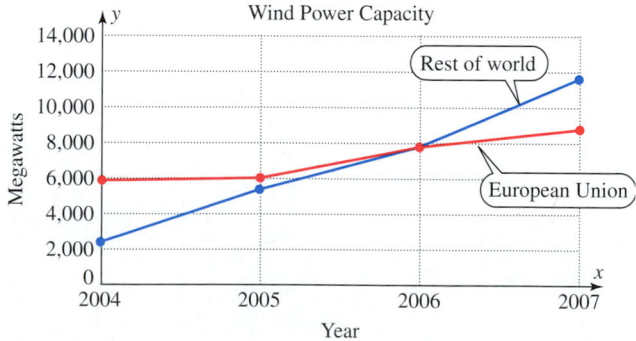

68. Phone Service The number of cellular phone subscribers and the number of land lines in the United States from 2001 to 2006 are displayed in the accompanying graph. Determine the point of intersection of these two line graphs. Then interpret the meaning of the *x*- and *y*-coordinates of this point.

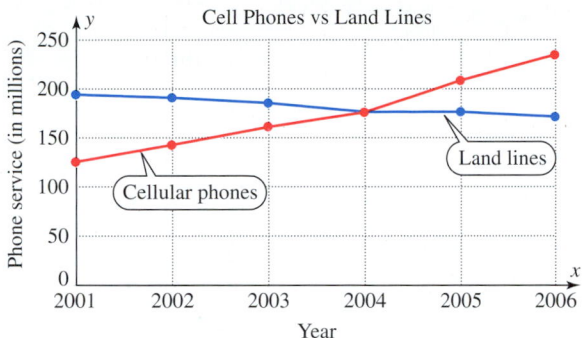

Cell Phones vs Land Lines

Service A			Service B	
Students *x*	Cost *y*		Students *x*	Cost *y*
10	45		10	205
20	90		20	210
30	135		30	215
40	180		40	220
50	225		50	225
60	270		60	230

71. Angles in Two Parts The two angles shown represent parts that should be assembled so that the angles are supplementary and the larger angle is 5 more than 3 times the smaller angle. Determine the number of degrees in each angle.

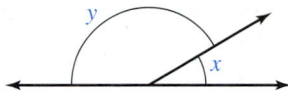

69. A Table of Flight Distances A small single-engine plane that travels 150 mi/h leaves Chicago's O'Hare Airport at noon. Two hours later, a larger commercial jet that travels 450 mi/h takes off and flies in the same direction. The tables display the distance in miles flown by each plane, where *x* is the number of hours past 2 P.M. Give the solution to the corresponding system of equations. Then interpret the meaning of the *x*- and *y*-coordinates of this solution.

72. Angles in Two Parts The two angles shown in Exercise 71 represent parts that should be assembled so that the angles are supplementary and the larger angle is 6 less than twice the smaller angle. Determine the number of degrees in each angle.

73. Fixed and Variable Costs One machine has a fixed daily cost of $75 and a variable cost of $3 per item produced, whereas a second machine has a fixed daily cost of $60 and a variable cost of $4.50 per item produced. Using *y* to represent the total daily costs of these items, determine the number of items *x* for which the total daily costs will be the same.

Small Plane			Large Jet	
Hours *x*	Distance *y*		Hours *x*	Distance *y*
0	300		0	0
0.5	375		0.5	225
1.0	450		1.0	450
1.5	525		1.5	675
2.0	600		2.0	900
2.5	675		2.5	1,125

70. A Table of Bus Fares When planning a student trip, the student council has to choose between two busing services. Service A charges $4.50 per person, while service B charges a fee of $200 plus $0.50 per person. The tables display the charges in dollars for each service based on the number of students on the trip. Give the solution to the corresponding system of equations. Then interpret the meaning of the *x*- and *y*-coordinates of this solution.

74. Fixed and Variable Costs A small manufacturing plant makes custom replacement parts for plastic molding equipment. The maximum production is limited to 100 units per day. The company has two machines available to make this part. One machine has a fixed daily cost of $450 and a variable cost of $12.50 per unit of this part produced. A second machine has a fixed daily cost of $200 and a variable cost of $17.50 per unit of this part produced. Using *y* to represent the total daily cost of producing *x* units of this part, determine the number of units for which the total daily cost of producing this part will be the same for both machines.

Multiple Representations

In Exercises 75 and 76, complete the numerical and graphical descriptions, and then write a full sentence that answers the question.

75. Word Problem

One number is seven more than twice another number. The sum of these numbers is thirteen. Find these numbers.

Algebraic Equations

$y = 2x + 7$
$x + y = 13$

Table

x	$y_1 = 2x + 7$	$y_2 = 13 - x$
−3	1	16
−2		
−1		
0		
1		
2		
3	13	10

Graph

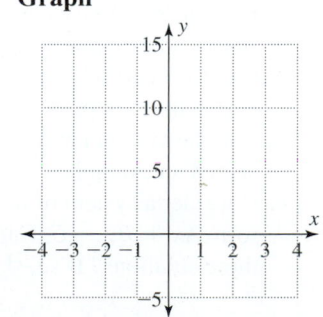

76. Word Problem

One number is five less than three times another number. The sum of these numbers is eleven. Find these numbers.

Algebraic Equations

$y = 3x - 5$

$x + y = 11$

Table

x	$y_1 = 3x - 5$	$y_2 = 11 - x$
-1	-8	12
-0		
1		
2		
3		
4		
5	10	6

Graph

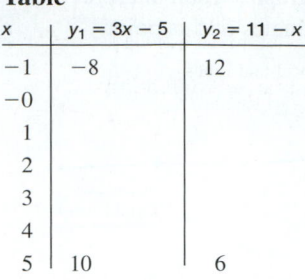

Review and Concept Development

77. Write a system of two linear equations with a solution of $(3, 5)$ so that one line is horizontal and the other is vertical.

78. Write a system of two linear equations with a solution of $(-4, -2)$ so that one line is horizontal and the other is vertical.

79. Write a system of two linear equations with a solution of $(0, 3)$ so that one line has a slope of $\frac{1}{2}$ and the other line has a slope of -4.

80. Write a system of two linear equations with a solution of $(0, -2)$ so that one line has a slope of $\frac{1}{3}$ and the other line has a slope of -1.

Group discussion questions

81. Discovery Question

a. Consider a system of two linear equations of the form $Ax + By = C$. Can the system have zero solutions? If so, sketch an example.

b. Can the system have exactly one solution? If so, sketch an example.

c. Can the system have exactly two solutions? If so, sketch an example.

d. Can the system have exactly three solutions? If so, sketch an example.

e. Can the system have more than three solutions? If so, sketch an example.

82. Discovery Question

a. Consider a system of three linear equations of the form $Ax + By = C$. Can the system have zero solutions? If so, sketch an example.

b. Can the system have exactly one solution? If so, sketch an example.

c. Can the system have exactly two solutions? If so, sketch an example.

d. Can the system have exactly three solutions? If so, sketch an example.

e. Consider a system of three linear equations of the form $Ax + By = C$. Can the system have more than three solutions? If so, sketch an example.

83. Error Analysis A student graphed the system shown here and concluded the lines were parallel and the system was inconsistent with no solution. Describe the error the student has made, and then solve this system of equations.

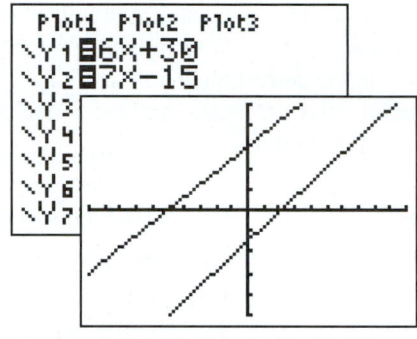

$[-10, 10, 1]$ by $[-50, 50, 10]$

84. Error Analysis A student graphed the following system and concluded that the system had no solution. Describe the error the student has made, and then solve this system of equations.

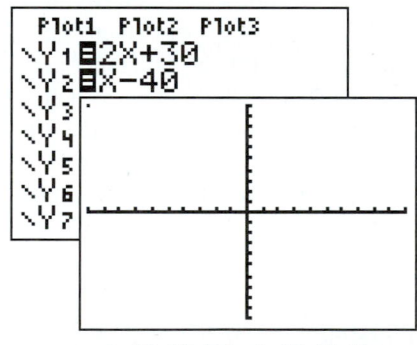

$[-10, 10, 1]$ by $[-10, 10, 1]$

85. Challenge Question Use the given graph to determine the solution of the system of linear equations, and write the equation of both lines in general form.

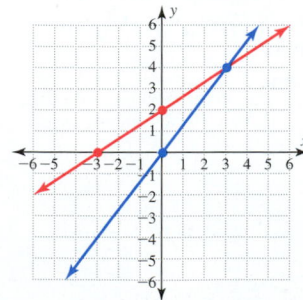

3.3	Cumulative Review

In Exercises 1 and 2, perform the indicated operations.

1. $-24 \div 4 \cdot 2$

2. $-24 \div (4 \cdot 2)$

In Exercises 3–5, write each inequality in interval notation.

3. $x > 5$

4. $x \le -2$

5. $-4 \le x < 10$

Section 3.4	Solving Systems of Linear Equations in Two Variables by the Substitution Method

Objective:

1. Solve a system of linear equations by the substitution method.

Although the graphical method is an excellent means of visualizing the solution of a system of equations, it has some limitations. Even with a computer or graphing calculator, this method can be time-consuming. This method also can produce some error because of limitations in estimating the point of intersection. Thus algebraic methods that are quicker and yield exact solutions often are preferred.

1. Solve a System of Linear Equations by the Substitution Method

What Is the Logic Behind One of the Algebraic Methods?

The first algebraic method presented in this chapter is based on the **substitution principle,** which states that a quantity may be substituted for its equal. The substitution method is particularly appropriate when it is easy to solve one equation for either x or y.

Substitution Method

Step 1. Solve one of the equations for one variable in terms of the other variable.

Step 2. Substitute the expression obtained in step 1 into the other equation (eliminating one of the variables), and solve the resulting equation.

Step 3. Substitute the value obtained in step 2 into the equation obtained in step 1 (back-substitution) to find the value of the other variable.

The ordered pair obtained in steps 2 and 3 is the solution.

The substitution method, which is shown here for systems of linear equations, can also be used to solve some nonlinear systems. The addition method for solving linear systems is covered in Section 3.5.

A Mathematical Note

Grace Hopper (1906–1992) was one of the foremost pioneers in the field of computer programming. She was one of the developers of the COBOL programming language. She retired from the U.S. Navy in 1986 as a rear admiral 43 years after joining as a lieutenant. In 1977, the author James Hall had the pleasure of hearing her humorously describe what she called the first computer bug—a moth that on September 9, 1947, flew into a relay of the early Mark II computer and caused it to malfunction. For a photo of this moth, do a web search for "first computer bug."

Example 1	Solving a Linear System by the Substitution Method

Solve $\left\{ \begin{array}{r} 3x + 4y = 8 \\ x + 2y = 10 \end{array} \right\}$ by the substitution method.

Solution

1. $x + 2y = 10$ Solve the second equation for x. (Note that this is relatively easy because x has a coefficient of 1 in this equation.)

$x = 10 - 2y$

2. $3x + 4y = 8$

$3(10 - 2y) + 4y = 8$ Substitute for x in the first equation. (Note that x has been eliminated.) Then solve this equation for y.

$30 - 6y + 4y = 8$

$30 - 2y = 8$

$-2y = 8 - 30$

$-2y = -22$

$y = 11$ This is the y-coordinate of the solution.

3. $x = 10 - 2y$

$x = 10 - 2(11)$ Back-substitute 11 for y into the equation that was solved for x in step 1.

$x = 10 - 22$

$x = -12$ This is the x-coordinate of the solution.

Check:

First Equation

$3x + 4y = 8$

$3(-12) + 4(11) \overset{?}{=} 8$

$-36 + 44 \overset{?}{=} 8$

$8 \overset{?}{=} 8$ checks

Second Equation

$x + 2y = 10$

$-12 + 2(11) \overset{?}{=} 10$

$-12 + 22 \overset{?}{=} 10$

$10 \overset{?}{=} 10$ checks

Answer: $(-12, 11)$

Self-Check 1

Solve $\begin{Bmatrix} 3x + y = 6 \\ 2x + y = 2 \end{Bmatrix}$ by the substitution method.

The answer to a consistent system of independent linear equations should be written in ordered-pair notation. The notation $(-12, 11)$ emphasizes that the system in Example 1 has only one solution and that the x- and y-coordinates must be used in the correct order.

Example 2 revisits Example 3 from Section 3.3. Solving even a relatively simple system of linear equations by using a table or graph can take some time as we decide what input values would be appropriate. Fortunately, the substitution method provides an algebraic solution that is quick and exact.

Example 2 Determining the Number of Degrees in Two Supplementary Angles

A metal fabrication shop must assemble a part so that the angles shown in the figure are supplementary. For it to function properly, the specifications for the part require that the larger angle be twice the smaller angle. Determine the number of degrees in each angle.

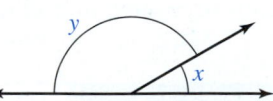

Solution

Let x = number of degrees in smaller angle   Select a variable to represent each unknown.

y = number of degrees in larger angle

Word Equations	Algebraic Equations	

1. The angles are supplementary.

2. The larger angle is twice the smaller angle.

$$x + y = 180$$
$$y = 2x$$

Angles are supplementary if the total of their measures is 180°. Solve the system using the substitution method.

Algebraic Solution

$$x + 2x = 180$$
$$3x = 180$$
$$x = 60$$
$$y = 2x$$
$$y = 2(60)$$
$$y = 120$$

Substitute $2x$ for y in the first equation and then solve for x.

Back-substitute 60 for x in the second equation.

Answer: The smaller angle is 60° and the larger angle is 120°.

Does this answer check?

Self-Check 2

Rework Example 2 with the same conditions, except that the angles must be complementary (sum is 90°).

What Is the Best Strategy to Use if the Equations Contain Fractional Coefficients?

A system of equations that contains fractional coefficients may be easier to solve if we first convert the coefficients to integers by multiplying through by the LCD. For example, multiplying both sides of $\dfrac{x}{4} + \dfrac{y}{3} = 1$ by 12 yields $3x + 4y = 12$, which has integer coefficients.

Example 3 Solving a Linear System with Fractional Coefficients

Solve $\begin{cases} \dfrac{x}{2} + \dfrac{y}{6} = \dfrac{2}{3} \\ \dfrac{x}{4} - \dfrac{y}{5} = \dfrac{7}{4} \end{cases}$.

Solution

First Equation

$$\frac{x}{2} + \frac{y}{6} = \frac{2}{3}$$

$$6\left(\frac{x}{2} + \frac{y}{6}\right) = 6\left(\frac{2}{3}\right)$$

$$6\left(\frac{x}{2}\right) + 6\left(\frac{y}{6}\right) = 6\left(\frac{2}{3}\right)$$

(1) $\quad 3x + y = 4$

Second Equation

$$\frac{x}{4} - \frac{y}{5} = \frac{7}{4}$$

$$20\left(\frac{x}{4} - \frac{y}{5}\right) = 20\left(\frac{7}{4}\right)$$

$$20\left(\frac{x}{4}\right) - 20\left(\frac{y}{5}\right) = 20\left(\frac{7}{4}\right)$$

(2) $\quad 5x - 4y = 35$

We start by producing a simpler system of equations with integer coefficients.

Multiply both sides of the first equation by the LCD of 6 and both sides of the second equation by the LCD of 20.

(1) $\begin{cases} 3x + y = 4 \\ 5x - 4y = 35 \end{cases}$
(2)

This system with integer coefficients has the same solution as the original system of equations.

(1) $3x + y = 4$ Solve equation (1) for y since y has a coefficient of 1.

$y = 4 - 3x$

(2) $5x - 4y = 35$ Substitute for y in equation (2). Then solve this equation for x.

$5x - 4(4 - 3x) = 35$

$5x - 16 + 12x = 35$

$17x - 16 = 35$

$17x = 51$

$x = 3$ This is the x-coordinate of the solution.

(3) $y = 4 - 3x$

$y = 4 - 3(3)$ Back-substitute 3 for x in the equation that was solved for y in step 1.

$y = 4 - 9$

$y = -5$ This is the y-coordinate of the solution.

Answer: $(3, -5)$ Does this solution check in the original equations?

Self-Check 3

Solve $\begin{cases} \dfrac{x}{2} + \dfrac{y}{3} = -4 \\ \dfrac{x}{6} - \dfrac{y}{6} = -3 \end{cases}$

If I Am Using the Substitution Method, How Will I Know When a System Is Inconsistent?

Examples 4 and 5 illustrate what will happen when the substitution method is used to solve an inconsistent system of linear equations or a consistent system of dependent linear equations. For an inconsistent system, the algebraic steps will produce a contradiction.

Example 4 Solving an Inconsistent System

Solve $\begin{cases} y = 3x - 6 \\ 6x - 2y = 9 \end{cases}$ by the substitution method.

Solution

$y = 3x - 6$ The first equation is already solved for y.

$6x - 2y = 9$

$6x - 2(3x - 6) = 9$ Substitute for y in the second equation.

$6x - 6x + 12 = 9$ Note that the simplification of this equation eliminates all the variables and produces a contradiction.

$12 = 9$ (a contradiction)   The conditions given by the two equations are contradictory. Thus this system has no solution; it is an inconsistent system.

Answer: There is no solution.

Self-Check 4

Solve $\begin{cases} y = 5x + 2 \\ 20x - 4y = 8 \end{cases}$ by the substitution method.

If the substitution method is used to solve a system of dependent equations, the algebraic steps will produce an identity.

| **Example 5** | Solving a Consistent System of Dependent Equations |

Solve $\begin{cases} 2x = 4y + 6 \\ 3x - 6y = 9 \end{cases}$ by the substitution method.

Solution

$$2x = 4y + 6 \qquad \text{Solve the first equation for } x.$$
$$x = 2y + 3$$
$$3x - 6y = 9$$
$$3(2y + 3) - 6y = 9 \qquad \text{Then substitute this value for } x \text{ into the second equation.}$$
$$6y + 9 - 6y = 9 \qquad \text{Note that the simplification of this equation eliminates all variables and produces an identity.}$$

$$9 = 9 \text{ (an identity)} \qquad$$ This identity means that the system contains dependent equations and has an infinite number of solutions.

Points $(3, 0)$, $\left(0, -\dfrac{3}{2}\right)$, and $(-1, -2)$ are three of the points that lie on the common line and satisfy both equations. Points $(0, 0)$, $(1, 1)$, and $(2, 3)$ are three points that do not lie on the common line and do not satisfy either equation.

Answer: There is an infinite number of solutions.

Self-Check 5

Solve $\begin{cases} 5y = 4x - 10 \\ 12x - 15y = 30 \end{cases}$.

The accompanying box summarizes what happens when each of the three types of linear systems is solved by the substitution method.

Algebraic Solution of the Three Types of Linear Systems

1. **Consistent system of independent equations:** The solution process will produce unique x- and y-values.
2. **Inconsistent system:** The solution process will produce a contradiction.
3. **Consistent system of dependent equations:** The solution process will produce an identity.

Example 6 compares the algebraic, numerical, and graphical solutions of a consistent system of dependent equations.

| **Example 6** | Using Multiple Perspectives to Examine a System of Dependent Equations |

Solve $\begin{cases} y = 2x - 3 \\ 6x - 3y = 9 \end{cases}$ algebraically. Then examine this solution numerically and graphically, and describe the answer verbally.

Solution

Algebraically

Substitution Method

(1) $\qquad\qquad y = 2x - 3$

(2) $\qquad 6x - 3y = 9$

$\qquad 6x - 3(2x - 3) = 9$

$\qquad\quad 6x - 6x + 9 = 9$

$\qquad\qquad\qquad 9 = 9$

This identity signals a system of dependent equations.

*Writing Both Equations in
Slope-Intercept Form*

(1) $\qquad\quad y = 2x - 3$

(2) $6x - 3y = 9$

$\qquad -3y = -6x + 9$

$\qquad\quad y = 2x - 3$

Equation (2) simplified to exactly the same form as equation (1). Thus this is a system of dependent equations.

Numerically

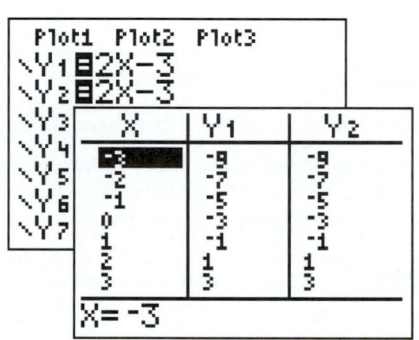

Graphically

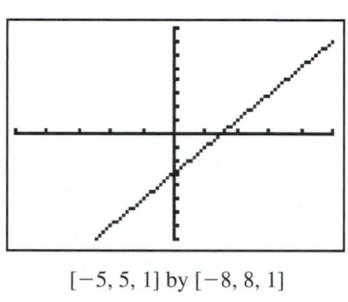

$[-5, 5, 1]$ by $[-8, 8, 1]$

Verbally

Answer: There is an infinite number of solutions that satisfy both equations. Because both equations are just different forms of the same equation, they produce exactly the same output values, the same table of values, and the same graph.

Self-Check 6

Solve $\begin{Bmatrix} 4x - 2y = 0 \\ 8x - 5y = -1 \end{Bmatrix}$ by the substitution method. Then check your solution.

The mean of a set of test scores is one type of average for these scores. The range is the difference between the highest and lowest scores. Example 7 involves both the mean and the range of two test scores.

Example 7 — Determining Test Scores with a Given Mean and Range

A student has two test scores in an algebra class. The mean of these test scores is 82, and their range is 12. Use this information to determine both test scores.

Solution

Let x = higher test score

$\quad\; y$ = lower test score

Word Equations	Algebraic Equations
(1) The mean of the two test scores is 82.	$\dfrac{x + y}{2} = 82$
(2) The range of the two test scores is 12.	$x - y = 12$

$\boxed{1}$ $x - y = 12$ Solve the second equation for x.

    $x = y + 12$

$\boxed{2}$ $\dfrac{x + y}{2} = 82$ Simplify the first equation by multiplying both sides by the LCD of 2.

    $x + y = 164$

  $(y + 12) + y = 164$ Then substitute for x in equation (2).

    $2y + 12 = 164$ Simplify this equation and solve for y.

      $2y = 152$

        $y = 76$ This is the lower of the two test scores.

$\boxed{3}$ $x = y + 12$

    $x = 76 + 12$ Back-substitute 76 for y into the equation that was solved for x in step 1.

    $x = 88$ This is the higher of the two test scores.

Answer: The two scores are 88 and 76.

Self-Check 7

Two test scores have a mean of 82 and a range of 20. Determine both test scores.

The answer to a word problem should be written as a full sentence. The question in Example 7 did not mention either x or y, so the answer should not use these symbols since they were introduced in the solution process.

Example 8 is solved by using the substitution method. The solution process includes the use of fractions, which many students would prefer to avoid. This is one reason that we will examine the addition method for solving this system in Section 3.5.

Example 8   Solving a Linear System by the Substitution Method

Solve $\begin{cases} 3x - 2y = 5 \\ 9x + 4y = -10 \end{cases}$ by the substitution method.

Solution

$\boxed{1}$ $3x - 2y = 5$ The coefficient of y is the smaller in magnitude, so we elect to solve for y.

    $-2y = -3x + 5$

      $y = \dfrac{3}{2}x - \dfrac{5}{2}$

$\boxed{2}$ $9x + 4y = -10$ Substitute this expression for y in the second equation. Then solve this equation for x.

  $9x + 4\left(\dfrac{3}{2}x - \dfrac{5}{2}\right) = -10$

    $9x + 6x - 10 = -10$

        $15x = 0$

         $x = \dfrac{0}{15}$

         $x = 0$ This is the x-coordinate of the solution.

3. $y = \dfrac{3}{2}x - \dfrac{5}{2}$

$y = \dfrac{3}{2}(0) - \dfrac{5}{2}$ Back-substitute 0 for x in the equation that was solved for y in step 1.

$y = 0 - \dfrac{5}{2}$

$y = -\dfrac{5}{2}$ This is the y-coordinate of the solution.

Answer: $\left(0, -\dfrac{5}{2}\right)$ Does this answer check?

Self-Check 8

Solve $\begin{cases} 4x - 3y = 0 \\ 8x + 9y = 10 \end{cases}$ by the substitution method.

Self-Check Answers

1. $(4, -6)$
2. The smaller angle is 30° and the larger angle is 60°.
3. $(-12, 6)$
4. No solution
5. An infinite number of solutions; three of these solutions are $(2.5, 0)$, $(0, -2)$, and $(1.25, -1)$.
6. $\left(\dfrac{1}{2}, 1\right)$
7. The two scores are 92 and 72.
8. $\left(\dfrac{1}{2}, \dfrac{2}{3}\right)$

3.4   Using the Language and Symbolism of Mathematics

1. The substitution principle states that a quantity may be substituted for its _____.

2. The solution for a system of two linear equations in two variables should be written in _____ _____ notation.

3. A contradiction is an equation that is always _____.

4. An identity is an equation that is always _____.

5. The solution of a consistent system of independent linear equations by the substitution method will produce **a conditional equation/a contradiction/an identity.** (Select the correct choice.)

6. The solution of a consistent system of dependent linear equations by the substitution method will produce **a conditional equation/a contradiction/an identity.** (Select the correct choice.)

7. The solution of an inconsistent system of independent linear equations by the substitution method will produce **a conditional equation/a contradiction/an identity.** (Select the correct choice.)

8. Using the substitution method produced the equations $x = 3$ and $y = 4$. Thus the system of two linear equations is a _____ system of _____ linear equations with one solution that is the ordered pair _____.

9. Using the substitution method produced the equation $3 = 4$. Thus the system of two linear equations is an _____ system of _____ linear equations with _____ solution(s).

10. Using the substitution method produced the equation $3 = 3$. Thus the system of two linear equations is a _____ system of _____ linear equations with an _____ number of solutions.

3.4 Quick Review

In Exercises 1–3, match each equation with the most appropriate description.

1. $x + x = 2$ **A.** Conditional equation

2. $x + x = 2x$ **B.** Contradiction

3. $2x = 2(x + 1)$ **C.** Identity

4. Two angles are complementary if the sum of their measures is _____.

5. Write in slope-intercept form the equation of the line through the points $(5, 4)$ and $(-5, 0)$.

3.4 Exercises

Objective 1 Solve a System of Linear Equations by the Substitution Method

In Exercises 1–42, solve each system of linear equations by the substitution method.

1. $y = 2x - 3$
   $x + y = 9$

2. $y = 3x - 4$
   $x + 2y = 6$

3. $y = 5x - 10$
   $2x + 3y = 4$

4. $y = 9 - 3x$
   $x + 2y = 8$

5. $x = 6 - 5y$
   $2x + 9y = 4$

6. $x = 4y - 9$
   $2x + 3y = 15$

7. $3x - y = 1$
   $x + 2y = 2$

8. $x - 3y = -14$
   $2x + y = 7$

9. $2x - y - 6 = 0$
   $x - y - 6 = 0$

10. $2x + y - 11 = 0$
    $x + 2y - 13 = 0$

11. $2x + 3y = -7$
    $x + 4y = -6$

12. $3x + 4y = -2$
    $x - 4y = -6$

13. $2x - 5y = 9$
    $x - 3 = 0$

14. $4x - 7y = -1$
    $x + 2 = 0$

15. $3x - 2y = 1$
    $y + 7 = 0$

16. $5x - 4y = 3$
    $y - 4 = 0$

17. $y = -2x$
    $5x - y = -7$

18. $y = 6x$
    $8x - 5y = -11$

19. $3x - 2y = 0$
    $11x - 9y = 5$

20. $5x + 4y = 0$
    $6x + 5y = 2$

21. $x + y = 21$
    $x - y = 3$

22. $2x + y = 3$
    $x + 2y = 9$

23. $2x - 7y = 42$
    $x = 0$

24. $5x - 9y = 35$
    $y = 0$

25. $5x - 2y = 11$
    $3x + 3y = 15$

26. $3x + 4y + 2 = 0$
    $5x - 20y + 30 = 0$

27. $5x - 4y + 12 = 0$
    $2x - 3y + 2 = 0$

28. $3x + 4y + 5 = 0$
    $-2x + 5y - 11 = 0$

29. $y = x + 3$
    $\dfrac{x}{2} - \dfrac{y}{5} = 3$

30. $y = x + 4$
    $\dfrac{x}{4} - \dfrac{y}{2} = 0$

31. $x = 2y + 4$
    $3x - 6y = 12$

32. $y = 4 - 2x$
    $6x + 3y = 12$

33. $y = 5x + 3$
    $10x - 2y = 6$

34. $x = 5y - 3$
    $4x - 20y = -1$

35. $4x + 7y = 0$
    $7x - 4y = 0$

36. $9x - 5y = 0$
    $5x + 9y = 0$

37. $x = 2y - 5$
    $\dfrac{x}{6} + \dfrac{y}{8} = 1$

38. $\dfrac{x}{6} + \dfrac{y}{12} = \dfrac{1}{6}$
    $\dfrac{x}{8} - \dfrac{y}{4} = -2$

39. $\dfrac{x}{2} + \dfrac{y}{5} = \dfrac{4}{5}$
    $\dfrac{x}{6} - \dfrac{y}{2} = \dfrac{5}{6}$

40. $\dfrac{x}{8} - \dfrac{y}{5} = \dfrac{1}{10}$
    $y = \dfrac{5}{8}x - \dfrac{1}{4}$

41. $\dfrac{x}{4} - \dfrac{y}{2} = \dfrac{7}{24}$
    $\dfrac{x}{3} + \dfrac{y}{2} = 0$

42. $\dfrac{x}{2} - \dfrac{y}{3} = -\dfrac{7}{12}$
    $\dfrac{x}{8} + \dfrac{y}{9} = 0$

Review and Concept Development

43. **a.** Write in slope-intercept form the equation of the line through the points $(2, -2)$ and $(-1, 4)$.
    b. Write in slope-intercept form the equation of the line through the points $(-1, 1)$ and $(5, -2)$.
    c. Find the point that is on both lines from parts **a** and **b**.

44. **a.** Write in slope-intercept form the equation of the line through the points $(-1, 4)$ and $(-3, 7)$.
    b. Write in slope-intercept form the equation of the line through the points $(4, -1)$ and $(-2, 3)$.
    c. Find the point that is on both lines from parts **a** and **b**.

In Exercises 45–48, select the system of linear equations (choice A, B, C, or D) that represents each word problem. Then solve this system by the substitution method and answer the problem, using a full sentence.

Word Problem	**Algebraic Equations**
45. Find two numbers whose sum is 120 and whose difference is 30.	**A.** $x + y = 20$ $\quad x - y = 30$
46. Find two numbers whose sum is 30 and the first number is 120 more than the second number.	**B.** $x + y = 120$ $\quad x - y = 30$
47. Find two numbers whose sum is 30 and whose difference is 30.	**C.** $x + y = 30$ $\quad x - y = 30$
48. Find two numbers whose sum is 20 and whose difference is 30.	**D.** $x + y = 30$ $\quad x - y = 120$

Multiple Representations In Exercises 49 and 50, complete the numerical and graphical descriptions for each problem and then write a full sentence that answers the question.

Word Problem | **Algebraic Equations** | **Table** | **Graph**

49. The sum of two numbers is 8 and their difference is 2. Find these numbers.

$x + y = 8$
$x - y = 2$

x	$y_1 = 8 - x$	$y_2 = x - 2$
2	6	0
3		
4		
5		
6		
7		
8	0	6

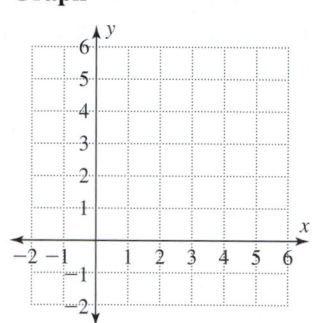

Word Problem | **Algebraic Equations** | **Table** | **Graph**

50. The sum of two numbers is 2 and their difference is 6. Find these numbers.

$x + y = 2$
$x - y = 6$

x	$y_1 = 2 - x$	$y_2 = x - 6$
0	2	-6
1		
3		
4		
5		
6	-4	0

Connecting Concepts to Applications

In Exercises 51–56, write a system of linear equations using the variables x and y, and use this system of equations to solve the problem.

51. Angles in Two Parts To function properly, two parts must be assembled so that the two angles are complementary, as shown in the figure, and one angle is 18° larger than the other angle. Determine the number of degrees in each angle.

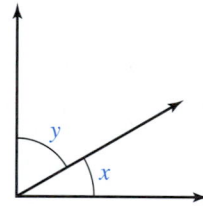

52. Angles in Two Parts To function properly, two parts must be assembled so that the two angles are complementary and one angle is 12° more than twice the other angle. Determine the number of degrees in each angle. (See Exercise 51.)

53. Mean and Range A student has two test scores in a psychology class. The mean of these scores is 76 and their range is 28. Use this information to determine both test scores.

54. Mean and Range A student has two test scores in a history class. The mean of these scores is 90 and their range is 12. Use this information to determine both test scores.

55. Mean and Range Tiger Woods had identical scores for three of the first four rounds at a PGA golf tournament. After he shot a tournament record in the final round, his mean score was 66 and the range of his scores was 8. Use this information to determine his score in each of the four rounds.

56. Mean and Range An amateur playing in a pro-am golf tournament in Scottsdale, Arizona, had identical scores for three of the first four rounds to enter the final day of play in first place. However, her score on the final day was her highest of the four rounds, and she finished fifth for the tournament. Her mean score was 73 and the range of her scores was 8. Use this information to determine her score in each of the four rounds.

57. Dosages for Children When a doctor knows the adult dosage of a medication but not a child's dosage, there are several formulas available for calculating an appropriate dosage for the child. Two formulas for calculating a child's dosage are Fried's rule and Cowling's rule. We use x to represent the age of the child and y to represent the child's dosage in milliliters. If the adult dosage of a medication is 48 mL, then these two formulas simplify to

Fried's rule: $y = 3.84x$

Cowling's rule: $y = 2x + 2$

a. Will these two formulas ever give the same dosage?

b. If so, at what age will the dosages be equal?

58. Tuition at Two Colleges The college tuition at one state university is $95 per credit-hour plus a student activity fee of $200. The tuition at the archrival across the state is $100 per credit-hour plus a student activity fee of $125. Use x to represent the number of credit-hours and y to represent the total cost of tuition and fees.

a. Write a linear equation to model the cost of a semester of tuition and fees at the state university.

b. Write a linear equation to model the cost of a semester of tuition and fees at the rival university.

c. Use this system of equations to determine the conditions for which these costs will be the same.

d. What is this cost?

59. Taxi Fares The accompanying graph compares the cost of a taxi ride for two different taxicab companies. The cost is based on the number of miles driven.

a. Use the y-intercept and an additional point to determine the equation of the line that models the cost for company A.

b. Use the y-intercept and an additional point to determine the equation of the line that models the cost for company B.

c. Solve this system of equations.

d. Does this algebraic solution seem consistent with the point of intersection shown on the graph?

e. Interpret the meaning of the x- and y-coordinates of this solution.

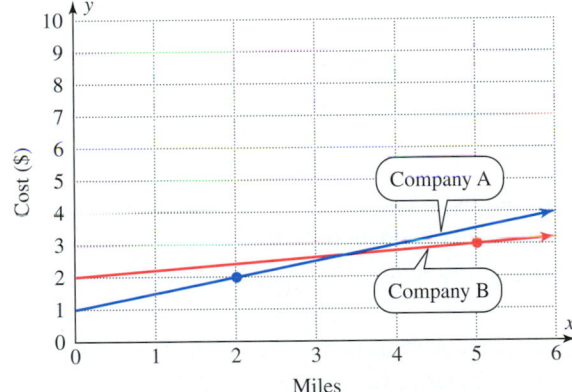

60. Population Growth Plainsville and Springfield were both founded in 1900. The accompanying graph compares the populations of each city based on the number of years since 1900.

a. Use the y-intercept and an additional point to determine the equation of the line that models the population for Plainsville.

b. Use the y-intercept and an additional point to determine the equation of the line that models the population for Springfield.

c. Solve the system of equations from parts **a** and **b**.

d. Does this algebraic solution seem consistent with the point of intersection shown on the graph?

e. Interpret the meaning of the x- and y-coordinates of this solution.

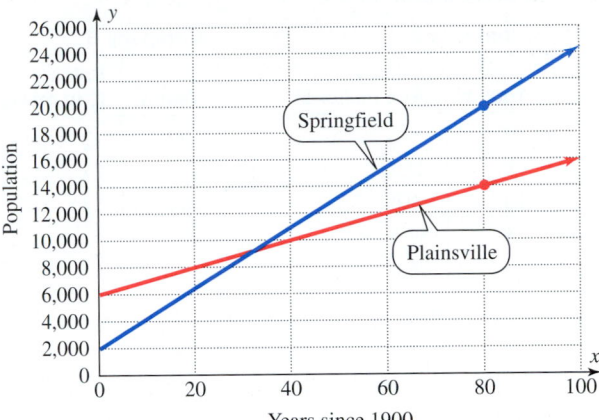

61. Investment Growth The tables display the values of two different investments made in 1985. Investment A was an initial deposit of $1,500 and has earned a simple interest rate of 12%. Investment B was an initial deposit of $2,000 and has earned a simple interest rate of 7%. The variable x represents the number of years since 1985.

a. Use the y-intercept and an additional point to determine the equation of the line that models the value of investment A.

b. Use the y-intercept and an additional point to determine the equation of the line that models the value of investment B.

c. Solve the system of equations from parts **a** and **b**.

d. Does this algebraic solution seem consistent with the values shown in the tables?

e. Interpret the meaning of the x- and y-coordinates of this solution.

Investment A			Investment B	
Year x	Value y		Year x	Value y
0	1,500		0	2,000
5	2,400		5	2,700
10	3,300		10	3,400
15	4,200		15	4,100
20	5,100		20	4,800
25	6,000		25	5,500

62. Sales Growth The tables display the sales in millions of dollars of two companies founded in 1990. The variable x represents the number of years since 1990.

a. Use the y-intercept and an additional point to determine the equation of the line that models the sales for company A.

b. Use the y-intercept and an additional point to determine the equation of the line that models the sales for company B.

c. Solve the system of equations from parts **a** and **b**.
d. Does this algebraic solution seem consistent with the values shown in the tables?
e. Interpret the meaning of the x- and y-coordinates of this solution.

Company A	
Year x	Sales y ($ in millions)
0	5.0
5	22.5
10	40.0
15	57.5
20	75.0

Company B	
Year x	Sales y ($ in millions)
0	17
5	27
10	37
15	47
20	57

63. The table illustrates that the x-coordinate of the solution of the system $\begin{cases} x + 2y = 9 \\ x - y = -2 \end{cases}$ is between the integers _____ and _____ because Y_2 switches from less than Y_1 to more than Y_1 in this x-interval. This indicates that the lines cross in this x-interval and signals a solution of the system of equations. Use the substitution method to confirm this fact and to find the exact coordinates of this solution.

x	$y_1 = -0.5x + 4.5$	$y_2 = x + 2$
-2	5.5	0
-1	5.0	1
0	4.5	2
1	4.0	3
2	3.5	4
3	3.0	5
4	2.5	6

64. The table illustrates that the x-coordinate of the solution of the system $\begin{cases} 7x + y = 1 \\ 7x - y = -9 \end{cases}$ is between the integers _____ and _____ because Y_2 switches from less than Y_1 to more than Y_1 in this x-interval. This indicates that the lines cross in this x-interval and signals a solution of the system of equations. Use the substitution method to confirm this fact and to find the exact coordinates of this solution.

x	$y_1 = -7x + 1$	$y_2 = 7x + 9$
-2	15	-5
-1	8	2
0	1	9
1	-6	16
2	-13	23
3	-20	30
4	-27	37

Applying Technology

In Exercises 65–68, use the substitution method and a calculator to solve each system of linear equations.

65.
$$y = 1.05x - 2.1325$$
$$7.14x - 8.37y = 10.1835$$

66.
$$y = 4.55x - 10.696$$
$$2.08x - 1.17y = 1.0972$$

67.
$$x = 4.91y - 5.899$$
$$2.1x - 9.9y = -12.84$$

68. $2.409x + 2.409y = 4.818$
$$3.56x - 3.56y = 14.24$$

Group discussion questions

69. Discovery Question Given the system $\begin{cases} x - 5y = -1 \\ 3x - y = 11 \end{cases}$:
a. Solve this system by solving the first equation for x and then substituting for x in the second equation.
b. Solve this system by solving the second equation for y and then substituting for y in the first equation.
c. Compare these solutions and generalize about implementing the substitution method for other systems of linear equations.

70. Discovery Question Given the system $\begin{cases} x + 4y = -5 \\ 2x + 7y = 10 \end{cases}$, have part of your group solve this system by using a graphical method and have the rest of the group solve this system by using the substitution method. Time each group to determine which method seems more efficient for this problem. Did both groups get the same answer?

In Exercises 71 and 72, determine the value of B so that each system of equations will be an inconsistent system.

71. $4x - 8y = 12$ **72.** $4x + 2y = 10$
   $3x + By = 10$ $10x + By = 20$

In Exercises 73 and 74, determine the value of C so that each system of equations will have dependent equations.

73. $6x - 15y = 9$ **74.** $5x - 15y = 10$
   $4x - 10y = C$ $4x - 12y = C$

75. Challenge Question Solve each of these systems for (x, y) in terms of a and b for $a \neq 0$.
a. $x - ay = b$ b. $3ax + y = b$
   $x + ay = 2b$ $2ax - y = 4b$

3.4 Cumulative Review

In Exercises 1 and 2 determine whether either $x = -2$ or $x = 2$ is a solution of each equation.

1. $3x - 4 = 5x - 8$

2. $-4(2x + 7) = 3x - 6$

In Exercises 3–5 solve each equation.

3. $1 - 4(x + 3) = -5(x - 1)$

4. $\dfrac{x}{2} = 4 - \dfrac{x - 3}{3}$

5. $3.4x - 36.17 = -2.3(x - 4.1)$

Section 3.5   Solving Systems of Linear Equations in Two Variables by the Addition Method

Objective:

1. Solve a system of linear equations by the addition method.

As noted in Section 3.4, the substitution method is well suited to systems that contain at least one variable with a coefficient of 1 or -1. For other systems it may be easier to use the addition method, which is described in the following box.

1. Solve a System of Linear Equations by the Addition Method

What Is the Logic Behind the Addition Method?

The addition method is based on the **addition-subtraction principle of equality,** which states that equal values can be added to or subtracted from both sides of an equation to produce an equivalent equation. This method is also called the **elimination method** because the strategy is to eliminate a variable in one of the equations.

Addition Method

Step 1. Write both equations in the general form $Ax + By = C$.

Step 2. If necessary, multiply one or both of the equations by a constant so that the equations have one variable for which the coefficients are additive inverses.

Step 3. Add the new equations to eliminate a variable and then solve the resulting equation.

Step 4. Substitute this value into one of the original equations (back-substitution), and solve for the other variable.

The ordered pair obtained in steps 3 and 4 is the solution that should check in both equations.

If $3x - y = 5$ and $2x + y = 10$, then the balance scales in the figure illustrate what happens when we add equals to equals. The first balance scale represents that $2x + y$ equals 10. This scale is still in balance when $3x - y$ is added to one side and 5 is added to the other side.

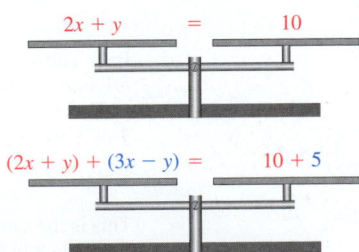

We will examine this pair of equations further in Example 1.

Example 1   Solving a Linear System by the Addition Method

Solve $\begin{cases} 2x + y = 10 \\ 3x - y = 5 \end{cases}$ by the addition method.

Solution

$$2x + y = 10$$
$$\underline{3x - y = 5}$$
$$5x = 15$$

The equations are already in the proper form, and the coefficients of y are additive inverses. Add these equations to eliminate y. Solve this equation for x.

$$x = 3$$ This is the x-coordinate of the solution.

$$2x + y = 10$$
$$2(3) + y = 10$$ Back-substitute 3 for x in the first equation. Then solve for y.
$$6 + y = 10$$
$$y = 4$$ This is the y-coordinate of the solution.

Check:

First Equation	Second Equation
$2x + y = 10$	$3x - y = 5$
$2(3) + 4 \overset{?}{=} 10$	$3(3) - 4 \overset{?}{=} 5$
$6 + 4 \overset{?}{=} 10$	$9 - 4 \overset{?}{=} 5$
$10 \overset{?}{=} 10$ checks.	$5 \overset{?}{=} 5$ checks.

Answer: (3, 4)   Remember to write the solution using ordered-pair notation.

Self-Check 1

Solve $\begin{cases} 4x + 3y = 7 \\ -2x - 3y = -11 \end{cases}$.

Example 2 revisits Example 8 from Section 3.4. When you compare these two solutions, you will likely prefer the addition method shown next.

Example 2   Solving a Linear System by the Addition Method

Solve $\begin{cases} 3x - 2y = 5 \\ 9x + 4y = -10 \end{cases}$.

Solution

$$3x - 2y = 5 \qquad\qquad 6x - 4y = 10$$
$$9x + 4y = -10 \qquad\qquad \underline{9x + 4y = -10}$$
$$15x = 0$$
$$x = 0$$

Multiply both sides of the first equation by 2 to obtain coefficients of y that are additive inverses. Add these equations to eliminate y.
Divide both sides of the equation by 15 to solve for x. This is the x-coordinate of the solution.

$$3x - 2y = 5$$
$$3(0) - 2y = 5$$
$$-2y = 5$$

Back-substitute 0 for x in the first equation of the original system. Then solve for y.

$$y = -\frac{5}{2}$$

This is the y-coordinate of the solution.

Answer: $\left(0, -\dfrac{5}{2}\right)$

This is the same solution we obtained in Example 8 in Section 3.4.

Self-Check 2

Solve $\begin{cases} 2x - 7y = 1 \\ x + 5y = 9 \end{cases}$.

When I Use the Addition Method, Does It Make Any Difference Whether I Eliminate *x* or Eliminate *y*?

In Example 3, both equations are multiplied by a constant so that the coefficients of y will be additive inverses. The choice to eliminate y is arbitrary since it would be just as easy to eliminate x. Although any common multiple of the coefficients of y can be used to eliminate the y-variable, we encourage you to use the least common multiple of the coefficients.

Example 3   Solving a Linear System by the Addition Method

Solve $\begin{cases} 5x + 3y = -6 \\ 3x + 2y = -5 \end{cases}$ by the addition method.

Solution

$5x + 3y = -6$	$10x + 6y = -12$	Multiply both sides by 2.
$3x + 2y = -5$	$-9x - 6y = 15$	Multiply both sides by -3.
	$x = 3$	Add these equations to eliminate y.

$$5x + 3y = -6$$
$$5(3) + 3y = -6$$ — Back-substitute 3 for x in the first equation of the original system.
$$15 + 3y = -6$$
$$3y = -21$$
$$y = -7$$ — Then solve for y.

Answer: $(3, -7)$ Does this solution check?

Self-Check 3

Solve $\begin{cases} 8x - 3y = 5 \\ -5x + 7y = 43 \end{cases}$.

In Example 4, the first step is the decision to eliminate the variable y. The coefficients of y are -3 and 4. Use this information to guide the steps taken to produce the least common multiple of 12.

Example 4 Solving a Linear System by the Addition Method

Solve $\begin{cases} 7x = 3y + 8 \\ 4y = -5(x + 5) \end{cases}$ by the addition method.

Solution

$$7x = 3y + 8 \qquad 7x - 3y = 8 \qquad 28x - 12y = 32 \qquad \text{First write both equations in the form } Ax + By = C.$$

$$4y = -5(x + 5) \qquad 5x + 4y = -25 \qquad \underline{15x + 12y = -75} \qquad \text{To produce the coefficients of } -12 \text{ and } 12 \text{ for } y,$$

$$43x \qquad = -43 \qquad \text{multiply the first equation by 4 and the second equation by 3. Add these equations to eliminate } y.$$

$$x = -1 \qquad \text{Solve for } x.$$

$$4y = -5(x + 5)$$
$$4y = -5(-1 + 5) \qquad \qquad \text{Back-substitute } -1 \text{ for } x \text{ in the second equation of the}$$
$$4y = -5(4) \qquad \qquad \text{original system.}$$
$$4y = -20$$
$$y = -5 \qquad \qquad \text{Then solve for } y.$$

Answer: $(-1, -5)$ \qquad\qquad Does this solution check?

Self-Check 4

Solve $\begin{cases} 2x = 6y - 1 \\ 9y + 1 = 2(2x + 1) \end{cases}$ by the addition method.

The word problem strategy introduced in Section 2.8 is illustrated again in Example 5. Note the emphasis on the use of word equations as an aid to forming the algebraic equations.

Example 5 Factory Defects

A quality control worker in a factory weighed eight soda cans. The weights of the first seven were all equal, but the eighth can had a defect and was too light. The mean of these weights was 15.65 oz, and the range was 2.8 oz. Use this information to determine each weight.

Solution

Definition of Variables

Let x = weight in ounces of each of first seven cans \qquad Select a variable to represent each unknown.

y = weight in ounces of eighth can

Word Equations

(1) The mean weight is 15.65 oz. \qquad First form word equations, and then translate them into algebraic

(2) The range of the weights is 2.8 oz. \qquad equations.

Algebraic Equations

(1) $\dfrac{7x + y}{8} = 15.65$ \qquad The mean of a set of measurements is found by dividing their sum by the number of measurements.

(2) $x - y = 2.8$ \qquad The range of a set of measurements is found by subtracting the smallest from the largest.

Solving Algebraically by the Addition Method

(1) $7x + y = 125.2$

(2) $\underline{x - y = \quad 2.8}$

$\quad 8x \qquad = 128$

$\quad\; x \qquad\;\; = 16$

Multiply both sides of equation (1) above by 8. Then add the two equations to eliminate y.

Solve for x.

$$x - y = 2.8$$
$$16 - y = 2.8$$
$$-y = -13.2$$
$$y = 13.2$$

Back-substitute 16 for x in the second equation and then solve for y.

Numerical Check

$$\frac{7(16) + 13.2}{8} = \frac{125.2}{8} = 15.65$$

$$16 - 13.2 = 2.8$$

These values produce a mean of 15.65 and a range of 2.8.

Answer: The first seven cans each weigh 16 oz, and the eighth can weighs 13.2 oz.

Self-Check 5

A quality control worker in a factory weighed 12 golf balls. The total weight of these golf balls was 19.6 ounces. The weights of 11 of the golf balls were equal, but the 12th had a defect and was too heavy and weighed 0.16 oz more than the others. Use this information to determine each weight.

If the Equations Involve Fractional Coefficients, Do I Multiply Both Sides by the LCD Just As I Did with the Substitution Method?

Yes, the first step is to multiply through by the LCD to produce equivalent equations with integer coefficients. This is illustrated in Example 6.

Example 6   Solving a Linear System with Fractional Coefficients

Solve $\begin{cases} \dfrac{x}{6} + \dfrac{y}{9} = 2 \\[2mm] \dfrac{x}{8} - \dfrac{y}{3} = 9 \end{cases}$ by the addition method.

Solution

(1) $\dfrac{x}{6} + \dfrac{y}{9} = 2 \qquad 18\left(\dfrac{x}{6} + \dfrac{y}{9}\right) = 18(2)$

Multiply both sides by 18, the LCD. Distribute the factor of 18 and simplify this equation.

(2) $\dfrac{x}{8} - \dfrac{y}{3} = 9 \qquad -24\left(\dfrac{x}{8} - \dfrac{y}{3}\right) = -24(9)$

Multiply both sides by -24, the opposite of the LCD. Distribute the factor of -24 and simplify this equation.

(1)	$3x + 2y = 36$	Add these equations to eliminate x.
(2)	$\underline{-3x + 8y = -216}$	
	$10y = -180$	
	$y = -18$	Solve for y.

$$3x + 2y = 36$$
$$3x + 2(-18) = 36$$
$$3x - 36 = 36$$
$$3x = 72$$
$$x = 24$$

Back-substitute -18 for y in the first equation with integer coefficients.

Then solve for x.

Answer: $(24, -18)$

Does this answer check in the original system?

Self-Check 6

Solve $\left\{\begin{array}{l} \dfrac{x}{2} - \dfrac{y}{8} = 3 \\ \dfrac{x}{4} + \dfrac{y}{2} = -3 \end{array}\right\}$.

If I Am Using the Addition Method, How Will I Know When a System Is Inconsistent?

Remember that the algebraic solution of an inconsistent system of linear equations will produce a contradiction. The algebraic solution of a consistent system of dependent linear equations will produce an identity. In Section 3.4, we examined these two possibilities by using the substitution method. Now we will see what happens when the addition method is used to solve an inconsistent system of linear equations or a consistent system of dependent linear equations.

Example 7 Solving an Inconsistent System

Solve $\left\{\begin{array}{l} 5x + 10y = 11 \\ x + 2y = 3 \end{array}\right\}$ by the addition method.

Solution

$5x + 10y = 11$	$5x + 10y = 11$	Multiply both sides of the second equation by -5. Adding these equations eliminates both variables and produces a contradiction.
$x + 2y = 3$	$\underline{-5x - 10y = -15}$	
	$0 = -4$, a contradiction	

Answer: There is no solution.

Since the result is a contradiction, the original system is inconsistent and has no solution.

Self-Check 7

Solve $\left\{\begin{array}{l} x - \dfrac{y}{2} = \dfrac{3}{2} \\ 6x - 3y = 5 \end{array}\right\}$ by the addition method.

Example 8   Solving a Consistent System of Dependent Equations

Solve $\begin{cases} 0.6x - 1.5y = 0.3 \\ -0.4x + 1.0y = -0.2 \end{cases}$ by the addition method.

Solution

$0.6x - 1.5y = 0.3$	$12x - 30y = 6$	Multiply both sides of the first equation by 20, and then multiply both sides of the second equation by 30.
$-0.4x + 1.0y = -0.2$	$\underline{-12x + 30y = -6}$	
	$0 = 0$, an identity	

Adding these equations eliminates both variables and produces an identity.

Answer: There is an infinite number of solutions.

Since the result is an identity, the original equations are dependent, and the system has an infinite number of solutions.

Self-Check 8

Solve $\begin{cases} 0.4x - 0.8y = 2.0 \\ 0.3x - 0.6y = 1.5 \end{cases}$.

Example 9 involves two unknown numbers. Variables x and y are used to represent these numbers so that a system of two linear equations can be formed to describe this problem.

Example 9   Solving a Numeric Word Problem

Find two numbers whose sum is 60 and whose difference is 14.

Solution

Let x = larger number

y = smaller number

Word Equations	Algebraic Equations	
(1) The sum of the two numbers is sixty	$x + y = 60$	
(2) The difference between the two numbers is fourteen	$\underline{x - y = 14}$	Add the two equations to eliminate y.
	$2x \qquad = 74$	
	$x = 37$	Divide both sides of the equation by 2 to solve for x.
$x + y = 60$		Back-substitute 37 for x in the first equation and solve for y.
$37 + y = 60$		
$y = 23$		

Answer: The numbers are 37 and 23.

Write the answer as a full sentence. Does this answer check?

Self-Check 9

Find two numbers whose sum is 81 and whose difference is 19.

3.5   Using the Language and Symbolism of Mathematics

1. The addition-subtraction principle of equality states that _____ values can be added to or subtracted from both sides of an equation to produce an equivalent equation.

2. The addition method for solving a system of linear equations is also sometimes referred to as the _____ method because the strategy is to _____ one of the variables when the two equations are added.

3. To eliminate the y-variable when adding two linear equations, we first must make the coefficients of y in the two equations _____ of each other.

4. Using the addition method to solve a system of two linear equations produced the equation $0 = 5$. Thus the system of equations is an _____ system with _____ solution(s).

5. Using the addition method to solve a system of two linear equations produced the equation $0 = 0$. Thus the system is a _____ system of _____ equations with an _____ number of solutions.

3.5   Quick Review

1. The LCD of $\dfrac{7}{15}$ and $\dfrac{6}{35}$ is _____.

2. Simplify $4(x - 2y - 3) - 5(3x + y - 2)$.

3. An even integer is represented by n.
   a. Represent the next three consecutive integers.
   b. Represent the next three consecutive even integers.
   c. Represent the next three consecutive odd integers.

4. An algebra student had exam scores of 84, 76, 93, and 87. Determine the mean of these scores.

5. A line shows the average weight of a U.S. eighth-grader over the last 20 years. Because this average weight has been trending higher, you would expect this line to have a **positive/negative** slope. (Select the correct choice.)

3.5   Exercises

Objective 1 Solve a System of Linear Equations by the Addition Method

In Exercises 1–30, solve each system of linear equations by the addition method.

1. $x + 2y = 6$
   $-x + 3y = 4$

2. $5x - y = 10$
   $2x + y = 4$

3. $5x + 2y = -26$
   $3x - 2y = -38$

4. $-5x + 3y = 27$
   $5x + 2y = -7$

5. $6x + 2y = -1$
   $12x - y = 3$

6. $5x - y = -1$
   $15x + 2y = 7$

7. $x + 2y = 1$
   $3x + 4y = 0$

8. $x + 6y = 0$
   $3x + 8y = 5$

9. $2x + 3y = -9$
   $-4x + 5y = -37$

10. $3x + 7y = 25$
    $-6x + 3y = 18$

11. $5x - 3y = 5$
    $4x + 6y = 46$

12. $2x + 15y = 3$
    $3x - 5y = -1$

13. $2x + 5y = -3$
    $3x + 8y = -5$

14. $2x - 3y = 4$
    $11x - 5y = -1$

15. $2x + 3y = 15$
    $5x + 4y = -1$

16. $4x + 3y = 11$
    $5x + 2y = -9$

17. $2x - 13y = 38$
    $5x + 27y = 95$

18. $23x - 2y = -30$
    $47x + 3y = 45$

19. $2x = 11y$
    $5x = 19y$

20. $3x = 17y$
    $4x = 9y$

21. $2x + 6 = 0$
$3x + 2y = 1$

22. $2y + 14 = 0$
$2x + 3y = -13$

23. $\dfrac{x}{2} + \dfrac{y}{3} = 5$
$\dfrac{x}{3} - \dfrac{y}{2} = -1$

24. $\dfrac{x}{8} + \dfrac{y}{8} = 0$
$\dfrac{x}{2} - \dfrac{y}{4} = -3$

25. $6x - 8y = 10$
$-15x + 20y = -20$

26. $2x - 6y = 8$
$-3x + 9y = -12$

27. $6x = 9 - 3y$
$4y = 12 - 8x$

28. $3x = 6 - 12y$
$7 - 4x = 16y$

29. $\dfrac{x}{2} - \dfrac{y}{2} = \dfrac{3}{4}$
$\dfrac{x}{2} + \dfrac{y}{4} = \dfrac{5}{8}$

30. $\dfrac{x}{3} + \dfrac{y}{3} = 0$
$10x + 5y = -1$

Review and Concept Development

In Exercises 31–42, solve each system of linear equations by either the substitution method or the addition method.

31. $y = 2x - 1$
$7x - 4y = -1$

32. $x = 3y - 5$
$-5x + 6y = -11$

33. $3x + 7y = 20$
$5x - 7y = -4$

34. $4x + 3y = 26$
$5x - 2y = -25$

35. $y = -x$
$5x - 7y = 6$

36. $y = \dfrac{5}{3}x$
$x + y = 2$

37. $\dfrac{x}{4} - \dfrac{y}{3} = \dfrac{5}{12}$
$\dfrac{x}{2} - \dfrac{2y}{3} = 1$

38. $\dfrac{x}{4} - \dfrac{y}{6} = 6$
$\dfrac{x}{6} + \dfrac{y}{3} = -4$

39. $2x + 3y = 4x + 5y - 6$
$x - 5y = 7x + 2y - 21$

40. $3x + 4(2y - 1) = x + 8y - 2$
$5x + 7(y + 5) = 4(x + y + 5) + 1$

41. $0.5x + 1.2y = 0.3$
$0.4x + 0.9y = 0.3$

42. $0.05x - 0.13y = 0.01$
$0.03x - 0.07y = 0.03$

43. a. Write in slope-intercept form the equation of the line through the points $(11.5, -3.0)$ and $(0.5, 1.0)$.
  b. Write in slope-intercept form the equation of the line through the points $(40, -7)$ and $(-11, 2)$.
  c. Find the point that is on both lines from parts **a** and **b**.

44. a. Write in slope-intercept form the equation of the line through the points $(-3.0, 2.5)$ and $(3, -2)$.
  b. Write in slope-intercept form the equation of the line through the points $(0, 0)$ and $(5, -4)$.
  c. Find the point that is on both lines from parts **a** and **b**.

Connecting Concepts to Applications

In Exercises 45–48, select the system of linear equations (choice A, B, C, or D) that represents the word problem. Then solve this system by the addition method and answer the problem, using a full sentence.

Algebraic Equations

A. $2x + y = 14$
$x - y = 4$

B. $2x + y = 14$
$x + 3y = 22$

C. $3x + y = 1$
$x - y = 7$

D. $5x + 2y = 2$
$2x + 5y = 26$

Word Problem

45. The sum of three times one number and a second number is one. The difference of the first number minus the second number is seven. Find these numbers.

46. The sum of twice one number plus a second number is fourteen. The first number plus three times the second number is twenty-two. Find these numbers.

47. The sum of five times one number plus twice a second number is two. The sum of twice the first number plus five times the second number is twenty-six. Find the numbers.

48. The sum of twice one number plus a second number is fourteen. The difference of the first number minus the second number is four. Find the numbers.

In Exercises 49–52, write a system of linear equations using the variables x and y, and use this system to solve the problem.

49. Find two numbers whose sum is 88 and whose difference is 28.

50. Find two numbers whose sum is 133 and whose difference is 11.

51. Find two numbers whose sum is 102 if one number is twice the other number.

52. Find two numbers whose sum is 213 if one number is twice the other number.

Mean and Range In Exercises 53 and 54, write a system of linear equations using the variables x and y and use this system of equations to solve the problem.

53. Basketball Scoring A basketball player had four games with identical scores and a fifth game that was his high for the whole season. His mean score for the five games was 25 and the range was 15. Use this information to determine his score in each of the five games.

54. Factory Defects A quality control worker in a factory measured ten wheel rims. The diameters of the first nine were all equal, but the tenth had a defect and was too small. The mean of these measurements was 24.9 cm and the range was 1 cm. Use this information to determine each measurement.

55. College Enrollment Patterns In 1995, 38% of all graduates at Mason High School attended a 4-year college and 18% of all graduates attended a community college. In 2011, of all graduates, 30% attended a 4-year college and 42% attended a community college.

a. Use the two points (1995, 38) and (2011, 30) to write an equation for the line showing the percent of graduates attending a 4-year college.

b. Use the two points (1995, 18) and (2011, 42) to write an equation for the line showing the percent of graduates attending a community college.

c. Solve the system of equations from parts **a** and **b**.

d. Interpret the meaning of the x- and y-coordinates of the solution from part **c**.

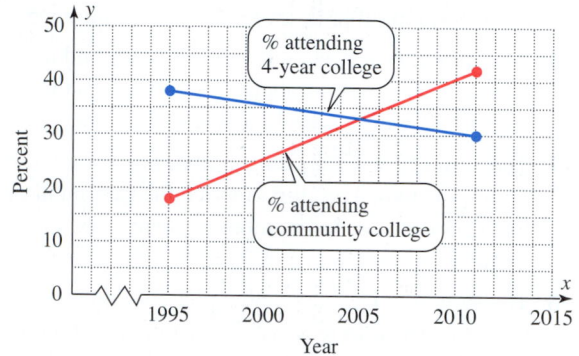

56. Generation of Electricity from the Wind In 2000, only 3% of all electricity needed in one community was generated by wind, and 39% was generated by coal. By 2010, of the electricity needed, 20% was generated by wind and 16% was generated by coal.

a. Use the two points (2000, 3) and (2010, 20) to write an equation for the line showing the percent of electricity generated by wind.

b. Use the two points (2000, 39) and (2010, 16) to write an equation for the line showing the percent of electricity generated by coal.

c. Solve the system of equations from parts **a** and **b**.

d. Interpret the meaning of the x- and y-coordinates of the solution from part **c**.

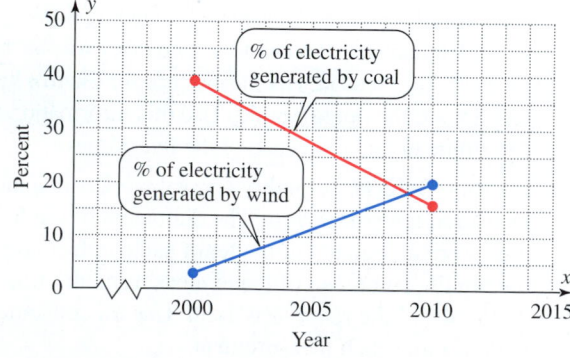

57. The table illustrates that the x-coordinate of the solution of the system $\begin{cases} 11x - 2y = 1 \\ 11x + 3y = 11 \end{cases}$ is between the integers _____ and _____ because Y_2 switches from more than Y_1 to less than Y_1 in this x-interval. This indicates that the lines cross in this x-interval and signals a solution of the system of equations. Use the addition method to confirm this fact and to find the exact coordinates of this solution.

x	$Y_1 = \dfrac{11x}{2} - \dfrac{1}{2}$	$Y_2 = \dfrac{-11x}{3} + \dfrac{11}{3}$
-2	-11.5	11.0000
-1	-6.0	7.3333
0	-0.5	3.6667
1	5.0	0.0000
2	10.5	-3.6667
3	16.0	-7.3333
4	21.5	-11.0000

58. The table illustrates that the x-coordinate of the solution of the system $\begin{cases} 3x + 4y = 8 \\ -3x + 2y = -20 \end{cases}$ is between the integers _____ and _____ because Y_2 switches from less than Y_1 to more than Y_1 in this x-interval. This indicates that the lines cross in this x-interval and signals a solution of the system of equations. Use the addition method to confirm this fact and to find the exact coordinates of this solution.

x	$Y_1 = -0.75x + 2$	$Y_2 = 1.5x - 10$
0	2.00	-10.0
1	1.25	-8.5
2	0.50	-7.0
3	-0.25	-5.5
4	-1.00	-4.0
5	-1.75	-2.5
6	-2.50	-1.0

Applying Technology

In Exercises 59–62, use a calculator to assist you in solving each system of linear equations.

59. $75x - 45y = -150$
$25x + 15y = 550$

60. $22x + 19y = 29$
$33x + 14y = -29$

61. $8.93x - 7.21y = -0.363$
$6.04x + 7.21y = 66.231$

62. $2.14x - 3.15y = -4.131$
$3.07x + 4.21y = 0.184$

Group discussion questions

63. Review and Concept Development Using x to represent the number of years after 2000, an equation that models the percent of College of Lake County students enrolling by phone is $y_1 = -2.054x + 18.804$.

An equation that models the percent of students enrolling by the Web is $y_2 = 7.839x - 0.732$.

a. Determine the point of intersection of these two lines.

b. Interpret the meaning of the x- and y-coordinates of this point of intersection.

c. Which method of enrollment is increasing?

d. Which method of enrollment is decreasing?

e. The phone system and the Web-based system each cost the college over $100,000 to operate each year. What would you advise the college to do? Why?

64. Review and Concept Development

a. Two lines both have a positive slope. Can these lines intersect in a single point? If so, illustrate your answer with a sketch.

b. Two lines both have a negative slope. Can these lines intersect in a single point? If so, illustrate your answer with a sketch.

c. One line has a positive slope and a second line has a negative slope. Can these lines intersect in a single point? If so, illustrate your answer with a sketch.

d. Two lines both have a positive slope. Is it possible for these lines to have no point of intersection? If so, illustrate your answer with a sketch.

e. Two lines both have a negative slope. Is it possible for these lines to have no point of intersection? If so, illustrate your answer with a sketch.

f. One line has a positive slope and a second line has a negative slope. Is it possible for these lines to have no point of intersection? If so, illustrate your answer with a sketch.

65. Communicating Mathematically

a. Two linear equations produce the same line when they are graphed. Write a sentence describing what would happen if you used the addition method to solve this system.

b. Two linear equations produce distinct parallel lines when they are graphed. Write a sentence describing what would happen if you used the addition method to solve this system.

66. Communicating Mathematically Write a word problem for each system of linear equations. (*Hint:* See Exercises 45–48.)

a. $3x - 4y = 5$
$4x + y = 13$

b. $6x + 2y = -16$
$5x + y = -18$

67. Challenge Question Solve this system for (x, y) in terms of a, b, and c. Assume a, b, and c are all nonzero.

$ax + by = c$
$2ax - by = 2c$

68. Discovery Question Given the system

$$\left\{ \begin{array}{l} 7x - 11y = -2 \\ -7x + 22y = 5 \end{array} \right\}, \text{ have part of your group solve}$$

this system by using a graphical method and the rest of the group solve this system by using the addition method. Time each group to determine which method seems more efficient for this problem. Did both groups get exactly the same answer? If not, which group was more accurate?

3.5 | Cumulative Review

In Exercises 1–3 write the given interval in inequality notation.

1. $[-3, \infty)$

2. $(-\infty, -2)$

3. $(-2, 6]$

In Exercises 4 and 5 translate each graph into inequality notation.

4.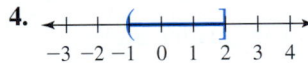

5.
```
←+--+--(--+--+--+--]--+--+→
 -4 -2  0  2  4  6  8
```

Section 3.6 | More Applications of Linear Systems

Objective:

1. Use systems of linear equations to solve word problems.

1. Use Systems of Linear Equations to Solve Word Problems

As we noted in Section 2.8, the mathematics that most students will encounter after college will be stated in words. To help prepare you for this, we use this section to further develop the word problem strategy, which is restated here for your reference.

Strategy for Solving Word Problems

Step 1. Read the problem carefully to determine what you are being asked to find.

Step 2. Select a variable to represent each unknown quantity. Specify precisely what each variable represents, and note any restrictions on each variable.

Step 3. If necessary, make a sketch and translate the problem into a word equation or a system of word equations. Then translate each word equation into an algebraic equation.

Step 4. Solve the equation or the system of equations, and answer the question completely in the form of a sentence.

Step 5. Check the reasonableness of your answer.

We suggest that you continue to use the steps in this strategy as a checklist when you work word problems. Writing word equations is a key step that many students prefer to skip. However, students who practice this step often make major gains in their problem-solving skills. We strongly urge you to write word equations for the problems in this section.

The major new feature of the word problems in this section is that they contain two unknowns rather than just one unknown. Thus we will use two variables and a system of two equations to solve these problems. As a rule of thumb, we usually use as many variables as we have unknowns. For a problem with two unknown quantities, it is often easier to solve the problem by using two variables than it is by using just one variable.

If a word problem involves any shape or design, it is wise to make a sketch of this shape. Observing the visual relationship of the variables can assist you in writing equations for the problem. Example 1 illustrates the steps in the strategy for solving word problems with a problem involving two supplementary angles.

Example 1   Number of Degrees in Two Supplementary Angles

A metal brace on the platform on the side of a crane is designed so that the two angles shown in the figure are supplementary. The specifications require that the larger angle be 40° more than the smaller angle. Determine the number of degrees in each angle.

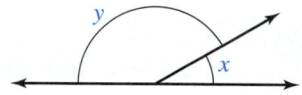

Solution

Definition of Variables

Let x = number of degrees in smaller angle

y = number of degrees in larger angle

Select a variable to represent each unknown. (The restrictions on the variables are considered when the system is solved graphically.)

Word Equations

(1) The total of the measures of the two angles is 180°.

(2) The larger angle is 40° more than the smaller angle.

Angles are supplementary if the total of their measures is 180°. Specifications require that the larger angle be 40° more than the smaller angle.

Algebraic Equations

(1) $x + y = 180$

(2) $y = x + 40$

Translate each word equation into an algebraic equation.

Algebraic Solution

(1) $x + (x + 40) = 180$ **(2)** $y = x + 40$

$2x + 40 = 180$ $y = 70 + 40$

$2x = 140$ $y = 110$

$x = 70$

To solve this system algebraically, use the substitution method, and substitute $x + 40$ for y in equation (1). Then solve this equation for x. Back-substitute 70 for x in equation (2).

Answer: The smaller angle is 70°, and the larger angle is 110°.   Does this answer check?

Self-Check 1

Two parts that brace a portion of a playground slide are joined at a common point. The two angles that they form at this common vertex must be complementary (sum is 90°). If the larger angle is 20° more than the smaller angle, determine the number of degrees in each angle.

One important measure of your academic success is your **grade-point average (GPA)**. Your GPA is calculated by dividing your grade points by the number of academic hours you have completed. Your grade points are calculated by multiplying the number of academic hours of a course by the numeric grade (e.g., 4 for an A) for the course. We will use Example 2 to assist one student with some academic planning.

Example 2   Computing a GPA

A student who has already completed 30 semester-hours with a GPA of 2.50 at Montgomery College needs to raise her GPA to 3.00 to qualify for a scholarship. How many semester-hours with a grade of A would be required to meet her goal? How many grade points will she have if she meets her goal? (Assume an A = 4 points.)

Solution

Definition of Variables

Let x = number of semester-hours that student will need to take with straight A's

y = grade points earned

Word Equations

(1) Actual grade points earned

= Current grade points + New grade points

(2) Desired grade points = Grade points with a 3.00 GPA

If we have two variables, then we will need to write two word equations.

Algebraic Equations

(1) $y = 2.50(30) + 4x$

To translate these word equations into algebraic form, multiply the current GPA by 30 semester-hours to obtain the current grade points. Multiply a numerical grade of 4 by the number of hours x to obtain the new grade points.

(2) $y = 3.00(30 + x)$

The desired GPA is 3.00 when $(30 + x)$ semester-hours have been completed.

Algebraic Solution

$3.00(30 + x) = 2.50(30) + 4x$

$90.00 + 3.00x = 75.00 + 4x$

$15 = x$

To determine when the desired grade points will equal the actual grade points, solve this system of equations. Substitute $3.00(30 + x)$ for y in the first equation.

(1) $y = 2.50(30) + 4x$

$y = 75 + 4(15)$

$y = 75 + 60$

$y = 135$

Back-substitute 15 for x in equation (1). Then solve this equation for y.

Answer: She would need to earn all A's in 15 semester-hours to raise the cumulative GPA to 3.00. She would then have 135 grade points.

Does this seem reasonable? Does this answer check?

Self-Check 2

The first semester a student enrolled in college, he earned the following grades: C in a 3-hour English course, A in a 4-hour mathematics class, B in a 3-hour speech class, and C in a 4-hour psychology class. (Assume an A = 4 points.)

a. Determine his grade points for this semester.

b. Determine his GPA for this semester.

c. The student planned to take a course during an intersession to try to raise his GPA to 3.00. How many semester-hours with a grade of A would be required to meet this goal?

In Section 2.8, we examined a variety of applications of the mixture principle, which is restated here:

Mixture principle: Amount in first + Amount in second = Amount in mixture

Are There Any Other General Principles That I Can Use to Set Up Word Problems?

Yes, one other general principle is the rate principle, which is given in the following box.

Rate Principle
Amount = Rate × Base $A = R \cdot B$

Applications of the Rate Principle

1. Variable cost = Cost per item × Number of items
2. Interest = Principal invested × Rate × Time
3. Distance = Rate × Time
4. Amount of active ingredient = Rate of concentration × Amount of mixture
5. Work = Rate × Time

Example 3 uses both the mixture principle and the rate principle. The mixture principle is used to express the total cost as the sum of the variable cost plus the fixed cost. Then the rate principle is used to determine the variable cost. The rate in this problem is the variable cost per pizza, and the base for this problem is the number of pizzas made.

Example 3   Fixed and Variable Costs for a Pizza Business

A family pizza business opened a new store. The daily cost for making pizzas includes a fixed daily cost and a variable cost per pizza. The total cost for making 200 pizzas on Monday was $750. The total cost for making 250 pizzas on Tuesday was $900. Determine the fixed daily cost and variable cost per pizza for this business.

Solution

Definition of Variables

Let x = variable cost per pizza   Select a variable to represent each unknown.
y = fixed dollar cost per day

Word Equations

(1) Monday: Variable cost + Fixed cost = Total cost   Note that each of these word equations is based on the
(2) Tuesday: Variable cost + Fixed cost = Total cost   mixture principle.

Algebraic Equations

(1) $200x + y = 750$ Using the rate principle, the variable cost for each day is the
(2) $250x + y = 900$ number of pizzas times the cost per pizza.

Algebraic Solution

(1) $-200x - y = -750$

(2) $\underline{\quad 250x + y = \quad 900}$

$\qquad\qquad 50x \quad = \quad 150$

$\qquad\qquad\quad x = \qquad 3$

(2) $250x + y = 900$

$250(3) + y = 900$

$750 + y = 900$

$\qquad\quad y = 150$

To solve this system algebraically, multiply each side of equation (1) by -1 and then use the addition method to eliminate y. Solve this equation for x. Then back-substitute 3 for x in equation (2) and solve for y.

Answer: The fixed cost per day is $150, and $3 is the variable cost per pizza.

Does this answer check?

Self-Check 3

An umbrella manufacturer has daily costs that are both fixed and variable. The company produced 500 umbrellas on Monday at a cost of $1,450 and 600 umbrellas Tuesday at a cost of $1,700. Determine the fixed daily cost and the variable cost per umbrella.

Example 4 uses the mixture principle to form both equations. The total principal is composed of two separate investments, and the total interest is the combined interest from the two investments. The rate principle is also used in Example 4 to calculate the interest earned on each investment. The rate in this application is the interest rate, and the formula is $I = PRT$.

Example 4   Income from Two Investments

A student saving for college was given $6,000 by her grandparents. She invested part of this money in a savings account that earned interest at the rate of 3% per year. The rest was invested in a bond that paid interest at the rate of 5.5% per year. If the combined interest at the end of 1 year was $305, how much was invested at each rate?

Solution

Definition of Variables

Let x = principal invested in savings account

$\quad\;\; y$ = principal invested in bond

Select a variable to represent each unknown. (The restrictions on the variables are considered when the system is solved graphically.)

Word Equations

Both word equations are based on the mixture principle.

(1) $\begin{pmatrix} \text{Principal in} \\ \text{savings account} \end{pmatrix} + \begin{pmatrix} \text{Principal in} \\ \text{bond} \end{pmatrix} = \begin{pmatrix} \text{Total} \\ \text{principal} \end{pmatrix}$

(2) $\begin{pmatrix} \text{Interest on} \\ \text{savings account} \end{pmatrix} + \begin{pmatrix} \text{Interest on} \\ \text{bond} \end{pmatrix} = \begin{pmatrix} \text{Total} \\ \text{interest} \end{pmatrix}$

Algebraic Equations

(1) $x + y = 6,000$

(2) $0.03x + 0.055y = 305$

Algebraic Solution

(1) $-0.03x - 0.030y = -180$

(2) $\underline{\quad 0.03x + 0.055y = \quad 305}$

$\qquad\qquad\quad 0.025y = \quad 125$

$\qquad\qquad\qquad\quad y = \dfrac{125}{0.025}$

$\qquad\qquad\qquad\quad y = 5,000$

By using $I = PRT$ and 1 year for time, the interest on the savings account is $I = (x)(0.03)(1) = 0.03x$, and the interest on the bond is $I = (y)(0.055)(1) = 0.055y$.

(1) $\qquad x + y = 6,000$

$\qquad x + 5,000 = 6,000$

$\qquad\qquad\quad x = 1,000$

To solve this system algebraically, multiply each side of equation (1) by -0.03 and then use the addition method to solve this system. Divide both sides by 0.025 to solve for y. Back-substitute 5,000 for y in equation (1) and solve for x.

Answer: She invested $1,000 in the savings account and $5,000 in the bond.

Does this answer check?

Example 5   Determining the Speed Flown by Two Airplanes

Two airplanes depart from an airport simultaneously, one flying 100 km/h faster than the other. These planes travel in opposite directions, and after 1.5 hours they are 1,275 km apart. Determine the speed of each plane. (*Hint: $D = RT$*; distance equals rate times the time.)

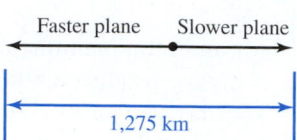

Solution

Definition of Variables

Let r_1 = rate of slower plane, km/h

r_2 = rate of faster plane, km/h

Select a variable to represent each unknown. (The restrictions on the variables are considered when the system is solved graphically.)

Word Equations

(1) $\left(\begin{array}{c}\text{Rate of}\\\text{faster plane}\end{array}\right) = \left(\begin{array}{c}\text{Rate of}\\\text{slower plane}\end{array}\right) + 100 \text{ km/h}$

(2) $\left(\begin{array}{c}\text{Distance by}\\\text{slower plane}\end{array}\right) + \left(\begin{array}{c}\text{Distance by}\\\text{faster plane}\end{array}\right) = \left(\begin{array}{c}\text{Total distance}\\\text{between planes}\end{array}\right)$

Note that the second word equation is based on the mixture principle.

Algebraic Equations

(1) $r_2 = r_1 + 100$

(2) $1.5r_1 + 1.5r_2 = 1,275$

By using $D = RT$ and a time of 1.5 hours, the distance is $1.5r_1$ for the slower plane and $1.5r_2$ for the faster plane.

Algebraic Solution

(1) $r_1 - r_2 = -100$

(2) $\underline{r_1 + r_2 = \;\;850}$

$2r_1 = \;\;750$

$r_1 = \;\;375$

(1) $r_2 = r_1 + 100$

$r_2 = 375 + 100$

$r_2 = 475$

To solve this system algebraically, rearrange the terms in equation (1) and simplify equation (2) by dividing both sides of the equation by 1.5.

Then use the addition method to solve this system. Back-substitute 375 for r_1 in equation (1).

Answer: The planes traveled at 375 and 475 km/h.

Does this seem reasonable? Does the answer check?

Is There Any Strategy That I Can Use When Solving Word Problems Involving Two Mixtures?

Yes, for many problems you can write one equation for the total volume (or amount). Then you can write a second equation based on the amount of active ingredients involved. For example, when two liquids of different concentrations are mixed, the total volume is found by adding the volumes of the liquids that are mixed. The total of the chemicals found in the mixture is calculated by adding the chemicals found in each of the individual solutions. This is illustrated in Example 6, where we also use the rate principle to determine the amount of chlorine in each solution.

The Mixture Principle

Example 6   Calculating the Amount of Two Solutions to Mix

A water treatment plant uses batch deliveries of two levels of chlorine concentration to mix to create the level of concentration needed on any given day. Determine how many liters of a 3% chlorine solution and how many liters of a 5% chlorine solution should be mixed to produce 500 L of a 4.4% chlorine solution.

Solution

Definition of Variables

Let x = number of liters of 3% chlorine solution

y = number of liters of 5% chlorine solution

Select a variable to represent each unknown. (The restrictions on the variables are considered when the system is solved graphically.)

Word Equations

(1) $\left(\begin{array}{c}\text{Volume of}\\\text{first solution}\end{array}\right) + \left(\begin{array}{c}\text{Volume of}\\\text{second solution}\end{array}\right) = \left(\begin{array}{c}\text{Total volume}\\\text{of mixture}\end{array}\right)$

(2) $\left(\begin{array}{c}\text{Volume of chlorine}\\\text{in first solution}\end{array}\right) + \left(\begin{array}{c}\text{Volume of chlorine}\\\text{in second solution}\end{array}\right) = \left(\begin{array}{c}\text{Total chlorine}\\\text{in mixture}\end{array}\right)$

Note that both word equations are based on the mixture principle.

The volume of chlorine in each solution is calculated by using the rate principle.

Algebraic Equations

(1) $x + y = 500$

(2) $0.03x + 0.05y = 0.044(500)$

The volume of chlorine in the x L of 3% solution is $0.03x$. The volume of chlorine in the y L of the 5% solution is $0.05y$. The volume of chlorine in 500 L of 4.4% solution is $0.044(500)$.

Algebraic Solution

(1) $-0.03x - 0.03y = -15$

(2) $\quad 0.03x + 0.05y = \quad 22$

$$0.02y = \quad 7$$

$$\frac{0.02y}{0.02} = \frac{7}{0.02}$$

$$y = 350$$

(1) $x + y = 500$

$$x + 350 = 500$$

$$x = 150$$

To solve this system algebraically, multiply both sides of equation (1) by -0.03 and then use the addition method to solve this system.

Divide both sides of this equation by 0.02. Back-substitute 350 for y in equation (1) and solve for x.

Answer: Use 150 L of the 3% solution and 350 L of the 5% solution to produce 500 L of a 4.4% chlorine solution.

Does this answer check?

Self-Check 6

A water treatment plant uses batch deliveries of two levels of chlorine concentration to mix to create the level of concentration needed on any given day. Determine how many liters of a 3% chlorine solution and how many liters of a 5% chlorine solution should be mixed to produce 500 L of a 4.5% chlorine solution.

Example 7 illustrates another application of the rate principle. The amount of work done is equal to the rate of work times the time worked, or $W = RT$. Each of the examples in this section can be checked by using either graphs or tables. This is illustrated in Example 7.

Example 7 | Determining the Rates of Two Workers

A small building contractor plans to add a bricklayer to his full-time crew. He has two bricklayers on a current job that he is considering for this position. On Monday he observed that these two bricklayers each worked 7 hours and laid a total of 3,150 bricks. On Tuesday the older bricklayer worked 6 hours and the younger bricklayer worked 5 hours, and they laid a total of 2,500 bricks. Determine for the contractor the rate of work for each bricklayer, assuming that both bricklayers work at a fairly consistent rate.

Solution

Definition of Variables

Let r_1 = rate of work for younger bricklayer, bricks/h

$\quad r_2$ = rate of work for older bricklayer, bricks/h

Select a variable to represent each unknown. (The restrictions on the variables are considered when the system is solved graphically.)

Word Equations

(1) Monday: $\left(\begin{array}{c}\text{Work by}\\\text{younger worker}\end{array}\right) + \left(\begin{array}{c}\text{Work by}\\\text{older worker}\end{array}\right) = $ Total work

(2) Tuesday: $\left(\begin{array}{c}\text{Work by}\\\text{younger worker}\end{array}\right) + \left(\begin{array}{c}\text{Work by}\\\text{older worker}\end{array}\right) = $ Total work

The mixture principle is used to form the word equation for Monday and another word equation for Tuesday.

Algebraic Equations

(1) $7r_1 + 7r_2 = 3{,}150$

(2) $5r_1 + 6r_2 = 2{,}500$

The amount of work done by each bricklayer is determined by using the rate principle in the form $W = RT$.

Algebraic Solution

(1) $r_1 + r_2 = 450$

(2) $5r_1 + 6r_2 = 2{,}500$

(1)	$-5r_1 - 5r_2 = -2{,}250$	**(1)** $r_1 + r_2 = 450$
(2)	$\underline{5r_1 + 6r_2 = 2{,}500}$	$r_1 + 250 = 450$
	$r_2 = 250$	$r_1 = 200$

To solve this system algebraically, divide both sides of equation (1) by 7. Then multiply both sides of this equation by -5 and use the addition method to solve this system. Back-substitute 250 for r_2 in equation (1) and solve for r_1.

Graphical Check

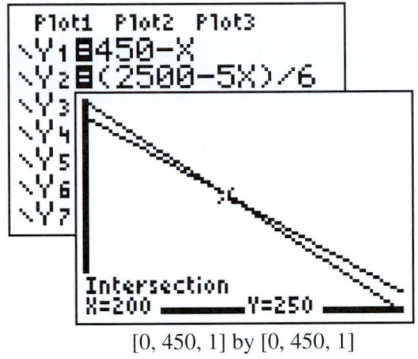

[0, 450, 1] by [0, 450, 1]

On a calculator, use x to represent r_1 and use y to represent r_2.

To solve this system using a graphing calculator, solve each equation for the y-variable.

(1) $y = 450 - x$

(2) $y = \dfrac{2{,}500 - 5x}{6}$

To establish restrictions on x, consider what r_1 would be if the younger bricklayer laid all 3,150 bricks in 7 hours (3,150 bricks/7 h = 450 bricks/h). Therefore this rate must be at least 0 bricks/h and at most 450 bricks/h. Thus, we select the x-values [0, 450, 1] for the viewing window. The **ZoomFit** option selects the y-values and graphs these equations on this window. Use the **Intersect** feature to find that these lines intersect where $x = 200$ and $y = 250$.

Numerical Check

X	Y₁	Y₂
0	450	416.67
50	400	375
100	350	333.33
150	300	291.67
200	250	250
250	200	208.33
300	150	166.67

X=200

The table verifies that the x-value of 200 produces y-values of 250. Also note that other values of x do not check.

Answer: The younger worker lays 200 bricks/h, and the older worker lays 250 bricks/h.

Does this seem reasonable?

Self-Check 7

Two machines are used to produce pencils. During a 2-hour period in the morning, the two machines produced a total of 1,560 pencils. In the afternoon, machine 1 operated for 4 hours and machine 2 operated for 3 hours, and they produced 2,640 pencils. Determine the rate of production for each machine.

Self-Check Answers

1. The smaller angle is 35° and the larger angle is 55°.
2. **a.** He would have 39 grade points.
   b. His GPA would be approximately 2.79.
   c. He would need to earn an A in 3 semester-hours.
3. The fixed cost per day is $200, and the variable cost per umbrella is $2.50.
4. The first investment was for $20,000, and the second investment was for $80,000.
5. The first train travels at 80 km/h and the second at 100 km/h.
6. Use 125 L of the 3% solution and 375 L of the 5% solution to produce 500 L of a 4.5% chlorine solution.
7. Machine 1 produces 300 pencils/h, and machine 2 produces 480 pencils/h.

3.6 Using the Language and Symbolism of Mathematics

1. The first step in the word problem strategy given in this book is to read the problem carefully to determine what you are being asked to _____.

2. The second step in the word problem strategy given in this book is to select a _____ to represent each unknown quantity.

3. The third step in the word problem strategy given in this book is to model the problem verbally with word equations and then to translate these word equations into _____ equations.

4. The fourth step in the word problem strategy given in this book is to _____ the equation or system of equations and to answer the question completely in the form of a sentence.

5. The fifth step in the word problem strategy given in this book is to check your answer to make sure the answer is _____.

6. The _____ principle states that the amount obtained by combining two parts is equal to the amount obtained from the first part plus the amount obtained from the second part.

7. The _____ principle states that the amount obtained is equal to the rate times the base to which this rate is applied.

8. In the formula $D = RT$, D represents distance, R represents _____, and T represents _____.

9. In the formula $I = PRT$, I represents interest, P represents _____, R represents _____, and T represents _____.

10. In the formula $W = RT$, W represents the amount of _____, R represents the rate of work, and T represents the _____ worked.

3.6 Quick Review

1. The cost that does not change with the number of items produced is called the _____ cost.

2. The cost of making each unit that can change with the number of items produced is called the _____ cost.

3. The phrase "_____ on the variable" refers to the values of the variable that are permissible in an application.

4. _____ equations have the same solution set.

5. Two angles are _____ if the sum of their measures is 180°.

3.6 Exercises

Objective 1 Use Systems of Linear Equations to Solve Word Problems

In Exercises 1 and 2, solve each problem by completing each of the given steps.

1. **Dimensions of a Rectangular Poster** The perimeter of a rectangular poster is 204 cm. Find the dimensions of the poster if the length is 6 cm more than the width.
   a. Select a variable to represent each of the unknown quantities, and identify each variable, including the units of measurement.
   b. Write a pair of word equations that models this problem.
   c. Translate the word equations from part **b** into algebraic equations.
   d. Solve the system of equations from part **c**.
   e. Is the solution from part **d** within the restrictions on the variables, and does it seem reasonable?
   f. Write a sentence that answers the problem.

2. Dimensions of a Rectangular Pane of Glass The perimeter of a rectangular pane of glass is 68 cm. Find the dimensions of the pane of glass if the length is 2 cm less than 3 times the width.

   a. Select a variable to represent each of the unknown quantities, and identify each variable, including the units of measurement.

   b. Write a pair of word equations that models this problem.

   c. Translate the word equations from part **b** into algebraic equations.

   d. Solve the system of equations from part **c.**

   e. Is the solution from part **d** within the restrictions on the variables, and does it seem reasonable?

   f. Write a sentence that answers the problem.

In Exercises 3–40, solve each problem by using a system of equations and the word problem strategy developed in this section.

3. Numeric Word Problem Find two numbers whose sum is 100 if one number is 16 more than twice the smaller number.

4. Numeric Word Problem Find two numbers whose sum is 100 if one number is 16 more than three times the smaller number.

5. Dimensions of an Isosceles Triangle An isosceles triangle has two equal sides. The triangle shown here has a perimeter of 72 cm. The base is 12 cm longer than the two equal sides. Find the dimensions of this triangle.

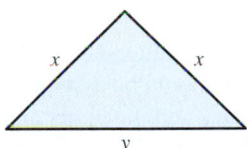

6. Angles of an Isosceles Triangle The sum of the interior angles of a triangle is 180°. An isosceles triangle has two equal angles. Find the angles of the isosceles triangle shown here if angle y is 33° larger than angle x.

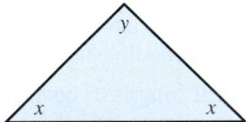

7. Coded Messages To receive a message, a spy first must send a pair of authorization numbers that satisfies the conditions of the two independent supervisors on duty that day. Supervisor A must receive the code and confirm that the sum of seven times the first number plus four times the second number is two hundred ninety-four. Supervisor B must receive the code and confirm that six times the first number minus the second number is ninety-seven. What two numbers should be sent when supervisors A and B are on duty?

8. Coded Messages To receive a message, an engineer traveling abroad first must send a pair of authorization numbers to his home office that satisfies the conditions of the two independent supervisors on duty that day. Supervisor A must receive the code and confirm that the sum of five times the first number plus three times the second number is eighty-four. Supervisor B must receive the code and confirm that four times the first number minus the second number is twenty-three. What two numbers should be sent when supervisors A and B are on duty?

9. Mean and Range A golfer recorded five scores for five rounds of golf one week. Four of these rounds were the same, and one was 10 strokes higher than the others. If the average score for the week was 83, what was the score for each of the five rounds?

10. Mean and Range A company has six employees. Five of these employees make exactly the same amount each month, which is less than the amount their supervisor makes. The mean (average) hourly salary of all employees is $26/h, and the range of these salaries is $12/h. Determine the hourly salary of each employee.

11. GPA A student enrolled in college for the first semester in the spring of 2010. She earned the following grades: C in a 3-hour English course, B in a 4-hour mathematics class, C in a 3-hour speech class, and C in a 3-hour Spanish class.

   a. Determine her grade points for this semester.

   b. Determine her GPA for this semester.

   c. The student planned to take courses during the summer to try to raise her GPA to 3.00. How many semester-hours with a grade of A would be required to meet this goal? (Assume an A = 4 points.)

12. GPA A student enrolled in college for the first semester in the spring of 2010. He earned the following grades: B in a 3-hour English course, C in a 4-hour business class, C in a 3-hour speech class, and B in a 3-hour French class.

   a. Determine his grade points for this semester.

   b. Determine his GPA for this semester.

   c. The student planned to take courses during the summer to try to raise his GPA to 3.00. How many semester-hours with a grade of A would be required to meet this goal? (Assume an A = 4 points.)

13. GPA A student who has already completed 30 semester-hours with a GPA of 2.50 at Harper College needs to raise her GPA to 2.75 to meet transfer requirements into one college program. How many semester-hours with a grade of A would be required to meet this goal? How many grade points will the student have if this goal is met? (Assume an A = 4 points.)

14. **GPA** A student who has already completed 36 semester-hours with a GPA of 2.75 at Parkland College needs to raise his GPA to 3.00 to meet transfer requirements into one college program. How many semester-hours with a grade of A would be required to meet this goal? How many grade points will the student have if this goal is met? (Assume an A = 4 points.)

15. **Cable TV Bills** A cable TV bill consists of a fixed monthly charge and a variable charge, which depends on the number of pay-per-view movies ordered. In January the $44 bill included four pay-per-view movies. In February the $54.50 bill included seven pay-per-view movies. Determine the fixed monthly charge and the charge for each pay-per-view movie.

16. **Rental Truck Costs** The cost of a rental truck used to move furniture includes a fixed daily cost and a variable cost based on miles driven. For a customer driving 180 mi, the cost for 1 day will be $75. For a customer driving 150 mi, the cost for 1 day will be $67.50. Determine the fixed daily cost and the cost per mile for this truck.

17. **Costs for Union Employees** Regular union employees earn $120/day, and union stewards earn $144/day. A company has a need for a total of 50 employees and has a daily payroll budget of $6,072. How many of each type of employee will meet the company's needs and consume all the budget?

18. **Equipment Costs** A bulldozer can move 25 tons/h of material, and an end loader can move 18 tons/h. The bulldozer costs $75/h to operate, and the end loader costs $50/h to operate. During 1 hour, a construction company has moved 204 tons at a cost of $600. How many bulldozers and how many end loaders were operating during this hour?

19. **Number of Basketball Tickets** The total receipts for a basketball game are $1,400 for 788 tickets sold. Adults paid $2.50 for admission and students paid $1.25. The ticket takers counted only the number of tickets, not the type of tickets. Determine the number of each type of ticket sold.

20. **Loads of Topsoil** Topsoil sells for $55 per truckload, and fill dirt sells for $40 per truckload. A landscape architect estimates that 20 truckloads will be needed for a certain job. The estimated cost is $920. How many loads of topsoil and how many loads of fill dirt are planned for this job?

21. **Complementary Angles** The angles shown in the figure are complementary, and one angle is 28° larger than the other angle. Determine the number of degrees in each angle.

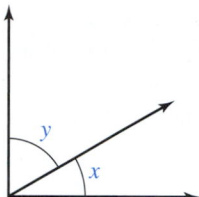

22. **Complementary Angles** The angles shown in Exercise 21 are complementary, and one angle is 15° more than twice the other angle. Determine the number of degrees in each angle.

23. **Supplementary Angles** The two angles shown in the figure represent parts that should be assembled so that the angles are supplementary and the larger angle is 5° more than 4 times the smaller angle. Determine the number of degrees in each angle.

24. **Supplementary Angles** The two angles shown in Exercise 23 represent parts that should be assembled so that the angles are supplementary and the larger angle is 9° less than twice the smaller angle. Determine the number of degrees in each angle.

25. **Order for Left- and Right-Handed Desks** When ordering new desks for classrooms, the business manager for a college uses the fact that there are approximately 9 times as many right-handed people as left-handed people. How many desks of each type must be ordered if 600 new desks are needed?

26. **Length of a Board** A board 12 ft long is to be cut into two pieces so that one piece is 3 ft longer than the other piece. Determine the length of each piece.

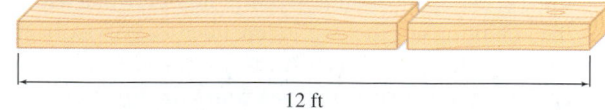

12 ft

27. Costs for Lightbulbs An incandescent 60-W lightbulb costs $0.50 to buy and $0.12 per day to use (24 hours of usage). A compact fluorescent lamp (CFL) that produces the equivalent amount of light costs $2.50 to buy and $0.02 per day to use. Determine the number of days of usage when these costs are equivalent. What is this equivalent cost? (Source: Salt River Project Power Utility.)

28. Fixed and Variable Costs A family business makes custom rocking chairs, which it markets as retirement gifts. One month the total fixed and variable costs for producing 250 rockers were $32,250. The next month the total fixed and variable costs for producing 220 rockers were $28,500. Determine the fixed cost and the variable cost per rocking chair.

29. Interest on Two Investments Part of a $12,000 investment earned interest at a rate of 7%, and the rest earned interest at a rate of 9%. The combined interest earned at the end of 1 year was $890. How much was invested at each rate?

30. Interest on Two Investments An investment of $10,000 earned a net income of $345 in 1 year. Part of the investment was in bonds and earned income at a rate of 8%. The rest of the investment was in stocks and lost money at the rate of 5%. How much was invested in bonds, and how much was invested in stocks?

31. Interest on Two Investments A broker made two separate investments on behalf of a client. The first investment earned 8% and the second earned 5%, for a total gain of $600. If these investments had earned 4% and 7%, respectively, then the gain would have been $480. How much was actually invested at each rate?

32. Interest on Two Investments A broker made two separate investments on behalf of a client. The first investment earned 12% and the second investment lost 5%, for a total gain of $1,040. If the amounts in these two investments had been switched, then the result would have been a gain of $360. Determine how much was actually invested at each rate.

33. Rates of Two Trains Two trains depart simultaneously from a station, traveling in opposite directions. One averages 10 km/h more than the other. After $\frac{1}{2}$ hour they are 89 km apart. Determine the speed of each train.

34. Rates of Two Airplanes A jet plane and a tanker that are 525 mi apart head toward each other so that the jet can refuel. The jet flies 200 mi/h faster than the tanker. Determine the speed of each aircraft if they meet in 45 minutes. (*Hint:* Use consistent units of measurement.)

35. Rates of Two Trains Two trains depart from a station, traveling in opposite directions. The train that departs at 6:30 A.M. travels 15 km/h faster than the train that departs at 6 A.M. At 7 A.M., they are 135 km apart. Determine the rate of each train.

36. Rates of Two Buses Two buses that are 90 mi apart travel toward each other on an interstate highway. The slower bus, which departs at 9:30 A.M., travels 5 mi/h slower than the bus that departs at 9:45 A.M. At 10:30 A.M., the buses pass each other. Determine the rate of each bus.

37. Rate of a River Current When a boat travels downstream with a river current, the rate of the boat in still water and the rate of the current are added. When a boat travels upstream, the rate of the current is subtracted from the rate of the boat.

A paddlewheel riverboat takes 1 hour to go 24 km downstream and another 4 hours to return upstream. Determine the rate of the boat and the rate of the current.

38. Rate of a River Current A small boat can go 40 km downstream in 1 hour but only 10 km upstream in 1 hour. Determine the rate of the boat and the rate of the current. (*Hint:* See Exercise 37.)

39. Mixture of Disinfectant A hospital needs 80 L of a 12% solution of disinfectant. How many liters of a 33% solution and a 5% solution should be mixed to obtain this 12% solution?

40. Mixture of Medicine The dosage of a medicine ordered by a doctor is 40 mL of a 16% solution. A nurse has available both a 20% solution and a 4% solution of this medicine. How many milliliters of each could be mixed to prepare this 40-mL dosage?

41. Gold Alloy A goldsmith has 80 g of an alloy that is 50% pure gold. How many grams of an alloy that is 80% pure gold must be combined with the 80 g of alloy that is 50% pure gold to form an alloy that will be 72% pure gold?

42. Saline Mixture A 30% salt solution is prepared by mixing a 20% salt solution and a 45% salt solution. How many liters of each must be used to produce 60 L of the 30% salt solution?

43. Mixture of Fruit Drinks A fruit drink concentrate is 15% water. How many liters of pure water should be added to 12 L of concentrate to produce a mixture that is 50% water?

44. Insecticide Mixture A nursery owner is preparing an insecticide by mixing a 95% solution with water (0% solution). How much solution and how much water are needed to fill a 500-gal tank with 3.8% solution?

45. Rates of Two Workers A small building contractor plans to add a bricklayer to his full-time crew. He has two bricklayers on a current job that he is considering for this position. On Monday, he observed that these two bricklayers each worked 8 hours and laid a total of 4,000 bricks. On Tuesday, the older bricklayer worked 6 hours, the younger bricklayer worked 7 hours, and they laid a total of 3,220 bricks. Determine for the contractor the rate of work for each bricklayer, assuming that both bricklayers work at a fairly consistent rate.

46. Rates of Two Machines A small machine shop is considering buying one of two metal presses at an auction. The owner visited the company selling these presses on two different days. On Monday, she observed that these two presses were used 8 hours each and they produced a total of 88 stamped parts. On Tuesday, the older machine worked 4 hours and the newer machine worked 6 hours, and they produced a total of 56 stamped parts. Determine for the owner of the machine shop the rate of work for each metal press, assuming that both presses work at a fairly consistent rate.

47. Scientific Articles An important indicator of the innovation and the health of a country's economy is the publication of articles in scholarly journals. These articles are often followed by patents and new products.

Based on data from the National Science Foundation, the following equations can be used:

(1) United States $y = 200$
(2) Europe $y = 6.8x + 160$

where x is the number of years since 1990 and y is the number of publications in thousands.
a. Solve this system of equations.
b. Interpret the meaning of this solution.

48. International Travel By using x to represent the number of years since 1985, the following equations can be used to model the number in millions of U.S. travelers to other countries and the number of foreign visitors to the United States.

(1) U.S. travelers $y = 1.8x + 30$
(2) Foreign visitors to United States $y = 0.64x + 39.4$

a. Solve this system of equations.
b. Interpret the meaning of this solution.

49. Computer Lease Options The graph compares the costs of two options for leasing the computer equipment for a small engineering firm. The cost is based on the number of months the computer equipment is used.

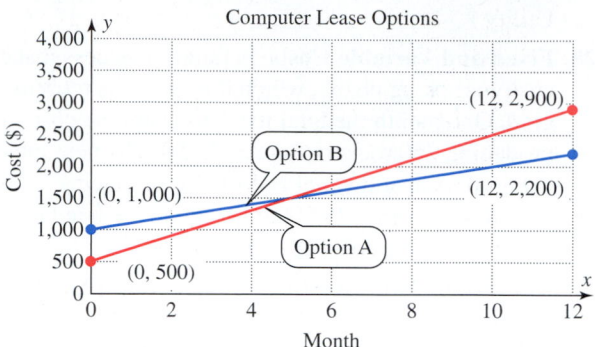

a. Use the y-intercept and an additional point to determine the equation of the line that models the cost for option A.
b. Use the y-intercept and an additional point to determine the equation of the line that models the cost for option B.
c. Solve the system of equations from parts **a** and **b**.
d. Does the algebraic solution from part **c** seem consistent with the point of intersection shown on the graph?
e. Interpret the meaning of the x- and y-coordinates of the solution from part **d**.

50. Fish Populations The graph compares the populations of bass and catfish in Lake Shelbyville after 1990.

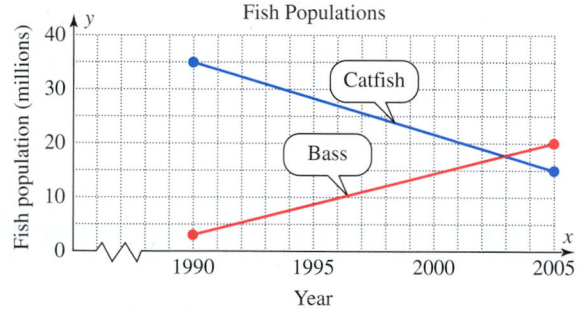

a. Use the two points (1990, 3) and (2005, 20) to write an equation for the line showing the bass population.
b. Use the two points (1990, 35) and (2005, 15) to write an equation for the line showing the catfish population.
c. Solve the system of equations from parts **a** and **b**.
d. Interpret the meaning of the x- and y-coordinates of the solution from part **c**.

In Exercises 51 and 52, use the given information to solve each part of the problem.

51. **Fixed and Variable Costs** A tennis shoe manufacturer has daily costs that are both fixed and variable. The variable costs depend on the number of shoes produced that day. Determine the fixed daily cost and the variable cost per pair of shoes for each of the following situations. Check to confirm that each situation is possible and that the proposed answer meets the restrictions on the variables.
   a. The company produced 400 pairs of shoes on Monday at a cost of $5,700 and 525 pairs on Tuesday at a cost of $6,700.
   b. The company produced 400 pairs of shoes on Monday at a cost of $9,100 and 525 pairs on Tuesday at a cost of $12,100.

52. **Investment Choices** A broker invested a total of $25,000 in two separate investments on behalf of a client. For the first year, the rate of return was 5% on the first investment and 7% on the second investment. Determine how much was invested in each of these investments for each of the following situations. Check to confirm that each situation is possible and that the proposed answer meets the restrictions on the variables.
   a. The total income from the two investments was $1,450.
   b. The total income from the two investments was $1,770.

53. **Point of No Return** A rescue helicopter is flying from an aircraft carrier 350 mi to a Pearl Harbor hospital. The point of no return is the point where it would take as much time to fly back to the aircraft carrier as it would to fly on to Pearl Harbor. The airspeed of the helicopter is 175 mi/h, and it is flying into a 25-mi/h head wind.
   a. Calculate the total time for the helicopter to fly to Pearl Harbor.

   b. Once the helicopter is at the point of no return, determine the time it will take to complete the trip.
   c. Use the results from parts **a** and **b** to determine the time for the helicopter to reach the point of no return.

54. **Point of No Return** Rework Exercise 53, assuming the helicopter is flying with a 25-mi/h tailwind to Pearl Harbor.

Group discussion questions

55. **Communicating Mathematically** Write a word problem that can be solved by using the system of equations
$$\begin{cases} x + y = 500 \\ 5x + 8y = 3{,}475 \end{cases}$$ and that models
   a. A numeric problem (see Example 9 in Section 3.5)
   b. A problem involving ticket sales (see Exercise 19)
   c. A rate-of-work problem (see Exercise 45)

56. **Communicating Mathematically** Write a word problem that can be solved by using the system of equations
$$\begin{cases} x + y = 500 \\ 0.05x + 0.08y = 34.75 \end{cases}$$ and that models
   a. A combined investment problem (see Exercise 29)
   b. A mixture of two piles of copper ore (see Exercise 41)

57. **Discovery Question**
   a. Compare the two systems of equations in Exercises 55 and 56. Are these systems equivalent; that is, can you convert one system to the other? If so, does this mean that this one system can represent all five of the word problems created in Exercises 55 and 56?
   b. Discuss the use of one general mathematical principle, such as the mixture principle, to solve many seemingly unrelated problems.
   c. Although pure numeric-based word problems have limited direct application, they are widely used in textbooks. Discuss a rationale of why this is helpful for students.

3.6 **Cumulative Review**

1. Simplify $2(x - 3) + 4(x - 3)$.
2. Solve $2(x - 3) + 4(x - 3) = 0$.

In Exercises 3 and 4, determine whether the ordered pair $(-1, 2)$ is a solution of each linear equation.

3. $3x + y = -1$

4. $y = -2x + 4$
5. Determine the x- and y-intercepts of $3x - 4y = 12$.

Chapter 3 | Key Concepts

1. Slope of a Line
- The slope m of a line through (x_1, y_1) and (x_2, y_2) with $x_1 \neq x_2$ is $m = \dfrac{\text{change in } y}{\text{change in } x} = \dfrac{y_2 - y_1}{x_2 - x_1} = \dfrac{\text{rise}}{\text{run}}$.

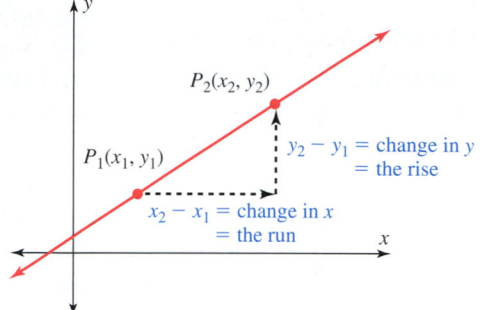

- Δx, read "delta x," denotes a change in x; and Δy, read "delta y," denotes a change in y; $m = \dfrac{\Delta y}{\Delta x}$.
- The slope of a line gives the change in y for each 1-unit change in x.
- A line with positive slope goes upward to the right.
- A line with negative slope goes downward to the right.
- The slope of a horizontal line is 0.
- The slope of a vertical line is undefined.
- The slopes of parallel lines are the same.
- The slopes of perpendicular lines are opposite reciprocals.
- The product of the slopes of perpendicular lines is -1.
- The slope of $y = mx + b$ is m.
- The slope of $y - y_1 = m(x - x_1)$ is m.

2. Slope Represents a Rate of Change
Rates of change compare how one variable is changing with respect to another variable. Examples of rates of change are:
- miles per gallon—how distance is changing as fuel is being used
- inches per hour—how the depth of a pool is changing as water flows in
- grade of a highway—how elevation changes over a length of highway
- common difference of an arithmetic sequence—change from one term to the next as the term number increases by 1

3. Forms of Linear Equations: A linear equation in x and y is first degree in both x and y.
- Slope-intercept form: $y = mx + b$ or $f(x) = mx + b$ with slope m and y-intercept $(0, b)$
- Vertical line: $x = a$ for a real constant a
- Horizontal line: $y = b$ for a real constant b
- General form: $Ax + By = C$
- Point-slope form: $y - y_1 = m(x - x_1)$ through (x_1, y_1) with slope m

4. System of Equations: When we consider two or more equations at the same time, we refer to this as a system of equations.

5. Solution of a System of Linear Equations in Two Variables
- A solution is an ordered pair that satisfies each equation in the system.
- A solution of a system of linear equations is represented graphically by a point that is on both lines in the system.
- A solution is represented in a table of values as an x-value for which both y-values are the same.

6. Terminology Describing Systems of Linear Equations
- Consistent system: The system of equations has at least one solution.
- Inconsistent system: The system of equations has no solution.
- Independent equations: The equations cannot be simplified and rewritten as the same equation.
- Dependent equations: The equations can both be written in exactly the same form.

7. Classifications of Systems of Linear Equations
- Exactly one solution: A consistent system of independent equations
- No solution: An inconsistent system of independent equations
- An infinite number of solutions: A consistent system of dependent equations

8. Using the Slope-Intercept Form to Classify Systems of Linear Equations: For the linear system
$$\left\{ \begin{array}{l} y = m_1 x + b_1 \\ y = m_2 x + b_2 \end{array} \right\} \text{ with}$$
- $m_1 \neq m_2$
  The system is consistent with independent equations and exactly one solution.
- $m_1 = m_2$ and $b_1 \neq b_2$
  The system is inconsistent with independent equations and no solution.
- $m_1 = m_2$ and $b_1 = b_2$
  The system is consistent with dependent equations and an infinite number of solutions.

9. Substitution Principle: A quantity may be substituted for its equal.

10. Addition-Subtraction Principle of Equality: Equal values can be added to or subtracted from both sides of an equation to produce an equivalent equation.

11. Methods for Solving Systems of Linear Equations in Two Variables
- Graphical method; the point(s) on both lines
- Numerical method; for the same x-value, the two y-values are equal.

- Algebraically by the substitution method
- Algebraically by the addition method

12. Algebraic Solution of the Three Types of Systems of Linear Equations in Two Variables
- Consistent system of independent equations: The solution process will produce unique x- and y-coordinates.
- Inconsistent system: The solution process will produce a contradiction.
- Consistent system of dependent equations: The solution process will produce an identity.

13. Solving a System of Linear Equations Containing Fractions: It may be easier to start by multiplying both sides of the equation by the least common denominator (LCD) of all the fractions in the equation.

14. Strategy for Solving Word Problems
Step 1 Read the problem carefully to determine what you are being asked to find.

Step 2 Select a variable to represent each unknown quantity. Specify precisely what each variable represents, and note any restrictions on each variable.

Step 3 If necessary, make a sketch and translate the problem into a word equation or a system of word equations. Then translate each word equation into an algebraic equation.

Step 4 Solve the equation or system of equations, and answer the question completely in the form of a sentence.

Step 5 Check the reasonableness of your answer.

15. Mixture Principle

$$\left(\begin{array}{c}\text{Amount}\\\text{in first}\end{array}\right) + \left(\begin{array}{c}\text{Amount}\\\text{in second}\end{array}\right) = \left(\begin{array}{c}\text{Amount}\\\text{in mixture}\end{array}\right)$$

16. Rate Principle: The amount obtained is equal to the rate times the base to which this rate is applied.

$$\text{Amount} = \text{Rate} \times \text{Base} \qquad A = RB$$

Chapter 3 | Review Exercises

Slope of a Line

In Exercises 1–10, calculate the slope of each line.

1. The line through $(1, -3)$ and $(4, 3)$

2. The line through $(1, 4)$ and $(6, 1)$

3. $f(x) = \dfrac{4}{7}x + 3$

4. $y - 2 = -(x - 8)$

5. $y = -7$

6.

(graph with line labeled $2x - 5y = 10$)

7.

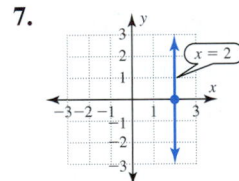

(graph with line labeled $x = 2$)

8. The line that contains the points in this table

x	y
-3	4
-2	4
-1	4
0	4
1	4
2	4
3	4

9. The line that contains the points in this table

x	y
4	-3
4	-2
4	-1
4	0
4	1
4	2
4	3

10. The line that contains the points in this table

x	y
-3	-0.5
-2	0.0
-1	0.5
0	1.0
1	1.5
2	2.0
3	2.5

11. Complete the following table involving the change in x, the change in y, and the slope of the line $f(x) = mx + b$.

Change in x	Change in y	Slope
a. 5		$\dfrac{4}{5}$
b.	4	$\dfrac{4}{5}$
c. -5		$\dfrac{4}{5}$
d. 1		$\dfrac{4}{5}$
e. -1		$\dfrac{4}{5}$

12. Complete each table, using only pencil and paper, so that the points all lie on a line with a slope of $m = \dfrac{3}{5}$.

a.

x	y
0	6
5	
10	
15	
20	

b.

x	y
0	6
10	
20	
30	
40	

c.

x	y
0	6
-5	
-10	
-15	
-20	

13. Slope of a Wheelchair Ramp Determine the slope of the wheelchair ramp shown in the figure.

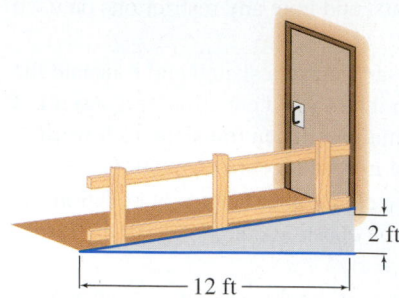

14. Determine whether the first line l_1 is parallel to, perpendicular to, or neither parallel nor perpendicular to the second line l_2.

a. Line l_1 passes through $(4, -2)$ and $(6, 2)$.
   Line l_2 passes through $(5, 3)$ and $(9, 11)$.
b. Line l_1 defined by $y = 2$
   Line l_2 defined by $x = -2$
c. Line l_1 defined by $y = 5x - 4$
   Line l_2 defined by $y - 4 = -5(x + 1)$

In Exercises 15 and 16, mentally estimate the slope of each line and then use a calculator to determine the slope.

15.

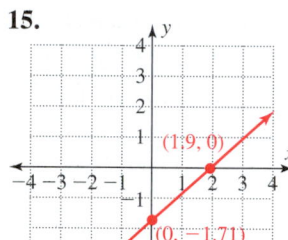

16.

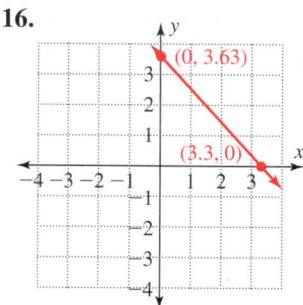

Graphing Lines

In Exercises 17–22, use pencil and paper to graph the line satisfying the given conditions.

17. Through $(-2, -1)$ and $(2, 4)$

18. With x-intercept $(-4, 0)$ and y-intercept $(0, 3)$

19. With y-intercept $(0, 1)$ and $m = -\dfrac{3}{4}$

20. $y = \dfrac{2}{5}x - 4$ **21.** $y - 1 = \dfrac{2}{3}(x + 1)$

22. $2x - 3y = 6$

Equations of Lines

In Exercises 23–30, write the equation of each line in slope-intercept form.

23. The line with y-intercept $(0, 6)$ and a slope of $-\dfrac{1}{2}$

24. The line with x-intercept $(3, 0)$ and y-intercept $(0, -2)$

25. The line that passes through $(2, 5)$ and $(4, -1)$

26. The line that passes through $(0, -4)$ and goes up 3 units for every 2-unit increase in x

27. The line that passes through the origin and is parallel to $y = 5x - 9$

28. The line that passes through $(2, 5)$ and is perpendicular to $y = -2x + 7$

29. Write the equation of the horizontal line that passes through $(4, 9)$.

30. Write the equation of the vertical line that passes through $(4, 9)$.

31. The graphing calculator display shows a table of values for a linear function. Use this table to determine the following:
a. y-intercept of the line
b. Value of Δx shown in the table
c. Value of Δy shown in the table
d. Slope of the line
e. Equation of the line in slope-intercept form

X	Y1	
-6	-13	
-4	-10	
-2	-7	
0	-4	
2	-1	
4	2	
6	5	

$X = -6$

32. Determine which points satisfy the equation $2x - 5y = 10$.
A. $(5, 0)$ **B.** $(10, 2)$
C. $(2, 10)$ **D.** $(0, -2)$

Solutions of Linear Systems

33. Determine which points satisfy the system of equations $\begin{cases} 3x + y = -2 \\ x + 2y = 6 \end{cases}$.
A. $(1, -5)$ **B.** $(2, 2)$
C. $(0, 1)$ **D.** $(-2, 4)$

In Exercises 34–36, use the graphs to solve the system of linear equations represented by each graph.

34.

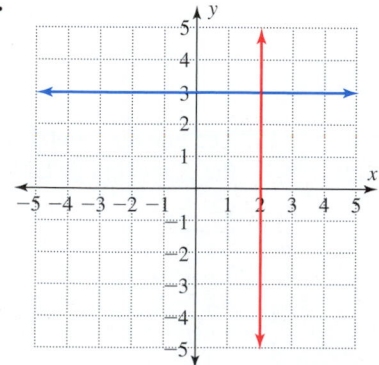

35.

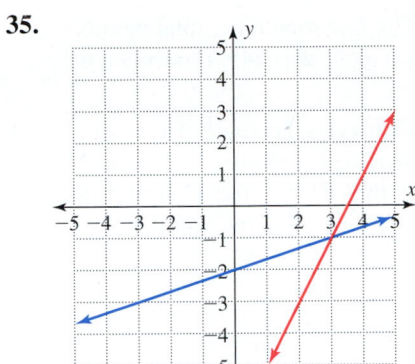

36.

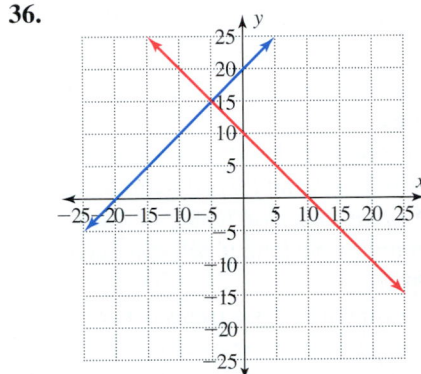

In Exercises 37 and 38, use the graphs to identify each system of linear equations as either an inconsistent system or a consistent system of dependent equations. Assume that each graph shows enough information to make this decision.

37.

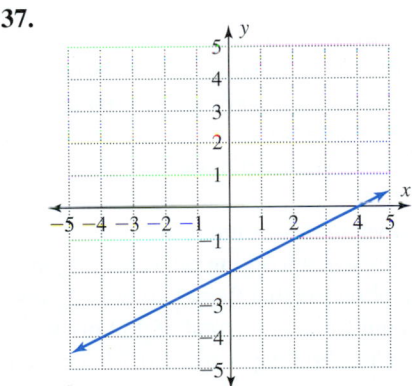

38.

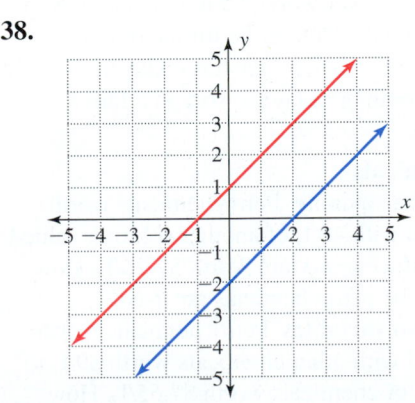

Exercises 39–42 give tables of values for each linear equation in a system of linear equations. Use these tables to solve each system.

39.

x	y_1	y_2
0	1.0000	−4.0
1	1.6667	−2.5
2	2.3333	−1.0
3	3.0000	0.5
4	3.6667	2.0
5	4.3333	3.5
6	5.0000	5.0

40.

x	y_1	y_2
0	−11	−11
1	−4	−4
2	3	3
3	10	10
4	17	17
5	24	24
6	31	31

41.

x	y_1	y_2
0	−11	−9
1	−4	−2
2	3	5
3	10	12
4	17	19
5	24	26
6	31	33

42.

x	y_1	y_2
−2.0	−3.5	−4.25
−1.5	−2.25	−2.75
−1.0	−1	−1.25
−0.5	.25	.25
0	1.5	1.75
0.5	2.75	3.25
1.0	4	4.75

In Exercises 43–46, solve each system of linear equations by using the substitution method.

43. $2x + y = 15$
$3x + 5y = -2$

44. $5x + 15y = 11$
$x + 3y = 11$

45. $x = 3y - 4$
$0.8x + 2.4y = 3.2$

46. $x = 4y + 3$
$0.8x + 3.2y = -4.0$

In Exercises 47–52, solve each system of linear equations by using the addition method.

47. $2x - y = -1$
$3x - 2y = -7$

48. $3x + 2y = 5$
$5x - 4y = 8$

49. $\dfrac{x}{2} + \dfrac{y}{3} = 1$
$\dfrac{x}{4} + \dfrac{y}{5} = 8$

50. $1.2x + 2.3y = 168$
$2.5x - 3.1y = -123.5$

51. $4x - 7y = 3$
$4x - 7y = 8$

52. $7x + 10y = 4$
$14x - 5y = -7$

In Exercises 53–56, use the slope-intercept form of each line to determine the number of solutions of the system. Then classify each system as a consistent system of independent equations, an inconsistent system, or a consistent system of dependent equations. Do not solve these equations.

53. $y = \dfrac{2}{9}x - 7$
$y = -\dfrac{2}{9}x + 7$

54. $y = \dfrac{2}{9}x - 7$
$y = \dfrac{2}{9}x + 7$

55. $6x + 2y = 4$
$15x = 10 - 5y$

56. $\dfrac{x}{6} + \dfrac{y}{4} = \dfrac{1}{3}$
$\dfrac{x}{3} + \dfrac{y}{2} = \dfrac{1}{4}$

Connecting Concepts to Applications

In Exercises 57–60, select the system of linear equations (choice A, B, C, or D) that represents the word problem. Then solve this system by the substitution method and answer the problem, using a full sentence.

Word Problem	Algebraic Equations
57. Find two numbers whose sum is 100 and whose difference is 20.	**A.** $x + y = 120$ $x - y = 20$
58. Find two numbers whose sum is 20 if the first number is 100 more than the second number.	**B.** $x + y = 100$ $x - y = 20$ **C.** $x + y = 20$ $x - y = 20$
59. Find two numbers whose sum is 20 and whose difference is 20.	**D.** $x + y = 20$ $x - y = 100$
60. Find two numbers whose sum is 120 and whose difference is 20.	

61. GPA A student who has already completed 45 credit-hours with a GPA of 2.80 at Chemeketa Community College needs to raise her GPA to 3.00 to qualify for a scholarship. How many credit-hours with a grade of A would be required to meet her goal? How many grade points will she have if she meets her goal? (Assume an $A = 4$ points.)

In Exercises 62–69, solve each problem by using a system of equations and each step of the word problem strategy outlined in this chapter.

62. Fixed and Variable Printing Costs A mathematics department is preparing a color supplement for its beginning algebra class. Two options are presented to the department chair. Option 1 is to pay 35 cents per page to a local printer. Option 2 is to buy a color printer for $2,200 and print the supplement in the department for a variable cost of 13 cents per page. What total page count would produce the same cost for both options?

63. Supplementary Angles The angles labeled x and y in the figure are supplementary. (Their sum is 180°.) The larger angle is 42° more than twice the smaller angle. Determine the number of degrees in each angle.

64. Mean and Range of Factory Measurements A quality-control worker in a factory measured ten wheels for shopping carts. The diameters of the first nine were all equal, but the tenth had a defect and was too large. The mean of these measurements was 25.05 cm and the range was 0.75 cm. Use this information to determine each measurement.

65. Interest on Two Loans A student borrowed a total of $9,000 on two loans—an 8% car loan and a 5%

educational loan. The first month her total monthly interest on these two loans was $45. Determine the amount of each loan.

66. Dimensions of an Isosceles Triangle The perimeter of an isosceles triangle (two equal sides) is 37 cm. The base is 5 cm shorter than each of the other two sides. Find the length of the base. (See the figure.)

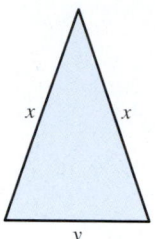

67. Rates of Two Trains
   a. Solve using one variable Two trains depart simultaneously from a station, traveling in opposite directions. One averages 80 km/h and the other 100 km/h. Determine how long it will take the trains to be 450 km apart.
   b. Solve using two variables Two trains depart simultaneously from a station, traveling in opposite directions. One averages 8 km/h more than the other, and after $\frac{1}{2}$ hour they are 100 km apart. Determine the speed of each train.

68. Rates of Two Workers
   a. Solve using one variable Two bricklayers are working together to complete a fireplace in a new home. The older bricklayer can lay 25 bricks per hour and the younger bricklayer can lay 20 bricks per hour. Assuming both bricklayers work at a consistent rate, how long it will take them to lay 1,170 bricks?
   b. Solve using two variables A union steward observed two bricklayers working on a current job. On Monday she observed that these two bricklayers each worked 8 hours and laid a total of 384 bricks. On Tuesday the older bricklayer worked 7 hours and the younger bricklayer worked 5 hours, and they laid a total of 292 bricks. Determine for the union steward the rate of work for each bricklayer, assuming that both bricklayers work at a fairly consistent rate.

69. Mixture of Chemicals
   a. Solve using one variable Petrochemicals worth $6/L are mixed with 40 L of another solution valued at $9/L to produce a mixture worth $7.20/L. How many liters of the petrochemicals are used?
   b. Solve using two variables Petrochemicals worth $6/L are mixed with other chemicals worth $9/L to produce 100 L of chemicals worth $7.65/L. How many liters of each are used?

70. Costs of Two Music Clubs Two popular music clubs have an introductory offer. Club A charges a $6 initiation fee plus $0.95 per CD. Club B does not charge any initiation fee but charges $1.95 per CD. The following graph compares the cost of membership in each club based on the number of CDs purchased.
  a. Give the equation in slope-intercept form for the line representing the costs for club A.
  b. Interpret the meaning of the slope and the y-intercept of the line for club A.
  c. Give the equation in slope-intercept form for the line representing the costs for club B.
  d. Interpret the meaning of the slope and the y-intercept of the line for club B.
  e. Determine the exact solution to the corresponding system of equations.
  f. Interpret the meaning of the x- and y-coordinates of this solution.

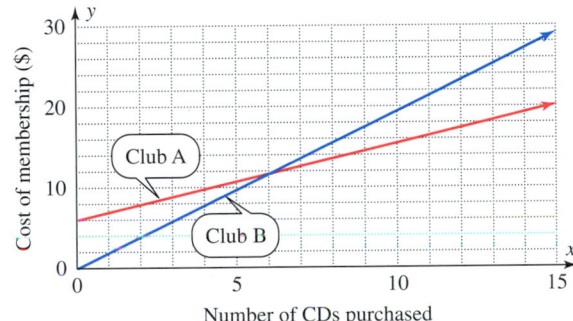

71. Automotive Service Charges The tables display the charges by two automotive repair shops based on the number of hours required for a repair.
  a. Give the equation in slope-intercept form for the line representing the costs for shop A.
  b. Interpret the meaning of the slope and the y-intercept of the line for shop A.
  c. Give the equation in slope-intercept form for the line representing the costs for shop B.
  d. Interpret the meaning of the slope and the y-intercept of the line for shop B.
  e. Give the solution to the corresponding system of equations.
  f. Interpret the meaning of the x- and y-coordinates of this solution.

Shop A

Hours x	Cost y ($)
1	50
2	80
3	110
4	140

Shop B

Hours x	Cost y ($)
1	56
2	82
3	108
4	134

Chapter 3 | Mastery Test

Objective 3.1.1 Determine the Slope of a Line

1. Calculate the slope of the line through the given points.
  a. $(3, -1)$ and $(5, 3)$
  b. $(3, -1)$ and $(1, 5)$
  c. Determine the slope of the line graphed.
  d. Determine the slope of the line containing all the points in the table.

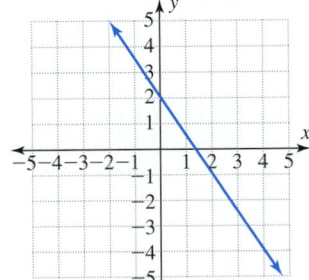

x	y
-5	1
5	5
10	7
20	11
35	17
50	23

  e. Determine the slope of the line defined by $y = -3x + 6$.
  f. Determine the slope of the line defined by $y - 4 = 7(x + 1)$.

Objective 3.1.2 Use Slopes to Determine Whether Two Lines Are Parallel, Perpendicular, or Neither

2. Determine whether the line defined by the first equation is parallel to, perpendicular to, or neither parallel nor perpendicular to the line defined by the second equation.

  a. $y = \frac{1}{2}x - 3$
     $y = -2x + 3$

  b. $y = \frac{1}{2}x - 3$
     $y = \frac{1}{2}x + 3$

  c. $y = 3$
     $x = 4$

  d. $y = 2x + 3$
     $y = -2x + 3$

Objective 3.1.3 Calculate and Interpret Rates of Change

3. Rate of Ascent The altitude of a model airplane is recorded every 5 seconds after liftoff. These data are displayed in the given table.

	A	B
1	Time (s)	Altitude (ft)
2	0	0
3	5	20
4	10	40
5	15	60

a. What is the value of Δx shown in column A of the table?
b. What is the value of Δy shown in column B of the table?
c. Determine the slope of the line containing these points.
d. Interpret the meaning of this rate of change.
e. At this rate, how long after the plane begins its ascent will it reach an altitude of 1,200 ft?

Objective 3.2.1 Use the Slope-Intercept Form to Write and Graph Linear Equations

4. Write in slope-intercept form the equation of the line satisfying the given conditions. Then sketch the graph of the line.

a. y-intercept $(0, 5)$ and slope $\dfrac{2}{3}$

b. y-intercept $(0, -2)$ and slope $-\dfrac{5}{3}$

Objective 3.2.2 Use the Point-Slope Form to Write and Graph Linear Equations

5. Write in point-slope form the equation of the line satisfying the given conditions. Then sketch the graph of the line and convert the equation to slope-intercept form.
a. Through $(-1, 3)$ with slope 4
b. Through $(4, -1)$ and $(1, -2)$

Objective 3.2.3 Use the Special Forms of Equations for Horizontal and Vertical Lines

6. a. Write the equation of the horizontal line through $(-4, 3)$.
b. Write the equation of the vertical line through $(-4, 3)$.
c. Write the equation of the line through $(-4, 3)$ and parallel to $x = 5$.
d. Write the equation of the line through $(-4, 3)$ and perpendicular to $x = 5$.

Objective 3.3.1 Check Possible Solutions of a Linear System

7. Determine whether each ordered pair is a solution of the linear system $\begin{cases} x + 2y = 4 \\ 3x - 5y = 1 \end{cases}$.

a. $(2, 1)$
b. $(-3, -2)$
c. $(-2, 3)$

Objective 3.3.2 Solve a System of Linear Equations by Using Graphs and Tables

8. Use graphs to determine the simultaneous solution for each system of linear equations.

a.

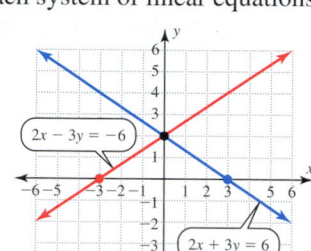

b. $f(x) = 2x + 1$
$f(x) = x + 3$

Use a numerical table to determine the simultaneous solution for each system of linear equations.

c.

x	$y_1 = x + 1$	$y_2 = -x - 3$
-4	-3	1
-3	-2	0
-2	-1	-1
-1	0	-2
0	1	-3
1	2	-4
2	3	-5

d. $y = 2x - 19$
$y = 3x - 31$

Objective 3.3.3 Identify Inconsistent Systems and Systems of Dependent Linear Equations

9. Select the choice A, B, or C that best describes each system of linear equations.

a. $f(x) = 2x - 3$
$f(x) = 3x + 7$

b. $f(x) = 2x - 3$
$f(x) = 2x + 7$

c. $f(x) = 2x - 3$
$f(x) = 2x - 3$

A. A consistent system of dependent equations

B. A consistent system of independent equations

C. An inconsistent system of independent equations

Objective 3.4.1 Solve a System of Linear Equations by the Substitution Method

10. Solve each system of linear equations by the substitution method.

a. $y = -4x - 1$
     $2x + y = 3$

b. $y = 4x - 1$
     $2x + 3y = -2$

c. $x + 3y = 1$
     $\dfrac{x}{3} - \dfrac{y}{2} = \dfrac{5}{6}$

d. $x - 2y = 5$
     $5x + 10y = 12$

Objective 3.5.1 Solve a System of Linear Equations by the Addition Method

11. Solve each system of linear equations by the addition method.

a. $3x - 7y = -13$
     $2x + 7y = 3$

b. $0.3x + 1.9y = 1.6$
     $0.6x - 1.1y = -1.7$

c. $3x - 2y = 6$
     $-6x + 4y = -12$

d. $y = \dfrac{3}{5}x + 2$
     $9x - 15y = 20$

Objective 3.6.1 Use Systems of Linear Equations to Solve Word Problems

12. Use systems of linear equations to solve each word problem.

a. Complementary angles: Two angles are complementary. If the larger angle is 12° more than 5 times the smaller, determine the number of degrees in each angle.

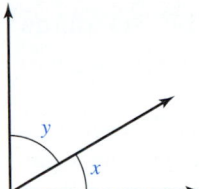

b. Value of two investments: An investment of $10,000 earned a net income of $625 in 1 year. Part of the investment was in bonds and earned income at a rate of 7%. The rest of the investment was in a savings account that earned income at a rate of 4%. How much was invested in bonds and how much in the savings account?

c. Mixture of two medicines: The dosage of medicine ordered by a doctor is 25 mL of a 50% solution. A nurse has available a 30% solution and an 80% solution of this medicine. How many milliliters of each should be mixed to produce this 25-mL dosage?

d. Rates of two planes: A jet plane and a tanker plane are 350 mi apart. They head toward each other so the jet can refuel. The jet flies 250 mi/h faster than the tanker. Determine the speed of each aircraft if they meet in 30 minutes.

Chapter 3 | Group Project

OSHA* Standards for Stair-Step Rise and Run

This project requires a tape measure, a ruler, or a yardstick.

The OSHA standards for stairs for single-family residences vary from the standards for other buildings. The standards for single-family residences set a maximum on the rise of 8 in and a minimum on the run of 9 in.

OSHA Standard 1910.24

Definitions:

Rise: *The vertical distance from the top of a tread to the top of the next-higher tread.*

Run: *The horizontal distance from the leading edge of a tread to the leading edge of an adjacent tread.*

1. There are 10 steps to the first landing in the stairs of the home of the author James Hall.

   Each of these steps has a rise of 8 in and a run of $9\frac{3}{4}$ in.

   a. Do these stairs meet the stated standard for single-family residences?
   b. What is the total rise in feet and inches of these 10 steps?
   c. What is the total run in feet and inches of these 10 steps?
   d. Calculate to the nearest hundredth the slope of the line from the leading edge of one step to the leading edge of the next step.

2. **a.** Using the OSHA definitions, measure the rise and run of a set of stairs on your campus.
   b. How many steps are on this set of stairs?
   c. What is the total rise of these steps?
   d. What is the total run of these steps?
   e. Calculate the slope of the line from the leading edge of one step to the leading edge of the next step.
   f. Which stairs are steeper, those given for the Hall house or those in your school?

*Occupational Safety and Health Act.

Linear Inequalities

Chapter Outline

Cable Strength

We rely on the strength of cables more than we may realize. Cranes, elevators, ski lifts, and many theme park rides use cables to move cargo or occupants safely from place to place. Mathematics plays a key role in designing cables for required strengths.

Gigantic cables support many of the most magnificent bridges in the world. Anyone driving across the graceful span of the Golden Gate Bridge places a vital trust in the two huge cables that suspend the bridge high over San Francisco Bay. These cables had to be designed and built to hold the enormous weight of the bridge as well as the traffic moving across it. Some bridge cables are composed of several strands with each strand containing several wires. The strength of a standard galvanized steel bridge strand depends on the number and size of the individual wires in the strand.

Exercise 34 in Section 4.1 and other exercises in this chapter will use inequalities to examine the number of strands required to meet strength requirements on a cable.

Golden Gate Bridge:
Width: 90 ft
Total length: 8,981 ft
(1.7 mi)
Weight of bridge:
894,500 tons

Supporting cables: 2
Wires in each cable:
27,572
Diameter of each cable:
36.375 in

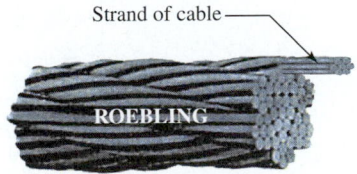

Strand of cable

ROEBLING

| **Section 4.1** | Solving Linear Inequalities Using the Addition-Subtraction Principle |

The essence of mathematics is not to make simple things complicated, but to make complicated things simple. —Stan Gudder

Objectives:

1. Identify linear inequalities and check a possible solution of an inequality.
2. Solve linear inequalities in one variable using the addition-subtraction principle for inequalities.
3. Use graphs and tables to solve linear inequalities in one variable.

A Mathematical Note

The symbols $>$ and $<$ for greater than and less than are due to Thomas Harriot (1631). These symbols were not immediately accepted, as many mathematicians preferred the symbols ⊏ and ⊐.

In Exercise 34 at the end of this section, we examine the fiber cable strength needed for a sailboat. The manufacturer specifies that each cable must support a minimum of 200 lb. We will use linear inequalities to examine this problem.

1. Identify Linear Inequalities and Check a Possible Solution of an Inequality

How Do Linear Inequalities Differ from the Linear Equalities That I Have Already Studied?

A linear equation in one variable can be written in the form $Ax + B = C$. If we replace this equality symbol with an inequality symbol, the result is a **linear inequality in one variable** of the form $Ax + B < C$, $Ax + B \leq C$, $Ax + B > C$, or $Ax + B \geq C$. In this section, we will solve linear inequalities algebraically and graphically and use tables to check these solutions. Although the steps used to solve linear inequalities are very similar to the steps used to solve linear equations, there are some important distinctions. We will build on these similarities and carefully point out these distinctions.

In the accompanying box, remember that a parenthesis indicates that an endpoint is not included in the interval. A bracket indicates that the endpoint is included in the interval.

Linear Inequalities

Verbally	Algebraically	Algebraic Examples	Graphical Examples
A linear inequality in one variable is an inequality that is first-degree in that variable.	For real constants A, B, and C, with $A \neq 0$.		
	$Ax + B > C$	$x > 2$	−1 0 1 2 3 4 5
	$Ax + B \geq C$	$x \geq 2$	−1 0 1 2 3 4 5
	$Ax + B < C$	$x < 2$	−1 0 1 2 3 4 5
	$Ax + B \leq C$	$x \leq 2$	−1 0 1 2 3 4 5

Example 1   Identifying Linear Inequalities

Determine which of the following are linear inequalities in one variable.

Solution

(a) $4x - 5$ Not a linear inequality   This is a first-degree algebraic expression, but it does not contain an inequality symbol.

(b) $4x - 5 = 11$ Not a linear inequality   This is a linear equation, not a linear inequality.

(c) $4x - 5 < 11$ Linear inequality in one variable   This is a linear inequality in the variable x.

(d) $4x^2 - 5 < 11$ Not a linear inequality   The exponent on the variable is 2.

Self-Check 1

Determine which of the following are linear inequalities in one variable.

a. $3x + 7 \geq 34$ **b.** $3x + 7 = 34$ **c.** $3x^2 + 7 = 34$ **d.** $3x + 7 - 34$

How Many Solutions Does a Linear Inequality Have?

A **conditional inequality** contains a variable and is true for some, but not all, real values of the variable. A value that makes an inequality a true statement is said to satisfy the inequality. The **solution** of a linear inequality consists of all values that satisfy the inequality. The solution of a conditional inequality will be an interval that contains an infinite set of values.

Example 2   Checking Possible Solutions of an Inequality

Determine whether either (a) $x = -3$ or (b) $x = -1.5$ satisfies $6x - 2 < 5x - 4$.

Solution

(a) $6x - 2 < 5x - 4$ Substitute the given value for x in the inequality and determine whether this makes the inequality a true statement.
    $6(-3) - 2 \overset{?}{<} 5(-3) - 4$
        $-18 - 2 \overset{?}{<} -15 - 4$
            $-20 \overset{?}{<} -19$ is true

    Answer: -3 satisfies   The solution of this inequality will include -3.
            $6x - 2 < 5x - 4.$

(b) $6x - 2 < 5x - 4$
    $6(-1.5) - 2 \overset{?}{<} 5(-1.5) - 4$
        $-9 - 2 \overset{?}{<} -7.5 - 4$ The solution of this inequality will not include -1.5.
            $-11 \overset{?}{<} -11.5$ is false

    Answer: -1.5 does not satisfy
            $6x - 2 < 5x - 4.$

Self-Check 2

Determine whether either $x = -2.4$ or $x = 3.2$ satisfies $6x - 2 < 5x - 4$.

Since conditional linear inequalities have an infinite interval of solutions, it is common to represent these solutions with a graph or by using the interval notation introduced in Section 1.2. It is very useful to be able to determine whether individual values satisfy an inequality by inspecting either a graph or the interval representation of the solution of an inequality. This is illustrated in Example 3.

Example 3 — Identifying Solutions of an Inequality from a Graph

Determine which of the values $-3, -2, -1, 0, 1, 2,$ and 3 satisfy the inequality whose solution is represented by this graph.

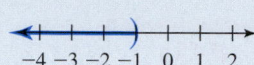

Solution

2 and 3 satisfy the inequality.   This graph represents the inequality $x \geq 2$. The graph includes 2 and all real numbers to the right of 2. The values $-3, -2, -1, 0,$ and 1 are not shaded and do not satisfy the inequality.

Self-Check 3

Determine which of the values $-6, -4, -2, 0, 2, 4,$ and 6 satisfy the inequality $x < -1$.

2. Solve Linear Inequalities in One Variable Using the Addition-Subtraction Principle for Inequalities

How Can I Solve Linear Inequalities Algebraically?

The procedure for solving a linear inequality algebraically is very similar to the procedure we already used to solve linear equations. The basic idea is to isolate the variable terms on one side of the inequality and the constant terms on the other side. Inequalities that have the same solution are called **equivalent inequalities.** We use the addition-subtraction principle to produce equivalent inequalities that are simpler than the original inequality.

Addition-Subtraction Principle for Inequalities

Verbally	Algebraically*	Numerical Example
If the same number is added to or subtracted from both sides of an inequality, the result is an equivalent inequality.	If a, b, and c are real numbers, then $a < b$ is equivalent to $a + c < b + c$ and to $a - c < b - c$.	$x - 2 < 5$ is equivalent to $x - 2 + 2 < 5 + 2$ and to $x < 7$.

*Similar statements can be made for the inequalities $\leq$, $>$, and $\geq$.

The logic behind the addition-subtraction principle can be explained by examining the two graphs in Figure 4.1.1. Adding 2 to both sides of an inequality shifts both points to the right 2 units. Thus their position relative to each other is preserved. Likewise, subtracting 3 from both sides of an inequality shifts both points to the left 3 units. Thus their position relative to each other is also preserved.

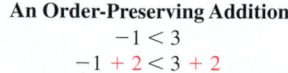

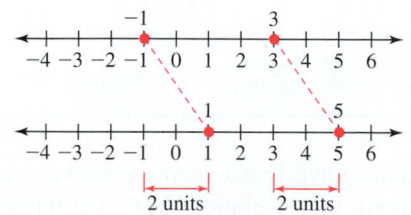

Figure 4.1.1

Example 4   Solving a Linear Inequality with Variables on Both Sides

Solve $6v - 2 < 5v - 3$, and graph the solution set.

Solution

$$6v - 2 < 5v - 3$$

$$6v - 5v - 2 < 5v - 5v - 3 \qquad \text{Subtract } 5v \text{ from both sides of the inequality.}$$

$$v - 2 < -3$$

$$v - 2 + 2 < -3 + 2 \qquad \text{Add 2 to both sides of the inequality.}$$

$$v < -1$$

Answer: $(-\infty, -1)$ whose graph is

$$-4 \ -3 \ -2 \ -1 \ \ 0 \ \ 1 \ \ 2$$

Self-Check 4

Solve $2x + 4 > x$, and graph the solution set.

The strategy for solving linear inequalities is much the same as the one used to solve linear equations. At each step, try to produce simpler expressions than on the previous step. In Example 5, we start by using the distributive property to remove parentheses.

Example 5   Solving a Linear Inequality Containing Parentheses

Solve $5(x + 2) \geq 3(2x + 3) - 1$, and graph the solution set.

Solution

$$5(x + 2) \geq 3(2x + 3) - 1$$

$$5x + 10 \geq 6x + 9 - 1 \qquad \text{Use the distributive property to remove the parentheses,}$$
$$\text{and then combine like terms.}$$

$$5x + 10 \geq 6x + 8$$

$$5x + 10 - 5x \geq 6x + 8 - 5x \qquad \text{Subtract } 5x \text{ from both sides of the inequality.}$$

$$10 \geq x + 8$$

$$10 - 8 \geq x + 8 - 8 \qquad \text{Subtract 8 from both sides of the inequality.}$$

$$2 \geq x \quad \text{or} \qquad \text{The statements } a \geq b \text{ and } b \leq a \text{ have the same meaning.}$$

$$x \leq 2$$

Answer: $(-\infty, 2]$ whose graph is

$$-1 \ \ 0 \ \ 1 \ \ 2 \ \ 3 \ \ 4 \ \ 5$$

Self-Check 5

Solve $3(x - 1.5) - 2 < 2(x - 2)$, and graph the solution set.

Note in Example 5 that the last step of the solution is written as $x \leq 2$. It is common to write answers so that x is the subject of the sentence, as in "x is less than or equal to 2."

3. Use Graphs and Tables to Solve Linear Inequalities in One Variable

Can I Solve Linear Inequalities Using Graphs?

Yes, the solution of a linear inequality in one variable can be determined graphically by letting y_1 represent the left side of the inequality and y_2 represent the right side of the inequality. In Example 6, we will solve $2x - 5 \leq 1$ by letting y_1 represent $2x - 5$ and y_2 represent 1.

Examine the graphs of y_1 and y_2 on the same coordinate system.

- If (x, y_1) is below (x, y_2), then $y_1 < y_2$. This x-value will satisfy $2x - 5 < 1$.
- If (x, y_1) is above (x, y_2), then $y_1 > y_2$. This x-value will satisfy $2x - 5 > 1$.

Example 6   Using a Graph to Solve a Linear Inequality

Solve $2x - 5 \leq 1$ by graphing $y_1 = 2x - 5$ and $y_2 = 1$.

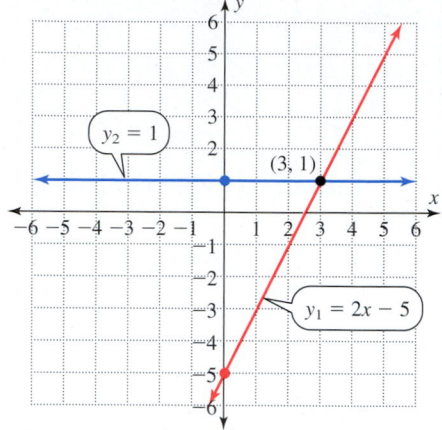

Solution

First graph both $y_1 = 2x - 5$ and $y_2 = 1$. Then determine the point of intersection of these two lines—in this case, the point $(3, 1)$. Observe from the graph that the graph of y_1 is below the graph of y_2 to the left of $(3, 1)$. Thus for $x \leq 3$, $y_1 \leq y_2$.

Answer: $(-\infty, 3]$ The x-values to the left of $(3, 1)$ are those for which $x \leq 3$. This interval can also be represented by $(-\infty, 3]$.

Self-Check 6

Solve $1.5x + 1 \geq -2$ by graphing $y_1 = 1.5x + 1$ and $y_2 = -2$.

Can I Use a Table to Check the Solution of a Linear Inequality?

Yes, start by letting y_1 represent the left side of an equation and y_2 represent the right side. Then examine x-values to the right and left of the point where y_1 and y_2 are equal. On one side $y_1 < y_2$ and on the other side $y_1 > y_2$. This is illustrated in the table, where we check the solution of $2x - 5 \leq 1$ from Example 6.

	A	B	C	D
	X	Y1 = 2x - 5	<, =, or >	Y2 = 1
1				
2	-1	-7	<	1
3	0	-5	<	1
4	1	-3	<	1
5	2	-1	<	1
6	3	1	=	1
7	4	3	>	1
8	5	5	>	1

You can create the table of values using either pencil and paper, a calculator, or a spreadsheet. Then inspect the table to determine the appropriate order relation between the y_1 and y_2 values. This order relation is inserted in this table in column C.

In Example 6, the graphs of y_1 and y_2 crossed at $(3, 1)$. This is confirmed by the table, which also shows that $y_1 \leq y_2$ for $x \leq 3$.

In Example 7, we will solve an inequality algebraically and graphically and then check the solution using a table.

Example 7   Using Multiple Perspectives to Solve a Linear Inequality

Solve $\dfrac{x}{2} - 1 \geq 1 - \dfrac{x}{2}$ algebraically and graphically. Then check this solution with a table and describe the solution verbally.

Solution

Algebraic Solution	Graphical Solution	Numerical Check

Algebraic Solution

$$\frac{x}{2} - 1 \geq 1 - \frac{x}{2}$$

$$\frac{x}{2} - 1 + \frac{x}{2} \geq 1 - \frac{x}{2} + \frac{x}{2}$$

$$x - 1 \geq 1$$

$$x - 1 + 1 \geq 1 + 1$$

$$x \geq 2$$

Graphical Solution

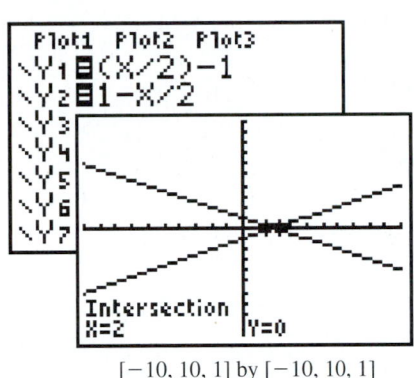

$[-10, 10, 1]$ by $[-10, 10, 1]$

Numerical Check

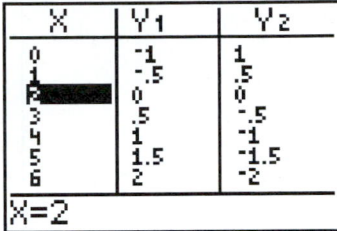

Verbal Description

All values of x greater than or equal to 2 satisfy this inequality. From the graph, Y_1 is on or above Y_2 for $x \geq 2$. In the table, $Y_1 \geq Y_2$ for $x \geq 2$.

Answer: $[2, \infty)$ whose graph is

$\xleftarrow{\quad}\!\!+\!\!+\!\!+\![\!\!+\!\!+\!\!+\!\!\xrightarrow{\quad}$
$\quad -1 \; 0 \; 1 \; 2 \; 3 \; 4 \; 5$

Self-Check 7

Solve $\dfrac{2x}{3} + 2 \leq 1 - \dfrac{x}{3}$ algebraically and use a table to check your solution.

How Will I Recognize When Word Problems Involve Inequalities?

Some of the more common phrases used to indicate inequalities are given in the box.

Phrases Used to Indicate Inequalities

Phrase	Verbal Meaning	Inequality Notation	Interval Notation	Graphical Notation
• "x is at least a" • "x is a minimum of a"	x is greater than or equal to a	$x \geq a$	$[a, \infty)$	$\xleftarrow{}\;\;[\!\!\xrightarrow{}$ a
• "x is at most a" • "x is a maximum of a" • "x never exceeds a"	x is less than or equal to a	$x \leq a$	$(-\infty, a]$	$\xleftarrow{}\!\!]\;\;\xrightarrow{}$ a
• "x exceeds a"	x is greater than a	$x > a$	(a, ∞)	$\xleftarrow{}\;\;(\!\!\xrightarrow{}$ a
• "x is smaller than a"	x is less than a	$x < a$	$(-\infty, a)$	$\xleftarrow{}\!\!)\;\;\xrightarrow{}$ a

Remember that correct usage of interval notation requires the smaller value on the left and the larger value on the right. For example, the notation $(2, 7]$ is correct while $[7, 2)$ is incorrect.

Example 8   Translating Verbal Statements into Algebraic Inequalities

Write each of these statements in both inequality notation and interval notation, and then graph the solution of this inequality.

Solution

	Inequality	Interval	Graph
(a) x is at least 4	$x \geq 4$	$[4, \infty)$	
(b) x is at most 3	$x \leq 3$	$(-\infty, 3]$	
(c) x exceeds 2	$x > 2$	$(2, \infty)$	
(d) x never exceeds 2	$x \leq 2$	$(-\infty, 2]$	
(e) x is a minimum of 5	$x \geq 5$	$[5, \infty)$	

Self-Check 8

Match each phrase with the corresponding algebraic inequality. (Some choices can be used more than once, and others may not be used.)

a. x exceeds 8   **A.** $x < 8$

b. x is at most 8   **B.** $x \leq 8$

c. x is at least 8   **C.** $x > 8$

d. x is a maximum of 8   **D.** $x \geq 8$

In Example 9, first we translate the verbally stated inequality to algebraic form and then we solve this inequality. Note that the answer is written as a full sentence.

Example 9   Solving a Verbally Stated Inequality

Determine all values of x so that four times the quantity x plus three exceeds eleven more than three times x.

Solution

| 4 times the quantity x plus 3 | exceeds | 11 more than 3 times x | Write this inequality in words and then translate these words into an algebraic inequality. |

$$4(x + 3) \qquad > \qquad 3x + 11$$
Use the distributive property to remove the parentheses on the left side of the inequality.

$$4x + 12 > 3x + 11$$

$$4x + 12 - 3x > 3x + 11 - 3x$$
Subtract $3x$ from both sides of the inequality.

$$x + 12 > 11$$

$$x + 12 - 12 > 11 - 12$$
Subtract 12 from both sides of the inequality.

$$x > -1$$

Answer: The values of x are greater than -1.

It might be wise to check a couple of values. For example, -2 doesn't check and 2 does check.

Self-Check 9

Determine all values of x so that five times the quantity x minus four is at most seven more than four times x.

Summary: The appearance of linear inequalities is very similar to the appearance of linear equations. Like a linear equation, a linear inequality in x is first degree in this variable. That is, the exponent on x is understood to be 1 even though it is not written. The algebraic method for solving these inequalities is similar to the method used to solve linear equations.

We limited the problems examined in this section to problems that could be solved by the addition-subtraction principle. In the next section, we will examine the multiplication-division principle. The primary distinction between solving equations and solving inequalities is that we must be very careful to reverse the inequality symbol when we multiply or divide both sides of the inequality by a negative number.

Self-Check Answers

1. **a.** A linear inequality
    b. Not a linear inequality
    c. Not a linear inequality
    d. Not a linear inequality
2. -2.4 satisfies the inequality but 3.2 does not.
3. -6, -4, and -2
4. $(-4, \infty)$ whose graph is

    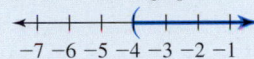
    $$-7\ -6\ -5\ -4\ -3\ -2\ -1$$
5. $(-\infty, 2.5)$ whose graph is

    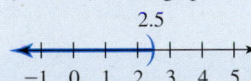
    2.5
    $$-1\ \ 0\ \ 1\ \ 2\ \ 3\ \ 4\ \ 5$$
6. $x \geq -2$
    The graph of $y = 1.5x + 1$ is above the graph of $y_2 = -2$ for $x \geq -2$.

    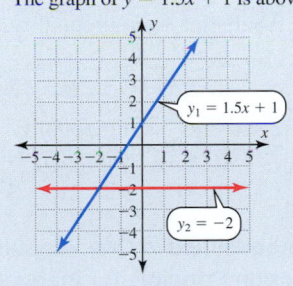
    $y_1 = 1.5x + 1$
    $y_2 = -2$

7. $(-\infty, -1]$
    In the table, Y_1 is less than or equal to Y_2 for $x \leq -1$.

    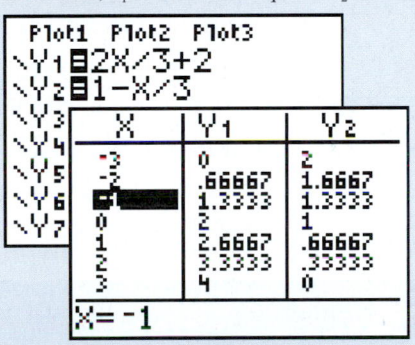

8. **a.** C   **b.** B   **c.** D   **d.** B
9. The values of x are less than or equal to 27.

4.1   Using the Language and Symbolism of Mathematics

1. An inequality that is first degree in x is called a _____ inequality in x.

2. An inequality is first degree in x if the exponent on x is _____.

3. Inequalities with the same solution are called _____ inequalities.

4. A _____ inequality contains a variable and is true for some, but not all, real values of the variable.

5. A value that makes an inequality a _____ statement satisfies the inequality.

6. By the _____-subtraction principle, $a < b$ is equivalent to $a + c < b + c$.

7. By the addition-_____ principle, $a < b$ is equivalent to $a - c < b - c$.

8. **a.** If $x > y$, then $x + 2$ _____ $y + 2$.
   b. If $x > y$, then $x - 2$ _____ $y - 2$.
   c. If $x < y$, then y _____ x.

9. The _____ property is often used to help us remove parentheses from expressions in an inequality.

10. The statement "x is at least five" is represented by the inequality x _____ 5.

11. The statement "x is at most five" is represented by the inequality x _____ 5.

12. The statement "x exceeds five" is represented by the inequality x _____ 5.

13. The statement "x is a maximum of five" is represented by the inequality x _____ 5.

14. The statement "x is a minimum of five" is represented by the inequality x _____ 5.

4.1 | Quick Review

1. What number is in the interval $[-2, 5)$ but not in the interval $(-2, 5)$?

2. In interval notation, a _____ indicates that an endpoint is not included in the interval.

3. Solve the equation $4(2x - 7) + 8 = 11 - 3(x - 8)$ algebraically.

4. Use the graphs of $y_1 = \frac{1}{2}x - 3$ and $y_2 = -x + 3$ to

solve the equation $\frac{1}{2}x - 3 = -x + 3$.

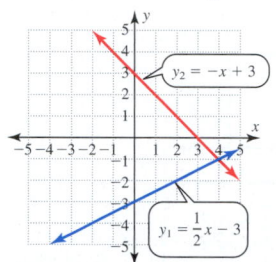

5. Use the table for $y_1 = 3(x - 2)$ and $y_2 = 7(x + 2)$ to solve the equation $3(x - 2) = 7(x + 2)$.

	A	B	C
1	X	Y1 = 3(x - 2)	Y2 = 7(x + 2)
2	-9	-33	-49
3	-7	-27	-35
4	-5	-21	-21
5	-3	-15	-7
6	-1	-9	7
7	1	-3	21
8	3	3	35

4.1 | Exercises

Objective 1 Identify Linear Inequalities and Check a Possible Solution of an Inequality

In Exercises 1 and 2, determine which of the following are linear inequalities in one variable.

1. A. $6x - 3$ B. $6x - 3 \le 4x$
   C. $y = 6x - 3$ D. $x^2 \ge 36$

2. A. $4x + 1$ B. $x^2 > 4$
   C. $4x + 1 > 9$ D. $y \le 4x + 1$

In Exercises 3 and 4, express each interval by using interval notation.

3. a. $x \le -4$ b. $x > 5$
   c. ⟵┼┼┼┼)┼┼┼⟶ d. x is at least -1
           3

4. a. $x \ge 1.5$ b. $x < -3$
   c. ⟵┼┼┼┼(┼┼┼⟶ d. x is at most 5
          -2

5. Determine whether $x = 3$ satisfies each inequality.
   a. $x < 3$ b. $x \le 3$
   c. $x > 3$ d. $x \ge 3$

6. Determine whether $x = 2$ satisfies each inequality.
   a. $x \le 2$ b. $x < 2$
   c. $x \ge 2$ d. $x > 2$

7. Determine whether $x = 3$ satisfies each inequality.
   a. $x > -3$ b. $x > 5$
   c. $3x - 5 \ge 2(x - 1)$ d. $5x - 3 \le 2x + 3$

8. Determine whether $x = 2$ satisfies each inequality.
   a. $x > 3$ b. $x \ge -1$
   c. $3x - 5 \ge 2(x - 1)$ d. $5x - 3 \le 2x + 3$

Objective 2 Solve Linear Inequalities in One Variable Using the Addition-Subtraction Principle for Inequalities

In Exercises 9–28, solve each inequality algebraically.

9. $x + 7 > 11$ 10. $x - 7 < -11$

11. $2x + 3 \le x + 4$ 12. $5x - 3 \ge 4x - 9$

13. $x - 7 \ge -3$ 14. $x - 3 \le -7$

15. $5x - 8 < 4x - 8$ 16. $11x + 12 > 10x + 12$

17. $7 \le x$ 18. $-4 \ge x$

19. $x - 4 > 2x - 3$ 20. $4x + 9 < 5x + 3$

21. $5 - 9x < 6 - 10x$ 22. $12 - 11x < 10 - 12x$

23. $5(x - 2) \ge 4(x - 2)$ 24. $7(x - 3) \ge 6(x - 4)$

25. $5(2 - x) > 4(3 - x)$ 26. $2(3 - 2x) > 3(2 - x)$

27. $9(x - 5) \ge 4(2x - 2) + 3$

28. $11(x + 4) \ge 5(2x + 3) - 7$

Objective 3 Use Graphs and Tables to Solve Linear Inequalities in One Variable

In Exercises 29–32, use the graph to determine the x-values that satisfy each equation and inequality.

29. a. $y_1 = y_2$
    b. $y_1 < y_2$
    c. $y_1 > y_2$

30. a. $y_1 = y_2$
    b. $y_1 < y_2$
    c. $y_1 > y_2$

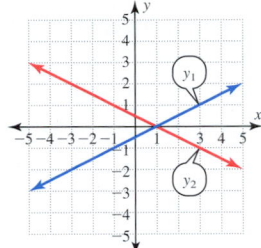

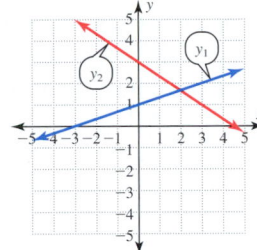

31. a. $y_1 = y_2$
    b. $y_1 \leq y_2$
    c. $y_1 \geq y_2$

32. a. $y_1 = y_2$
    b. $y_1 \leq y_2$
    c. $y_1 \geq y_2$

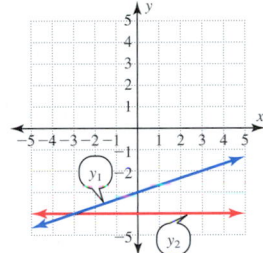

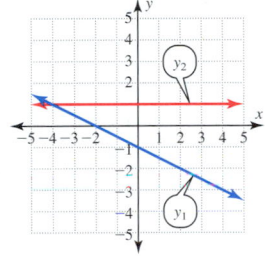

33. Using a Graph That Models Overtime Spending Limits Robinson Construction Company has budgeted a spending limit of $15,000 per year on overtime for its foremen. The graph of $y_1 = 15,000$ displays the spending limit, and the graph of $y_2 = f(x)$ displays the cost for overtime based on the number of overtime hours worked by the foremen. The maximum possible number of hours of overtime by all the foremen is 800 hours. Use the graphs to solve:
a. $y_1 = y_2$ **b.** $y_1 < y_2$ **c.** $y_1 > y_2$
d. Interpret the meaning of each of these solutions.

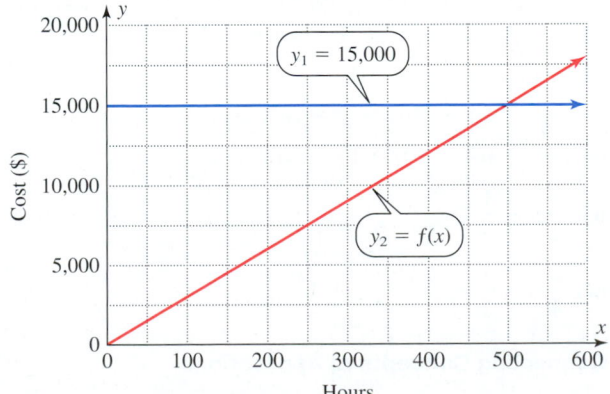

34. Using a Graph That Models Strength Limits for a Cable The contract for a composite fiber cable ordered for a sailboat from a cable manufacturer specifies that

each cable must support a minimum load of 200 lb. The graph of $y_1 = 200$ displays the strength limit, and the graph of $y_2 = f(x)$ displays the strength of a cable based on the number of separate fiber strands braided into the cable. Use the graphs to solve:
a. $y_1 = y_2$ **b.** $y_1 < y_2$ **c.** $y_1 > y_2$
d. Interpret the meaning of each of these solutions.

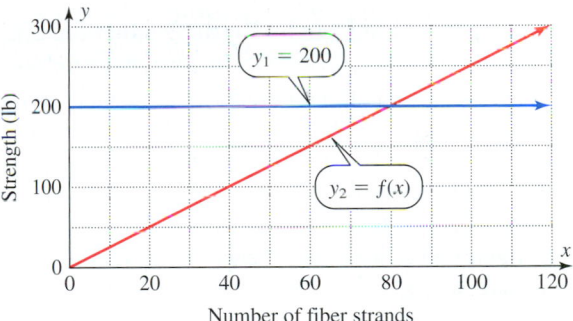

In Exercises 35–38, use the table of values to determine the x-values that satisfy each equation and inequality.

35. a. $y_1 = y_2$
    b. $y_1 \leq y_2$
    c. $y_1 \geq y_2$

36. a. $y_1 = y_2$
    b. $y_1 \leq y_2$
    c. $y_1 \geq y_2$

x	y_1	y_2
0	2	12
1	7	15
2	12	18
3	17	21
4	22	24
5	27	27
6	32	30

x	y_1	y_2
0	-5	-9
1	-1	-3
2	3	3
3	7	9
4	11	15
5	15	21
6	19	27

37. a. $y_1 = y_2$
b. $y_1 < y_2$
c. $y_1 > y_2$

x	y_1	y_2
-20	-67	-97
-15	-52	-72
-10	-37	-47
-5	-22	-22
0	-7	3
5	8	28
10	23	53

38. a. $y_1 = y_2$
b. $y_1 < y_2$
c. $y_1 > y_2$

x	y_1	y_2
1	-3.0	0.8
1.2	-2.4	-0.4
1.4	-1.8	-0.8
1.6	-1.2	-1.2
1.8	-0.6	-1.6
2	0.0	-2.0
2.2	0.6	-2.4

39. Using Tables That Model Supply Costs The accompanying two tables give the costs, including delivery, for ordering copier paper from two different office supply firms. Use these tables to solve:
a. $y_1 = y_2$ **b.** $y_1 < y_2$ **c.** $y_1 > y_2$
d. Interpret the meaning of each of the solutions from parts **a** to **c.**

Ace Office Supply

Boxes of Paper x	Cost y_1 ($)
4	112
8	224
12	336
16	448
20	560
24	672

Max Office Supply

Boxes of Paper x	Cost y_2 ($)
4	136
8	224
12	312
16	400
20	488
24	576

40. Using Tables That Model Bottle Production The accompanying two tables give the number of glass bottles produced after a number of minutes by two different machines. For each machine, there is an initial start-up time before any bottles will come out. Use these tables to solve:
a. $y_1 = y_2$ **b.** $y_1 < y_2$ **c.** $y_1 > y_2$
d. Interpret the meaning of each of the solutions from parts **a** to **c.**

Machine 1

Minutes After Start-Up x	Bottles Produced y_1
4	40
6	80
8	120
10	160
12	200
14	240
16	280

Machine 2

Minutes After Start-Up x	Bottles Produced y_2
4	20
6	70
8	120
10	170
12	220
14	270
16	320

Estimate Then Calculate

In Exercises 41–44, mentally estimate the solution of each inequality to the nearest integer, and then calculate the exact solution.

Problem	Mental Estimate	Exact Solution
41. $x - 0.727 > 0.385$		
42. $x + 29.783 \leq 70.098$		
43. $2x + 25.87 < x + 3.94$		
44. $5x - 1.61 \geq 4x + 5.42$		

Multiple Representations and Concept Development

In Exercises 45–48, solve each inequality algebraically and graphically. Then check this solution with a table and describe this solution verbally. (*Hint:* See Example 7.)

45. $2x - 1 \geq x + 2$ **46.** $0.5x + 2 > -0.5x - 2$
47. $-x + 3 < -2x + 5$ **48.** $3(x - 1) \leq 2(x - 2)$

In Exercises 49–54, first write an algebraic inequality for each verbal statement, and then solve this inequality.

49. Five minus two times x is less than or equal to seven minus three times x.

50. Six plus four times x is greater than or equal to three times x plus five.

51. Twice w plus three results in a minimum of eleven more than w.

52. Five times w never exceeds three more than four times w.

53. Three times the quantity m plus five exceeds twelve more than twice m.

54. Six times the quantity m minus eleven is a maximum of ninety-six more than five times m.

Skill and Concept Development

In Exercises 55–66, solve each inequality either algebraically or graphically.

55. $5(w - 2) + 14 > 4(w + 3)$
56. $7(x - 3) + 16 \geq 6(x - 2)$
57. $3(5n - 2) \leq 7(2n - 1)$
58. $13(2n - 1) < 5(5n - 3)$
59. $4(6m + 7) - 2m \geq 7(3m + 1) + 1$
60. $5(2m + 8) + 3m > 3(4m + 5) - 4$
61. $-3(2 - m) < -4(3 - m) + 4.5$
62. $-11(2 - w) - 10.5 \leq -12(3 - w)$
63. $4y - 4(5 + 2y) > y - 6(3 + y)$
64. $19 - 3(4 - y) \geq 11 - 4(1 - y)$
65. $\dfrac{1}{2}x - \dfrac{1}{3} \geq \dfrac{5}{3} - \dfrac{1}{2}x$
66. $\dfrac{2}{3}x - \dfrac{4}{7} < \dfrac{3}{7} - \dfrac{1}{3}x$

Connecting Concepts to Applications

67. Total Points To earn an A in an algebra class, a student must earn at least 630 points. He has 535 points going into the last 100-point test. How many points must he earn on the last test?

68. Total Grade Points To earn a grade of B, a student needs her exam points to total at least 320 points. Her first three test grades are 71, 79, and 78. What score must she earn on the fourth test to obtain a B?

69. Perimeter of a Rectangular Pen A farmer has 84 ft of woven wire fence available to enclose three sides of the rectangular pen shown here. What are the possible values for the length x?

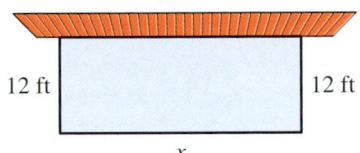

12 ft   12 ft

x

70. Perimeter of a Triangular Pen A pet security system for a yard is buried in a triangular shape with two sides of length 16 m. The pet owner can afford only 54 m of the buried wire. What are the possible values for the length x?

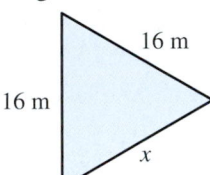

16 m

16 m

x

71. Costs and Revenue The daily cost of producing x units of computer mice includes a fixed cost of $450 per day and a variable cost of $4 per unit. The income produced by selling x units is $5 per unit. Letting y_1 represent the income and y_2 represent the cost, graph y_1 and y_2. Determine the values of x for which $y_1 < y_2$, the loss interval for this company. Also determine the values of x for which $y_1 > y_2$, the profit interval for this company.

72. Costs and Revenue The daily cost of producing x units of cellular phones includes a fixed cost of $600 per day and a variable cost of $14 per unit. The income produced by selling x units is $15 per unit. Letting y_1 represent the income and y_2 represent the cost, graph y_1 and y_2. Determine the values of x for which $y_1 < y_2$, the loss interval for this company. Also determine the values of x for which $y_1 > y_2$, the profit interval for this company.

Group discussion questions

73. Communicating Mathematically The sum of the lengths of any two sides of a triangle is greater than the length of the third side. Write three different inequalities expressing this concept for the triangle given here.

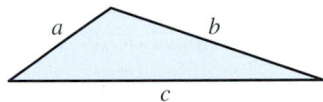

a b

c

74. Communicating Mathematically Write an algebraic expression for each verbal expression. Then explain the different meanings of these expressions.
a. 3 more than x
b. 3 is more than x
c. 2 less than x
d. 2 is less than x

75. Challenge Question You have solved an inequality and tested one value from your solution. This value checked, yet your solution is incorrect. Give two distinct examples that illustrate how this is possible.

4.1   Cumulative Review

1. The numerical coefficient of $-7x^2$ is _____.

2. The numerical coefficient of $-x^3$ is _____.

3. The exponent on x in $-x^3$ is _____.

4. The exponent on x in $-9x$ is _____.

5. Evaluate $|2x - 3|$ for $x = \dfrac{1}{2}$.

Section 4.2   Solving Linear Inequalities Using the Multiplication-Division Principle

Objective:

1. Solve linear inequalities using the multiplication-division principle for inequalities.

The specifications for many manufactured products place limits on the length, volume, weight, or strength of the product. For a tent, it is important that each supporting cable meet minimum strength requirements. Engineers frequently use inequality notation to indicate these strength requirements. In Example 8, we use inequality notation to determine the number of synthetic fibers to put into a tent cable to meet the minimum strength requirements.

1. Solve Linear Inequalities Using the Multiplication-Division Principle for Inequalities

Multiplication-Division Principle for Inequalities		
Verbally	**Algebraically***	**Numerical Examples**
Order-Preserving: If both sides of an inequality are multiplied or divided by a *positive number,* the result is an inequality that has the same solution as the original inequality.	If a, b, and c are real numbers and $c > 0$, then $a > b$ is equivalent to $ac > bc$.	$\dfrac{x}{2} > 3$ is equivalent to $2\left(\dfrac{x}{2}\right) > 2(3)$ and to $x > 6$
Order-Reversing: If both sides of an inequality are multiplied or divided by a *negative number and the order of inequality is reversed,* the result is an inequality that has the same solution as the original inequality.	If a, b, and c are real numbers and $c < 0$, then $a > b$ is equivalent to $ac < bc$.	$-\dfrac{x}{3} > 5$ is equivalent to $(-3)\left(-\dfrac{x}{3}\right) < (-3)(5)$ and to $x < -15$

*Similar statements can be made for division and for the inequalities $<$, $\leq$, and $\geq$.

The logic behind the multiplication-division principle can be understood by examining the two graphs in Figures 4.2.1 and 4.2.2.

An Order-Preserving Multiplication

$1 < 3$
$1(2) < 3(2)$
$2 < 6$

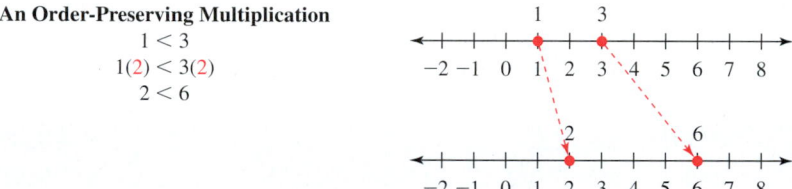

Figure 4.2.1 Multiplication-division principle with positive numbers.

Multiplying both sides of the inequality by $+2$ doubles the distance between the points, while their position relative to each other is preserved. Likewise, dividing by $+2$ would preserve the order relation, although the distance between the points is halved.

An Order-Reversing Multiplication

$-1 < 3$
$-1(-2) > 3(-2)$
$2 > -6$

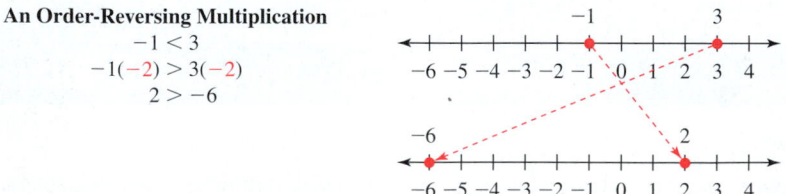

Figure 4.2.2 Multiplication-division principle with negative numbers.

Multiplying both sides of the inequality by -2 changes the sign of each side of the inequality and thus reverses the order of their positions relative to each other. Likewise, dividing by -2 would reverse the order relation.

One way to visually illustrate the order-reversing case is with a meter stick. Label the meter stick with $-$, 0, and $+$ and points a and b with a to the left of b. Multiplying by -1 changes the $-$ side to $+$, and the $+$ side to $-$. Illustrate this by rotating the meter stick $180°$ about the 0 point. This reverses the order that points a and b now occur on the meter stick.

How Is the Multiplication-Division Principle Used with the Addition-Subtraction Principle?

We use the addition-subtraction principle to isolate the variable terms on one side of an inequality and the constant terms on the other side. Then we use the multiplication-division principle to obtain a coefficient of 1 for the variable.

Example 1   Solving a Linear Inequality with a Positive Coefficient

Solve $3x \leq 6$, and graph the solution set.

Solution

$3x \leq 6$

$\dfrac{3x}{3} \leq \dfrac{6}{3}$ Dividing both sides of the inequality by $+3$ preserves the order.

$x \leq 2$

Answer: $(-\infty, 2]$ whose graph is

Self-Check 1

Solve $6w \geq -12$ and graph the solution.

A linear inequality is not solved until the coefficient of the variable is 1. If the coefficient is -1, the inequality is not yet solved. In Example 2, we solve for w by multiplying both sides of the inequality by -1. Note that this step reverses the order of the inequality.

Example 2   Solving a Linear Inequality with a Coefficient of -1

Solve $-w < 15$, and graph the solution set.

Solution

$-w < 15$

$(-1)(-w) > (-1)(15)$ Multiplying both sides of the inequality by -1 reverses the order of the inequality. You can obtain the same result by dividing both sides of the inequality by -1.

$w > -15$

Answer: $(-15, \infty)$ whose graph is

Self-Check 2

Solve $-\dfrac{1}{2}y < 2$, and graph the solution.

What Is the Basic Strategy Used to Solve Linear Inequalities?

If an inequality contains variables on both sides of the inequality, first we want to isolate the variable terms on one side of the inequality and the constant terms on the other side of the inequality. This is illustrated in Example 3.

Example 3 Solving a Linear Inequality with Variables on Both Sides

Solve $5m - 7 < 3m - 13$.

Solution

$$5m - 7 < 3m - 13$$

$$5m - 7 - 3m < 3m - 13 - 3m \qquad \text{Subtracting } 3m \text{ preserves the order.}$$

$$2m - 7 < -13$$

$$2m - 7 + 7 < -13 + 7 \qquad \text{Adding 7 preserves the order.}$$

$$2m < -6$$

$$\frac{2m}{2} < \frac{-6}{2} \qquad \text{Dividing by } +2 \text{ preserves the order.}$$

$$m < -3$$

Answer: $(-\infty, -3)$ whose graph is

-6 -5 -4 -3 -2 -1 0

Self-Check 3

Solve $7m + 8 \geq 10m - 4$, and graph the solution.

Example 4 Solving a Linear Inequality with Variables on Both Sides

Solve $6x - 2 \geq 8x + 12$.

Solution

$$6x - 2 \geq 8x + 12$$

$$6x - 2 - 8x \geq 8x + 12 - 8x \qquad \text{Subtracting } 8x \text{ preserves the order.}$$

$$-2x - 2 \geq 12$$

$$-2x - 2 + 2 \geq 12 + 2 \qquad \text{Adding 2 preserves the order.}$$

$$-2x \geq 14$$

$$\frac{-2x}{-2} \leq \frac{14}{-2} \qquad \text{Dividing by } -2 \text{ reverses the order.}$$

$$x \leq -7$$

Answer: $(-\infty, -7]$ whose graph is

-10 -9 -8 -7 -6 -5 -4

Self-Check 4

Solve $4y - 20 \leq 9y + 15$, and graph the solution.

We encourage you to use the steps as shown in Example 4 to avoid careless errors. In more advanced courses, you may see some of these steps combined as shown by the Alternative Solution.

Alternative Solution

A shortened version of the solution for Example 4 follows.

$$6x - 2 \geq 8x + 12$$

$$-2x \geq 14 \qquad \text{Subtracting } 8x \text{ and adding 2 preserve the order.}$$

$$x \leq -7 \qquad \text{Dividing by } -2 \text{ reverses the order.}$$

How Does the Procedure for Solving Linear Inequalities Compare to the Procedure for Solving Linear Equations?

They are much the same. At each step, we try to produce simpler expressions than on the previous step. This is illustrated in Example 5, which begins by using the distributive property to remove parentheses.

Example 5   Solving a Linear Inequality Containing Parentheses

Solve $6(3 - 4n) < 5(2 - 3n) - 28$.

Solution

$$6(3 - 4n) < 5(2 - 3n) - 28$$

$$18 - 24n < 10 - 15n - 28 \qquad \text{Use the distributive property to remove the parentheses, and}$$
$$18 - 24n < -15n - 18 \qquad\quad \text{then combine like terms.}$$

$$-24n < -15n - 36 \qquad\qquad \text{Subtracting 18 from both sides preserves the order.}$$

$$-9n < -36 \qquad\qquad\qquad\quad \text{Adding } 15n \text{ to both sides preserves the order.}$$

$$n > 4 \qquad\qquad\qquad\qquad\quad \text{Dividing both sides by } -9 \text{ reverses the order.}$$
$$\qquad\qquad\qquad\qquad\qquad\qquad \text{The solution is all values of } n \text{ greater than 4.}$$

Answer: $(4, \infty)$ whose graph is

$$\xleftarrow{\quad} +\!\!+\!\!+\!\!(+\!\!+\!\!+\!\!+\!\!+\!\!+\!\!\xrightarrow{\quad}$$
$$\;\;-2\;\;0\;\;2\;\;\;4\;\;6\;\;8\;\;10\;\;12$$

Self-Check 5

Solve $3(2m - 5) \geq 4(1 - 2m) - 5$, and give the answer using interval notation.

Linear inequalities with fractions, similar to linear equations with fractions, can be simplified by converting the inequality to an equivalent inequality that does not involve fractions. This can be accomplished by multiplying both sides of the inequality by the LCD (least common denominator) of all the terms.

Example 6   Solving a Linear Inequality Containing Fractions

Solve $\dfrac{x}{15} \leq \dfrac{x}{10} - \dfrac{5}{6}$.

Solution

$$\dfrac{x}{15} \leq \dfrac{x}{10} - \dfrac{5}{6} \qquad\qquad \begin{aligned} &15 = 3 \cdot 5 \\ &10 = 2 \cdot 5 \\ &6 = 2 \cdot 3 \end{aligned}$$

$$30\left(\dfrac{x}{15}\right) \leq 30\left(\dfrac{x}{10}\right) - 30\left(\dfrac{5}{6}\right) \qquad \begin{aligned} &\text{LCD} = 2 \cdot 3 \cdot 5 = 30 \\ &\text{Multiplying by 30, the LCD, preserves the order.} \end{aligned}$$

$$2x \leq 3x - 25$$

$$-x \leq -25 \qquad\qquad\qquad \text{Subtracting } 3x \text{ from both sides of the inequality preserves the order.}$$

$$(-1)(-x) \geq (-1)(-25) \qquad\quad \text{Multiplying by } -1 \text{ reverses the order.}$$

$$x \geq 25$$

Answer: $[25, \infty)$ whose graph is

$$\xleftarrow{\quad} +\!\!+\!\!+\!\![+\!\!+\!\!+\!\!+\!\!\xrightarrow{\quad}$$
$$-50\;-25\;\;\;0\;\;\;25\;\;50\;\;75\;\;100$$

Self-Check 6

Solve $-\dfrac{14}{15} > -\dfrac{2}{3}x$, and give the answer using interval notation.

The ability to examine problems graphically and numerically can deepen your overall understanding of many concepts, including the solutions of inequalities. Note that in Example 7 all perspectives yield the same result.

Example 7 — Using Multiple Perspectives to Solve a Linear Inequality

Solve $\dfrac{x}{2} + \dfrac{1}{6} \le \dfrac{x}{6} + \dfrac{1}{2}$ algebraically and graphically. Then check this solution with a table and write the solution using interval notation.

Solution

Algebraic Solution

$$\frac{x}{2} + \frac{1}{6} \le \frac{x}{6} + \frac{1}{2}$$

$$6\left(\frac{x}{2}\right) + 6\left(\frac{1}{6}\right) \le 6\left(\frac{x}{6}\right) + 6\left(\frac{1}{2}\right)$$

$$3x + 1 \le x + 3$$

$$3x \le x + 2$$

$$2x \le 2$$

$$x \le 1$$

Graphical Solution

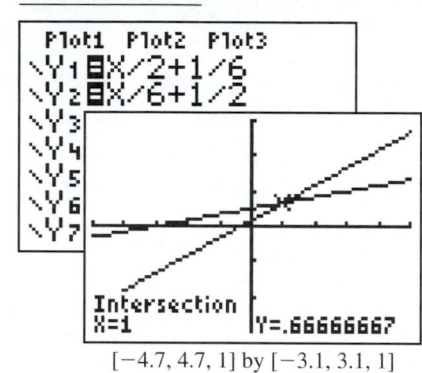

[−4.7, 4.7, 1] by [−3.1, 3.1, 1]

Numerical Check

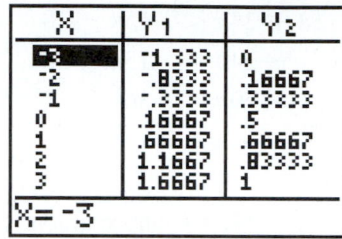

Verbal Description

All values of x less than or equal to 1 satisfy this inequality. On the graph Y_1 is below or on Y_2 for $x \le 1$. Thus $Y_1 \le Y_2$ for $x \le 1$. From the table, Y_1 is less than or equal to Y_2 for $x \le 1$.

Answer: $(-\infty, 1]$ whose graph is

-3 -2 -1 0 1 2 3 4

Self-Check 7

Use a graph to solve $6\left(\dfrac{x}{4} + 1\right) > -\dfrac{x}{2} + 1$. Then check this answer using a table.

Example 8 examines the limits placed on an order for a new cable.

Example 8 — Strength Limits on a Cable

The contract for a cable used to support a large tent requires that the cable be able to support a minimum load of 750 lb. The composite cable consists of a synthetic fiber wound around a hemp core that can support a load of 20 lb. Each synthetic fiber added to the cable adds 25 lb to the load that the cable can support. Determine the number of synthetic fibers to put in this cable.

Solution

Let n = number of synthetic fibers to put in this cable. The restrictions on n, the number of fibers in the cable, are $\{0, 1, 2, \ldots\}$. The number of fibers must be an integer and cannot be less than 0 fibers.

Word Inequality

| Load that synthetic fibers can support | + | Load that hemp core can support | ≥ | 750 |

The total load that the cable can support must be 750 lb or more.

Algebraic Inequality

$$25n + 20 \geq 750$$
$$25n \geq 730$$
$$\frac{25n}{25} \geq \frac{730}{25}$$
$$n \geq 29.2$$

First subtract 20 from both sides of the inequality. Then divide both sides of the inequality by 25.

Because the number of fibers must be an integer greater than 29, the number of fibers must be 30 or greater.

Answer: The cable must contain at least 30 synthetic fibers.

Self-Check 8

Rework Example 8 assuming the hemp core can support 80 lb and that each synthetic fiber can support 30 lb.

Self-Check Answers

1. $[-2, \infty)$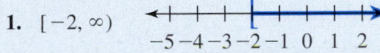
$-5 -4 -3 -2 -1 \ 0 \ 1 \ 2$

2. $(-4, \infty)$
$-5 -4 -3 -2 -1 \ 0 \ 1 \ 2$

3. $(-\infty, 4]$
$-2 -1 \ 0 \ 1 \ 2 \ 3 \ 4 \ 5$

4. $[-7, \infty)$
$-10 -9 -8 -7 -6 -5 -4$

5. $[1, \infty)$

6. $\left(\dfrac{7}{5}, \infty\right)$

7. $(-2.5, \infty)$

The graph of $Y_1 = 6\left(\dfrac{x}{4} + 1\right)$ is above

$Y_2 = -\dfrac{x}{2} + 1$ for $x > -2.5$.

The table verifies $Y_1 > Y_2$ for $x > -2.5$.

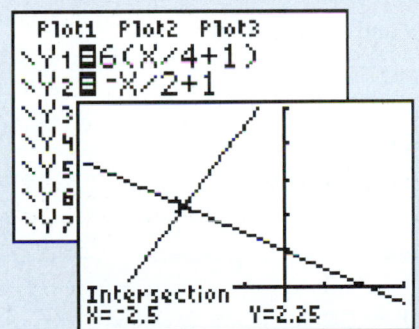

8. The cable must contain at least 23 synthetic fibers.

4.2 | Using the Language and Symbolism of Mathematics

1. By the _____-division principle for inequalities, $x > y$ is equivalent to $2x > 2y$.

2. By the _____-division principle for inequalities, $x > y$ is equivalent to $-2x < -2y$.

3. If a, x, and y are real numbers and $a > 0$, then $x > y$ is equivalent to ax _____ ay.

4. If a, x, and y are real numbers and $a < 0$, then $x > y$ is equivalent to ax _____ ay.

5. To clear an inequality of fractions, we multiply both sides of the inequality by the _____ _____ _____ of all the terms in the inequality.

6. The _____ property is often used to help us remove parentheses from expressions in an inequality.

4.2 Quick Review

1. Solve $1.5x + 1.2 = 1.3(x - 1)$.

2. Solve $\dfrac{x - 3}{2} = \dfrac{2x + 3}{5}$.

3. Solve $5(4x - 1) + 3 = 3(6x - 5) - 7$.

4. Two equations are equivalent if they have the _____ solution set.

5. The equation $2x - 6 = 0$ is equivalent to:
   A. $2x = 6$
   B. $-2x + 6 = 0$
   C. $2x + 6 = 0$
   D. Both **A** and **B** are correct.
   E. **A**, **B**, and **C** are all correct.

4.2 Exercises

Objective 1 Solve Linear Inequalities Using the Multiplication-Division Principle for Inequalities

In Exercises 1 and 2, fill in each blank with the correct inequality symbol.

1. a. If $x > y$, then $x + 2$___$y + 2$.
   b. If $x > y$, then $x - 2$___$y - 2$.
   c. If $x > y$, then $\dfrac{x}{2}$___$\dfrac{y}{2}$.
   d. If $x > y$, then $\dfrac{x}{-2}$___$\dfrac{y}{-2}$.

2. a. If $x < y$, then $x + 3$___$y + 3$.
   b. If $x < y$, then $x - 3$___$y - 3$.
   c. If $x < y$, then $3x$___$3y$.
   d. If $x < y$, then $-3x$___$-3y$.

In Exercises 3–6, solve each inequality and show each step of your solution process.

3. $-2x < 4$
4. $-3x \geq -12$
5. $-3x + 4 \geq 7$
6. $-5x - 7 \leq 32$

In Exercises 7–44, solve each inequality and give the answer using interval notation.

7. $2x \geq 4$
8. $4x < 2$
9. $\dfrac{1}{2}v < -4$
10. $\dfrac{1}{3}v > -\dfrac{2}{3}$
11. $-\dfrac{3}{4}w < -12$
12. $-\dfrac{4}{3}w \leq -12$
13. $-x \geq 0$
14. $-x \leq 5$
15. $1.1m < 2.53$
16. $2.1m \leq 1.68$
17. $-\dfrac{n}{4} > \dfrac{1}{2}$
18. $-\dfrac{n}{6} \geq \dfrac{2}{3}$
19. $-8z \geq 40$
20. $-9z < -36$
21. $-\dfrac{t}{3} \geq 0$
22. $-\dfrac{2t}{5} > 0$
23. $-10 > -5x$
24. $-21 > -7x$
25. $5t \leq 21 - 2t$
26. $8t < 3t + 35$
27. $9x + 7 > 5x - 13$
28. $23x - 14 > 18x + 56$
29. $11y - 8 > 14y + 7$
30. $8y + 19 \geq 2y - 11$

31. $5 - 9y \leq 19 - 2y$
32. $13 - 7y \leq 1 - y$
33. $2(3t - 4) > 2(t - 2)$
34. $7(2t - 4) \geq 3(3t + 2)$
35. $4(3 - x) \geq 7(2 - x)$
36. $5(1 - 2x) < 9(3 - x)$
37. $\dfrac{x}{6} + \dfrac{1}{2} \leq \dfrac{x}{4} + \dfrac{1}{3}$
38. $\dfrac{2x}{5} + \dfrac{1}{2} > \dfrac{3x}{4} + \dfrac{3}{20}$
39. $-2(3w + 7) < 3(5 - w) + 2$
40. $-5(4w - 2) > 3(3w + 5) - 5$
41. $-3(5 - 2x) > -2(5x + 1) + 3$
42. $-4(2 - 3x) \geq -7(3 - x) - 2$
43. $6(7x - 8) - 4(3x + 2) \leq 8(5x + 3)$
44. $-5y - 7(2y - 3) < 3y - 2(5 - y) - 5$

In Exercises 45–48, use the graph to solve each equation and inequality.

45. a. $y_1 = y_2$
    b. $y_1 < y_2$
    c. $y_1 > y_2$

46. a. $y_1 = y_2$
    b. $y_1 < y_2$
    c. $y_1 > y_2$

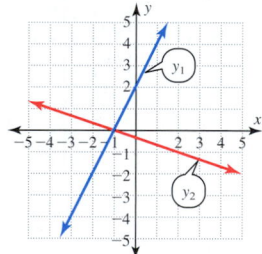

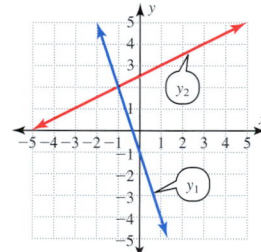

47. a. $y_1 = y_2$
    b. $y_1 \leq y_2$
    c. $y_1 \geq y_2$

48. a. $y_1 = y_2$
    b. $y_1 \leq y_2$
    c. $y_1 \geq y_2$

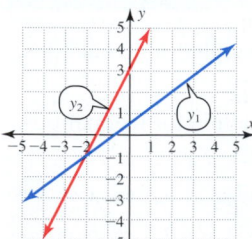

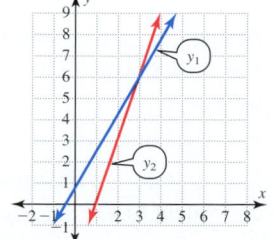

In Exercises 49–52, use the table of values to solve each equation and inequality.

49. a. $y_1 = y_2$
b. $y_1 < y_2$
c. $y_1 > y_2$

x	y_1	y_2
0	−3	3
1	2	6
2	7	9
3	12	12
4	17	15
5	22	18
6	27	21

50. a. $y_1 = y_2$
b. $y_1 < y_2$
c. $y_1 > y_2$

x	y_1	y_2
−3	−11	−14
−2	−7	−7
−1	−3	0
0	1	7
1	5	14
2	9	21
3	13	28

51. a. $y_1 = y_2$
b. $y_1 \leq y_2$
c. $y_1 \geq y_2$

x	y_1	y_2
−20	−67	−97
−15	−52	−72
−10	−37	−47
−5	−22	−22
0	−7	3
5	8	28
10	23	53

52. a. $y_1 = y_2$
b. $y_1 \leq y_2$
c. $y_1 \geq y_2$

x	y_1	y_2
−2	7	−23
−1	5	−19
0	3	−15
1	1	−11
2	−1	−7
3	−3	−3
4	−5	1

Estimate Then Calculate

In Exercises 53–56, mentally estimate the solution of each inequality, and then calculate the exact solution.

Problem	Mental Estimate	Exact Solution
53. $5.02x \leq 35.642$		
54. $-4.95x \geq 31.185$		
55. $0.47x > -4.183$		
56. $-0.99x < 118.8$		

Multiple Representations

In Exercises 57–60, solve each inequality algebraically and graphically. Then check this solution with a table and describe this solution verbally. (*Hint:* See Example 7.)

57. $2x < 5x + 6$
58. $3x + 1 < -x + 5$
59. $3(x - 5) \leq 1 - x$
60. $2(x - 1) \geq 4 - x$

In Exercises 61–64, write an algebraic inequality for each verbal statement, and solve this inequality.

61. Twice the quantity x plus five is less than six.

62. Three times the quantity x minus four is greater than fifteen.

63. Four times the quantity y minus seven is greater than or equal to six times the quantity y plus three.

64. Five times the quantity y plus nine is less than or equal to eight times the quantity y minus three.

Connecting Concepts to Applications

65. Rental Costs The costs for renting a rug-shampooing machine from two different rental companies are given by the graphs. The graph of $y_1 = f_1(x)$ gives the cost by Dependable Rental Company based on the number of hours of use. The graph of $y_2 = f_2(x)$ gives the cost by Anytime Rental Company based on the number of hours of use. Use these graphs to solve
a. $y_1 = y_2$ **b.** $y_1 < y_2$ **c.** $y_1 > y_2$
d. Interpret the meaning of each of these solutions.

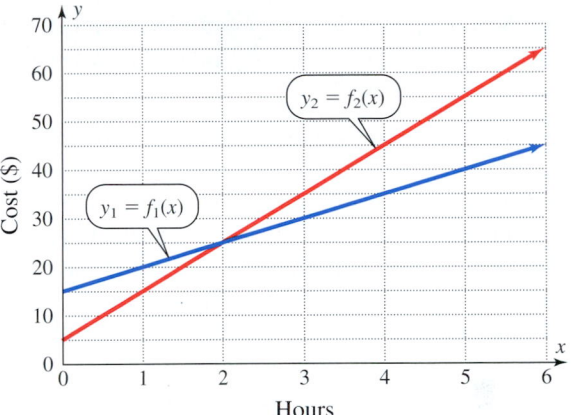

66. Strength of Two Cables A cable manufacturer makes cables that are wound around two different cores and braided from different fibers. Part of the tensile strength comes from the core, and the rest of the strength comes from the fiber strands. The graph of $y_1 = f_1(x)$ gives the strength of the first cable based on the number of separate fiber strands braided into this cable. The graph of $y_2 = f_2(x)$ gives the strength of the second cable based on the number of separate fiber strands braided into this cable. Use these graphs to solve
a. $y_1 = y_2$ **b.** $y_1 < y_2$ **c.** $y_1 > y_2$
d. Interpret the meaning of each of these solutions.

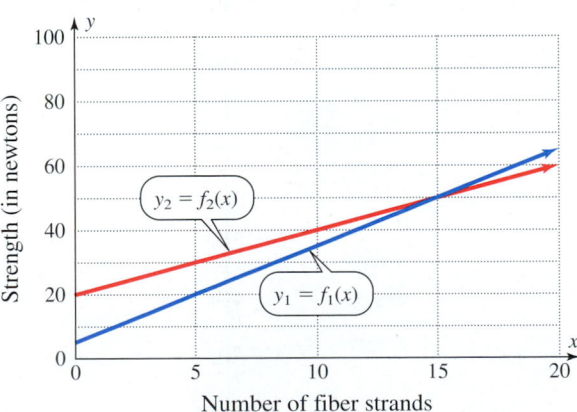

67. Sweatshirt Costs The accompanying two tables give the costs for ordering college logo sweatshirts from two sporting goods stores. In each case, there is a setup charge for designing the logo as well as a cost per sweatshirt. Use these tables to solve

a. $y_1 = y_2$ **b.** $y_1 < y_2$ **c.** $y_1 > y_2$

d. Interpret the meaning of each of these solutions.

Mel's Sports		Michael's Sporting Goods	
Number of Sweatshirts x	Cost y_1 ($)	Number of Sweatshirts x	Cost y_2 ($)
10	310	10	320
20	560	20	565
30	810	30	810
40	1,060	40	1,055
50	1,310	50	1,300
60	1,560	60	1,545

68. Using Tables That Model Bottle Production The accompanying two tables give the number of glass bottles produced after a number of minutes by two different machines. For each machine there is an initial start-up time before any bottles will come out. Use these tables to solve

a. $y_1 = y_2$ **b.** $y_1 < y_2$ **c.** $y_1 > y_2$

d. Interpret the meaning of each of these solutions.

Machine 1		Machine 2	
Minutes After Start-Up x	Bottles Produced y_1	Minutes After Start-Up x	Bottles Produced y_2
4	40	4	20
6	80	6	70
8	120	8	120
10	160	10	170
12	200	12	220
14	240	14	270
16	280	16	320

69. Temperature Limit The temperature in a computer room must be at least 41° Fahrenheit (41°F). Find the allowable temperature in degrees Celsius. The relationship between Fahrenheit and Celsius temperatures is $F = \dfrac{9}{5}C + 32$.

70. Budgetary Limit A company has set a limit of at most $450 to be spent on its monthly electric bill. If electricity costs $0.12/kW, find the number of kilowatts the company can use each month.

71. Interest Payment The most a family can budget each month for an interest payment is $450. Their interest payment for one month will be approximately 0.75% of the amount they can borrow. How much can the family borrow?

72. Taxi Fare The minimum charge for a taxi ride is $3.50, plus an additional $0.25 per 0.1 mi. How far can you travel if you have at most $20.50 to spend on taxi fare?

73. Costs and Revenue The cost of producing an order of bricks includes a fixed cost of $250 and a variable cost of $0.50 per brick. All bricks are custom-ordered, and the charge for manufacturing these bricks includes a setup fee of $150 plus a charge of $1 per brick. Let y_1 represent the income received from an order for x bricks and y_2 represent the cost of producing x bricks.

a. Determine the values of x for which $y_1 < y_2$, the loss interval for this order.

b. Determine the values of x for which $y_1 > y_2$, the profit interval for this order.

74. Costs and Revenue The cost of producing x flooring tiles includes a fixed cost of $400 and a variable cost of $1.25 per tile. All tiles are custom-ordered, and the charge for manufacturing these tiles includes a setup fee of $300 plus a charge of $2 per tile. Let y_1 represent the income received from an order for x tiles and y_2 represent the cost of producing x tiles.

a. Determine the values of x for which $y_1 < y_2$, the loss interval for this order.

b. Determine the values of x for which $y_1 > y_2$, the profit interval for this order.

Group discussion questions

75. Error Analysis The final solution to this problem is correct, but one step has an error. Find and fix this error.

Solve $-\dfrac{2x}{3} + 1 > 7$.

Solution: $-\dfrac{2x}{3} + 1 > 7$

Step 1. $-\dfrac{2x}{3} > 6$

Step 2. $\left(-\dfrac{3}{2}\right)\left(-\dfrac{2x}{3}\right) > \left(-\dfrac{3}{2}\right)(6)$

Step 3. $x < -9$

Step 4. Answer $(-\infty, -9)$

76. Error Analysis Examine the following argument and explain the error in reasoning.

$$x < 0$$

$x + x < x$ Add x to both sides.

$$2x < x$$

$2 < 1$ Divide both sides by x.

77. Challenge Question If x and y are real numbers and $x > y$, is there a real number a such that $ax > ay$ is false and $ax < ay$ is also false? If this is possible, list all values of a that make this possible.

78. Challenge Question Solve each inequality for x.
a. $2x + 3y < 5y - 2x$
b. $8x + 5y > 4x - 3y$
c. $2(x + 3y) \leq -3(x + 3y)$
d. $5(x - 2y) > 2(7x - 5y)$

4.2 | **Cumulative Review**

In Exercises 1 and 2, use this table to evaluate each expression.

x	$f(x)$
-2	1
0	3
2	6
3	-2

1. $f(-2)$

2. $f(x) = -2; x = \underline{?}$

A 16-ft board has two pieces each of length x cut off.

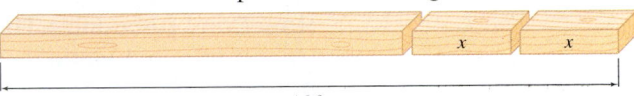

16 ft

3. Write a function f for the length of the remaining piece in terms of x.

4. What restrictions should be placed on the variable x?

5. Evaluate and interpret $f(3)$.

Section 4.3 | Solving Compound Inequalities

Objectives:

1. Identify an inequality that is a contradiction or an unconditional inequality.
2. Write the intersection or union of two intervals.
3. Solve compound inequalities.

1. Identify an Inequality That Is a Contradiction or an Unconditional Inequality

Most of the inequalities that we consider are conditional inequalities. **Conditional inequalities** contain a variable and are true for some, but not all, real values of the variable.

Are All Inequalities Conditional Inequalities?

No, an inequality that is always true is called an **unconditional inequality,** whereas an inequality that is always false is called a **contradiction.**

Example 1 | Classifying Inequalities

Identify each inequality as a conditional inequality, an unconditional inequality, or a contradiction.

Solution

(a) $x < x + 1$ This inequality is true for every value of x and therefore is an unconditional inequality.

Every real number is less than 1 more than that number. Thus the solution set is the set of all real numbers $\mathbb{R}$. The graph of the solution set is the entire real number line.

(b) $y < y - 1$ This inequality is false for every value of y and therefore is a contradiction.

No real number is less than the number 1 unit less than itself. The solution set is the empty set, a set with no elements. There are no solutions to plot on the number line.

(c) $z > 4$ This inequality is a conditional inequality since it is true for real numbers greater than 4 but is false if 4 or any value less than 4 is substituted for z.

The graph of the solution set is

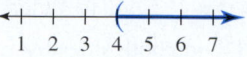

Self-Check 1

Identify each inequality as a conditional inequality, an unconditional inequality, or a contradiction.

a. $2x \le x + x$ **b.** $2x < x + x$ **c.** $2x < x + 1$

How Can I Recognize That an Inequality Is a Contradiction or an Unconditional Inequality?

If solving an inequality results in an inequality with only constants and no variable, then the inequality is either a contradiction or an unconditional inequality. This is illustrated in Example 2.

Example 2   Solving an Inequality When the Variable Is Eliminated

Solve each inequality.

(a) $3x + 4 < 3x + 1$ **(b)** $4v + 3 \ge 4v + 1$

Solution

(a) $3x + 4 < 3x + 1$

     $3x + 4 - 3x < 3x + 1 - 3x$

              $4 < 1$ is a false statement.

Subtracting $3x$ preserves the order. Because the last inequality is a false statement, the original inequality is also a contradiction.

   Answer: There is no solution.

Since the inequality is a contradiction, there are no points to graph.

(b) $4v + 3 \ge 4v + 1$

     $4v + 3 - 4v \ge 4v + 1 - 4v$

              $3 \ge 1$ is a true statement.

Subtracting $4v$ preserves the order. Since the last inequality is an unconditional inequality, the original inequality is also an unconditional inequality.

   Answer: $\mathbb{R}$ (the set of all real numbers).

Because the inequality is an unconditional inequality, the graph of the solution set is the set of all real numbers.

Self-Check 2

Solve each inequality.

a. $2x + 5 \le 2x - 1$ **b.** $2x + 5 \le 3x - 1$ **c.** $2x + 5 \ge 2x - 1$

2. Write the Intersection or Union of Two Intervals

A compound inequality is formed when we place more than one restriction on a variable—for example, on the golf hole shown in the figure, a golfer must hit a drive more than 153 yd but less than 195 yd to clear the water and to land safely on this island green.

How Do I Read and Write Compound Inequalities?

The compound inequality $153 < x < 195$ can be read, "x is between 153 and 195," or it can be read "x is greater than 153 *and x* is less than 195." The key word connecting these two inequalities is the word *and*. The word *and* indicates the intersection of these two sets. The intersection of two high-ways consists of the road common to both highways. The intersection of two sets consists of the elements common to both sets.

Another way to combine two sets is to form their union. The union of sets A and B consists of the elements in either set A *or* set B. The United States is a union of states, the set consisting of all these states.

> **A Mathematical Note**
>
> The set symbols ∩ and ∪ were introduced by Giuseppe Peano in 1888.

Intersection and Union of Two Sets			
Algebraic Notation	Verbally	Graphical Example	Algebraic Example
$A \cap B$	The intersection of A and B is the set that contains the elements that are in *both A and B*.	The intersection of (number line graph) −3−2−1 0 1 2 3 4 5 6 and (number line graph) −3−2−1 0 1 2 3 4 5 6 is (number line graph) −3−2−1 0 1 2 3 4 5 6	$(-3, 4) \cap [0, 6]$ $= [0, 4)$
$A \cup B$	The union of A and B is the set that contains the elements in either A or B (or both).	The union of (number line graph) −3−2−1 0 1 2 3 4 5 6 and (number line graph) −3−2−1 0 1 2 3 4 5 6 is (number line graph) −3−2−1 0 1 2 3 4 5 6	$(-3, 4) \cup [0, 6]$ $= (-3, 6]$

The word *or* can be misunderstood if you think of this word only in the exclusive sense (as in "black *or* white") rather than in the inclusive sense used in the definition of the union of two sets.

| **Example 3** | Determining the Intersection and the Union of Two Sets |

Determine the intersection and union of $\{-3, -0.8, 1.7, \pi\}$ and $\{-5, -0.8, \pi, 11\}$.

Solution

Intersection

$\{-3, -0.8, 1.7, \pi\} \cap \{-5, -0.8, \pi, 11\}$
$= \{-0.8, \pi\}$

The intersection contains only the elements in both of these sets.

Union

$\{-3, -0.8, 1.7, \pi\} \cup \{-5, -0.8, \pi, 11\}$
$= \{-5, -3, -0.8, 1.7, \pi, 11\}$

The union contains all of the elements in one set or the other (or both).

Self-Check 3

Determine the intersection and union of $\left\{ -\dfrac{10}{3}, -2, 0, 3, 4.6 \right\}$ and $\{-2, -1, 0, 1, 2, 3\}$.

| **Example 4** | Determining the Intersection and the Union of Two Intervals |

Determine the intersection and union of the intervals $(-3, 2]$ and $[-1, 4)$.

Solution

Intersection

$(-3, 2] \cap [-1, 4) = [-1, 2]$

Only the numbers from -1 to 2 are included in both intervals.

Union

$(-3, 2] \cup [-1, 4) = (-3, 4)$

The union contains all of the numbers that are in the first interval or the second interval (or both).

Self-Check 4

Determine the intersection and union of $[-4, 2]$ and $(-2, 7)$.

---

3. Solve Compound Inequalities

The compound inequality $a \leq x \leq b$ is equivalent to $x \geq a$ and $x \leq b$. In Example 5, the compound inequality $-3 < 2x - 1 \leq 5$ is equivalent to $2x - 1 > -3$ and $2x - 1 \leq 5$. Compound inequalities can be solved by considering each of the component inequalities.

| **Example 5** | Solving a Compound Inequality |

Solve $-3 < 2x - 1 \leq 5$.

Solution

$$-3 < 2x - 1 \leq 5$$

$\quad -3 < 2x - 1 \qquad$ and $\qquad 2x - 1 \leq 5$

$\quad -2 < 2x \qquad\quad$ and $\qquad\quad 2x \leq 6$

$\quad -1 < x \qquad\qquad$ and $\qquad\qquad x \leq 3$

$\qquad\qquad -1 < x \leq 3$

$-3 < 2x - 1 \leq 5$ is equivalent to $-3 < 2x - 1$ and $2x - 1 \leq 5$. Solve each of these inequalities separately and then form their intersection.

Answer: $(-1, 3]$ whose graph is

Self-Check 5

Solve $-7 \le 1 - 2x < 7$.

Can the Solution in Example 5 Be Condensed?

Yes, note that the same steps are performed on each member of the compound inequality solved in Example 5. Thus this solution can be condensed as illustrated in the Alternate Solution given here.

Alternative Solution

$$-3 < 2x - 1 \le 5$$

$$-3 + 1 < 2x - 1 + 1 \le 5 + 1 \qquad \text{Adding 1 to each member of the compound inequality preserves the order of the inequalities.}$$

$$-2 < 2x \le 6$$

$$\frac{-2}{2} < \frac{2x}{2} \le \frac{6}{2} \qquad \text{Dividing each member of the compound inequality by 2 preserves the order of the inequalities.}$$

$$-1 < x \le 3$$

Can the Alternate Solution Used for Example 5 Be Used to Solve All Compound Inequalities?

No, for some compound inequalities, the variable cannot be isolated in the middle of the compound inequality. In such cases, the compound inequality must be split into two simple inequalities that can be solved individually. The final solution is then formed by the intersection of these individual sets. This procedure is illustrated in Example 6.

Example 6   Solving an Inequality Whose Solution Is the Intersection of Two Sets

Solve $5v - 8 \le 2v + 4 < 4v - 2$.

Solution

This compound inequality is equivalent to

$5v - 8 \le 2v + 4$	and	$2v + 4 < 4v - 2$	Solve each of the inequalities separately.
$3v - 8 \le 4$		$-2v + 4 < -2$	
$3v \le 12$		$-2v < -6$	
$v \le 4$		$v > 3$	

The intersection of $v \le 4$ and $v > 3$ is $3 < v \le 4$.

Answer: $(3, 4]$ whose graph is

Self-Check 6

Solve $3v - 7 \le 2v + 5 \le 6v - 3$.

Remember that it is customary to write the smaller number as the left member of an inequality and the larger value as the right member—in the same order that these numbers occur on the number line. Thus in Example 6, the solution was written in the form $3 < v \le 4$ to match the correct interval notation $(3, 4]$.

In Example 7, we solve an inequality involving the union of two intervals. Note that the connecting word *or* indicates that we are forming the union of two sets.

Example 7   Solving an Inequality Whose Solution Is the Union of Two Sets

Solve $2(w - 5) \geq 3w - 8$ or $3(w + 2) > w + 8$.

Solution

$$2(w - 5) \geq 3w - 8 \qquad \text{or} \qquad 3(w + 2) > w + 8$$
$$2w - 10 \geq 3w - 8 \qquad\qquad\qquad 3w + 6 > w + 8$$
$$2w \geq 3w + 2 \qquad\qquad\qquad\quad 3w > w + 2$$
$$-w \geq 2 \qquad\qquad\qquad\qquad\quad 2w > 2$$
$$w \leq -2 \qquad\qquad\qquad\qquad\quad w > 1$$

Solve each of these inequalities individually. Then graph all the points that are in one set or the other set.

Answer: $(-\infty, -2] \cup (1, \infty)$ whose graph is

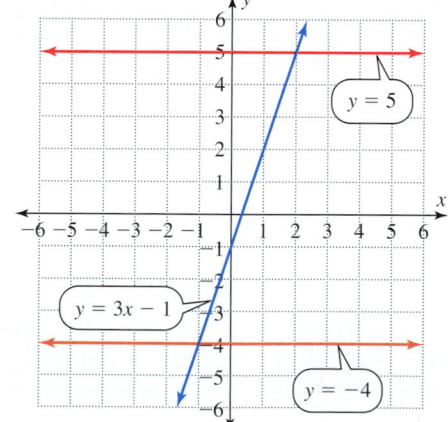

The union symbol $\cup$ is used to indicate that the numbers can be in either the first interval or the second interval.

Self-Check 7

Solve $3v + 7 \leq 2v + 5$ or $4v + 1 \leq 6v - 5$.

In Example 8, we will examine the solution of a compound inequality using multiple perspectives. We will solve this inequality algebraically and graphically and use a table to confirm the solution.

Example 8   Using Multiple Perspectives to Solve a Compound Inequality

Solve $-4 < 3x - 1 \leq 5$ algebraically and graphically. Then check the solution with a table and describe the solution verbally.

Solution

Algebraic Solution

$$-4 < 3x - 1 \leq 5$$
$$-3 < 3x \leq 6$$
$$-1 < x \leq 2$$

Graphical Solution

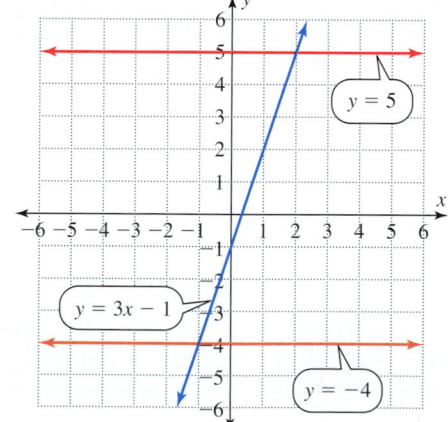

Numerical Check

x	$y = 3x - 1$
-3	-10
-2	-7
-1	-4
0	-1
1	2
2	5
3	8

Verbal Description

The graph of $y = 3x - 1$ is between the lines $y = -4$ and $y = 5$ for values of x in the interval $(-1, 2]$. This agrees with the algebraic solution. The values in the table confirm that $3x - 1$ is greater than -4 and less than or equal to 5 for values of x in the interval $(-1, 2]$.

Answer: $(-1, 2]$ whose graph is

Self-Check 8

Use the given graph and table to solve $-3 \le 2x + 1 < 5$.

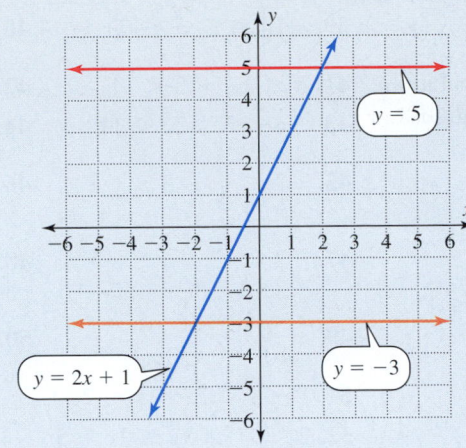

x	y = 2x + 1
−3	−5
−2	−3
−1	−1
0	1
1	3
2	5
3	7
4	9

Self-Check Answers

1. **a.** Unconditional inequality   **b.** Contradiction
   c. Conditional inequality
2. **a.** No solution   **b.** $[6, \infty)$ **c.** All real numbers, $\mathbb{R}$
3. Intersection: $\{-2, 0, 3\}$; Union: $\left\{-\dfrac{10}{3}, -2, -1, 0, 1, 2, 3, 4.6\right\}$
4. Intersection: $(-2, 2]$; Union: $[-4, 7)$

5. $(-3, 4]$
6. $[2, 12]$
7. $(-\infty, -2] \cup [3, \infty)$ whose
   graph is
   $-3 \ -2 \ -1 \ \ 0 \ \ 1 \ \ 2 \ \ 3 \ \ 4$
8. $[-2, 2)$

4.3   Using the Language and Symbolism of Mathematics

1. A _____ inequality is true for some values of the variable in the inequality but false for other values.

2. An inequality that is always true is an _____ inequality.

3. An inequality that is always false is a _____.

4. The _____ inequality $a \le x \le b$ is equivalent to $x \ge a$ _____ $x \le b$.

5. The _____ of sets A and B is the set of elements that are in both A and B.

6. The _____ of sets A and B is the set of elements that are either in set A or in set B.

4.3   Quick Review

Identify each equation as a conditional equation, a contradiction, or an identity.

1. $2(x - 2) + 3 = 3(x - 2) - (x - 5)$
2. $3(x - 2) + 1 = 2(x - 2) + x$
3. $2(x - 2) + 3 = 3(x - 2)$

Solve each equation.

4. $\dfrac{2}{5}(x + 3) + 2 = \dfrac{4}{7}(x + 2)$
5. $0.24(x + 10) + 1.2 = 0.76x - 1.6$

4.3   Exercises

Objective 1 Identify an Inequality That Is a Contradiction or an Unconditional Inequality

In Exercises 1–3, match each inequality with the choice that best describes this inequality.

1. $x > x + 5$ **A.** A conditional inequality
2. $x > 5$ **B.** An unconditional inequality
3. $x < x + 5$ **C.** A contradiction

In Exercises 4–9, each inequality is a conditional inequality, an unconditional inequality, or a contradiction. Identify the type of each inequality and solve it.

4. $2x + 3 < 2(x + 1) + 1$ 5. $2(x + 3) \ge 3(x - 2)$
6. $3x + 2 \le 3(x + 1)$ 7. $2(x + 3) < 2(x - 2)$
8. $4x + 1 \le 3(x + 1)$ 9. $2(x + 3) < 2(x + 4)$

Objective 2 Write the Intersection or Union of Two Intervals

In Exercises 10–13, rewrite each inequality as two separate inequalities, using the word *and* to connect the inequalities.

10. $-3 \le x < 17$ **11.** $2 < x \le 8$

12. $2 \le 3x - 1 \le 14$ **13.** $-13 < 5x + 2 < 7$

In Exercises 14–16, write each inequality expression as a single compound inequality.

14. $x > -2$ and $x \le 5$ **15.** $x \ge 0$ and $x \le 4$

16. $x \ge -1$ and $x < \pi$

In Exercises 17 and 18 determine $A \cap B$ and $A \cup B$.

17. $A = \{-4, 0, 1, 3\}; B = \{-5, 0, 1, 2\}$

18. $A = \{-3, -2, 0, 2\}; B = \{-2, 1, 2, 3\}$

In Exercises 19–22, represent each union of intervals by two separate inequalities, using the word *or* to connect the inequalities.

19. $(-\infty, 5] \cup (7, \infty)$ **20.** $(-\infty, -2) \cup (3, \infty)$

21. $(-2, 5] \cup (7, 10)$ **22.** $[-4, 2] \cup (5, 9]$

In Exercises 23–28, determine $A \cap B$ and $A \cup B$ for the given intervals A and B.

23. $A = (-4, 7], B = [5, 11)$

24. $A = (-3, 6), B = [2, 9]$

25. $A = (-4, \infty), B = [5, 11)$

26. $A = (-3, \infty), B = [2, 9]$

27. $A = (-\infty, -2], B = [-4, \infty)$

28. $A = (-\infty, 5), B = (1, \infty)$

In Exercises 29 and 30, complete the table:

Inequality Notation	Interval Notation	Graph	Verbally
29. $-1 \le x < 6$			
30. $-4 < x \le 2$			

In Exercises 31–34, write an inequality for each statement.

31. a. x is greater than or equal to 1 and less than 3.
   b. x is less than 1 or greater than 3.

32. a. x is greater than -5 and less than or equal to 1.
   b. x is less than -5 or greater than 1.

33. a. x is greater than or equal to 4 or is less than or equal to -2.
   b. x is between -2 and 4.

34. a. x is greater than 0 or is less than -3.
   b. x is between -3 and 0.

Objective 3 Solve Compound Inequalities

In Exercises 35–60, solve each inequality.

35. $-30 < 5x < -20$ **36.** $21 \le 7x \le 42$

37. $-4 < -4x \le 8$ **38.** $-15 \le -5x < 5$

39. $-14 \le -\dfrac{7v}{2} \le 0$ **40.** $-15 \le -\dfrac{5v}{3} \le 0$

41. $3 \le 2w + 3 < 11$ **42.** $1 < 3w - 2 \le 13$

43. $-4 < 6 - 5t < 11$ **44.** $-1 \le 5 - 6t < 17$

45. $-2 \le \dfrac{m}{2} - 3 \le -1$ **46.** $-3 < \dfrac{m}{2} - 2 < -1$

47. $-2 < \dfrac{m - 3}{2} \le -1$ **48.** $-3 \le \dfrac{m - 2}{3} < -1$

49. $\dfrac{x - 3}{10} \ge \dfrac{x - 2}{3}$ **50.** $\dfrac{x - 3}{3} < \dfrac{x + 1}{9}$

51. $x - 1 < 2x < x + 2$

52. $2x - 1 \le 3x \le 2x + 5$

53. $3x - 4 \le 4x + 1 \le 3x + 2$

54. $x - 1 \le 3x + 1 \le x + 5$

55. $5(x + 4) < 28$ and $3(x - 4) > -11$

56. $x - 1 < 2$ and $2 > x + 1$

57. $2x - 1 < -5$ or $3x + 1 > 4$

58. $x - 1 < \dfrac{2x + 1}{3}$ or $\dfrac{4x - 1}{3} \ge x - 1$

59. $2x - 1 < 3x + 1$ or $5x - 6 < 3x - 12$

60. $7(4x - 5) - 16 < 5(8x - 3)$ or
    $11 - 4(2x - 3) \le -3(2 - 7x)$

Review and Concept Development

In Exercises 61–64, use the graph to solve each compound inequality.

61. $2 \le x + 3 < 5$ **62.** $-4 \le x - 2 \le 2$

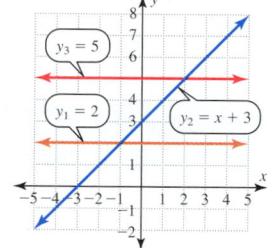

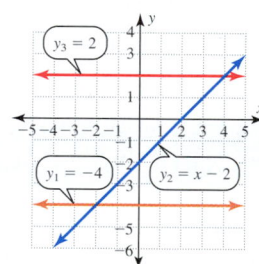

63. $x - 2 < 2x - 1 < x + 3$

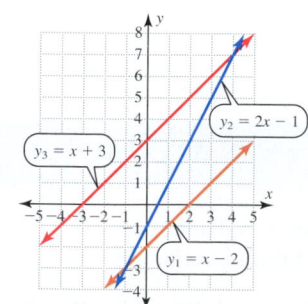

64. $-\dfrac{x}{2} - 3 < \dfrac{x}{2} + 2 < -\dfrac{x}{2} + 4$

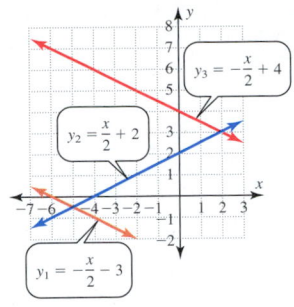

In Exercises 65–68, use the table to solve each inequality.

65. $-1 < 4 - x < 3$

x	y = 4 − x
0	4
1	3
2	2
3	1
4	0
5	−1
6	−2

66. $0 < 3 - x < 4$

x	y = 3 − x
−3	6
−2	5
−1	4
0	3
1	2
2	1
3	0

67. $-3 < 2x + 1 \le 7$

x	y = 2x + 1
−3	−5
−2	−3
−1	−1
0	1
1	3
2	5
3	7

68. $0 \le \dfrac{x}{2} + 1 \le 2$

x	$y = \dfrac{x}{2} + 1$
−3	−0.5
−2	0.0
−1	0.5
0	1.0
1	1.5
2	2.0
3	2.5

In Exercises 69–72, solve each inequality algebraically and graphically. Then check the solution with a table and describe the solution verbally.

69. $0 \le x + 5 < 6$ **70.** $-1 < x - 3 \le 1$

71. $-7 \le x - 8 \le -6$ **72.** $6 < x + 7 < 9$

Estimate Then Calculate

In Exercises 73–76, mentally estimate the solution of each inequality and then determine the exact solution.

Problem	Mental Estimate	Exact Solution
73. $5.719 < 3.01x < 9.331$		
74. $2.01 \le x + 0.99 \le 5.08$		
75. $-4.179 \le -1.99x < 5.771$		
76. $7.81 < -7.1x \le 13.49$		

Connecting Concepts to Applications

77. Length of a Golf Shot Write a compound inequality that expresses the length of the tee shot that a golfer

must make to clear the water and to land safely on the green of the golf hole shown here.

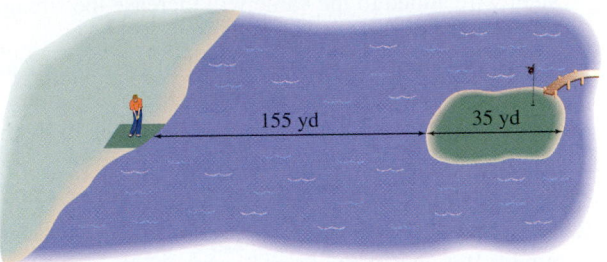

155 yd   35 yd

78. Length of a Golf Shot A golfer must hit a shot more than 170 yd to clear a lake in front of a green. The depth of the green is only 20 yd. Write a compound inequality that expresses the length of the tee shot that a golfer must make to clear the water and to land safely on the green of this golf hole. (*Hint:* See the figure in Exercise 77.)

79. Length of a Golf Shot A golfer is faced with a decision on the 15th fairway of the Lake of the Woods golf course. A safe shot less than 150 yd will land short of a lake surrounding the green. A more aggressive shot must travel more than 220 yd and less than 270 yd to clear the lake and to stay on the green. Write a union of two intervals that gives the lengths of shots that will not land in the lake.

80. Length of a Golf Shot A golfer is faced with a decision on the third fairway of the Springfield golf course. A safe shot less than 100 yd will land short of a lake surrounding the green. A more aggressive shot must travel more than 195 yd and less than 220 yd to clear the lake and to stay on the green. Write a union of two intervals that gives the lengths of shots that will not land in the lake.

81. Elevator Cable Strength The engineering specifications for many products require a safety factor that exceeds the expected value ever needed for the product to perform safely. The strength needed for one of the cables used on an elevator is 2,000 lb. Using a safety factor of 2 would require the cable to have a strength of 4,000 lb, and a safety factor of 3 would require the cable to have a strength of 6,000 lb. Each steel wire added to the cable will increase the strength of the cable by approximately 250 lb.

a. Using x to represent the number of wires, write a single compound inequality that represents the number of wires that should be put in the cable to produce a safety factor from 2 to 3.

b. Solve this inequality for x and interpret this solution.

82. Shoe Sizes A shoe store that caters to long-distance runners imports shoes from Europe for its customers. The store plans to stock men's U.S. sizes 7 through 13. If the European size is represented by x, then the U.S. size is given by $0.8x - 24.5$.

a. Solve the inequality $7 \le 0.8x - 24.5 \le 13$ to determine all European shoe sizes the store must stock in order to fit U.S. sizes from 7 to 13.

b. Shaquille O'Neal's European shoe size is 58. Does the store plan to stock his size?

83. Strength of a Belt The drive belt for a wood chipper needs a minimum strength of 200 lb. However, there is also a required maximum strength of 400 lb as the equipment is designed for the belt to shear to prevent major damage to other components if the machine encounters a foreign object such as a steel bar. The graphs of $y_1 = 200$ and $y_2 = 400$ display the strength limits, and the graph of $y_3 = f(x)$ displays the strength of the belt based on the number of separate fiber strands embedded into the belt. Use the graphs to solve $200 \le y_3 \le 400$ and to interpret the meaning of this solution.

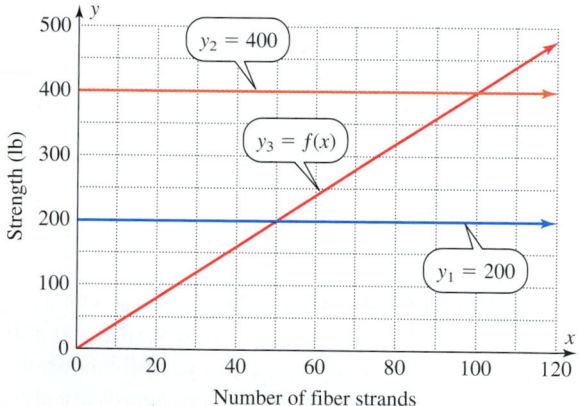

84. Company Bonuses A local automobile dealership calculated a holiday bonus for each salesperson based on the number of automobiles that the person sold during the year. The bonus records show that each salesperson received at least $400 and that the most anyone received was $1,500. The graphs of $y_1 = 400$ and $y_2 = 1,500$ display the minimum and maximum bonus amounts, and the graph of $y_3 = f(x)$ displays the bonus paid based on the number of cars sold. Use the

graph to solve $400 \le y_3 \le 1,500$ and to interpret the meaning of this solution.

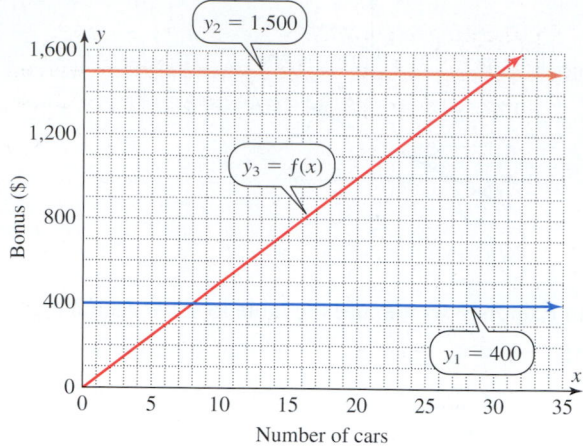

In Exercises 85–88, translate each problem into a linear inequality and then solve this inequality.

85. Temperature Limits The warranty for a computer specifies that the temperature of its environment must be maintained between 41° and 95°F. Find the temperatures that are within this range, expressed in degrees Celsius. Solve $41 < \dfrac{9}{5}C + 32 < 95$ to find the acceptable range of Celsius temperatures.

86. Glass Blowing Temperatures The range of acceptable Celsius temperatures for working one type of glass by a glass blower is from 650° to 700°C. Solve
$650 \le \dfrac{5}{9}(F - 32) \le 700$ to find the acceptable range of Fahrenheit temperatures.

87. Perimeter of a Triangle Two sides of a triangle must be 8 and 12 m. The perimeter must be greater than 24 m and less than 40 m. Determine the length that can be used for the third side.

88. Perimeter of a Rectangle The width of a rectangle must be 9 m. If the perimeter must be between 44 and 64 m, determine the length that can be used for the rectangle.

Group discussion questions

89. Communicating Mathematically If possible, list all values of x for which $0.9 < x < 1.0$. If this is not possible, explain why.

90. Challenge Question Can you determine, without direct calculations, which is larger, $25(\pi - 2)$ or $25(4 - \pi)$?

91. Challenge Question Can you determine, without direct calculation, which is greater, the height or the circumference of the can of three tennis balls shown? (Assume that one ball is touching the top of the can and one ball is touching the bottom with no extra space between the balls.)

92. Communicating Mathematically
a. Explain why this graph indicates that $1.4(x + 0.5) \geq 1.4x + 1.9$ is a contradiction.

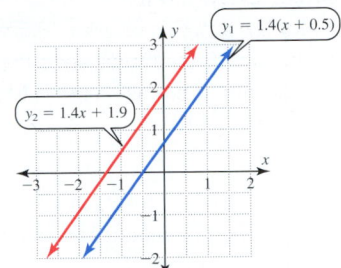

b. Explain why this table indicates that $2.5(2x - 4) \leq 5(x - 1)$ is an unconditional inequality.

x	$y = 2.5(2x - 4)$	$y = 5(x - 1)$
-2	-20	-15
-1	-15	-10
0	-10	-5
1	-5	0
2	0	5
3	5	10

4.3 Cumulative Review

1. Evaluate $5|2x - 11| - 8$ for $x = -4$.
2. Evaluate $|-5| + |3| + |-5 + 3|$.

Fill in each blank with either $<$, $=$, or $>$.

3. 2^3 __ 3^2 **4.** 2^4 __ 4^2 **5.** 2^5 __ 5^2

Section 4.4 Solving Absolute Value Equations and Inequalities

Objectives:

1. Use absolute value notation to represent intervals.
2. Solve absolute value equations and inequalities.

In industrial applications, there is generally an allowance for a small variation, or leeway, between the standard size for a part or component and the actual size. This acceptable variation is called the **tolerance.** For a 42-cm steel rod, for example, the tolerance may be 0.05 cm. A worker on an assembly line might merely lay the rod on a table that is marked to indicate the upper and lower limits of tolerance, as shown in the figure. However, an engineer doing calculations would need to describe this tolerance algebraically.

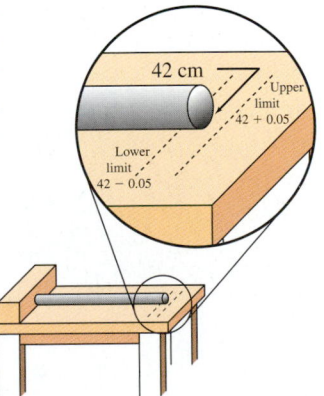

1. Use Absolute Value Notation to Represent Intervals

How Can I Use Mathematical Notation to Express Tolerances?

One way to write a tolerance algebraically is to use absolute value notation. The absolute value of x, denoted by $|x|$, is the distance on the number line between 0 and x. Thus the absolute value inequality $|x| < 4$ represents the interval of points for which the distance from the origin is less than 4 units. Likewise, the absolute value inequality $|x| > 4$ represents the two intervals containing points for which the distance from the origin is more than 4 units. The accompanying box illustrates these concepts.

Absolute Value Expressions

For any real number x and any nonnegative real number d:

Algebraically	Verbally	Algebraic Example	Graphical Example
$\lvert x \rvert = d$	The distance from 0 to x is d units.	$\lvert x \rvert = 4$ if $x = -4$ or $x = 4$	4 units left of 0 4 units right of 0 −4−3−2−1 0 1 2 3 4
$\lvert x \rvert < d$	The distance from 0 to x is less than d units.	$\lvert x \rvert < 4$ if $-4 < x < 4$ $x > -4$ and $x < 4$	Points less than 4 units from 0 −4−3−2−1 0 1 2 3 4
$\lvert x \rvert > d$	The distance from 0 to x is more than d units.	$\lvert x \rvert > 4$ if $x < -4$ or $x > 4$	Points more than 4 units left of 0 Points more than 4 units right of 0 −4−3−2−1 0 1 2 3 4

Example 1   Writing the Solution for an Absolute Value Inequality Using Interval Notation

Use interval notation to represent the real numbers that are solutions of each inequality.

Solution

(a) $\lvert x \rvert \leq 8$ $[-8, 8]$ The distance from 0 to x is less than or equal to 8 units.

(b) $\lvert x \rvert > 7$ $(-\infty, -7) \cup (7, \infty)$ The distance from 0 to x is 7 or more units.

Self-Check 1

Use interval notation to represent the real numbers that are solutions of these inequalities.

a. $\lvert x \rvert < 5$ **b.** $\lvert x \rvert \geq 5$

The distance between two real numbers on the number line can be determined by subtraction.

Questions: Can you determine the distance between these real numbers on the number line?

    A: The distance from 2 to 5.
    B: The distance from -5 to -2.
    C: The distance from -2 to -5.

Graphically   **Absolute Value of Difference**

Answers: **A:**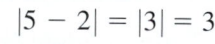

3 units
−2 −1 0 1 2 3 4 5 6

$\lvert 5 - 2 \rvert = \lvert 3 \rvert = 3$

B: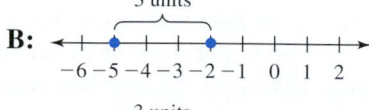

3 units
−6 −5 −4 −3 −2 −1 0 1 2

$\lvert -2 - (-5) \rvert = \lvert 3 \rvert = 3$

C: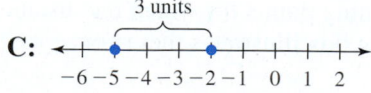

3 units
−6 −5 −4 −3 −2 −1 0 1 2

$\lvert -5 - (-2) \rvert = \lvert -3 \rvert = 3$

If a is larger than b, then the distance from a to b is given by the difference $a - b$. To indicate that the distance between a and b is always nonnegative, we can denote this distance by $|a - b|$. In particular, $|x - 0|$, or $|x|$, can be interpreted as the distance between x and the origin. Likewise, $|a + b| = |a - (-b)|$ equals the distance between a and $-b$. This concept of distance is examined in Examples 2 and 3.

Example 2   Representing Distance Using Absolute Value Notation

Use absolute value notation to represent the distance between each pair of points.

Solution

(a) -5 and 1 $\quad |1 - (-5)| = |6| = 6$ or

$\qquad\qquad\qquad |-5 - 1| = |-6| = 6$

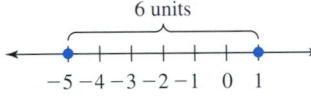

(b) v and w $\quad |v - w|$ or $|w - v|$ $\quad$ Note that $|v - w| = |w - v|$.

Self-Check 2

Use absolute value notation to represent the distance between these real numbers.

a. 8 and x **b.** x and y **c.** v and $-w$

When you see a simple absolute value equation or inequality, try to think of this expression in terms of distance.

Example 3   Using Distance to Interpret an Absolute Value Expression

Using the geometric concept of distance, interpret **(a)** $|x - 3| = 4$, **(b)** $|x - 3| < 4$, and **(c)** $|x - 3| > 4$.

Solution

(a) $|x - 3| = 4$ indicates that the distance between x and 3 is 4 units. This means that x is either 4 units to the left of 3 or 4 units to the right of 3.

Algebraically

Left 4 Units	*or*	*Right 4 Units*
$x - 3 = -4$	or	$x - 3 = +4$
$x = -1$	or	$x = 7$

Answer: $x = -1$ or $x = 7$

Graphically

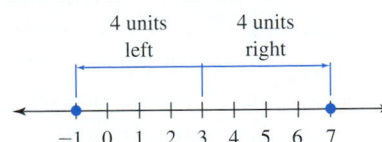

Do these values check?

(b) $|x - 3| < 4$ indicates that the distance between x and 3 is less than 4 units. This means that x is less than 4 units to the left of 3 and less than 4 units to the right of 3.

Algebraically

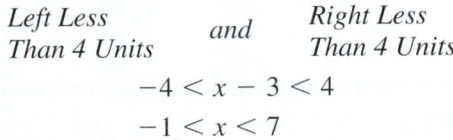

$$-4 < x - 3 < 4$$
$$-1 < x < 7$$

Answer: $(-1, 7)$

Graphically

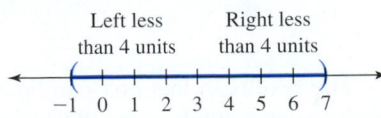

Can you check this interval?

(c) $|x - 3| > 4$ indicates that the distance between x and 3 is more than 4 units. This means that x is either more than 4 units to the left of 3 or more than 4 units to the right of 3.

Algebraically

Left More	*or*	*Right More*
Than 4 Units		*Than 4 Units*
$x - 3 < -4$	or	$x - 3 > 4$
$x < -1$	or	$x > 7$

Answer: $(-\infty, -1) \cup (7, \infty)$

Graphically

Left more than 4 units   Right more than 4 units

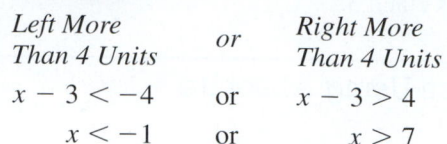

$$-1 \quad 0 \quad 1 \quad 2 \quad 3 \quad 4 \quad 5 \quad 6 \quad 7$$

Can you check the union of these two intervals?

Self-Check 3

Match each absolute value expression with its solution.

a. $\|x - 7\| > 4$	**A.** $x = 3$ or $x = 11$
b. $\|x - 7\| = 4$	**B.** $(3, 11)$
c. $\|x - 7\| < 4$	**C.** $(-\infty, 3) \cup (11, \infty)$

---

2. Solve Absolute Value Equations and Inequalities

The results observed in Example 3 are summarized in the box. Note that $|x - a| < d$ corresponds to the intersection of two separate intervals. This requires the word *and* to describe $x - a > -d$ and $x - a < d$. Also note that $|x - a| > d$ corresponds to the union of two separate intervals. This requires the word *or* to describe $x - a < -d$ or $x - a > d$. Again the visual aid of a graph and a description in terms of distance should help you to use the correct notation.

Solving Absolute Value Equations and Inequalities*

For any real numbers x and a and positive real number d:

Absolute Value Expression	Verbally	Graphically	Equivalent Expression
$\|x - a\| = d$	x is d units either left or right of a.	$a - d \quad a \quad a + d$	$x - a = -d$ or $x - a = +d$
$\|x - a\| < d$	x is less than d units from a.	$a - d \quad a \quad a + d$	$x - a > -d$ and $x - a < d$, $-d < x - a < +d$
$\|x - a\| > d$	x is more than d units from a.	$a - d \quad a \quad a + d$	$x - a < -d$ or $x - a > +d$

*Similar statements can also be made about the order relations less than or equal to ($\leq$) and greater than or equal to ($\geq$). Expressions with d negative are examined in the group exercises at the end of this section.

   The results in this box can be applied to solve more complicated expressions within absolute value symbols, as illustrated in Example 4.

Example 4	Solving an Absolute Value Equation

Solve $|2x - 3| = 31$.

Solution

Left 31 Units	*or*	*Right 31 Units*	
$2x - 3 = -31$	or	$2x - 3 = +31$	To obtain all solutions for this equation, we must consider both of these equations.
$2x = -28$		$2x = 34$	
$x = -14$		$x = 17$	

Answer: $x = -14$ or $x = 17$ Do these values check?

Self-Check 4

Solve $|4x - 6| = 10$.

What Procedure Do I Use to Solve a More Complicated Absolute Value Inequality?

First isolate the absolute value expression on the left side of the expression, as illustrated in Example 5.

Example 5	Solving an Absolute Value Inequality

Solve $\left|\dfrac{x + 2}{2}\right| + 5 > 6$.

Solution

$$\left|\dfrac{x + 2}{2}\right| + 5 > 6$$

$$\left|\dfrac{x + 2}{2}\right| + 5 - 5 > 6 - 5$$

Subtract 5 from both sides of the inequality to isolate the absolute value expression on the left side of the inequality. Subtracting 5 preserves the order of the inequality.

$$\left|\dfrac{x + 2}{2}\right| > 1$$

Left More Than 1 Unit	*or*	*Right More Than 1 Unit*	
$\dfrac{x + 2}{2} < -1$	or	$\dfrac{x + 2}{2} > 1$	Multiplying both sides of an inequality by 2 preserves the order of the inequality. Subtracting 2 from both sides of an inequality preserves the order of the inequality.
$x + 2 < -2$	or	$x + 2 > 2$	
$x < -4$	or	$x > 0$	

Answer: $(-\infty, -4) \cup (0, \infty)$

Self-Check 5

Solve $\left|\dfrac{2x - 3}{5}\right| - 7 \geq -6$.

What Does a Graph or a Table for an Absolute Value Inequality Reveal?

We can compare y-values in two-dimensional graphs to solve an absolute value inequality in one variable. We can also compare values in a table to solve an absolute value inequality. This is illustrated in Example 6.

Example 6 Using Multiple Perspectives to Solve an Absolute Value Inequality

Solve $|4x - 6| < 10$ algebraically and graphically. Then check the solution with a table.

Solution

Algebraic Solution

Left Less Than 10 Units **and** *Right Less Than 10 Units*

$$-10 < 4x - 6 < 10$$
$$-4 < 4x < 16$$
$$-1 < x < 4$$

$|x - a| < d$ is equivalent to $-d < x - a < d$. Adding 6 to each member of an inequality preserves the order of the inequality. Dividing each member of an inequality by 4 preserves the order of the inequality.

Graphical Solution

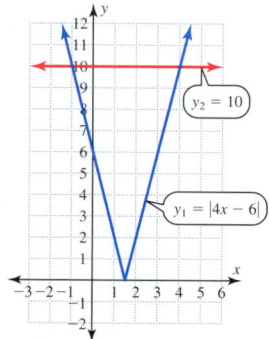

The graph of $y_1 = |4x - 6|$ is below $y_2 = 10$ for x between -1 and 4.

Numerical Check

x	$y_1 = |4x - 6|$	<, =, or >	$y_2 = 10$
-2	14	>	10
-1	10	=	10
0	6	<	10
1	2	<	10
2	2	<	10
3	6	<	10
4	10	=	10
5	14	>	10

You may wish to use a calculator or a spreadsheet to fill in the numerical values and then insert by hand either <, =, or >.

Note that $y_1 < y_2$ for x between -1 and 4.

Answer: $(-1, 4)$

Self-Check 6

Use this table to solve $|2x - 1| \geq 3$.

x	$y_1 = |2x - 1|$	<, =, or >	$y_2 = 3$
-2	5		3
-1	3		3
0	1		3
1	1		3
2	3		3
3	5		3

If $|a| = |b|$, then a and b are equal in magnitude but their signs can either agree or disagree. Thus $|a| = |b|$ is equivalent to $a = b$ or $a = -b$. In Example 7, we will use a table to reveal the x-values for which the two sides of an equation are equal. Similarly, we will use the point of intersection of two graphs to reveal the x-value for which the two sides of an equation are equal.

Example 7   Using Multiple Perspectives to Solve an Equation Involving Two Absolute Value Expressions

Solve $|3x - 5| = |5x - 7|$.

Solution

Algebraic Solution

$$3x - 5 = 5x - 7 \quad \text{or} \quad 3x - 5 = -(5x - 7)$$
$$-2x = -2 \qquad\qquad\qquad 3x - 5 = -5x + 7$$
$$x = 1 \qquad\qquad\qquad\qquad 8x = 12$$
$$x = \frac{3}{2}$$

Graphical Solution

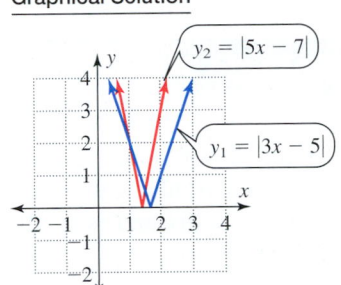

Numerical Check

x	$y_1 = |3x - 5|$	$y_2 = |5x - 7|$
0.0	5.0	7.0
0.5	3.5	4.5
1.0	2.0	2.0
1.5	0.5	0.5
2.0	1.0	3.0
2.5	2.5	5.5
3.0	4.0	8.0

Verbal Description

From the table, $y_1 = y_2$ for $x = 1$ and $x = 1.5$, and the graphs of y_1 and y_2 intersect at $x = 1$ and $x = 1.5$.

Answer: $x = 1$ or $x = \dfrac{3}{2}$

Self-Check 7

Solve $|x - 3| = |2x|$.

Some of the earlier examples in this section start with an absolute value inequality and then produce the interval of values that satisfy this inequality. Example 8 goes in the other direction. It starts with an interval of values and then produces an absolute value inequality that represents this interval.

Example 8   Representing an Interval Using Absolute Value Notation

Write an absolute value inequality to represent the interval $(-\infty, -3] \cup [7, \infty)$.

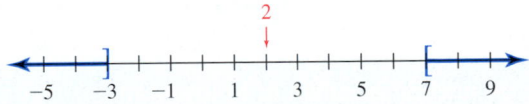

Solution

Step 1. Distance between endpoints:
$$|7 - (-3)| = |10| = 10$$

Determine the distance from one interval endpoint to the other.

Step 2. Midpoint between endpoints:
$$-3 + 5 = 2 \quad \text{and} \quad 7 - 5 = 2$$

Determine the midpoint of the interval. The midpoint of 2 is 5 units (one-half of 10 units) from each endpoint.

Step 3. Absolute value inequality:
$$|x - 2| \geq 5$$

Write the absolute value inequality using 2 as the midpoint and 5 as the distance to each endpoint.

Answer: $|x - 2| \geq 5$ represents the interval $(-\infty, -3] \cup [7, \infty)$.

Self-Check 8

Write an absolute value inequality to represent the interval $[-8, 4]$.

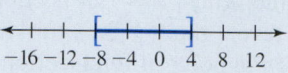

How Can I Use Absolute Value Notation to Represent Tolerances?

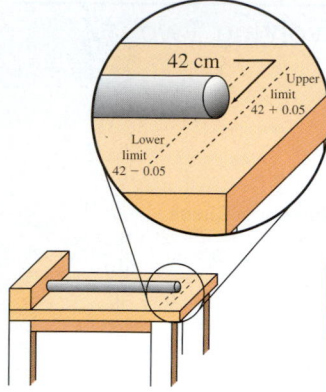

At the beginning of this section we introduced **tolerance,** the acceptable variation between the standard size for a part or component and the actual size. For example, the tolerance may be 0.05 cm for a 42-cm steel rod. An engineer can describe this tolerance algebraically as $|r - 42| \leq 0.05$. This is an algebraic statement that the length of the rod r and the desired length of 42 cm can differ by at most 0.05 cm. In Example 9, we examine the tolerance for a drug dosage.

Example 9 Representing a Tolerance Interval Using Absolute Value Notation

The amount of a drug that is placed in a bag for intravenous administration to a patient should be 20 mL, with a tolerance of ± 0.5 mL. Express this interval as a compound linear inequality and as an absolute value inequality, using d to represent the volume of the drug.

Solution

Algebraically

$$-0.5 \leq d - 20 \leq 0.5$$
$$|d - 20| \leq 0.5$$

Graphically

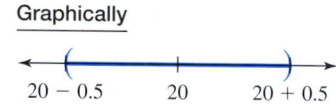

$20 - 0.5 \quad 20 \quad 20 + 0.5$

Verbally

The drug can be from 0.5 mL below 20 mL to 0.5 mL above 20 mL.

Self-Check 9

The weight of one of the camera lenses for a Mars rover should be 7 grams with a tolerance of 0.003 gram. Express this tolerance interval using an absolute value inequality.

Self-Check Answers

1. **a.** $(-5, 5)$ **b.** $(-\infty, -5] \cup [5, \infty)$
2. **a.** $|x - 8|$ **b.** $|x - y|$ **c.** $|v + w|$
3. **a.** C **b.** A **c.** B
4. $x = -1$ or $x = 4$
5. $(-\infty, -1] \cup [4, \infty)$
6. $(-\infty, -1] \cup [2, \infty)$
7. $x = -3$ or $x = 1$
8. $|x + 2| \leq 6$
9. $|x - 7| \leq 0.003$

4.4 Using the Language and Symbolism of Mathematics

1. The absolute value of x is the _____ on the number line between 0 and x.

2. Fill in these blanks with $=$, $<$, or $>$.
    a. The distance from 0 to x is less than d units.
    $|x|$ _____ d
    b. The distance from 0 to x is d units. $|x|$ _____ d
    c. The distance from 0 to x is greater than d units.
    $|x|$ _____ d

3. Write an absolute value equation that indicates x is d units left or right of a. _____

4. The acceptable variation between the standard size for a part and the actual size of the part is called the

    _____.

5. To solve an absolute value equation or inequality, first _____ the absolute value expression on the left side of the expression.

4.4 | Quick Review

Evaluate each absolute value expression.

1. $|34|$ **2.** $|-34|$ **3.** $|-3 + 4|$ **4.** $|-3| + |4|$ **5.** $|-3| - |4|$

4.4 | Exercises

Objective 1 Use Absolute Value Notation to Represent Intervals

In Exercises 1–22, write an absolute value equation or inequality to represent each set of points.

1.
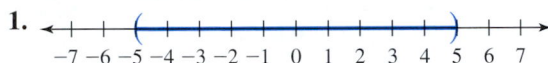

2.

3.

4.

5. The points between -4 and 4.

6. The numbers that are at least -3 and at most 3.

7. The numbers that are at least -5 and at most 5.

8. The points between -8 and 8.

9.

10.

11.

12.

13.

14.

15.

16.

17. $[-3, 3]$ **18.** $(-9, 9)$

19. $(-\infty, -3) \cup (3, \infty)$ **20.** $(-\infty, -9] \cup [9, \infty)$

21. $(-7, 7)$ **22.** $[-\pi, \pi]$

In Exercises 23 and 24, use interval notation to represent the real numbers that are solutions of each inequality.

23. a. $|x| < 3$ **b.** $|x| \le 3$ **c.** $|x| \ge 3$

24. a. $|x| < 10$ **b.** $|x| \le 10$ **c.** $|x| > 10$

Objective 2 Solve Absolute Value Equations and Inequalities

In Exercises 25–52, solve each equation and inequality.

25. $|x| = 6$ **26.** $|x| = 13$

27. $|x| = 0$ **28.** $|2x| = 0$

29. $|x - 2| = 5$ **30.** $|x - 7| = 8$

31. $|x + 2| = 5$ **32.** $|x + 7| = 8$

33. $|3x + 4| \ge 2$ **34.** $|3x - 5| \ge 1$

35. $\left|\dfrac{-7x}{2}\right| \le 14$ **36.** $\left|\dfrac{-5y}{11}\right| \le 55$

37. $|2x + 3| - 4 < 1$ **38.** $|2x - 5| - 2 < 3$

39. $|2(x + 3) - 4| < 1$ **40.** $|2(x - 5) - 2| < 3$

41. $|2x + 1| + 4 > 11$ **42.** $|2x - 1| + 6 > 9$

43. $\left|\dfrac{x - 1}{7}\right| \ge 14$ **44.** $\left|\dfrac{2x - 1}{3}\right| \ge 6$

45. $|2(2x - 1) - (x - 3)| < 11$

46. $|3(2x + 1) - 4(3x - 2)| < 5$

47. $|2x| = |x - 1|$ **48.** $|3x| = |4 - x|$

49. $|x - 1| = \left|\dfrac{x + 1}{2}\right|$ **50.** $|x + 3| = \left|\dfrac{x - 1}{2}\right|$

51. $|2x + 3| = |3x - 1|$ **52.** $|4x + 2| = |5x - 1|$

Review and Concept Development

In Exercises 53–56, use the graph to solve each equation and inequality.

53. a. $|x - 2| = 4$
    b. $|x - 2| < 4$
    c. $|x - 2| > 4$

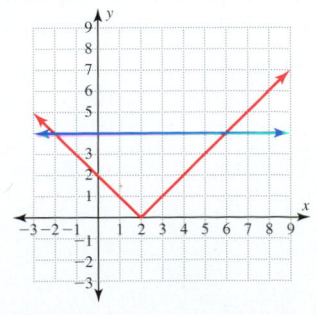

54. a. $|x + 3| = 2$
 b. $|x + 3| \le 2$
 c. $|x + 3| > 2$

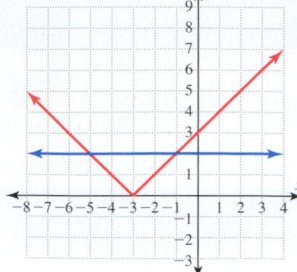

55. a. $\left|\dfrac{x - 4}{2}\right| = 1$

 b. $\left|\dfrac{x - 4}{2}\right| < 1$

 c. $\left|\dfrac{x - 4}{2}\right| \ge 1$

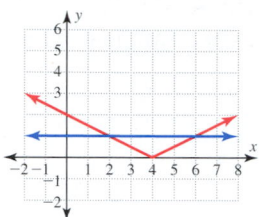

56. a. $|3x - 6| = 3$
 b. $|3x - 6| \le 3$
 c. $|3x - 6| \ge 3$

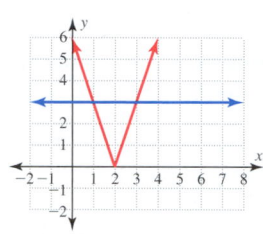

In Exercises 57 and 58, complete the table by placing $<$, $=$, or $>$ in the third column. Then determine the x-values that satisfy each equation and inequality. In each case y_1 represents the left side of the expression and $\dfrac{1}{2}$ represents the right side of the expression.

57. a. $|x + 2| = 2$
 b. $|x + 2| \le 2$
 c. $|x + 2| \ge 2$

x	y_1	$<, =, $ or $>$	y_2
-5	3		2
-4	2		2
-3	1		2
-2	0		2
-1	1		2
0	2		2
1	3		2

58. a. $|2x + 1| = 3$
 b. $|2x + 1| < 3$
 c. $|2x + 1| > 3$

x	y_1	$<, =, $ or $>$	y_2
-3	5		3
-2	3		3
-1	1		3
0	1		3
1	3		3
2	5		3
3	7		3

In Exercises 59 and 60, use the table of values to determine the x-values that satisfy each equation and inequality.

59. a. $|x - 5| = 10$
 b. $|x - 5| < 10$
 c. $|x - 5| \ge 10$

x	y_1
-10	15
-5	10
0	5
5	0
10	5
15	10
20	15

60. a. $|x + 25| = 50$
 b. $|x + 25| \le 50$
 c. $|x + 25| > 50$

x	y_1
-100	75
-75	50
-50	25
-25	0
0	25
25	50
50	75

Multiple Representations

In Exercises 61–64, solve each inequality algebraically and graphically. Then use a table to check your solution.

61. $|x + 1| < 1$ **62.** $|x + 2| < 1$

63. $|2x - 1| < 3$ **64.** $|2x + 3| < 5$

In Exercises 65–70, write an absolute value inequality to represent each interval.

65. $(-12, 6)$ **66.** $(-3, 11)$

67. $[-4, 26]$ **68.** $[12, 32]$

69. $(-\infty, -2) \cup (6, \infty)$ **70.** $(-\infty, -6] \cup [4, \infty)$

Estimate Then Calculate

In Exercises 71–74, mentally estimate the solution of each equation to the nearest integer, and then calculate the exact solution.

Problem	Mental Estimate	Exact Solution
71. $|2.01x| = 9.849$		
72. $|3.9x| = 39.39$		
73. $|x + 7.93| = 15.03$		
74. $|x - 9.01| = 29.13$		

Connecting Concepts to Applications

75. Length of a Truck Spring The engineering specifications for a truck specify that one of the springs in the suspension system should compress to absorb shocks and extend when the wheels go into holes. The distance x in inches between the bottom of the spring and the ground is given by $|x - 22| < 6$. Write an interval that gives all possible distances satisfying the engineering specifications.

76. Length of a Transmission Cable The engineering specifications for an electrical transmission cable specify that the cable that spans the distance between two towers and crosses an interstate highway must be able to contract during cold weather and expand during warm weather. The distance x in feet between the bottom of the cable over the interstate and the highway below is given by $|x - 35| < 3$. Write an interval that gives all possible distances satisfying the engineering specifications.

In Exercises 77–82, express the tolerance interval as an absolute value inequality, and determine the lower and upper limits of the interval.

77. Tolerance of a Piston Rod The length of a piston rod in an automobile engine is specified by contract to the manufacturer to be 15 cm with a tolerance of ± 0.001 cm.

78. Tolerance of a False Tooth A dental clinic has received an order for a dental implant that is 1.3 cm long with a tolerance of ± 0.05 cm.

79. Tolerance of a Soft Drink Bottle The specifications for a bottle-filling machine in a soft drink plant call for the machine to dispense 16 oz into a bottle with a tolerance of ± 0.25 oz.

80. Tolerance of a Fuel Tank A military contract for an aircraft fuel tank specifies a capacity of 420 L with a tolerance of ± 1 L.

81. Tolerance of a Temperature Control A temperature control at a pharmaceutical manufacturing plant is set to keep the temperature at 25°C with a tolerance of ± 2°C.

82. Tolerance of a Voltage Control The electrical supply to an important Internet provider is monitored to keep the line feed at 120 V with a tolerance of ± 1 V.

83. Strength Limits on a Drive Belt The drive belt for a mulching machine must be able to handle a load of at least 3,000 lb. However, the design for the machine is for the belt to break on any load of 9,000 lb or greater. This is done so the belt is sacrificed to prevent damage to the rest of the machine if foreign material is input into the mulching machine. The belt consists of synthetic strands attached to a nylon core that can support a load of 250 lb. Each synthetic strand added to the cable adds 500 lb to the load the belt can handle.

a. Write a compound inequality to model this problem.
b. Determine the number of synthetic strands to put in this belt.

84. Strength Limits on a Drive Belt Rework Exercise 83, assuming that the drive belt for a mulching machine must be able to handle a load of at least 4,000 lb and that the design for the machine is for the belt to break on any load of 10,000 lb or greater.

a. Write a compound inequality to model this problem.
b. Determine the number of synthetic strands to put in this belt.

85. Safety Factor for a Bridge Cable Some bridge cables are composed of several strands with each strand containing several wires. The design specifications for a bridge call for using a cable that can handle a load of 50,000 lb. Using a safety factor of 2 would require the cable to have a strength of 100,000 lb, and a safety factor of 5 would require the cable to have a strength of 250,000 lb. For one type of cable in this load range, the equation $y = 250x - 150$, where x is the strand diameter in inches, can be used to approximate the load strength y in thousands of pounds.

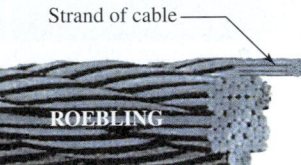
Strand of cable

a. Write a compound linear inequality to represent strand diameters that would satisfy safety factors between 2 and 5 for this bridge cable.
b. Solve this inequality and interpret your answer.

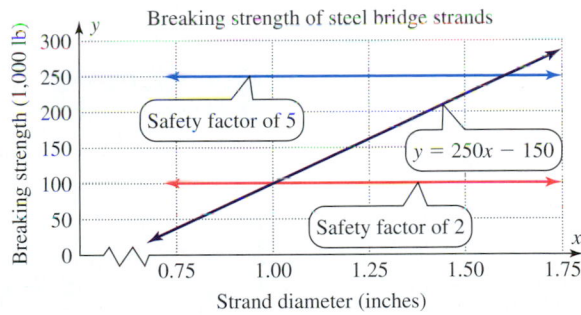
Breaking strength of steel bridge strands

Group discussion questions

86. Discovery Question Solve these special-case absolute value equations and inequalities.
a. $|x| = -1$ **b.** $|x| < -1$ **c.** $|x| > -1$

Then generalize these results to give the solution of each of these inequalities for real numbers x and a and any negative real number d.
d. $|x - a| = d$ **e.** $|x - a| < d$ **f.** $|x - a| > d$

87. Error Analysis A student solved $|3x - 5| = 26$ by using the following steps:

$$|3x - 5| = 26$$
$$3x + 5 = 26$$
$$3x = 21$$
$$x = 7$$

Identify any steps where this student made an error, and then determine the correct solution to this equation.

4.4 | **Cumulative Review**

Use the given information to graph each line.

1. The line through $(-2, -1)$ and $(3, 2)$.

2. A line with intercepts $(3, 0)$ and $(0, 4)$.

3. The line through $(-2, -1)$ with slope $\dfrac{3}{5}$.

4. The horizontal line through $(-2, -1)$.

5. The line defined by $2x + 3y = 6$.

Section 4.5 | Graphing Systems of Linear Inequalities in Two Variables

Objectives:

1. Identify a solution of a linear inequality in two variables.
2. Graph a linear inequality in two variables.
3. Graph a system of linear inequalities.

A Mathematical Note

Winifred Edgerton Merrill (1862–1951) was the first American woman to receive a Ph.D. in mathematics. She studied mathematical astronomy and was later instrumental in the foundation of Barnard College for Women.

In this section, we examine linear inequalities in two variables and systems of these inequalities. This topic is a natural extension of systems of linear equations covered in Chapter 3 and our work in this chapter with inequalities involving only one variable. In Exercise 62, we will examine how inequalities can be used to describe possible options for businesses.

1. Identify a Solution of a Linear Inequality in Two Variables

What Is a Solution of a Linear Inequality in Two Variables?

The **solution to a linear inequality** in two variables is the set of all ordered pairs that, when substituted into the inequality, make a true statement. Each of these solutions is said to satisfy the inequality.

Example 1 | Checking Possible Solutions of an Inequality

Determine whether each ordered pair satisfies $2x + 3y \leq 6$.

(a) $(5, 0)$ **(b)** $(0, 2)$ **(c)** $(0, 0)$

Solution

(a) $2x + 3y \leq 6$

$2(5) + 3(0) \stackrel{?}{\leq} 6$

$10 \stackrel{?}{\leq} 6$ is false.

Substitute the given coordinates into the inequality and determine whether a true statement results. The solution of $2x + 3y \leq 6$ does not contain this point.

Answer: $(5, 0)$ does not satisfy $2x + 3y \leq 6$.

(b) $2x + 3y \leq 6$

$2(0) + 3(2) \stackrel{?}{\leq} 6$

$6 \stackrel{?}{\leq} 6$

The inequality $\leq$, less than or equal to, includes the equality, so the point $(0, 2)$ is a solution. The solution of $2x + 3y \leq 6$ contains this point.

Answer: $(0, 2)$ satisfies $2x + 3y \leq 6$.

(c) $2x + 3y \leq 6$

$2(0) + 3(0) \stackrel{?}{\leq} 6$

$0 \stackrel{?}{\leq} 6$

The solution of $2x + 3y \leq 6$ contains this point.

Answer: $(0, 0)$ satisfies $2x + 3y \leq 6$.

Self-Check 1

Determine whether each ordered pair satisfies $5x + y > 10$.

a. $(2, 1)$ **b.** $(2, 0)$ **c.** $(0, 0)$

The line $Ax + By = C$ separates the plane into two regions called half-planes. (See the accompanying figure.) One half-plane will satisfy $Ax + By < C$, and the other half-plane will satisfy $Ax + By > C$. If the inequality is $\leq$ or $\geq$, the line will be part of the solution. If the inequality is $<$ or $>$, the line will be the boundary separating the half-planes, but it will not be part of the solution. Standard convention is to use a solid line for $\leq$ or $\geq$ when the line is part of the solution and to use a dashed line for $<$ or $>$ when the line is not part of the solution.

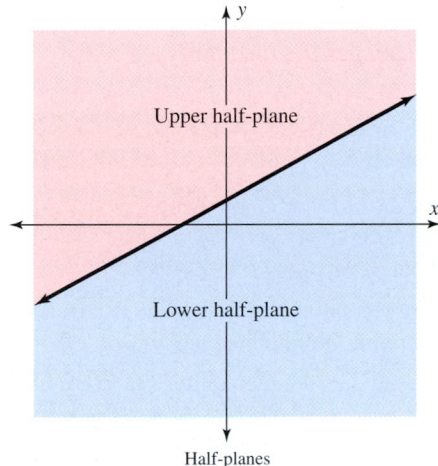

Half-planes

Example 2   Identifying Solutions of an Inequality from a Graph

Determine whether either point A or point B satisfies the inequality $y > \dfrac{x}{2} - 1$.

Solution

$A: (1, 1)$

$$y > \frac{x}{2} - 1$$

$$1 \overset{?}{>} \frac{1}{2} - 1$$

$$1 \overset{?}{>} -\frac{1}{2} \quad \text{is true}$$

$B: (2, -3)$

$$y > \frac{x}{2} - 1$$

$$-3 \overset{?}{>} \frac{2}{2} - 1$$

$$-3 \overset{?}{>} 0 \quad \text{is false}$$

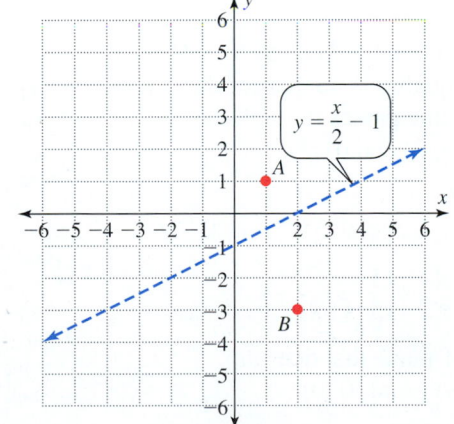

Note that we could determine this by inspection because point A is in the upper half-plane where $y > \dfrac{x}{2} - 1$ and point B is in the lower half-plane where $y < \dfrac{x}{2} - 1$.

Answer: Point A satisfies $y > \dfrac{x}{2} - 1$ but point B does not.

Self-Check 2

Plot the points $(-3, -1)$ and $(5, 1)$ on the graph in Example 2. Determine whether each point satisfies $y < \dfrac{x}{2} - 1$.

2. Graph a Linear Inequality in Two Variables

In Example 2, you could never list all the solutions of the inequality $y > \dfrac{x}{2} - 1$. Thus we must represent these points algebraically or graphically.

How Do I Graph a Linear Inequality in Two Variables?

First graph the line that forms the boundary for the half-planes. Then use a test point to determine which half-plane satisfies the inequality.

Graphing a Linear Inequality in Two Variables

Step 1. Graph the equality $Ax + By = C$, by using
  a. A solid line for $\leq$ or $\geq$
  b. A dashed line for $<$ or $>$

Step 2. Choose an arbitrary test point not on the line; $(0, 0)$ is often convenient. Substitute this test point into the inequality.

Step 3. a. If the test point satisfies the inequality, shade the half-plane containing this point.
  b. If the test point does not satisfy the inequality, shade the other half-plane.

Examples 3 and 4 illustrate each step of this procedure.

Example 3 Graphing a Linear Inequality

Graph the solution of $3x - 5y < -15$.

Solution

1. Draw a dashed line for $3x - 5y = -15$ because the equality is not part of the solution. The line passes through the intercepts $(-5, 0)$ and $(0, 3)$.

2. Test the origin:

$$3x - 5y < -15$$
$$3(0) - 5(0) \not< -15$$
$$0 \not< -15 \text{ is false.}$$

3. Shade the half-plane that does *not* include the test point $(0, 0)$.

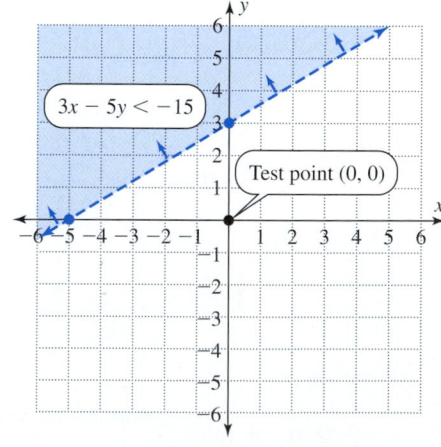

Self-Check 3

Graph $2x + y \leq 10$.

Note that in Example 3 we could solve the inequality for y as shown to the right. The greater than inequality in $y > \dfrac{3}{5}x + 3$ indicates that the solution is the upper half-plane.

$$3x - 5y < -15$$
$$-5y < -3x - 15$$
$$y > \frac{3}{5}x + 3$$

This is also the solution that we obtained in Example 3.

In Example 3 we used $(0, 0)$ as a test point but this is not appropriate in Example 4 because this point is on the line separating the half-planes. We must always select a test point that is not on the line separating the half-planes.

Example 4   Graphing a Linear Inequality

Graph the solution of $5x \geq 3y$.

Solution

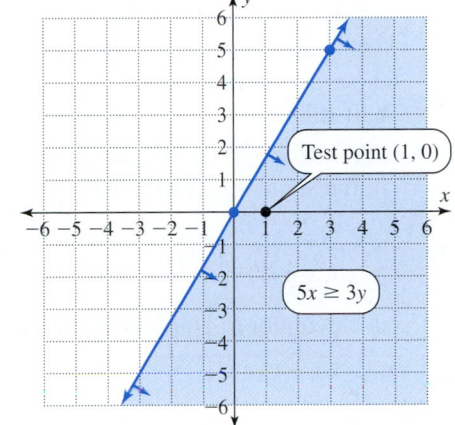

1 Draw a solid line for $5x = 3y$ since the original statement includes the equality. The line passes through $(0, 0)$ and $(3, 5)$.

2 Test the point $(1, 0)$. (Do not test $(0, 0)$ since the origin lies on the line.)

$$5x \geq 3y$$
$$5(1) \overset{?}{\geq} 3(0)$$
$$5 \overset{?}{\geq} 0 \text{ is true.}$$

3 Shade the half-plane that includes the test point $(1, 0)$.

Test point $(1, 0)$

$5x \geq 3y$

Self-Check 4

Graph $2x \leq 3y$.

Note that in Example 4 we could solve the inequality for y as shown to the right. The less than or equal inequality in $y \leq \dfrac{5}{3}x$ indicates that the solution is the lower half-plane. This is also the solution that we obtained in Example 4.

$$5x \geq 3y$$
$$3y \leq 5x$$
$$y \leq \frac{5}{3}x$$

To graph a linear inequality on a TI-84 Plus calculator, first solve the inequality for y. Then enter the right side of the inequality next to Y_1, and select the shading that represents the appropriate half-plane. This is illustrated in Technology Perspective 4.5.1. Note that on a TI-84 Plus calculator, the symbol ◥ denotes that the upper half-plane should be shaded and the symbol ◢ denotes that the lower half-plane should be shaded.

Technology Perspective 4.5.1   Graphing a Linear Inequality

Graph the solution of $3x - 5y < -15$ from Example 3.

TI-84 Plus Keystrokes

1. First solve $3x - 5y < -15$ for y to rewrite this inequality as
    $$y > \frac{3}{5}x + 3.$$

2. Press [Y=] and enter the equation.

3. Use the arrow keys to move the cursor to the left of Y_1 and press [ENTER] until the marker indicates the graph will shade the upper half-plane.

4. Press [ZOOM] [6] to display the graph in the standard viewing window.

TI-84 Plus Calculator

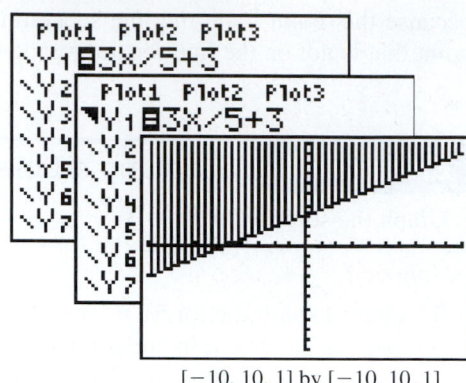

$[-10, 10, 1]$ by $[-10, 10, 1]$

Technology Self-Check 1

Use a calculator to graph the solution of $5x \ge 3y$ from Example 4.

Note: For $\ge$ and $>$ the graph will be shaded above the line, and for $\le$ and $<$ the graph will be shaded below the line. The calculator does not distinguish between $\ge$ and $>$, nor does it distinguish between $\le$ and $<$.

Example 5   Using a Graphing Calculator to Solve an Inequality

Use a graphing calculator to graph the solution of $2x + 3y \le 6$.

Solution

$$2x + 3y \le 6$$
$$3y \le -2x + 6$$
$$y \le -\frac{2}{3}x + 2$$

First solve for y. Then enter the equation into Y_1 and select the appropriate shading for the inequality. The solution of the inequality is the lower half-plane.

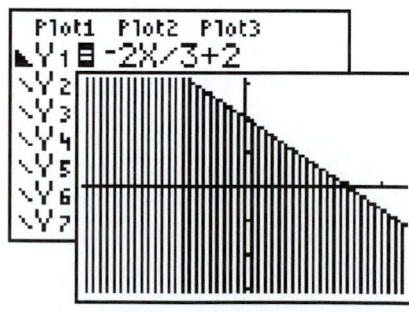

$[-4.7, 4.7, 1]$ by $[-4.7, 4.7, 1]$

Self-Check 5

Use a graphing calculator to graph the solution of $2x + 3y \ge 6$.

3. Graph a System of Linear Inequalities

What Is a System of Linear Inequalities?

When we consider two or more linear inequalities at the same time, we refer to this as a **system of linear inequalities.**

How Do I Graph a System of Linear Inequalities in Two Variables?

First graph each inequality on the same coordinate system. The solution of the system is the intersection of these two individual regions. To clarify which points satisfy each inequality, the TI-84 Plus calculator uses vertical and horizontal lines to indicate the solutions of the individual inequalities. The solution of the system is the crosshatched region where these lines intersect. When we graph these inequalities by using pencil and paper, we use small arrows to indicate each individual half-plane. Then we shade the intersection of these half-planes as illustrated in Example 6.

Example 6 Graphing a System of Linear Inequalities

Graph the solution of $\begin{Bmatrix} x \geq -3 \\ x \leq 2 \end{Bmatrix}$.

Solution

Inspection reveals that both $x = -3$ and $x = 2$ represent vertical lines. The region containing $(0, 0)$ satisfies both inequalities. Thus the solution set is the violet strip between these solid lines.

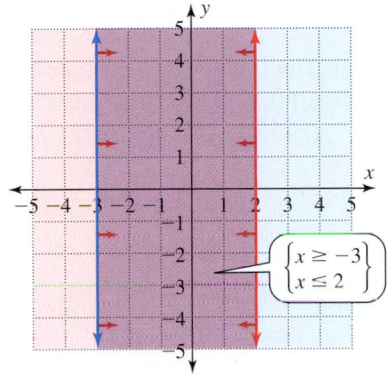

$\begin{Bmatrix} x \geq -3 \\ x \leq 2 \end{Bmatrix}$

Self-Check 6

Graph the solution of $\begin{Bmatrix} y \leq 4 \\ y > -3 \end{Bmatrix}$.

The solution to Example 7 proceeds step by step. First one inequality is graphed, and then the second inequality is graphed. Finally the intersection of these regions is shaded as the solution for the system.

Example 7 Graphing a System of Linear Inequalities

Graph the solution of $\begin{Bmatrix} x + y \geq 2 \\ 3x - 2y < 6 \end{Bmatrix}$.

Solution

(a) $\boxed{1}$ To graph $x + y \geq 2$, draw a solid line for
$x + y = 2$. Use the intercepts $(2, 0)$ and $(0, 2)$
to draw the line.

$\boxed{2}$ Test the point $(0, 0)$ in $x + y \geq 2$
$0 + 0 \not\geq 2$ is false.

$\boxed{3}$ Shade the half-plane that does *not* contain $(0, 0)$
(shown here in light red).

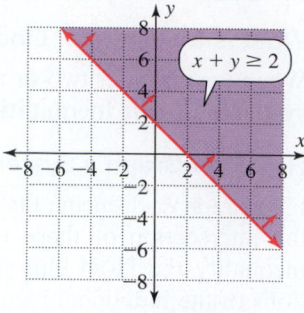

(b) $\boxed{1}$ To graph $3x - 2y < 6$, draw a dashed line for
$3x - 2y = 6$. Use the intercepts $(2, 0)$ and
$(0, -3)$ to draw the line.

$\boxed{2}$ Test the point $(0, 0)$ in $3x - 2y < 6$
$3(0) - 2(0) \not< 6$
$0 \not< 6$ is true.

$\boxed{3}$ Shade the half-plane that contains $(0, 0)$
(shown here in light blue).

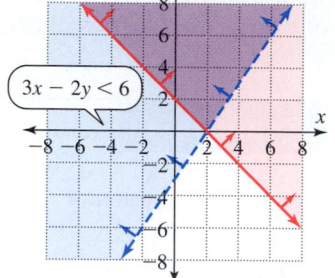

(c) The intersection of these two regions is the set of
all the points satisfying this system of inequalities
(shown here in violet).

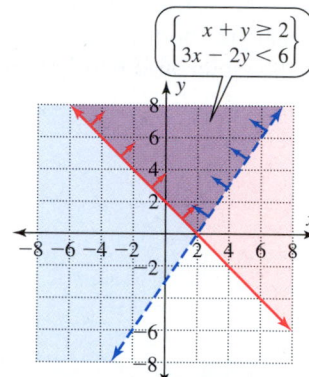

Self-Check 7

Graph the solution of $\begin{Bmatrix} x + y \leq 3 \\ 2x - 3y \leq 6 \end{Bmatrix}$.

A graphing calculator displays the solution of a system of inequalities as the intersection of the different shading patterns. This is illustrated in Example 8.

Example 8   Using a Graphing Calculator to Solve a
System of Inequalities

Use a graphing calculator to graph the solution of $\begin{Bmatrix} x + y \geq 2 \\ 3x - 2y < 6 \end{Bmatrix}$. This is the same
system as in Example 7.

Solution

$$x + y \geq 2 \qquad\qquad 3x - 2y < 6$$

$$y \geq 2 - x \qquad\qquad -2y < -3x + 6$$

$$y > \frac{3}{2}x - 3$$

To graph these inequalities on a calculator, first solve each inequality for y.

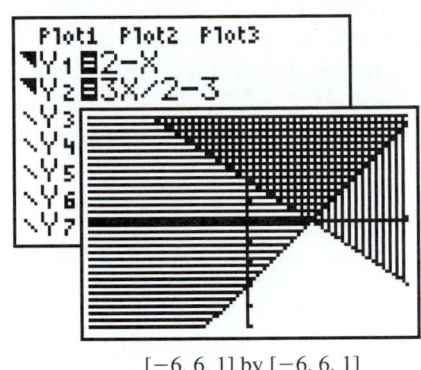

$[-6, 6, 1]$ by $[-6, 6, 1]$

Graphing calculators are powerful tools, but you must interpret properly the information they display. For inequalities, be careful to determine from the given inequalities whether the boundary lines on the display should be included in the solution set.

Answer: The solution set is represented by the crosshatched area. To interpret this solution properly, note that the line representing $y = \frac{3}{2}x - 3$ is not part of the solution.

Self-Check 8

Use a graphing calculator to graph the solution of $\begin{cases} x + y \leq 3 \\ 2x - 3y \leq 6 \end{cases}$. (This is the same system as in Self-Check 7.)

Example 9 presents an application that is modeled by a system of three linear inequalities. The solution for this system is graphed using pencil-and-paper methods.

Example 9   Length of Two Pieces of Optic Fiber

An installer for a telecommunications company used two pieces of optic fiber from a spool containing 100 m. Lengths x and y must both be positive, and their total is at most 100 m. Write a system of linear inequalities to model this problem, and then graph the solution of this system.

Solution

Word Inequality	Algebraic Inequality	
x is positive.	$x > 0$	Translate each word inequality into an algebraic inequality.
y is positive.	$y > 0$	
The total of x and y is at most 100.	$x + y \leq 100$	

Graphically

First graph each inequality on the same coordinate system. Then test the point $(25, 25)$ in each inequality.

$$x > 0 \qquad\qquad y > 0 \qquad\qquad x + y \leq 100$$

$$25 \overset{?}{>} 0 \text{ is true.} \qquad 25 \overset{?}{>} 0 \text{ is true.} \qquad 25 + 25 \overset{?}{\leq} 100$$

$$50 \overset{?}{\leq} 100 \text{ is true.}$$

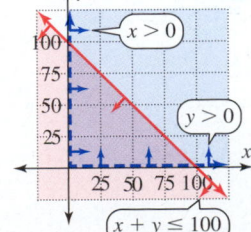

Use arrows to indicate each individual region formed by the linear boundaries, and then use violet shading to indicate the triangular region that is the solution of the system.

All points within the violet shaded region are solutions of this system of inequalities.

Self-Check 9

The owner of 45,000 time-share points can spend these points on either of two vacation options. The points x and y that are spent on each of these two options must be nonnegative. Write a system of inequalities to model this problem, and then graph the solution of this system.

Self-Check Answers

1. $(2, 1)$ satisfies the inequality but $(2, 0)$ and $(0, 0)$ do not.
2. $(-3, -1)$ does not satisfy the inequality but $(5, 1)$ does.

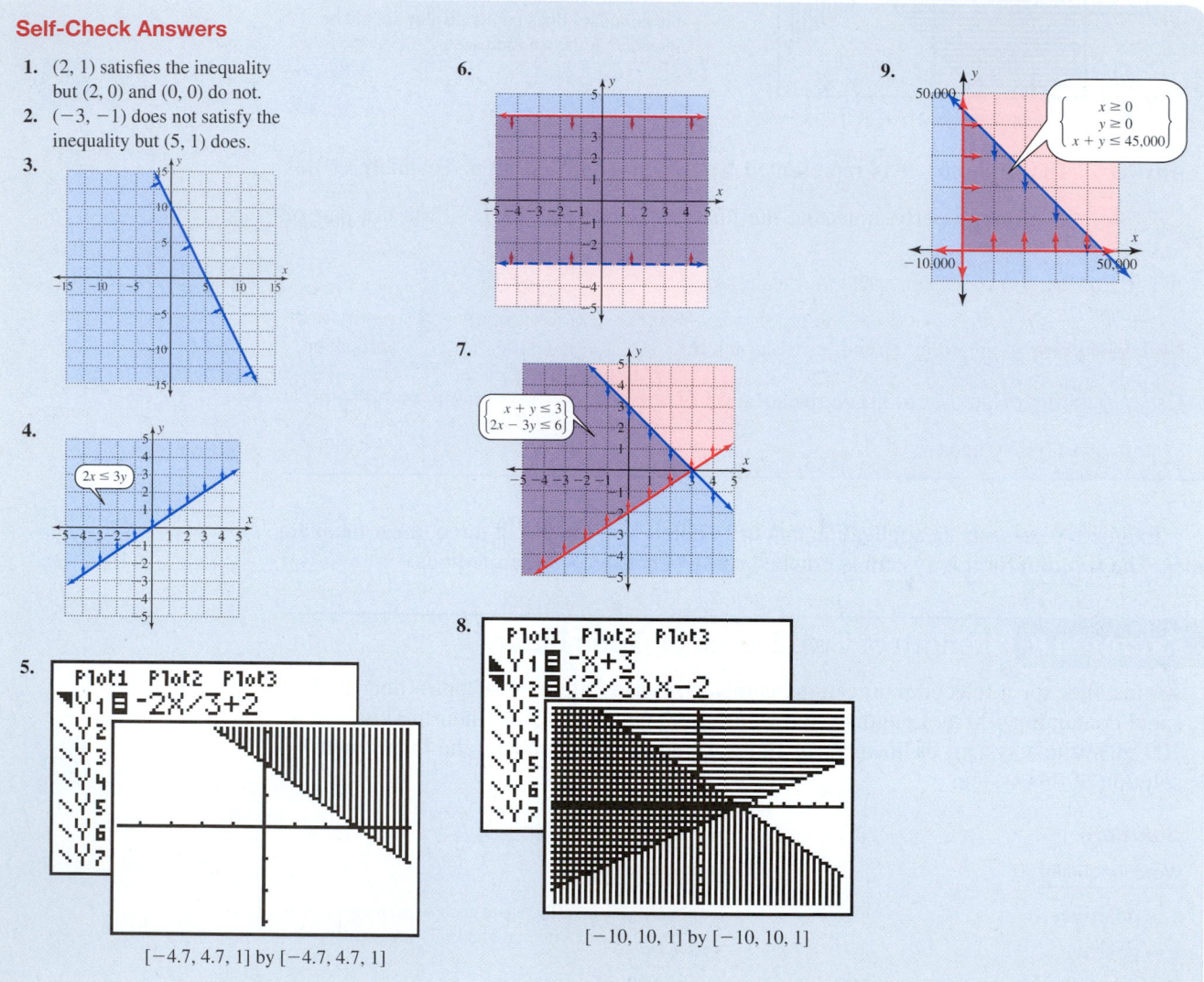

$$\begin{cases} x + y \le 3 \\ 2x - 3y \le 6 \end{cases}$$

$$\begin{cases} x \ge 0 \\ y \ge 0 \\ x + y \le 45{,}000 \end{cases}$$

$2x \le 3y$

Y1 = -2X/3+2

$[-4.7, 4.7, 1]$ by $[-4.7, 4.7, 1]$

Y1 = -X+3
Y2 = (2/3)X-2

$[-10, 10, 1]$ by $[-10, 10, 1]$

Technology Self-Check Answers

1.

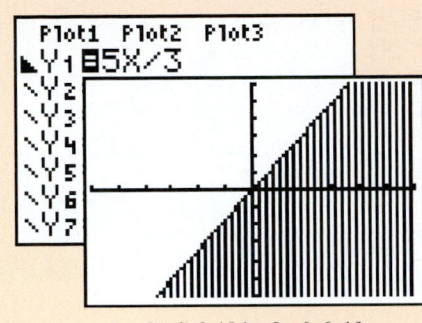

$[-6, 6, 1]$ by $[-6, 6, 1]$

4.5   Using the Language and Symbolism of Mathematics

1. A solution to a linear inequality in two variables is an _____ _____ of values that, when substituted into the inequality, makes a _____ statement.

2. The line $Ax + By = C$ separates the plane into two regions called _____-_____.

3. For the linear inequalities $\leq$ and $\geq$, we graph the boundary equation by using a _____ line.

4. For the linear inequalities $<$ and $>$, we graph the boundary equation by using a _____ line.

5. If a test point satisfies an inequality, shade the half-plane that _____ this point.

6. If a test point does not satisfy an inequality, shade the half-plane that does _____ _____ this point.

7. If ◥ $y_1 = 3x - 7$ is used to graph an inequality on a graphing calculator, then the calculator will shade the _____ half-plane.

8. If ◣ $y_1 = 3x - 7$ is used to graph an inequality on a graphing calculator, then the calculator will shade the _____ half-plane.

4.5   Quick Review

1. Determine the x- and y-intercepts of the graph of $2x + 5y = 10$.

2. Write $2x + 5y = 10$ in slope-intercept form.

3. Graph $2x + 5y = 10$.

4. Graph $x = 4$.

5. Graph $y = 4$.

4.5   Exercises

Objective 1   Identify a Solution of a Linear Inequality in Two Variables

In Exercises 1–4, determine whether the given point is a solution of the inequality.

1. Check $(0, 0)$ in these inequalities.
   a. $2x + 3y < 1$ b. $2x + 3y \leq 1$
   c. $2x + 3y > 1$ d. $2x + 3y \geq 1$

2. Check $(1, 2)$ in these inequalities.
   a. $2x + 3y < 8$ b. $2x + 3y \leq 8$
   c. $2x + 3y > 8$ d. $2x + 3y \geq 8$

3. Check $(2, -3)$ in these inequalities.
   a. $3x - y < 9$ b. $3x - y \leq 9$
   c. $3x - y > 9$ d. $3x - y \geq 9$

4. Check $(-4, 1)$ in these inequalities.
   a. $x + 5y < 2$ b. $x + 5y \leq 2$
   c. $x + 5y > 2$ d. $x + 5y \geq 2$

5. Determine whether points A through D are solutions of the inequality graphed here.

A (0, 0)
B (4, −2)
C (0, 2)
D (−5, 0)

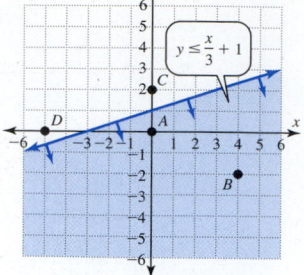

6. Determine whether points A through D are solutions of the inequality graphed here.

A (4, 2)
B (0, 4)
C (−2, 0)
D (−4, −4)

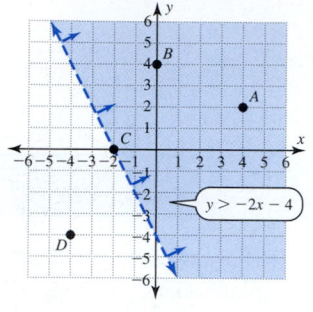

7. Determine whether points A through D are solutions of the inequality graphed here.

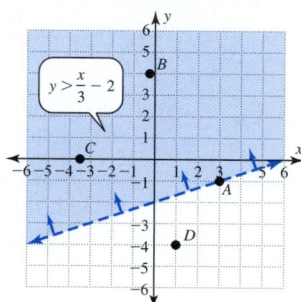

8. Determine whether points A through D are solutions of the inequality graphed here.

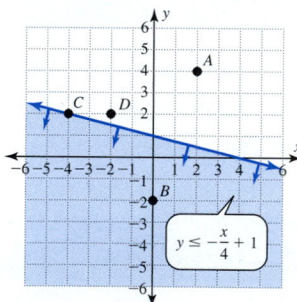

9. Which of the following inequalities corresponds to the graph in the figure?
A. $x - 3y \leq 3$
B. $x - 3y < 3$
C. $x - 3y \geq 3$
D. $x - 3y > 3$

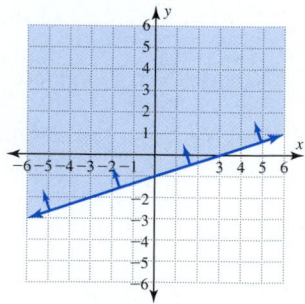

10. Which of the following inequalities corresponds to the graph in the figure?
A. $x + 2y \leq 4$
B. $x + 2y < 4$
C. $x + 2y \geq 4$
D. $x + 2y > 4$

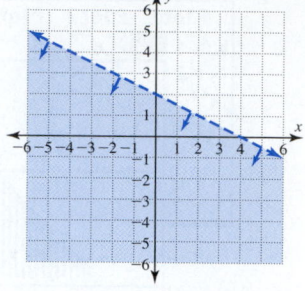

Objective 2 Graph a Linear Inequality in Two Variables

In Exercises 11–18, graph each linear inequality.

11. $x - y \geq 5$
12. $2x + y \leq 6$
13. $3x - 2y - 12 < 0$
14. $-5x + 2y + 10 < 0$
15. $x > 3y$
16. $5x < -y$
17. $\frac{1}{2}x + \frac{1}{3}y \leq 1$
18. $\frac{1}{4}x - \frac{1}{5}y \geq 1$

Objective 3 Graph a System of Linear Inequalities

19. Determine which of points A through D satisfy the system of inequalities $\begin{cases} -x + 2y > 4 \\ 2x - y \geq -2 \end{cases}$.

A (4, 5)   B (0, 2)
C (2, 6)   D (−3, −2)

20. Determine which of points A through D satisfy the system of inequalities $\begin{cases} x - y \geq -4 \\ x + y < 4 \end{cases}$.

A (0, 4)   B (−4, 0)
C (0, −4)   D (0, 0)

In Exercises 21–24, shade the regions on the graphs of $y_1 = m_1 x + b_1$ and $y_2 = m_2 x + b_2$ that satisfy each system of inequalities. In Exercises 23 and 24, $m_1 = m_2$.

a. $\begin{cases} y_1 \geq m_1 x + b_1 \\ y_2 \geq m_2 x + b_2 \end{cases}$

b. $\begin{cases} y_1 \geq m_1 x + b_1 \\ y_2 \leq m_2 x + b_2 \end{cases}$

c. $\begin{cases} y_1 \leq m_1 x + b_1 \\ y_2 \leq m_2 x + b_2 \end{cases}$

d. $\begin{cases} y_1 \leq m_1 x + b_1 \\ y_2 \geq m_2 x + b_2 \end{cases}$

21.

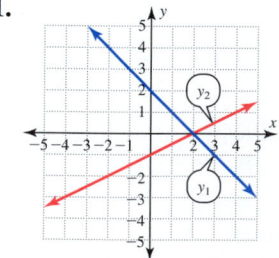

22.

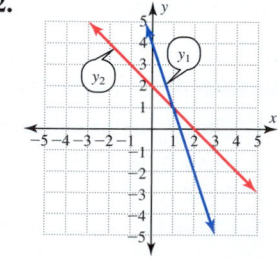

23.

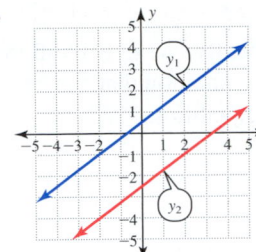

24.

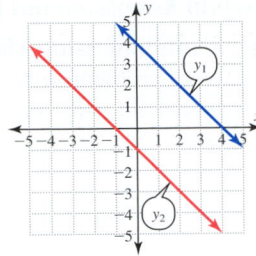

47.

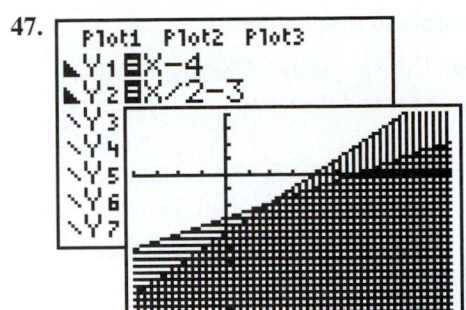

$[-4, 10, 1]$ by $[-10, 4, 1]$

48.

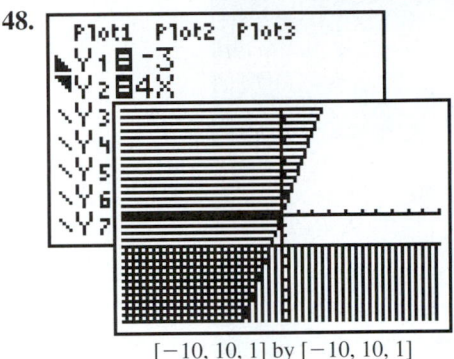

$[-10, 10, 1]$ by $[-10, 10, 1]$

In Exercises 25–46, graph the solution of each system of linear inequalities.

25. $x - y \geq 4$
$2x - y < 6$

26. $x + 2y \leq 2$
$2x - y \geq 4$

27. $2x - 5y - 10 < 0$
$2x - y - 7 \geq 0$

28. $3x + 2y - 12 > 0$
$2x + 5y - 10 < 0$

29. $\dfrac{x}{2} - \dfrac{y}{2} < 1$
$\dfrac{x}{2} + \dfrac{y}{2} > -1$

30. $\dfrac{x}{3} - \dfrac{y}{3} \geq -1$
$\dfrac{x}{4} + \dfrac{y}{4} \geq -1$

31. $x \geq 1$
$x < 4$

32. $x > -5$
$x \leq -2$

33. $y < -2$
$y \geq -6$

34. $y > 3$
$y < 4$

35. $x > -2$
$x \leq 3$

36. $y < 5$
$y > 3$

37. $y \geq -2$
$y \leq -4$

38. $x \leq -4$
$x \geq -1$

39. $\quad\quad x \geq 0$
$\quad\quad y \geq 0$
$2x + 3y < 6$

40. $\quad\quad x \geq 0$
$\quad\quad y \geq 0$
$5x + 2y < 10$

41. $x + y > 0$
$x - y < 0$
$\quad\quad y < 4$

42. $2x + 3y \leq 0$
$3x - 2y \geq 0$
$\quad\quad y \geq -4$

43. $x \geq -2$
$x \leq 2$
$y \geq -3$
$y \leq 3$

44. $x > 1$
$x < 3$
$y > -5$
$y < -2$

45. $3x + 3y - 15 \leq 0$
$6x + 2y - 18 \leq 0$
$\quad\quad\quad x \geq 0$
$\quad\quad\quad y \geq 0$

46. $\quad x - y - 2 \leq 0$
$2x + 2y - 8 \leq 0$
$\quad\quad\quad x \geq 0$
$\quad\quad\quad y \geq 0$

Applying Technology

In Exercises 47 and 48, write the system of inequalities that is graphed on the given graphing calculator display. (Assume that the inequalities are either $\leq$ or $\geq$ and are not $<$ or $>$.)

Multiple Representations

In Exercises 49–52, write an algebraic inequality for each verbal statement.

49. The x-coordinate is at least two more than the y-coordinate.

50. The y-coordinate is at least three more than the x-coordinate.

51. The y-coordinate is at most four more than the x-coordinate.

52. The x-coordinate is at most one more than the y-coordinate.

In Exercises 53–56, write a system of algebraic inequalities for these verbal statements.

53. Both the x- and y-coordinates are positive, and the sum of the two coordinates does not exceed ten.

54. Both the x- and y-coordinates are positive, and the sum of the two coordinates does not exceed eight.

55. Both the x- and y-coordinates are nonnegative, and the sum of x and twice y is at most five.

56. Both the x- and y-coordinates are nonnegative, and the sum of twice x and three times y is at most six.

Connecting Concepts to Applications

In Exercises 57–60, write a system of inequalities that represents each situation and then graph this system of inequalities.

57. Length of Two Pieces of Rope A store sold two pieces of rope from a spool containing 150 m. Lengths x and y must both be positive, and their total is at most 150 m.

58. Theater Tickets A theater can seat at most 2,500 customers. Neither the number of adult tickets x nor the number of child tickets y can be negative, and their total is at most 2,500.

59. Production Units A company makes a profit of $40 for each of the x stereos it ships and $50 for each of the y televisions it ships. The number of units of each item shipped is nonnegative, and the profit per day for the factory has never exceeded $4,400.

60. Factory Production A printer makes a profit of $2 for each of the x books produced and $0.25 for each of the y magazines produced. The number produced of each is nonnegative, and the profit per day for the printer has never exceeded $15,000.

Group discussion questions

61. Communicating Mathematically Write two different word problems that describe a situation that can be modeled by this system of inequalities.

$$\begin{cases} x \geq 0 \\ y \geq 0 \\ x + y \leq 100 \end{cases}$$

62. Challenge Question: Production Choices The following system of inequalities models the restrictions on the number x of wooden table chairs and the number y of wooden rocking chairs made at a furniture factory each week. These restrictions are caused by limitations on the factory's equipment and labor supply.

$$\begin{cases} x \geq 0 \\ y \geq 0 \\ x + y \leq 50 \\ 3x + y \leq 90 \end{cases}$$

a. Determine each corner point of the region formed by graphing this system of inequalities.

b. The factory makes $20 for each table chair and $25 for each rocking chair. Write an expression for the profit involving x and y.

c. Evaluate the profit expression from part **b** at each of the corner points from part **a**. Which of these points produces the greatest profit for the factory?

63. Challenge Question Write a system of linear inequalities that is represented by the following graph.

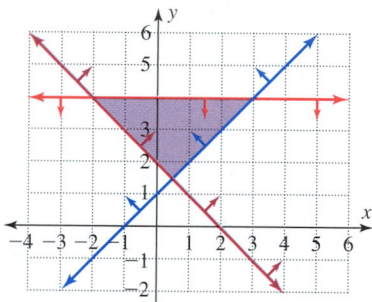

4.5   Cumulative Review

Evaluate each expression.

1. $(1 + 2 + 3)^2$

2. $1^2 + 2^2 + 3^2$

3. $(7 - 4)^2$

4. $7^2 - 4^2$

5. $\dfrac{3^2}{5} - \left(\dfrac{3}{5}\right)^2$

Chapter 4 | Key Concepts

1. Solution of an Inequality
- The solution of an inequality in one variable is the set of all values that make the inequality a true statement.
- The solution of an inequality in two variables is the set of all ordered pairs that make the inequality a true statement.

2. Types of Inequalities
- Conditional inequality: An inequality that contains a variable and is true for some, but not all, real values of the variable(s).
- Unconditional inequality: An inequality that is always true.
- Contradiction: An inequality that is always false.
- Equivalent inequalities: Inequalities that have the same solution are called equivalent inequalities.
- Linear inequality: An inequality that is first degree in each variable.

3. Principles Used to Solve Inequalities*
- **Addition-Subtraction Principle:** If a, b, and c are real numbers, then $a < b$ is equivalent to $a + c < b + c$ and to $a - c < b - c$.
- **Multiplication-Division Principle:** If a, b, and c are real numbers and $c > 0$, then $a < b$ is equivalent to $ac < bc$. If $c < 0$, then $a < b$ is equivalent to $ac > bc$.[†]

4. Intersection and Union of Two Sets
- Intersection: $A \cap B$ is the set of points in both A and B.
- Union: $A \cup B$ is the set of points in either set A *or* B (or both).

5. Compound Inequality: $a \leq x \leq b$ is equivalent to $x \geq a$ and $x \leq b$.

6. Absolute Value Equations and Inequalities: For any real numbers x and a and positive real number d:
- $|x - a| = d$ is equivalent to $x - a = -d$ or $x - a = d$.
- $|x - a| < d$ is equivalent to $-d < x - a < d$.
- $|x - a| > d$ is equivalent to $x - a < -d$ or $x - a > d$.
- $|x - a| = -d$ is a contradiction and has no solution.
- $|x - a| < -d$ is a contradiction and has no solution.
- $|x - a| > -d$ is an unconditional inequality, and the solution set is the set of all real numbers.

7. Solving an Absolute Value Equation or Inequality: To solve an absolute value equation or inequality such as $|ax - b| + c < d$, first isolate the absolute value expression on the left side of the expression.

8. Tolerance: The acceptable variation between the standard size for a part or component and the actual size is called the tolerance. For example, $|x - L| \leq t$ expresses the tolerance t between the desired length of L and the actual length x.

9. Half-Planes: The line $Ax + By = C$ separates the plane into two regions called half-planes. One half-plane satisfies $Ax + By < C$, and the other half-plane satisfies $Ax + By > C$.

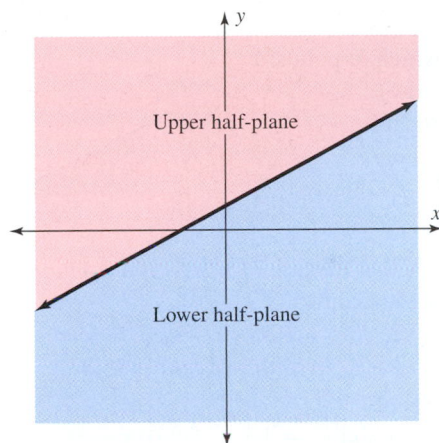

10. Graphing a Linear Inequality in Two Variables:
Step 1 Graph the equality $Ax + By = C$ by using
  a. A solid line for $\leq$ or $\geq$.
  b. A dashed line for $<$ or $>$.
Step 2 Choose an arbitrary test point not on the line; $(0, 0)$ is often convenient. Substitute this test point into the inequality.
Step 3 a. If the test point satisfies the inequality, shade the half-plane containing this point.
  b. If the test point does not satisfy the inequality, shade the other half-plane.

11. Graphing a System of Linear Inequalities: Graph each inequality in the system on the same coordinate system. Use arrows to indicate each individual region formed by these lines. Then use shading to indicate the intersection of these regions. The solution of the system is represented by this intersection.

---

*Similar statements can be made for the inequalities $\leq$, $>$, and $\geq$.
[†]Similar statements can be made for division.

Chapter 4 | Review Exercises

Solutions of Linear Inequalities and Systems of Inequalities

1. Determine if $x = -4$ is a solution of each of these inequalities.
   a. $2x > -8$
   b. $x + 3 \le -1$
   c. $5x + 3 < 2x - 3$
   d. $5(x - 2) \ge 3(2x - 4) + 7$

2. Determine which of points A through D satisfy the inequality $y < \dfrac{x}{3} - 2$

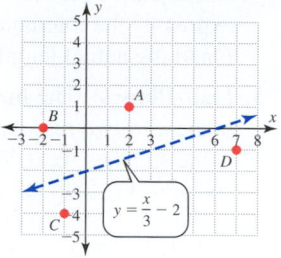

3. Determine which of points A through D satisfy:
   $x \ge 0, y \ge 0$, and
   $y \le -0.5x + 3$.

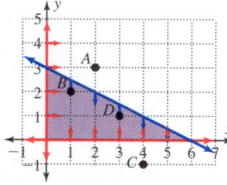

4. Determine whether the point $(-2, 5)$ is a solution of each of these inequalities.
   a. $y \le 2x - 12$
   b. $y \ge 3x - 12$
   c. $3x + 5y < 15$
   d. $-2x + 4y > 22$

5. Use the graphs of $y_1 = 0.2x + 0.4$ and $y_2 = 1$ to solve each equation and inequality.
   a. $0.2x + 0.4 = 1$
   b. $0.2x + 0.4 > 1$
   c. $0.2x + 0.4 < 1$

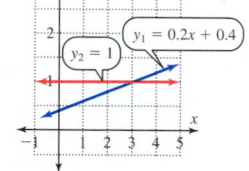

6. Complete the table for $y_1 = 5(x + 2)$ and $y_2 = 2x + 7$ by placing $<, =, >$ in the third column. Then determine the x-values that satisfy each equation and inequality.
   a. $5(x + 2) = 2x + 7$
   b. $5(x + 2) < 2x + 7$
   c. $5(x + 2) > 2x + 7$

x	y_1	<, =, OR >	y_2
−3	−5		1
−2	0		3
−1	5		5
0	10		7
1	15		9
2	20		11
3	25		13

In Exercises 7–30, solve each inequality.

7. $x + 7 < 5$
8. $x - 11 > -10$
9. $-10 \ge 2x$
10. $-3x \le 12$
11. $3x + 7 > 5x + 13$
12. $7 \ge 7 - 9y$

13. $-\dfrac{x}{2} < \dfrac{1}{4}$

14. $\dfrac{3x}{7} + \dfrac{4}{5} > \dfrac{3x}{5} + \dfrac{2}{7}$

15. $7y + 14 \ge 2(3y + 8)$

16. $-5(2y - 2) \le 7 - 11y$

17. $2(3t - 4) > 3(t - 6) + 1$

18. $2(11t - 3) < 5(3t + 2) - 20$

19. $5(x - 4) + 6 < (x + 4) - 30$

20. $7(x - 3) + 6 \ge 12(x + 4) + 2$

21. $4(y - 1) - 7(y + 1) \le -3(y + 2) - 5$

22. $1 - 4(y - 2) \ge 3(y - 7) - 5(4 - y)$

23. $\dfrac{v}{2} - \dfrac{3v + 4}{4} > \dfrac{v + 40}{4}$

24. $\dfrac{v}{8} + 6 > \dfrac{v}{12} + 5$

25. $9 < x + 7 \le 13$

26. $0 \le \dfrac{4}{5}x < 20$

27. $42 \le \dfrac{-3m}{7} < 60$

28. $-5 \le 9 - x \le 5$

29. $-7 \le 5x + 3 \le 13$

30. $-10 < 5 - 3x < 8$

In Exercises 31–34, each inequality is a conditional inequality, an unconditional inequality, or a contradiction. Identify the type of each inequality and solve it.

31. $3v + 3v > 5v$
32. $3v + 3v \le 5v$
33. $3(v + 1) < 3v + 4$
34. $5(3v - 1) > 3(5v - 1)$

35. Use the given graph to solve $-1 \le \dfrac{2 - x}{3} \le 1$.

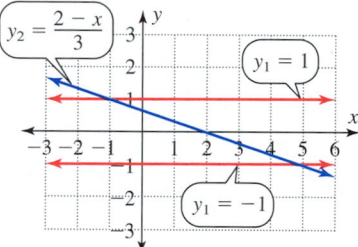

36. Use the following table for $y = \dfrac{x - 1}{2}$ to solve $-1 \le \dfrac{x - 1}{2} < 1$.

x	$y = \dfrac{x - 1}{2}$
−2	−1.5
−1	−1.0
0	−0.5
1	0.0
2	0.5
3	1.0
4	1.5

In Exercises 37–42, graph the solution of each system of inequalities.

37. $x \ge -1$
    $x \le 3$

38. $y \le 4$
    $y \ge 1$

39. $2x - 3y < 6$
    $2x + 5y > 10$

40. $2x + y < 4$
    $2x + y > 1$

41. $3x + 4y \le 12$
$3x - 4y \le 12$

42. $2x + y \le 4$
$x \ge 0$
$y \ge 0$

Absolute Value Equations and Inequalities

In Exercises 43–46, solve each absolute value equation or inequality.

43. $|x - 3| = 4$

44. $|2x - 5| + 6 = 9$

45. $|5x - 2| > 8$

46. $|x - 5| < 3$

In Exercises 47–50, write an absolute value inequality to represent each interval.

47. $(-5, 5)$

48. $[-4, 4]$

49. $(-2, 10)$

50. $[-4, 20]$

Multiple Representations

51. Complete the table.

Verbally	Inequality Notation	Interval Notation	Graph
a. x exceeds 3			
b.	$x \le 4$		
c.		$[2, +\infty)$	
d.			⟵─┼──┼──┼──)──┼──┼──┼─⟶ 2 3 4 5 6 7 8

52. Complete the following table.

Verbally	Inequality Notation	Interval Notation	Graph
a. x is greater than 2 and less than 6			
b.	$-3 < x \le 4$		
c.		$[-2, 0)$	
d.			⟵─┼──[──┼──┼──)──┼──┼─⟶ −2 −1 0 1 2 3 4

53. Write an algebraic inequality for each word inequality.
   a. x plus seven is at most eleven.
   b. Twice the quantity x minus one exceeds thirteen.
   c. Three x minus five is at least seven.
   d. Four x plus nine never exceeds twenty-one.

In Exercises 54–56, write an algebraic inequality for each word inequality, and then solve this inequality.

54. Four times the quantity x plus three is greater than eight. Solve for x.

55. Three times the quantity x minus seventeen is less than or equal to twice the quantity x plus eleven. Solve for x.

56. Three x minus two is greater than or equal to seven and is less than nineteen. Solve for x.

In Exercises 57 and 58, write each inequality expression as a compound inequality.

57. $x > -3$ and $x \le 4$

58. $x \ge 0$ and $x < \pi$

In Exercises 59–64, solve each inequality algebraically and graphically. Then use a table to check this solution and describe the solution verbally.

59. $x + 3 > 1$

60. $\dfrac{x}{2} - 1 \le 1$

61. $\dfrac{x}{2} + 4 \ge 1 - \dfrac{x}{2}$

62. $2x - 1 < x + 2$

63. $2(x + 2) \le 3(x + 1)$

64. $|x - 2| < 2$

Intersections and Unions

In Exercises 65–68, determine $A \cap B$ and $A \cup B$ for the given intervals A and B.

65. $A = (1, 4), B = (2, 7)$

66. $A = [2, 5), B = [3, 6]$

67. $A = (-\infty, 2], B = [-3, \infty)$

68. $A = (-\infty, 5), B = [-2, 6)$

In Exercises 69 and 70, solve each inequality.

69. $x - 2 < 3$ and $3 > x + 2$

70. $x - 2 < 4x + 2$ or $6x - 6 < 4x - 12$

Connecting Concepts to Applications

71. Electricity Production from Wind and Coal In 1990, only 3% of all electricity needed in one community was generated by wind and 43% was generated by coal. By 2000, 30% of the electricity needed was generated by wind and 20% was generated by coal. The graph of y_1 represents the percent of electricity generated by wind. The graph of y_2 represents the percent of electricity generated by coal. Use the graphs that follow to solve the following.
   a. $y_1 = y_2$
   b. $y_1 < y_2$
   c. $y_1 > y_2$
   d. Interpret the meaning of each of these solutions.

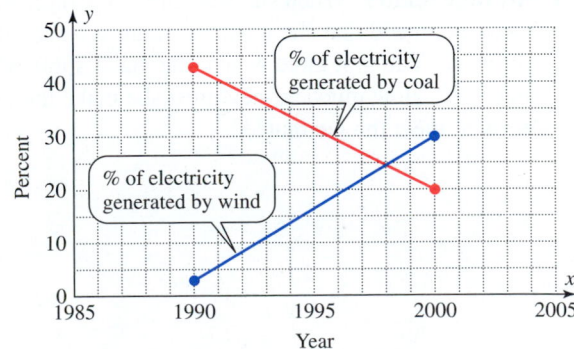

72. Repair Shop Charges The accompanying tables display the charges by two TV repair shops based on the number of hours required for a repair. Use these tables to solve the following.

   a. $y_1 = y_2$
   b. $y_1 < y_2$
   c. $y_1 > y_2$
   d. Interpret the meaning of each of these solutions.

Shop A

Hours x	Cost y_1 ($)
1	50
2	85
3	120
4	155

Shop B

Hours x	Cost y_2 ($)
1	55
2	85
3	115
4	145

In Exercises 73–78, solve each word problem.

73. Basketball Average A basketball player scored 17 points, 27 points, and 18 points in the first three games of the season. How many points will she have to score in the fourth game to average at least 20 points for the first four games?

74. Average Salary When she was hired, an employee was guaranteed that she would average at least $500 per week. The first three weeks she made $560, $450, and $480. How much money must she make the fourth week to meet the guarantee?

75. Perimeter of a Rectangle The width of a rectangle must be exactly 12 cm. If the perimeter must be between 44 and 64 cm, determine the length that can be used for the rectangle.

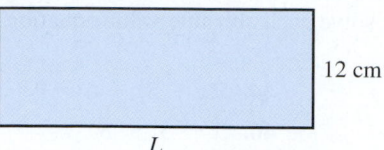

12 cm

L

76. The Length of a Golf Shot Write a compound inequality that expresses the length of the tee shot that a golfer must make for the ball to clear the water and to land safely on the green of the golf hole shown here.

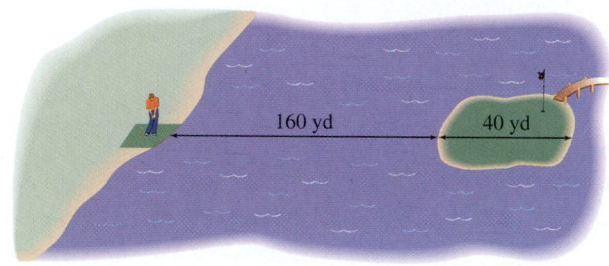

160 yd   40 yd

77. Tolerance of Rocket Fuel To adjust the flight of a Mars probe, a rocket is allowed to burn 7.2 g of fuel with a tolerance of 0.25 g. Express this tolerance interval as an absolute value inequality, and determine the lower and upper limits of the interval.

78. Costs and Revenue The cost of printing x advertising posters includes a fixed cost of $75 and a variable cost of $0.50 per poster. All posters are custom-ordered, and the charge for printing these posters includes a setup fee of $50 plus a charge of $1 per poster. Let y_1 represent the income received from an order for x posters and y_2 represent the cost of printing x posters. Determine the values of x for which $y_1 < y_2$, the loss interval for this order. Also determine the values of x for which $y_1 > y_2$, the profit interval for this order.

In Exercises 79–81, mentally estimate the solution of each inequality and then calculate the exact solution.

Problem	Mental Estimate	Exact Solution
79. $x - 2.53 < 7.52$		
80. $-2.53x \geq 5.0853$		
81. $10.89 < 9.9x \leq 29.601$		

Chapter 4 | Mastery Test

Objective 4.1.1 Identify Linear Inequalities and Check a Possible Solution of an Inequality

1. Determine whether $x = 10$ is a solution of each inequality.
   a. $4x - 1 \geq x + 29$
   b. $2(x + 15) > -2(x + 5)$
   c. $5(x + 2) < 2(x + 20)$
   d. $5(x - 2) \leq 3(x - 10)$

Objective 4.1.2 Solve Linear Inequalities in One Variable Using the Addition-Subtraction Principle for Inequalities

2. Solve each inequality algebraically.
   a. $x + 12 \leq 17$
   b. $x - 9 < -11$
   c. $4x + 3 > 3x - 8$
   d. $-2x - 5 \geq -3x - 1$

Objective 4.1.3 Use Graphs and Tables to Solve Linear Inequalities in One Variable

3. In parts **a** and **b**, use the graph to solve each inequality.
   a. $\dfrac{x}{3} - 2 < -\dfrac{1}{2}(x - 1)$
   b. $\dfrac{x}{3} - 2 \geq -\dfrac{1}{2}(x - 1)$

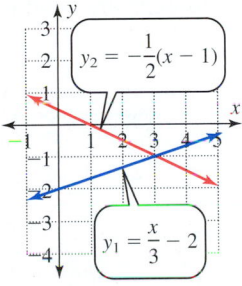

In parts **c** and **d**, use the table to solve each inequality.
   c. $4(x + 1) \leq 2(x + 6)$
   d. $4(x + 1) > 2(x + 6)$

x	$y_1 = 4(x + 1)$	$<, =,$ or $>$	$y_2 = 2(x + 6)$
0	4		12
1	8		14
2	12		16
3	16		18
4	20		20
5	24		22
6	28		24

Objective 4.2.1 Solve Linear Inequalities Using the Multiplication-Division Principle for Inequalities

4. Solve each inequality algebraically.
   a. $-11x \leq 165$
   b. $\dfrac{3x}{7} > 105$
   c. $8(x + 3) < 4(x + 4) - (10 - 2x)$
   d. $-5 \leq \dfrac{4x - 9}{3}$

Objective 4.3.1 Identify an Inequality That Is a Contradiction or an Unconditional Inequality

5. Identify each of these inequalities as a conditional inequality, an unconditional inequality, or a contradiction, and then solve the inequality.
   a. $x + x < x$
   b. $2(3x + 5) \geq 3(2x - 6)$
   c. $3(8x + 1) > 4(6x + 5)$
   d. $2x + 3 \leq 3x + 3$

Objective 4.3.2 Write the Intersection or Union of Two Intervals

6. Determine each intersection and union.
   a. $[-3, 7) \cup [2, 9]$ b. $[-3, 7) \cap [2, 9]$
   c. $(-3, 3) \cup (-1, 5)$ d. $(-\infty, 5] \cap (-2, \infty)$

Objective 4.3.3 Solve Compound Inequalities

7. Solve each of these inequalities.
   a. $-30 < -6x \leq 48$
   b. $-30 \leq x - 6 < 48$
   c. $-1 \leq 4x + 3 \leq 19$
   d. $2x + 1 \leq -7$ or $3x - 2 \geq 7$

Objective 4.4.1 Use Absolute Value Notation to Represent Intervals

8. Write an absolute value equation or inequality to represent each set of points.
   a. -2 and 2 b. $(-2, 2)$ c. $(-\infty, -2] \cup [2, \infty)$

Objective 4.4.2 Solve Absolute Value Equations and Inequalities

9. Solve each equation and inequality algebraically.
   a. $|x| = 8$ b. $|2x + 5| = 49$
   c. $|x - 5| > 4$ d. $|5x - 3| + 2 \leq 10$

In parts **e** and **f**, use the graph to solve each inequality.
   e. $|2x - 3| < 3$
   f. $|2x - 3| > 3$

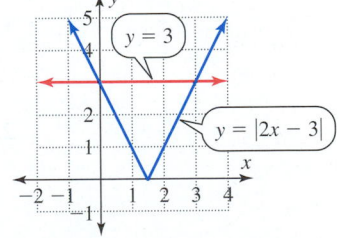

In parts **g** and **h**, use the table to solve each inequality.

g. $\left| \dfrac{x}{2} + 1 \right| \leq 1$

h. $\left| \dfrac{x}{2} + 1 \right| \geq 1$

x	$y = \left|\dfrac{x}{2} + 1\right|$
-5	1.5
-4	1.0
-3	0.5
-2	0.0
-1	0.5
0	1.0
1	1.5

Objective 4.5.1 Identify a Solution of a Linear Inequality in Two Variables

10. Determine whether $(3, -5)$ or $\left(\dfrac{2}{3}, -\dfrac{4}{5}\right)$ satisfies

$6x - 5y \leq 40$.

Objective 4.5.2 Graph a Linear Inequality in Two Variables

11. Graph each of these linear inequalities.

   a. $2x - 3y < 6$ **b.** $y \geq \dfrac{x}{2}$

   c. $x \leq -2$ **d.** $y > -2$

Objective 4.5.3 Graph a System of Linear Inequalities

12. Graph the solution of each system of inequalities.

   a. $x \geq 2$ **b.** $y > -2$
       $x \leq 4$ $y < 3$

   c. $x + 2y \geq 4$ **d.** $x \geq 0$
      $-2x + y < 4$ $y \geq 0$
                        $2x + y \leq 4$

Chapter 4 | **Group Project**

Tolerance Intervals and Business Decisions

An engineer is responsible for the installation of the machinery that produces rollers for laser jet printers. Part of his job is to ensure that the rollers produced meet the specifications in a contract with one of the largest customers for these rollers. These rollers are supposed to be 2.50 cm in diameter, and 98.5% or more of the delivered rollers must satisfy the tolerance interval $|x - 2.50| < 0.015$, where the units are centimeters.

1. **a.** Express this tolerance interval in inequality notation. _____

     b. What is the lower limit of this tolerance interval? _____

     c. What is the upper limit of this tolerance interval? _____

2. As part of the daily quality control check, workers randomly select 20 rollers each hour to carefully measure. The table shown here gives the measurements taken between 8:00 and 9:00 on a Monday morning.

a. Using check marks, complete the table to indicate whether each measurement is acceptable or unacceptable.

Roller No.	Measurement (cm)	Acceptable	Unacceptable
1	2.51		
2	2.48		
3	2.48		
4	2.49		
5	2.50		
6	2.50		
7	2.49		
8	2.50		
9	2.51		
10	2.52		
11	2.50		
12	2.49		
13	2.48		
14	2.50		
15	2.50		
16	2.49		
17	2.47		
18	2.50		
19	2.51		
20	2.49		

b. How many of these 20 rollers were acceptable? _____

c. How many of these 20 rollers were unacceptable? _____

d. What percent of these 20 rollers was acceptable? _____

3. Based on data collected over several days, the engineer in charge of this machinery actually expects 99.5% of the rollers produced to be within the tolerance interval $|x - 2.50| < 0.015$. Suppose that data collected from 100 rollers during the first 5 hours of operation on Monday suggest trouble with the machinery. The samples yield 95 rollers that are acceptable and 5 rollers that are unacceptable. The next routine downtime for maintenance is not scheduled until midnight on Friday night.

The current plant schedule is to produce 20,000 rollers on the Monday through Friday shifts. All this production will be shipped the next day to the plant's primary customer for use in its production on the following Monday.

The engineer who has received this information must recommend a course of action for the plant manager. There are three options:

Option A: *Continue to operate the plant as it is now and ship the next 20,000 rollers to the customer, requiring that 98.5% or more of the delivered rollers must satisfy the tolerance interval $|x - 2.50| < 0.015$ where the units are centimeters. If these rollers fail this customer's inspections, the order may be rejected and a $13,000,000 contract can be canceled.*

Option B: *Continue production but use a backup plan of putting quality control inspectors on the packaging line to examine the product. Past data show that 99% of the rollers screened by inspectors that are marked as acceptable are actually acceptable. (Even the inspectors make some errors.) Hiring a full crew of inspectors will cost the company $2,500 to pay the inspectors and an additional $3,500 in lost materials because of the discarded rejects that are produced. The cost of discarded rollers could increase if the problem with the machinery gets worse instead of staying the same.*

Option C: *Stop production immediately and readjust the machinery to bring it back to specifications. Estimated time to readjust the equipment is 4 hours. This will mean one shift is sent home 3 hours early and another shift will be paid for 1 hour of work before production can resume. This will also result in another 4 hours of overtime to make up for lost production. The cost for the added maintenance work is estimated to be $1,250, and the cost for the extra production labor is $5,000.*

Which option would you recommend? _____

Describe why you would select this option rather than one of the other two. _____

_____

_____

_____

Exponents and Operations with Polynomials

Chapter Outline

The Ozone Layer

The *1987 Montreal Protocol on Substances That Deplete the Ozone Layer* restricted the use of chlorofluorocarbons (CFCs), which damage the ozone layer above the earth. The accompanying table gives the world production of CFCs for selected years after 1950 as measured in thousands of tons. In Exercises 75 and 76 of Section 5.4, these data are modeled by a third-degree polynomial equation.

x Years after 1950	C(x) CFCs (thousands of tons)
0	60
10	100
20	700
30	900
40	1,250
50	800

| Section 5.1 | Product and Power Rules for Exponents |

Mathematics knows no races or geographic boundaries; for mathematics, the cultural world is one country. —DAVID HILBERT (1862–1943), GERMAN MATHEMATICIAN

Objectives:

1. Convert between exponential form and expanded form.
2. Use the product rule for exponents.
3. Use the power rule for exponents.

A Mathematical Note

In his authoritative *History of Mathematics*, David Eugene Smith attributes our present use of integral exponents to René Descartes (1596–1650).

One of the themes of this book is that algebra presents the language and symbolism of mathematics—a language and symbolism invented by humans to make things easier to represent and examine. Obviously, a student of algebra must study this notation in order to use it effectively—but please keep in mind that without algebra, mathematical concepts will be very cumbersome and lengthy to describe. Exponential notation, first examined in Section 1.6, is one example of a compact algebraic notation. This notation is reviewed in the next box.

1. Convert Between Exponential Form and Expanded Form

Exponential Notation

Algebraically	Verbally	Numerical Examples
For any natural number n: $b^n = \underbrace{b \cdot b \cdot \cdots \cdot b}_{n \text{ factors of } b}$ with base b and exponent n.	For any natural number n: b^n is the product of b used as a factor n times. The expression b^n is read as "b to the nth power."	$7^4 = 7 \cdot 7 \cdot 7 \cdot 7$ $(-3)^2 = (-3)(-3)$

What Is Meant by Saying That Exponential Notation Is a Compact Algebraic Notation?

To answer this, consider the expanded form and the equivalent exponential form for 1.07 used as a factor 10 times.

Expanded Form   **Exponential Form**

$(1.07)(1.07)(1.07)(1.07)(1.07)(1.07)(1.07)(1.07)(1.07)(1.07)$ $(1.07)^{10}$

The calculator display shown confirms that both results are identical. It also illustrates that the exponential form makes it easier to see that there are exactly 10 factors of 1.07. Most students prefer to enter the compact exponential form into a calculator or spreadsheet. This is an important consideration because expressions like $(1.07)^{10}$ arise in many applications in the workplace. For example, $(1.07)^{10}$ can be used to determine the value of each $1 invested at a 7% annual rate of return for a period of 10 years (1.07 is 100% plus 7%).

```
1.07*1.07*1.07*1
.07*1.07*1.07*1.
07*1.07*1.07*1.0
7
           1.967151357
1.07^10
           1.967151357
```

Before examining two rules for simplifying exponential expressions, we will practice interpreting exponential notation. The most common error made in evaluating exponential expressions is using the wrong base. If there are no symbols of grouping, only the constant or variable immediately to the left of the exponent is the base. Recall from Section 1.6 that $(-3)^2 = (-3)(-3) = 9$, whereas $-3^2 = -(3 \cdot 3) = -9$. The subtle but significant differences in the expressions in each part of Example 1 are very important for you to understand.

Example 1   Identifying the Base of an Exponential Expression

Write each exponential expression in expanded form.

Solution

(a) x^4	$x \cdot x \cdot x \cdot x$	The base is x, and the exponent is 4.
(b) $-x^4$	$-x \cdot x \cdot x \cdot x$	The base is x, not $-x$, and the exponent is 4.
(c) $(-x)^4$	$(-x)(-x)(-x)(-x)$	The base is $-x$, and the exponent is 4.
(d) $(xy)^3$	$(xy)(xy)(xy)$	The base is xy, and the exponent is 3.
(e) xy^3	$x \cdot y \cdot y \cdot y$	The base of the exponent 3 is y, not xy.

Self-Check 1

Write each exponential expression in expanded form.

a. $-5a^3$ **b.** $(-5a)^3$ **c.** $a^3 b^2$

To understand the properties of exponents, it is important that you note the base of the exponent in each part of Examples 1 and 2. You can also practice this skill by completing Self-Checks 1 and 2.

Example 2   Identifying Bases and Exponents

Write each expression in exponential form.

Solution

(a) $x \cdot x \cdot y \cdot y$	$x^2 y^2$
(b) $x \cdot x + y \cdot y$	$x^2 + y^2$
(c) $(x + y)(x + y)$	$(x + y)^2$
(d) $-3 \cdot a \cdot a \cdot a \cdot a$	$-3a^4$
(e) $(-3a)(-3a)(-3a)(-3a)$	$(-3a)^4$

Self-Check 2

Write each expression in exponential form.

a. $-x \cdot x$ **b.** $(-x)(-x)$ **c.** $x \cdot y \cdot y \cdot y$

2. Use the Product Rule for Exponents

Is There a Shortcut Rule for Multiplying Exponential Expressions?

Yes, if these expressions have the same base. The product rule for exponents can simplify our work with exponential expressions. This property of exponents follows directly from the definition of natural number exponents. We start by examining the logic behind the product rule for exponents. The product $x^3 \cdot x^4$ is written here in expanded form. Note that three factors of x plus four more factors of x results in seven factors of x.

$$x^3 x^4 = \overbrace{x \cdot x \cdot x}^{\substack{\text{Three factors} \\ \text{of } x}} \cdot \overbrace{x \cdot x \cdot x \cdot x}^{\substack{\text{Four factors} \\ \text{of } x}} = \overbrace{x \cdot x \cdot x \cdot x \cdot x \cdot x \cdot x}^{\substack{\text{Seven factors} \\ \text{of } x}} = x^7$$

Thus, $x^3 \cdot x^4 = x^{3+4} = x^7$. The fact that x^3x^4 and x^7 are equivalent expressions means they have the same value for all values of x. This is confirmed by comparing the values of Y_1 and Y_2 in the accompanying table.

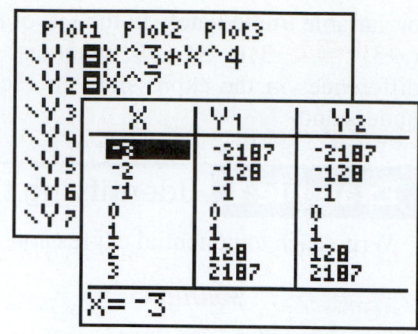

In general, the product $x^mx^n = x^{m+n}$ has the base x used as a factor a total of $m + n$ times. Note that m factors of x plus n more factors of x results in $m + n$ factors of x.

$$x^mx^n = \overbrace{(x \cdot x \cdots x)}^{\substack{m \text{ factors} \\ \text{of } x}} \cdot \overbrace{(x \cdot x \cdots x)}^{\substack{n \text{ factors} \\ \text{of } x}} = \overbrace{x \cdot x \cdots x}^{\substack{m+n \text{ factors} \\ \text{of } x}} = x^{m+n}$$

Product Rule for Exponents

For any real number x and natural numbers m and n:

Algebraically	Verbally	Algebraic Example
$x^m \cdot x^n = x^{m+n}$	To multiply two factors with the same base, use the common base and add the exponents.	$x^3 \cdot x^4 = x^7$

Although it is useful to memorize "add the exponents" for multiplying two factors, it is also important to remember that the bases must be the same.

Example 3   Using the Product Rule

Use the product rule to simplify each expression.

(a) z^3z^{10} (b) $v^2v^4v^5$ (c) $-x^7x^8$ (d) $(3y^4)(5y^5)$ (e) y^3x^2

Solution

(a) $z^3z^{10} = z^{3+10}$ Add the exponents with the common base z.

$\quad = z^{13}$

(b) $v^2v^4v^5 = v^{2+4+5}$ Add the exponents with the common base v.

$\quad = v^{11}$ The product rule can be extended to any number of factors that have the same base.

(c) $-x^7x^8 = -x^{7+8}$ $-x^7x^8 = -(x^7x^8)$; the base for each exponent is x, not $-x$.

$\quad = -x^{15}$

(d) $(3y^4)(5y^5) = (3)(5)y^{4+5}$ Add the exponents with the common base y.

$\quad = 15y^9$

(e) $y^3x^2 = x^2y^3$ Since these factors do not have the same base, this expression cannot be simplified. However, factors are usually written in alphabetical order.

Self-Check 3

Simplify each expression.

a. $x^2 x^3$ **b.** $(2y^9)(7y^{12})$ **c.** $m^2 n^4$ **d.** $-z^2 z^9$

3. Use the Power Rule for Exponents

Is There a Shortcut Rule for Raising a Power to a Power?

Yes, the power rule for exponents allows us to quickly simplify expressions like $(x^2)^4$. We start by examining the logic for this rule. The expression $(x^2)^4$ is expanded here.

$$(x^2)^4 = \overbrace{x^2 \cdot x^2 \cdot x^2 \cdot x^2}^{\text{Four factors of } x^2}$$

$$= \overbrace{(x \cdot x)(x \cdot x)(x \cdot x)(x \cdot x)}^{\text{Four groups of two factors of } x}$$

$$= \overbrace{x \cdot x \cdot x \cdot x \cdot x \cdot x \cdot x \cdot x}^{2 \cdot 4 \text{ factors of } x}$$

$$= x^8$$

Thus, $(x^2)^4 = x^8$. This fact is confirmed by comparing the values of Y_1 and Y_2 in the accompanying table.

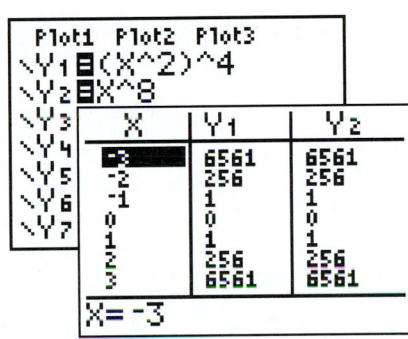

In general,

$$(x^m)^n = \overbrace{x^m \cdot x^m \cdots x^m}^{n \text{ factors of } x^m}$$

$$= \overbrace{(x \cdot x \cdots x)(x \cdot x \cdots x) \cdots (x \cdot x \cdots x)}^{n \text{ groups of } m \text{ factors of } x}$$

$$= \overbrace{x \cdot x \cdot x \cdots x \cdot x \cdot x}^{m \cdot n \text{ factors of } x}$$

$$= x^{mn}$$

Power Rule for Exponents

For any real number x and natural numbers m and n:

Algebraically	Verbally	Algebraic Example
$(x^m)^n = x^{mn}$	To raise a power to a power, multiply the exponents.	$(x^2)^4 = x^8$

Example 4 Using the Power Rule

Simplify each expression.

(a) $(b^3)^4$ **(b)** $-(x^3)^2$

Solution

(a) $(b^3)^4 = b^{3 \cdot 4}$ To raise a power to a power, multiply the exponents.

$\qquad\quad = b^{12}$

(b) $-(x^3)^2 = -x^{3 \cdot 2}$ Multiply the exponents. The base is x, not $-x$.

$\qquad\qquad\quad = -x^6$

Self-Check 4

Simplify each expression.

a. $(x^2)^5$ **b.** $(x^5)^2$ **c.** $(-x^5)^2$ **d.** $(-x^2)^5$ **e.** $-(x^5)^2$

Can the Power Rule for Exponents Be Applied to Products and Quotients?

Yes; this is described in the accompanying box. We develop the logic for these rules immediately following this box.

Raising Products and Quotients to a Power

For any real numbers x and y and natural number m:

Algebraically	Verbally	Algebraic Example
$(xy)^m = x^m y^m$	To raise a product to a power, raise each factor to this power.	$(xy)^3 = x^3 y^3$
$\left(\dfrac{x}{y}\right)^m = \dfrac{x^m}{y^m}$ for $y \neq 0$	To raise a quotient to a power, raise both the numerator and the denominator to this power.	$\left(\dfrac{x}{y}\right)^3 = \dfrac{x^3}{y^3}$

A Product to a Power: As an example of raising a product to a power, we use the expression $(xy)^3$. It can be expanded as

$$(xy)^3 = \overbrace{(xy)(xy)(xy)}^{\text{Three factors of } xy}$$

$$= \overbrace{(x \cdot x \cdot x)}^{\text{Three factors of } x} \cdot \overbrace{(y \cdot y \cdot y)}^{\text{Three factors of } y}$$

$$= x^3 y^3$$

Thus, $(xy)^3 = x^3 y^3$

In general,

$$(xy)^m = \overbrace{(xy)(xy) \cdots (xy)}^{m \text{ factors of } xy}$$

$$= \overbrace{(x \cdot x \cdots x)}^{m \text{ factors of } x}\overbrace{(y \cdot y \cdots y)}^{m \text{ factors of } y}$$

$$= x^m y^m$$

A Quotient to a Power: As an example of raising a quotient to a power, we use the expression $\left(\dfrac{x}{y}\right)^3$. It can be expanded as

$$\left(\frac{x}{y}\right)^3 = \overbrace{\left(\frac{x}{y}\right)\left(\frac{x}{y}\right)\left(\frac{x}{y}\right)}^{\text{Three factors of } \frac{x}{y}}$$

$$= \frac{\overbrace{x \cdot x \cdot x}^{\text{Three factors of } x}}{\underbrace{y \cdot y \cdot y}_{\text{Three factors of } y}}$$

$$= \frac{x^3}{y^3}$$

Thus, $\left(\dfrac{x}{y}\right)^3 = \dfrac{x^3}{y^3}$.

In general,

$$\left(\frac{x}{y}\right)^m = \overbrace{\left(\frac{x}{y}\right)\left(\frac{x}{y}\right)\cdots\left(\frac{x}{y}\right)}^{m \text{ factors of } \frac{x}{y}}$$

$$= \frac{\overbrace{x \cdot x \cdots x}^{m \text{ factors of } x}}{\underbrace{y \cdot y \cdots y}_{m \text{ factors of } y}}$$

$$= \frac{x^m}{y^m}$$

In Example 5, we use the rule for raising a product to a power. Then in Example 6, we use the rule for raising a quotient to a power.

Example 5   Simplifying a Product to a Power

Simplify each expression.

(a) $(ab)^5$ (b) $(a^2b)^5$ (c) $(2t)^3$

(d) $(-3t)^4$ (e) $-(3t)^4$ (f) $-3t^4$

Solution

(a) $(ab)^5 = a^5b^5$ For a product to a power, $(xy)^m = x^my^m$.

(b) $(a^2b)^5 = (a^2)^5b^5$ For a product to a power, $(xy)^m = x^my^m$.

$\qquad = a^{10}b^5$ Use the power rule $(x^m)^n = x^{m \cdot n}$.

(c) $(2t)^3 = 2^3t^3$ For a product to a power, $(xy)^m = x^my^m$.

$\qquad = 8t^3$

(d) $(-3t)^4 = (-3)^4t^4$ Notice the subtle distinctions among parts (d), (e), and (f). In part (d), the base is $-3t$.

$\qquad = 81t^4$

(e) $-(3t)^4 = -(3^4 \cdot t^4)$ In part (e), the base is $3t$.

$\qquad = -(81t^4)$

$\qquad = -81t^4$

(f) $-3t^4 = -3t^4$ In part (f), the exponent applies to only the base of t; thus this expression is already in simplified form.

Example 6   Simplifying a Quotient to a Power

Simplify each expression, assuming each variable is nonzero.

(a) $\left(\dfrac{a}{b}\right)^5$ **(b)** $\left(\dfrac{3}{v}\right)^4$ **(c)** $\left(-\dfrac{5}{y}\right)^2$ **(d)** $-\left(\dfrac{5}{y}\right)^2$ **(e)** $\left(\dfrac{2v^2}{5w^4}\right)^3$

Solution

(a) $\left(\dfrac{a}{b}\right)^5 = \dfrac{a^5}{b^5}$ Use the quotient to a power rule $\left(\dfrac{x}{y}\right)^m = \dfrac{x^m}{y^m}$.

(b) $\left(\dfrac{3}{v}\right)^4 = \dfrac{3^4}{v^4}$ Use the quotient to a power rule $\left(\dfrac{x}{y}\right)^m = \dfrac{x^m}{y^m}$.

$\qquad = \dfrac{81}{v^4}$

(c) $\left(-\dfrac{5}{y}\right)^2 = \left(\dfrac{-5}{y}\right)^2$ The fraction $-\dfrac{a}{b}$ equals $\dfrac{-a}{b}$.

$\qquad = \dfrac{(-5)^2}{y^2}$ Use the quotient to a power rule $\left(\dfrac{x}{y}\right)^m = \dfrac{x^m}{y^m}$.

$\qquad = \dfrac{25}{y^2}$

(d) $-\left(\dfrac{5}{y}\right)^2 = -\left(\dfrac{5^2}{y^2}\right)$ Use the quotient to a power rule $\left(\dfrac{x}{y}\right)^m = \dfrac{x^m}{y^m}$.

$\qquad = -\dfrac{25}{y^2}$ The exponent applies to only the base $\dfrac{5}{y}$, not $-\dfrac{5}{y}$.

(e) $\left(\dfrac{2v^2}{5w^4}\right)^3 = \dfrac{(2v^2)^3}{(5w^4)^3}$ Use the quotient to a power rule.

$\qquad = \dfrac{(2)^3(v^2)^3}{(5)^3(w^4)^3}$ Use the product to a power rule.

$\qquad = \dfrac{8v^6}{125w^{12}}$

Example 7 reviews some problems that were covered when order of operations was discussed in Section 1.6. Although these problems are not difficult, they can easily be misinterpreted. It is important to pay careful attention to the small changes in notation that dictate the order of operations.

Example 7   Evaluating Algebraic Expressions

Evaluate each expression for $x = -3$ and $y = -5$.

(a) $(-x)^2$ **(b)** $-x^2$ **(c)** $x^2 + y^2$ **(d)** $(x + y)^2$

Solution

(a) $(-x)^2 = [-(-3)]^2$

$\qquad\quad = (3)^2$

$\qquad\quad = 9$

Note in parts (a) and (b) that the difference in notation leads to different answers. In part (a) the base for the exponent is $-x$, but in part (b) the base is just x.

(b) $-x^2 = -(x^2)$

$\qquad\quad = -[(-3)^2]$

$\qquad\quad = -(9)$

$\qquad\quad = -9$

(c) $x^2 + y^2 = (-3)^2 + (-5)^2$

$\qquad\qquad = 9 + 25$

$\qquad\qquad = 34$

In parts (c) and (d), the different notations indicate distinctive orders of operation. This produces different answers.

(d) $(x + y)^2 = [-3 + (-5)]^2$

$\qquad\qquad = (-8)^2$

$\qquad\qquad = 64$

Self-Check 7

Evaluate each expression for $x = -3$ and $y = -5$.

a. xy^2 **b.** $(xy)^2$ **c.** $x + y^2$ **d.** $(x - y)^2$

In Section 1.6, we used the distributive property to add like terms. In Example 8, we review this topic and compare it to multiplying factors by using the product rule.

Example 8   Comparing Addition and Multiplication

(a) Add $x + x$ and multiply $x \cdot x$.

(b) Add $v^3 + v^3$ and multiply $v^3 \cdot v^3$.

(c) Add $3x^2 + 5x^2$ and multiply $(3x^2)(5x^2)$.

(d) Add $x^2 + x^4$ and multiply $x^2 \cdot x^4$.

Solution

Sum	Product

(a) $x + x = 2x$ $x \cdot x = x^2$

(b) $v^3 + v^3 = 2v^3$ $v^3 \cdot v^3 = v^6$

(c) $3x^2 + 5x^2 = 8x^2$ $(3x^2)(5x^2) = (3)(5)(x^2)(x^2)$

$\qquad\qquad\qquad\qquad\qquad\qquad\qquad\qquad = 15x^4$

(d) $x^2 + x^4$ cannot be simplified further as these terms are unlike. $x^2 \cdot x^4 = x^6$

Self-Check 8

Add the like terms and simplify the products.

	Add	Multiply
a.	$y + y + y$	$y \cdot y \cdot y$
b.	$m^2 + m^2$	$m^2 \cdot m^2$
c.	$4a^3 + 7a^3$	$(4a^3)(7a^3)$
d.	$w^2 + w^3$	$w^2 \cdot w^3$

Self-Check Answers

1. **a.** $-5 \cdot a \cdot a \cdot a$
   b. $(-5a)(-5a)(-5a)$
   c. $a \cdot a \cdot a \cdot b \cdot b$
2. **a.** $-x^2$ **b.** $(-x)^2$
   c. xy^3
3. **a.** x^5 **b.** $14y^{21}$
   c. m^2n^4 **d.** $-z^{11}$

4. **a.** x^{10} **b.** x^{10}
   c. x^{10} **d.** $-x^{10}$
   e. $-x^{10}$
5. **a.** x^5y^5 **b.** $x^{15}y^5$
6. **a.** $\dfrac{x^5}{y^5}$ **b.** $\dfrac{x^5}{y^{20}}$
7. **a.** -75 **b.** 225
   c. 22 **d.** 4

	Sum	Product
8. a.	$3y$	y^3
b.	$2m^2$	m^4
c.	$11a^3$	$28a^6$
d.	$w^2 + w^3$ cannot be simplified further as these terms are unlike.	w^5

5.1   Using the Language and Symbolism of Mathematics

1. In the exponential expression x^n, x is the _____ and n is the _____.

2. In the product $x \cdot x$, both x and x are called _____.

3. In the sum $x + x$, both x and x are called _____ or _____.

4. In the exponential expression $-x^4$ the base is _____.

5. In the exponential expression $(-x)^4$ the base is _____.

6. In the exponential expression $(xy)^4$ the base is _____.

7. In the exponential expression xy^4 the base of the exponent 4 is _____.

8. In the expression $x^5 = x \cdot x \cdot x \cdot x \cdot x$, x^5 is referred to as the _____ form and $x \cdot x \cdot x \cdot x \cdot x$ is referred to as the _____ form.

9. The product rule for exponents states that $x^m \cdot x^n =$ _____ for any real number x and natural numbers m and n.

10. The power rule for exponents states that $(x^m)^n =$ _____ for any real number x and natural numbers m and n.

11. For any real numbers x and y and any natural number m, $(xy)^m =$ _____.

12. For any real numbers x and y, $y \neq 0$, and any natural number m, $\left(\dfrac{x}{y}\right)^m =$ _____.

5.1   Quick Review

1. Evaluate -5^2.

2. Evaluate $(-5)^2$.

3. Evaluate $2(3 + 4)$.

4. Evaluate $(3 + 4)^2$.

5. Evaluate $3^2 + 4^2$.

5.1 Exercises

Objective 1 Convert Between Exponential Form and Expanded Form

In Exercises 1–4, write each expression in exponential form.

1. a. $x \cdot x \cdot x \cdot x \cdot x$ **b.** $x \cdot y \cdot y \cdot y \cdot y$
c. $(x \cdot y)(x \cdot y)(x \cdot y)$

2. a. $-3 \cdot a \cdot a \cdot a \cdot a$ **b.** $(-3a)(-3a)(-3a)(-3a)$
c. $-(3a)(3a)(3a)(3a)$

3. a. $(a + b)(a + b)(a + b)$ **b.** $a \cdot a \cdot a + b \cdot b \cdot b$
c. $a + b \cdot b \cdot b$

4. a. $a \cdot a \cdot a + b$ **b.** $(a + b)(a + b)$
c. $(ab)(ab)(ab)(ab)$

In Exercises 5–10, write each exponential expression in expanded form.

5. a. m^6 **b.** $(-m)^6$ **c.** $-m^6$

6. a. mn^6 **b.** $(mn)^6$ **c.** $6m^2$

7. a. $(m + n)^2$ **b.** $m^2 + n^2$ **c.** $m + n^2$

8. a. $m^2 + n^3$ **b.** m^2n^3 **c.** $(m + n)^3$

9. a. $\dfrac{a^2}{b^3}$ **b.** $\left(\dfrac{a}{b}\right)^4$ **c.** $\dfrac{a}{b^4}$

10. a. $\dfrac{x^3}{y^2}$ **b.** $\left(\dfrac{x}{y}\right)^3$ **c.** $\dfrac{x^3}{y}$

Objective 2 Use the Product Rule for Exponents

In Exercises 11 and 12, simplify each expression. Do this by first converting the expression to expanded form and counting the factors of x. Then write your answer as a power of x.

11. x^5x^3 **12.** x^6x^2

In Exercises 13–18, simplify each expression.

13. x^9x^{11} **14.** y^7y^3

15. $(3m^4)(5m^6)$ **16.** $(6m^5)(4m^3)$

17. $-x(2x^2)(3x^3)$ **18.** $-y(3y^3)(5y^5)$

Objective 3 Use the Power Rule for Exponents

In Exercises 19 and 20, simplify each expression. Do this by first converting the expression to expanded form and counting the factors of x. Then write your answer as a power of x.

19. $(x^3)^2$ **20.** $(x^2)^4$

In Exercises 21–36, simplify each expression. Assume that variables are restricted to values that prevent division by zero.

21. $(x^5)^2$ **22.** $(x^3)^7$

23. $(3xy)^2$ **24.** $(5xy)^2$

25. $(-n^2)^6$ **26.** $(-3m)^4$

27. $5(abc)^2$ **28.** $-5(xyz)^2$

29. $(xy^2)^3$ **30.** $(m^2n^3)^4$

31. $\left(\dfrac{3}{4}\right)^2$ **32.** $\left(\dfrac{4}{5}\right)^2$

33. $\left(\dfrac{4}{w}\right)^3$ **34.** $-\left(\dfrac{3}{y^3}\right)^2$

35. $\left(\dfrac{x^2}{y^3}\right)^{11}$ **36.** $\left(\dfrac{m}{n^5}\right)^7$

Skill and Concept Development

In Exercises 37–50, simplify each expression. Assume that $b \neq 0$ and $y \neq 0$.

37. a. -1^{10} **b.** $(-1)^{10}$

38. a. -1^{12} **b.** $(-1)^{12}$

39. a. $(-1)^{11}$ **b.** -1^{11}

40. a. $(-1)^{13}$ **b.** $(-1)^{13}$

41. $(-5x^2y^4)^3$ **42.** $(-5x^2y^7)^2$

43. $(-2xy^2)(-3x^2y^4)$ **44.** $(-6x^2y^3)(7x^4y^5)$

45. $(5x^2)^2(2x^3)^3$ **46.** $(2x^5)^2(3x^4)^3$

47. $\left(\dfrac{3x}{2y^2}\right)^3$ **48.** $\left(\dfrac{-2x^3}{3y}\right)^4$

49. $\left(-\dfrac{2x^2}{3y^3}\right)^4$ **50.** $\left(-\dfrac{5a^5}{3b^3}\right)^2$

In Exercises 51–60, evaluate each expression for $x = 2$ and $y = 3$.

51. a. $-x^2$ **b.** $(-x)^2$

52. a. $-y^2$ **b.** $(-y)^2$

53. a. $(xy)^2$ **b.** xy^2

54. a. $(xy)^3$ **b.** xy^3

55. a. $x^3 + y^3$ **b.** $(x + y)^3$

56. a. $x^3 - y^3$ **b.** $(x - y)^3$

57. a. $5x^2y^2$ **b.** $(5xy)^2$

58. a. $(3xy)^2$ **b.** $3x^2y^2$

59. a. $\left(-\dfrac{x^2}{y}\right)$ **b.** $\left(-\dfrac{x}{y}\right)^2$

60. a. $\left(-\dfrac{x^4}{y}\right)$ **b.** $\left(-\dfrac{x}{y}\right)^4$

Comparing Addition and Multiplication

In Exercises 61–66, add the like terms and simplify the products.

Add	Multiply
61. a. $v + v + v$	**b.** $v \cdot v \cdot v$
62. a. $w^2 + w^2$	**b.** $w^2 \cdot w^2$
63. a. $4m^3 + 6m^3$	**b.** $(4m^3)(6m^3)$
64. a. $2x + 3x + 10x$	**b.** $(2x)(3x)(10x)$
65. a. $3x^2 + 2x$	**b.** $(3x^2)(2x)$
66. a. $4x^3 + 4x^2$	**b.** $(4x^3)(4x^2)$

Estimate Then Calculate

In Exercises 67–72, mentally estimate the value of each expression and then calculate the exact value.

Problem	Mental Estimate	Exact Value
67. -6.98^2		
68. $(-9.9)^4$		
69. $(1.01 + 1.01)^2$		
70. $1.01^2 + 1.01^2$		
71. $7.99^2 - 5.9^2$		
72. $(7.99 - 5.9)^2$		

Connecting Concepts to Applications

Compound interest

In Exercises 73 and 74, use the formula $A = P(1 + r)^t$ to determine the value of a principal P that is compounded for t years at an interest rate r.

73. What is the value of a principal P of \$1 compounded for 10 years at a rate of 8%?

74. What is the value of a principal P of \$1 compounded for 12 years at a rate of 7%?

Generation of electricity by a windmill

Use the following information to work Exercises 75–78.

There are several environmental reasons to promote greater generation of electricity by windmills. There are also several concerns with this generation method: noise, visual pollution, reliability, efficiency, cost, and the variability of the wind from hour to hour. To make sound business decisions, a company needs the facts related to all these concerns. One manufacturer of windmills has prepared the accompanying table that gives the number of kilowatt-hours generated from a given speed of wind. The output of the windmill actually varies directly as the cube of the speed of the wind. Thus the windmill is rather inefficient at low wind speeds and very efficient at higher wind speeds that are within its safe operating range. The function the company used to create the table is $f(x) = 0.01x^3$.

Wind Speed x (mi/h)	$f(x)$ (kW of Electricity)
0	0.00
5	1.25
10	10.00
15	33.75
20	80.00
25	156.25
30	270.00

75. a. Evaluate and interpret $f(20)$.
  b. Evaluate and interpret $f(27)$.

76. What value of x will produce $f(x) = 270$? Interpret this result.

77. Compare the results of a company installing one of these windmills in an area with a wind speed of 10 mi/h with a company that installs the same type of windmill in an area with a wind speed of 20 mi/h. How many times greater will be the power generated by the higher 20-mi/h wind?

78. Suppose the company made an error and the function it should have used to produce this table is $f(x) = 0.01x^2$. Use this formula to produce a new table of values for the same input values of x.

Connecting Algebra to Geometry

79. Use the figure to determine the area of each region.

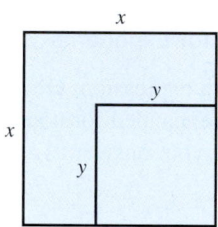

  a. The area of the larger square.
  b. The area of the smaller square.
  c. The area inside the larger square and outside the smaller square.

80. Each side of the larger square is double that of the smaller square. Compare their areas.

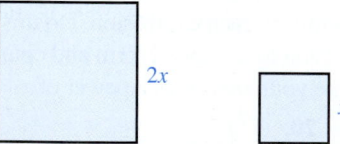

81. Each side of the larger cube is double that of the smaller cube. Compare their volumes.

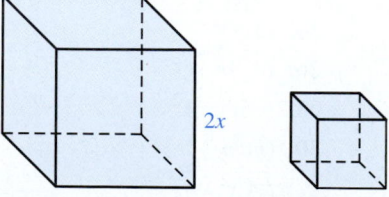

Applying Technology

An important technology skill is the ability to use calculators or spreadsheets to compare options. For example, you can test whether it is appropriate to add exponents or multiply exponents. In Exercises 82 and 83, use a calculator or a spreadsheet to complete the given table of values. Then use the values in the table to determine whether $Y_1 = Y_2$ or $Y_1 = Y_3$.

x	Y_1	Y_2	Y_3
-3			
-2			
-1			
0			
1			
2			
3			

82. $Y_1 = x^2 x^3$
$Y_2 = x^5$
$Y_3 = x^6$

83. $Y_1 = (x^2)^3$
$Y_2 = x^5$
$Y_3 = x^6$

Group discussion questions

84. Challenge Question Use the product and power rules for exponents to simplify each expression, assuming m is a natural number.
a. $x^m x^2$ **b.** $(x^m)^2$ **c.** $x^{m+1} x^{m+2}$ **d.** $(x^{m+1})^2$

85. Challenge Question One person becomes infected with a new strain of bacteria. Over the course of the next 24 hours, the victim will infect 4 more persons. On the following day, the original victim and the new victims will each infect 4 more people. Assume this infection continues in this manner for 8 days. Determine the total number of people infected on each

of the 8 days in this table. Can you determine a formula for the number infected on the nth day?

Day n	Population Infected P
0	1
1	$1 + 4 = 5$
2	
3	
4	
5	
6	
7	
8	

86. Discovery Question Use a calculator or a spreadsheet to complete the given table of values for each part. Based on your observations in parts **a** and **b**, make a conjecture for a rule for simplifying $\dfrac{x^m}{x^n}$.

x	Y_1	Y_2
-3		
-2		
-1		
0		
1		
2		
3		

a. $Y_1 = \dfrac{x^5}{x^3}$
$Y_2 = x^2$

b. $Y_1 = \dfrac{x^7}{x^4}$
$Y_2 = x^3$

5.1 | Cumulative Review

1. Use a pencil and paper to evaluate $\dfrac{3(4) - 6 \div 2}{3(4)}$.

2. Evaluate each expression.
a. $9 + 0$ **b.** $9 - 0$ **c.** $9 \cdot 0$
d. $0 \cdot 9$ **e.** $0 \div 9$ **f.** $9 \div 0$

3. Solve $5x - 3(2x + 1) = 5 - 2(x - 5)$ for x.

4. Solve $3x - 4 \le 3 - 5(x - 2)$ and give your answer in interval notation.

5. Solve $|2x - 1| \ge 9$ and give your answer in interval notation.

Section 5.2 | Quotient Rule and Zero Exponents

Objectives:

1. Use the quotient rule for exponents.
2. Simplify expressions with zero exponents.
3. Combine the properties of exponents to simplify expressions.

1. Use the Quotient Rule for Exponents

Is There a Shortcut Rule for Dividing Exponential Expressions?

Yes, the quotient rule can be used if the exponential expressions have the same base. The logic behind the quotient rule is now examined. The quotient of $\dfrac{x^5}{x^3}$ is written below in expanded form.

For $x \neq 0$, the quotient is

$$\frac{x^5}{x^3} = \frac{\overbrace{x \cdot x \cdot x \cdot x \cdot x}^{\text{Five factors of } x}}{\underbrace{x \cdot x \cdot x}_{\text{Three factors of } x}}$$

The fraction is reduced by dividing both the numerator and the denominator by three factors of x.

$$= \overbrace{x \cdot x}^{\text{Two factors of } x}$$

$$= x^2$$

Thus $\frac{x^5}{x^3} = x^2$. The fact that these two expressions are equivalent for $x \neq 0$ is confirmed by comparing the values of Y_1 and Y_2 in the accompanying table.

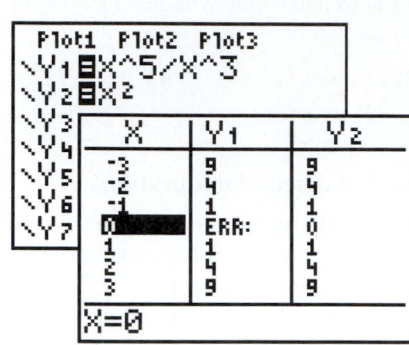

In the screen shot, note that $\frac{x^5}{x^3}$ is undefined for $x = 0$; for all other values, $\frac{x^5}{x^3} = x^2$.

In general, if m and n are natural numbers and $m > n$, then the quotient $\frac{x^m}{x^n}$ can be reduced by dividing out each of the n common factors of x. This will leave $m - n$ factors of x in the numerator; that is, $\frac{x^m}{x^n} = x^{m-n}$ for $x \neq 0$ and $m > n$. (We will examine the case where $m < n$ in Example 2 and the case where $m = n$ after Example 2.) As we note in the box, $\frac{x^m}{x^n}$ is undefined for $x = 0$.

Quotient Rule for Exponents

For any real number x and natural numbers m and n with $m > n$:*

Algebraically	Verbally	Algebraic Example
$\dfrac{x^m}{x^n} = x^{m-n}$ for $x \neq 0$ $\dfrac{x^m}{x^n}$ is undefined for $x = 0$	To divide two expressions with the same base, use the common base and subtract the exponents.	$\dfrac{x^5}{x^3} = x^2$

*We will soon note that the restriction $m > n$ is unnecessary, when we extend the quotient rule.

Example 1   Using the Quotient Rule

Use the quotient rule to simplify each expression. Assume the denominators are nonzero.

(a) $\dfrac{z^{10}}{z^3}$ **(b)** $\dfrac{m^4 n^7}{m^2 n^6}$ **(c)** $-\dfrac{7^8}{7^6}$ **(d)** $\dfrac{x^5}{y^3}$

Solution

(a) $\dfrac{z^{10}}{z^3} = z^{10-3}$ Subtract the exponents with the same base. Dividing both the numerator and denominator by 3 factors of z leaves 7 factors of z in the numerator.

$\qquad = z^7$

(b) $\dfrac{m^4 n^7}{m^2 n^6} = m^{4-2} n^{7-6}$ Subtract the exponents on the common bases m and n.

$\qquad = m^2 n^1$

$\qquad = m^2 n$

(c) $-\dfrac{7^8}{7^6} = -(7^{8-6})$ Note how much easier it is to apply the quotient rule than it is to evaluate 7^8 and 7^6 and then divide.

$\qquad = -(7^2)$

$\qquad = -49$

(d) $\dfrac{x^5}{y^3} = \dfrac{x^5}{y^3}$ Because the numerator and the denominator do not have the same base, this expression cannot be simplified.

Self-Check 1

Simplify each expression.

a. $\dfrac{m^{14}}{m^9}$ **b.** $\dfrac{x^5 y^5}{x^3 y^2}$ **c.** $\dfrac{5^8}{5^6}$ **d.** $-\dfrac{v^4}{w^2}$

The quotient rule, as developed so far, allows us to simplify $\dfrac{x^m}{x^n}$ when $m > n$. The expression $\dfrac{x^m}{x^n}$ can also be simplified when $m < n$. The reasoning is similar to that used when $m > n$. If $x \neq 0$ and m and n are natural numbers, then $\dfrac{x^m}{x^n} = \dfrac{1}{x^{n-m}}$ for $m < n$. The example $\dfrac{x^4}{x^7}$ is examined first. This example is closely related to negative exponents, which are covered in Section 5.3.

$$\frac{x^4}{x^7} = \frac{\overset{1}{\cancel{x} \cdot \cancel{x} \cdot \cancel{x} \cdot \cancel{x}}}{\underset{1}{\cancel{x} \cdot \cancel{x} \cdot \cancel{x} \cdot \cancel{x}} \cdot x \cdot x \cdot x}$$

$$= \frac{1}{x \cdot x \cdot x}$$

$$= \frac{1}{x^3}$$

Thus $\dfrac{x^4}{x^7} = \dfrac{1}{x^3}$. The fact that these two expressions are equivalent for $x \neq 0$ is confirmed by comparing the values of Y_1 and Y_2 in the accompanying table.

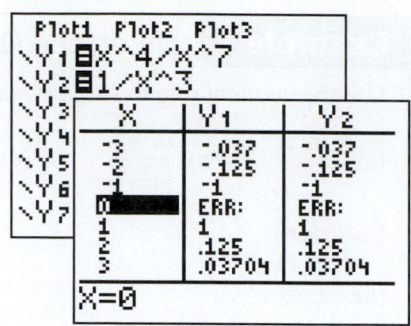

In the screen shot, note that both expressions are undefined for $x = 0$.

The properties of exponents are so important that many exercises are given on each individual property. These properties will be revisited extensively throughout the book. In Example 2 we use the quotient rule to simplify $\dfrac{x^m}{x^n}$. In each case, this is accomplished by subtracting the smaller exponent from the larger exponent.

Example 2   Simplifying Exponential Expressions

Simplify each of the following expressions.

(a) $\dfrac{x^3}{x^5}$ (b) $\dfrac{x^4 y^4}{x^6 y}$ (c) $\dfrac{x^8}{y^9}$ (d) $\dfrac{11^8}{11^{10}}$

Solution

(a) $\dfrac{x^3}{x^5} = \dfrac{1}{x^{5-3}}$ Reduce the fraction by dividing both the numerator and the denominator by the three common factors of x. This is accomplished by subtracting the smaller exponent from the larger exponent. This leaves two factors of x in the denominator.

$= \dfrac{1}{x^2}$

(b) $\dfrac{x^4 y^4}{x^6 y} = \dfrac{y^{4-1}}{x^{6-4}}$ For each base, subtract the smaller exponent from the larger exponent.

$= \dfrac{y^3}{x^2}$

(c) $\dfrac{x^8}{y^9} = \dfrac{x^8}{y^9}$ The bases are not the same, so this expression cannot be simplified.

(d) $\dfrac{11^8}{11^{10}} = \dfrac{1}{11^{10-8}}$ Note that it is easier to first apply the properties of exponents than to start by evaluating 11^8.

$= \dfrac{1}{11^2}$

$= \dfrac{1}{121}$

Self-Check 2

Simplify each expression.

a. $\dfrac{m^3}{m^{11}}$ b. $\dfrac{x^3 y^5}{x^4 y^4}$ c. $\dfrac{10^9}{10^{12}}$ d. $\dfrac{v^7}{w^8}$

2. Simplify Expressions with Zero Exponents

Can I Use Exponents Other Than the Natural Numbers 1, 2, 3, 4, ...?

Yes, we will now develop a definition of a zero exponent. Consider two approaches to simplifying $\dfrac{5^2}{5^2}$.

Option 1

$$\frac{5^2}{5^2} = \frac{25}{25} = 1$$

For option 1, reduce by dividing both the numerator and denominator by 25.

Option 2

$$\frac{5^2}{5^2} = 5^{2-2} = 5^0$$

For option 2, examine what would happen if we try to use the quotient rule and subtract the exponents.

For these two options to obtain the same result, we must define $5^0 = 1$. In general, we define $x^0 = 1$ for $x \neq 0$. This is further confirmed by examining Y_1 in the accompanying table.

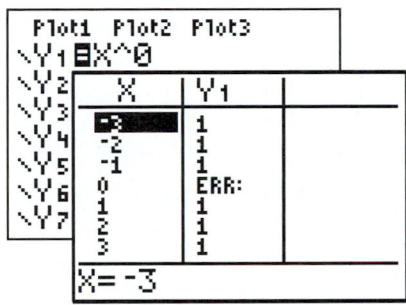

In the screen shot, note that 0^0 is undefined. This is also emphasized in the box.

Definition of Zero Exponent		
For any real number x:		
Algebraically	**Verbally**	**Numerical Example**
$x^0 = 1$ for $x \neq 0$ 0^0 is undefined	Any nonzero real number raised to the 0 power is 1.	$17^0 = 1$

Example 3 illustrates why we must be very careful to correctly identify the base when we use exponential expressions.

Example 3 Using Zero Exponents

Simplify each expression, assuming all bases are nonzero.

(a) 3^0 **(b)** $(-3)^0$ **(c)** -3^0 **(d)** $3x^0$

(e) $(3x)^0$ **(f)** $(x + y)^0$ **(g)** $x^0 + y^0$

Solution

(a) $3^0 = 1$ The base is 3.

(b) $(-3)^0 = 1$ The base is -3.

(c) $-3^0 = -(3)^0 = -1$ The base is 3 (not -3).

(d) $3x^0 = 3(1) = 3$ The base is x.

(e) $(3x)^0 = 1$ The base is $3x$.

(f) $(x + y)^0 = 1$ The base is $x + y$.

(g) $x^0 + y^0 = 1 + 1 = 2$ The different results in parts (f) and (g) are a result of the different order of operations.

Self-Check 3

Simplify each expression, assuming all bases are nonzero.

a. $(5v)^0$ **b.** $5v^0$ **c.** $(5v + w)^0$ **d.** $5v^0 + w^0$

3. Combine the Properties of Exponents to Simplify Expressions

Do the Product, Quotient, and Power Rules Also Apply for Zero Exponents?

Yes, these rules for exponents are summarized in the accompanying box. In Section 5.3, these rules will be applied to all integral exponents: positive, zero, and negative. So far, we have developed these rules only for whole-number exponents. Remember the whole numbers include 0, 1, 2, 3, 4, ...

Summary of the Properties of Exponents

For any nonzero real numbers x and y and whole-number exponents m and n:

Product rule: $x^m \cdot x^n = x^{m+n}$

Power rules: $(x^m)^n = x^{mn}$

$(xy)^m = x^m y^m$

$\left(\dfrac{x}{y}\right)^m = \dfrac{x^m}{y^m}$

Quotient rule: $\dfrac{x^m}{x^n} = x^{m-n}$ for $m > n$

$\dfrac{x^m}{x^n} = \dfrac{1}{x^{n-m}}$ for $m < n$

Zero exponent: $x^0 = 1$ for $x \neq 0$

0^0 is undefined

Problems that involve several of these properties can sometimes be simplified correctly by using more than one sequence of steps. It is often wise to simplify the expressions inside grouping symbols first. Study the examples in the text and those given by your instructor. Try to use a sequence of steps that will minimize your effort as well as produce a correct result.

Example 4 Simplifying Exponential Expressions

Simplify each expression, assuming $x \neq 0$ and $y \neq 0$.

(a) $\left(\dfrac{x^{12}}{x^4}\right)^2$ **(b)** $[(2a^7)(5a^4)]^3$ **(c)** $\dfrac{12x^7y^7}{18x^5y^8}$

Solution

(a) $\left(\dfrac{x^{12}}{x^4}\right)^2 = (x^{12-4})^2$ Use the quotient rule; subtract the exponents.

$= (x^8)^2$ Use the power rule; multiply the exponents.

$= x^{16}$

(b) $[(2a^7)(5a^4)]^3 = [(2)(5)a^{7+4}]^3$ Use the product rule; add the exponents.

$$= (10a^{11})^3$$

$$= 10^3 a^{(3)(11)}$$ Use the power rule; multiply the exponents.

$$= 1{,}000a^{33}$$

(c) $\dfrac{12x^7 y^7}{18x^5 y^8} = \dfrac{2x^{7-5}}{3y^{8-7}}$ Divide both the numerator and the denominator by 6. Also use the quotient rule and subtract the exponents on the common bases.

$$= \dfrac{2x^2}{3y}$$

Self-Check 4

Simplify each expression for $x \neq 0$.

a. $\left(\dfrac{x^7}{x^5}\right)^3$ **b.** $[(2x^3)(3x^2)]^2$

Example 5 compares the subtraction of like terms to the division of expressions with a common base. Please note the distinction between these operations.

Example 5   Comparing Subtraction and Division

Assuming $x \neq 0$:

(a) Subtract $10x^2 - 2x^2$ and divide $\dfrac{10x^2}{2x^2}$.

(b) Subtract $8x^2 - 4x$ and divide $\dfrac{8x^2}{4x}$.

Solution

Difference	Quotient
(a) $10x^2 - 2x^2 = 8x^2$	$\dfrac{10x^2}{2x^2} = 5$
(b) $8x^2 - 4x$ cannot be simplified further as these terms are unlike.	$\dfrac{8x^2}{4x} = 2x$

Self-Check 5

Subtract the like terms and simplify the quotients. Assume $x \neq 0$.

Subtract	Divide
a. $6x^6 - 2x^2$	$\dfrac{6x^6}{2x^2}$
b. $8x^3 - 2x^3$	$\dfrac{8x^3}{2x^3}$

The compound interest formula $A = P(1 + r)^t$ can be used to find the value A of an initial investment of principal P at an annual interest rate r compounded for t years. In Example 6, we use this formula and a spreadsheet to compute the value of an investment over a 10-year period. We could use the table feature on a calculator to determine these values, but a spreadsheet can handle a larger number of values with greater ease. Note that the expression $P(1 + r)^t$ involves an exponent that is a variable. Example 6 illustrates one way to evaluate this expression for different values of t. We will examine expressions with variable exponents further in Chapter 11.

Example 6 Calculating the Value of an Investment

Use the formula $A = P(1 + r)^t$ to find the value of a $7,500 investment compounded at 7% at the end of each year over a 10-year period.

Solution

$A = P(1 + r)^t$

$A = 7,500(1 + 0.07)^t$

$A = 7,500(1.07)^t$

Using the compound interest formula, substitute $7,500 for the principal P and 0.07 for the interest rate of 7%.

B2	▾	f_x =7500*(1.07)^A2

	A	B
1	Years, t	Value, $A = 7500(1.07)^t$
2	0	$7,500.00
3	1	$8,025.00
4	2	$8,586.75
5	3	$9,187.82
6	4	$9,830.97
7	5	$10,519.14
8	6	$11,255.48
9	7	$12,043.36
10	8	$12,886.40
11	9	$13,788.44
12	10	$14,753.64

Enter the input values for t in column A and enter the formula in the appropriate cells in column B. See Technology Perspective 2.2.1 on Creating a Table of Values for assistance.

Note for a zero exponent $(1.07)^0 = 1$ and the value of $P = \$7,500$. This is the value at $t = 0$, the initial time of the investment.

The last row shows that the value of the investment after 10 years is $14,753.64.

Self-Check 6

Use the formula $A = P(1 + r)^t$ to find the value of a $7,500 investment compounded at 8% at the end of each year over a 10-year period.

Self-Check Answers

1. a. m^5 b. x^2y^3 c. 25   d. $-\dfrac{v^4}{w^2}$

2. a. $\dfrac{1}{m^8}$ b. $\dfrac{y}{x}$ c. $\dfrac{1}{1,000}$ d. $\dfrac{v^7}{w^8}$

3. a. 1   b. 5   c. 1   d. 6

4. a. x^6 b. $36x^{10}$

5. a. **Difference:** $6x^6 - 2x^2$ cannot be simplified further as these terms are unlike;
   Quotient: $3x^4$
   b. **Difference:** $6x^3$;
   Quotient: 4

6. The last row shows that the value of the investment after 10 years is $16,191.94.

B2	▾	f_x =7500*(1.08)^A2

	A	B
1	Years, t	Value, $A = 7500(1.08)^t$
2	0	$7,500.00
3	1	$8,100.00
4	2	$8,748.00
5	3	$9,447.84
6	4	$10,203.67
7	5	$11,019.96
8	6	$11,901.56
9	7	$12,853.68
10	8	$13,881.98
11	9	$14,992.53
12	10	$16,191.94

5.2 | Using the Language and Symbolism of Mathematics

1. Two numbers that are added are called _____ or _____.

2. Two numbers that are multiplied are called _____.

3. The answer to a division of two numbers is called the _____.

4. The quotient rule for exponents states that
$$\frac{x^m}{x^n} = \text{_____}$$
for any real number x for
$x \neq$ _____ and for natural numbers m
and n with $m > n$.

5. If $m < n$, then we can use the quotient rule to write
$$\frac{x^m}{x^n} = \text{_____}.$$

6. To divide two expressions with the same base, use the common base and _____ the exponents.

7. The value of x^0 is _____ for $x \neq$ _____.

8. 0^0 is _____.

5.2 | Quick Review

1. Evaluate $-3x^2$ for $x = -2$.

2. Evaluate $(-3x)^2$ for $x = -2$.

3. Simplify 6^2.

4. Simplify 2^6.

5. Simplify $\dfrac{45}{60}$.

5.2 | Exercises

Objective 1 Use the Quotient Rule for Exponents

In Exercises 1 and 2, simplify each expression. Do this by first converting the expression to expanded form, reducing the fraction, and counting the remaining factors of x. Then write your answer as a power of x.

1. $\dfrac{x^6}{x^3}$

2. $\dfrac{x^6}{x^2}$

In Exercises 3–26, simplify each expression. Assume all bases are nonzero.

3. $\dfrac{x^{15}}{x^5}$

4. $\dfrac{y^{18}}{y^6}$

5. $\dfrac{m^{36}}{m^{12}}$

6. $\dfrac{m^{40}}{m^{10}}$

7. $\dfrac{-x^{11}}{x^7}$

8. $\dfrac{-y^{23}}{y^{17}}$

9. $\dfrac{t^8}{t^{12}}$

10. $\dfrac{w^5}{w^{11}}$

11. $\dfrac{v^{10}}{v^{15}}$

12. $\dfrac{n^{15}}{n^{18}}$

13. $-\dfrac{a^6}{a^9}$

14. $\dfrac{-b^6}{b^8}$

15. $\dfrac{a^3}{b^4}$

16. $\dfrac{a^4}{b^3}$

17. $\dfrac{14m^5}{21m^2}$

18. $\dfrac{35m^8}{42m^3}$

19. $\dfrac{36v^5}{66v^8}$

20. $\dfrac{20v^6}{12v^2}$

21. $\dfrac{15a^2b^3}{20a^2b^2}$

22. $\dfrac{33a^5b^5}{22a^3b^8}$

23. $\dfrac{8m^7n^7}{6m^9n^2}$

24. $\dfrac{6x^6y^6}{3x^3y^{10}}$

25. $-\dfrac{27v^4w^9}{45v^6w^6}$

26. $-\dfrac{35a^9b^7}{28a^8b^9}$

Objective 2 Simplify Expressions with Zero Exponents

In Exercises 27–44, simplify each expression. Assume all bases are nonzero.

27. **a.** 7^0 **b.** -7^0 **c.** $(-7)^0$

28. **a.** 10^0 **b.** -10^0 **c.** $(-10)^0$

29. **a.** $2^0 + 8^0$ **b.** $(2 + 8)^0$ **c.** $2^0 - 8^0$

30. **a.** $4^0 + 6^0$ **b.** $(4 + 6)^0$ **c.** $4^0 - 6^0$

31. **a.** 3^0 **b.** 0^3 **c.** -3^0

32. **a.** 5^0 **b.** 0^5 **c.** -5^0

33. **a.** x^0 for $x \neq 0$ **b.** x^0 for $x = 0$

34. **a.** $5y^0$ for $y \neq 0$ **b.** $(5y)^0$ for $y = 0$

35. **a.** $-3x^0$ for $x \neq 0$ **b.** $(-3x)^0$ for $x = 0$

36. **a.** $-9y^0$ for $y \neq 0$ **b.** $(-9y)^0$ for $y = 0$

37. **a.** $(4x)^0$ **b.** $4x^0$ **c.** $4 + x^0$

38. **a.** $(-5y)^0$ **b.** $-5y^0$ **c.** $-5 + y^0$

39. **a.** $(4x - 3y)^0$ **b.** $(4x)^0 - (3y)^0$ **c.** $4x^0 - 3y^0$

40. **a.** $(-5y + 2z)^0$ **b.** $(-5y)^0 + (2z)^0$ **c.** $-5y^0 + 2z^0$

41. $(5a + 3b)^0 + (5a)^0 + (3b)^0 + 5a^0 + 3b^0$

42. $(7v - 8w)^0 + (7v)^0 + (-8w)^0 + 7v^0 - 8w^0$

43. $\dfrac{y^{48}}{y^{48}}$

44. $\dfrac{v^{113}}{v^{113}}$

Objective 3 Combine the Properties of Exponents to Simplify Expressions

In Exercises 45–64, simplify each expression. Assume the variables are nonzero.

45. $\left(\dfrac{x^5}{x^2}\right)^2$

46. $\left(\dfrac{a^7}{a^4}\right)^2$

47. $\left(\dfrac{m^3}{m^9}\right)^2$

48. $\left(\dfrac{b^2}{b^6}\right)^3$

49. $[(2x^5)(3x^4)]^2$

50. $[(5x^2)(2x^4)]^3$

51. $[(3x^2)(2x^3)]^2$

52. $[(2b^4)(5b^3)]^3$

53. $\left(\dfrac{10x^7}{5x^4}\right)^5$

54. $\left(\dfrac{21v^9}{7v^4}\right)^3$

55. $[(6m^4)(2m^5)]^2$

56. $[(5m^7)(3m^5)]^2$

57. $(5x^3)^2(2x^5)^3$

58. $(10v^4)^3(6v^5)^2$

59. $\left(\dfrac{12x^5}{6x^3}\right)\left(\dfrac{15x^7}{5x^3}\right)$

60. $\left(\dfrac{24a^9}{8a^3}\right)\left(\dfrac{42a^7}{14a^6}\right)$

61. $\dfrac{36a^7b^8}{12a^3b^3}$

62. $\dfrac{64m^9n^9}{16mn^7}$

63. $\dfrac{(4x^2y)^3}{(8xy^2)^2}$

64. $\dfrac{(2x^3y^2)^5}{(4x^2y^3)^3}$

Skill and Concept Development

In Exercises 65–70, simplify each expression.

65. a. $\dfrac{2^{45}}{2^{43}}$ **b.** $\dfrac{2^{43}}{2^{45}}$ **c.** $\dfrac{2^{43}}{2^{43}}$

66. a. $\dfrac{5^{87}}{5^{84}}$ **b.** $\dfrac{5^{84}}{5^{87}}$ **c.** $\dfrac{5^{87}}{5^{87}}$

67. a. $0 - 5$ **b.** $0 \cdot 5$ **c.** 5^0
d. 0^5 **e.** $5 \div 0$

68 a. $0 + 6$ **b.** $6 \cdot 0$ **c.** 6^0
d. 0^6 **e.** $6 \div 0$

69. a. $\dfrac{10^8}{10^5}$ **b.** $\dfrac{10^5}{10^5}$ **c.** $\dfrac{10^5}{10^8}$

70. a. $\dfrac{5^6}{5^4}$ **b.** $\dfrac{5^6}{5^6}$ **c.** $\dfrac{5^4}{5^6}$

Comparing Subtraction and Division

In Exercises 71–74, subtract the like terms and simplify the quotients. Assume the variables are nonzero.

Subtract	Divide
71. a. $38x - 2x$	**b.** $\dfrac{38x}{2x}$
72. a. $x^3 - x^3$	**b.** $\dfrac{x^3}{x^3}$
73. a. $3x^2 - 3x$	**b.** $\dfrac{3x^2}{3x}$
74. a. $8x^3 - 2x^2$	**b.** $\dfrac{8x^3}{2x^2}$

Applying Technology

Estimate Then Calculate

In Exercises 75–78, mentally estimate the value of each expression. Then approximate this value to the nearest hundredth.

Problem	Mental Estimate	Approximation
75. $\dfrac{8.014^2}{7.99}$		
76. $\dfrac{1.99^2}{2.01^4}$		
77. $(1.99)^3(5.02)^2$		
78. $[(1.99)(5.02)]^4$		

Compound Interest

79. Use the formula $A = P(1 + r)^t$ to find the value of an $8,500 investment compounded at 5.5% at the end of a 10-year period.

80. Use the formula $A = P(1 + r)^t$ to find the value of an $8,500 investment compounded at 6.5% at the end of a 10-year period.

In Exercises 81 and 82, use a calculator or a spreadsheet to complete the given table of values. Then use the values in the table to determine whether $Y_1 = Y_2$ or $Y_1 = Y_3$.

x	Y_1	Y_2	Y_3
-2			
-1			
0			
1			
2			

81. $Y_1 = \dfrac{x^6}{x^2}$

$Y_2 = x^4$

$Y_3 = x^3$

82. $Y_1 = \dfrac{x^3}{x^{12}}$

$Y_2 = \dfrac{1}{x^9}$

$Y_3 = \dfrac{1}{x^4}$

b. Evaluate $0^6, 0^5, 0^4, 0^3, 0^2$, and 0^1. If you were defining 0^0 based on these observations, what value would you give for 0^0?

c. Use a calculator to evaluate 0^0. What result did you get?

d. 0^0 is undefined. Based on your observations in parts **a** and **b**, explain why you think 0^0 is undefined.

Group discussion questions

83. Challenge Question Use the properties of exponents to simplify each expression, assuming m is a natural number and x is a nonzero real number.

a. $\dfrac{x^{m+3}}{x^{m+1}}$

b. $\dfrac{x^{3m}}{x^m}$

c. $\left(\dfrac{x^{2m+1}}{x^m}\right)^3$

d. $[(x^{m+1})(x^{m+2})]^4$

e. $\dfrac{(-1)^{m+2}}{(-1)^m}$

84. Challenge Question Insert parentheses in the expression $3 + 4 \cdot 5^2$ to create an order of operations that produces a result of

a. 175   **b.** 403   **c.** 529   **d.** 1,225

85. Discovery Question

a. Evaluate $6^0, 5^0, 4^0, 3^0, 2^0$, and 1^0. If you were defining 0^0 based on these observations, what value would you give for 0^0?

86. Discovery Question Use a calculator or a spreadsheet to complete the given table of values for $Y_1 = x^{-1}$. Then discuss what you think this notation represents.

x	Y_1
1	
2	
3	
4	
5	
6	
7	

5.2   **Cumulative Review**

1. Use a pencil and paper to add $\dfrac{2}{15} + \dfrac{5}{12}$.

2. Determine in slope-intercept form the equation of the line that passes through $(-2, 5)$ and $(4, -3)$.

3. Solve the system of equations $\begin{cases} 2x - y = 1 \\ 3x + 4y = 18 \end{cases}$.

4. Use a pencil and paper to sketch the graph of the line $y = -\dfrac{2}{3}x + 5$.

5. Solve $-3 < 2x + 7 \le 11$ and give your answer in interval notation.

Section 5.3   Negative Exponents and Scientific Notation

Objectives:

1. Simplify expressions with negative exponents.

2. Use the properties of exponents to simplify expressions.

3. Use scientific notation.

Planning for space flights to Mars and other planets must include many factors and solve many technical problems. If a flight crew has an emergency and needs to communicate with their ground crew on Earth, how long will it take for an emergency message to reach Earth from Mars?

Due to the vast distance and the speed of the signal, we will use scientific notation to examine this problem. In Example 8 we will represent the distance the signal must travel as 3.78×10^{11} m.

To use scientific notation, we must first examine negative exponents.

1. Simplify Expressions with Negative Exponents

A Positive Exponent Counts the Number of Times a Base Is Used As a Factor; What Is the Meaning of a Negative Exponent?

A Mathematical Note

John Wallis, in *Arithmetica Infinitorum* (1655), was the first writer to explain the use of zero and negative exponents.

We will answer this question by exploring the logic for the definition of a negative exponent. Consider two approaches to simplifying $\dfrac{5^2}{5^3}$.

Option 1

$$\frac{5^2}{5^3} = \frac{25}{125} = \frac{1}{5}$$

For option 1, reduce by dividing both the numerator and denominator by 25.

Option 2

$$\frac{5^2}{5^3} = 5^{2-3} = 5^{-1}$$

For option 2, examine what happens if we try to use the quotient rule and subtract the exponents.

For these two options to obtain the same result, we must define $5^{-1} = \dfrac{1}{5}$. In general, we define $x^{-1} = \dfrac{1}{x}$ for $x \neq 0$. This is further confirmed by examining Y_1 and Y_2 in the accompanying table:

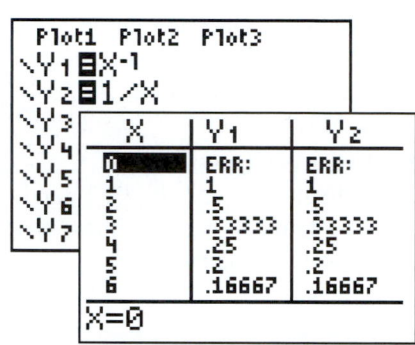

Note in the screen shot that x^{-1} and $\dfrac{1}{x}$ are both undefined for $x = 0$.

Similarly, $\dfrac{x^3}{x^5} = \dfrac{1}{x^2}$ or $\dfrac{x^3}{x^5} = x^{-2}$. Thus, we define $x^{-2} = \dfrac{1}{x^2}$ for $x \neq 0$. This suggests the general definition of negative exponents given in the box.

Definition of Negative Exponents

For any real number x and natural number n:

Algebraically	Verbally	Algebraic Example
$x^{-n} = \dfrac{1}{x^n}$ for $x \neq 0$ x^{-n} is undefined for $x = 0$	A nonzero base with a negative exponent can be rewritten by using the reciprocal of the base and the corresponding positive exponent.	$x^{-3} = \dfrac{1}{x^3}$

Example 1   Simplifying Expressions with Negative Exponents

Simplify each of the following expressions.

(a) 3^{-4} (b) $(-3)^{-4}$ (c) -3^{-4}

(d) $2^{-1} + 3^{-1}$ (e) $(2 + 3)^{-1}$

Solution

(a) $3^{-4} = \dfrac{1}{3^4}$

Take the reciprocal of the base. Note that the answer is positive.

$= \dfrac{1}{81}$

(b) $(-3)^{-4} = \dfrac{1}{(-3)^4}$

The base is -3. Take the reciprocal of the base.

$= \dfrac{1}{81}$

(c) $-3^{-4} = -(3^{-4})$

The base is 3 (not -3). Take the reciprocal of the base. The negative coefficient makes the answer negative.

$= -\dfrac{1}{3^4}$

$= -\dfrac{1}{81}$

(d) $2^{-1} + 3^{-1} = \dfrac{1}{2} + \dfrac{1}{3}$

Simplify each term, convert to a common denominator, and then add these fractions.

$= \dfrac{3}{6} + \dfrac{2}{6}$

$= \dfrac{5}{6}$

(e) $(2 + 3)^{-1} = (5)^{-1} = \dfrac{1}{5}$

First simplify inside the parentheses and then take the reciprocal of the base.

Self-Check 1

Simplify each expression.

a. 2^{-3} **b.** -2^3 **c.** $3^{-1} + 4^{-1}$ **d.** $(3 + 4)^{-1}$

A special case of the definition of negative exponents is useful for working with fractions. The fact that

$$\left(\frac{x}{y}\right)^{-1} = \frac{y}{x} \text{ for } x \neq 0 \text{ and } y \neq 0 \text{ is shown as follows:}$$

$$\left(\frac{x}{y}\right)^{-1} = \frac{1}{\dfrac{x}{y}} = 1 \div \frac{x}{y} = \frac{1}{1} \cdot \frac{y}{x} = \frac{y}{x}$$

The general case is given in the following box.

Fraction to a Negative Power

For any real numbers x and y and natural number n:

Algebraically	Verbally	Numerical Example
$\left(\dfrac{x}{y}\right)^{-n} = \left(\dfrac{y}{x}\right)^{n}$ for $x \neq 0$ and $y \neq 0$	A nonzero fraction to a negative exponent can be rewritten by taking the reciprocal of the fraction and using the corresponding positive exponent.	$\left(\dfrac{3}{5}\right)^{-2} = \left(\dfrac{5}{3}\right)^{2}$ $= \dfrac{25}{9}$

We encourage you to rewrite $\left(\dfrac{5}{3}\right)^{-2}$ as $\left(\dfrac{3}{5}\right)^{2}$ rather than as $\dfrac{5^{-2}}{3^{-2}}$. Although both are correct, the first simplification is easier to use.

Example 2   Simplifying Expressions with Negative Exponents

Simplify the following expressions for $x \neq 0$ and $y \neq 0$.

(a) $\left(\dfrac{2}{3}\right)^{-1}$ 　　(b) $\left(\dfrac{5}{3}\right)^{-2}$ 　　(c) $\left(\dfrac{x}{y}\right)^{-3}$ 　　(d) $\dfrac{x^{-3}}{y}$

Solution

(a) $\left(\dfrac{2}{3}\right)^{-1} = \dfrac{3}{2}$ 　　　　　Take the reciprocal of the base.

(b) $\left(\dfrac{5}{3}\right)^{-2} = \left(\dfrac{3}{5}\right)^{2}$ 　　　　Take the reciprocal of the base, and then square this fraction.

$\qquad\quad = \left(\dfrac{3}{5}\right)\left(\dfrac{3}{5}\right)$

$\qquad\quad = \dfrac{9}{25}$

(c) $\left(\dfrac{x}{y}\right)^{-3} = \left(\dfrac{y}{x}\right)^{3}$ 　　　　Take the reciprocal of the base and then apply the power rule.

$\qquad\quad = \dfrac{y^{3}}{x^{3}}$

(d) $\dfrac{x^{-3}}{y} = \dfrac{x^{-3}}{1}\dfrac{1}{y}$ 　　　　The base for the exponent -3 is just x.

$\qquad\quad = \left(\dfrac{1}{x^{3}}\right)\left(\dfrac{1}{y}\right)$ 　　Take the reciprocal of this base but not of the y.

$\qquad\quad = \dfrac{1}{x^{3}y}$

Self-Check 2

Simplify each expression. Assume $x \neq 0$.

a. $\left(\dfrac{10}{3}\right)^{-1}$ 　　b. $\left(\dfrac{2}{7}\right)^{-2}$ 　　c. $\dfrac{y}{x^{-3}}$

One of the most common errors made is to confuse the meaning of negative coefficients with the meaning of negative exponents. Negative coefficients designate negative numbers. Negative exponents designate reciprocals, *not* negative numbers.

Example 3   Evaluating Algebraic Expressions

Evaluate each expression for $x = 3$ and $y = 5$.

(a) $-x^4$ **(b)** $\dfrac{1}{x^{-4}}$ **(c)** $(x + y)^{-1}$ **(d)** $x^{-1} + y^{-1}$

Solution

(a) $\begin{aligned}-x^4 &= -(3)^4 \\ &= -81\end{aligned}$ Note that part (a) has a negative coefficient while part (b) has a positive coefficient.

(b) $\begin{aligned}\dfrac{1}{x^{-4}} &= \dfrac{1}{3^{-4}} \\ &= 3^4 \\ &= 81\end{aligned}$

(c) $\begin{aligned}(x + y)^{-1} &= (3 + 5)^{-1} \\ &= 8^{-1} \\ &= \dfrac{1}{8}\end{aligned}$ Using the correct order of operations, first simplify inside the parentheses and then take the reciprocal of the base of 8.

(d) $\begin{aligned}x^{-1} + y^{-1} &= 3^{-1} + 5^{-1} \\ &= \dfrac{1}{3} + \dfrac{1}{5} \\ &= \dfrac{5}{15} + \dfrac{3}{15} \\ &= \dfrac{8}{15}\end{aligned}$ Simplify each term and then add these fractions, using the LCD of 15.

Self-Check 3

Evaluate each expression for $x = 2$ and $y = 4$.

a. $-x^5$ **b.** x^{-5} **c.** $\dfrac{1}{y^{-2}}$ **d.** $x^{-2} + y^{-2}$ **e.** $(x + y)^{-2}$

---

2. Use the Properties of Exponents to Simplify Expressions

Do the Rules Given Earlier for Positive Exponents Also Apply to Zero and Negative Exponents?

Yes, the product, power, and quotient rules for exponents are restated here in algebraic form. Fractional exponents will be examined in Chapter 9.

Properties of Integer Exponents

For any nonzero real number x and integers m and n:

Product rule: $x^m \cdot x^n = x^{m+n}$

Power rule: $(x^m)^n = x^{mn}$

Quotient rule: $\dfrac{x^m}{x^n} = x^{m-n}$

Use the properties of exponents whenever possible instead of applying the definitions. The properties of exponents are shortcuts that can save you time. When more than one property of exponents is involved, there may be several ways to simplify the expression. As you practice, try to gain efficiency by studying the examples given by your instructor and presented in this book.

Example 4   Using the Properties of Exponents

Simplify each expression to a form containing only positive exponents. Assume each variable is nonzero.

(a) $\dfrac{x^4}{x^{-7}}$ (b) $(x^{-4})^{-2}$ (c) $(2a^{-2}b^4)^{-3}$ (d) $\left(\dfrac{12x^{-2}y^4}{15x^5y^{-6}}\right)^{-2}$

Solution

(a) $\dfrac{x^4}{x^{-7}} = x^{4-(-7)}$ Use the quotient rule. Subtract the smaller exponent from the larger exponent to obtain an expression with a positive exponent.

$\phantom{(a)\ \dfrac{x^4}{x^{-7}}} = x^{11}$

(b) $(x^{-4})^{-2} = x^{(-4)(-2)}$ Use the power rule. Multiply the exponents.

$\phantom{(b)\ (x^{-4})^{-2}} = x^8$

(c) $(2a^{-2}b^4)^{-3} = 2^{-3}(a^{-2})^{-3}(b^4)^{-3}$ Product to a power rule.

$\phantom{(c)\ (2a^{-2}b^4)^{-3}} = 2^{-3}a^6b^{-12}$ Use the power rule: Multiply the exponents.

$\phantom{(c)\ (2a^{-2}b^4)^{-3}} = \dfrac{a^6}{2^3b^{12}}$ Express in terms of positive exponents by taking the reciprocal of the bases.

$\phantom{(c)\ (2a^{-2}b^4)^{-3}} = \dfrac{a^6}{8b^{12}}$

(d) $\left(\dfrac{12x^{-2}y^4}{15x^5y^{-6}}\right)^{-2} = \left(\dfrac{4y^{10}}{5x^7}\right)^{-2}$ First simplify the expression inside the parentheses, observing carefully the order of operations. Use the quotient rule on each base by subtracting the smaller exponent from the larger exponent. Then take the reciprocal of the base to remove the negative exponent.

$\phantom{(d)\ \left(\dfrac{12x^{-2}y^4}{15x^5y^{-6}}\right)^{-2}} = \left(\dfrac{5x^7}{4y^{10}}\right)^{2}$

$\phantom{(d)\ \left(\dfrac{12x^{-2}y^4}{15x^5y^{-6}}\right)^{-2}} = \dfrac{5^2(x^7)^2}{4^2(y^{10})^2}$ Raise each factor to the second power.

$\phantom{(d)\ \left(\dfrac{12x^{-2}y^4}{15x^5y^{-6}}\right)^{-2}} = \dfrac{25x^{14}}{16y^{20}}$

Self-Check 4

Simplify each expression to a form using only positive exponents. Assume $x \neq 0$ and $y \neq 0$.

a. x^4x^{-7} b. $\dfrac{x^2y^{-3}}{x^{-1}y^4}$ c. $\left(\dfrac{14x^{-3}y^2}{35x^2y^{-4}}\right)^{-2}$

3. Use Scientific Notation

Why Should I Learn About Scientific Notation?

One reason is that you may see scientific notation appear in the results of calculator or spreadsheet calculations. Another reason is that scientific notation is used extensively in the sciences to represent numbers of either macroscopic or microscopic proportions. For example, the mean distance between Earth and the planet Uranus is 2,870,000,000,000 m, whereas the width of one line etched on an Intel computer chip is 0.000000032 m.

Our decimal system of representing numbers is based on powers of 10. Scientific notation uses exponents to represent powers of 10 and thus provides an alternative format for representing numbers. Each of the powers of 10 from 10,000 to 0.001 is written here in exponential notation:

$$10,000 = 10^4$$
$$1,000 = 10^3$$
$$100 = 10^2$$
$$10 = 10^1$$
$$1 = 10^0$$
$$0.1 = 10^{-1}$$
$$0.01 = 10^{-2}$$
$$0.001 = 10^{-3}$$

As these numbers illustrate, the exponent on 10 determines the position of the decimal point. **Scientific notation** uses this fact to express any decimal number as a product of a number between 1 and 10 (or between -1 and -10 if the number is negative) and an appropriate power of 10. Table 5.3.1 shows some representative numbers written in scientific notation.

Scientific Notation Format

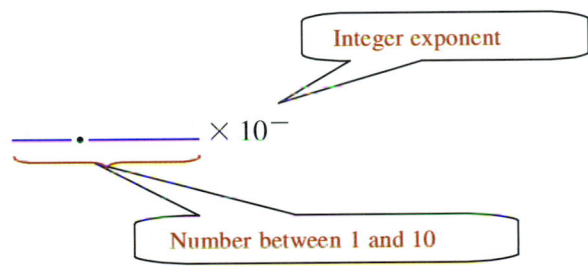

Integer exponent

$\underline{\qquad} . \underline{\qquad} \times 10^{-}$

Number between 1 and 10

Table 5.3.1

Number	Scientific Notation
$12,345 = 1.2345 \times 10,000$	1.2345×10^4
$123.45 = 1.2345 \times 100$	1.2345×10^2
$12.345 = 1.2345 \times 10$	1.2345×10^1
$0.12345 = 1.2345 \times 0.1$	1.2345×10^{-1}
$0.0012345 = 1.2345 \times 0.001$	1.2345×10^{-3}
$-1,234.5 = -1.2345 \times 1,000$	-1.2345×10^3

A Mathematical Note

Extremely small numbers and extremely large numbers are difficult to comprehend. For example, try to grasp the value of each of the following amounts: a million dollars, a billion dollars, and a trillion dollars. Spent at the rate of a dollar a second, a million dollars would last almost 12 days, a billion dollars would last almost 32 years, and a trillion dollars would last almost 32,000 years.

If a number is written in scientific notation, then it can be rewritten in standard decimal notation by multiplying by the power of 10. This multiplication is easily accomplished by shifting the decimal point the number of places indicated by the exponent on the 10. Positive exponents indicate larger magnitudes and shift the decimal point to the right. Negative exponents indicate smaller magnitudes and shift the decimal point to the left.

Writing a Number in Standard Decimal Notation

Verbally	Numerical Examples
Multiply the two factors by using the given power of 10.	
1. If the exponent on 10 is positive, move the decimal point to the right.	1. $3.456 \times 10^2 = 3.456 \times 100 = 345.6$ The decimal point is moved two places to the right.
2. If the exponent on 10 is zero, do not move the decimal point.	2. $3.456 \times 10^0 = 3.456 \times 1 = 3.456$ The decimal point is not moved.
3. If the exponent on 10 is negative, move the decimal point to the left.	3. $3.456 \times 10^{-2} = 3.456 \times 0.01 = 0.03456$ The decimal point is moved two places to the left.

Example 5   Writing Numbers in Standard Decimal Notation

Write each of the following numbers in standard decimal notation.

(a) 5.789×10^2 **(b)** 8.6×10^4 **(c)** 4.61×10^{-4} **(d)** -1.4×10^7

Solution

(a) $5.789 \times 10^2 = 578.9$ Move the decimal point two places to the right.
Two places right

(b) $8.6 \times 10^4 = 86,000$ Fill in zeros so that the decimal point can be moved four places to the right.
Four places right

(c) $4.61 \times 10^{-4} = 0.000461$ Fill in zeros so that the decimal point can be moved four places to the left.
Four places left

(d) $-1.4 \times 10^7 = -14,000,000$ Move the decimal point seven places to the right.
Seven places right

Self-Check 5

Write each of these numbers in standard decimal format.

a. 7.23×10^5 **b.** 7.23×10^{-5} **c.** 1.4×10^{-7}

The steps for writing a number in scientific notation are given in the next box. Remember that positive exponents indicate larger magnitudes and negative exponents indicate smaller magnitudes.

Writing a Number in Scientific Notation

Verbally	Numerical Examples
1. Move the decimal point immediately to the right of the first nonzero digit of the number.	
2. Multiply by a power of 10 determined by counting the number of places the decimal point has been moved.	
a. The exponent on 10 is zero or positive if the magnitude of the original number is 1 or greater.	$3.456 = 3.456 \times 10^0$
	$345.6 = 3.456 \times 10^2$
b. The exponent on 10 is negative if the magnitude of the original number is between 0 and 1.	$0.03456 = 3.456 \times 10^{-2}$

Example 6   Writing Numbers in Scientific Notation

Write each of the following numbers in scientific notation.

(a) 9,876   **(b)** 325.49   **(c)** 0.0009876

Solution

(a) $9,876. = 9.876 \times 10^3$

Three places left

The decimal point is moved three places to position it behind the first nonzero digit, 9. The exponent 3 is positive, since the original number is greater than 1.

(b) $325.49 = 3.2549 \times 10^2$

Two places left

The exponent on 10 is positive 2, since the decimal point was moved two places and the original number is greater than 1.

(c) $0.0009876 = 9.876 \times 10^{-4}$

Four places right

The exponent on 10 is -4, since the decimal point was moved four places and the original number is less than 1.

Self-Check 6

Write each of these numbers in scientific notation.

a. 80,000   **b.** 0.008   **c.** 0.723   **d.** 72.3

Calculators often use a special syntax to represent scientific notation. Many calculators use E to precede the power of 10 that would be used to represent a number in scientific notation, as illustrated in the following table.

Standard Decimal Notation	Scientific Notation	Calculator Syntax
12,345	1.2345×10^4	1.2345E4
1.2345	1.2345×10^0	1.2345E0
0.00567	5.67×10^{-3}	5.67E−3

The TI-84 Plus calculator and Excel spreadsheets can both display results and accept entries in scientific notation. If the magnitude of a number is sufficiently large or small, then the number displayed can be in scientific notation even if the input is in standard decimal notation. The TI-84 Plus calculator can also be set to display all numbers using scientific notation. This is done by using the **SCI** option in the **MODE** menu. In an Excel spreadsheet, a similar result can be obtained by using the **Scientific** option in the **Number** menu to format a cell.

In Technology Perspective 5.3.1, observe that under the same formatting, some of these products are displayed in scientific notation and others are not. This determination is made automatically based on the magnitude of the number. Also observe how input can be made using scientific notation.

| **Technology Perspective 5.3.1** | Using Scientific Notation |

Evaluate the products 99×125, $990{,}000 \times 125{,}000$, and $(1.5 \times 10^{12})(3.2 \times 10^{-8})$. Then reevaluate the product 99×125 and display the result in scientific notation.

TI-84 Plus Calculator

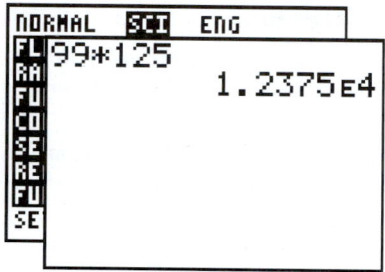

```
99*125
               12375
990000*125000
          1.2375E11
(1.5E12)(3.2E-8)
               48000
```

```
NORMAL  SCI  ENG
FL 99*125
RA         1.2375E4
FU
CO
SE
RE
FU
SE
```

Excel Spreadsheet

C1 ▾	f_x	=A1*B1	
	A	B	C
1	99	125	12375
2	990000	125000	1.24E+11
3	1.50E+12	3.20E-08	48000
4			
5	99	125	1.24E+04

1. The first three products are displayed using scientific notation only if the magnitude is sufficiently large or small.

2. To enter the number 1.5×10^{12}, use the [EE] feature, which is the secondary function of the [,] key.

3. Press [MODE] [>] [ENTER] to set the calculator to scientific mode.

4. In this mode, the product 99×125 is displayed using scientific notation.

1. The first three products are displayed using scientific notation only if the magnitude is sufficiently large or small.

2. The entry in cell **A3** was made by typing 1.5E12. Cell **B3** was entered in the same manner.

3. Format cell **C5** using the **Scientific** option in the **Number** menu.

4. The product in cell **C5** will then be displayed using scientific notation.

Technology Self-Check 1

Compute each product and give the result in scientific notation.

a. 65×783 **b.** $(4.5 \times 10^{-14})(3.7 \times 10^{-10})$

Scientific notation can also be used to perform pencil-and-paper calculations or to perform quick estimates of otherwise lengthy calculations.

Example 7   Using Scientific Notation to Estimate a Product

Use scientific notation to perform a pencil-and-paper estimate of (149,736)(0.003999).

Solution

	Estimated Value	Calculator Value

$149{,}736 \approx 150{,}000$

$\qquad = 1.5 \times 10^5$

$0.003999 \approx 0.004$

$\qquad = 4.0 \times 10^{-3}$

Estimated Value:

$(1.5 \times 10^5)(4.0 \times 10^{-3})$

$= (1.5)(4)(10^5)(10^{-3})$

$= 6 \times 10^2$

$= 600$

Calculator Value:

```
149736*.003999
         598.794264
```

Answer: The calculator product of 598.794264 seems reasonable based on our estimate of 600.

Self-Check 7

Use scientific notation to estimate (4,997,483)(0.030147).

The results of calculations involving very small or very large numbers often need to be rounded to avoid reporting answers with an unrealistic number of significant digits. (Optional material on significant digits and precision is given in Appendix A.) This is illustrated in Example 8.

Example 8   Calculating Time for an Emergency Message from Mars

If a flight crew orbiting Mars has an emergency and needs to communicate with the ground crew on Earth, how long will it take for an emergency message to reach Earth from Mars? Assume that the distance the signal must travel is 3.78×10^{11} m and the speed of the signal is approximately 2.998×10^8 m/s.

Solution

Let t = time in seconds for the message to reach Earth.

$\boxed{\text{Time}} = \boxed{\text{Distance}} \div \boxed{\text{Rate}}$

$t = \dfrac{3.78 \times 10^{11}}{2.998 \times 10^8}$

$t \approx 1{,}260 \text{ seconds}$

$t \approx \dfrac{1{,}260}{60} \text{ minutes}$

$t \approx 21 \text{ minutes}$

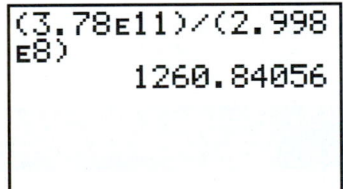

```
(3.78E11)/(2.998
E8)
        1260.84056
```

Rearrange the formula $D = RT$ to form the word equation.

Then substitute in the given values, and use a calculator to approximate t.

The answer in seconds has been rounded to three significant digits.

Answer: It will take an emergency message approximately 21 minutes for the signal to travel from Mars to Earth.

Self-Check 8

Determine how long it will take a message to reach Earth if the signal must travel a distance of 2.31×10^{10} m and the speed of the signal is approximately 2.998×10^8 m/s.

Self-Check Answers

1. **a.** $\frac{1}{8}$ **b.** -8 **c.** $\frac{7}{12}$ **d.** $\frac{1}{7}$

2. **a.** $\frac{3}{10}$ **b.** $\frac{49}{4}$ **c.** x^3y

3. **a.** -32 **b.** $\frac{1}{32}$ **c.** 16 **d.** $\frac{5}{16}$ **e.** $\frac{1}{36}$

4. **a.** $\frac{1}{x^3}$ **b.** $\frac{x^3}{y^7}$ **c.** $\frac{25x^{10}}{4y^{12}}$

5. **a.** 723,000   **b.** 0.0000723   **c.** 0.00000014
6. **a.** 8.0×10^4
   b. 8.0×10^{-3}
   c. 7.23×10^{-1}
   d. 7.23×10^1
7. 150,000
8. It will take the signal approximately 77 seconds to reach Earth.

Technology Self-Check Answers

1. **a.** 5.0895×10^4 **b.** 1.665×10^{-23}

5.3 Using the Language and Symbolism of Mathematics

1. For any real number x, $x \neq 0$, and natural number n,
   $x^{-n} = $ _____.

2. For any nonzero real number x and integers m and n,
   $x^m \cdot x^n = $ _____.

3. For any nonzero real number x and integers m and n,
   $(x^m)^n = $ _____.

4. For any nonzero real number x and integers m and n,
   $\dfrac{x^m}{x^n} = $ _____.

5. Scientific notation is used to express a decimal number as a product of a number between 1 and 10 (or between -1 and -10 if the number is negative) and an appropriate power of _____.

6. In scientific notation, the exponent on _____ determines the position of the decimal point.

5.3 Quick Review

Evaluate each expression.

1. 10^4

2. $\left(\dfrac{1}{10}\right)^3$

3. $10^3 + 10^2 + 10^1$

Simplify each expression. Assume $y \neq 0$.

4. $(-2x^3)(3x^2)$

5. $\left(\dfrac{-2x^3}{y^5}\right)^2$

5.3 Exercises

Objective 1 Simplify Expressions with Negative Exponents

In Exercises 1–16, simplify each expression by writing it in a form free of negative exponents. Assume that all bases are nonzero real numbers.

1. **a.** 2^{-5} **b.** -2^5 **c.** $\dfrac{1}{5^{-2}}$

2. **a.** 4^{-3} **b.** 3^{-4} **c.** $\dfrac{-1}{4^{-3}}$

3. **a.** $\left(\dfrac{6}{5}\right)^{-1}$ **b.** $\left(\dfrac{6}{5}\right)^{-2}$ **c.** $\left(\dfrac{6}{5}\right)^{0}$

4. **a.** $\left(\dfrac{3}{4}\right)^{-1}$ **b.** $\left(\dfrac{3}{4}\right)^{-2}$ **c.** $\left(\dfrac{3}{4}\right)^{0}$

5. **a.** 10^{-2} **b.** 10^{-3} **c.** -10^{-2}

6. **a.** 2^{-4} **b.** -2^{-4} **c.** -4^2

7. **a.** $(2+5)^{-1}$ **b.** $2^{-1}+5^{-1}$ **c.** $-2+5^{-1}$

8. **a.** $(5+10)^{-1}$ **b.** $5^{-1}+10^{-1}$ **c.** $-5+10^{-1}$

9. a. $\left(\dfrac{1}{2} + \dfrac{1}{5}\right)^{-1}$ **b.** $\left(\dfrac{1}{2}\right)^{-1} + \left(\dfrac{1}{5}\right)^{-1}$ **c.** $\left(\dfrac{1}{2}\right)^{0} + \left(\dfrac{1}{5}\right)^{0}$

10. a. $\left(\dfrac{1}{3} + \dfrac{1}{4}\right)^{-1}$ **b.** $\left(\dfrac{1}{3}\right)^{-1} + \left(\dfrac{1}{4}\right)^{-1}$ **c.** $\left(\dfrac{1}{3} + \dfrac{1}{4}\right)^{0}$

11. a. $\left(\dfrac{x}{y}\right)^{-1}$ **b.** $\left(\dfrac{x}{y}\right)^{-2}$ **c.** $\dfrac{x^{-2}}{y}$

12. a. $\left(\dfrac{v}{w}\right)^{-1}$ **b.** $\left(\dfrac{v}{w}\right)^{-2}$ **c.** $\dfrac{v}{w^{-2}}$

13. a. $(3x)^{-2}$ **b.** $3x^{-2}$ **c.** $\dfrac{-3}{x^{-2}}$

14. a. $(2x)^{-3}$ **b.** $2x^{-3}$ **c.** $\dfrac{-2}{x^{-3}}$

15. a. $(m + n)^{-1}$ **b.** $m^{-1} + n^{-1}$ **c.** $m + n^{-1}$

16. a. $(2x + y)^{-1}$ **b.** $2x + y^{-1}$ **c.** $2x^{-1} + y^{-1}$

Objective 2 Use the Properties of Exponents to Simplify Expressions

In Exercises 17–38, simplify each expression, writing each in a form free of negative exponents. Assume all variables are nonzero.

17. a. $v^{-3}v^{12}$ **b.** $\dfrac{v^{12}}{v^{-3}}$ **c.** $\dfrac{v^{-12}}{v^{3}}$

18. a. $w^{21}w^{-7}$ **b.** $\dfrac{w^{21}}{w^{-7}}$ **c.** $\dfrac{w^{-21}}{w^{7}}$

19. a. $x^{3}x^{0}x^{-7}$ **b.** $(x^{3}x^{-7})^{0}$ **c.** $x^{0}x^{-3}x^{-7}$

20. a. $y^{0}yy^{-9}$ **b.** $(y^{9}y^{-1})^{0}$ **c.** $y^{0}y^{-1}y^{-9}$

21. a. $\dfrac{6x^{-4}}{2x^{-3}}$ **b.** $(2x^{-3})(6x^{4})$ **c.** $(2x^{-3})(6x^{-4})$

22. a. $\dfrac{15x^{7}}{5x^{-2}}$ **b.** $(15x^{7})(5x^{-2})$ **c.** $(15x^{-7})(5x^{-2})$

23. a. $\left(\dfrac{3x}{5y}\right)^{2}$ **b.** $\left(\dfrac{3x}{5y}\right)^{-2}$ **c.** $[(3x)(5y)]^{2}$

24. a. $\left(\dfrac{5v}{7w}\right)^{2}$ **b.** $\left(\dfrac{5v}{7w}\right)^{-2}$ **c.** $[(5v)(7w)]^{2}$

25. $[(5x^{3})(-4x^{-2})]^{-1}$ **26.** $[(6x^{-4})(5x^{7})]^{-1}$

27. $(2x^{-3}y^{4})^{3}(3x^{4}y^{-2})^{2}$ **28.** $(5x^{3}y^{-4})(2x^{-5}y^{3})^{4}$

29. $\dfrac{-6x^{6}}{4x^{-4}}$ **30.** $\dfrac{-12x^{-3}}{-3x^{-12}}$

31. $\dfrac{24x^{3}y^{-8}}{-6x^{-12}y^{24}}$ **32.** $\dfrac{-16x^{-8}y^{-5}}{8x^{4}y^{-10}}$

33. $\left(\dfrac{12x^{3}}{6x^{-2}}\right)^{2}$ **34.** $\left(\dfrac{24x^{5}}{8x^{-3}}\right)^{3}$

35. $\left(\dfrac{m^{3}n^{-7}}{m^{7}n^{-11}}\right)^{-3}$ **36.** $\left(\dfrac{v^{-6}w^{4}}{v^{-9}w^{-7}}\right)^{-5}$

37. $\dfrac{(2x^{-1}y^{2})(4x^{2}y^{-3})^{-2}}{(12x^{-2}y^{-2})^{-1}}$ **38.** $\left[\dfrac{(5x^{-3}y^{4})^{-2}(6x^{2}y^{-5})}{15x^{2}y^{-4}}\right]^{-2}$

Objective 3 Use Scientific Notation

In Exercises 39–42, write each number in standard decimal notation.

39. a. 4.58×10^{4} **b.** 4.58×10^{-4}
   c. 4.58×10^{6} **d.** 4.58×10^{-6}

40. a. 1.7×10^{1} **b.** 1.7×10^{-1}
   c. 1.7×10^{5} **d.** 1.7×10^{-5}

41. a. 8.1×10^{3} **b.** 8.1×10^{-3}
   c. -8.1×10^{3} **d.** -8.1×10^{-7}

42. a. 3.6×10^{7} **b.** 3.6×10^{-7}
   c. -3.6×10^{7} **d.** -3.6×10^{-7}

In Exercises 43–46, write each number in scientific notation.

43. a. 9,700 **b.** 97,000,000
   c. 0.97 **d.** 0.00097

44. a. 18,900 **b.** 189
   c. 0.0000189 **d.** 0.00189

45. a. 35,000,000,000 **b.** 0.0000000035
   c. -3,500 **d.** -0.035

46. a. 470,000,000,000 **b.** 0.000000047
   c. -47,000 **d.** -0.0047

Connecting Concepts to Applications

In Exercises 47–50, write the numbers in each statement in standard decimal notation.

47. Distance to Neptune The mean distance from our sun to the planet Neptune, which was examined by the *Voyager* 2 spacecraft in 1989, is 4.493×10^{9} km.

48. Operations by a Computer A 1990 advertisement for one computer claimed that the computer could perform 4.5×10^{8} floating-point operations per second.

49. Computer Switching Speed The switching speeds of some of the most sophisticated computers are measured in picoseconds. A picosecond is one-trillionth of a second, or 1.0×10^{-12} second.

50. Wavelength of Red Light The wavelength of red light is 7,000 angstroms, which is 7.0×10^{-7} m.

In Exercises 51–54, write the numbers in each statement in scientific notation.

51. Temperature of the Sun Temperatures inside the sun are estimated to be 14,000,000°C.

52. Speed of Light The speed of light is approximately 299,790,000 m/s.

53. Pain Signals A burn to a human finger will trigger a pain message to the brain. This pain signal will travel approximately 1 ft in 0.0000568 second.

54. Tidal Friction The friction between the ocean and the ocean floor due to tides is causing the earth's rotation to slow down by about 0.00000002 second/day.

55. Moore's Law Gordon Moore gave a talk in 1965, four years after the first integrated circuit was developed, which predicted that the number of transistors per integrated circuit would double every 18 months. Remarkably, this doubling prediction has proved accurate for 40 plus years, yielding fantastic advances in computers. This prediction has become known as Moore's law. Use the accompanying table to answer each part of this question.

Year	Number of Transistors
1971	2.3×10^3
1978	2.9×10^4
1985	2.8×10^5
1993	3.1×10^6
2000	4.2×10^7
2003	2.2×10^8
2006	1.1×10^9

a. Write in standard decimal notation the number of transistors on an integrated circuit in 1971.

b. Write in standard decimal notation the number of transistors on an integrated circuit in 2003.

c. Newspaper articles have claimed that there were 100,000 times as many transistors on an integrated circuit in 2003 as in 1971. Use the data in this table to determine whether this is an accurate claim.

56. Time with a Cesium Clock The NIST (National Institute of Standards & Technology) F-1 Cesium fountain clock built in 1999 is accurate to 5.0×10^{-8} second/year. Accuracy of this level is important for many scientific processes such as global positioning systems on airplanes and the design of microelectronic devices.

How many seconds does the clock vary from the actual time in

a. A century   b. 10,000 years   c. 20,000,000 years

57. Time for an Emergency Message If a flight crew on the way to Mars has an emergency and needs to communicate with the ground crew on Earth, how long will it take for an emergency message to reach Earth? Assume that the distance the signal must travel is 2.85×10^{11} m and the speed of the signal is approximately 2.998×10^8 m/s.

58. Storage Capacity of a Personal Music Player Each song on a personal music player requires approximately 4 megabytes (MB) (4.0×10^6 bytes) of memory. If you purchase a personal music player with 60 gigabytes (GB) (6.0×10^{10} bytes) of memory, approximately how many songs can you store on this player?

Applying Technology

Estimate Then Calculate

In Exercises 59–64, use scientific notation and pencil and paper to estimate the value of each expression. Then use a calculator or spreadsheet to approximate the value of each expression.

Problem	Pencil-and-Paper Estimate	Approximation
59. $(10{,}013)(0.00007943)$		
60. $(9{,}853{,}493)(0.00061)$		
61. $\dfrac{0.000005034}{0.00009893}$		
62. $\dfrac{901{,}053{,}792}{0.02987}$		
63. $(0.02973)^3$		
64. $(0.0005012)^2$		

Review and Concept Development

In Exercises 65 and 66, use the given values of x and y to evaluate each expression.

	x^{-3}	$-x^3$
65. $x = 3$		
66. $x = 4$		

In Exercises 67 and 68, use the given values of x and y to evaluate each expression.

	$x^2 + y^2$	$(x + y)^2$	$x^{-1} + y^{-1}$	$(x + y)^{-1}$
67. $x = -3, y = 4$				
68. $x = 2, y = -4$				

In Exercises 69 and 70, write each result on the calculator screen in standard decimal notation.

69.
```
2.8E6*6.3E-11
            1.764E-4
```

70.
```
2.3E5*1.5E3
            3.45E8
```

In Exercises 71 and 72, column C displays the product of A and B.

C1	▾	f_x	=A1*B1	
	A	B	C	
1	1.30E+06	2.70E-09	3.51E-03	
2	8.40E-04	3.20E+09	2.69E+06	

71. Write the value in cell **C1** in standard decimal form.

72. Write the value in cell **C2** in standard decimal form.

In Exercises 73 and 74, use a calculator or spreadsheet to approximate each expression accurate to three significant digits. Write the answer in standard decimal notation. (*Hint:* See Appendix A on significant digits.)

73. a. $(4.32 \times 10^{15})(8.49 \times 10^{-17})$

   b. $(7.16 \times 10^3)^2$

74. a. $\dfrac{9.71 \times 10^{13}}{6.53 \times 10^{11}}$

   b. $[(6.7)(4.98 \times 10^2)]^2$

In Exercises 75 and 76, use a calculator or a spreadsheet to complete the given table of values. Then use the values in the table to determine whether $Y_1 = Y_2$ or $Y_1 = Y_3$.

x	Y_1	Y_2	Y_3
-3			
-2			
-1			
0			
1			
2			
3			

75. $Y_1 = x^{-2}$

$Y_2 = \dfrac{1}{x^2}$

$Y_3 = \dfrac{2}{x}$

76. $Y_1 = x^{-3}$

$Y_2 = \dfrac{1}{x^3}$

$Y_3 = \dfrac{3}{x}$

Group discussion questions

77. Risks and Choices

   a. If you live to be 1 billion seconds old, how many years will you have lived?

   b. If you live to be 80 years old, how many seconds will you have lived?

   c. If smoking cigarettes decreases one individual's lifetime from 80 years to 65, how many seconds has this individual lost of the potential lifetime?

   d. Assume in part **c** that this person smoked $20 \times 365 \times 50$ cigarettes. How many seconds of life has each cigarette cost this person?

78. Multiple Representations Engineering notation is similar to scientific notation. Engineering notation expresses a decimal number as a product of a number between 1 and 1,000 (or between -1 and $-1,000$ if the number is negative) and a power of 10 that is a multiple of 3, as in 12.4×10^3 or 453.7×10^6. Set a calculator first to scientific notation (**SCI**), then to engineering notation (**ENG**), and complete this table.

Standard Decimal Notation	Scientific Notation	Engineering Notation
a. 45		
b. 45,678		
c. 456,789		
d. 4,567,890		
e. 0.045		
f. 0.00045		

79. Discovery Question If x is negative and m is an integer:

   a. Can x^m be positive? If so, give the values of m for which x^m is positive.

   b. Can x^m be negative? If so, give the values of m for which x^m is negative.

   c. Can x^{-2} represent a negative value? If so, give an example.

   d. Can x^{-3} represent a negative value? If so, give an example.

5.3   Cumulative Review

1. Write each expression in terms of multiplication.
   a. $y + y + y + y$ **b.** y^4

2. Simplify $8x + 5 - 2(x - (3x + 5))$ to the form $ax + b$.

3. The dosage of a medicine ordered by a doctor is 40 milliliters of a 16% solution. A nurse has available a 20% solution and a 4% solution of this medicine. How many milliliters of each could be mixed to prepare this 40-milliliter dosage?

4. Determine the slope of the line that is perpendicular to $4x - 3y = 24$.

5. Determine $A \cup B$ and $A \cap B$ for $A = [-3, 8)$ and $B = (-\infty, 6]$.

Section 5.4 | Adding and Subtracting Polynomials

Objectives:

1. Use the terminology associated with polynomials.
2. Add and subtract polynomials.

This section examines important algebraic expressions called polynomials. We have already used polynomials such as $2x + 3$, $\frac{5}{9}(F - 32)$, $2\pi r$, and πr^2 throughout the previous sections of this book. We now extend our earlier work on operations with polynomials and define some key terminology.

1. Use the Terminology Associated with Polynomials

Recall that factors are constants or variables that are multiplied together to form a product. A single number or an indicated product of factors is called a **term.** The terms in an algebraic expression are separated from one another by plus or minus symbols.

The expression $7x^2 + 9xy - 3y^2$, which also can be written as $7x^2 + 9xy + (-3y^2)$, has three terms: $7x^2$, $9xy$, and $-3y^2$. The second term, $9xy$, has three factors: 9, x, and y.

What Is a Polynomial?

A **monomial** is a real number, a variable, or a product of real numbers and variables. Because a variable can be used repeatedly as a factor, whole-number exponents can occur in monomials. A **polynomial** is a monomial or a sum of a finite number of monomials.

Monomials and Polynomials

	Verbally	Algebraic Examples
Monomials	A monomial is a real number, a variable, or a product of real numbers and variables with whole-number exponents.	-5, π, x, A, $5x$, $7xy$, and πr^2 are monomials.
Polynomials	A polynomial is a monomial or a sum of a finite number of monomials.	-5, $7xy$, and $7x^2 + 9xy - 3y^2$ are polynomials.

A monomial in the variable x has the form ax^n, where a is the constant coefficient. Although the word *coefficient* is not usually applied to a constant term, it can be. In Example 1, we note that the coefficient of 5 is 5. The exponent n in the monomial ax^n can be any whole number 0, 1, 2, 3, ... Because the exponent n cannot be negative, a monomial cannot have a variable in the denominator.

Example 1 | Identifying Monomials

Determine which of the following expressions are monomials.

(a) $-7x^4$

(b) $\frac{4}{7}x^3y^2$

Solution

Monomial   The coefficient of x^4 is -7.

Monomial   The numerical coefficient is the constant $\frac{4}{7}$. The exponent on x is $+3$ and on y is $+2$.

(c) $\dfrac{4x^3}{7y^2}$ or $\dfrac{4}{7}x^3y^{-2}$ Not a monomial   This expression contains a variable in the denominator; in the optional form the exponent on y is not a whole number.

(d) 5   Monomial   All constants are monomials. Also 5 can be written as $5x^0$. So the numerical coefficient is 5.

(e) $5x^{1/2}$ Not a monomial $5x^{1/2}$ (or $5\sqrt{x}$)* has an exponent that is not a whole number.

*The relationship between exponential and radical notation will be examined in Section 10.5.

Self-Check 1

Determine which of the following expressions are monomials.

a. $4x^2$ **b.** $\dfrac{3}{x^2}$ **c.** $\dfrac{x^2}{3}$ **d.** $4x^{1/2}$

Polynomials containing one, two, and three terms are called **monomials, binomials,** and **trinomials,** respectively. Although the prefix *poly* means many, polynomials can have a single term. Example 2 illustrates these classifications.

Example 2   Identifying Polynomials

Determine whether each expression is a polynomial. Classify each polynomial according to the number of terms it contains.

Solution

(a) $5x^2 + 3x - 7$ Trinomial   The three terms are $5x^2$, $3x$, and -7.

(b) $x^5y - 4x^3y^2$ Binomial   The two terms are x^5y and $-4x^3y^2$.

(c) $\dfrac{x+5}{x-5}$ Not a polynomial   A polynomial cannot contain a variable in the denominator.

(d) $\dfrac{7xy}{9}$ Monomial   The polynomial has only one term, with a coefficient of $\dfrac{7}{9}$.

(e) 18.93   Monomial   All constants are monomials.

(f) $3x^4 - 9x^3 + 7x^2 + 8x - 1$ Polynomial with five terms   Polynomials with more than three terms are not assigned special names.

Self-Check 2

Determine whether each expression is a polynomial. Classify each polynomial according to the number of terms it contains.

a. $7x + 3$ **b.** $\dfrac{7x}{3}$ **c.** $\dfrac{3}{7x}$ **d.** $7x^3 + x - 9$

The **degree of a monomial** is the sum of the exponents for all the variables in this term. A nonzero constant is understood to have degree zero ($4 = 4x^0$ with exponent 0), but no degree is assigned to the monomial 0. The degree of a polynomial plays a key role in determining the number of possible solutions to an nth-degree polynomial equation. For example, a first-degree linear equation of the form $ax + b = 0$ has one solution and a second-degree equation of the form $ax^2 + bx + c = 0$ has two solutions.

Example 3 Determining the Degree of a Monomial

Determine the coefficient and the degree of each monomial.

Solution

	Coefficient	Degree	
(a) $-5x^3$	-5	3	The numerical coefficient is -5, not 5.
(b) $5x^3y^7$	5	10	The sum of the exponents is $3 + 7 = 10$.
(c) 5	5	0	$5 = 5x^0$ with exponent 0.
(d) $-x$	-1	1	$-x = -1x^1$ with coefficient of -1 and exponent of 1 (usually not written).
(e) mn^3	1	4	The coefficient is understood to be 1. The sum of the exponents is $1 + 3 = 4$.

Self-Check 3

Determine the coefficient and the degree of each monomial.

a. $-3x^4$ b. xyz c. π

The **degree of a polynomial** is the same as the degree of the term with the highest degree. To find this highest degree, examine each term individually—do **not** sum the degrees of the terms.

Example 4 Determining the Degree of a Polynomial

Determine the degree of each of the following polynomials.

Solution

(a) $5x^3 + 7x^2$	3	The degrees of the individual terms are 3 and 2.
(b) $-11x^2 + 7x + 8$	2	The degrees of the individual terms are 2, 1, and 0.
(c) $4x^6y - 3x^3y^5$	8	The degrees of the individual terms are 7 and 8.
(d) $a^3 + 5a^2b - 3ab^2 + b^3$	3	Each of these terms is of degree 3.

Self-Check 4

Give the degree of each polynomial.

a. $3x^2 + 4x + 5$ b. $4a^4b^2 - 7a^3b^5 + 5$

Is There Any Preference for How I Should Write a Polynomial?

Yes, polynomials often are easier to compare and manipulate if they are written in a standard form.

A polynomial is in **standard form** if (1) the variables in each term are written in alphabetical order and (2) the terms are arranged in descending powers of the first variable.

A polynomial in x is in **descending order** if the exponents on x decrease from left to right. For example, $3x^2 - 7x^3 + 4 - 9x$ can be written in descending order as $-7x^3 + 3x^2 - 9x + 4$.

The **leading term** is the first term of a polynomial in standard form and the term of highest degree. In Section 5.7, it is definitely best to write polynomials in standard form before performing long division of polynomials.

Example 5 Writing Polynomials in Standard Form

Write each of the following polynomials in standard form.

Solution

(a) $8y^3z^2x$ $8xy^3z^2$ Write the factors in alphabetical order.

(b) $2x^3 + 7 + x^4 - 5x^2$ $x^4 + 2x^3 - 5x^2 + 7$ Arrange the terms in descending order. The leading term is x^4.

(c) $y^2 + 4yx + x^2$ $x^2 + 4xy + y^2$ Write each term in alphabetical order, and then arrange the terms in decreasing powers of x.

Self-Check 5

Write each polynomial in standard form.

a. $-9 + 3x^2 + 8x$ b. $-9y^2zx^5$ c. $-4v + 9v^5 + 3v^2 - v^3 + 1$

A polynomial whose only variable is x is called a **polynomial in x**. The standard form of an nth-degree polynomial in x is $a_nx^n + a_{n-1}x^{n-1} + \cdots + a_1x + a_0$. If each of the coefficients of this polynomial is a real number, then this polynomial is called a **real polynomial**. A polynomial function in x can be represented by the function notation $f(x)$. For example, $f(x) = 5x^2 - 13$ represents a polynomial function. Sometimes we use letters other than f to represent functions. For example, we can let $P(x) = 2x^2 - 4x + 11$. To evaluate this polynomial for $x = 3$, we then substitute 3 for x in the polynomial: $P(3) = 2(3)^2 - 4(3) + 11 = 17$.

Example 6 Evaluating a Profit Polynomial

The profit in dollars made by producing and selling x units is given by the polynomial $P(x) = -x^2 + 14x - 33$. Evaluate and interpret each expression.

(a) $P(0)$ (b) $P(3)$ (c) $P(10)$

Solution

(a) $P(0) = -(0)^2 + 14(0) - 33$ Substitute 0 for x and then evaluate this expression.

$= -33$

Selling 0 units results in a loss of $33.

(b) $P(3) = -(3)^2 + 14(3) - 33$

$= -9 + 42 - 33$

$= 0$

The seller breaks even if 3 units are sold.

(c) $P(10) = -(10)^2 + 14(10) - 33$

$= -100 + 140 - 33$

$= 7$

The seller makes $7 if 10 units are sold.

Self-Check 6

The profit in dollars made by producing and selling x units is given by the polynomial $P(x) = -x^2 + 17x - 30$. Evaluate and interpret each expression.

a. $P(0)$ b. $P(5)$

2. Add and Subtract Polynomials

How Do I Add or Subtract Polynomials?

We add or subtract polynomials by combining like terms. **Like terms** or **similar terms** have exactly the same variable factors. For example, $5x^2y$ and $-11x^2y$ have the same variable factors and are like terms, whereas $5x^2y$ and $11xy$ have different variable factors and are unlike terms. We first used the distributive property in Section 1.6 to add like terms and to remove parentheses from a group of terms. We will use the distributive property extensively to add and subtract polynomials.

Some individuals find a vertical format helpful to organize the work for adding or subtracting two polynomials. If you use a vertical format, align only like terms in the same column. To do this, you may need to leave blanks in some rows (or insert zero coefficients if you write terms in these locations). Example 7 illustrates both the horizontal and the vertical format.

Example 7   Adding Polynomials

Add $5x^2 - 7x + 9$ and $3x^2 + 6x - 8$.

Solution

Horizontal Format

$(5x^2 - 7x + 9) + (3x^2 + 6x - 8)$	Use a plus symbol to indicate the addition.
$= 5x^2 - 7x + 9 + 3x^2 + 6x - 8$	Remove the parentheses.
$= 5x^2 + 3x^2 - 7x + 6x + 9 - 8$	Use the commutative and associative properties of addition to reorder the terms.
$= (5 + 3)x^2 + (-7 + 6)x + (9 - 8)$	Then use the distributive property to combine like terms.
$= 8x^2 - x + 1$	The answer is written in standard form.

Vertical Format

$$5x^2 - 7x + 9$$
$$\underline{3x^2 + 6x - 8}$$
$$8x^2 -\ \ x + 1$$

Align only like terms in the same column.

Although use of the distributive property is not as apparent in this format, it is still the key justification for combining the like terms. The answer is written in standard form.

Answer: $(5x^2 - 7x + 9) + (3x^2 + 6x - 8) = 8x^2 - x + 1$

Self-Check 7

a. Add $5x - 9$ and $8x + 7$.

b. Add $-9x^4 + 8x^3 - 9$ and $6x^4 + 8x^2 - 3x + 1$.

To subtract polynomials, we also combine like terms. Subtraction can be defined as the addition of an additive inverse or opposite. The opposite of a polynomial is formed by taking the opposite of each term of the polynomial. For example, the opposite of $7x^2 - 5x + 4$ is denoted by $-(7x^2 - 5x + 4)$, and $-(7x^2 - 5x + 4) = -7x^2 + 5x - 4$. Example 8 illustrates how to subtract the polynomial $7x^2 - 5x + 4$ by adding its opposite.

Example 8   Subtracting Polynomials

Subtract $7x^2 - 5x + 4$ from $3x^2 + 6x - 8$.

Solution

Horizontal Format

$(3x^2 + 6x - 8) - (7x^2 - 5x + 4)$ Use a minus symbol to indicate the subtraction.

$= 3x^2 + 6x - 8 - 7x^2 + 5x - 4$ Remove the parentheses by using the distributive property.

$= 3x^2 - 7x^2 + 6x + 5x - 8 - 4$ Reorder the terms. Then use the distributive property to combine like terms.

$= (3 - 7)x^2 + (6 + 5)x + (-8 - 4)$

$= -4x^2 + 11x - 12$ The answer is written in standard form.

Vertical Format

$$\begin{array}{r} 3x^2 + 6x - 8 \\ +(-7x^2 + 5x - 4) \\ \hline -4x^2 + 11x - 12 \end{array}$$

To subtract using the vertical format, align like terms. To subtract, add the opposite of $7x^2 - 5x + 4$. Note that this is accomplished by changing the sign of each term of $7x^2 - 5x + 4$.

Answer: $(3x^2 + 6x - 8) - (7x^2 - 5x + 4) = -4x^2 + 11x - 12$

Self-Check 8

a. Subtract $2x^2 - 4$ from $5x^2 + 7x$.

b. Subtract $-3x^4 - 4x^3 + 2x^2 - 4$ from $x^3 - 5x^2 + 7x$.

If two polynomials are equal, then a table of values will show identical values for the two polynomials and their graphs will be identical. In Example 9 we use a table as a check on the difference two polynomials. We will examine the graphs of polynomials in Chapter 6.

The table in Example 9 is created using a calculator. This table could also be created using pencil and paper or a spreadsheet. The main point is not how the table is created but that the table can be another way to check our work.

Example 9   Using Tables to Compare Two Polynomials

Subtract $2(3x + 5)$ from $4(2x - 3)$ and check the difference by using a table of values.

Solution

Algebraically

$4(2x - 3) - 2(3x + 5)$ Indicate the subtraction and then use the distributive property to remove the parentheses.

$= 8x - 12 - 6x - 10$

$= (8 - 6)x + (-12 - 10)$ Reorder the terms.

$= 2x - 22$ Then use the distributive property to combine like terms.

Numerical Check

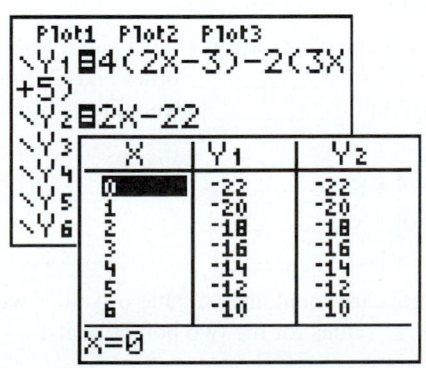

Let Y_1 represent the indicated difference of the two terms and Y_2 represent the calculated difference. Since the table values of Y_1 and Y_2 are identical, this indicates that $4(2x - 3) - 2(3x + 5) = 2x - 22$.

Answer: $4(2x - 3) - 2(3x + 5) = 2x - 22$

Self-Check 9

A student subtracted $4x^2 - 5x + 2$ from $3x^2 - 2x + 6$ and obtained $x^2 - 3x - 4$. Check this result by using a table of values.

We must use caution when using tables or graphs to compare two polynomials. Two polynomials can share several points without being the same polynomial. In general, two nth-degree polynomials can share up to n points and still be distinct polynomials. If these two nth-degree polynomials share $n + 1$ points, then the polynomials are equal. In Self-Check 9, these two distinct second-degree polynomials share exactly two common points—they have different values for all other values of x.

Self-Check Answers

1. **a.** Monomial
   b. Not a monomial
   c. Monomial
   d. Not a monomial
2. **a.** Binomial
   b. Monomial
   c. Not a polynomial
   d. Trinomial
3. **a.** -3; degree 4
   b. 1; degree 3
   c. π; degree 0
4. **a.** 2
   b. 8
5. **a.** $3x^2 + 8x - 9$
   b. $-9x^5y^2z$
   c. $9v^5 - v^3 + 3v^2 - 4v + 1$
6. **a.** $P(0) = -30$; Selling 0 units results in a loss of $30.
   b. $P(5) = 30$; The seller makes $30 if 5 units are sold.
7. **a.** $13x - 2$
   b. $-3x^4 + 8x^3 + 8x^2 - 3x - 8$

8. **a.** $3x^2 + 7x + 4$
   b. $3x^4 + 5x^3 - 7x^2 + 7x + 4$
9.

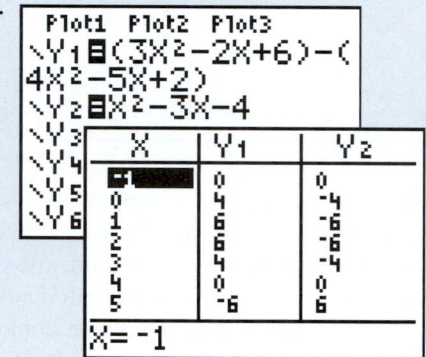

Because $Y_2 = -Y_1$, this indicates the student subtracted the polynomials in the wrong order.

5.4   Using the Language and Symbolism of Mathematics

1. _____ are constants or variables that are multiplied together to form a product.

2. Terms in an algebraic expression are separated from each other by _____ or _____ symbols.

3. The whole numbers are _____.

4. A _____ is a real number, a variable, or a product of real numbers and variables with whole-number exponents.

5. A _____ is a monomial or a sum of monomials.

6. In the monomial ax^n, a is a _____ of x^n.

7. A polynomial containing exactly one term is a _____.

8. A polynomial containing exactly two terms is a _____.

9. A polynomial containing exactly three terms is a _____.

10. The degree of a monomial is the _____ of the exponents for all the variables in the monomial.

11. The degree of a polynomial is the same as the degree of the term with the _____ degree.

12. The degree of a nonzero constant is _____.

13. No degree is assigned to the constant _____.

14. A polynomial is in standard form if (1) the variables in each term are written in _____ order and (2) the terms are arranged in _____ powers of the first variable.

15. The coefficient of x is _____.

16. The coefficient of $-x$ is _____.

17. The degree of $-x$ is _____.

18. If two polynomials are equal, then a table of values will show _____ values for the two polynomials.

5.4 Quick Review

1. The result of the addition of two numbers is called their _____.

2. The result of the subtraction of two numbers is called their _____.

3. The additive inverse of 15 is _____.

Evaluate each expression for $f(x) = 3x - 7$.

4. $f(-4)$ **5.** $f(0)$

5.4 Exercises

Objective 1 Use the Terminology Associated with Polynomials

In Exercises 1–6, determine whether each expression is a monomial. If the expression is a monomial, give its coefficient and its degree.

1. a. -12 **b.** $-12x$ **c.** $-x^{12}$ **d.** x^{-12}

2. a. 3 **b.** $3x$ **c.** x^3 **d.** x^{-3}

3. a. $\frac{2}{5}x^4$ **b.** $\frac{5x^4}{2}$ **c.** $-\frac{5x^4}{2}$ **d.** $\frac{2}{5x^4}$

4. a. $\sqrt{2}$ **b.** $\sqrt{2}x$ **c.** $2\sqrt{x}$ **d.** $\frac{x}{2}$

5. a. $5x^2y^2$ **b.** $-5x^2y^2$ **c.** $\frac{5x^2}{y^2}$ **d.** $5x^2 + y^2$

6. a. $9a^2b^2c^2$ **b.** $\frac{a^2b^2c^2}{9}$ **c.** $\frac{9a^2b^2}{c^2}$ **d.** $9a^2 + b^2c^2$

In Exercises 7–10, determine whether each expression is a polynomial. Classify each polynomial according to the number of terms it contains, and give its degree.

7. a. $3x^2 - 7x$ **b.** $x^3 - 8x^2 + 9$
  c. $-9x^4$ **d.** -9

8. a. $\frac{-4x^5}{7}$ **b.** $x^4 - 5$
  c. $x^2 - 5x + 7$ **d.** 0

9. a. $11xy$ **b.** $11x + y$
  c. $\frac{11x}{y}$ **d.** $x^3 + y^2 - 11$

10. a. $4x^2 - 3xy + y^2$ **b.** $-3xy$
  c. $-3x + y$ **d.** $\frac{3x}{y}$

In Exercises 11 and 12, write each polynomial in standard form.

11. a. $8yx^2z^3$ **b.** $-9 + x^2 - 5x$
  c. $-2x^2 + 7 - 4x^3 + 9x$

12. a. $-11c^3a^4b$ **b.** $-7x + 8 - 3x^2$
  c. $13x + 7x^2 - 9x^3 - 11$

In Exercises 13 and 14, identify each pair of terms as like or unlike.

13. a. $7xy$ and $-xy$
  b. $-4x^3y^2$ and $6x^2y^3$
  c. $5xy^2z^3$ and $5xy^2z^3$

14. a. $5v^4w^3$ and $8v^3w^4$
  b. $3vw^2$ and πvw^2
  c. $\frac{3}{5}xy^2z^3$ and $-0.9xy^2z^3$

Objective 2 Add and Subtract Polynomials

In Exercises 15 and 16, write the opposite or additive inverse of each polynomial.

15. a. $7xy$
  b. $7x - y$
  c. $-2x^2 + 3xy - 7y^2$

16. a. $-5vw$
  b. $-5v + w$
  c. $5v^2 - 6vw - w^2$

In Exercises 17–54, determine each sum or difference, and write the result in standard form.

17. a. $7x + (5x + 9x)$ **b.** $7x - (5x + 9x)$
18. a. $8y + (3y + 12y)$ **b.** $8y - (3y + 12y)$
19. a. $(3x^2 - 7x) + (4x^2 - 5x)$
  b. $(3x^2 - 7x) - (4x^2 - 5x)$
20. a. $(-5x + 3) + (8x - 11)$
  b. $(-5x + 3) - (8x - 11)$
21. a. $(5x^2 - 7x + 9) + (4x^2 + 6x - 3)$
  b. $(5x^2 - 7x + 9) - (4x^2 + 6x - 3)$
22. a. $(7x^2 + 6x - 13) + (3x^2 - 9x - 4)$
  b. $(7x^2 + 6x - 13) - (3x^2 - 9x - 4)$
23. $(6x + 5) + 2(8x - 7)$
24. $(-5x + 3) + 3(4x - 11)$
25. $(-7v + 3) - 5(2v - 11)$
26. $(-4v - 13) - 3(-7v + 11)$
27. $(5a^2 - 7a + 9) + (4a^2 + 6a - 3)$
28. $(7a^2 + 6a - 13) + (3a^2 - 9a - 4)$
29. $(8x^2 - 3x - 4) - (2x^2 - 2x - 9)$
30. $(11x^2 + 7x - 1) - (3x^2 - 12x + 8)$
31. $(-w^2 + 4) - (-4w^2 + 7w - 6)$
32. $(17w - 13) - (3w^2 - 4w + 6)$
33. $(-2m^4 + 3m^3 + m^2 - 1) + (m^4 - m^2 - 2m + 5)$
34. $(4m^3 + 7m^2 - 9m - 1) + (2m^3 - m^2 + 7)$
35. $(-3n^4 + 7n^3 + n - 8) - (n^4 - 2n^3 - n^2 + 4)$
36. $(-2n^4 - n^2 + 3n - 5) - (5n^4 + 7n^3 - 2n + 5)$

37. $(-7y^4 + 3y^6 - y + 4y^5 - 11 + 2y^2) -$
    $2(3y^5 - 5y + 7y^6 - 9y^2 + 7 - y^4)$

38. $(8y^3 + 3 - 5y^2 + y^6 - 4y^4 + 9y) -$
    $3(5 + 3y^2 - 2y^6 + y - 2y^4 - y^5)$

39. $2(x^2 - 3x - 5) - 7(x^2 + 2x - 1)$

40. $8(x^2 + 5x - 2) - 5(2x^2 - 3x - 4)$

41. $3(2a^2 - 5a + 1) - 2(3a^2 - a - 5)$

42. $(7a^2 - a + 3) - 5(5a^2 + a - 2)$

43. $(5x + 7) + (4x - 8) + (3x - 11)$

44. $(6x - 4) + (2x - 3) + (5x + 9)$

45. $(13x - 8) - 4(7x - 9)$

46. $(17x + 5) - 2(14x - 7)$

47. $(5x^2 + 7xy - 9y^2) + (6x^2 - 3xy + y^2)$

48. $(11x^2 - 9xy - 4y^2) + (8x^2 + xy - 5y^2)$

49. $(-7x^2 + 6xy + 8y^2) - (13x^2 - xy - 3y^2)$

50. $(4x^2 - 5xy + 12y^2) - (7x^2 - 8xy - 11y^2)$

51. $(5x^3 - 7x + 9 + 5x^2) + (2x^2 + 13 + x^3 - x) -$
    $(4x - 3x^3 + x^2 - 8)$

52. $(6x - 11x^3 + 7 - x^2) - (14 - 2x^2 - x^3 - 5x) +$
    $(9x - 12 - x^3 + 8x^2)$

53. $(x^4 - 2x^3y + x^2y^2 + xy^3 - 3y^4) -$
    $(2x^4 + x^3y - 5x^2y^2 - 7y^4) - (7x^4 + 3x^2y^2 + 2y^4)$

54. $(3x^4 - 9x^3y + 7x^2y^2 - y^4) -$
    $(7x^4 + 4x^2y^2 - 11y^4) - (-11x^4 - 9x^3y + y^4)$

Review and Concept Development

In Exercises 55 and 56, evaluate each expression for
$P(x) = 5x^2 + 3x - 2$.

55. a. $P(0)$ **b.** $P(2)$ **c.** $P(10)$ **d.** $P(8)$

56. a. $P(1)$ **b.** $P(5)$ **c.** $P(20)$ **d.** $P(100)$

In Exercises 57–62, use the given table to answer each
question. The profit in dollars made by producing and
selling x units is given by $P(x) = -x^2 + 22x - 40$.

B2		f_x	=-1*(A2^2)+22*A2-40	
	A	**B**		**C**
1	x	$P(x) = -x^2 + 22x - 40$		
2	0	-$40.00		
3	1	-$19.00		
4	2	$0.00		
5	3	$17.00		
6	4	$32.00		
7	5	$45.00		
8	6	$56.00		
9	7	$65.00		
10	8	$72.00		
11	9	$77.00		
12	10	$80.00		
13	11	$81.00		
14	12	$80.00		
15	13	$77.00		

57. Evaluate and interpret $P(0)$.

58. Evaluate and interpret $P(2)$.

59. Evaluate and interpret $P(4)$.

60. Evaluate and interpret $P(7)$.

61. Determine the x-values for which $P(x) = 77$. Interpret the meaning of these x-values.

62. Determine the x-values for which $P(x) = 80$. Interpret the meaning of these x-values.

Connecting Algebra to Geometry

In Exercises 63–66, write a polynomial for the perimeter of each polygon.

63. Quadrilateral

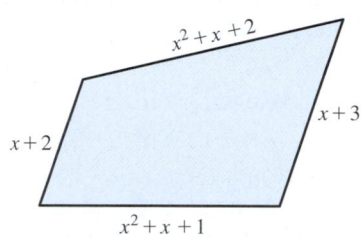

64. Trapezoid

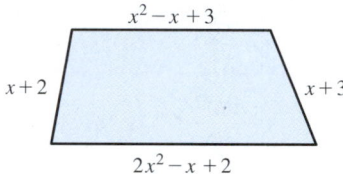

65. Pentagon

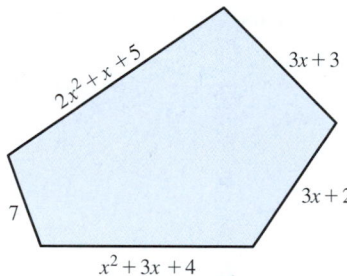

66. Hexagon

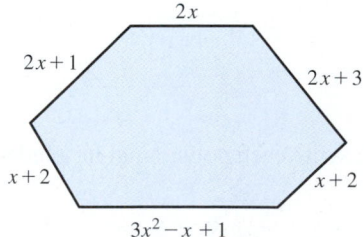

Multiple Representations

In Exercises 67–74, write a polynomial for each verbal expression.

67. Two times x cubed minus five x plus eleven.

68. Nine times y squared plus seven y minus nine.

69. Three times the sixth power of w plus five times the fourth power of w.

70. Eight times the fifth power of v minus v cubed plus eleven v.

71. x squared minus y squared.

72. a cubed plus b cubed.

73. Seven more than twice w.

74. Eight less than the opposite of four y.

Connecting Concepts to Applications

World production of chlorofluorocarbons

The 1987 Montreal Protocol on Substances That Deplete the Ozone Layer restricted the use of CFCs which damage the ozone layer above the earth. The accompanying table gives the world production of CFCs for selected years after 1950 as measured in thousands of tons.

x (Years After 1950)	C(x) (Thousands of Tons of CFCs)
0	60
10	100
20	700
30	900
40	1,250
50	800

75. These data can be modeled by the function $C(x) = -0.05x^3 + 3.0x^2 - 13.5x + 47.7$. Use this function to approximate the tons of CFC produced in 1965.

76. Use the function in Exercise 75 to predict the tons of CFC produced in 2005.

Applying Technology

In Exercises 77 and 78, use a calculator or a spreadsheet to complete the given table of values. Then use the values in the table to determine whether $Y_1 = Y_2$ or $Y_1 = Y_3$.

x	Y₁	Y₂	Y₃
−3			
−2			
−1			
0			
1			
2			
3			

77. $Y_1 = (3x^2 - 2x - 5) - (x^2 + 3x + 10)$
$Y_2 = 2x^2 - 5x + 5$
$Y_3 = 2x^2 - 5x - 15$

78. $Y_1 = (5x^2 + 7x - 9) - 2(x^2 - 5x - 6)$
$Y_2 = 3x^2 + 17x + 3$
$Y_3 = 3x^2 + 17x - 3$

Group discussion questions

79. Challenge Question Give an example that satisfies the given conditions.
  a. A fifth-degree monomial in x with a coefficient of 2
  b. A monomial of degree 0
  c. A first-degree binomial in x with a constant term of -3
  d. A second-degree binomial in x with a constant term of -3
  e. Two first-degree monomials in x whose sum is a constant
  f. Two first-degree monomials in x whose difference is a constant
  g. Two fourth-degree binomials in x whose sum is a third-degree monomial.

80. Error Analysis A classmate subtracted $3x^2 - x - 8$ from $7x^2 + 10x + 9$ by writing the expression $7x^2 + 10x + 9 - 3x^2 - x - 8$ to obtain $4x^2 + 9x + 1$. Explain the error in this work, and make a suggestion as to how the classmate could organize the work to avoid this error.

1. Write the equation of the line that passes through $(-3, 4)$ with an undefined slope.

2. Write the equation of the line that passes through $(-3, 4)$ with a slope of zero.

Use the given graph and table to determine the following input and output values.

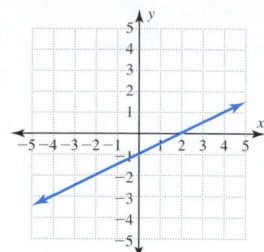

x	$f(x)$
-4	-3
-2	-2
0	-1
2	0
4	1

3. $f(4) = ?$ 4. $f(2) = ?$ 5. $f(x) = 0; x = ?$

Section 5.5   Multiplying Polynomials

Objectives:

1. Multiply polynomials.
2. Multiply binomials by inspection.

Polynomials are used to describe many applications, especially those involving geometric shapes. Exercise 84 examines a polynomial expression that represents the area of concrete around a rectangular swimming pool. Each factor of this expression has a practical meaning that may be important to those planning the pool area. On the other hand, the expanded form might be preferred by those calculating the amount of concrete that will be needed.

1. Multiply Polynomials

The product of two monomials was considered when the product rule for exponents was discussed in Section 5.1. Recall that the product rule for exponents states that $x^m x^n = x^{m+n}$.

Example 1   Multiplying Monomials

Simplify $(7x^3y^2)(-8x^4y^6)$.

Solution

$$(7x^3y^2)(-8x^4y^6) = (7)(-8)(x^3x^4)(y^2y^6)$$ Regroup and reorder the factors, using the associative and commutative properties of multiplication.

$$= -56x^{3+4}y^{2+6}$$ Using the product rule for exponents, add the
$$= -56x^7y^8$$ exponents on the common bases.

Self-Check 1

Simplify $(-9a^2bc)(-4ab^2c^3)$

How Do I Multiply a Monomial by a Polynomial with Two or More Terms?

The distributive property is used to find the product of a monomial and a polynomial. This use of the distributive property is illustrated in Example 2 by the product of a monomial and a binomial.

Example 2   Multiplying a Monomial by a Polynomial

Multiply these polynomials.

(a) $5x^3(7x - 9)$ **(b)** $(4x^2 - 5xy + 11y^2)(2xy)$

Solution

(a) $5x^3(7x - 9) = 5x^3(7x) - 5x^3(9)$

$= (5)(7)(x^3x) - (5)(9)x^3$

$= 35x^4 - 45x^3$

Distribute the multiplication of $5x^3$ to each term of the binomial. Then simplify each term, using the product rule for exponents.

(b) $(4x^2 - 5xy + 11y^2)(2xy)$

$= 4x^2(2xy) - 5xy(2xy) + 11y^2(2xy)$

$= 8x^3y - 10x^2y^2 + 22xy^3$

Distribute the multiplication of $2xy$ to each term of the trinomial. Then simplify, using the product rule for exponents.

Self-Check 2

Multiply these polynomials.

a. $9x^3(8x - 5)$ **b.** $-3x^2(7x^2 - 2xy - 3y^2)$ **c.** $(7x^2 - 2xy - 3y^2)(4y)$

The procedure for multiplying a monomial by a polynomial is summarized in the following box.

Multiplying a Monomial by a Polynomial

Verbally	Algebraic Example
To multiply a monomial by a polynomial, use the distributive property to multiply the monomial by each term of the polynomial.	$5x(2x^2 - 3x + 8) = 5x(2x^2) - 5x(3x) + 5x(8)$ $= 10x^3 - 15x^2 + 40x$

Multiplying two polynomials with more than one term each requires the repeated use of the distributive property. An illustration is provided in Example 3.

Example 3   Multiplying by a Binomial

Multiply the following factors.

(a) $(7x + 3)(2x + 5)$ **(b)** $(3x - 2)(4x^2 - 5x + 6)$

Solution

(a) $(7x + 3)(2x + 5)$

$= 7x(2x + 5) + 3(2x + 5)$

$= 7x(2x) + 7x(5) + 3(2x) + 3(5)$

$= 14x^2 + 35x + 6x + 15$

$= 14x^2 + 41x + 15$

First distribute the multiplication of $2x + 5$ to each term of $7x + 3$. Then distribute $7x$ times each term of $2x + 5$; also distribute 3 times each term of $2x + 5$.

Finally, simplify by combining like terms.

(b) $(3x - 2)(4x^2 - 5x + 6)$

$\quad = 3x(4x^2 - 5x + 6) - 2(4x^2 - 5x + 6)$

$\quad = 3x(4x^2) - 3x(5x) + 3x(6) - 2(4x^2) - (-2)(5x) - 2(6)$

$\quad = 12x^3 - 15x^2 + 18x - 8x^2 + 10x - 12$

$\quad = 12x^3 - 23x^2 + 28x - 12$

The distributive property justifies forming the product of each term of $4x^2 - 5x + 6$ by each term of $3x - 2$.

Simplify by combining like terms.

Self-Check 3

Multiply the polynomial factors.

a. $(2x + 3y)(2x - 3y)$ **b.** $(x - 2)(x^2 + 2x + 4)$

Fortunately, all the steps shown in Example 3 to illustrate the use of the distributive property can be shortened by using the procedure in the following box.

Multiplying Two Polynomials

To multiply one polynomial by another, multiply each term of the first polynomial by each term of the second polynomial and then combine the like terms.

Next we examine the product of two trinomials. Again we start by multiplying each term of the first trinomial by each term of the second trinomial. It is generally easier to keep our steps organized and to obtain accurate results if we write each polynomial factor in standard form. Example 4 illustrates an optional vertical format for organizing the steps used to multiply these polynomials.

Example 4 Multiplying Two Trinomials

Multiply $(x + 1 + x^2)$ by $(4 + x^2 - 2x)$.

Solution

$(x + 1 + x^2)(4 + x^2 - 2x) = (x^2 + x + 1)(x^2 - 2x + 4)$

$\quad\quad x^2 - 2x + 4$

$\underline{\times\ x^2 +\ x + 1}$

$\quad\quad x^4 - 2x^3 + 4x^2$

$\quad\quad\quad\quad x^3 - 2x^2 + 4x$

$\underline{\quad\quad\quad\quad\quad\quad x^2 - 2x + 4}$

$\quad x^4 -\ x^3 + 3x^2 + 2x + 4$

First write each factor in standard form.

Then write one factor on the first row and the second factor on the next row.

Multiply each term on the second row by every term on the first row. First multiply by x^2, then by x, and then by 1.

Note that only similar terms are aligned in the same column. Add the similar terms in each column, and write the product in standard form.

Answer: $(x + 1 + x^2)(4 + x^2 - 2x) = x^4 - x^3 + 3x^2 + 2x + 4$

Self-Check 4

Multiply $3y + y^2 - 1$ by $2 + y^2 - 2y$.

We have examined the multiplication of polynomials by using algebraic properties. Now we give a geometric viewpoint of the multiplication of two binomials. The area of the rectangle is considered in two different ways. First the area of the whole rectangle is calculated, and then the area is computed by adding the areas of the four parts. Since the area is the same either way, these two polynomials must be equal.

A Mathematical Note

Plato (ca. 427–348 B.C.) was a pupil of Socrates. In 388 B.C., he formed a school called the Academy in Athens. Inscribed above the gate to this academy was the following motto: "Let no man ignorant of geometry enter." Many of our algebraic formulas are based on early geometric discoveries.

Area of Whole Rectangle

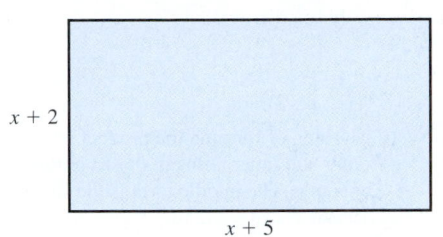

$x + 2$

$x + 5$

Areas of Four Parts

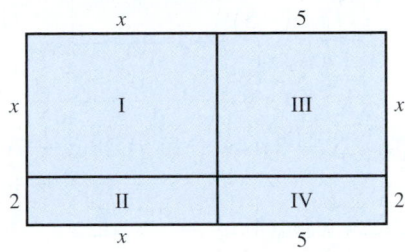

Area = Length × Width   Total area = Area I + Area II + Area III + Area IV

Area = $(x + 5)(x + 2)$ Total area = $x^2 + 2x + 5x + 2(5)$

Total area = $x^2 + 7x + 10$

Thus $(x + 5)(x + 2) = x^2 + 7x + 10$.

2. Multiply Binomials by Inspection

Is There a Way That I Can Multiply Binomials More Efficiently?

Yes, with practice, the procedure of multiplying each term of the first polynomial by each term of the second polynomial can be used to multiply two binomials by inspection. One way to remember the steps for performing this mental multiplication is to use the **FOIL** method. FOIL is an acronym for **F**irst, **O**uter, **I**nner, and **L**ast. The logic of the FOIL method is illustrated here for the product $(2x + 5)(x - 6)$.

$$(2x + 5)(x - 6) = 2x(x - 6) + 5(x - 6)$$
$$= 2x^2 - 12x + 5x - 30$$

Using the distributive property results in multiplying each term of $2x + 5$ by each term of $x - 6$.

Think:

F irst terms, $2x(x)$ ────────

O uter terms, $2x(-6)$ ────────

I nner terms, $5(x)$ ────────

L ast terms, $5(-6)$ ────────

When the middle two terms are like terms, we combine these like terms. In the current example, the sum of the middle terms is $-12x + 5x = -7x$. Thus $(2x + 5)(x - 6) = 2x^2 - 7x - 30$.

An extra step is shown in parts (a) and (b) of Example 5 to illustrate the FOIL method. Usually the inner and outer products are added mentally and the product is written in its final form. Once you feel comfortable omitting this extra step, you should do so.

Example 5   Multiplying Binomials by Inspection

Use the FOIL method to multiply each pair of binomials.

(a) $(x + 9)(x - 5)$ (b) $(7v - 3)(4v + 1)$

(c) $(5w + 4)(w + 2)$ (d) $(3v - 8w)(2v - 5w)$

Solution

(a) $(x + 9)(x - 5) = x^2 - 5x + 9x - 45$
$$= x^2 + 4x - 45$$

F	$(x)(x)$
O	$(x)(-5)$
I	$(9)(x)$
L	$(9)(-5)$

The step of forming the product of the outer and inner terms is shown in this example. The middle term is the sum of $-5x$ and $9x$.

(b) $(7v - 3)(4v + 1) = 28v^2 + 7v - 12v - 3$
$$= 28v^2 - 5v - 3$$

F	$(7v)(4v)$
O	$(7v)(1)$
I	$(-3)(4v)$
L	$(-3)(1)$

The step of forming the product of the outer and inner terms is shown in this example. The middle term is the sum of $7v$ and $-12v$.

(c) $(5w + 4)(w + 2) = 5w^2 + 14w + 8$

The middle term is the sum of $10w$ and $4w$.

(d) $(3v - 8w)(2v - 5w) = 6v^2 - 31vw + 40w^2$

The middle term is the sum of $-15vw$ and $-16vw$.

Self-Check 5

Use the FOIL method to multiply these binomials.

a. $(x + 4)(x + 7)$ **b.** $(3v + 5)(4v - 1)$ **c.** $(2v - 5w)(3v - 4w)$

The ability to multiply binomials by inspection not only facilitates your work with multiplication, but also gives you a strong preparation to factor trinomials in Chapter 6.

Example 6 illustrates the multiplication of three binomial factors. We start by multiplying the first two factors, and then we multiply this product by the third factor.

Example 6   Finding the Product of Three Binomials

Multiply $(x - 3)(x + 2)(x - 5)$.

Solution

$$(x - 3)(x + 2)(x - 5) = [(x - 3)(x + 2)](x - 5)$$
$$= [x^2 + 2x - 3x - 6](x - 5)$$
$$= (x^2 - x - 6)(x - 5)$$
$$= x^2(x - 5) - x(x - 5) - 6(x - 5)$$
$$= x^3 - 5x^2 - x^2 + 5x - 6x + 30$$
$$= x^3 - 6x^2 - x + 30$$

First find the product of $(x - 3)(x + 2)$ by using FOIL.

Then distribute $x - 5$ to each term of $x^2 - x - 6$.

Answer: $(x - 3)(x + 2)(x - 5) = x^3 - 6x^2 - x + 30$

Self-Check 6

Multiply $(x - 1)(x - 4)(x + 5)$.

The number of units that can be sold for some products depends on the price of the item. For example, lowering the price of new cars will usually generate more sales. Lowering interest rates (the cost of borrowing money) is a standard tool of the Federal Reserve to promote more borrowing in the economy. The revenue generated by selling a number of units is found by multiplying the price per unit by the number of units sold. Thus some companies can exert control over their revenue by controlling their prices. Example 7 expresses the revenue for automobile sales as a function of the price of the car.

Example 7 Revenue for Automobile Sales

For a price of x dollars per car (x from $20,000 to $30,000), an automobile manufacturer estimates that it can sell $10,000 - 0.2x$ cars.

(a) Write a revenue polynomial $R(x)$ based on this estimate.

(b) Create a table of values using input values of $x = 20,000, 22,000, 24,000, 26,000, 28,000, 30,000,$ and $32,000.

(c) Use this table to evaluate and interpret $R(20,000)$.

Solution

(a)   Let x = price per car in dollars

$10,000 - 0.2x$ = number of cars that will sell

$$\boxed{\text{Revenue}} = \boxed{\text{Number of items sold}} \cdot \boxed{\text{Price per item}}$$

$$R(x) = (10,000 - 0.2x)(x)$$

Write a word equation as the basis of an algebraic equation. This is the revenue polynomial.

(b)

B2 ▾ f_x =(10000-0.2*A2)*A2

	A	B
1	Price per car, x	Revenue, (10000-0.2x)(x)
2	$20,000	$120,000,000
3	$22,000	$123,200,000
4	$24,000	$124,800,000
5	$26,000	$124,800,000
6	$28,000	$123,200,000
7	$30,000	$120,000,000
8	$32,000	$115,200,000

A spreadsheet makes a good choice to examine tables with monetary values. These cells have been formatted to display currency.

You can obtain the values in this table using a graphing calculator, but the revenue values would be displayed in scientific notation due to their size.

(c) From the table, $R(20,000) = 120,000,000$. A price of $20,000 per car will produce a revenue of $120,000,000.

Self-Check 7

For a price of x dollars per car (x from $25,000 to $40,000), an automobile manufacturer estimates that it can sell $9,000 - 0.15x$ cars.

a. Write a revenue polynomial $R(x)$ based on this estimate.

b. Create a table of values using input values of $x = 25,000, 30,000, 35,000,$ and $40,000.

c. Use this table to evaluate and interpret $R(30,000)$.

Equal polynomials will have identical output values for all input values. In Example 8, we use a table to compare two polynomials as a check that these polynomials are equal. This table could have been created with a spreadsheet or by pencil-and-paper calculations. The main point is that a table with sufficient values can be used to check that two polynomials are equal.

Example 8   Using a Table to Compare Two Polynomials

Multiply $(2x + 3)(2x - 3)$ and check this product using a table of values.

Solution

Algebraically

$$(2x + 3)(2x - 3) = 4x^2 - 9$$

Numerical Check

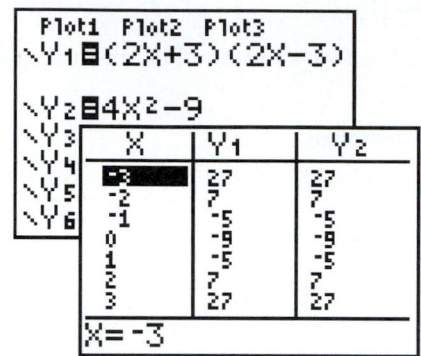

Multiplying these binomials by inspection, we note that the middle term is the sum of $-6x$ and $6x$. Thus the product is $4x^2 + 0x - 9 = 4x^2 - 9$.

Let Y_1 represent the indicated product of the two factors and Y_2 represent the calculated product. Because the values of Y_1 and Y_2 in the table are identical, this indicates that the polynomials $(2x + 3)(2x - 3)$ and $4x^2 - 9$ are equal.

Answer: $(2x + 3)(2x - 3) = 4x^2 - 9$

Self-Check 8

Multiply $(3x + 4)(3x - 4)$ and check this product using a table of values.

The distributive property is one of the most used properties of algebra, a property used both to multiply and expand expressions and to factor expressions. In Example 8 of Section 1.6, we used this property to expand the expression $9(x + 4) = 9x + 36$. We also used the distributive property to rewrite $13x + 8x$ as $(13 + 8)x = 21x$. Both parts of Example 9 use the distributive property, first to expand $2x(5x - 4) + 3(5x - 4)$ and then to factor this expression. In this section we have concentrated on multiplying polynomials and will examine factoring in greater detail in Chapter 6.

Example 9   Comparison of Expanding and Factoring a Polynomial

Use the distributive property first to expand $2x(5x - 4) + 3(5x - 4)$ and then to factor this expression.

Solution

Expand

$$2x(5x - 4) + 3(5x - 4)$$
$$= 2x(5x) - 2x(4) + 3(5x) - 3(4)$$
$$= 10x^2 - 8x + 15x - 12$$
$$= 10x^2 + 7x - 12$$

Factor

$$2x(5x - 4) + 3(5x - 4)$$
$$= (2x + 3)(5x - 4)$$

To expand, distribute the factor $2x$ to each term of $5x - 4$ and the factor 3 to each term of $5x - 4$. Use the product rule for exponents and then add like terms. To factor, use the distributive property to factor out the common factor of $5x - 4$.

Self-Check 9

Use the distributive property to:

a. Expand $7x(2x - 5) - 3(2x - 5)$.

b. Factor $7x(2x - 5) - 3(2x - 5)$.

Self-Check Answers

1. $36a^3b^3c^4$
2. **a.** $72x^4 - 45x^3$
   b. $-21x^4 + 6x^3y + 9x^2y^2$
   c. $28x^2y - 8xy^2 - 12y^3$
3. **a.** $4x^2 - 9y^2$
   b. $x^3 - 8$
4. $y^4 + y^3 - 5y^2 + 8y - 2$
5. **a.** $x^2 + 11x + 28$
   b. $12v^2 + 17v - 5$
   c. $6v^2 - 23vw + 20w^2$
6. $x^3 - 21x + 20$
7. **a.** $R(x) = (9,000 - 0.15x)(x)$
   b.

	A	B
	B2 ▾ f_x =(9000-0.15*A2)*A2	
1	Price per car, x	Revenue, (9000-0.15x)(x)
2	$25,000	$131,250,000
3	$30,000	$135,000,000
4	$35,000	$131,250,000
5	$40,000	$120,000,000

   c. $R(30,000) = 135,000,000$; A price of $30,000 will generate a revenue of $135,000,000.

8. $9x^2 - 16$
   Check: Because all values of Y_1 and Y_2 in the table are equal, the product checks.

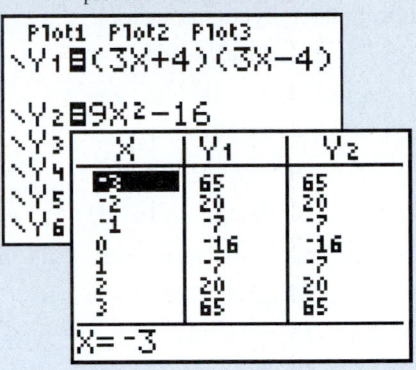

9. **a.** $14x^2 - 41x + 15$
   b. $(7x - 3)(2x - 5)$

5.5 Using the Language and Symbolism of Mathematics

1. The degree of $5x^2$ is _____, and its coefficient is _____.

2. The product rule for exponents says $x^m x^n =$ _____.

3. In the product $(5x^3)(8x^4) = 40x^7$, the coefficients are **added/multiplied** and the exponents are **added/multiplied.** (Select the correct choices.)

4. To multiply a monomial by a polynomial, use the _____ property to multiply the monomial by each term of the polynomial.

5. To multiply one polynomial by another, multiply each term of the first polynomial by _____ term of the second polynomial and then combine like terms.

6. A polynomial with two terms is called a _____.

7. A polynomial with three terms is called a _____.

8. The acronym FOIL represents _____, _____, _____, _____.

9. **a.** The polynomial $4x - 5$ has _____ terms.
   b. The polynomial $3x + 2$ has _____ terms.
   c. The product of $4x - 5$ and $3x + 2$ is the polynomial $12x^2 - 7x - 10$ which has _____ terms.

10. **a.** The binomial $4x - 5$ is of degree _____.
    b. The binomial $3x + 2$ is of degree _____.
    c. The product of $4x - 5$ and $3x + 2$ is the trinomial $12x^2 - 7x - 10$ of degree _____.

11. The polynomial $P(x) = x^2 - x + 3$ is read "P of _____ equals x squared minus x plus 3."

5.5 Quick Review

Simplify each expression.

1. $-3(4x - 7)$

2. $5(2x + 9)$

3. $-6(3x - 5y + 8)$

If possible, simplify each expression by performing the indicated operation. If it is not possible to simplify the expression, explain why.

4. **a.** $(3x^2)(4x)$ **b.** $3x^2 + 4x$

5. **a.** $(-2x^2)(5x^2)$ **b.** $-2x^2 + 5x^2$

5.5 | Exercises

Objective 1 Multiply Polynomials

In Exercises 1–38, multiply the polynomial factors.

1. $-9v^2(-7v^3)$ **2.** $11w^3(-w^5)$

3. $(a^3b^4)(4a^6b)$ **4.** $(-11x^2y^3)(-8xy^6)$

5. $4x^2(6x - 5)$ **6.** $7x^3(9x + 4)$

7. $-4v^2(8v + 3)$ **8.** $-9v^4(2v - 11)$

9. $2x(7x^2 - 9x - 4)$ **10.** $3x(8x^2 + 4x - 5)$

11. $-11x^2(4x^2 - 5x - 8)$ **12.** $-14x^2(2x^2 - 3x + 5)$

13. $(2x^2 - xy - y^2)(5xy)$ **14.** $(3x^2 + xy - y^2)(2xy)$

15. $3a^2b(2a^2 + 4ab - b^2)$ **16.** $5v^2w^2(8v^2 - vw + 3w^2)$

17. $(y + 2)(y + 3)$ **18.** $(y - 4)(y - 6)$

19. $(2v + 5)(3v - 4)$ **20.** $(2v - 3)(5v + 2)$

21. $(6x + 5)(9x + 3)$ **22.** $(7x + 11)(4x + 6)$

23. $(x + y)(2x - y)$ **24.** $(x - y)(3x + y)$

25. $(2x + 4y)(5x - 7y)$ **26.** $(2x - 5y)(3x + 4y)$

27. $(x + 2)(x^2 - 3x - 4)$ **28.** $(2x - 1)(x^2 + x - 3)$

29.
$$\begin{array}{r} m^2 + 3m + 2 \\ \times \quad 2m - 5 \\ \hline \end{array}$$
30.
$$\begin{array}{r} m^2 - 2m - 7 \\ \times \quad 3m + 4 \\ \hline \end{array}$$

31.
$$\begin{array}{r} x^2 - xy + y^2 \\ \times \quad x + y \\ \hline \end{array}$$
32.
$$\begin{array}{r} x^2 + xy + y^2 \\ \times \quad x - y \\ \hline \end{array}$$

33.
$$\begin{array}{r} 2x^2 - 5x + 11 \\ \times \ 3x^2 + 4x - \ 9 \\ \hline \end{array}$$
34.
$$\begin{array}{r} -6x^2 + 3x - 8 \\ \times \ 5x^2 - 2x + 7 \\ \hline \end{array}$$

35. $5x(x + 3)(x - 2)$ **36.** $7x(x - 4)(x + 2)$

37. $(2x - 1)(x + 3)(x + 4)$

38. $(3x + 2)(x - 1)(x + 5)$

Objective 2 Multiply Binomials by Inspection

In Exercises 39–50, multiply the binomials by inspection. (*Hint:* See Example 5, which illustrates the FOIL method.)

39. $(m + 3)(m + 4)$ **40.** $(m + 7)(m + 11)$

41. $(n - 5)(n + 8)$ **42.** $(n + 9)(n - 4)$

43. $(5x + 7)(4x - 3)$ **44.** $(4x - 5)(3x + 8)$

45. $(9y - 1)(y - 7)$ **46.** $(8y - 3)(y - 4)$

47. $(3a - 2b)(4a + 7b)$ **48.** $(5a + 9b)(2a - b)$

49. $(9x - 7y)(10x + 3y)$ **50.** $(5x + 7y)(7x - 5y)$

Connecting Algebra to Geometry

In Exercises 51–56, write a polynomial for the area of each figure. See the formulas on the inside back cover of this book.

51.

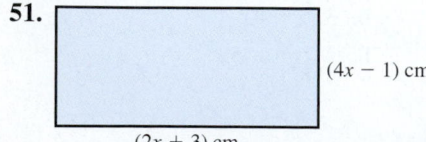

$(4x - 1)$ cm

$(2x + 3)$ cm

52.

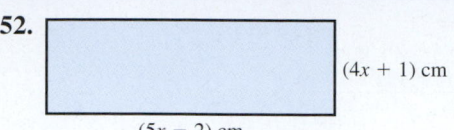

$(4x + 1)$ cm

$(5x - 2)$ cm

53.

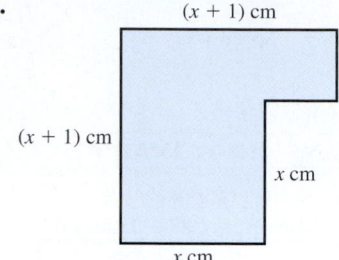

$(x + 1)$ cm

$(x + 1)$ cm

x cm

x cm

54.

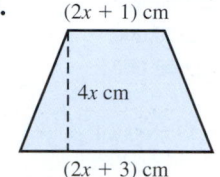

$(2x + 1)$ cm

$4x$ cm

$(2x + 3)$ cm

55.

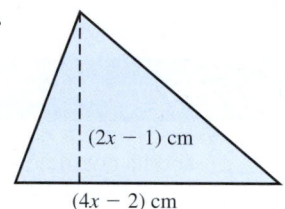

$(2x - 1)$ cm

$(4x - 2)$ cm

56.

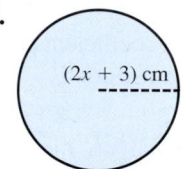

$(2x + 3)$ cm

In Exercises 57–59, write a polynomial for the area of the shaded region. Assume the curved portions are semicircles of the same size.

57.

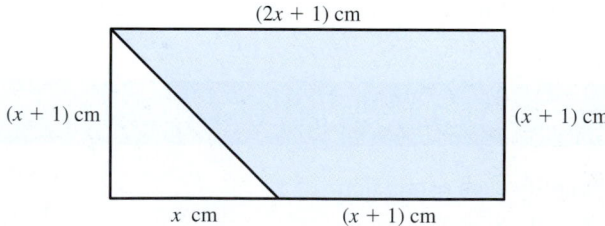

$(2x + 1)$ cm

$(x + 1)$ cm

$(x + 1)$ cm

x cm

$(x + 1)$ cm

58.

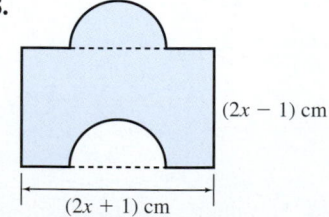

$(2x - 1)$ cm

$(2x + 1)$ cm

59.

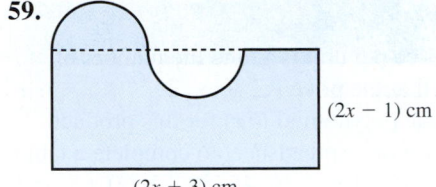

(2x − 1) cm

(2x + 3) cm

Volume of a solid

In Exercises 60 and 61, write a polynomial for the volume of each solid.

60.

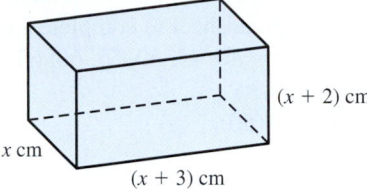

(x + 2) cm

x cm

(x + 3) cm

61.

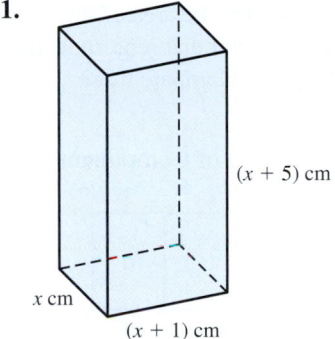

(x + 5) cm

x cm

(x + 1) cm

Applying Technology

In Exercises 62–65, use a calculator or a spreadsheet to complete the given table of values for each pair of expressions. Then use the values in the table to determine whether $Y_1 = Y_2$.

x	Y_1	Y_2
−3		
−2		
−1		
0		
1		
2		
3		

62. $Y_1 = (4x − 5)(2x + 3)$
$Y_2 = 8x^2 + 2x − 15$

63. $Y_1 = (3x − 4)(4x + 3)$
$Y_2 = 12x^2 − 12$

64. $Y_1 = (x + 3)(2x − 1)$
$Y_2 = 2x^2 − 3$

65. $Y_1 = (x + 2)(x − 2)$
$Y_2 = x^2 − 4$

Comparison of Expanding and Factoring Polynomials

In Exercises 66–74, use the distributive property to expand the expression in the first column and to factor the greatest common factor out of the polynomial in the second column.

Expand	**Factor**
66. a. $4x(9x + 2)$	**b.** $(4x)(9x) + (4x)(2)$
67. a. $5v(8v − 7)$	**b.** $(5v)(8v) − (5v)(7)$
68. a. $8xy^2(3x^2 − 2y)$	**b.** $(8xy^2)(3x^2) − (8xy^2)(2y)$
69. a. $2mn(m^2 − mn + 5n^2)$	**b.** $(2mn)(m^2) − (2mn)(mn) + (2mn)(5n^2)$
70. a. $3ab(a^2 + 2ab − 6b^2)$	**b.** $(3ab)(a^2) + (3ab)(2ab) − (3ab)(6b^2)$
71. a. $x(x − 3) + 2(x − 3)$	**b.** $x(x − 3) + 2(x − 3)$
72. a. $x(x + 4) − 5(x + 4)$	**b.** $x(x + 4) − 5(x + 4)$
73. a. $3x(5x − 2) − 4(5x − 2)$	**b.** $3x(5x − 2) − 4(5x − 2)$
74. a. $(2x + 7)(5x) + (2x + 7)(3)$	**b.** $(2x + 7)(5x) + (2x + 7)(3)$

Skill Development

In Exercises 75–82, multiply the polynomial factors.

75. $(5 − 2x^2)(3x^2 + 4)$

76. $(2 − x)(x − x^2 + 3)$

77. $(x + 4 − x^2)(5 − x + x^2)$

78. $(3 + 2x − x^2)(4x − x^2 + 1)$

79. $(v + 5)^2$ **80.** $(v − 6)^2$

81. $(x + 2)^3$ **82.** $(x − 5)^3$

Connecting Concepts to Applications

83. Volume of a Tray A 24-in by 12-in metal sheet has squares x units on a side cut from each corner. Then the sides are bent upward and perpendicular to the base to form a metal tray. From the formula $V = LWH$, the volume of the metal tray will be $V = (24 − 2x)(12 − 2x)(x)$. Write an expanded form of this polynomial.

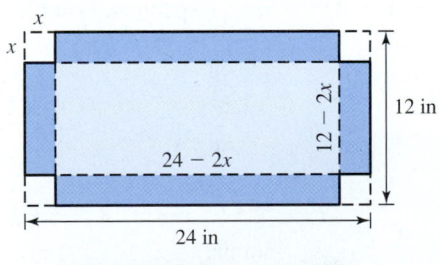

x

x

12 − 2x

12 in

24 − 2x

24 in

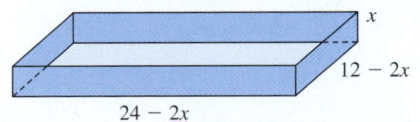

x

12 − 2x

24 − 2x

84. Area of a Concrete Pad Concrete of width x ft is poured around all four sides of a 20-ft by 40-ft pool.

   a. One way to calculate the area covered by the concrete is to take the total area occupied by the pool and concrete and subtract the area of the pool. A polynomial representing this area is $A = (40 + 2x)(20 + 2x) - (40)(20)$. Expand and simplify this polynomial.

   b. Another way to calculate the area covered by the concrete is to add the areas covered by concrete on each end and on the two sides. A polynomial representing this area is $A = 2[x(20 + 2x) + 40x]$. Expand and simplify this polynomial.

   c. Compare the polynomials in parts **a** and **b**.

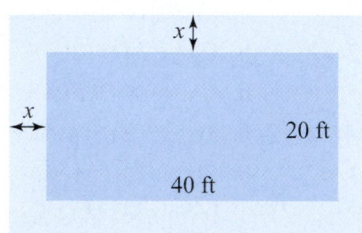

85. Area of a Concrete Pad A concrete sidewalk of width x ft is poured around a circular pool of radius 15 ft. From the formula $A = \pi r^2$, the area covered by the concrete will be $A = \pi(x + 15)^2 - \pi(15)^2$. Expand and simplify this polynomial.

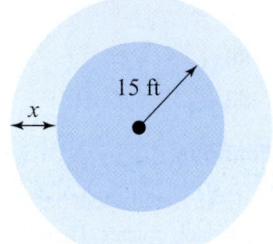

86. Area of a Concrete Pad A concrete sidewalk of width 4 ft is poured around a circular pool of radius x ft. From the formula $A = \pi r^2$, the area covered by the concrete will be $A = \pi(x + 4)^2 - \pi x^2$. Expand and simplify this polynomial.

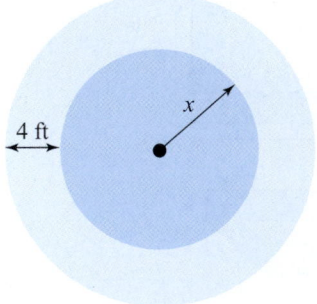

Applying Technology

87. Revenue The price per unit is x, and the number of units that will sell at the price is $280 - 2x$.

   a. Write a revenue polynomial $R(x)$ for this product.

   b. Use a calculator or a spreadsheet to complete a table using input values of $x = 40, 45, 50, 55, 60, 65$, and 70. (*Hint:* See Example 7.)

   c. Evaluate and interpret $R(50)$.

88. Revenue The price per unit is x, and the number of units that will sell at that price is $540 - 3x$.

   a. Write a revenue polynomial $R(x)$ for this product.

   b. Use a calculator or a spreadsheet to complete a table using input values of $x = 40, 45, 50, 55, 60, 65$, and 70. (*Hint:* See Example 7.)

   c. Evaluate and interpret $R(50)$.

Group discussion questions

89. Challenge Question For each rectangular region, first write a polynomial that represents the area of the whole rectangle. Then write a polynomial that represents the sum of the areas of all the parts. Compare these polynomials. Are they equal?

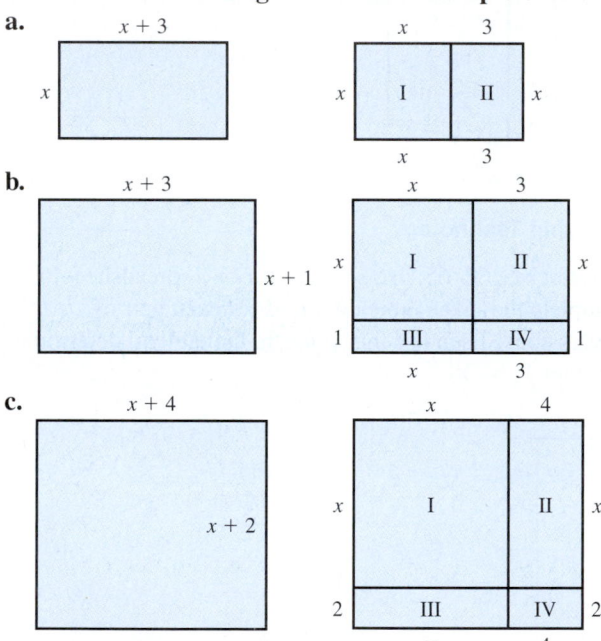

90. Challenge Question

   a. Give two binomials whose sum is a monomial.

   b. Give two binomials whose sum is a binomial.

   c. Give two binomials whose sum is a trinomial.

   d. Give two binomials whose product is a trinomial.

   e. Give two binomials whose product is a binomial.

91. Discovery Question

    a. Complete each product.

        i. $(x + 1)(x - 1) =$ _____

        ii. $(x + 2)(x - 2) =$ _____

        iii. $(x + 3)(x - 3) =$ _____

        iv. $(x + 4)(x - 4) =$ _____

        v. $(x + 5)(x - 5) =$ _____

    b. Make a general conjecture based on this work. State this conjecture in algebraic form and be prepared to describe this conjecture verbally to your teacher.

    c. Test this conjecture on $(x + 12)(x - 12)$.

    d. Test this conjecture on $(2v + 3)(2v - 3)$.

92. Challenge Question Expand the product of the factors $(x - a)(x - b)(x - c) \cdots (x - y)(x - z)$.

5.5 | Cumulative Review

1. Use a pencil and paper to perform the addition $\dfrac{5}{12} + \dfrac{3}{10}$.

2. Use a pencil and paper to perform the multiplication $\dfrac{5}{12} \cdot \dfrac{3}{10}$.

3. Write in slope-intercept form $f(x) = mx + b$ the equation of the line graphed here.

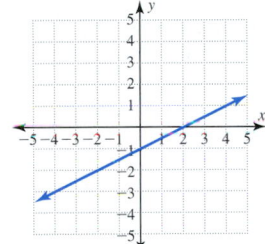

4. Solve each equation and inequality.

    a. $|x - 5| = 9$

    b. $|x - 5| \le 9$

    c. $|x - 5| > 9$

5. Write the simplified form of $\left(\dfrac{-12x^8y^{-3}}{20x^{-2}y^6}\right)^{-2}$ using only positive exponents.

Section 5.6 | Special Products of Binomials

Objectives:

1. Determine the product of a sum and a difference of binomials by inspection.

2. Determine the square of a binomial by inspection.

3. Use the order of operations to simplify polynomials.

The recognition of patterns algebraically, numerically, and graphically is an important part of mathematics. Once patterns are recognized, we can often take advantage of these patterns to shorten our work or to gain insights on the problems we are examining. We will now examine two patterns that occur frequently in the multiplication of binomials. The first is the product of a sum and a difference.

1. Determine the Product of a Sum and a Difference of Binomials by Inspection

I Can Already Multiply Polynomials; Why Should I Learn How to Form Special Products?

Learning to use these special products will improve your efficiency and give you a head start when we factor polynomials. We will now examine three products illustrating a sum times a difference. Note that each of these products produces a binomial with no middle term. Thus these special products can be formed even more quickly than with the FOIL method.

$$\textbf{(Sum)} \cdot \textbf{(Difference)} \quad \textbf{Product}$$
$$(x + 2)(x - 2) = x^2 - 4$$
$$(x + 3)(x - 3) = x^2 - 9$$
$$(x + y)(x - y) = x^2 - y^2$$

Product of a Sum and a Difference

Algebraically	Verbally	Algebraic Example
$(A + B)(A - B) = A^2 - B^2$	This product of a sum and a difference is the difference of their squares.	$(x + 11)(x - 11) = x^2 - 121$

The product of a sum and a difference is illustrated geometrically by the areas of the figures shown next.

The difference between the areas of two squares with sides of x and y is $x^2 - y^2$. This is illustrated by the L-shaped region at the end of the first row of figures. This region consists of two rectangular regions, both of width $x - y$. One has length x, and the other has length y. These two rectangles can be combined to form a rectangle with width $x - y$ and length $x + y$. The area of this rectangle is $(x - y)(x + y)$.

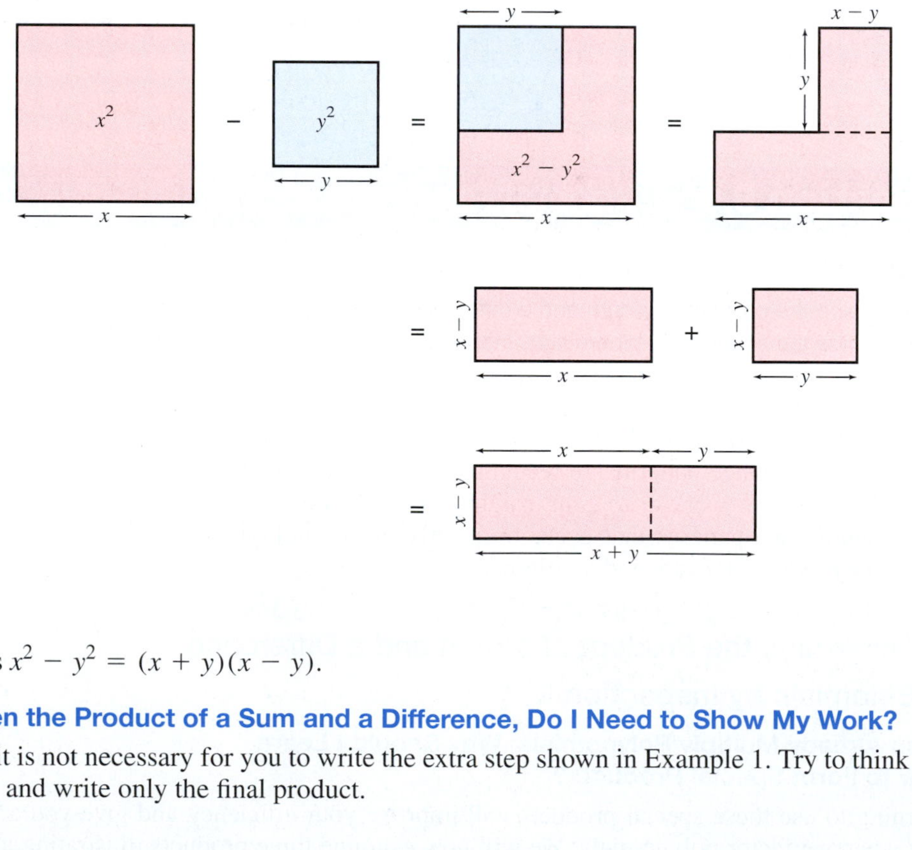

Thus $x^2 - y^2 = (x + y)(x - y)$.

Given the Product of a Sum and a Difference, Do I Need to Show My Work?

No, it is not necessary for you to write the extra step shown in Example 1. Try to think this step, and write only the final product.

Example 1　Multiplying a Sum by a Difference

Multiply these binomial factors by inspection.

(a) $(x + 9)(x - 9)$　　　　**(b)** $(3v - 8)(3v + 8)$　　　　**(c)** $(5x + 7y)(5x - 7y)$

Solution

(a) $(x + 9)(x - 9) = (x)^2 - (9)^2$　　　The product is the difference of the squares of x
$$= x^2 - 81$$　　　and 9.

(b) $(3v - 8)(3v + 8) = (3v)^2 - 8^2$　　　Since multiplication is commutative, this product
$$= 9v^2 - 64$$　　　equals $(3v + 8)(3v - 8)$. The product is the
　　　difference of the squares.

(c) $(5x + 7y)(5x - 7y) = (5x)^2 - (7y)^2$
$$= 25x^2 - 49y^2$$

Self-Check 1

Multiply these binomial factors.

a. $(w + 12)(w - 12)$　　　　**b.** $(8v - 1)(8v + 1)$　　　　**c.** $(2a + 3b)(2a - 3b)$

As a look ahead to Chapter 6, we will emphasize the strong connection between multiplying and factoring polynomials. Expanding and factoring are reverse processes. These two processes are two different uses of the same algebraic information. Note that the steps used in Example 2 to factor $25v^2 - 1$ are the same steps in reverse order that we use to expand $(5v + 1)(5v - 1)$.

Example 2　Comparison of Expanding and Factoring

Use the special product $(A + B)(A - B) = A^2 - B^2$ to expand $(5v + 1)(5v - 1)$ and to factor $25v^2 - 1$.

Solution

Expand

$$(5v + 1)(5v - 1) = (5v)^2 - (1)^2$$
$$= 25v^2 - 1$$

Factor

$$25v^2 - 1 = (5v)^2 - (1)^2$$
$$= (5v + 1)(5v - 1)$$

Self-Check 2

a. Expand $(x + 11)(x - 11)$.　　　　**b.** Factor $x^2 - 121$.

c. Expand $(2a + 7b)(2a - 7b)$.　　　　**d.** Factor $x^2 - 36y^2$.

2. Determine the Square of a Binomial by Inspection

The second special pattern that we examine is the square of a binomial, the product of a binomial multiplied by itself. Each of the products listed below is the square of a binomial. For example, $(x + 2)^2 = (x + 2)(x + 2) = x^2 + 4x + 4$. Note the pattern formed by each of these products. By taking advantage of this pattern we can square binomials by inspection.

(Binomial)² **Product**

$$(x + 2)^2 = x^2 + 4x + 4$$
$$(x + 5)^2 = x^2 + 10x + 25$$
$$(x + y)^2 = x^2 + 2xy + y^2$$
$$(x - 2)^2 = x^2 - 4x + 4$$
$$(x - 5)^2 = x^2 - 10x + 25$$
$$(x - y)^2 = x^2 - 2xy + y^2$$

This pattern is summarized in the box.

Square of a Binomial

Algebraically	Verbally	Algebraic Example
Square of a sum: $(A + B)^2 = A^2 + 2AB + B^2$	The square of a binomial is a trinomial that has:	
	1. A first term that is the square of the first term of the binomial.	$(x + 1)^2 = x^2 + 2x + 1$
Square of a difference: $(A - B)^2 = A^2 - 2AB + B^2$	**2.** A middle term that is twice the product of the two terms of the binomial.	$(x - 1)^2 = x^2 - 2x + 1$
	3. A last term that is the square of the last term of the binomial.	

The formula for the square of a sum is illustrated geometrically by the areas of the following figures. The total area inside the square is equal to the sum of the areas of the four parts shown.

Area of a Whole Square

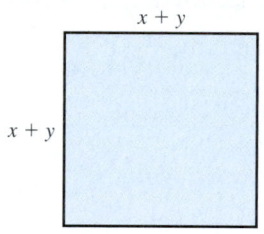

Areas of Four Parts

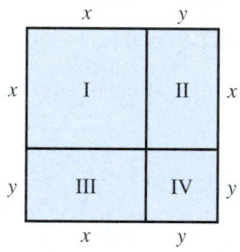

Area $= (x + y)(x + y)$ Total area = Area I + Area II + Area III + Area IV

Area $= (x + y)^2$ Total area $= x^2 + xy + xy + y^2$

                             Total area $= x^2 + 2xy + y^2$

Thus $(x + y)^2 = x^2 + 2xy + y^2$.

Example 3   Squaring Binomials

Write the expanded form for each square of a binomial.

(a) $(x + 7)^2$ **(b)** $(v - 4)^2$ **(c)** $(2x + 3y)^2$

Solution

(a) $(x + 7)^2 = x^2 + 2(7)(x) + (7)^2$ Use the form $(A + B)^2 = A^2 + 2AB + B^2$ to square this binomial.

The square of x
Twice the product of x and 7
The square of 7

                     $= x^2 + 14x + 49$

(b) $(v - 4)^2 = v^2 + 2(-4)(v) + (4)^2$ Use the form $(A - B)^2 = A^2 - 2AB + B^2$ to square this binomial.

The square of v

Twice the product of v and (-4)

The square of 4

$$= v^2 - 8v + 16$$

(c) $(2x + 3y)^2 = (2x)^2 + 2(2x)(3y) + (3y)^2$ Use the form $(A + B)^2 = A^2 + 2AB + B^2$ to square this binomial. The first term is the square of $2x$. The second term is twice the product of $2x$ and $3y$. The third term is the square of $3y$.

$$= 4x^2 + 12xy + 9y^2$$

Self-Check 3

Write the expanded form for each square of a binomial.

a. $(x - 3)^2$ **b.** $(7m + n)^2$ **c.** $(3a - 4b)^2$

We will examine the factoring of polynomials in detail in Chapter 6. The main purpose of Example 4 is to continue our emphasis that expanding and factoring are reverse processes that both rely heavily on the distributive property.

Example 4   Comparison of Expanding and Factoring

Use the special product $(A + B)^2 = A^2 + 2AB + B^2$ to expand $(x + 8)^2$ and to factor $x^2 + 16x + 64$.

Solution

Expand

$$(x + 8)^2 = (x)^2 + 2(x)(8) + (8)^2$$
$$= x^2 + 16x + 64$$

Factor

$$x^2 + 16x + 64 = (x)^2 + 2(x)(8) + (8)^2$$
$$= (x + 8)^2$$

Self-Check 4

a. Expand $(x - 11)^2$.   **b.** Factor $x^2 - 22x + 121$.

c. Expand $(2a + 7b)^2$.   **d.** Factor $x^2 - 12xy + 36y^2$.

3. Use the Order of Operations to Simplify Polynomials

Problems involving operations with polynomials are often written using parentheses. Thus it is extremely important to follow the order of operations first given in Section 1.6 and restated here.

Order of Operations

Step 1. Start with the expression within the innermost pair of grouping symbols.

Step 2. Perform all exponentiations.

Step 3. Perform all multiplications and divisions as they appear from left to right.

Step 4. Perform all additions and subtractions as they appear from left to right.

In Example 5, note the distinct meanings of $(x + y)^2$ and $x^2 + y^2$.

Example 5   Using Order of Operations to Simplify a Polynomial

Simplify $(x + y)^2 + (x^2 + y^2)$.

Solution

$$(x + y)^2 + (x^2 + y^2) = (x^2 + 2xy + y^2) + (x^2 + y^2)$$ First square the binomial to obtain $x^2 + 2xy + y^2$. Then add $x^2 + y^2$.

$$= 2x^2 + 2xy + 2y^2$$

Self-Check 5

Simplify $(3x - 5)^2 - [(3x)^2 - (5)^2]$.

Example 6 presents two polynomial expressions that share a similar appearance but indicate a distinct order of operations.

Example 6   Using Order of Operations to Simplify Polynomials

Simplify each expression.

(a) $3x - 4(3x + 4)$ **(b)** $(3x - 4)(3x + 4)$

Solution

(a) $3x - 4(3x + 4) = 3x - 12x - 16$ Multiplication by -4 is distributed to each term of $3x + 4$. Then like terms are added.

$$= -9x - 16$$

(b) $(3x - 4)(3x + 4) = (3x)^2 - (4)^2$ The parentheses indicate the product of two binomials. This product is a special form that can be multiplied by inspection.

$$= 9x^2 - 16$$

Self-Check 6

Simplify each expression.

a. $6x + 5(6x - 5)$ **b.** $(6x + 5)(6x - 5)$

Should I Try to Work All These Problems Mentally?

Even if you can, we advise against this. The risk of order-of-operation errors outweighs the possible gain in speed. It is important to write your intermediate steps clearly and carefully, as illustrated in Example 7.

Example 7   Using Order of Operations to Simplify a Polynomial

Simplify $3(4x - 5y)^2 - 2(4x + 5y)^2$.

Solution

$$3(4x - 5y)^2 - 2(4x + 5y)^2$$
$$= 3(16x^2 - 40xy + 25y^2) - 2(16x^2 + 40xy + 25y^2)$$ First square each of the binomials by inspection.

$$= 48x^2 - 120xy + 75y^2 - 32x^2 - 80xy - 50y^2$$ Next distribute the factor of 3 and the factor of -2.

$$= 16x^2 - 200xy + 25y^2$$ Then combine like terms.

Self-Check 7

Simplify $5(2x + 3)^2 - 3(2x - 3)^2$.

5.6 Using the Language and Symbolism of Mathematics

1. Match each of these polynomials with the description that best fits each polynomial,
   a. $A^2 - B^2$ **A.** The square of a sum
   b. $(A - B)^2$ **B.** The sum of two squares
   c. $(A + B)^2$ **C.** The difference of two squares
   d. $A^2 + B^2$ **D.** The square of a difference

2. A binomial is a polynomial with _____ terms.

3. A trinomial is a polynomial with _____ terms.

4. Expanding and _____ are reverse processes.

5. The property that allows us to rewrite $2x(3x + 5)$ as $2x(3x) + 2x(5)$ is the _____ property.

6. Expanding $(x + 2)(x - 2)$ produces _____.

7. _____ $x^2 - 4$ produces $(x + 2)(x - 2)$.

5.6 Quick Review

Use a pencil and paper to evaluate each expression.

1. $(3 + 7)^2$
2. $3^2 + 7^2$
3. $5^2 - 3^2 + (5 - 3)^2$

Multiply these binomials by inspection.

4. $(x + 4)(x - 5)$
5. $(2x + 3)(2x + 5)$

5.6 Exercises

Objective 1 Determine the Product of a Sum and a Difference of Binomials by Inspection

In Exercises 1–12, multiply each sum by a difference by inspection.

1. $(7a + 1)(7a - 1)$
2. $(v + 9)(v - 9)$
3. $(z - 10)(z + 10)$
4. $(w - 13)(w + 13)$
5. $(2w - 3)(2w + 3)$
6. $(3w + 5)(3w - 5)$
7. $(9a + 4b)(9a - 4b)$
8. $(7a + 3b)(7a - 3b)$
9. $(4x - 11y)(4x + 11y)$
10. $(5x - 12y)(5x + 12y)$
11. $(x^2 - 2)(x^2 + 2)$
12. $(x^2 + 3)(x^2 - 3)$

Objective 2 Determine the Square of a Binomial by Inspection

In Exercises 13–30, write the expanded form for each square of a binomial.

13. $(4m + 1)^2$
14. $(7m + 1)^2$
15. $(n - 9)^2$
16. $(n - 10)^2$
17. $(3t + 2)^2$
18. $(8t + 3)^2$
19. $(5v - 8)^2$
20. $(7v - 3)^2$
21. $(4x + 5y)^2$
22. $(7x + 2y)^2$
23. $(6a - 11b)^2$
24. $(8a - 7b)^2$

25. $(a + bc)^2$
26. $(a - bc)^2$
27. $(x^2 - 3)^2$
28. $(x^2 + 5)^2$
29. $(x^2 + y)^2$
30. $(x^2 - 3y)^2$

Objective 3 Use the Order of Operations to Simplify Polynomials

In Exercises 31–44, use the order of operations to simplify each expression.

31. **a.** $5x - 6(5x + 6)$ **b.** $(5x - 6)(5x + 6)$
32. **a.** $4x + 7(4x - 7)$ **b.** $(4x + 7)(4x - 7)$
33. **a.** $(4x - 9)^2$ **b.** $(4x)^2 - (9)^2$
34. **a.** $(7x + 10)^2$ **b.** $(7x)^2 + (10)^2$
35. $(v + 7)^2 - (v^2 - 7^2)$
36. $(w - 5)^2 - (w^2 - 5^2)$
37. $(2x + y)^2 - [(2x)^2 + y^2]$
38. $(x + y)^2 - (x - y)^2$
39. $(v + w)(v - w) - (v - w)^2$
40. $(3v - w)(3v + w) - (3v - w)^2$
41. $(a - b)^2 - (b - a)^2$
42. $(a - b)^2 + (b - a)^2$

43. $2(3x - 2y)^2 - (3x + 2y)^2$

44. $5(2x + 7y)^2 - 3(2x - 7y)^2$

Review and Concept Development

Multiple representations

In Exercises 45–50, write each verbal statement in algebraic form and then perform the indicated operation.

45. Square the quantity $3x + 7y$.

46. Square the quantity $8x - 9y$.

47. Subtract the square of $x - 6$ from the square of $x + 8$.

48. Subtract the square of $x + 7$ from the square of $x - 9$.

49. Multiply $a^2 + 1$ by the product of $a + 1$ and $a - 1$.

50. Multiply $a^2 + 4$ by the product of $a - 2$ and $a + 2$.

Applying Technology

In Exercises 51–54, use a calculator or a spreadsheet to complete the given table of values for each pair of expressions. Then use the values in the table to determine whether $Y_1 = Y_2$.

x	Y_1	Y_2
-3		
-2		
-1		
0		
1		
2		
3		

51. $Y_1 = (2x + 1)(2x - 1)$
   $Y_2 = 4x^2 - 1$

52. $Y_1 = x^2 + 9$
   $Y_2 = (x + 3)(x - 3)$

53. $Y_1 = (2x + 3)^2$
   $Y_2 = 4x^2 + 9$

54. $Y_1 = 4x^2 - 28x + 49$
   $Y_2 = (2x - 7)^2$

Comparison of Expanding and Factoring Polynomials

In Exercises 55–58, expand each expression in the first column and factor each expression in the second column.

Expand	**Factor**
55. $(9x - 1)(9x + 1)$	$81x^2 - 1$
56. $(7x - 11)(7x + 11)$	$49x^2 - 121$
57. $(10w - 3x)^2$	$100w^2 - 60wx + 9x^2$
58. $(8m + 5n)^2$	$64m^2 + 80mn + 25n^2$

Group discussion questions

59. Mental Arithmetic Sometimes impressive feats of mental arithmetic have a very simple algebraic basis.

   a. Use the fact that $(103)(97) = (100 + 3)(100 - 3)$ to compute this product mentally.

   b. Use the fact that $(96)(104) = (100 - 4)(100 + 4)$ to compute this product mentally.

   c. Use the fact that $(99)(99) = (100 - 1)^2$ to compute this product mentally.

   d. Use the fact that $(101)(101) = (100 + 1)^2$ to compute this product mentally.

   e. Make up two problems of your own that you can compute mentally.

60. Discovery Question Substitute $-y$ for y in the equation $(x + y)^2 = x^2 + 2xy + y^2$ and simplify the result. What do you observe?

61. Discovery Question

   a. Let Y_1 represent each of these polynomials, and then graph Y_1, using the viewing window $[-6, 6, 1]$ by $[-20, 10, 5]$. What relationship do you observe between the factors of Y_1 and attributes of the graph?

   i. $(x + 1)(x - 1)$
   ii. $(x + 2)(x - 2)$
   iii. $(x + 2)(x - 3)$
   iv. $(x + 4)(x - 1)$

   b. Given the graph of $y = P(x)$, can you predict the factors of $P(x)$? Test your prediction with a graphing calculator.

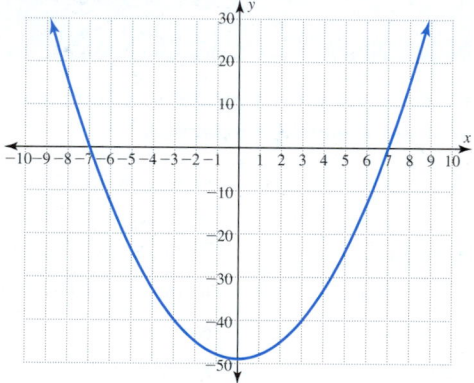

5.6   Cumulative Review

Match each equation with its classification.

1. $3x + 3 = 2x + 3$

2. $3x + 3 = 3x + 2$

3. $3x + 3 = 3(x + 1)$

   A. Conditional equation
   B. Identity
   C. Contradiction

4. Write in slope-intercept form the equation of a line passing through $(-2, 3)$ and parallel to $y = -\frac{2}{5}x + 7$.

5. Write in slope-intercept form the equation of a line passing through $(-2, 3)$ and perpendicular to $y = -\frac{2}{5}x + 7$.

Section 5.7 | Dividing Polynomials

Objectives:

1. Divide a polynomial by a monomial.
2. Use long division of polynomials.

This section examines the division of polynomials. We build on our familiarity with the division of real numbers and monomials. Recall that $\frac{15}{3} = 5$ because 15 equals the product of the factors 3 and 5. Thus we can check the division of polynomials by multiplying two factors (the divisor by the quotient). This also helps us prepare for factoring polynomials in Chapter 6. We will start by reviewing the quotient of two monomials, which was first considered in Section 5.2.

1. Divide a Polynomial by a Monomial

Is the Quotient Rule for Exponents Useful When We Divide a Polynomial by a Monomial?

Yes, recall that the quotient rule for exponents states that $\frac{x^m}{x^n} = x^{m-n}$ for $x \neq 0$. We assume throughout this section that the variables are restricted to values that will avoid division by zero.

Example 1   Dividing a Monomial by a Monomial

Simplify $\dfrac{8x^2y^5}{4xy^3}$ assuming that $x \neq 0$ and $y \neq 0$. Then check your answer by multiplying the divisor by the quotient.

Solution

$$\frac{8x^2y^5}{4xy^3} = \frac{2x^{2-1}y^{5-3}}{1}$$

Divide both the numerator and the denominator by 4. Using the quotient rule for exponents, subtract the exponents on the common bases. Note that the coefficients are divided and the exponents are subtracted.

$$= 2xy^2$$

This is the quotient.

Check: $(4xy^3)(2xy^2) = 8x^2y^5$ The answer checks because $8x^2y^5$ equals the product of two factors, the divisor and the quotient.

Self-Check 1

Divide $21x^3y^2$ by $3x^2y^2$ and use multiplication to check your answer.

The quotient of a polynomial divided by a monomial is often written in fractional form. Since $\dfrac{a+b}{c} = \dfrac{a}{c} + \dfrac{b}{c}$ $(c \neq 0)$, each term of the polynomial in the numerator is divided by the monomial in the denominator. Then each fraction is written in reduced form.

Dividing a Polynomial by a Monomial

Verbally	Algebraic Example
To divide a polynomial by a monomial, divide each term of the polynomial by the monomial.	$\dfrac{6x^2 + 4x}{2x} = \dfrac{6x^2}{2x} + \dfrac{4x}{2x}$ $= 3x + 2$

Example 2   Dividing a Binomial by a Monomial

Divide $15x^3 + 12x^2$ by $3x^2$ assuming that $x \neq 0$. Then check your answer by multiplying the divisor by the quotient.

Solution

$$\frac{15x^3 + 12x^2}{3x^2} = \frac{15x^3}{3x^2} + \frac{12x^2}{3x^2}$$ Divide each term of the numerator by $3x^2$, and then simplify.

$$= 5x + 4$$ This is the quotient.

Check: $3x^2(5x + 4) = (3x^2)(5x) + (3x^2)(4)$

$$= 15x^3 + 12x^2$$ This quotient checks.

Self-Check 2

Simplify $\dfrac{10x^3y + 4x^2y^2}{2x^2y}$ and use multiplication to check your answer.

Does Each Division by a Polynomial Yield a Zero Remainder?

No, some divisions do not have a zero remainder. In fact, the quotient $-6y^2 + 9y - \dfrac{4}{y}$ in Example 3 is not even a polynomial because $-\dfrac{4}{y}$ has a variable in the denominator. Nonetheless, the answers to all division problems can be checked by multiplication.

Example 3   Dividing a Trinomial by a Monomial

Find $(20y + 30y^4 - 45y^3) \div (-5y^2)$ and check your result. Assume that $y \neq 0$.

Solution

$$\frac{30y^4 - 45y^3 + 20y}{-5y^2} = \frac{30y^4}{-5y^2} - \frac{45y^3}{-5y^2} + \frac{20y}{-5y^2}$$ First write the division in fractional form, expressing the numerator in descending order.

$$= -6y^2 + 9y - \frac{4}{y}$$ Divide each term of the numerator by $-5y^2$, and then simplify.

Check: $(-5y^2)\left(-6y^2 + 9y - \dfrac{4}{y}\right)$ The answer to a division problem can be checked by multiplication.

$$= (-5y^2)(-6y^2) + (-5y^2)(9y) - (-5y^2)\left(\frac{4}{y}\right)$$ Distribute the multiplication of $-5y^2$ to each term, and then simplify.

$$= 30y^4 - 45y^3 + 20y$$ This answer checks.

Self-Check 3

Find $(12x^9 - 24x^8 + 10x^6) \div (2x^6)$.

2. Use Long Division of Polynomials

Is the Procedure for Dividing Polynomials Similar to the Procedure for Dividing Integers?

Yes, this similarity will be demonstrated in a side-by-side comparison of the long division of integers and the long division of polynomials. This procedure is outlined first.

Long Division of Polynomials

Step 1. Write the polynomials in long-division format, expressing each in standard form.

Step 2. Divide the first term of the divisor into the first term of the dividend. The result is the first term of the quotient.

Step 3. Multiply the first term of the quotient by every term in the divisor, and write this product under the dividend, aligning like terms.

Step 4. Subtract this product from the dividend, and bring down the next term.

Step 5. Use the result of step 4 as a new dividend, and repeat steps 2 through 4 until either the remainder is 0 or the degree of the remainder is less than the degree of the divisor.

Long Division of Integers
Problem: Divide 672 by 21.

Long Division of Polynomials
Problem: Divide $6x^2 + 7x + 2$ by $2x + 1$.

Step 1. Write the division in the long-division format.

$$21\overline{)672}$$

$$2x + 1\overline{)6x^2 + 7x + 2}$$

Step 2. Divide to obtain the first term in the quotient.

$$\begin{array}{r} 3 \\ 21\overline{)672} \end{array}$$

$$\begin{array}{r} 3x \\ 2x + 1\overline{)6x^2 + 7x + 2} \end{array}$$

$6x^2$ divided by $2x$ is $3x$.

Step 3. Multiply the first term in the quotient by the divisor.

$$\begin{array}{r} 3 \\ 21\overline{)672} \\ 63 \end{array}$$

$$\begin{array}{r} 3x \\ 2x + 1\overline{)6x^2 + 7x + 2} \\ 6x^2 + 3x \end{array}$$

Multiply $3x$ by $2x + 1$ and align similar terms.

Step 4. Subtract this product from the dividend, and bring down the next term.

$$\begin{array}{r} 3 \\ 21\overline{)672} \\ 63 \\ \hline 42 \end{array}$$

$$\begin{array}{r} 3x \\ 2x + 1\overline{)6x^2 + 7x + 2} \\ -(6x^2 + 3x) \\ \hline 4x + 2 \end{array}$$

Subtract $6x^2 + 3x$.

Step 5. Repeat steps 2–4 to obtain the next term in the quotient.

$$\begin{array}{r} 32 \\ 21\overline{)672} \\ 63 \\ \hline 42 \\ 42 \\ \hline 0 \end{array}$$

$$\begin{array}{r} 3x + 2 \\ 2x + 1\overline{)6x^2 + 7x + 2} \\ 6x^2 + 3x \\ \hline 4x + 2 \\ -(4x + 2) \\ \hline 0 \end{array}$$

$4x$ divided by $2x$ is 2. Multiply 2 by $2x + 1$ to obtain $4x + 2$. Then subtract $4x + 2$.

Because the remainder is 0, the division is finished.

Answer: $672 \div 21 = 32$ $(6x^2 + 7x + 2) \div (2x + 1) = 3x + 2$

Note that if 10 is substituted for x in $(6x^2 + 7x + 2) \div (2x + 1)$, we obtain $[6(10)^2 + 7(10) + 2] \div [2(10) + 1]$, or $(600 + 70 + 2) \div (20 + 1) = 672 \div 21$. Thus both of the division problems shown above represent the same thing for $x = 10$.

Example 4 illustrates the use of long division of polynomials to divide a trinomial by a binomial. The unstated assumption in Example 4 and all the remaining examples in this section is that all values that would cause division by zero are excluded. In Example 4, this means we are assuming $x \neq -3$.

Example 4   Using Long Division to Divide a Trinomial by a Binomial

Divide $2x^2 + x - 15$ by $x + 3$.

Solution

Step 1. $x + 3 \overline{)2x^2 + x - 15}$

Set up the format for long division, writing both polynomials in standard form.

Step 2.
$$\begin{array}{r} 2x \\ x + 3 \overline{)2x^2 + x - 15} \end{array}$$

Divide $2x^2$ (the first term of the dividend) by x (the first term of the divisor) to obtain $2x$ (the first term of the quotient). Align similar terms.

Step 3.
$$\begin{array}{r} 2x \\ x + 3 \overline{)2x^2 + x - 15} \\ \underline{2x^2 + 6x} \end{array}$$

Multiply $2x$ by every term in the divisor, aligning under similar terms in the dividend.

Step 4.
$$\begin{array}{r} 2x \\ x + 3 \overline{)2x^2 + x - 15} \\ \underline{-(2x^2 + 6x)} \\ -5x - 15 \end{array}$$

Subtract $2x^2 + 6x$ from the dividend.

Step 5.
$$\begin{array}{r} 2x - 5 \\ x + 3 \overline{)2x^2 + x - 15} \\ \underline{2x^2 + 6x} \\ -5x - 15 \\ \underline{-(-5x - 15)} \\ 0 \end{array}$$

Divide $-5x$ (the first term in the last row) by x (the first term of the divisor) to obtain -5 (the next term in the quotient). Then multiply each term of the divisor by -5, aligning under similar terms in the dividend. Subtract to obtain the remainder of 0.

Answer: $\underbrace{(2x^2 + x - 15)}_{\text{Dividend}} \div \underbrace{(x + 3)}_{\text{Divisor}} = \underbrace{2x - 5}_{\text{Quotient}}$

To check your answer, multiply $(x + 3)(2x - 5)$.

Self-Check 4

Divide $12x^2 + x - 35$ by $3x - 5$.

What Is the First Step in the Long-Division Procedure?

Write the polynomials in standard form. In Example 5, we write $7v + 6v^3 + 2 - 19v^2$ in standard form before setting up the long-division format.

Example 5   Using Long Division of Polynomials

Divide $(7v + 6v^3 + 2 - 19v^2)$ by $3v - 2$.

Solution

$$\begin{array}{r} 2v^2 \\ 3v - 2 \overline{)6v^3 - 19v^2 + 7v + 2} \\ \underline{- \,(6v^3 - 4v^2)} \\ -15v^2 + 7v \end{array}$$

Write both polynomials in standard form. Divide to obtain the first term in the quotient, $\dfrac{6v^3}{3v} = 2v^2$. Multiply $2v^2$ by $3v - 2$ to obtain $6v^3 - 4v^2$, and then subtract.

$$\begin{array}{r} 2v^2 - 5v \\ 3v - 2 \overline{)6v^3 - 19v^2 + 7v + 2} \\ \underline{6v^3 - 4v^2} \\ -15v^2 + 7v \\ \underline{- \,(-15v^2 + 10v)} \\ -3v + 2 \end{array}$$

Divide to obtain the second term in the quotient,

$$-\frac{15v^2}{3v} = -5v.$$

Multiply $-5v$ by $3v - 2$ to obtain $-15v^2 + 10v$, and then subtract.

$$\begin{array}{r} 2v^2 - 5v - 1 \\ 3v - 2 \overline{)6v^3 - 19v^2 + 7v + 2} \\ \underline{6v^3 - 4v^2} \\ -15v^2 + 7v \\ \underline{-15v^2 + 10v} \\ -3v + 2 \\ \underline{-(-3v + 2)} \\ 0 \end{array}$$

Divide to obtain the third term in the quotient, $\dfrac{-3v}{3v} = -1$.

Multiply -1 by $3v - 2$ to obtain $-3v + 2$, and then subtract.

Answer: $\dfrac{6v^3 - 19v^2 + 7v + 2}{3v - 2}$

$= 2v^2 - 5v - 1$

To check your answer, multiply $(3v - 2)(2v^2 - 5v - 1)$.

Self-Check 5

Divide $15x - 31x^2 + 12x^3 - 2$ by $4x - 1$.

If we divide 22 by 5, we get 4 with a remainder of 2. Thus,

$$\frac{22}{5} = 4 + \frac{2}{5}$$

Because 5 does not divide 22 evenly, we say that 5 is not a factor of 22. In Example 6, we divide two polynomials that produce a nonzero remainder.

Example 6   Performing a Division That Has a Nonzero Remainder

Divide $x^2 + 2x - 13$ by $x + 5$.

Solution

$$
\begin{array}{r}
x - 3 \\
x + 5 \overline{)\, x^2 + 2x - 13} \\
-\,\underline{(x^2 + 5x)} \\
-3x - 13 \\
-\,\underline{(-3x - 15)} \\
2
\end{array}
$$

Divide x^2 by x to obtain x.

Multiply x by $x + 5$ to obtain $x^2 + 5x$ and then subtract.

Divide $-3x$ by x to obtain -3.

Multiply -3 by $x + 5$ to obtain $-3x - 15$, and then subtract to obtain the remainder of 2.

Answer:
$$\frac{x^2 + 2x - 13}{x + 5} = x - 3 + \frac{2}{x + 5}$$

The quotient is expressed as $x - 3$ plus the remainder 2, divided by $x + 5$.

Check: $(x + 5)\left(x - 3 + \dfrac{2}{x + 5}\right)$

To check, we multiply the divisor $x + 5$ by the quotient $x - 3 + \dfrac{2}{x + 5}$.

$$= (x + 5)(x - 3) + (x + 5)\left(\frac{2}{x + 5}\right)$$
$$= x^2 - 3x + 5x - 15 + 2$$
$$= x^2 + 2x - 13$$

The answer checks.

Self-Check 6

Divide $x^2 - 3x - 14$ by $x - 2$.

The first step in the long-division procedure is to write the polynomials in standard form. Some polynomials have missing terms. In Example 7, the polynomial $x^3 - 5x - 14$ is missing the x^2 term. This polynomial can be rewritten as $x^3 + 0x^2 - 5x - 14$. This format can help prevent careless errors in the long-division procedure.

Example 7 Dividing a Polynomial with Missing Terms

Divide $x^3 - 5x - 14$ by $x - 3$.

Solution

$$
\begin{array}{r}
x^2 + 3x + 4 \\
x - 3 \overline{)\, x^3 + 0x^2 - 5x - 14} \\
-\,\underline{(x^3 - 3x^2)} \\
3x^2 - 5x \\
-\,\underline{(3x^2 - 9x)} \\
4x - 14 \\
-\,\underline{(4x - 12)} \\
-2
\end{array}
$$

Use 0 as the coefficient of the missing x^2 term.

Divide x^3 by x to obtain x^2.

Multiply x^2 by $x - 3$ to obtain $x^3 - 3x^2$, and then subtract.

Divide $3x^2$ by x to obtain $3x$.

Multiply $3x$ by $x - 3$ to obtain $3x^2 - 9x$, and then subtract.

Divide $4x$ by x to obtain 4. Multiply 4 by $x - 3$ to obtain $4x - 12$. Subtract to obtain the remainder of -2.

Answer:
$$\frac{x^3 - 5x - 14}{x - 3} = x^2 + 3x + 4 - \frac{2}{x - 3}$$

Because the remainder is not zero, $x - 3$ is not a factor of $x^3 - 5x - 14$.

Self-Check 7

Divide $8x^3 - 27$ by $2x - 3$.

Example 8 examines a problem with a second-degree trinomial as a divisor.

Example 8 Dividing by a Second-Degree Trinomial

Divide $x^4 - x^3 + 6x^2 - 7x + 15$ by $x^2 - 2x + 3$.

Solution

$$
\begin{array}{r}
x^2 + x + 5 \\
x^2 - 2x + 3 \overline{)x^4 - x^3 + 6x^2 - 7x + 15} \\
-(x^4 - 2x^3 + 3x^2) \\
\hline
x^3 + 3x^2 - 7x \\
-(x^3 - 2x^2 + 3x) \\
\hline
5x^2 - 10x + 15 \\
-(5x^2 - 10x + 15) \\
\hline
0
\end{array}
$$

Divide x^4 by x^2 to obtain the first term of x^2 in the quotient.

Multiply x^2 by $x^2 - 2x + 3$ to obtain $x^4 - 2x^3 + 3x^2$, and then subtract.

Divide x^3 by x^2 to obtain x.

Multiply x by $x^2 - 2x + 3$ to obtain $x^3 - 2x^2 + 3x$, and then subtract.

Divide $5x^2$ by x^2 to obtain 5.

Multiply 5 by $x^2 - 2x + 3$ to obtain $5x^2 - 10x + 15$, and then subtract to produce the remainder of 0.

Answer: $\dfrac{x^4 - x^3 + 6x^2 - 7x + 15}{x^2 - 2x + 3} = x^2 + x + 5$

Self-Check 8

Divide $2x^4 + 6x^3 + 7x^2 - 3x - 4$ by $x^2 + 3x + 4$.

The **average cost per unit** for units produced in a factory is the total cost of the units divided by the number of units. In Example 9, we determine the average cost per unit for the production of portable music players.

Example 9 Determining the Average Cost per Unit

A company recorded the number of portable music players produced and the cost for producing these units for different periods of time. Using x to represent the number of units produced, the company determined that the cost function for this music player is $C(x) = 35x + 12,000$.

(a) Write an algebraic expression $A(x)$ for the average cost of producing x units.

(b) Evaluate and interpret $C(1,000)$.

(c) Evaluate and interpret $A(1,000)$.

Solution

(a) $\boxed{\text{Average cost}} = \boxed{\text{Total cost}} \div \boxed{\text{Number of units}}$

$A(x) = (35x + 12,000) \div x$

$A(x) = \dfrac{35x + 12,000}{x}$

$A(x) = 35 + \dfrac{12,000}{x}$

Translate this word equation into an algebraic representation, using the given information.

(b) $C(x) = 35x + 12,000$

$C(1,000) = 35(1,000) + 12,000$

$C(1,000) = 35,000 + 12,000$

$C(1,000) = 47,000$

Substitute 1,000 into the cost function to determine the cost of 1,000 units.

The cost of producing 1,000 portable music players is $47,000.

(c) $A(x) = 35 + \dfrac{12{,}000}{x}$

$A(1{,}000) = 35 + \dfrac{12{,}000}{1{,}000}$

$A(1{,}000) = 35 + 12$

$A(1{,}000) = 47$

Substitute 1,000 into the average cost function to determine the average cost of 1,000 units.

The average cost of producing 1,000 portable music players is $47 per unit.

Self-Check 9

Rework all parts of Example 9 assuming the cost function for this music player is $C(x) = 40x + 10{,}000$.

Self-Check Answers

1. $7x$; $(3x^2 y^2)(7x) = 21x^3 y^2$
2. $5x + 2y$; $(2x^2 y)(5x + 2y) = 10x^3 y + 4x^2 y^2$
3. $6x^3 - 12x^2 + 5$
4. $4x + 7$
5. $3x^2 - 7x + 2$
6. $x - 1 - \dfrac{16}{x - 2}$

7. $4x^2 + 6x + 9$
8. $2x^2 - 1$
9. **a.** $A(x) = 40 + \dfrac{10{,}000}{x}$
    b. $C(1{,}000) = 50{,}000$; The cost of producing 1,000 portable music players is $50,000.

c. $A(1{,}000) = 50$; The average cost of producing 1,000 portable music players is $50 per unit.

5.7   Using the Language and Symbolism of Mathematics

1. In the division $\dfrac{15}{3} = 5$, the dividend is 15, the divisor is 3, and the _____ is 5.

2. In the division $\dfrac{x^2 - x - 6}{x - 3} = x + 2$, the dividend is $x^2 - x - 6$, the divisor is _____, and the quotient is $x + 2$.

3. The division $\dfrac{35}{5} = 7$ can be checked by writing $5 \cdot 7 = 35$. In the equation $5 \cdot 7 = 35$, both 5 and 7 are factors and 35 is their _____.

4. The division $\dfrac{x^2 - x - 6}{x - 3} = x + 2$ can be checked by writing $(x - 3)(x + 2) = x^2 - x - 6$. In the equation

$(x - 3)(x + 2) = x^2 - x - 6$, the factors are _____ and _____, and the product is _____.

5. The first step in the long-division procedure for dividing polynomials is to write both the divisor and the dividend in _____ form.

6. After starting the long-division procedure for dividing polynomials, we continue the steps until either the remainder is _____ or the degree of the remainder is less than the degree of the _____.

7. The polynomial $3x^4 + 5x^2 - 7x + 2$ is said to have a missing x^3 term. If this polynomial is rewritten with an x^3 term, the coefficient of x^3 will be _____.

8. The _____ cost per unit is the total cost of the units divided by the number of units.

5.7   Quick Review

In Exercises 1–4, simplify each expression.

1. $\dfrac{56}{24}$

2. $\dfrac{25x^{10}}{10x^5}$

3. $-3x^2 y^5 (4x^2 - 5xy + 7y^2)$

4. $(4x + 3)(2x - 7) - 9$

5. Write $\dfrac{621}{57}$ as a mixed number.

All exercises are assumed to restrict the variables to values that avoid division by zero.

Objective 1 Divide a Polynomial by a Monomial

In Exercises 1–18, find each quotient.

1. $\dfrac{18x^3}{6x}$

2. $\dfrac{24x^5}{3x^2}$

3. $\dfrac{35a^3b^2}{-7ab}$

4. $\dfrac{48a^4b^4}{-8ab^2}$

5. $\dfrac{15a^2 - 20a}{5a}$

6. $\dfrac{24a^5 - 16a^3}{4a}$

7. $\dfrac{9a^3 - 15a^2 + 6a}{3a^2}$

8. $\dfrac{24a^5 - 6a^4 - 12a^2}{6a^3}$

9. $\dfrac{16m^4 - 8m^3 + 10m^2 + 6m}{2m}$

10. $\dfrac{24m^4 - 18m^3 + 36m^2 - 6m}{3m}$

11. $\dfrac{45x^5y^2 - 54x^4y^5 + 99x^2y^4}{9x^2y^2}$

12. $\dfrac{-77x^7y^2 + 55x^5y^4 - 33x^3y^6}{11x^3y^2}$

13. $\dfrac{54v^2 - 36v^4 + 12 - 18v^3}{-6v}$

14. $\dfrac{15v^3 - 6 + 9v^4 + 27v^2}{-3v}$

15. $(35v^3w - 7v^2w - 28vw^2) \div 7vw$

16. $(54m^3n - 18m^2n + 27mn^2) \div 9mn$

17. $(100x^{20} - 50x^{12} + 30x^9) \div 10x^9$

18. $(66x^{21} - 48x^{14} - 18x^7) \div 6x^7$

Objective 2 Use Long Division of Polynomials

In Exercises 19–22, find each quotient, and then use multiplication to check your answer.

19. $x + 2 \overline{)\, x^2 + 9x + 14}$

20. $x + 5 \overline{)\, x^2 + 8x + 15}$

21. $\dfrac{v^2 + 2v - 24}{v - 4}$

22. $\dfrac{v^2 - 2v - 35}{v - 7}$

In Exercises 23 and 24, perform the indicated division and then use multiplication to check your answer. (*Hint:* See Example 6.)

23. $\dfrac{4m^3 - 9m^2 + 10m + 7}{m - 3}$

24. $\dfrac{6m^3 - 19m^2 - 23m + 15}{m - 4}$

In Exercises 25–44, find each quotient.

25. $\dfrac{6w^2 + w - 12}{2w + 3}$

26. $\dfrac{12w^2 + 14w - 10}{3w + 5}$

27. $\dfrac{20m^2 - 43m + 14}{5m - 2}$

28. $\dfrac{28m^2 - 27m + 5}{4m - 1}$

29. $\dfrac{63y^2 - 130y + 63}{7y - 9}$

30. $\dfrac{32y^2 + 36y - 35}{8y - 5}$

31. $\dfrac{30x^2 - 45x^3 + 10x + 35x^4}{5x}$

32. $\dfrac{14x^2 - 35x^3 + 28x^4 - 42x}{7x}$

33. $(x^2 - 12x + 35) \div (x - 5)$

34. $(x^2 - 16x + 60) \div (x - 6)$

35. $(6a^2 - 11a - 10) \div (2a - 5)$

36. $(15a^2 - 19a - 10) \div (5a + 2)$

37. $(3v^3 - 4v^2 - 8) \div (v - 2)$

38. $(4v^3 - 6v^2 - 54) \div (v - 3)$

39. $(x^3 - 8) \div (x - 2)$

40. $(x^3 + 125) \div (x + 5)$

41. $(20y^3 + 16y^2 - 3y - 54) \div (5y - 6)$

42. $(18y^3 - 63y^2 + 106y - 55) \div (6y - 5)$

43. $\dfrac{21x^4 - 62x^2 - 7x^3 + 9x + 45}{3x^2 - x - 5}$

44. $\dfrac{7x^3 - 19x - 20 + 6x^4 - x^2}{3x + 4 + 2x^2}$

Connecting Algebra to Geometry

45. **Area of a Rectangle** The area of the rectangle shown is $(4x^2 + 4x - 3)$ cm^2. Find the width of this rectangle.

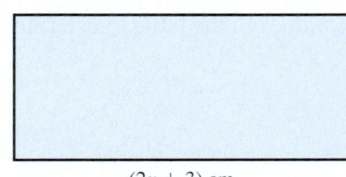

$(2x + 3)$ cm

46. **Area of a Rectangle** The area of the rectangle shown is $(9x^2 - 1)$ cm^2. Find the width of this rectangle.

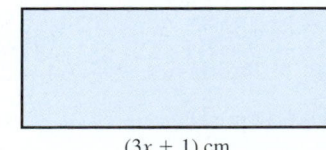

$(3x + 1)$ cm

47. Area of a Triangle The area of the triangle shown is $(12x^2 - 17x - 7)$ cm². Find the altitude of this triangle as shown by the dashed line.

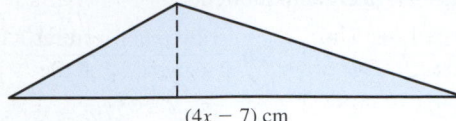

(4x − 7) cm

48. Area of a Triangle The area of the triangle shown is $(5x^2 + 13x + 6)$ cm². Find the altitude of this triangle as shown by the dashed line.

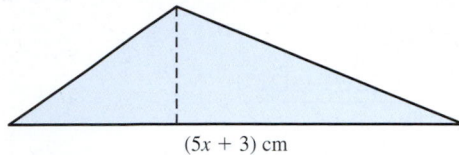

(5x + 3) cm

49. Volume of a Box The volume of the solid shown is $(8x^3 + 14x^2 + 3x)$ cm³. Find the height of this solid.

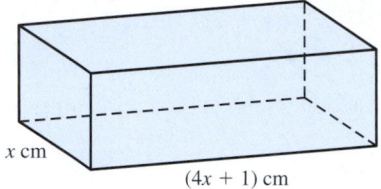

x cm

(4x + 1) cm

50. Volume of a Box The volume of the solid shown is $(18x^3 + 33x^2 + 5x)$ cm³. Find the height of this solid.

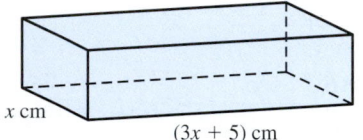

x cm

(3x + 5) cm

Applying Technology

In Exercises 51–54, use a calculator or a spreadsheet to complete the given table of values for each pair of expressions. Then use the values in the table to determine whether $Y_1 = Y_2$.

x	Y_1	Y_2
−3		
−2		
−1		
0		
1		
2		
3		

51. $Y_1 = (2x^2 - 11x + 15)/(-x + 3)$
$Y_2 = -2x + 5$

52. $Y_1 = (3x^2 + x - 14)/(x + 2)$
$Y_2 = 3x - 7$

53. $Y_1 = (x^2 - 9)/(x - 3)$
$Y_2 = x - 3$

54. $Y_1 = (x^3 - 8)/(x^2 + 2x + 4)$
$Y_2 = x - 2$

Concept Development

55. What polynomial, when divided by $x + 3$, yields a quotient of $x^2 - 3x - 5$?

56. What polynomial, when divided by $x - 7$, yields a quotient of $2x^2 + x - 6$?

57. What polynomial, when divided into $6x^2 - 2x - 20$, produces a quotient of $2x - 4$?

58. What polynomial, when divided into $20x^2 + 7x - 6$, produces a quotient of $5x - 2$?

59. a. If a fourth-degree trinomial in x is added to a second-degree binomial in x, the degree of the sum is _____.
   b. If a fourth-degree trinomial in x is multiplied by a second-degree binomial in x, the degree of the product is _____.
   c. If a fourth-degree trinomial in x is divided by a second-degree binomial in x, the degree of the quotient is _____.

60. a. If a third-degree trinomial in x is added to a second-degree binomial in x, the degree of the sum is _____.
   b. If a third-degree trinomial in x is multiplied by a second-degree binomial in x, the degree of the product is _____.
   c. If a third-degree trinomial in x is divided by a second-degree binomial in x, the degree of the quotient is _____.

61. When a polynomial is divided by $2x + 3$, the quotient is $x + 2 + \dfrac{4}{2x + 3}$. Determine this polynomial.

62. When a polynomial is divided by $3x - 2$, the quotient is $2x + 1 + \dfrac{2}{3x - 2}$. Determine this polynomial.

63. One factor of $28x^2 + 11x - 30$ is $7x - 6$. Determine the other factor.

64. One factor of $40x^2 + 63x + 18$ is $8x + 3$. Determine the other factor.

65. a. One factor of $6x^3 - 29x^2 - 17x + 60$ is $2x + 3$. Determine the other factor.
   b. Another factor of $6x^3 - 29x^2 - 17x + 60$ is $3x - 4$. Complete this equation:
   $6x^3 - 29x^2 - 17x + 60 = (2x + 3)(3x - 4)(?)$.

66. a. One factor of $20x^3 + 8x^2 - 125x - 50$ is $5x + 2$. Determine the other factor.
b. Another factor of $20x^3 + 8x^2 - 125x - 50$ is $2x - 5$. Complete this equation:
$20x^3 + 8x^2 - 125x - 50 = (5x + 2)(2x - 5)(?)$.

Connecting Concepts to Applications

67. Entries in a Spreadsheet Table A two-dimensional table in an Excel spreadsheet has $30k^2 + 39k + 12$ entries. If there are $6k + 3$ rows in the table, how many columns are in the table?

68. Entries in a Spreadsheet Table A two-dimensional table in an Excel spreadsheet has $4k^2 + 81k + 20$ entries. If there are $k + 20$ columns in the table, how many rows are in the table?

69. Average Cost per Unit Using x to represent the number of units produced, a factory determined that the cost in dollars to produce these units was $C(x) = 125x + 960$.
a. Write an algebraic expression for $A(x)$, the average cost of producing x units.
b. Evaluate and interpret $A(50)$.

70. Average Cost per Unit Using x to represent the number of units produced, a factory determined that the cost in dollars to produce these units was $C(x) = 60x + 720$.
a. Write an algebraic expression for $A(x)$, the average cost of producing x units.
b. Evaluate and interpret $A(90)$.

71. Average Cost per Unit A company manufactures lawn chairs. The company determined that the cost in dollars to produce x chairs is $C(x) = 8x + 1,200$.

a. Write an algebraic expression for $A(x)$, the average cost of producing x chairs.
b. Evaluate and interpret $C(200)$.
c. Evaluate and interpret $A(200)$.
d. Evaluate and interpret $C(250)$.
e. Evaluate and interpret $A(250)$.

f. What is the change in cost when production is increased from 200 to 250 chairs?
g. What is the change in the average cost per unit when production is increased from 200 to 250 chairs?

72. Average Cost per Unit A company manufactures baseball gloves. The company determined that the cost in dollars to produce x baseball gloves is $C(x) = 11x + 800$.
a. Write an algebraic expression for $A(x)$, the average cost of producing x gloves.
b. Evaluate and interpret $C(100)$.
c. Evaluate and interpret $A(100)$.
d. Evaluate and interpret $C(200)$.
e. Evaluate and interpret $A(200)$.
f. What is the change in cost when production is increased from 100 to 200 gloves?
g. What is the change in the average cost per unit when production is increased from 100 to 200 gloves?

Group discussion questions

73. Challenge Question
a. Can you give two binomials whose quotient is a monomial?
b. Can you give two binomials whose quotient is a binomial?
c. Can you give two binomials whose quotient is a trinomial?

74. Discovery Question Complete each of these divisions.
a. $(x^2 - 1) \div (x - 1)$
b. $(x^3 - 1) \div (x - 1)$
c. $(x^4 - 1) \div (x - 1)$
d. Based on parts **a** through **c,** predict the quotient when $x^5 - 1$ is divided by $x - 1$. Then perform this division to check your prediction.
e. Use inspection and this pattern to divide $x^8 - 1$ by $x - 1$.

75. Discovery Question

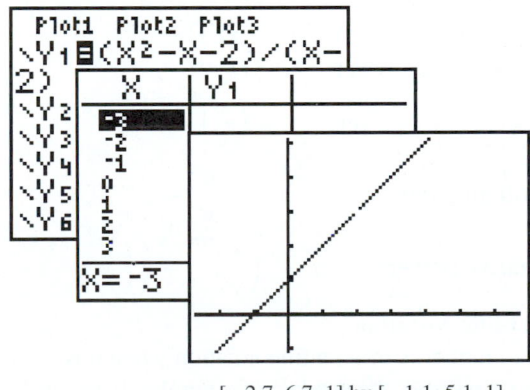

$[-2.7, 6.7, 1]$ by $[-1.1, 5.1, 1]$

a. Use a graphing calculator to complete the table of values shown here.
b. Then use the **Trace** feature on a graphing calculator to examine this graph on the window indicated.

c. From the table, what is the value of Y_1 when $x = 2$?

d. Describe what happens to the graph when $x = 2$.

e. This graph is not a straight line. Explain why it is not a straight line.

f. Compare $Y_1 = \dfrac{x^2 - x - 2}{x - 2}$ to $Y_2 = x + 1$.

How are they the same? How are they different?

5.7   Cumulative Review

Match each system of equations with its correct classification.

1. $y = 3x + 2$
$y = 2x + 3$

2. $y = 3x + 2$
$y = 3x + 3$

3. $y = 3x + 2$
$y = 3x + 2$

A. Consistent system of independent equations
B. Consistent system of dependent equations
C. Inconsistent system

4. Graph the solution of $3x - 4y > 16$.

5. Expand the expression $(2x - 5)^2$.

Chapter 5 | Key Concepts

1. Exponential Notation: For any real base x and natural number n,

- $x^n = x \cdot x \cdot \cdots \cdot x$ (n factors of x)
- $x^0 = 1$ for $x \neq 0$; 0^0 is undefined.
- $x^{-n} = \dfrac{1}{x^n}$ for $x \neq 0$.

2. Base of an Exponent: If there are no symbols of grouping, only the constant or variable immediately to the left of an exponent is the base. Thus $-x^2$ and $(-x)^2$ have distinct meanings.

3. Summary of the Properties of Exponents: For any nonzero real numbers x and y and integral exponents m and n,

Product rule: $x^m \cdot x^n = x^{m+n}$

Power rule: $(x^m)^n = x^{mn}$

Product to a power: $(xy)^m = x^m y^m$

Quotient to a power: $\left(\dfrac{x}{y}\right)^m = \dfrac{x^m}{y^m}$

Quotient rule: $\dfrac{x^m}{x^n} = x^{m-n}$

Negative power: $\left(\dfrac{x}{y}\right)^{-n} = \left(\dfrac{y}{x}\right)^n$

4. Scientific Notation

- A number is in scientific notation when it is expressed as the product of a number between 1 and 10 (or between -1 and -10 if negative) and an appropriate power of 10.
- On many calculators the **EE** feature is used to enter the power of 10.

- If a number such as 6.89E5 appears on a calculator display, this represents 6.89×10^5 in scientific notation.

5. Converting from Scientific Notation to Decimal Notation: Multiply the two factors by using the given power of 10.

- If the exponent on 10 is positive, move the decimal point to the right.
- If the exponent on 10 is zero, do not move the decimal point.
- If the exponent on 10 is negative, move the decimal point to the left.

6. Converting from Decimal Notation to Scientific Notation

- Move the decimal point immediately to the right of the first nonzero digit of the number.
- Multiply by a power of 10 determined by counting the number of places the decimal point has been moved.
  - **i.** The exponent on 10 is 0 or positive if the magnitude of the original number is 1 or greater.
  - **ii.** The exponent on 10 is negative if the magnitude of the original number is between 0 and 1.

7. Monomial: A monomial is a real number, a variable, or a product of real numbers and variables.

8. Polynomial: A polynomial is a monomial or a sum of a finite number of monomials.

9. Classification of Polynomials

- Monomials contain one term.
- Binomials contain two terms.
- Trinomials contain three terms.

10. Degree of a Monomial: The degree of a monomial is the sum of the exponents on all the variables in the term.

11. Degree of a Polynomial: The degree of a polynomial is the same as the degree of the term of highest degree.

12. A Polynomial Is in Standard Form if
- The variables in each term are written in alphabetical order.
- The terms are arranged in descending powers of the first variable.

13. Real Polynomial in x: An nth-degree real polynomial in x can be written in the standard form $a_nx^n + a_{n-1}x^{n-1} + \cdots + a_1x + a_0$, where each of the coefficients of this polynomial is a real number.

14. Equal Polynomials: Equal polynomials in x have exactly the same values for each value of x. The tables and graphs for two equal polynomials are identical.

15. Adding and Subtracting Polynomials: To add or subtract polynomials, combine like terms.

16. Product of Polynomials
- To multiply a monomial by a polynomial, use the distributive property to multiply the monomial by each term of the polynomial.

- To multiply two polynomials, use the distributive property to multiply each term of the first polynomial by each term of the second polynomial, and then combine like terms.
- To multiply binomials by inspection, you can use FOIL to keep track of the steps. FOIL is an acronym for **F**irst, **O**uter, **I**nner, and **L**ast.

17. Special Products
- Product of a sum and a difference $(A + B)(A - B) = A^2 - B^2$
- The square of a sum $(A + B)^2 = A^2 + 2AB + B^2$
- The square of a difference $(A - B)^2 = A^2 - 2AB + B^2$

18. Quotient of Polynomials
- To divide a polynomial by a monomial, divide each term of the polynomial by the monomial.
- To divide a polynomial by a polynomial of more than one term, use long division of polynomials.

19. Average Cost per Unit: The average cost per unit for units that are manufactured is the total cost of the units divided by the number of units.

Chapter 5 | Review Exercises

Properties of Exponents

All exercises are assumed to restrict the variables to values that avoid division by 0 and avoid 0^0.

1. Write each expression in exponential form.
   a. $xyyy$ **b.** $(xy)(xy)(xy)$
   c. $xx + yy$ **d.** $(x + y)(x + y)$

2. Write each exponential expression in expanded form.
   a. $x^2 + y^2$ **b.** $(x + y)^2$
   c. $-x^4$ **d.** $(-x)^4$

In Exercises 3–12, simplify each expression.

3. **a.** 3^2 **b.** 3^{-2}
   c. -3^2 **d.** $(-3)^{-2}$

4. **a.** 3^0 **b.** 0^3
   c. $3^0 + 4^0$ **d.** $(3 + 4)^0$

5. **a.** $\left(\frac{1}{2}\right)^{-1} + \left(\frac{1}{3}\right)^{-1}$ **b.** $\left(\frac{1}{2} + \frac{1}{3}\right)^{-1}$
   c. $\left(\frac{2}{3}\right)^{-1}$ **d.** $\frac{2^{-1}}{3}$

6. **a.** -1^6 **b.** $(-1)^6$
   c. 6^{-1} **d.** -6^{-1}

7. **a.** $x^3 + x^3$ **b.** x^3x^3
   c. $\frac{x^3}{x^3}$ **d.** $x^3 - x^3$

8. **a.** 10^2 **b.** $(-10)^2$
   c. 10^{-2} **d.** -10^{-2}

9. **a.** $5^0 + 6^0 + 7^0$ **b.** $(5 + 6)^0 + 7^0$
   c. $(5 + 6 + 7)^0$ **d.** $0^5 + 0^6 + 0^7$

10. **a.** $2x^0 + 3y^0 + 4z^0$ **b.** $(2x + 3y + 4z)^0$
    c. $(2x)^0 + (3y)^0 + (4z)^0$ **d.** $2x^0 + (3y + 4z)^0$

11. **a.** x^2x^5 **b.** $\frac{x^5}{x^2}$
    c. $(x^2)^5$ **d.** $\left(\frac{x^5}{x}\right)^2$

12. **a.** $(4x^3y^2)(12x^5y^4)$ **b.** $\frac{12x^5y^4}{4x^3y^2}$
    c. $(4x^3y^2)^2$ **d.** $\left(\frac{12x^5y^4}{4x^3y^2}\right)^3$

In Exercises 13–18, evaluate each expression for $x = 3$ and $y = 4$.

13. **a.** $-x^2$ **b.** $(-x)^2$
14. **a.** xy^2 **b.** $(xy)^2$
15. **a.** $x^2 + y^2$ **b.** $(x + y)^2$
16. **a.** $x^{-1} + y^{-1}$ **b.** $(x + y)^{-1}$
17. **a.** $x^0 + y^0$ **b.** $(x + y)^0$
18. **a.** $\frac{x}{y^{-2}}$ **b.** $\left(\frac{x}{y}\right)^{-2}$

In Exercises 19–32, simplify each expression.

19. $(5x^2y^3z^4)(6xy^4z^7)$ **20.** $(5x^2y^3z^4)^2$

21. $\dfrac{24a^2b^4c^5}{40ab^7c^3}$ **22.** $\left(\dfrac{12m^3}{4n^2}\right)^{-2}$

23. $(3x^2)(4x^3)(5x^{11})$ **24.** $\dfrac{(3x^2)^4}{9x^5}$

25. $(-2x^5)^3(-5x^3)^2$ **26.** $(7y^8)(8y^{-6})$

27. $\dfrac{16w^{-4}}{8w^{-6}}$ **28.** $(5m^{-6})^{-4}$

29. $(6x)^{-3}\left(\dfrac{x}{6}\right)^{-4}$ **30.** $\left(\dfrac{x^{188}}{x^{-439}}\right)^0$

31. $\dfrac{(2a^3b^3)(8a^2b^5)}{(4ab^2)^2}$ **32.** $(x^2y^{-3}z^{-4})^{-2}(x^4y^{-1}z^{-2})^{-1}$

Scientific Notation

33. Size of Ink Drops An inkjet printer delivers ink droplets as tiny as one picoliter. One picoliter is 1.0×10^{-12} liter. Write this number in standard decimal notation.

34. Number of Ink Drops An inkjet printer can deliver 2.2×10^4 droplets per second. Write this number in standard decimal notation.

35. Express the number shown on the calculator display in standard decimal notation.

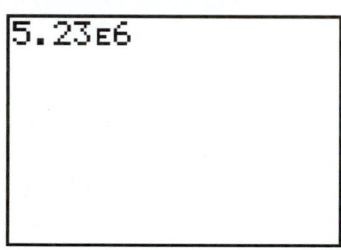

36. Express the number shown on the calculator display in standard decimal notation.

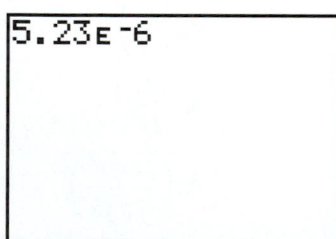

37. Time for a Satellite Signal If the distance from *Voyager 2* to Earth on its 1989 flyby of Neptune was 4.5×10^{12} m, determine the time for a *Voyager 2* signal traveling at 2.99×10^8 m/s to reach Earth.

38. Estimation Skills Use scientific notation and pencil and paper to select the best approximation of $(41,010)(0.0000000001989)$.
A. 8.0×10^{-8} **B.** 8.0×10^{-6}
C. 8.0×10^{-4} **D.** 6.0×10^{-5}
E. 6.0×10^5

Polynomials

In Exercises 39–42, classify each polynomial according to the number of terms it contains, and give its degree.

39. π **40.** $3x^6 - 17x^2$

41. $-9x^3 + 7x^2 + 8x - 11$ **42.** $x^5y - 7x^4y^2 + 23x^3y^3$

43. Write $11x^4 - 3x^2 + 9x^3 + 7x^5 + 4 - 8x$ in standard form.

44. Write a fifth-degree monomial in x with a coefficient of negative seven.

45. Write a second-degree binomial in x whose leading coefficient is one and with a constant term of negative three.

In Exercises 46–64, perform the indicated operations.

46. $(7x^2 - 9x + 13) + (4x^2 + 6x - 11)$

47. $(9x^3 - 5x^2 - 7) - (4x^3 + 8x - 11)$

48. $(3x^4 - 8x^3 + 7x^2 + 9x - 4) + 2(2x^4 + 6x^2 + 9)$

49. $7x^5 + 9x^3 + 6x - 3$
$\quad - 3(4x^5 - 3x^4 - 7x^3 - x^2 - x + 8)$

50. $(x^2 - 8x + 7) - (2x^2 + 7x + 11) + (3x^2 + 4x - 8)$

51. $5x^2(7x^3 - 9x^2 + 3x + 1)$

52. $(5v + 1)(7v - 1)$ **53.** $(5y - 7)^2$

54. $(9y + 5)^2$ **55.** $(3a + 5b)(3a - 5b)$

56. $(3m + 5)(2m^2 - 6m + 7)$ **57.** $\dfrac{-36a^3b^7}{9a^4b^4}$

58. $\dfrac{15m^5n^4 - 21m^4n^5 - 3m^3n^6}{3m^2n^3}$ **59.** $\dfrac{v^2 - 6v + 8}{v - 4}$

60. $\dfrac{21w^2 - 40w - 21}{3w - 7}$ **61.** $\dfrac{(a + b)^7}{(a + b)^6}$

62. $(3x + 4y)^2 - (3x + 4y)(3x - 4y)$

63. $(2x + 3y)^2 - (2x - 3y)^2$

64. $\dfrac{6x^2 + x - 2}{2x - 1} - \dfrac{8x^2 + 18x - 35}{4x - 5}$

65. The answer to $\dfrac{x^3 + 64}{x + 4}$ is $x^2 - 4x + 16$. Check this answer.

66. The answer to $\dfrac{x^2 + 4x - 5}{x - 2}$ is $x + 6 + \dfrac{7}{x - 2}$. Check this answer.

In Exercises 67–70, use the distributive property and the special products to expand the first expression and to factor the second expression.

Expand	**Factor**
67. a. $3x(x - 4)$	**b.** $(3x)(x) - (3x)(4)$
68. a. $5x(x + 3) - 7(x + 3)$	**b.** $5x(x + 3) - 7(x + 3)$
69. a. $(4x + 3)(4x - 3)$	**b.** $16x^2 - 9$
70. a. $(6x - 7)^2$	**b.** $36x^2 - 84x + 49$

In Exercises 71–76, multiply these binomial factors by inspection.

71. $(9x - 2)(5x + 3)$ **72.** $(4x - 3y)(5x + 6y)$

73. $(5x - 6y)(5x + 6y)$ **74.** $(5x - 6y)^2$

75. $(7x + 2y)^2$ **76.** $(x + 3)(x - 3)(x^2 + 9)$

In Exercises 77–80, use the order of operations to simplify each expression.

77. a. $3x - 7(4x + 1)$ **b.** $(3x - 7)(4x + 1)$

78. a. $(8x)^2 - (5y)^2$ **b.** $(8x - 5y)^2$

79. $(7a - 2b)^2 + (7a + 2b)^2$

80. $2(a - 3b)^2 + 3(a + 3b)^2 + 4a^2 + 5(3b)^2$

81. Divide $8x^4 - 2x^3 + 12x^2 - 31x + 7$ by $2x^3 + 3x - 7$.

82. Divide $81x^4 - 16$ by $3x - 2$.

Connecting Concepts to Applications

83. Area of a Rectangle Determine the area of the rectangle shown here.

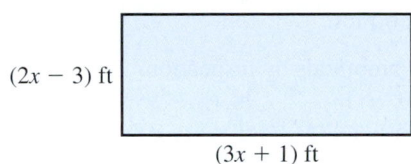

84. Height of a Box The volume of the box shown here is $(2x^3 + 2x^2)$ cm^3. Determine the height of this box.

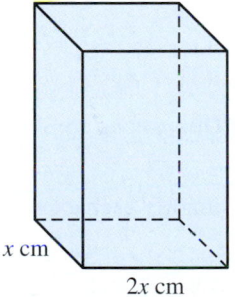

85. Revenue The price per unit of an item is x dollars, and the number of units that will sell at this price is estimated to be $350 - 4x$.
   a. Write $R(x)$, a revenue polynomial for this item.
   b. Evaluate and interpret $R(0)$.
   c. Evaluate and interpret $R(40)$.
   d. Evaluate and interpret $R(80)$.

86. Value of an Investment Use the formula $A = P(1 + r)^t$ to find the value of a $10,000 investment compounded at 7.5% for 8 years.

87. Average Cost A company manufactures range finders for golfers. The company determined that the cost in dollars to produce x range finders is $C(x) = 15x + 1,000$.
   a. Write an algebraic expression for $A(x)$, the average cost of producing x range finders.
   b. Evaluate and interpret $C(100)$.
   c. Evaluate and interpret $A(100)$.
   d. Evaluate and interpret $C(200)$.
   e. Evaluate and interpret $A(200)$.
   f. What is the change in cost when production is increased from 100 to 200 range finders?
   g. What is the change in the average cost per unit when production is increased from 100 to 200 range finders?

Applying Technology

88. Use a calculator or a spreadsheet to complete the given table of values. Then use the values in the table to determine which, if any, of the polynomials are equal.

x	Y_1	Y_2	Y_3	Y_4
-3				
-2				
-1				
0				
1				
2				
3				

$Y_1 = x^3 - 1$
$Y_2 = x^3 + 1$
$Y_3 = (x + 1)(x^2 - x + 1)$
$Y_4 = (x - 1)(x^2 + x + 1)$

Chapter 5 | Mastery Test

All exercises are assumed to restrict the variables to values that avoid division by 0 and avoid 0^0.

Objective 5.1.1 Convert Between Exponential Form and Expanded Form

1. Write each expression in exponential form.
   a. $x \cdot x \cdot x \cdot x$ **b.** $x \cdot x \cdot x \cdot y \cdot y \cdot y \cdot y$
   Write each expression in expanded form.
   c. $-2x^3y^2$ **d.** $(-2x)^3y^2$

Objective 5.1.2 Use the Product Rule for Exponents

2. Simplify each of these expressions.
   a. x^5x^8 **b.** $(5y^4)(8y^3)$
   c. $-v^4v^5v^6$ **d.** $(mn^2)(m^3n)$

Objective 5.1.3 Use the Power Rule for Exponents

3. Simplify each of these expressions.
   a. $(w^5)^8$ **b.** $(2y^4)^5$
   c. $(-v^2w^3)^4$ **d.** $\left(\dfrac{-3m}{2n}\right)^3$

Objective 5.2.1 Use the Quotient Rule for Exponents

4. Simplify each of these expressions.

a. $\dfrac{z^8}{z^2}$

b. $\dfrac{4x^9}{2x^3}$

c. $-\dfrac{v^{23}}{v^{23}}$

d. $\dfrac{-6a^4b^6}{3a^2b^2}$

Objective 5.2.2 Simplify Expressions with Zero Exponents

5. Simplify each of these expressions.

a. 1^0

b. $\left(\dfrac{2}{5}\right)^0$

c. $(3x + 5y)^0$

d. $3x^0 + 5y^0$

Objective 5.2.3 Combine the Properties of Exponents to Simplify Expressions

6. Simplify each of these expressions.

a. $(x^3y^5)(x^2y^6)$

b. $\left(\dfrac{x^2y^3}{xy}\right)^3$

c. $\dfrac{(6x^2)^2}{(3x^3)^3}$

d. $[(2x^2)(3x^4)]^2$

Objective 5.3.1 Simplify Expressions with Negative Exponents

7. Simplify each of these expressions.

a. 3^{-1}

b. 7^{-2}

c. $\left(\dfrac{2}{3}\right)^{-2}$

d. $\left(\dfrac{1}{3}\right)^{-1} + \left(\dfrac{1}{6}\right)^{-1}$

Objective 5.3.2 Use the Properties of Exponents to Simplify Expressions

8. Simplify each expression to a form containing only positive exponents.

a. $\dfrac{x^6y^{-2}}{x^{-3}y^4}$

b. $\left(\dfrac{-15x^{10}y^{-12}}{12x^{15}y^4}\right)^{-2}$

Objective 5.3.3 Use Scientific Notation

9. Write these numbers in standard decimal notation.

a. 3.57×10^5

b. 7.35×10^{-5}

Write these numbers in scientific notation.

c. 0.000509

d. 93,050,000

Objective 5.4.1 Use the Terminology Associated with Polynomials

10. Classify each of these polynomials according to the number of terms it contains, and give its degree.

a. $-5y^3 - 13y$

b. 273

c. $2x^2 - 7x + 1$

d. $17x^5 - 4x^3 + 9x + 8$

Objective 5.4.2 Add and Subtract Polynomials

11. Simplify each of these expressions.

a. $(4x - 9y) + (3x + 8y)$

b. $(2x^2 - 3x + 7) - (5x^2 - 9x - 11)$

c. $(5x^4 - 9x^3 + 7x^2 + 13) + (4x^4 + 12x^2 + 9x - 5)$

d. $(3x^3 - 7x + 1) - (12x^2 + 9x - 5)$

Objective 5.5.1 Multiply Polynomials

12. Find each of the indicated products.

a. $(5x^2y^3)(11x^4y^7)$

b. $-3x^2(5x^3 - 2x^2 + 7x - 9)$

c. $(x + 5)(x^2 + 3x + 1)$

d. $(x - 3y)(x^2 + xy + y^2)$

Objective 5.5.2 Multiply Binomials by Inspection

13. Multiply these binomials by inspection.

a. $(x + 8)(x + 11)$ b. $(x - 9)(x - 5)$

c. $(2x + 5y)(3x - 4y)$ d. $(5x - 3y)(2x + 7y)$

Objective 5.6.1 Determine the Product of a Sum and a Difference by Inspection

14. Multiply these binomials by inspection.

a. $(6x + 1)(6x - 1)$ b. $(x - 9y)(x + 9y)$

c. $(5x + 11)(5x - 11)$ d. $(3x - 8y)(3x + 8y)$

Objective 5.6.2 Determine the Square of a Binomial by Inspection

15. Write the expanded form of each square of a binomial by inspection.

a. $(6x + 1)^2$ b. $(x - 9y)^2$

c. $(5x - 11)^2$ d. $(3x + 8y)^2$

Objective 5.6.3 Use the Order of Operations to Simplify Polynomials

16. Use the order of operations to simplify each expression.

a. $(x - 3)^2 - (x^2 - 3^2)$

b. $(x - 5y)^2 - (x + 5y)^2$

Objective 5.7.1 Divide a Polynomial by a Monomial

17. Find each of the indicated quotients.

a. $\dfrac{15a^4b^2}{5a^2b^2}$

b. $\dfrac{36m^4n^3 - 48m^2n^5}{12m^2n^2}$

Objective 5.7.2 Use Long Division of Polynomials

18. Find each of the indicated quotients.

a. $x - 2\overline{)x^2 + 9x - 22}$

b. $\dfrac{6x^3 + 10x^2 - 32}{3x - 4}$

| Chapter 5 | Group Project |

A Polynomial Model for the Construction of a Box

Supplies Needed for Each Group

- Heavy-weight paper (120-lb weight is suggested) for constructing a box
- Metric rulers with centimeters marked for measuring and to use as an aid in folding the paper
- Scissors for cutting out the corners of the paper
- Masking tape to tape up the sides of the box at the corners
- Rice to place into the box and to be measured for volume
- Graduated cylinders marked in milliliters ($1 \text{ mL} = 1 \text{ cm}^3$)
- Funnels to be used to pour rice into the graduated cylinders

I. Lab: Constructing a Box and Measuring the Volume of Rice That the Box Contains

1. Measure the width of your sheet of paper to the nearest tenth of a centimeter.
   _____

2. Measure the length of your sheet of paper to the nearest tenth of a centimeter.
   _____

3. Assign different groups different-sized squares to cut out (e.g., 4 cm, 4.5 cm, 5 cm). Mark off equal squares to be cut from each corner, and record to the nearest tenth of a centimeter the width of each square. _____

4. Carefully cut the squares from each corner of your sheet of paper as illustrated.

5. Using a ruler to maintain a straight fold, fold the paper along lines that you have drawn between the corners.

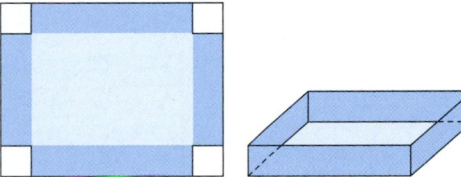

6. Using the masking tape, tape up the corners as shown in the illustration. Take care to form vertical sides.

7. Using graduated cylinders, measure the rice in milliliters, and carefully pour the rice into your box until it is full (level the top of the rice with your ruler as needed). Record the volume of rice your box contains in cubic centimeters. _____

8. Calculate the volume of the box, using the formula $V = l \cdot w \cdot h$ and the measurements of the length, width, and height recorded above. _____

9. Compare the volume of rice measured to the calculated volume of the box. What is the relative error of the rice measured to that predicted by the formula? _____ Is your relative error acceptable? Be prepared to defend your answer or to explain any excessive error.*

10. Create a table of values using the results from the box of each group. Use x to represent the width of the square cut from each corner of the sheet of paper. Use y to represent the volume of rice contained in each box. (Discard any values with excessive error.)

II. Creating a Mathematical Model for the Volume of a Box

1. Use x to represent the height of each box. Write an interval to express the restrictions on x for the sheet of paper used by your group. _____

2. Write an equation that gives the width of the box in terms of the height x.
   $W =$ _____

*Relative error is the error of the measurement divided by the value predicted by the formula. See Appendix A for more information on relative error.

3. Write an equation that gives the length of the box in terms of the height x.

$L =$ _____

4. Write an equation that gives the volume of the box in terms of the height x.

$V =$ _____

5. Use the volume equation to create a table of values over the set of x-values listed in part **1** above.

6. Use the volume equation to create a graph over the set of x-values listed in part **1** above.

7. Do all values of x produce the same volume? If so, why? If not, is there any value that seems to produce greater volume than other values of x?

This review is intended to help you make a realistic assessment of your areas of strength and weakness of the material covered in Chapters 1–5. A few questions in this review may cover topics not discussed at your school. You may wish to ask your instructor for topics that your school stresses. Some of these questions require you to interpret graphs or calculator screens.

Review of Chapter 1: Operations with Real Numbers and a Review of Geometry

Arithmetic and Order of Operations

In Exercises 1–6, calculate the value of each expression using pencil and paper.

1. a. $15 + (-3)$ b. $15 - (-3)$
   c. $15(-3)$ d. $15 \div (-3)$

2. a. $-0.6 + 0.02$ b. $-0.6 - 0.02$
   c. $-0.6(0.02)$ d. $-0.6 \div 0.02$

3. a. $\dfrac{1}{6} + \dfrac{3}{4}$ b. $\dfrac{1}{6} - \dfrac{3}{4}$
   c. $\dfrac{1}{6}\left(\dfrac{3}{4}\right)$ d. $\dfrac{1}{6} \div \dfrac{3}{4}$

4. a. $3 + 0$ b. $3 - 0$
   c. $3(0)$ d. $3 \div 0$

5. a. $6 - 12 \div 3 + 4^2$ b. $(6 - 12) \div 3 + 4^2$

6. a. $\dfrac{2(4) \div 4(-2)}{8 \div 2(-4)}$ b. $\dfrac{5 - 2(3 - 7)}{(10 + 3)^2}$

Properties and Subsets of the Real Numbers

7. Identify all of the numbers from the set
   $$\left\{-5, -3.28, -\sqrt{3}, -\frac{1}{3}, 0, \sqrt{9}, 11\right\}$$ that are:
   a. Rational numbers   b. Irrational numbers
   c. Integers   d. Natural numbers

8. Complete the table to represent each rational number in fractional form, in decimal form, and as a percent.

	Fraction	Decimal	Percent
a.	$\frac{1}{5}$		
b.		0.35	
c.			60%

9. Name the property that justifies each statement.
   a. $2(x + y) = 2x + 2y$
   b. $2(x + y) = 2(y + x)$
   c. $2(x + y) = (x + y)(2)$
   d. $2 + (x + y) = (2 + x) + y$

10. a. The additive identity is _____.
    b. The additive inverse of 5 is _____.

11. a. The multiplicative identity is _____.
    b. The multiplicative inverse of 5 is _____.

Interval Notation

12. Write the interval notation for each inequality.
    a. $x \le 5$ b. $x > -2$
    c. $-3 \le x < 0$ d. $3 < x \le 9$

Geometry

13. Determine the area and perimeter of each figure.

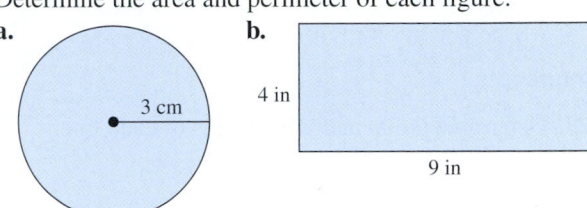

a. 3 cm   b. 4 in   9 in

14. Given that $m\angle A = 28°$, $\angle A$ and $\angle B$ are complementary, and $\angle A$ and $\angle C$ are supplementary, determine:
    a. $m\angle B$
    b. $m\angle C$

Review of Chapter 2: Linear Equations and Patterns

Solving Equations

15. Check $x = 5$ to determine whether it is a solution of each equation.
    a. $2x - 3 = x + 1$ b. $2(x - 4) = x - 3$

In Exercises 16 and 17, simplify each expression in part **a** and solve each equation in part **b**.

Simplify	Solve
16. a. $3(x - 2) - 5(x - 4)$	b. $3(x - 2) - 5(x - 4) = 0$
17. a. $2(x + 5) - (x + 3)$	b. $2(x + 5) - (x + 3) = 0$

Solve each linear equation.

18. a. $3(4x + 1) - 7x = 2x - 21$ b. $5(x - 3) = 3(x + 1)$

19. a. $\dfrac{x}{4} - 1 = \dfrac{2x}{5}$ b. $2x - \dfrac{3x + 1}{2} = \dfrac{1}{3}$

20. Solve each equation for y.
    a. $2x - 3y = 12$ b. $\dfrac{x}{3} - \dfrac{y}{5} = 1$

21. Use the table to solve
    $0.712x - 0.328 = -0.671x + 5.204$.

x	$y_1 = 0.712x - 0.328$	$y_2 = -0.671x + 5.204$
0	-0.328	5.204
1	0.384	4.533
2	1.096	3.862
3	1.808	3.191
4	2.520	2.520
5	3.232	1.849
6	3.944	1.178

22. Use the graph to solve $\dfrac{2x}{3} + 3 = -x - 7$.

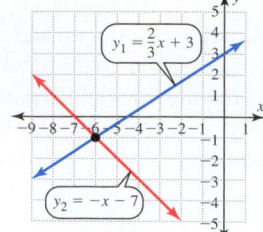

Intercepts

23. Determine the x- and y-intercepts of each line.

a.

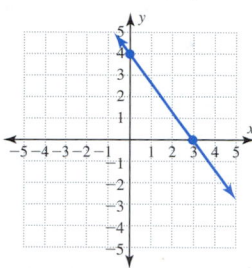

b. $3x - 2y = 18$

Proportions and Direct Variation

24. The amount of concession sales at a new ballpark varies directly as the number of fans in attendance. Concession sales are $1,000,000 at a game where 40,000 people attend.

a. Write an equation relating the concession sales at a game to attendance at the game.

b. What does the constant of variation mean?

c. Use this equation to complete the table of values.

Attendance	Sales ($)
34,000	
38,000	
42,000	

25. Use similar triangles and proportional sides to find the value of x.

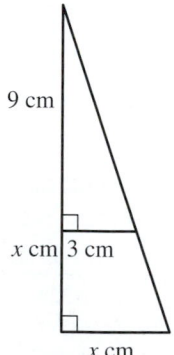

26. Determine whether each sequence is an arithmetic sequence. If the sequence is arithmetic, write the common difference.

a. 0, 4, 8, 12, 16   **b.** 12, 9, 6, 3, 0
c. 1, 2, 4, 8, 16   **d.** 5, 5, 5, 5, 5

Function Notation

27. Evaluate each expression given $f(x) = 5x - 2$.
a. $f(0)$ **b.** $f(-3)$ **c.** $f(4)$ **d.** $f(10)$

28. Use the given graph to determine the missing input and output values.
a. $f(-4) = $ _____   **b.** $f(0) = $ _____
c. $f(x) = -1; x = $ _____   **d.** $f(x) = 0; x = $ _____

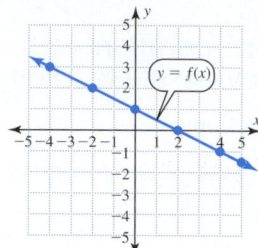

29. Price Discount A discount store had a holiday special with a 20% discount on the price of all sporting goods.

a. Write a function for the new price of an item with an original price of x dollars.

b. Complete the table for the new price of each item whose original price is given.

x	f(x)
20	16.00
48	
75	
90	
110	

Applications of Linear Equations

30. Travel Time Two boats leave a dock at the same time. One travels downstream at a rate of 15 mi/h and the other travels upstream 6 mi/h slower than the first boat. Determine the number of hours it will take the boats to be 72 miles apart.

31. Amount of Acid in a Solution A chemist needs to create a 20% acid solution. She has 5 L of a 25% acid solution in a 10-L container. How much water should she mix with this solution to decrease the concentration to the desired 20%?

Review of Chapter 3: Lines and Systems of Linear Equations in Two Variables

Graphs of Linear Equations

32. Determine whether the ordered pair $(1, -3)$ is a solution of each linear equation.
a. $5x + 2y = -1$ **b.** $y = -2x + 4$
c. $x = 1$ **d.** $y = 2$

33. Graph each of these lines.
a. The line through $(3, 1)$ and $(-2, -4)$

b. The line with slope $m = \dfrac{2}{3}$ and with a y-intercept $(0, -1)$

c. The line with slope $m = -\dfrac{3}{4}$ and through $(-1, 2)$

d. The line defined by $2x - y = 8$

e. The line defined by $y - 2 = \dfrac{1}{2}(x + 3)$

34. Match each equation with the best description of the graph of this linear equation.

a. $y = \dfrac{-3}{2}x + 4$ **A.** A horizontal line

b. $y = \dfrac{2}{3}x + 7$ **B.** A vertical line

c. $y = 4$ **C.** A line parallel to $y = \dfrac{-3}{2}x + 1$

d. $x = -1$ **D.** A line perpendicular to $y = \dfrac{-3}{2}x + 1$

35. Determine the slope of each line.
   a. The line through $(3, 7)$ and $(-2, 1)$
   b. $y = \dfrac{3}{4}x - 2$ **c.** $y = 2$ **d.** $x = 2$

Writing Equations of Lines

36. Write in slope-intercept form the equation of the line with slope $m = 3$ and y-intercept $(0, -2)$.

37. Write in slope-intercept form the equation of the line with slope $m = -2$ through $(3, 1)$.

38. Write in slope-intercept form the equation of the line through $(-1, 1)$ and $(2, 3)$.

39. Write the equation of a horizontal line through $(3, 4)$.

40. Write the equation of a vertical line through $(3, 4)$.

41. Write in slope-intercept form the equation of the line through $(0, 2)$ and parallel to $y = \dfrac{-2}{5}x + 7$.

Systems of Linear Equations

42. Select the choice that best describes each system of linear equations.
   a. $y = 3x - 5$ **A.** A consistent system of
          $y = 2x + 1$ independent equations
   b. $y = 3x - 2$ **B.** A consistent system of
          $y = 3x + 2$ dependent equations
   c. $y = 3x - 2$ **C.** An inconsistent system
          $3x = y + 2$ of independent equations

43. Solve this system of linear equations by the substitution method.
$$y = 3x + 5$$
$$3x + 2y = 28$$

44. Solve this system of linear equations by the addition method.
$$4x - 7y = 40$$
$$4x + 7y = -16$$

45. Solve this system of linear equations by the addition method.
$$-3x + 4y = 19$$
$$5x + 7y = -18$$

Applications Systems of Linear Equations

46. Rates of Two Trains Two trains depart simultaneously from a station, traveling in opposite directions. One averages 8 km/h less than the other. After $\dfrac{1}{2}$ hour they are 99 km apart. Determine the speed of each train.

47. Amount of Acid in a Solution A chemist needs 6 L of a 20% acid solution. He has available a 25% acid solution and a 10% acid solution. How much of each solution should be mixed in order to create this 20% solution?

48. Loads of Topsoil Topsoil sells for $60 per truckload, and fill dirt sells for $35 per truckload. A landscape architect estimates that 24 truckloads will be needed for a certain job. The estimated cost is $1,090. How many loads of topsoil and how many loads of fill dirt are planned for this job?

Review of Chapter 4: Linear Inequalities

Linear Inequalities

49. Use the table to solve each inequality.
   a. $4x - 1 \geq 3x + 1$ **b.** $4x - 1 \leq 3x + 1$

x	$y_1 = 4x - 1$	$y_2 = 3x + 1$
-3	-13	-8
-2	-9	-5
-1	-5	-2
0	-1	1
1	3	4
2	7	7
3	11	10

50. Use the graph to solve each inequality.
   a. $2x + 1 \geq -x - 2$
   b. $2x + 1 \leq -x - 2$

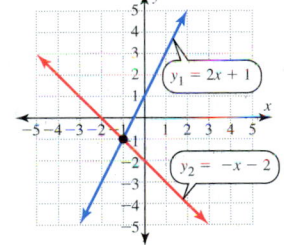

51. a. Solve $5(x - 3) \geq 2x + 9$.
   b. Solve $2(x + 3) < 4(x - 1) - 3(x + 2)$.

52. Identify each of these inequalities as either a conditional inequality, an unconditional inequality, or a contradiction. Then solve each inequality.
   a. $x + x < 0$ **b.** $x + x < 2x$ **c.** $x + x \leq 2x$

53. Solve each equation and inequality.
   a. $|2x - 1| = 7$ **b.** $|2x + 1| < 5$ **c.** $|x - 3| \geq 5$

54. Solve $2 < 3x - 4 \leq 5$.

55. Solve $2x + 1 \leq -5$ or $3x - 2 \geq 4$.

56. Graph the solution of $3x - 4y \leq 12$.

57. Graph the solution of this system of inequalities.
$$x + 2y \geq 2$$
$$-x + 2y < 4$$

58. Graph the solution of this system of inequalities.
$$x \geq 0$$
$$y \geq 0$$
$$3x + y \leq 3$$

Review of Chapter 5: Exponents and Operations with Polynomials

Using Properties of Exponents

In Exercises 59–67, simplify each expression assuming $x \neq 0$ and $y \neq 0$.

59. a. $-5^0 + 3^0 + 4^0$ **b.** $(-5)^0 + 3^0 + 4^0$
    c. $(-5 + 3 + 4)^0$ **d.** $-(5 + 3 + 4)^0$

60. a. $\left(\dfrac{1+1}{2+3}\right)^{-1}$ **b.** $\left(\dfrac{1}{2+3}\right)^{-1}$

    c. $\left(\dfrac{1}{2} + \dfrac{1}{3}\right)^{-1}$ **d.** $\left(\dfrac{1}{2}\right)^{-1} + \left(\dfrac{1}{3}\right)^{-1}$

61. a. 5^0 **b.** 0^5
    c. 6^{-1} **d.** $(-1)^6$

62. a. $\left(\dfrac{2}{3}\right)^0$ **b.** $\left(\dfrac{2}{3}\right)^{-2}$

    c. $\left(\dfrac{2}{3}\right)^{-1}$ **d.** $\left(\dfrac{2}{3}\right)^2$

63. a. 10^2 **b.** 10^4
    c. 10^0 **d.** 10^{-1}

64. a. $(3x + 7y)^0$ **b.** $(3x)^0 + (7y)^0$
    c. $3(x + 7y)^0$ **d.** $3x^0 + 7y^0$

65. a. $x^3 + x^3$ **b.** $x^2 \cdot x^6$
    c. $\dfrac{x^6}{x^2}$ **d.** $(x^2)^6$

66. a. $(8x^2y^3)(4xy)$ **b.** $\dfrac{8x^2y^3}{4xy}$

    c. $\dfrac{(2x^2)^3}{4x^3}$ **d.** $(8x^2y^3)^2$

67. a. $x^{-2}x^5$ **b.** $\left(\dfrac{x^{-3}}{x^{-4}}\right)^{-2}$

    c. $(3x^{-3}y^4)^{-2}$ **d.** $\dfrac{-6x^{-6}y^2}{12x^{12}y^{-8}}$

68. Write each number in standard decimal notation.

    a. 1.23×10^{-2} **b.** 1.23×10^4

69. Write each number in scientific notation.

    a. 0.0000435   **b.** 1,870,000

Operations with Polynomials

In Exercises 70–74, simplify each expression.

70. a. $(3x + 7y) + (2x - 4y)$
    b. $(3x + 7y) - (2x - 4y)$
    c. $(3x + 7y) - 2(x - 4y)$

71. a. $(4x^2 - 5x + 3) + (3x^2 + 4x - 9)$
    b. $(4x^2 - 5x + 3) - (3x^2 + 4x - 9)$
    c. $3(5x^2 - 4x + 2) - 4(3x^2 + x - 5)$

72. a. $(4x + 7)(4x - 7)$ **b.** $(2x + 3)^2$ **c.** $(5x - 6)^2$

73. a. $(x + 4)(x^2 - 2x - 3)$
    b. $(x - 3)(x + 2)(x - 4)$
    c. $(3x^2 + x - 2)(2x^2 - 4x - 1)$

74. a. $\dfrac{6x^3 + 4x^2}{2x}$ **b.** $\dfrac{x^2 - x - 12}{x + 3}$ **c.** $\dfrac{2x^4 - 4x^3 + 2x^2 - 8}{2x - 4}$

Review of Technology Perspectives

75. 1.2.1 Representing Additive Inverses
Approximate the additive inverse of 3π to the nearest thousandth.

76. 1.2.2 Evaluating Absolute Value Expressions
Evaluate $|35.2 - 44.35|$.

77. 1.2.3 Approximating Square Roots
Approximate $\sqrt{35}$ to the nearest thousandth.

78. 1.3.1 Reducing a Fraction to Lowest Terms
Reduce $\dfrac{756}{840}$ to lowest terms.

79. 1.6.1 Using Exponents
Evaluate -6^4.

80. 1.7.1 Storing Values Under Variable Names
Evaluate $2x^3 - 3x + 5$ for $x = 1.2$.

81. 2.2.1 Creating a Table of Values
Create a table of values for $x = -2, -1, 0, 1, 2, 3, 4$ using the equation $y = 5x - 7$.

82. 2.2.2 Creating a Graph
Graph $y = -2x + 6$ using the standard viewing window $[-10, 10, 1]$ by $[-10, 10, 1]$ on a graphing calculator.

83. 2.2.3 Changing the Table Settings
Create a table of values for $x = 0, 4, 8, 12, 16, 20, 24$ using the equation $y = -3x + 10$.

84. 2.3.1 Using a Trace Feature
Use the equation $y = 3x - 7$ and a graphing calculator to determine whether the points $(-1, 10)$ and $(4, 5)$ are on the graph of this line.

85. 2.3.2 Approximating a Point of Intersection
Use the **Intersect** feature on a graphing calculator to solve the system of linear equations $\begin{Bmatrix} y = 2x - 7 \\ y = -x + 5 \end{Bmatrix}$.

86. 3.3.1 Determining Rational Solutions of a System of Linear Equations
Use the **Intersect** feature on a graphing calculator to solve the system of linear equations $\begin{Bmatrix} y = 4x + 2 \\ y = -3x - 6 \end{Bmatrix}$ and convert the decimal coordinates of x and y to fractions.

87. 3.3.2 Using the ZoomFit Option to Select an Appropriate Viewing Window
Restrict the x-values from 0 to 90, and use the **ZoomFit** option to set an appropriate viewing window for the system of linear equations $\begin{Bmatrix} y = 90 - x \\ y = 3x \end{Bmatrix}$. Then solve this system graphically.

88. 4.5.1 Graphing a Linear Inequality
Use a calculator to graph the solution of $y \leq \dfrac{2}{3}x - 4$ in the standard viewing window $[-10, 10, 1]$ by $[-10, 10, 1]$.

89. 5.3.1 Using Scientific Notation
Compute the product $(3.0 \times 10^{12})(6.022 \times 10^{-23})$ and give the result in scientific notation.

Preparing for Intermediate Algebra—A Review of Key Concepts of Beginning Algebra

This review of Chapters 1–5 emphasizes a limited number of key concepts. It is designed to focus on some of the important concepts and skills needed to start an Intermediate Algebra course and is not meant to be a comprehensive review of Beginning Algebra. An outline of these concepts is given below. More information on each of the topics highlighted in this review can be found in Chapters 1–5 or by using **ALEKS,** the artificial intelligence tutor that accompanies this textbook.

- Evaluating Algebraic Expressions
- Simplifying Algebraic Expressions
- Solving Equations and Inequalities
- Solving Systems of Equations
- Using Equations to Solve Word Problems
- Using Linear Functions: Graphing Lines and Writing the Equation of a Line

Evaluating Algebraic Expressions

One skill that is needed throughout algebra is the ability to evaluate algebraic expressions. To **evaluate an algebraic expression** for given values of the variables means to replace each variable by the specific value given for that variable and then to simplify the expression. For everyone to get the same value when we evaluate an expression, we all must follow the same order of operations. Mathematicians developed the following order of operations so that each expression will have a unique interpretation and value. You should always verify that a calculator or a computer program is following this order before undertaking any important calculation. Technology Perspective 1.7.1 on page 81 illustrates how to evaluate an algebraic expression using a graphing calculator or a spreadsheet.

Order of Operations

Step 1. Start with the expression within the innermost pair of grouping symbols.

Step 2. Perform all exponentiations.

Step 3. Perform all multiplications and divisions as they appear from left to right.

Step 4. Perform all additions and subtractions as they appear from left to right.

When a constant is substituted for a variable, it is often wise to use parentheses to avoid a careless error in sign or an incorrect order of operations.

Example 1   Evaluating Algebraic Expressions

Given $x = 3$, $y = -5$, and $z = 4$, evaluate:

(a) $2x - 3(y - z)$ **(b)** $-2x^2 + (2y)^2$

Solution

(a) $2x - 3(y - z) = 2(3) - 3[(-5) - (4)]$

$\qquad\qquad\qquad = 6 - 3[-9]$

$\qquad\qquad\qquad = 6 + 27$

$\qquad\qquad\qquad = 33$

Use parentheses and substitute in the given values for each variable. Then start inside the innermost grouping symbols. Multiplication has priority over subtraction.

(b) $-2x^2 + (2y)^2 = -2(3)^2 + [2(-5)]^2$

$\qquad\qquad\qquad = -2(9) + (-10)^2$

$\qquad\qquad\qquad = -18 + 100$

$\qquad\qquad\qquad = 82$

Substitute in the given values for each variable. Then start inside the innermost grouping symbols. Exponentiation has priority over multiplication, and multiplication has priority over addition.

Self-Check 1

Evaluate each expression for $x = 12$, $y = -4$, and $z = -3$.

a. $-x + 5yz$ **b.** $-x \div y(z)$ **c.** $-x \div (yz)$

d. $\sqrt{y^2} + \sqrt{z^2}$ **e.** $\sqrt{y^2 + z^2}$ **f.** $\dfrac{2x - 5yz}{x^2 - x - 114}$

We will need to evaluate algebraic expressions whenever we apply formulas to applications. Formulas are used to state many algebraic definitions and properties. Several common geometric formulas are listed inside the back jacket of this book. Formulas are also widely used in business and in spreadsheets.

Example 2   Using Formulas

Solution

(a) Use $I = PRT$ to determine the interest on a $5,000 loan at 6% for 1 month.

$I = PRT$

$I = (5{,}000)(0.06)\left(\dfrac{1}{12}\right)$

$ = \dfrac{5{,}000(0.01)}{2}$

$ = 2{,}500(0.01)$

$ = 25$

The interest is $25.

Substitute in the given values for each variable. Note that $6\% = 0.06$ and that 1 month $= \dfrac{1}{12}$ year. Then simplify this expression.

(b) Use $m = \dfrac{y_2 - y_1}{x_2 - x_1}$ to determine the slope of the line through $(-2, 5)$ and $(3, -3)$.

$m = \dfrac{y_2 - y_1}{x_2 - x_1}$

$ = \dfrac{-3 - (5)}{3 - (-2)}$

$ = \dfrac{-8}{5}$

The slope m is $-\dfrac{8}{5}$.

Substitute in the given x- and y-values for the first point (x_1, y_1) and the second point (x_2, y_2). Then simplify this expression.

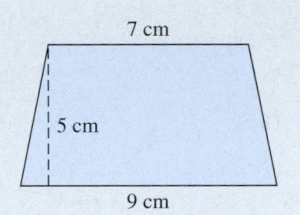

Self-Check 2

a. Use $A = \dfrac{1}{2}h(a + b)$ to determine the area of this trapezoid.

b. Use $d = \sqrt{(x_2 - x_1)^2 + (y_2 - y_1)^2}$ to determine the distance between $(-2, 5)$ and $(3, -7)$.

7 cm

5 cm

9 cm

Simplifying Algebraic Expressions

The properties of real numbers frequently are used to rewrite algebraic expressions in a more convenient form. Two reasons to rewrite algebraic expressions are to put them in as simple a form as possible or to put them in a standard form that reveals important information. We will first review some of the properties of the real number system and then we will simplify some algebraic expressions.

Properties of the Real Numbers

Commutative Properties:

of addition $a + b = b + a$

of multiplication $ab = ba$

Associative Properties:

of addition $(a + b) + c = a + (b + c)$

of multiplication $(ab)(c) = a(bc)$

Identities:

additive $a + 0 = 0 + a = a$

multiplicative $a \cdot 1 = 1 \cdot a = a$

Inverses:

additive $a + (-a) = (-a) + a = 0$

multiplicative $a \cdot \dfrac{1}{a} = \dfrac{1}{a} \cdot a = 1$ for $a \neq 0$

Distributive Properties of Multiplication Over Addition:

$$a(b + c) = ab + ac$$

$$(b + c)a = ba + ca$$

One way to simplify algebraic expressions is by adding or combining like terms. **Like terms** have exactly the same variable factors. The distributive property plays a key role in adding like terms. For example, we can rewrite $5x + 7x$ as $(5 + 7)x$ or $12x$. The constant factor in a term is called the **numerical coefficient** of the term. The numerical coefficient of $5x$ is 5 and the numerical coefficient of $7x$ is 7. Thus we combine like terms by adding or subtracting their coefficients.

Example 3   Simplifying Algebraic Expressions

Use the distributive property to combine the like terms and to simplify each expression.

(a) $-7w + 4w$ **(b)** $-2a - 5(3b + 4a)$

Solution

(a) $-7w + 4w = (-7 + 4)w$

$\qquad\qquad\quad = -3w$

Usually we skip writing the middle step and perform this addition mentally. It is the distributive property that justifies what we are doing.

(b) $-2a - 5(3b + 4a) = -2a - 5(3b) - 5(4a)$

$= -2a - 15b - 20a$

$= -2a - 20a - 15b$

$= (-2 - 20)a - 15b$

$= -22a - 15b$

Distribute the factor of -5 to each term inside the parentheses. Use the associative property of addition to reorder the terms; without rewriting, we think of this expression as $-2a + (-15b) + (-20a)$. Then use the distributive property again to add the coefficients of the like terms.

Self-Check 3

Simplify each expression by using the distributive property to combine the like terms.

a. $5x - 4y - 3x + 11y$

b. $5x - 4(y - 3x) + 11y$

c. $-(2x^2 - 5x - 9) + 3(5x^2 + 4x - 8)$

d. $4x(x - 3) - 3x(2x - 5)$

The properties of exponents also play an important role in evaluating and simplifying many algebraic expressions.

Properties of Exponents

For real numbers m, n, x, x^m, x^n, y^m, and y^n:

Product rule: $x^m \cdot x^n = x^{m+n}$

Power rule: $(x^m)^n = x^{mn}$

Product to a power: $(xy)^m = x^m y^m$

Quotient to a power: $\left(\dfrac{x}{y}\right)^m = \dfrac{x^m}{y^m}$ for $y \neq 0$

Quotient rule: $\dfrac{x^m}{x^n} = x^{m-n}$ for $x \neq 0$

Negative power: $\left(\dfrac{x}{y}\right)^{-n} = \left(\dfrac{y}{x}\right)^n$ for $x \neq 0, y \neq 0$

Special identities: $x^1 = x,$ $x^0 = 1$ for $x \neq 0,$ $x^{-1} = \dfrac{1}{x}$ for $x \neq 0$

In Example 4, we use the properties of zero and negative exponents to simplify each expression.

Example 4   Simplifying Algebraic Expressions Containing Exponents

Simplify each expression assuming the base for each exponent is nonzero.

(a) $3x^0 + 5y^0 + (3x)^0 + (5y)^0$ **(b)** $\left(\dfrac{1}{2}\right)^{-3} + \left(\dfrac{1}{5}\right)^{-1}$

Solution

(a) $3x^0 + 5y^0 + (3x)^0 + (5y)^0$

$= 3(1) + 5(1) + 1 + 1$

$= 3 + 5 + 1 + 1$

$= 10$

Note the different bases for the exponents of 0 in each of these terms.

(b) $\left(\dfrac{1}{2}\right)^{-3} + \left(\dfrac{1}{5}\right)^{-1} = \left(\dfrac{2}{1}\right)^{3} + \left(\dfrac{5}{1}\right)^{1}$ Convert each expression to one involving positive exponents, and then apply the positive exponents and add the terms.

$$= 8 + 5$$
$$= 13$$

Self-Check 4

Simplify each expression assuming the base for each exponent is nonzero.

a. $2^3 + 3^2 + (5 - 3)^4$

b. $-2x^0 + 6y^0 + (-2x)^0 + (6y)^0$

c. $\left(\dfrac{1}{3}\right)^{-2} + \left(\dfrac{1}{4}\right)^{-1}$

In Example 5, we use the product, quotient, and power rules to simplify each expression.

Example 5   Simplifying Algebraic Expressions Containing Exponents

Use the product, quotient, and power rules for exponents to simplify each expression.

(a) $(2x^3y^4)(5x^6y^7)$ **(b)** $\dfrac{12x^{10}y^8}{6x^2y^{-2}}$ **(c)** $(-2x^4y^5)^3$

Solution

(a) $(2x^3y^4)(5x^6y^7) = 2(5)x^{3+6}y^{4+7}$ Multiply the coefficients and add the exponents.
$$= 10x^9y^{11}$$

(b) $\dfrac{12x^{10}y^8}{6x^2y^{-2}} = 2x^{10-2}y^{8-(-2)}$ Divide the coefficients and subtract the exponents.

$$= 2x^8y^{10}$$

(c) $(-2x^4y^5)^3 = (-2)^3(x^4)^3(y^5)^3$ Raise each factor to the 3rd power.
$$= -8x^{4(3)}y^{5(3)}$$ Using the power rule, multiply the exponents.
$$= -8x^{12}y^{15}$$

Self-Check 5

Simplify each expression using the product, quotient, and power rules for exponents.

a. $(-3x^2y^5)(4x^3y^8)$ **b.** $\dfrac{15x^{12}y^9}{3x^{-3}y^3}$ **c.** $(-5x^5y^6)^2$ **d.** $2x^2(3xy^2)^2$

Many of the algebraic expressions that we simplify are polynomials. An nth-degree **real polynomial in x** is an algebraic expression of the form $a_nx^n + a_{n-1}x^{n-1} + \cdots + a_1x + a_0$, where n is a natural number, and each coefficient is a real number. We add and subtract polynomials by using the distributive property to combine like terms. When we use the distributive property to multiply polynomials, this results in the product of each term from the first factor multiplied by each term of the second factor. This is illustrated in Example 6.

Example 6   Multiplying Polynomials

Find the product of these polynomials.

(a) $3x^2y(4x^2 - 2xy - y^2)$ **(b)** $(2x - 3y)(5x + 4y)$

Solution

(a) $3x^2y(4x^2 - 2xy - y^2)$

$= 3x^2y(4x^2) - 3x^2y(2xy) - 3x^2y(y^2)$ Distribute the first factor of $3x^2y$ to each term of the

$= 12x^4y - 6x^3y^2 - 3x^2y^3$ second factor.

(b) $(2x - 3y)(5x + 4y)$ Use the distributive property to multiply each term

$= 2x(5x + 4y) - 3y(5x + 4y)$ from the first factor by each term of the second

$= 10x^2 + 8xy - 15xy - 12y^2$ factor. Then combine like terms.

$= 10x^2 - 7xy - 12y^2$

Self-Check 6

Find the product of these polynomials.

a. $5xy(3x^2 - 4xy)$ **b.** $(x + 7y)(8x - 3y)$ **c.** $(2x - 3y)^2$

Sometimes we can use the properties of exponents and inspection to divide one polynomial by another. When this is not possible, one option is to use long division of polynomials, as illustrated in Example 7.

Example 7   Dividing Polynomials

Divide $6x^2 - x - 2$ by $3x - 2$.

Solution

$$
\begin{array}{r}
2x + 1 \\
3x - 2 \overline{)6x^2 - x - 2} \\
\underline{-(6x^2 - 4x)} \\
3x - 2 \\
\underline{-(3x - 2)} \\
0
\end{array}
$$

Divide $6x^2$ by $3x$ to obtain $2x$.

Multiply $2x$ by $3x - 2$ to obtain $6x^2 - 4x$, and then subtract.

Divide $3x$ by $3x$ to obtain 1. Multiply 1 by $3x - 2$ to obtain $3x - 2$. Then subtract to obtain the remainder of 0.

Answer: $\dfrac{6x^2 - x - 2}{3x - 2} = 2x + 1$

You can check this answer by multiplying $(3x - 2)(2x + 1)$.

Self-Check 7

a. Divide $6x^2y^2 - 4xy$ by $2xy$.   **b.** Divide $35x^2 + 13x - 4$ by $7x + 4$.

Solving Equations

One topic that almost everyone associates with algebra is solving equations. Solving equations is analogous to solving a puzzle that has a prize. To get to the prize or unlock the solution to the equation, you must go through a series of steps. You can then check this solution by substituting back into the original equation.

In algebra, recognizing the type of equation is the key to knowing how many solutions to expect and to selecting the steps to produce the solution(s). A **linear equation** in x is a first-degree equation in x. The general strategy for solving a linear equation, regardless of its complexity, is to isolate the variable whose value is to be determined on one side of the equation and place all other terms on the other side. A good rule of thumb is to try to produce simpler equivalent equations at each step of the solution process. The principles used to accomplish this are given in the following box and are then illustrated in Example 8.

Addition-Subtraction Principle of Equality

If a, b, and c are real numbers, then:

- $a = b$ is equivalent to $a + c = b + c$
- $a = b$ is equivalent to $a - c = b - c$

Multiplication-Division Principle of Equality

If a, b, and c are real numbers, then:

- $a = b$ is equivalent to $ac = bc$ for $c \neq 0$
- $a = b$ is equivalent to $\dfrac{a}{c} = \dfrac{b}{c}$ for $c \neq 0$

Example 8   Solving a Linear Equation

Solve $\dfrac{3v + 2}{5} = \dfrac{4v - 2}{7} + 1$.

Solution

$$\frac{3v + 2}{5} = \frac{4v - 2}{7} + 1$$

$$35\left(\frac{3v + 2}{5}\right) = 35\left(\frac{4v - 2}{7} + 1\right) \qquad \text{Multiply both sides of the equation by the LCD, 35.}$$

$$\frac{35}{5}(3v + 2) = \frac{35}{7}(4v - 2) + 35(1) \qquad \text{Use the distributive property to remove parentheses, and then combine like terms.}$$

$$7(3v + 2) = 5(4v - 2) + 35$$

$$7(3v) + 7(2) = 5(4v) - 5(2) + 35$$

$$21v + 14 = 20v - 10 + 35$$

$$21v + 14 = 20v + 25$$

$$21v + 14 - 14 = 20v + 25 - 14 \qquad \text{Subtract 14 from both sides of the equation.}$$

$$21v = 20v + 11$$

$$21v - 20v = 20v - 20v + 11 \qquad \text{Then subtract } 20v \text{ from both sides.}$$

$$v = 11$$

Answer: $v = 11$ Does this answer check?

Self-Check 8

Solve each equation.

a. $3(2a + 1) = 2(1 - a) + 9$ **b.** $\dfrac{w}{2} = \dfrac{w}{3} - 2$ **c.** $\dfrac{4v - 3}{7} + 2 = \dfrac{3v + 2}{4}$

d. $9(2x - 3) - 13(5x - 1) = 14(3x - 1)$

The solution of many equations can be obtained algebraically, graphically, or numerically. Examining a problem from these different perspectives and then describing it verbally is often referred to as the use of multiple perspectives. In Example 9, we will illustrate the use of multiple representations to solve a linear equation.

Example 9   Using Multiple Representations to Solve a Linear Equation

Solve $2(x + 4) = 4(x + 3)$.

Solution

Algebraically

$$2(x + 4) = 4(x + 3)$$
$$2x + 8 = 4x + 12$$
$$2x + 8 - 8 = 4x + 12 - 8$$
$$2x = 4x + 4$$
$$2x - 4x = 4x - 4x + 4$$
$$-2x = 4$$
$$\frac{-2x}{-2} = \frac{4}{-2}$$
$$x = -2$$

Graphically

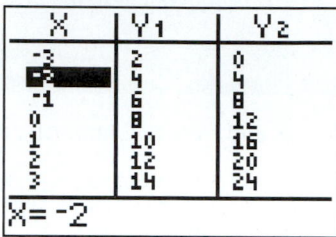

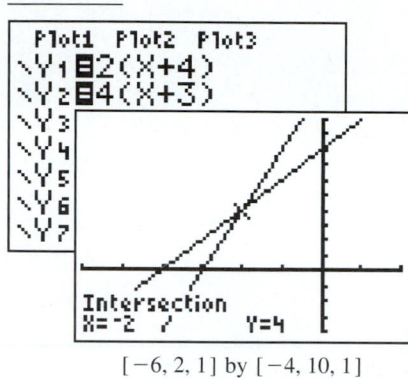

$[-6, 2, 1]$ by $[-4, 10, 1]$

Numerically

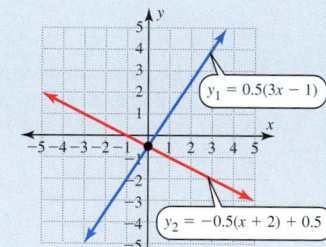

Verbally

The value of x that satisfies the equation is $x = -2$. From the graph, the x-coordinate of the point of intersection is $x = -2$. From the table, $y_1 = y_2$ when $x = -2$.

Self-Check 9

a. Use the table to solve
$2.5(3x - 1) = 3.5(x + 2) - 1.5$.

x	$y_1 = 2.5(3x - 1)$	$y_2 = 3.5(x + 2) - 1.5$
-3	-25.0	-5.0
-2	-17.5	-1.5
-1	-10.0	2.0
0	-2.5	5.5
1	5.0	9.0
2	12.5	12.5
3	20.0	16.0

b. Use the graph to solve
$0.5(3x - 1) = -0.5(x + 2) + 0.5$.

$y_1 = 0.5(3x - 1)$

$y_2 = -0.5(x + 2) + 0.5$

The linear equations given in Examples 8 and 9 are first degree in x and each has one solution. Most of the linear equations that you encounter in textbooks will have one solution. An equation with exactly one solution is an example of a **conditional equation**—an equation that is true for some values of the variable and false for other values. There are two other types of equations. A **contradiction** is an equation that is false for all values of the variable. The equation $x = x + 1$ is a contradiction. An **identity** is an equation that is true for all values of the variable. The equation $3(x + 2) = 3x + 6$ is an identity.

Some equations can produce exactly two solutions. An example of this is the absolute value equation $|2x - 1| + 5 = 8$. In Example 10, we will solve this equation algebraically. You could also examine this equation using the multiple perspectives shown in Example 9.

The absolute value of a number can be interpreted as the distance from this number to 0 on the number line. For example, both $|24| = 24$ and $|-24| = 24$. Thus when we solve an absolute value equation we must allow for the possibility that the expression inside the absolute value symbols can be either positive or negative.

Example 10 Solving an Absolute Value Equation

Solve $|2x - 1| + 5 = 8$.

Solution

$\begin{aligned} |2x - 1| + 5 &= 8 \\ |2x - 1| &= 3 \end{aligned}$	First subtract 5 from both sides of the equation to isolate the absolute value expression on the left side. The expression inside the absolute value symbols can be either -3 or 3.
$\begin{aligned} 2x - 1 &= -3 \quad\text{or}\quad & 2x - 1 &= 3 \end{aligned}$	
$\begin{aligned} 2x &= -2 & 2x &= 4 \end{aligned}$	Add 1 to each side of both equations.
$\begin{aligned} x &= -1 & x &= 2 \end{aligned}$	Divide both sides of each equation by 2.

Answer: $x = -1$ or $x = 2$ Do both of these solutions check?

Self-Check 10

Solve each equation.

a. $|x| = 13$ **b.** $|2x - 1| = 7$ **c.** $|2x + 11| - 9 = 6$

Solving Inequalities

Given two real numbers x and a, there are three ways that x and a can be related. Either $x < a$ or $x = a$ or $x > a$.

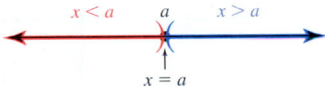

Whereas a conditional linear equation will have only one solution, a conditional linear inequality will often have a solution that consists of a whole interval of values. Although the steps used to solve linear inequalities are very similar to the steps used to solve linear equations, there are some important distinctions. Multiplying or dividing both sides of an inequality by a negative number will reverse the order of the inequality, as illustrated in Example 11.

Addition-Subtraction Principle for Inequalities

If a, b, and c are real numbers, then:

- $a > b$ is equivalent to $a + c > b + c$ and
- $a > b$ is equivalent to $a - c > b - c$

Multiplication-Division Principle for Inequalities

If a, b, and c are real numbers, then:

- $a > b$ is equivalent to $ac > bc$ for $c > 0$ but
- $a > b$ is equivalent to $ac < bc$ for $c < 0$

- $a > b$ is equivalent to $\dfrac{a}{c} > \dfrac{b}{c}$ for $c > 0$ but

- $a > b$ is equivalent to $\dfrac{a}{c} < \dfrac{b}{c}$ for $c < 0$

Example 11   Solving a Linear Inequality

Solve $5(2 - 3n) < 6(4 - n) + 2$.

Solution

$$5(2 - 3n) < 6(4 - n) + 4$$

$$5(2) - 5(3n) < 6(4) - 6(n) + 4 \qquad$$ Use the distributive property to remove the parentheses. Then combine like terms.

$$10 - 15n < 24 - 6n + 4$$

$$10 - 15n < 28 - 6n$$

$$10 - 15n + 6n < 28 - 6n + 6n \qquad$$ Adding $6n$ to both sides of the inequality preserves the order relation.

$$10 - 9n < 28$$

$$10 - 10 - 9n < 28 - 10 \qquad$$ Subtracting 10 from both sides of the inequality preserves the order relation.

$$-9n < 18$$

$$\frac{-9n}{-9} > \frac{18}{-9} \qquad$$ Dividing both sides of the inequality by -9 reverses the order relation.

$$n > -2$$

Answer: $(-2, \infty)$ whose graph is

Self-Check 11

Solve each inequality.

a. $2x + 1 \geq -x - 2$ **b.** $2x + 9 \leq 5(x - 3)$

c. $4(x - 1) - 3(x + 2) > 2(x + 3)$

The compound inequality $a \leq x \leq b$ is equivalent to $x \geq a$ *and* $x \leq b$. The compound inequality $-5 \leq 2x + 3 \leq 5$ is examined in Example 12.

Example 12   Solving a Compound Inequality

Solve $-5 \leq 2x + 3 \leq 5$.

Solution

$$-5 \leq 2x + 3 \leq 5$$

$$-5 - 3 \leq 2x + 3 - 3 \leq 5 - 3 \qquad$$ Subtracting 3 from each member of the inequality preserves the order relation.

$$-8 \leq 2x \leq 2$$

$$\frac{-8}{2} \leq \frac{2x}{2} \leq \frac{2}{2} \qquad$$ Dividing each member of the inequality by $+2$ preserves the order relation.

$$-4 \leq x \leq 1$$

Answer: $[-4, 1]$ whose graph is

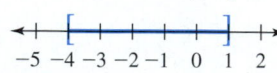

Self-Check 12

Solve each inequality.

a. $-1 \leq x - 2 \leq 5$ **b.** $-4 \leq -2x < 6$ **c.** $-8 < 3x - 2 < 7$

Another type of compound inequality uses the logical connector *or*. For example, the compound inequality $2x - 1 < -3$ or $2x - 1 > 3$ has as its solution the union of two distinct intervals.

Example 13 Solving a Compound Inequality

Solve $2x - 1 < -3$ or $2x - 1 > 3$.

Solution

$2x - 1 < -3$	or	$2x - 1 > 3$
$2x < -2$	or	$2x > 4$
$x < -1$	or	$x > 2$

Solve each inequality separately.

Answer: $(-\infty, -1) \cup (2, \infty)$ whose graph is

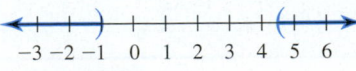

$$-3 \quad -2 \quad -1 \quad 0 \quad 1 \quad 2 \quad 3 \quad 4$$

The union symbol is used to indicate that the solutions can be in either the first or the second interval.

Self-Check 13

Solve each inequality.

a. $4x - 1 < -13$ or $2x - 1 > 9$ **b.** $5x - 2 < 3$ or $6x - 1 > 23$

As we noted before Example 10, the absolute value of a number can be interpreted as the distance from this number to 0 on the number line. Absolute value inequalities can also be interpreted in terms of distance. For example, the inequality $|x - a| < d$ can be interpreted as the set of all points x that are less than d units from a. This inequality can be rewritten as the compound inequality $-d < x - a < d$.

Likewise, the inequality $|x - a| > d$ can be interpreted as the set of all points x that are more than d units from a. This inequality can be rewritten as the compound inequality $x - a < -d$ or $x - a > d$. Note that the solution of $|2x + 3| \le 5$ was obtained in Example 12 by solving $-5 \le 2x + 3 \le 5$. Also the solution of $|2x - 1| > 3$ was obtained in Example 13 by solving $2x - 1 < -3$ or $2x - 1 > 3$. A summary of equivalent forms for absolute value equations and inequalities is given in the following table.

Absolute Value Equations and Inequalities

If x and a are real numbers and d is a positive real number, then:

- $|x - a| = d$ is equivalent to $x - a = -d$ or $x - a = d$
- $|x - a| < d$ is equivalent to $-d < x - a < d$
- $|x - a| > d$ is equivalent to $x - a < -d$ or $x - a > d$

Example 14 Solving an Absolute Value Inequality

Solve $|7 - 4x| > 11$.

Solution

$7 - 4x < -11$	or	$7 - 4x > 11$
$-4x < -18$	or	$-4x > 4$
$x > 4.5$	or	$x < -1$

Solve each inequality separately. Note that the inequality is reversed when each side of the inequalities is divided by -4.

Answer: $(-\infty, -1) \cup (4.5, \infty)$ whose graph is

$$-3 \quad -2 \quad -1 \quad 0 \quad 1 \quad 2 \quad 3 \quad 4 \quad 5 \quad 6$$

The union symbol is used to indicate that the solutions can be in either the first or the second interval.

Self-Check 14

Solve each inequality.

a. $|x| \le 5$ **b.** $|x| > 5$ **c.** $|7 - 2x| < 5$ **d.** $|5x + 1| \ge 4$

Solving Systems of Linear Equations

When we are solving an application that has only one quantity or variable, then a single linear equation may be the algebraic model to use. When we are solving an application that involves two different quantities, then a system of linear equations with two variables may be a better algebraic model. Two principles are used to solve systems of equations.

- **Substitution principle:** A quantity may be substituted for its equal.
- **Addition-subtraction principle:** Equal values can be added to or subtracted from both sides of an equation to produce an equivalent equation.

In Example 15, we use the addition method to solve a system of two linear equations with two variables. Note that there is only one answer—an ordered pair with both an x-coordinate and a y-coordinate. You can also solve systems of equations using multiple perspectives nearly identical to those shown in Example 9.

| **Example 15** | Solving a System of Linear Equations |

Solve $\begin{cases} 5x + 2y = 3 \\ 3x - 4y = 33 \end{cases}$.

Solution

$$
\begin{array}{ll}
5x + 2y = 3 & 10x + 4y = 6 \\
3x - 4y = 33 & \underline{3x - 4y = 33} \\
 & 13x \quad\;\; = 39 \\
 & x \quad\;\; = 3
\end{array}
$$

Multiply both sides of the first equation by 2.

Add these equations to eliminate y and then solve for x.

$$
\begin{aligned}
5x + 2y &= 3 \\
5(3) + 2y &= 3 \\
15 + 2y &= 3 \\
2y &= -12 \\
y &= -6
\end{aligned}
$$

Back-substitute 3 for x in the first equation of the original system. Then solve for y.

Answer: $(3, -6)$

This answer is a single point—not an interval of values. Does this solution check?

Self-Check 15

a. Solve $\begin{cases} 4x - 7y = 40 \\ 4x + 7y = -16 \end{cases}$.

b. Solve $\begin{cases} -3x + 4y = 19 \\ 5x + 7y = -18 \end{cases}$.

c. Use the table to solve $\begin{cases} 7x + 3y = 1 \\ 5x - y = 7 \end{cases}$.

d. Use the graph to solve $\begin{cases} x - y = 1 \\ x + 2y = 7 \end{cases}$.

x	$y_1 = (1 - 7x)/3$	$y_2 = 5x - 7$
-3	7.333	-22.000
-2	5.000	-17.000
-1	2.667	-12.000
0	0.333	-7.000
1	-2.000	-2.000
2	-4.333	3.000
3	-6.667	8.000

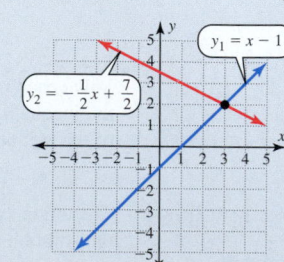

Solving Word Problems

Almost all the mathematics that you encounter outside the classroom will likely be stated in words. Thus to profit from algebra you must be able to work with word problems. A basic strategy for working with word problems is given below.

Strategy for Solving Word Problems

Step 1. Read the problem carefully to determine what you are being asked to find.

Step 2. Select a variable to represent each unknown quantity. Specify precisely what each variable represents and note any restrictions on each variable.

Step 3. If necessary, make a sketch and translate the problem into a word equation or a system of word equations. Then translate each word equation into an algebraic equation.

Step 4. Solve the equation or the system of equations, and answer the question completely in the form of a sentence.

Step 5. Check the reasonableness of your answer.

Two principles that are used frequently to translate word problems into equations are the mixture principle and the rate principle.

- **Mixture principle:**

$$\begin{pmatrix} \text{Amount} \\ \text{in first} \end{pmatrix} + \begin{pmatrix} \text{Amount} \\ \text{in second} \end{pmatrix} = \begin{pmatrix} \text{Amount} \\ \text{in mixture} \end{pmatrix}$$

- **Rate principle:** Amount = Rate × Base $A = R \cdot B$

In Example 16, the mixture principle is used to write each daily cost as the fixed cost plus the variable cost. Also the rate principle is used to write variable cost as the cost per pizza times the number of pizzas.

Example 16 | Solving a Word Problem

A family pizza business opened a new store. The daily cost for making pizzas includes a fixed daily cost and a variable cost per pizza. The total cost for making 200 pizzas on Monday was $1,020. The total cost for making 250 pizzas on Tuesday was $1,220. Determine the fixed daily cost and variable cost per pizza for this business.

Solution

Let x = the variable cost per pizza

Let y = the fixed dollar cost per day

	Word Equations		Algebraical Equations
(1) Mon:	Variable cost + Fixed cost = Total cost	**(1)**	$200x + y = 1{,}020$
(2) Tu:	Variable cost + Fixed cost = Total cost	**(2)**	$250x + y = 1{,}220$

$$\begin{array}{rcl} -200x - y &=& -1{,}020 \\ \underline{250x + y} &=& \underline{1{,}220} \\ 50x &=& 200 \\ x &=& 4 \end{array}$$ Multiply both sides of the first equation by -1.

Add these equations to eliminate y and then solve for x.

$$\begin{array}{rcl} 200x + y &=& 1{,}020 \\ 200(4) + y &=& 1{,}020 \\ 800 + y &=& 1{,}020 \\ y &=& 220 \end{array}$$ Back-substitute 4 for x in the first equation and then solve for y.

Answer: The fixed cost per day is $220 and the variable cost per pizza is $4.

Self-Check 16

a. A student invested $2,000 of summer income to save for college. A portion was deposited in a savings account that earned only 4%. The rest was invested in a local electrical utility and earned 8%. The total income for 1 year was $136. How much did the student put in the savings account?

b. A chemist needs 6 L of a 20% acid solution. She has available a 25% acid solution and a 10% acid solution. How much of each solution should she mix to create this 20% solution?

Using Linear Functions

The graph of a linear function is a line. Two points are needed to graph a line. The equation for a line can be written in the slope-intercept form $f(x) = mx + b$. In this form, the letter m represents the slope: $m = \dfrac{y_2 - y_1}{x_2 - x_1} = \dfrac{\Delta y}{\Delta x}$. The letter b represents the y-coordinate of the y-intercept $(0, b)$.

In Example 17, we will first graph a line through given points and then we will graph a line given an equation of the line.

Example 17   Graphing a Line

(a) All the points listed in this table lie on the same line. Graph this line.

(b) Graph $y = \dfrac{2}{5}x - 1$.

x	y
-4	-2
-2	-1
0	0
2	1
4	2

Solution

(a)

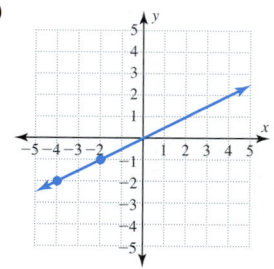

Plot any two points and draw the line through these points.

Then check that all points do lie on this line.

(b)

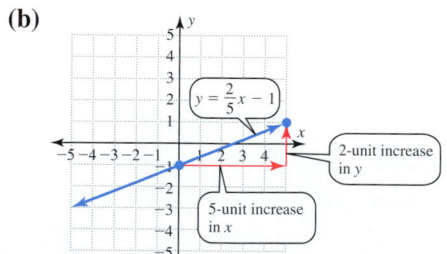

First plot the y-intercept $(0, -1)$. Using the slope $m = \dfrac{\Delta y}{\Delta x} = \dfrac{2}{5}$, determine a second point. Do this by moving up 2 units for every 5-unit move to the right.

Self-Check 17

a. Graph the line containing the points in this table.

b. Graph $y = -\dfrac{3}{2}x + 1$.

x	y
-4	3
-2	1
0	-1
2	-3
4	-5

Although we can always write a linear function in the form $y = mx + b$, other forms sometimes conveniently fit the given information. The most common forms of linear equations are:

- **Slope-intercept form:** $y = mx + b$, with slope m and y-intercept $(0, b)$
- **Point-slope form:** $y - y_1 = m(x - x_1)$ through (x_1, y_1) with slope m
- **General form:** $Ax + By = C$
- **Vertical line:** $x = a$ for a real constant a
- **Horizontal line:** $y = b$ for a real constant b

It is useful to remember that parallel lines have the same slope and that perpendicular lines have slopes whose product is -1.

Example 18   Writing the Equation of a Line

(a) Write in slope-intercept form the equation of the line containing the listed points.

x	y
-4	-17
-2	-11
0	-5
2	1
4	7

(b) Write in slope-intercept form the equation of the line containing the listed points.

x	y
-3	9
-1	5
1	1
3	-3
5	-7

(c) Write in slope-intercept form the equation of the line perpendicular to $y = \dfrac{2}{5}x + 4$ and through $(4, -2)$.

Solution

(a) $y = mx + b$

$$m = \frac{y_2 - y_1}{x_2 - x_1} = \frac{7 - 1}{4 - 2}$$

$$m = 3$$

$b = -5$

$y = 3x - 5$

Use any two points that are on the line to determine the slope. The points $(2, 1)$ and $(4, 7)$ are used here.

By inspection, the y-intercept is $(0, -5)$.

Substitute m and b into the slope-intercept form to write an equation of the line.

(b) $y = mx + b$

$$m = \frac{y_2 - y_1}{x_2 - x_1} = \frac{5 - 9}{-1 - (-3)}$$

$$m = \frac{-4}{2}$$

$$m = -2$$

$y = -2x + b$

$9 = -2(-3) + b$

$9 = 6 + b$

$3 = b$

$y = -2x + 3$

Use the points $(-3, 9)$ and $(-1, 5)$ to determine the slope.

Note that the y-intercept is not listed in the table so we will calculate b. Substitute the slope -2 and the point $(-3, 9)$ to determine b. Now use m and b to write an equation of the line.

(c) $y = mx + b$

$$m = -\frac{5}{2}$$

$$-2 = -\frac{5}{2}(4) + b$$

$$-2 = -10 + b$$

$$b = 8$$

$$y = -\frac{5}{2}x + 8$$

By inspection, the line defined by $y = \frac{2}{5}x + 4$ has slope $\frac{2}{5}$.

A perpendicular line must have a slope of $-\frac{5}{2}$. Substitute the slope $-\frac{5}{2}$ and the point $(4, -2)$ to determine b. Now use m and b to write an equation of the line.

Self-Check 18

Write an equation of the line containing the given points.

a. Write in slope-intercept form the equation of the line containing the listed points.

x	y
−4	2
−2	3
0	4
2	5
4	6

b. Write in slope-intercept form the equation of the line containing the listed points.

x	y
−3	−11
−1	−7
1	−3
3	1
5	5

c. Write in slope-intercept form the equation of the line parallel to $2x + 3y = 6$ with y-intercept $(0, 5)$.

Self-Check Answers

1. **a.** 48   **b.** −9   **c.** −1   **d.** 7   **e.** 5   **f.** −2
2. **a.** 40 cm²   **b.** 13
3. **a.** $2x + 7y$ **b.** $17x + 7y$ **c.** $13x^2 + 17x − 15$ **d.** $−2x^2 + 3x$
4. **a.** 33   **b.** 6   **c.** 13
5. **a.** $−12x^5y^{13}$ **b.** $5x^{15}y^6$ **c.** $25x^{10}y^{12}$ **d.** $18x^4y^4$
6. **a.** $15x^3y − 20x^2y^2$ **b.** $8x^2 + 53xy − 21y^2$ **c.** $4x^2 − 12xy + 9y^2$
7. **a.** $3xy − 2$ **b.** $5x − 1$
8. **a.** $a = 1$ **b.** $w = −12$ **c.** $v = 6$ **d.** $x = 0$
9. **a.** $x = 2$ **b.** $x = 0$
10. **a.** $x = −13$ or $x = 13$ **b.** $x = −3$ or $x = 4$ **c.** $x = −13$ or $x = 2$
11. **a.** $[−1, \infty)$ **b.** $[8, \infty)$ **c.** $(−\infty, −16)$
12. **a.** $[1, 7]$ **b.** $(−3, 2]$ **c.** $(−2, 3)$
13. **a.** $(−\infty, −3) \cup (5, \infty)$ **b.** $(−\infty, 1) \cup (4, \infty)$
14. **a.** $[−5, 5]$ **b.** $(−\infty, −5) \cup (5, \infty)$ **c.** $(1, 6)$
    d. $(−\infty, −1] \cup [0.6, \infty)$
15. **a.** $(3, −4)$ **b.** $(−5, 1)$ **c.** $(1, −2)$ **d.** $(3, 2)$
16. **a.** $600 was put in the savings account.
    b. Mix 4 L of the 25% solution and 2 L of the 10% solution.

17. **a.**

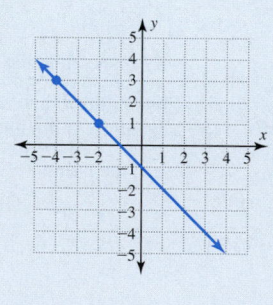

b.

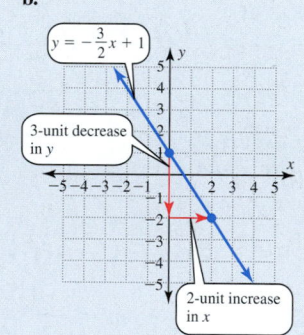

18. **a.** $y = 0.5x + 4$ **b.** $y = 2x − 5$ **c.** $y = -\frac{2}{3}x + 5$

Factoring Polynomials

Chapter Outline

Roadway Design

Polynomials are used to describe many applications, especially those involving geometric shapes. The volume of dirt that must be moved to create the base of a new roadway is very important to contractors who bid on a construction project. The cross section of this roadway can be approximated by a trapezoid. Using the desired dimensions of the roadway and design limitations that include safety considerations, engineers can form each factor in the formula for the volume of dirt that must be moved.

Exercise 75 in Section 6.1 and the Chapter 6 Group Project both examine the trapezoidal cross section of a roadway.

| Section 6.1 | An Introduction to Factoring Polynomials |

For the things of this world cannot be made known without a knowledge of mathematics   —Roger Bacon (13th-century philosopher)

A Mathematical Note

The word *factor* comes from the Latin root *facere* meaning "to do." Things that are necessary in order to do something are known as its factors. A *factory* is a place where things are put together to make a product. Factors of a polynomial are put together using multiplication to form a product. In mathematics, the word *factor* is associated with multiplication.

Objectives:

1. Factor the GCF out of a polynomial.
2. Factor by grouping.
3. Use the zeros of a polynomial $P(x)$ and the x-intercepts of the graph of $y = P(x)$ to factor the polynomial.

What Does It Mean to Factor a Polynomial?

Factoring is the process of rewriting an expression as a product of two or more factors. Multiplication uses factors to produce a product. Factoring reverses this process to produce the factors, as illustrated here.

Multiplication	**Factorization**	
$3 \cdot 5 = 15$	$15 = 3 \cdot 5$	The factors are 3 and 5.
$2x(x + 3) = 2x^2 + 6x$	$2x^2 + 6x = 2x(x + 3)$	The factors are $2x$ and $x + 3$.
$(x + 4)(x - 1) = x^2 + 3x - 4$	$x^2 + 3x - 4 = (x + 4)(x - 1)$	The factors are $x + 4$ and $x - 1$.

Why Do I Need to Factor Polynomials?

One reason is that the factored form of a polynomial consists of simpler pieces that may each reveal an important fact about this polynomial. Other reasons that we factor polynomials are that factors can be used to:

- Determine key values for a polynomial   The zeros of a polynomial
- Determine key points on the graph of a function   The x-intercepts of a graph
- Solve equations   In Section 6.6 and Chapter 7
- Reduce rational expressions   In Chapter 9

1. Factor the GCF Out of a Polynomial

We will begin factoring polynomials by determining the greatest common factor (**GCF**) of each term of a polynomial. The GCF of a set of integers is the largest integer that will exactly divide each integer in the set. For example, the integers 12 and 18 have positive common factors of 1, 2, 3, and 6. The GCF of 12 and 18 is 6. The GCF of a polynomial is the "largest" factor common to each term. This means that the **GCF of a polynomial** is the common factor of each term that contains both:

- the largest possible numerical coefficient, and
- the largest possible exponent on each variable factor

Will I Be Able to Determine the GCF of a Polynomial by Inspection?

The GCF can be determined by inspection for many polynomials. For those that you cannot determine by inspection, you can use the steps illustrated in Example 1.

Example 1 Determining the GCF of Two Monomials

Determine the GCF of each pair of monomials.

(a) 40 and 60   **(b)** $24x^2y^3$ and $36xy^5$

Solution

(a) $40 = 2^3 \cdot 5$ Write each monomial in factored form using exponents to denote
        $60 = 2^2 \cdot 3 \cdot 5$ repeated factors.

Common Factor	
2^2	2 occurs twice as a factor in each term
5	5 occurs once as a factor in each term

        GCF $= 2^2 \cdot 5$
        GCF $= 20$

The GCF is the product of the common factors of 2^2 and 5.

(b) $24x^2y^3 = 2^3 \cdot 3 \cdot x^2 \cdot y^3$ Write each monomial in factored form using exponents to denote
        $36xy^5 = 2^2 \cdot 3^2 \cdot x \cdot y^5$ repeated factors.

Common Factor	
2^2	2 occurs twice as a factor in each term
3	3 occurs once as a factor in each term
x	x occurs once as a factor in each term
y^3	y occurs 3 times as a factor in each term

        GCF $= 2^2 \cdot 3 \cdot x \cdot y^3$
        GCF $= 12xy^3$

The GCF is the product of the common factors of 2^2, 3, x, and y^3.

Self-Check 1

Determine the GCF of each pair of monomials.

a. 56 and 70   **b.** $40x^3y^5$ and $100x^4y^2$

Example 2 also illustrates some of the steps that you can take to determine the GCF of a polynomial. It is acceptable for you to skip some of these steps if you can determine the GCF of the polynomial by inspection.

Example 2 Determining the GCF of a Polynomial

Determine the GCF of each polynomial.

(a) $6x^2 + 9x$ **(b)** $35x^3y^3 + 50x^2y^5 - 5x^2y^3$

Solution

(a) $6x^2 = 2 \cdot 3 \cdot x^2$ Each term in the expression $6x^2 + 9x$ contains 3 and x
        $9x = 3^2 \cdot x$ as factors.
        GCF $= 3x$

(b) $35x^3y^3 = 5 \cdot 7 \cdot x^3 \cdot y^3$ Each term in the expression $35x^3y^3 + 50x^2y^5 - 5x^2y^3$ contains 5, x^2,
        $50x^2y^5 = 2 \cdot 5^2 \cdot x^2 \cdot y^5$ and y^3 as factors.
        $5x^2y^3 = 5 \cdot x^2 \cdot y^3$
        GCF $= 5x^2y^3$

Self-Check 2

Determine the GCF of each polynomial.

a. $12x^2 - 20x$ **b.** $36x^3y - 4x^2y^2 + 10xy^3$

What Do I Do First to Factor a Polynomial?

The best way to start factoring polynomials is to determine the GCF and then factor out any GCF other than 1. The distributive property can be used to write the expression $3x(2x + 5)$ as $3x(2x) + 3x(5)$, which equals $6x^2 + 15x$. The distributive property can also be used to write $6x^2 + 15x$ in the factored form $3x(2x + 5)$. In this case, the GCF $3x$ was factored out of the polynomial $6x^2 + 15x$. This process is known as **factoring out the greatest common factor.** If the GCF of a polynomial is known, then the other factor can be determined by division. The result of this division can often be determined by inspection.

Example 3   Factoring Out the Greatest Common Factor

Factor the GCF out of each polynomial.

(a) $16x + 24$ **(b)** $20x^2y - 30xy^2$ **(c)** $12a^3b - 18a^2b^2 + 30ab^3$

Solution

(a) $16x + 24 = 8(2x) + 8(3)$

$= 8(2x + 3)$

The GCF is 8. Use the distributive property $ab + ac = a(b + c)$ to factor the GCF of 8 out of this binomial.

(b) $20x^2y - 30xy^2 = 10xy(2x) - 10xy(3y)$

$= 10xy(2x - 3y)$

The GCF is $10xy$. Use the distributive property $ab - ac = a(b - c)$ to factor the GCF out of this binomial.

(c) $12a^3b - 18a^2b^2 + 30ab^3$

$= 6ab(2a^2) - 6ab(3ab) + 6ab(5b^2)$

$= 6ab(2a^2 - 3ab + 5b^2)$

The GCF is $6ab$. Use the distributive property to factor the GCF out of this trinomial.

You can reverse these steps to multiply these factors and check this factorization.

Self-Check 3

Factor the GCF out of each polynomial.

a. $6x + 10$ **b.** $6x^3 - 9x^2 + 12x$

Sometimes the GCF is a polynomial with two or more terms. Note the similarity of the two parts of Example 4. By observing this similarity, you may find it easier to factor out a binomial GCF.

Example 4   Factoring Out a Binomial GCF

Factor the GCF out of each polynomial.

(a) $xz + 2z$ **(b)** $x(x - 3) + 2(x - 3)$

Solution

(a) $xz + 2z = (x + 2)(z)$

Factor out the GCF of z.

(b) $x(x - 3) + 2(x - 3) = (x + 2)(x - 3)$

Factor out the GCF of $x - 3$.

Note that if $z = x - 3$, these two expressions represent the same polynomial.

Self-Check 4

Factor the GCF out of each polynomial.

a. $xy - 7y$ **b.** $x(x + 5) - 7(x + 5)$

2. Factor by Grouping

The GCF of $x(x - 3) + 2(x - 3)$ from Example 4(b) is the binomial factor $x - 3$. Some polynomials with four or more terms can be factored by strategically grouping the terms to produce a common binomial factor. This is called **factoring by grouping.**

To factor a polynomial by grouping pairs of terms, we group terms that we already know how to factor with the hope of producing a common factor of each group. This is described in the following box. Section 6.5 examines other grouping possibilities.

Factoring a Four-Term Polynomial by Grouping Pairs of Terms

1. Be sure you have factored out the GCF if it is not 1.

2. Pair the terms and factor the GCF out of each pair of terms.

3. If there is a common binomial factor of these two groups, factor out this GCF. If there is no common binomial factor, try to use step 2 again with a different pairing of terms. If all possible pairs fail to produce a common binomial factor, the polynomial will not factor by this method.

A key characterization of the polynomials in Example 5 is that each has four terms. This indicates that factoring by grouping is an appropriate strategy to use to factor these polynomials.

Example 5 Factoring by Grouping

Factor each polynomial by using the grouping method.

(a) $ax + bx + 3a + 3b$ **(b)** $5x^2 - 20x + 2x - 8$ **(c)** $x^2 + 4x - 6x - 24$

Solution

(a) $ax + bx + 3a + 3b = (ax + bx) + (3a + 3b)$ Group the first two terms and the last two terms. Factor x out of the first pair
$\qquad = x(a + b) + 3(a + b)$ of terms and 3 out of the second pair
$\qquad = (x + 3)(a + b)$ of terms. Then factor the common factor $(a + b)$ out of each group.

(b) $5x^2 - 20x + 2x - 8 = (5x^2 - 20x) + (2x - 8)$ Group the first two terms and the last two terms. Factor $5x$ out of the first
$\qquad = 5x(x - 4) + 2(x - 4)$ pair of terms and 2 out of the second
$\qquad = (5x + 2)(x - 4)$ pair of terms. Then factor the common factor $(x - 4)$ out of each group.

(c) $x^2 + 4x - 6x - 24 = x(x + 4) - 6(x + 4)$ Factor x out of the first pair of terms and -6 out of the last two terms.
$\qquad = (x - 6)(x + 4)$ Then factor the common factor $(x + 4)$ out of each group.

Self-Check 5

Factor each polynomial by using the grouping method.

a. $ax - ay + bx - by$ **b.** $x^2 + 3x - 7x - 21$

The factorization of $ax + bx + 3a + 3b$ in Example 5(a) can be written as either $(x + 3)(a + b)$ or as $(a + b)(x + 3)$. This is justified by the commutative property of multiplication. In fact, this property can be used to reorder the factors of any polynomial.

It is important to read directions before "working" a math problem. You could be given the expression $x(x - 3) + 2(x - 3)$ and asked to do two different tasks: to expand or to factor. Both of these options are displayed here.

Expand   **Factor**

$x(x - 3) + 2(x - 3) = x^2 - 3x + 2x - 6$ $x(x - 3) + 2(x - 3) = (x + 2)(x - 3)$
$ = x^2 - x - 6$

Note that all three forms shown above, $x^2 - x - 6$, $x(x - 3) + 2(x - 3)$, and $(x + 2)(x - 3)$, are different representations of the same polynomial. Equal polynomials have identical tables of values and identical graphs. Example 6 uses a table to compare $x^2 - x - 6$ to its factored form.

The table for Example 6 was created using a calculator. It could have been created using a spreadsheet or pencil and paper. The key concept is that different forms of the same polynomial will produce exactly the same table of values. In Example 6, we use Y_1 to represent the first polynomial and Y_2 to represent the second polynomial. If the values for Y_1 and Y_2 do not agree for each value in the table, then this confirms that Y_1 and Y_2 are different polynomials.

Example 6   Using a Table to Compare Two Polynomials

Use a table to compare $x^2 - x - 6$ to $(x + 2)(x - 3)$.

Solution

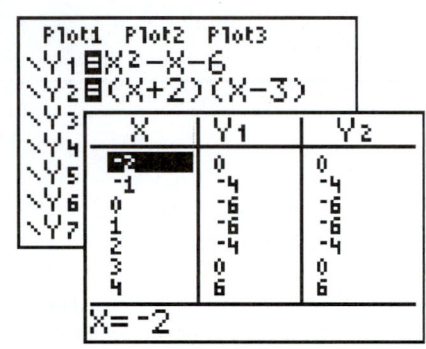

To enter the polynomials into a calculator, we use $Y_1 = x^2 - x - 6$ and $Y_2 = (x + 2)(x - 3)$.

The values for Y_1 and Y_2 in the table are identical. Thus Y_1 and Y_2 represent equal polynomials.

Also note for both $x = -2$ and $x = 3$ that $x^2 - x - 6 = 0$ and $(x + 2)(x - 3) = 0$. This relationship between the factors and the values making the polynomial 0 is examined in the following paragraph.

Answer: $x^2 - x - 6 = (x + 2)(x - 3)$

Self-Check 6

Use a table to compare $x^2 + 3x - 10$ to $(x + 5)(x - 2)$.

---

3. Use the Zeros of a Polynomial *P(x)* and the *x*-intercepts of the Graph of *y* = *P(x)* to Factor the Polynomial

An examination of the table in Example 6 reveals that both Y_1 and Y_2 are 0 for $x = -2$ and $x = 3$. If $P(x)$ represents a polynomial, we call an x-value that produces $P(x) = 0$ a **zero of the polynomial**. The second-degree polynomial $P(x) = x^2 - x - 6$ has both -2 and 3 as zeros, as shown here.

Zeros of $P(x) = x^2 - x - 6$

$P(-2) = (-2)^2 - (-2) - 6$ $\qquad\qquad$ $P(3) = (3)^2 - (3) - 6$
$ = 4 + 2 - 6$ $\qquad\qquad\qquad$ $ = 9 - 3 - 6$
$ = 0$ $\qquad\qquad\qquad\qquad\quad$ $ = 0$

Because $P(-2) = 0$ and $P(3) = 0$, we also know that $(-2, 0)$ and $(3, 0)$ are points on the graph of $y = x^2 - x - 6$. In fact, these points are the x-intercepts. These intercepts are shown in the accompanying graph. As we continue to examine second-degree (quadratic) polynomials in Chapters 7 and 8, we will see that graphs of quadratic functions will have at most two x-intercepts.

Graph of $y = x^2 - x - 6$

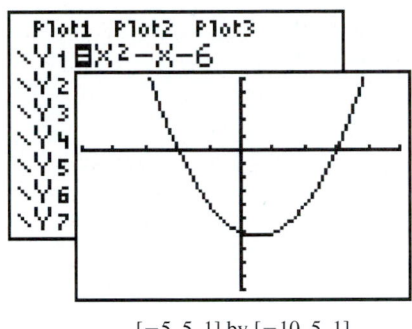

$[-5, 5, 1]$ by $[-10, 5, 1]$

Relationship among x-intercepts, zeros, and factors

x-intercepts:	$(-2, 0)$ and $(3, 0)$ are x-intercepts of the graph of $y = x^2 - x - 6$.
Zeros:	-2 and 3 are zeros of $x^2 - x - 6$.
Factors:	$x^2 - x - 6 = (x + 2)(x - 3)$

What Is the Relationship Between the Factors and the Zeros of P(x)?

In the examination just given of $y = x^2 - x - 6$, note that for the factor $x + 2$ to equal 0, x must equal -2; and likewise for the factor $x - 3$: if $x - 3 = 0$, then $x = +3$. What if $x - c$ is a factor of a polynomial? This is examined in the following display.

Factor	Zero
$x + 2$	$x + 2 = 0$, thus $x = -2$
$x - 3$	$x - 3 = 0$, thus $x = 3$
$x - c$	$x - c = 0$, thus $x = c$

A generalization of the relationship among the x-intercepts, zeros, and factors of a polynomial is given in the table. The fact that these statements are equivalent means that any time we know any one of these three pieces of information, we also know the other two pieces of information.

Equivalent Statements About Linear Factors of a Polynomial

For a real constant c and a real polynomial $P(x)$, the following statements are equivalent.

Graphically	Numerically	Algebraically
$(c, 0)$ is an x-intercept of the graph of $y = P(x)$	$P(c) = 0$; that is, c is a zero of $P(x)$	$x - c$ is a factor of $P(x)$

A polynomial that has real numbers as the coefficients for each term is a real polynomial. Each real polynomial of degree n has exactly n zeros. What can vary from polynomial to polynomial is how many of these zeros are real numbers. The zeros that are not real numbers will be imaginary numbers. In Chapter 7 we examine complex numbers and imaginary numbers.

In Example 7, note that the second-degree trinomial $x^2 - 7x + 12$ has two zeros. (A carefully constructed table displays these zeros.) These zeros are used to obtain its factored form. We will examine algebraic techniques for factoring trinomials of this type in Section 6.2. The key concept here is to make the connection between the zeros and the factors of a polynomial.

Example 7	Using Zeros to Factor a Polynomial

(a) Use the given table to determine the
    zeros of $x^2 - 7x + 12$.

(b) Then use these zeros to factor $x^2 - 7x + 12$.

x	$y = x^2 - 7x + 12$
-1	20
0	12
1	6
2	2
3	0
4	0
5	2

Solution

(a) The zeros of $x^2 - 7x + 12$ are 3 and 4.

The table shows $y = 0$ for $x = 3$ and $x = 4$.

(b) The factors of $x^2 - 7x + 12$ are
    $x - 3$ and $x - 4$.

Because $y = 0$ for $x = 3$, we have $x - 3$ as a
factor of $x^2 - 7x + 12$.

Because $y = 0$ for $x = 4$, we have $x - 4$ as a
factor of $x^2 - 7x + 12$.

Thus $x^2 - 7x + 12 = (x - 3)(x - 4)$.

You can check this factorization by multiplying
$x - 3$ by $x - 4$.

Self-Check 7

a. Use the given table to determine the zeros of $x^2 + 2x - 8$.

b. Then use these zeros to factor $x^2 + 2x - 8$.

x	$y = x^2 + 2x - 8$
-4	0
-3	-5
-2	-8
-1	-9
0	-8
1	-5
2	0

As a follow-up to Example 7, evaluate the polynomial expression $x^2 - 7x + 12$ for
$x = 3$. We can do this by substituting 3 for x either into the original expression
$x^2 - 7x + 12$ or into the factored form $(x - 3)(x - 4)$.

Using the form $x^2 - 7x + 12$ **Using the form $(x - 3)(x - 4)$**

$$(3)^2 - 7(3) + 12 = 9 - 21 + 12 \qquad (3 - 3)(3 - 4) = (0)(-1)$$
$$= 0 \qquad\qquad\qquad\qquad = 0$$

Note that the factored form makes it much easier to recognize the zeros of a polynomial.

In Example 8, we use the x-intercepts of the graph of $y = x^2 - 3x - 10$ to obtain the
zeros and the factored form of $x^2 - 3x - 10$. We will examine algebraic techniques for
factoring trinomials of this type in Section 6.2. The key concept here is to make the connection between the x-intercepts of a graph and the zeros and factors of a polynomial.

Example 8   Using *x*-intercepts to Factor a Polynomial

(a) Use the given graph to determine the *x*-intercepts of $y = x^2 - 3x - 10$.

(b) Determine the zeros of $x^2 - 3x - 10$.

(c) Factor $x^2 - 3x - 10$.

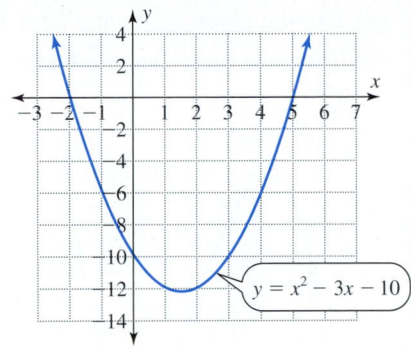

Solution

(a) The *x*-intercepts of $y = x^2 - 3x - 10$ are $(-2, 0)$ and $(5, 0)$.

The best estimate from the graph is that it crosses the *x*-axis at $x = -2$ and $x = 5$. This can be checked by verifying that the zeros check or that the factorization checks.

(b) The zeros of $x^2 - 3x - 10$ are -2 and 5.

An *x*-intercept of $(-2, 0)$ means that -2 is a zero of $x^2 - 3x - 10$.

An *x*-intercept of $(5, 0)$ means that 5 is a zero of $x^2 - 3x - 10$.

(c) The factors of $x^2 - 3x - 10$ are $x + 2$ and $x - 5$.

Because $y = 0$ for $x = -2$, $x - (-2)$ or $x + 2$ is a factor of $x^2 - 3x - 10$. Because $y = 0$ for $x = 5$, $x - 5$ is a factor of $x^2 - 3x - 10$.

Thus $x^2 - 3x - 10 = (x + 2)(x - 5)$.

You can check this factorization by multiplying $(x + 2)$ by $(x - 5)$.

Self-Check 8

a. Use the given graph to determine the *x*-intercepts of $y = x^2 - 3x - 4$.

b. Determine the zeros of $x^2 - 3x - 4$.

c. Factor $x^2 - 3x - 4$.

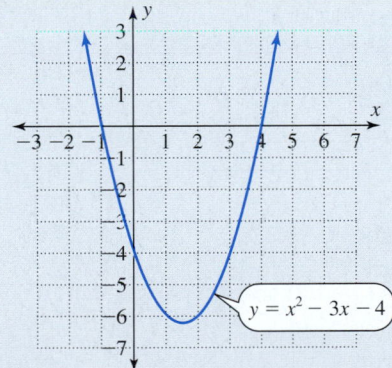

In Example 9, we use the concepts from this section to factor a third-degree polynomial. It is worth pointing out that this approach of using zeros and/or *x*-intercepts to obtain the factored form can assist us in factoring some polynomials that are difficult to factor algebraically.

Example 9 Using Zeros to Factor a Third-Degree Polynomial

(a) Use the given graph and table to determine the zeros of $x^3 - 3x^2 - 10x + 24$.

(b) Then use these zeros to factor $x^3 - 3x^2 - 10x + 24$.

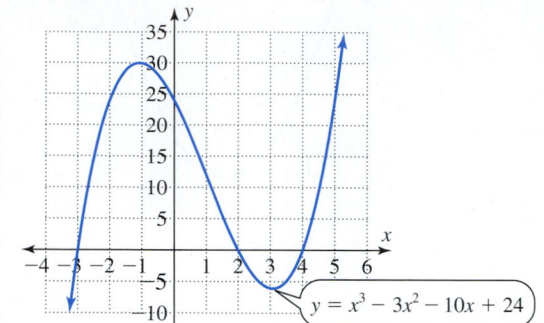

x	y
-3	0
-2	24
-1	30
0	24
1	12
2	0
3	-6
4	0

$y = x^3 - 3x^2 - 10x + 24$

Solution

(a) The zeros of $x^3 - 3x^2 - 10x + 24$ are -3, 2, and 4.

By inspection, it appears that the graph has x-intercepts of $(-3, 0)$, $(2, 0)$, and $(4, 0)$. This is confirmed by the table, which shows $y = 0$ for $x = -3$, $x = 2$, and $x = 4$.

(b) The factors of $x^3 - 3x^2 - 10x + 24$ are $x + 3$, $x - 2$, and $x - 4$.

Because $y = 0$ for $x = -3$, $x + 3$ is a factor of $x^3 - 3x^2 - 10x + 24$.

Because $y = 0$ for $x = 2$, $x - 2$ is a factor of $x^3 - 3x^2 - 10x + 24$.

Thus $x^3 - 3x^2 - 10x + 24$
$$= (x + 3)(x - 2)(x - 4).$$

Because $y = 0$ for $x = 4$, $x - 4$ is a factor of $x^3 - 3x^2 - 10x + 24$.

Self-Check 9

a. Use the given graph and table to determine the zeros of $x^3 + 5x^2 + 6x$.

b. Then use these zeros to factor $x^3 + 5x^2 + 6x$.

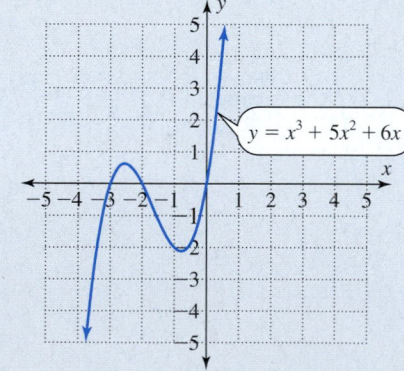

$y = x^3 + 5x^2 + 6x$

x	y
-4	-8
-3	0
-2	0
-1	-2
0	0
1	12
2	40

6.1 Using the Language and Symbolism of Mathematics

1. The GCF of a polynomial is the _____ _____ factor of the polynomial.

2. The GCF of a polynomial is a common factor of each term of the polynomial that has the largest possible numerical _____ and the largest possible _____ on each variable factor.

3. When we factor out a common factor from each term of a polynomial, we are using the _____ property.

4. Some polynomials with four or more terms can be factored by strategically _____ the terms to produce a common binomial factor.

5. An x-intercept of a graph is a point on the graph whose y-coordinate is _____.

6. If $P(x)$ is a polynomial and $P(5) = 0$, then 5 is called a _____ of the polynomial.

7. If $x + 4$ is a factor of the polynomial $P(x)$, then _____ is an x-intercept of the graph of $y = P(x)$.

8. If $P(x)$ is a polynomial and $(-2, 0)$ is an x-intercept of the graph of $y = P(x)$, then _____ is a zero of $P(x)$.

9. If $P(x)$ is a polynomial and $(8, 0)$ is an x-intercept of the graph of $y = P(x)$, then _____ is a factor of $P(x)$.

10. Each coefficient of a real polynomial $P(x)$ is a _____ number.

11. Each real polynomial function of degree n has exactly _____ zeros. Some (perhaps all) of these zeros can be real numbers. If some of the zeros are not real numbers, then they will be imaginary numbers. (We will examine imaginary numbers and complex numbers in Section 7.5.)

6.1 Quick Review

1. Factor 77 into its prime factors.

2. Factor 90 into its prime factors.

3. Determine the x-intercept of the line graphed here.

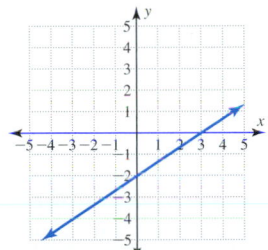

4. All the points in this table lie on the same line. What is the x-intercept of this line?

x	y
-1.0	-7.5
0.0	-5.0
1.0	-2.5
2.0	0.0
3.0	2.5

5. Multiply $(x - 4)(x + 5)$.

6.1 Exercises

Objective 1 Factor the GCF Out of a Polynomial

In Exercises 1–6, determine the GCF of each pair of monomials.

1. 24 and 54
2. 30 and 48
3. $20x^2$ and $28x$
4. $15y^2$ and $20y$
5. $18x^4y^3$ and $45x^2y$
6. $50a^2b^5$ and $44a^3b^2$

In Exercises 7–16, fill in the missing information. Then write your answer as the complete factorization of the given polynomial.

7. $12x^3 - 18x^2 = 6x^2(2x - \underline{\quad})$
8. $15x^4 + 20x^3 = 5x^3(3x + \underline{\quad})$
9. $10a^2b + 6ab^2 = 2ab(\underline{\quad} + \underline{\quad})$
10. $21a^3b - 15a^2b^2 = 3a^2b(\underline{\quad} - \underline{\quad})$
11. $10x^3 - 15x^2y + 20xy^2 = 5x(\underline{\quad} - \underline{\quad} + \underline{\quad})$
12. $12x^3 + 16x^2y + 20xy^2 = 4x(\underline{\quad} + \underline{\quad} + \underline{\quad})$
13. $2x(x - 2y) + y(x - 2y) = (\underline{\quad} + y)(x - 2y)$
14. $5x(x + 3y) + 4y(x + 3y) = (\underline{\quad} + 4y)(x + 3y)$
15. $7x(2x - 5) - 9(2x - 5) = (7x - \underline{\quad})(2x - 5)$
16. $(3x + 4)(5v) - (3x + 4)(4w) = (3x + 4)(5v - \underline{\quad})$

In Exercises 17–28, factor the GCF out of each polynomial.

17. $14x^2 + 77x$
18. $22x^3 - 33x^2$
19. $10x^3 - 15x^2$
20. $8x^4 + 20x^2$
21. $8x^2y - 20xy^2$
22. $18x^2y^2 - 30xy^2$
23. $6x^3 - 8x^2 + 10x$
24. $6x^3 - 9x^2 - 21x$
25. $(5x - 2)(x) + (5x - 2)(3)$
26. $(3x + 4)(7x) - (3x + 4)(5)$
27. $(x + 2y)(6x) - (x + 2y)(7y)$
28. $(5x - 2y)(3x) + (5x - 2y)(4y)$

Objective 2 Factor by Grouping

In Exercises 29–38, factor each polynomial using the grouping method.

29. $ax + bx + 2a + 2b$

30. $ax + bx - 3a - 3b$

31. $5ax - 2a + 15x - 6$

32. $3ax + a + 6x + 2$

33. $2ax - 7a - 2bx + 7b$

34. $12ax - 8a - 21bx + 14b$

35. $3x^2 - 5x + 12x - 20$

36. $2x^2 - 8x + 7x - 28$

37. $12x^2 + 4x - 15x - 5$

38. $10x^2 - 15x - 14x + 21$

Objective 3 Use the Zeros of a Polynomial $P(x)$ and the x-intercepts of the Graph of $y = P(x)$ to Factor the Polynomial

In Exercises 39–42, use the polynomial $P(x)$ and the table for $y = P(x)$ to:

a. List the zeros of $P(x)$.   **b.** Factor each polynomial.

39.

x	$y = x^2 - 7x + 10$
0	10
1	4
2	0
3	-2
4	-2
5	0
6	4

40.

x	$y = x^2 - 2x - 24$
-6	24
-4	0
-2	-16
0	-24
2	-24
4	-16
6	0

41.

x	$y = x^2 - 4x - 32$
-4	0
-2	-20
0	-32
2	-36
4	-32
6	-20
8	0

42.

x	$y = x^2 + 12x + 35$
-9	8
-8	3
-7	0
-6	-1
-5	0
-4	3
-3	8

In Exercises 43–46, use the graph of $y = P(x)$ to:

a. Determine the x-intercepts of the graph.

b. List the zeros of $P(x)$.

c. Factor the polynomial $P(x)$.

43. $P(x) = x^2 - 7x + 10$ **44.** $P(x) = x^2 - 5x - 24$

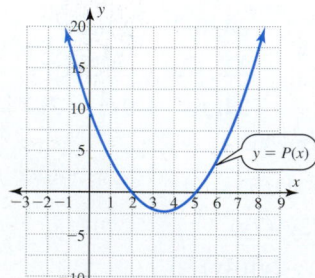

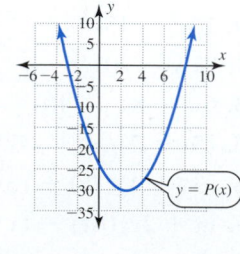

45. $P(x) = x^3 - x^2 - 6x$ **46.** $P(x) = x^3 - x^2 - 12x$

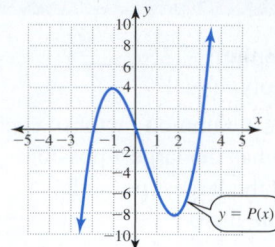

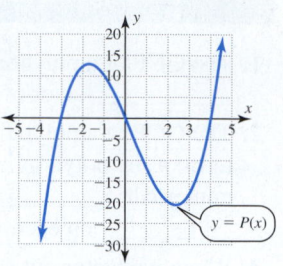

In Exercises 47–52, use the factored form of each polynomial $P(x)$ to:

a. List the zeros of $P(x)$.

b. List the x-intercepts of the graph of $y = P(x)$.

47. $P(x) = (x + 5)(x - 9)$

48. $P(x) = (x + 7)(x - 11)$

49. $P(x) = (x + 3)(x + 1)(x - 8)$

50. $P(x) = (x + 5)(x - 6)(x - 12)$

51. $P(x) = x(x + 3) - 17(x + 3)$ (*Hint:* Complete the factorization first.)

52. $P(x) = x(x + 4) - 15(x + 4)$ (*Hint:* Complete the factorization first.)

Applying Technology

In Exercises 53–56:

a. Use a calculator or spreadsheet to complete each table for $y = P(x)$.

b. Determine the zeros of $P(x)$.

c. Factor the polynomial $P(x)$.

53.

x	$y = x^2 - 35x + 300$
0	
5	
10	
15	
20	
25	
30	

54.

x	$y = x^2 + 26x + 168$
-15	
-14	
-13	
-12	
-11	
-10	
-9	

55.

x	$y = x^3 - 10x^2 + 24x$
0	
1	
2	
3	
4	
5	
6	

56.

x	$y = x^3 - 35x^2 + 350x - 1,000$
0	
5	
10	
15	
20	
25	
30	

In Exercises 57–62, use a calculator or a graphing utility to graph each polynomial. Then use the x-intercepts of the graph of $y = P(x)$ to factor $P(x)$. A window of $[-10, 10, 1]$ by $[-100, 100, 1]$ will work for each polynomial.

57. $P(x) = x^2 - x - 56$

58. $P(x) = x^2 - 2x - 48$

59. $P(x) = x^3 + 3x^2 - 49x + 45$

60. $P(x) = x^3 + 8x^2 - 4x - 32$

61. $P(x) = x^4 - 17x^2 + 16$

62. $P(x) = x^4 - 10x^2 + 9$

Review and Concept Development

In Exercises 63–66, complete the following table for each polynomial.

Polynomial $P(x)$	Factored Form of $P(x)$	Zeros of $P(x)$	x-intercepts of the Graph of $y = P(x)$
63. $x^2 + 2x - 63$	$(x - 7)(x + 9)$		
64. $x^2 - 17x + 66$		6 and 11	
65. $x^2 - 2x - 80$			$(-8, 0), (10, 0)$
66. $x^3 - x^2 - 30x$		$-5, 0,$ and 6	

Comparison of Factoring and Expanding

In Exercises 67–70, factor the polynomial in the left column. Then expand and simplify the polynomial in the right column.

Factor	Expand
67. $x(x + 4) + 2(x + 4)$	$x(x + 4) + 2(x + 4)$
68. $x(x - 5) + 3(x - 5)$	$x(x - 5) + 3(x - 5)$
69. $2x(5x - 2) - 7(5x - 2)$	$2x(5x - 2) - 7(5x - 2)$
70. $3x(2x + 5) - 8(2x + 5)$	$3x(2x + 5) - 8(2x + 5)$

Connecting Concepts to Applications

71. Investment Value If a principal of P dollars earns interest at an annual interest rate r, then the value of this investment (in dollars) at the end of the year is given by $P + Pr$.
   a. Factor this polynomial expression.
   b. Evaluate this factored form using $P = 2,000$ and $r = 0.04$.

72. Number of Games A Little League consists of n teams. The league director plans to create a schedule so that each team will play each of the other teams 4 times during the season. The number of games required to accomplish this is given by $2n^2 - 2n$.

a. Factor this polynomial expression.
b. If there are only two teams in the league, evaluate $2n^2 - 2n$ using the factored form.
c. If there are six teams in the league, evaluate $2n^2 - 2n$ using the factored form.

Connecting Algebra to Geometry

73. Surface Area The surface area of a cylindrical soda can is given by $A = 2\pi r^2 + 2\pi rh$ for a can with radius r and height h. Factor this polynomial expression.

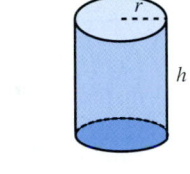

74. Area A concrete pad near a small storage building is in the shape of a trapezoid. The surface area of this pad is $x^2 + 3x$, where the area of the square portion of the figure is x^2 and the area of the triangular portion is $3x$. Factor this polynomial expression.

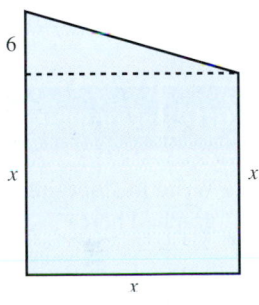

Group discussion questions

75. Challenge Question The area of the trapezoidal cross section of a roadway shown here is given by either

$$A = 2x^2 + 20x \text{ or by } A = \frac{1}{2}(x)(4x + 40).$$

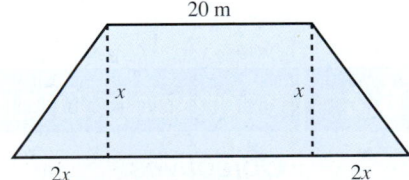

20 m

x x

$2x$ $2x$

a. Verify that these two equations are equivalent.
b. Explain the relationship of each term of the first equation to the sketch of the cross section of the roadway.
c. Explain the relationship of each factor of the second equation to the sketch of the cross section of the roadway.

76. Discovery Question Two forms of a polynomial $P(x)$ are given here.

Factored Form	Expanded Form
$(x + 5)(x - 7)$	$x^2 - 2x - 35$

a. What is the degree of this polynomial?

b. Which of these forms did you use to determine the answer in part **a**?

c. What are the x-intercepts of the graph of $y = x^2 - 2x - 35$? (Be sure to write your answers as ordered pairs.)

d. Which of these forms did you use to determine the answer in part **c**?

e. What is the y-intercept of the graph of $y = x^2 - 2x - 35$? (Be sure to write your answer as an ordered pair.)

f. Which of these forms did you use to determine the answer in part **e**?

g. What are the zeros of $x^2 - 2x - 35$?

h. Which of these forms did you use to determine the answer in part **g**?

77. Error Analysis A student graphed the equation as displayed and concluded that the zeros of $(x + 4)(x - 6)$ are -2 and 3. Describe the error the student has made, and give the correct zeros.

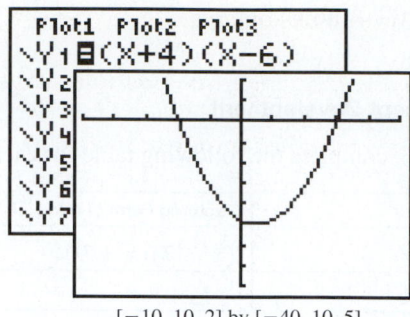

[−10, 10, 2] by [−40, 10, 5]

78. Discovery Question

a. Graph $y = x^2 - x - 42$, and use this graph to factor $x^2 - x - 42$.

b. Graph $y = -x^2 + x + 42$, and use this graph to factor $-x^2 + x + 42$.

c. Graph $y = x^2 + 5x - 24$, and use this graph to factor $x^2 + 5x - 24$.

d. Graph $y = -x^2 - 5x + 24$, and use this graph to factor $-x^2 - 5x + 24$.

e. What is the relationship between the graph of $y = P(x)$ and the graph of $y = -P(x)$?

f. What is the relationship between the zeros of $P(x)$ and of $-P(x)$?

g. What is the relationship between the factored form of $P(x)$ and of $-P(x)$?

6.1   Cumulative Review

1. Write in slope-intercept form the equation of the line graphed here.

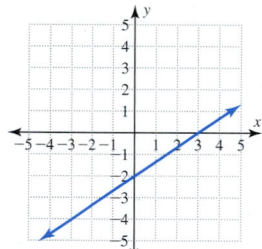

2. Solve the inequality $-9 < 2x + 3 \leq 11$. Write the answer in interval notation.

3. Simplify the expression $\left(\dfrac{6x^{-2}y^8}{2x^{-4}y^{-2}}\right)^{-2}$ and write the answer using only positive exponents. Assume all variables are nonzero.

4. Use long division to divide $6x^2 + 11x - 10$ by $3x - 2$.

5. The factor pairs of 15 are $1 \cdot 15$ and $3 \cdot 5$. Determine all factor pairs of:

    a. 21   b. 20

Section 6.2   Factoring Trinomials of the Form $x^2 + bx + c$

Objectives:

1. Factor trinomials of the form $x^2 + bx + c$ by the trial-and-error method or by inspection.
2. Identify a prime trinomial of the form $x^2 + bx + c$.
3. Factor trinomials of the form $x^2 + bxy + cy^2$ by the trial-and-error method or by inspection.

In Section 6.1, we learned how to:

  i. Factor the GCF out of a polynomial.

  ii. Factor some polynomials with four terms by grouping.

  iii. Use the zeros shown in a table or the x-intercepts displayed on a graph to factor a polynomial.

In this section, we will focus on an algebraic method for factoring trinomials of the form $x^2 + bx + c$.

The trinomial $ax^2 + bx + c$ is referred to as a **quadratic trinomial.** The term ax^2 is called the **second-degree term,** or the **quadratic term;** bx is the **first-degree term** or the **linear term;** and c is the **constant term.** For example, $6x^2 - 7x - 5$ is a quadratic trinomial with $a = 6$, $b = -7$, and $c = -5$. The quadratic term is $6x^2$, the linear term is $-7x$, and the constant term is -5. A polynomial is **prime over the integers** if its only factorization must involve either 1 or -1 as one of the factors.

1. Factor Trinomials of the Form $x^2 + bx + c$ by the Trial-and-Error Method or by Inspection

The easiest quadratic trinomials to factor are those in which the leading coefficient a is equal to 1, so we will begin with these trinomials. Either $x^2 + bx + c$ is prime or it can be factored as $(x + c_1)(x + c_2)$.

Where Do I Start to Factor $x^2 + bx + c$?

Start by looking at the factors of the constant term c. This logic is developed further by examining the following four products.

Factors		F	O	I	L		Product
$(x + 1)(x + 24)$	=	x^2	$+24x$	$+x$	$+24$	=	$x^2 + 25x + 24$
$(x + 2)(x + 12)$	=	x^2	$+12x$	$+2x$	$+24$	=	$x^2 + 14x + 24$
$(x + 3)(x + 8)$	=	x^2	$+8x$	$+3x$	$+24$	=	$x^2 + 11x + 24$
$(x + 4)(x + 6)$	=	x^2	$+6x$	$+4x$	$+24$	=	$x^2 + 10x + 24$

Product is x^2

Product is 24

Sum is the linear term

Thus, a factorization of $x^2 + bx + c$ must have a pair of factors of c whose sum is b.

Factoring Trinomials of the Form $x^2 + bx + c$

Algebraically	Verbally	Examples
$x^2 + bx + c = (x + c_1)(x + c_2)$ where $c_1 c_2 = c$ and $c_1 + c_2 = b$.	A trinomial $x^2 + bx + c$ is factorable into a pair of binomial factors with integer coefficients if and only if there are two integers whose product is c and whose sum is b. Otherwise, the trinomial is prime over the integers.	$x^2 + 14x + 24 = (x + 2)(x + 12)$ where $2 \cdot 12 = 24$ and $2 + 12 = 14$. $x^2 + 50x + 24$ is prime because there there are no factor pairs of 24 whose sum is 50.

Since several possible factors of c may have to be examined before the correct factors are found, this method is referred to as the **trial-and-error method.** Once you gain some proficiency with this process, you will be able to do many problems by inspection. Nonetheless, you should always examine the possibilities in a systematic manner rather than through random guessing.

In Example 1, we use the trial-and-error method to factor $x^2 + 9x + 18$, a trinomial with a positive constant term of 18 and a positive linear coefficient of 9. Because both of these terms are positive, we start by using factor pairs of 18 that fit the sign pattern $(x + \underline{})(x + \underline{})$.

Example 1 Factoring a Trinomial with a Positive Constant Term

Factor $x^2 + 9x + 18$.

Solution

$$x^2 + 9x + 18 = (x + \underline{\quad})(x + \underline{\quad})$$

Product is x^2 Product is 18

First set up a form with the correct sign pattern.

Then use the factor pairs of 18 to examine all possible factors of this form. The correct factorization will result in a linear term of $9x$.

Possible Factors	Resulting Linear Term	Factors of 18		Sum of Factors
$(x + 1)(x + 18)$	$19x$	1	18	19
$(x + 2)(x + 9)$	$11x$	2	9	11
$(x + 3)(x + 6)$	$9x$	3	6	9

Answer: $x^2 + 9x + 18 = (x + 3)(x + 6)$ **Check:** $(x + 3)(x + 6) = x^2 + 6x + 3x + 18$
$$= x^2 + 9x + 18$$

Self-Check 1

Factor $x^2 + 11x + 18$.

The factored form of $x^2 + 9x + 18$ in Example 1 could also be expressed as $(x + 6)(x + 3)$ since multiplication is commutative. It is also worth observing that the factored form of $x^2 - 9x + 18$ is closely related to the factored form of $x^2 + 9x + 18$. Note that $x^2 - 9x + 18 = (x - 3)(x - 6)$ is very similar to the factored form of $x^2 + 9x + 18 = (x + 3)(x + 6)$.

An important first step in factoring a quadratic trinomial is to use the signs of b and c from $x^2 + bx + c$ to determine the correct sign pattern of the binomial factors.

Sign Pattern for the Factors of $x^2 + bx + c$

Verbally	Algebraically	Examples
If the constant term is positive, the factors of this term must have the same sign. These factors will share the same sign as the linear term.	$x^2 + bx + c = (x + \underline{\quad})(x + \underline{\quad})$ $x^2 - bx + c = (x - \underline{\quad})(x - \underline{\quad})$	$x^2 + 5x + 6 = (x + 2)(x + 3)$ $x^2 - 5x + 6 = (x - 2)(x - 3)$
If the constant term is negative, the factors of this term must be opposite in sign. The sign of the constant factor with the larger absolute value will be the same as that of the linear term.	$x^2 + bx - c = (x + \underline{\quad})(x - \underline{\quad})$ $x^2 - bx - c = (x + \underline{\quad})(x - \underline{\quad})$	$x^2 + 5x - 6 = (x + 6)(x - 1)$ $x^2 - 5x - 6 = (x + 1)(x - 6)$

Example 2 condenses some of the work that was shown in Example 1. The goal of these examples is to reinforce the understanding needed to correctly apply any shortcuts that you may know. With practice, there are many shortcuts that you can use to factor these trinomials. In Example 2, $m^2 + 8m - 20$ is a trinomial with a negative constant term of -20 and a positive linear coefficient of 8. Thus the sign pattern of the factors is of the form $(x + \underline{\quad})(x - \underline{\quad})$.

Example 2	Factoring a Trinomial with a Negative Constant Term

Factor $m^2 + 8m - 20$.

Solution

$$m^2 + 8m - 20 = (m + \underline{\quad})(m - \underline{\quad})$$

Product is m^2 Product is -20

First set up a form with the correct sign pattern.

Then use the factor pairs of -20 to examine possible factors of this form. The correct factorization will result in a linear term of $8m$. Note that we did not list all possible factor pairs; we stopped once we found the correct pair of factors.

Factors of -20		Sum of Factors
-1	20	19
-2	10	8

Answer: $m^2 + 8m - 20 = (m + 10)(m - 2)$ **Check:** $(m + 10)(m - 2) = m^2 - 2m + 10m - 20$
$$= m^2 + 8m - 20$$

Self-Check 2

Factor $n^2 - n - 20$.

Note in Example 2 that switching the signs on the last term of each factor would produce $(m - 10)(m + 2)$. This is the factorization of $m^2 - 8m - 20$.

2. Identify a Prime Trinomial of the Form $x^2 + bx + c$

In Example 3, we use the trial-and-error method to confirm that a trinomial is prime and to factor a similar trinomial. Both trinomials in Example 3 are of the form $v^2 - \underline{\quad}v + 12$, so the potential factored form of both trinomials has the same sign pattern. In each case, we consider only options for which the product of the first terms is v^2 and the product of the constant terms is 12.

Example 3	Identifying a Prime Trinomial

Determine which trinomial is prime, and factor the trinomial that is factorable over the integers.

(a) $v^2 - 9v + 12$ **(b)** $v^2 - 8v + 12$

Solution

(a) $v^2 - 9v + 12 = (v - \underline{\quad})(v - \underline{\quad})$
(b) $v^2 - 8v + 12 = (v - \underline{\quad})(v - \underline{\quad})$

Note that both trinomials have factors with the same sign pattern. Use the factor pairs of 12 to examine all possible factors of this form.

The correct factorization for part (a) will result in a linear term of $-9v$, and the correct factorization for part (b) will result in a linear term of $-8v$.

Factors of 12		Sum of Factors
-1	-12	-13
-2	-6	-8
-3	-4	-7

(a) Answer: $v^2 - 9v + 12$ is prime.

None of the factor pairs of 12 has a sum of -9. Since we have systematically examined and eliminated all integer factor pairs of 12, we know this polynomial is prime over the integers.

(b) Answer: $v^2 - 8v + 12 = (v - 2)(v - 6)$ **Check:** $(v - 2)(v - 6) = v^2 - 6v - 2v + 12$
$$= v^2 - 8v + 12$$

Self-Check 3

Determine which trinomial is prime, and factor the trinomial that is factorable over the integers.

a. $w^2 - 7w + 12$ **b.** $w^2 - 11w + 12$

Example 3 shows us that if we have created a table of all possible factors and corresponding sums, then we know all trinomials that can factor with the given quadratic and constant terms. In Example 3, the quadratic term is v^2 and the constant term is 12. The table of possible factors and resulting linear terms shows that the only linear terms that will lead to a factorization are $-13v$, $-8v$, and $-7v$ (or their opposites). With any other linear term, the trinomial will be prime.

If you compare Example 1 and Example 4, you will note that we are showing less work now. In fact, we encourage you to try to work toward factoring $x^2 + bx + c$ without writing a table. Before you look at the table in the solution for Example 4, try to mentally determine the correct pair of factors. Then you can examine the table that is shown to validate the steps that you may have done mentally.

Example 4 **Factoring a Trinomial with a Negative Quadratic Term**

Factor $-x^2 + 12x - 32$.

Solution

$$-x^2 + 12x - 32 = -1(x^2 - 12x + 32)$$

First factor out -1 from each term.

$$= -(x - \underline{\quad})(x - \underline{\quad})$$

Then set up a form with the correct sign pattern for the factors of the remaining trinomial.

Factors of 32		Sum of Factors
-1	-32	-33
-2	-16	-18
-4	-8	-12

Then examine the factor pairs of 32 for a pair with a sum of -12.

Answer: $-x^2 + 12x - 32 = -(x - 4)(x - 8)$

Check: $-(x - 4)(x - 8) = -(x^2 - 8x - 4x + 32)$
$$= -1(x^2 - 12x + 32)$$
$$= -x^2 + 12x - 32$$

Self-Check 4

Factor $-y^2 - 13y - 30$.

In the paragraph before Example 4, we asked you to try to mentally determine the correct pair of factors. This is called **factoring by inspection.** In Example 5, we use factoring by inspection to factor two trinomials that differ only by the sign of the linear term.

Example 5 Understanding the Significance of the Sign of the Linear Term

Factor each trinomial.

(a) $x^2 + 4x - 60$ **(b)** $x^2 - 4x - 60$

Solution

(a) $x^2 + 4x - 60 = (x + \underline{\quad})(x - \underline{\quad})$ First set up a form with the correct sign pattern. Form a table only if necessary.

Answer: $x^2 + 4x - 60 = (x + 10)(x - 6)$ The pair of factors of -60 that has a sum of 4 is the pair of 10 and -6.

$$\textbf{Check:} \quad (x + 10)(x - 6) = x^2 - 6x + 10x - 60$$
$$= x^2 + 4x - 60$$

(b) $x^2 - 4x - 60 = (x + \underline{\quad})(x - \underline{\quad})$ First set up a form with the correct sign pattern.

Answer: $x^2 - 4x - 60 = (x + 6)(x - 10)$ The pair of factors of -60 that has a sum of -4 is the pair of 6 and -10.

$$\textbf{Check:} \quad (x + 6)(x - 10) = x^2 - 10x + 6x - 60$$
$$= x^2 - 4x - 60$$

Self-Check 5

Factor each trinomial.

a. $x^2 + 11x + 28$ **b.** $x^2 - 11x + 28$

In Example 5, note that the only difference between $x^2 + 4x - 60$ and $x^2 - 4x - 60$ is the sign of the linear term. Thus, the correct factorization of $x^2 - 4x - 60$ can be obtained by simply changing signs of the constants in the factors of $x^2 + 4x - 60$.

In Example 6, we factor a trinomial with a GCF of $2x$. We first factor out this GCF and then factor the remaining trinomial by inspection.

Example 6 Completely Factoring a Trinomial

Factor $2x^3 + 6x^2 - 80x$.

Solution

$$2x^3 + 6x^2 - 80x = 2x(x^2 + 3x - 40)$$ First factor out the GCF of $2x$ from each term.
$$= 2x(x + \underline{\quad})(x - \underline{\quad})$$ Then set up a form with the correct sign pattern for the factors of the trinomial factor. Form a table only if necessary.

Answer: $2x^3 + 6x^2 - 80x = 2x(x + 8)(x - 5)$ The pair of factors of -40 that has a sum of 3 is -5 and 8. Note that $(-5)(8) = -40$ and $-5 + 8 = 3$.

$$\textbf{Check:} \quad 2x(x + 8)(x - 5) = 2x(x^2 - 5x + 8x - 40)$$
$$= 2x(x^2 + 3x - 40)$$
$$= 2x^3 + 6x^2 - 80x$$

Self-Check 6

Factor $3ax^3 - 12ax^2 - 135ax$.

3. Factor Trinomials of the Form $x^2 + bxy + cy^2$ by the Trial-and-Error Method or by Inspection

Observe the pattern illustrated by the following factorizations. We can use this pattern to factor polynomials that contain more than one variable.

Factorizations of Polynomials in x	Factorizations of Polynomials in x *and* y
$x^2 + 7x + 12 = (x + 3)(x + 4)$	$x^2 + 7xy + 12y^2 = (x + 3y)(x + 4y)$
$a^2 - 5a - 24 = (a + 3)(a - 8)$	$a^2 - 5ab - 24b^2 = (a + 3b)(a - 8b)$
$v^2 - 9v + 12$ is prime.	$v^2 - 9vw + 12w^2$ is prime.

In Example 7, we factor a quadratic trinomial in two variables.

Example 7 Factoring a Trinomial in Two Variables

Factor $x^2 + 19xy + 48y^2$.

Solution

$x^2 + 19xy + 48y^2 = (x + \underline{\quad} y)(x + \underline{\quad} y)$ First set up a form with the correct sign pattern.

Answer: $x^2 + 19xy + 48y^2 = (x + 3y)(x + 16y)$ Mentally determine that the factor pair of 48 that has a sum of 19 is the pair of 3 and 16. Form a table only if necessary.

$$\textbf{\textit{Check:}} \quad (x + 3y)(x + 16y) = x^2 + 16xy + 3xy + 48y^2$$
$$= x^2 + 19xy + 48y^2$$

Self-Check 7

Factor $m^2 - 12mn + 35n^2$.

In Example 8, we will examine the graph of $y = x^2 + 4x - 32$, a table of values for $x^2 + 4x - 32$, and the factorization of $x^2 + 4x - 32$. Continuing to examine topics graphically, numerically, and algebraically not only deepens your understanding but it also improves your ability to apply mathematical concepts in the workplace.

Example 8 Using Multiple Perspectives to Factor a Trinomial

(a) Graph $y = x^2 + 4x - 32$ and determine the x-intercepts.

(b) Complete a table of values and determine the zeros of $x^2 + 4x - 32$.

(c) Factor $x^2 + 4x - 32$.

Solution

(a)

$[-10, 10, 1]$ by $[-40, 10, 5]$

The graph can provide an excellent clue as to the correct pair of numerical factors of the constant term.

The x-intercepts of $(-8, 0)$ and $(4, 0)$ are determined by inspection.

The x-intercepts are $(-8, 0)$ and $(4, 0)$.

(b)

X	Y1	
-12	64	
-8	0	
-4	-32	
0	-32	
4	0	
8	64	
12	160	

X= -12

The graph indicates that the x-values to check are $x = -8$ and $x = 4$. Select input values of x that will include these values. (Sometimes you may need to scroll down through the values to see all the values that are of interest.) The table shows that $x = -8$ and $x = 4$ are zeros of the polynomial and also confirms that the x-intercepts are $(-8, 0)$ and $(4, 0)$.

The zeros of $x^2 + 4x - 32$ are $x = -8$ and $x = 4$.

(c) $x^2 + 4x - 32 = (x + 8)(x - 4)$

Because $x = -8$ is a zero of $x^2 + 4x - 32$, $x + 8$ is a factor. Because $x = 4$ is a zero of $x^2 + 4x - 32$, $x - 4$ is a factor. You could also obtain this factorization by the trial-and-error method.

Self-Check 8

a. Use the given graph of $y = x^2 + 2x - 35$ to determine the x-intercepts.

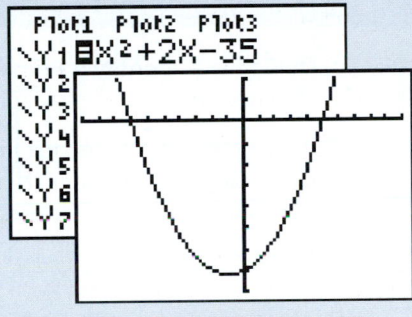

$[-10, 10, 1]$ by $[-40, 10, 5]$

b. Complete a table of values and determine the zeros of $x^2 + 2x - 35$.

c. Factor $x^2 + 2x - 35$.

Self-Check Answers

1. $(x + 2)(x + 9)$
2. $(n + 4)(n - 5)$
3. **a.** $(w - 3)(w - 4)$ **b.** $w^2 - 11w + 12$ is prime.
4. $-(y + 3)(y + 10)$
5. **a.** $(x + 4)(x + 7)$ **b.** $(x - 4)(x - 7)$
6. $3ax(x + 5)(x - 9)$
7. $(m - 5n)(m - 7n)$

8. **a.** The x-intercepts are $(-7, 0)$ and $(5, 0)$.

   b.

X	Y1	
-10	45	
-7	0	
-4	-27	
-1	-36	
2	-27	
5	0	
8	45	

   X= -10

   The zeros are $x = -7$ and $x = 5$.

   c. $x^2 + 2x - 35 = (x + 7)(x - 5)$

6.2 **Using the Language and Symbolism of Mathematics**

1. A polynomial is _____ over the integers if its only factorization must involve either 1 or -1 as one of the factors.

2. A trinomial of the form $ax^2 + bx + c$ is called a _____ trinomial.

3. In the trinomial $ax^2 + bx + c$:
   a. The second-degree term ax^2 is called the _____ term.
   b. The first-degree term bx is called the _____ term.
   c. The term c is called the _____ term.

4. A trinomial $x^2 + bx + c$ is factorable over the integers if there are two integers whose _____ is c and whose _____ is b.

5. If the constant c is positive in the trinomial $x^2 + bx + c$, the factors of c must have the _____ sign.

6. If the constant c is negative in the trinomial $x^2 + bx + c$, the factors of c must have _____ signs.

7. Factoring $x^2 + bx + c$ directly without writing any steps to determine the factors of c whose sum is b is called factoring by _____.

6.2 Quick Review

1. Multiply and simplify $(x + 3)(x - 5)$.

2. Multiply and simplify $(x + 3y)(x - 5y)$.

3. Multiply and simplify $x(x - 5)(x - 2)$.

The factor pairs of 33 are $1 \cdot 33$ and $3 \cdot 11$. In Exercises 4 and 5, write all possible factor pairs of each integer.

4. 35

5. 48

6.2 Exercises

Objective 1 Factor Trinomials of the Form $x^2 + bx + c$ by the Trial-and-Error Method or by Inspection

In Exercises 1–4, match each trinomial with the appropriate sign pattern for the factors of this trinomial.

1. $x^2 - 10x - 11$ **A.** $(x + \underline{})(x + \underline{})$

2. $x^2 - 18x + 17$ **B.** $(x + \underline{})(x - \underline{})$

3. $x^2 + 6x + 5$ **C.** $(x - \underline{})(x - \underline{})$

4. $x^2 + 3x - 4$

In Exercises 5–8, use the given sign pattern for the binomial factors to complete the factorization of each trinomial.

5. $x^2 - 6x + 8 = (x - \underline{})(x - \underline{})$

6. $x^2 + 9x + 8 = (x + \underline{})(x + \underline{})$

7. $x^2 - 11x - 12 = (x + \underline{})(x - \underline{})$

8. $x^2 + 4x - 12 = (x + \underline{})(x - \underline{})$

In Exercises 9–34, factor each trinomial.

9. $x^2 + 3x + 2$ **10.** $x^2 + 4x + 3$

11. $x^2 - x - 2$ **12.** $x^2 - 2x - 3$

13. $x^2 + x - 20$ **14.** $x^2 + x - 6$

15. $x^2 - 9x + 20$ **16.** $x^2 - 5x + 6$

17. $x^2 - 7x - 18$ **18.** $x^2 - 3x - 18$

19. $x^2 + 2x - 24$ **20.** $x^2 + 10x - 24$

21. $a^2 - 10a - 11$ **22.** $b^2 + 6b + 5$

23. $y^2 - 18y + 17$ **24.** $x^2 + 3x - 4$

25. $x^2 - 8x + 7$ **26.** $x^2 + 14x + 13$

27. $y^2 - 2y - 99$ **28.** $y^2 + 30y - 99$

29. $p^2 - 29p + 100$ **30.** $p^2 + 20p + 100$

31. $t^2 + 14t + 48$ **32.** $t^2 - 26t + 48$

33. $n^2 + 5n - 36$ **34.** $n^2 - 9n - 36$

Objective 2 Identify a Prime Trinomial of the Form $x^2 + bx + c$

In Exercises 35–40, use the given tables to assist you in identifying which trinomials are prime and factoring the trinomials that can be factored over the integers.

35.

Factors of 12		Sum of Factors
1	12	13
2	6	8
3	4	7

a. $x^2 + 7x + 12$ **b.** $x^2 + 8x + 12$

c. $x^2 + 10x + 12$ **d.** $x^2 + 13x + 12$

36.

Factors of −12		Sum of Factors
−1	12	11
−2	6	4
−3	4	1

a. $x^2 + x - 12$ **b.** $x^2 + 2x - 12$

c. $x^2 + 4x - 12$ **d.** $x^2 + 11x - 12$

37.

Factors of −28		Sum of Factors
1	−28	−27
2	−14	−12
4	−7	−3

a. $x^2 - 3x - 28$ **b.** $x^2 - 12x - 28$

c. $x^2 - 15x - 28$ **d.** $x^2 - 27x - 28$

38.

Factors of 28		Sum of Factors
−1	−28	−29
−2	−14	−16
−4	−7	−11

a. $x^2 - 10x + 28$ **b.** $x^2 - 11x + 28$

c. $x^2 - 16x + 28$ **d.** $x^2 - 29x + 28$

39.

Factors of −36		Sum of Factors
−1	36	35
−2	18	16
−3	12	9
−4	9	5
−6	6	0

a. $x^2 - 36$ **b.** $x^2 + 9x - 36$

c. $x^2 + 16x - 36$ **d.** $x^2 + 18x - 36$

40.

Factors of 36		Sum of Factors
1	36	37
2	18	20
3	12	15
4	9	13
6	6	12

a. $x^2 + 36$ **b.** $x^2 + 12x + 36$

c. $x^2 + 13x + 36$ **d.** $x^2 + 20x + 36$

Objective 3 Factor Trinomials of the Form $x^2 + bxy + cy^2$ by the Trial-and-Error Method or by Inspection

In Exercises 41–48, factor each trinomial.

41. $x^2 + 5xy + 6y^2$ **42.** $x^2 - 7xy + 6y^2$

43. $x^2 + xy - 6y^2$ **44.** $x^2 - 3xy - 10y^2$

45. $x^2 - 10xy + 25y^2$ **46.** $x^2 - 14xy + 49y^2$

47. $a^2 - 16ab - 36b^2$ **48.** $a^2 + 5ab - 36b^2$

Review and Concept Development

In Exercises 49–54, first factor −1 out of each polynomial and then complete the factorization.

49. $-x^2 + 2x + 35$ **50.** $-x^2 + 34x + 35$

51. $-m^2 - 3m + 18$ **52.** $-m^2 + m + 12$

53. $-x^2 + 12xy + 45y^2$ **54.** $-x^2 - xy + 20y^2$

In Exercises 55–64, first factor out the GCF and then complete the factorization.

55. $5x^3 + 15x^2 + 10x$ **56.** $7y^3 - 14y^2 - 21y$

57. $4ax^3 - 20ax^2 + 24ax$ **58.** $7bx^3 - 56bx^2 - 63bx$

59. $10az^2 + 290az + 1{,}000a$ **60.** $4xy^2 + 8xy - 192x$

61. $2abx^2 + 4abx - 96ab$ **62.** $33v^2w - 99vw + 66w$

63. $(a + b)x^2 - 6(a + b)x + 6(a + b)$

64. $(2a - 3)y^2 + 9(2a - 3)y - 22(2a - 3)$

In Exercises 65–68, some of the polynomials are factorable over the integers and others are prime over the integers. Factor those that are factorable, and label the others as prime.

65. a. $x^2 - 3x - 18$ **66. a.** $x^2 + 4x + 45$
   b. $x^2 - 3x + 18$ **b.** $x^2 + 4x - 45$

67. a. $x^2 + 13xy + 12y^2$ **68. a.** $x^2 + 8xy - 15y^2$
   b. $x^2 + 13xy - 12y^2$ **b.** $x^2 + 8xy + 15y^2$

In Exercises 69 and 70, use the given table for $y = P(x)$ to:
a. Give the zeros of $P(x)$.
b. Factor the polynomial $P(x)$.

69.

x	$y = x^2 + 14x + 45$
−10	5
−9	0
−8	−3
−7	−4
−6	−3
−5	0
−4	5

70.

x	$y = x^2 - 16x + 63$
5	8
6	3
7	0
8	−1
9	0
10	3
11	8

In Exercises 71 and 72, use the given graph of $y = P(x)$ to:
a. Give the x-intercepts of the graph of $y = P(x)$.
b. Factor the polynomial $P(x)$.

71.

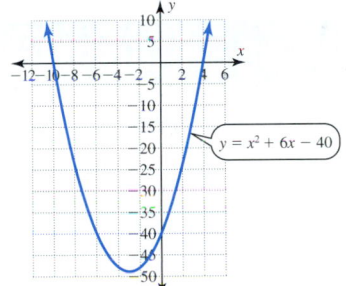

$y = x^2 + 6x - 40$

72.

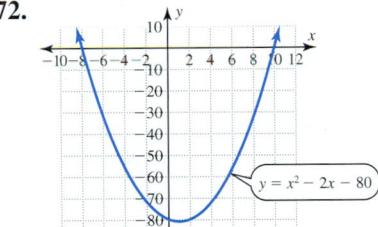

$y = x^2 - 2x - 80$

In Exercises 73 and 74, complete the following table for each polynomial.

Polynomial $P(x)$	Factored Form	Zeros of $P(x)$	x-intercepts of the Graph of $y = P(x)$
73. $x^2 + 2x - 15$			
74. $x^2 - 17x + 60$			

Comparison of Expanding and Factoring

In Exercises 75 and 76, expand and simplify the expression in the left column and factor the expression in the right column.

Expand	Factor
75. $x(x-3) - 2(x-3)$	$x(x-3) - 2(x-3)$
76. $x(x+1) - 8(x+1)$	$x(x+1) - 8(x+1)$

Group discussion questions

77. Discovery Question Determine the zeros of each polynomial function.

a. $y = (2x+1)(3x-1)$ b. $y = (4x+1)(3x-2)$

c. $y = (4x-3)(5x+2)$

Use your observations from parts **a** to **c** to write in factored form a polynomial function with the given zeros.

d. $\dfrac{2}{3}$ and $\dfrac{1}{5}$ e. $-\dfrac{1}{2}$ and $-\dfrac{3}{5}$

f. $\dfrac{2}{5}$ and $-\dfrac{1}{7}$

78. Challenge Question List all polynomials of the form $x^2 + ax + 48$ that have a as an integer and can be factored over the integers.

79. Challenge Question List all polynomials of the form $x^2 + ax - 36$ that have a as an integer and can be factored over the integers.

6.2	Cumulative Review

1. Solve the equation $\dfrac{2x-7}{3} = \dfrac{x+4}{2}$.

2. Write in slope-intercept form the equation of the line that passes through $(-2, 5)$ and $(-5, 9)$.

3. Solve the inequality $|x+8| \le 9$. Give the answer in interval notation.

4. Simplify $(3x-2)^2 - (2x-3)^2$.

5. Factor the GCF out of $12x^4y^3 + 9x^3y^4 + 3x^2y^2$.

Section 6.3	Factoring Trinomials of the Form $ax^2 + bx + c$

Objectives:

1. Factor trinomials of the form $ax^2 + bx + c$ by the trial-and-error method.
2. Factor trinomials of the form $ax^2 + bx + c$ by the AC method.
3. Identify a prime trinomial of the form $ax^2 + bx + c$.

In Section 6.2, we factored some trinomials of the form $x^2 + bx + c$ by the trial-and-error method or by inspection. We now focus on methods for factoring trinomials of the form $ax^2 + bx + c$ where the leading coefficient, a, is not 1. Note that $x^2 + bx + c$ also fits the form $ax^2 + bx + c$ with $a = 1$. Therefore, any of the methods we discuss here will also work for $x^2 + bx + c$. Recall that a polynomial is prime over the integers if its only factorization must involve either 1 or -1 as one of the factors.

In this section we will continue to develop the trial-and-error method. Then we will present a method known as the AC method or the AC-grouping method.

Why Are Two Algebraic Methods Presented for Factoring $ax^2 + bx + c$?

Both methods are widely used and each has advantages. Since your instructor knows you and your curriculum better than anyone else, we suggest that you follow your instructor's advice on which method to use. The important thing is that you develop a method that works for you as efficiently as possible.

1. Factor Trinomials of the Form $ax^2 + bx + c$ by the Trial-and-Error Method

Where Do I Start to Factor $ax^2 + bx + c$ Using the Trial-and-Error Method?

After you have factored out the GCF, start by looking at the first term and the last term. This logic is developed further by examining the following four products.

		Product is $6x^2$			Product is 5		
Factors		F	O	I	L		Product
$(6x + 1)(x + 5)$	$=$	$6x^2$	$+30x$	$+x$	$+5$	$=$	$6x^2 + 31x + 5$
$(6x + 5)(x + 1)$	$=$	$6x^2$	$+6x$	$+5x$	$+5$	$=$	$6x^2 + 11x + 5$
$(2x + 5)(3x + 1)$	$=$	$6x^2$	$+2x$	$+15x$	$+5$	$=$	$6x^2 + 17x + 5$
$(2x + 1)(3x + 5)$	$=$	$6x^2$	$+10x$	$+3x$	$+5$	$=$	$6x^2 + 13x + 5$

Sum is the linear term

Thus, a factorization of $ax^2 + bx + c$ must match a pair of factors of a with a pair of factors of c in order to produce a middle term with a coefficient of b. Selecting the proper matching of these factors is where the trial and error comes in.

Factoring Trinomials of the Form $ax^2 + bx + c$

Algebraically	Verbally	Examples
$ax^2 + bx + c = (a_1x + c_1)(a_2x + c_2)$ where $a_1a_2 = a$, $c_1c_2 = c$, and $a_1c_2 + a_2c_1 = b$.	A trinomial $ax^2 + bx + c$ is factorable into a pair of binomial factors with integer coefficients if and only if there is a pair of factors of a that can be matched with a pair of factors of c to produce a middle term with a coefficient of b. Otherwise, the trinomial is prime over the integers.	$6x^2 + 13x + 5 = (2x + 1)(3x + 5)$ where $2 \cdot 3 = 6$, $1 \cdot 5 = 5$, and $2 \cdot 5 + 3 \cdot 1 = 13$. $6x^2 + 100x + 5$ is prime because there are no factor pairs of 6 and 5 that can be matched to produce a middle term of $100x$.

In Example 1, we factor a trinomial with a positive constant term. When considering possible factors of $5x^2 + 36x + 7$, we consider only possibilities for which the product of the first terms is $5x^2$ and the product of the last terms is 7. To select the factors that produce the correct middle term may require some trial and error. Using the FOIL procedure from Section 5.5 may allow you to select the factors more efficiently. Note that we list the factors of 5 only once but the factors of 7 twice so that we can create all possible matches between the pairs of factors.

Example 1 Factoring a Trinomial by the Trial-and-Error Method

Factor $5x^2 + 36x + 7$.

Solution

$5x^2 + 36x + 7 = (\underline{\quad}x + \underline{\quad})(\underline{\quad}x + \underline{\quad})$

Product is $5x^2$ Product is 7

Since both the linear term and the constant term are positive, set up a form with the sign pattern shown. Then examine all possible factor pairs of 5 and 7. Be sure to list all possible pairs in the second column so a combination is not overlooked. The correct factorization will result in a linear term of $36x$.

Possible Factors	Resulting Linear Term
$(x + 1)(5x + 7)$	$12x$
$(x + 7)(5x + 1)$	$36x$

Factors of 5		Factors of 7	
☞ 1	5	1	7
		☞ 7	1

Answer: $5x^2 + 36x + 7 = (x + 7)(5x + 1)$

Check: $(x + 7)(5x + 1) = 5x^2 + x + 35x + 7$
$= 5x^2 + 36x + 7$

The factored form of $5x^2 + 36x + 7$ from Example 1 could also be expressed as $(5x + 1)(x + 7)$ since multiplication is commutative. Example 1 also verifies that $5x^2 + 2x + 7$ is prime because the middle term of $2x$ is not among the possibilities listed for resulting linear terms.

Can I Also Use the Sign Pattern to Factor $ax^2 + bx + c$?

Yes; after factoring out the GCF, use the signs of b and c to determine the correct sign pattern of the binomial factors.

Sign Pattern for the Factors of $ax^2 + bx + c$, $a < 0$

Verbally	Algebraically	Examples
If the constant term is positive, the factors of this term must have the same sign. These factors will share the same sign as the linear term.	$ax^2 + bx + c = (\_\_x + \_\_)(\_\_x + \_\_)$ or $ax^2 - bx + c = (\_\_x - \_\_)(\_\_x - \_\_)$	$2x^2 + 7x + 6 = (2x + 3)(x + 2)$ $2x^2 - 7x + 6 = (2x - 3)(x - 2)$
If the constant term is negative, the factors of this term must be opposite in sign. The factor with the negative sign will depend on the right combination of factors of a and c.	$ax^2 + bx - c = (\_\_x + \_\_)(\_\_x - \_\_)$ or $ax^2 - bx - c = (\_\_x + \_\_)(\_\_x - \_\_)$	$2x^2 + 11x - 6 = (2x - 1)(x + 6)$ $2x^2 - 11x - 6 = (2x + 1)(x - 6)$

Do I Need to List All of the Factors in a Table If I Can Determine the Factors Without Doing This?

No, in Example 2 we stop the list as soon as the correct factors are determined.

Example 2 — Factoring a Trinomial by the Trial-and-Error Method

Factor $4a^2 - 5a - 6$.

Solution

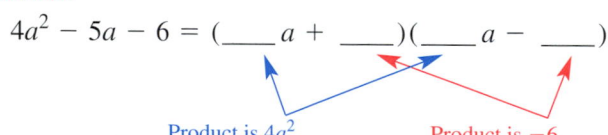

$$4a^2 - 5a - 6 = (\_\_\_a + \_\_\_)(\_\_\_a - \_\_\_)$$

Product is $4a^2$ Product is -6

First set up a form with the correct sign pattern.

Then examine all possible factor pairs of 4 and -6. The correct factorization will result in a linear term of $-5a$. Be sure to list all possible pairs in the second column so a combination is not overlooked.

Possible Factors	Resulting Linear Term
$(a + 1)(4a - 6)$	$-2a$
$(a - 1)(4a + 6)$	$2a$
$(a + 2)(4a - 3)$	$5a$
$(a - 2)(4a + 3)$	$-5a$

Factors of 4	
☞ 1	4
2	2

Factors of -6	
1	-6
-1	6
2	-3
☞ -2	3
3	-2
-3	2

Answer: $4a^2 - 5a - 6 = (a - 2)(4a + 3)$

Check: $(a - 2)(4a + 3) = 4a^2 + 3a - 8a - 6$
$= 4a^2 - 5a - 6$

Self-Check 2

Factor $9b^2 + 5b - 4$.

Because the GCF of $4a^2 - 5a - 6$ in Example 2 is 1, the GCF of each binomial factor must also be 1. Therefore you can omit all possibilities that have a binomial factor with a GCF other than 1. The first two possible factorizations of $4a^2 - 5a - 6$, $(a + 1)(4a - 6)$ and $(a - 1)(4a + 6)$, could have been ruled out before computing the linear term because each contains a factor with a GCF other than 1.

In Example 3, we shorten the process of factoring by listing only the factors of a and c. In this example, the constant term is positive, but the linear term is negative.

Example 3   Factoring a Trinomial by the Trial-and-Error Method

Factor $12m^2 - 29m + 15$.

Solution

$12m^2 - 29m + 15 = (\underline{}m - \underline{})(\underline{}m - \underline{})$ First set up a form with the correct sign pattern.

Factors of 12		Factors of 15	
1	12	-1	-15
2	6	-15	-1
☞ 3	4	-3	-5
		☞ -5	-3

The possibilities that can be used to fill in this form will come from the factor pairs of 12 and 15. Examine these possibilities until we find the combination that will result in a linear term of $-29m$.

With practice, you can perform the FOIL procedure mentally to find the middle term that would result from each combination.

Answer: $12m^2 - 29m + 15 = (3m - 5)(4m - 3)$

Check:

$(3m - 5)(4m - 3) = 12m^2 - 9m - 20m + 15$
$ = 12m^2 - 29m + 15$

Self-Check 3

Factor $10n^2 - 53n + 15$.

2. Factor Trinomials of the Form $ax^2 + bx + c$ by the AC Method

An alternative to the trial-and-error method for factoring trinomials of the form $ax^2 + bx + c$ is a method known as the **AC-grouping method** or the **AC method.** The steps used to factor a trinomial by the AC method are the same steps used to multiply two binomials—but in reverse order. This is illustrated below.

Multiplying Two Binomials

$(2x + 3)(6x - 5) = 2x(6x - 5) + 3(6x - 5)$
$ = 12x^2 - 10x + 18x - 15$
$ = 12x^2 + 8x - 15$

Factoring a Trinomial

$12x^2 + 8x - 15 = 12x^2 - 10x + 18x - 15$
$ = 2x(6x - 5) + 3(6x - 5)$
$ = (2x + 3)(6x - 5)$

Note that in the factorization of $12x^2 + 8x - 15$, the first step is to rewrite $8x$ as $-10x + 18x$. This is the key step in the AC method. When we rewrite $ax^2 + bx + c$ to split bx into two linear terms, the product of the coefficients of these two new linear terms must be ac. The fact that the product of these coefficients must be ac is the crucial clue to using the AC method.

Verification that the product of the linear coefficients must be *ac*

Examine $ax^2 + bx + c = (a_1x + c_1)(a_2x + c_2)$ where $a_1a_2 = a$, $c_1c_2 = c$, and $a_1c_2 + a_2c_1 = b$.

If the linear term $bx = a_1c_2x + a_2c_1x$, then the product of the coefficients of these two linear terms is $a_1c_2a_2c_1 = ac$.

How Does Knowing That the Product of the Coefficients of the Two Linear Terms Is *ac* Help Us to Factor $ax^2 + bx + c$?

It provides a complete list of possibilities to check. In Example 4, the trinomial $6x^2 + 17x + 7$ has $6 \cdot 7 = 42$. Thus examining all possible factor pairs of 42 will enable us either to factor this trinomial or to determine that it is prime.

Example 4 Factoring a Trinomial Using the AC Method

Factor $6x^2 + 17x + 7$.

Solution

Factors of 42		Sum of Factors
1	42	43
2	21	23
3	14	17

Note that $ac = 6 \cdot 7 = 42$. Then examine the factor pairs of 42 for a pair of factors whose sum is 17.

The only pair of factors of 42 with a sum of 17 is the pair of 3 and 14. The list was stopped as soon as the correct pair of factors was found.

$$6x^2 + 17x + 7 = 6x^2 + 3x + 14x + 7$$
$$= 3x(2x + 1) + 7(2x + 1)$$
$$= (3x + 7)(2x + 1)$$

Rewrite the linear term $17x$ as $3x + 14x$.

Use the distributive property to factor $3x$ out of the first pair of terms and 7 out of the last pair of terms.

Then factor out the GCF of $2x + 1$.

Does this factorization check?

Answer: $6x^2 + 17x + 7 = (3x + 7)(2x + 1)$

Self-Check 4

Factor $15x^2 + 13x + 2$.

The AC method used to factor the trinomial in Example 4 is summarized in the box.

Factoring Trinomials of the Form $ax^2 + bx + c$ by the AC Method

Procedure	Example
Step 1. Factor out the GCF. If a is negative, factor out -1.	Factor $6x^2 - x - 12$.
Step 2. Find a pair of factors of ac whose sum is b. If there is not a pair of factors whose sum is b, the trinomial is prime over the integers.	**Factors of -72** **Sum of Factors**

Procedure	Factors of -72	Sum of Factors
	1 -72	-71
	2 -36	-34
	3 -24	-21
	4 -18	-14
	6 -12	-6
	8 -9	-1

- If the constant c is positive, the factors of ac must have the same sign. These factors will share the same sign as the linear coefficient b.
- If the constant c is negative, the factors of ac must be opposite in sign. The sign of b will determine the sign of the factor with the larger absolute value.

Step 3. Rewrite the linear term of $ax^2 + bx + c$ so that b is the sum of the pair of factors from step 2.

Step 4. Factor the polynomial from step 3 by grouping the terms and factoring the GCF out of each pair of terms.

$$6x^2 - x - 12 = 6x^2 + 8x - 9x - 12$$
$$= 2x(3x + 4) - 3(3x + 4)$$
$$= (2x - 3)(3x + 4)$$

The AC method will work for the trinomials we factored in Section 6.2. The trinomials in that section were of the form $ax^2 + bx + c$ with $a = 1$.

Example 5 provides further practice using the AC method to factor trinomials of the form $ax^2 + bx + c$. In $15x^2 - 41x + 14$, b is negative and ac is positive, so both factors of ac must be negative.

Example 5   Factoring a Trinomial Using the AC Method

Factor $15x^2 - 41x + 14$.

Solution

Factors of 210		Sum of Factors
-1	-210	-211
-2	-105	-107
-3	-70	-73
-5	-42	-47
-6	-35	-41

Note that $ac = 15 \cdot 14 = 210$. The constant term 14 is positive and the linear coefficient is -41. Thus both factors of 210 must be negative. Examine the factor pairs of 210 for a pair of factors whose sum is -41.

The only pair of factors of 210 with a sum of -41 is the pair of -6 and -35. The list was stopped as soon as the correct pair of factors was found.

$$15x^2 - 41x + 14 = 15x^2 - 6x - 35x + 14$$
$$= 3x(5x - 2) - 7(5x - 2)$$
$$= (3x - 7)(5x - 2)$$

Rewrite the linear term $-41x$ as $-6x - 35x$.

Use the distributive property to factor $3x$ out of the first pair of terms and -7 out of the last pair of terms.

Then factor out the GCF of $5x - 2$.

Answer: $15x^2 - 41x + 14 = (3x - 7)(5x - 2)$ Does this factorization check?

Self-Check 5

Factor $15x^2 - 31x + 14$.

---

3. Identify a Prime Trinomial of the Form $ax^2 + bx + c$

In Example 6, we use the AC method to confirm that $6x^2 + 15x + 5$ is prime. In the trinomial $6x^2 + 15x + 5$, the product ac is 30, so we are in search of a pair of factors of 30 whose sum is 15. An alternative would be to use the trial-and-error method to examine each of the possible pairs of binomial factors.

Example 6   Identifying a Prime Trinomial by Using the AC Method

Verify that $6x^2 + 15x + 5$ is prime.

Solution

Factors of 30		Sum of Factors
1	30	31
2	15	17
3	10	13
5	6	11

First note that $ac = 6 \cdot 5 = 30$.

Then examine all of the factor pairs of 30 for a pair whose sum is 15.

Answer: $6x^2 + 15x + 5$ is prime.   Since there is not a pair of factors of 30 that has a sum of 15, this trinomial is prime.

> **Self-Check 6**
>
> Given $4x^2 + 9x + 5$ and $4x^2 + 10x + 5$, which one of these trinomials is prime?

In Example 6, we confirmed that $6x^2 + 15x + 5$ is prime. The table from Example 6 also indicates that if the linear term had a coefficient of 11, 13, 17, or 31 instead of 15, then we would be able to factor this trinomial. In fact, a quadratic trinomial with a first term of $6x^2$ and a constant term of 5 is factorable over the integers only if the middle term has one of these coefficients or their opposites.

Any time the leading coefficient is negative, it is a good idea to factor out -1 or -1 times the GCF so the leading coefficient will be positive. In Example 7, we factor -1 out of a trinomial with a leading coefficient of -5.

Example 7   Factoring a Trinomial Using the AC Method

Factor $-5a^2 + 3a + 2$.

Solution

$$-5a^2 + 3a + 2 = -1(5a^2 - 3a - 2)$$ First factor out -1.

Factors of -10		Sum of Factors
1	-10	-9
2	-5	-3

For the trinomial $5a^2 - 3a - 2$ note that $ac = 5(-2) = -10$.

Examine the factor pairs of -10 for a pair of factors whose sum is -3. Because the linear term has a negative coefficient, we use a negative for the factor with the larger absolute value.

$$
\begin{aligned}
-5a^2 + 3a + 2 &= -(5a^2 - 3a - 2) \\
&= -(5a^2 + 2a - 5a - 2) \\
&= -[a(5a + 2) - 1(5a + 2)] \\
&= -(a - 1)(5a + 2)
\end{aligned}
$$

Rewrite the linear term $-3a$ as $2a - 5a$.

Use the distributive property to factor a out of the first pair of terms and -1 out of the last pair of terms. Then factor out the GCF of $5a + 2$.

Answer: $-5a^2 + 3a + 2 = -(a - 1)(5a + 2)$ **Check:**
$$
\begin{aligned}
-(a - 1)(5a + 2) &= -(5a^2 + 2a - 5a - 2) \\
&= -(5a^2 - 3a - 2) \\
&= -5a^2 + 3a + 2
\end{aligned}
$$

Self-Check 7

Factor $-7b^2 - 12b + 4$.

If you want additional practice with the trial-and-error method, we suggest that you reexamine Examples 4–7 using this method. On the other hand, if you want additional practice with the AC method, rework Examples 8 and 9 using this method. In Example 8, we factor a trinomial with two variables by the trial-and-error method.

Example 8 Factoring a Trinomial with Two Variables by the Trial-and-Error Method

Factor $6v^2 - 19vw - 7w^2$.

Solution

$6v^2 - 19vw - 7w^2 = (\underline{} v + \underline{} w)(\underline{} v - \underline{} w)$ Set up the correct sign pattern.

Factors of 6		Factors of −7	
1	6	☞ 1	−7
2	3	7	−1
☞ 3	2		
6	1		

The possibilities that can be used to fill in this form will come from the factors of 6 and −7. Note that we include all combinations of the factors that can occur in the sign pattern. Examine these possibilities until we find the combination that will result in a linear term of $-19vw$.

Answer:

$6v^2 - 19vw - 7w^2 = (3v + w)(2v - 7w)$

Check: $(3v + w)(2v - 7w) = 6v^2 - 21vw + 2vw - 7w^2$
$= 6v^2 - 19vw - 7w^2$

Self-Check 8

Factor $11v^2 - 20vw - 4w^2$.

If the terms of a trinomial have a GCF other than 1, then the first step in factoring this trinomial is to factor out the GCF. In Example 9, we factor a trinomial with two variables by trial and error.

Example 9 Completely Factoring a Trinomial by the Trial-and-Error Method

Factor $22ax^2 - 28axy + 6ay^2$.

Solution

$22ax^2 - 28axy + 6ay^2 = 2a(11x^2 - 14xy + 3y^2)$ First factor out $2a$ from each term.

$= 2a(\underline{} x - \underline{} y)(\underline{} x - \underline{} y)$ Then set up a form with the correct sign pattern for the remaining trinomial.

Factors of 11		Factors of 3	
☞ 1	11	☞ −1	−3
		−3	−1

The possibilities that can be used to fill in this form will come from the factors of 11 and 3. We examine these possibilities until we find the combination that will result in a linear term of $-14xy$.

Answer:

$22ax^2 - 28axy + 6ay^2 = 2a(x - y)(11x - 3y)$

Check:

$2a(x - y)(11x - 3y) = 2a(11x^2 - 3xy - 11xy + 3y^2)$
$= 2a(11x^2 - 14xy + 3y^2)$
$= 22ax^2 - 28axy + 6ay^2$

Self-Check 9

Factor $60x^3 - 35x^2 - 60x$.

Summary: Sections 6.1–6.3 have covered a variety of methods for factoring trinomials of the form $ax^2 + bx + c$. The accompanying box lists some of the advantages and disadvantages of each method.

Methods for Factoring $ax^2 + bx + c$	Advantages	Disadvantages
Graphs	Visually displays the x-intercepts that correspond to the factors of the trinomial. If there are no x-intercepts, this guarantees that the polynomial is prime over the integers.	Can be time-consuming to select the appropriate viewing window and to approximate the x-intercepts. Because the x-intercepts are approximated, these factors should be checked.
Tables	Can see the zeros of the trinomial and can observe numerical patterns that are important in many applications. Spreadsheets allow us to use the power of computers to exploit this method.	Often requires insight or some trial and error to select the most appropriate table.
AC Method	This is a precise step-by-step process that can factor any trinomial of the form $ax^2 + bx + c$ or can identify the trinomial as prime. This method has the same steps used to multiply binomial factors but the steps are reversed.	Many trinomials with small integer coefficients can be factored by inspection, and it is not necessary to write the table and all the steps of this method.
Trial-and-Error Method	Takes advantage of patterns and insights to quickly factor trinomials with small integer coefficients. Observing the mathematical patterns in these trinomials can improve your foundation for other topics.	For novices, this process can be very challenging. It is important to examine all possibilities before deciding the trinomial is prime.

Self-Check Answers

1. $(3x + 11)(x + 1)$
2. $(9b - 4)(b + 1)$
3. $(10n - 3)(n - 5)$
4. $(5x + 1)(3x + 2)$
5. $(3x - 2)(5x - 7)$
6. $4x^2 + 10x + 5$ is prime.
7. $-(7b - 2)(b + 2)$
8. $(11v + 2w)(v - 2w)$
9. $5x(3x - 4)(4x + 3)$

6.3   Using the Language and Symbolism of Mathematics

1. The first step in factoring $ax^2 + bx + c$ is to factor out the _____ if it is not 1.

2. When factoring the trinomial $ax^2 + bx + c$ by trial and error, the sign pattern will be
(___ x + ___)(___ x + ___) or
(___ x − ___)(___ x − ___) when c is _____.

3. When factoring the trinomial $ax^2 + bx + c$ by trial and error, the sign pattern will be
(___ x + ___)(___ x − ___) when c is _____.

4. When factoring the trinomial $ax^2 + bx + c$ by the AC method, we use a pair of factors of ac whose sum is _____.

5. When we factor the trinomial $ax^2 + bx + c$, the factors of ac are of the _____ sign when $ac > 0$.

6. When we factor the trinomial $ax^2 + bx + c$, the factors of ac are of _____ sign when $ac < 0$.

7. A polynomial is prime over the integers if its only factorization must involve either _____ or _____ as one of the factors.

| **6.3** | **Quick Review** |

1. Multiply and simplify $(2x + 7)(3x - 1)$.

2. Multiply and simplify $(5x + 4y)(2x + 7y)$.

3. Multiply and simplify $(2x + 5)(3x - 4)$.

4. Multiply and simplify $-3xy(3x + y)(6x - 5y)$.

5. Factor out the GCF of $2x(3x - 5) + 7(3x - 5)$.

| **6.3** | **Exercises** |

Objective 1 Factor Trinomials of the Form $ax^2 + bx + c$ by the Trial-and-Error Method

In Exercises 1–4, match each trinomial with the appropriate sign pattern for the factors of this trinomial.

1. $16x^2 - 50x - 21$ **A.** $(\underline{\quad}x + \underline{\quad})(\underline{\quad}x + \underline{\quad})$

2. $16x^2 + 10x - 21$ **B.** $(\underline{\quad}x + \underline{\quad})(\underline{\quad}x - \underline{\quad})$

3. $16x^2 + 38x + 21$ **C.** $(\underline{\quad}x - \underline{\quad})(\underline{\quad}x - \underline{\quad})$

4. $16x^2 - 50x + 21$

In Exercises 5–8, fill in the missing $+$ or $-$ symbols. Then write your answer as the complete factorization of each trinomial.

5. a. $3x^2 + 4x + 1 = (3x\underline{\quad}1)(x\underline{\quad}1)$
   b. $3x^2 - 4x + 1 = (3x\underline{\quad}1)(x\underline{\quad}1)$

6. a. $3x^2 + 7x + 2 = (3x\underline{\quad}1)(x\underline{\quad}2)$
   b. $3x^2 - 7x + 2 = (3x\underline{\quad}1)(x\underline{\quad}2)$

7. a. $2x^2 + x - 1 = (2x\underline{\quad}1)(x\underline{\quad}1)$
   b. $2x^2 - x - 1 = (2x\underline{\quad}1)(x\underline{\quad}1)$

8. a. $3x^2 + 5x - 2 = (3x\underline{\quad}1)(x\underline{\quad}2)$
   b. $3x^2 - x - 2 = (3x\underline{\quad}2)(x\underline{\quad}1)$

In Exercises 9–16, fill in the missing information. Then write your answer as the complete factorization of each trinomial.

9. $5x^2 + 9x - 2 = (5x - \underline{\quad})(x + \underline{\quad})$

10. $5x^2 - 3x - 2 = (5x + \underline{\quad})(x - \underline{\quad})$

11. $5x^2 - 7x + 2 = (5x - \underline{\quad})(x - \underline{\quad})$

12. $5x^2 - 11x + 2 = (5x - \underline{\quad})(x - \underline{\quad})$

13. $6v^2 - 5v - 6 = (3v + 2)(\underline{\quad} - \underline{\quad})$

14. $6v^2 + 5v - 6 = (3v - 2)(\underline{\quad} + \underline{\quad})$

15. $6v^2 - 37v + 6 = (6v - 1)(\underline{\quad} - \underline{\quad})$

16. $6v^2 - 35v - 6 = (6v + 1)(\underline{\quad} - \underline{\quad})$

In Exercises 17–26, factor each trinomial by the trial-and-error method.

17. $5x^2 + 6x + 1$ **18.** $11x^2 + 12x + 1$

19. $7x^2 - 8x + 1$ **20.** $2x^2 - 3x + 1$

21. $7x^2 - 6x - 1$ **22.** $5x^2 - 4x - 1$

23. $11x^2 + 10x - 1$ **24.** $7x^2 + 6x - 1$

25. $2x^2 + 5x + 3$ **26.** $2x^2 + 7x + 3$

Objective 2 Factor Trinomials of the Form $ax^2 + bx + c$ by the AC Method

In Exercises 27–30, write the factorization of each trinomial.

27. $3a^2 + 17a + 10 = 3a^2 + 15a + 2a + 10$

28. $3a^2 - 11a + 10 = 3a^2 - 6a - 5a + 10$

29. $3m^2 - 10m - 8 = 3m^2 - 12m + 2m - 8$

30. $3m^2 + 23m - 8 = 3m^2 - m + 24m - 8$

In Exercises 31 and 32, use the given table to assist you in factoring each trinomial by the AC method.

31.

Factors of 24		Sum of Factors
1	24	25
2	12	14
3	8	11
4	6	10

a. $3m^2 + 10m + 8$ **b.** $3m^2 + 11m + 8$

c. $3m^2 + 14m + 8$ **d.** $3m^2 + 25m + 8$

32.

Factors of 24		Sum of Factors
-1	-24	-25
-2	-12	-14
-3	-8	-11
-4	-6	-10

a. $8v^2 - 14v + 3$ **b.** $6v^2 - 11v + 4$

c. $12v^2 - 25v + 2$ **d.** $6v^2 - 25v + 4$

In Exercises 33–54, factor each trinomial by the AC method.

33. $2x^2 + 5x - 7$ **34.** $11m^2 + 100m + 9$

35. $4w^2 - 9w + 5$ **36.** $9w^2 + 9w - 10$

37. $9z^2 - 9z - 10$ **38.** $3z^2 - 35z + 50$

39. $10b^2 + 29b + 10$ **40.** $15b^2 - 22b + 8$

41. $12m^2 - mn - 35n^2$ **42.** $12m^2 + 13mn - 35n^2$

43. $18y^2 + 55y - 28$ **44.** $18y^2 + 3y - 28$

45. $55y^2 - 29y - 12$ **46.** $24v^2 - 43v + 18$

47. $20x^2 + 37xy + 15y^2$ **48.** $15x^2 - 37xy + 20y^2$

49. $14x^2 - 39xy + 10y^2$ **50.** $15x^2 + 61xy + 22y^2$

51. $33m^2 + 13mn - 6n^2$ **52.** $35m^2 + 4mn - 4n^2$

53. $6v^2 + 13vw + 6w^2$ **54.** $10v^2 - 29vw + 10w^2$

Objective 3 Identify a Prime Trinomial of the Form $ax^2 + bx + c$

In Exercises 55–58, use the given table to assist you in identifying whether each trinomial is prime or can be factored. Factor the trinomials that can be factored over the integers.

55.

Factors of −40		Sum of Factors
−1	40	39
−2	20	18
−4	10	6
−5	8	3

a. $4n^2 + 39n - 10$ **b.** $2n^2 + 3n - 20$
c. $5n^2 + 6n - 8$ **d.** $40n^2 + 8n - 1$

56.

Factors of −40		Sum of Factors
1	−40	−39
2	−20	−18
4	−10	−6
5	−8	−3

a. $4n^2 - 3n - 10$ **b.** $4n^2 - 39n - 10$
c. $5n^2 - 18n - 8$ **d.** $5n^2 - 12n - 8$

57.

Factors of 36		Sum of Factors
−1	−36	−37
−2	−18	−20
−3	−12	−15
−4	−9	−13
−6	−6	−12

a. $3x^2 - 25xy + 12y^2$ **b.** $4x^2 - 15xy + 9y^2$
c. $2x^2 - 13xy + 18y^2$ **d.** $6x^2 - 37xy + 6y^2$

58.

Factors of −36		Sum of Factors
−1	36	35
−2	18	16
−3	12	9
−4	9	5
−6	6	0

a. $3x^2 + 16xy - 12y^2$ **b.** $18x^2 + 9xy - 2y^2$
c. $6x^2 + 11xy - 6y^2$ **d.** $9x^2 + 35xy - 4y^2$

In Exercises 59 and 60, determine whether each trinomial is prime. Factor the trinomials that can be factored over the integers.

59. a. $4x^2 - 23x + 15$ **b.** $4x^2 - 18x + 15$
60. a. $16x^2 + 23x - 9$ **b.** $16x^2 + 18x - 9$

Review and Skill Development

In Exercises 61–66, first factor −1 out of each polynomial and then complete the factorization.

61. $-12x^2 + 20x - 3$ **62.** $-12x^2 - 20x - 3$
63. $-35y^2 + 4y + 4$ **64.** $-35y^2 + 24y - 4$
65. $-12z^2 - 47z - 40$ **66.** $-12z^2 + 17z + 40$

In Exercises 67–76, first factor out the GCF and then completely factor the polynomial.

67. $-38ax^2 + 32ax + 6a$
68. $-51ax^2 - 99ax + 6a$
69. $60x^5 - 145x^4 + 75x^3$
70. $48x^4 + 112x^3 + 60x^2$
71. $15m^3n - 6m^2n^2 - 21mn^3$
72. $44m^3n - 8m^2n^2 - 52mn^3$
73. $210av^3 + 77av^2w - 210avw^2$
74. $350av^2w + 240avw^2 - 350aw^3$
75. $35x^2(a + b) - 29x(a + b) + 6(a + b)$
76. $21x^2(a - b) + 41x(a - b) + 10(a - b)$

In Exercises 77–80, complete the following table for each polynomial.

Polynomial $P(x)$	Factored Form	Zeros of $P(x)$	x-intercepts of the Graph of $y = P(x)$
Example: $2x^2 + 3x - 5$	$(2x + 5)(x - 1)$	$-\dfrac{5}{2}$ and 1	$\left(-\dfrac{5}{2}, 0\right)$, $(1, 0)$
77. $3x^2 - 16x + 5$	$(3x - 1)(x - 5)$		
78. $5x^2 + 9x + 4$	$(5x + 4)(x + 1)$		
79. $3x^2 + 10x - 8$		$\dfrac{2}{3}$ and -4	
80. $4x^2 - 5x - 6$			$\left(-\dfrac{3}{4}, 0\right)$, $(2, 0)$

Group discussion questions

81. Discovery Question Write each of these polynomials as a trinomial in the form $ax^2 + 0x + c$ and then factor this trinomial by inspection.

a. $x^2 - 25$ **b.** $x^2 - 64$
c. $4x^2 - 1$ **d.** $4x^2 - 25$

Use your observations from these problems to write the factors of the following polynomials by inspection.

e. $x^2 - 49$ **f.** $9x^2 - 1$
g. $9x^2 - 16$

Explain what you have observed to your instructor.

82. Discovery Question
a. Use $2x^2 + 3x + 8x + 12$ to factor $2x^2 + 11x + 12$.
b. Use $2x^2 + 8x + 3x + 12$ to factor $2x^2 + 11x + 12$ and compare the results to part **a.**
c. Use $-2x^2 + 3x - 4x + 6$ to factor $-2x^2 - x + 6$.
d. Use $-2x^2 - 4x + 3x + 6$ to factor $-2x^2 - x + 6$ and compare the results to part **c.**
e. What conclusion can you reach from this work?

Explain your conclusion to your instructor.

6.3 | Cumulative Review

Given $x = -2$, $y = 3$, and $z = -5$, evaluate each expression.

1. $2x - 3(y - z)$

2. $-3x^2 + (2y)^2 + 2z^2$

3. Simplify the given expression assuming the base for each exponent is nonzero. $2x^0 + 3y^0 + (2x)^0 + (3y)^0$

4. Solve $|2x + 9| - 3 = 11$ for x.

5. Multiply and simplify $(2x - 3)(4x^2 + 6x + 9)$.

Section 6.4 | Factoring Special Forms

Objectives:

1. Factor perfect square trinomials.
2. Factor the difference of two squares.
3. Factor the sum or difference of two cubes.

The connection of factoring to other topics in mathematics is one of the primary reasons we study factoring. The special forms examined in this section occur frequently in many different contexts throughout mathematics. You can save considerable time and effort by using these special forms. Although you can continue to factor quadratic trinomials by either the trial-and-error method or the AC method, you can gain insight and efficiency by recognizing these patterns.

1. Factor Perfect Square Trinomials

We start by reexamining the special forms first introduced in Section 5.6. A perfect square trinomial factors as the square of a binomial.

Perfect Square Trinomials

Algebraically	Verbally	Algebraic Example
Square of a sum $A^2 + 2AB + B^2 = (A + B)^2$ **Square of a difference** $A^2 - 2AB + B^2 = (A - B)^2$	A trinomial that is the square of a binomial has **1.** A first term that is the square of the first term of the binomial. **2.** A middle term that is twice the product of the two terms of the binomial. **3.** A last term that is the square of the last term of the binomial.	$x^2 + 12x + 36 = (x + 6)^2$ $x^2 - 12x + 36 = (x - 6)^2$

Why Is It Useful to Be Able to Determine Whether a Trinomial Is a Perfect Square Trinomial?

If a trinomial fits this form, then it can be factored quickly using one of the special patterns in the preceding box. If it does not fit this form, then we can still factor it using either the trial-and-error method or the AC method. Example 1 illustrates how to identify perfect square trinomials.

Example 1 Factoring Perfect Square Trinomials

Determine which of these polynomials are perfect square trinomials, and factor those that fit this form.

(a) $25v^2 + 10v + 1$ **(b)** $m^2 - 12m + 36$ **(c)** $m^2 - 13m + 36$
(d) $x^2 - 10x - 25$ **(e)** $x^2 + 8x + 10$

Solution

(a) $25v^2 + 10v + 1 = (5v)^2 + 10v + (1)^2$

Write the first and last terms as perfect squares. Then check to see whether the middle term fits the form $A^2 + 2AB + B^2$.

$\quad\quad\quad\quad\quad\quad\quad = (5v)^2 + 2(5v)(1) + (1)^2$

$\quad\quad\quad\quad\quad\quad\quad = (5v + 1)^2$

Use the form $A^2 + 2AB + B^2 = (A + B)^2$ to write the factored form.

(b) $m^2 - 12m + 36 = m^2 - 12m + (6)^2$

This trinomial fits the form $A^2 - 2AB + B^2 = (A - B)^2$ with $A = m$ and $B = 6$.

$\quad\quad\quad\quad\quad\quad\quad\quad = m^2 - 2(m)(6) + (6)^2$

$\quad\quad\quad\quad\quad\quad\quad\quad = (m - 6)^2$

(c) This trinomial is not a perfect square.

Although both the first and last terms are perfect squares, this trinomial does not fit the form $A^2 - 2AB + B^2$.

To be a perfect square, the middle term would have to be $12m$ or $-12m$. [See part (b).]

(d) This trinomial is not a perfect square.

Because the last term has a negative coefficient, it cannot be a perfect square of an integer.

(e) This trinomial is not a perfect square.

The last term 10 is not a perfect square of an integer.

Self-Check 1

Determine which of these polynomials are perfect square trinomials, and factor those that fit this form.

a. $x^2 - 18x + 81$ **b.** $121w^2 + 22w + 1$
c. $25v^2 + 60v + 36$ **d.** $x^2 - 10x + 16$

---

2. Factor the Difference of Two Squares

Another special form that is encountered frequently in subsequent mathematics courses is the difference of two squares. This form was first examined in Section 5.6.

Difference of Two Squares

Algebraically	Verbally	Algebraic Example
$A^2 - B^2 = (A + B)(A - B)$	The difference of the squares of two terms factors as the sum of these terms times their difference.	$x^2 - 49 = (x + 7)(x - 7)$

In Example 2, we practice identifying and factoring the difference of two squares.

Example 2 Factoring the Difference of Two Squares

Determine which of these polynomials are the difference of two squares, and factor those that fit this form.

(a) $100a^2 - 9$ **(b)** $9v^2 - 16w^2$ **(c)** $9v^2 + w^2$ **(d)** $x^2 - 5$

Solution

(a) $100a^2 - 9 = (10a)^2 - (3)^2$
$\qquad\qquad\quad = (10a + 3)(10a - 3)$

First write this binomial in the form $A^2 - B^2$, and then factor as a sum times a difference.

(b) $9v^2 - 16w^2 = (3v)^2 - (4w)^2$
$\qquad\qquad\quad = (3v + 4w)(3v - 4w)$

(c) This binomial is not the difference of two squares.

This binomial is the sum of two perfect squares, not their difference.

(d) This binomial is not the difference of two squares.

5 is not a perfect square integer. However, you could factor this binomial by using irrational factors as $x^2 - 5 = x^2 - (\sqrt{5})^2 = (x + \sqrt{5})(x - \sqrt{5})$

Self-Check 2

Determine which of these polynomials are the difference of two squares, and factor those that fit this form.

a. $36y^2 - 25$ **b.** $100a^2 - 121b^2$ **c.** $16a^2b^2 - 1$ **d.** $x^2 - 35$

Is the Sum of Two Squares As Useful to Recognize As the Difference of Two Squares?

Yes; consider the binomial $A^2 + B^2$ in the trinomial form $A^2 + 0AB + B^2$. An attempt to factor this trinomial over the integers cannot produce a middle term of 0. Since the coefficient of the last term is positive, the sign of the last term in each binomial factor must be the same. Thus, the middle term cannot be 0. Therefore, the sum of two squares is prime over the integers, and we do not need to spend time trying to factor quadratic binomials with this form.

Sum of Two Squares

Algebraically	Verbally	Algebraic Example
$A^2 + B^2$ is prime*	The sum of two squares is prime.	$16x^2 + 9$ is prime.

*For binomials of degree greater than 2, this statement is not always true.

A Mathematical Note

Pierre de Fermat's last theorem is one of the most famous theorems in mathematics. It states that there do not exist positive integers x, y, z, and n such that $x^n + y^n = z^n$ for $n > 2$. Two of the features that make this theorem so interesting are as follows: (1) It is easily understood—it even resembles the Pythagorean theorem $x^2 + y^2 = z^2$ and thus shows that the Pythagorean theorem does not generalize to a higher-degree equation. (2) It resisted the best attempts of mathematicians to prove until 1993 when Andrew Wiles, a British-American mathematician, proved this result.

Each of the polynomials in Example 3 can be factored by the trial-and-error method or by the AC method. (We can consider $4a^2 - 25b^2$ as $4a^2 + 0ab - 25b^2$.) However, note how much more efficient it is to use the special forms to quickly select the correct factors.

Example 3 | **Factoring Special Forms**

Factor each of these polynomials.

(a) $4a^2 - 20ab + 25b^2$ **(b)** $4a^2 - 25b^2$ **(c)** $4a^2 + 25b^2$

Solution

(a) $4a^2 - 20ab + 25b^2$ First write this trinomial in the form $A^2 - 2AB + B^2$ and
$= (2a)^2 - 2(2a)(5b) + (5b)^2$ then factor this perfect square trinomial.
$= (2a - 5b)^2$

(b) $4a^2 - 25b^2$ First write this binomial in the form $A^2 - B^2$ and then factor
$= (2a)^2 - (5b)^2$ this difference of two squares.
$= (2a + 5b)(2a - 5b)$

(c) $4a^2 + 25b^2$ is prime over the   This binomial can be written as $(2a)^2 + (5b)^2$. As the sum
integers.   of two squares, it must be prime over the integers.

Self-Check 3

Factor each of these polynomials.

a. $169v^2 - 100w^2$ **b.** $169v^2 + 100w^2$ **c.** $169v^2 + 260vw + 100w^2$

---

3. Factor the Sum or Difference of Two Cubes

Two other special forms for binomials are $A^3 + B^3$ and $A^3 - B^3$. These forms are known as the sum of two cubes and the difference of two cubes, respectively. We start by examining $x^3 - 8$, which factors as $x^3 - 8 = (x - 2)(x^2 + 2x + 4)$. This factorization of $x^3 - 8$ can be checked by multiplication.

$$(x - 2)(x^2 + 2x + 4) = x(x^2 + 2x + 4) - 2(x^2 + 2x + 4)$$
$$= x^3 + 2x^2 + 4x - 2x^2 - 4x - 8$$
$$= x^3 - 8$$

Note in the graph of $y = x^3 - 8$ that $(2, 0)$ is an x-intercept of the graph. This is visual confirmation that $x - 2$ is a linear factor of $x^3 - 8$. The graph also suggests that there are no other x-intercepts corresponding to the factor $x^2 + 2x + 4$. Is $x^2 + 2x + 4$ factorable, or is it prime? We use the AC method to determine this.

Does $x^2 + 2x + 4$ factor?

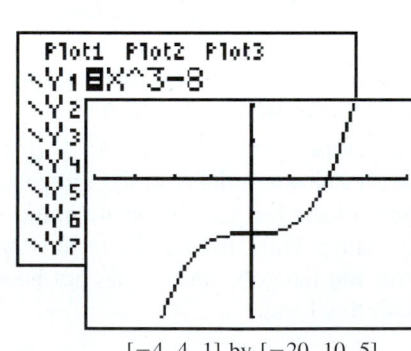

$[-4, 4, 1]$ by $[-20, 10, 5]$

Factors of 4		Sum of Factors
1	4	5
2	2	4

In the trinomial $x^2 + 2x + 4$, $a = 1$ and $c = 4$. Thus $ac = 1 \cdot 4 = 4$.

Examine the factors of 4 for a pair of factors whose sum is 2. None of the pairs of factors of 4 have a sum of 2. Therefore $x^2 + 2x + 4$ is prime over the integers.

Since $x^2 + 2x + 4$ is prime over the integers, the complete factorization of $x^3 - 8$ is $x^3 - 8 = (x - 2)(x^2 + 2x + 4)$.

We now list some other products and factors that fit the form of the sum or difference of two cubes. You may wish to multiply these factors to verify these products.

Products

$(x + 1)(x^2 - x + 1) = x^3 + 1$

$(x - 1)(x^2 + x + 1) = x^3 - 1$

$(x + 2)(x^2 - 2x + 4) = x^3 + 8$

$(x - 2)(x^2 + 2x + 4) = x^3 - 8$

Factors

$x^3 + 1 = (x + 1)(x^2 - x + 1)$

$x^3 - 1 = (x - 1)(x^2 + x + 1)$

$x^3 + 8 = (x + 2)(x^2 - 2x + 4)$

$x^3 - 8 = (x - 2)(x^2 + 2x + 4)$

The pattern exhibited by these polynomials is summarized in the box.

Sum or Difference of Two Cubes

Algebraically	Verbally	Algebraic Examples
$A^3 + B^3 = (A + B)(A^2 - AB + B^2)$ $A^3 - B^3 = (A - B)(A^2 + AB + B^2)$	Write the binomial factor as the sum (or difference) of the two cube roots. Then use the binomial factor to obtain each term of the trinomial: 1. The square of the first term of the binomial is the first term of the trinomial factor. 2. The opposite of the product of the two terms of the binomial is the second term of the trinomial factor. 3. The square of the last term of the binomial is the third term of the trinomial factor.	$x^3 + 8 = (x + 2)(x^2 - 2x + 4)$ $x^3 - 8 = (x - 2)(x^2 + 2x + 4)$

Are There Any Clues That Will Help Me Recognize a Sum or Difference of Two Cubes?

It is helpful to be able to recognize the perfect squares 1, 4, 9, 16, 25, and so on, when you are examining a binomial to determine whether it is a sum or a difference of two squares. Likewise, it is helpful to be able to recognize the perfect cubes 1, 8, 27, 64, 125, and so on, when you are examining a binomial to determine whether it is a sum or a difference of two cubes. We practice this skill in Example 4.

Example 4   Identifying the Sum or Difference of Two Cubes

Examine each binomial to determine whether it is a sum or difference of two cubes or neither.

(a) $x^3 + 64$ **(b)** $x^3 + 16$ **(c)** $27x^3 - 1$ **(d)** $8x^3 - 125$

Solution

(a) This binomial is the sum of two cubes.   In the form $A^3 + B^3$, $x^3 + 64 = x^3 + 4^3$.

(b) This binomial is neither the sum nor the   The last term, 16, is not a perfect cube.
difference of two cubes.

(c) This binomial is the difference of two cubes. $27x^3 - 1 = (3x)^3 - 1^3$

(d) This binomial is the difference of two cubes. $8x^3 - 125 = (2x)^3 - 5^3$

In Examples 5 and 6, note that we first write the binomial factor and then use this binomial to obtain each term of the trinomial factor.

Example 5 Factoring the Sum of Two Cubes

Factor $x^3 + 64$.

Solution

$$x^3 + 64 = x^3 + 4^3$$
$$= (x + 4)[(\underline{\quad})^2 - (\underline{\quad})(\underline{\quad}) + (\underline{\quad})^2]$$
$$= (x + 4)[x^2 - (x)(4) + 4^2]$$
$$= (x + 4)(x^2 - 4x + 16)$$

First express this binomial in the form $A^3 + B^3$. Write the binomial factor and then use this factor and the form $A^3 + B^3 = (A + B)(A^2 - AB + B^2)$ to complete the factorization.

The trinomial $x^2 - 4x + 16$ is prime over the integers.

Square of the first term, x. ⟶
Opposite of the product of the two terms. ⟶
Square of the last term, 4. ⟶

Self-Check 5

Factor $x^3 + 125$.

Example 6 Factoring the Difference of Two Cubes

Factor $27x^3 - 1$.

Solution

$$27x^3 - 1 = (3x)^3 - (1)^3$$
$$= (3x - 1)[(\underline{\quad})^2 + (\underline{\quad})(\underline{\quad}) + (\underline{\quad})^2]$$
$$= (3x - 1)[(3x)^2 + (3x)(1) + (1)^2]$$
$$= (3x - 1)(9x^2 + 3x + 1)$$

First express this binomial in the form $A^3 - B^3$. Write the binomial factor and then use this factor and the form $A^3 - B^3 = (A - B)(A^2 + AB + B^2)$ to complete the factorization.

The trinomial $9x^2 + 3x + 1$ is prime over the integers.

Square of the first term, $3x$. ⟶
Opposite of the product of the two terms. ⟶
Square of the last term, 1. ⟶

Self-Check 6

Factor $64x^3 - 1$.

Example 7 contains two variables and fits the form of the difference of two cubes.

Example 7 Factoring the Difference of Two Cubes

Factor $125a^3 - 64b^3$.

Solution

$$125a^3 - 64b^3 = (5a)^3 - (4b)^3$$
$$= (5a - 4b)[(\underline{\quad})^2 + (\underline{\quad})(\underline{\quad}) + (\underline{\quad})^2]$$
$$= (5a - 4b)[(5a)^2 + (5a)(4b) + (4b)^2]$$
$$= (5a - 4b)(25a^2 + 20ab + 16b^2)$$

First express this binomial in the form $A^3 - B^3$.
Write the binomial factor and then use this factor and the form $A^3 - B^3 = (A - B)(A^2 + AB + B^2)$ to complete the factorization.

Square of the first term, $5a$. ———
Opposite of the product of the two terms. ———
Square of the last term, $4b$. ———

Self-Check 7

Factor $1{,}000x^3 + 27y^3$.

Some polynomials with degree greater than 2 can also be factored as special forms. When you factor polynomials of higher degree, be sure to continue your work until the polynomial is factored completely. A polynomial is factored completely if each factor other than a constant factor is prime over the integers. This is illustrated in Example 8 with a fourth-degree binomial.

Example 8 Factoring a Fourth-Degree Binomial

Factor $16y^4 - 1$.

Solution

$$16y^4 - 1 = (4y^2)^2 - 1^2$$

First write this binomial in the form $A^2 - B^2$ and then factor as a sum times a difference.

$$= (4y^2 + 1)(4y^2 - 1)$$

$4y^2 + 1$ is the sum of two squares and is prime.

$$= (4y^2 + 1)[(2y)^2 - 1^2]$$

$4y^2 - 1$ is the difference of two squares and factors as a sum times a difference.

$$= (4y^2 + 1)(2y + 1)(2y - 1)$$

Self-Check 8

Factor $x^4 - 81$.

Self-Check Answers

1. **a.** $(x - 9)^2$
   b. $(11w + 1)^2$
   c. $(5v + 6)^2$
   d. This trinomial is not a perfect square.
2. **a.** $(6y + 5)(6y - 5)$
   b. $(10a + 11b)(10a - 11b)$
   c. $(4ab + 1)(4ab - 1)$
   d. This binomial is not the difference of two squares.
3. **a.** $(13v + 10w)(13v - 10w)$
   b. Prime
   c. $(13v + 10w)^2$

4. **a.** Neither
   b. Sum of two cubes
   c. Difference of two cubes
5. $(x + 5)(x^2 - 5x + 25)$
6. $(4x - 1)(16x^2 + 4x + 1)$
7. $(10x + 3y)(100x^2 - 30xy + 9y^2)$
8. $(x^2 + 9)(x + 3)(x - 3)$

6.4 Using the Language and Symbolism of Mathematics

Match each of these forms with the most appropriate description.

1. $A^2 + 2AB + B^2 = (A + B)^2$
2. $A^2 - 2AB + B^2 = (A - B)^2$
3. $A^2 - B^2 = (A + B)(A - B)$
4. $A^3 + B^3 = (A + B)(A^2 - AB + B^2)$
5. $A^3 - B^3 = (A - B)(A^2 + AB + B^2)$

A. Difference of two squares
B. Difference of two cubes
C. Sum of two cubes
D. Square of a sum
E. Square of a difference

6.4 Quick Review

1. Multiply and simplify $(x + 7)(x - 7)$.
2. Multiply and simplify $(5x + 2y)(5x + 2y)$.
3. Multiply and simplify $(6x - 5)(6x - 5)$.
4. Multiply and simplify $(3x - 2)(9x^2 + 6x + 4)$.

5. Complete the following tables using only pencil and paper.

x	x^2		x	x^3
1			1	
2			2	
3			3	
4			4	
5			5	
6				
7				
8				
9				
10				

6.4 Exercises

Objective 1 Factor Perfect Square Trinomials

In Exercises 1–4, fill in the missing information. Then write your answer as the factorization of the trinomial.

1. $a^2 + 6a + 9 = a^2 + 2(\underline{})(a) + (\underline{})^2$
   $= (a + \underline{})^2$
2. $a^2 - 10a + 25 = a^2 - 2(\underline{})(a) + (\underline{})^2$
   $= (a - \underline{})^2$
3. $4w^2 - 12w + 9 = (\underline{})^2 - 2(\underline{})(3) + 3^2$
   $= (\underline{} - 3)^2$
4. $9w^2 + 30w + 25 = (\underline{})^2 + 2(\underline{})(5) + 5^2$
   $= (\underline{} + 5)^2$

In Exercises 5–18, completely factor each trinomial, using the forms for perfect square trinomials. If necessary, first factor out the GCF.

5. $m^2 - 2m + 1$
6. $m^2 + 2m + 1$
7. $36v^2 + 12v + 1$
8. $100v^2 - 20v + 1$
9. $25x^2 - 20xy + 4y^2$
10. $81x^2 + 180xy + 100y^2$
11. $121m^2 + 88mn + 16n^2$
12. $169m^2 - 78mn + 9n^2$
13. $-x^2 + 16xy - 64y^2$
14. $-25x^2 + 60xy - 36y^2$
15. $9 + 16x^2 - 24x$
16. $121 + x^2 - 22x$
17. $2x^3 + 16x^2 + 32x$
18. $3x^3 + 42x^2 + 147x$

Objective 2 Factor the Difference of Two Squares

In Exercises 19–32, completely factor each binomial, using the form for the difference of two squares. If necessary, first factor out the GCF.

19. $w^2 - 49$
20. $w^2 - 36$
21. $9v^2 - 1$
22. $16v^2 - 1$
23. $81m^2 - 25$
24. $36m^2 - 49$
25. $4a^2 - 9b^2$
26. $100a^2 - 49b^2$
27. $16v^2 - 121w^2$
28. $25v^2 - 144w^2$
29. $36 - m^2$
30. $81 - m^2$
31. $20x^2 - 45y^2$
32. $75x^2 - 48y^2$

Objective 3 Factor the Sum or Difference of Two Cubes

In Exercises 33–40, completely factor each binomial, using the form for the sum or difference of two cubes.

33. $x^3 + 27$

34. $x^3 - 27$

35. $m^3 - 125$

36. $m^3 + 125$

37. $64a^3 - b^3$

38. $64a^3 + b^3$

39. $125x^3 + 8y^3$

40. $64a^3 - 125b^3$

Skill Development

In Exercises 41–48, some of the polynomials are factorable over the integers, and others are prime over the integers. Factor those that are factorable, and label the others as prime. If necessary, first factor out the GCF.

41. a. $x^2 - 6x + 9$
   b. $x^2 + 9$
   c. $x^2 - 10x + 9$

42. a. $x^2 + 4x + 4$
   b. $x^2 + 5x + 4$
   c. $x^2 + 4$

43. a. $x^2 - 12xy + 36y^2$
   b. $x^2 - 36y^2$
   c. $x^2 - 13xy + 36y^2$

44. a. $64x^2 - y^2$
   b. $64x^2 + 20xy + y^2$
   c. $64x^2 + 16xy + y^2$

45. a. $4x^2 + 9y^2$
   b. $4x^2 - 12xy + 9y^2$
   c. $4x^2 + 15xy + 9y^2$

46. a. $16x^2 + 56xy + 49y^2$
   b. $16x^2 - 70xy + 49y^2$
   c. $16x^2 + 49y^2$

47. a. $25x^2 - 64y^2$
   b. $25x^2 - 80xy + 64y^2$
   c. $25x^2 - 100xy + 64y^2$

48. a. $100x^2 - 81y^2$
   b. $100x^2 + 225xy + 81y^2$
   c. $100x^2 + 180xy + 81y^2$

In Exercises 49–52, determine the missing term so each trinomial will be a perfect square trinomial.

49. a. $x^2 + \underline{\quad} + 36$
   b. $x^2 + 12x + \underline{\quad}$

50. a. $x^2 - \underline{\quad} + 25$
   b. $x^2 - 10x + \underline{\quad}$

51. a. $x^2 - \underline{\quad} + 100$
   b. $x^2 - 20x + \underline{\quad}$

52. a. $x^2 + \underline{\quad} + 49$
   b. $x^2 + 14x + \underline{\quad}$

In Exercises 53–72, completely factor each polynomial.

53. $-16t^2 + 96t - 144$

54. $-18x^2 - 60x - 50$

55. $x^4 - 9$

56. $x^6 - 36$

57. $25y^4 - 1$

58. $x^4 - 16$

59. $81y^4 - 1$

60. $m^2n^2 - 25$

61. $(x - 5)^2 - 9y^2$

62. $(x + 7)^2 - 16y^2$

63. $4x^2 - (y - 3)^2$

64. $9x^2 - (y + 2)^2$

65. $(a + 2b)^2 - (2a + b)^2$

66. $(a - 3b)^2 - (3a - b)^2$

67. $(a - 5b)x^2 - (a - 5b)y^2$

68. $(x^2 - y^2)a^2 - (x^2 - y^2)b^2$

69. $(a + b)x^2 + 6(a + b)x + 9(a + b)$

70. $(a - b)x^2 - 10(a - b)x + 25(a - b)$

71. $(x^2 + 20xy + 100y^2)a^2 - 16(x^2 + 20xy + 100y^2)$

72. $(x^2 - 6xy + 9y^2)m^2 - 25n^2(x^2 - 6xy + 9y^2)$

Connecting Algebra to Geometry

In Exercises 73–75, completely factor each polynomial.

73. $x_1^2 - x_2^2$ (the area between two squares)

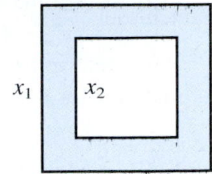

74. $\pi r_1^2 - \pi r_2^2$ (the area between two concentric circles)

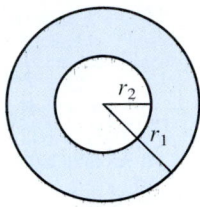

75. $\pi r_1^2 h - \pi r_2^2 h$ (the volume of steel in a steel pipe)

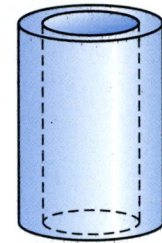

76. The area of a window is given by the function $A(x) = 7x - x^2$, where x represents the width of the window.
   a. Write this function in factored form.
   b. Explain the meaning of each factor. (*Hint: A = l · w.*)
   c. Is this function more meaningful in expanded form or in factored form? Explain.

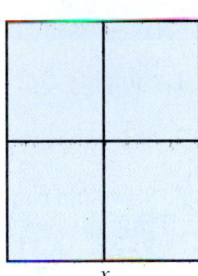

77. A steel sheet 7 in wide by 9 in long has squares x cm on a side cut from each corner, and then the sides are turned up to form the steel tray shown. The volume of this tray is given by the function $V(x) = 63x - 32x^2 + 4x^3$.
   a. Write this function in factored form. (*Hint:* Do not reorder the terms.)
   b. Explain the meaning of each factor. (*Hint: V = l · w · h.*)
   c. Is this function more meaningful in expanded form or in factored form? Explain.

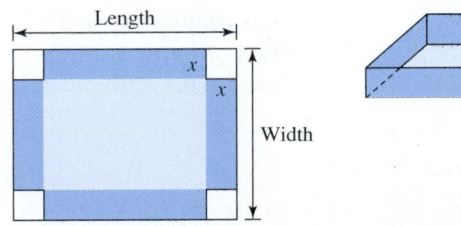

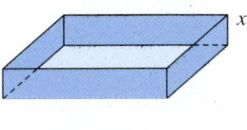

Review and Concept Development

78. Use the graph and table to:

(a) Determine the x-intercepts of the graph of $y = x^3 - 4x$.

(b) Determine the zeros of the polynomial $x^3 - 4x$.

(c) Factor the polynomial $x^3 - 4x$.

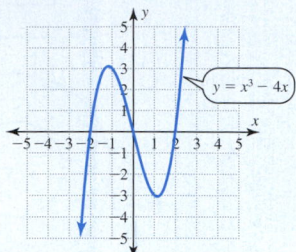

x	y
-3	-15
-2	0
-1	3
0	0
1	-3
2	0
3	15

79. Use the graph and table to:

(a) Determine the x-intercepts of the graph of $y = x^3 - x$.

(b) Determine the zeros of the polynomial $x^3 - x$.

(c) Factor the polynomial $x^3 - x$.

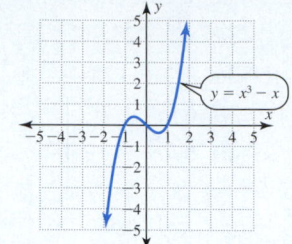

x	y
-3	-24
-2	-6
-1	0
0	0
1	0
2	6
3	24

Group discussion questions

80. Challenge Question Factor each polynomial completely, assuming that m and n are natural numbers.

a. $81y^{2n} - 16$
b. $x^{2m} - y^{2n}$
c. $4x^{2m} + 20x^my^n + 25y^{2n}$
d. $10x^{2m} + 4x^my^n - 6y^{2n}$

81. Challenge Question The binomial $a^6 + b^6$ can be written as the sum of two squares $(a^3)^2 + (b^3)^2$. Yet $a^6 + b^6$ is not prime. Factor $a^6 + b^6$.

6.4 Cumulative Review

1. Simplify $5x + 6 - 2(3x + 7)$.

2. Solve the system of equations $\begin{cases} 3x + 4y = 32 \\ 5x - 2y = -3 \end{cases}$.

3. Solve the compound inequality $3x - 5 < 10$ or $7 - x < -1$ and give your answer in interval notation.

4. Using pencil and paper, simplify $\left(\dfrac{1}{3}\right)^{-2} + \left(\dfrac{2}{5}\right)^{-1}$. Give your answer as a fraction.

5. Simplify $\dfrac{18x^{10}y^8}{3x^{-2}y^4}$. Give your answer in terms of positive exponents.

Section 6.5 Factoring by Grouping and a General Strategy for Factoring Polynomials

Objectives:

1. Factor polynomials by grouping.

2. Determine the most appropriate method for factoring a polynomial.

1. Factor Polynomials by Grouping

The factoring methods developed in previous sections of this chapter are used primarily with binomials and trinomials. For polynomials with four or more terms, **factoring by grouping** is often used. This is an extension of the grouping method introduced in Section 6.1 with greatest common factors.

How Do I Determine How to Group the Terms?

Instead of examining the entire polynomial at one time, we start by grouping terms that we already know how to factor. Thus factoring by grouping relies heavily on the ability to spot common factors and special forms.

Example 1   Factoring by Grouping

Factor $3ac - bc + 6a - 2b$.

Solution

$$\underline{3ac - bc} \;\; \underline{+\, 6a - 2b} = c(3a - b) \;\; + 2(3a - b)$$
$$= (c + 2)(3a - b)$$

There is no common factor other than 1 for all four terms. Thus we group the first two terms, which have a GCF of c. Also we group the last two terms, which have a GCF of 2. Then we factor $3a - b$ out of both terms.

Self-Check 1

Factor $rt - st + rv - sv$.

To determine an advantageous grouping for a polynomial, examine the polynomial for groups that are special forms. This is illustrated in Example 2, where the first two terms form the difference of two squares.

Example 2   Factoring by Grouping

Factor $x^2 - y^2 + x - y$.

Solution

$$\underline{x^2 - y^2} \;\; \underline{+\, x - y} = (x - y)(x + y) \;\; + (x - y)(1)$$
$$= (x - y)(x + y + 1)$$

Note that the first two terms form the difference of two squares. Group them and factor $x^2 - y^2$. Then factor $x - y$ out of both terms.

Caution: This term of 1 must be included to account for the term $(x - y)(1)$ in the previous step.

Self-Check 2

Factor $x^2 - y^2 + 7x + 7y$.

Will All Groupings of the Terms Lead to the Same Factorization?

No, it is possible that one grouping will lead to a factorization, whereas other groupings will prove useless. For example, it is tempting to factor $z^2 - a^2 + 4ab - 4b^2$ by grouping the first two terms and the last two terms, which would result in $(z + a)(z - a) + 4b(a - b)$. However, this grouping, which has a GCF of 1, is not useful. A more useful grouping of this polynomial is shown in Example 3.

Example 3 Factoring by Grouping

Factor $z^2 - a^2 + 4ab - 4b^2$.

Solution

$$
\begin{aligned}
z^2 - a^2 + 4ab - 4b^2 &= z^2 - (a^2 - 4ab + 4b^2) \\
&= z^2 - (a - 2b)^2 \\
&= [z + (a - 2b)][z - (a - 2b)] \\
&= (z + a - 2b)(z - a + 2b)
\end{aligned}
$$

Group the last three terms, noting the sign change of each term within the parentheses. Then factor the perfect square trinomial within the parentheses. This polynomial fits the form of the difference of two squares. Factor this special form as a sum times a difference and then simplify.

Self-Check 3

Factor $x^2 - 10x + 25 - 36y^2$.

Sometimes it is advantageous to reorder the terms of a polynomial to facilitate a useful grouping, as illustrated in Example 4.

Example 4 Factoring by Grouping

Factor $y^3 + x + x^3 + y$.

Solution

$$
\begin{aligned}
y^3 + x + x^3 + y &= (x^3 + y^3) + (x + y) \\
&= (x + y)(x^2 - xy + y^2) + (x + y)(1) \\
&= (x + y)[(x^2 - xy + y^2) + 1] \\
&= (x + y)(x^2 - xy + y^2 + 1)
\end{aligned}
$$

Reorder the terms and group the sum of the two perfect cubes. Then factor this special form. Now factor out the GCF of $x + y$. Be sure to include the last term of 1.

Self-Check 4

Factor $x^3 - 4x + y^3 - 4y$.

2. Determine the Most Appropriate Method for Factoring a Polynomial

It is important to practice factoring a variety of polynomials so that you can quickly select the appropriate method. The exercises at the end of this section contain a mix of problems intended to promote the development of a general factoring strategy. The following box outlines the methods of factoring covered earlier in this chapter.

Is There a Key to Using This Strategy?

Yes, a key is to determine the number of terms in the polynomial.

Strategy for Factoring a Polynomial Over the Integers

After factoring out the GCF (greatest common factor), proceed as follows.

Binomials

Factor special forms:

$A^2 - B^2 = (A + B)(A - B)$	Difference of two squares
$A^3 - B^3 = (A - B)(A^2 + AB + B^2)$	Difference of two cubes
$A^3 + B^3 = (A + B)(A^2 - AB + B^2)$	Sum of two cubes
$A^2 + B^2$ is prime	The sum of two squares is prime if x^2 and y^2 are only second-degree terms and have no common factor other than 1.

Trinomials

Factor the forms that are perfect squares:

$A^2 + 2AB + B^2 = (A + B)^2$	Perfect square trinomial; a square of a sum
$A^2 - 2AB + B^2 = (A - B)^2$	Perfect square trinomial; a square of a difference

Factor trinomials that are not perfect squares by inspection if possible; otherwise, use the trial-and-error method or the AC method.

Polynomials of Four or More Terms

Factor by grouping.

The first step in Example 5 is to factor out the GCF of $5xy$ from each term.

Example 5   Completely Factoring a Binomial

Factor $5x^3y - 5xy^3$.

Solution

$$5x^3y - 5xy^3 = 5xy(x^2 - y^2) \qquad \text{Factor out the GCF of } 5xy.$$
$$= 5xy(x + y)(x - y) \qquad \text{Noting that } x^2 - y^2 \text{ is the difference of two perfect squares, factor this special form.}$$

Self-Check 5

Factor $12ax^3 - 75axy^2$.

The first step in Example 6 is to factor out the GCF of $6ab^2$ from each term.

Example 6   Completely Factoring a Trinomial

Factor $6ab^2x^2 + 30ab^2xy + 36ab^2y^2$.

Solution

$$6ab^2x^2 + 30ab^2xy + 36ab^2y^2 = 6ab^2(x^2 + 5xy + 6y^2) \qquad \text{Factor out the GCF of } 6ab^2.$$
$$= 6ab^2(x + 2y)(x + 3y) \qquad \text{Next factor the trinomial by inspection. Two factors of 6 with a sum of 5 are 2 and 3.}$$

Self-Check 6

Factor $8x^4y - 12x^3y^2 - 20x^2y^3$.

The polynomial in Example 7 contains four terms. Factor out the GCF, and then use the grouping method to factor the polynomial.

Example 7 Completely Factoring a Polynomial

Factor $3x^3y^2 - 3xy^4 - 3x^2y^2 + 3xy^3$.

Solution

$3x^3y^2 - 3xy^4 - 3x^2y^2 + 3xy^3$

$= 3xy^2(\underline{x^2 - y^2} \ - x + y)$ Factor out the GCF of $3xy^2$.

$= 3xy^2[(x + y)(x - y) - 1(x - y)]$ Group the first two terms that fit the form of a difference

$= 3xy^2(x + y - 1)(x - y)$ of two squares. Factor the difference of two squares as a
sum times a difference. Also factor -1 out of the last
two terms $-x + y$ to rewrite them as $-1(x - y)$. Then
factor out the common factor of $x - y$. Be sure to
include the last term of -1 in the last factor.

Self-Check 7

Factor $5x^5y - 5x^3y^3 - 10x^4y - 10x^3y^2$.

The substitution shown in Example 8 is not necessary and may be skipped. However, the substitution is shown to emphasize that this polynomial can be considered a trinomial of the form $a^2 - 8ab + 15b^2$. To apply the special forms and the methods of factoring presented here, it is important to recognize the variety of expressions that fit these forms.

Example 8 Using Substitution to Factor a Polynomial

Factor $(x + 3y)^2 - 8b(x + 3y) + 15b^2$.

Solution

The polynomial $(x + 3y)^2 - 8b(x + 3y) + 15b^2$ is of the form $a^2 - 8ab + 15b^2$.

Substitute a for the quantity $x + 3y$.

To factor $a^2 - 8ab + 15b^2$, notice that

$a^2 - 8ab + 15b^2 = (a - \underline{\quad} b)(a - \underline{\quad} b)$

$= (a - 5b)(a - 3b)$

The coefficient of $15b^2$ is positive, so the factors of 15 must be of the same sign. The middle term has a coefficient of -8, so both factors of 15 are negative and the sign pattern is $(a - \underline{\quad} b)(a - \underline{\quad} b)$. By inspection you can determine that the factors of 15 with a sum of -8 are -3 and -5.

$a^2 - 8ab + 15b^2 = (a - 5b)(a - 3b)$

$(x + 3y)^2 - 8(x + 3y)b + 15b^2 = [(x + 3y) - 5b][(x + 3y) - 3b]$

Substitute $x + 3y$ back for a. Then simplify each factor.

$= (x + 3y - 5b)(x + 3y - 3b)$

Answer: $(x + 3y)^2 - 8b(x + 3y) + 15b^2 = (x + 3y - 5b)(x + 3y - 3b)$

Self-Check 8

Factor $(a + 3b)^2 - 6(a + 3b) + 8$.

A polynomial is prime over the integers if its only factorization must involve 1 or -1 as one of the factors. You should continue all factorizations until each factor, other than a monomial factor, is prime. The polynomials presented in the examples and exercises in this section either are prime or can be factored by the strategy given in this section. Other methods of factoring are usually considered in college algebra courses.

6.5   Using the Language and Symbolism of Mathematics

1. The method often used to factor polynomials with four or more terms is called factoring by _____.

2. GCF stands for _____ _____ _____.

3. The binomial $x^2 + y^2$ is a _____ polynomial.

4. The perfect square trinomial $x^2 + 2xy + y^2$ is the square of a _____.

5. The perfect square trinomial $x^2 - 2xy + y^2$ is the square of a _____.

6. The form $x^3 + y^3$ is the sum of two _____.

7. The form $x^3 - y^3$ is the _____ of two cubes.

6.5   Quick Review

In Exercises 1–3, factor the GCF out of each polynomial.

1. $a(x - y) - 3a$

2. $2bx - 6b(y - 1)$

3. $(3a - b)x^2 - 3(3a - b)x + 9(3a - b)$

In Exercises 4 and 5, factor each polynomial as the difference of two squares.

4. $(a + 2b)^2 - 9y^2$

5. $25x^2 - (a - 3b)^2$

6.5   Exercises

Objective 1 Factor Polynomials by Grouping

In Exercises 1–4, fill in the missing information. Then write your answer as the factorization of the polynomial.

1. $x^2 - xy + 5x - 5y = x(\_\_ - \_\_) + 5(\_\_ - \_\_)$
$= (x + 5)(\_\_ - \_\_)$

2. $xy + xz - 2y - 2z = x(\_\_ + \_\_) - 2(\_\_ + \_\_)$
$= (x - 2)(\_\_ + \_\_)$

3. $x^2 - y^2 + 2y - 1 = x^2 - (\_\_ - \_\_ + \_\_)$
$= x^2 - (\_\_ - \_\_)^2$
$= (x + y - 1)(\_\_ - \_\_ + \_\_)$

4. $x^2 - 5xy - 6y^2 + x - 6y$
$= (x^2 - 5xy - 6y^2) + (x - 6y)$
$= (x - 6y)(\_\_ + \_\_) + (x - 6y)(\_\_)$
$= (x - 6y)(\_\_ + \_\_ + \_\_)$

In Exercises 5–22, factor each polynomial completely using the grouping method.

5. $ac + bc + ad + bd$

6. $xy + xz + 2y + 2z$

7. $3a - 6b + 5ac - 10bc$

8. $6a^2 + 3ab + 2a + b$

9. $ab + bc - ad - cd$

10. $ac + bc + a + b$

11. $v^2 - vw - 7v + 7w$

12. $mn - 7m - n + 7$

13. $4a^2 + 12a + 9 - 16b^2$

14. $16x^2 - a^2 - 2a - 1$

15. $3mn + 15m - kn - 5k$

16. $az^2 + bz^2 + aw^2 + bw^2$

17. $az^3 + bz^3 + aw^2 + bw^2$

18. $s^3 + 11s^2 + s + 11$

19. $9b^2 - 24b + 16 - a^2$

20. $ax - ay - az + bx - by - bz$

21. $ay^2 + 2ay - y + a - 1$

22. $x^2 - 14x + 49 - 16y^2$

Objective 2 Determine the Most Appropriate Method for Factoring a Polynomial

In Exercises 23–72, completely factor each polynomial using the strategy outlined in this section.

23. $64y^2 - 9z^2$

24. $25a^2 - 144b^2$

25. $16x^2 + 49y^2$

26. $3ax^2 + 33ax + 72a$

27. $12x^2 - 27x + 15$

28. $121m^2 + 49n^2$

29. $49a^2 - 28a + 4$

30. $25a^2 - 10a + 1$

31. $x(a - b) + y(a - b)$

32. $a(x - y) - b(x - y)$

33. $10w^2 - 6w - 21$

34. $3s^2 + 3s + 3t - 3t^2$

35. $25v^2 - vw + 36w^2$

36. $32x^3 + 12x^2 - 20x$

37. $4x^{10} + 12x^5y^3 + 9y^6$ **38.** $9s^2 - 63$

39. $12x^3y - 12xy^3$

40. $25y^2 - 30yz + 9z^2$

41. $cx + cy + dx + dy$ **42.** $5a^2bc - 5b^3c$

43. $ax^2 + ax + bxy + by$ **44.** $20x^3y - 245xy^3$

45. $x^6 + 4x^3y + 4y^2$ **46.** $3ax^2 + 3ay^2$

47. $64ax^2 - 104ax + 40a$ **48.** $100s^4 + 120s^3t + 36s^2t^2$

49. $5x^2 - 55$ **50.** $200x^2 + 2$

51. $63a^3b - 175ab$ **52.** $49b^2 + 126bc + 81c^2$

53. $-12x^2 + 12xy - 3y^2$ **54.** $-71ax^4 + 71a$

55. $144x^2 + 81$ **56.** $x^2 + 2xy + y^2 - 16z^2$

57. $9x^2 - 6x + 1 - 25y^2$ **58.** $2x^2 + 2x + 2y - 2y^2$

59. $12ax^2 - 10axy - 12ay^2$ **60.** $8ax^2 - 648ay^4$

61. $7s^5t - 7st^5$ **62.** $-6ax^3 + 24ax$

63. $3ax^2 - 3ay^2 + 6ay - 3a$ **64.** $x^3 + 4x^2y + 4xy^2$

65. $2ax^3 + 10ax^2 - 28ax$ **66.** $30ax^3 + 5ax^2 - 10ax$

67. $a^2 + 2a + 1 + ab + b$ **68.** $30x^3 - 57x^2 + 18x$

69. $(a - 2b)x^2 - 2(a - 2b)x - 24(a - 2b)$

70. $36(a - b)x^2 - 6(a - b)xy - 20(a - b)y^2$

71. $(4x^2 - 12xy + 9y^2) + (72ay - 48ax) - 25a^2$

72. $(25x^2 - 10xy + y^2) + (10xz - 2yz) - 24z^2$

Using Technology

In Exercises 73 and 74, use a graphing calculator or a graphing utility to graph $y = P(x)$. Use the graph to determine the x-intercepts of this graph, the zeros of $P(x)$, and the factors of this polynomial.

Polynomial P(x)	x-intercepts of the Graph of y = P(x)	Zeros of P(x)	Factored Form
73. $x^3 - 3x^2 - 10x + 24$			
74. $(x^3 - 4x) + (5x^2 - 20)$			

Skill Development

In Exercises 75–78, completely factor each polynomial using the strategy outlined in this section.

75. $5ab^3 - 5a$

76. $bw^3 + b$

77. $x^3 + y^3 + x^2 - y^2$

78. $x^3 + y^3 - 3x - 3y$

Group discussion questions

79. Challenge Question Factor each polynomial completely, assuming that m and n are natural numbers.
  a. $x^{m+2} + 4x^m$
  b. $x^{3m} + 6x^{2m} + 9x^m$

80. Challenge Question Use the distributive property to expand the first expression and to factor the second expression.

Expand	Factor
a. $x^{-4}(x^2 - 3)$	$x^{-2} - 3x^{-4} = x^{-4}(\underline{} - \underline{})$
b. $2x^{-5}(x^2 - 2)$	$2x^{-3} - 4x^{-5} = 2x^{-5}(\underline{} - \underline{})$
c. $x^{-4}(2x - 1)(x - 3)$	$2x^{-2} - 7x^{-3} + 3x^{-4}$
	$= x^{-4}(\underline{} - \underline{} + \underline{})$
	$= x^{-4}(\underline{} - \underline{})(\underline{} - \underline{})$

81. Challenge Question First factor $2{,}025x^4 + 2{,}700x^2y + 900y^2$ by factoring this perfect square trinomial as the square of a sum. Then start over and factor the GCF from this trinomial. Complete each of these factorizations and compare the results, stating which factorization you find easier.

6.5 Cumulative Review

1. If the factored form of a polynomial is $P(x) = (x + 5)(x - 8)$, then what are the x-intercepts of the graph of $y = P(x)$?

2. If the factored form of a polynomial is $P(x) = (x - 3)(x + 9)$, then what are the zeros of $P(x)$?

In Exercises 3–5, solve each equation and classify the equation as a contradiction, a conditional equation, or an identity.

3. $2(x + 3) = x + 6$

4. $2(x + 3) = 2x + 6$

5. $2(x + 3) = 2x + 5$

Section 6.6 | Solving Equations by Factoring

Objectives:

1. Identify a quadratic equation and write it in standard form.
2. Use factoring to solve selected quadratic equations.
3. Construct a quadratic equation with given solutions.
4. Solve a quadratic inequality by using the *x*-intercepts of the corresponding graph.

The factored form of an expression is often more useful to a person designing a product or setting up a problem. Later work with this design may require one to expand these factors. In Exercise 79 at the end of this section, the factors that represent the length and width of a rectangular area can be obtained by examining the figure. After writing the expression for the area in factored form, you will need to expand this expression to solve a second-degree equation for a given area.

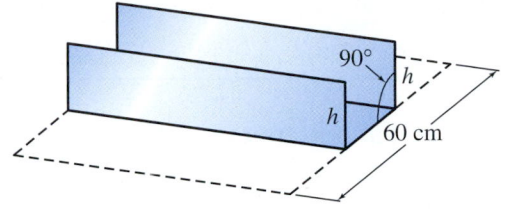

> **A Mathematical Note**
>
> The importance of quadratic equations has been noted by many ancient civilizations. A surviving document from Egypt (ca. 2000 B.C.) contains quadratic equations. The Hindu mathematician Brahmagupta (ca. 628) included a version of the quadratic formula for solving quadratic equations in his works.

1. Identify a Quadratic Equation and Write It in Standard Form

A **quadratic equation** in *x* is a second-degree equation that can be written in the standard form $ax^2 + bx + c = 0$, where *a*, *b*, and *c* represent real constants and $a \neq 0$. (If we allowed *a* to equal zero, the equation would not be quadratic; instead it would be the linear equation $bx + c = 0$.) In the quadratic equation $ax^2 + bx + c = 0$, ax^2 is called the **quadratic term,** bx is called the **linear term,** and *c* is called the **constant term.**

Quadratic Equation

Algebraically	Verbally	Algebraic Example
If *a*, *b*, and *c* are real constants and $a \neq 0$, then $ax^2 + bx + c = 0$ is the standard form of a quadratic equation in *x*.	A quadratic equation in *x* is a second-degree equation in *x*.	$3x^2 - 4x + 1 = 0$ is a quadratic equation with Quadratic term: $3x^2$ Linear term: $-4x$ Constant term: 1

Example 1 | Identifying Quadratic Equations

Determine which of these are quadratic equations in one variable. Write each quadratic equation in standard form and identify *a*, *b*, and *c*.

Solution

(a) $5y - 6 = -y^2$ — This equation is a quadratic equation in *y* and can be written in standard form as $y^2 + 5y - 6 = 0$.

$a = 1$, $b = 5$, and $c = -6$.

It is customary to write the equation so that a is positive.

(b) $5w = 12$ — This equation is a linear equation in *w*; it is not a quadratic equation.

Quadratic equations are second degree. Linear equations are first degree.

(c) $5v^2 = 3v - 4$ — This equation is a quadratic equation in *v* and can be written in standard form as $5v^2 - 3v + 4 = 0$.

$a = 5$, $b = -3$, and $c = 4$.

The quadratic term is $5v^2$, the linear term is $-3v$, and the constant term is 4.

(d) $x^2 + 7x = x^3 - 5$ — This equation is a third-degree equation; it is not a quadratic equation.

(e) $x^2 + 3x + 7$ This is not an equation, just a polynomial expression.

(f) $x^2 = 6x$ This equation is a quadratic equation and can be written in standard form as $x^2 - 6x = 0$.

$a = 1, b = -6$, and $c = 0$.

Self-Check 1

Determine which of these are quadratic equations in one variable. Write each quadratic equation in standard form.

a. $3x^2 - 2 = x$ **b.** $3x^2 - 2$ **c.** $3x - 2 = 0$

Why Is It Important to Identify Linear and Quadratic Equations?

If we decide to solve these equations algebraically, we use different methods to solve linear equations than we do to solve quadratic equations.

2. Use Factoring to Solve Selected Quadratic Equations

The easiest method for solving some quadratic equations with small integer coefficients is to use factoring and the zero-factor principle. Because some quadratic equations are not factorable over the integers, some quadratic equations cannot be solved by factoring over the integers. We will cover methods for solving those quadratic equations in Section 7.3, but first we state the zero-factor principle.

Zero-Factor Principle

Algebraically	Verbally	Algebraic Example
For real numbers a and b: If $a = 0$ or $b = 0$, then $ab = 0$.	The product of zero and any other factor is zero.	$0 \cdot x = 0$
If $ab = 0$, then $a = 0$ or $b = 0$.	If the product of two or more factors is zero, at least one of the factors must be zero.	If $(x - 3)(x + 4) = 0$, then either $x - 3 = 0$ or $x + 4 = 0$.

Example 2   Solving a Quadratic Equation in Factored Form

Solve $(v - 2)(5v + 3) = 0$.

Solution

$(v - 2)(5v + 3) = 0$ By the zero-factor principle, either the first factor or the second factor is zero.

$v - 2 = 0$ or $5v + 3 = 0$

$v = 2$ $5v = -3$ Notice that there are two solutions.

$v = -\dfrac{3}{5}$

Answer: $v = 2, v = -\dfrac{3}{5}$ Do both of these values check?

Self-Check 2

Solve $(2x - 9)(x + 5) = 0$.

The quadratic equation in Example 2 was carefully selected so we could easily apply the zero-factor principle. We now examine a wider variety of quadratic equations by using the strategy described in the following box. The steps in this procedure are illustrated in Example 3.

Solving Quadratic Equations by Factoring

Verbally	Algebraic Example
1. Write the equation in standard form, with the right side zero.	$x^2 + x = 6$ $x^2 + x - 6 = 0$
2. Factor the left side of the equation.	$(x - 2)(x + 3) = 0$
3. Set each factor equal to zero.	$x - 2 = 0$ or $x + 3 = 0$ $x = 2$ $x = -3$
4. Solve the resulting first-degree equations.	

Example 3   Solving a Quadratic Equation by Factoring

Solve $6x^2 + 7x = 5$.

Solution

$$6x^2 + 7x = 5$$ Write the equation in standard form, with the right side 0.

$$6x^2 + 7x - 5 = 0$$ Factor the left side of the equation.

$$(3x + 5)(2x - 1) = 0$$ Use the sign pattern (___ x + ___)(___ x − ___) to factor $6x^2 + 7x - 5$ by inspection. Note that the factors of -5 must have opposite signs. Also because 5 is prime, the factors of 5 must be 1 and 5. The possible factor pairs of 6 are 1 and 6 or 2 and 3.

$3x + 5 = 0$ or $2x - 1 = 0$ Set each factor equal to 0.

$3x = -5$ $2x = 1$ Solve the resulting first-degree equations.

$x = -\dfrac{5}{3}$ $x = \dfrac{1}{2}$ Notice that there are two solutions.

Answer: $x = -\dfrac{5}{3}, x = \dfrac{1}{2}$ Do both of these values check?

Self-Check 3

Solve $6x^2 - 37x + 22 = 0$.

What Do I Do If the Quadratic Equation Has Fractions for the Coefficients?

Simplify the equation by converting it to an equivalent equation that does not involve fractions. To do this, multiply both sides of the equation by the LCD of all the terms. The LCD in Example 4 is 8.

Example 4   Solving a Quadratic Equation Containing Fractions

Solve $\dfrac{1}{8}x^2 - \dfrac{1}{4}x = 1$.

Solution

$$\frac{1}{8}x^2 - \frac{1}{4}x = 1$$

$\frac{1}{8}x^2 - \frac{1}{4}x - 1 = 0$	First write the equation in standard form.
$8\left(\frac{1}{8}x^2 - \frac{1}{4}x - 1\right) = 8(0)$	Multiply both sides by the LCD of 8 to obtain integer coefficients.
$x^2 - 2x - 8 = 0$	Use the sign pattern
$(x - 4)(x + 2) = 0$	$(x - \underline{\quad})(x + \underline{\quad})$ to factor $x^2 - 2x - 8$ by inspection.
$x - 4 = 0 \qquad$ or $\qquad x + 2 = 0$	Set each factor equal to 0.
$x = 4 \qquad\qquad\qquad x = -2$	Solve each of these linear equations.
Answer: $x = 4, x = -2$	Do both of these values check?

Self-Check 4

Solve $x^2 - \frac{5}{2}x - \frac{3}{2} = 0$.

It is important that one side of a quadratic equation be 0 before you try to apply the zero-factor principle. Failure to observe this is the source of many student errors.

Example 5 Solving a Quadratic Equation Not in Standard Form

Solve $(6x + 1)(x - 2) = 8$.

Solution

$(6x + 1)(x - 2) = 8$	The first step is to multiply the factors on the left side of the equation and write the equation in standard form with the right side equal to 0. Then factor the left side of the equation.
$6x^2 - 11x - 2 = 8$	
$6x^2 - 11x - 10 = 0$	
$(2x - 5)(3x + 2) = 0$	Use the sign pattern $(\underline{\quad} x - \underline{\quad})(\underline{\quad} x + \underline{\quad})$ to factor $6x^2 - 11x - 10$ by inspection. Note that the factors of -10 must have opposite signs. The possible factor pairs of 6 are 1 and 6 or 2 and 3. The possible factor pairs of 10 are 1 and 10 or 2 and 5.
$2x - 5 = 0 \qquad$ or $\qquad 3x + 2 = 0$	
$2x = 5 \qquad\qquad\qquad 3x = -2$	
$x = \frac{5}{2} \qquad\qquad\qquad x = -\frac{2}{3}$	Set each factor equal to 0. Then solve each of the linear equations for x.
Answer: $x = \frac{5}{2}, x = -\frac{2}{3}$	Do both of these solutions check?

Self-Check 5

Solve $(x - 4)(x - 3) = 2$.

3. Construct a Quadratic Equation with Given Solutions

How Is It Possible to Construct a Quadratic Equation with Given Solutions?

If we can factor a quadratic equation in standard form, then we can find its solutions. If we reverse this procedure, we can create a quadratic equation whose solutions are given. This process is illustrated in Example 6.

Example 6 Constructing a Quadratic Equation Whose Solutions Are Given

Construct a quadratic equation in x with integer coefficients with solutions of $x = -1$ and $x = \dfrac{2}{3}$.

Solution

$x = -1$	and	$x = \dfrac{2}{3}$	The solutions are $x = -1$ and $x = \dfrac{2}{3}$.

$x + 1 = 0$

$3x = 2$

$3x - 2 = 0$

Rewrite each equation so that the right side is zero and free of fractions. Note that $x + 1 = 0$ is a linear equation whose solution is $x = -1$, and $3x - 2 = 0$ is a linear equation whose solution is $x = \dfrac{2}{3}$.

$(x + 1)(3x - 2) = 0$

$3x^2 + x - 2 = 0$

These factors equal zero, so their product is zero.

Answer: $3x^2 + x - 2 = 0$ is a quadratic equation with solutions of $x = -1$ and $x = \dfrac{2}{3}$.

Self-Check 6

Construct a quadratic equation in x with integer coefficients with solutions of $x = 4$ and $x = -\dfrac{3}{7}$.

Earlier we used factoring to find the solutions of a quadratic equation $P(x) = 0$. Now we have reversed this process and used the solutions of $P(x) = 0$ to determine the factors of $P(x)$.

What Other Information Can I Determine from the Solutions of $P(x) = 0$?

The following box gives the relationship among the factors of $P(x) = 3x^2 + x - 2$, the solutions of $P(x) = 0$, the zeros of $P(x)$, and the x-intercepts of the graph of $y = P(x)$. This is an extension of the topic first examined in Section 6.1.

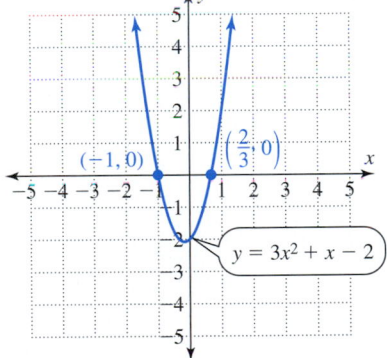

Factor of $3x^2 + x - 2$	Solutions of $3x^2 + x - 2 = 0$	Zeros of $3x^2 + x - 2$	x-intercepts of the Graph of $y = 3x^2 + x - 2$
$x + 1$	$x = -1$	-1	$(-1, 0)$
$3x - 2$	$x = \dfrac{2}{3}$	$\dfrac{2}{3}$	$\left(\dfrac{2}{3}, 0\right)$

A generalization of this relationship is given in the following table. We will examine this relationship further in Section 7.3.

Equivalent Statements About Linear Factors of a Polynomial

For a real constant c and a real polynomial $P(x)$, the following statements are equivalent.

Factor	Solution	Zero	x-intercept
$x - c$ is a factor of $P(x)$.	$x = c$ is a solution of $P(x) = 0$.	$P(c) = 0$, that is, c is a zero of $P(x)$.	$(c, 0)$ is an x-intercept of the graph of $y = P(x)$.

Can I Use the Information in This Box to Solve Quadratic Equations?

Yes, in Example 7, we will show how the algebraic solution for $x^2 + 2x - 35 = 0$ can also be obtained from a graph or a table.

Example 7   Using Multiple Perspectives to Solve a Quadratic Equation

(a) Solve $x^2 + 2x - 35 = 0$ algebraically.
(b) Solve $x^2 + 2x - 35 = 0$ graphically.
(c) Solve $x^2 + 2x - 35 = 0$ numerically.

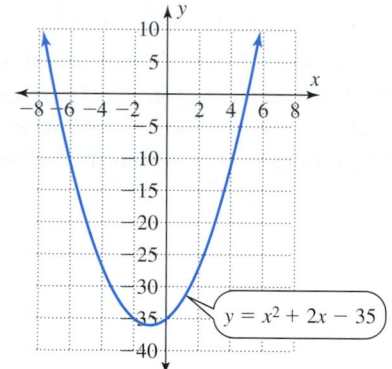

x	y
−11	64
−7	0
−3	−32
1	−32
5	0
9	64
13	160

$y = x^2 + 2x - 35$

Solution

(a) Algebraically

$$x^2 + 2x - 35 = 0$$
$$(x + 7)(x - 5) = 0$$
$$x + 7 = 0 \qquad \text{or} \qquad x - 5 = 0$$
$$x = -7 \qquad\qquad\qquad x = 5$$

Use the sign pattern $(x + \underline{\quad})(x - \underline{\quad})$ to factor $x^2 + 2x - 35$ by inspection.

This illustrates the relationship of the factors of $x^2 + 2x - 35$ to the solutions of $x^2 + 2x - 35 = 0$.

(b) Graphically

The x-intercepts of the graph are $(-7, 0)$ and $(5, 0)$. Thus, the values of x for which $x^2 + 2x - 35 = 0$ are $x = -7$ and $x = 5$.

This illustrates the relationship of the x-intercepts of the graph of $y = x^2 + 2x - 35$ to the solutions of $x^2 + 2x - 35 = 0$.

(c) Numerically

The y-value in the table equals 0 for $x = -7$ and $x = 5$. Thus, the values of x for which $x^2 + 2x - 35 = 0$ are $x = -7$ and $x = 5$.

This illustrates the relationship of the zeros of $x^2 + 2x - 35$ to the solutions of $x^2 + 2x - 35 = 0$.

Answer: $x = -7, x = 5$

Self-Check 7

a. Solve $x^2 - 5x - 36 = 0$ algebraically.

b. Determine the x-intercepts of the graph of $y = x^2 - 5x - 36$.

c. Determine the zeros of $x^2 - 5x - 36$.

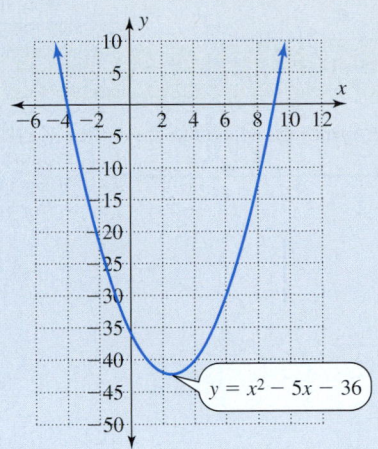

x	y
-5	14
-4	0
-3	-12
0	-36
8	-12
9	0
10	14

4. Solve a Quadratic Inequality by Using the x-intercepts of the Corresponding Graph

The graph of $y = x^2 + 2x - 35$ in Example 7 actually reveals much more than we used to solve $x^2 + 2x - 35 = 0$. By examining where the y-coordinate is positive, we can solve $x^2 + 2x - 35 > 0$. Likewise, by examining where the y-coordinate is negative, we can solve $x^2 + 2x - 35 < 0$. This information is summarized in the following box.

Solutions of Equations and Inequalities

If $P(x)$ is a real polynomial and c is a real number, then:

Verbally	Algebraically	Graphically
• c is a solution of $P(x) = 0$	$P(c) = 0$	$(c, 0)$ is an x-intercept of the graph of $y = P(x)$.
• c is a solution of $P(x) < 0$	$P(c) < 0$	At $x = c$, the graph is below the x-axis.
• c is a solution of $P(x) > 0$	$P(c) > 0$	At $x = c$, the graph is above the x-axis.

In Example 8 we use the graph of $y = x^2 + 2x - 35$ to solve two quadratic inequalities.

Example 8   Using x-intercepts to Solve Quadratic Inequalities

Use the graph and table for $y = x^2 + 2x - 35$ to solve

(a) $x^2 + 2x - 35 < 0$

(b) $x^2 + 2x - 35 \geq 0$

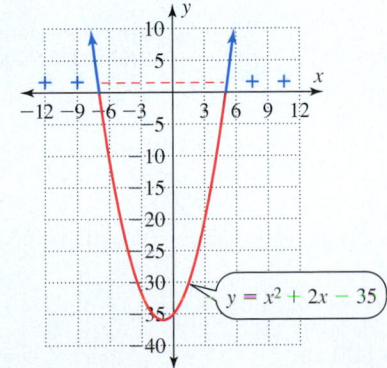

x	y	$<, =,$ or $>$	0
-11	64	$>$	0
-7	0	$=$	0
-3	-32	$<$	0
1	-32	$<$	0
5	0	$=$	0
9	64	$>$	0
13	160	$>$	0

Solution

(a) $(-7, 5)$

x-values for which $y < 0$

←+—+—(—+—+—)—+—+→
-15 -11 -7 -3 1 5 9 13

The graph of $y = x^2 + 2x - 35$ is below the *x*-axis for *x*-values in the interval $(-7, 5)$. In the table, the value of y is less than 0 for *x*-values between -7 and 5.

(b) $(-\infty, -7] \cup [5, \infty)$

x-values for which $y \geq 0$

←+—+—]—+—+—[—+—+→
-15 -11 -7 -3 1 5 9 13

The graph of $y = x^2 + 2x - 35$ is above or on the *x*-axis for *x*-values in the interval $(-\infty, -7] \cup [5, \infty)$. In the table, the value of y is greater than or equal to 0 for *x*-values less than or equal to -7 or greater than or equal to 5.

Self-Check 8

Use the graph and table in Self-Check 7 to solve:

a. $x^2 - 5x - 36 \leq 0$

b. $x^2 - 5x - 36 > 0$

In Section 7.4, we will examine several different applications of quadratic equations. One application illustrated in Example 9 involves a rectangular area.

Example 9   Determining the Height of a Rectangular Area

A rectangular steel sheet that is 100 cm wide has *h* cm turned down on both the left and right sides. This is done at right angles to form three sides of a rectangular frame for the opening for a storm gutter on the side of a street. Determine the height of the gutter if the cross-sectional area is 1,200 cm^2.

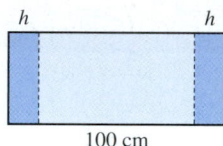

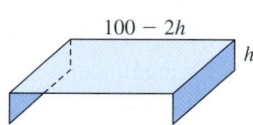

Solution

$$\text{Let } h = \text{the height of the rectangular opening in cm}$$
$$100 - 2h = \text{the width of the rectangular opening in cm}$$

If *h* cm is turned down on each side, this leaves $(100 - 2h)$ cm for the width.

Word Equation

$$(\text{Height})(\text{Width}) = \text{Area}$$

The area of the rectangular opening is the product of its height times its width.

Algebraic Equation

Substitute the expressions identified above for the height, width, and area.

$$h(100 - 2h) = 1{,}200$$
$$100h - 2h^2 = 1{,}200$$
$$-2h^2 + 100h - 1{,}200 = 0$$
$$h^2 - 50h + 600 = 0$$
$$(h - 20)(h - 30) = 0$$
$$h - 20 = 0 \quad \text{or} \quad h - 30 = 0$$
$$h = 20 \qquad\qquad h = 30$$

Write this quadratic equation in standard from. Simplify by dividing both sides of the equation by -2.

Solve this quadratic equation by factoring.

Answer: The height of the gutter can be either 20 cm or 30 cm.   Do both answers check?

Self-Check 9

Rework Example 9 if the cross-sectional area is 800 cm^2.

In Chapter 7, we examine other algebraic methods for solving quadratic equations. The factoring method used in this chapter has been used to find real solutions that are rational numbers. In Chapter 7, we examine quadratic equations with irrational solutions and quadratic equations with solutions that are not real numbers.

Self-Check Answers

1. **a.** This is a quadratic equation in x.
   Standard form: $3x^2 - x - 2 = 0$
   b. Not an equation. This is a quadratic expression.
   c. Not a quadratic equation. This is a linear equation.

2. $x = \dfrac{9}{2}, x = -5$

3. $x = \dfrac{2}{3}, x = \dfrac{11}{2}$

4. $x = -\dfrac{1}{2}, x = 3$

5. $x = 2, x = 5$

6. $7x^2 - 25x - 12 = 0$
7. **a.** $x = -4, x = 9$ **b.** $(-4, 0), (9, 0)$
   c. $-4, 9$
8. **a.** $[-4, 9]$ **b.** $(-\infty, -4) \cup (9, \infty)$
9. The height of the gutter can be either 10 cm or 40 cm.

6.6 Using the Language and Symbolism of Mathematics

1. A second-degree equation in x that can be written as $ax^2 + bx + c = 0$ is called a _____ equation.

2. In the equation $ax^2 + bx + c = 0$, ax^2 is called the _____ term, bx is called the _____ term, and c is called the _____ term.

3. By the zero-factor principle, if $(x - 3)(x + 6) = 0$, then _____ $= 0$ or _____ $= 0$.

4. To clear an equation of fractions, multiply each term by the _____ of all the terms.

In Exercises 5–9, assume that $P(x)$ represents a real polynomial in x.

5. If c is a real solution of $P(x) = 0$, then _____ is an x-intercept of the graph of $y = P(x)$.

6. If c is a real solution of $P(x) = 0$, then _____ is a factor of $P(x)$.

7. If the graph of $y = P(x)$ crosses the x-axis at $x = c$, then $P(c)$ _____ 0.

8. If the graph of $y = P(x)$ is above the x-axis at $x = c$, then $P(c)$ _____ 0.

9. If the graph of $y = P(x)$ is below the x-axis at $x = c$, then $P(c)$ _____ 0.

6.6 Quick Review

1. Solve $3x - 45 = 0$.

2. Solve $2x + 14 = 0$.

3. Solve $5x - 7 = 0$.

4. Determine whether $x = \dfrac{2}{3}$ is a solution of
   $(3x - 2)(x + 5) = 0$.

5. Determine whether $x = 3$ is a solution of
   $(x - 3)(x + 5) = 2$.

6.6 Exercises

Objective 1 Identify a Quadratic Equation and Write It in Standard Form

In Exercises 1 and 2, match each algebraic expression with the most appropriate description.

1. **a.** $2x - 3 = 0$ **A.** A quadratic equation with two distinct real solutions
   b. $2x - 3 = 2x + 3$ **B.** A quadratic polynomial but not an equation
   c. $(2x - 3)(x + 4) = 0$ **C.** A linear equation with no real solution
   d. $2x^2 + 5x - 12$ **D.** A linear equation with one real solution

2. **a.** $x^2 - 25 = 0$ **A.** A linear equation with one real solution
   b. $-10x + 25 = 0$ **B.** A quadratic equation with two distinct real solutions
   c. $x^2 - 10x + 25$ **C.** A cubic equation with three solutions
   d. $x^3 - 25x = 0$ **D.** A quadratic polynomial but not an equation

In Exercises 3 and 4, write each quadratic equation in the standard form $ax^2 + bx + c = 0$ and identify a, b, and c.

3. **a.** $2x^2 - 7x + 3 = 0$ **b.** $8x^2 = 3x$
   c. $7x^2 = -5$ **d.** $(x - 1)(2x + 1) = 3$

4. a. $3x^2 = 5x - 2$ **b.** $3x^2 = 17$
  c. $2x^2 = 17x$ **d.** $(3x + 1)(x - 2) = 3$

Objective 2 Use Factoring to Solve Selected Quadratic Equations

In Exercises 5–50, solve each equation.

5. $(m - 8)(m + 17) = 0$ **6.** $(m + 15)(m - 3) = 0$

7. $(2n - 5)(3n + 1) = 0$ **8.** $(5n + 2)(n - 6) = 0$

9. $z(2z + 7) = 0$ **10.** $z(6z + 1) = 0$

11. $v^2 - 121 = 0$ **12.** $v^2 - 169 = 0$

13. $x^2 + 3x + 2 = 0$ **14.** $x^2 + 5x + 4 = 0$

15. $y^2 - 3y = 18$ **16.** $t^2 + 4t = 12$

17. $3v^2 = -v$ **18.** $5v^2 = v$

19. $2w^2 = 7w + 15$ **20.** $3w^2 = 17w + 6$

21. $6x^2 + 19x + 10 = 0$ **22.** $10x^2 - 17x + 3 = 0$

23. $x^2 = 11x - 24$ **24.** $x^2 = 2x + 48$

25. $70z^2 = 5z + 15$ **26.** $8v^2 + 24v = 0$

27. $9x^2 = 25$ **28.** $4x^2 = 9$

29. $r(r + 3) = 10$ **30.** $(r + 6)(r - 1) = -10$

31. $\dfrac{x^2}{2} - \dfrac{x}{2} - 1 = 0$ **32.** $\dfrac{x^2}{3} - \dfrac{x}{3} - 2 = 0$

33. $\dfrac{x^2}{10} - \dfrac{x}{5} = \dfrac{3}{10}$ **34.** $\dfrac{x^2}{4} + \dfrac{x}{2} = \dfrac{15}{4}$

35. $\dfrac{m^2}{18} - \dfrac{m}{6} - 1 = 0$ **36.** $\dfrac{m^2}{20} - \dfrac{m}{4} + \dfrac{1}{5} = 0$

37. $\dfrac{x^2}{12} - \dfrac{x}{3} - 1 = 0$ **38.** $\dfrac{x^2}{18} + \dfrac{x}{6} - 1 = 0$

39. $(v - 12)(v + 1) = -40$

40. $(v - 5)(v + 6) = 12$

41. $(v + 2)(v + 1) = 8v + 2$

42. $(v + 5)(v + 3) = 5v + 25$

43. $(2w - 3)^2 = 25$

44. $(3w + 2)^2 = 16$

45. $(3x - 8)(x + 1) = (x + 1)(x - 3)$

46. $(x + 1)(5x + 1) = (x + 2)(x + 3)$

47. $(4x + 5)(x + 5) = 45x$

48. $(3x + 2)(3x - 4) = -3(4x + 3)$

49. $9m^2 = 42m - 49$

50. $25m^2 = 90m - 81$

Objective 3 Construct a Quadratic Equation with Given Solutions

In Exercises 51–58, construct a quadratic equation in x with integer coefficients with the given solutions.

51. $x = 1$ and $x = 5$ **52.** $x = 2$ and $x = 6$

53. $x = -3$ and $x = 6$ **54.** $x = 4$ and $x = -5$

55. $x = -\dfrac{1}{2}$ and $x = \dfrac{3}{4}$ **56.** $x = \dfrac{1}{3}$ and $x = -\dfrac{2}{5}$

57. $x = -\dfrac{7}{5}$ and $x = -\dfrac{2}{3}$ **58.** $x = -\dfrac{6}{5}$ and $x = -\dfrac{2}{7}$

Objective 4 Solve a Quadratic Inequality by Using the x-intercepts of the Corresponding Graph

In Exercises 59–62, use the graph to determine the x-values that satisfy each equation and inequality.

59. $P(x) = x^2 - 2x - 15$
  a. $x^2 - 2x - 15 = 0$
  b. $x^2 - 2x - 15 < 0$
  c. $x^2 - 2x - 15 > 0$

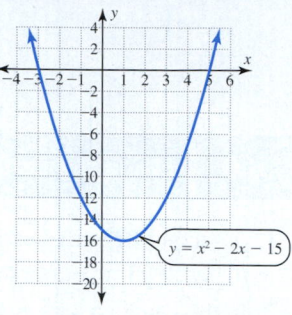

$y = x^2 - 2x - 15$

60. $P(x) = x^2 + 2x - 8$
  a. $x^2 + 2x - 8 = 0$
  b. $x^2 + 2x - 8 < 0$
  c. $x^2 + 2x - 8 > 0$

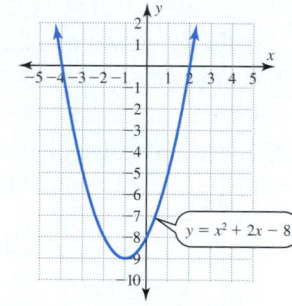

$y = x^2 + 2x - 8$

61. $P(x) = -x^2 - 3x + 10$
  a. $-x^2 - 3x + 10 = 0$
  b. $-x^2 - 3x + 10 < 0$
  c. $-x^2 - 3x + 10 > 0$

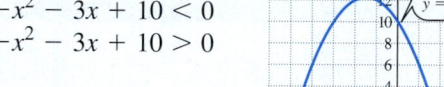
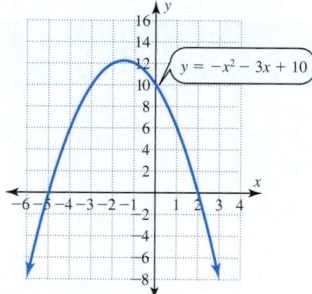

$y = -x^2 - 3x + 10$

62. $P(x) = -x^2 + 3x + 4$
  a. $-x^2 + 3x + 4 = 0$
  b. $-x^2 + 3x + 4 < 0$
  c. $-x^2 + 3x + 4 > 0$

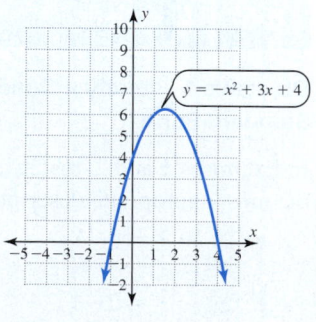

$y = -x^2 + 3x + 4$

Review and Concept Development

63. Use the table for $y = x^2 + 6x - 40$ to:
  a. Determine the zeros of $x^2 + 6x - 40$.
  b. Determine the x-intercepts of the graph of $y = x^2 + 6x - 40$.
  c. Factor $x^2 + 6x - 40$.
  d. Give the solutions of $x^2 + 6x - 40 = 0$.

x	$y = x^2 + 6x - 40$
-10	0
-8	-24
-6	-40
-4	-48
-2	-48
0	-40
2	-24
4	0

64. Use the table for $y = x^2 - 8x - 48$ to:
  a. Determine the zeros of $x^2 - 8x - 48$.
  b. Determine the x-intercepts of the graph of $y = x^2 - 8x - 48$.
  c. Factor $x^2 - 8x - 48$.
  d. Give the solutions of $x^2 - 8x - 48 = 0$.

x	$y = x^2 - 8x - 48$
-8	80
-4	0
0	-48
4	-64
8	-48
12	0
16	80

65. Use the graph to:
  a. Determine the x-intercepts of the graph of $y = x^3 - x^2 - 2x$.
  b. Determine the zeros of the polynomial $x^3 - x^2 - 2x$.
  c. Factor the polynomial $x^3 - x^2 - 2x$.
  d. Give the solutions of $x^3 - x^2 - 2x = 0$.

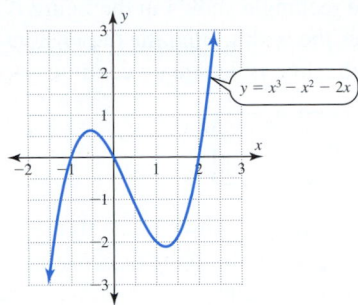

66. Use the graph to:
  a. Determine the x-intercepts of the graph of $y = x^3 - 3x^2 + 2x$.
  b. Determine the zeros of the polynomial $x^3 - 3x^2 + 2x$.
  c. Factor the polynomial $x^3 - 3x^2 + 2x$.
  d. Give the solutions of $x^3 - 3x^2 + 2x = 0$.

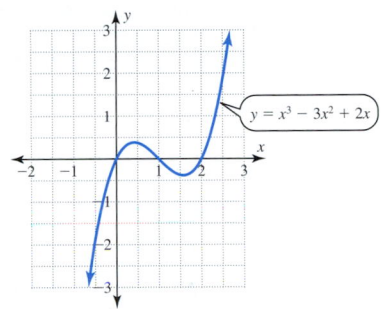

Comparison of Solving and Simplifying

Exercises 67–70 have two parts. In part **a**, solve the equation, and in part **b**, perform the indicated operations and simplify the result.

Solve   **Simplify**

67. a. $(5m - 3)(m - 2) = 0$ **b.** $(5m - 3)(m - 2)$

68. a. $(4x - 3)(5x + 2) = 0$ **b.** $(4x - 3)(5x + 2)$

69. a. $4(x - 3)(5x + 2) = 0$ **b.** $4(x - 3)(5x + 2)$

70. a. $x(x - 3)(5x + 2) = 0$ **b.** $x(x - 3)(5x + 2)$

Applying Technology

In Exercises 71–74, use a graphing calculator or a graphing utility to graph $y = P(x)$ and complete this table.

Polynomial $P(x)$	Solutions of $P(x) = 0$	Solutions of $P(x) < 0$	Solutions of $P(x) > 0$
71. $(x - 2)(x + 7)$			
72. $-2(x + 4)(x - 2)$			
73. $-3x^2 - 9x + 12$			
74. $x^2 + 7x - 8$			

Review and Concept Development (top right continued)

d. Give the solutions of $x^3 - x^2 - 2x = 0$.

Connecting Algebra to Geometry

75. The length of the rectangle shown in the figure is 2 cm more than 3 times the width. Find the dimensions of this rectangle if the area is 33 square centimeters (cm^2).

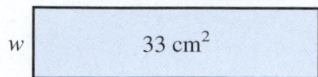

w 33 cm^2

76. The length of the rectangle shown in the figure is 5 cm more than twice the width. Find the dimensions of this rectangle if the area is 52 cm^2.

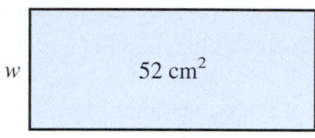

w 52 cm^2

77. The base of a triangle in the figure is 2 m longer than the height. Find the base if the area of this triangle is 24 m^2.

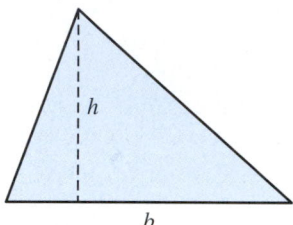

h

b

78. The base of the triangle shown in the figure is 10 cm longer than the height. Determine the height of the triangle if its area is 48 cm^2.

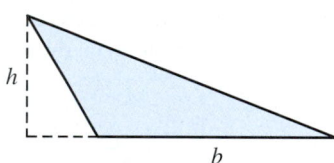

h

b

79. A metal sheet 60 cm wide is used to form a trough by bending up each side as illustrated in the figure. Determine the height of each side if the cross-sectional area is 450 cm^2.

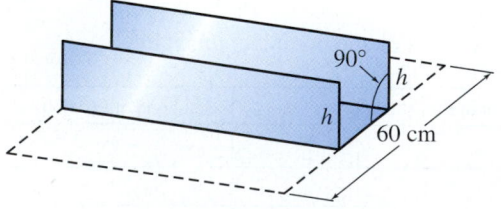

90° h

h

60 cm

80. Determine the height of each side of the metal sheet in Exercise 79 if the cross-sectional area is 432 cm^2.

Group discussion questions

81. Discovery Question Each of these polynomials is prime over the integers, but can be factored over the irrational numbers. Factor these polynomials and solve each equation.

$P(x)$	Factors of $P(x)$	Solutions of $P(x) = 0$
Example: $x^2 - 3$	$(x + \sqrt{3})(x - \sqrt{3})$	$x = -\sqrt{3}, x = \sqrt{3}$
a. $x^2 - 5$	$(x + \sqrt{5})(x - \sqrt{5})$	
b. $x^2 - 7$		$x = -\sqrt{7}, x = \sqrt{7}$
c. $x^2 - 11$		

82. Error Analysis A student solved the equation $(x - 3)(x + 5) = 9$ as follows:

$$(x - 3)(x + 5) = 9$$

$$x - 3 = 9 \qquad x + 5 = 9$$
$$x = 12 \qquad\quad x = 4$$

Find the error in this solution and determine the correct solution.

83. Challenge Question Solve each equation for x.
   a. $(x + y)(x - 3y) = 0$
   b. $x^2 - 5xy - 6y^2 = 0$
   c. $6x^2 - xy = 2y^2$
   d. $x^2 + 2xy = 15y^2$

84. Discovery Question
   a. Graph each of these equations, and then describe the relationship of the graph to the x-axis:
      $y = x^2 - 2x + 1, y = x^2 + 2x + 1,$
      $y = x^2 - 4x + 4,$ and $y = x^2 + 4x + 4.$
   b. Factor each of these polynomials, and then describe any relationship that you observe between the factors and the relationship of the corresponding graph to the x-axis: $x^2 - 2x + 1, x^2 + 2x + 1,$ $x^2 - 4x + 4,$ and $x^2 + 4x + 4.$
   c. Factor $x^2 + 18x + 81$, and predict the relationship of the graph of $y = x^2 + 18x + 81$ to the x-axis.

Match each algebraic equation with the property it illustrates.

1. $(xy)z = x(yz)$ **A.** Associative property of addition
2. $(x + y) + z = x + (y + z)$ **B.** Associative property of multiplication
3. $w(xy + z) = w(yx + z)$ **C.** Commutative property of addition
4. $w(xy + z) = wxy + wz$ **D.** Commutative property of multiplication
5. $w(xy + z) = w(z + xy)$ **E.** Distributive property of multiplication over addition

Chapter 6   Key Concepts

1. **Quadratic Trinomial:** $ax^2 + bx + c$ is a quadratic trinomial; ax^2 is called the second-degree term or the quadratic term, bx is the first-degree term or the linear term, and c is the constant term.

2. **Factors and Terms:** Factors are constants or variables that are multiplied together to form a product. A single number or an indicated product of factors is called a term. The terms in an algebraic expression are separated from each other by plus or minus symbols.

3. **Factored Form of a Polynomial:** The factored form of a polynomial $P(x)$ is useful for quickly determining the solutions of $P(x) = 0$, the zeros of $P(x)$, and the x-intercepts of the graph of $y = P(x)$.

4. **Prime Polynomial:** A polynomial is prime over the integers if its only factorization with integer coefficients must involve 1 or -1 as one of the factors.

5. **Completely Factoring a Polynomial:** A polynomial is factored completely if each factor other than a common monomial factor is prime.

6. **Factorable Trinomials:** A trinomial $ax^2 + bx + c$ is factorable into a pair of binomial factors with integer coefficients if and only if there are two integers whose product is ac and whose sum is b. Otherwise, the trinomial is prime over the integers.

7. **GCF of a Polynomial:** The greatest common factor of a polynomial is the factor of each term that contains both of the following:
   • Largest possible numerical coefficient
   • Largest possible exponent on each variable factor

8. **Factoring $ax^2 + bx + c$ by the trial-and-error method:**
   Step 1 Factor out the GCF. If a is negative, factor out -1.
   Step 2 Make a form for the two binomial factors, and fill in the obvious information, such as the sign pattern.
   Step 3 List the possible factors of the first and last terms that fit this pattern.
   Step 4 Select the factors that yield the correct middle term. If all possibilities fail, the polynomial is prime.

9. **Factoring $ax^2 + bx + c$ by the AC method**
   Step 1 Factor out the GCF. If a is negative, factor out -1.
   Step 2 Find two factors of ac whose sum is b. If all possibilities fail, the polynomial is prime.
   Step 3 Rewrite the linear term of $ax^2 + bx + c$ so that b is the sum of the factors from step 2.
   Step 4 Factor the polynomial from step 3 by grouping the terms and factoring the GCF out of each pair of terms.

10. **Factoring by Grouping:** To factor polynomials of four or more terms, we can first group terms together that we already know how to factor. This method relies on the ability to spot common factors and special forms.

11. **Strategy for Factoring a Polynomial Over the Integers:** After factoring out the GCF, proceed as follows:
   Binomials: Factor special forms:
   • Difference of two squares
     $A^2 - B^2 = (A + B)(A - B)$
   • Difference of two cubes
     $A^3 - B^3 = (A - B)(A^2 + AB + B^2)$
   • Sum of two cubes
     $A^3 + B^3 = (A + B)(A^2 - AB + B^2)$
   • $A^2 + B^2$ is prime

     The sum of two squares is prime if x^2 and y^2 are only second-degree terms and have no common factor other than 1.

   Trinomials: Factor the forms that are perfect squares:
   • Perfect square trinomial; a square of a sum
     $A^2 + 2AB + B^2 = (A + B)^2$
   • Perfect square trinomial; a square of a difference
     $A^2 - 2AB + B^2 = (A - B)^2$
   • Factor trinomials that are not perfect squares by inspection if possible; otherwise, use the trial-and-error method or the AC method.

   Polynomials of Four or More Terms: Factor by grouping.

12. Quadratic Equation: If a, b, and c are real constants and $a \neq 0$, then $ax^2 + bx + c = 0$ is the standard form of a quadratic equation in x; ax^2 is called the quadratic term, bx is called the linear term, and c is called the constant term.

13. Zero-Factor Principle: If a and b are real algebraic expressions, then $ab = 0$ if and only if $a = 0$ or $b = 0$.

14. Solving Quadratic Equations by Factoring

Step 1 Write the equation in standard form, with the right side 0.

Step 2 Factor the left side of the equation.

Step 3 Set each factor equal to 0.

Step 4 Solve the resulting first-degree equations.

15. Equivalent Statements About Linear Factors of a Polynomial: For a real constant c and a real polynomial $P(x)$, the following statements are equivalent:
- $x - c$ is a factor of $P(x)$.
- $x = c$ is a solution of $P(x) = 0$.

- $P(c) = 0$; that is, c is a zero of $P(x)$.
- $(c, 0)$ is an x-intercept of the graph of $y = P(x)$.

16. Solutions of a Quadratic Inequality
- The solution of $ax^2 + bx + c > 0$ is the set of x-values for which the graph of $y = ax^2 + bx + c$ is above the x-axis.
- The solution of $ax^2 + bx + c < 0$ is the set of x-values for which the graph of $y = ax^2 + bx + c$ is below the x-axis.

17. Solutions of Equations and Inequalities: If $P(x)$ is a real polynomial and c is a real number, then

Verbally	Algebraically	Graphically
c is a solution of $P(x) = 0$.	$P(c) = 0$	$(c, 0)$ is an x-intercept of the graph of $y = P(x)$.
c is a solution of $P(x) < 0$.	$P(c) < 0$	At $x = c$, the graph is below the x-axis.
c is a solution of $P(x) > 0$.	$P(c) > 0$	At $x = c$, the graph is above the x-axis.

Chapter 6 | Review Exercises

Factoring

In Exercises 1–4, factor the GCF out of each polynomial.

1. $4x - 36$
2. $2x^2 - 10x$
3. $12ax^2 - 24ax$
4. $5x^3 - 15x$

In Exercises 5–8, use grouping to factor each polynomial.

5. $ax + bx + ay + by$
6. $ax + 2ay - 3x - 6y$
7. $6x^2 + 21x - 10x - 35$
8. $24x^2 - 20x + 18x - 15$

In Exercises 9–40, factor each polynomial completely over the integers.

9. $x^2 - 4$
10. $4x^2 - 1$
11. $7x^2 - 28$
12. $2m^2 - 128$
13. $x^2 - 11x + 18$
14. $m^2 + 10m + 21$
15. $10x^2 + 90x + 140$
16. $5x^2 - 45x + 40$
17. $x^2 + 4x + 4$
18. $m^2 + 10m + 25$
19. $3x^2 - 18x + 27$
20. $2m^2 - 16m + 32$
21. $-4x^2 + 12xy - 9y^2$
22. $-9x^2 + 30xy - 25y^2$
23. $ax + 2ay - 7x - 14y$
24. $x^3y - 5x^2y + 4xy$
25. $x^2(a + b) - 25(a + b)$
26. $11x^2 - 11y^2 + 33x + 33y$
27. $v^2 + 2v + 1 - w^2$
28. $5a(2x - 3y) + 3b(3y - 2x)$
29. $v^2 + 9w^2$
30. $v^2 - 9w^2$
31. $100x^2 - 49y^2$
32. $100x^2 + 49y^2$
33. $4x^2 + 2xy - 30y^2$
34. $6x^2 - 7xy - 20y^2$

35. $20x^2 + 50x - 30$
36. $6x^2 + 61xy + 10y^2$
37. $v^4 - 1$
38. $81w^4 - 16$
39. $a^2(x + 5y) + 2ab(x + 5y) + b^2(x + 5y)$
40. $9x^2 - y^2 + 3x - y$

Identifying Prime Polynomials

In Exercises 41–44, use the given table to factor each polynomial completely over the integers. (*Hint:* Some of these polynomials are prime.)

41.

Factors of −100		Sum of Factors
1	−100	−99
2	−50	−48
4	−25	−21
5	−20	−15
10	−10	0

a. $x^2 - 48x - 100$
b. $x^2 - 15x - 100$
c. $x^2 - 100$
d. $x^2 - 20x - 100$

42.

Factors of −48		Sum of Factors
−1	48	47
−2	24	22
−3	16	13
−4	12	8
−6	8	2

a. $x^2 + 22x - 48$
b. $x^2 + 13x - 48$
c. $x^2 + 2x - 48$
d. $x^2 + x - 48$

43.

Factors of 100		Sum of Factors
1	100	101
2	50	52
4	25	29
5	20	25
10	10	20

a. $100x^2 + 52x + 1$ **b.** $4x^2 + 52x + 25$
c. $5x^2 + 29xy + 20y^2$ **d.** $10x^2 + 29xy + 10y^2$

44.

Factors of 48		Sum of Factors
-1	-48	-49
-2	-24	-26
-3	-16	-19
-4	-12	-16
-6	-8	-14

a. $48x^2 - 19x + 1$ **b.** $3x^2 - 26x + 16$
c. $2x^2 - 49xy + 24y^2$ **d.** $16x^2 - 16xy + 3y^2$

Sum and Difference of Two Cubes

In Exercises 45–48, factor each polynomial completely over the integers.

45. $8x^3 - y^3$ **46.** $x^3 + 64$

47. $2x^4 - 16x$ **48.** $v^4 - v^3 - v + 1$

Using Factoring to Solve Equations

In Exercises 49–62, solve each equation.

49. $(x - 5)(x + 1) = 0$ **50.** $(2x - 3)(3x + 2) = 0$

51. $x(x - 7)(7x - 2) = 0$ **52.** $v^2 - 4v - 21 = 0$

53. $10y^2 + 13y - 3 = 0$ **54.** $6w^2 = 11w + 21$

55. $\dfrac{v^2}{30} = \dfrac{v}{15} + \dfrac{1}{2}$ **56.** $(x + 6)(x - 2) = 9$

57. $x(x^2 - 36) = 0$ **58.** $2x(x^2 - 10x + 25) = 0$

59. $\dfrac{z^2}{12} = \dfrac{z + 3}{3}$ **60.** $\dfrac{x^2}{9} + \dfrac{9}{4} = -x$

61. $(3w + 2)(w + 1) = (2w + 3)(w - 2)$

62. $(x - 3)(x + 2) = (x + 9)(2x - 9)$

Comparison of Solving and Simplifying

Exercises 63 and 64 have two parts. In part **a,** solve the equation, and in part **b,** perform the indicated operations and simplify the expression.

Solve	Simplify
63. a. $(x - 3)(x + 12) = 0$	**b.** $(x - 3)(x + 12)$
64. a. $(3x - 1)(2x + 5) = 0$	**b.** $(3x - 1)(2x + 5)$

Connecting Concepts

65. Use the table and graph for $y = x^2 + 2x - 24$ to:
a. List the zeros of $x^2 + 2x - 24$.
b. List the x-intercepts of the graph of
$y = x^2 + 2x - 24$.
c. Factor the polynomial $x^2 + 2x - 24$.

d. Give the solutions of $x^2 + 2x - 24 = 0$.

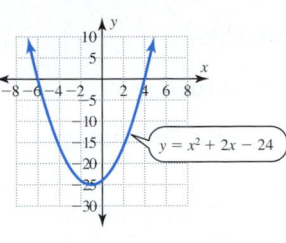

x	$y = x^2 + 2x - 24$
-6	0
-4	-16
-2	-24
0	-24
2	-16
4	0
6	24

66. Use the table and graph for $y = x^3 + 3x^2 - x - 3$ to:
a. List the zeros of $x^3 + 3x^2 - x - 3$.
b. List the x-intercepts of the graph of
$y = x^3 + 3x^2 - x - 3$.
c. Factor the polynomial $x^3 + 3x^2 - x - 3$.
d. Give the solutions of $x^3 + 3x^2 - x - 3 = 0$.

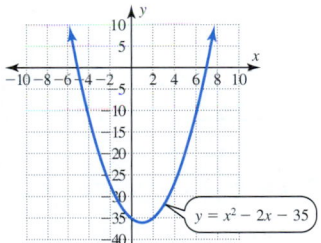

x	$y = x^3 + 3x^2 - x - 3$
-4	-15
-3	0
-2	3
-1	0
0	-3
1	0
2	15

Using x-intercepts to Solve Inequalities

67. Use the graph to determine the x-values that satisfy each equation and inequality.
a. $x^2 - 2x - 35 = 0$
b. $x^2 - 2x - 35 \le 0$
c. $x^2 - 2x - 35 > 0$

$y = x^2 - 2x - 35$

68. Use the graph to determine the x-values that satisfy each equation and inequality.
a. $-x^2 - x + 20 = 0$
b. $-x^2 - x + 20 < 0$
c. $-x^2 - x + 20 \ge 0$

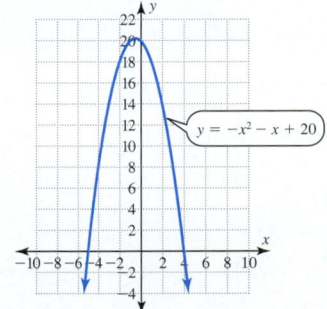

$y = -x^2 - x + 20$

Constructing Equations

69. Construct a quadratic equation with $x = 3$ and $x = -\dfrac{4}{7}$ as solutions.

70. Construct a quadratic equation with $x = -2$ and $x = \dfrac{5}{2}$ as solutions.

Applying Technology

In Exercises 71 and 72, use a graphing calculator or a graphing utility to graph $y = P(x)$ and then use this graph to assist you in completing this table.

Polynomial $P(x)$	x-intercepts of the Graph	Factored Form of $P(x)$	Zeros of $P(x)$	Solutions of $P(x) = 0$
71. $x^2 + 4x - 77$				
72. $x^2 - 20x - 525$				

Connecting Concepts to Applications

73. The difference in the distance traveled in 1 revolution by two bicycle tires of radii r_1 and r_2 is $2\pi r_1 - 2\pi r_2$. Factor this polynomial.

74. The lengths of two booms on circular irrigation systems are L_1 and L_2. The difference in the areas covered by these irrigation systems is $\pi L_1^2 - \pi L_2^2$. Factor this polynomial.

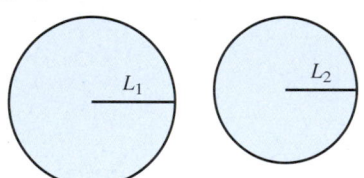

75. The difference in the volumes of two spherical oxygen tanks is $\dfrac{4}{3}\pi r_1^3 - \dfrac{4}{3}\pi r_2^3$. Factor this polynomial.

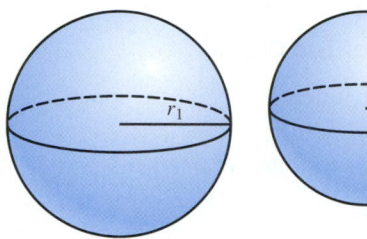

76. The base of the triangle shown in the figure is 3 cm longer than the height. Determine the height of the triangle if its area is 14 cm^2.

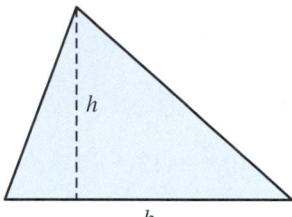

77. A metal sheet 60 cm wide is used to form a trough by bending up each side, as illustrated in the figure. Determine the height of each side if the cross-sectional area is 400 cm^2.

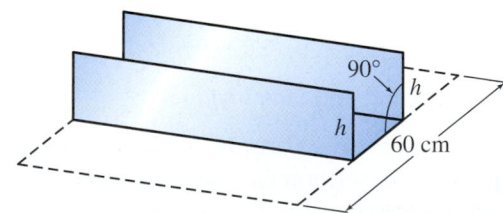

Chapter 6 | Mastery Test

Objective 6.1.1 Factor the GCF out of a Polynomial

1. Factor the GCF out of each polynomial.
  a. $18x^3 - 99x^2$
  b. $4x^3y^2 - 6x^2y^3 + 10xy^4$
  c. $3x(7x - 2) + 4(7x - 2)$
  d. $x(2x + 3y) - 5y(2x + 3y)$

Objective 6.1.2 Factor by Grouping

2. Factor each polynomial using the grouping method.
  a. $7ax + a + 21x + 3$
  b. $12ax - 8a - 15bx + 10b$
  c. $2x^2 - 6x + 4x - 12$
  d. $10x^2 - 15x - 12x + 18$

Objective 6.1.3 Use the Zeros of a Polynomial $P(x)$ and the x-intercepts of the Graph of $y = P(x)$ to Factor the Polynomial

3. Use the table and graph for $y = x^2 + 10x - 24$ to:
   a. List the zeros of $x^2 + 10x - 24$.
   b. List the x-intercepts of the graph of $y = x^2 + 10x - 24$.
   c. Factor the polynomial $x^2 + 10x - 24$.

x	$y = x^2 + 10x - 24$
-33	735
-26	392
-19	147
-12	0
-5	-49
2	0
9	147

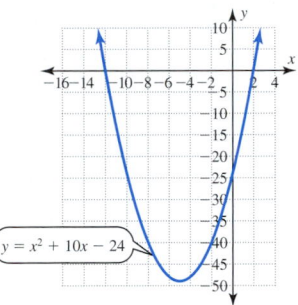

Objective 6.2.1 Factor Trinomials of the Form $x^2 + bx + c$ by the Trial-and-Error Method or by Inspection

4. Factor each trinomial.
   a. $w^2 - 4w - 45$ b. $w^2 + 14w + 45$
   c. $v^2 - 10v + 24$ d. $v^2 + 5v - 36$

Objective 6.2.2 Identify a Prime Trinomial of the Form $x^2 + bx + c$

5. Determine which trinomials are prime, and factor the trinomials that are factorable over the integers.
   a. $x^2 - 5x - 36$ b. $x^2 - 5x + 36$
   c. $x^2 + 17x + 30$ d. $x^2 + 17x - 30$

Objective 6.2.3 Factor Trinomials of the Form $x^2 + bxy + cy^2$ by the Trial-and-Error Method or by Inspection

6. Factor each trinomial.
   a. $x^2 - xy - 12y^2$ b. $x^2 - 13xy + 12y^2$
   c. $a^2 + 30ab + 144b^2$ d. $a^2 + 2ab - 48b^2$

Objective 6.3.1 Factor Trinomials of the Form $ax^2 + bx + c$ by the Trial-and-Error Method

7. Factor each trinomial.
   a. $5x^2 - 33x - 14$ b. $6x^2 - 73x + 12$
   c. $9x^2 + 21x + 10$ d. $12x^2 - 7xy - 12y^2$

Objective 6.3.2 Factor Trinomials of the Form $ax^2 + bx + c$ by the AC Method

8. Use the given table to factor each trinomial using the AC method.

Factors of -72		Sum of Factors
-1	72	71
-2	36	34
-3	24	21
-4	18	14
-6	12	6
-8	9	1

   a. $9x^2 + 21x - 8$ b. $12x^2 + x - 6$
   c. $3x^2 + x - 24$ d. $18x^2 + 71xy - 4y^2$

Objective 6.3.3 Identify a Prime Trinomial of the Form $ax^2 + bx + c$

9. Determine which trinomials are prime, and factor the trinomials that are factorable over the integers.
   a. $8x^2 - 14x - 3$ b. $8x^2 - 14x + 3$
   c. $9x^2 + 9x - 4$ d. $9x^2 + 9x + 4$

Objective 6.4.1 Factor Perfect Square Trinomials

10. Factor each perfect square trinomial.
   a. $x^2 + 14xy + 49y^2$ b. $x^2 - 16xy + 64y^2$
   c. $9x^2 + 60xy + 100y^2$ d. $25x^2 - 110xy + 121y^2$

Objective 6.4.2 Factor the Difference of Two Squares

11. Factor each difference of two squares.
   a. $x^2 - 4y^2$ b. $400a^2 - b^2$
   c. $16v^2 - 49w^2$ d. $36x^2 - 25y^2$

Objective 6.4.3 Factor the Sum or Difference of Two Cubes

12. Factor each sum or difference of two cubes.
   a. $64v^3 - 1$ b. $v^3 + 125$
   c. $8x^3 + 125y^3$ d. $27a^3 - 1{,}000b^3$

Objective 6.5.1 Factor Polynomials by Grouping

13. Use grouping to factor each polynomial.
   a. $2ax + 3a + 2bx + 3b$
   b. $14ax - 6bx - 35ay + 15by$
   c. $a^2 - 4b^2 + a - 2b$
   d. $x^2 + 10xy + 25y^2 - 4$

Objective 6.5.2 Determine the Most Appropriate Method for Factoring a Polynomial

14. Completely factor each polynomial.
   a. $5x^2 - 245$
   b. $2ax^2 + 20ax + 50a$
   c. $5av^2 - 5ax^2 + 20axy - 20ay^2$
   d. $19x^3 - 19$

Objective 6.6.1 Identify a Quadratic Equation and Write It in Standard Form

15. Determine which of these are quadratic equations in one variable. Write each quadratic equation in standard form, and identify a, b, and c.
   a. $2x - 5 = x^2$ b. $2x - 5 = x^3$
   c. $2x - 5$ d. $2x^2 - 5 = 0$

Objective 6.6.2 Use Factoring to Solve Selected Quadratic Equations

16. Solve each equation.
   a. $(2x + 1)(x - 3) = 0$
   b. $x^2 - 2x - 99 = 0$
   c. $(2x - 1)(x + 5) = 6$
   d. $(3x + 1)(2x + 7) = 4(x + 1)$

Objective 6.6.3 Construct a Quadratic Equation with Given Solutions

17. Construct a quadratic equation with the given solutions.

 a. -7 and 9

 b. -13 and 13

 c. $\dfrac{1}{7}$ and -4

 d. $\dfrac{3}{7}$ and $\dfrac{4}{3}$

Objective 6.6.4 Solve a Quadratic Inequality by Using the x-intercepts of the Corresponding Graph

18. Use the graph to determine the x-values that satisfy each equation and inequality.

 a. $x^2 + 2x - 15 = 0$

 b. $x^2 + 2x - 15 < 0$

 c. $x^2 + 2x - 15 > 0$

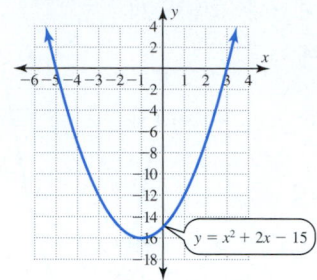

$y = x^2 + 2x - 15$

Chapter 6 | Group Project

Estimating the Cost for a Highway Construction Project

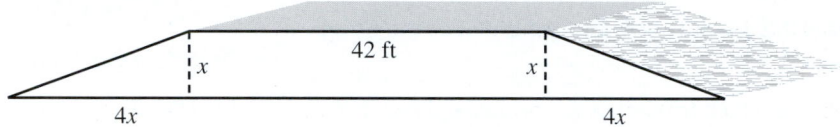

An engineer preparing a bid for a highway construction project made this sketch. The project will require dirt to be moved and filled in to a depth of x ft on a 1-mile section of roadway that is currently too low. The model has a trapezoidal cross section.

1. Determine the slope of the embankment of the left side of the roadway.

2. Using the formula for the area of a trapezoid, $A = \dfrac{1}{2}h(a + b)$, write an equation in terms of x for a cross-sectional area (in ft^2) of the roadway.

3. Write an equation in terms of x for the volume (in yd^3) of dirt that must be filled in for a 1-mile section of this roadway. (*Hint:* 1 mile = 5,280 ft; 1 yd^3 = 27 ft^3.)

4. Write an equation in terms of x for the cost (in $) to fill in this dirt for a 1-mile section of this roadway if it costs \$12 per yd^3 to move the dirt from a nearby location.

5. Write an equation in terms of x for the cost (in $) to fill in this dirt for a 1-mile section of this roadway if it costs \$22 per yd^3 to move the dirt from a more distant location.

6. Using either a calculator or a spreadsheet, complete this table for fill depths of 5, 10, 15, and 20 ft.

	A	B	C	D	E
1	Height (ft)	Area (sq ft)	Vol (cubic yd)	Cost ($) @ $12	Cost ($) @ $22
2	5	310	60622	$727,467	$1,333,689
3	10				
4	15				
5	20				

Solving Quadratic Equations

Chapter Outline

Brake Design

The engineering team designing a car must make many important decisions. Team members must first consider the type of driving for which the car will be used. Will it run on the highway, off road, or on a racetrack? This information will determine not only the engine to be used but also the steering, suspension, and braking systems.

To design the braking system, the team must consider all the components: the pistons, cylinders, levers, friction pads and surfaces, and tires. The car's ability to stop depends on how these components have been designed and put together. To calculate the predicted stopping distance of the car, the design team uses the specifications of the various parts of the braking system to set up a quadratic equation. A sample equation might look like

$$d = 0.01v^2 + 0.03v$$

The stopping distance d is given in meters and the velocity of the car in kilometers per hour is represented by v. The coefficients 0.01 and 0.03 are based on the braking system's design and the typical driver's reaction time (usually, a little less than 1 second). Exercise 39 in Section 7.3 considers $100 = 0.01v^2 + 0.03v$ to determine the speed v in km/h that a car is going when it requires 100 m to stop.

Section 7.1 | Extraction of Roots and Properties of Square Roots

When algebra and geometry united, they lent each other mutual support and have marched together towards perfection —JOSEPH-LOUIS LAGRANGE

Objectives:

1. Solve quadratic equations by extraction of roots.
2. Use the product rule for radicals to simplify square roots.
3. Use the quotient rule for radicals to simplify square roots.
4. Simplify expressions of the form $\dfrac{\sqrt{a}}{\sqrt{b}}$ by rationalizing the denominator.

Quadratic equations are equations that can be written in the standard form $ax^2 + bx + c = 0$. In Section 6.6 we solved equations like $x^2 - 4 = 0$ by factoring the left side of the equation over the integers.

Why Do I Need to Learn Another Method for Solving Quadratic Equations?

Not all quadratic equations can be solved by factoring over the integers. When the left side of a quadratic equation cannot be factored over the integers, we need another method to solve these equations. This is illustrated below.

Quadratic Equation	**Solution by Factoring**
$x^2 - 4 = 0$	$x^2 - 4 = 0$
	$(x + 2)(x - 2) = 0$
	$x + 2 = 0 \quad$ or $\quad x - 2 = 0$
	$x = -2 \quad$ or $\quad x = 2$
$x^2 - 3 = 0$	This equation cannot be solved by factoring over the integers.
$x^2 - \dfrac{1}{5} = 0$	This equation cannot be solved by factoring over the integers.

1. Solve Quadratic Equations by Extraction of Roots

The equation $x^2 - 3 = 0$ can be rewritten as $x^2 = 3$. The two solutions of this equation are the two numbers whose square is 3. One of these numbers is positive and the other is negative, and we can denote them by $x = \sqrt{3}$ and $x = -\sqrt{3}$. This is sometime condensed to $x = \pm\sqrt{3}$. This process of solving quadratic equations is called **extraction of roots.**

Extraction of Roots	Example
If k is a positive real number, then the equation $x^2 = k$ has two real solutions, $x = \pm\sqrt{k}$: $x = \sqrt{k} \quad$ and $\quad x = -\sqrt{k}$.	$x^2 = 3$ $x = \sqrt{3} \quad$ or $\quad x = -\sqrt{3}$

If $k = 0$ in $x^2 = k$, then the only solution of the equation $x^2 = 0$ will be $x = 0$. If $k < 0$, the two solutions of $x^2 = k$ will not be real numbers. We will examine solutions of this type in Section 7.5.

What Is the First Step in Solving a Quadratic Equation by Extraction of Roots?

The first step is to write the left side of the equation as a perfect square. This step is needed in Example 1(c).

Example 1   Solving Quadratic Equations by Extraction of Roots

Solve each quadratic equation by using extraction of roots to obtain exact solutions. Then approximate each irrational solution to the nearest hundredth.

(a) $x^2 = 49$ **(b)** $v^2 = 7$ **(c)** $6w^2 - 30 = 0$

Solution

(a) $x^2 = 49$

$\qquad x = \pm\sqrt{49}$

$\qquad x = 7 \quad \text{or} \quad x = -7$

This solution uses extraction of roots. The equation could also be solved by factoring $x^2 - 49 = 0$ as $(x - 7)(x + 7) = 0$.

(b) $v^2 = 7$

$\qquad v = \pm\sqrt{7}$

$\qquad v = \sqrt{7} \quad \text{or} \quad v = -\sqrt{7}$

$\qquad v \approx 2.65 \quad \text{or} \quad v \approx -2.65$

Use extraction of roots to determine the exact solutions of $v = \pm\sqrt{7}$. These solutions cannot be determined by factoring over the integers and they cannot be simplified. They can be approximated on a calculator as illustrated here.

(c) $6w^2 - 30 = 0$

$\qquad 6w^2 = 30$

$\qquad w^2 = 5$

$\qquad w = \pm\sqrt{5}$

$\qquad w = \sqrt{5} \quad \text{or} \quad w = -\sqrt{5}$

$\qquad w \approx 2.24 \quad \text{or} \quad w \approx -2.24$

First add 30 to both sides of the equation. Then divide both sides of the equation by 6 to isolate the perfect square on the left side of the equation. Then solve this equation by extraction of roots.

Self-Check 1

Solve each quadratic equation by using extraction of roots to obtain exact solutions. Then approximate each irrational solution to the nearest hundredth.

a. $m^2 = 16$ **b.** $y^2 = 17$ **c.** $5x^2 - 15 = 0$

2. Use the Product Rule for Radicals to Simplify Square Roots

Do I Need to Use the Product Rule for Radicals to Solve Quadratic Equations?

The solution of quadratic equations often produces answers that contain square roots. If all you need is decimal approximations of these values, then you can do this directly with a calculator without simplifying the radicals. However, for some situations it is desirable to write the radical expressions in as simple a form as possible. If you need a simplified form, then you will need to use the product rule for radicals. This rule is stated below for square roots.

Product Rule for Square Roots	Numerical Example	
If $\sqrt{x}$ and $\sqrt{y}$ are both real numbers, then $\sqrt{xy} = \sqrt{x}\,\sqrt{y}$.	$\sqrt{12} = \sqrt{4 \cdot 3}$ $\qquad = \sqrt{4}\,\sqrt{3}$ $\qquad = 2\sqrt{3}$	`√(12)` `         3.464101615` `2√(3)` `         3.464101615`

Example 2   Using the Product Rule for Square Roots

Simplify each square root using the product rule for square roots.

(a) $\sqrt{9 \cdot 5}$ **(b)** $\sqrt{18}$ **(c)** $\sqrt{125}$

Solution

(a) $\sqrt{9 \cdot 5} = \sqrt{9}\,\sqrt{5}$ Use the product rule for square roots.

$\quad\quad\quad = 3\sqrt{5}$ Then simplify $\sqrt{9}$.

(b) $\sqrt{18} = \sqrt{9 \cdot 2}$ First determine the largest perfect square that is a

$\quad\quad = \sqrt{9}\sqrt{2}$ factor of 18. Then use the factored form to simplify

$\quad\quad = 3\sqrt{2}$ this expression.

(c) $\sqrt{125} = \sqrt{25 \cdot 5}$ First determine the largest perfect square that is a

$\quad\quad\quad = \sqrt{25}\sqrt{5}$ factor of 125. Then use the factored form to simplify

$\quad\quad\quad = 5\sqrt{5}$ this expression.

Self-Check 2

Simplify each expression using the product rule for square roots.

a. $\sqrt{16 \cdot 3}$ **b.** $\sqrt{50}$

Example 3 illustrates a quadratic equation whose solutions can be simplified using the product rule for square roots.

Example 3   Solving a Quadratic Equation by Extraction of Roots

Solve $x^2 - 12 = 0$ by using extraction of roots to obtain exact solutions. Then approximate each irrational solution to the nearest thousandth.

Solution

$x^2 - 12 = 0$ Add 12 to both sides of the equation to isolate the

$x^2 = 12$ perfect square on the left side.

$x = \pm\sqrt{12}$ Solve this equation using extraction of roots.

$x = \pm\sqrt{4 \cdot 3}$

$x = \pm\sqrt{4}\sqrt{3}$ The decimal approximation can be obtained directly

$x = \pm 2\sqrt{3}$ from $\sqrt{12}$ or from the simplified form $2\sqrt{3}$.

$x \approx 3.464$ or $x \approx -3.464$ $\sqrt{12} \approx 3.464$ and $2\sqrt{3} \approx 3.464$

Self-Check 3

Solve each quadratic equation by using extraction of roots to obtain an exact solution. Then approximate each irrational solution to the nearest thousandth.

a. $x^2 = 8$ **b.** $2y^2 - 56 = 0$

The ability to check your work is important not only on tests but also to maintain the quality of production in the workplace. There are different ways to check your work other than simply proofreading the steps that have already been performed. To check the solutions of an equation, we can substitute these solutions back into the original equation. We can do this either by using pencil and paper or by using a calculator or spreadsheet. We can also work the same problem a different way to see whether the two solutions agree.

This is not proof that the solutions are correct, but it is still a very useful practice. In the following illustration, we check our work in Example 3 by substituting the solutions back into the equation $x^2 - 12 = 0$. Then we use the **TRACE** feature to verify that the solutions correspond to the x-intercepts of the graph of $y = x^2 - 12$. This was first shown in Technology Perspective 2.3.1: Using the **TRACE** Feature.

Checking Solutions by Substitution

To check $2\sqrt{3}$:

$$x^2 - 12 = 0$$
$$(2\sqrt{3})^2 - 12 \overset{?}{=} 0$$
$$4(3) - 12 \overset{?}{=} 0$$
$$12 - 12 \overset{?}{=} 0 \text{ checks}$$

To check $-2\sqrt{3}$:

$$x^2 - 12 = 0$$
$$(-2\sqrt{3})^2 - 12 \overset{?}{=} 0$$
$$4(3) - 12 \overset{?}{=} 0$$
$$12 - 12 \overset{?}{=} 0 \text{ checks}$$

Checking Using Calculator

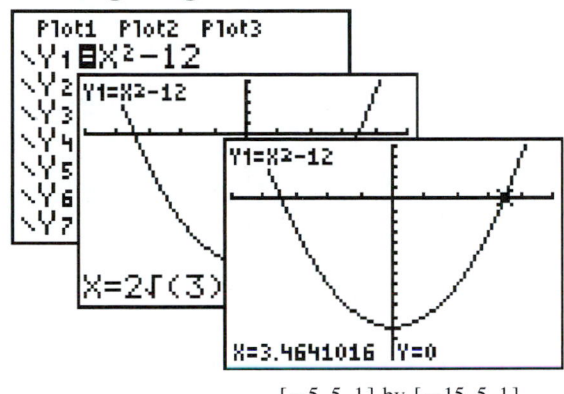

$[-5, 5, 1]$ by $[-15, 5, 1]$

To check $2\sqrt{3}$ using a calculator, graph the equation and use the **TRACE** feature to enter $x = 2\sqrt{3}$ and check this solution. The solution $-2\sqrt{3}$ can be checked using the same procedure.

3. Use the Quotient Rule for Radicals to Simplify Square Roots

Another rule that can be used to simplify square roots is the quotient rule for radicals. This rule is stated below for square roots.

Quotient Rule for Square Roots	Numerical Example

If $\sqrt{x}$ and $\sqrt{y}$ are both real numbers and $y > 0$, then $\sqrt{\dfrac{x}{y}} = \dfrac{\sqrt{x}}{\sqrt{y}}$.	$\sqrt{\dfrac{25}{36}} = \dfrac{\sqrt{25}}{\sqrt{36}}$ $= \dfrac{5}{6}$
	
	 ```√(25/36)►Frac```
	``` 5/6```
	```√(25)/√(36)►Frac```
	``` 5/6```

Example 4 Using the Quotient Rule for Square Roots

Simplify each square root using the quotient rule for square roots.

(a) $\sqrt{\dfrac{16}{81}}$ **(b)** $\dfrac{\sqrt{18}}{\sqrt{2}}$

Solution

(a) $\sqrt{\dfrac{16}{81}} = \dfrac{\sqrt{16}}{\sqrt{81}}$ Use the quotient rule for square roots to examine the square root of the numerator and denominator separately.

$$= \dfrac{4}{9}$$

(b) $\dfrac{\sqrt{18}}{\sqrt{2}} = \sqrt{\dfrac{18}{2}}$ Use the quotient rule for square roots to rewrite the quotient of these two radicals as a single radical. Reduce the fraction under this square root symbol.

$= \sqrt{9}$ Then evaluate $\sqrt{9}$.

$= 3$

Self-Check 4

Simplify each expression using the quotient rule for square roots.

a. $\sqrt{\dfrac{9}{25}}$ **b.** $\dfrac{\sqrt{75}}{\sqrt{3}}$

Example 5   Solving a Quadratic Equation by Extraction of Roots

Solve $4v^2 = 5$ by using extraction of roots to obtain exact solutions. Then approximate each irrational solution to the nearest thousandth.

Solution

$4v^2 = 5$

$v^2 = \dfrac{5}{4}$ First isolate v^2 on the left side of the equation.

$v = \pm\sqrt{\dfrac{5}{4}}$ Then solve this equation by extraction of roots.

$v = \pm\dfrac{\sqrt{5}}{\sqrt{4}}$ The decimal approximation can be obtained directly from

$v = \pm\dfrac{\sqrt{5}}{2}$ $\sqrt{\dfrac{5}{4}}$ or from the simplified form $\dfrac{\sqrt{5}}{2}$.

$v \approx 1.118$ or $v \approx -1.118$ $\sqrt{\dfrac{5}{4}} \approx 1.118$ and $\dfrac{\sqrt{5}}{2} \approx 1.118$

Self-Check 5

Solve each quadratic equation by using extraction of roots to obtain exact solutions. Then approximate each irrational solution to the nearest thousandth.

a. $9z^2 = 2$ **b.** $16x^2 - 17 = 0$

---

4. Simplify Expressions of the Form $\dfrac{\sqrt{a}}{\sqrt{b}}$ by Rationalizing the Denominator

The radical expression $\dfrac{2}{\sqrt{5}}$ has an irrational number $\sqrt{5}$ in the denominator and is not

considered to be in simplest form. The equivalent expression $\dfrac{2\sqrt{5}}{5}$ has a rational number 5

in the denominator and is considered to be in simplest form. The process of rewriting a radical expression so that the denominator does not contain any radicals is called **rationalizing**

the denominator. Example 6 illustrates the procedure for rewriting $\dfrac{2}{\sqrt{5}}$ as $\dfrac{2\sqrt{5}}{5}$. Note that

multiplying $\sqrt{x}$ by itself produces $\sqrt{x^2} = x$ for $x \geq 0$.

Example 6   Rationalizing the Denominator

Simplify each expression by rationalizing the denominator.

(a) $\dfrac{2}{\sqrt{5}}$ **(b)** $\sqrt{\dfrac{7}{11}}$

Solution

(a) $\dfrac{2}{\sqrt{5}} = \dfrac{2}{\sqrt{5}} \cdot \dfrac{\sqrt{5}}{\sqrt{5}}$ Multiply both the numerator and the denominator by $\sqrt{5}$. This does not change the value of the expression because $\dfrac{\sqrt{5}}{\sqrt{5}} = 1$.

$\qquad\;\; = \dfrac{2\sqrt{5}}{5}$

(b) $\sqrt{\dfrac{7}{11}} = \dfrac{\sqrt{7}}{\sqrt{11}}$ Multiply both the numerator and the denominator by $\sqrt{11}$. This does not change the value of the expression because $\dfrac{\sqrt{11}}{\sqrt{11}} = 1$.

$\qquad\;\;\; = \dfrac{\sqrt{7}}{\sqrt{11}} \cdot \dfrac{\sqrt{11}}{\sqrt{11}}$

$\qquad\;\;\; = \dfrac{\sqrt{7}\sqrt{11}}{\sqrt{11}\sqrt{11}}$

$\qquad\;\;\; = \dfrac{\sqrt{77}}{11}$ You can check this result on a calculator by approximating both $\sqrt{\dfrac{7}{11}}$ and $\dfrac{\sqrt{77}}{11}$.

Self-Check 6

Simplify each expression by rationalizing the denominator.

a. $\dfrac{6}{\sqrt{7}}$ **b.** $\sqrt{\dfrac{11}{7}}$

Example 7   Solving a Quadratic Equation by Extraction of Roots

Solve $6x^2 = 14$ by using extraction of roots to obtain exact solutions. Then approximate each irrational solution to the nearest thousandth.

Solution

$6x^2 = 14$

$x^2 = \dfrac{14}{6}$ First isolate x^2 on the left side of the equation.

$x^2 = \dfrac{7}{3}$ Then reduce the fraction $\dfrac{14}{6}$ to lowest terms.

$x = \pm\sqrt{\dfrac{7}{3}}$ Solve this equation by extraction of roots.

$x = \pm\dfrac{\sqrt{7}}{\sqrt{3}} \cdot \dfrac{\sqrt{3}}{\sqrt{3}}$ Rationalize the denominator.

$x = \pm\dfrac{\sqrt{21}}{3}$ The decimal approximation can be obtained directly from $\sqrt{\dfrac{7}{3}}$ or from the simplified form $\dfrac{\sqrt{21}}{3}$.

$x \approx 1.528$ or $x \approx -1.528$ $\sqrt{\dfrac{7}{3}} \approx 1.528$ and $\dfrac{\sqrt{21}}{3} \approx 1.528$

Self-Check 7

Solve each quadratic equation by using extraction of roots to obtain exact solutions. Then approximate each irrational solution to the nearest thousandth.

a. $10w^2 = 65$ **b.** $6x^2 - 15 = 0$

In the following box, we extend the method of extraction of roots to a quadratic equation that contains the square of a binomial on the left side of the equation.

Extraction of Roots	Example
For a positive real number k and nonzero real numbers a and b: If $(ax + b)^2 = k$, then $ax + b = \pm\sqrt{k}$.	$(2x + 1)^2 = 3$ $2x + 1 = \sqrt{3}$ or $2x + 1 = -\sqrt{3}$

The result of performing extraction of roots will produce two linear equations. We then need to solve each of these linear equations, as illustrated in Example 8.

Example 8   Solving a Quadratic Equation by Extraction of Roots

Solve $(2t - 3)^2 = 25$ by extraction of roots.

Solution

$$(2t - 3)^2 = 25$$
$$2t - 3 = \pm\sqrt{25} \qquad \text{Solve the quadratic equation using extraction of roots.}$$
$$2t - 3 = \pm 5$$
$$2t = 3 \pm 5 \qquad \text{Simplify this expression and solve for } t.$$
$$t = \frac{3 \pm 5}{2}$$
$$t = \frac{3 + 5}{2} \quad \text{or} \quad t = \frac{3 - 5}{2} \qquad \text{Find the two solutions.}$$
$$t = \frac{8}{2} \quad \text{or} \quad t = \frac{-2}{2} \qquad \text{To check these two solutions, substitute each value back into the original equation.}$$
$$t = 4 \quad \text{or} \quad t = -1$$

Self-Check 8

Solve each quadratic equation by extraction of roots.

a. $(2m + 1)^2 = 9$ **b.** $(2v - 1)^2 = 25$

When Will I Use Quadratic Equations in Applications?

Many problems involving geometrical shapes will require quadratic equations. Quadratic equations are also required for some business applications. Example 9 uses a quadratic

equation to determine an interest rate on an investment. The formula $A = P\left(1 + \dfrac{r}{n}\right)^n$

gives the amount of an investment that is compounded n times a year for 1 year at an annual interest rate of r. If the investment is compounded twice a year, this formula

becomes the quadratic equation $A = P\left(1 + \dfrac{r}{2}\right)^2$.

If the left side of a quadratic equation is written as a perfect square, then extraction of roots may be the most efficient way to solve the equation. If the equation is not in this form, then some steps may be needed to put the equation in this form. Example 9 will allow us to review the main steps that are needed to apply the method of extraction of roots:

- Rewrite the equation so that the coefficient of the perfect square term is 1.
- Isolate the perfect square term on the left side of the equation.
- Extract the roots and then simplify.

Example 9   Determining the Interest Rate on an Investment

An advertisement stated, "Invest \$1,000 with us and see your investment grow to \$1,100 by the end of the year." If this investment is compounded twice during the year, use the

formula $A = P\left(1 + \dfrac{r}{2}\right)^2$ to determine the annual interest rate to the nearest hundredth of a percent.

Solution

$$A = P\left(1 + \frac{r}{2}\right)^2$$

Substitute \$1,100 for A, the amount of the investment at the end of the year. Substitute \$1,000 for P, the principal invested at the beginning of the year.

$$1{,}100 = 1{,}000\left(1 + \frac{r}{2}\right)^2$$

$$\left(1 + \frac{r}{2}\right)^2 = \frac{11}{10}$$

Simplify and then solve this quadratic equation using extraction of roots. Because we want a decimal approximation for r, it is not necessary to rationalize the denominator. Instead we enter the form shown here into a calculator.

$$1 + \frac{r}{2} = \pm\sqrt{\frac{11}{10}}$$

$$\frac{r}{2} = -1 \pm \sqrt{\frac{11}{10}}$$

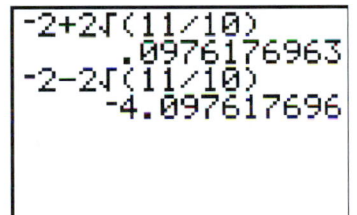

$$r = -2 \pm 2\sqrt{\frac{11}{10}}$$

$$r \approx 0.0976 \quad \text{or} \quad r \approx -4.0976$$

Answer: The annual interest rate is approximately 9.76%.

A negative interest rate is not meaningful in this application. Does this value check?

Self-Check 9

Rework Example 9 assuming that the \$1,000 investment grows to \$1,075 by the end of 1 year.

Self-Check Answers

1. **a.** $m = 4$ or $m = -4$ **b.** $y = \sqrt{17}$ or $y = -\sqrt{17}$; $y \approx \pm 4.12$
   c. $x = \sqrt{3}$ or $x = -\sqrt{3}$; $x \approx \pm 1.73$
2. **a.** $4\sqrt{3}$ **b.** $5\sqrt{2}$
3. **a.** $x = 2\sqrt{2}$ or $x = -2\sqrt{2}$; $x \approx \pm 2.828$
   b. $y = 2\sqrt{7}$ or $y = -2\sqrt{7}$; $y \approx \pm 5.292$
4. **a.** $\dfrac{3}{5}$ **b.** 5
5. **a.** $z = \dfrac{\sqrt{2}}{3}$ or $z = -\dfrac{\sqrt{2}}{3}$; $z \approx \pm 0.471$
   b. $x = \dfrac{\sqrt{17}}{4}$ or $x = -\dfrac{\sqrt{17}}{4}$; $x \approx \pm 1.031$
6. **a.** $\dfrac{6\sqrt{7}}{7}$ **b.** $\dfrac{\sqrt{77}}{7}$
7. **a.** $w = \dfrac{\sqrt{26}}{2}$ or $w = -\dfrac{\sqrt{26}}{2}$; $w \approx \pm 2.550$
   b. $x = \dfrac{\sqrt{10}}{2}$ or $x = -\dfrac{\sqrt{10}}{2}$; $x \approx \pm 1.581$
8. **a.** $m = 1$ or $m = -2$ **b.** $v = 3$ or $v = -2$
9. The annual interest rate is approximately 7.36%.

7.1 Using the Language and Symbolism of Mathematics

1. An equation that can be written in the standard form $ax^2 + bx + c = 0$ is called a _____ equation.

2. The notation _____ is read plus or minus.

3. By extraction of roots, the solutions of $x^2 = k$ are the real numbers _____ for $k > 0$.

4. To use extraction of roots to solve a quadratic equation, we first write the left side of the equation as a perfect _____.

5. If $\sqrt{x}$ and $\sqrt{y}$ are both real numbers, then the product rule for square roots says $\sqrt{xy} = $ _____.

6. If $\sqrt{x}$ and $\sqrt{y}$ are both real numbers and $y > 0$, then the quotient rule for square roots says $\sqrt{\dfrac{x}{y}} = $ _____.

7. The process of rewriting a radical expression so that the denominator does contain any radicals is called _____ the denominator.

7.1 Quick Review

1. Evaluate $\sqrt{36}$ by inspection.

2. Use a calculator or a spreadsheet to approximate $\sqrt{39}$ to the nearest thousandth.

3. Use a calculator or a spreadsheet to approximate $\dfrac{\sqrt{3}}{5}$ to the nearest thousandth.

4. Reduce $\dfrac{35}{77}$ to lowest terms.

5. Solve $1 + \dfrac{x}{2} = 9$.

7.1 Exercises

Objective 1 Solve Quadratic Equations by Extraction of Roots

In Exercises 1–12, solve each quadratic equation by using extraction of roots to obtain exact solutions. Then approximate any irrational solutions to the nearest hundredth.

1. $m^2 = 4$ 2. $m^2 = 25$

3. $x^2 = 2$ 4. $x^2 = 11$

5. $5y^2 = 45$ 6. $11y^2 = 1{,}100$

7. $3v^2 = 39$ 8. $2v^2 = 38$

9. $2w^2 - 72 = 0$ 10. $10w^2 - 90 = 0$

11. $10z^2 - 50 = 0$ 12. $3z^2 - 21 = 0$

Objective 2 Use the Product Rule for Radicals to Simplify Square Roots

In Exercises 13–18, simplify each expression using the product rule for square roots.

13. $\sqrt{16 \cdot 5}$ 14. $\sqrt{81 \cdot 17}$

15. $\sqrt{20}$ 16. $\sqrt{12}$

17. $\sqrt{200}$ 18. $\sqrt{75}$

In Exercises 19–24, solve each quadratic equation by using extraction of roots to obtain exact solutions. Then approximate any irrational solutions to the nearest hundredth.

19. $x^2 = 18$ 20. $x^2 = 44$

21. $3y^2 = 81$ 22. $5y^2 = 40$

23. $7a^2 - 56 = 0$ 24. $3a^2 - 150 = 0$

Objective 3 Use the Quotient Rule for Radicals to Simplify Square Roots

In Exercises 25–32, simplify each expression using the quotient rule for square roots.

25. $\sqrt{\dfrac{4}{9}}$ **26.** $\sqrt{\dfrac{25}{16}}$ **27.** $\sqrt{\dfrac{6}{49}}$ **28.** $\sqrt{\dfrac{13}{64}}$

29. $\dfrac{\sqrt{20}}{\sqrt{5}}$ **30.** $\dfrac{\sqrt{99}}{\sqrt{11}}$ **31.** $\dfrac{\sqrt{18}}{\sqrt{50}}$ **32.** $\dfrac{\sqrt{20}}{\sqrt{245}}$

In Exercises 33–36, solve each quadratic equation by using extraction of roots to obtain exact solutions. Then approximate any irrational solutions to the nearest hundredth.

33. $4v^2 = 81$ **34.** $25v^2 = 121$

35. $36x^2 - 7 = 0$ **36.** $81x^2 - 5 = 0$

Objective 4 Simplify Expressions of the Form $\dfrac{\sqrt{a}}{\sqrt{b}}$ by Rationalizing the Denominator

In Exercises 37–44, simplify each expression by rationalizing the denominator.

37. $\dfrac{3}{\sqrt{5}}$ **38.** $\dfrac{6}{\sqrt{13}}$ **39.** $\dfrac{6}{\sqrt{10}}$ **40.** $\dfrac{15}{\sqrt{6}}$

41. $\dfrac{\sqrt{2}}{\sqrt{3}}$ **42.** $\dfrac{\sqrt{5}}{\sqrt{2}}$ **43.** $\sqrt{\dfrac{5}{17}}$ **44.** $\sqrt{\dfrac{2}{11}}$

In Exercises 45–48, solve each quadratic equation by using extraction of roots to obtain exact solutions. Then approximate any irrational solutions to the nearest hundredth.

45. $8x^2 = 20$ **46.** $15x^2 = 55$

47. $10w^2 - 14 = 0$ **48.** $70w^2 - 110 = 0$

Skill Development

In Exercises 49–54, solve each quadratic equation by extraction of roots.

49. $(n - 1)^2 = 36$ **50.** $(n + 2)^2 = 64$

51. $(2b + 3)^2 = 49$ **52.** $(2b - 3)^2 = 169$

53. $(3x - 1)^2 + 7 = 32$ **54.** $(5x + 1)^2 - 2 = 79$

Comparison of Linear Equations and Quadratic Equations

In Exercises 55–58, solve each equation.

Linear Equation	Quadratic Equation
55. a. $2x = 36$	**b.** $x^2 = 36$
56. a. $w - 2 = 25$	**b.** $(w - 2)^2 = 25$
57. a. $2m - 1 = 9$	**b.** $(2m - 1)^2 = 9$
58. a. $2(y - 1) - 3 = 1$	**b.** $(y - 1)^2 - 3 = 1$

Review and Concept Development

59. Use the graph of $y = (2x - 1)^2 - 1$ to:
a. Solve $(2x - 1)^2 - 1 = 0$.
b. Solve $(2x - 1)^2 - 1 < 0$.
c. Solve $(2x - 1)^2 - 1 > 0$.

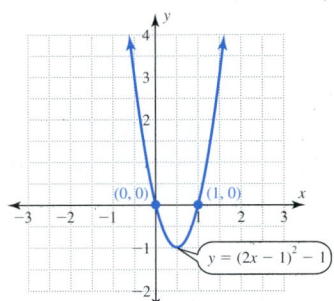

60. Use the graph of $y = (x - 2)^2 - 3$ to:
a. Solve $(x - 2)^2 - 3 = 0$.
b. Solve $(x - 2)^2 - 3 < 0$.
c. Solve $(x - 2)^2 - 3 > 0$.

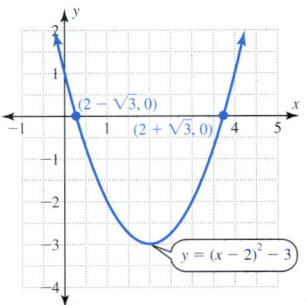

61. a. Determine the exact solutions of $x^2 - 6 = 0$.
b. Determine the exact zeros of $x^2 - 6$.
c. Determine the exact x-intercepts of the parabola whose equation is $y = x^2 - 6$.

62. a. Determine the exact solutions of $(2x - 1)^2 - 5 = 0$.
b. Determine the exact zeros of $(2x - 1)^2 - 5$.
c. Determine the exact x-intercepts of the parabola whose equation is $y = (2x - 1)^2 - 5$.

63. Construct a quadratic equation in x that has the given solutions.
a. -7 and 7 **b.** $-\sqrt{7}$ and $\sqrt{7}$

64. Construct a quadratic equation in x that has the given solutions.
a. -11 and 11 **b.** $-\sqrt{11}$ and $\sqrt{11}$

Connecting Algebra to Geometry

In Exercises 65 and 66, use the given lengths of legs a and b and the equation $a^2 + b^2 = c^2$ to approximate the length of the hypotenuse c to the nearest tenth of a centimeter.

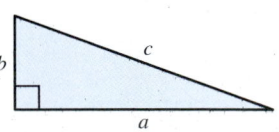

65. $a = 29.5$ cm, $b = 35.7$ cm

66. $a = 43.6$ cm, $b = 57.3$ cm

Connecting Concepts to Applications

67. Interest Rate A certificate of deposit earns an annual rate of interest r that is compounded twice a year. An investment of \$850 grows to \$923.79 by the end of 1 year. Use the formula $A = P\left(1 + \dfrac{r}{2}\right)^2$ to approximate the annual interest rate to the nearest tenth of a percent.

68. Interest Rate A certificate of deposit earns an annual rate of interest r that is compounded twice a year. An investment of \$2,500 grows to \$2,678.06 by the end of 1 year. Use the formula $A = P\left(1 + \dfrac{r}{2}\right)^2$ to approximate the annual interest rate to the nearest tenth of a percent.

69. Free-Fall Time If a hammer is dropped from the 90th floor of a construction project, the distance in feet that it will fall after t seconds is given by $d = 16t^2$. Approximate to the nearest tenth of a second the time it takes the hammer to fall 50 ft.

70. Free-Fall Time If a hammer is dropped from the 90th floor of a construction project, the distance in feet that it will fall after t seconds is given by $d = 16t^2$. Approximate to the nearest tenth of a second the time it takes the hammer to fall 100 ft.

Group discussion questions

71. Error Analysis A student used a calculator to approximate $\dfrac{4 + \sqrt{7}}{3}$ and produced the work shown.

```
4+√(7)/3
          4.881917104
```

a. What is the error?

b. Use a calculator or a spreadsheet to approximate the value of $\dfrac{-2 + \sqrt{5}}{3}$ to the nearest thousandth.

c. Use a calculator or a spreadsheet to approximate the value of $\dfrac{5 - \sqrt{3}}{7}$ to the nearest thousandth.

72. Constructing an Equation with Given Solutions

a. Construct a quadratic equation that has solutions $x = 1 - \sqrt{7}$ and $x = 1 + \sqrt{7}$.

b. Construct a cubic equation that has solutions $x = 0$, $x = \sqrt{5}$, and $x = -\sqrt{5}$.

c. Construct a cubic equation that has solutions $x = 0$, $x = \dfrac{1 - \sqrt{5}}{2}$, and $x = \dfrac{1 + \sqrt{5}}{2}$.

73. Checking a Solution A student solved $x^2 - 5x - 1 = 0$ and obtained $x = \dfrac{5 - \sqrt{29}}{2}$ as one of the solutions.

a. Using only pencil and paper, substitute this value into the equation to check this solution.

b. Using a calculator or a spreadsheet, substitute this value into the equation to check this solution.

| 7.1 | Cumulative Review |

1. Write 4,718.5 in scientific notation.

2. Write 2.3×10^{-4} in standard decimal notation.

3. Simplify $(5x^2y^3)(4x^3y^4)$.

4. Simplify $(5x^2y^3)^2$.

5. List all natural numbers between -17.3 and 4.3.

Section 7.2 | Solving Quadratic Equations by Completing the Square

Objectives:

1. Determine the constant term in a perfect square trinomial.

2. Solve quadratic equations by completing the square.

What Is the Purpose of This Section?

This section extends the method of extraction of roots that was covered in Section 7.1 so that we can solve any quadratic equation. This material also allows us to develop the quadratic formula which is given in Section 7.3. We can apply extraction of roots when the left side of a quadratic equation is written as a perfect square. By using the skills developed in this section, we will be able to write the left side of any quadratic equation as a perfect square. In Example 1 we start by examining a quadratic equation whose left member is already the square of a binomial.

Example 1 | **Writing the Left Side of a Quadratic Equation as a Perfect Square**

Write each equation so that the left side is expressed as the square of a binomial.

(a) $x^2 + 6x + 9 = 1$ **(b)** $y^2 - 10y + 25 = 9$

Solution

(a) $x^2 + 6x + 9 = 1$ The left side factors as the perfect square of a sum.

$(x + 3)^2 = 1$

(b) $y^2 - 10y + 25 = 9$ The left side factors as the perfect square of a difference.

$(y - 5)^2 = 9$

Self-Check 1

Write each of these equations so that the left side is the square of a binomial.

a. $x^2 - 4x + 4 = 3$ **b.** $y^2 + 12y + 36 = 4$

1. Determine the Constant Term in a Perfect Square Trinomial

Most quadratic equations will not appear in the convenient form given in Example 1. However, even when the left side is not the square of a binomial, we can rewrite the equation so that this form will occur. Recall from Section 6.4 that a trinomial that is the perfect square of a binomial will have one of these two forms.

Square of a Sum

$(x + k)^2 = x^2 + 2kx + k^2$

Constant term: $k^2 = \left(\dfrac{2k}{2}\right)^2$

Square of a Difference

$(x - k)^2 = x^2 - 2kx + k^2$

Constant term: $k^2 = \left(\dfrac{-2k}{2}\right)^2$

In either case, the coefficient of x^2 is 1 and the constant term k^2 is the square of one-half the coefficient of x. We can use this observation to determine the constant term that must occur in a perfect square trinomial.

Example 2 | **Determining the Constant Term in a Perfect Square Trinomial**

Fill in the missing constant term that is needed to make each expression a perfect square trinomial. Then factor this trinomial.

(a) $x^2 + 8x +$ ____ **(b)** $v^2 - 6v +$ ____ **(c)** $x^2 + 2ax +$ ____

Solution

(a) $x^2 + 8x + \left(\dfrac{8}{2}\right)^2 = x^2 + 8x + 16$ The constant term needed is the square of one-half the coefficient of x. In this case, the missing constant term is 16.

$= (x + 4)^2$

(b) $v^2 - 6v + \left(\dfrac{-6}{2}\right)^2 = v^2 - 6v + 9$ The coefficient of v is -6 so the missing constant term must be 9.

$= (v - 3)^2$

(c) $x^2 + 2ax + \left(\dfrac{2a}{2}\right)^2 = x^2 + 2ax + a^2$ The coefficient of x is $2a$ so the missing constant term must be a^2.

$= (x + a)^2$

A Mathematical Note

Many methods that we now use quickly and efficiently in algebraic form have their roots in ancient geometric methods. In particular, the method of completing the square was used by both Greek and Arab mathematicians in the geometric form shown in the following geometric viewpoint. Al-Khowârizmi (ca. 825) was a noted Arab mathematician, astronomer, and author who illustrated this method in his writings.

Is There a Name for the Process Used to Fill in the Missing Constant Term to Produce a Perfect Square Trinomial?

Yes, this process is called **completing the square.** This terminology has a very natural geometrical basis, as illustrated in the following geometrical viewpoint. In this box, we will fill in the missing constant term that is needed to make $x^2 + 6x +$ ____ a perfect square trinomial.

Completing the Square for $x^2 + 6x +$ ____

Algebraically $x^2 + 6x + 9$ $= (x + 3)^2$

Geometrically

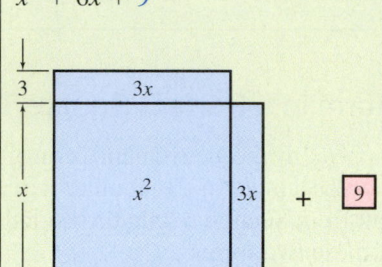   + $\boxed{9}$ =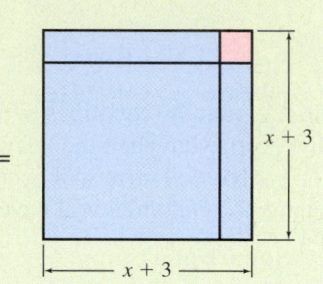

---

2. Solve Quadratic Equations by Completing the Square

In Example 3, we will use the process of completing the square to put a quadratic equation in a form that allows us to use extraction of roots to solve the equation. Remember that adding or subtracting the same number from both sides of an equation will produce an equivalent equation.

Example 3   Completing the Square to Find Two Integer Solutions

Rewrite the left side of $x^2 + 6x = -5$ so that it is a perfect square, and then solve this equation by extraction of roots.

Solution

$$x^2 + 6x = -5$$
$$x^2 + 6x + 9 = -5 + 9$$ Complete the square by adding 9 to each side of the equation.
$$(x + 3)^2 = 4$$
$$x + 3 = \pm 2$$ Then solve this equation by extraction of roots.
$$x = -3 \pm 2$$
$$x = -3 - 2 \quad \text{or} \quad x = -3 + 2$$ Find the two solutions.
$$x = -5 \quad\quad \text{or} \quad x = -1$$ Do these solutions check?

Self-Check 3

Solve $y^2 + 2y = 8$ by completing the square.

In Example 4, we divide both sides of the equation by the coefficient of w^2 to obtain a coefficient of 1.

Example 4   Completing the Square to Find Two Irrational Solutions

Rewrite the left side of $2w^2 - 16w = -22$ so that it is a perfect square, and then solve this equation by extraction of roots. Give exact solutions and then approximate the solutions to the nearest thousandth.

Solution

$$2w^2 - 16w = -22$$

The coefficient of w^2 is 2, so divide both sides of the equation by 2.

$$w^2 - 8w = -11$$

$$w^2 - 8w + 16 = -11 + 16$$

Complete the square by adding 16 to each side of the equation.

$$(w - 4)^2 = 5$$

$$w - 4 = \pm\sqrt{5}$$

Then solve this equation by extraction of roots.

$$w = 4 \pm\sqrt{5}$$

$$w = 4 - \sqrt{5} \quad \text{or} \quad w = 4 + \sqrt{5}$$

The exact solutions have been approximated to the nearest thousandth using a calculator.

$$w \approx 1.764 \quad \text{or} \quad w = 6.236$$

Self-Check 4

Solve $3v^2 - 12v = 18$ by completing the square to obtain the exact solutions. Then approximate the solutions to the nearest thousandth.

The process of completing the square can be used to solve any quadratic equation. The key steps in this process are shown in the following box.

Solving Quadratic Equations by Completing the Square

Verbally	Algebraic Example
Step 1. Write the equation with the constant term on the right side.	$x^2 + 3x - 4 = 0$ $x^2 + 3x = 4$
Step 2. Divide both sides of the equation by the coefficient of x^2 to obtain a coefficient of 1 for x^2.	$x^2 + 3x + \left(\dfrac{3}{2}\right)^2 = 4 + \left(\dfrac{3}{2}\right)^2$
Step 3. Take one-half the coefficient of x, square this number, and add the result to both sides of the equation.	$\left(x + \dfrac{3}{2}\right)^2 = \dfrac{16}{4} + \dfrac{9}{4}$
Step 4. Write the left side of the equation as a perfect square.	$\left(x + \dfrac{3}{2}\right)^2 = \dfrac{25}{4}$ $x + \dfrac{3}{2} = \pm\dfrac{5}{2}$
Step 5. Solve this equation by extraction of roots.	$x = -\dfrac{3}{2} \pm \dfrac{5}{2}$ $x = -\dfrac{3}{2} - \dfrac{5}{2} \quad \text{or} \quad x = -\dfrac{3}{2} + \dfrac{5}{2}$ $x = -4 \qquad\qquad\qquad x = 1$

The steps outlined in the box will be used to solve the equation in Example 5. Then we use this procedure in Section 7.3 to create a formula that can be used to solve any quadratic equation.

Example 5 Completing the Square to Find Two Irrational Solutions

Solve $4x^2 - 20x + 7 = 0$ by completing the square to obtain the exact solutions. Then approximate the solutions to the nearest hundredth.

Solution

$$4x^2 - 20x + 7 = 0$$

$$4x^2 - 20x = -7 \qquad \text{Shift the constant to the right side of the equation.}$$

$$x^2 - 5x = -\frac{7}{4} \qquad \text{Divide both sides of the equation by the coefficient of } x^2, \text{ which is 4.}$$

$$x^2 - 5x + \left(-\frac{5}{2}\right)^2 = -\frac{7}{4} + \left(-\frac{5}{2}\right)^2 \qquad \text{Take one-half the coefficient of } x: \ \frac{1}{2}(-5) = -\frac{5}{2}.$$

$$\left(x - \frac{5}{2}\right)^2 = -\frac{7}{4} + \frac{25}{4} \qquad \text{Square this number: } \left(-\frac{5}{2}\right)^2 = \frac{25}{4}. \text{ Then add this result to both sides of the equation. Write the left side as a perfect square.}$$

$$\left(x - \frac{5}{2}\right)^2 = \frac{9}{2} \qquad \text{Extract the roots.}$$

$$x - \frac{5}{2} = \pm\sqrt{\frac{9}{2}} \qquad \text{Add } \frac{5}{2} \text{ to both sides of the equation.}$$

$$x = \frac{5}{2} \pm \frac{3\sqrt{2}}{2} \qquad \text{Simplify, noting that } \sqrt{\frac{9}{2}} = \frac{3}{\sqrt{2}} \cdot \frac{\sqrt{2}}{\sqrt{2}} = \frac{3\sqrt{2}}{2}.$$

$$x = \frac{5 - 3\sqrt{2}}{2}, \quad x = \frac{5 + 3\sqrt{2}}{2} \qquad \text{The exact answers can be approximated to the nearest hundredth with a calculator.}$$

$$x \approx 0.38, \quad x \approx 4.62$$

Self-Check 5

Solve $2w^2 - 4w + 1 = 0$ by completing the square to obtain the exact solutions. Then approximate the solutions to the nearest thousandth.

From our earlier work in Chapter 6, we know that the real zeros of a function and the solutions of the corresponding equation are directly related. In Technology Perspective 7.2.1, we illustrate how to use the **ZERO** feature on the TI-84 Plus calculator to solve the quadratic equation from Example 5. As equations and the corresponding algebraic solutions become more difficult, it is a very handy feature. However, this feature is not available in most spreadsheet programs.

Technology Perspective 7.2.1   Calculating the Zeros of a Function

Calculate the smaller of the two zeros of $Y_1 = 4x^2 - 20x + 7$ (from Example 5).

TI-84 Plus Keystrokes

1. Press Y= and enter the function for Y_1.

2. Press 2nd CALC 2 to select the **zero** option from the menu.

3. Use the arrow keys to move the cursor until it is located slightly to the left side of the first zero. Then press ENTER.

4. Use the right arrow key to move the cursor until it is located slightly on the right side of the zero. Then press ENTER.

5. When the screen requests a guess, use the left arrow key to move the cursor until it is close to the zero. Then press ENTER.

TI-84 Plus Screens

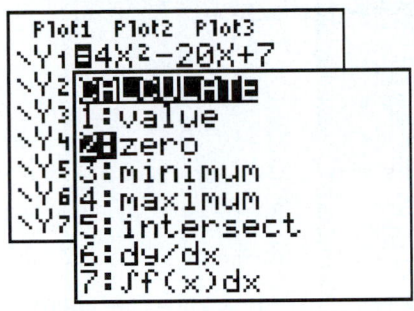

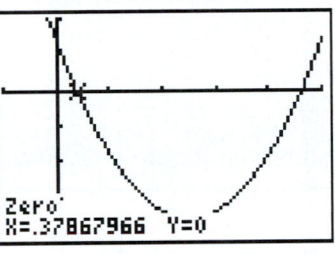

$[-1, 5, 1]$ by $[-20, 10, 5]$

Answer: Because 0.38 is approximately a zero of $4x^2 - 20x + 7$, we also know that one solution of $4x^2 - 20x + 7 = 0$ is $x \approx 0.38$.

Note: We will use the Self-Check to approximate the other zero.

Technology Self-Check 1

Calculate the larger of the two zeros of $Y_1 = 4x^2 - 20x + 7$ (from Example 5).

It is important to be able to solve quadratic equations as efficiently as possible, as such equations arise in many applications, especially those from geometrically based problems. Example 6 involves the area of a rectangular opening.

Example 6   Area of a Rectangular Opening

A rectangular opening for the ductwork installation in a shopping center must have the dimensions as shown in the figure. This opening must have a cross-sectional area of 30 ft² for the airflow through the ductwork. Calculate the dimensions of this rectangular opening.

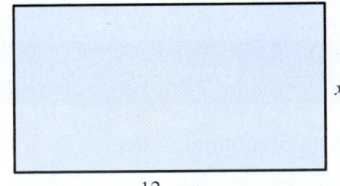

Solution

$$A = LW$$

$$30 = x(12 - x)$$

$$30 = 12x - x^2$$

$$x^2 - 12x = -30$$

$$x^2 - 12x + 36 = -30 + 36$$

$$(x - 6)^2 = 6$$

$$x - 6 = \pm\sqrt{6}$$

$$x = 6 \pm \sqrt{6}$$

$$x = 6 - \sqrt{6} \quad \text{or} \quad x = 6 + \sqrt{6}$$

$$x \approx 3.55 \qquad \text{or} \quad x \approx 8.45$$

$$12 - x \approx 8.45 \quad \text{or} \quad 12 - x \approx 3.55$$

Substitute the given area, width, and length into the formula for the area of a rectangle.

Rewrite the equation so the quadratic term has a positive coefficient and the constant term is on the right side of the equation. Then add 36 to both sides of the equation to complete the square; finally, solve by extracting the roots.

Answer: The dimensions are approximately 3.55 ft by 8.45 ft.

If the width is 3.55 ft, then the length is 8.45 ft.
If the width is 8.45 ft, then the length is 3.55 ft.

Self-Check 6

Solve Example 6 if the cross-sectional area of the opening must be 32 ft^2.

Self-Check Answers

1. **a.** $(x - 2)^2 = 3$ **b.** $(y + 6)^2 = 4$

2. **a.** $m^2 + 10m + 25 = (m + 5)^2$ **b.** $v^2 - 12v + 36 = (v - 6)^2$
   c. $x^2 - 2ax + a^2 = (x - a)^2$

3. $y = -4$ or $y = 2$

4. $v = 2 - \sqrt{10}$ or $v = 2 + \sqrt{10}$; $v \approx -1.162$ or $v \approx 5.162$

5. $w = 1 - \dfrac{\sqrt{2}}{2}$ or $w = 1 + \dfrac{\sqrt{2}}{2}$; $w \approx 0.293$ or $w \approx 1.707$

6. The dimensions are 4 ft by 8 ft.

Technology Self-Check Answer

1. The other zero is approximately 4.62.

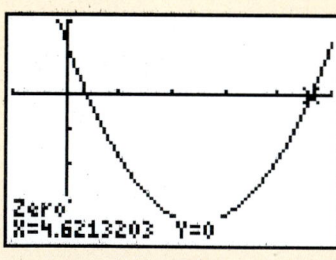

Zero
X=4.6213203   Y=0

$[-1, 5, 1]$ by $[-20, 10, 5]$

7.2   **Using the Language and Symbolism of Mathematics**

1. A _____ trinomial is a polynomial of degree 2 with three terms.

2. The polynomial $x^2 + 2kx + k^2$ can be written as $(x + k)^2$ and is called a perfect _____ trinomial.

3. Filling in the missing constant term that is needed to produce an expression that is a perfect square trinomial is called _____ the square.

4. Adding the same number to both sides of an equation will produce an _____ equation.

7.2 Quick Review

1. Simplify $\left(\dfrac{7}{2}\right)^2$.

2. Expand $(x - 7)^2$.

3. Expand and simplify $(1 - \sqrt{7})^2$.

4. Factor $x^2 + 14x + 49$.

5. Simplify $\dfrac{6 - 4\sqrt{5}}{2}$.

7.2 Exercises

Objective 1 Determine the Constant Term in a Perfect Square Trinomial

In Exercises 1–4, write each equation so that the left side is expressed as the square of a binomial.

1. $m^2 - 2m + 1 = 4$ **2.** $m^2 - 8m + 16 = 9$

3. $y^2 + 14y + 49 = 5$ **4.** $y^2 + 20y + 100 = 7$

In Exercises 5–10, fill in the missing constant term that is needed to make each expression a perfect square trinomial.

5. $x^2 + 12x +$ _____   **6.** $x^2 + 16x +$ _____

7. $v^2 - 18v +$ _____   **8.** $v^2 - 22v +$ _____

9. $m^2 + 4bm +$ _____   **10.** $m^2 - 6bm +$ _____

Objective 2 Solve Quadratic Equations by Completing the Square

In Exercises 11–18, rewrite the left side of each equation so that it is a perfect square, and then solve this equation by using extraction of roots.

11. $w^2 - 4w = 0$ **12.** $v^2 - 2v = 0$

13. $w^2 - 4w = 12$ **14.** $v^2 - 2v = 24$

15. $y^2 + 6y = 0$ **16.** $x^2 + 8x = 0$

17. $y^2 + 6y = -5$ **18.** $x^2 + 8x = -7$

In Exercises 19–34, solve each equation by completing the square to obtain the exact solutions. Then approximate any irrational solutions to the nearest hundredth.

19. $t^2 + 2t + 2 = 6$ **20.** $t^2 - 2t - 2 = 2$

21. $w^2 + 10w + 25 = \dfrac{1}{4}$ **22.** $w^2 + 12w + 36 = \dfrac{1}{9}$

23. $2y^2 - 8y + 4 = 2$ **24.** $2y^2 + 8y + 6 = 4$

25. $3x^2 + 6x + 2 = 0$ **26.** $5x^2 + 30x + 20 = 0$

27. $m^2 - \dfrac{3}{2}m = 1$ **28.** $m^2 + \dfrac{5}{2}m = \dfrac{3}{2}$

29. $4v^2 - 4v - 1 = 0$ **30.** $4v^2 - 12v + 7 = 0$

31. $5v^2 + v + 1 = 3v + 2$ **32.** $v^2 - 2v + 3 = 2v + 4$

33. $(x - 3)(x + 2) = 1$ **34.** $(x - 1)(x + 4) = 7$

Comparing Linear Equations and Quadratic Equations

In Exercises 35 and 36, solve each equation. Give exact solutions and decimal approximations rounded to the nearest hundredth.

Linear Equation	Quadratic Equation
35. a. $2x + x = 3$	**b.** $x^2 + 2x = 3$
36. a. $2v + 8v = -10$	**b.** $v^2 + 8v = -10$

Review and Concept Development

37. Use the graph of $y = x^2 - 3$ to:
  a. Solve $x^2 - 3 = 0$.
  b. Solve $x^2 - 3 \leq 0$.
  c. Solve $x^2 - 3 \geq 0$.

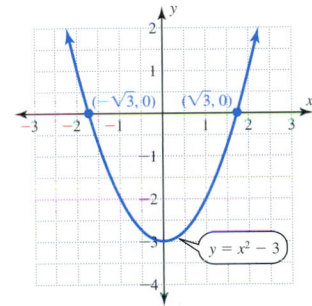

38. Use the graph of $y = x^2 + 4x + 2$ to:
  a. Solve $x^2 + 4x + 2 = 0$.
  b. Solve $x^2 + 4x + 2 \leq 0$.
  c. Solve $x^2 + 4x + 2 \geq 0$.

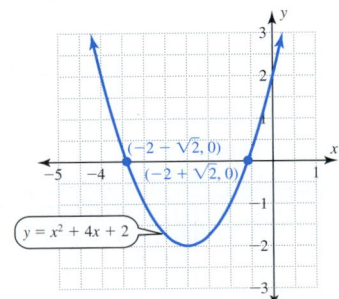

39. a. Determine the exact solutions of $2x^2 - 2x - 1 = 0$.
  b. Determine the exact zeros of $2x^2 - 2x - 1$.
  c. Determine the exact x-intercepts of the parabola whose equation is $y = 2x^2 - 2x - 1$.

40. a. Determine the exact solutions of $x^2 - 2x - 5 = 0$.
  b. Determine the exact zeros of $x^2 - 2x - 5$.
  c. Determine the exact x-intercepts of the parabola whose equation is $y = x^2 - 2x - 5$.

41. Construct a quadratic equation in x that has the given solutions.
  a. $-\dfrac{2}{3}$ and $\dfrac{2}{3}$ **b.** $-\dfrac{\sqrt{2}}{3}$ and $\dfrac{\sqrt{2}}{3}$

42. Construct a quadratic equation in x that has the given solutions.
  a. $-\dfrac{3}{5}$ and $\dfrac{3}{5}$ **b.** $-\dfrac{\sqrt{3}}{5}$ and $\dfrac{\sqrt{3}}{5}$

Connecting Algebra to Geometry

43. Dimensions of a Photo
A rectangular photograph in a magazine is 1.5 times as wide as it is tall. The photo occupies an area of 24 cm^2 on the page. Determine the dimensions of the photo.

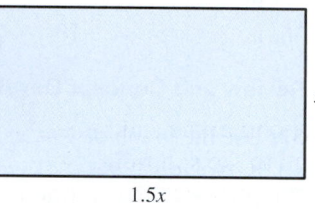
x
$1.5x$

44. Manufacturing of a Disk A blank circular disk is cut from the square metal stock shown. Determine to the nearest tenth of a cm the radius of the disk if the area of metal wasted is 124 cm^2.

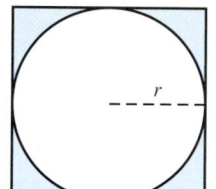
r

45. Manufacturing of a Ring
A metal ring is manufactured by cutting a circle from the center of a larger circular blank disk as shown. The radius of the blank disk is twice the radius of the circle that is removed, and the area of the metal ring remaining is 15 cm^2. Determine to the nearest tenth of a cm the radius of the circle that is removed.

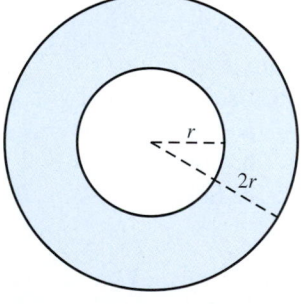
r
$2r$

46. Framing a Print A square print is surrounded by a mat and then framed. The width of the square mat is twice that of the print. If the area covered by the mat is 432 in^2, determine the width of the print.

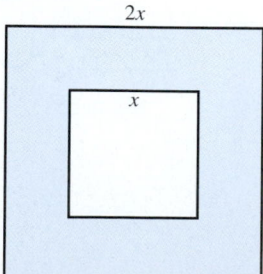
$2x$
x

Applying Technology

In Exercises 47–50, use a graphing calculator to approximate the solutions of each equation to the nearest hundredth. (*Hint:* See Technology Perspective 7.2.1 on Calculating the Zeros of a Function.)

47. $x^2 - 2x - 2 = 0$ **48.** $5v^2 - 2v - 1 = 0$
49. $6w^2 - w - 35 = 0$ **50.** $3w^2 + w - 3 = 0$

Group discussion questions

51. Comparing Methods of Solving Quadratic Equations Solve $3x^2 + 13x - 10 = 0$ by each of the following methods if this is possible. If any of these methods cannot be used to solve the equation, state why not. Then discuss which method you thought was easiest and which was the most difficult.
  a. Factoring
  b. Completing the square
  c. Using a graphing calculator to approximate the solutions to the nearest hundredth

52. Comparing Methods of Solving Quadratic Equations Solve $(16x^2 + 8x + 1) - 25 = 0$ by each of the following methods if this is possible. If any of these methods cannot be used to solve the equation, state why not. Then discuss which method you thought was easiest and which was the most difficult.
  a. Factoring
  b. Completing the square
  c. Using a graphing calculator to approximate the solutions to the nearest hundredth

53. Comparing Methods of Solving Quadratic Equations Solve $7x^2 + 19x + 11 = 0$ by each of the following methods if this is possible. If any of these methods cannot be used to solve the equation, state why not. Then discuss which method you thought was easiest and which was the most difficult.
  a. Factoring
  b. Completing the square
  c. Using a graphing calculator to approximate the solutions to the nearest hundredth

54. Discovery Question Complete the square of $x^2 + 10x$ both algebraically and geometrically.

Algebraically	Geometrically
$x^2 + 10x$	

5    $5x$

x x^2 $5x$

x 5

7.2	**Cumulative Review**

1. Simplify $12 - 5(11 - 8 \div 2)$.

2. Simplify $(12 - 5)(11 - 8 \div 2)$.

3. If a line has a slope of $\dfrac{3}{5}$, then a parallel line has a slope of _____.

4. If a line has a slope of $\dfrac{3}{5}$, then a perpendicular line has a slope of _____.

5. Simplify $\dfrac{36x^5 y^{10}}{2x^2 y^2}$.

Section 7.3   Using the Quadratic Formula to Find Real Solutions

Objectives:

1. Use the quadratic formula to solve quadratic equations with real solutions.

2. Use the discriminant to determine the nature of the solutions of a quadratic equation.

In this section, we will continue to solve only quadratic equations with real solutions. We will examine quadratic equations that have complex solutions after introducing imaginary numbers in Section 7.5. The methods that we have already covered for finding real solutions of quadratic equations are listed here.

- Tables   The zeros in a table correspond to the solutions.
- Graphs   The *x*-intercepts correspond to the solutions.
- Factoring   The factors of the polynomial produce the solutions.
- Extraction of roots
- Completing the square

The steps used to solve a quadratic equation in the form $ax^2 + bx + c = 0$ by completing the square can be generalized for any coefficients a, b, and c. The end result of this process is known as the **quadratic formula.**

Why Should I Learn the Quadratic Formula?

This formula allows you to skip all the steps of completing the square and just apply a shortcut formula to produce the solutions. This method can solve any quadratic equation; it produces exact answers; and it can be programmed on computers or calculators. We now show the development of the quadratic formula.

Development of the Quadratic Formula

$ax^2 + bx + c = 0$ Start with a quadratic equation in standard form with a, b, and c real numbers and $a > 0$.

$ax^2 + bx = -c$ Shift the constant to the right side of the equation.

$x^2 + \dfrac{b}{a}x = -\dfrac{c}{a}$ Divide both sides by a, which is the coefficient of x^2. To divide by a, we assume $a \neq 0$.

$x^2 + \dfrac{b}{a}x + \left(\dfrac{b}{2a}\right)^2 = -\dfrac{c}{a} + \left(\dfrac{b}{2a}\right)^2$ Add the square of one-half the coefficient of x to both sides of the equation.

$\left(x + \dfrac{b}{2a}\right)^2 = -\dfrac{4ac}{4a^2} + \dfrac{b^2}{4a^2}$ Write the left side as a perfect square, and the right side in terms of a common denominator.

$\left(x + \dfrac{b}{2a}\right)^2 = \dfrac{b^2 - 4ac}{4a^2}$ Simplify by combining the terms on the right side of the equation.

$$x + \frac{b}{2a} = \pm \sqrt{\frac{b^2 - 4ac}{4a^2}}$$ Extract the roots.

$$x = -\frac{b}{2a} \pm \frac{\sqrt{b^2 - 4ac}}{2a}$$ Simplify the radical, and subtract $\frac{b}{2a}$ from both sides.

$$x = \frac{-b \pm \sqrt{b^2 - 4ac}}{2a}$$ This is the quadratic formula.

1. Use the Quadratic Formula to Solve Quadratic Equations with Real Solutions

The quadratic formula is given in the following box. You should memorize this formula. Note that the symbol $\pm$ is used to express the two solutions concisely. Also note that if $a = 0$, the equation $ax^2 + bx + c = 0$ becomes the linear equation $bx + c = 0$; and if $a < 0$, it is customary to divide each term by -1 so the coefficient of x^2 is positive.

Quadratic Formula

The solutions of the quadratic equation $ax^2 + bx + c = 0$ with real coefficients a, b, and c, and $a \neq 0$, are

$$x = \frac{-b \pm \sqrt{b^2 - 4ac}}{2a}$$

Example 1   Using the Quadratic Formula to Determine Two Rational Solutions

Use the quadratic formula to solve $x^2 - 5x - 6 = 0$.

Solution

$x^2 - 5x - 6 = 0$ Substitute $a = 1$, $b = -5$, and $c = -6$ into

$$x = \frac{-(-5) \pm \sqrt{(-5)^2 - 4(1)(-6)}}{2(1)}$$ $x = \dfrac{-b \pm \sqrt{b^2 - 4ac}}{2a}$.

$$x = \frac{5 \pm \sqrt{25 + 24}}{2}$$ Simplify, and find both solutions.

$$x = \frac{5 \pm \sqrt{49}}{2}$$

$$x = \frac{5 \pm 7}{2}$$

$$x = \frac{5 - 7}{2} \quad \text{or} \quad x = \frac{5 + 7}{2}$$

$$x = -1 \qquad\qquad x = 6$$

Answer: The solutions are $x = -1$ and $x = 6$.

Self-Check 1

Use the quadratic formula to solve $2x^2 - 13x + 15 = 0$.

Note that we could also have solved the equation in Example 1 by factoring it as $(x + 1)(x - 6) = 0$. The primary advantage of the quadratic formula is that it can be used to solve problems that cannot be solved by factoring over the integers. This is illustrated in Example 2. Before you use the quadratic formula, be sure to write the equation in the standard form $ax^2 + bx + c = 0$. This will help prevent making an error in the sign of a, b, or c.

Example 2   Using the Quadratic Formula to Determine Two Irrational Solutions

Use the quadratic formula to determine the exact solutions of $4x^2 - 4x = 1$, and approximate these solutions to the nearest hundredth.

Solution

$$4x^2 - 4x = 1$$

$$4x^2 - 4x - 1 = 0$$

First write the equation in standard form.

$$x = \frac{-(-4) \pm \sqrt{(-4)^2 - 4(4)(-1)}}{2(4)}$$

Substitute $a = 4$, $b = -4$, and $c = -1$ into the quadratic formula.

$$x = \frac{4 \pm \sqrt{16 + 16}}{8}$$

$$x = \frac{4 \pm \sqrt{32}}{8}$$

$$x = \frac{4 \pm 4\sqrt{2}}{8}$$

Note that $\sqrt{32} = \sqrt{16} \cdot \sqrt{2} = 4\sqrt{2}$

$$x = \frac{4(1 \pm \sqrt{2})}{8}$$

Simplify, and write both solutions separately.

$$x = \frac{1 \pm \sqrt{2}}{2}$$

Answer: $x = \dfrac{1 - \sqrt{2}}{2}, x = \dfrac{1 + \sqrt{2}}{2}$

$$x \approx -0.21, x \approx 1.21$$

The exact solutions are approximated to the nearest hundredth with a calculator.

Self-Check 2

a. Use the quadratic formula to determine the exact solutions of $x^2 - 2x = 5$.

b. Approximate these solutions to the nearest hundredth.

The x-intercepts of a graph have a y-coordinate of 0. In Example 3, we use the solutions of $ax^2 + bx + c = 0$ to determine the exact x-intercepts of the graph of $y = ax^2 + bx + c$. We start by setting $y = 0$ and then solve $ax^2 + bx + c = 0$.

Example 3   Determining the Exact Intercepts of a Parabola

Algebraically determine the exact y-intercept and x-intercepts of the parabola defined by $y = x^2 - 4x + 1$. Then approximate these intercepts to the nearest tenth.

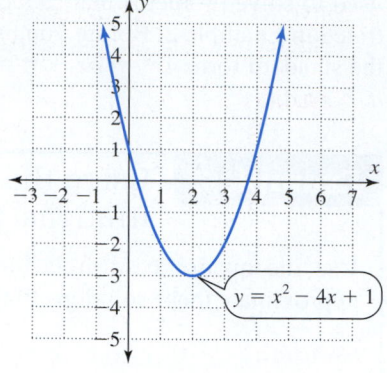

$y = x^2 - 4x + 1$

Solution

To find the y-intercept, set $x = 0$ and determine y.

$y = x^2 - 4x + 1$

$y = (0)^2 - 4(0) + 1$

$y = 1$

y-intercept: $(0, 1)$

To find the x-intercepts, set $y = 0$ and then solve for x.

$0 = x^2 - 4x + 1$ or $x^2 - 4x + 1 = 0$

$x = \dfrac{-(-4) \pm \sqrt{(-4)^2 - 4(1)(1)}}{2(1)}$

$x = \dfrac{4 \pm \sqrt{16 - 4}}{2}$

$x = \dfrac{4 \pm \sqrt{12}}{2}$

$x = \dfrac{4 \pm 2\sqrt{3}}{2}$

$x = 2 \pm \sqrt{3}$

Exact x-intercepts: $(2 - \sqrt{3}, 0)$ and $(2 + \sqrt{3}, 0)$
Approximate x-intercepts: $(0.3, 0)$ and $(3.7, 0)$

The graph crosses the y-axis if the x-coordinate is 0.

The graph crosses the x-axis if the y-coordinate is zero.

Use the quadratic formula
$x = \dfrac{-b \pm \sqrt{b^2 - 4ac}}{2a}$ *to solve*
$x^2 - 4x + 1 = 0$. *In this equation* $a = 1, b = -4,$ *and* $c = 1$.

$\sqrt{12} = \sqrt{4}\sqrt{3} = 2\sqrt{3}$

The exact values are approximated to the nearest tenth with a calculator.

Self-Check 3

a. Determine the exact intercepts of the parabola defined by $y = 2x^2 - x - 2$.

b. Approximate the x-intercepts to the nearest tenth.

The parabola in Example 3 has two distinct x-intercepts. It is possible that a parabola can have zero, one, or two x-intercepts. We will now begin to examine each of these possibilities. The parabola $y = (x - 3)^2$ in Example 4 has exactly one x-intercept, $(3, 0)$. If a parabola has exactly one x-intercept, we say that the corresponding quadratic equation has a **double real solution** or a **solution of multiplicity 2**. In Example 4, $x = 3$ is a double real solution of $(x - 3)^2 = 0$.

Example 4 Solving for a Double Real Solution

Solve $(x - 3)^2 = 0$ and determine the
x-intercept(s) of $y = (x - 3)^2$.

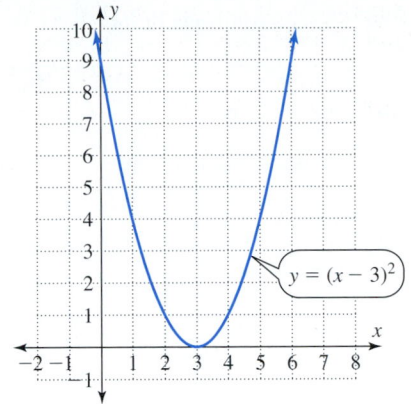

$y = (x - 3)^2$

Solution

$(x - 3)^2 = 0$

$x - 3 = \pm\sqrt{0}$

$x = 3 \pm 0$

$x = 3 - 0$ or $x = 3 + 0$

$x = 3$ or $x = 3$

Both solutions are the same. That means
that $x = 3$ is a double solution.
Note that the parabola is tangent to
the x-axis at the x-intercept of $(3, 0)$.

Answer: $x = 3$

Self-Check 4

Solve $(x + 2)^2 = 0$ and determine the x-intercept(s) of $y = (x + 2)^2$.

2. Use the Discriminant to Determine the Nature of the Solutions of a Quadratic Equation

What Happens If a Parabola Has No x-intercepts?

In this case, the corresponding quadratic equation will not have any real solutions. Both
solutions will be complex numbers with imaginary parts. We will examine this case in
Section 7.5.

Can I Identify a Quadratic Equation That Has Imaginary Solutions Without Actually Solving the Equation?

Yes, there are two ways. First, the graph of the corresponding parabola will not have any
x-intercepts. Second, we can determine the nature of the solutions algebraically by examining the radicand $b^2 - 4ac$ of the quadratic formula

$$x = \frac{-b \pm \sqrt{b^2 - 4ac}}{2a}$$

Because $b^2 - 4ac$ can be used to discriminate between real solutions and complex solutions with imaginary parts, it is called the **discriminant.**

A Mathematical Note

James Joseph Sylvester
(1814–1897) was born in
England as James Joseph. He
changed his last name to
Sylvester when he moved to
the United States. At Johns
Hopkins University, he led
efforts to establish graduate
work in mathematics in the
United States. He also founded
the *American Journal of
Mathematics.* Among his
lasting contributions to
mathematics are the many new
terms he introduced, including
the term *discriminant.*

The Nature of the Solutions of a Quadratic Equation

There are three possibilities for the types of solutions of $ax^2 + bx + c = 0$.

Value of the Discriminant	Solutions of $ax^2 + bx + c = 0$	The Parabola $y = ax^2 + bx + c$	Graphical Example
1. $b^2 - 4ac > 0$	Two distinct real solutions	Two x-intercepts	$y = x^2 - x - 2$
2. $b^2 - 4ac = 0$	A double real solution	One x-intercept; the parabola is tangent to the x-axis.	$y = x^2 - 4x + 4$
3. $b^2 - 4ac < 0$	Neither solution is real; both solutions are complex numbers with imaginary parts. These solutions will be complex conjugates. (Complex numbers are covered in Section 7.5.)	No x-intercepts	$y = x^2 + 1$

Example 5   Determining the Nature of the Solutions of a Quadratic Equation

Use the discriminant and a graph to determine the nature of the solutions of each quadratic equation.

(a) $x^2 - 6x + 8 = 0$ **(b)** $x^2 - 6x + 9 = 0$ **(c)** $x^2 - 6x + 10 = 0$

Solution

Discriminant

(a) $x^2 - 6x + 8 = 0$

$$b^2 - 4ac = (-6)^2 - 4(1)(8)$$
$$= 36 - 32$$
$$= 4$$

Because $b^2 - 4ac > 0$, there are two distinct real solutions. On the parabola this is illustrated by the two x-intercepts.

Graph

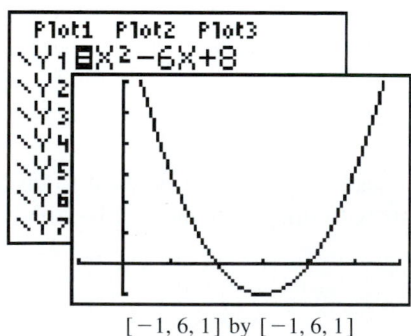

$[-1, 6, 1]$ by $[-1, 6, 1]$

(b) $x^2 - 6x + 9 = 0$

$$b^2 - 4ac = (-6)^2 - 4(1)(9)$$
$$= 36 - 36$$
$$= 0$$

Because $b^2 - 4ac = 0$, there is a double real solution. This is confirmed by the parabola that is tangent to the x-axis.

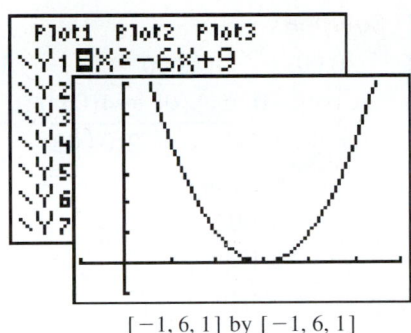

$[-1, 6, 1]$ by $[-1, 6, 1]$

(c) $x^2 - 6x + 10 = 0$

$$b^2 - 4ac = (-6)^2 - 4(1)(10)$$
$$= 36 - 40$$
$$= -4$$

Because $b^2 - 4ac < 0$, the solutions are complex numbers with imaginary parts. On the graph this is illustrated by the parabola with no x-intercepts.

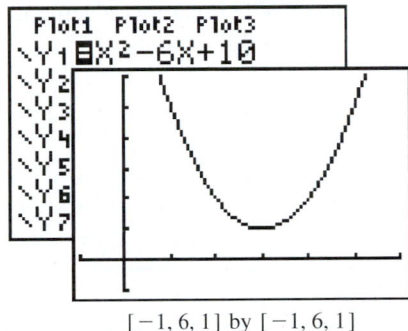

$[-1, 6, 1]$ by $[-1, 6, 1]$

Self-Check 5

Determine the nature of the solutions of each quadratic equation.

a. $x^2 + 4x + 4 = 0$ **b.** $x^2 + 4x + 5 = 0$ **c.** $x^2 + 4x + 3 = 0$

A quadratic equation written in the standard form $ax^2 + bx + c = 0$ will sometimes have either the constant term c or the linear term bx missing. In this case, it may be easier to solve the equation by factoring or by extraction or roots. If we decide to use the quadratic formula, we must substitute zero in as the coefficient for the missing term. This is illustrated in Example 6, in which we calculate the break-even value for an Alaskan fishing boat. The break-even value is the number of units that must be produced and sold to pay for the overhead costs and the production costs. The break-even value produces a profit of $0.

Example 6   Calculating a Break-Even Value

An Alaskan fishing boat has a permit to catch up to 10,000 pounds of Alaskan crab on one fishing trip. The profit in dollars for this trip is approximated by the function $P(x) = 0.002x^2 - 18,000$, where x is the number of pounds of Alaskan crab that is caught. Determine the break-even value for this trip.

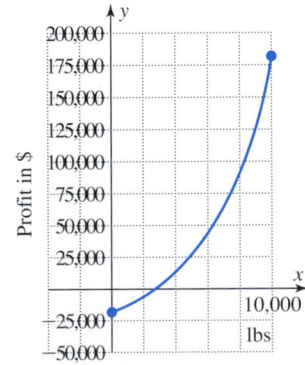

Solution

$$0.002x^2 - 18,000 = 0$$

$$x = \frac{-0 \pm \sqrt{0^2 - 4(0.002)(-18,000)}}{2(0.002)}$$

$$x = \frac{\pm\sqrt{144}}{0.004}$$

$$x = \frac{\pm 12}{0.004}$$

$$x = \pm 3,000$$

To determine the break-even value, set the profit to zero. In the standard form $ax^2 + bx + c = 0$, $a = 0.002$, $b = 0$, and $c = -18,000$.

Substitute these values into the quadratic formula.

Answer: The break-even value is 3,000 pounds.

A value of $-3,000$ is not an appropriate value for this problem.

Self-Check 6

A company can produce and sell from 0 to 500 units per month. The monthly profit function for this company is approximated by $P(x) = 0.3x^2 - 15,052.8$. Determine the break-even value for this company.

Can I Create Quadratic Equations with Specific Irrational Solutions?

Actually, we have already created equations with given solutions in the previous sections. We will review this strategy now to show how we can create the equation given in Example 2.

Example 7   Constructing a Quadratic Equation with Given Solutions

Construct a quadratic equation with solutions of $x = \dfrac{1 - \sqrt{2}}{2}$ and $x = \dfrac{1 + \sqrt{2}}{2}$.

Solution

$$x = \frac{1 - \sqrt{2}}{2} \qquad \text{or} \qquad x = \frac{1 + \sqrt{2}}{2}$$

$$2x = 1 - \sqrt{2} \qquad \text{or} \qquad 2x = 1 + \sqrt{2}$$

$$2x - 1 + \sqrt{2} = 0 \quad \text{or} \quad 2x - 1 - \sqrt{2} = 0$$

$$(2x - 1 + \sqrt{2})(2x - 1 - \sqrt{2}) = 0$$

$$[(2x - 1) + \sqrt{2}][(2x - 1) - \sqrt{2}] = 0$$

$$(2x - 1)^2 - (\sqrt{2})^2 = 0$$

$$4x^2 - 4x + 1 - 2 = 0$$

$$4x^2 - 4x - 1 = 0$$

Clear each equation of fractions. Rewrite each linear equation so that the right side of the equation is 0. If the factors of a quadratic equation are 0, then their product is 0. Multiply these factors by taking advantage of the special form $(A + B)(A - B) = A^2 - B^2$, and then combine like terms.

Answer: $4x^2 - 4x - 1 = 0$ has solutions $x = \dfrac{1 - \sqrt{2}}{2}$ and $x = \dfrac{1 + \sqrt{2}}{2}$.

Compare this result to Example 2.

Self-Check 7

Construct a quadratic equation with solutions of $x = \dfrac{1 - \sqrt{3}}{2}$ and $x = \dfrac{1 + \sqrt{3}}{2}$.

Summary: We have now covered a variety of methods for solving quadratic equations. The following box lists some of the advantages and disadvantages of each method. Graphical methods have the advantage of helping us "see" the solutions but often lack some of the precision and speed of algebraic methods. It is worth noting that both graphs and tables can be used not only for quadratic equations but also for many other types of equations. Other than for very simple quadratic equations, the algebraic approach that we recommend is the quadratic formula.

Most business applications and many in the social sciences produce quadratic equations with only real solutions. Applications with solutions that are complex numbers with imaginary parts are more likely to be found in the sciences. The quadratic formula will also be used to solve these equations in Section 7.5.

Methods for Solving $ax^2 + bx + c = 0$	Advantages	Disadvantages
Graphs	Visually displays the x-intercepts that correspond to real solutions. If there are no x-intercepts, this indicates the solutions are imaginary.	It can be time-consuming to select the appropriate viewing window and to approximate the real solutions. This method cannot determine complex solutions with imaginary parts.
Tables	Tables allow us to observe numerical patterns that are important in many applications. Spreadsheets allow us to use the power of computers to exploit this method.	Often it requires insight or some trial and error to select the most appropriate table.
Factoring	Factoring is easiest to use when the coefficients a, b, and c are small integers. Factoring also makes it easy to identify the x-intercepts of the corresponding parabola.	Most quadratic equations that arise from applications do not have small integers for the coefficients a, b, and c. This method works best only for real solutions that are rational.
Extraction of roots	This method is very easy to use if the left side of the quadratic equation is written as a perfect square.	If the left side of the quadratic equation is not written as a perfect square, then one must first complete the square.
Completing the square	This method can be used to solve any quadratic equation. This method is used to develop the quadratic formula and is the basis for work done in later mathematics courses.	It requires one to repeat all the steps used to develop the quadratic formula.
Quadratic formula	This method can be used to solve any quadratic equation. It is routine to substitute a, b, and c into the quadratic formula. This formula can easily be programmed into calculators and computers.	For some easy equations, factoring may be faster.

Self-Check Answers

1. $x = 1.5, x = 5$

2. **a.** $x = 1 - \sqrt{6}, x = 1 + \sqrt{6}$
   b. $x \approx -1.45, x \approx 3.45$

3. **a.** y-intercept: $(0, -2)$; x-intercepts: $\left(\dfrac{1 - \sqrt{17}}{4}, 0\right)$ and $\left(\dfrac{1 + \sqrt{17}}{4}, 0\right)$
   b. $(-0.8, 0)$ and $(1.3, 0)$

4. $x = -2$; $(-2, 0)$

5. **a.** A double real solution
   b. Complex solutions with imaginary parts
   c. Two distinct real solutions

6. The break-even value is 224 units.

7. $2x^2 - 2x - 1 = 0$

7.3 Using the Language and Symbolism of Mathematics

1. The formula $x = \dfrac{-b \pm \sqrt{b^2 - 4ac}}{2a}$ is called the _____ formula.

2. Before you solve a quadratic equation using the quadratic formula, you should write the equation in _____ form.

3. The standard form of $5x = 7 - 3x^2$ is _____.

4. The symbol that is read "plus or minus" is _____.

5. From the quadratic formula, $b^2 - 4ac$ is called the _____.

6. If a quadratic equation $ax^2 + bx + c = 0$ has a double real solution, then $b^2 - 4ac =$ _____.

7. If a quadratic equation $ax^2 + bx + c = 0$ has $b^2 - 4ac > 0$, then this equation has _____ real solutions.

8. If a quadratic equation $ax^2 + bx + c = 0$ has $b^2 - 4ac < 0$, then this equation has _____ real solutions.

7.3 Quick Review

1. Simplify $\sqrt{20}$.

2. Simplify $\dfrac{6 + 4\sqrt{5}}{2}$.

3. Evaluate $y = 3x^2 + 2x - 1$ for $x = 0$.

4. Evaluate $y = 3x^2 + 2x - 1$ for $x = \dfrac{1}{3}$.

5. If the parabola $y = 2x^2 + 19x - 60$ has x-intercepts of $(-12, 0)$ and $(2.5, 0)$, determine the solutions of $2x^2 + 19x - 60 = 0$.

7.3 Exercises

Objective 1 Use the Quadratic Formula to Solve Quadratic Equations with Real Solutions

In Exercises 1–20, use the quadratic formula to determine the exact solutions of each equation. Then approximate each irrational solution to the nearest hundredth.

1. $m^2 - 2m - 3 = 0$
2. $m^2 + 6m - 16 = 0$
3. $8y^2 = 2y + 1$
4. $2y^2 + 5y = -3$
5. $4x^2 = 4x + 35$
6. $4x^2 + 4x = 8$
7. $5w^2 - 2w - 1 = 0$
8. $3w^2 + 7w + 3 = 0$
9. $9v^2 + 1 = 12v$
10. $9v^2 + 12v = 1$
11. $6t^2 + 5t = 0$
12. $5t^2 = 6t$
13. $z^2 = 8$
14. $z^2 = 14$
15. $4x^2 - 12x + 9 = 0$
16. $9x^2 + 12x + 4 = 0$
17. $(w + 4)(w + 5) = 12$
18. $(w + 1)(w - 1) = 15$
19. $x(x + 2) = 2x + 3$
20. $x^2 + x + 18 = x(2x + 1)$

Objective 2 Use the Discriminant to Determine the Nature of the Solutions of a Quadratic Equation

In Exercises 21 and 22, match the description of the discriminant of each quadratic equation with the most appropriate graph.

21. a. A quadratic equation whose discriminant is negative
    b. A quadratic equation whose discriminant is zero
    c. A quadratic equation whose discriminant is positive

A.

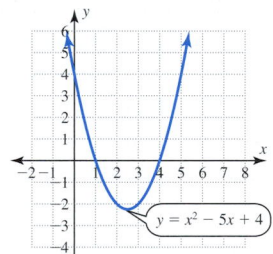

B.

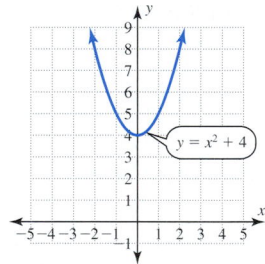

C.

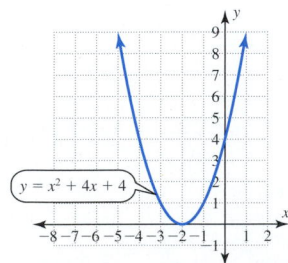

22. a. A quadratic equation whose discriminant is negative
   b. A quadratic equation whose discriminant is zero
   c. A quadratic equation whose discriminant is positive

A.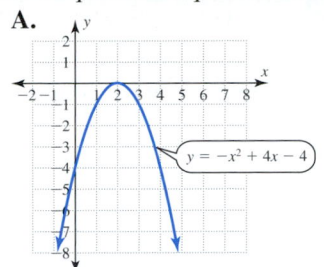

B.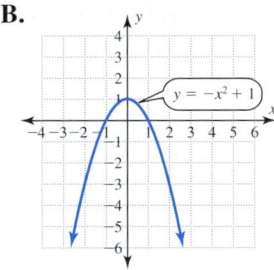

C.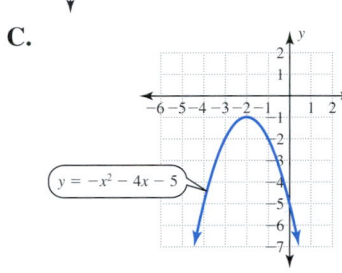

In Exercises 23–28, calculate the discriminant of each quadratic equation and determine the nature of the solutions of each equation.

23. $w^2 - 20w + 100 = 0$ **24.** $w^2 - 5w + 7 = 0$

25. $v^2 - 5v + 5 = 0$ **26.** $v^2 - 15 = 0$

27. $w^2 + 6 = 0$ **28.** $y^2 + 8y = -16$

Connecting Algebra to Geometry

In Exercises 29–32, determine the exact y-intercept and x-intercepts of each parabola. Then approximate these intercepts to the nearest hundredth.

29.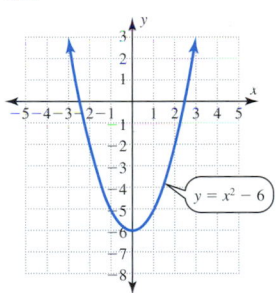

30.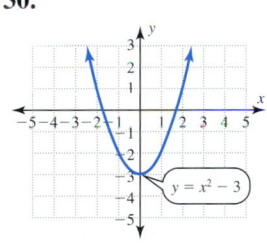

31.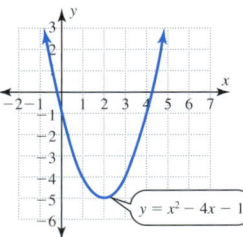

32.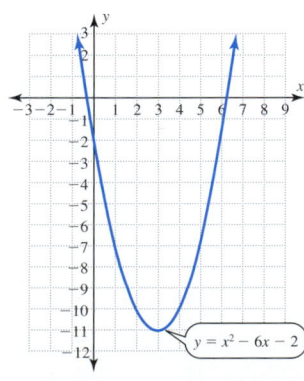

Connecting Concepts to Applications

33. One number is 6 more than another number. Find these numbers if their product is 8.

34. The product of a number and twice this number is 98. Find this number.

35. The rectangle shown in the figure is 4 cm longer than it is wide. Find the dimensions if the area is 8 cm².

36. The perimeter of the square shown in the figure is numerically 2 more than the area. Find the length in meters of one side of the square.

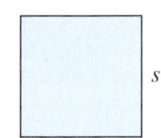

37. Break-Even Values Given the profit function $P(x) = -80x^2 + 5,280x - 61,200$, where x represents the number of units produced and $P(x)$ represents the profit in dollars, determine the following.
   a. Overhead costs [evaluate $P(0)$]
   b. Break-even values [determine to the nearest unit when $P(x) = 0$]

38. Break-Even Values Given the profit function $P(x) = -75x^2 + 5,925x - 47,250$ where x represents the number of units produced and $P(x)$ represents the profit in dollars, determine the following.
   a. Overhead costs [evaluate $P(0)$]
   b. Break-even values [determine to the nearest unit when $P(x) = 0$]

39. Speed of an Automobile A design team uses the specifications of the various parts of the braking system to approximate the stopping distance in meters of an automobile by $d = 0.01v^2 + 0.03v$. Solve the equation $100 = 0.01v^2 + 0.03v$ to determine the speed v to the nearest km/h that an automobile is going when it requires 100 m to stop.

40. Speed of an Automobile Determine the speed v to the nearest km/h that the automobile in Exercise 39 is going when it requires 115 m to stop.

Review and Concept Development

In Exercises 41–44, solve each cubic equation.

41. $v(v - 4)(v + 7) = 0$ **42.** $v(v + 5)(v - 9) = 0$

43. $w(w^2 - 10) = 0$

44. $w(w^2 - 10w + 20) = 0$

45. The solutions of the equation $-x^2 + 2x + 6 = 0$ are $x = 1 - \sqrt{7}$ and $x = 1 + \sqrt{7}$. Use the graph of $y = -x^2 + 2x + 6$ to:

a. Solve $-x^2 + 2x + 6 < 0$.

b. Solve $-x^2 + 2x + 6 > 0$.

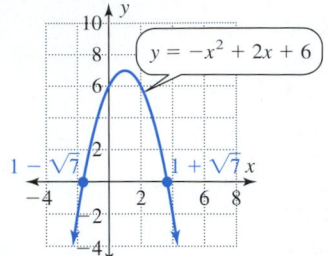

46. The solutions of the equation $4x^2 - 11 = 0$ are

$x = -\dfrac{\sqrt{11}}{2}$ and $x = \dfrac{\sqrt{11}}{2}$.

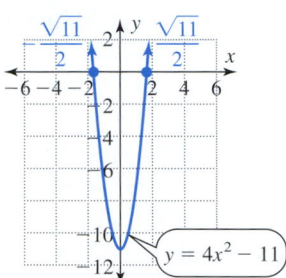

Use the graph of $y = 4x^2 - 11$ to:

a. Solve $4x^2 - 11 < 0$.

b. Solve $4x^2 - 11 > 0$.

47. Path of a Model Rocket The height of a model rocket is given by $f(x) = -16x^2 + 48x - 18$, where $f(x)$ represents the height of the rocket in feet and x is the number of seconds that have elapsed since the rocket was launched from a pit below ground level.

a. Determine the times at which the rocket is at ground level.

b. Determine the interval of times for which the rocket is above ground level.

48. Rework Exercise 47 assuming the height of the rocket is given by $f(x) = -16x^2 + 64x - 12$.

49. Construct a quadratic equation in x that has solutions $\dfrac{3 - \sqrt{5}}{2}$ and $\dfrac{3 + \sqrt{5}}{2}$.

50. Construct a quadratic equation in x that has solutions $\dfrac{5 - \sqrt{3}}{2}$ and $\dfrac{5 + \sqrt{3}}{2}$.

Group discussion questions

51. Comparing Methods of Solving Quadratic Equations Solve $x^2 - 36 = 0$ by each of the following methods if this is possible. Then discuss which method you prefer and why you prefer this method.

a. Factoring

b. Extraction of roots

c. The quadratic formula

d. Using a graphing calculator to determine the x-intercepts of $y = x^2 - 36$

52. Comparing Methods of Solving Quadratic Equations Match each method of solving quadratic equations with the most appropriate description and discuss the limitations noted with your group.

a. Can be used only to solve quadratic equations

b. Can be used only to find rational solutions of equations

c. Can be used only to find approximate real solutions of equations

A. Factoring polynomials over the integers

B. Graphing functions and using their x-intercepts

C. The quadratic formula

53. Challenge Question Solve each equation for x.

a. $x^2 - 4y^2 = 0$ **b.** $x^2 - 3y^2 = 0$

c. $x^2 + 4xy + 3y^2 = 0$ **d.** $x^2 + 5xy + 3y^2 = 0$

7.3 | **Cumulative Review**

1. Write $-3y^3 + 2xy^2 + 2x^3 - 5x^2y$ in standard form.

2. Write the interval notation for the solutions of $|x| < 4$.

3. Simplify $(-1)^{333}$.

4. Simplify $(-4)(-3)(-2)(-1)(0)(1)(2)(3)(4)$.

5. Division by _____ is undefined.

Section 7.4 | Applications of Quadratic Equations

Objectives:

1. Use quadratic equations to solve word problems.
2. Use the Pythagorean theorem.

Why Is There a Whole Section on Word Problems?

The context in which most college graduates encounter mathematics is verbally stated problems—problems that need to be translated into algebra to be solved. It is rare that anyone will ask you to solve a quadratic equation. More likely, someone will ask, "How do you . . ." or "What is the cheapest way to . . .?" These are the real-world problems that will challenge your problem-solving skills.

Do I Need to Learn a New Strategy for Solving Word Problems?

No, the word problem strategy repeated here for reference is the same strategy that we used in Section 2.8. The only difference is that now we will examine problems that will translate into quadratic equations.

1. Use Quadratic Equations to Solve Word Problems

Strategy for Solving Word Problems

Step 1. Read the problem carefully to determine what you are being asked to find.

Step 2. Select a variable to represent each unknown quantity. Specify precisely what each variable represents and note any restrictions on each variable.

Step 3. If necessary, make a sketch and translate the problem into a word equation or a system of word equations. Then translate each word equation into an algebraic equation.

Step 4. Solve the equation or system of equations, and answer the question completely in the form of a sentence.

Step 5. Check the reasonableness of your answer.

Each sport or profession requires practice time on the basics in order to execute efficiently in a real performance. Even Tiger Woods practices chipping and putting to hone these skills for tournaments. Likewise, we will include a few basic drill exercises like Example 1. Sometimes it is easy to determine an answer to a word problem like Example 1 without taking the time to write the equation, and the answer may even check. What you risk by not using algebra to form an equation is to overlook other possible solutions. In a practical setting, this may mean an engineer or businessperson overlooks another solution that may be cheaper or has preferable attributes over the obvious solution.

Example 1 | Solving a Numeric Word Problem

Find two consecutive even integers whose product is 48.

Solution

Let n = the smaller even integer

$n + 2$ = the next consecutive even integer

Consecutive even integers (such as 2, 4, and 6) differ by 2 units.

Word Equation

(First integer) · (Second integer) = 48

Form a word equation and then translate it into an algebraic equation.

Algebraic Equation

$$n(n + 2) = 48$$

$$n^2 + 2n - 48 = 0$$

$$(n + 8)(n - 6) = 0$$

$n + 8 = 0$	or	$n - 6 = 0$
$n = -8$	or	$n = 6$
$n + 2 = -6$	or	$n + 2 = 8$

Write the quadratic equation in standard form. The form $(n + \underline{})(n - \underline{})$ is used to factor the left side of the equation by inspection.

Note that there are two pairs of answers.

Answer: Both the consecutive even integers −8 and −6 and the integers 6 and 8 have a product of 48.

Self-Check 1

Find two consecutive odd integers whose product is 143.

In Section 2.8, we noted that it is important to pay attention to the restrictions on the variable—the values of the variable that are permissible in a given application. After solving the equations that are used to model the problem, always inspect the solutions of the equation(s) to see if they are appropriate for the original problem. For example, check for negative lengths, negative interest rates, negative speeds for vehicles, more than 24 hours in a day, or more than 31 days in a month.

Example 2   Finding the Dimensions of Two Square Tarps

A square tarp is used to shield the pitcher's mound of a baseball diamond from the rain. A new design for the tarp is 1 m longer on each side. The combined coverage area of the two tarps is 61 m². Find the dimensions of each tarp.

Solution

$x = $ length of each side of smaller tarp, m

$x + 1 = $ length of each side of the larger tarp, m

The new tarp is 1 m longer on each side.

Word Equation

$$\left(\begin{array}{c}\text{Area of}\\\text{1st tarp}\end{array}\right) + \left(\begin{array}{c}\text{Area of}\\\text{2nd tarp}\end{array}\right) = 61 \text{ m}^2$$

Form a word equation. Then use the formula $A = s^2$, the formula for the area of a square, to translate this statement into an algebraic equation.

Algebraic Equation

$$x^2 + (x + 1)^2 = 61$$

$$x^2 + x^2 + 2x + 1 = 61$$

Write the quadratic equation in standard form and simplify by dividing both sides of the equation by 2.

$$2x^2 + 2x - 60 = 0$$

$$x^2 + x - 30 = 0$$

$$(x - 5)(x + 6) = 0$$

Then solve this equation by factoring. The form $(x - \underline{})(x + \underline{})$ is used to factor the left side of the equation by inspection.

$x - 5 = 0$	or $x + 6 = 0$
$x = 5$	$x = -6$
$x + 1 = 6$	$x + 1 = -5$

Be sure to find $x + 1$ as well as x.

The negative values −6 and −5 are not meaningful dimensions.

Answer: The dimensions of the tarps are, respectively, 5 m and 6 m on each side.

Check:

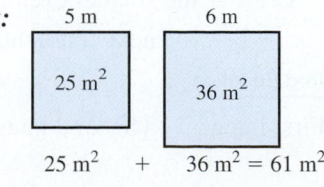

$$25 \text{ m}^2 \quad + \quad 36 \text{ m}^2 = 61 \text{ m}^2$$

Self-Check 2

If each side of a square is increased by 4 cm, the total area of the both the new square and the old square will be 250 cm^2. Determine the length of each side of both squares.

   Example 2 illustrated that we can solve an application involving two unknowns using only one variable. We could also use two variables to solve this problem. If we used y to represent the length of the larger tarp, then $y = x + 1$. We could then substitute $x + 1$ for y throughout the rest of the problem.

   In Example 3, we combine the calculations from an algebraic approach with the insight gained from a graph to solve a problem.

Example 3   Determining the Height of Fireworks

A faulty fireworks rocket is launched vertically with an initial velocity of 96 ft/s and then falls to the earth unexploded. Its height h in feet after t seconds is given by $h(t) = -16t^2 + 96t$. During what time interval after launch will the height of the rocket exceed 90 ft? The given figure shows the relationship of the height to the time. It does not show the path of the rocket, which goes straight up and then down.

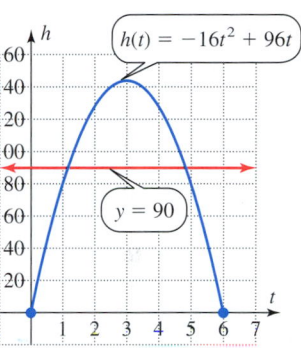

Solution

$$h(t) = -16t^2 + 96t$$

$$90 = -16t^2 + 96t$$

First determine the times when the height is 90 ft.

$$16t^2 - 96t + 90 = 0$$

$$8t^2 - 48t + 45 = 0$$

Write the quadratic equation in standard form, and then divide both sides of the equation by 2.

$$t = \frac{48 \pm \sqrt{(-48)^2 - 4(8)(45)}}{2(8)}$$

Solve this equation by using the quadratic formula.

$$t = \frac{48 \pm \sqrt{864}}{16}$$

$$t = \frac{48 - \sqrt{864}}{16} \quad \text{or} \quad t = \frac{48 + \sqrt{864}}{16}$$

These are the exact values of t.

$$t \approx 1.2 \qquad\qquad t \approx 4.8$$

Use a calculator to approximate these values to the nearest tenth of a second. At these times, the height of the rocket is 90 ft.

Answer: The height of the rocket will exceed 90 ft between 1.2 and 4.8 seconds after launch.

From the graph, note that the parabola is above $y = 90$ for t values between 1.2 and 4.8. The height will be over 90 ft from 1.2 to 4.8 seconds after launch.

Self-Check 3

Rework Example 3 to determine the time interval for which the rocket is above 80 ft.

   Another way to approximate the solution in Example 3 is to use the **Intersect** feature on a graphing calculator. (See Technology Perspective 2.3.2 on Approximating a Point of Intersection.) We can approximate the points of intersection of the graphs of $Y_1 = -16x^2 + 96x$ and $Y_2 = 90$. Then the interval of points between these values will represent the times when the rocket is above 90 ft.

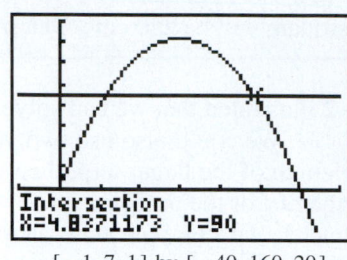

[−1, 7, 1] by [−40, 160, 20]   [−1, 7, 1] by [−40, 160, 20]

The rocket first reaches a height of 90 ft after approximately 1.2 seconds.

The rocket returns to a height of 90 ft after approximately 4.8 seconds.

Thus the rocket is above 90 ft from 1.2 seconds to 4.8 seconds.

A Mathematical Note

The Greek mathematician Pythagoras (ca. 500 B.C.) taught orally and required secrecy of his initiates. Thus, records of the society formed by Pythagoras are anecdotal. His society produced a theory of numbers that was part science and part numerology. It assigned numbers to many abstract concepts, for example, 1 for reason, 2 for opinion, 3 for harmony, and 4 for justice. The star pentagon was the secret symbol of the Pythagoreans.

2. Use the Pythagorean Theorem

The triangle is one of the most basic shapes in geometry and in the world around us. A **right triangle** is a triangle containing a 90° angle; that is, two sides of the triangle are perpendicular. In a right triangle, the two shorter sides are called the **legs** and the longest side is called the **hypotenuse.** The lengths of the three sides can be denoted, respectively, by a, b, and c. (See the figure.) The Pythagorean theorem states an important relationship among the sides of a right triangle.

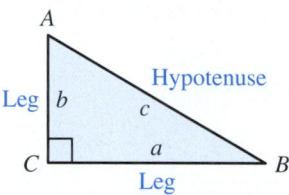

Pythagorean Theorem

If triangle ABC is a right triangle, then:

Algebraically	Geometric Example	Verbally
$a^2 + b^2 = c^2$		The sum of the areas of the squares formed on the legs of a right triangle is equal to the area of the square formed on the hypotenuse.
(The converse of this theorem is also true. If $a^2 + b^2 = c^2$, then triangle ABC is a right triangle.)	$3^2 + 4^2 = 5^2$ $9 + 16 = 25$	

Will I Ever Need to Use the Pythagorean Theorem If I Am Not an Engineer?

This is quite possible because many applications involve straight lines and triangles. Examples 4, 5, and 6 all use the Pythagorean theorem and quadratic equations.

Example 4   Using the Pythagorean Theorem to Solve an Application

An actress has a spotlight located 8 ft above her head. If the limit of the light's effective
range from her head is 17 ft, how far can she move from underneath this light?

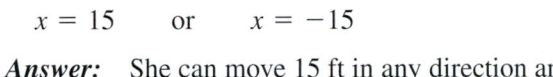

Solution

Let x = number of feet she can move across the floor

   d = number of feet from spotlight to top of her head

$d^2 = x^2 + 8^2$ Use the Pythagorean theorem
$d^2 = x^2 + 64$ and the values identified in the
              given sketch.

$17^2 = x^2 + 64$ Substitute 17 for d.

$289 = x^2 + 64$ Simplify this quadratic equation, and use
              extraction of roots to solve this equation.
$225 = x^2$

$x^2 = 225$

$x = \pm\sqrt{225}$

$x = 15$ or $x = -15$ Distance moved is interpreted to be positive.

Answer: She can move 15 ft in any direction and stay
         within the effective range of the light.

Self-Check 4

a. Two legs of a right triangle are 5 and 12 cm. Determine the length of the hypotenuse.

b. The hypotenuse of a right triangle is 25 cm, and one leg is 7 cm. How long is the other leg?

Example 5 is based on the analysis of structural components for a building. The shape
that results may not be exactly a right triangle. However, this mathematical model is close
enough to the actual result to yield useful information.

Example 5   Determining the Deflection of a Beam

A steel beam that is 20 ft long is placed in a hydraulic press for testing. Under
extreme pressure, the beam deflects upward so that the distance between the
hydraulic rams decreases by 2 in. Assume the deflected shape is approximated by
the figure shown. Determine the height of the bulge in the middle of the beam.

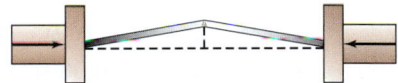

Solution

Let x = height of bulge, in inches   First identify the unknown value with a variable.

Then note that the shape in the sketch is a right triangle, so you can apply
the Pythagorean theorem.

Word Equation

$$\left(\begin{array}{c}\text{Length to}\\\text{center point}\end{array}\right)^2 + \left(\begin{array}{c}\text{Height of}\\\text{bulge}\end{array}\right)^2 = \left(\begin{array}{c}\text{Length of left}\\\text{half of beam}\end{array}\right)^2$$

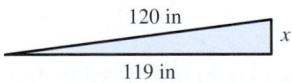

Algebraic Equation

$119^2 + x^2 = 120^2$ One-half the length of the beam is 10 ft or 120 in. The distance between the
$14{,}161 + x^2 = 14{,}400$ rams has decreased from 240 in to 238 in. One-half the distance between
              the presses is 119 in. Substitute these values into the word equation.
$x^2 = 239$

$x = \pm\sqrt{239}$

$x = \sqrt{239}$ or $x = -\sqrt{239}$ Solve this quadratic equation by extraction of roots. Then approximate
              these solutions.
$x \approx 15.5$ $x \approx -15.5$

The beam could deflect either up or down 15.46 in.

Answer: The bulge would be approximately 15.5 in.   This large a bulge is not tolerable in most building projects. Therefore
engineers plan for expansion and contraction spaces in their designs.

Example 6 uses both the Pythagorean theorem and the formula $D = R \cdot T$ for calculating the distance for a given rate and time. We will assume for simplicity that the rates and distances are given for flat ground measurements so that the planes' altitude changes are not considered.

Example 6   Applying the Pythagorean Theorem

Two airplanes depart simultaneously from an airport. One flies due south; the other flies due east at a rate 50 mi/h faster than that of the first airplane. After 2 hours, radar indicates that the airplanes are 500 mi apart. What is the ground speed of each airplane?

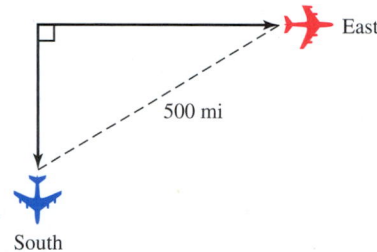

East

500 mi   Make a sketch that models the problem.

South

Solution

Let r = ground speed of first plane, mi/h

$r + 50$ = ground speed of second plane, mi/h

Represent each unknown quantity using a variable. To be consistent with the formula $D = R \cdot T$, we use the variable r to represent speed.

Word Equation

$$\left(\begin{array}{c}\text{Distance first} \\ \text{plane travels}\end{array}\right)^2 + \left(\begin{array}{c}\text{Distance second} \\ \text{plane travels}\end{array}\right)^2 = \left(\begin{array}{c}\text{Distance between} \\ \text{two planes}\end{array}\right)^2$$

The word equation is based on the Pythagorean theorem.

Algebraic Equation

$$(2r)^2 + [2(r + 50)]^2 = (500)^2$$
$$(2r)^2 + (2r + 100)^2 = (500)^2$$
$$4r^2 + (4r^2 + 400r + 10{,}000) = 250{,}000$$
$$8r^2 + 400r - 240{,}000 = 0$$
$$r^2 + 50r - 30{,}000 = 0$$
$$(r - 150)(r + 200) = 0$$

$r - 150 = 0$ or $r + 200 = 0$

$\qquad r = 150 \qquad\qquad\qquad r = -200$

$r + 50 = 200$

Using $D = R \cdot T$ and a time of 2 hours, we find that the first plane travels $2r$ mi and the second travels $2(r + 50)$ mi.

Substitute the distances into the word equation and simplify.

Write the quadratic equation in standard form.
Divide both sides of the equation by 8.

Solve this quadratic equation by factoring.

A negative rate is not meaningful for this problem.

Find both r and $r + 50$.

Answer: The first plane is flying south at 150 mi/h, and the second is flying east at 200 mi/h.

Do these rates check, and are they reasonable?

The most common description for the size of a computer monitor is the diagonal measurement of the screen. Based on that specification, two different monitors could have the same diagonal dimension yet have different viewable areas. Another technical specification typically listed for a computer monitor is the aspect ratio. The aspect ratio is the ratio of the width to the height of the screen. In Example 7, the listed specifications for a monitor are used to calculate the dimensions of the screen.

Example 7   Using the Aspect Ratio and Screen Size to Determine Screen Dimensions

A computer monitor is advertised as a 21-inch monitor with an 8:5 aspect ratio. Approximate the height, width, and viewable area of this monitor.

Solution

Let x = height of screen

$\dfrac{8x}{5}$ = width of screen

An aspect ratio of 8:5 indicates the width is $\dfrac{8}{5}$ as large as the height.

Word Equation

$$\left(\begin{array}{c}\text{Screen}\\ \text{width}\end{array}\right)^2 + \left(\begin{array}{c}\text{Screen}\\ \text{height}\end{array}\right)^2 = \left(\begin{array}{c}\text{Screen}\\ \text{diagonal}\end{array}\right)^2$$

The word equation is based on the Pythagorean theorem.

Algebraic Equation

$$\left(\frac{8x}{5}\right)^2 + x^2 = 21^2$$

Substitute the dimensions into the word equation and simplify.

$$\frac{64x^2}{25} + x^2 = 441$$

$$25\left(\frac{64x^2}{25} + x^2\right) = 25(441)$$

Multiply both sides of the equation by 25 to clear the equation of fractions.

$$64x^2 + 25x^2 = 11{,}025$$

$$89x^2 = 11{,}025$$

$$x^2 = \frac{11{,}025}{89}$$

$$x = \pm\sqrt{\frac{11{,}025}{89}}$$

Combine like terms and use extraction of roots to solve for x.

$$x = \sqrt{\frac{11{,}025}{89}} \qquad \text{or} \qquad x = -\sqrt{\frac{11{,}025}{89}}$$

$$x \approx 11.1 \qquad\qquad\qquad x \approx -11.1$$

A negative value for the screen height is not meaningful. Find both x and $\dfrac{8x}{5}$.

$$\frac{8x}{5} \approx 17.8$$

Answer: The height of the screen is 11.1 in, and the width of the screen is 17.8 in. The viewable area of this monitor is approximately 198 in².

The viewable area is found by multiplying the screen width by the screen height.

Self-Check 7

A computer monitor is advertised as a 21-inch monitor with a 4:3 aspect ratio. Approximate the height, width, and viewable area of this monitor.

7.4   Using the Language and Symbolism of Mathematics

1. The first step in the word problem strategy given in this book is to read the problem carefully to determine what you are being asked to _____ .

2. The second step in the word problem strategy given in this book is to select a _____ to represent each unknown quantity.

3. The third step in the word problem strategy given in this book is to model the problem verbally with word equations and then to translate these word equations into _____ equations.

4. The fourth step in the word problem strategy given in this book is to _____ the equation or system of equations and answer the question completely in the form of a sentence.

5. The fifth step in the word problem strategy given in this book is to check your answer to make sure the answer is _____ .

6. In the formula $D = RT$, D represents distance, R represents _____ , and T represents _____ .

7. In a right triangle, the two shorter sides are called the _____ and the longest side is called the _____ .

8. The two legs of a right triangle are _____ to each other.

9. The Pythagorean theorem states that for a right triangle with hypotenuse c and sides of length a and b, _____ .

7.4   Quick Review

1. If n represents an even integer, represent the next two integers in terms of n.

2. If n represents an even integer, represent the next two even integers in terms of n.

3. Solve $x^2 = 64$ by extraction of roots.

4. Use a calculator or spreadsheet to approximate $\sqrt{\dfrac{567}{13}}$ to the nearest tenth.

5. The length of wire remaining on a 100-ft roll after an electrician has cut off n pieces is given by $L = 100 - 5n$. There are certain implied restrictions on the variable n.
   a. What is smallest possible value of n?
   b. What is largest possible value of n?
   c. List all possible values of n from the smallest to the largest.

7.4   Exercises

Objective 1 Use Quadratic Equations to Solve Word Problems

In Exercises 1–52, solve each problem.

Numeric Word Problems

1. Find two consecutive integers whose product is 156.
2. Find two consecutive integers whose product is 240.
3. Find two consecutive even integers whose product is 288.

4. Find two consecutive odd integers whose product is 99.
5. The sum of the squares of two consecutive integers is 113. Find these integers.
6. The sum of the squares of three consecutive integers is 50. Find these integers.
7. One number is 4 more than three times another number. Find these numbers if their product is 175.
8. One number is 3 less than four times another number. Find these numbers if their product is 10.

Dimensions of a Rectangle

9. The length of a rectangle (see the figure) is 2 cm more than three times the width. Find the dimensions of this rectangle if the area is 21 cm².

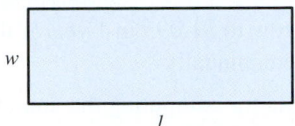

10. The length of a rectangle (see the figure) is 10 m less than twice the width. Find the dimensions of this rectangle if the area is 72 m².

Dimensions of a Triangle

11. The base of a triangle (see the figure) is 3 m longer than the height. Find the base if the area of this triangle is 77 m².

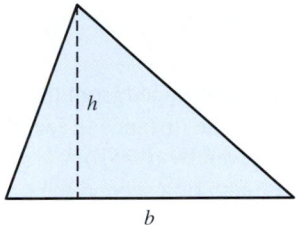

12. The base of the triangle shown in the figure is 5 cm longer than the height. Determine the height of the triangle if its area is 12 cm².

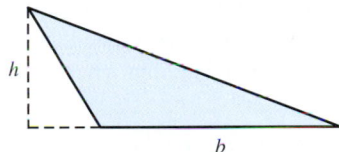

Dimensions of a Square

13. If each side of a square is increased by 5 cm, the total area of both the new square and the original square will be 200 cm². Approximate to the nearest tenth of a centimeter the length of each side of the original square.

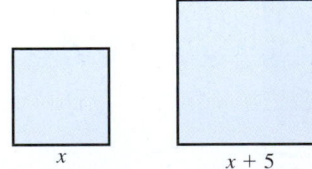

14. If each side of a square is increased by 2 cm, the total area of both the new square and the original square will be 400 cm². Approximate to the nearest tenth of a centimeter the length of each side of the original square.

15. Trajectory of a Projectile The height in meters of a ball released from a ramp is given by the function $h(t) = -4.9t^2 + 29.4t + 34.3$, where t represents the time in seconds since the ball was released from the end of the ramp. Determine the time interval that the ball is above 50 m.

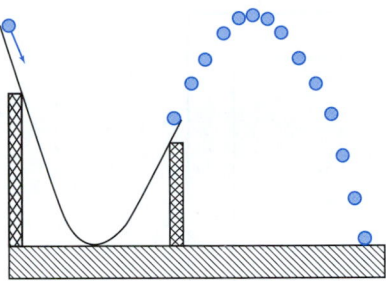

16. Height of a Golf Ball The height in feet of a golf ball hit from an elevated tee box is given by $h(t) = -16t^2 + 60t + 25$, where t represents the time in seconds the ball is in flight.

   a. Determine the number of seconds before the golf ball hits the ground at a height level of 0 ft.

   b. Determine the time interval for which the ball is above 25 ft.

17. Profit Polynomial The business manager of a small windmill manufacturer projects that the profit in dollars from making x windmills per week will be $P(x) = -x^2 + 70x - 600$. How many windmills should be produced each week in order to generate a profit? (*Hint:* Only an integral number of units can be produced.)

18. Average Cost The president of a company producing water pumps has gathered data suggesting that the average cost in dollars of producing x units per hour of a new style of pump will be $C(x) = x^2 - 22x + 166$. Determine the number of pumps per hour to produce to keep the average cost below $75 per pump.

19. Radius of a Circle The area enclosed between two concentric circles is 56π cm². The radius of the larger circle is 1 cm less than twice the radius of the smaller circle. Determine the length of the shorter radius. (See the figure.)

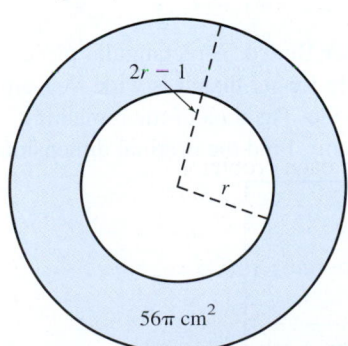

20. Dimensions of a Mat A square mat used for athletic exercises has a uniform red border on all four sides. The rest of the mat is blue. The width of the blue square is three-fourths the width of the entire square. If the area colored red is 28 m², determine the length of each side of the mat.

21. Pizza Size A couple plans to order a 10-in pizza when another couple they know joins them. Approximately what size pizza should they order to double the amount in a 10-in pizza?

22. Spaghetti Recipe A recipe for spaghetti for four people suggests using enough spaghetti to fill a 1-in-diameter opening on a spaghetti measuring device. What size hole would be recommended for two people?

23. Rope Length The length of one piece of rope is 8 m less than twice the length of another piece of rope. Each rope is used to enclose a square region. The area of the region enclosed by the longer rope is 279 m² more than the area of the region enclosed by the shorter rope. Determine the length of the shorter rope.

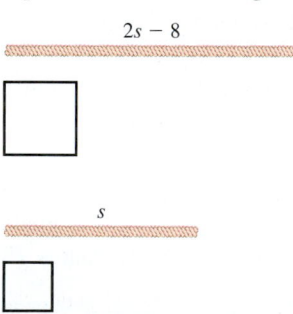

24. Dimensions of Poster Board A rectangular piece of poster board is 6.0 cm longer than it is wide. A 1-cm strip is cut off each side. The area of the remaining poster board is 616 cm². Find the original dimensions.

25. Interest Rate The formula for computing the amount A of an investment of principal P invested at interest rate r for 1 year and compounded semiannually is $A = P\left(1 + \dfrac{r}{2}\right)^2$. Approximately what interest rate is necessary for $1,000 to grow to $1,095 in 1 year if the interest is compounded semiannually?

26. Interest Rate The formula for computing the amount A of an investment of principal P invested at interest rate r for 1 year and compounded semiannually is $A = P\left(1 + \dfrac{r}{2}\right)^2$. Approximately what interest rate is necessary for $1,000 to grow to $1,075 in 1 year if the interest is compounded semiannually?

27. Free-Fall Time If a hammer is dropped from the 33rd floor of a construction project, the distance in feet that it will fall after t seconds is given by $d = 16t^2$. Approximate to the nearest tenth of a second the time it takes the hammer to fall 75 ft.

28. Free-Fall Time If a hammer is dropped from the 33rd floor of a construction project, the distance in feet that it will fall after t seconds is given by $d = 16t^2$. Approximate to the nearest tenth of a second the time it takes the hammer to fall 110 ft.

29. Speed of an Automobile A design team uses the specifications of the various parts of the braking system to approximate the stopping distance in meters of an automobile by $d = 0.01v^2 + 0.02v$. Solve the equation $100 = 0.01v^2 + 0.02v$ to determine the speed v to the nearest km/h that an automobile is going when it requires 100 m to stop.

30. Speed of an Automobile Determine the speed v to the nearest km/h that the automobile in Exercise 29 is going when it requires 115 m to stop.

Objective 2 Use the Pythagorean Theorem

31. Ladder Position If the bottom of a 17-ft ladder is 8 ft from the base of the chimney, how far is it from the bottom of the chimney to the top of the ladder?

32. Length of a Guy Wire A television tower has a guy wire attached 40 ft above the base of the tower. If the anchor point of the guy wire is 42 ft from the base of the tower, how long is the guy wire?

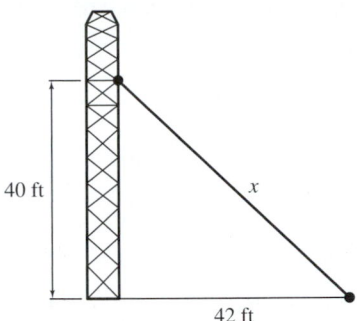

40 ft

x

42 ft

33. Positioning Effective Lighting The lighting system planned for a basketball court has bulbs that should be placed at least 30 ft above the floor for safety reasons. One bulb is located directly above the center of the court at a height of 30 ft. A ball is located at a position *x* ft from center court. If the limit of the light's effective range is 35 ft, how far can the ball be positioned from center court and still receive adequate light from this bulb? Approximate this distance to the nearest tenth of a foot.

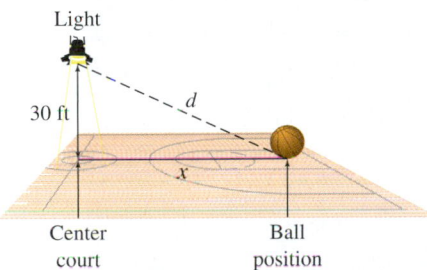

Light

30 ft

d

x

Center court

Ball position

34. Positioning an Effective Sound System The sound system planned for an airport concourse has speakers that should be placed at least 18 ft above the floor and thus approximately 12 ft above the heads of most customers. A customer is standing at a position *x* ft from a reference point directly under one speaker that is 18 ft above the floor. If the limit of the speaker's effective range is 20 ft, approximate to the nearest foot the distance a customer can walk from the reference point and stay within the effective range of the speaker.

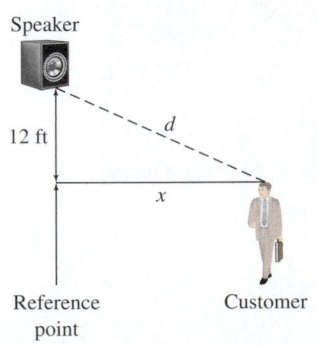

Speaker

12 ft

d

x

Reference point

Customer

35. Deflection of a Beam A 40-ft concrete beam is fixed between two rigid anchor points. If this beam expands by 0.5 inch due to a 50° increase in temperature, determine the bulge in the middle of the beam. (Assume for simplicity that the beam deflects as shown in this diagram.)

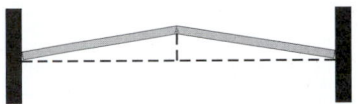

36. Deflection of a Beam A 28-ft aluminum beam is tested between two hydraulic rams. If the distance between the rams is decreased by 0.25 in, determine the bulge in the middle of the beam. (Assume for simplicity that the beam deflects as shown in this diagram.)

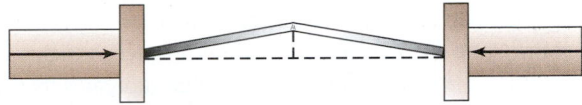

37. Screen Size The size of a computer monitor is usually given as the length of a diagonal of the screen. A new computer comes with a 17-in monitor. Give the height of the screen if the width is 13 in.

38. Screen Size A handheld organizer offers a 3.5-in screen (diagonal length). Approximate its width to the nearest hundredth if its height is 3.0 in.

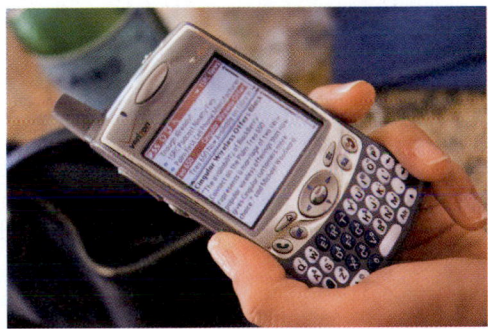

39. Screen Size A computer monitor is advertised as a 17-in monitor with a 4:3 aspect ratio. Determine the actual dimensions and viewable area of this monitor. (*Hint:* See Example 7.)

40. Screen Size A computer monitor is advertised as a 19-in monitor with a 5:4 aspect ratio. Determine the actual dimensions and viewable area of this monitor. (*Hint:* See Example 7.)

41. Distance by Plane Upon leaving an airport, an airplane flew due south and then due east. After it had flown 17 mi farther east than it had flown south, it was 25 mi from the airport. How far south had it flown?

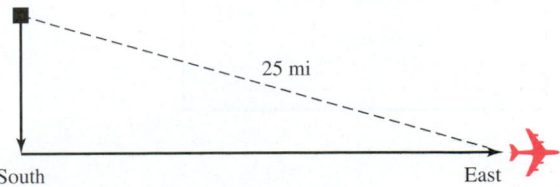

25 mi

South

East

42. Distance by Plane Upon leaving an airport, an airplane flew due west and then due north. After it had flown 1 mi farther north than it had flown west, it was 29 mi from the airport. How far west had it flown?

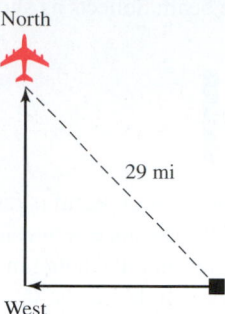

North

29 mi

West

43. Speed of an Airplane Two airplanes depart simultaneously from an airport. One flies due south; the other flies due east at a rate 30 mi/h faster than that of the first airplane. After 3 hours, radar indicates that the airplanes are 450 mi apart. What is the ground speed of each airplane?

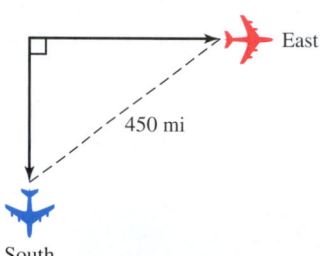

East

450 mi

South

44. Speed of an Airplane Two airplanes depart simultaneously from an airport. One flies due south; the other flies due east at a rate 10 mi/h faster than that of the first airplane. After 1 hour, radar indicates that the airplanes are 290 mi apart. What is the ground speed of each airplane?

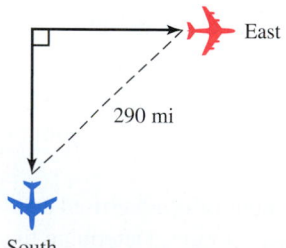

East

290 mi

South

45. Room Width Examining the blueprints for a rectangular room that is 17 ft longer than it is wide, an electrician determines that a wire run diagonally across this room will be 53 ft long. What is the width of the room?

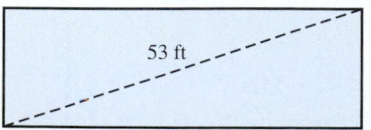

53 ft

46. Diameter of a Storage Bin The length of the diagonal brace in the cylindrical storage bin shown in the figure is 17 m. If the height of the cylindrical portion of the bin is 7 m more than the diameter, determine the diameter.

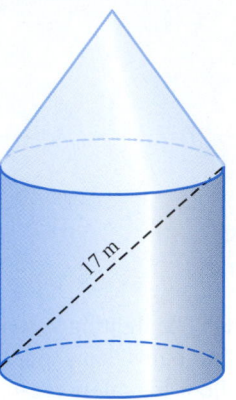

17 m

47. Distances on a Baseball Diamond The bases on a baseball diamond are placed at the corners of a square whose sides are 90 ft long. How much farther does a catcher have to throw the ball to get it from home plate to second base than from home plate to third base?

48. Speed of a Baseball If a catcher on a baseball team throws a baseball at 120 ft/s (over 80 mi/h), approximately how long will it take his throw to go from home plate to second base? Approximately how long will it take his throw to go from home plate to third base? (*Hint:* See Exercise 47.)

49. Distance to the Horizon The radius of the earth is approximately 4,000 mi. Approximate to within 10 mi the distance from the horizon to a plane flying at an altitude of 4 mi.

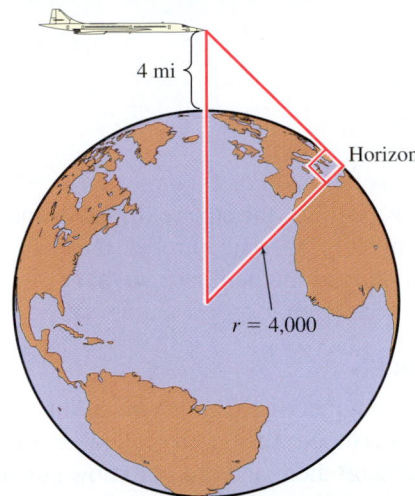

4 mi

Horizon

$r = 4,000$

50. Distance to the Horizon The radius of the earth is approximately 4,000 mi. Approximate to within 10 mi the distance from the horizon to a plane flying at an altitude of 5 mi. (See the figure for Exercise 49.)

51. Metal Machining A round stock of metal that is $18\sqrt{2}$ cm in diameter is milled into a square piece of stock. How long are the sides of the largest square that can be milled?

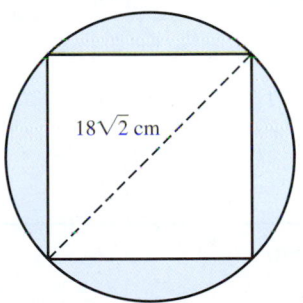

$18\sqrt{2}$ cm

52. Metal Machining A round stock of metal that is 30 cm in diameter is milled into a square piece of stock. How long are the sides of the largest square that can be milled? (See the figure in Exercise 51.)

Group discussion questions

53. Challenge Question Given the equation $x^2 + (x + 2)^2 = 244$: (*Hint:* See Exercise 2.)
   a. Write a numeric word problem that is modeled by this equation.
   b. Write a word problem that models the area of two squares. (*Hint:* See Exercise 31.)
   c. Write a word problem that models the distance between two airplanes. (*Hint:* See Exercise 43.)

54. Challenge Question
   a. Use the Pythagorean theorem to determine the length of diagonal d_1.
   b. Then determine the length of diagonal d_2.

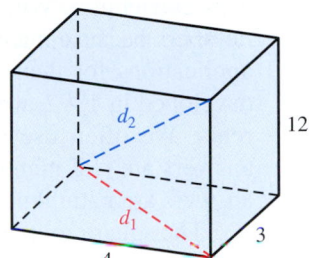

d_2

12

d_1

4

3

---

1. The slope of the line defined by $y = -3x + 5$ is _____.

2. The y-intercept of the line defined by $y = -3x + 5$ is _____.

3. Write the equation of a line with slope $m = 2$ and y-intercept $(0, -3)$.

4. The commutative property of addition allows us to rewrite $2x + 3y$ as _____.

5. Write an inequality using the variable x for the interval $(-2, 4]$.

<table>
<tr><td>**Section 7.5**</td><td>Complex Numbers and Solving Quadratic Equations with Complex Solutions</td></tr>
</table>

Objectives:

1. Express complex numbers in standard form.
2. Add, subtract, multiply, and divide complex numbers.
3. Solve a quadratic equation with imaginary solutions.

A Mathematical Note

Jean-Victor Poncelet (1788–1867) was a lieutenant in Napoleon's army. After being left for dead on the Russian front, he was captured and marched for months across frozen plains in subzero weather. To survive the boredom of captivity, he decided to reproduce as much mathematics as he could. When he returned to France after the war, he had developed projective geometry and a geometric interpretation of imaginary numbers.

Some quadratic equations do not have real-number solutions. For example, $x^2 = -1$ has no real solutions. The square of a negative number is positive, the square of 0 is 0, and the square of a positive number is positive. Thus there is no real number that satisfies $x^2 = -1$. In the 17th century, mathematicians first defined the number i so that $i^2 = -1$. We use i to represent $\sqrt{-1}$ and $-i$ to represent $-\sqrt{-1}$. Both $i^2 = -1$ and $(-i)^2 = -1$.

Will I Ever Need to Use Numbers That Contain *i*?

This is more likely if you study electrical circuits or engineering than if you are in a business curriculum. When these numbers were first developed, they were called imaginary numbers because mathematicians were not familiar with them and did not know concrete applications for them. One of the first concrete applications of imaginary numbers was developed in 1892, when Charles P. Steinmetz used them in his theory of alternating currents. We often use imaginary numbers in the computation of problems whose final answers are real numbers, just as we use fractions in the computation of problems whose answers are natural numbers.

The square root of any negative number is imaginary and can be expressed in terms of the imaginary unit i. For $x > 0$, we define $\sqrt{-x}$ to be $i\sqrt{x}$.

The Imaginary Number *i*

$i = \sqrt{-1}$ so $i^2 = -1$

$-i = -\sqrt{-1}$ so $(-i)^2 = i^2 = -1$

For any positive real number x, $\sqrt{-x} = i\sqrt{x}$.

Example 1   Writing Imaginary Numbers by Using the *i* Notation

Write each imaginary number in terms of i.

(a) $\sqrt{-25}$ (b) $\sqrt{-3}$

Solution

(a) $\sqrt{-25} = i\sqrt{25}$ $\sqrt{-x} = i\sqrt{x}$

$\qquad\quad = i(5)$ We usually write $5i$, not $i5$, just as we usually write $7x$, not $x7$.

$\qquad\quad = 5i$

(b) $\sqrt{-3} = i\sqrt{3}$ We usually write $i\sqrt{3}$ instead of $\sqrt{3}i$ so we do not accidentally interpret this expression as $\sqrt{3i}$.

Self-Check 1

Write each imaginary number in terms of i.

a. $\sqrt{-4}$ b. $\sqrt{-144}$

1. Express Complex Numbers in Standard Form

Using the real numbers, the imaginary number i, and the operations of addition, subtraction, multiplication, and division, we obtain numbers that can be written in the form $a + bi$, where a and b are real numbers.

Any number that can be written in the **standard form** $a + bi$ is called a **complex number.** If $b = 0$, then $a + bi$ is just the real number a. If $b \neq 0$, then $a + bi$ is called **imaginary.** If $a = 0$ and $b \neq 0$, then bi is called **pure imaginary.** Thus the complex numbers include both the real numbers and the pure imaginary numbers.

Complex Numbers

If a and b are real numbers and $i = \sqrt{-1}$, then:

Algebraic Form	Numerical Example
$a + bi$ is a complex number with a **real term** a and an **imaginary term** bi.	$5 + 6i$ has a real term 5 and an imaginary term $6i$.

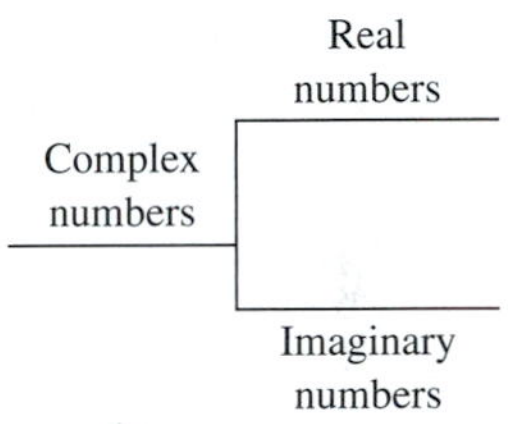

Every real number is a complex number, and every imaginary number is a complex number. The real numbers and the imaginary numbers are distinct subsets of the complex numbers with no elements in common. The relationships among the real, complex, and imaginary numbers are shown in Figure 7.5.1 and illustrated by the examples in Table 7.5.1. Every number in this table is a complex number.

Figure 7.5.1

Table 7.5.1   Classifying Complex Numbers

Complex Number	Standard Form	Real Term	Coefficient of the Imaginary Term	Classification
6	$6 + 0i$	6	0	Real
$-7i$	$0 - 7i$	0	-7	Pure imaginary
$3 - 4i$	$3 - 4i$	3	-4	Imaginary
0	$0 + 0i$	0	0	Real
$-\sqrt{25}$	$-5 + 0i$	-5	0	Real
$\sqrt{-25}$	$0 + 5i$	0	5	Pure imaginary

Many graphing calculators have the ability to work with complex numbers. The TI-84 Plus calculator has a **REAL** mode and an **$a + bi$** (complex) mode. Attempting to evaluate $\sqrt{-25}$ in **REAL** mode will generate an error because $\sqrt{-25}$ is not a real number. When the calculator is set to complex mode, the calculator correctly evaluates $\sqrt{-25}$ as $5i$. This is illustrated in Technology Perspective 7.5.1. This is one of several features built into the TI-84 Plus calculator that is not available in most spreadsheets.

Technology Perspective 7.5.1 | Using the COMPLEX Mode on a Calculator

Evaluate $\sqrt{-25}$ both in REAL mode and in COMPLEX mode.

TI-84 Plus Keystrokes

1. Enter $\sqrt{-25}$ into the calculator and press 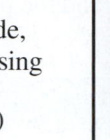. If the calculator is in **REAL** mode, an error message will be displayed. (Pressing again will clear the error message.)

TI-84 Plus Calculator

2. To switch to **COMPLEX** mode, press MODE and then use the arrow keys to move the cursor over $a + bi$ and press ENTER. Press 2nd QUIT to exit the **MODE** menu.

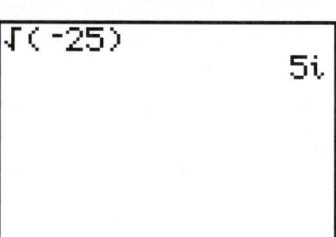

3. Enter $\sqrt{-25}$ into the calculator and press ENTER. With the calculator in **COMPLEX** mode, $\sqrt{-25}$ will display as $5i$.

Technology Self-Check 1

Use the **COMPLEX** mode on a calculator to evaluate $\sqrt{36} + \sqrt{-49}$.

2. Add, Subtract, Multiply, and Divide Complex Numbers

Since complex numbers consist of two terms—a real term and an imaginary term—the arithmetic of complex numbers is very similar to the arithmetic of binomials. The similarity is illustrated in the following examples by using the operation of addition.

Addition of Binomials

$(2x + 3y) + (4x + 7y)$
$= (2x + 4x) + (3y + 7y)$
$= 6x + 10y$

Addition of Complex Numbers

$(2 + 3i) + (4 + 7i)$
$= (2 + 4) + (3i + 7i)$
$= 6 + 10i$

Add complex numbers by adding the real terms and the imaginary terms separately. That is,

$$(a + bi) + (c + di) = (a + c) + (b + d)i$$

Subtraction is performed similarly.

Example 2   Adding and Subtracting Complex Numbers

Perform the indicated operations.

(a) $(2 + 5i) + (8 + 4i)$ **(b)** $6 + (11 - 4i)$

(c) $(8 - 5i) - 6i$ **(d)** $(-7 + 6i) - (4 + 10i)$

Solution

(a) $(2 + 5i) + (8 + 4i) = 2 + 5i + 8 + 4i$

$= (2 + 8) + (5 + 4)i$

$= 10 + 9i$

Group like terms together, and then add like terms. Add the real terms, and then add the imaginary terms.

(b) $6 + (11 - 4i) = (6 + 11) + (0 - 4)i$

$= 17 - 4i$

Also 6 can be written in the form $6 + 0i$.

(c) $(8 - 5i) - 6i = (8 + 0) + (-5 - 6)i$

$= 8 - 11i$

Also $-6i$ can be written in the form $0 - 6i$.

(d) $(-7 + 6i) - (4 + 10i) = -7 + 6i - 4 - 10i$

$= (-7 - 4) + (6 - 10)i$

$= -11 - 4i$

Use the distributive property to remove the second pair of parentheses. Then add like terms.

Self-Check 2

Simplify $(21 + 4i) - (9 - 7i)$.

Do I Need Symbols Other Than *i* to Represent Other Complex Numbers?

No, since $i^2 = -1$, higher powers of i can always be simplified to $i, -1, -i$, or 1. The first four powers of i are keys to simplifying higher powers to standard form and therefore should be memorized.

First Four Powers of *i*

$i^1 = i$

$i^2 = i \cdot i = -1$

$i^3 = i^2 \cdot i = (-1)i = -i$

$i^4 = i^2 \cdot i^2 = (-1)(-1) = 1$

The powers of i repeat in cycles of four: i^4, i^8, i^{12}, and so on are all equal to 1. We will use this fact to simplify i^n, where n is any integer exponent, to $i, -1, -i$, or 1.

Example 3   Powers of *i*

Simplify each power of i.

(a) i^7 **(b)** i^{13} **(c)** i^{406}

Solution

(a) $i^7 = i^4 \cdot i^3$

$= (1)(-i)$

$= -i$

First extract the largest multiple of 4 from each exponent. Then simplify, replacing i^4, i^8, i^{12}, etc., by 1.

(b) $i^{13} = i^{12}i^1$

$= (1)(i)$

$= i$

(c) $i^{406} = i^{404} \cdot i^2$

$\qquad = (i^4)^{101} \cdot (-1)$

$\qquad = (1)^{101}(-1)$

$\qquad = (1)(-1)$

$\qquad = -1$

$$\begin{array}{r} 101 \\ 4\overline{)406} \\ \underline{404} \\ 2 \end{array}$$

Self-Check 3

Simplify i^{33}.

We used some of the properties of radicals in Section 7.1. For now, please recall that for positive values of x and y, $\sqrt{x}\sqrt{y} = \sqrt{xy}$. However, for negative values of x and y, $\sqrt{x}\sqrt{y} \neq \sqrt{xy}$. For example,

$$\begin{aligned} \sqrt{-2}\sqrt{-3} &= (i\sqrt{2})(i\sqrt{3}) \\ &= i^2\sqrt{6} \\ &= -\sqrt{6} \end{aligned} \qquad \text{whereas} \qquad \begin{aligned} \sqrt{(-2)(-3)} &= \sqrt{+6} \\ &= \sqrt{6} \end{aligned}$$

Note that, in the previous equation, we wrote $i\sqrt{2}$ rather than $\sqrt{2}i$. The latter form could easily be confused with $\sqrt{2i}$, which has i under the radical symbol. To make it clear that the factor i is not in the radicand, it is best to put the i in front of the radical.

Another example illustrating $\sqrt{x}\sqrt{y} \neq \sqrt{xy}$ for negative values of x and y is given here.

$$\begin{aligned} (\sqrt{-5})^2 &= (i\sqrt{5})^2 \\ &= 5i^2 \\ &= -5 \end{aligned} \qquad \text{whereas} \qquad \begin{aligned} \sqrt{(-5)^2} &= \sqrt{25} \\ &= 5 \end{aligned}$$

To avoid the potential misinterpretations just noted, we recommend that you write every complex number in standard form before proceeding with any operations. We can multiply complex numbers as if they are binomials with a real term and an imaginary term. We use the distributive property extensively to form these products. This is illustrated in Example 4.

Example 4   Multiplying Complex Numbers

Calculate each product.

(a) $3(4 - 5i)$　　　　**(b)** $\sqrt{-2}(\sqrt{3} - \sqrt{-2})$　　　　**(c)** $(5 - 6i)(5 + 6i)$

(d) $(2 - 7i)(5 + 3i)$　　**(e)** $(a + bi)(a - bi)$

Solution

(a) $3(4 - 5i) = 3(4) - 3(5i)$ Distribute the factor 3.

$\qquad\qquad\quad = 12 - 15i$

(b) $\sqrt{-2}(\sqrt{3} - \sqrt{-2}) = i\sqrt{2}(\sqrt{3} - i\sqrt{2})$ First write each factor in the standard $a + bi$ form.

$\qquad\qquad\qquad\quad = i\sqrt{2}(\sqrt{3}) - i\sqrt{2}(i\sqrt{2})$ Then distribute the factor $i\sqrt{2}$.

$\qquad\qquad\qquad\quad = i\sqrt{6} - 2i^2$ Simplify each term, noting that $\sqrt{2}\sqrt{2} = 2$.

$\qquad\qquad\qquad\quad = i\sqrt{6} - 2(-1)$ Replace i^2 by -1.

$\qquad\qquad\qquad\quad = 2 + i\sqrt{6}$ Reorder the terms to write the answer in the standard form.

(c) $(5 - 6i)(5 + 6i) = (5)^2 - (6i)^2$ Note that this expression is of the form $(A - B)(A + B) = A^2 - B^2$.

$\qquad\qquad\qquad\quad = 25 - 36i^2$ Replace i^2 by -1.

$\qquad\qquad\qquad\quad = 25 - 36(-1)$

$\qquad\qquad\qquad\quad = 61$

(d) $(2 - 7i)(5 + 3i) = 2(5 + 3i) - 7i(5 + 3i)$ Distribute the factor of $5 + 3i$. Then

$$= 10 + 6i - 35i - 21i^2$$ distribute the factors of 2 and of $-7i$.

$$= 10 - 29i - 21(-1)$$ Combine like terms and replace i^2 by -1.

$$= 10 - 29i + 21$$

$$= 31 - 29i$$

(e) $(a + bi)(a - bi) = a^2 - (bi)^2$ Note that this expression is of the form

$$= a^2 - b^2i^2$$ $(A + B)(A - B) = A^2 - B^2$.

$$= a^2 + b^2$$ Replace i^2 by -1.

Self-Check 4

Calculate the product $(5 - 3i)(6 + 2i)$.

The **conjugate** of any complex number $a + bi$ is $a - bi$. This definition is illustrated in the following table.

Complex Number	Complex Number in Standard Form	Complex Conjugate in Standard Form	Complex Conjugate
$5 - 6i$	$5 - 6i$	$5 + 6i$	$5 + 6i$
$\sqrt{2} + i$	$\sqrt{2} + i$	$\sqrt{2} - i$	$\sqrt{2} - i$
$7i$	$0 + 7i$	$0 - 7i$	$-7i$
8	$8 + 0i$	$8 - 0i$	8

The product of a complex number and its conjugate uses the special product of a sum and a difference to yield the difference of two squares. This product is always a real number, as we illustrated in Example 4(e). We will now look at some specific examples of the product of conjugates.

Example 5   Multiplying Complex Conjugates

Multiply each complex number by its conjugate and simplify the result.

(a) $3 + 4i$ **(b)** $5 - 2i$ **(c)** $7i$ **(d)** 8

Solution

(a) $(3 + 4i)(3 - 4i) = 9 - 16i^2$ The conjugate of $3 + 4i$ is $3 - 4i$. Multiply,

$$= 9 + 16$$ using the fact that this expression is of the form

$$= 25$$ $(A + B)(A - B) = A^2 - B^2$.

(b) $(5 - 2i)(5 + 2i) = 25 - 4i^2$ The conjugate of $5 - 2i$ is $5 + 2i$.

$$= 25 + 4$$

$$= 29$$

(c) $(7i)(-7i) = -49i^2$ The conjugate of $7i$ is $-7i$.

$$= 49$$

(d) $8 \cdot 8 = 64$ The conjugate of 8 is 8.

Self-Check 5

Multiply $11 + 10i$ by its conjugate.

The fact that the product of a complex number and its conjugate is always a real number plays a key role in the division of complex numbers, as outlined in the following box.

Division of Complex Numbers

Verbally	Numerical Example
Step 1. Write the division problem as a fraction.	$20 \div (1 + 3i) = \dfrac{20}{1 + 3i}$
Step 2. Multiply both the numerator and the denominator by the conjugate of the denominator.	$= \dfrac{20}{1 + 3i} \cdot \dfrac{1 - 3i}{1 - 3i}$
Step 3. Simplify the result, and express it in standard $a + bi$ form.	$= \dfrac{20(1 - 3i)}{1 + 9}$
	$= 2(1 - 3i)$
	$= 2 - 6i$

Example 6 Dividing Complex Numbers

Simplify $(8 - i) \div (1 - 2i)$.

Solution

$$(8 - i) \div (1 - 2i) = \frac{8 - i}{1 - 2i}$$ Write the division problem as a fraction.

$$= \frac{8 - i}{1 - 2i} \cdot \frac{1 + 2i}{1 + 2i}$$ Multiply the numerator and the denominator by $1 + 2i$, the conjugate of the denominator.

$$= \frac{8 + 15i - 2i^2}{1 - 4i^2}$$ Multiply the numerators and multiply the conjugates in the denominators.

Simplify, replacing i^2 by -1.

$$= \frac{10 + 15i}{5}$$

$$= \frac{10}{5} + \frac{15}{5}i$$

$$= 2 + 3i$$ Write the result in standard $a + bi$ form. You can check this answer by multiplying $2 + 3i$ by $1 - 2i$. Is the product $8 - i$?

Self-Check 6

Calculate the quotient $\dfrac{3 + 2i}{2 - 3i}$.

3. Solve a Quadratic Equation with Imaginary Solutions

Where Am I Most Likely to Encounter Imaginary Numbers?

One of the uses you are most likely to have for imaginary numbers is as solutions of quadratic equations. Every quadratic equation has two complex solutions (a double real solution is counted as two solutions). Some of these complex solutions are real numbers and others are imaginary numbers. Problems with imaginary solutions occur more in engineering and the sciences than they do in business or the social sciences.

Example 7 gives a quadratic equation with imaginary solutions. It is not possible to obtain these solutions by factoring over the integers or by graphing the corresponding parabola on the real coordinate plane. However, we can obtain imaginary solutions by extraction of roots, by completing the square, or by using the quadratic formula. Example 7 illustrates the use of the quadratic formula to find imaginary solutions. Because of the $\pm$ symbol in the quadratic formula, imaginary solutions will always appear as complex conjugates.

Example 7   Using the Quadratic Formula to Determine Two Imaginary Solutions

Use the quadratic formula to solve $x^2 - 4x + 5 = 0$.

Solution

$x^2 - 4x + 5 = 0$

$x = \dfrac{-(-4) \pm \sqrt{(-4)^2 - 4(1)(5)}}{2(1)}$

$x = \dfrac{4 \pm \sqrt{16 - 20}}{2}$

$x = \dfrac{4 \pm \sqrt{-4}}{2}$

$x = \dfrac{4 \pm 2i}{2}$

$x = \dfrac{4}{2} \pm \dfrac{2i}{2}$

$x = 2 \pm i$

The equation is in the standard form $ax^2 + bx + c = 0$ with $a = 1$, $b = -4$, and $c = 5$. Substitute these values into the quadratic formula:

$x = \dfrac{-b \pm \sqrt{b^2 - 4ac}}{2a}$

Because the discriminant $b^2 - 4ac$ is negative, these solutions will be imaginary. Simplify and replace $\sqrt{-4}$ by $2i$.

Write each of these complex numbers in standard $a + bi$ form. Note that these imaginary solutions are complex conjugates. Do these values check?

Answer: $x = 2 - i$ or $x = 2 + i$

Self-Check 7

Solve $x^2 - 6x + 34 = 0$.

The solutions in Example 7 can be checked by substituting them back into the original equation. We can do this using either pencil and paper or a calculator. We will illustrate how to check $x = 2 - i$ using a TI-84 Plus calculator. First we will store the value of $2 - i$ under the variable name x. (See Technology Perspective 1.7.1 on Storing Values Under Variable Names.) Then we will evaluate the expression $x^2 - 4x + 5$ to check whether this expression is 0 for $x = 2 - i$. Note that the ⬜i key is the secondary function of the ⬤ key.

```
2-i→X
                2-i
X²-4X+5
                  0
```

Because $x^2 - 4x + 5 = 0$ when $x = 2 - i$, this value is a solution of the equation.

Although calculators are powerful tools that allow us to leverage our mathematical power, it is important for us to know some of the limitations of these tools. One of the limitations is that calculators sometimes introduce small errors into calculations. Often these errors are not significant or go unnoticed. In Example 8, the small error is noticeable because the answer is shown using scientific notation.

Example 8 Evaluating Powers of i and Interpreting Calculator Results

Evaluate i^{27} using pencil and paper and using a calculator.

Solution

By Pencil and Paper

$$i^{27} = (i^{24})(i^3)$$
$$= (1)(-i)$$
$$= -i$$

Answer: $i^{27} = -i$

This answer is exact and is supported by the calculator approximation.

By Calculator

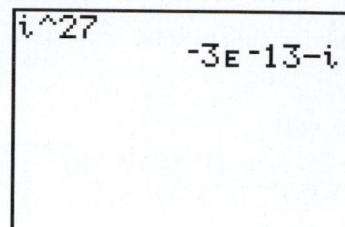

In $a + bi$ form, the display reads $(-3.0 \times 10^{-13}) - i$.

Because $-3.0 \times 10^{-13} = -0.0000000000003 \approx 0$, the calculator has introduced a small error. The approximation on the display is that $i^{27} \approx 0 - i$.

Self-Check 8

Evaluate i^{44} using pencil and paper and using a calculator.

Complex numbers have concrete meanings in electrical circuits, and electrical engineers can create equations that help them design products with desired characteristics. Example 9 shows how to create an equation with given complex solutions. Because imaginary solutions of quadratic equations occur in conjugate pairs, we must start with solutions that are complex conjugates if we want a quadratic equation with real coefficients.

Example 9 Constructing a Quadratic Equation with Given Complex Solutions

Construct a quadratic equation in x with solutions of $3 - i$ and $3 + i$.

Solution

$$x = 3 - i \quad \text{or} \quad x = 3 + i$$
$$x - (3 - i) = 0 \qquad x - (3 + i) = 0$$

Rewrite each linear equation so that the right side of the equation is 0. If the factors of a quadratic equation are 0, then their product is 0.

$$(x - 3 + i)(x - 3 - i) = 0$$
$$[(x - 3) + i][(x - 3) - i] = 0$$
$$(x - 3)^2 - i^2 = 0$$

Multiply these factors, taking advantage of products of special forms.

$$x^2 - 6x + 9 - (-1) = 0$$

Then combine like terms.

$$x^2 - 6x + 10 = 0$$

This quadratic equation has the given solutions.

Self-Check 9

Construct a quadratic equation in x with solutions of $-3i$ and $3i$.

7.5   Using the Language and Symbolism of Mathematics

1. The imaginary number i is used to represent _____.

2. The complex number $a + bi$ has a real term _____, and an imaginary term _____.

3. A complex number is **always/sometimes/never** a real number.

4. A real number is **always/sometimes/never** a complex number.

5. An imaginary number is **always/sometimes/never** a complex number.

6. A complex number is **always/sometimes/never** an imaginary number.

7. A real number is **always/sometimes/never** an imaginary number.

8. The standard form of a complex number is _____.

9. Powers of i repeat in cycles of _____.

10. The complex _____ of $3 + 5i$ is $3 - 5i$.

11. Every quadratic equation has _____ complex solutions (a double real solution is counted as two solutions).

12. If $7 + 2i$ is a solution of $ax^2 + bx + c = 0$ and a, b, and c are real numbers, then _____ is also a solution of this equation.

7.5   Quick Review

1. Write 37,469.21 by using scientific notation.

2. Write 4.06×10^{-5} in standard decimal notation.

3. Use the distributive property to expand and simplify $5x(2x - 3) + 4(2x - 3)$.

4. Use the distributive property to factor $5x(2x - 3) + 4(2x - 3)$.

5. Multiply $(5x + 3y)(5x - 3y)$.

7.5   Exercises

Objective 1 Express Complex Numbers in Standard Form

In Exercises 1–4, write each complex number in standard $a + bi$ form.

1. **a.** $-\sqrt{36}$ **b.** $\sqrt{-36}$
2. **a.** $-\sqrt{49}$ **b.** $\sqrt{-49}$
3. **a.** $\sqrt{-9} + \sqrt{16}$ **b.** $\sqrt{9} + \sqrt{-16}$
4. **a.** $\sqrt{4} + \sqrt{-25}$ **b.** $\sqrt{-4} - \sqrt{25}$

Objective 2 Add, Subtract, Multiply, and Divide Complex Numbers

In Exercises 5–42, perform the indicated operations and express the result in the standard $a + bi$ form.

5. $(1 + 2i) + (8 - 3i)$
6. $(6 - 7i) + (13 - 4i)$
7. $(5 + 3i) - (2 + 2i)$
8. $(7 + i) - (5 - 2i)$
9. $6(10 + 4i)$
10. $-9(3 - 12i)$

11. $(3 + 3i) + \dfrac{1}{2}(8 - 6i)$ **12.** $(3 + 3i) - \dfrac{1}{3}(6 - 9i)$

13. $\sqrt{-9 - 16}$ **14.** $\sqrt{-25 - 144}$

15. $\sqrt{-9} + \sqrt{-16}$ **16.** $\sqrt{-25} + \sqrt{-144}$

17. $(2i)(3i)$ **18.** $(5i)(7i)$

19. $2i(3 - 5i)$ **20.** $-3i(4 - 2i)$

21. $i(3 + 2i) + 2(5 - 3i)$ **22.** $4(3 - 2i) - i(5 + 3i)$

23. $(2 - 7i)(2 + 7i)$ **24.** $(6 + 9i)(6 - 9i)$

25. $(5 - 2i)(4 + 7i)$ **26.** $(6 + i)(3 - 5i)$

27. $\sqrt{-4}\sqrt{-25}$ **28.** $\sqrt{-9}\sqrt{-100}$

29. $\sqrt{(-4)(-25)}$ **30.** $\sqrt{(-9)(-100)}$

31. $(5 + i)^2$ **32.** $(6 - i)^2$

33. $(4 - 3i)^2$ **34.** $(3 + 5i)^2$

35. $(4 + 7i)(7 - 4i)$ **36.** $(11 - 3i)(2 + 5i)$

37. $\sqrt{-3}(\sqrt{2} + \sqrt{-3})$

38. $(\sqrt{2} - \sqrt{-5})(\sqrt{2} - 3\sqrt{-5})$

39. $2\sqrt{-75} + \sqrt{-27}$

40. $3\sqrt{-8} - 5\sqrt{-98}$

41. $\sqrt{4} + \sqrt{-9} - \sqrt{9} - \sqrt{-25}$

42. $\sqrt{64} + \sqrt{-36} - \sqrt{9} - \sqrt{-1}$

In Exercises 43–46, simplify each power of i.

43. i^9 **44.** i^{11}

45. i^{58} **46.** i^{81}

In Exercises 47–50, multiply each complex number by its conjugate.

47. $2 + 5i$ **48.** $3 - 8i$

49. $13i$ **50.** 13

In Exercises 51–64, perform the indicated operations and express the result in standard $a + bi$ form.

51. $\dfrac{4}{1 + i}$ **52.** $\dfrac{6}{1 - i}$

53. $\dfrac{5 + i}{5 - i}$ **54.** $\dfrac{4 - i}{4 + i}$

55. $\dfrac{3 - 2i}{i}$ **56.** $\dfrac{-2 + 5i}{i}$

57. $\dfrac{2 + 3i}{4 - 5i}$ **58.** $\dfrac{5 - 2i}{4 + 3i}$

59. $85 \div (7 - 6i)$ **60.** $185 \div (11 + 8i)$

61. $\sqrt{-\dfrac{25}{9}}$ **62.** $\dfrac{\sqrt{-36}}{\sqrt{49}}$

63. $\dfrac{\sqrt{25}}{\sqrt{-9}}$ **64.** $\dfrac{\sqrt{36}}{\sqrt{-49}}$

Objective 3 Solve a Quadratic Equation with Imaginary Solutions

In Exercises 65–70, solve each quadratic equation.

65. $x^2 - 4x + 29 = 0$ **66.** $x^2 - 8x + 25 = 0$

67. $x^2 = 6x - 13$ **68.** $x^2 = 4x - 6$

69. $-2w(w - 3) = 5$

70. $3w(w + 1) = -(w + 2)$

Review and Concept Development

In Exercises 71–74, solve each equation by extraction of roots.

71. $x^2 = -16$ **72.** $x^2 = -25$

73. $(2x + 3)^2 = -9$ **74.** $(2x - 3)^2 = -4$

In Exercises 75–80, calculate the discriminant of each quadratic equation and determine the nature of the solutions of this equation. Do not solve the equation.

75. $x^2 - 10x + 25 = 0$ **76.** $x^2 - 10x + 24 = 0$

77. $x^2 - 10x + 26 = 0$ **78.** $y^2 - 2 = 0$

79. $y^2 + 2 = 0$ **80.** $y^2 + 2y = 0$

81. Substitute $3 - 2i$ for x to determine whether this value is a solution of $x^2 - 6x + 13 = 0$.

82. Substitute $2 + 3i$ for x to determine whether this value is a solution of $x^2 - 4x + 13 = 0$.

In Exercises 83–86, construct a quadratic equation in x that has the given solutions.

83. $-4i$ and $4i$ **84.** $-5i$ and $5i$

85. $4 - i$ and $4 + i$ **86.** $5 - i$ and $5 + i$

In Exercises 87 and 88, solve each cubic equation.

87. $(x - 2)(x^2 + x + 1) = 0$

88. $(x + 2)(x^2 - x + 1) = 0$

89. Given $z = 3 - i$:
   a. What is the conjugate of z?
   b. What is the additive inverse of z?
   c. What is the multiplicative inverse of z?

Applying Technology

In Exercises 90 and 91, interpret the calculator display and give the exact value of the power of i.

90.

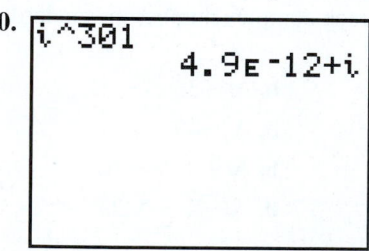

91.

```
i^46
        -1+4ᴇ-13i
```

Group discussion questions

92. Discussion Question

  a. Is $\sqrt{3}$ a real number?

  b. Is $\sqrt{3}$ an imaginary number?

  c. Is $\sqrt{3}$ a complex number?

  d. Is $\sqrt{-3}$ a real number?

  e. Is $\sqrt{-3}$ an imaginary number?

  f. Is $\sqrt{-3}$ a complex number?

93. Discussion Question

  a. What can you say about a complex number that is equal to its conjugate?

  b. Give an example of two imaginary numbers whose sum is a real number.

  c. Give an example of two imaginary numbers whose product is a real number.

  d. Name a complex number that cannot be used as a divisor.

94. Discovery Question Complete the following diagram for i to i^{12} to illustrate that powers of i repeat in cycles of four.

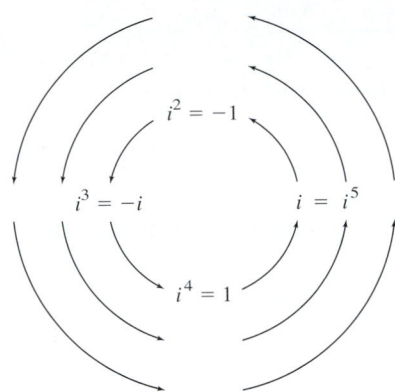

95. Challenge Question Use a calculator to answer these questions.

  a. Is 2 a solution of $x^3 - 8 = 0$?

  b. Is $-1 + i\sqrt{3}$ a solution of $x^3 - 8 = 0$?

  c. Is $-1 - i\sqrt{3}$ a solution of $x^3 - 8 = 0$?

  d. How many cube roots does 8 have? What are these cube roots, and which is the principal cube root of 8?

96. Challenge Question

  a. Simplify $i + i^2 + i^3 + i^4 + \cdots + i^{99} + i^{100}$.

  b. Explain how this sum can be computed mentally.

7.5   Cumulative Review

1. Solve $4x - 1 = 7$.

2. Solve $4(x - 1) = 7$.

3. Solve $|x - 1| = 7$.

4. Determine the perimeter of a square if one side is of length 5 cm.

5. Determine the area of a square if one side is of length 5 cm.

Chapter 7   Key Concepts

1. Standard Form of a Quadratic Equation: If x is a real variable and a, b, and c are real constants with $a \neq 0$, then $ax^2 + bx + c = 0$ is the standard form of a quadratic equation in x with ax^2 the quadratic term, bx the linear term, and c the constant term.

2. Methods of Solving Quadratic Equations

  • Tables

  • Graphs

  • Factoring

  • Extraction of roots

  • Completing the square

  • The quadratic formula

3. Product Rule for Square Roots: If $\sqrt{x}$ and $\sqrt{y}$ are both real numbers, then $\sqrt{xy} = \sqrt{x}\sqrt{y}$.

4. Quotient Rule for Square Roots: If $\sqrt{x}$ and $\sqrt{y}$ are both real numbers, then $\sqrt{\dfrac{x}{x}} = \dfrac{\sqrt{x}}{\sqrt{y}}$ for $y \neq 0$.

5. Rationalizing the Denominator: If an expression contains a radical in the denominator, the process of rewriting the expression so that there is no radical in the denominator is called rationalizing the denominator.

6. Perfect Square Trinomials

  • Square of a sum: $A^2 + 2AB + B^2 = (A + B)^2$

  • Square of a difference: $A^2 - 2AB + B^2 = (A - B)^2$

7. Extraction of Roots: The solutions of $x^2 = k$ are $x = \pm\sqrt{k}$.

8. Completing the Square: The process of completing the square can be used to set up quadratic equations for extraction of roots. It is also used to develop the quadratic formula.

9. Quadratic Formula: The solutions of $ax^2 + bx + c = 0$ with real coefficients a, b, and c and $a \neq 0$ are $x = \dfrac{-b \pm \sqrt{b^2 - 4ac}}{2a}$.

10. Discriminant: In the quadratic formula $x = \dfrac{-b \pm \sqrt{b^2 - 4ac}}{2a}$, the expression $b^2 - 4ac$ is called the discriminant. The discriminant, which is the expression under the radical symbol, can be used to discriminate between the real solutions and imaginary solutions of a quadratic equation.

11. Solution of Multiplicity 2: If both solutions of a quadratic equation are the same, we call the solution a double solution or a solution of multiplicity 2.

12. Nature of the Solutions of a Quadratic Equation: There are three possibilities for the solutions of $ax^2 + bx + c = 0$.

Value of the Discriminant	Solutions of $ax^2 + bx + c = 0$	The Parabola $y = ax^2 + bx + c$	Graphical Example
$b^2 - 4ac > 0$	Two distinct real solutions	Two x-intercepts	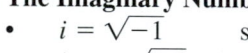
$b^2 - 4ac = 0$	A double real solution	One x-intercept with the vertex on the x-axis	
$b^2 - 4ac < 0$	Neither solution is real; both solutions are complex numbers with imaginary parts. These solutions will be complex conjugates.	No x-intercepts	

13. Constructing a Quadratic Equation with Given Solutions: If $x = r_1$ and $x = r_2$ are two solutions of a quadratic equation, then $(x - r_1)(x - r_2) = 0$ is a factored form of this quadratic equation.

14. Pythagorean Theorem and Converse
- A right triangle is a triangle containing a 90° angle.
- Triangle ABC is a right triangle if, and only if, $a^2 + b^2 = c^2$.

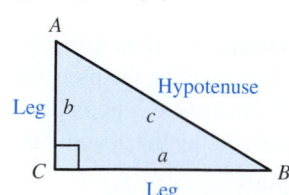

15. The Imaginary Number i
- $i = \sqrt{-1}$ so $i^2 = -1$
  $-i = -\sqrt{-1}$ so $(-i)^2 = i^2 = -1$
- For any positive real number x, $\sqrt{-x} = i\sqrt{x}$.
- The first four powers of i are $i^1 = i$, $i^2 = -1$, $i^3 = -i$, $i^4 = 1$.

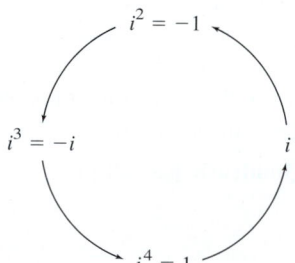

16. Complex Numbers
- If a and b are real numbers and $i = \sqrt{-1}$, then $a + bi$ is a complex number with real term a and imaginary term bi.
- $a + bi$ and $a - bi$ are complex conjugates.

17. Operations with Complex Numbers: The addition, subtraction, and multiplication of complex numbers are similar to the corresponding operations with binomials. Division by a complex number is accomplished by multiplying both the numerator and the denominator by the conjugate of the denominator and then simplifying the result.

18. Subsets of the Complex Numbers: The relationships of important subsets of the set of complex numbers are summarized in the tree diagram, where a and b are real numbers and $i = \sqrt{-1}$.

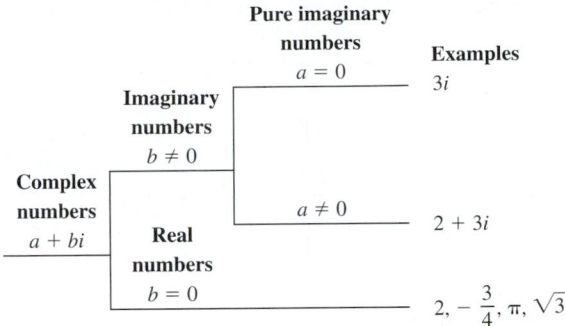

Chapter 7 | Review Exercises

Simplifying Expressions with Square Roots

In Exercises 1–8, simplify each radical expression.

1. $\sqrt{25 \cdot 7}$ **2.** $\sqrt{99}$

3. $\sqrt{\dfrac{25}{49}}$ **4.** $\sqrt{\dfrac{6}{25}}$

5. $\dfrac{\sqrt{45}}{\sqrt{20}}$ **6.** $\dfrac{4}{\sqrt{6}}$

7. $\dfrac{\sqrt{5}}{\sqrt{6}}$ **8.** $\sqrt{\dfrac{3}{11}}$

Completing the Square

9. Write the equation $x^2 - 12x + 36 = 0$ so that the left side is expressed as the square of a binomial.

10. Fill in the missing constant term that is needed to make $x^2 + 30x +$ _____ a perfect square trinomial.

11. Fill in the missing constant term that is needed to make $x^2 - 20x +$ _____ a perfect square trinomial.

12. Rewrite the equation $v^2 - 8v = -7$ so that the left side is expressed as the square of a binomial.

Solving Quadratic Equations

13. Use the graph of $y = -x^2 + 2x + 8$ to solve $-x^2 + 2x + 8 = 0$.

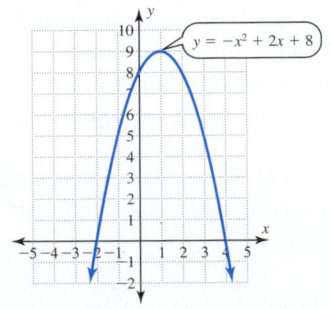

14. Use the table for $y = x^2 + 12x - 64$ to solve $x^2 + 12x - 64 = 0$.

x	$y = x^2 + 12x - 64$
-20	96
-16	0
-12	-64
-8	-96
-4	-96
0	-64
4	0
8	96

In Exercises 15 and 16, solve each quadratic equation by factoring.

15. $x^2 - 7x + 12 = 0$

16. $(2x - 1)(4x - 5) = 20$

In Exercises 17–20, solve each quadratic equation by using the quadratic formula.

17. $v^2 - 9v + 20 = 0$

18. $w^2 = 4w - 2$

19. $10(m^2 - 1) = 21m$

20. $\dfrac{2}{3}x^2 = 2x + 3$

In Exercises 21 and 22, solve each quadratic equation by extraction of roots.

21. $m^2 = 64$

22. $(2y - 3)^2 = 1$

In Exercises 23 and 24, solve each quadratic equation by completing the square.

23. $w^2 + 10w = 24$

24. $v^2 - 18v = -65$

25. Use the graph of $y = x^2 + 6x - 7$ to:
  a. Solve $x^2 + 6x - 7 = 0$.
  b. Solve $x^2 + 6x - 7 \leq 0$.
  c. Solve $x^2 + 6x - 7 \geq 0$.

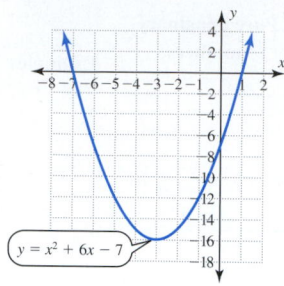

In Exercises 26–30, construct a quadratic equation in x with integer coefficients that has the given solutions.

26. -3 and 5

27. $-\dfrac{2}{5}$ and $\dfrac{1}{4}$

28. $-\sqrt{5}$ and $\sqrt{5}$

29. A double solution of $\dfrac{5}{7}$

30. $3 - \sqrt{5}$ and $3 + \sqrt{5}$

Applying Technology

In Exercises 31 and 32, use a graphing calculator or a graphing utility to graph $y = P(x)$, to determine the x-intercepts of this graph, and to complete this table.

Polynomial $P(x)$	x-intercepts of the Graph of $y = P(x)$	Factored Form of $P(x)$	Zeros of $P(x)$	Solutions of $P(x) = 0$
31. $-3x^2 + 8x + 3$				
32. $x^3 + x^2 - 2x$				

Connecting Concepts to Applications

33. Compound Interest The formula for computing the amount A of an investment of principal P at interest rate r for 1 year compounded semiannually is $A = P\left(1 + \dfrac{r}{2}\right)^2$. Approximately what interest rate is necessary for \$1,000 to grow to \$1,050 in 1 year if the interest is compounded semiannually?

34. Pasta Recipe A recipe for pasta for eight people suggests using enough angel-hair pasta to fill a 6-cm-diameter opening on a pasta-measuring device. Approximately what size hole would be recommended for four people?

35. Profit The net income in dollars produced by selling x units of a product is given by $P(x) = -x^2 + 45x - 200$. If $P(x) > 0$, there is a profit. Determine the profit interval, the values of x that will generate a profit.

36. Height of a Baseball The height h in feet of a baseball t seconds after being hit by a batter is given by $h(t) = -16t^2 + 80t + 3$.
  a. Determine to the nearest hundredth of a second when the ball will hit the ground.
  b. During what time interval after the baseball is hit will its height exceed 67 ft?

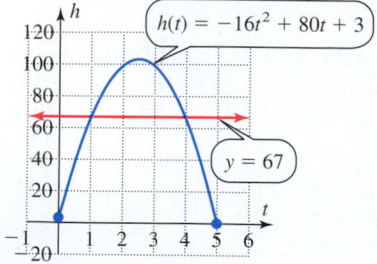

37. Dimensions of a Right Triangle The length of the hypotenuse of the right triangle shown in the figure is 2 cm more than the length of the longer leg. If the longer leg is 7 cm longer than the short leg, determine the length of each side.

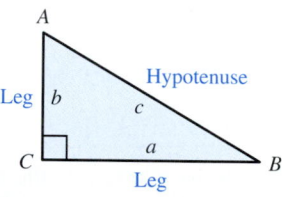

38. Deflection of a Beam A 40-ft concrete beam is fixed between two rigid anchor points. If this beam expands by 0.4 inch due to a 45° increase in temperature, determine the bulge in the middle of the beam. (Assume for simplicity that the beam deflects as shown in this diagram.)

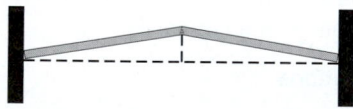

39. Positioning an Effective Sound System The sound system planned for an airport concourse has speakers that should be placed at least 14 ft above the floor and thus approximately 8 ft above the heads of most customers. A customer is standing at a position x ft from a reference point directly under one speaker that is 14 ft above the floor.
  a. If a customer is standing 16 ft from the reference point, how far is it from the speaker to the customer's ears?

b. If the limit of the speaker's effective range is 20 ft, approximate to the nearest foot the distance a customer can walk from the reference point and stay within the effective range of the speaker.

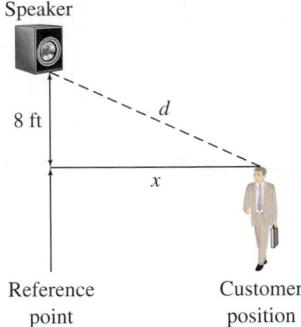

40. Speed of an Airplane Two airplanes depart simultaneously from an airport. One flies due south; the other flies due east at a rate 20 mi/h faster than that of the first airplane. After 1 hour, radar indicates that the airplanes are approximately 450 mi apart. What is the ground speed of each airplane?

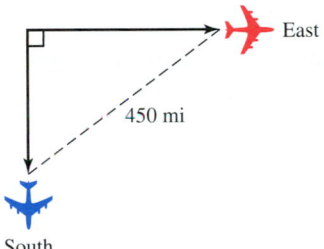

Complex Numbers

In Exercises 41–54, simplify each expression and write the result in standard $a + bi$ form.

41. $\sqrt{-100}$

42. $\sqrt{64} - \sqrt{-64}$

43. $(5 - 6i) - (3 - 2i)$

44. $2(4 - 3i) - 5(2 + 6i)$

45. $2i(5 - 6i)$

46. $(5 - 7i)(6 + 3i)$

47. $(7 - 3i)^2$

48. $(5 - 2i)(5 + 2i)$

49. i^5

50. $i^6 + i^7$

51. $\dfrac{1 + i}{1 - i}$

52. $\dfrac{58}{2 + 5i}$

53. i^{-11}

54. i^{32}

55. Use the parabola defined by $y = ax^2 + bx + c$ to determine whether the discriminant for $ax^2 + bx + c = 0$ is negative, zero, or positive. Then identify the nature of the solutions of the quadratic equation as distinct real solutions, a double real solution, or complex solutions with imaginary parts.

a.

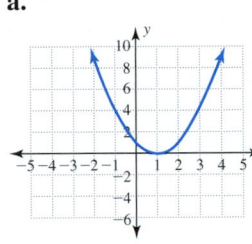

b.

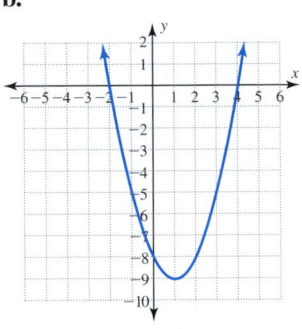

c.

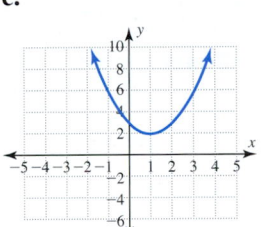

In Exercises 56–58, use the discriminant to determine the nature of the solutions of each quadratic equation.

56. $-3x^2 = 75 - 30x$

57. $4v^2 + 2v + 1 = 0$

58. $x^2 + 3x = 2$

In Exercises 59–62, solve each quadratic equation.

59. $z^2 = -49$

60. $(2z + 1)^2 = -9$

61. $x^2 = 6x - 10$

62. $(x - 1)(x - 5) = -6$

In Exercises 63 and 64, write a quadratic equation with the given solutions.

63. $-11i$ and $11i$

64. $5 - i$ and $5 + i$

65. a. Write the conjugate of $2 - 5i$.
   b. Write the additive inverse of $2 - 5i$.
   c. Write the multiplicative inverse of $2 - 5i$.

66. a. If possible, write a complex number that is not a real number. If this is not possible, write a reason why this is impossible.
   b. If possible, write a real number that is not a complex number. If this is not possible, write a reason why this is impossible.

c. If possible, write a complex number that is not an imaginary number. If this is not possible, write a reason why this is impossible.

d. If possible, write an imaginary number that is not a complex number. If this is not possible, write a reason why this is impossible.

67. Interpret the calculator display and give the exact value of the power of i.

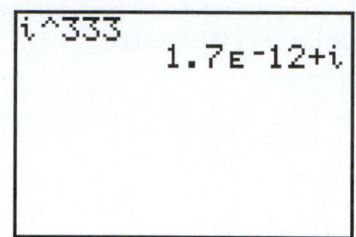

68. Solve each of these equations.
a. $2x + 1 = 9$
b. $(2x + 1)^2 = 9$
c. $|2x + 1| = 9$

Chapter 7 | Mastery Test

Objective 7.1.1 Solve Quadratic Equations by Extraction of Roots

1. Solve each quadratic equation by extraction of roots.
a. $x^2 = 400$
b. $(2v - 5)^2 - 81 = 0$

Objective 7.1.2 Use the Product Rule for Radicals to Simplify Square Roots

2. Simplify each radical expression.
a. $\sqrt{36 \cdot 2}$
b. $\sqrt{28}$

Objective 7.1.3 Use the Quotient Rule for Radicals to Simplify Square Roots

3. Simplify each radical expression.
a. $\sqrt{\dfrac{49}{64}}$
b. $\sqrt{\dfrac{11}{25}}$
c. $\dfrac{\sqrt{48}}{\sqrt{3}}$
d. $\dfrac{\sqrt{18}}{\sqrt{98}}$

Objective 7.1.4 Simplify Expressions of the Form $\dfrac{\sqrt{a}}{\sqrt{b}}$ by Rationalizing the Denominator

4. Simplify each radical expression.
a. $\dfrac{6}{\sqrt{2}}$
b. $\sqrt{\dfrac{3}{7}}$

Objective 7.2.1 Determine the Constant Term in a Perfect Square Trinomial

5. Fill in the missing constant term that is needed to make each expression a perfect square trinomial.
a. $w^2 - 20w + \underline{\hspace{0.5cm}}$
b. $w^2 + 8w + \underline{\hspace{0.5cm}}$

Objective 7.2.2 Solve Quadratic Equations by Completing the Square

6. Solve each equation by completing the square to determine exact solutions. Then approximate any irrational solutions to the nearest hundredth.
a. $y^2 - 6y = 16$
b. $w^2 - 3w = \dfrac{7}{4}$

Objective 7.3.1 Use the Quadratic Formula to Solve Quadratic Equations with Real Solutions

7. Solve each equation by using the quadratic formula to determine exact solutions. Then approximate any irrational solutions to the nearest hundredth.
a. $x^2 + 6x - 5 = 0$
b. $6m^2 - 11m - 4 = 6$
c. $4z^2 - 28z = -49$
d. $(x - 1)^2 = 2(x + 1)$

Objective 7.3.2 Use the Discriminant to Determine the Nature of the Solutions of a Quadratic Equation

8. Use the discriminant to determine the nature of the solutions of each quadratic equation as distinct real solutions, a double real solution, or complex solutions with imaginary parts.
a. $5w^2 + 5w + 1 = 0$
b. $7y^2 = 84y - 252$
c. $3x^2 + 2x - 1 = 0$
d. $(z + 1)(z - 2) = -3$

Objective 7.4.1 Use Quadratic Equations to Solve Word Problems

9. Break-Even Values $P(x) = -x^2 + 190x - 925$ is a profit polynomial that gives the profit in dollars made by selling x units of a product. Determine the break-even values of x.

10. Length of a Side of a Square The side of one square is 5 cm longer than the side of another square. Their total area is 193 cm². Determine the length of a side of each square.

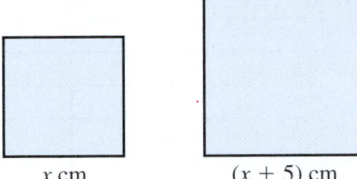

x cm   (x + 5) cm

Objective 7.4.2 Use the Pythagorean Theorem

11. Length of a Rafter What length rafter is needed to span a horizontal distance of 24 feet if the roof must rise 7 feet?

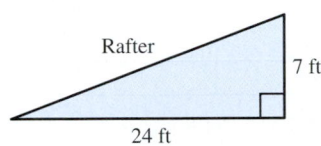

Rafter   7 ft

24 ft

12. Ladder Position What is the distance of the base of the ladder from the wall in the figure?

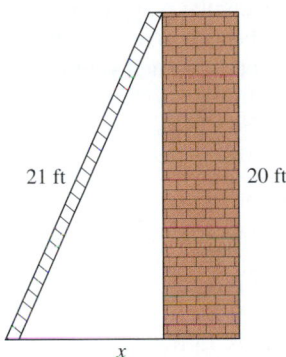

21 ft   20 ft

x

Objective 7.5.1 Express Complex Numbers in Standard Form

13. Write each complex number in standard $a + bi$ form.
  a. $\sqrt{-81}$ **b.** $\sqrt{16} - 25$
  c. $\sqrt{-16} - \sqrt{25}$ **d.** $i^2 + i^3$

Objective 7.5.2 Add, Subtract, Multiply, and Divide Complex Numbers

14. Simplify each expression and write your answer in standard $a + bi$ form.
  a. $2(4 - 5i) - 3(3 - 4i)$ **b.** $(4 - 5i)(2 - 4i)$
  c. $\dfrac{4 - 5i}{2 - 4i}$ **d.** $(3 - i)^2$

Objective 7.5.3 Solve a Quadratic Equation with Imaginary Solutions

15. Solve each quadratic equation.
  a. $(y - 1)^2 = -36$
  b. $x^2 + 29 = 10x$
  c. $(v - 2)(v - 4) = -26$

Chapter 7 | Group Project

Choices: Creating a Mathematical Model and Using This Model to Find an Optimal Solution

A fencing contractor in Sacramento, California, was contracted to install 150 ft of chain-link fencing to surround a playground area next to an existing grade-school building. This is illustrated in the figure. For security reasons, the specifications indicate that the only entrances to this area come through the school. The length of the wall along the school where no fencing is needed is 180 ft. The question that we will explore is: What dimensions should the contractor use to enclose the maximum possible rectangular area inside this fence?

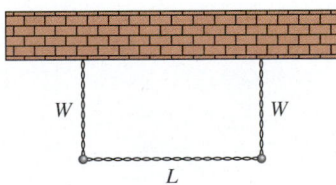

W   W

L

Exploration: Examining the relationship between the width of the playground and the area of the playground.

I. **Creating Numerical, Graphical, and Algebraic Models for This Problem**

Width W (ft)	Length L (ft)	Area A (ft^2)
5.0	140	700
10.0		
15.0		
20.0		
25.0		
30.0		
35.0		
37.5		
40.0		
45.0		
50.0		
55.0		
60.0		
65.0		
70.0		

1. Complete the table to form a numerical model for this problem.
2. **a.** Using the first two columns of this table, plot these data on a scatter diagram, using an appropriate scale for each axis. This will produce a graphical model for the length as a function of the width.
   b. Now write an equation that gives the length L of the rectangular playground area in terms of the width W. This will produce an algebraic model for the length in terms of the width. (*Suggestion:* Use your algebraic model to double-check your table and your graph.)
   c. What are the restrictions on the variable W based on your equation in part 2(b)? (*Hint:* There is only so much fencing.)
   d. What are the restrictions on the variable L based on the input values listed in part 2(c)?
3. **a.** Using the first and third columns of this table, plot these data on a scatter diagram, using an appropriate scale for each axis. This will produce a graphical model for the area in terms of the width.
   b. Now write an equation for A that gives the area of the rectangular playground in terms of the width W. This will produce an algebraic model for the area in terms of the width. (*Suggestion:* Use your algebraic model to double-check your table and your graph.)
   c. What are the restrictions on the variable W based on your equation in part 3(b)?
   d. What is the interval of values of A that corresponds to the values of W listed in part 3(c)?

II. **Using the Models to Answer Questions About This Problem**

1. **a.** Determine the length of the playground when the width is 32 ft.
   b. Determine the area of the playground when the width is 32 ft.
2. **a.** Determine the length of the playground when the width is 0 ft.
   b. Determine the area of the playground when the width is 0 ft.
3. **a.** Determine the length of the playground when the width is 75 ft.
   b. Determine the area of the playground when the width is 75 ft.
4. Determine the exact value of W for which the length is 105 ft.
5. Determine the exact value of W for which the area is 2,772 ft^2.
6. Determine the width and the length of the fence that will enclose the maximum possible playground area. What is the maximum possible area?
7. Explain how you determined the dimensions that yield the maximum area. Which of the three types of mathematical models (numerical, graphical, or algebraic) did you use to answer this question? Why did you use this model?

Functions: Linear, Absolute Value, and Quadratic

CHAPTER 8

Chapter Outline

How Far Did That Home Run Travel?

This question is commonly asked at baseball parks. Like anybody else who saw Albert Pujols' blast to the fourth story of a building outside of San Diego's stadium, Greg Rybarczyk felt it had to be a mammoth home run, one with the distance to match the reaction of the players.

"When I saw that ball hit, I thought it was a huge homer—something about the angle off the bat, the speed of the swing, it just looked like a real bomb," Rybarczyk said. "But when I analyzed it on my website Hit Tracker Online, surprisingly it turned out to be more of a regular fly ball than a screaming liner."

Baseballs travel a parabolic path that we can analyze using quadratic functions. We can determine how far the ball traveled, how high it went, and how long it was in the air once we have the correct function to model its path. Exercises 27 and 28 in Section 8.6 will use real data to determine an equation to model the path of a baseball.

| **Section 8.1** | Functions and Representations of Functions |

"Mathematics has a beauty of its own, a symmetry and proportion in its results that is found only in works of greatest beauty."   —Jacob William Young (19th century mathematics author)

Objectives:

1. Identify a function and determine its domain and range.
2. Use function notation.

1. Identify a Function and Determine Its Domain and Range

A Mathematical Note

The word *function* was introduced by the German mathematician Gottfried Wilhelm Freiherr von Leibniz (ca. 1682). Leibniz is credited as a developer of calculus, together with the English mathematician Sir Isaac Newton (ca. 1680). Leibniz contributed both original terminology and notation as well as his results in the areas of algebra and calculus.

Function notation and linear functions were first examined in Section 2.2. Many mathematicians view the concept of a function as the single most important concept in mathematics.[*] Because of the importance of this function concept, we will define a function, reexamine function notation, and examine functions from multiple perspectives:

- Mapping representation of a function
- Ordered-pair representation of a function
- Table representation of a function
- Graphical representation of a function
- Function notation representation of a function

What Is a Function?

In mathematics, we use the word *function* to designate a correspondence that has very specific properties. A function must match each input value with exactly one value of an output variable.

> **Function**
>
> A **function** is a correspondence that matches each input value with exactly one value of the output variable. The set of all possible input values is called the **domain** of the function or the set of **independent** values. The set of all output values is called the **range** of the function or the set of **dependent** values.

A key point is to understand the importance of the words *exactly one*. Each input value is paired with one and only one output value. When you first begin studying the function concept, you may want to think about three different interrelated parts of a function:

1. The domain of the function, the set of all possible input values

2. The range of the function, the set of all possible output values

3. The correspondence (rule or formula), which pairs each domain element with exactly one range element

Functions can be denoted in a variety of ways. Example 1 uses mapping notation.

Why Should I Learn to Use Mapping Notation to Represent a Function?

The mapping or arrow notation clarifies the pairing that a function creates between the domain elements and the range elements. Mapping notation also helps emphasize that a

*See Ed Dubinsky and Gvershon Harel, "The Concept of Function: Aspects of Epistemology and Pedagogy," *MAA Notes*, Vol. 25, 1992.

function can be viewed as a process that takes input values and produces unique output values. Compare parts (b) and (c) of Example 1. It is important to understand the distinction between these examples.

Example 1   Identifying Functions by Using Mapping Notation

Classify each correspondence as either a function or a correspondence that is not a function. For each function, identify the domain D and range R.

Solution

(a)

D		R
3	→	15
9	→	39
13	→	55

This correspondence is a function.
Domain $D = \{3, 9, 13\}$
Range $R = \{15, 39, 55\}$

Each domain element is paired with exactly one output value in the range.

(b)

D		R
16	↗↘	4
		−4
0	→	0

This correspondence is not a function.

The domain element 16 is not paired with exactly one element in the range; it is paired with both 4 and −4.

(c)

D		R
−4	↘	16
4	↗	
0	→	0

This correspondence is a function.
Domain $D = \{-4, 0, 4\}$
Range $R = \{0, 16\}$

−4 is paired only with 16; 4 is paired only with 16; and 0 is paired only with 0. Thus each element in D is paired with exactly one element in R.

Self-Check 1

Classify each correspondence as either a function or a correspondence that is not a function. Give the domain and range of each function.

a.

D		R
0	→	0
3	↗↗	3
−3		

b.

D		R
0	→	0
3	↗↘	3
		−3

The arrows in the mapping notation for functions have the advantage of clearly stressing the active nature of a function in pairing input values with output values. Conceptually, this is a great notation. Practically, mapping notation has some limitations. If there are many domain values, it can take a long time to draw all the arrows for the function. Arrows are also more awkward to type than other notations. The ordered-pair notation shown in Example 2 can convey the same information more concisely.

What Is the Advantage of Representing a Function Using Ordered-Pair Notation?

The first coordinate of an ordered pair is the input value, the second coordinate is the output value, and the parentheses establish how these elements are paired.

A **relation** is defined as any set of ordered pairs. Thus some relations are functions and others aren't. To be a function, a relation must pair each input value with exactly one output value.

Example 2 | Comparing Mapping Notation and Ordered-Pair Notation

(a) Convert this function, given in mapping notation, to ordered-pair notation.

Solution

Mapping Notation | Ordered-Pair Notation

D	R
$-2 \rightarrow -1$	
$-1 \rightarrow 3$	
$0 \rightarrow 5$	
$1 \rightarrow 7$	
$2 \rightarrow 9$	

$(-2, -1)$
$(-1, 3)$
$(0, 5)$
$(1, 7)$
$(2, 9)$

Both notations indicate a function whose domain is $D = \{-2, -1, 0, 1, 2\}$ and whose range is $R = \{-1, 3, 5, 7, 9\}$. The two notations also pair the elements exactly the same way.

This function can be written as the set of ordered pairs
$\{(-2, -1), (-1, 3), (0, 5), (1, 7), (2, 9)\}$.

(b) Convert this function, given as a set of ordered pairs, to mapping notation.

Solution

Ordered-Pair Notation

$\{(-5, 5), (5, 5), (-7, 7), (7, 7)\}$

Mapping Notation

D	R
$-5 \rightarrow 5$	
$5 \nearrow$	
$-7 \rightarrow 7$	
$7 \nearrow$	

Both notations indicate a function whose domain is $D = \{-7, -5, 5, 7\}$ and whose range is $R = \{5, 7\}$.

Self-Check 2

a. Convert this function, given in mapping notation, to ordered-pair notation.

D	R
$-2 \rightarrow 5$	
$-1 \rightarrow 2$	
$0 \rightarrow -1$	
$1 \rightarrow -4$	

b. Convert this function, given as a set of ordered pairs, to mapping notation.
$\{(1, -1), (2, 5), (3, 5), (5, 2)\}$

In an ordered pair (x, y), it is crucial to give the coordinates in the correct order. In contrast, a set is just a collection of elements—elements that we can list in any order. Thus $\{-5, 5, -7, 7\} = \{-7, -5, 5, 7\}$.

How Can I Use a Graph to Represent a Function?

The Cartesian coordinate system provides a pictorial means of presenting the relationship between two variables. Each point in a plane can be uniquely identified by an ordered pair (x, y). Thus graphs can be used to represent mathematical relations. The x-coordinate of the ordered pair (x, y) is called the **independent variable,** and the y-coordinate is called the **dependent variable.** In an experiment, the experimenter has the freedom to select different input values. To reflect this freedom, we refer to the variable representing these input values as the independent variable. When the experiment is run, the corresponding output values are determined by the input values that were used. Thus we refer to the variable representing the output values as the dependent variable.

Example 3   Graphing Relations

Graph each relation and determine whether the relation is a function.

Solution

(a) $\{(2, -2), (2, -1), (2, 0), (2, 1), (2, 2)\}$

Not a function

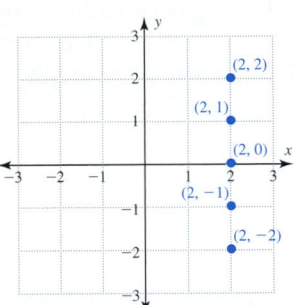

This relation is not a function. The input of 2 has more than one output paired with it.

(b)

D		R
-2	$\rightarrow$	-3
-1	$\rightarrow$	-1
0	$\rightarrow$	1
1	$\rightarrow$	3
2	$\rightarrow$	5

Function

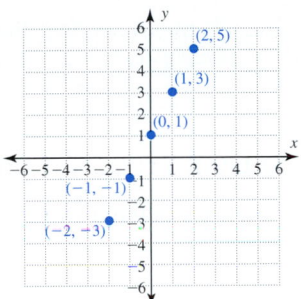

This relation is a function with domain $D = \{-2, -1, 0, 1, 2\}$ and range $R = \{-3, -1, 1, 3, 5\}$.

(c)

x	y
-2	2
-1	2
0	2
1	2
2	2

Function

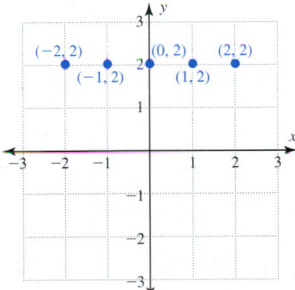

This relation is a function with domain $D = \{-2, -1, 0, 1, 2\}$ and range $R = \{2\}$. Each input value has exactly one output paired with it. Each input is paired with 2.

Self-Check 3

Graph each relation and determine whether the relation is a function.

a. $\{(-2, 3), (-1, 3), (0, 3), (1, 3), (2, 3)\}$

b.

D		R
-2	$\rightarrow$	3
-1	$\rightarrow$	1
0	$\rightarrow$	0
1	$\rightrightarrows$	-2
		-4

How Can I Determine the Domain and Range of a Function From Its Graph?

If a function is defined by a graph, then the ordered pairs can be determined by examining the points that form the graph. The domain can be found by projecting these points onto the *x*-axis, and the range can be found by projecting these points onto the *y*-axis. This is illustrated by the point (5, 3) shown in the graph.

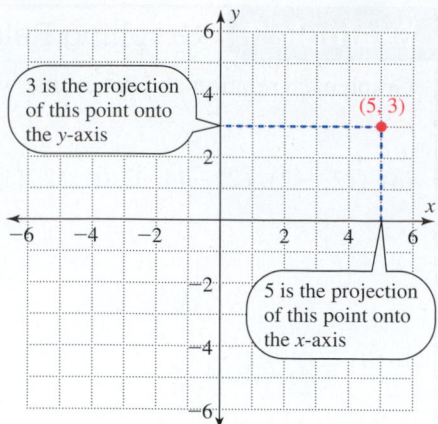

3 is the projection of this point onto the *y*-axis

5 is the projection of this point onto the *x*-axis

Domain and Range from the Graph of a Function

Domain: The domain of a function is the projection of its graph onto the *x*-axis.

Range: The range of a function is the projection of its graph onto the *y*-axis.

Example 4 Determining the Domain and Range from the Graph of a Function

Write the domain and the range of the functions defined by these graphs.

Solution

Domain: Projection onto *x*-axis   Range: Projection onto *y*-axis

(a)

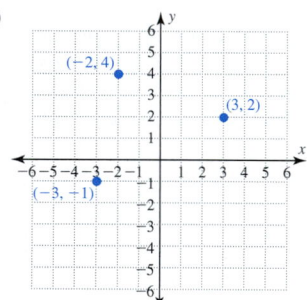
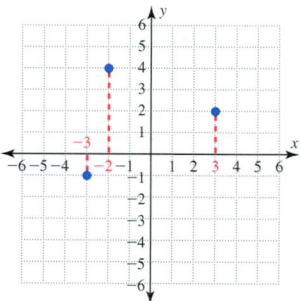
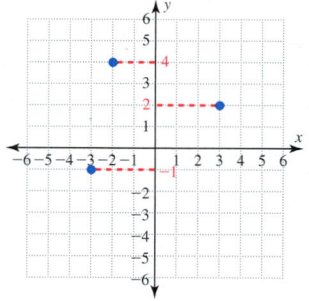

Domain = $\{-3, -2, 3\}$ Range = $\{-1, 2, 4\}$

This function consists of the ordered pairs $\{(-3, -1), (-2, 4), (3, 2)\}$.

(b)

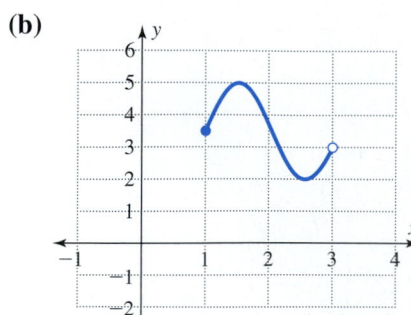
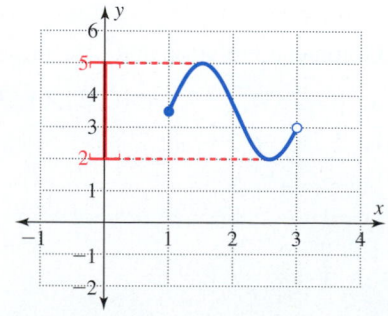

Domain = $[1, 3)$ Range = $[2, 5]$

The domain includes the endpoint 1 but does not include the endpoint 3.

(c)

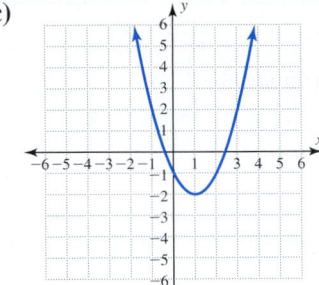   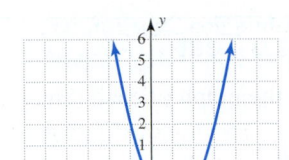   

Domain = $\mathbb{R}$, the set of all real numbers

Range = $[-2, \infty)$

Self-Check 4

Write the domain and the range of the functions defined by these graphs.

a.

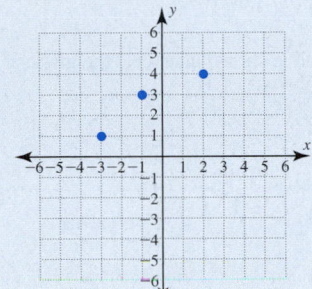

b.

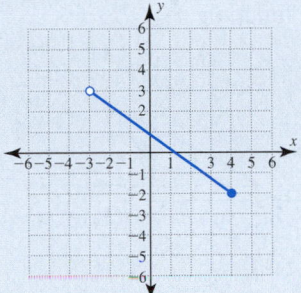

Is There an Easy Way to Determine Whether a Graph Represents a Function?

Yes, imagine a vertical line sweeping across the graph. A vertical line hitting a graph in exactly one point pairs that input value of x with exactly one output value of y. A vertical line hitting a graph in two points pairs that input value of x with two output values of y—thus the graph does not represent a function.

A Graph of a Function

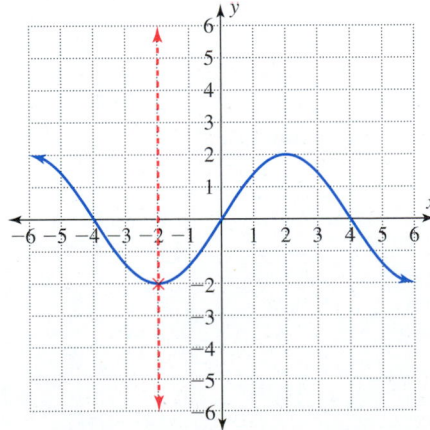

A Graph That Is Not a Function

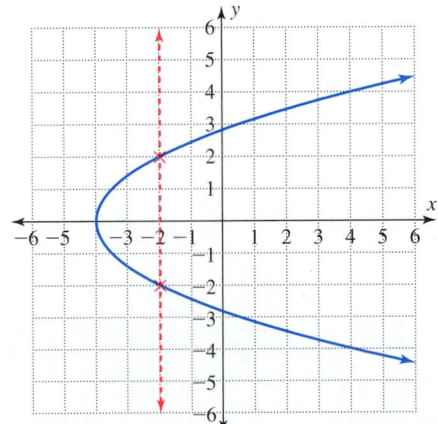

The Vertical Line Test

A graph represents a function if it is impossible to have any vertical line intersect the graph at more than one point.

Example 5   Using the Vertical Line Test

Use the vertical line test to determine whether the graph of each relation represents a function.

(a)

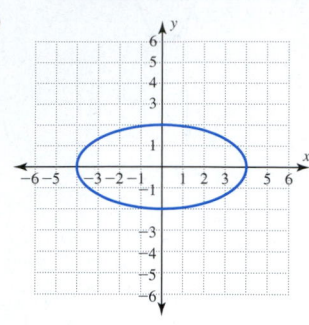

Solution

Not a function

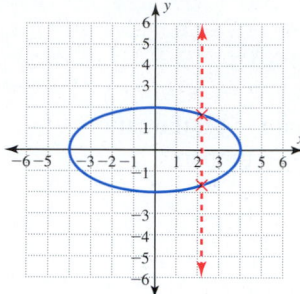

It is possible to draw a vertical line, as the one shown here, that intersects the graph in two points.

(b)

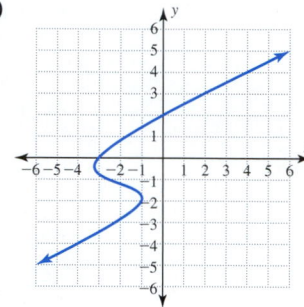

Not a function

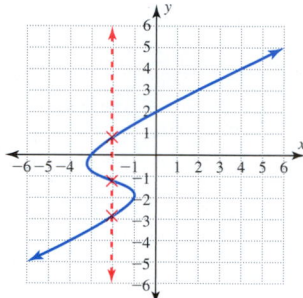

The vertical line shown intersects the graph at three points.

(c)

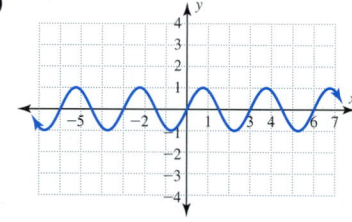

A function

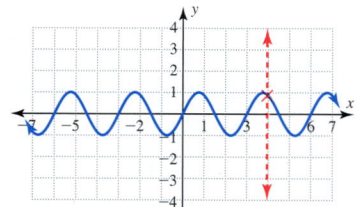

Every vertical line will intersect this graph in exactly one point. This function pairs each input value with exactly one output value.

Self-Check 5

Use the vertical line test to determine whether the graph of each relation represents a function.

a.

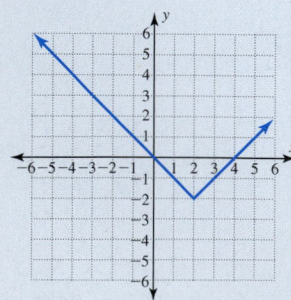

b.

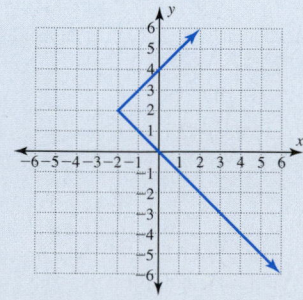

2. Use Function Notation

What Is the Advantage of Using Function Notation to Represent a Function?

A table of values or a graph of a few points can each give us valuable representations for showing the relationship between a set of x-y pairs. For many x-y pairs, it may be inconvenient or impossible to put all these points in a table. Thus it is useful to be able to express the relationship between x and y by using an equation. The notation $f(x)$ is referred to as **function notation** and is read **"f of x"** or **"$f(x)$ is the output value for an input value of x."** Function notation, first introduced in Section 2.2, is used to give an equation that describes a unique output that we can calculate for each input of x. To evaluate $f(x)$ for a specific value of x, replace x on both sides of the equation by this specific value.

Function Notation

The expression is $f(x)$ read as "f of x" or "the value of f at x." The letter f names the function, and the variable x represents an input value from the domain. The notation $f(x)$ represents a specific output value corresponding to x.

Caution: $f(x)$ does not mean f times x in this context. It represents an output value for a specific input value of x.

It is common to use the letter f to represent the function, although any letter can be used. Example 6 illustrates how function notation is used to represent a linear function.

Example 6   Evaluating Function Notation

Evaluate $f(x) = 2x - 4$ for each input value, and then graph this linear function.

Solution

(a) $x = 0$ $f(x) = 2x - 4$

$f(0) = 2(0) - 4$ Substitute each input value of x into the formula $f(x) = 2x - 4$.

$f(0) = 0 - 4$

$f(0) = -4$ $(0, -4)$ is a point on the graph.

(b) $x = 2$ $f(x) = 2x - 4$

$f(2) = 2(2) - 4$

$f(2) = 4 - 4$

$f(2) = 0$ $(2, 0)$ is a point on the graph.

(c) $x = 3$ $f(x) = 2x - 4$

$f(3) = 2(3) - 4$

$f(3) = 6 - 4$

$f(3) = 2$ $(3, 2)$ is a point on the graph.

The equation $f(x) = 2x - 4$ is in the slope-intercept form $(y = mx + b)$ of a line with slope $m = 2$ and y-intercept $(0, -4)$.

The graph of $f(x) = 2x - 4$ is one continuous line, not just the three distinct points calculated here. These points allow us to represent the graph of all the points on this line.

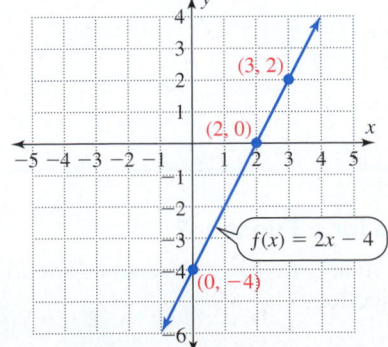

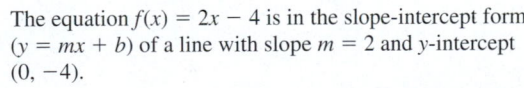

In Example 7, the equation that produces this graph is given by $y = f(x)$. For each point (x, y) on the graph, x represents an input value and y represents $f(x)$; $f(x)$ is the output of the function f for the input x.

Example 7   Using a Graph to Evaluate Input and Output Values

Use the given graph to complete the following.

(a) Evaluate $f(-3)$.

(b) Evaluate $f(1)$.

(c) Determine the value of x for which $f(x) = 1$.

(d) Determine the value of x for which $f(x) = 3$.

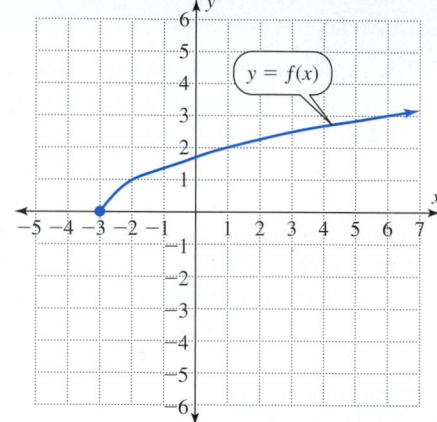

Solution

(a) $f(-3) = 0$ $(-3, 0)$ is a point on the graph. Thus $f(-3) = 0$.

(b) $f(1) = 2$ $(1, 2)$ is a point on the graph. Thus $f(1) = 2$.

(c) $x = -2$ $(-2, 1)$ is a point on the graph. Thus $f(-2) = 1$.

(d) $x = 6$ $(6, 3)$ is a point on the graph. Thus $f(6) = 3$.

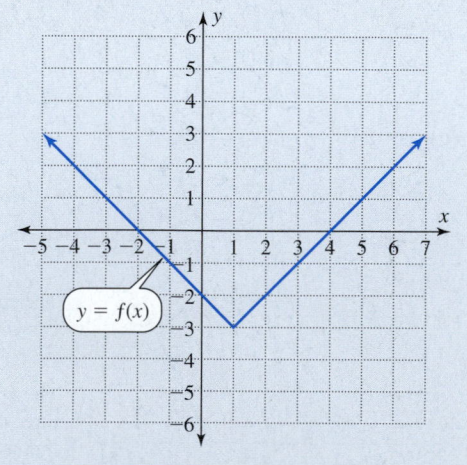

What Are the Multiple Representations I Can Use for a Function?

As we noted at the beginning of this section, there are many ways to represent a function. Example 8 compares the representations given in this section.

Example 8   Using Multiple Representations for a Function

An investment in a bond tripled in value over a 25-year period. Four individuals invested, respectively, $50, $100, $200, and $500. Using these four values as the input values, represent this function by using function notation, mapping notation, a table, ordered pairs, and a graph.

Solution

Verbal Representation	Function Notation	Mapping Notation

Triple each input value to obtain the output value.

Function Notation

$f(x) = 3x$ for each x in the domain

$D = \{50, 100, 200, 500\}$

Mapping Notation

x		$f(x)$
50	→	150
100	→	300
200	→	600
500	→	1,500

Table

x	$f(x)$
50	150
100	300
200	600
500	1,500

Ordered-Pair Notation

$\{(50, 150), (100, 300),$
$(200, 600), (500, 1,500)\}$

Graph

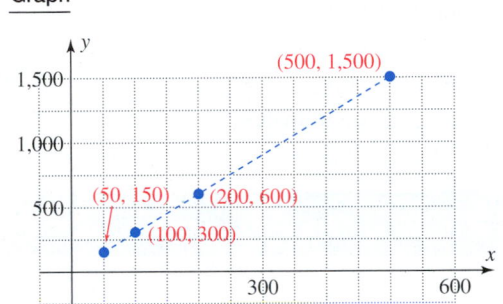

Self-Check 8

Use the domain $D = \{-2, 0, 1, 3\}$ and $f(x) = -2x + 5$.

a. Give a table to represent this function.

b. Represent this function as a set of ordered pairs.

Example 8 contains only selected discrete points. Thus the graph contains only these discrete points. The dashed line is not part of the graph but does help show the linear pattern.

Although there are many representations for a function, the key concept is that one value uniquely determines another value. For example, the show that you watch on TV is determined by the channel that you select, and your monthly mortgage payment depends on the interest rate.

Self-Check Answers

1. a. Function with $D = \{-3, 0, 3\}$ and $R = \{0, 3\}$
   b. Not a function

2. a. $\{(-2, 5), (-1, 2), (0, -1), (1, -4)\}$
   b.

D		R
1	→	-1
2	→	5
3	↗	
5	→	2

3. a. A function   **b.** Not a function

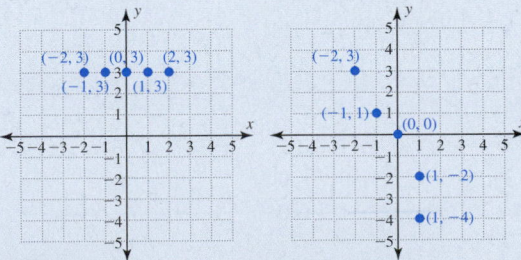

4. a. $D = \{-3, -1, 2\}$, $R = \{1, 3, 4\}$
   b. $D = (-3, 4]$, $R = [-2, 3)$

5. a. A function   **b.** Not a function

6. a. $f(-2) = 13$
   b. $f(0) = 7$
   c. $f(4) = -5$

7. a. $f(-3) = 1$ **b.** $f(2) = -2$
   c. $x = 1$
   d. $x = -4$ and $x = 6$

8. a.

x	$f(x)$
-2	9
0	5
1	3
3	-1

   b. $\{(-2, 9), (0, 5),$
   $(1, 3), (3, -1)\}$

8.1   Using the Language and Symbolism of Mathematics

1. A function is a correspondence that matches each input value with exactly _____ value of the _____ variable.

2. The set of all input values of a function is called the _____ of the function.

3. The set of all output values of a function is called the _____ of the function.

4. The independent values are represented by the _____ variable.

5. The dependent values are represented by the _____ variable.

6. In ordered-pair notation, the domain of a function is represented by the set of _____-coordinates.

7. In ordered-pair notation, the range of a function is represented by the set of _____-coordinates.

8. If a function is defined by a graph, the domain of the function is represented by the projection of the graph onto the _____-axis.

9. If a function is defined by a graph, the range of the function is represented by the projection of the graph onto the _____-axis.

10. The _____ line test can be used to visually inspect a graph to determine whether it represents a function.

11. The notation $f(x)$ is called _____ notation.

12. The notation $f(x) = 3x^2$ is read "_____ of _____ equals three x squared." In this notation the input variable is represented by _____, and $f(x)$ represents the _____ variable. The letter _____ names the function.

8.1   Quick Review

1. Evaluate $3x^2 - 5x - 1$ for $x = 10$.

2. Graph the ordered pair $(-1, 3)$.

3. Graph the interval $(-1, 3)$.

4. Represent the points in this graph using interval notation.

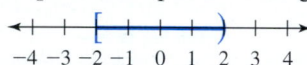

$$-4 \; -3 \; -2 \; -1 \quad 0 \quad 1 \quad 2 \quad 3 \quad 4$$

5. Represent the points in this graph using interval notation.

$$-4 \; -3 \; -2 \; -1 \quad 0 \quad 1 \quad 2 \quad 3 \quad 4$$

8.1   Exercises

Objective 1   Identify a Function and Determine Its Domain and Range

In Exercises 1–4, determine whether each relation is a function. For each function, identify the domain and range.

1. a.

D		R
1	↗↘	−1
		1
2	→	2
3	→	3

b.

D		R
−1	→	1
1	→	1
2	→	2
−3	→	3

c.

D		R
3	→	−1
2	→	1
1	→	3
0	→	0

2. a.

D		R
		−5
3	↗↘	0
		4

b.

D		R
−5	→	3
0	→	3
4	→	3

c.

D		R
4	→	2
1	→	1
0	→	0
−1	→	0

3. a. $\{(1, 1), (-1, 1), (2, 0)\}$

b. $\{(1, 1), (1, -1), (0, 2)\}$

c.

x	y
−3	−1
−2	0
−1	1
0	2
1	3

4. a. $\{(2, 3), (2, -3), (0, 13), (13, 0)\}$

b. $\{(3, 2), (-3, 2), (0, 13), (13, 0)\}$

c.

x	y
−2	−8
−1	−1
0	0
1	1
2	8

In Exercises 5–10, use the vertical line test to determine whether each graph represents a function.

5. a.

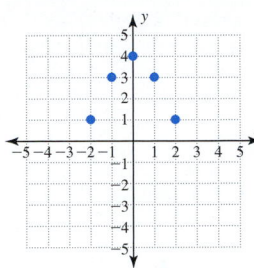

b.

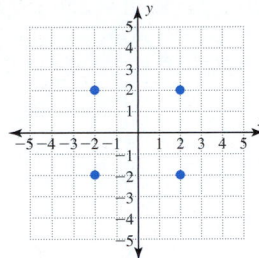

c.

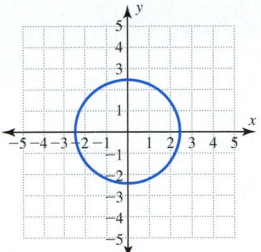

c.

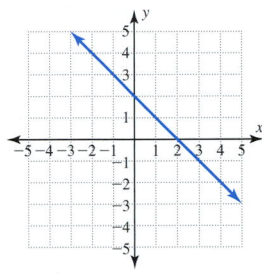

6. a.

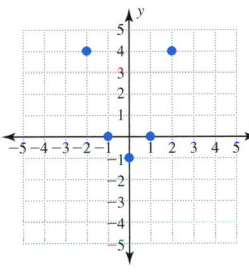

b.

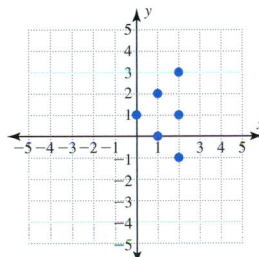

8. a.

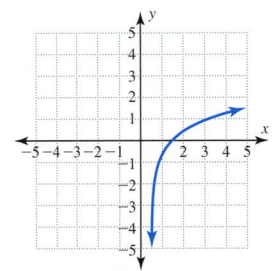

b.

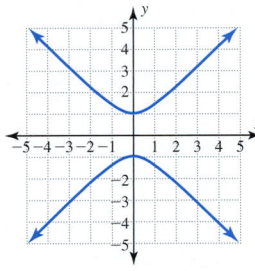

c.

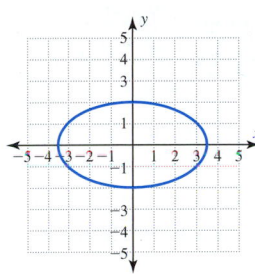

c.

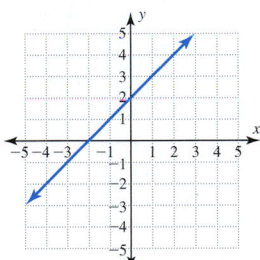

9. a.

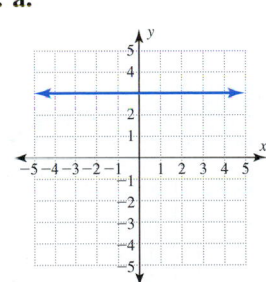

b.

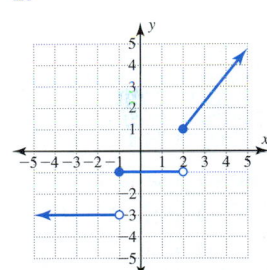

c.

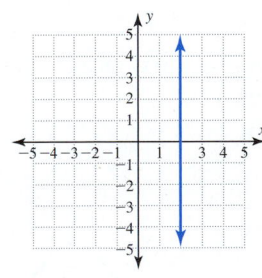

7. a.

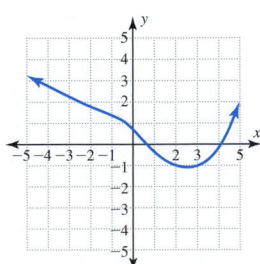

b.

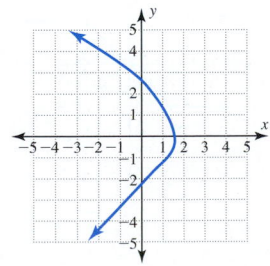

10. a.

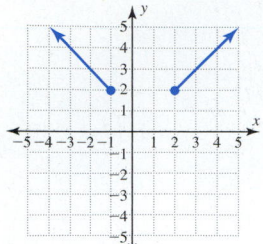

b.

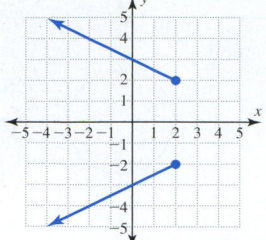

c.

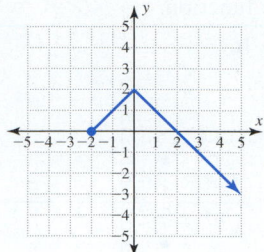

c.

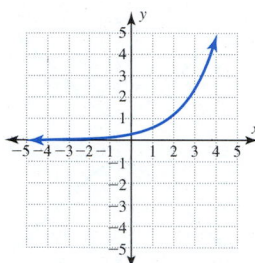

In Exercises 11 and 12, use the given set of ordered pairs to determine the domain and range of each function.

11. a. $\{(-4, 8), (-3, 6), (-2, 4), (-1, 2), (0, 0)\}$
   b. $\{(1, 5), (2, 5), (3, 5), (4, 5), (5, 5)\}$

12. a. $\{(0, 0), (1, 1), (2, 4), (3, 9), (4, 16)\}$
   b. $\{(-2, -1), (-1, -1), (0, -1), (1, -1), (2, -1)\}$

In Exercises 13 and 14, use the given table of values to determine the domain and range of each function.

13. a.

x	y
2	-3
5	1
8	5
11	9
14	13

b.

x	y
5	9
6	5
8	2
12	4
16	5

14. a.

x	y
-4	16
-2	4
0	0
2	4
4	16

b.

x	y
-4	5
-1	4
2	3
5	2
8	1

In Exercises 15–18, use the given graph to determine the domain and range of each function.

15. a.

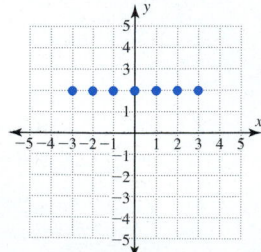

b.

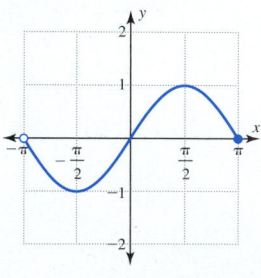

16. a.

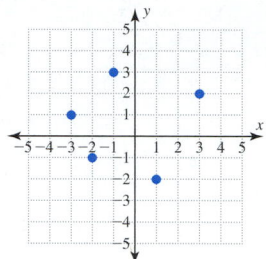

b.

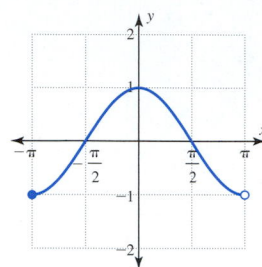

c.

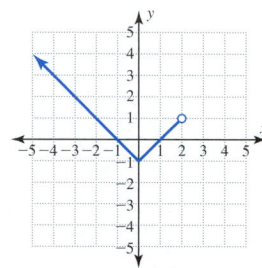

17. a.

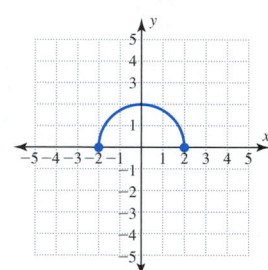

b.

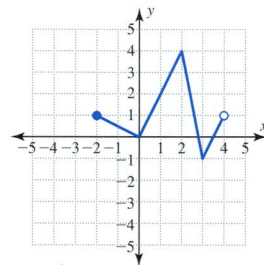

c.

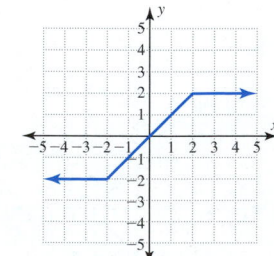

18. a.

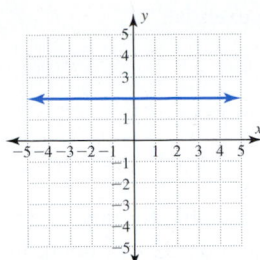

b.

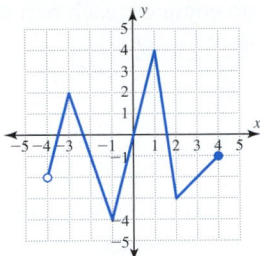

c.

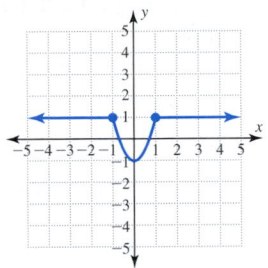

Objective 2 Use Function Notation

Given $f(x) = 2x - 7$, evaluate each expression in Exercises 19 and 20.

19. a. $f(-10)$ **b.** $f(0)$ **c.** $f(5)$

20. a. $f(-4)$ **b.** $f(-1)$ **c.** $f(10)$

Given $f(x) = \sqrt{x}$, evaluate each expression in Exercises 21 and 22.

21. a. $f(0)$ **b.** $f(9)$ **c.** $f\left(\dfrac{1}{4}\right)$

22. a. $f(16)$ **b.** $f(100)$ **c.** $f\left(\dfrac{9}{25}\right)$

Given $f(x) = -|x - 5| + 7$, evaluate each expression in Exercises 23 and 24.

23. a. $f(0)$ **b.** $f(5)$ **c.** $f(8)$

24. a. $f(-1)$ **b.** $f(4)$ **c.** $f(10)$

Given $f(x) = x^2 - 4x - 5$, evaluate each expression in Exercises 25 and 26.

25. a. $f(-1)$ **b.** $f(2)$ **c.** $f(7)$

26. a. $f(-2)$ **b.** $f(0)$ **c.** $f(5)$

27. Use the table to answer each question.

x	y
−1	1
2	3
5	5
8	7

 a. Is 3 an input value or an output value?
 b. Is −1 an input value or an output value?
 c. If 2 is the input value, what is the output value?
 d. If the output value is 7, what was the input value?

28. Use the set of ordered pairs below to answer each question. {(0, 4), (3, 5), (5, 1), (2, 0)}
 a. Is 3 an input value or an output value?
 b. Is 4 an input value or an output value?
 c. If 5 is the input value, what is the output value?
 d. If the output value is 5, what was the input value?

29. Use the graph to answer each question.

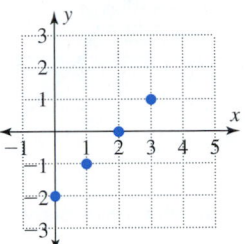

 a. Is 3 an input value or an output value?
 b. Is −1 an input value or an output value?
 c. If 2 is the input value, what is the output value?
 d. If the output value is 1, what was the input value?

30. Use the graph to answer each question.

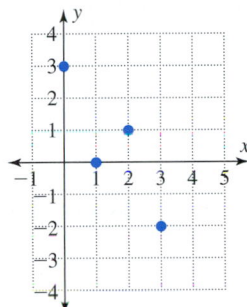

 a. Is −2 an input value or an output value?
 b. Is 2 an input value or an output value?
 c. If 2 is the input value, what is the output value?
 d. If the output value is 0, what was the input value?

In Exercises 31 and 32, use the given table to evaluate each value of $f(x)$.

31. a. $f(-3)$ **b.** $f(1)$
 c. $f(2)$

32. a. $f(-2)$ **b.** $f(0)$
 c. $f(3)$

In Exercises 33 and 34, use the given table to determine the value of x that produces the given value of $f(x)$.

33. a. $f(x) = -1$ **b.** $f(x) = 0$
 c. $f(x) = 8$

34. a. $f(x) = 3$ **b.** $f(x) = -2$
 c. $f(x) = -10$

x	f(x)
−3	8
−2	3
−1	0
0	−1
1	−2
2	−5
3	−10

Table for Exercises 31–34

In Exercises 35 and 36, use the given graph to evaluate each value of $f(x)$.

35. a. $f(-2)$ **b.** $f(0)$ **c.** $f(3)$

36. a. $f(-1)$ **b.** $f(1)$ **c.** $f(2)$

In Exercises 37 and 38, use the given graph to determine the value of x that produces the given value of $f(x)$.

37. a. $f(x) = 4$ **b.** $f(x) = 0$
    c. $f(x) = -2$

38. a. $f(x) = 2$ **b.** $f(x) = -4$
    c. $f(x) = -3$

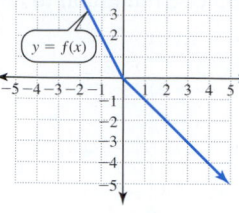

Graph for Exercises 35–38

Multiple Representations

In Exercises 39–46, represent each function as described in each exercise.

39. Use the function defined by the table to complete each part of this exercise.

x	-5	-3	-2	0	1	4
y	4	2	0	-2	-3	-4

  a. Express this function using mapping notation.
  b. Express this function using ordered-pair notation.
  c. Graph this function.

40. Use the function defined by the given table to complete each part of this exercise.

x	-5	-3	-2	0	1	4
y	3	3	3	3	3	3

  a. Express this function using mapping notation.
  b. Express this function using ordered-pair notation.
  c. Graph this function.

41. Use the function defined by the graph to complete each part of this exercise.
  a. Express this function using mapping notation.
  b. Express this function using ordered-pair notation.
  c. Express this function using a table format.

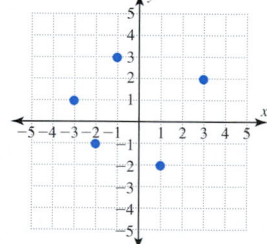

42. Use the function defined by the graph to complete each part of this exercise.
  a. Express this function using mapping notation.
  b. Express this function using ordered-pair notation.
  c. Express this function using a table format.

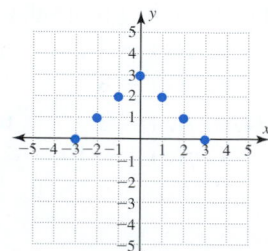

43. Use the function defined by the given mapping notation to complete each part of this exercise.

D		R
-1	$\rightarrow$	1
1	$\rightarrow$	3
2	$\rightarrow$	-1
4	$\rightarrow$	-2

  a. Express this function using a table format.
  b. Express this function using ordered-pair notation.
  c. Graph this function.

44. Use the function defined by the given mapping notation to complete each part of this exercise.

D		R
-4	$\rightarrow$	-4
-2	$\rightarrow$	-2
0	$\rightarrow$	0
1	$\rightarrow$	1
3	$\rightarrow$	3

  a. Express this function using a table format.
  b. Express this function using ordered-pair notation.
  c. Graph this function.

45. Use the function $\{(-5, 4), (-3, 4), (1, 4), (2, 4), (3, 4)\}$ to complete each part of this exercise.
  a. Express this function using a table format.
  b. Express this function using mapping notation.
  c. Graph this function.

46. Use the function $\{(-4, -2), (-2, 4), (4, -3), (3, 1)\}$ to complete each part of this exercise.
  a. Express this function using a table format.
  b. Express this function using mapping notation.
  c. Graph this function.

Connecting Concepts to Applications

47. Value of an Investment An investor purchased four properties for $25,000, $40,000, $50,000, and $65,000. Over a 7-year period the value of each property doubled. Use these four values as the input values for a function to calculate the current value of each property.
  a. Represent this function using function notation.
  b. Represent this function using mapping notation.
  c. Represent this function using a table.
  d. Represent this function using ordered pairs.

48. Diet Plan Predictions A diet plan promoted by a nationwide business promised a 5% weight loss for its customers during the first 2 weeks. One local franchise of this business had five new customers sign contracts in 1 week. The sign-in weights in pounds of these customers were 140, 160, 190, 230, and 280.
  a. Using function notation, represent the promised new weight of each customer after 2 weeks on this plan.
  b. Represent this function by using mapping notation.
  c. Represent this function by using a table.
  d. Represent this function by using ordered pairs.

49. Pressure on a Gas (Boyle's Law) Boyle's law gives the relationship between the pressure and the volume of a gas at a constant temperature. The accompanying histogram shows the pressure and volume of a gas. Use the function defined by this histogram to complete each part of this exercise.

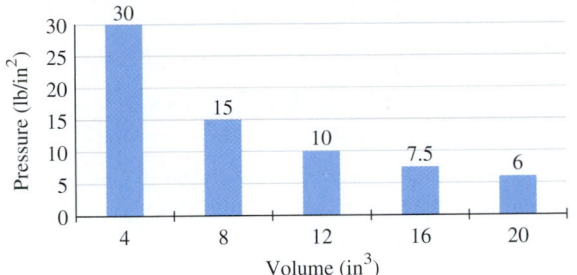

a. If the input x represents the volume of gas, the output y represents _____.

b. What is the volume of gas if the pressure is 7.5 lb/in^3?

c. What is the pressure of a gas that has a volume of 8 in^3?

50. World Series Scores The Philadelphia Phillies defeated the Tampa Bay Rays in five games to win the 2008 World Series. The given histogram shows the number of runs scored by the Phillies in each of the five games. Use the function defined by the histogram to complete each part of this exercise.

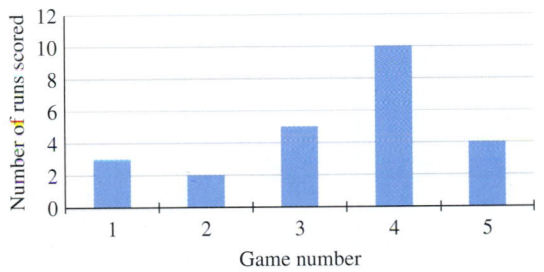

a. If the input x represents the game number, the output y represents _____.

b. How many runs did the Phillies score in game 4?

c. In which game(s) did the Phillies score only 2 runs?

51. Smoking Trends The table shows the percent of U.S. adults who smoke.

Year x	Percent of U.S. Adults Who Smoke y
1965	41.9
1974	37.0
1985	29.9
1990	25.3
1995	24.6
2000	24.1

Source: Center for Disease Control.

a. Graph these points.

b. What percent of U.S. adults smoked in 1985?

c. In what year did 37% of U.S. adults smoke?

52. Salaries by Gender The table shows women's annual earnings as a percent of men's annual earnings in the United States.

Year x	Annual Earnings for Women As a Percent of Men's y
1979	59.7
1984	63.7
1989	68.7
1994	72.0
1999	72.2
2002	76.0

Source: U.S. Department of Labor.

a. Graph these points.

b. What were the earnings for women as a percent of men's in 1989?

c. In what year did women earn 76% of what men earned in the United States?

53. College Tuition and Fees The cost of tuition and fees at 2-year colleges in the United States can be approximated by the function $C(x) = 200x + 1,000$, where x represents the number of years since 1970. (*Source:* U.S. Department of Education.)

a. Use this function to calculate the tuition for years 1970, 1980, 1990, and 2000. Then graph these four points.

b. Estimate the cost of tuition and fees in 2010.

c. Determine the approximate year that the cost of tuition and fees was $5,800.

54. Cost of a Car Lease The total cost to lease a car for x months is given by the function $C(x) = 300x + 1,500$.

a. Use this function to calculate the total cost to lease a car for each of the first 6 months. Then graph these six points.

b. Give the cost to lease the car for 24 months.

c. The lessee has already spent $4,800 on this car lease. How many months of the lease has this person paid?

55. Sales Tax The sales tax in Houston, Texas, is 8.25% (6.25% state tax, 1% city tax, and 1% for the Houston MTA).

a. If x represents the amount of a purchase at a Houston clothing store, write a function T so that $T(x)$ represents the tax on this purchase.

b. Use the function from part **a** to complete the table shown.

x	$T(x)$
0	
100	
200	
300	
400	
500	
600	

56. Jet Fuel Use According to Captain Michael Ragsdale, the number of pounds of jet fuel used in an hour by an MD88 passenger jet at cruising altitude is 9,000 lb.
  a. If x represents the number of hours the plane has been at cruising altitude, write a function F so that $F(x)$ represents the number of pounds of jet fuel that the plane will use during this portion of its flight.
  b. Use the function from part **a** to complete the table shown.

x	$F(x)$
0.00	
.25	
.50	
.75	
1.00	
1.25	
1.50	

In Exercises 57–60, match each graph with its domain and range.

57.

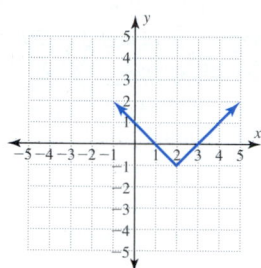

58.

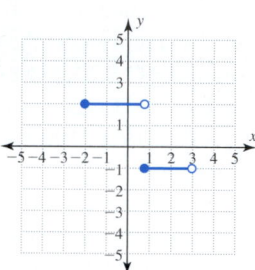

59.

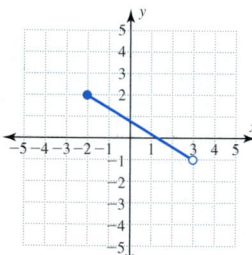

60.

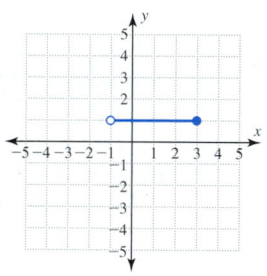

A. $D = (-1, 3], R = \{1\}$ **B.** $D = [-2, 3), R = (-1, 2]$
C. $D = [-2, 3), R = \{-1, 2\}$
D. $D = \mathbb{R}$ (all real numbers), $R = [-1, \infty)$

61. a. Draw a semicircle that represents a function.
  b. Draw a semicircle that does not represent a function.

62. Can a circle ever be a function? Explain your answer.

Group discussion questions

63. Error Analysis A student examined this graph and concluded that it passed the vertical line test and therefore the graph represents a function. Correct this answer and explain the error in this reasoning.

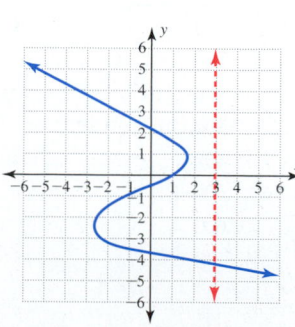

64. Challenge Question Graph a function with the given domain D and range R. For some parts there may be several correct graphs.
  a. $D = \{2\}, R = \{3\}$ **b.** $D = \{-3\}, R = \{2\}$
  c. $D = (-3, 2], R = \{-2\}$ **d.** $D = [-4, 4), R = \{3\}$
  e. $D = (-3, 3), R = \{-1, 2\}$
  f. $D = [-2, 2], R = \{-4, 4\}$
  g. $D = [-4, 3), R = (-3, 4]$
  h. $D = (-3, 4], R = [-2, 2)$
  i. $D = \mathbb{R}$ (all real numbers), $R = [-2, \infty)$
  j. $D = \mathbb{R}$ (all real numbers), $R = (-\infty, 2]$

65. Calculator Discovery Use a graphing calculator to graph each function and to determine the domain and range of each function.
  a. $f(x) = \sqrt{x}$ **b.** $f(x) = -\sqrt{x}$
  c. $f(x) = -x^2 + 1$ **d.** $f(x) = (x - 1)^2$
  e. $f(x) = \sqrt{-x}$ **f.** $f(x) = \sqrt{2 - x}$

66. Calculator Discovery Each of these graphs is produced by an equation of the form $y = |x| + c$ for a specific value of c. Write the equation for each graph, and use a graphing calculator to test each function.

a.

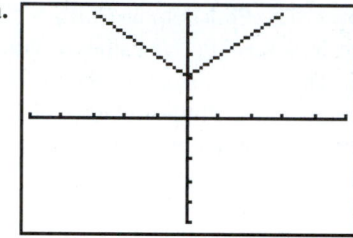

$[-5, 5, 1]$ by $[-5, 5, 1]$

b.

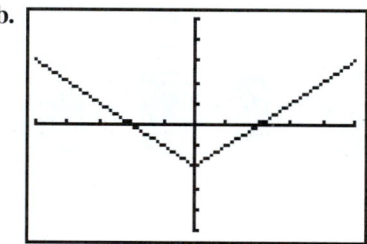

$[-5, 5, 1]$ by $[-5, 5, 1]$

c.

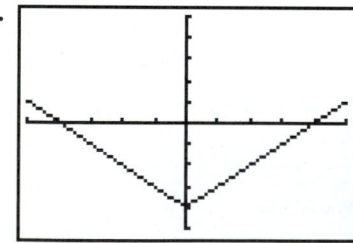

$[-5, 5, 1]$ by $[-5, 5, 1]$

d.

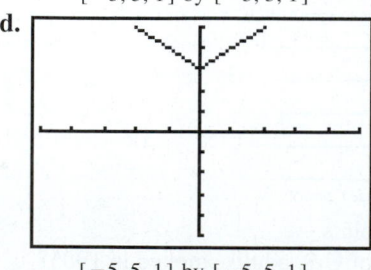

$[-5, 5, 1]$ by $[-5, 5, 1]$

67. Calculator Discovery Use each set of data to write a function $y = f(x)$ that will produce the given table of values. Use a graphing calculator to test each function.

a.

X	Y1
-3	-12
-2	-8
-1	-4
0	0
1	4
2	8
3	12

X=-3

b.

X	Y1
-3	9
-2	4
-1	1
0	0
1	1
2	4
3	9

X=-3

c.

X	Y1
-3	-11
-2	-7
-1	-3
0	1
1	5
2	9
3	13

X=-3

d.

X	Y1
-3	8
-2	3
-1	0
0	-1
1	0
2	3
3	8

X=-3

8.1 Cumulative Review

1. Solve $\dfrac{x}{3} + \dfrac{1}{2} = \dfrac{2x - 5}{4}$.

2. Simplify $\dfrac{-5x^3}{2y^7} \cdot \dfrac{12x^2y^{-2}}{25x^{-1}y^{-4}}$. Write your answer using only positive exponents.

3. Factor $25x^2 + 60xy + 36y^2$.

4. Factor $25x^2 - 36y^2$.

5. Factor $-5x^2 + 20$.

Section 8.2 | Linear Functions

Objectives:

1. Determine the slope of a line.
2. Sketch the graph of a linear function.
3. Write the equation of the line through given points.
4. Determine the intercepts of a line.
5. Determine the *x*-values for which a linear function is positive and the *x*-values for which a linear function is negative.

Although the variety of functions that you can encounter in the workplace is considerable, it is remarkable that most of these functions belong to a few classic families of functions. Each member of a family shares many important characteristics with other members of that family. Thus by examining only a few families of functions, you can be well prepared for many contingencies that you may encounter. We will examine some of the key characteristics of these functions from algebraic, graphical, and numerical perspectives.

We examined linear functions in Sections 2.2, 3.1, and 3.2. At that point of the textbook, our emphasis was more on the equations, their graphs, and notation, and less on the formal function concept. Now that we have defined the function concept more formally, we will revisit linear functions before introducing other families of functions.

What Are the Key Characteristics of a Linear Function?

- The graph of a linear function is a straight line.
- The equation is first degree and can be written in the slope-intercept form $f(x) = mx + b$ with slope m and y-intercept $(0, b)$.
- Linear functions have a constant rate of change. The slope is the same between any two points on the line.
- The domain of all linear functions is $\mathbb{R}$. The range of all linear functions is also $\mathbb{R}$ unless the function is a constant function, whose graph is a horizontal line.

Linear Function

Algebraically	Numerical Example	Graphical Example	Verbally
A function of the form $f(x) = mx + b$ is called a linear function.	$f(x) = 2x - 5$ $\begin{array}{c\|c} x & y = f(x) \\ \hline -1 & -7 \\ 0 & -5 \\ 1 & -3 \\ 2 & -1 \\ 3 & 1 \\ 4 & 3 \\ 5 & 5 \end{array}$	$f(x) = 2x - 5$ 	In the table for $y = 2x - 5$, each 1-unit increase in x produces a 2-unit increase in y. The graph of $y = 2x - 5$ is a straight line. Each point on this line satisfies this equation. This line rises 2 units for each 1-unit move to the right.

1. Determine the Slope of a Line

Recall from Section 3.1 that we define the slope of a line as the ratio of the change in y to the change in x. This rise over the run can be expressed algebraically by $m = \dfrac{y_2 - y_1}{x_2 - x_1}$ or $m = \dfrac{\Delta y}{\Delta x}$. In general, the slope is an indication of how the y-values are changing with respect to the x-values. One of the things that we can observe quickly from the slope of a line is whether the line is sloping upward to the right or downward to the right. A line with positive slope goes upward as it moves from left to right. A line with negative slope goes downward as it moves from left to right. A line with zero slope is a horizontal line that does not move up or down.

Interpreting Slopes

Positive Slope	Zero Slope	Negative Slope
Algebraically In $f(x) = mx + b$, m will be positive. *Example:* $f(x) = \dfrac{1}{2}x - 1$	**Algebraically** In $f(x) = mx + b$, m will be 0 and $f(x) = b$. *Example:* $f(x) = 4$	**Algebraically** In $f(x) = mx + b$, m will be negative. *Example:* $f(x) = -2x + 1$

Graphically	**Graphically**	**Graphically**
The line will slope upward to the right.	This is a horizontal line that does not slope upward or downward.	The line slopes downward to the right.
Example:	*Example:*	*Example:*

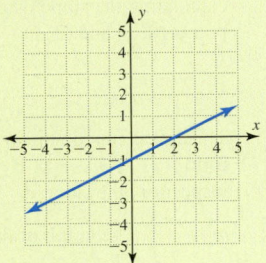
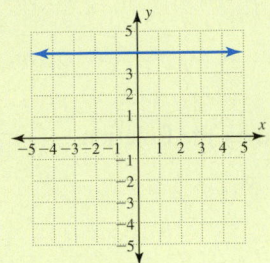
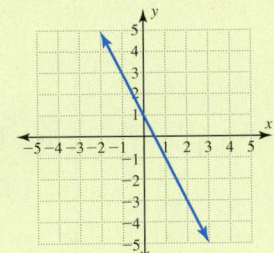

Numerically	**Numerically**	**Numerically**
The *y*-values will increase as the *x*-values increase.	The *y*-values will not change as the *x*-values change.	The *y*-values will decrease as the *x*-values increase.
Example:	*Example:*	*Example:*

x	y
-4	-3
-2	-2
0	-1
2	0
4	1

x	y
-2	4
-1	4
0	4
1	4
2	4

x	y
-2	5
-1	3
0	1
1	-1
2	-3

The equation of a vertical line is of the form $x = k$. This equation does not represent a function—it fails the vertical line test. The slope of a vertical line is undefined.

How Can I Determine the Slope of a Line?

To determine the slope of a line, we can substitute any two points on the line into the equation for the slope of a line. We can obtain these two points from a table, a graph, or the equation of a line. We can also determine the slope by inspection if the equation of the line is written in slope-intercept form. This is illustrated in Example 1.

Example 1   Determining the Slope of a Line

Determine the slope of each line.

Solution

(a) The line that passes through all the points listed in the table

x	y
-2	-3
1	-1
4	1
7	3
10	5

$$m = \frac{y_2 - y_1}{x_2 - x_1}$$

$$m = \frac{-1 - (-3)}{1 - (-2)}$$

$$m = \frac{2}{3}$$

Select the two points $(-2, -3)$ and $(1, -1)$ from the table and use the slope formula. Also note that as the *x*-values increase by 3, the *y*-values increase by 2. The slope of this line is positive so it will slope upward to the right.

(b) The line in the graph

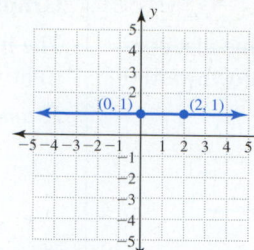

$$m = \frac{y_2 - y_1}{x_2 - x_1}$$

$$m = \frac{1 - 1}{2 - 0}$$

$$m = \frac{0}{2}$$

$$m = 0$$

Note that this is a horizontal line, so the slope is zero.

This can be confirmed by calculating the slope using the points $(0, 1)$ and $(2, 1)$.

(c) The line defined by the function

$$f(x) = -\frac{3}{2}x + 5$$

$$m = -\frac{3}{2}$$

This linear function is in the form $f(x) = mx + b$. By inspection, the slope is

$m = -\frac{3}{2}$. The graph of this line will slope downward to the right.

Self-Check 1

Determine the slope of each line.

a. The line that passes through all the points listed in the table

x	1	2	3	4	5
y	-2	-2	-2	-2	-2

b. The line in the accompanying graph

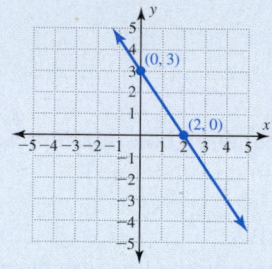

c. The line defined by the function $f(x) = 4x + 8$

2. Sketch the Graph of a Linear Function

What Is the Most Efficient Way to Graph a Line?

Two points are needed to graph a line. If we have the equation of a line, we can always select two arbitrary x-values and calculate the corresponding y-values. However, there is a quicker way to do this directly from the slope-intercept form $f(x) = mx + b$. By inspection, one point is the y-intercept $(0, b)$. The slope $m = \dfrac{\Delta y}{\Delta x}$ can be used to determine a second point. Start with the y-intercept $(0, b)$ and use the changes in x and y to form a new point $(\Delta x, b + \Delta y)$. This is illustrated in Example 2.

Example 2   Graphing a Line Using the Slope-Intercept Form

Use the slope and y-intercept to graph the line defined by $f(x) = \dfrac{2}{3}x - 1$.

Solution

$f(x) = \dfrac{2}{3}x - 1$

Slope: $m = \dfrac{2}{3}$

y-intercept: $(0, -1)$

Second point: $(0 + 3, -1 + 2)$

$= (3, 1)$

Use the slope-intercept form $f(x) = mx + b$ to determine the slope $m = \dfrac{2}{3}$ and the y-intercept $(0, -1)$.

Plot the y-intercept $(0, -1)$ and then use the slope to plot a second point. Do this by increasing y by 2 units for a 3-unit increase in x.

Sketch the line through these two points.

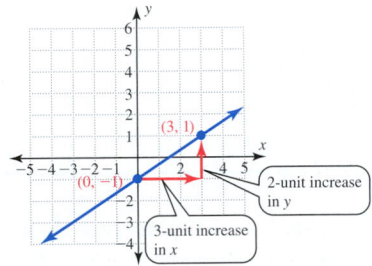

Self-Check 2

Use the slope and y-intercept to graph the line defined by $f(x) = \dfrac{5}{2}x - 3$.

The ability to write an equation for the line that passes through given points is an essential skill for an algebra student. The steps used to write an equation for the line will depend on the given information. For every linear function, it is possible to write the equation in the form $y = mx + b$.

3. Write the Equation of the Line Through Given Points

What Information Do I Need to Write the Equation of a Line in the Form $y = mx + b$?

In every case, the slope either must be given or be calculated. The other steps will depend on the given information. In Example 3, we are asked to write the equation of a line that passes through points listed in a table. If all the points in the table are on the same line, then we can use any two points in the table to determine the slope. *Caution:* Check to be sure that all the points produce the same slope or that all points satisfy the same equation. In Example 3, the y-intercept is listed in the table.

Example 3   Writing the Equation of the Line Through Given Points

All the points in this table lie on the same line. Write in slope-intercept form the equation of the line that passes through the points listed in this table.

x	y
-3	8
0	6
3	4
6	2
9	0

Solution

$$m = \frac{y_2 - y_1}{x_2 - x_1} = \frac{6 - 8}{0 - (-3)}$$

Calculate the slope m by using any two of the given points. The points $(-3, 8)$ and $(0, 6)$ are used here. Also note that as the x-values increase by 3, the y-values decrease by 2.

$$m = \frac{-2}{3}$$

$$b = 6$$

The y-intercept $(0, 6)$ is listed in the table.

$$y = mx + b$$

$$y = -\frac{2}{3}x + 6$$

Substitute $-\frac{2}{3}$ for m and 6 for b. You can use a graphing calculator to check that this is the correct equation. Do all of the points lie on this line?

Answer: $y = -\frac{2}{3}x + 6$

Self-Check 3

All the points in this table lie on the same line. Write in slope-intercept form the equation of the line that passes through the points listed in this table.

x	-4	-2	0	2	4
y	5	10	15	20	25

The main difference between Examples 3 and 4 is that the y-intercept is listed in the table in Example 3, but not in Example 4. Thus to use the slope-intercept form in Example 4 we must first calculate the y-intercept.

Example 4   Writing the Equation of the Line Through Given Points

All the points in this table lie on the same line. Write in slope-intercept form the equation of the line that passes through the points listed in this table.

x	y
-4	-1
1	3
6	7
11	11
16	15

Solution

$$m = \frac{4}{5}$$

Note that as the x-values increase by 5, the y-values increase by 4.

$$y = mx + b$$

Use the slope-intercept form to calculate the y-intercept $(0, b)$.

$$-1 = \frac{4}{5}(-4) + b$$

Substitute $(-4, -1)$ for (x, y) and $\frac{4}{5}$ for m. Solve this equation for b.

$$-1 = \frac{-16}{5} + b$$

$$b = \frac{11}{5}$$

$$y = mx + b$$

Substitute $\frac{4}{5}$ for m and $\frac{11}{5}$ for b into the slope-intercept form to write the equation satisfying the given conditions. You can use a graphing calculator to check that this is the correct equation. Do all of the points lie on this line?

$$y = \frac{4}{5}x + \frac{11}{5}$$

Answer: $y = \frac{4}{5}x + \frac{11}{5}$

Self-Check 4

All the points in this table lie on the same line. Write in slope-intercept form the equation of the line that passes through the points listed in this table.

x	-3	1	5	9	13
y	11	8	5	2	-1

Both Examples 3 and 4 used the slope-intercept form to write an equation of a line. One reason to do this is to stress the importance of the functional form of a linear equation, $f(x) = mx + b$. However, the point-slope form, $y - y_1 = m(x - x_1)$, used in Chapter 3, could also be used to write each of these equations.

4. Determine the Intercepts of a Line

In Example 5, we determine x- and y-intercepts of a linear function by solving linear equations in one variable.

Example 5 Finding the Intercepts of the Graph of a Linear Function

Determine the intercepts of the graph of $f(x) = \dfrac{3}{4}x - 2$. Then use these intercepts to sketch the graph of this function.

Solution

x-intercept:

$$y = \frac{3}{4}x - 2$$

First write the equation in the form $y = mx + b$ by replacing $f(x)$ with y.

$$0 = \frac{3}{4}x - 2$$

To determine the x-intercept, set $y = 0$ and solve for x.

$$4(0) = 4\left(\frac{3}{4}x\right) - 4(2)$$

$$0 = 3x - 8$$

$$8 = 3x$$

$$\frac{8}{3} = x$$

$\left(\dfrac{8}{3}, 0\right)$ is the x-intercept.

y-intercept:

$$y = \frac{3}{4}x - 2$$

From the form $y = mx + b$, the y-intercept is $(0, -2)$. You will obtain the same result if you substitute 0 for x and solve for y.

$(0, -2)$ is the y-intercept.

Graph:

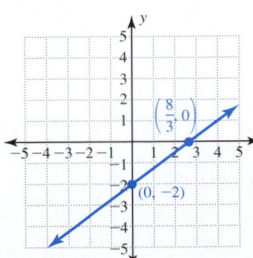

Plot the x-intercept $\left(\dfrac{8}{3}, 0\right)$, the y-intercept $(0, -2)$, and draw the line through these two points.

Would you obtain the same graph if you started with the y-intercept $(0, -2)$ and used the slope $m = \dfrac{3}{4}$ to obtain a second point, $(4, 1)$?

Self-Check 5

Determine the intercepts of the graph of $f(x) = -\dfrac{1}{2}x + 4$. Then use these intercepts to sketch the graph of this function.

In many linear applications, both the slope and the y-intercept have a meaningful interpretation, as illustrated below. Understanding the significance of these interpretations can help you as you try to write a linear model for a problem.

Cost Equation

$y = 2.50x + 3.00$

Table of Costs

Miles x	Cost y ($)
0	3.00
2	8.00
5	15.50
9	25.50

Meaning of m and b

$m = 2.50$; The cost is increasing by $2.50 per mile.
$b = 3.00$; The initial cost is $3.00.

Height Equation

$y = -50x + 400$

Graph of Heights

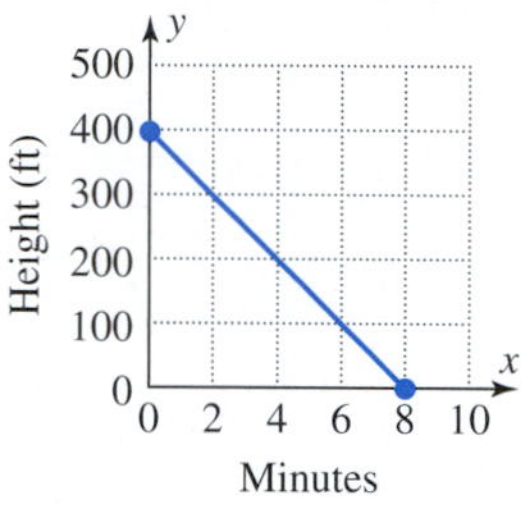

Meaning of m and b

$m = -50$; The height is decreasing by 50 ft per minute.
$b = 400$; The initial height is 400 ft.

To write a linear function to model a problem, it is helpful to keep in mind the meaning of the slope and y-intercept. It is also important to observe whether the y-values are increasing or decreasing as the x-values increase. This will indicate whether the slope is positive or negative. Example 6 examines data on engine performance taken from the September 2001 issue of *Popular Science,* where the horsepower generated by an engine is decreasing in a linear fashion as the elevation of the car increases.

Example 6 **Writing a Linear Function That Models Engine Performance**

The data in the table form a linear model and give the actual horsepower generated by a 100-hp engine at various elevations.

(a) Use this table to determine the linear function
$H(x) = mx + b$ for these data points.

(b) Interpret the meaning of the slope in this application.

(c) Interpret the meaning of the y-intercept in the application.

Elevation in feet, x	Horsepower, y
0	100
1,000	97
2,000	94
3,000	91
4,000	88

Solution

(a) $m = \dfrac{y_2 - y_1}{x_2 - x_1}$

$\quad = \dfrac{97 - 100}{1{,}000 - 0}$

$\quad = \dfrac{-3}{1{,}000}$

$\quad = -0.003$

Using the first two points in the table, (0, 100) and (1,000, 97), calculate the slope. Also note that as the x-values increase by 1,000, the y-values decrease by 3.

$b = 100$

The y-intercept is listed as (0, 100).

$H(x) = -0.003x + 100$

Substitute m and b into the slope-intercept form.

(b) $m = -0.003$

The actual horsepower is decreasing by 0.003 hp for each foot the elevation increases.

The actual horsepower is decreasing by 3 hp for each 1,000 feet the elevation increases. $-0.003 = \dfrac{-3}{1{,}000}$.

(c) $b = 100$

The power generated by the engine at sea level is 100 hp.

Sea level has an elevation of 0 ft.

Self-Check 6

Rework Example 6 assuming the table represents the actual horsepower generated by a 200-hp engine at various altitudes.

Altitude in feet, x	Horsepower, y
0	200
500	199
1,000	198
1,500	197
2,000	196

5. Determine the x-values for Which a Linear Function Is Positive and the x-values for Which a Linear Function Is Negative

What Is Meant by Saying That a Function Is Positive or Negative?

For a given point or an interval of points, this means that the output value of this function is positive or negative.

Classifying a Function as Positive or Negative

Verbally	Graphically	Numerically	Algebraically
The function is **positive** at (x, y).	(x, y) is above the x-axis.	(x, y) has a positive y-value.	$f(x) > 0$
The function is **negative** at (x, y).	(x, y) is below the x-axis.	(x, y) has a negative y-value.	$f(x) < 0$

These terms define when a function is positive or negative at a given point. Often a real-valued function will be either positive or negative over an interval of x-values. In Example 7, we will determine the interval of x-values for which a function is positive and the interval of x-values for which a function is negative.

Example 7   Determining Where a Function Is Positive and Where a Function Is Negative

Determine the x-values for which $f(x) = -\dfrac{1}{2}x + 1$ is positive and the x-values for which this function is negative.

(a) **Graphically:** Use the graph to solve this problem.

(b) **Numerically:** Use the table to solve this problem.

(c) **Algebraically:** Use linear inequalities to solve this problem.

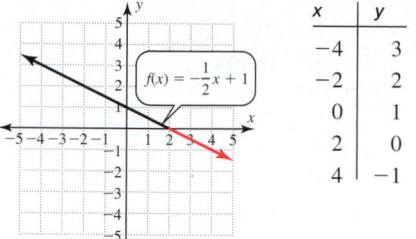

x	y
-4	3
-2	2
0	1
2	0
4	-1

Solution

Positive:

(a) Graphically

The graph is above the x-axis (in the black) to the left of the x-intercept $(2, 0)$.

(b) Numerically

The y-values in the table are positive for x-values less than 2.

(c) Algebraically

Solve $f(x) > 0$.

$$-\frac{1}{2}x + 1 > 0$$

$$-x + 2 > 0$$

$$-x > -2$$

$$x < 2$$

Negative:

The graph is below the x-axis (in the red) to the right of the x-intercept $(2, 0)$.

The y-values in the table are negative for x-values greater than 2.

Solve $f(x) < 0$.

$$-\frac{1}{2}x + 1 < 0$$

$$-x + 2 < 0$$

$$-x < -2$$

$$x > 2$$

Answer: The function is positive for x-values in the interval $(-\infty, 2)$ and negative for x-values in the interval $(2, \infty)$. This is confirmed by all three perspectives examined.

Self-Check 7

Determine the x-values for which $f(x) = \dfrac{2}{3}x + 2$ is positive and the x-values for which this function is negative.

a. **Graphically:** Use the graph to solve this problem.

b. **Numerically:** Use the table to solve this problem.

c. **Algebraically:** Use linear inequalities to solve this problem.

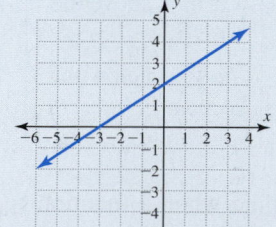

x	y
-6	-2
-3	0
0	2
3	4
6	6

Although graphical and numerical methods can provide insight into a problem that we are solving, they often lack the conciseness and the exactness of algebraic methods. The *x*-values for which a linear function is positive or negative can be approximated by graphical and numerical methods. However, if the *x*-intercept is not an integer value as it is in Example 7, then the algebraic method may be preferable.

In Section 2.3, we looked at the significance of intercepts in an application involving the overhead costs and the break-even value for production using a machine. The overhead costs are the costs involved if no units are produced. The break-even value is the number of units a machine must make before a profit is realized. That is, the break-even value will result in a profit of $0, not a loss or a gain. In Example 8, if production is less than the break-even number of units, this will result in a loss. The set of input values that will produce a loss is called the **loss interval.** Likewise, if production is more than the break-even number of units, this will result in a profit. The set of input values that will produce a profit is called the **profit interval.**

Example 8   Determining Profit and Loss Intervals for Machine Production

In the given graph, the input value of *x* is the number of tons of refined ore that a machine produces in 1 day. The output *y* is the profit generated by the sale of this ore when it is produced. The machine is capable of producing a maximum of 125 tons in 1 day, but it is not always able to produce this maximum number. The break-even value for this machine is 50 tons produced in 1 day. Determine **(a)** the loss interval and **(b)** the profit interval for this machine.

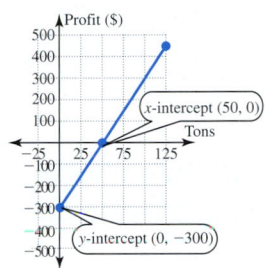

Solution

(a) The loss interval is [0, 50).

The machine could produce 0 tons. The bracket is used to indicate that 0 is included in the loss interval. Producing up to but not including 50 tons will result in a loss. Producing

(b) The profit interval is (50, 125].

more than 50 tons will result in a profit. The maximum this machine can produce is 125 tons. The bracket indicates that 125 is included in the profit interval.

Self-Check 8

Use the given graph for a machine similar to the one in Example 8. This machine is capable of producing a maximum of 100 tons in 1 day, but it is not always able to produce this maximum number. The break-even value for this machine is 60 tons produced in 1 day. Determine **(a)** the loss interval and **(b)** the profit interval for this machine.

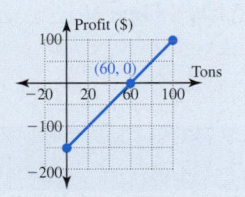

Some production or manufacturing processes result in output that can be counted—that is, output that corresponds to the whole numbers 0, 1, 2, 3, ... For example, a company manufacturing windows can produce 0, 1, 2, 3, . . . windows. Other companies produce or manufacture a product that can be measured—that is, output can correspond to an interval of positive real numbers. In Example 8, ore production in tons per day was the interval [0, 125]. Depending on the product involved, the volume of production can be represented by either just whole numbers or an interval of real numbers. Nonetheless, it is customary in business and economics to refer to profit and loss intervals in both situations.

Self-Check Answers

1. a. $m = 0$ **b.** $m = -\dfrac{3}{2}$ **c.** $m = 4$

2.

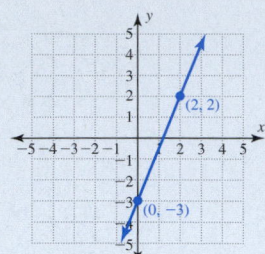

Graph:

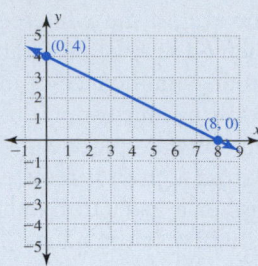

3. $y = \dfrac{5}{2}x + 15$

4. $y = -\dfrac{3}{4}x + \dfrac{35}{4}$

5. x-intercept: $(8, 0)$
   y-intercept: $(0, 4)$

6. a. $H(x) = -0.002x + 200$
   b. $m = -0.002$: The actual horsepower is decreasing by 0.002 hp for each foot the altitude increases.
   c. $b = 200$: The power generated by the engine at sea level is 200 hp.
7. Examining this function graphically, numerically, and algebraically, we see that $f(x) = \dfrac{2}{3}x + 2$ is positive for x-values in the interval $(-3, \infty)$ and this function is negative for x-values in the interval $(-\infty, -3)$.
8. a. The loss interval is $[0, 60)$.
   b. The profit interval is $(60, 100]$.

8.2   Using the Language and Symbolism of Mathematics

1. The slope of a line that passes through (x_1, y_1) and (x_2, y_2) is $m = $ _____ .

2. A line with _____ slope goes upward as it moves from left to right.

3. A line with _____ slope goes downward as it moves from left to right.

4. A line with _____ slope is a horizontal line.

5. A function of the form $f(x) = mx + b$ is called a _____ function.

6. The graph of $f(x) = 2x + 5$ has a slope of _____ and a y-intercept _____ .

7. To determine the x-intercept of a line, set _____ = 0 and solve for _____ .

8. To determine the y-intercept of a line, set _____ = 0 and solve for _____ .

9. A function $y = f(x)$ is positive when the output value $f(x)$ is _____ .

10. A function $y = f(x)$ is negative when the output value $f(x)$ is _____ .

8.2   Quick Review

1. Use interval notation to represent the real numbers that satisfy $x < 6$.

2. Use interval notation to represent the real numbers that satisfy $x \geq 5$.

3. Solve $-\dfrac{1}{3}x + 1 < 9$. Give your answer in interval notation.

4. Sketch the graph of the line through $(-2, 4)$ and $(3, 6)$.

5. Complete this table so that a line through these points will have a slope of $\dfrac{3}{7}$.

x	y
-14	5
-7	
0	
7	
14	

8.2   Exercises

Objective 1 Determine the Slope of a Line

In Exercises 1–4, determine the slope of each line.

1. $f(x) = 5x + 3$ **2.** $f(x) = -1$

3. $f(x) = 2$ **4.** $f(x) = -3x + 7$

In Exercises 5–8, determine the slope of the line that passes through all the points in the table.

5.

x	y
0	3
1	5
2	7
3	9
4	11

6.

x	y
-2	1
-1	4
0	7
1	10
2	13

7.

x	y
3	8
6	4
9	0
12	-4
15	-8

8.

x	y
3	-7
5	-7
7	-7
9	-7
11	-7

In Exercises 9–12, determine the slope of each line.

9.
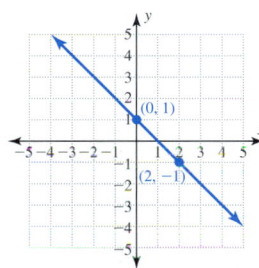

10.
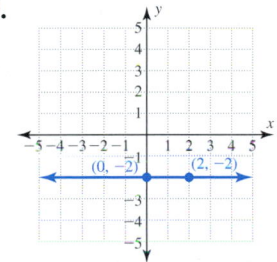

11.
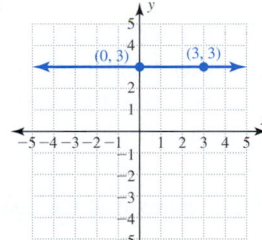

12.
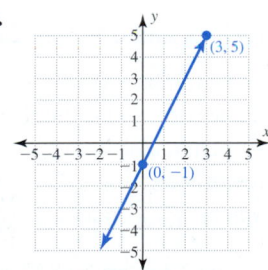

Objective 2 Sketch the Graph of a Linear Function

13. Use the function $f(x) = -x + 2$ to:
  a. Complete a table for this function with input values of $x = -2, -1, 0, 1, 2$ using only a pencil and paper.
  b. Describe the change in y for each 1-unit increase in x.

c. Plot these points and sketch the line through these points.
d. Describe the behavior of the line for every 1-unit move to the right.

14. Use the function $f(x) = \dfrac{x}{2} + 1$ to:
  a. Complete a table for this function with input values of $x = -2, -1, 0, 1, 2$ using only a pencil and paper.
  b. Describe the change in y for each 1-unit increase in x.
  c. Plot these points and sketch the line through these points.
  d. Describe the behavior of the line for every 1-unit move to the right.

In Exercises 15–18, sketch the graph of each linear function.

15. $f(x) = -2x + 5$ **16.** $f(x) = 3x - 5$

17. $f(x) = \dfrac{5}{3}x - 4$ **18.** $f(x) = -\dfrac{2}{5}x + 3$

Objective 3 Write the Equation of the Line Through Given Points

In Exercises 19–22, determine the equation of the line in the form $f(x) = mx + b$ that passes through each pair of points.

19. $(0, -2)$ and $(4, 7)$ **20.** $(0, 3)$ and $(-2, 5)$

21. $(5, -1)$ and $(2, -6)$ **22.** $(-2, -1)$ and $(5, -4)$

In Exercises 23–26, all the points listed in the table lie on the same line. Determine:

a. The slope of the line.
b. The y-intercept of the line.
c. The equation of the line in the form $f(x) = mx + b$.

23.

x	y
-1	5
0	3
1	1
2	-1
3	-3

24.

x	y
-3	-2
-2	1
-1	4
0	7
1	10

25.

x	y
7	-6
11	1
15	8
19	15
23	22

26.

x	y
-9	9
-7	6
-5	3
-3	0
-1	-3

In Exercises 27–30, use the information displayed in the graph to determine:

a. The slope of the line.
b. The y-intercept of the line.
c. The equation of the line in the form $f(x) = mx + b$.

27.

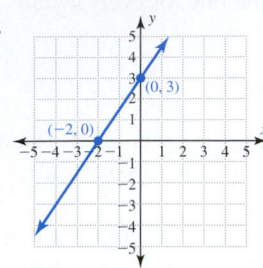

28.

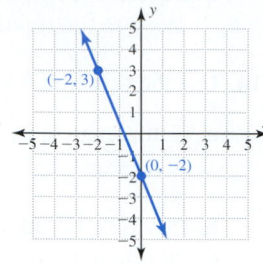

29.

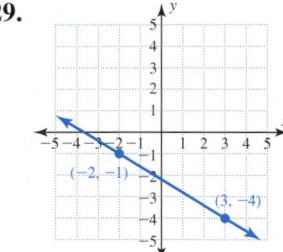

30.

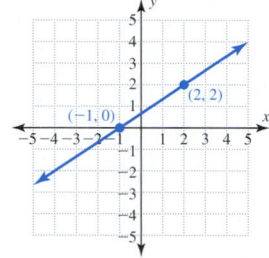

Objective 4 Determine the Intercepts of a Line

In Exercises 31 and 32, use the graph to determine the intercepts of each linear function.

31.

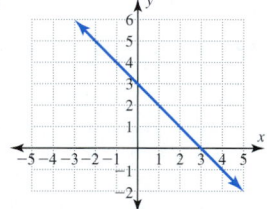

32.

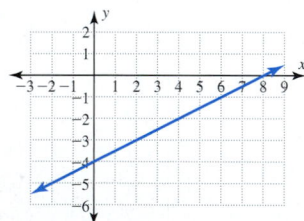

In Exercises 33 and 34, use the table to determine the intercepts of the line through these points.

33.

x	y
−2	12
−1	8
0	4
1	0
2	−4
3	−8
4	−12

34.

x	y
−3	0
−2	5
−1	10
0	15
1	20
2	25
3	30

In Exercises 35–38, determine the intercepts of the graph of each linear function. Then use the intercepts to sketch the graph. Select a scale for your graph that will display both intercepts.

35. $f(x) = 2x - 4$

36. $f(x) = -3x + 6$

37. $f(x) = -\dfrac{2}{3}x + 12$

38. $f(x) = \dfrac{5}{3}x - 10$

Objective 5 Determine the x-values for Which a Linear Function Is Positive and the x-values for Which a Linear Function Is Negative

In Exercises 39–44, determine the x-values for which each linear function is positive and the x-values for which it is negative.

39.

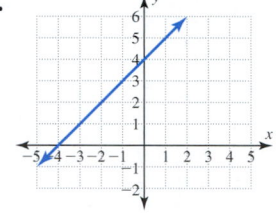

40.

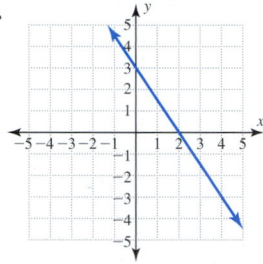

41.

x	y
−3	−6
0	−4
3	−2
6	0
9	2
12	4
15	6

42.

x	y
−4	3
−2	2
0	1
2	0
4	−1
6	−2
8	−3

43. $f(x) = -10x + 500$

44. $f(x) = 20x - 300$

Review and Concept Development

For Exercises 45–48, recall from Section 3.1 that for nonvertical lines, parallel lines have the same slope and perpendicular lines have slopes that are opposite reciprocals.

45. Write in slope-intercept form the equation of the line passing through $(2, -3)$ and parallel to $y = \dfrac{3}{5}x + 1$.

46. Write in slope-intercept form the equation of the line passing through $(2, -3)$ and parallel to $y = -\dfrac{2}{3}x - 5$.

47. Write in slope-intercept form the equation of the line passing through $(2, -3)$ and perpendicular to
$y = -\dfrac{2}{3}x - 5$.

48. Write in slope-intercept form the equation of the line passing through $(2, -3)$ and perpendicular to
$y = \dfrac{3}{5}x + 1$.

In Exercises 49–56, sketch the graph of each line.

49. $x = 3$ **50.** $x = -2$

51. $y = -4$ **52.** $y = 5$

53. $3x + 5y = 30$ **54.** $4x - 3y = 12$

55. $y - 2 = -3(x + 4)$ **56.** $y + 1 = 5(x - 2)$

57. a. Graph the first five terms of $a_n = -n + 5$.
(*Hint:* See Example 6 in Section 1.7.)
   b. For the linear function $f(x) = -x + 5$, plot the points with x-coordinates of 1, 2, 3, 4, and 5, and then sketch the line through these points.

58. a. Graph the first five terms of $a_n = n - 4$. (*Hint:* See Example 6 in Section 1.7.)
   b. For the linear function $f(x) = x - 4$, plot the points with x-coordinates of 1, 2, 3, 4, and 5, and then sketch the line through these points.

59. Use the function $f(x) = 5x - 8$ to determine the missing input and output values.
   a. $f(12) = $ ____
   b. $f(x) = 12; x = $ ____

60. Use the function $f(x) = 3x + 7$ to determine the missing input and output values.
   a. $f(25) = $ ____
   b. $f(x) = 25; x = $ ____

61. Use the given table of values to determine the missing input and output values.

x	$f(x)$
0	-10
1	-8
6	2
8	6

   a. $f(6) = $ ____
   b. $f(x) = 6; x = $ ____

62. Use the given table of values to determine the missing input and output values.

x	$f(x)$
0	2
2	3
6	5
8	6

   a. $f(2) = $ ____
   b. $f(x) = 2; x = $ ____

63. Use the given graph to determine the missing input and output values.

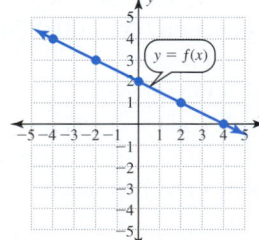

   a. $f(2) = $ ____
   b. $f(x) = 2; x = $ ____

64. Use the given graph to determine the missing input and output values.

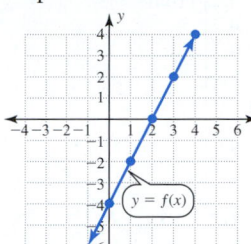

   a. $f(0) = $ ____
   b. $f(x) = 0; x = $ ____

Connecting Concepts to Applications

65. Health Care Costs A patient has an insurance policy that requires a 20% copayment by the patient for all medical expenses. Letting x represent the dollar cost of a medical bill, the function $C(x) = 0.20x$ gives the cost to the patient.
   a. Use this function to complete the table.

x	$C(x)$
0	
500	
1,000	
1,500	
2,000	
2,500	
3,000	

   b. Determine the cost to a patient if the medical bill is $1,000.
   c. Determine the cost of a medical bill if the cost to the patient is $425.

66. Health Care Costs A patient has an insurance policy that requires a 20% copayment by the patient for all medical expenses. Letting x represent the dollar cost of a medical bill, the function $C(x) = 0.80x$ gives the cost to the insurance company.
   a. Use this function to complete the table.

x	$C(x)$
0	
500	
1,000	
1,500	
2,000	
2,500	
3,000	

   b. Determine the cost to the insurance company if the medical bill is $1,000.
   c. Determine the cost of a medical bill if the cost to the insurance company is $2,250.

Profit and Loss Intervals

In Exercises 67 and 68, determine the profit and loss intervals for each of these profit functions. The x-variable represents the number of units of production, and the y-variable represents the profit generated by the sale of this production.

67.

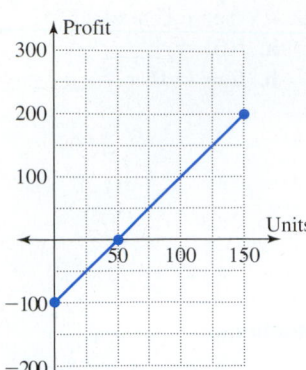

68.

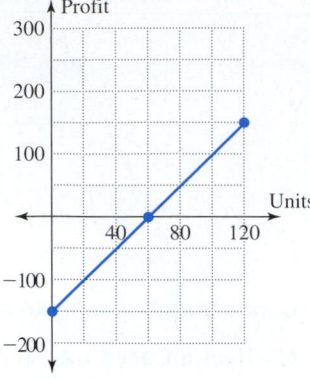

a. Use this table to determine the linear function $S(x) = mx + b$ for these data points.

b. Interpret the meaning of m in this problem.

c. Interpret the meaning of b in this problem.

d. Evaluate and interpret $S(23)$.

73. Depreciation of Carpentry Tools The value of a \$500 set of tools depreciates by \$50 per year.

a. Write a linear function V so that $V(x)$ gives the dollar value of these tools after x years.

b. Use this function to calculate the value of these tools after 7 years.

c. Determine the number of years before the value of the tools becomes \$50.

d. Use this function to complete the table.

x	V(x)
0	
1	
2	
3	
4	
5	
6	

69. Monthly Profit The function $f(x) = 50x - 500$ represents the monthly profit by a small business producing and selling x units of an item.

a. Algebraically determine the intercepts and interpret the meaning of these points.

b. Determine the profit and loss intervals for this profit function. Assume the practical domain is $[0, 30]$.

70. Monthly Profit The function $f(x) = 40x - 200$ represents the monthly profit by a small business producing and selling x units of an item.

a. Algebraically determine the intercepts and interpret the meaning of these points.

b. Determine the profit and loss intervals for this profit function. Assume the practical domain is $[0, 20]$.

71. Interest on a Savings Account The data points given in the table all lie on a line.

x	y = V(x)
Savings Account (\$)	Yearly Interest (\$)
500	20
750	30
1,000	40
1,250	50
1,500	60
1,750	70
2,000	80

a. Use this table to determine the linear function $V(x) = mx + b$ for these data points.

b. Interpret the meaning of m in this problem.

c. Interpret the meaning of b in this problem.

d. Evaluate and interpret $V(1,100)$.

72. Stretch by a Steel Spring The data points given in the table all lie on a line. Write the equation of this line.

x	y = S(x)
Mass on the Spring (kg)	Stretch by the Spring (cm)
5	0.4
10	0.8
15	1.2
20	1.6
25	2.0
30	2.4
35	2.8

74. Length of Hair The length of a human hair increases by approximately 1.3 cm per month. A young girl has a ponytail 20 cm long and wants a ponytail much longer.

a. Write a linear function L so that $L(x)$ gives the length in cm of her ponytail after x months.

b. Determine the length of her ponytail after 6 months.

c. Use this function to determine how many months it will take her to have a ponytail that is 50 cm long.

d. Use this function to complete the table.

x	L(x)
0	
6	
12	
18	
24	
30	
36	

Group discussion questions

75. Communicating Mathematically The monthly interest charge of 1.5% of the unpaid balance on a credit card is modeled by the function $f(x) = 0.015x$.

a. What does the input value x represent?

b. What does the output value $f(x)$ represent?

c. Describe the meaning of both x and $f(x)$ when $x = 100$.

d. Describe the meaning of both x and $f(x)$ when $f(x) = 100$.

76. Communicating Mathematically The function

$$f(x) = \frac{x}{4}$$ models the test average for a student after four exams.

a. What does the input value x represent?

b. What does the output value $f(x)$ represent?

c. Describe the meaning of both x and $f(x)$ when $x = 280$.

d. Describe the meaning of both x and $f(x)$ when $f(x) = 96$.

77. Communicating Mathematically

a. The function $f(t) = 100t + 300$ models the cost to rent and operate a bulldozer for t hours in a day. As you read this function, the story that it is telling you is

that it costs _____ dollars to rent this bulldozer even if it is not used this day. When it is used, each hour of usage costs an additional _____ dollars.

b. The function $P(x) = 25x - 200$ models the profit made by producing x glass tabletops per day in a factory. As you read this function, the story that it is telling you is that _____.

78. Discovery Question Use a graphing calculator to graph $f(x) = x^2$ and $f(x) = |x|$. Describe the shape of each of these graphs.

8.2 Cumulative Review

1. Add $\dfrac{5}{12} + \dfrac{1}{3}$ using pencil and paper.

2. Simplify $\dfrac{4x^8}{8x^4}$.

3. Multiply $-3(x + 4)(x - 1)$.

4. Factor $-2x^2 + 8x + 24$ completely.

5. Solve $(x - 2)(x + 3) = 6$.

Section 8.3 | Absolute Value Functions

Objectives:

1. Sketch the graph of an absolute value function.

2. Determine the intercepts of the graph of an absolute value function.

3. Determine the x-values for which an absolute value function is positive and the x-values for which an absolute value function is negative.

1. Sketch the Graph of an Absolute Value Function

We will now examine the family of absolute value functions. These functions are named for the algebraic equation that defines the function. Every absolute value function of the form $f(x) = |ax + b| + c$ or $f(x) = -|ax + b| + c$, where a, b, and c are real numbers, has a characteristic V-shape.

The **vertex** of the graph of an absolute value function is the point at the tip of the V-shape. If the V-shape opens upward, the vertex will be the lowest point on the graph and the y-value there will be the minimum y-value. If the V-shape opens downward, the vertex will be the highest point on the graph and the y-value there will be the maximum y-value. For an absolute value function, the points in a table and on a graph will form a symmetric pattern on both sides of the vertex.

Absolute Value Function

Algebraically	Numerically	Graphically	Verbally
$f(x) = |x|$	$\begin{array}{c|c} x & f(x) = |x| \\ \hline -3 & 3 \\ -2 & 2 \\ -1 & 1 \\ 0 & 0 \\ 1 & 1 \\ 2 & 2 \\ 3 & 3 \end{array}$		This V-shaped graph opens upward with the vertex of the V-shape at $(0, 0)$. The minimum y-value is 0. The domain, the projection of this graph onto the x-axis, is $D = \mathbb{R}$. The range, the projection of this graph onto the y-axis, is $R = [0, \infty)$. The points in the table and on the graph form a symmetric pattern on both sides of the vertex.

In Example 1, we analyze the behavior of the graph of the absolute value function $f(x) = -|x|$. We will first create a table of values for this function, and then we will graph the function by plotting these points.

Example 1   Analyzing the Graph of an Absolute Value Function

Use the function $f(x) = -|x|$ to complete each part of this example.

(a) Create a table using input values of $x = -3, -2, -1, 0, 1, 2, 3$ for this function.

(b) Plot these points and sketch a graph of $y = f(x)$.

(c) Does this graph open upward or downward?

(d) Determine the vertex.

(e) Determine the minimum or maximum y-value.

(f) Determine the domain of this function.

(g) Determine the range of this function.

Solution

(a)

x	$f(x) = -|x|$
-3	-3
-2	-2
-1	-1
0	0
1	-1
2	-2
3	-3

(b)

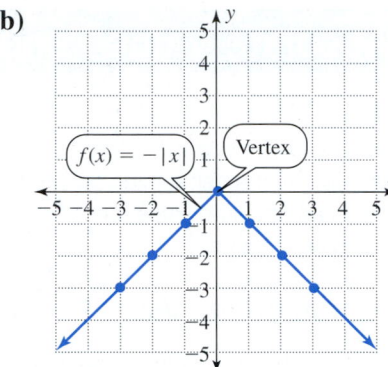

(c) This V-shaped graph opens downward.

(d) The vertex of the V-shape is at $(0, 0)$.

(e) The maximum y-value is 0. This value is located at the vertex.

(f) The domain, the projection of this graph onto the x-axis, is $D = \mathbb{R}$.

(g) The range, the projection of this graph onto the y-axis, is $R = (-\infty, 0]$.

Self-Check 1

Use the function $f(x) = -|x| + 3$ to complete each part.

a. Create a table using input values of $x = -3, -2, -1, 0, 1, 2, 3$ for this function.

b. Plot these points and sketch a graph of $y = f(x)$.

c. Does this graph open upward or downward?

d. Determine the vertex.

e. Determine the minimum or maximum y-value.

f. Determine the domain of this function.

g. Determine the range of this function.

Example 2 is very similar to Example 1. Although the function in Example 2 is slightly more complicated, we will analyze it in the same way.

Example 2	Analyzing the Graph of an Absolute Value Function

Use the function $f(x) = |x + 1| - 2$ to complete each part of this example.

(a) Create a table using input values of $x = -3, -2, -1, 0, 1, 2, 3$ for this function.

(b) Plot these points and sketch a graph of $y = f(x)$.

(c) Does this graph open upward or downward?

(d) Determine the vertex.

(e) Determine the minimum or maximum y-value.

(f) Determine the domain of this function.

(g) Determine the range of this function.

Solution

(a)

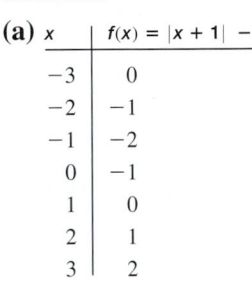

x	$f(x) = |x + 1| - 2$
-3	0
-2	-1
-1	-2
0	-1
1	0
2	1
3	2

(b)

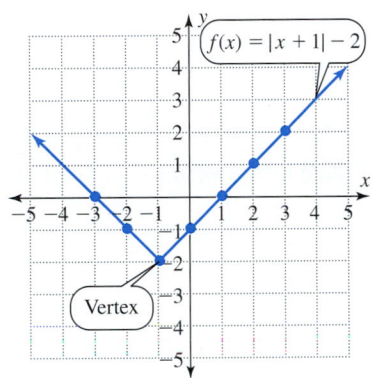

(c) This V-shaped graph opens upward.

(d) The vertex of the V-shape is at $(-1, -2)$.

(e) The minimum y-value is -2. This value is located at the vertex.

(f) The domain, the projection of this graph onto the x-axis, is $D = \mathbb{R}$.

(g) The range, the projection of this graph onto the y-axis, is $R = [-2, \infty)$.

Self-Check 2

Use the function $f(x) = -|x - 2| + 1$ to complete each part.

a. Create a table using input values of $x = -1, 0, 1, 2, 3, 4, 5$ for this function.

b. Plot these points and sketch a graph of $y = f(x)$.

c. Does this graph open upward or downward?

d. Determine the vertex.

e. Determine the minimum or maximum y-value.

f. Determine the domain of this function.

g. Determine the range of this function.

What Causes the Graph of an Absolute Value Function to Open Downward?

One observation we can make about Examples 1 and 2 is that if there is a negative sign in front of the absolute value symbols, then the graph will open downward. In general:

- Absolute value functions of the form $f(x) = |ax + b| + c$ will open upward.
- Absolute value functions of the form $f(x) = -|ax + b| + c$ will open downward.

This is a good opportunity to use a calculator or computer to graph several absolute value functions to verify this observation.

How Can I Quickly Determine the Location of the Vertex of an Absolute Value Function?

Another observation we can make about Examples 1 and 2 is that the vertex occurs at the x-value for which the expression inside the absolute value is equal to zero. In Example 1, the function is $f(x) = -|x|$. The vertex occurs at $x = 0$. In Example 2, the function is $f(x) = |x + 1| - 2$. The vertex occurs at $x = -1$, the x-value for which $x + 1 = 0$. We use this observation to determine the x-coordinate of the vertex in Example 3.

Example 3   Sketching the Graph of an Absolute Value Function

Sketch the graph of $f(x) = -|2x + 5| + 3$.

Solution

Determine the x-coordinate of the vertex:

$$2x + 5 = 0$$

$$x = -\frac{5}{2}$$

The x-coordinate of the vertex occurs at the x-value for which $2x + 5 = 0$.

Determine the y-coordinate of the vertex:

$$f\left(-\frac{5}{2}\right) = 3$$

To determine the y-coordinate of the vertex, evaluate the function for $x = -\frac{5}{2}$.

The vertex is located at $\left(-\frac{5}{2}, 3\right)$.

Complete a table of values using x-values on both sides of the vertex.

x	$f(x) = -\lvert 2x + 5 \rvert + 3$
-4	0
-3	2
$-\dfrac{5}{2}$	3
-2	2
-1	0

This can be done using pencil and paper or by using a calculator or a spreadsheet.

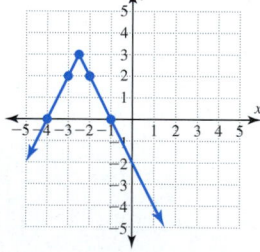

Plot the points from the table. Note the symmetric pattern formed by these points.

Self-Check 3

Sketch the graph of $f(x) = |2x - 3| - 3$.

A summary of the procedure we use to sketch the graph of an absolute value function is given in the following box.

Sketching the Graph of an Absolute Value Function

1. Determine whether the graph opens upward or opens downward.

2. Determine the location of the vertex.

3. Complete a table of values using x-values on both sides of the vertex.

To evaluate a function for a specific value of x, recall that we substitute the input value for each occurrence of the variable. In Technology Perspective 2.2.3 on **Changing the Table Settings,** we showed how to adjust the table settings to evaluate a function for input values having a constant change from one term to the next. A calculator or a spreadsheet can also be used to evaluate functions for input values not separated by a constant change. On a TI-84 Plus calculator, this is done by changing the calculator to the **ASK** mode. Spreadsheets are automatically set up to accept input values that are not evenly spaced. This is illustrated in Technology Perspective 8.3.1.

Technology Perspective 8.3.1 | **Selecting Input Values for a Function**

Evaluate the function $f(x) = -|2x + 5| + 3$ from Example 3 using $-4, -3, -\dfrac{5}{2}, -2,$ and -1 as input values.

TI-84 Plus Calculator

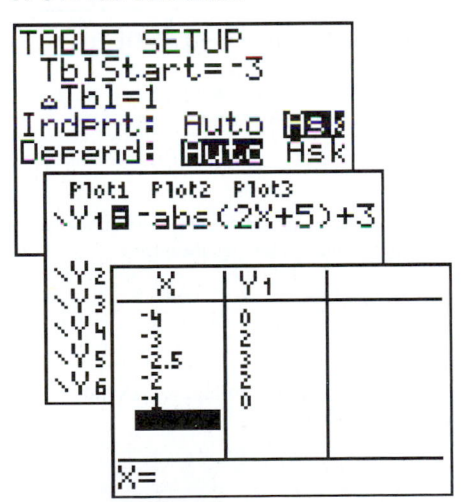

Excel Spreadsheet

	A	B	C
		fx	=-ABS(2*A2+5)+3
1	x	f(x) = -abs(2x+5)+3	
2	-4	0	
3	-3	2	
4	-2.5	3	
5	-2	2	
6	-1	0	

1. Press [2nd] [TBLSET] to edit the table setup as shown so that the Indpnt: option is set to **ASK.**

2. Press [Y=] and enter the function Y_1.

3. Press [2nd] [TABLE] and input the x-values to be evaluated.

1. Enter the desired x-values in column A.

2. Enter the formula "$= -\text{abs}(2*A2+5)+3$" in cell **B2** and copy this formula through column B.

3. This formula is used to determine the y-values shown in column B.

Technology Self-Check 1

Evaluate the function $f(x) = |5x + 3| - 2$ using $-2, -1, -\dfrac{3}{5}, 0,$ and 1 as input values.

2. Determine the Intercepts of the Graph of an Absolute Value Function

The following discussion and examples review absolute value equations and inequalities and enrich these topics by examining them from a graphical perspective. We will start by restating some key points from Section 4.4.

Solving Absolute Value Equations and Inequalities

For any real numbers x and a and positive real number d:

Absolute Value Expression	Verbally	Graphically	Equivalent Expression
$\lvert x - a \rvert = d$	x is either d units left or right of a.	(number line: $a-d$, a, $a+d$)	$x - a = -d$ or $x - a = +d$
$\lvert x - a \rvert < d$	x is less than d units from a.	(number line: $a-d$, a, $a+d$)	$-d < x - a < +d$
$\lvert x - a \rvert > d$	x is more than d units from a.	(number line: $a-d$, a, $a+d$)	$x - a > d$ or $x - a < -d$

Similar statements can also be made about the order relations less than or equal to ($\leq$) and greater than or equal to ($\geq$).

Can I Determine the Intercepts of an Absolute Value Function Without First Graphing the Function?

Yes, we can determine x- and y-intercepts using the same strategy used with linear functions. To find the x-intercept(s), substitute 0 for y and solve for x. To find the y-intercept, substitute 0 for x and solve for y. In Example 4, we algebraically determine the intercepts of an absolute value function.

Example 4   Determining the Intercepts of the Graph of an Absolute Value Function

Algebraically determine the intercepts of the graph of $f(x) = \lvert x - 3 \rvert - 5$.

Solution

x-intercepts:

$y = \lvert x - 3 \rvert - 5$

$0 = \lvert x - 3 \rvert - 5$

$5 = \lvert x - 3 \rvert$

$x - 3 = -5$ or $x - 3 = 5$

$x = -2$ or $x = 8$

To find the x-intercepts, substitute 0 for y and solve the resulting absolute value equation for x.

The expression inside the absolute value symbols can be either -5 or $+5$.

The x-intercepts are $(-2, 0)$ and $(8, 0)$.

y-intercept:

$y = \lvert x - 3 \rvert - 5$

$y = \lvert 0 - 3 \rvert - 5$

$y = \lvert -3 \rvert - 5$

$y = 3 - 5$

$y = -2$

To find the y-intercept, substitute 0 for x and solve for y.

The y-intercept is $(0, -2)$.

Self-Check 4

Determine the intercepts of the graph of $f(x) = \lvert x + 4 \rvert - 2$.

Examples of the graphs of some absolute value functions are shown here. It is worth pointing out that all absolute value functions will have a y-intercept, but the number of x-intercepts will vary. Some absolute value functions will have two x-intercepts, some will have one x-intercept, and some will have no x-intercepts.

Two x-intercepts   **One x-intercept**   **Zero x-intercepts**

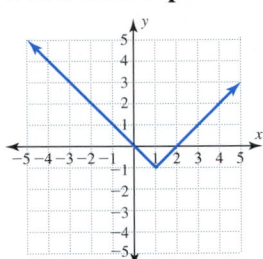   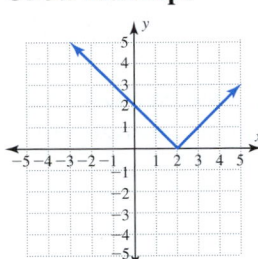   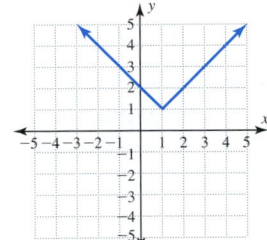

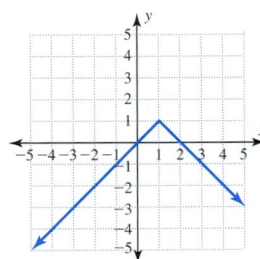   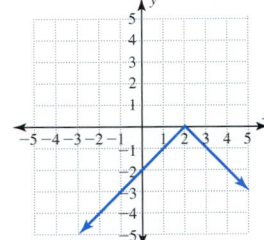   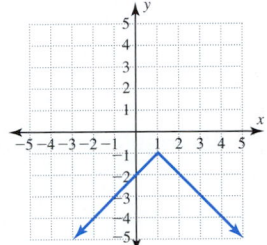

3. Determine the x-values for Which an Absolute Value Function Is Positive and the x-values for Which an Absolute Value Function Is Negative

Recall from Section 8.2 that a function $y = f(x)$ is positive when the output value $f(x)$ is positive, and a function $y = f(x)$ is negative when the output value $f(x)$ is negative.

Example 5	Determining Where a Function Is Positive and Where a Function Is Negative

Determine the x-values for which $f(x) = |x - 1| - 2$ is positive and the x-values for which this function is negative.

(a) **Graphically:** Use the graph to solve this problem.

(b) **Numerically:** Use the table to solve this problem.

(c) **Algebraically:** Use inequalities to solve this problem.

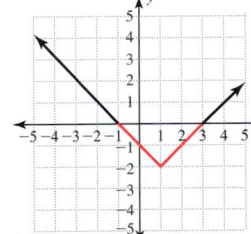

x	y
-3	2
-1	0
1	-2
3	0
5	2

Solution

Positive | Negative

(a) Graphically

The function is above the x-axis (in the black) to the left of the x-intercept $(-1, 0)$ or to the right of the x-intercept $(3, 0)$.

The function is below the x-axis (in the red) between the x-intercepts $(-1, 0)$ and $(3, 0)$.

(b) Numerically

The y-values in the table are positive for $x < -1$ or $x > 3$.

The y-values in the table are negative for $-1 < x < 3$.

(c) Algebraically

Solve $f(x) > 0$.

$$|x - 1| - 2 > 0$$
$$|x - 1| > 2$$
$$x - 1 < -2 \quad \text{or} \quad x - 1 > 2$$
$$x < -1 \quad \text{or} \quad x > 3$$

Solve $f(x) < 0$.

$$|x - 1| - 2 < 0$$
$$|x - 1| < 2$$
$$-2 < x - 1 < 2$$
$$-1 < x < 3$$

Answer: The function is positive for x-values in the interval $(-\infty, -1) \cup (3, \infty)$ and negative for x-values in the interval $(-1, 3)$. This is confirmed in all three perspectives given above.

Self-Check 5

Determine the x-values for which $f(x) = |x + 2| - 3$ is positive and the x-values for which this function is negative.

a. Graphically: Use the graph to solve this problem.

b. Numerically: Use the table to solve this problem.

c. Algebraically: Use inequalities to solve this problem.

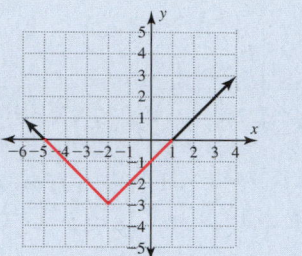

x	y
−6	1
−5	0
−2	−3
1	0
2	1

Self-Check Answers

1. a.

x	$f(x) = -|x| + 3$
−3	0
−2	1
−1	2
0	3
1	2
2	1
3	0

b.

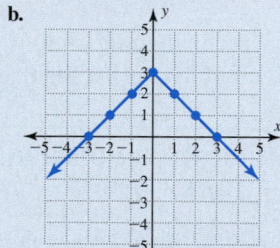

c. The graph opens downward.
d. The vertex is $(0, 3)$.
e. The maximum y-value is 3.
f. The domain is $D = \mathbb{R}$.
g. The range is $R = (-\infty, 3]$.

2. a.

x	$f(x) = -|x - 2| + 1$
−1	−2
0	−1
1	0
2	1
3	0
4	−1
5	−2

b.

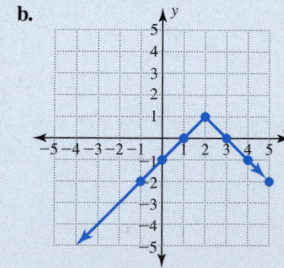

c. The graph opens downward.
d. The vertex is $(2, 1)$.
e. The maximum y-value is 1.
f. The domain is $D = \mathbb{R}$.
g. The range is $R = (-\infty, 1]$.

3.

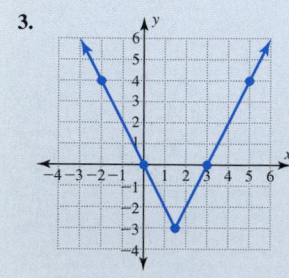

4. The x-intercepts are $(-6, 0)$ and $(-2, 0)$. The y-intercept is $(0, 2)$.

5. The function is positive for x-values in the interval $(-\infty, -5) \cup (1, \infty)$ and negative for x-values in the interval $(-5, 1)$.

Technology Self-Check Answer

1.

B2 ▾		*fx*	=ABS(5*A2+3)-2	
	A	B	C	
1	x	f(x)=abs(5x+3)-2		
2	-2	5		
3	-1	0		
4	-0.6	-2		
5	0	1		
6	1	6		

8.3 Using the Language and Symbolism of Mathematics

1. The graph of an absolute value function has a _____-shape.

2. The _____ of the graph of an absolute value function is the point at the tip of the V-shape.

3. The y-value of the vertex of an absolute value function is the _____ y-value if the graph opens downward.

4. The y-value of the vertex of an absolute value function is the _____ y-value if the graph opens upward.

5. The graph of an absolute value function of the form $f(x) = |ax + b| + c$ will open _____.

6. The graph of an absolute value function of the form $f(x) = -|ax + b| + c$ will open _____.

8.3 Quick Review

1. Evaluate $-|3 - 8|$.

2. Evaluate $-|8 - 3|$.

3. Use interval notation to represent the real numbers that satisfy $-3 < x < 3$.

4. Solve $-9 < 2x + 1 < 9$. Give your answer using interval notation.

5. Solve the compound inequality $5x - 2 \leq -10$ or $5x - 2 \geq 10$. Give your answer using interval notation.

8.3 Exercises

Objective 1 Sketch the Graph of an Absolute Value Function

In Exercises 1–4, use the absolute value function $f(x) = |2x - 1| - 3$ to evaluate each expression.

1. $f(0)$
2. $f(-3)$
3. $f(-4)$
4. $f(5)$

In Exercises 5–8, determine whether the graph of each absolute value function opens upward or opens downward.

5. $f(x) = |x - 1| - 3$
6. $f(x) = -|x + 3| - 1$
7. $f(x) = -|x - 4| + 2$
8. $f(x) = |x + 5| + 1$

In Exercises 9–12, match each function with its graph.

9. $f(x) = |x + 1| - 2$
10. $f(x) = |x - 2| + 3$
11. $f(x) = |x - 1| + 2$
12. $f(x) = |x + 2| - 3$

A.

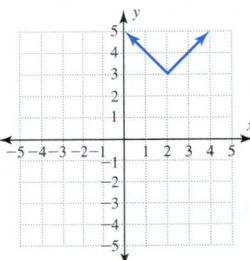

B.

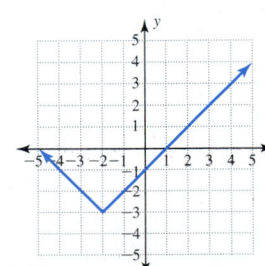

C.

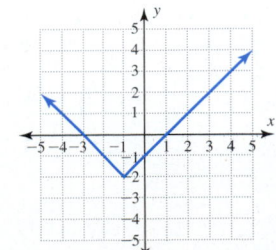

D.

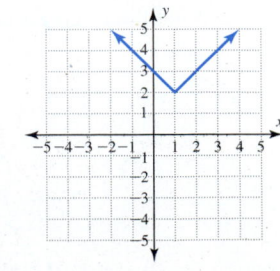

In Exercises 13–16, complete the given table of values and then use these points to sketch the graph of each absolute value function.

x	−3	−2	−1	0	1	2	3
y							

13. $f(x) = |x + 2|$ **14.** $f(x) = |x| + 2$

15. $f(x) = |x| - 1$ **16.** $f(x) = |x - 1|$

In Exercises 17 and 18, use the graph of each absolute value function to determine the following.

  a. Vertex of the graph
  b. Minimum y-value of the function
  c. The x-value at which the minimum y-value occurs

17.   **18.**

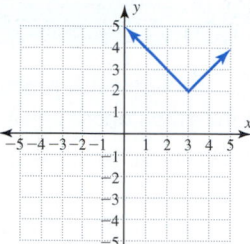   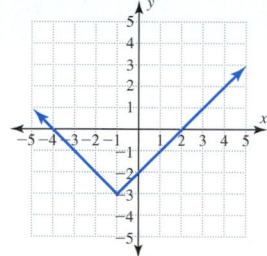

In Exercises 19 and 20, use the graph of each absolute value function to determine:

  a. Vertex of the graph
  b. Maximum y-value of the function
  c. The x-value at which the maximum y-value occurs

19.   **20.**

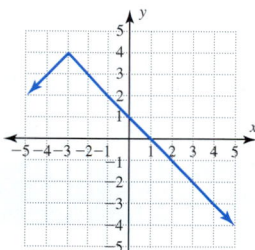   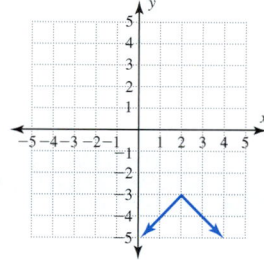

In Exercises 21–24, use the given absolute value function to complete each part.

  a. Complete the table of values for this function.

x	−3	−2	−1	0	1	2	3
y							

  b. Plot these points, and sketch a graph of $y = f(x)$.
  c. Does this graph open upward or downward?
  d. Determine the vertex.
  e. Determine the minimum or maximum y-value.
  f. Determine the domain of this function.
  g. Determine the range of this function.

21. $f(x) = -|x - 1| + 3$ **22.** $f(x) = |x + 1| - 3$

23. $f(x) = |x + 2| - 4$ **24.** $f(x) = -|x - 2| + 4$

In Exercises 25–28, sketch a graph of each absolute value function and then determine the domain and range of the function.

25. $f(x) = |2x - 3| - 1$ **26.** $f(x) = |4x + 3| - 3$

27. $f(x) = -|3x + 2| + 5$ **28.** $f(x) = -|2x + 1| + 4$

Objective 2 Determine the Intercepts of the Graph of an Absolute Value Function

In Exercises 29 and 30, determine the x- and y-intercepts of the graph of each absolute value function.

29.   **30.**

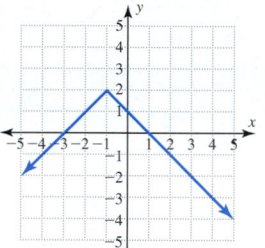   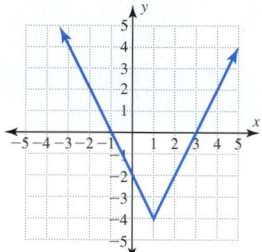

In Exercises 31 and 32, use the table for each absolute value function to determine the x- and y-intercepts of the graph of the function.

31.

x	y
−3	0
−2	−1
−1	−2
0	−1
1	0
2	1
3	2

32.

x	y
−2	−2
0	−1
2	0
4	1
6	0
8	−1
10	−2

In Exercises 33–36, algebraically determine the x- and y-intercepts of the graph of each absolute value function.

33. $f(x) = |x - 3| - 7$ **34.** $f(x) = |x + 2| - 5$

35. $f(x) = -|x + 1| + 7$ **36.** $f(x) = -|x - 3| + 4$

Objective 3 Determine the x-values for Which an Absolute Value Function Is Positive and the x-values for Which an Absolute Value Function Is Negative

In Exercises 37–40, use the graph of each absolute value function to determine the interval of x-values for which the function is positive and the interval of x-values for which the function is negative.

37.   **38.**

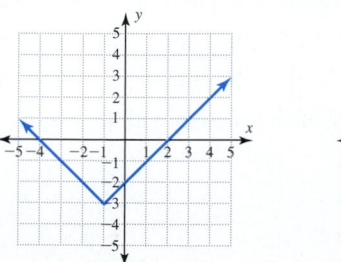   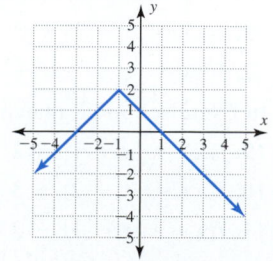

39.

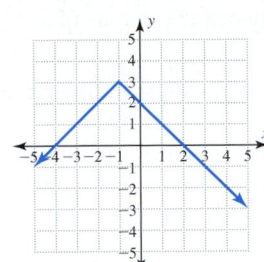

40.

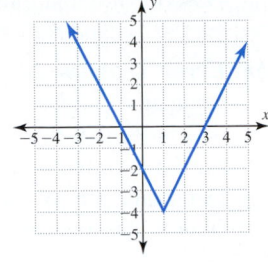

In Exercises 41–44, use the table for each absolute value function to determine the interval of x-values for which the function is positive and the interval of x-values for which the function is negative.

41.

x	y
-5	-1
-4	0
-3	1
-2	0
-1	-1
0	-2
1	-3

42.

x	y
-1	2
0	1
1	0
2	-1
3	0
4	1
5	2

43.

x	y
-4	-1
-3	0
-2	1
-1	2
0	1
1	0
2	-1

44.

x	y
-7	1
-6	0
-5	-1
-4	-2
-3	-1
-2	0
-1	1

In Exercises 45–48, algebraically determine the interval of x-values for which the function is positive and the interval of x-values for which the function is negative.

45. $f(x) = |x + 5| - 7$ **46.** $f(x) = |x - 3| - 5$

47. $f(x) = -|x + 2| + 3$ **48.** $f(x) = -|x - 2| + 4$

Review and Skill Development

In Exercises 49–52, solve each equation and inequality.

49. a. $|x - 2| = 8$ **b.** $|x - 2| < 8$ **c.** $|x - 2| \geq 8$

50. a. $|x + 4| = 7$ **b.** $|x + 4| \leq 7$ **c.** $|x + 4| > 7$

51. a. $|3x - 1| = 5$ **b.** $|3x - 1| \leq 5$ **c.** $|3x - 1| > 5$

52. a. $|2x + 3| = 7$ **b.** $|2x + 3| \leq 7$ **c.** $|2x + 3| > 7$

In Exercises 53 and 54, use each function to solve each equation and inequality.

53. $f(x) = |x + 5| - 1$

   a. $f(x) = 10$ **b.** $f(x) < 10$ **c.** $f(x) \geq 10$

54. $f(x) = |x - 4| + 2$

   a. $f(x) = 6$ **b.** $f(x) \leq 6$ **c.** $f(x) > 6$

55. Use the given table of values to determine the missing input and output values.

x	$f(x)$
-4	-2
-3	-1
-2	0
-1	1
0	0
1	-1
2	-2

   a. $f(-3) = $ _____
   b. $f(-2) = $ _____
   c. $f(x) = -2; x = $ _____ and $x = $ _____
   d. $f(x) = 0; x = $ _____ and $x = $ _____

56. Use the given table of values to determine the missing input and output values.

x	$f(x)$
1	5
2	4
3	3
4	2
5	3
6	4
7	5

   a. $f(4) = $ _____
   b. $f(5) = $ _____
   c. $f(x) = 4; x = $ _____ and $x = $ _____
   d. $f(x) = 5; x = $ _____ and $x = $ _____

57. Use the given graph to determine the missing input and output values.

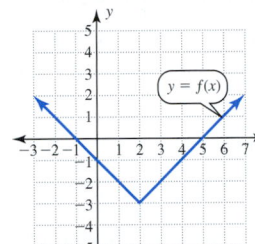

   a. $f(-2) = $ _____
   b. $f(2) = $ _____
   c. $f(x) = -2; x = $ _____ and $x = $ _____
   d. $f(x) = 2; x = $ _____ and $x = $ _____

58. Use the given graph to determine the missing input and output values.

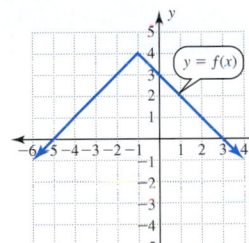

   a. $f(0) = $ _____
   b. $f(2) = $ _____
   c. $f(x) = 0; x = $ _____ and $x = $ _____
   d. $f(x) = 2; x = $ _____ and $x = $ _____

59. Use the function $f(x) = |x - 8| - 4$ to determine the missing input and output values.

   a. $f(0) = $ _____
   b. $f(3) = $ _____
   c. $f(x) = 3; x = $ _____ and $x = $ _____
   d. $f(x) = 12; x = $ _____ and $x = $ _____

60. Use the function $f(x) = |x + 2| - 5$ to determine the missing input and output values.

   a. $f(0) = $ _____
   b. $f(3) = $ _____
   c. $f(x) = -4; x = $ _____ and $x = $ _____
   d. $f(x) = 3; x = $ _____ and $x = $ _____

Connecting Concepts to Applications

61. Rolling a Pair of Dice When rolling a pair of dice, the probability that the sum of the numbers on those dice is x is given by $P(x) = \dfrac{6 - |7 - x|}{36}$, where x can be any integer from 2 to 12. Evaluate and interpret each of these expressions.

   a. $P(2)$ **b.** $P(7)$ **c.** $P(12)$

d. List each value in the domain of this function.

e. Which has a greater probability, a sum of 2 or a sum of 7?

62. Electrical Voltage The voltage in an electrical circuit designed by a beginning electronics class is a function of time. This function is given by $V(t) = -120|2t - 1| + 120$, where t is the time in seconds and $V(t)$ is the voltage. Evaluate and interpret each of these expressions.

a. $V(0)$ **b.** $V(0.5)$ **c.** $V(1.0)$

Group discussion questions

63. Discovery Question Each of these equations and inequalities has either no solution or all real numbers as its solution. Explain how you can determine each solution by inspection.

a. $|x - 3| = -2$

b. $|x + 1| < -5$

c. $|x - 3| + 8 > 2$

64. Calculator Discovery Use a graphing calculator to graph each function: $f(x) = |x|, f(x) = |x - 1|,$

$f(x) = |x - 2|, f(x) = |x - 3|$, and $f(x) = |x - 4|$. Describe the relationship between the graphs of $f(x) = |x|$ and $f(x) = |x - c|$.

65. Error Analysis A student produced the following calculator window for the function shown and concluded that the function was a linear function and the range of the function is the set of all real numbers. Determine the correct analysis and the source of the error in this answer.

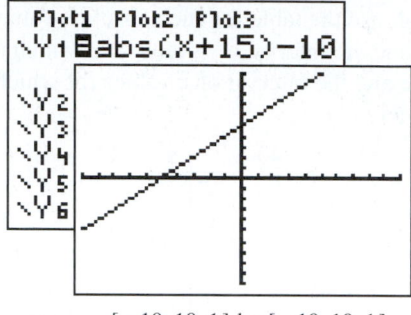

$[-10, 10, 1]$ by $[-10, 10, 1]$

8.3	**Cumulative Review**

1. Simplify $\sqrt{24}$ and give a decimal approximation for $\sqrt{24}$ to the nearest hundredth.

2. Solve $x^2 = 8$.

3. Solve $3x^2 - 5x - 7 = 0$.

4. Construct a quadratic equation with $x = -2$ and $x = 7$ as solutions.

5. Simplify $\dfrac{4 + 5i}{2 - i}$. Write your answer in $a + bi$ form.

Section 8.4	**Quadratic Functions**

Objectives:

1. Distinguish between linear and quadratic functions.

2. Determine the vertex of a parabola.

3. Sketch the graph of a quadratic function and determine key features of the resulting parabola.

4. Solve problems involving a maximum or minimum value.

Parabolic shapes are all around us in our daily lives. The photo illustrates the parabolic path of a stream of water. The path of a thrown ball will be parabolic (assuming that air resistance is negligible). To consider the design and behavior of ejection seats, the National Aeronautics and Space Administration examined parabolic paths for the seats after they blast away from an aircraft. Many manufactured products display a parabolic shape. Some cables supporting bridges hang in a parabolic shape, and parabolic reflectors are used for spotlights, on microphones at sporting events, and in satellite receiver dishes.

All of the functions graphed here are quadratic functions, and the graph of each is a parabola.

Examples of Quadratic Functions

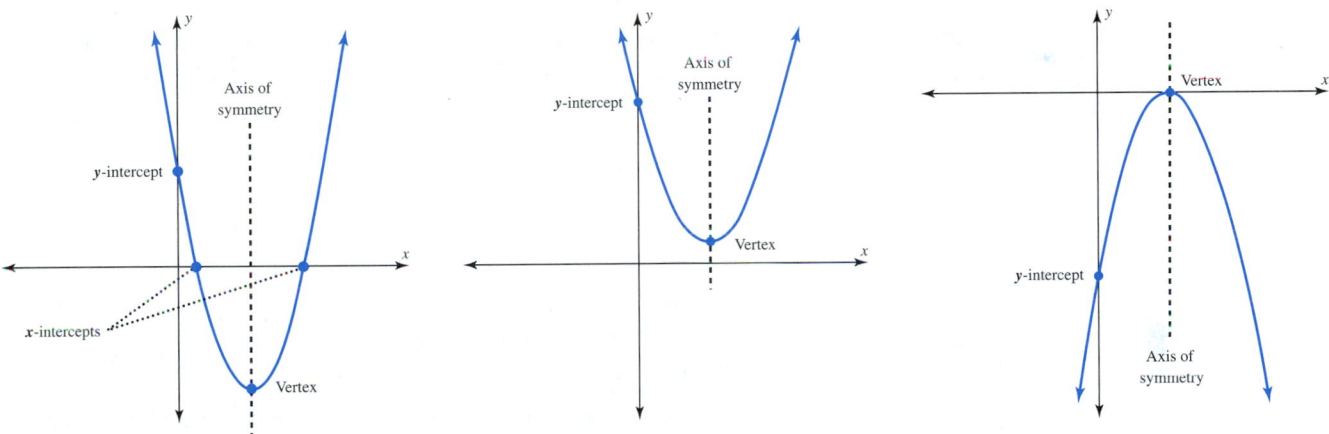

A second-degree polynomial function can be written in the form $f(x) = ax^2 + bx + c$ and is called a **quadratic function.** The term ax^2 is called the **quadratic** or the **second-degree term,** bx is called the **linear term,** and c is called the **constant term.** The graph of a quadratic function of this form is always a **parabola** that is either opening upward or opening downward. The **vertex** of a parabola opening upward is the lowest point on the parabola. The vertex of a parabola opening downward is the highest point on the parabola. Parabolas opening upward are called concave up and those opening downward are called concave down. The **axis of symmetry** of a parabola is the vertical line that passes through the vertex. The portion of the parabola to the left of the axis of symmetry is a mirror image of the portion to the right of this axis of symmetry. Quadratic functions have applications in many areas including the business applications given later in this chapter.

Both linear functions and quadratic functions are polynomial functions. Some characteristics of the graphs of polynomial functions are listed here. Note that the parabolas shown in the previous figure have these characteristics.

Characteristics of the Graphs of Polynomial Functions:

- Each polynomial function is continuous (there are no breaks or gaps between points on the graph).
- Each polynomial function is smooth (there are no sharp points or corners on the graph).
- The domain of each polynomial function is the set of all real numbers (the graph extends without limit both to the left and to the right).
- Some polynomial functions have relative (localized) maximum or minimum points (there are high or low points on the graph); there are more possibilities for these points for higher-degree polynomials.

1. Distinguish Between Linear and Quadratic Functions

In Section 8.2 we examined first-degree polynomial functions—functions that we refer to as linear functions. Linear functions have these characteristics that distinguish them from all other functions.

First-Degree Functions—Linear Functions

Algebraically	Numerically	Graphically
$f(x) = mx + b$	A fixed change in x produces a constant change in y.	A straight line
Example: $f(x) = 2x - 5$	*Example:* <table><tr><td>x</td><td>y</td></tr><tr><td>−1</td><td>−7</td></tr><tr><td>0</td><td>−5</td></tr><tr><td>1</td><td>−3</td></tr><tr><td>2</td><td>−1</td></tr><tr><td>3</td><td>1</td></tr></table>	*Example:*

In this section, we will examine second-degree polynomial functions and their distinctive characteristics.

Second-Degree Functions—Quadratic Functions

Algebraically	Numerically	Graphically
$f(x) = ax^2 + bx + c$	The y-values form a symmetric pattern about the vertex.	A parabola
Example: $f(x) = x^2 - 4x + 4$	*Example:* <table><tr><td>x</td><td>y</td></tr><tr><td>−1</td><td>9</td></tr><tr><td>0</td><td>4</td></tr><tr><td>1</td><td>1</td></tr><tr><td>2</td><td>0</td></tr><tr><td>3</td><td>1</td></tr><tr><td>4</td><td>4</td></tr><tr><td>5</td><td>9</td></tr></table>	*Example:*

What Causes a Parabola to Open Upward or Open Downward?

This is determined by the sign of a in the function $f(x) = ax^2 + bx + c$. If $a > 0$, the parabola will open upward as illustrated in Example 1 by $f(x) = x^2$. If $a < 0$, the parabola will open downward as illustrated in Example 1 by $f(x) = -x^2$. In Example 1, we use the defining equation to create a table of values, plot these points, and then smoothly connect them with a parabolic shape. This is a good procedure as long as you are able to locate the vertex in your table. We will develop a general procedure for graphing a parabola later in this section.

Example 1   Graphing a Quadratic Function

Create a table of values using inputs of $x = -3, -2, -1, 0, 1, 2, 3$ for each quadratic function, plot these points, and sketch the graph of the parabola through these points. Then determine the domain and the range of each function.

(a) $f(x) = x^2$ **(b)** $f(x) = -x^2$

Solution

Algebraically	Numerically	Graphically	Verbally

(a) $f(x) = x^2$

x	$f(x) = x^2$
-3	9
-2	4
-1	1
0	0
1	1
2	4
3	9

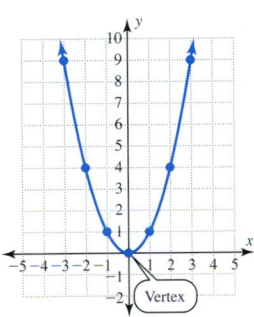

This parabola opens upward with a vertex of (0, 0). This function is defined for all real numbers. The domain $D = \mathbb{R}$. The range, the projection of the graph onto the y-axis, is $R = [0, \infty)$.

(b) $f(x) = -x^2$

x	$f(x) = -x^2$
-3	-9
-2	-4
-1	-1
0	0
1	-1
2	-4
3	-9

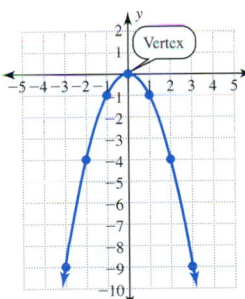

This parabola opens downward with a vertex of (0, 0). This function is defined for all real numbers. The domain $D = \mathbb{R}$. The range, the projection of the graph onto the y-axis, is $R = (-\infty, 0]$.

Self-Check 1

Create a table of values using inputs of $x = -3, -2, -1, 0, 1, 2, 3$ for $f(x) = x^2 - 2$, plot these points, and sketch the graph of the parabola through these points. Then determine the domain and the range of this function.

It is important to be able to predict the shape of a graph from the defining equation. Knowing the shape of a graph can also help us to determine an appropriate viewing window. Without this knowledge, we may select an inappropriate window that fails to show key features of the graph. When we are working with real applications, knowing the shape of the graph can help us select a model that fits the behavior that we are trying to describe and explain. In Example 2, we use the defining equation to describe the graph of each linear or quadratic function.

A Mathematical Note

Ian Stewart writes in *Nature's Numbers* (a 1995 book by Basic Books) that his "book ought to have been called *Nature's Numbers and Shapes*." Even though shapes can always be reduced to numbers, "it is often better to think of shapes that make use of our powerful visual capabilities."

Example 2 Identifying Lines or Parabolas

Identify the graph of each function as a line or a parabola. If the graph is a line, determine whether the slope is positive or negative. If the graph is a parabola, determine whether the parabola opens upward or downward.

Solution

(a) $f(x) = 4x - 11$ Line with positive slope   This first-degree function is a linear function in the slope-intercept form $f(x) = mx + b$. The slope $m = 4$ is positive.

(b) $f(x) = -5x + 6$ Line with negative slope   This first-degree function is a linear function in the slope-intercept form $f(x) = mx + b$. The slope $m = -5$ is negative.

(c) $f(x) = -2x^2 + x - 9$ Parabola that opens downward   This second-degree function is a quadratic function in the form $f(x) = ax^2 + bx + c$.

Because $a = -2$, this parabola opens downward.

(d) $f(x) = 3x^2 - 5x - 1$ Parabola that opens upward   This second-degree function is a quadratic function in the form $f(x) = ax^2 + bx + c$.

Because $a = 3$, this parabola opens upward.

(e) $f(x) = 7$ Line with a slope of 0   The graph of this constant function will be the horizontal line with a y-intercept of $(0, 7)$. Thus the slope is neither positive nor negative; it is 0.

Self-Check 2

Identify the graph of each function as a line or a parabola. If the graph is a line, determine whether the slope is positive or negative. If the graph is a parabola, determine whether the parabola opens upward or downward.

1. $f(x) = 2x^2 - 1$ **2.** $f(x) = 2x - 1$ **3.** $f(x) = -x + 2$ **4.** $f(x) = -2x^2 + x$

2. Determine the Vertex of a Parabola

If a problem is modeled by a quadratic function, then this function will have a maximum or minimum value that occurs at the vertex of the parabola defined by this function. The ability to determine relative minima and maxima is one of the most powerful applications of algebraic models of problems. This ability allows us to pick among many possible options and select the one that has optimal results. For example, these optimal results could be minimum cost, minimum production of pollutants, maximum size, or maximum profit.

Maximum Value   **Minimum Value**

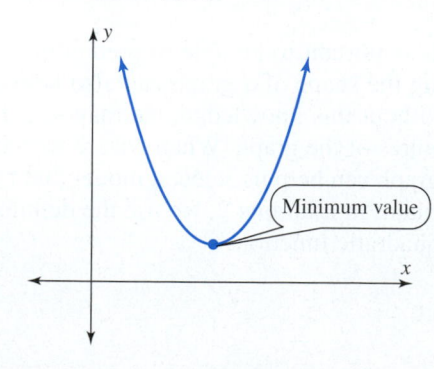

The symmetry of a parabola can be used to develop an algebraic method for determining its vertex. To start, note that the vertex lies on the axis of symmetry and that the x-intercepts (if there are any) are equally spaced on either side of this axis of symmetry.

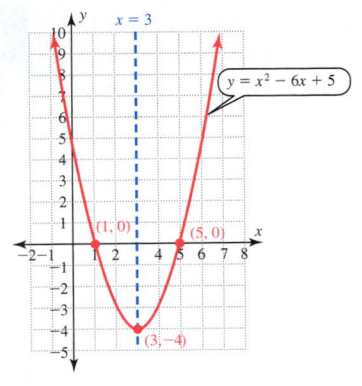

In the example shown, the x-intercepts $(1, 0)$ and $(5, 0)$ are both 2 units from the axis of symmetry. The axis of symmetry is the vertical line $x = 3$. The vertex $(3, -4)$ lies on the axis of symmetry.

The procedure that we just used to find the vertex of $y = x^2 - 6x + 5$ can be generalized to determine the vertex of any parabola, including those that do not have any x-intercepts. From the quadratic formula, a parabola with two x-intercepts will have these x-intercepts located at $\left(-\dfrac{b}{2a} - \dfrac{\sqrt{b^2 - 4ac}}{2a}, 0\right)$ and $\left(-\dfrac{b}{2a} + \dfrac{\sqrt{b^2 - 4ac}}{2a}, 0\right)$. The x-coordinate of the vertex is located midway between these x-intercepts. Adding $-\dfrac{b}{2a} - \dfrac{\sqrt{b^2 - 4ac}}{2a}$ and $-\dfrac{b}{2a} + \dfrac{\sqrt{b^2 - 4ac}}{2a}$

and dividing by 2, we obtain $-\dfrac{b}{2a}$. To determine the y-coordinate of the vertex, evaluate $f\left(-\dfrac{b}{2a}\right)$.

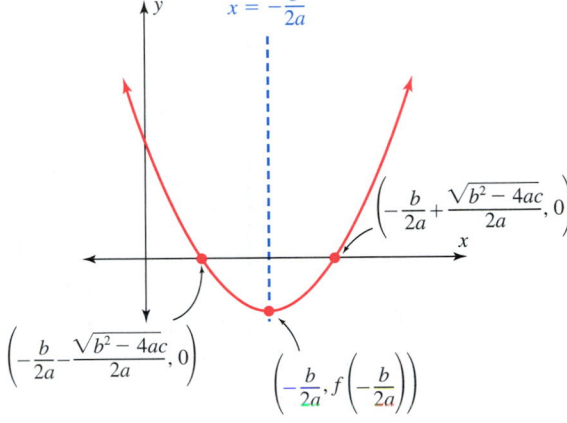

Vertex of the Parabola Defined by $f(x) = ax^2 + bx + c$

Algebraically	Numerically	Graphically
$\left(-\dfrac{b}{2a}, f\left(-\dfrac{b}{2a}\right)\right)$	The y-values form a symmetric pattern about the vertex. The y-coordinate of the vertex will be either the largest or the smallest y-value in the table.	The vertex is either the highest or the lowest point on the parabola.

Example:

$f(x) = -x^2 + 7x + 8$

$a = -1$ and $b = 7$

$x = -\dfrac{b}{2a} = -\dfrac{7}{2(-1)} = \dfrac{7}{2} = 3.5$

$f(3.5) = 20.25$

The vertex is $(3.5, 20.25)$.

Example:

x	y
-2	-10
-1	0
2	18
3.5	20.25
5	18
8	0
9	-10

Example:

A graph of $f(x) = -x^2 + 7x + 8$ with axis of symmetry $x = 3.5$, vertex $(3.5, 20.25)$, and x-intercepts $(-1, 0)$ and $(8, 0)$.

Example 3 — Determining the Vertex of a Parabola

Determine the vertex of the parabola defined by $f(x) = 2x^2 + 5x - 12$.

Solution

Calculate the x-coordinate of the vertex.

$$-\frac{b}{2a} = -\frac{5}{2(2)}$$

$$= -1.25$$

The x-coordinate of the vertex is $-\dfrac{b}{2a}$. For the function $f(x) = 2x^2 + 5x - 12$, $a = 2$ and $b = 5$.

Calculate the y-coordinate of the vertex.

$$f(x) = 2x^2 + 5x - 12$$

$$f\left(-\frac{b}{2a}\right) = f(-1.25)$$

$$= 2(-1.25)^2 + 5(-1.25) - 12$$

$$= -15.125$$

The y-coordinate of the vertex is $f\left(-\dfrac{b}{2a}\right)$.

Answer: Vertex is $(-1.25, -15.125)$.

You can use a graphing calculator to see if this answer seems reasonable.

Self-Check 3

Determine the vertex of the parabola defined by $f(x) = -2x^2 + 4x + 5$.

3. Sketch the Graph of a Quadratic Function and Determine Key Features of the Resulting Parabola

Can I Quickly Sketch the Graph of a Quadratic Function?

Yes. Just as we use the slope and y-intercept to quickly sketch a linear function, we can use key information to sketch the graph of a quadratic function. A summary of the procedure we use to sketch the graph of a quadratic function is given in the box. This procedure is illustrated in Example 4.

Sketching the Graph of a Quadratic Function

1. Determine whether the parabola opens upward or downward.
2. Determine the coordinates of the vertex.
3. Determine the intercepts.
4. Complete a table using points on both sides of the vertex.
5. Connect all points with a smooth parabolic shape.

Example 4 — Sketching the Graph of a Parabola

Determine the following information and use it to sketch the graph of the function $f(x) = 2x^2 + 5x - 12$.

(a) Does the parabola open upward or open downward?
(b) Determine the coordinates of the vertex.
(c) Determine the intercepts of the graph of this function.
(d) Complete a table of values using inputs on both sides of the vertex.
(e) Use this information to sketch the graph of this function.

Solution

(a) The graph of this function will open upward.   Since the coefficient of the quadratic term, $2x^2$, is positive, the graph will open upward.

(b) Vertex

The vertex is $(-1.25, -15.125)$.   The vertex of this parabola was found in Example 3.

(c) y-intercept

$$f(0) = -12$$

The y-intercept is $(0, -12)$.   The y-intercept is found by evaluating the function for $x = 0$.

x-intercept(s)

$$2x^2 + 5x - 12 = 0$$

$$(2x - 3)(x + 4) = 0$$

$$2x - 3 = 0 \quad \text{or} \quad x + 4 = 0$$

$$x = 1.5 \quad \text{or} \quad x = -4$$

The x-intercepts are $(1.5, 0)$ and $(-4, 0)$.   The x-intercepts are found by setting the function equal to zero and solving the resulting quadratic equation.

(d) Table

x	y
-4	0
-3	-9
-2	-14
-1.25	-15.125
-1	-15
0	-12
1.5	0

These input values can be evaluated by hand, but using a calculator or a spreadsheet to complete the table can make this process easier.

(e) Graph

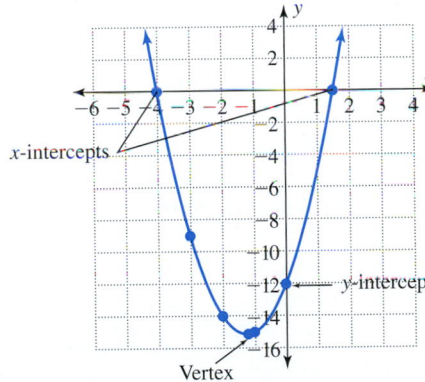

To sketch the graph, plot the intercepts and the points from the table, and connect them with a smooth parabolic shape.

Self-Check 4

Determine the following information and use it to sketch the graph of the function $f(x) = -5x^2 + 7x + 6$.

a. Does the parabola open upward or open downward?

b. Determine the coordinates of the vertex.

c. Determine the intercepts of the graph of this function.

d. Complete a table of values using inputs on both sides of the vertex.

e. Use this information to sketch the graph of this function.

4. Solve Problems Involving a Maximum or Minimum Value

Does the Vertex of a Parabola Have a Special Meaning?

Yes. As noted earlier in this section, the ability to determine the vertex of a parabola allows us to pick among many possible options and select the one that has optimal results. In Example 5, we examine a function that models the height of a ball as a function of time in order to determine the maximum height reached by the ball. This maximum height is represented by the vertex of the graph of this function.

Example 5	Maximum Height of a Ball

As part of a dynamic exhibit at a children's museum, hard plastic balls are released down a curved ramp. When the balls leave the end of the ramp, their height is then a function of time. The height of each ball in meters is given by $h(t) = -4.9t^2 + 10t + 4$, where t is given in seconds. Determine the maximum height of each ball and when this maximum height occurs.

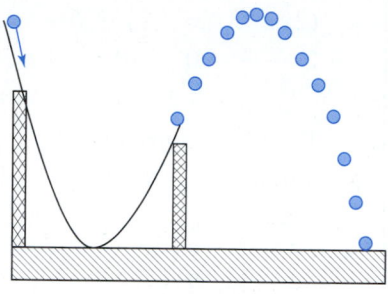

Solution

Calculate the t-coordinate of the vertex.

$$-\frac{b}{2a} = -\frac{10}{2(-4.9)}$$

$$\approx 1.02$$

Because the leading coefficient of $h(t) = -4.9t^2 + 10t + 4$ is negative, the parabola is concave downward. The maximum height will occur at the vertex of this parabola, $\left(-\frac{b}{2a}, h\left(-\frac{b}{2a}\right)\right)$. For this function $a = -4.9$ and $b = 10$.

Calculate the h-coordinate of the vertex (t, h).

$$h(t) = -4.9t^2 + 10t + 4$$

$$h\left(-\frac{b}{2a}\right) \approx h(1.02)$$

$$\approx -4.9(1.02)^2 + 10(1.02) + 4$$

$$\approx 9.10$$

The vertex is approximately $(1.02, 9.10)$.

Answer: The maximum height of approximately 9.1 m is reached after 1.02 sec.

Self-Check 5

The height of a golf ball in feet is given by $h(t) = -16t^2 + 104t$, where t is the number of seconds that have elapsed since the ball was hit. Determine the highest point that the ball reaches and when this maximum height occurs.

We have shown how to determine the coordinates of the vertex of a parabola using $\left(-\frac{b}{2a}, f\left(-\frac{b}{2a}\right)\right)$. Now we illustrate how to use a graphing calculator to approximate the coordinates of a relative maximum. The steps for approximating a relative minimum are nearly identical. Before reading Technology Perspective 8.4.1, enter $y = -4.9x^2 + 10x + 4$ from Example 5 into your calculator and select $[0, 3, 1]$ by $[-4, 12, 1]$ for the viewing window.

Technology Perspective 8.4.1	Calculating the Coordinates of a Relative Maximum

Approximate the coordinates of the vertex of $y = -4.9x^2 + 10x + 4$ from Example 5.

TI-84 Plus Keystrokes

1.

2. By inspection, the location of the maximum appears to be close to $x = 1$ (between $x = 0.5$ and $x = 1.5$).

3. The **maximum** feature under the **CALC** menu first asks for the left bound. Input 0.5.

4. We are then asked for the right bound. Input 1.5.

5. Finally, we are asked for a guess. Input 1.

Note: The x-inputs given above must be between 0 and 3 because of the window selected.

Answer: (1.02, 9.10)

The coordinates of the vertex have been rounded to the nearest hundredth.

TI-84 Plus Screens

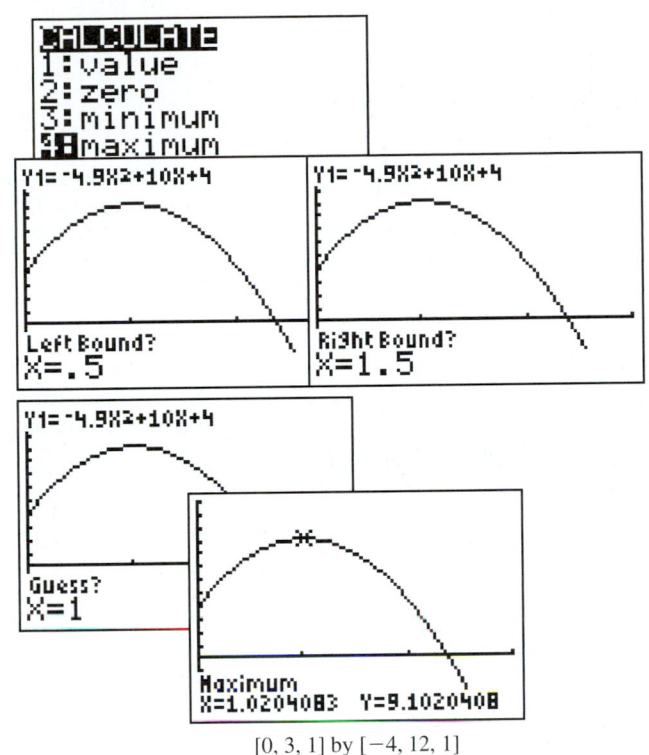

[0, 3, 1] by [−4, 12, 1]

Technology Self-Check 1

Use a graphing calculator to determine to the nearest hundredth the coordinates of the vertex of the parabola defined by $y = -1.28x^2 + 3.54x + 7.29$.

In Example 6, we use a graphing calculator to analyze a quadratic function. Although we have already developed the skills to perform each of these tasks using pencil and paper, we would usually use technology with a problem like this in the workplace.

Example 6   Analyzing a Quadratic Model for Profits

As part of his diversified production, a farmer can harvest and sell up to 5,000 pounds of catfish per week. Each week he has overhead costs, and when he increases his harvest, he must hire extra help and use additional equipment. All these factors are accounted for in the following function that gives his weekly profit in dollars as a function of the number of pounds x that he sells. His weekly profit is modeled by the function $P(x) = -0.0001x^2 + 0.6x - 500$.

a. Determine $P(0)$ and interpret this value.

b. Use a calculator to determine the x-intercepts of the graph of this function and interpret these values.

c. Determine the profit interval for this function.

d. Use a calculator to determine the vertex of this graph and interpret the meaning of both coordinates.

Solution

(a) $P(x) = -0.0001x^2 + 0.6x - 500$

$P(0) = -0.0001(0)^2 + 0.6(0) - 500$

$P(0) = -500$

If he sells 0 lb, he will lose $500.

Evaluate $P(0)$ by substituting 0 for x. Because of overhead costs, he will lose money if he does not sell any catfish.

The y-intercept of the graph is $(0, -500)$.

(b)

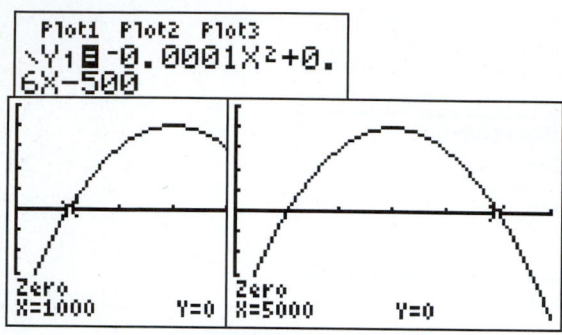

[0, 6,000, 1,000] by [−500, 500, 100]

Graph this function on a graphing calculator and then use the **Zero** feature on the **CALC** menu to calculate the x-value where y is zero in order to determine the x-intercepts of the function.

Note: We could have found these intercepts by solving $-0.0001x^2 + 0.6x - 500 = 0$.

The x-intercepts are $(1,000, 0)$ and $(5,000, 0)$. If the farmer sells 1,000 or 5,000 lb, he will break even.

(c) The farmer will make a profit with sales from 1,000 to 5,000 lb.

The graph is above the x-axis and the profit is positive for the x-values in the interval $(1,000, 5,000)$.

(d)

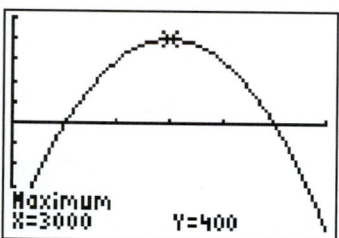

[0, 6,000, 1,000] by [−500, 500, 100]

Use the **Maximum** feature under the **CALC** menu on a graphing calculator to determine the maximum of this quadratic function.

Note: Since we already found the zeros to be 1,000 and 5,000, we could have found the x-coordinate of the vertex by determining the average of these zeros. We could also have used $x = -\dfrac{b}{2a}$ to find this x-coordinate.

The vertex is $(3,000, 400)$.

If the farmer sells 3,000 pounds of catfish, he will make $400. This is the maximum amount he can make in a week from his catfish sales.

Self-Check 6

Rework Example 6 assuming the weekly profit is modeled by the function $P(x) = -0.0001x^2 + 0.8x - 1,200$.

In Example 7, we will use a quadratic function to model a problem and then we will use this quadratic function to determine the maximum value of the function.

Example 7   Area Enclosed by a Fence

A homeowner plans to enclose a rectangular exercise area for her dog. One side of the playing area will be formed by the wall of an existing building, and the other three sides will be formed by using up to 16 meters of fencing that she has purchased. Write a function that models the area as a function of the width w. Then find the maximum area that she can enclose with this fencing.

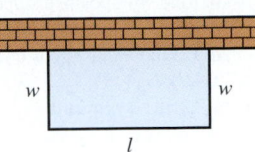

Solution

Let w = the width in meters of this rectangular area

$\qquad\quad l$ = the length in meters of this rectangular area

$16 - 2w$ = the length in meters of this rectangular area

The total fence available is 16 m. Thus

$$2w + l = 16$$
$$l = 16 - 2w$$

Area = (Length)(Width)

The word equation is based on the formula for the area of a rectangle.

$A = l \cdot w$

$A = (16 - 2w)(w)$

$A = -2w^2 + 16w$

Substitute the length and width into the word equation.

The area is a function of the width. To examine this function on a graphing calculator, use the variables x and y.

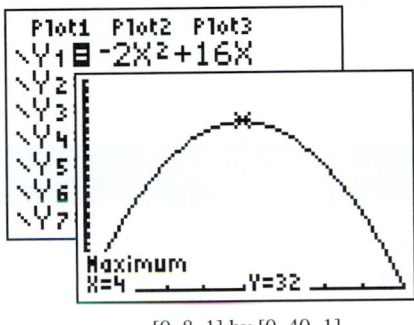

[0, 8, 1] by [0, 40, 1]

The vertex of the parabola defined by $y = -2x^2 + 16x$ is (4, 32).

Use a graphing calculator to determine the maximum of the function $y = -2x^2 + 16x$.

Note: We could also have calculated the vertex

$$\left(-\frac{b}{2a},\ f\left(-\frac{b}{2a}\right)\right)$$ to find this maximum.

The vertex identifies the maximum area that can be enclosed by 16 m of this fencing. In the ordered pair (4, 32), 4 is the width in meters and 32 is the area in square meters. When $w = 4$, the length is $l = 8$, as shown below.

$l = 16 - 2w$

$l = 16 - 2(4)$

$l = 16 - 8$

$l = 8$

Answer: The maximum area is 32 m². This area is obtained by using a width of 4 m and a length of 8 m.

Self-Check 7

Rework Example 7 assuming the homeowner has 20 m of fencing available.

Self-Check Answers

1. Numerically

x	$f(x) = x^2 - 2$
-3	7
-2	2
-1	-1
0	-2
1	-1
2	2
3	7

Graphically

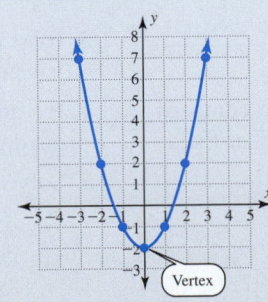

Verbally

This parabola opens upward with a vertex of (0, −2). This function is defined for all real numbers. The domain $D = \mathbb{R}$. The range, the projection of the graph onto the y-axis, is $R = [-2, \infty)$.

2. **a.** Parabola that opens upward
   b. Line with positive slope
   c. Line with negative slope
   d. Parabola that opens downward

3. Vertex is (1, 7).

4. **a.** Opens downward
   b. Vertex (0.7, 8.45)
   c. x-intercepts: (−0.6, 0) and (2, 0); y-intercept: (0, 6)
   d.

x	y
−1	−6
−0.6	0
0	6
0.7	8.45
1	8
2	0

   e.

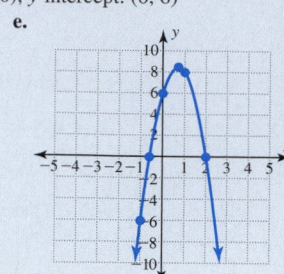

5. The maximum height of 169 ft is reached after 3.25 sec.

6. **a.** If he sells 0 lb, he will lose $1,200.
   b. The x-intercepts are (2,000, 0) and (6,000, 0). If the farmer sells 2,000 or 6,000 lb, he will break even.
   c. The farmer will make a profit with sales from 2,000 to 6,000 lb.
   d. If the farmer sells 4,000 pounds of catfish, he will make $400. This is the maximum amount he can make in a week from his catfish sales.

7. The maximum area is 50 m^2. This area is obtained by using a width of 5 m and a length of 10 m.

Technology Self-Check Answer

1. (1.38, 9.74)

8.4 Using the Language and Symbolism of Mathematics

1. A function of the form $f(x) = ax^2 + bx + c$ with $a \neq 0$ has a graph whose shape is a _____.

2. A function of the form $f(x) = ax^2 + bx + c$ with $a \neq 0$ is called a _____ function.

3. The highest point on a parabola that is concave downward is called the _____ of the parabola.

4. The vertex is the lowest point on a parabola that is concave _____.

5. A parabola defined by $f(x) = ax^2 + bx + c$ is symmetric about a vertical line passing through its vertex. This vertical line is called the _____ of _____.

6. In the function $f(x) = ax^2 + bx + c$, ax^2 is called the _____ term, bx is called the _____ term, and c is called the _____ term.

7. A function that is defined by a first-degree polynomial has a graph that is _____.

8. A parabola is defined by a function that is a _____-degree polynomial.

9. If a line is defined by $y = mx + b$ and $m > 0$, then the line slopes _____ to the right.

10. If a parabola is defined by $f(x) = ax^2 + bx + c$ and $a > 0$, then the parabola is concave _____.

11. If a parabola is defined by $f(x) = ax^2 + bx + c$ and $a < 0$, then the parabola is concave _____.

12. The x-coordinate of the vertex of a parabola defined by $f(x) = ax^2 + bx + c$ is given by _____.

13. The y-coordinate of the vertex of a parabola defined by $f(x) = ax^2 + bx + c$ is given by _____.

8.4 Quick Review

1. The equation $y = mx + b$ is known as the _____ _____ form of a line.

2. The fixed costs or the _____ costs for a company can include insurance, rent, electricity, and other fixed expenses when 0 units are produced.

3. If (x, y) lies in quadrant II, then x is _____ and y is positive.

4. If x and y have the same sign, then the point (x, y) lies in quadrant I or quadrant _____.

5. Solve $6x^2 - 13x + 6 = 0$.

8.4	Exercises

Objective 1 Distinguish Between Linear and Quadratic Functions

In Exercises 1–4, use the shape of each graph to match it with the function that defines this graph.

1.

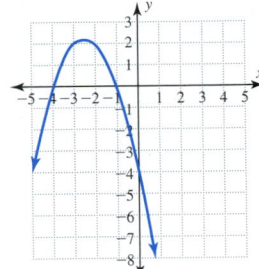

A. $f(x) = 3x - 4$

2.

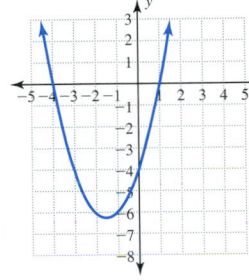

B. $f(x) = -x - 4$

3.

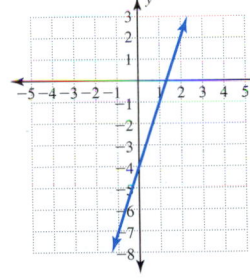

C. $f(x) = x^2 + 3x - 4$

4.

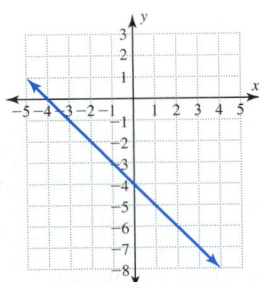

D. $f(x) = -x^2 - 5x - 4$

In Exercises 5–12, identify the graph of each function as a line or a parabola. If the graph is a line, determine whether the slope is negative, zero, or positive. If the graph is a

parabola, determine whether the parabola opens upward or downward.

5. $y = -3x + 4$

6. $y = -2x - 1$

7. $y = 4x - 3$

8. $y = 2x - 1$

9. $y = 8x^2$

10. $y = -4x^2$

11. $y = -3x^2 + 8x + 9$

12. $y = 2x^2 + 4x + 5$

Objective 2 Determine the Vertex of a Parabola

In Exercises 13 and 14, determine the vertex of each parabola.

13.

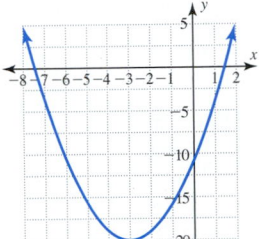

14.

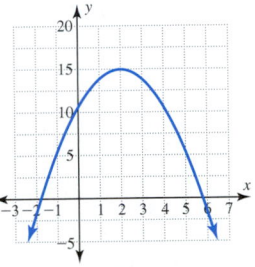

15. The x-intercepts of a parabola are $(6, 0)$ and $(-2, 0)$. Determine the x-coordinate of the vertex.

16. The x-intercepts of a parabola are $(-5, 0)$ and $(3, 0)$. Determine the x-coordinate of the vertex.

In Exercises 17–20, use the given function to calculate the vertex of the parabola.

17. $f(x) = x^2 - 6x + 11$

18. $f(x) = x^2 + 4x - 9$

19. $f(x) = -5x^2 + 10x + 6$

20. $f(x) = -3x^2 + 6x + 7$

In Exercises 21–24, use the given equation to calculate:

a. The y-intercept of the parabola

b. The x-intercepts of the parabola

c. The vertex of the parabola

21. $y = 2x^2 - 3x - 2$

22. $y = 5x^2 - x - 4$

23. $y = (2x - 3)(x + 4)$

24. $y = (5x - 2)(x + 3)$

Objective 3 Sketch the Graph of a Quadratic Function and Determine Key Features of the Resulting Parabola

In Exercises 25–28, use the given quadratic function to:

a. Complete the table to the right.

b. Plot these points on a graph.

c. Connect these points with a smooth parabolic curve.

25. $y = -x^2 + 3$

26. $y = x^2 - 4$

27. $y = x^2 + x - 6$

28. $y = -x^2 + x + 5$

x	y
-3	
-2	
-1	
0	
1	
2	
3	

In Exercises 29–34, use the given graphs to determine the following key features of each parabola.

a. Does the parabola open upward or downward?
b. Vertex
c. The y-intercept
d. The x-intercepts
e. Domain of f
f. Range of f

29.

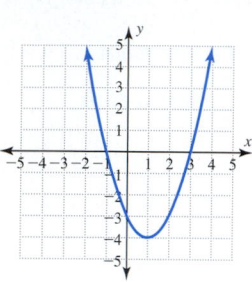

30.

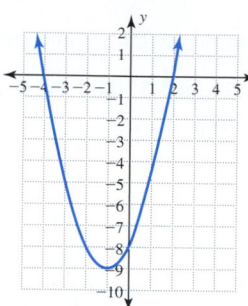

31.

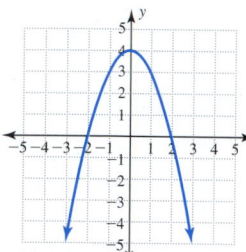

32.

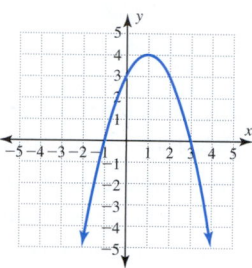

33.

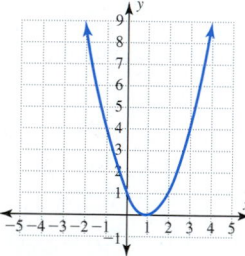

34.

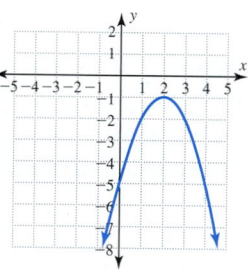

For each quadratic function in Exercises 35–38, determine the requested information and then use this information to sketch its graph.

a. Will the parabola open upward or open downward?
b. Determine the coordinates of the vertex.
c. Determine the intercepts of the graph of this function.
d. Complete a table of values using inputs on both sides of the vertex.
e. Use this information to sketch the graph of this function.

35. $f(x) = x^2 + 5x - 6$ **36.** $f(x) = -x^2 - x + 12$
37. $f(x) = -2x^2 + x + 15$ **38.** $f(x) = 2x^2 + 11x - 21$

Objective 4 Solve Problems Involving a Maximum or Minimum Value

In Exercises 39–42, $P(x)$ gives the profit in dollars when x units are produced and sold. Use the graph of the profit function to estimate the following.

a. Overhead costs [*Hint:* Evaluate $P(0)$.]
b. Break-even values [*Hint:* When does $P(x) = 0$?]
c. Maximum profit that can be made and the number of units to sell to create this profit

39.

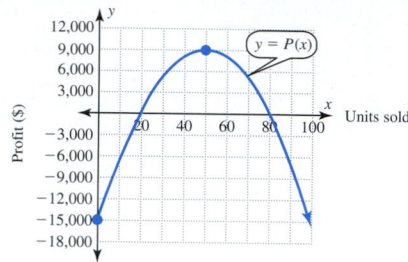

40.

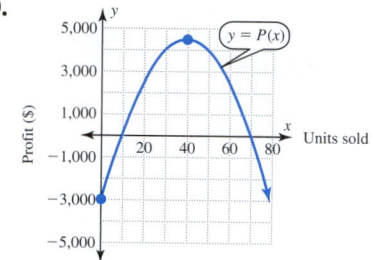

41.

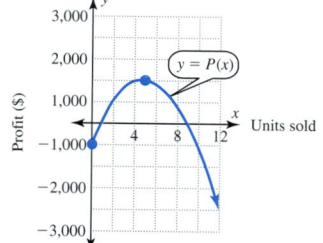

42.

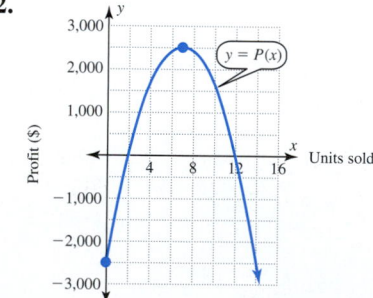

In Exercises 43–46, $P(x)$ gives the profit in dollars when x tons are produced and sold. Use the graph of the profit function to determine the following:

a. Overhead costs (*Hint:* Evaluate $P(0)$.)
b. Break-even values (*Hint:* When does $P(x) = 0$?)
c. Maximum profit that can be made and the number of tons to sell to create this profit

43. $P(x) = -100x^2 + 14,000x - 240,000$

44. $P(x) = -50x^2 + 7,000x - 65,000$

45. $P(x) = -140x^2 + 7,000x - 28,665$

46. $P(x) = -160x^2 + 8,000x - 39,160$

47. Height of a Baseball The height of a baseball is given by $y = -16x^2 + 96x + 3$, where y represents the height of the ball in feet and x is the number of seconds that have elapsed since the ball was released. Determine the highest point that the ball reaches. Determine how many seconds into the flight the maximum height is reached.

48. Height of a Softball The height of a softball is given by $y = -16x^2 + 48x + 8$, where y represents the height of the ball in feet and x is the number of seconds that have elapsed since the ball was released. Determine the highest point that the ball reaches. Determine how many seconds into the flight the maximum height is reached.

49. Maximum Grapefruit Production Research at several test plots has revealed that 50 grapefruit trees per plot will average 200 grapefruit per tree. Increasing the number of trees on the plot to $(50 + t)$ will result in reducing the number of grapefruit per tree to $(200 - 2t)$. The function $N(t) = (200 - 2t)(50 + t)$ gives the total number of grapefruit that will be produced on a plot as a function of t, the number of extra (in excess of 50) trees planted on the plot. Determine the number of trees that should be planted on a plot this size to maximize the number of grapefruit the plot will yield.

50. Optimal Selling Price A toy store can obtain a toy tractor from the manufacturer at a cost of $8 each and estimates that if it sells the tractors for x dollars apiece, approximately $12(18 - x)$ toy tractors will be sold each month. Under these conditions, the monthly profit from the sale of these toys as a function of price x will be $P(x) = 12(x - 8)(18 - x)$. Determine the selling price at which these toy tractors should be sold to maximize the profit.

51. Maximum Area
   a. Use the figure to help construct a function that gives the area of the rectangular region that can be enclosed by 40 m of fencing as a function of the width w.
   b. Determine the maximum area.

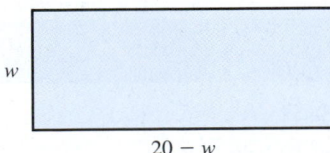

$20 - w$

52. Maximum Area
   a. Use the figure to help construct a function that gives the area of the rectangular region that can be enclosed by 90 m of fencing as a function of the width w.
   b. Determine the maximum area.

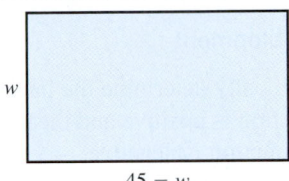

$45 - w$

Maximum Area A lumberyard plans to enclose a rectangular storage area for storing treated lumber. One side of the storage area will be formed by the wall of an existing building, and the other three sides will be formed by using a fixed amount of fencing. Use the information given in Exercises 53 and 54 to answer each part of these questions.

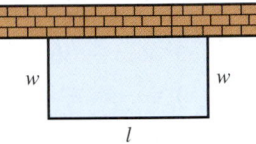

53. Assume the fixed amount of fencing available is 60 ft.
   a. Write a function that models the area as a function of the width w.
   b. Find the maximum area that can be enclosed with this fencing.

54. Assume the fixed amount of fencing available is 80 ft.
   a. Write a function that models the area as a function of the width w.
   b. Find the maximum area that can be enclosed with this fencing.

Maximum Area A rectangular steel sheet has x cm turned down on two sides. This is done at right angles to form three sides of a rectangular frame for the opening for a storm gutter on the side of a street. Use the dimensions given in Exercises 55 and 56 to answer each part of these questions.

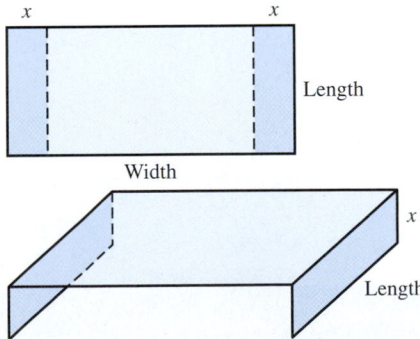

55. For a steel sheet 100 cm wide by 30 cm long:
   a. Write a polynomial function $A(x)$ to represent the cross-sectional area of the frame that is formed.
   b. Determine the value of x that will produce the form with the maximum cross-sectional area.
   c. What is this cross-sectional area?

56. For a steel sheet 80 cm wide by 20 cm long:
  a. Write a polynomial function $A(x)$ to represent the cross-sectional area of the frame that is formed.
  b. Determine the value of x that will produce the form with the maximum cross-sectional area.
  c. What is this cross-sectional area?

Review and Concept Development

In Exercises 57–64, algebraically determine the interval of x-values for which the function is positive and the interval of x-values for which the function is negative.

57. $f(x) = -2x + 8$ **58.** $f(x) = 5x + 20$

59. $f(x) = 3x - 10$ **60.** $f(x) = -4x + 5$

61. $f(x) = x^2 - 36$ **62.** $f(x) = -x^2 + 25$

63. $f(x) = -x^2 + 4x + 5$ **64.** $f(x) = x^2 + 5x + 6$

In Exercises 65 and 66, use each function to solve each equation and inequality.

65. $f(x) = x^2 - 2x - 8$
  a. $f(x) = 7$ **b.** $f(x) < 7$ **c.** $f(x) \geq 7$

66. $f(x) = -x^2 + 2x + 15$
  a. $f(x) = 12$ **b.** $f(x) \leq 12$ **c.** $f(x) > 12$

67. Use the given table of values to determine the missing input and output values.

x	$f(x)$
-4	-7
-3	-2
-2	1
-1	2
0	1
1	-2
2	-7

  a. $f(-2) =$ ____
  b. $f(x) = -2; x =$ ____ and $x =$ ____

68. Use the given table of values to determine the missing input and output values.

x	$f(x)$
0	11
1	6
2	3
3	2
4	3
5	6
6	11

  a. $f(6) =$ ____
  b. $f(x) = 6; x =$ ____ and $x =$ ____

69. Use the given graph to determine the missing input and output values.

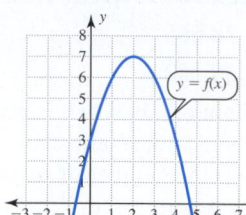

  a. $f(3) =$ ____
  b. $f(x) = 3; x =$ ____ and $x =$ ____

70. Use the given graph to determine the missing input and output values.

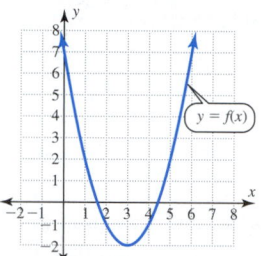

  a. $f(2) =$ ____
  b. $f(x) = 2; x =$ ____ and $x =$ ____

71. Use the function $f(x) = x^2 - 9x - 10$ to determine the missing input and output values.
  a. $f(12) =$ ____
  b. $f(x) = 12; x =$ ____ and $x =$ ____

72. Use the function $f(x) = -x^2 - 8x - 12$ to determine the missing input and output values.
  a. $f(-5) =$ ____
  b. $f(x) = -5; x =$ ____ and $x =$ ____

Group discussion questions

73. Discovery Question
  a. Graph $y = x^2 + 1$, $y = x^2 + 3$, and $y = x^2 + 4$, and verbally compare these graphs to the graph of $y = x^2$.
  b. On the basis of these observations, what would you predict about the graph of $y = x^2 + 6$ as compared with the graph of $y = x^2$?

74. Discovery Question
  a. Graph $y = (x + 1)^2$, $y = (x + 3)^2$, and $y = (x + 4)^2$, and verbally compare these graphs to the graph of $y = x^2$.
  b. On the basis of these observations, what would you predict about the graph of $y = (x + 6)^2$ as compared with the graph of $y = x^2$?

8.4   **Cumulative Review**

1. Sketch a graph of $f(x) = |2x - 3| - 2$.

2. Solve $|2x - 3| - 2 < 11$ and give your answer in interval notation.

3. Determine the slope-intercept form of the equation of the line that passes through $(-2, -5)$ and $(3, -4)$.

4. Factor $2x^2 - 11x - 21$.

5. Construct a quadratic equation in x with integer coefficients that has $x = 3$ and $x = -\dfrac{5}{2}$ as solutions.

Section 8.5 | Analyzing Graphs

Objectives:

1. Identify a function as an increasing or a decreasing function.
2. Analyze the graphs of linear, absolute value, and quadratic functions.

In this section, we take a summative look at the graphs of linear, absolute value, and quadratic functions. We have already examined the key properties of the graphs of each of these types of functions. After we introduce the concept of increasing or decreasing functions, we will review and further analyze the graphs of these three families of functions.

1. Identify a Function As an Increasing or a Decreasing Function

One of the ways we can analyze problems is by constructing a mathematical model of the problem and then using our mathematical language to describe this model. We often use the words *increasing* or *decreasing* to describe the behavior of these models.

What Does It Mean to Say That a Function Is Increasing?

A function is **increasing** if its graph rises as it moves from left to right. As the x-values increase, the y-values also increase. We say that a function is **decreasing** if its graph drops as it moves from left to right. As the x-values increase, the y-values decrease. A function is called an **increasing function** if it increases over its entire domain, and a function is called a **decreasing function** if it decreases over its entire domain.

Increasing and Decreasing Functions

	Graphical Example	Verbally
Increasing Function		A function is increasing over its entire domain if the graph rises as it moves from left to right. For all x-values, as the x-values increase, the y-values also increase.
Decreasing Function		A function is decreasing over its entire domain if the graph drops as it moves from left to right. For all x-values, as the x-values increase, the y-values decrease.

In Example 1, we identify increasing and decreasing functions by inspecting their graphs.

Example 1 Identifying Increasing and Decreasing Functions from Their Graphs

Use the graph of each function to identify the function as an increasing function or as a decreasing function.

Solutions

a.

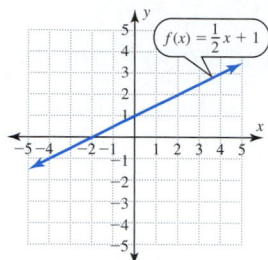

Increasing function   The graph of this linear function rises as the graph moves from left to right. Thus this function increases over its entire domain.

b.

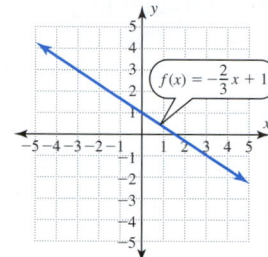

Decreasing function   The graph of this linear function drops as the graph moves from left to right. Thus this function decreases over its entire domain.

c.

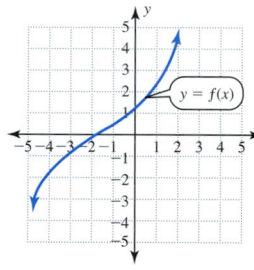

Increasing function   The graph of this function rises as the graph moves from left to right. Thus this function increases over its entire domain.

d.

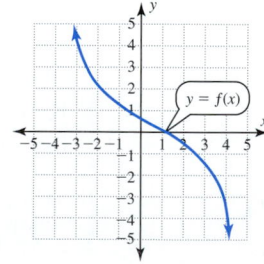

Decreasing function   The graph of this function drops as the graph moves from left to right. Thus this function decreases over its entire domain.

Self-Check 1

Identify each function defined by these graphs as either an increasing function or a decreasing function.

a.

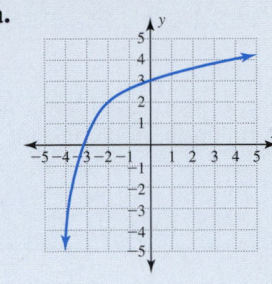

b.

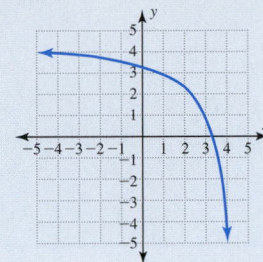

Can I Determine Whether a Function Is Increasing or Decreasing Without Graphing the Function?

If the function defines a line, the answer is yes. A line with positive slope rises as it moves from left to right. Thus a line with positive slope is an increasing function. A line with negative slope drops as it moves from left to right. Thus a line with negative slope is a decreasing function. A line with a slope of zero does not rise or drop as it moves from left to right. Thus a line with a slope of zero is a constant function.

Positive Slope:
An Increasing Linear Function

Negative Slope:
A Decreasing Linear Function

Zero Slope:
A Constant Linear Function

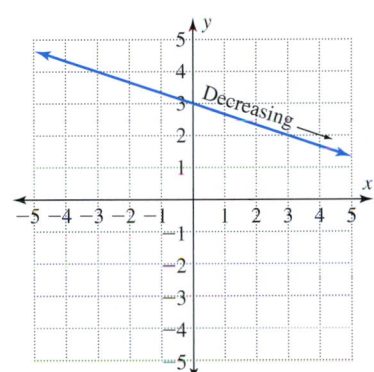

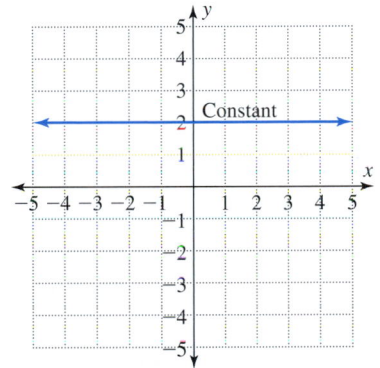

For an increasing function, a positive value of Δx produces a positive value of Δy.

For a decreasing function, a positive value of Δx produces a negative value of Δy.

For a constant function, any change in x produces no change in y.

In Example 2, we use the slope-intercept form of each line to determine the slope of the line. Then we identify the linear function as increasing, decreasing, or constant.

Example 2   Identifying Increasing, Decreasing, and Constant Linear Functions

Identify each function as an increasing function, a decreasing function, or a constant function.

(a) $f(x) = 3x - 17$ **(b)** $f(x) = -5x + 20$ **(c)** $f(x) = 7$

Solution

(a) $f(x) = 3x - 17$ is an increasing function.

This linear function is given in the slope-intercept form $f(x) = mx + b$ with a slope of 3. This is an increasing function because the slope is positive.

(b) $f(x) = -5x + 20$ is a decreasing function.

This linear function is decreasing because the slope -5 is negative.

(c) $f(x) = 7$ is a constant function.

This is a horizontal line with a slope of 0. The line is neither rising nor dropping.

Self-Check 2

Identify each function as an increasing function, a decreasing function, or a constant function.

a. $f(x) = -4x - 1$ **b.** $f(x) = 4x - 19$ **c.** $f(x) = -4$

Sometimes a function can be increasing over part of its domain, decreasing over another part of its domain, and constant over yet another part of its domain. This is illustrated in Example 3.

Example 3 | Identifying Intervals Where a Function Is Increasing, Decreasing, or Constant

The accompanying graph gives the altitude of a small airplane at a given time. The time x is given in minutes from the start of the flight and the altitude y is given in thousands of feet. Answer each question by examining this graph.

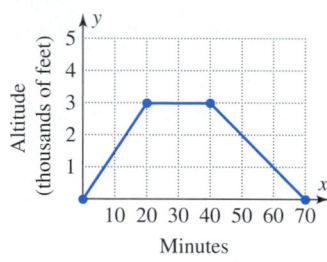

Solution

(a) During what time interval was the altitude increasing?

The altitude was increasing during the first 20 min of the flight—that is, during the interval (0, 20).

(b) During what time interval was the altitude constant?

The altitude was constant between 20 and 40 min after the flight began—that is, during the interval (20, 40).

(c) During what time interval was the altitude decreasing?

The altitude was decreasing between 40 and 70 min after the flight began—that is, during the interval (40, 70).

Self-Check 3

The graph gives the altitude y in thousands of feet in terms of the time x in minutes since the start of a flight in a small plane.

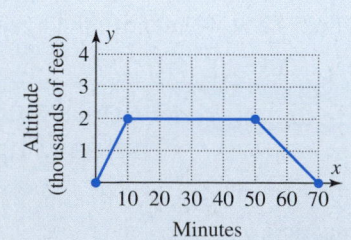

a. During what time interval was the altitude increasing?

b. During what time interval was the altitude constant?

c. During what time interval was the altitude decreasing?

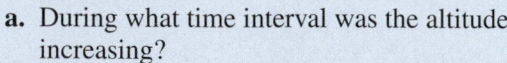

2. Analyze the Graphs of Linear, Absolute Value, and Quadratic Functions

The following box contains a checklist of key information that we are currently able to obtain about the graphs of linear, absolute value, and quadratic functions. We will use this list as a guide for analyzing these functions and their graphs.

Analyzing the Graphs of Linear, Absolute Value, and Quadratic Functions

1. Identify the type of function.
   a. For linear functions, determine the slope.
   b. For absolute value and quadratic functions, determine the vertex and whether the graph opens upward or downward.
2. Determine the y-intercept and the x-intercept(s).
3. Determine the domain and range.
4. Determine where the function is positive and where the function is negative.
5. Determine where the function is increasing and where the function is decreasing.

Consider the function $f(x) = \dfrac{3}{2}x - 3$, whose graph is shown.

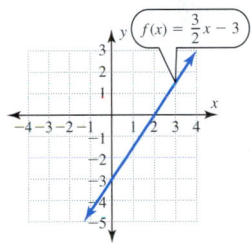

- This is a linear function with slope $m = \dfrac{3}{2}$.

- We can calculate the intercepts algebraically and confirm them on the graph.
  x-intercept: $(2, 0)$
  y-intercept: $(0, -3)$
- The domain and range of this function are both $\mathbb{R}$.
- This function is positive over the interval $(2, \infty)$ and negative over the interval $(-\infty, 2)$.
- Since the slope is positive, this function is increasing over its entire domain, $\mathbb{R}$.

We have seen that linear functions are always increasing, decreasing, or constant over their entire domain. As we will see in Examples 4 and 5, absolute value and quadratic functions will be increasing on one side of the vertex and decreasing on the other side. We analyze the graph of an absolute value function in Example 4.

Example 4   Analyzing the Graph of an Absolute Value Function

Use the graph of $f(x) = -|x - 2| + 4$ to determine the following information.

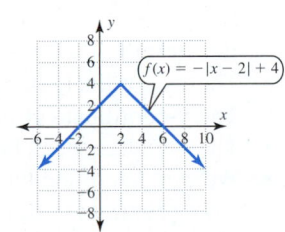

(a) Is the V-shape opening upward or downward?

(b) Vertex

(c) y-intercept

(d) x-intercepts

(e) Domain of f

(f) Range of f

(g) Interval where f is positive

(h) Interval where f is negative

(i) Interval where f is decreasing

(j) Interval where f is increasing

Solution

(a) Opens: Downward

(b) Vertex: (2, 4)   The vertex is the highest point on the graph.

(c) y-intercept: (0, 2)   The graph crosses the y-axis at (0, 2).

(d) x-intercepts: $(-2, 0)$ and $(6, 0)$ The graph crosses the x-axis at $(-2, 0)$ and $(6, 0)$.

(e) Domain $= \mathbb{R}$ The projection of this graph onto the x-axis is the set of all real numbers.

(f) Range $= (-\infty, 4]$ The projection of this graph onto the y-axis is the y-interval $(-\infty, 4]$.

(g) Positive on the interval $(-2, 6)$ The graph is above the x-axis between $x = -2$ and $x = 6$.

(h) Negative on the interval $(-\infty, -2) \cup (6, \infty)$.   The graph is below the x-axis for $x < -2$ or $x > 6$.

(i) Decreasing on the interval $(2, \infty)$ The graph is decreasing to the right of 2.

(j) Increasing on the interval $(-\infty, 2)$ The graph is increasing to the left of 2.

Self-Check 4

Use the graph of $f(x) = |x + 1| - 5$ to determine the following information.

a. Is the V-shape opening upward or downward?

b. Vertex

c. y-intercept

d. x-intercepts

e. Domain of f

f. Range of f

g. Interval where f is positive

h. Interval where f is negative

i. Interval where f is decreasing

j. Interval where f is increasing

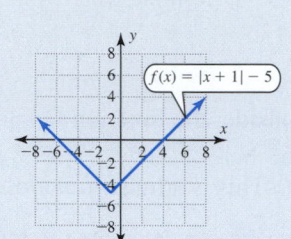

Can I Analyze the Graph of an Absolute Value Function Using Only Its Equation?

Yes, we can use algebra to determine some key information about the graph, but it is still wise to make a quick sketch of the graph. Then we can use knowledge of the basic properties of the graph to complete the rest of our analysis. This is illustrated now for $f(x) = -|x - 2| + 4$, the function from Example 4.

- $f(x) = -|x - 2| + 4$ has a factor of -1 in front of the absolute value symbol. Because this factor is negative, the graph will open downward.
- The x-coordinate of the vertex of this graph is the point where $x - 2 = 0$ or $x = 2$. The y-coordinate of the vertex is found by evaluating $f(2)$.
- The y-intercept can be calculated by evaluating $f(0)$.
- We can determine the x-intercepts algebraically by finding the x-values where the y-coordinate is 0. Solving $-|x - 2| + 4 = 0$, we get $x = -2$ and $x = 6$.
- To determine the domain and range, it is generally a good idea to quickly sketch a graph or use a graphing tool to create a graph if one is not provided.
- Once we have the x-intercepts, we use the graph (or an understanding of the graph) to determine where the function is positive or negative.
- The graph of an absolute value function will be increasing on one side of the vertex and decreasing on the other side.

In Example 5, we analyze the graph of a quadratic function. It is also possible to analyze this graph algebraically directly from its equation.

Example 5   Analyzing the Graph of a Quadratic Function

Use the graph of $f(x) = x^2 + 2x - 8$ to determine the following information.

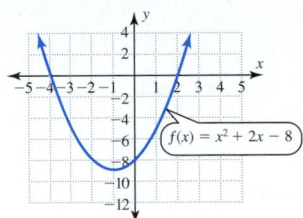

(a) Is the parabola opening upward or downward?

(b) Vertex

(c) y-intercept

(d) x-intercepts

(e) Domain of f

(f) Range of f

(g) Interval where f is positive

(h) Interval where f is negative

(i) Interval where f is decreasing

(j) Interval where f is increasing

Solution

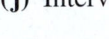

(a) Opens: Upward

(b) Vertex: $(-1, -9)$ The vertex is the lowest point on the parabola.

(c) y-intercept: $(0, -8)$ The graph crosses the y-axis at $(0, -8)$.

(d) x-intercepts: $(-4, 0)$ and $(2, 0)$ The graph crosses the x-axis at $(-4, 0)$ and $(2, 0)$.

(e) Domain $= \mathbb{R}$ The projection of this graph onto the x-axis is the set of all real numbers.

(f) Range $= [-9, \infty)$ The projection of this graph onto the y-axis is the y-interval $[-9, \infty)$.

(g) Positive on the interval $(-\infty, -4) \cup (2, \infty)$ The graph is above the x-axis for $x < -4$ or $x > 2$.

(h) Negative on the interval $(-4, 2)$ The graph is below the x-axis between $x = -4$ and $x = 2$.

(i) Decreasing on the interval $(-\infty, -1)$ The graph is decreasing to the left of -1.

(j) Increasing on the interval $(-1, \infty)$ The graph is increasing to the right of -1.

Self-Check 5

Use the graph of $f(x) = -x^2 + 2x + 3$ to determine the following information.

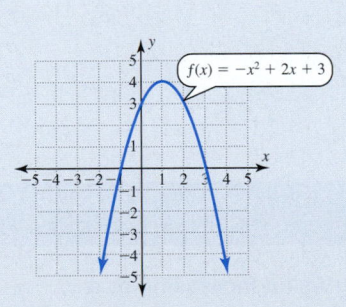

a. Is the parabola opening upward or downward?

b. Vertex

c. y-intercept

d. x-intercepts

e. Domain of f

f. Range of f

g. Interval where f is positive

h. Interval where f is negative

i. Interval where f is decreasing

j. Interval where f is increasing

Self-Check Answers

1. **a.** Increasing function   **b.** Decreasing function
2. **a.** Decreasing function   **b.** Increasing function
   c. Constant function
3. **a.** The altitude is increasing for the first 10 minutes—that is, during the interval (0, 10).
   b. The altitude is constant between 10 and 50 minutes—that is, during the interval (10, 50).
   c. The altitude is decreasing between 50 and 70 minutes—that is, during the interval (50, 70).
4. **a.** The V-shape is opening upward.
   b. Vertex: $(-1, -5)$
   c. y-intercept: $(0, -4)$
   d. x-intercepts: $(-6, 0)$ and $(4, 0)$
   e. Domain $= \mathbb{R}$

f. Range $= [-5, \infty)$
g. Positive on the interval $(-\infty, -6) \cup (4, \infty)$
h. Negative on the interval $(-6, 4)$
i. Decreasing on the interval $(-\infty, -1)$
j. Increasing on the interval $(-1, \infty)$
5. **a.** The parabola is opening downward.   **b.** Vertex: $(1, 4)$
   c. y-intercept: $(0, 3)$
   d. x-intercepts: $(-1, 0)$ and $(3, 0)$
   e. Domain $= \mathbb{R}$
   f. Range $= (-\infty, 4]$
   g. Positive on the interval $(-1, 3)$
   h. Negative on the interval $(-\infty, -1) \cup (3, \infty)$
   i. Decreasing on the interval $(1, \infty)$
   j. Increasing on the interval $(-\infty, 1)$

8.5   Using the Language and Symbolism of Mathematics

1. We say that a function is _____ if its graph rises as it moves from left to right.

2. We say that a function is _____ if its graph drops as it moves from left to right.

3. A function is called an increasing function if it _____ over its entire domain.

4. A function is called a decreasing function if it _____ over its entire domain.

5. A line with _____ slope is an increasing function.

6. A line with _____ slope is a decreasing function.

7. The checklist we use to analyze the graph of a linear, absolute value, or quadratic function is to:
   a. Identify the type of function.
      i. For linear functions, determine the _____.
      ii. For absolute value and quadratic functions, determine the _____ and whether the graph _____ upward or downward.
   b. Determine the y-intercept and the _____-intercept(s).
   c. Determine the _____ and range.
   d. Determine where the function is _____ and where the function is negative.
   e. Determine where the function is increasing and where the function is _____.

8.5   Quick Review

1. Sketch the graph of $f(x) = 4$.

2. Sketch the graph of $f(x) = -2x + 9$.

3. Sketch the graph of $f(x) = -|x - 2| + 5$.

4. Sketch the graph of $f(x) = x^2 - 4x - 5$.

5. Determine the domain and range of each function.
   a. $f(x) = 4$ **b.** $f(x) = -2x + 9$
   c. $f(x) = -|x - 2| + 5$ **d.** $f(x) = x^2 - 4x - 5$

8.5   Exercises

Objective 1 Identify a Function As an Increasing or a Decreasing Function

In Exercises 1–4, use the graph of each function to identify the function as an increasing function or as a decreasing function.

1.
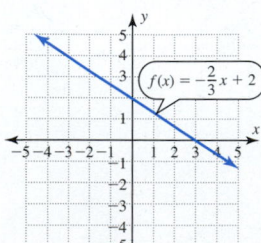

2.
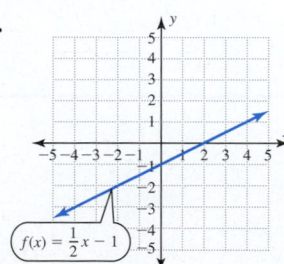

3.
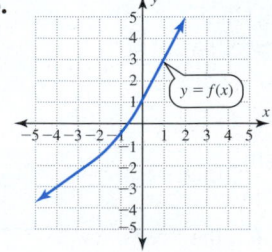

4.
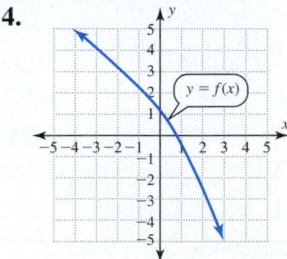

In Exercises 5–10, use the equation defining each function to identify the function as an increasing function, a decreasing function, or a constant function.

5. $f(x) = -2x + 5$ **6.** $f(x) = 4x - 2$

7. $f(x) = 2x - 5$ **8.** $f(x) = -4x + 2$

9. $f(x) = -3$ **10.** $f(x) = 3$

11. The accompanying graph corresponds to the price of a stock over a 6-month period. The time x is given in months and the price y is given in dollars. Use this graph to answer the following questions.
  a. During which months was the price increasing?
  b. During which months was the price decreasing?
  c. During which months was the price constant?

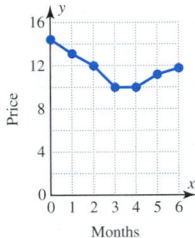

12. The given graph corresponds to the depth of water recorded over a 6-month period in a city reservoir. The time x is given in months and the depth y is given in feet. Use this graph to answer the following questions.
  a. During which months was the depth increasing?
  b. During which months was the depth decreasing?
  c. During which months was the depth constant?

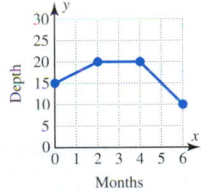

Objective 2 Analyze the Graphs of Linear, Absolute Value, and Quadratic Functions

In Exercises 13–20, use the graph or the equation of the linear function to complete each part of the exercise.

a. Determine the slope of this line.
b. Determine the x-intercept.
c. Determine the y-intercept.
d. Determine the domain of this function.
e. Determine the range of this function.
f. Determine the x-values for which the function is positive.
g. Determine the x-values for which the function is negative.
h. Determine the x-values for which the function is decreasing.
i. Determine the x-values for which the function is increasing.

13.

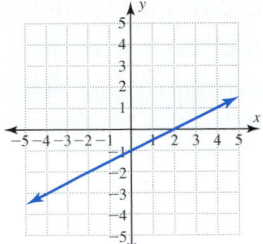

14.

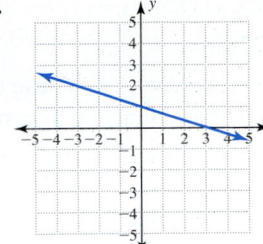

15.

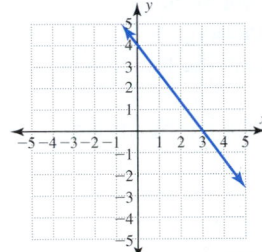

16.

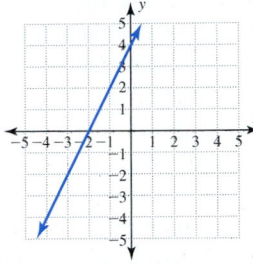

17. $f(x) = 5x - 20$ **18.** $f(x) = 4x + 40$

19. $f(x) = -150x + 300$ **20.** $f(x) = -200x - 1,000$

In Exercises 21–28, use the graph or the equation of the absolute value function to complete each part of the exercise.

a. Determine the vertex of this graph.
b. Does the graph open upward or downward?
c. Determine the x-intercept(s).
d. Determine the y-intercept.
e. Determine the domain of this function.
f. Determine the range of this function.
g. Determine the x-values for which the function is positive.
h. Determine the x-values for which the function is negative.
i. Determine the x-values for which the function is decreasing.
j. Determine the x-values for which the function is increasing.

21.

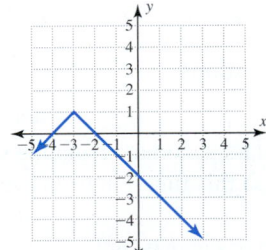

22.

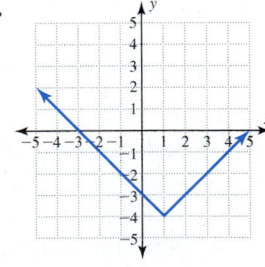

23.

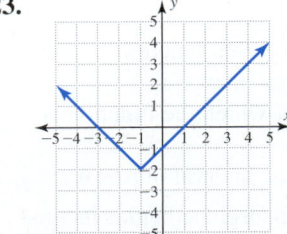

24.

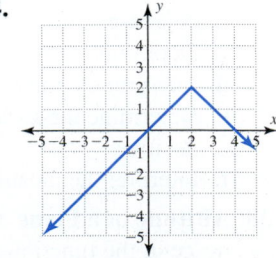

25. $f(x) = |x + 7| - 12$ **26.** $f(x) = |x - 10| - 11$

27. $f(x) = -|x - 15| + 20$ **28.** $f(x) = -|x + 10| + 12$

In Exercises 29–36, use the graph or the equation of the quadratic function to complete each part of the exercise.

a. Determine the vertex of this graph.

b. Does the graph open upward or downward?

c. Determine the x-intercept(s).

d. Determine the y-intercept.

e. Determine the domain of this function.

f. Determine the range of this function.

g. Determine the x-values for which the function is positive.

h. Determine the x-values for which the function is negative.

i. Determine the x-values for which the function is decreasing.

j. Determine the x-values for which the function is increasing.

29.

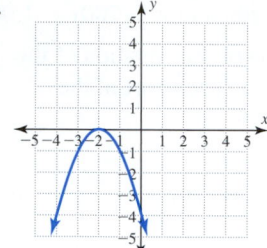

30.

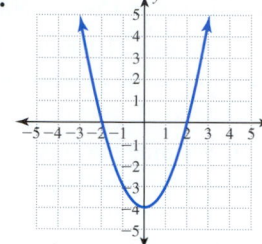

31.

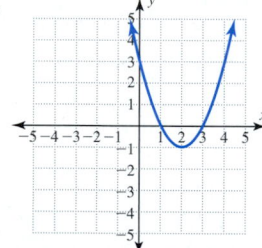

32.

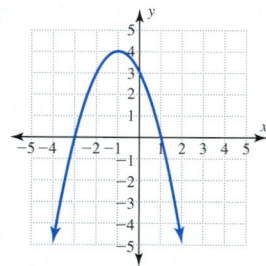

33. $f(x) = x^2 - 9x + 20$ **34.** $f(x) = -x^2 + x + 20$

35. $f(x) = -2x^2 + 9x + 26$ **36.** $f(x) = 2x^2 + 3x - 20$

Connecting Concepts to Applications

37. Profit Function The function $P(x) = 0.25x - 400$ gives the yearly profit in dollars generated from the sale of x snow cones in a year. Use this function to answer the following questions.

a. Is this function increasing or decreasing?

b. Interpret the meaning of the rate of increase or decrease.

c. Determine the intercepts for this function.

d. Interpret the meaning of the intercepts.

e. Determine the x-values for which the function is positive.

f. Interpret the meaning of the interval in part **e.**

38. Altitude of a Plane Once an airplane begins its final descent, the function $A(t) = -800t + 12,000$ gives the

altitude in feet as a function of time t in minutes after the start of the descent.

a. Is this function increasing or decreasing?

b. Interpret the meaning of the rate of increase or decrease.

c. Determine the intercepts for this function.

d. Interpret the meaning of the intercepts.

e. Determine the t-values for which the function is positive.

f. Interpret the meaning of the interval in part **e.**

39. Height of a Golf Ball The function $H(t) = -4.9t^2 + 24.5t$ gives the height of a golf ball in meters as a function of the time t in seconds after it was hit.

a. Determine the vertex of this parabolic function. Round both coordinates to the nearest tenth.

b. Interpret the meaning of both coordinates of the vertex.

c. Determine the intercepts for this function.

d. Interpret the meaning of the intercepts.

e. Determine the t-values for which the function is increasing.

f. Interpret the meaning of the interval in part **e.**

40. Profit Function The function $P(x) = -x^2 + 36x - 180$ gives the profit in dollars generated by the sale of x units of a product.

a. Determine the vertex of this parabolic function.

b. Interpret the meaning of both coordinates of the vertex.

c. Determine the intercepts for this function.

d. Interpret the meaning of the intercepts.

e. Determine the t-values for which the function is positive.

f. Interpret the meaning of the interval in part **e.**

Group discussion questions

41. Discovery Question Use a graphing calculator to graph

$f(x) = \dfrac{x}{2}, f(x) = \dfrac{x}{2} + 1, f(x) = \dfrac{x}{2} + 3$, and $f(x) = \dfrac{x}{2} - 3$. Describe the relationship that you observe among these graphs.

42. Error Analysis A student produced the following calculator window for the function shown and concluded that the function was an increasing function over its entire domain. Determine the correct analysis and the source of the error in this answer.

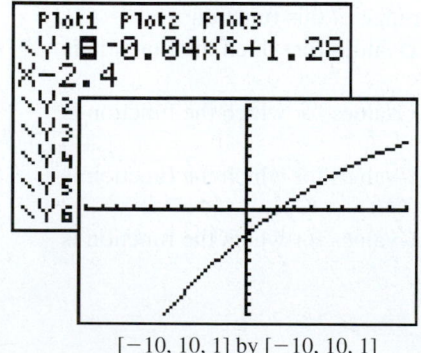

$[-10, 10, 1]$ by $[-10, 10, 1]$

43. Discovery Question In this exercise, you will compare the change in linear functions to the change in quadratic functions.

a. For a linear function like $y = 2x - 3$, the change in y for each 1-unit change in x is constant. In this example, this constant change in y is _____. This constant change represents the _____ of this line.

x	$y = 2x - 3$	Change in y
0	-3	—
1	-1	2
2	1	2
3	3	2
4	5	2
5	7	2
6	9	2

b. Before you complete this table, predict what each change in y will be for a 1-unit change in x. Then complete a table similar to the one in part **a** for $y = -3x + 1$. The constant change in y is _____.

x	$y = -3x + 1$	Change in y
0		—
1		
2		
3		
4		
5		
6		

c. For a quadratic function like $y = (x - 4)^2$, the changes in y vary for each 1-unit change in x. Thus the graph is not a line with a constant slope, but a parabola. The last column in the table gives the change in the changes for each 1-unit change in x. Complete a table similar to the previous table for $y = -x^2 + 5$. In the last column, what is the value of each change in the changes?

x	$y = (x - 4)^2$	Change in y	Change in the changes
0	16	—	—
1	9	-7	—
2	4	-5	2
3	1	-3	2
4	0	-1	2
5	1	1	2
6	4	3	2

x	$y = -x^2 + 5$	Change in y	Change in the changes
-3		—	—
-2			—
-1			
0			
1			
2			
3			

8.5 | Cumulative Review

1. Write 5,123 in scientific notation.
2. Write -0.0078 in scientific notation.
3. Write 4.5×10^4 in standard decimal notation.

4. Write -6.73×10^{-3} in standard decimal notation.
5. The coefficient of $7x^3$ is _____.

Section 8.6 | Curve Fitting—Selecting a Line or Parabola That Best Fits a Set of Data

Objectives:

1. Create a scatter diagram and select an appropriate model for the data.
2. Select a linear function that best fits a set of data.
3. Select a quadratic function that best fits a set of data.

In this chapter, we have examined some of the most basic functions that you are likely to encounter—linear, absolute value, and quadratic. We have noted that each family of functions produces a graph with a characteristic shape. If we have a formula, then we can produce a graph or predict outputs for any input value.

What we want to do now is much more sophisticated. We want to take a table of values and select the type of curve that fits the data. Then we want to select the one curve from that family that best fits the data. In this section, we will examine fitting straight lines and parabolas to data; we will examine other types of functions later in this book. The topic of selecting a curve that best fits a set of data points is called **curve fitting.**

1. Create a Scatter Diagram and Select an Appropriate Model for the Data

How Will I Determine What Type of Curve to Use for a Set of Data Points?

Start by examining the visual pattern formed by a scatter diagram of the points. (Scatter diagrams were first examined in Section 2.1.) The scatter diagram for a set of points often reveals a visual pattern, which can help us select the type of function that best describes the data. We can then use a graphing calculator or a spreadsheet to calculate the equation for the function that best fits the data. The data for a real application often do not lie exactly on a curve due to experimental error and other factors. However, these data points may all lie very close to a curve. The scatter diagrams that follow illustrate three possibilities. The first set of data displays a linear pattern; the second set of data displays a parabolic pattern typical of a second-degree function; and the third set of data displays the pattern of exponential growth. We will examine linear patterns and parabolic patterns in this section and exponential patterns in Chapter 11.

Linear Pattern **Parabolic Pattern** **Exponential Pattern**

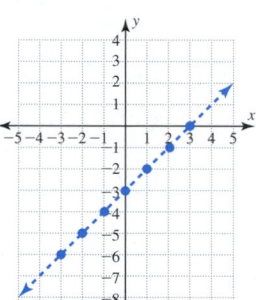 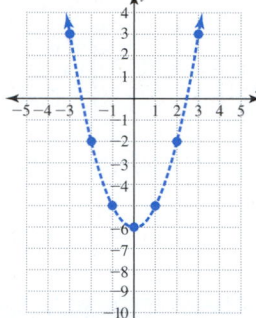 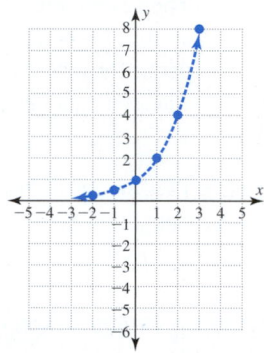

| **Example 1** | Selecting the Type of Graph that Best Fits a Set of Data |

Draw a scatter diagram for each set of data points and then specify the shape of the graph that describes the visual pattern formed by these points.

Solution

(a)

x	y
−3	−9
−2	−7
−1	−5
0	−3
1	−1
2	1
3	3

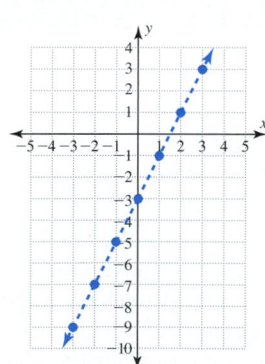

Shape: A line with a positive slope

(b)

x	y
−3	4
−2	−1
−1	−4
0	−5
1	−4
2	−1
3	4

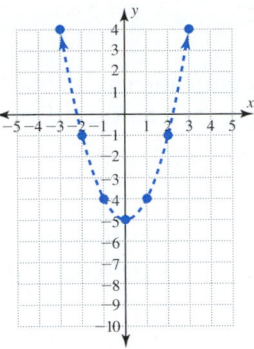

Shape: A parabola opening upward

Self-Check 1

Draw a scatter diagram for each set of data points and then specify the shape of the graph that describes the visual pattern formed by these points.

a.

x	−6	−4	−2	0	2	4	6
y	5	4	3	2	1	0	−1

b.

x	−3	−2	−1	0	1	2	3
y	−3	2	5	6	5	2	−3

Once I Have Selected the Type of Curve That Fits the Data Points, How Will I Determine the Best Equation?

In the real world, most curve fitting is done using a spreadsheet or other computer software. These computer programs can handle large sets of data and provide detailed analysis. To make it easier to focus on the concepts involved, we will illustrate only a graphing calculator. This will allow us to avoid the wide variety and complexity of the available computer programs and still allow us to cover all the main mathematical concepts. Technology Perspectives 8.6.1 and 8.6.2 illustrate how to use a graphing calculator to create a scatter diagram for the data points in the table from Example 1(a).

Technology Perspective 8.6.1	Using Lists to Store a Set of Data Points

Enter the data points from the table from Example 1(a) on a TI-84 Plus calculator.

TI-84 Plus Keystrokes

1. To **Edit** lists **L1** and **L2**, press  .

2. Enter the x-values under **L1**.

3. Press ⟩ to access **L2**. Enter the y-values under **L2**.

4. Press 2nd QUIT to return to the home screen.

TI-84 Plus Screens

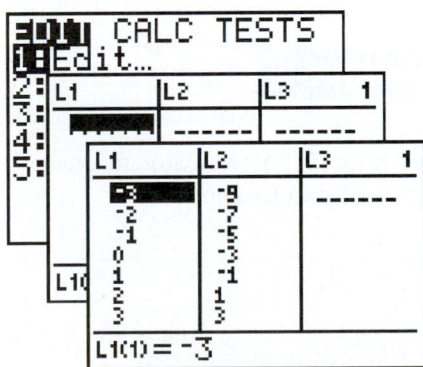

Technology Self-Check 1

Enter the data points from the table into a graphing calculator.

x	-3	-2	-1	0	1	2	3
y	-3	-1	1	5	7	9	12

Once data points are entered into a graphing calculator, we can then use the calculator to create the scatter diagram. This is illustrated in Technology Perspective 8.6.2.

Technology Perspective 8.6.2   Creating a Scatter Diagram

Create a scatter diagram for the data points from Example 1(a) on a TI-84 Plus calculator.

TI-84 Plus Keystrokes

1. Turn on **Plot1** by pressing [2nd] [STAT PLOT]. Set up the menu as shown here.

2. Press [Y=] and check to see that the **Y =** screen is blank.

3. Use **Zoom Stat** to automatically select a graphing window for these points. This is option **9** in the [ZOOM] menu.

Note: This calculator display corresponds to the linear pattern shown in Example 1(a).

TI-84 Plus Screens

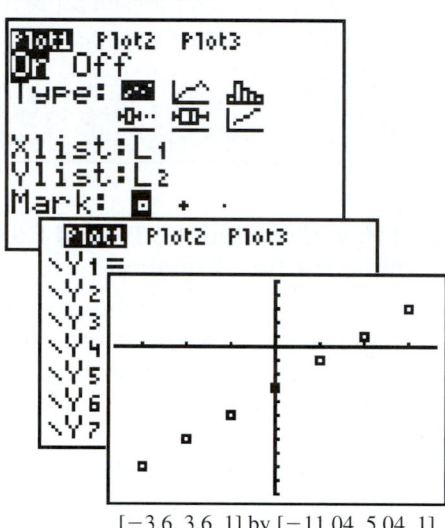

$[-3.6, 3.6, 1]$ by $[-11.04, 5.04, 1]$

Technology Self-Check 2

Use a graphing calculator to create a scatter diagram for the data points in the table.

x	-3	-2	-1	0	1	2	3
y	-3	-1	1	5	7	9	12

Example 2   Using a Graphing Calculator to Create a Scatter Diagram

Use a graphing calculator to create a scatter diagram for this set of data points as repeated from Example 1(b).

x	-3	-2	-1	0	1	2	3
y	4	-1	-4	-5	-4	-1	4

Solution

First enter the data points using **STAT-EDIT** feature.

Then use the **STAT-PLOT** feature to
create a scatter diagram of the
data points listed in **L1** and **L2.**

Note: This calculator display corresponds
to the parabolic pattern shown in
Example 1(b).

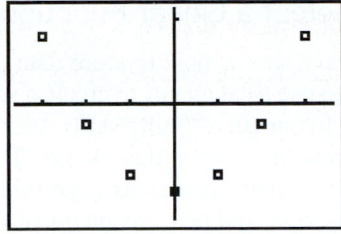

[−3.6, 3.6, 1] by [−6.53, 5.53, 1]

Self-Check 2

Use a graphing calculator to create a scatter diagram for the set of data points in the table.

x	−3	−2	1	0	1	2	3
y	−7	−2	−1	2	1	−2	−7

Example 3 examines a scatter diagram that illustrates the relationship between the speed
in miles per hour of an automobile and its rate of fuel consumption, measured in miles per
gallon. At higher speeds, there is greater wind resistance and less engine efficiency. Thus
greater speeds generally mean less fuel efficiency, as illustrated by the data in Example 3.

Example 3 Using a Graphing Calculator to Create a
Scatter Diagram

Use a graphing calculator to create a scatter diagram for the data
points in the given table. Then describe the visual pattern formed
by these points.

x mi/h	y mi/gal
20	23
25	22
30	21
35	20
40	19
45	18
50	17

Solution

First enter the data points, using the
STAT-EDIT feature.

Then use the **STAT-PLOT** feature to create a
scatter diagram of the data points listed
in **L1** and **L2.**

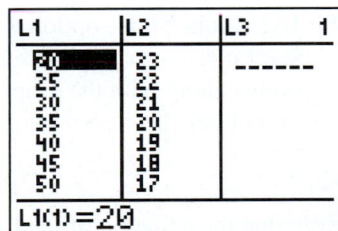

Answer: These points appear to form a linear
pattern with a negative slope.
The meaning of the negative slope
is that fuel efficiency is decreasing as
speed increases.

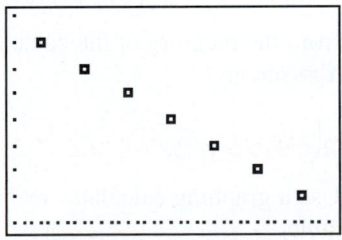

[17, 53, 1] by [15.98, 24.02, 1]

Self-Check 3

Use a graphing calculator to create a scatter
diagram for the set of data points in the table.

x	10	20	30	40	50	60	70
y	105	110	115	120	125	130	135

2. Select a Linear Function That Best Fits a Set of Data

We have shown how to store data points in a graphing calculator and we have examined how to use a calculator to create a scatter diagram of these points. We will now add the last piece to our curve-fitting skills. We will examine how to select a linear or a quadratic model that best fits a set of data points. The topic of fitting an equation to a set of data points is known in mathematics as **regression analysis.** We will limit our discussion of the details of this topic and take advantage of the capabilities of a graphing calculator to perform this work for us.

Technology Perspective 8.6.3 illustrates how to use a calculator to determine the line of best fit for the data in Example 3 and to then enter this equation into the ⬭Y= screen, using the slope and *y*-intercept just calculated. (Although there are shortcuts for transferring this equation directly to the ⬭Y= screen, we show the entry here to emphasize that the calculator has found the slope and *y*-intercept for the line best fitting these data.) Before you read Technology Perspective 8.6.3, enter the data from Example 3 in a graphing calculator and create the scatter diagram. Based on the linear pattern shown by the points, we calculate the line of best fit for the data. This is done by selecting option **4** under the **STAT-CALC** menu. We will use option **5** to select a quadratic model when the data exhibit a parabolic pattern.

Technology Perspective 8.6.3	Calculating and Graphing the Line of Best Fit

Calculate the line of best fit for the data in Example 3 and then graph this line.

TI-84 Plus Keystrokes

1. Select the **LinReg** option—this is option **4** in the **STAT-CALC** menu—by pressing ⬭STAT ⬭▷ ⬭4 .

2. Press ⬭ENTER to calculate the equation of the line of best fit.

3. Press ⬭Y= and enter the equation of the line of best fit as shown.

4. Use **Zoom Stat**—option **9** in the ⬭ZOOM menu—to display the data points along with the graph of the line of best fit.

TI-84 Plus Screens

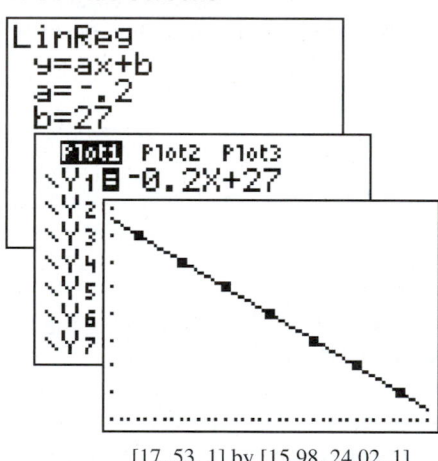

[17, 53, 1] by [15.98, 24.02, 1]

Note that the equation of the line of best fit can also be inserted into the ⬭Y= screen from the memory of the calculator by pressing ⬭VARS ⬭5 ⬭▷ ⬭▷ ⬭ENTER from the **Y=** screen.

Technology Self-Check 3

Use a graphing calculator to calculate and graph the line of best fit for the data in the table.

x	-3	-2	-1	0	1	2	3
y	-3	-1	1	5	7	9	12

Do All Data Fit an Equation Perfectly?

No, the data in Example 3 are phony in the sense that the authors made up the data set to exactly fit a line—it was manipulated to meet an ideal model. Textbooks are full of idealized data because it is easier to teach concepts piece-by-piece with nice clean data. It is important to realize that we can use the same procedures with real data—it just may be a little messier. The next examples address working with real data. Real data rarely lie perfectly on an idealized curve. Thus we try to select the curve that best fits the data, even when the curve will miss some of the data points.

Scatter Diagrams of Real Data

Data Forming a Linear Pattern   **Data Forming a Parabolic Pattern**

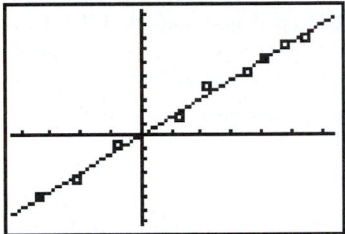   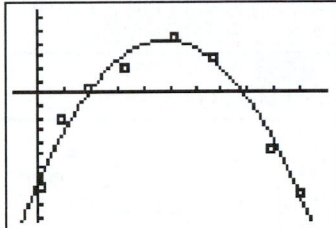

These data points do not lie exactly on a line or on a parabola, but they do exhibit, respectively, a strong linear and a strong parabolic pattern. The criteria for deciding which curve best fits a set of data is covered in most introductory statistics courses. We will use calculators to help us focus on the key steps rather than on the theoretical background. This is illustrated in the next two examples.

Example 4   Modeling Athletic Records Using a Line of Best Fit

Year	x	y (meters)
1948	48	5.69
1952	52	6.24
1956	56	6.35
1960	60	6.37
1964	64	6.76
1968	68	6.82
1976	76	6.72
1980	80	7.06
1984	84	6.96
1988	88	7.40
1992	92	7.14
1996	96	7.12
2000	100	6.99
2004	104	7.07
2008	108	7.04

The table displays the winning lengths for the Olympic women's long jump from 1948 through 2008. Each distance is given in meters. (The actual dates are not used in the calculations. The x-values represent the number of years after 1900. These smaller x-values allow us to round the coefficients of the equation of best fit without introducing an unacceptable amount of error.)

(a) Draw a scatter diagram for these data and then determine the line of best fit for the data.

(b) Interpret the meaning of the slope of this line.

(c) Use the equation of the line of best fit to estimate the missing entry in the table from 1972. How does this estimation compare to the actual 1972 winning jump of 6.78 m?

(d) Use the equation of the line of best fit to predict the women's winning long jump for the year 2012.

Solution

(a) Enter the data points using the **STAT-EDIT** feature.

L1	L2	L3	1
48	5.69	------	
52	6.24		
56	6.35		
60	6.37		
64	6.76		
68	6.82		
76	6.72		

L1(1)=48

Be sure to enter all the data points since these points will fill more than one display screen.

Use the **STAT-PLOT** feature to draw a scatter diagram of the data points listed in **L1** and **L2**.

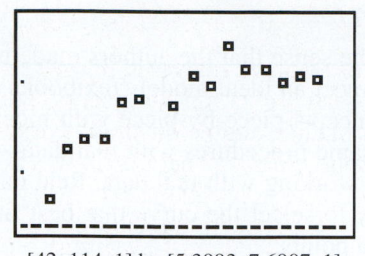

[42, 114, 1] by [5.3993, 7.6907, 1]

Note that the points in this scatter diagram do not lie on the same line. The upward trend to the right suggests that the slope of the line of best fit will be positive.

Use the **STAT-CALC** feature to calculate the line of best fit.

Enter the line of best fit on the (Y=) screen and then graph this line.

The line of best fit is approximately $y = 0.019x + 5.294$.

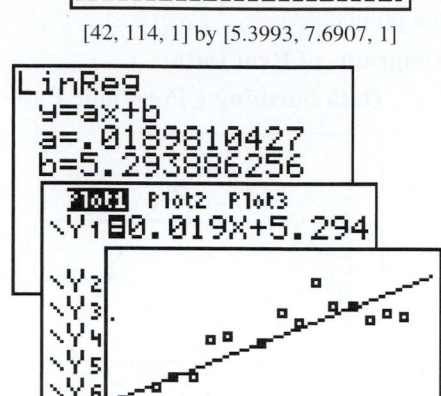

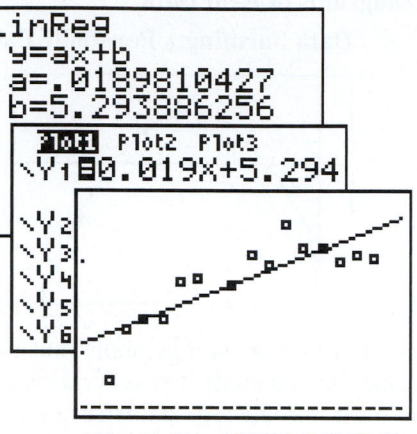

[42, 114, 1] by [5.3993, 7.6907, 1]

Round each coefficient in the equation to the nearest thousandth. The slope of the line of best fit is approximately 0.019, and the y-intercept is approximately 5.294. This line of best fit does not contain each data point but it fits these points better than any other line.

Is the y-intercept meaningful in this problem?

(b) The slope of the line of best fit is positive, which indicates that this linear function is increasing. The trend is for the winning distance to increase by about 1.9 cm per year.

Caution: Although the long-term trend is for the winning long jump to increase, this does not mean that the winning jump does increase at each Olympics. In fact, for some Olympics, the winning long jump is less than that at the previous Olympics.

0.019 m = 1.9 cm

(c) For 1972, the line of best fit yields an estimation of 6.66 meters. The actual 1972 winning distance of 6.78 meters is 12 cm longer than that given by this linear model.

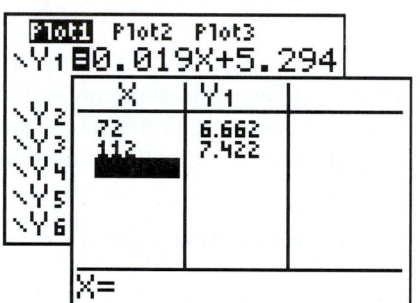

To estimate the winning long jump for 1972, substitute 72 for x in the equation for the line of best fit. To estimate the winning long jump distance for 2012, substitute 112 for x in the equation for the line of best fit.

The estimation error of 12 cm is about the same as the width of about 6 adult fingers.

Both calculations are made using the **ASK** mode in a table for this linear function.

(d) For 2012, the line of best fit yields an estimation of 7.42 meters.

Many factors could affect this prediction. Do you think a linear model is reasonable for this problem?

Self-Check 4

a. Use a graphing calculator to determine the line of best fit for these data points.

x	−3	−2	−1	0	1	2	3
y	−6	−3	−2	2	2	6	6

Write your answer in slope-intercept form with m and b rounded to the nearest thousandth.

b. Use this linear equation to estimate y to the nearest tenth when x = −1.3.

3. Select a Quadratic Function That Test Fits a Set of Data

If the scatter diagram for a set of data points displays an overall linear pattern, then we will likely examine linear curve fitting. If the scatter diagram for a set of data points displays an overall parabolic pattern, then we will likely examine quadratic curve fitting. The data points given in Example 5 form a parabolic pattern. Therefore a graphing calculator is used to determine the quadratic function of best fit. The keystrokes for calculating the parabola of best fit are identical to those for calculating the line of best fit except that we select the **QuadReg** option instead of the **LinReg** option under the **STAT-CALC** menu. Option **5** is the **QuadReg** option. The coefficients in this function will be rounded to the nearest thousandth. Rounding can sometimes create significant differences when we use the quadratic function of best fit.

Example 5	Using a Graphing Calculator to Determine a Parabola of Best Fit

Enter these points into a graphing calculator, draw the scatter diagram, and calculate the parabola of best fit. Use this quadratic equation to approximate y when $x = 6$.

x	y
0.1	-7.7
0.9	-2.0
1.9	0.4
3.3	2.1
5.2	4.5
6.7	2.9
8.9	-4.6
10.0	-8.2

Solution

1. Enter the data points using the **STAT-EDIT** feature.

Be sure to enter all points even though they cannot all be displayed on one screen.

2. Turn **Plot 1** on to draw a scatter diagram of the data points listed in **L1** and **L2.** Use the visual pattern to select the most appropriate family of curves. In this case, the visual pattern is parabolic.

$[-0.89, 10.99, 1]$ by $[-10.359, 6.659, 1]$

Since the points appear to form a parabola opening downward, try to find a quadratic function that best fits these data.

3. Use the **STAT-CALC** feature to calculate the parabola of best fit.

QuadReg
$y = ax^2 + bx + c$
$a = -.4874969798$
$b = 4.720802979$
$c = -7.171408121$

To calculate the parabola of best fit, select option 5 under the **STAT** **CALC** menu.

4. Enter the parabola of best fit on the Y= screen, and then graph this curve. Round each coefficient to the nearest thousandth.

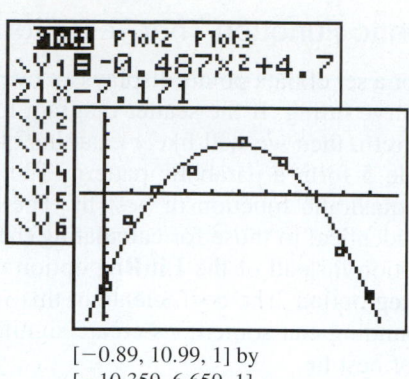

[−0.89, 10.99, 1] by
[−10.359, 6.659, 1]

This parabola of best fit does *not* contain each data point, but it does fit these points better than any other parabola.

Note that the equation of the parabola of best fit also can be inserted into the Y= screen from the memory of the calculator by pressing

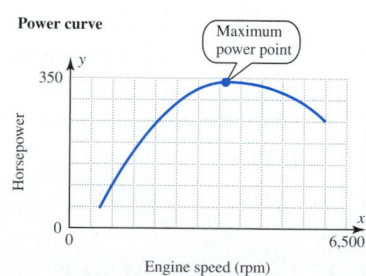

5. $y = -0.487x^2 + 4.721x - 7.171$

$y = -0.487(6)^2 + 4.721(6) - 7.171$

$y = 3.623$

$y \approx 3.6$

To approximate y for $y = 6$, substitute 6 into the equation of the parabola of best fit. Since the original data were accurate only to the tenths place, the estimate is rounded to the nearest tenth.

Answer: $y = -0.487x^2 + 4.721x - 7.171$ is approximately the parabola of best fit. For $x = 6$, $y \approx 3.6$.

The equation of the parabola of best fit is written in $y = ax^2 + bx + c$ form with each coefficient rounded to the nearest thousandth.

Self-Check 5

a. Use a graphing calculator to determine the parabola of best fit for these data points.

x	0	1	2	3	4	5	6
y	−4	0	4	4	2	0	−4

Write your answer in the form $y = ax^2 + bx + c$ with each coefficient rounded to the nearest thousandth.

b. Use this quadratic equation to approximate y to the nearest tenth when $x = 1.5$.

The amount of power an engine produces depends on the speed at which it is running. The graph relating speed to power is called a power curve. For some engines, this power curve is approximately parabolic. The point of maximum power is important because engines usually perform better when running faster than the maximum power point than they do when running slower than the maximum power point. Suppose the engine on a boat is temporarily put under extra load due to a wave or debris on the propeller. For an engine running faster than the maximum power level, this will briefly slow down the engine, increase the power, and smoothly overcome the extra load. We can use data collected on an engine to determine its power curve and its maximum power level. This is illustrated in Example 6.

Power curve

Maximum power point

Horsepower

350

0

0   6,500

Engine speed (rpm)

Example 6 — Modeling the Power Curve for the Engine on a Boat

Speed x (rpm)	Horsepower y = f(x) (hp)
0	0
500	130
1,000	240
2,000	340
3,000	280
3,500	180
4,000	50

(a) The data points shown in the table give the horsepower of one engine running at some tested speeds, in revolutions per minute (rpm). Create a power curve that is a quadratic function that best fits these data. (Round the coefficients of this quadratic function to three significant digits.)

(b) Use this power curve to determine the point of maximum power for this engine.

(c) Interpret both coordinates of this maximum power point.

Solution

(a)

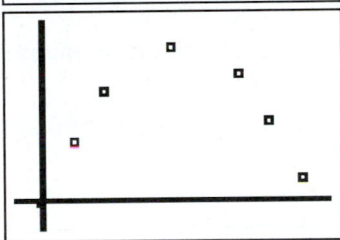

[−400, 4400, 1] by [−57.8, 397.8, 1]

Use a graphing calculator to plot the given points on a scatter diagram, and then use the **QuadReg** feature under the **CALC** submenu of the **STAT** menu to calculate the equation of the parabola of best fit.

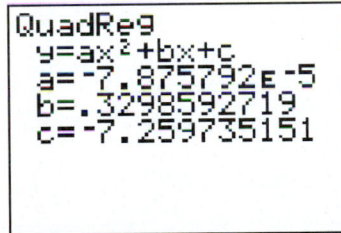

The equation of the parabola of best fit (with coefficients rounded to three significant digits) is
$f(x) = -0.0000788x^2 + 0.330x - 7.26$.

Power curve:
$$f(x) = -0.0000788x^2 + 0.330x - 7.26$$

(b)

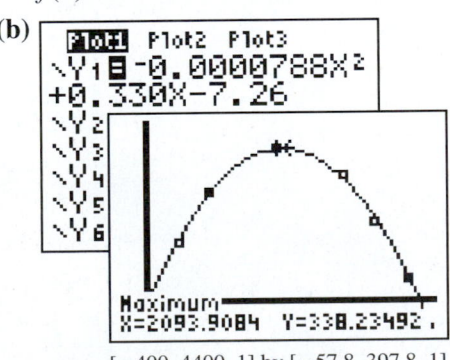

[−400, 4400, 1] by [−57.8, 397.8, 1]

Graph this parabola, and then use the **maximum** feature in the **CALC** menu to determine the vertex of the parabola.

Maximum power point: The vertex of this parabola of best fit is approximately at (2,100, 340).

The coordinates of the vertex have been rounded to two significant digits.

(c) To achieve the maximum possible horsepower of approximately 340 hp, run the engine at approximately 2,100 rev/min.

From the preceding paragraph it might be wise to actually run the engine at about 2,400 rev/min. The horsepower at this level is about 330 hp; and when an extra load is placed on the engine at this speed, the engine will slow down and respond with more power.

$$f(x) = -0.0000788x^2 + 0.330x - 7.26$$
$$f(2,400) = -0.0000788(2,400)^2 + 0.330(2,400) - 7.26$$
$$f(2,400) \approx 330$$

Self-Check 6

a. The data in the table were taken from a physics experiment where a ball was thrown from the top of a building and the distance from the ground to the ball was recorded as a function of the number of seconds after it was thrown. Use a graphing calculator to determine the parabola of best fit for these data points.

x (sec)	0	0.4	0.8	1.2	1.6	2.0	2.4
y (ft)	20	30	36	35	31	20	5

Write your answer in the form $y = ax^2 + bx + c$ with each coefficient rounded to the nearest hundredth.

b. Approximate the coordinates of the vertex of this function with each coordinate rounded to the nearest hundredth.

c. Interpret the meaning of the x- and y-coordinates of this vertex.

Self-Check Answers

1. a.

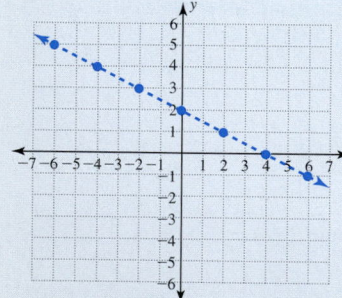

b.

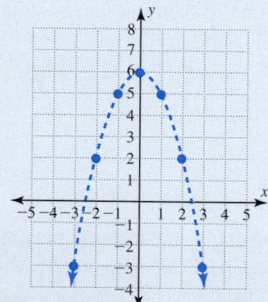

4. a. $y = 2.071x + 0.714$
   b. For $x = -1.3$, $y \approx -2.0$.
5. a. $y = -0.881x^2 + 5.214x - 3.905$
   b. For $x = 1.5$, $y \approx 1.9$.
6. a. $y = -16.07x^2 + 32.32x + 19.93$
   b. (1.00, 36.18)
   c. The maximum height is reached by the object 1 second after it is thrown. The maximum height is 36.18 feet.

Shape: A line with negative slope

Shape: A parabola opening downward

2.

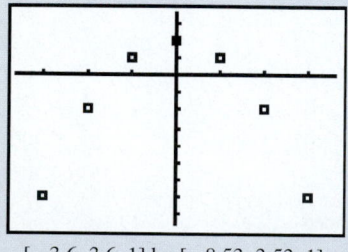

$[-3.6, 3.6, 1]$ by $[-8.53, 3.53, 1]$

3.

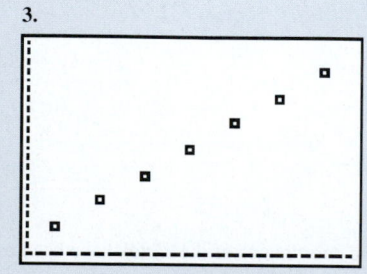

$[4, 76, 1]$ by $[99.9, 140.1, 1]$

Technology Self-Check Answers

1 and 2.

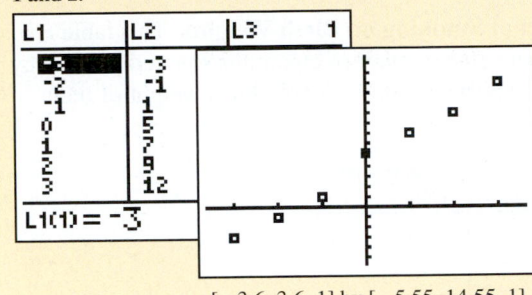

[−3.6, 3.6, 1] by [−5.55, 14.55, 1]

3.

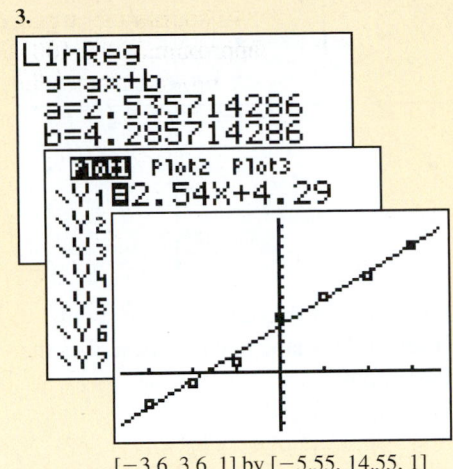

[−3.6, 3.6, 1] by [−5.55, 14.55, 1]

8.6   Using the Language and Symbolism of Mathematics

1. A _____ diagram for a set of data points is formed by graphing these points on a coordinate system.

2. If the pattern formed by the points of a scatter diagram is approximately a _____, then we will select an equation of the form $y = mx + b$.

3. If the pattern formed by the points of a scatter diagram is approximately a _____, then we will select an equation of the form $y = ax^2 + bx + c$.

4. The topic of selecting a curve that best fits a set of data points is called _____ fitting.

5. Selecting an equation that best fits a set of data points is called _____ analysis.

6. In the _____-_____ form $y = mx + b$, m represents the _____ of a line with y-intercept (_____, _____).

8.6   Quick Review

Use the function $f(x) = 15x + 20$ to answer each question.

1. Determine the slope.

2. Determine the intercepts.

Use the function $f(x) = -x^2 + 5x + 6$ to answer each question.

3. Determine the vertex.

4. Does the graph of this function open upward or downward?

5. Determine the intercepts.

8.6   Exercises

Objective 1 Create a Scatter Diagram and Select an Appropriate Model for the Data

Exercises 1–4 give a family of functions. Select the scatter diagram that best matches the corresponding description.

1. A line with positive slope

2. A line with negative slope

3. A parabola opening upward

4. A parabola opening downward

A.

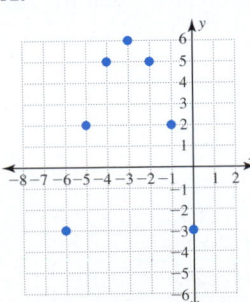

B.

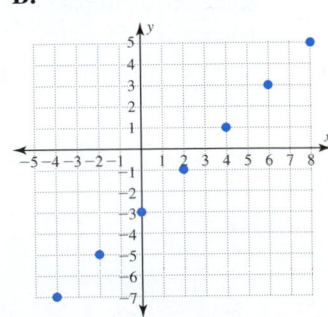

C.

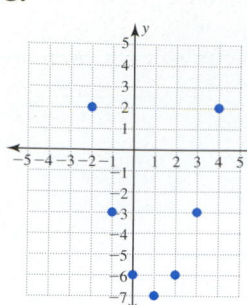

D.

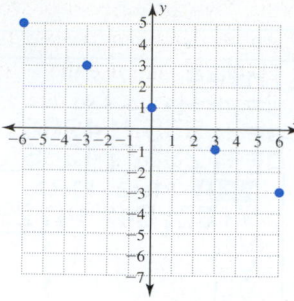

In Exercises 5–8, draw a scatter diagram for each set of data points and then select the shape that best describes the pattern formed by these points.

5.

x	−3	−2	−1	0	1	2	3
y	6	4	3	0	−1	−5	−6

6.

x	−3	−2	−1	0	1	2	3
y	−6	−2	1	2	1	−3	−8

7.

x	−3	−2	−1	0	1	2	3
y	10	4	0	−2	−2	0	4

8.

x	−3	−2	−1	0	1	2	3
y	16	21	23	27	30	34	37

A. A line with positive slope
B. A parabola opening upward
C. A line with negative slope
D. A parabola opening downward

Objective 2 Select a Linear Function That Best Fits a Set of Data

In Exercises 9–12, use a graphing calculator and the points given in the table.

a. Draw a scatter diagram of these points.
b. Determine the linear function that best fits these points. (Round the coefficients to the nearest thousandth.)

9.

x	y
1	3
2	5
3	8
4	9
5	11
6	12

10.

x	y
10	18
15	16
19	12
26	9
29	6
34	2

c. Use this linear equation to estimate y when x is 3.5.

c. Use this linear equation to estimate y when x is 22.

11.

x	y
−10	5
−7	1
−5	−3
−2	−7
0	−10
3	−14

12.

x	y
1	4
4	7
9	8
12	11
15	15
19	19

c. Use this linear equation to estimate y when x is 1.

c. Use this linear equation to estimate y when x is 10.

13. Effect of Smoking on Birth Weights This table compares the number of cigarettes smoked per day by eight expectant mothers to the birth weight of their children.

Number of Cigarettes x	Birth Weight y (kg)
5	3.20
10	3.15
15	3.10
20	3.06
25	3.09
30	3.05
35	3.01
40	3.00

a. Draw a scatter diagram for these data points, and determine the line of best fit. (Round the coefficients to the nearest thousandth.)
b. Interpret the meaning of the slope of the line of best fit.
c. Use this line of best fit to estimate the birth weight of a child whose mother smoked 18 cigarettes per day.
d. Use this line of best fit to estimate the birth weight of a child whose mother smoked 45 cigarettes per day.

14. Body Height Based on Femur Length Forensic scientists can estimate the height of a person based on the length of different bones of that person. This table of values compares the length of a man's femur (the thigh bone) to the height of that man.

Length of Femur x (cm)	Height y (cm)
43.0	164
43.5	166
44.0	166
44.5	168
45.0	170
45.5	169
46.0	172
46.5	173
47.0	175
47.5	175

a. Draw a scatter diagram for these data points and determine the line of best fit. (Round the coefficients to the nearest thousandth.)
b. Interpret the meaning of the slope of the line of best fit.
c. Use the line of best fit from part **a** to estimate the height of a man whose femur length is 44.7 cm.
d. Use the line of best fit from part **a** to estimate the height of a man whose femur length is 48.5 cm.

15. Olympic Men's Marathon Records The given table displays the winning times for the Olympic men's marathon for selected years from 1900 through 2008. Because each time is given as 2 hours and some number of minutes and seconds, the y-values in the table include only the time in seconds in excess of 2 hours. (Do not use the actual dates in your calculations. Use the x-values so that you can approximate the coefficients of the equation of best fit without introducing an unacceptable amount of error.)

Year	x	y Seconds After 2 Hours
1900	0	3,585
1912	12	2,214
1920	20	1,956
1932	32	1,896
1948	48	2,091
1960	60	916
1972	72	740
1980	80	663
1996	96	756
2000	100	611
2008	108	392

a. Draw a scatter diagram for these data points and then determine the line of best fit. (Round the coefficients to the nearest thousandth.)
b. Interpret the meaning of the slope of the line of best fit.
c. Use the equation of the line of best fit to estimate the missing entry in the table from 1984. How does this estimation compare to the actual 1984 record of 2 hours, 9 minutes, 21 seconds?
d. Use the equation of the line of best fit to predict the year in which the men's marathon record will become 2 hours. (Remember that the Olympics are held only every 4 years.)

16. Number of Women in the U.S. House The given table displays the number of women in the U.S. House of Representatives for selected years from 1980 through 2008. (Do not use the actual dates in your calculations. Use the x-values so that you can approximate the coefficients of the equation of best fit without introducing an unacceptable amount of error.)

Year	x	Women in House y
1980	80	19
1982	82	21
1984	84	22
1986	86	23
1988	88	25
1990	90	28
1992	92	47
1994	94	47
1996	96	51
1998	98	56
2000	100	59
2002	102	59
2004	104	68
2006	106	73
2008	108	78

a. Draw a scatter diagram for these data and then determine the line of best fit. (Round the coefficients to the nearest thousandth.)
b. Interpret the meaning of the slope of the line of best fit from part **a.**
c. Use the equation of the line of best fit to estimate the number of women Representatives in 1976. There were 18 women Representatives in 1976. How close was your estimate?
d. Use the equation of the line of best fit to predict the number of women Representatives in the year 2012.

Objective 3 Select a Quadratic Function That Best Fits a Set of Data

In Exercises 17–20, use a graphing calculator and the points given in the table.

a. Draw a scatter diagram of these points.
b. Determine the quadratic function that best fits these points. (Round the coefficients to three significant digits.)

17.

x	y
0	11
1	2
2	−4
3	−5
4	−3
5	1

18.

x	y
2	5.3
3	3.1
5	1.0
6	1.4
7	2.9
8	5.5

c. Use this quadratic equation to estimate y when x is 2.5.

c. Use this quadratic equation to estimate y when x is 4.

19.

x	y
−5	−18
−4	−12
−3	−9
−2	−8
−1	−9
0	−13

c. Use this quadratic equation to estimate y when x is 1.

20.

x	y
−5	−5.1
−3	−2.8
−1	−1.9
1	−2.6
3	−5.2
5	−9.1

c. Use this quadratic equation to estimate y when x is 6.

21. Load Strength of a Beam The given experimental data relate the maximum load in kilograms that a beam can support to its depth in centimeters.

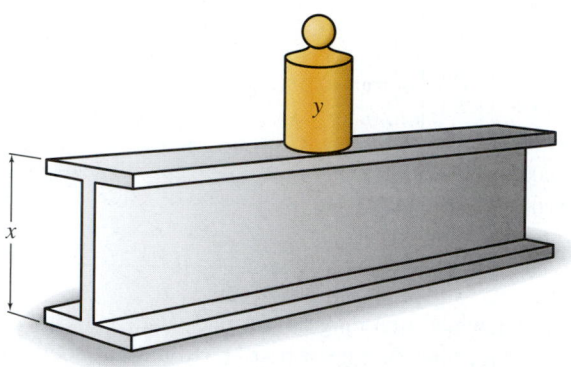

a. Use a calculator to determine the parabola of best fit for these data.

b. Use this quadratic equation to estimate the load that the beam can support if its depth is 15 cm.

Depth x (cm)	Load y (kg)
10	4,900
12	7,200
14	9,900
16	12,500
18	16,000
20	20,000
22	24,000
24	28,500

22. Modeling Distance of a Fall Based on Time
An experiment is run in a physics course to compare the distance an object falls to the time of the fall.

a. Draw a scatter diagram for these data and determine the parabola of best fit. (Round the coefficients of this quadratic function to three significant digits.)

b. Use this quadratic equation to estimate the distance an object would fall in 4.5 seconds.

x (sec)	y (ft)
0	0
1	15.9
2	62.5
3	146.0
4	257.1
5	400.9
6	580.2
7	783.6

23. Engine Power Curve

a. The data points shown in the table give the horsepower of a racing engine running at some tested revolutions per minute. Create a power curve that is a quadratic function that best fits these data. (Round the coefficients of this quadratic function to three significant digits.)

x (rev/min)	y (hp)
0	0
500	340
1,000	600
2,000	800
3,000	700
4,000	130

b. Then use this power curve to determine the point of maximum power for this engine.

c. Interpret both coordinates of this maximum power point.

24. Engine Power Curve

a. The data points shown in the table give the horsepower of one engine running at some tested value of revolutions per minute. Create a power curve that is a quadratic function that best fits these data. (Round the coefficients of this quadratic function to three significant digits.)

x (rev/min)	y (hp)
0	0
500	120
1,000	220
2,000	310
3,000	250
3,500	160
4,000	45

b. Use this power curve to determine the point of maximum power for this engine.

c. Interpret both coordinates of this maximum power point.

25. Profit as a Function of Production A business collected the data in the given table.

a. Use a graphing calculator to determine the quadratic function that gives the profit in dollars as a function of the number of units produced per day. Round the coefficients to three significant digits.

x Units	Profit y ($)
0	−350
10	−80
20	100
30	300
40	400
50	400
60	350
70	280

b. Use this function to approximate the break-even values for this business.

c. Use this function to approximate the maximum profit the business can make per day and the number of units it should produce to generate this profit.

26. Profit as a Function of Production A business collected the data in the given table.

a. Use a graphing calculator to determine the quadratic function that gives the profit in dollars as a function of the number of units produced per day. Round the coefficients to three significant digits.

x Units	Profit y ($)
0	−70
10	70
20	140
30	210
40	230
50	170
60	100
70	50

b. Use this function to approximate the break-even values for this business.

c. Use this function to approximate the maximum profit the business can make per day and the number of units it should produce to generate this profit.

Home Run Distance

The technology of tracking home run distances in baseball has improved greatly over the years. When a home run lands 40 ft above the ground in the bleachers 350 ft from home plate, it can be very difficult to determine exactly how far it would have traveled. Today special cameras are placed near home plate and at each of the foul poles to record data on the exact location of the ball. Then computers analyze the data to determine how far the ball would have traveled. In Exercises 27 and 28, x represents the horizontal distance the ball is from home plate, and y represents the height of the ball. Use a graphing calculator and the points given in the table to do the following.

a. Draw a scatter diagram of these points.
b. Determine the quadratic function that best fits these points. (Round the coefficients to three significant digits.)
c. Use this function f to approximate $f(325)$.
d. Interpret the meaning of the value found in part **c.**
e. Determine the x-intercepts of the graph of this function.
f. Interpret the meaning of the positive x-intercept found in part **e.** (The negative x-intercept has no practical meaning.)
g. Determine the vertex of this parabola.
h. Interpret the meaning of the vertex found in part **g.**

Group discussion questions

29. **Local Linearity** Over a small segment of a curve, almost any curve will appear linear.
    a. Determine the line of best fit for these data points:

x	0	1	2
y	4	3	0

    b. Determine the parabola of best fit for these data points:

x	−3	−2	−1	0	1	2	3
y	−5	0	3	4	3	0	−5

    c. Which curve do you think better fits the points in part **b,** a line or a parabola?
    d. Discuss your choice in part **c** with respect to "local linearity."

30. **Risks and Choices** For engineers and scientists, choosing an appropriate model for a set of data is extremely important. In Exercises 30 and 31, we compare a parabola of best fit to a line of best fit. Points that lie along one branch of a parabola may appear to be approximately linear, but a parabola will still be a better fit for these points.

x	−7	−6	−5	−4	−3	−2	−1	0	1
y	4	3	2	1	1	0.5	0.4	0.5	0.8

For the data points shown in the table, use a graphing calculator to:

    a. Draw a scatter diagram of these points.
    b. Determine the line of best fit in the form $y = mx + b$ with m and b rounded to the nearest thousandth.
    c. Graph this line of best fit on the scatter diagram.
    d. Determine the parabola of best fit in the form $y = ax^2 + bx + c$ with the coefficients rounded to the nearest thousandth.
    e. Graph this parabola of best fit on the scatter diagram.
    f. Explain why these graphs indicated that the parabola fits the data points better than the line.
    g. For $x = -2.5$, use the line of best fit to calculate y.
    h. For $x = -2.5$, use the parabola of best fit to calculate y.
    i. Compare the points calculated in parts **g** and **h.** Which of these points is closer to the given data points?

31.

x	−2	−1	0	1	2	3	4	5	7
y	0.5	0.4	0.5	0.8	1	2	3	4	7

For the data points shown in the table, use a graphing calculator to:

    a. Draw a scatter diagram of these points.
    b. Determine the line of best fit in the form $y = mx + b$ with m and b rounded to the nearest thousandth.
    c. Graph this line of best fit on the scatter diagram.
    d. Determine the parabola of best fit in the form $y = ax^2 + bx + c$ with the coefficients rounded to the nearest thousandth.
    e. Graph this parabola of best fit on the scatter diagram.

27.

x ft	y ft
0	4
100	75
200	90
300	49

28.

x ft	y ft
0	3
100	110
200	160
300	153

f. Explain why these graphs indicated that the parabola fits the data points better than the line.

g. For $x = 2.5$, use the line of best fit to calculate y.

h. For $x = 2.5$, use the parabola of best fit to calculate y.

i. Compare the points calculated in parts **g** and **h**. Which of these points is closer to the given data points?

32. Challenge Question Collect a pair of data items from approximately 15 people of the same gender and display these data in a table. For example, measure on

the same individual both the distance from the tip of an elbow to the end of an index finger and the length of the individual's right foot.

a. Before you collect these data, predict whether or not a line of best fit will be a good model.

b. Do you expect this line of best fit to have a positive or negative slope? Why?

c. Draw the scatter diagram, and calculate the line of best fit.

d. Did your predictions match the results you found?

8.6 Cumulative Review

Solve each equation.

1. $2x + 5 = 19$

2. $1 - \dfrac{2x - 5}{3} = \dfrac{x}{2}$

3. $|4x + 3| = 8$

4. $(3x + 1)^2 = 12$

5. $(x + 5)(x - 4) = 22$

Chapter 8 Key Concepts

1. Functions
- A function is a correspondence that matches each input value with exactly one value of the output variable.
- The input variable is also called the independent variable.
- The output variable is also called the dependent variable.
- The domain of a function is the set of all input values.
- The range of a function is the set of all output values.

2. Notations for Representing Functions
- Mapping notation
- Ordered-pair notation
- Table of values
- Graphs
- Function notation
- Verbally

3. Domain and Range from the Graph of a Function
- The domain of a function is the projection of its graph onto the x-axis.
- The range of a function is the projection of its graph onto the y-axis.

4. Vertical Line Test
- A graph represents a function if it is impossible to have any vertical line intersect the graph at more than one point.

5. Function Notation: The function notation $f(x)$ is read "f of x." $f(x)$ represents a unique output value for each input value of x.

6. Slope of a Line
- The slope m of a line through (x_1, y_1) and (x_2, y_2) with $x_1 \neq x_2$ is

$$m = \frac{\text{Change in } y}{\text{Change in } x} = \frac{y_2 - y_1}{x_2 - x_1} = \frac{\Delta y}{\Delta x} = \frac{\text{rise}}{\text{run}}$$

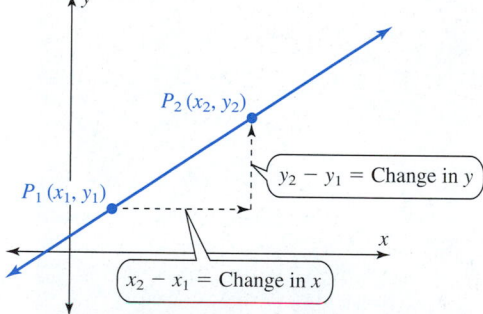

- The slope of a line gives the change in y for each 1-unit change in x.
- A line with positive slope goes upward to the right.
- A line with negative slope goes downward to the right.
- The slope of a horizontal line is 0.

- The slope of a vertical line is undefined.
- The slopes of parallel lines are the same.
- The slopes of perpendicular lines are opposite reciprocals.
- The product of the slopes of perpendicular lines is -1.
- The slope of $y = mx + b$ is m.
- The slope of $y - y_1 = m(x - x_1)$ is m.
- The common difference of an arithmetic sequence is the same as the slope of the line through the points formed by this sequence.

7. Linear Functions
- The graph of a linear function is a straight line.
- The equation is first degree and can be written in the slope-intercept form $f(x) = mx + b$ with slope m and y-intercept $(0, b)$.
- Linear functions have a constant rate of change. The slope is the same between any two points on the line.
- The domain of all linear functions is $\mathbb{R}$. The range of all linear functions is also $\mathbb{R}$ unless the function is a constant function, whose graph is a horizontal line.

8. Sketching the Graph of a Linear Function
- **Method 1:** Use the equation and any two input values and determine their corresponding output values. The graph of the linear function is the line through these two points.
- **Method 2:** Start with the y-intercept as one point on the graph and use the slope to produce a second point. The graph of the linear function is the line through these points.

9. Intercepts of a Line
- The x-intercept, often denoted by $(a, 0)$, is the point where the line crosses the x-axis.
- The y-intercept, often denoted by $(0, b)$, is the point where the line crosses the y-axis.

10. Positive and Negative Functions
- A function is positive when $f(x)$ is positive. On the graph of $y = f(x)$, this occurs at points above the x-axis.
- A function is negative when $f(x)$ is negative. On the graph of $y = f(x)$, this occurs at points below the x-axis.

11. Maximum and Minimum Values
- A maximum y-value of a function is at the highest point on the graph of this function.
- A minimum y-value of a function is at the lowest point on the graph of this function.

12. Absolute Value Functions
- The graph of an absolute value function is V-shaped.
- The graph will open upward if the sign in front of the absolute value symbol is positive.
- The graph will open downward if the sign in front of the absolute value symbol is negative.
- The vertex of an absolute value function is at the tip of the V-shape.
- If the graph of an absolute value function opens upward, the minimum y-value will be at the vertex.
- If the graph of an absolute value function opens downward, the maximum y-value will be at the vertex.
- The domain of an absolute value function is $\mathbb{R}$.

13. Sketching the Graph of an Absolute Value Function
- Determine whether the graph opens upward or opens downward.
- Determine the location of the vertex.
- Complete a table of values using x-values on both sides of the vertex.

14. Absolute-Value Equations and Inequalities: For any real numbers x and a and positive real number d:
- $|x - a| = d$ is equivalent to $x - a = -d$ or $x - a = d$.
- $|x - a| < d$ is equivalent to $-d < x - a < d$.
- $|x - a| > d$ is equivalent to $x - a < -d$ or $x - a > d$.
- $|x - a| = -d$ is a contradiction and has no solution.
- $|x - a| < -d$ is a contradiction and has no solution.
- $|x - a| > -d$ is an unconditional inequality and the solution set is the set of all real numbers.

15. Quadratic Functions
- A function of the form $f(x) = ax^2 + bx + c$ is a quadratic function.
- The graph of a quadratic function is a parabola.
- ax^2 is called the quadratic term or the second-degree term.
- bx is called the linear term or the first-degree term.
- c is called the constant term.
- For $a > 0$, the parabola opens upward.
- For $a < 0$, the parabola opens downward.
- The vertex is the lowest point on a parabola that opens upward.
- The vertex is the highest point on a parabola that opens downward.
- The axis of symmetry is a vertical line passing through the vertex.
- The vertex is located at $\left(-\dfrac{b}{2a}, f\left(-\dfrac{b}{2a} \right) \right)$.

16. Comparison of Linear Functions and Quadratic Functions

	Algebraically	Graphically	Numerically
Linear Function	$y = mx + b$	A straight line	A fixed change in x produces a constant change in y.
Quadratic Function	$y = ax^2 + bx + c$	A parabola	The y-values form a symmetric pattern about the vertex.

17. Sketching the Graph of a Quadratic Function
- Determine whether the parabola opens upward or downward.
- Determine the coordinates of the vertex.
- Determine the intercepts.
- Complete a table using points on both sides of the vertex.
- Connect all points with a smooth parabolic shape.

18. Increasing and Decreasing Functions
- A function is increasing on a portion of its graph if as the x-values increase, the y-values also increase.
- A function is increasing on a portion of its graph if the graph rises as it moves from left to right.
- A function is decreasing on a portion of its graph if as the x-values increase, the y-values decrease.
- A function is decreasing on a portion of its graph if the graph drops as it moves from left to right.

19. Analyzing the Graphs of Linear, Absolute Value, and Quadratic Functions
- Identify the type of function.
- For linear functions, determine the slope. For absolute value and quadratic functions, determine the vertex and whether the graph opens upward or downward.
- Determine the y-intercept and the x-intercept(s).
- Determine the domain and range.
- Determine where the function is positive and where the function is negative.
- Determine where the function is increasing and where the function is decreasing.

20. Curve Fitting
- A scatter diagram for a set of data points is a graph of these points.
- Selecting a curve that best fits a set of data is called curve fitting.
- On a graphing calculator, use linear regression, **LinReg,** to calculate a line of best fit for a set of data.
- On a graphing calculator, use quadratic regression, **QuadReg,** to calculate the parabola of best fit for a set of data.

Chapter 8 | Review Exercises

Functions, Domain, and Range

In Exercises 1 and 2, determine whether each relation is a function.

1. a.

x	y
9	1
8	2
6	4
3	5

b. x → y
9 → 1
8 →
3 →

c. $\{(1, 3), (1, 8), (1, 9)\}$ **d.** $\{(1, \pi), (2, 4\pi), (3, 9\pi)\}$

2. a.

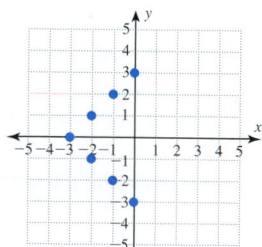

b.

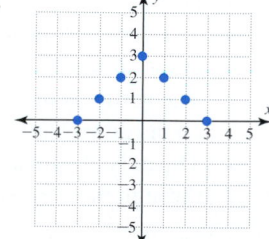

c.

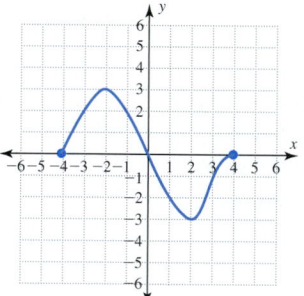

d.

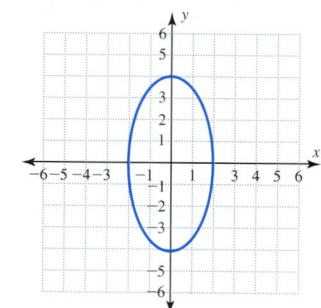

3. Determine the domain and range of each function.

a. $\{(5, 8), (6, 8), (7, 8)\}$

b.

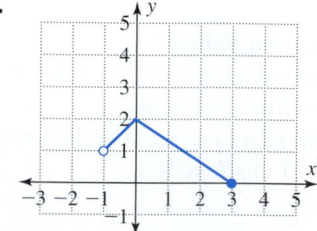

c. $f(x) = -|x + 3| + 4$

d. $f(x) = x^2 - 6x + 8$

4. Evaluate each expression for $f(x) = (x + 2)^2 + 3$.
  a. $f(-2)$ **b.** $f(0)$
  c. $f(1)$ **d.** $f(8)$

5. Evaluate each expression in parts **a** and **b**, given the table of values for $f(x)$.
  a. $f(-2)$
  b. $f(0)$
  c. Determine the input value of x for which $f(x) = 8$.
  d. Determine the input value of x for which $f(x) = 3$.

x	$f(x)$
-3	5
-2	4
-1	3
0	2
1	6
2	8
3	10

6. Evaluate each expression in parts **a** and **b**, given the graph of $y = f(x)$.
  a. $f(-4)$ **b.** $f(3)$
  c. Determine the input value of x for which $f(x) = 0$.
  d. Determine the input value of x for which $f(x) = -1$.

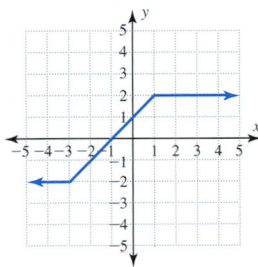

Linear Functions

7. Use the function defined by the given table to complete each part of this exercise.
  a. Express this function using mapping notation.
  b. Express this function using ordered-pair notation.
  c. Graph this function by making a scatter diagram of these data points.
  d. Write an equation for the line containing these points.

x	y
-10	4
-5	3
0	2
5	1
10	0
15	-1
20	-2

8. Use the function graphed in the given scatter diagram to complete each part of this exercise.
  a. Express this function using ordered-pair notation.
  b. Express this function using a table format.
  c. Express this function using mapping notation.
  d. Write the equation of the line containing these points.

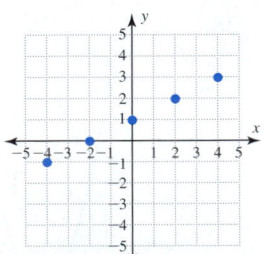

In Exercises 9 and 10, determine the slope of each line.

9.

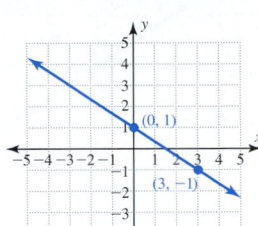

10.

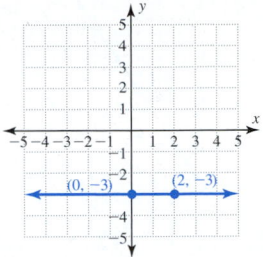

In Exercises 11 and 12, sketch the graph of each linear function.

11. $f(x) = -3x + 12$ **12.** $f(x) = \dfrac{1}{2}x + 3$

In Exercises 13 and 14, determine the equation of the line in the form $f(x) = mx + b$ that passes through each pair of points.

13. $(0, 2)$ and $(4, -3)$ **14.** $(-1, 3)$ and $(2, 5)$

In Exercises 15–18, identify the function as an increasing function or a decreasing function.

15. $f(x) = 2x - 3$ **16.** $f(x) = -\dfrac{3}{2}x - 1$

17.

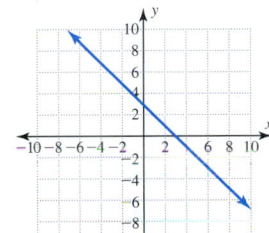

18.

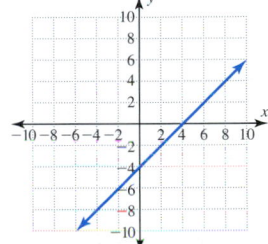

In Exercises 19–22, use the graph or the equation of the linear function to complete each part of the exercise.

  a. Determine the slope of this line.
  b. Determine the x-intercept.
  c. Determine the y-intercept.
  d. Determine the domain of this function.
  e. Determine the range of this function.
  f. Determine the x-values for which the function is positive.
  g. Determine the x-values for which the function is negative.
  h. Determine the x-values for which the function is decreasing.
  i. Determine the x-values for which the function is increasing.

19.

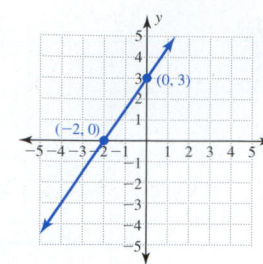

20.

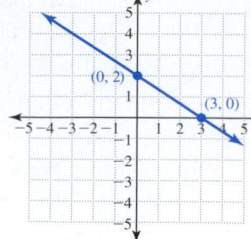

29.

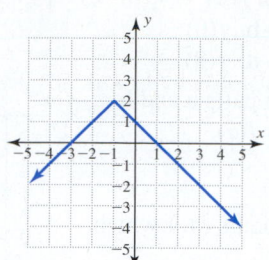

30.

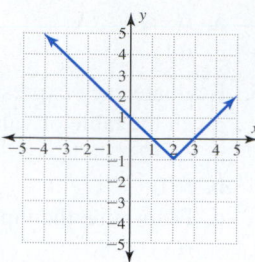

21. $f(x) = -12.5x + 125$ **22.** $f(x) = 35x - 210$

Absolute Value Functions

In Exercises 23 and 24, use the given absolute value function to complete each part of the exercise.

a. Complete the table of values for this function.

x	−3	−2	−1	0	1	2	3
y							

b. Plot these points, and sketch a graph of $y = f(x)$.
c. Does this graph open upward or downward?
d. Determine the vertex.
e. Determine the minimum or maximum y-value.
f. Determine the domain of this function.
g. Determine the range of this function.

23. $f(x) = -|x + 1| + 2$ **24.** $f(x) = |x - 1| - 3$

In Exercises 25 and 26, sketch the graph of each absolute value function.

25. $f(x) = |2x - 5| - 7$ **26.** $f(x) = -|2x - 1| + 3$

In Exercises 27 and 28, solve each equation and inequality.

27. a. $|x + 1| = 8$ **b.** $|x + 1| < 8$ **c.** $|x + 1| \geq 8$
28. a. $|2x + 3| = 7$ **b.** $|2x + 3| \leq 7$ **c.** $|2x + 3| > 7$

In Exercises 29–32, use the graph or the equation of the absolute value function to complete each part of the exercise.

a. Determine the vertex of this graph.
b. Does the graph open upward or downward?
c. Determine the x-intercept(s).
d. Determine the y-intercept.
e. Determine the domain of this function.
f. Determine the range of this function.
g. Determine the x-values for which the function is positive.
h. Determine the x-values for which the function is negative.
i. Determine the x-values for which the function is decreasing.
j. Determine the x-values for which the function is increasing.

31. $f(x) = |x - 8| - 9$ **32.** $f(x) = -|x + 7| + 12$

Quadratic Functions

In Exercises 33–36, use the shape of each graph to match it with the most appropriate description.

33.

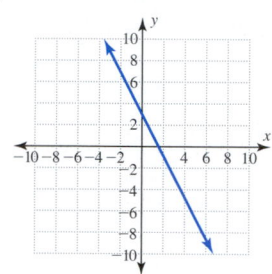

34.

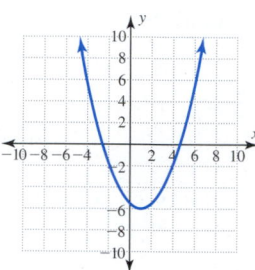

35.

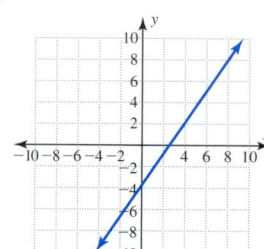

36.

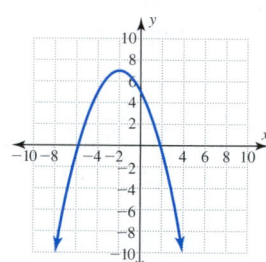

A. A linear function in the form $f(x) = mx + b$ with $m > 0$.
B. A linear function in the form $f(x) = mx + b$ with $m < 0$.
C. A quadratic function in the form $f(x) = ax^2 + bx + c$ with $a > 0$.
D. A quadratic function in the form $f(x) = ax^2 + bx + c$ with $a < 0$.

In Exercises 37–40, determine whether each parabola opens upward or opens downward, and determine its vertex.

37.

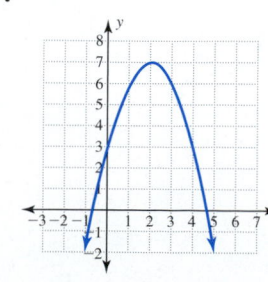

38.

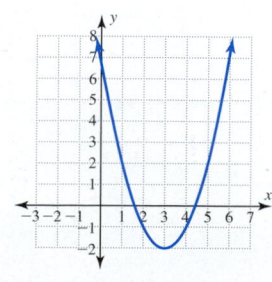

39. $f(x) = x^2 + 10x + 5$

40. $f(x) = -x^2 + 4x + 7$

In Exercises 41 and 42, sketch the graph of each function. Label the intercepts and the vertex on your graph.

41. $f(x) = -4x^2 + 4x + 3$

42. $f(x) = 2x^2 + 3x - 9$

In Exercises 43–46, use the graph or the equation of the quadratic function to complete each part of the exercise.

a. Determine the vertex of this graph.

b. Does the graph open upward or downward?

c. Determine the x-intercept(s).

d. Determine the y-intercept.

e. Determine the domain of this function.

f. Determine the range of this function.

g. Determine the x-values for which the function is positive.

h. Determine the x-values for which the function is negative.

i. Determine the x-values for which the function is decreasing.

j. Determine the x-values for which the function is increasing.

43. 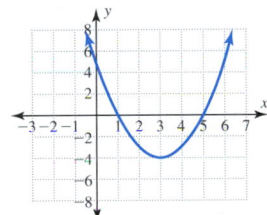   **44.**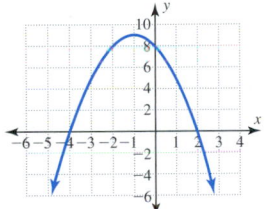

45. $f(x) = x^2 + 12x - 28$ **46.** $f(x) = -x^2 + 6x + 27$

Connecting Concepts to Applications

47. Cost of a Collect Call The accompanying table displays the dollar cost of a collect telephone call, based on a call of x minutes.

a. Use this table to determine the linear function $C(x) = mx + b$ for these data points.

b. Interpret the meaning of m in this problem.

c. Interpret the meaning of b in this problem.

d. Evaluate and interpret $C(15)$.

x Minutes	C $ Cost
3	3.80
7	6.20
10	8.00
16	11.60

48. Monthly Profit The function $P(x) = 4x - 640$ represents the monthly profit by a small business producing and selling x units of an item.

a. Algebraically determine the intercepts and interpret the meaning of these points.

b. Determine the profit and loss intervals for this profit function. Assume the practical domain is [0, 300].

49. Break-Even Values The function $P(x) = -100x^2 + 8,500x - 150,000$ gives the profit in dollars when x tons are produced and sold. Use the graph of the profit function to determine the following:

a. Overhead costs (*Hint:* Evaluate $P(0)$.)

b. Break-even values (*Hint:* When does $P(x) = 0$?)

c. Maximum profit that can be made and the number of tons to sell to create this profit

d. Determine the x-values for which $P(x) > 0$.

e. Interpret the meaning of the x-values found in part **d.**

50. Stretch of a Steel Spring The table displays data points taken from an experiment in which a mass was hung from a spring. The distance the spring stretched was measured for each mass.

a. Create a scatter diagram for these data and then determine the line of best fit. (Round the coefficients to the nearest thousandth.)

b. Interpret the meaning of the slope of the line of best fit.

c. Use the equation of the line of best fit to estimate the stretch for a mass of 12 kg.

d. Use the equation of the line of best fit to predict the number of kilograms of mass hung on the spring when the stretch is 2.4 cm.

x Mass on the Spring (kg)	y Stretch by the Spring (cm)
5	0.3
10	0.9
15	1.4
20	1.6
25	2.2
30	2.5
35	2.7

51. Modeling Profit Based on Sales A new company recorded its sales units versus its profit (or loss) for eight consecutive quarters.

a. Create a scatter diagram for these data and determine the parabola of best fit. Round the coefficients to three significant digits.

x Units sold	y Profit (hundreds of $)
0	−50
25	1,000
50	1,800
75	2,400
100	3,000
125	2,900
150	2,900
175	2,000

b. Use this quadratic equation to estimate the profit in dollars generated by the sale of 80 units in a quarter.

c. Determine the break-even values.

d. Determine the maximum profit that can be made and the number of units sold that generates this profit.

52. Height of a Softball The height of a softball is given by $H(t) = -16t^2 + 56t$, where $H(t)$ represents the height of the ball in feet and t is the number of seconds that have elapsed since the ball was hit. Determine the highest point that the ball reaches. Determine how many seconds into the flight the maximum height is reached.

53. Maximum Area A lumberyard plans to enclose a rectangular storage area for storing treated lumber. One side of the storage area will be formed by the wall of an existing building, and the other three sides will be formed by using 120 ft of fencing.

a. Write a function that models the area as a function of the width w.

b. Find the maximum area that can be enclosed with this fencing.

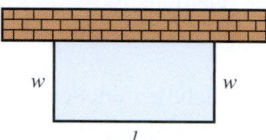

Chapter 8   Mastery Test

Objective 8.1.1 Identify a Function and Determine Its Domain and Range

1. Determine whether each relation is a function. If the relation is a function, determine its domain and range.

a.
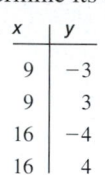
D   R

-3 ⟶ 9
3
-4 ⟶ 16
4

b.

x	y
9	−3
9	3
16	−4
16	4

c.
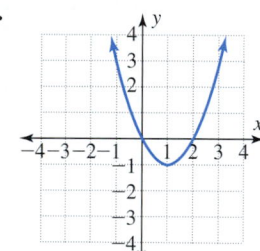

d.
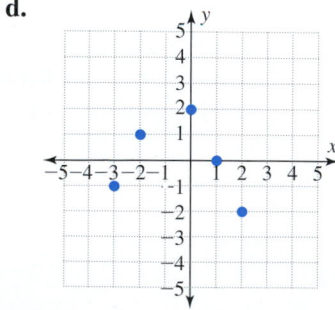

e.
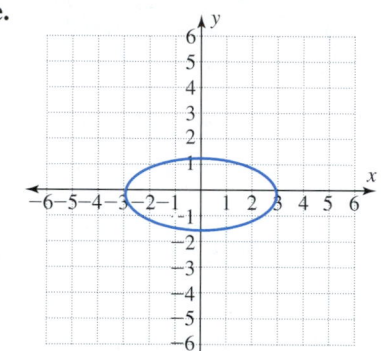

Objective 8.1.2 Use Function Notation

2. Evaluate each of the following expressions for $f(x) = 3x^2 + x - 12$.

a. $f(2)$ b. $f(-2)$

Use the given graph to determine the missing input and output values.

c. $f(0) = $ _____   d. $f(x) = 0; x = $ _____

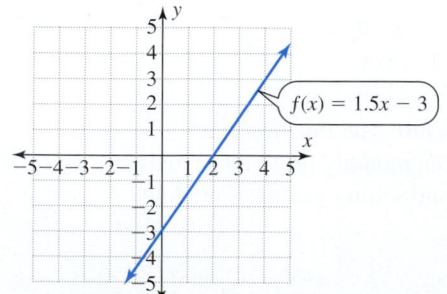

$f(x) = 1.5x - 3$

Use the given table of values to determine the missing input and output values.

e. $f(-10) = $ _____

f. $f(x) = 12; x = $ _____

x	f(x)
-10	-18.0
-5	-10.5
0	-3.0
5	4.5
10	12.0
15	19.5
20	27.0

Objective 8.2.1 Determine the Slope of a Line

3. Determine the slope of each line.

a. The line that passes through all the points listed in the table

x	y
2	-3
4	0
6	3
8	6
10	9

b. The line defined by the function $f(x) = x + 5$

Objective 8.2.2 Sketch the Graph of a Linear Function

4. Sketch the graph of each linear function.

a. $f(x) = 8$

b. $f(x) = -\dfrac{5}{3}x + 7$

c. $f(x) = \dfrac{1}{2}x - 7$

Objective 8.2.3 Write the Equation of the Line Through Given Points

5. Write in slope-intercept form the equation of the line that passes through the points listed in this table. All the points in each table lie on a single line.

a.

x	y
0	-3
2	-6
4	-9
6	-12
8	-15

b.

x	y
4	-1
9	3
14	7
19	11
24	15

Objective 8.2.4 Determine the Intercepts of a Line

6. Determine the intercepts of each linear function.

a.

x	y
-4	-9
0	-6
4	-3
8	0
12	3

b. $f(x) = -0.25x + 12$

Objective 8.2.5 Determine the x-values for Which a Linear Function Is Positive and the x-values for Which a Linear Function Is Negative

7. Determine x-values for which each linear function is positive and the x-values for which it is negative.

a.

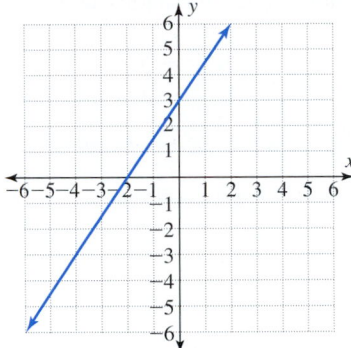

b. $f(x) = -8x + 24$

Objective 8.3.1 Sketch the Graph of an Absolute Value Function

8. Use the absolute value function $f(x) = |x + 2| - 5$ to complete each part of this exercise.

a. Does this graph open upward or downward?

b. Determine the vertex.

c. Sketch the graph of this function.

d. Determine the minimum or maximum y-value.

e. Determine the domain of this function.

f. Determine the range of this function.

Objective 8.3.2 Determine the Intercepts of the Graph of an Absolute Value Function

9. Use the absolute value function $f(x) = |x + 2| - 5$ to:

a. Determine the y-intercept.

b. Determine the x-intercepts.

Objective 8.3.3 Determine the x-values for Which an Absolute Value Function Is Positive and the x-values for Which an Absolute Value Function Is Negative

10. Use the absolute value function $f(x) = |x + 2| - 5$ to:

a. Determine the interval of x-values for which the function is positive.

b. Determine the interval of x-values for which the function is negative.

Objective 8.4.1 Distinguish Between Linear and Quadratic Functions

11. Identify the graph of each function as a line or a parabola. If the graph is a line, determine whether the slope is negative, zero, or positive. If the graph is a parabola, determine whether the parabola opens upward or downward.
   a. $y = -2x + 5$ b. $y = 2x^2 - 5$
   c. $y = 3x^2 - 4x + 5$ d. $y = 3x - 4$

Objective 8.4.2 Determine the Vertex of a Parabola

12. Determine the vertex of each parabola.
   a.

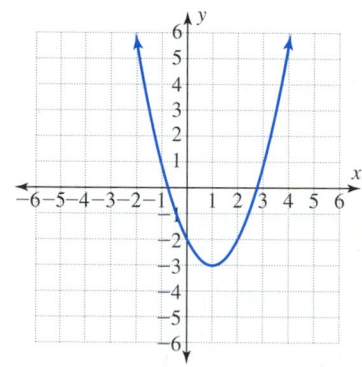

   b. $f(x) = -x^2 + 8x - 15$

Objective 8.4.3 Sketch the Graph of a Quadratic Function and Determine key Features of the Resulting Parabola

13. Use the quadratic function $f(x) = 2x^2 + 11x - 40$ to complete each part of this exercise.
   a. Will the parabola open upward or open downward?
   b. Determine the coordinates of the vertex.
   c. Determine the intercepts of the graph of this function.
   d. Complete a table of values using inputs on both sides of the vertex.
   e. Use this information to sketch the graph of this function.

Objective 8.4.4 Solve Problems Involving a Maximum or Minimum Value

14. **Height of a Softball** The height of a softball is given by $y = -16x^2 + 48x + 4$, where y represents the height of the ball in feet and x represents the number of seconds that have elapsed since the ball was hit.
   a. Determine the highest point the ball reaches.
   b. Determine the number of seconds into the flight the maximum height is reached.

Objective 8.5.1 Identify a Function As an Increasing or a Decreasing Function

15. Identify each function as an increasing function or a decreasing function.
   a.

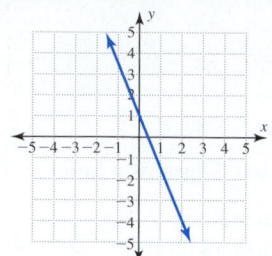

   b.

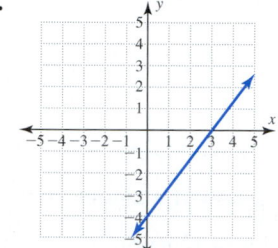

Objective 8.5.2 Analyze the Graphs of Linear, Absolute Value, and Quadratic Functions

16. Use either the graph or the equation of the quadratic function to complete each part of the exercise.

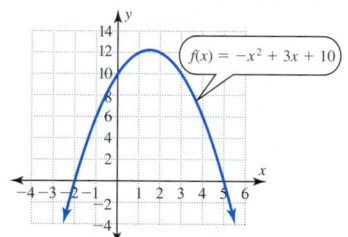

   a. Determine the vertex of this graph.
   b. Does the graph open upward or downward?
   c. Determine the x-intercept(s).
   d. Determine the y-intercept.
   e. Determine the domain of this function.
   f. Determine the range of this function.
   g. Determine the x-values for which the function is positive.
   h. Determine the x-values for which the function is negative.
   i. Determine the x-values for which the function is decreasing.
   j. Determine the x-values for which the function is increasing.

Objective 8.6.1 Create a Scatter Diagram and Select an Appropriate Model for the Data

17. Create a scatter diagram for each set of data points and then select the shape that best describes the pattern formed by these points.

a.
x	4	6	8	10	12	14	16
y	1	4	8	11	14	18	22

b.
x	3	6	9	12	15	18	21
y	−1	5	9	10	9	4	0

c.
x	−5	0	5	10	15	20	25
y	1	−1	−4	−2	2	6	11

d.
x	10	20	30	40	50	60	70
y	9	8	6	4	3	1	−1

A. A line with positive slope

B. A parabola opening upward

C. A line with negative slope

D. A parabola opening downward

Objective 8.6.2 Select a Linear Function That Best Fits a Set of Data

18. **Olympic Women's High-Jump Records**

The accompanying table displays the winning heights for the Olympic women's high jump from 1948 through 2008. Each distance is given in meters. (The actual dates are not used in the calculations. The x-values represent the number of years after 1900. These smaller x-values allow us to round the coefficients of the equation of best fit without introducing an unacceptable amount of error.)

Year	x	y (meters)
1948	48	1.68
1952	52	1.67
1956	56	1.76
1960	60	1.85
1964	64	1.90
1968	68	1.82
1976	76	1.93
1980	80	1.97
1984	84	2.02
1988	88	2.03
1992	92	2.02
1996	96	2.05
2000	100	2.01
2004	104	2.06
2008	108	2.05

a. Create a scatter diagram for these data and then determine the line of best fit. (Round the coefficients to the nearest thousandth.)
b. Interpret the meaning of the slope of this line.
c. Use the equation of the line of best fit to estimate the missing entry in the table from 1972. How does this estimation compare to the actual 1972 record of 1.92 m?
d. Use the equation of the line of best fit to predict the women's winning high-jump height for the year 2012.

Objective 8.6.3 Select a Quadratic Function That Best Fits a Set of Data

19. Create a scatter diagram of the points given in the table. Then determine the quadratic function that best fits these points. (Round the coefficients to three significant digits.) Use this function f to approximate $f(7)$.

x	−6	−4	−2	0	2	4	6	8
y	7	2	−2	−3	−4	−2	1	6

| Chapter 8 | Group Project |

An Algebraic Model for Real Data

Supplies needed for the lab:

I. A clear cylindrical container such as a 2-L soft drink container with a hole drilled near the bottom of the straight-walled portion and with the height labeled in centimeters on an attached paper strip or on the plastic with a permanent marker. Label the numerical scale from bottom to top, so that 0 is on the hole. *Suggestion:* Experiment with the hole size—approximately $\frac{1}{4}$ in—until the total drain time is less than 3 minutes.

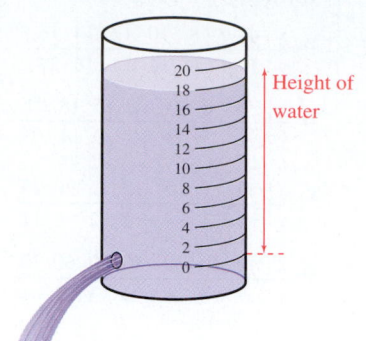

II. Water and a bucket to drain the water into.

III. A watch that displays seconds.

IV. A graphing calculator with a linear and a quadratic regression feature.

Lab: Recording the Time to Drain a Cylindrical Container

Place water into a cylindrical container up to the fill line of the cylindrical portion of this container while keeping the drain hole closed. Set your time to 0 seconds, and record this time and the initial height of the water. Then open the drain hole and remove the cap from the top of the container. As the water drains from the container, record the time for each height in the table. (You may decide to repeat this experiment by recording the height every 10 or more seconds.)

Time x (seconds)	Height y (cm)
0	20
	18
	16
	14
	12
	10
	8
	6
	4
	2
	0

1. Record your data in a table similar to the one shown.

2. Plot these data on a scatter diagram, using an appropriate scale for each axis.

Creating an Algebraic Model for the Data

3. a. Which of these models, a linear model or a quadratic model, do you think is the better model for this experiment?

  b. What information in the table or in the graph supports your choice in part **a**? Why do you believe this model is better?

  c. Without referring to any of the data that you collected, describe any observations that you made during the course of the experiment that you think support your choice in part **a**.

4. a. Using your graphing calculator, calculate either the line of best fit or the parabola of best fit (depending on which is the better choice). (*Hint:* Watch out for scientific notation.)

  b. Graph this curve on the scatter diagram.

5. Let $y = H(x)$ represent the curve of best fit from **4a**:

  a. Evaluate and interpret $H(50)$.

  b. Evaluate and interpret $H(70)$.

  c. Determine the value of x for which $H(x) = 11$. Interpret the meaning of this value.

Rational Functions

Chapter Outline

Weight Versus Mass

The weight of an object depends on the mass of the object, the distance from the object to the center of the nearest planet or moon, and the mass of this planet or moon. This is known as the universal law of gravitation. As the distance from the object to the nearest planet or moon increases, the weight of the object decreases.

Since the mass of Earth is much larger than that of the moon, the same object will weigh more on Earth than on the moon. The graph of the first function gives the weight in newtons (N) of an astronaut based on a distance in kilometers above the surface of the earth. The y-intercept of this graph indicates the weight of the astronaut on the surface of the earth. The graph of the second function gives the weight of this astronaut based on a distance above the surface of the moon. The y-intercept of this graph indicates the weight of the astronaut on the surface of the moon. Exercise 29 in Section 9.6 develops an equation for the weight of an astronaut above the earth.

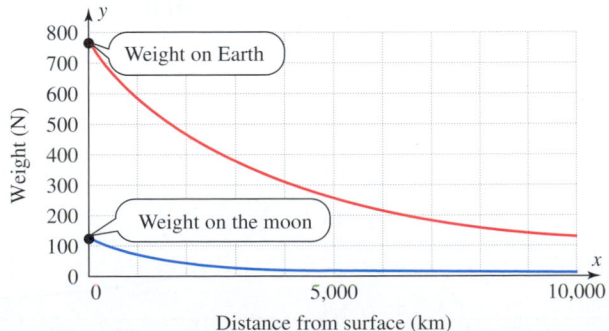

Section 9.1	**Graphs of Rational Functions and Reducing Rational Expressions**

"The most important thing we can do is inspire young minds and to advance the kind of science, math and technology education that will help youngsters take us to the next phase of space travel." —John Glenn (Astronaut and U.S. Senator)

Objectives:

1. Identify rational functions.
2. Determine the domain and identify the vertical asymptotes of a rational function.
3. Reduce a rational expression to lowest terms.

Each family of functions has its own distinctive numerical pattern, its own graph, and its own algebraic definition. Note the distinctive numerical patterns exhibited in the following two tables. The first table gives the distance that a spring stretches when a mass is attached to it, and the second table gives the time to travel a distance of 100 mi based on the rate of travel.

As x increases, y increases.	Distance a Spring Stretches		Time to Travel 100 mi		As x increases, y decreases.
	Mass x (kg)	Distance y (cm)	Speed x (mi/h)	Time y (h)	
	0	0	5	20.00	
	10	1.25	10	10.00	
	20	2.50	20	5.00	
	30	3.75	25	4.00	
	40	5.00	50	2.00	
	50	6.25	100	1.00	
	60	7.50	200	0.50	
	70	8.75	400	0.25	
	80	10.00	500	0.20	

In the first table, the distance that the spring stretches increases as the mass attached to the spring increases. In the second table, the time to cover 100 mi decreases as the speed increases. These two numerical patterns are vastly different and produce completely different shapes and algebraic equations. This is illustrated by the following two graphs.

Distance a Spring Stretches

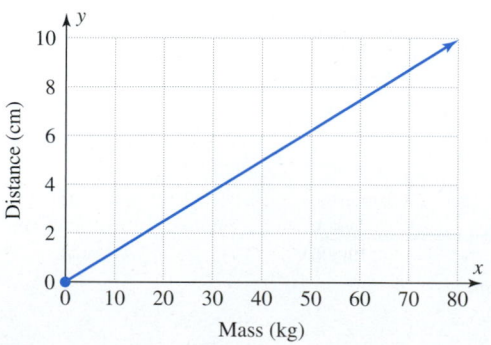

Time to Travel 100 mi

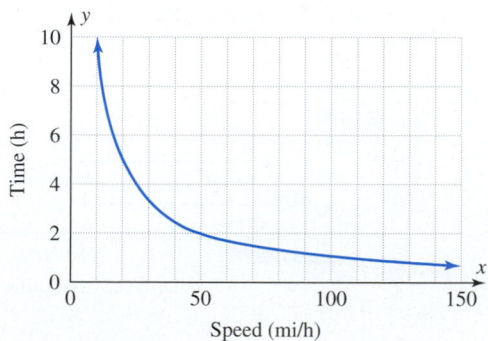

The graph on the left exhibits a linear pattern while the graph on the right does not have the characteristics of a polynomial function. Before we analyze the behavior of this graph, we first summarize some of the characteristics of polynomial functions. These are characteristics exemplified both by first-degree functions of the form $y = mx + b$ whose graphs have a characteristic linear shape and by second-degree functions of the form $y = ax^2 + bx + c$ whose graphs have a characteristic parabolic shape.

Characteristics of the Graphs of Polynomial Functions

- Each polynomial function is continuous (there are no breaks or gaps between points on the graph).
- Each polynomial function is smooth (there are no sharp points or corners on the graph).
- The domain of each polynomial function is the set of all real numbers (the graph extends without limit both to the left and to the right).
- Some polynomial functions have relative (localized) maximum or minimum points (there are high or low points on the graph)—there are more possibilities for these points for higher-degree polynomials.

1. Identify Rational Functions

What Is a Rational Function?

A **rational function** can be written as the ratio of two polynomials. By this definition, linear, quadratic, and other polynomial functions are rational functions with a denominator of 1. Rational functions containing variables in the denominator have graphs that are considerably more complicated than the graphs of polynomial functions. These functions can have breaks in the graph, and their domain may not be all real numbers.

Rational Function			
Algebraically	**Verbally**	**Algebraic Example**	**Graphical Example**
$f(x) = \dfrac{P(x)}{Q(x)}$ is a rational function if $P(x)$ and $Q(x)$ are polynomials and $Q(x) \neq 0$.	A rational function is defined by the ratio of two polynomials.	$f(x) = \dfrac{1}{x}$	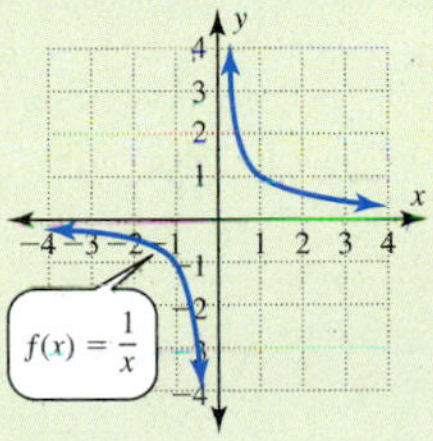

In Example 1, we use the algebraic forms of several functions to identify whether they are rational functions.

Example 1   Identifying Rational Functions

Determine whether each function defines a rational function.

(a) $f(x) = \dfrac{x + 5}{x - 3}$ **(b)** $f(x) = -5x + 2$ **(c)** $f(x) = 7x^2 - 5x + 12$

(d) $f(x) = \sqrt{x + 5}$ **(e)** $f(x) = |3x - 1| + 4$

Solution

(a) This is a rational function whose graph has a break in it.

Both the numerator $x + 5$ and the denominator $x - 3$ are polynomials. In Example 2, we examine the break in the graph of this rational function.

(b) This is a rational function whose graph is a line.

This function can be written in the form $f(x) = \dfrac{-5x + 2}{1}$. Both the numerator $-5x + 2$ and the denominator 1 are polynomials.

(c) This is a rational function whose graph is a parabola.

This function can be written in the form $f(x) = \dfrac{7x^2 - 5x + 12}{1}$. Both the numerator $7x^2 - 5x + 12$ and the denominator 1 are polynomials.

(d) This is not a rational function. This is a square root function.

The expression $\sqrt{x + 5}$ is not a polynomial. Thus this is not a rational function. We will explore graphs of these functions in Chapter 10.

(e) This is not a rational function. This is an absolute value function whose graph is a V-shape.

The expression $|3x - 1| + 4$ is not a polynomial. Thus this is not a rational function.

Self-Check 1

Determine whether each function defines a rational function.

a. $f(x) = 2x^2 - x + 1$ **b.** $f(x) = \dfrac{3x - 1}{x + 2}$ **c.** $f(x) = \dfrac{\sqrt{x}}{7}$ **d.** $f(x) = \dfrac{5}{x}$

---

2. Determine the Domain and Identify the Vertical Asymptotes of a Rational Function

What Is Meant by a Vertical Asymptote?

This is illustrated by the graph of $f(x) = \dfrac{1}{x}$. One of the classic behaviors possible with rational functions is that their graphs can approach and get very near to vertical and horizontal lines. A vertical line that a graph approaches but does not touch is called a **vertical asymptote**. A horizontal line that a graph approaches is called a **horizontal asymptote.**[*]

Characteristics of the Graph of $f(x) = \dfrac{1}{x}$

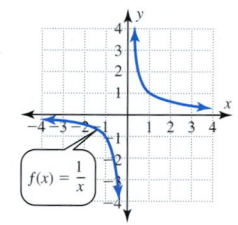

- The domain of this function does not include 0.
- The function is discontinuous; there is a break in the graph at $x = 0$. The graph consists of two unconnected branches.
- The graph approaches the vertical line $x = 0$ asymptotically.
- The graph approaches the horizontal line $y = 0$ asymptotically.

We can determine the domain of a rational function algebraically by solving for the values that make the denominator 0 and then excluding these values from the domain. In a numerical table, there will be error messages at the excluded values due to division by 0. Graphically there will be a break in the graph of a rational function at an excluded value of the domain, and the function can have a vertical asymptote at this x-value. To denote that the number 2 is excluded from the domain of a function, we can write the domain by using interval notation as $(-\infty, 2) \cup (2, \infty)$. Another notation to represent this domain is $\mathbb{R} \sim \{2\}$, which is read "the set of all real numbers except for 2."

---

[*]We limit our analysis of vertical and horizontal asymptotes to an intermediate algebra–level discussion. Rational functions are examined in greater depth in college algebra and calculus. For example, the graph of a rational function can have a hole at $x = c$ instead of a vertical asymptote (see Exercise 88 at the end of this section) if each factor of $x - c$ in the denominator is also in the numerator. Although the graph of a rational function can never touch a vertical asymptote, the behavior with respect to a horizontal asymptote is more complex—the graph can cross a horizontal asymptote before eventually approaching and staying very near to this horizontal line. The graphs of rational functions can also approach lines that are neither horizontal nor vertical.

Domain of a Rational Function

The domain of a rational function must exclude values that would cause division by zero.

Algebraic Example	Numerical Example		Graphical Example	Verbal Example
The domain of $f(x) = \dfrac{1}{x}$ is $\mathbb{R} \sim \{0\}$.	x	$y = \dfrac{1}{x}$		Only zero is excluded from the domain to prevent division by zero. Note that 1 divided by 0 is undefined. Also note that there is a break in the graph at the vertical asymptote $x = 0$.
	-3	-0.333		
	-2	-0.500		
	-1	-1.000		
	0	undefined		
	1	1.000		
	2	0.500		
	3	0.333		

Loosely speaking, the domain of a rational function consists of all real numbers that do not cause division by zero. Consider the function $f(x) = \dfrac{x + 5}{x - 3}$.

We can evaluate this function for $x = 1$ as follows:

$$f(1) = \frac{1 + 5}{1 - 3}$$

$$f(1) = -3$$

Observe what happens if we try to evaluate this same function for $x = 3$:

$$f(3) = \frac{3 + 5}{3 - 3}$$

$$f(3) = \frac{8}{0}$$

$f(3)$ is undefined.

As we will see for $f(x) = \dfrac{x + 5}{x - 3}$ in Example 2, $x = 3$ is the only value of x that causes division by 0. Thus the domain is all real numbers except 3, written as $\mathbb{R} \sim \{3\}$. To determine the domain of a rational function, we start by determining the values that are excluded from the domain because they cause division by 0. Algebraically, this is accomplished by setting the denominator equal to 0 and solving the resulting equation. The values that cause division by 0 in a rational function (see the previous footnote for information on holes) can also indicate where the graph of the function has vertical asymptotes. This is illustrated in Example 2.

Example 2   Determining the Domain and Vertical Asymptote of a Rational Function

Use the function $f(x) = \dfrac{x + 5}{x - 3}$ to:

(a) Determine the domain of this function.

(b) Determine the vertical asymptote of the graph of this function.

Solution

(a) The denominator is $x - 3$. $\quad$ Set the denominator equal to zero and solve this equation to
$x - 3 = 0$ when $x = 3$. $\quad$ determine the excluded values.

Domain: $\mathbb{R} \sim \{3\}$. $\quad$ $x = 3$ is excluded from the domain.

(b) The line $x = 3$ is a vertical asymptote of the graph of this function.

Notice the break in the graph at $x = 3$ on the graph of this function.

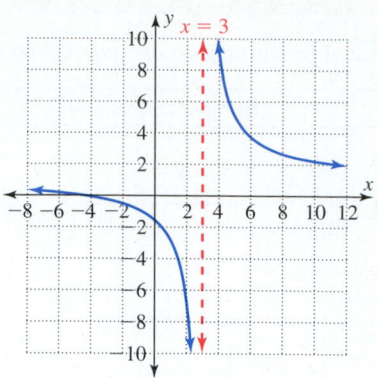

Self-Check 2

Use the function $f(x) = \dfrac{x - 1}{x + 4}$ to:

a. Determine the domain of this function.

b. Determine the vertical asymptote of the graph of this function.

A graphing calculator is an excellent tool to verify the algebraic results of Example 2. Note that the graph shows a break at $x = 3$, and the table shows that the function is undefined for $x = 3$. While it is wise to use your calculator to verify an algebraic result, it can be misleading to rely only on a calculator to find the domain or vertical asymptotes of a rational function.

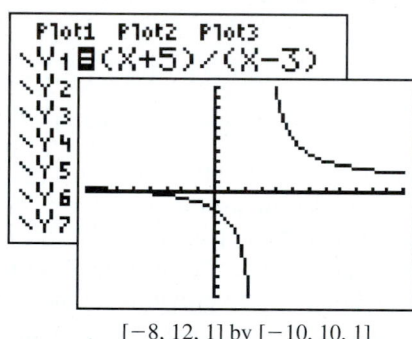
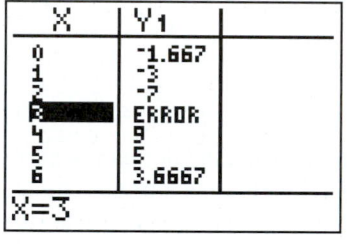

$[-8, 12, 1]$ by $[-10, 10, 1]$

Problems that involve ratios, rates, and averages can all produce rational expressions. Example 3 examines the asymptotic behavior of an average cost function.

Example 3 Analyzing the Asymptotic Behavior of a Function

To add an additional line to a cell-phone plan, a wireless company charges $25 for the phone and $10 per month for the additional line. The function $A(t) = \dfrac{25 + 10t}{t}$ gives the average monthly cost to own this additional phone as a function of time t in months.

(a) Describe what happens to the average monthly cost as the period of time the phone is owned becomes very small.

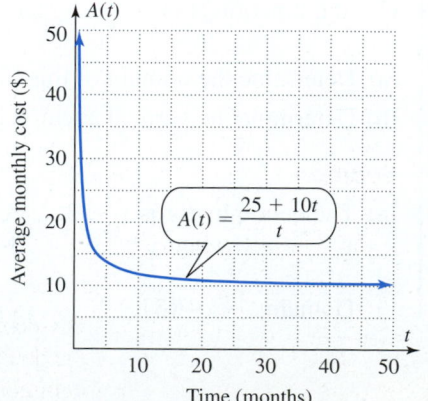

(b) Describe what happens to the average monthly cost as the period of time the phone is owned becomes very large.

Solution

(a) The asymptotic behavior near the vertical asymptote means that for a very short period of ownership, the average monthly cost is very large.

This graph has a vertical asymptote at $x = 0$. For x-values close to zero, the y-values become very large.

(b) The asymptotic behavior near the horizontal asymptote means that for a very long period of ownership, the average monthly cost approaches $10.

This graph has a horizontal asymptote at $y = 10$. For larger and larger x-values, the y-values appear to get close to 10.

Self-Check 3

Repeat Example 3 assuming the function $A(t) = \dfrac{15 + 20t}{t}$ gives the average monthly cost to own this phone as a function of time t in months.

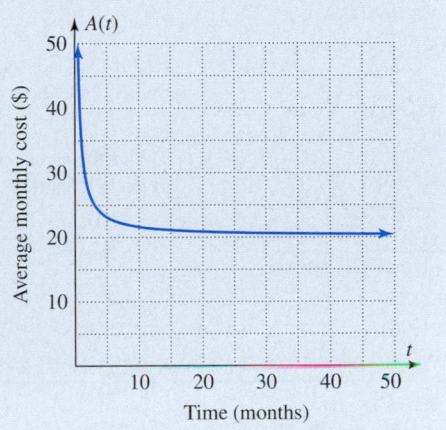

If the graph of a rational function has a vertical asymptote, it is important that the graph accurately reflects the behavior of the function close to this asymptote. If you graph a rational function using pencil and paper, you may want to sketch each asymptote using a dashed line. An asymptote can be an aid to sketching the graph as it approaches the asymptote. Whether you graph a rational function using pencil and paper or a graphing calculator, it is important to understand that the vertical asymptote is *not* part of the graph of the function.

3. Reduce a Rational Expression to Lowest Terms

To analyze rational functions, we need to be able to reduce rational expressions to lowest terms and to perform operations with rational expressions. We now examine the procedure for simplifying rational expressions. This procedure is identical to that used to simplify rational numbers in Chapter 1. A rational expression is in **lowest terms** when the numerator and the denominator have no common factor other than -1 or 1. The basic principle for reducing rational expressions is given in the following box.

Reducing a Rational Expression to Lowest Terms

Algebraically	Verbally	Algebraic Example
If A, B, and C are polynomials and $B \neq 0$ and $C \neq 0$, then $\dfrac{AC}{BC} = \dfrac{A}{B}$.	**1.** Factor both the numerator and the denominator of the rational expression. **2.** Divide the numerator and the denominator by any common nonzero factors.	$\dfrac{3x - 6}{x^2 - 2x} = \dfrac{3\overset{1}{\cancel{(x - 2)}}}{x\underset{1}{\cancel{(x - 2)}}}$ $= \dfrac{3}{x}$ for $x \neq 2$

To reduce a rational expression, the numerator and denominator must have a common factor. Note that $\dfrac{C}{C} = 1$, the multiplicative identity for $C \neq 0$.

To divide the numerator and denominator by a common factor, it is helpful to first write each as a product. Example 4 illustrates this method and reviews the rules for exponents from Chapter 5.

Example 4   Reducing a Rational Expression with a Monomial Denominator

Reduce $\dfrac{15x^3y^2}{25x^4y^2}$ to lowest terms. Assume that x and y are nonzero.

Solution

$$\frac{15x^3y^2}{25x^4y^2} = \frac{5x^3y^2(3)}{5x^3y^2(5x)}$$

Factor the GCF out of the numerator and the denominator. Note that x and y must both be nonzero to avoid division by zero.

$$= \frac{3}{5x}$$

This is the same answer you would get by using the properties of exponents given in Chapter 5.

Self-Check 4

Reduce $\dfrac{22a^2b^3}{33a^3b^2}$ to lowest terms.

Remember that equal algebraic expressions have the same outputs for all input values of the variables. Looking at a limited set of values in a table does not confirm that the expressions are equal for all values, but it does serve as a good check that is likely to catch potential errors from an algebraic computation. This is illustrated in Example 5.

Example 5   Reducing a Rational Expression and Using a Table as a Check

Reduce $\dfrac{x^2 + 3x}{x^2 + x - 6}$ and compare this result to the original expression using a table of values.

Solution

Algebraically

$$\frac{x^2 + 3x}{x^2 + x - 6} = \frac{x(x + 3)}{(x - 2)(x + 3)}$$

Factor both the numerator and the denominator. Note that the excluded values are -3 and 2.

$$= \frac{x}{x - 2} \qquad \text{for } x \neq 2,\, x \neq -3$$

Divide both the numerator and the denominator by the common factor $x + 3$. For $x \neq -3$, $x + 3$ is not zero.

Numerical Check

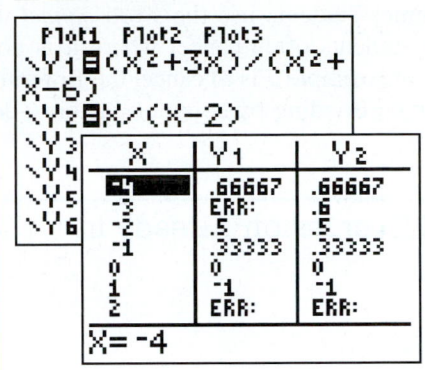

Let Y_1 represent the original expression and Y_2 represent the reduced expression.

The expressions are identical except for $x = -3$. The original expression has excluded values of -3 and 2. The reduced expression was obtained under these restrictions.

It is understood that this equality means that these expressions are identical for all values for which they are both defined. For $x \neq -3$ and $x \neq 2$ these expressions are identical.

Answer: $\dfrac{x^2 + 3x}{x^2 + x - 6} = \dfrac{x}{x - 2}$ for $x \neq 2, x \neq -3$

Self-Check 5

Reduce $\dfrac{x^2 - 9}{4x - 12}$ and compare this result to the original expression using a table of values.

We will assume that the simplified form of all rational expressions excludes values that cause division by zero. Then you will not need to write these restrictions with each of your answers.

Two polynomials are **opposites** if every term of the first polynomial is the opposite of the corresponding term in the second polynomial. The ratio of opposites is -1.

Example 6 **Reducing a Rational Expression When the Denominator Is the Opposite of the Numerator**

Reduce each of these rational expressions to lowest terms.

(a) $\dfrac{2x - 3y}{3y - 2x}$ (b) $\dfrac{-4a + 3b + c}{4a - 3b - c}$

Solution

(a) $\dfrac{2x - 3y}{3y - 2x} = \dfrac{\overset{1}{\cancel{2x - 3y}}}{-\underset{1}{(\cancel{2x - 3y})}}$ The denominator is the opposite of the numerator. The ratio of these opposites is -1.

$= -1$

(b) $\dfrac{-4a + 3b + c}{4a - 3b - c} = \dfrac{-\overset{1}{\cancel{(4a - 3b - c)}}}{\underset{1}{\cancel{4a - 3b - c}}}$ This expression equals -1, because the numerator and the denominator are opposites.

$= -1$

Self-Check 6

Reduce each rational expression to lowest terms.

a. $\dfrac{2x - 3y}{3y - 2x}$ b. $\dfrac{5r - 15t}{3t - r}$

To reduce a rational expression to lowest terms, we divide both the numerator and the denominator by a common nonzero factor. Sometimes students use the word *cancel* to describe this step. This can be dangerous because students often cancel where it is not applicable: You cannot cancel terms. A classic error in Example 7 is to cancel the $3x$ terms in the numerator and the denominator. Reduce only by dividing by common factors—do *not* cancel terms.

Example 7   Recognizing a Rational Expression Already in Reduced Form

Reduce $\dfrac{3x - 7}{3x + 11}$ to lowest terms.

Solution

Answer: $\dfrac{3x - 7}{3x + 11}$ This expression is already in reduced form, because the greatest common factor of the numerator and the denominator is 1.

Self-Check 7

Reduce each expression to lowest terms.

a. $\dfrac{2x + 5}{2x - 5}$ b. $\dfrac{2x - 5}{5 - 2x}$

The fundamental principle of fractions, the property that $\dfrac{AC}{BC} = \dfrac{A}{B}$ for $B \neq 0$ and $C \neq 0$, is used both to reduce fractions to lowest terms and to express fractions so that they have a common denominator. In Example 8, both the numerator and the denominator are multiplied by the same nonzero value, called the **building factor.** This step is important in adding rational expressions.

Example 8   Converting a Rational Expression to a Given Denominator

Convert $\dfrac{x + 3}{x - 2}$ to an equivalent expression with a denominator of $x^2 - 3x + 2$.

Solution

$\dfrac{x + 3}{x - 2} = \dfrac{?}{x^2 - 3x + 2}$

$\dfrac{x + 3}{x - 2} = \dfrac{?}{(x - 2)(x - 1)}$ To determine the building factor, first factor $x^2 - 3x + 2$. The other factor needed to build the denominator that we want is $x - 1$.

$\dfrac{x + 3}{x - 2} = \dfrac{x + 3}{x - 2} \cdot \dfrac{x - 1}{x - 1}$ Multiply $\dfrac{x + 3}{x - 2}$ by the multiplicative identity 1 in the form $\dfrac{x - 1}{x - 1}$.

$\dfrac{x + 3}{x - 2} = \dfrac{x^2 + 2x - 3}{x^2 - 3x + 2}$ This expression is equivalent to $\dfrac{x + 3}{x - 2}$ and has a denominator of $x^2 - 3x + 2$.

Self-Check 8

Determine the missing numerator in $\dfrac{3x - 2}{2x - 3} = \dfrac{?}{10x^2 - 7x - 12}$.

Self-Check Answers

1. **a.** A rational function whose graph is a parabola
   b. A rational function
   c. Not a rational function
   d. A rational function
2. **a.** Domain: $\mathbb{R} \sim \{-4\}$
   b. The line $x = -4$ will be a vertical asymptote of the graph.
3. **a.** For a very short period of ownership, the average monthly cost is very large.
   b. For a very long period of ownership, the average monthly cost approaches $20.
4. $\dfrac{2b}{3a}$
5. $\dfrac{x + 3}{4}$

Check:

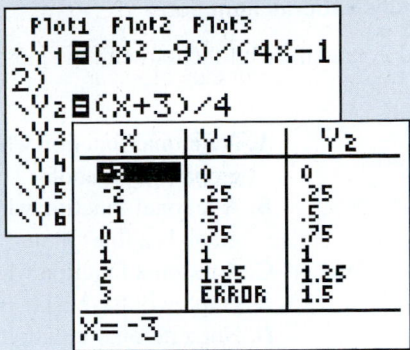

The two expressions agree for all values of x for which they are both defined.

6. **a.** -1 **b.** -5
7. **a.** $\dfrac{2x + 5}{2x - 5}$ **b.** -1
8. $\dfrac{3x - 2}{2x - 3} = \dfrac{15x^2 + 2x - 8}{10x^2 - 7x - 12}$

9.1   Using the Language and Symbolism of Mathematics

1. A rational expression is defined by the _____ of two polynomials.

2. Division by 0 is _____.

3. A function defined by $f(x) = \dfrac{P(x)}{Q(x)}$, where $P(x)$ and $Q(x)$ are polynomials and $Q(x) \neq 0$, is called a

   _____ function.

4. To determine the values excluded from the domain of a rational function, we look at the values for which the _____ is equal to _____.

5. A vertical line that a graph approaches but does not touch is called a vertical _____.

6. A horizontal line that a graph approaches is called a horizontal _____.

7. The graph of the rational function $f(x) = \dfrac{5}{x - 4}$ will have the line $x = 4$ as a vertical _____.

8. A rational expression is in _____ _____ when the numerator and denominator have no common factor other than -1 or 1.

9. If A, B, and C are polynomials and $B \neq 0$ and $C \neq 0$, then $\dfrac{AC}{BC} =$ _____.

10. If two polynomials are opposites, their ratio is _____.

9.1   Quick Review

1. Factor 56 as the product of prime integers.

2. Reduce $\dfrac{36}{90}$ to lowest terms.

Factor each expression.

3. $-3x^3 + 4x^2$

4. $x^3 - 6x^2 - 40x$

5. $6x^2 + xy - 35y^2$

9.1 | Exercises

Objective 1 Identify Rational Functions

In Exercises 1 and 2, match each function with the most appropriate description.

1. a. $f(x) = \dfrac{x}{x - 6}$ **A.** A rational function whose graph is a parabola

  b. $f(x) = |5x - 2|$ **B.** A rational function whose graph is a line

  c. $f(x) = 2x^2 - 3x - 4$ **C.** A rational function whose graph will have a break

  d. $f(x) = 3x - 7$ **D.** Not a rational function; an absolute value function whose graph has a V-shape

2. a. $f(x) = \dfrac{2}{3}x - 1$ **A.** A rational function whose graph is a parabola

  b. $f(x) = x^2 - 5$ **B.** A rational function whose graph is a line

  c. $f(x) = \dfrac{2}{x + 1}$ **C.** A rational function whose graph will have a break

  d. $f(x) = \sqrt{x + 2}$ **D.** Not a rational function; a square root function

Objective 2 Determine the Domain and Identify the Vertical Asymptotes of a Rational Function

3. If $f(x) = \dfrac{2x + 1}{2x - 1}$, evaluate each expression.

  a. $f(0)$ **b.** $f(1)$ **c.** $f\left(\dfrac{1}{2}\right)$

4. If $f(x) = \dfrac{5x - 11}{7x - 14}$, evaluate each expression.

  a. $f(0)$ **b.** $f(1)$ **c.** $f(2)$

5. Given the rational function $f(x) = \dfrac{5}{x - 4}$:

  a. What value is excluded from the domain of this function?

  b. What is the domain of this function?

  c. What is the equation of the vertical asymptote of the graph of this function?

6. Given the rational function $f(x) = \dfrac{x + 1}{x + 6}$:

  a. What value is excluded from the domain of this function?

  b. What is the domain of this function?

  c. What is the equation of the vertical asymptote of the graph of this function?

In Exercises 7–10, match the values excluded from the domain of the rational function $y = f(x)$ with the domain of this function.

Excluded Values	Domain
7. -3	**A.** $\mathbb{R} \sim \{0\}$ or equivalently $(-\infty, 0) \cup (0, \infty)$
8. 0	**B.** $\mathbb{R} \sim \{0, 3\}$ or equivalently $(-\infty, 0) \cup (0, 3) \cup (3, \infty)$
9. -3 and 0	**C.** $\mathbb{R} \sim \{-3\}$ or equivalently $(-\infty, -3) \cup (-3, \infty)$
10. 0 and 3	**D.** $\mathbb{R} \sim \{-3, 0\}$ or equivalently $(-\infty, -3) \cup (-3, 0) \cup (0, \infty)$

In Exercises 11–14, match the rational expressions with the corresponding excluded values.

11. $f(x) = \dfrac{2x - 1}{x + 1}$ **A.** Excluded value 0.5

12. $f(x) = \dfrac{x + 1}{2x - 1}$ **B.** Excluded values 0.5 and 1

13. $f(x) = \dfrac{2x - 1}{(x + 1)(2x + 1)}$ **C.** Excluded values -0.5 and -1

14. $f(x) = \dfrac{2x + 1}{(x - 1)(2x - 1)}$ **D.** Excluded value -1

In Exercises 15 and 16, determine the domain of each rational function.

15. a. $f(x) = \dfrac{2x + 1}{2x - 1}$ **b.** $f(x) = \dfrac{3x - 7}{x^2 - 81}$

16. a. $f(x) = \dfrac{5x - 11}{7x - 14}$ **b.** $f(x) = \dfrac{x^2 - 9}{x^2 - x - 42}$

In Exercises 17 and 18, use the given table to determine the domain of each rational function.

17. $f(x) = \dfrac{x + 4}{2x - 3}$ **18.** $f(x) = \dfrac{2x - 3}{x + 4}$

x	$f(x) = \dfrac{x + 4}{2x - 3}$	x	$f(x) = \dfrac{2x - 3}{x + 4}$
-3.0	$-.1111$	-6	7.500
-1.5	$-.4167$	-5	13.000
0.0	-1.333	-4	undefined
1.5	undefined	-3	-9.000
3.0	2.3333	-2	-3.500
4.5	1.4167	-1	-1.667
6.0	1.1111	0	$-.750$

In Exercises 19–22, match each function with the graph of the function.

19. $f(x) = \dfrac{1}{x + 2}$ **20.** $f(x) = \dfrac{1}{x - 2}$

21. $f(x) = \dfrac{1}{x - 4}$ **22.** $f(x) = \dfrac{1}{x + 4}$

A.

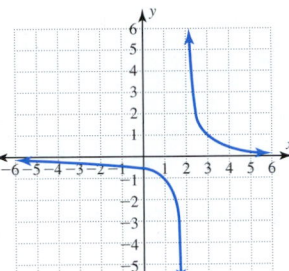

B.

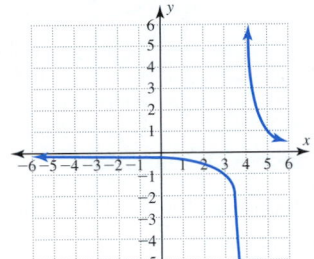

C.

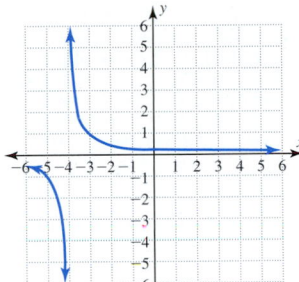

D.

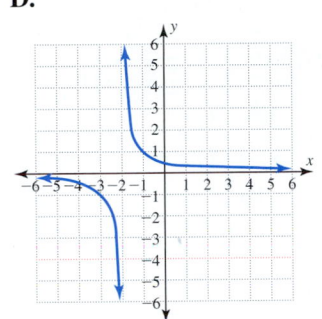

In Exercises 23–26, match each graph with the vertical asymptote(s) for this function.

23. $x = -1$ **24.** $x = -3$

25. $x = -1$ and $x = 1$ **26.** $x = -3$ and $x = 2$

A.

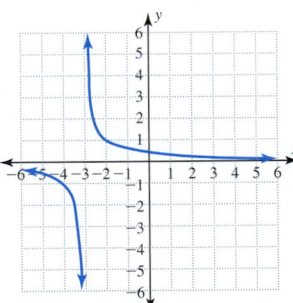

B.

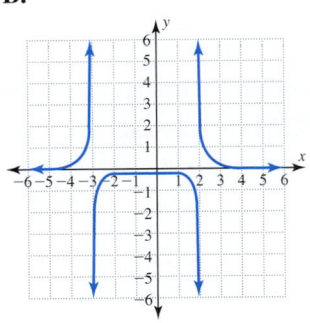

C.

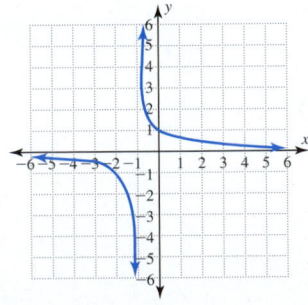

D.

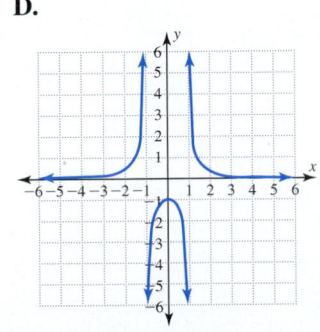

Objective 3 Reduce a Rational Expression to Lowest Terms

In Exercises 27–70, reduce each rational expression to lowest terms. Assume that the variables are restricted to values that prevent division by 0.

27. $\dfrac{22a^2b^3}{33a^3b}$ **28.** $\dfrac{21x^7y^5}{35x^4y^8}$

29. $\dfrac{30x^2 - 45x}{5x}$ **30.** $\dfrac{46x^4 - 69x^3}{23x^2}$

31. $\dfrac{7x}{14x^2 - 21x}$ **32.** $\dfrac{6m^2}{9m^4 - 15m^3}$

33. $\dfrac{a^2b(2x - 3y)}{-ab^2(2x - 3y)}$ **34.** $\dfrac{ab(2x - 3y)^2}{ab(2x - 3y)^3}$

35. $\dfrac{7x - 8y}{8y - 7x}$ **36.** $\dfrac{3a - 5b}{5b - 3a}$

37. $\dfrac{ax - ay}{by - bx}$ **38.** $\dfrac{5x - 10}{8 - 4x}$

39. $\dfrac{(x - 2y)(x + 5y)}{(x + 2y)(x + 5y)}$ **40.** $\dfrac{(x - 4y)(x + 4y)}{(x - 4y)(x + 7y)}$

41. $\dfrac{x^2 - y^2}{3x + 3y}$ **42.** $\dfrac{v^2 - w^2}{7v - 7w}$

43. $\dfrac{25x^2 - 4}{14 - 35x}$ **44.** $\dfrac{4x^2 - 9}{9 - 6x}$

45. $\dfrac{3x^2 - 2xy}{3x^2 - 5xy + 2y^2}$ **46.** $\dfrac{x^2 - 2xy + y^2}{x^2 - y^2}$

47. $\dfrac{4x^2 + 12xy + 9y^2}{14x + 21y}$ **48.** $\dfrac{x^2 + 9x + 14}{x^2 + 5x - 14}$

49. $\dfrac{14ax + 21a}{35ay + 42a}$ **50.** $\dfrac{3ax - 3ay}{3ax + 3ay}$

51. $\dfrac{ax - y - z}{y + z - ax}$ **52.** $\dfrac{x^2 - x - 3}{x + 3 - x^2}$

53. $\dfrac{2a^2 - ab - b^2}{a^2 - b^2}$ **54.** $\dfrac{ax + bx - ay - by}{5x - 5y}$

55. $\dfrac{vx + vy - wx - wy}{v^2 - w^2}$ **56.** $\dfrac{2x^2 - 9x - 5}{4x^2 - 1}$

57. $\dfrac{x^2 - 25}{3x^2 + 14x - 5}$ **58.** $\dfrac{2x^2 + xy - 3y^2}{3x^2 - 5xy + 2y^2}$

59. $\dfrac{5a^2 + 4ab - b^2}{5a^2 - 6ab + b^2}$ **60.** $\dfrac{4a^2 - 4ab + b^2}{2a^2 + ab - b^2}$

61. $\dfrac{(b^2 - 2b + 1) - a^2}{5ab - 5a + 5a^2}$ **62.** $\dfrac{(a^2 - 4ab + 4b^2) - z^2}{5a - 10b + 5z}$

63. $\dfrac{4z^2 - 5w}{5w - 4z^2}$

64. $\dfrac{x^4 - y^3}{y^3 - x^4}$

65. $\dfrac{12x^2 + 24xy + 12y^2}{16x^2 - 16y^2}$

66. $\dfrac{6y^2 + 11yz - 7z^2}{9y^2 + 42yz + 49z^2}$

67. $\dfrac{14x^2 - 9xy + y^2}{y^2 - 7xy}$

68. $\dfrac{x^2 - 25y^2}{x^2 - 5xy}$

69. $\dfrac{b^4 - 1}{5b + 5}$

70. $\dfrac{w^4 - 1}{w^3 + w}$

Connecting Concepts to Applications

71. Canoeing Time Upstream The time it takes a camper to paddle 12 mi upstream in a river that flows 2 mi/h is given by $T(x) = \dfrac{12}{x - 2}$, where x is the speed the camper can paddle in still water. Evaluate and interpret each of these expressions.

   a. $T(6)$ **b.** $T(8)$ **c.** $T(2)$

72. Canoeing Time Downstream The time it takes a camper to paddle 12 mi downstream in a river that flows 3 mi/h is given by $T(x) = \dfrac{12}{x + 3}$, where x is the speed the camper can paddle in still water. Evaluate and interpret each of these expressions.

   a. $T(0)$ **b.** $T(1)$ **c.** $T(3)$

73. Average Cost The average cost of producing x m of copper cable is displayed in the graph.

   a. Describe what happens to the average cost when the number of meters of cable produced approaches 0 m.

   b. Describe what happens to the average cost when the number of meters of cable produced becomes very large.

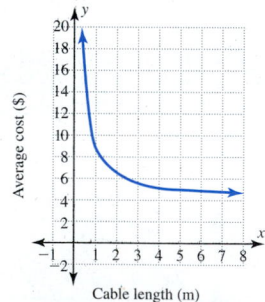

Cable length (m)

74. Average Cost The average cost of producing x kg of detergent is displayed in the graph.

   a. Describe what happens to the average cost when the number of kilograms of detergent produced approaches 0 kg.

b. Describe what happens to the average cost when the number of kilograms of detergent produced becomes very large.

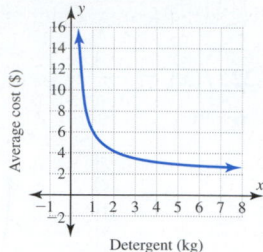

Detergent (kg)

Concept Development

In Exercises 75–86, fill in the missing numerator or denominator.

75. $\dfrac{7}{12} = \dfrac{?}{24}$

76. $\dfrac{3}{5} = \dfrac{?}{20}$

77. $\dfrac{2x}{x - 5} = \dfrac{?}{7(x - 5)}$

78. $\dfrac{3x}{x + 4} = \dfrac{?}{5(x + 4)}$

79. $\dfrac{7x - 8y}{3a - 5b} = \dfrac{?}{10b - 6a}$

80. $\dfrac{3x - 2y}{b - 3a} = \dfrac{?}{9a - 3b}$

81. $\dfrac{10}{a + b} = \dfrac{?}{a^2 - b^2}$

82. $\dfrac{6}{x - y} = \dfrac{?}{x^2 - y^2}$

83. $\dfrac{5}{x - 6} = \dfrac{?}{x^2 - 7x + 6}$

84. $\dfrac{2x - y}{x + 3y} = \dfrac{?}{x^2 + 5xy + 6y^2}$

85. $\dfrac{2x - y}{x + 3y} = \dfrac{2x^2 + xy - y^2}{?}$

86. $\dfrac{x - 2y}{x + 3y} = \dfrac{x^2 - 4y^2}{?}$

Group discussion questions

87. Error Analysis Examine each student's work, and correct any mistakes you find.

   a. Student A

$$\dfrac{x - 3}{x + 5} = \dfrac{-3}{5}$$

   b. Student B

$$\dfrac{x^2 - 9}{x^2 - 4x + 3} = \dfrac{(x - 3)(x + 3)}{(x - 3)(x - 1)}$$

$$= \dfrac{x + 3}{x - 1}$$

$$= \dfrac{3}{-1}$$

$$= -3$$

88. Discovery Question

   a. Reduce $\dfrac{x^2 - 5x + 6}{x - 2}$ to lowest terms.

   b. Determine the value excluded from the domain of
$$f(x) = \dfrac{x^2 - 5x + 6}{x - 2}.$$

   c. Graph $f(x) = \dfrac{x^2 - 5x + 6}{x - 2}.$

   d. Graph $g(x) = x - 3.$

   e. What is the domain of $f(x)$?

   f. What is the domain of $g(x)$?

   g. The graphs of $y = f(x)$ and $y = g(x)$ are different (although you may not see the difference on a graphing calculator). Describe the difference between these graphs.

89. Challenge Question

   a. Have each person in your group create a different rational function whose domain is all real numbers except for 2 and 5.

   b. Compare these functions algebraically.

   c. Use a graphing calculator to compare the graphs of these functions.

90. Communicating Mathematically Complete the following table.

x	$y = \dfrac{x + 3}{x - 2}$
1.00	
1.50	
1.90	
1.99	

   a. Describe the trend you observe in the x-values in the table.

   b. Describe the trend you observe in the y-values after you complete the table.

91. Communicating Mathematically Complete the following table.

x	$y = \dfrac{x + 3}{x - 2}$
1	
10	
100	
1,000	

   a. Describe the trend you observe in the x-values in the table.

   b. Describe the trend you observe in the y-values after you complete the table.

| **9.1** | **Cumulative Review** |

Sketch the graph of each function.

1. $h(x) = 7$

2. $f(x) = -\dfrac{1}{3}x + 2$

3. $g(x) = |x - 1|$

4. $p(x) = -|x - 1| + 5$

5. $f(x) = x^2 + 2x - 8$

| **Section 9.2** | Multiplying and Dividing Rational Expressions |

Objective:

1. Multiply and divide rational expressions.

Since the variables in a rational expression represent real numbers, the rules and procedures for performing operations with rational expressions are the same as those used in Chapter 1 for performing these operations with arithmetic fractions.

1. Multiply and Divide Rational Expressions

Why Are Multiplication and Division Covered Before Addition and Subtraction?

This is done so that we can develop some of the skills needed to add rational expressions with different denominators. When you are multiplying arithmetic fractions, there are two methods that will produce the correct result. Method 1 involves simply multiplying the numerators and multiplying the denominators and then reducing the resulting fraction. Method 2 involves first factoring each numerator and denominator. The product is still formed by multiplying the numerators and multiplying the denominators, but leaving these products in factored form makes it easier to recognize factors common to the numerator and denominator. Both methods are now illustrated to multiply $\frac{15}{8} \cdot \frac{12}{25}$.

Method 1

$$\frac{15}{8} \cdot \frac{12}{25} = \frac{180}{200}$$

$$= \frac{9 \cdot \cancel{20}}{10 \cdot \cancel{20}}$$

$$= \frac{9}{10}$$

Method 2

$$\frac{15}{8} \cdot \frac{12}{25} = \frac{3 \cdot 5}{2 \cdot 4} \cdot \frac{3 \cdot 4}{5 \cdot 5}$$

$$= \frac{3 \cdot \cancel{5} \cdot 3 \cdot \cancel{4}}{2 \cdot \cancel{4} \cdot 5 \cdot \cancel{5}}$$

$$= \frac{9}{10}$$

Which of These Methods Is Better to Use to Multiply Rational Expressions?

Method 2 is more efficient because reducing before multiplying the factors produces simpler expressions to work with. The procedure for multiplying rational expressions is given in the following box. Throughout this section, we will assume that the variables are restricted to values that prevent division by 0.

Multiplying Rational Expressions

Algebraically	Verbally	Algebraic Example
If A, B, C, and D are real polynomials and $B \neq 0$ and $D \neq 0$, then $$\frac{A}{B} \cdot \frac{C}{D} = \frac{AC}{BD}.$$	1. Factor the numerators and the denominators. Then write the product as a single fraction, indicating the product of the numerators and the product of the denominators. 2. Reduce this fraction by dividing the numerator and the denominator by any common nonzero factors.	$$\frac{x^2 - xy}{4a - 12b} \cdot \frac{3a - 9b}{5x - 5y}$$ $$= \frac{x(x - y)}{4(a - 3b)} \cdot \frac{3(a - 3b)}{5(x - y)}$$ $$= \frac{3x\cancel{(x - y)}\cancel{(a - 3b)}}{20\cancel{(x - y)}\cancel{(a - 3b)}}$$ $$= \frac{3x}{20}$$

A Mathematical Note

The Babylonians left considerable evidence of their mathematical abilities. Early in the 19th century, approximately 500,000 clay tablets were excavated. All these tablets were written in *cuneiform* script. About 400 of these tablets contain mathematical problems and tables. Tablet number 322 of the G. A. Plimpton collection at Columbia University is famous because it contains a list of Pythagorean triples for the lengths of the three sides of a right triangle.

Example 1   Multiplying Rational Expressions

Find the product of $\dfrac{2x - 2y}{35}$ and $\dfrac{15}{3x - 3y}$.

Solution

$$\frac{2x - 2y}{35} \cdot \frac{15}{3x - 3y} = \frac{2(x - y)}{35} \cdot \frac{15}{3(x - y)}$$

Factor the numerators and the denominators. (Usually you do not factor the constant coefficients.)

$$= \frac{2\cancel{(x - y)}(\cancel{3})(\cancel{5})}{\cancel{5}(7)(\cancel{3})\cancel{(x - y)}}$$

Then write the product as a single fraction, indicating the product of these factors.

Reduce by dividing out the common factors.

$$= \frac{2}{7}$$

Self-Check 1

Multiply $\dfrac{2x - 5y}{7xy} \cdot \dfrac{35x^2}{25y - 10x}$.

As Example 2 illustrates, once the polynomials in rational expressions are in factored form, the simplification of a product is easy.

Example 2   Multiplying Rational Expressions

Multiply $\dfrac{(x - 2)(x - 3)}{(x + 1)(x - 2)} \cdot \dfrac{(x + 1)(x - 4)}{(x - 3)(x + 4)}$.

Solution

$$\frac{(x - 2)(x - 3)}{(x + 1)(x - 2)} \cdot \frac{(x + 1)(x - 4)}{(x - 3)(x + 4)} = \frac{\cancel{(x - 2)}\cancel{(x - 3)}\cancel{(x + 1)}(x - 4)}{\cancel{(x + 1)}\cancel{(x - 2)}\cancel{(x - 3)}(x + 4)}$$

The numerators and denominators are already in factored form. Thus write the product as a single fraction and then reduce by dividing out the common factors.

$$= \frac{x - 4}{x + 4}$$

Self-Check 2

Multiply $\dfrac{(x - 1)(x + 3)}{(x + 3)(x + 5)} \cdot \dfrac{2x(x + 5)}{(x - 2)(x - 1)}$.

Example 2 illustrates how easy it is to multiply rational expressions once all the polynomials in the numerator and the denominator are factored. Thus the majority of work in multiplying rational expressions is in factoring the numerators and the denominators.

In Example 3, note that each factor in the numerator also occurs in the denominator. Thus the reduced form has only a factor of 1 remaining in the numerator.

Example 3   Multiplying Rational Expressions

Multiply $\dfrac{2x - y}{4x^2 - 4y^2} \cdot \dfrac{x^2 + 2xy + y^2}{2x^2 + xy - y^2}$.

Solution

$$\frac{2x - y}{4x^2 - 4y^2} \cdot \frac{x^2 + 2xy + y^2}{2x^2 + xy - y^2} = \frac{2x - y}{4(x^2 - y^2)} \cdot \frac{(x + y)(x + y)}{(2x - y)(x + y)}$$

Factor the numerators and the denominators. Then write the product as a single fraction.

$$= \frac{\cancel{(2x - y)}\cancel{(x + y)}\cancel{(x + y)}}{4(x - y)\cancel{(x + y)}\cancel{(2x - y)}\cancel{(x + y)}}$$

Reduce this fraction by dividing out the common factors.

$$= \frac{1}{4(x - y)}$$

Self-Check 3

Multiply $\dfrac{10x^2 - 6x}{7xy - 7y} \cdot \dfrac{5x^2 - 2x - 3}{25x^2 - 9}$.

If two rational expressions are equal, then tables for these two expressions will be equal for all values for which both expressions are defined. Looking at a limited set of values in a table does not confirm that the expressions are equal for all values, but it does serve as a good check that is likely to catch potential errors from an algebraic computation. This is illustrated in Example 4.

Example 4   Using Tables to Compare Two Rational Expressions

Multiply $\dfrac{x^3 + x}{x^3 - 3x^2} \cdot \dfrac{x^2 - 4x + 3}{5x^2 + 5}$ and compare the simplified form of the product to the original expression by using a table of values.

Solution

Algebraically

$$\frac{x^3 + x}{x^3 - 3x^2} \cdot \frac{x^2 - 4x + 3}{5x^2 + 5} = \frac{x(x^2 + 1)}{x^2(x - 3)} \cdot \frac{(x - 1)(x - 3)}{5(x^2 + 1)}$$

Factor the numerators and the denominators.

$$= \frac{\overset{1}{\cancel{x}}\overset{1}{\cancel{(x^2 + 1)}}(x - 1)\cancel{(x - 3)}}{\underset{x}{\cancel{x^2}}\underset{1}{\cancel{(x - 3)}}(5)\underset{1}{\cancel{(x^2 + 1)}}}$$

Then write the product as a single fraction. Reduce this fraction by dividing out the common factors.

$$= \frac{x - 1}{5x}$$

Numerical Check

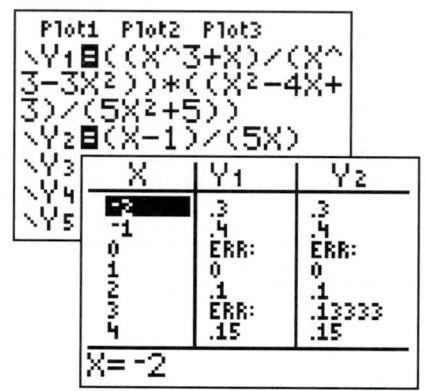

Let Y_1 represent the original expression and Y_2 represent the simplified product.

Both expressions are undefined for $x = 0$.

The expressions are identical except for $x = 3$.

The original expression has excluded values of 0 and 3. The reduced expression has an excluded value of 0.

It is understood that this equality means that these expressions are identical for all values for which they are both defined. For $x \neq 0$ and $x \neq 3$, these expressions are identical.

Answer: $\dfrac{x^3 + x}{x^3 - 3x^2} \cdot \dfrac{x^2 - 4x + 3}{5x^2 + 5} = \dfrac{x - 1}{5x}$

Self-Check 4

Multiply $\dfrac{x^2 + 3x}{x^3 + x^2} \cdot \dfrac{x^2 - 2x - 3}{5x^2 - 45}$ and compare the simplified product to the original expression by using a table of values.

We now examine the division of rational expressions. The procedure for dividing rational expressions is given in the box.

Dividing Rational Expressions

Algebraically	Verbally	Algebraic Example
If A, B, C, and D are real polynomials and $B \neq 0$, $C \neq 0$, and $D \neq 0$, then $$\frac{A}{B} \div \frac{C}{D} = \frac{A}{B} \cdot \frac{D}{C}$$ $$= \frac{AD}{BC}$$	1. Rewrite the division problem as the product of the dividend and the reciprocal of the divisor. 2. Perform the multiplication by using the procedure for multiplying rational expressions.	$$\frac{12x}{10x - 35} \div \frac{15x}{4x - 14}$$ $$= \frac{12x}{5(2x - 7)} \cdot \frac{2(2x - 7)}{15x}$$ $$= \frac{\overset{4}{\cancel{(12x)}}(2)\cancel{(2x - 7)}}{5\cancel{(2x - 7)}\underset{5}{\cancel{(15x)}}}\overset{1}{}$$ $$= \frac{8}{25}$$

Example 5 Dividing Rational Expressions

Divide $\dfrac{10ab}{3y - x} \div \dfrac{5a^3}{7x - 21y}$.

Solution

$$\frac{10ab}{3y - x} \div \frac{5a^3}{7x - 21y} = \frac{10ab}{3y - x} \cdot \frac{7x - 21y}{5a^3}$$

Rewrite the division problem as the product of the dividend and the reciprocal of the divisor.

$$= \frac{10ab}{(-1)(x - 3y)} \cdot \frac{7(x - 3y)}{5a^3}$$

Factor the binomials in the numerator and in the denominator. Note that -1 is factored out of $3y - x$.

$$= \frac{\overset{2b}{\cancel{(10ab)}}(7)\,\overset{1}{\cancel{(x - 3y)}}}{(-1)\cancel{(x - 3y)}\underset{a^2}{\cancel{(5a^3)}}}\underset{1}{}$$

Write this product as a single fraction. Then reduce by dividing out the common factors.

$$= \frac{(2b)(7)}{(-1)a^2}$$

$$= -\frac{14b}{a^2}$$

Self-Check 5

Divide $\dfrac{10x - 15}{x^2 - 5x} \div \dfrac{14x - 21}{10 - 2x}$.

Example 6   Dividing Rational Expressions

Divide $\dfrac{x^2 - 9}{x^2 - 25} \div \dfrac{x - 3}{x + 5}$.

Solution

$$\frac{x^2 - 9}{x^2 - 25} \div \frac{x - 3}{x + 5} = \frac{x^2 - 9}{x^2 - 25} \cdot \frac{x + 5}{x - 3}$$

Rewrite the division problem as the product of the dividend and the reciprocal of the divisor.

$$= \frac{(x + 3)(x - 3)}{(x + 5)(x - 5)} \cdot \frac{x + 5}{x - 3}$$

Factor the numerator and denominator of the first fraction.

$$= \frac{(x + 3)\cancel{(x - 3)}\cancel{(x + 5)}}{\cancel{(x + 5)}(x - 5)\cancel{(x - 3)}}$$

Then reduce by dividing out the common factors.

$$= \frac{x + 3}{x - 5}$$

Self-Check 6

Divide $\dfrac{x^2 - 36}{x^2 - 4x} \div \dfrac{x + 6}{x - 4}$.

Can I Combine Some of the Steps Illustrated in Example 6?

Yes, but be sure that your proficiency is well developed before you try to perform too many steps mentally. The steps that are skipped are the source of many student errors—especially errors in sign.

Examples 7 and 8 involve both multiplication and division. Compare these two examples to note the correct order of operations in each example.

Example 7   Simplifying a Rational Expression Involving Both Multiplication and Division

Simplify $\dfrac{12x}{x + y} \div \left(\dfrac{5x - 5}{x^2 - y^2} \cdot \dfrac{3xy}{xy - y} \right)$.

Solution

$$\frac{12x}{x + y} \div \left(\frac{5x - 5}{x^2 - y^2} \cdot \frac{3xy}{xy - y} \right) = \frac{12x}{x + y} \div \left[\frac{(5x - 5)(3xy)}{(x^2 - y^2)(xy - y)} \right]$$

The grouping symbols indicate that the multiplication inside the brackets has priority over the division. Multiply the numerators and the denominators within the brackets first.

$$= \frac{12x}{x + y} \cdot \frac{(x^2 - y^2)(xy - y)}{(5x - 5)(3xy)}$$

Then invert the divisor within the brackets and multiply.

$$= \frac{12x}{x + y} \cdot \frac{(x - y)(x + y)(y)(x - 1)}{(5)(x - 1)(3xy)}$$

Factor the numerator and the denominator.

$$= \frac{\overset{4}{\cancel{(12x)}}(x - y)\cancel{(x + y)}\cancel{(y)}\cancel{(x - 1)}}{\cancel{(x + y)}(5)\cancel{(x - 1)}\cancel{(3xy)}}$$

Then reduce by dividing out the common factors.

$$= \frac{4(x - y)}{5}$$

Self-Check 7

Simplify $\left(\dfrac{12x}{x + y} \div \dfrac{5x - 5}{x^2 - y^2} \right)\left(\dfrac{3xy}{xy - y} \right)$.

Example 8   Simplifying a Rational Expression Involving Both Multiplication and Division

Simplify $\dfrac{x^2 - 25}{x^2 - x - 12} \div \dfrac{x^2 - x - 20}{3x - 3} \cdot \dfrac{x^2 - 16}{x^2 + 4x - 5}$.

Solution

$$\dfrac{x^2 - 25}{x^2 - x - 12} \div \dfrac{x^2 - x - 20}{3x - 3} \cdot \dfrac{x^2 - 16}{x^2 + 4x - 5}$$

Following the correct order of operations from left to right, invert only the fraction in the middle.

$$= \dfrac{x^2 - 25}{x^2 - x - 12} \cdot \dfrac{3x - 3}{x^2 - x - 20} \cdot \dfrac{x^2 - 16}{x^2 + 4x - 5}$$

$$= \dfrac{(x + 5)(x - 5)}{(x + 3)(x - 4)} \cdot \dfrac{3(x - 1)}{(x - 5)(x + 4)} \cdot \dfrac{(x + 4)(x - 4)}{(x + 5)(x - 1)}$$

Factor each numerator and denominator and then write this product as a single fraction.

$$= \dfrac{\overset{1}{\cancel{(x + 5)}}\overset{1}{\cancel{(x - 5)}}(3)\overset{1}{\cancel{(x - 1)}}\overset{1}{\cancel{(x + 4)}}\overset{1}{\cancel{(x - 4)}}}{(x + 3)\underset{1}{\cancel{(x - 4)}}\underset{1}{\cancel{(x - 5)}}\underset{1}{\cancel{(x + 4)}}\underset{1}{\cancel{(x + 5)}}\underset{1}{\cancel{(x - 1)}}}$$

Reduce by dividing out the common factors.

$$= \dfrac{3}{x + 3}$$

Self-Check 8

Simplify $\dfrac{5x^2}{x^2 - 3x + 2} \cdot \dfrac{x^2 - 4x + 4}{x^2 - x} \div \dfrac{x^2 - 2x}{x^2 - 2x + 1}$.

Self-Check Answers

1. $-\dfrac{x}{y}$

2. $\dfrac{2x}{x - 2}$

3. $\dfrac{2x}{7y}$

4. $\dfrac{1}{5x}$

Check:

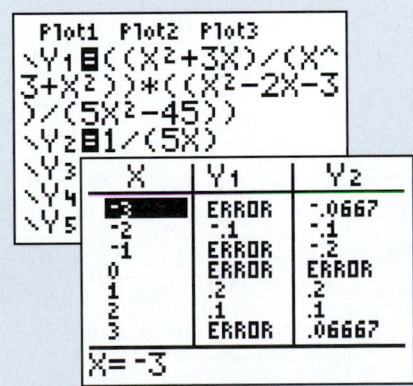

The two expressions agree for all values of x for which they are both defined.

5. $-\dfrac{10}{7x}$

6. $\dfrac{x - 6}{x}$

7. $\dfrac{36x^2(x - y)}{5(x - 1)^2}$

8. 5

9.2 Using the Language and Symbolism of Mathematics

1. If A, B, C, and D are real polynomials and $B \neq 0$ and $D \neq 0$, then $\dfrac{A}{B} \cdot \dfrac{C}{D} = $ _____.

2. To multiply two rational expressions:
   a. Factor the numerators and the denominators. Then write the product as a single _____, indicating the product of the numerators and the product of the denominators.
   b. Reduce this fraction by dividing the numerator and the denominator by any common _____ factors.

3. If A, B, C, and D are real polynomials and $B \neq 0$, $C \neq 0$, and $D \neq 0$, then $\dfrac{A}{B} \div \dfrac{C}{D} = $ _____.

4. To divide two rational expressions:
   a. Rewrite the division problem as the product of the dividend and the _____ of the divisor.
   b. Perform the multiplication by using the procedure for multiplying rational expressions.

5. Writing an equal symbol between two rational expressions means that these expressions are identical for all values for which _____ expressions are defined.

9.2 Quick Review

Perform the indicated operations using only pencil and paper.

1. $\dfrac{5}{3} \cdot \dfrac{2}{5} \cdot \dfrac{8}{5}$

2. $\dfrac{5}{3} \div \dfrac{2}{5} \cdot \dfrac{8}{5}$

3. $\dfrac{5}{3} \div \left(\dfrac{2}{5} \cdot \dfrac{8}{5} \right)$

Determine the missing fraction without using a calculator.

4. $\dfrac{3}{10} \cdot (\underline{}) = \dfrac{6}{20}$

5. $\dfrac{5}{8} \cdot (\underline{}) = \dfrac{35}{56}$

9.2 Exercises

Objective 1 Multiply and Divide Rational Expressions

In Exercises 1–46, perform the indicated operations and reduce all results to lowest terms. Assume the variables are restricted to values that prevent division by 0.

1. $\dfrac{14}{15} \cdot \dfrac{55}{42}$

2. $\dfrac{33}{10} \cdot \dfrac{28}{77}$

3. $\dfrac{9}{35} \div \dfrac{27}{55}$

4. $\dfrac{36}{49} \div \dfrac{9}{4}$

5. $\dfrac{7m}{9} \cdot \dfrac{6}{14m^2}$

6. $\dfrac{5m^3}{21} \cdot \dfrac{14}{15m^9}$

7. $\dfrac{66x^2}{30y} \div \dfrac{77x^8}{45y^3}$

8. $\dfrac{36x^5}{143} \div \dfrac{45x^7}{22}$

9. $\dfrac{x - 2y}{6} \cdot \dfrac{3}{2y - x}$

10. $\dfrac{15}{3a - 2b} \cdot \dfrac{2b - 3a}{12}$

11. $\dfrac{2x(x - 1)}{6(x + 3)} \cdot \dfrac{3(x + 3)}{x^2(x + 1)}$

12. $\dfrac{x^2(x + 5)}{12(x - 7)} \cdot \dfrac{36(x + 7)}{5x(x + 5)}$

13. $\dfrac{(x - 2)(x - 3)}{10(x - 2)} \cdot \dfrac{5(x - 3)}{(x + 3)(x - 3)}$

14. $\dfrac{3(x^2 - 4)}{14(x - 2)} \cdot \dfrac{7x}{11(x + 2)}$

15. $\dfrac{x^2 - y^2}{x^2 - 2xy + y^2} \div \dfrac{3x + 3y}{7x - 21}$

16. $\dfrac{4x^2 + 12x + 9}{5x^3 - 2x^2} \div \dfrac{14x^2 + 21x}{10x^2y^2}$

17. $\dfrac{x^2 - 3x + 2}{x^2 - 4x + 4} \div \dfrac{x^2 - 2x + 1}{3x^2 - 12}$

18. $\dfrac{a^2 - 9b^2}{4a^2 - b^2} \cdot \dfrac{4a^2 - 4ab + b^2}{2a^2 - 7ab + 3b^2}$

19. $\dfrac{7(c - y) - a(c - y)}{2c^2 - 7cy + 5y^2} \cdot \dfrac{2c - 5y}{c - y}$

20. $\dfrac{5(a + b) - x(a + b)}{2a^2 - 2b^2} \cdot \dfrac{6ax}{15 - 3x}$

21. $\dfrac{10y - 14x}{x - 1} \cdot \dfrac{x^2 - 2x + 1}{21x - 15y}$

22. $\dfrac{3w - 5z}{6w^2z} \cdot \dfrac{18w^2z + 30wz^2}{9w^2 - 25z^2}$

23. $\dfrac{9x^2 - 9xy + 9y^2}{5x^2y + 5xy^2} \cdot \dfrac{2x^3y - 2xy^3}{3x^2 - 3xy + 3y^2}$

24. $\dfrac{4x^2 - 1}{18xy} \div \dfrac{6x - 3}{16x^2 + 8x}$

25. $\dfrac{4x^2 - 4}{18xy} \cdot \dfrac{30x^2y}{5 - 5x^2}$

26. $\dfrac{2x^2 - 4xy + 2y^2}{6xy} \cdot \dfrac{25xy^2}{2xy - x^2 - y^2}$

27. $(3x^2 - 14x - 5) \cdot \dfrac{x^2 - 2x - 35}{3x^2 - 20x - 7}$

28. $(6x^2 - 19x + 10) \div \dfrac{6x^2 - 11x - 10}{9x^2 + 12x + 4}$

29. $\dfrac{20x^2 + 3x - 9}{21x - 35x^2} \div (12x^2 - 11x - 15)$

30. $\dfrac{4a^2 - ab - 5b^2}{ax + by + ay + bx} \div (8a - 10b)$

31. $\dfrac{-4}{7} \cdot \left(\dfrac{15}{6} \div \dfrac{-10}{14}\right)$ **32.** $-\dfrac{2}{9} \div \left(\dfrac{22}{12} \cdot \dfrac{-20}{33}\right)$

33. $\dfrac{18x^2}{5a} \cdot \dfrac{15ax}{81a^2} \cdot \dfrac{44ax}{24x^3}$ **34.** $\dfrac{14ab^2}{18x} \cdot \dfrac{27x^3}{15a^3b^3} \div \dfrac{7a^2x}{25ab}$

35. $\dfrac{x^2 + x - y^2 - y}{3x^2 - 3y^2} \div \dfrac{5x + 5y + 5}{7x^2y + 7xy^2}$

36. $\dfrac{a(2x - y) - b(2x - y)}{a^2 - b^2} \div \dfrac{6x - 3y}{4a + 4b}$

37. $\dfrac{2x - 5}{2x^2 - x - 15} \cdot \dfrac{-2x^2 - x + 10}{3x + 4} \cdot \dfrac{-15x - 20}{2x^2 - 9x + 10}$

38. $\dfrac{x^4 - y^4}{-2x - 3y} \cdot \dfrac{2x + 3y}{7x^2 + 7y^2} \cdot \dfrac{12xy}{8x^2 + 8xy}$

39. $\dfrac{5a - b}{a^2 - 5ab + 4b^2} \div \left(\dfrac{6ab}{3a - 12b} \cdot \dfrac{b^2 - 5ab}{4a - 4b}\right)$

40. $\dfrac{x - y}{x^2 - y^2} \div \dfrac{6x}{x + y} \cdot \dfrac{15x^2}{4}$ **41.** $\dfrac{x^2 - xy}{5x^2y^2} \div \dfrac{4x - 4y}{3xy} \cdot \dfrac{20y}{11}$

42. $\dfrac{4v^2 + 11v + 6}{2v^2 - v - 10} \div$

$\left(\dfrac{4v^2 - 21v - 18}{2v^2 - 3v - 5} \cdot \dfrac{4v^2 - 7v + 3}{4v^2 - 27v + 18}\right)$

43. $\dfrac{x(a - b + 2c) - 2y(a - b + 2c)}{x^2 - 4xy + 4y^2} \div$

$\dfrac{2a^2 - 2ab + 4ac}{7xy^2 - 14y^3}$

44. $\dfrac{a(2x - y) - b(2x - y)}{a^2 - b^2} \div \dfrac{6x - 3y}{4a + 4b}$

45. $\dfrac{5x^2 - 5xy}{6x^2 - 6y^2} \div \dfrac{10xy}{x^2 + 2xy + y^2}$

46. $\dfrac{45x^2 - 45xy}{18x^2y^3} \cdot \dfrac{x^2 + 2xy + y^2}{x^2 - y^2} \div \dfrac{9x^2y + 9xy^2}{36y^4}$

Estimate Then Calculate

In Exercises 47–50, mentally estimate the value of each expression for $x = 10.01$, and then use a calculator or a spreadsheet to approximate each value to the nearest hundredth.

Problem	Mental Estimate	Approximation
47. $\dfrac{(x + 6)(x - 7)}{(x - 7)(x - 6)}$		
48. $\dfrac{(x - 2)(x + 5)}{(x - 2)(x - 5)}$		
49. $\dfrac{x^2 + 50}{x^2 - 50}$		
50. $\dfrac{x^2 - 80}{x - 6}$		

Applying Technology

In Exercises 51–54, use the tables of values for each function to match each rational expression with its reduced form.

51. $Y_1 = \dfrac{3x^2 + 6x + 3}{(x + 1)^2}$

X	Y1
-3	3
-2	3
-1	ERR:
0	3
1	3
2	3
3	3

X= -3

52. $Y_1 = \dfrac{-x^3 + 4x^2 - 4x}{(x - 2)^2}$

X	Y1
-3	3
-2	2
-1	1
0	0
1	-1
2	ERR:
3	-3

X= -3

53. $Y_1 = \dfrac{x^3 + 4x^2 + 4x}{(x + 2)^2}$

X	Y1
-3	-3
-2	ERR:
-1	-1
0	0
1	1
2	2
3	3

X= -3

54. $Y_1 = \dfrac{6x^2 - 6}{2 - 2x^2}$

X	Y1
-3	-3
-2	-3
-1	ERR:
0	-3
1	ERR:
2	-3
3	-3

X= -3

A. $Y_2 = -3$

X	Y2
-3	-3
-2	-3
-1	-3
0	-3
1	-3
2	-3
3	-3

X= -3

B. $Y_2 = x$

X	Y2
-3	-3
-2	-2
-1	-1
0	0
1	1
2	2
3	3

X= -3

C. $Y_2 = -x$

X	Y2
-3	3
-2	2
-1	1
0	0
1	-1
2	-2
3	-3

X= -3

D. $Y_2 = 3$

X	Y₂
-3	3
-2	3
-1	3
0	3
1	3
2	3
3	3

X=-3

Skill Development

In Exercises 55–64, determine the unknown expression in each equation.

55. $12 \cdot (\underline{}) = 60$

56. $12 \div (\underline{}) = 4$

57. $(x^2y^3) \div (\underline{}) = xy$

58. $(x^2y^3) \cdot (\underline{}) = x^4y^4$

59. $(x^2 - 7x - 8) \div (\underline{}) = x + 1$

60. $(2x - 3) \cdot (\underline{}) = 4x^2 - 9$

61. $\left(\dfrac{x + 2}{x - 3}\right) \cdot (\underline{}) = \dfrac{x^2 - x - 6}{x^2 - 6x + 9}$

62. $\left(\dfrac{x + 5}{x - 5}\right) \cdot (\underline{}) = \dfrac{x^2 + 7x + 10}{x^2 - 3x - 10}$

63. $\left(\dfrac{2x^2 - 5x + 2}{3x^2 - 4x + 1}\right) \div (\underline{}) = \dfrac{2x - 1}{3x - 1}$

64. $\dfrac{x^2 - 2x - 15}{x^2 + x - 12} \div (\underline{}) = \dfrac{x + 3}{x - 3}$

Connecting Concepts to Applications

65. Average Cost of Lamps The number of lamps that a factory can assemble in a day is approximated by the function $N(t) = 2t^2 - 2t$, where t is the number of hours the factory operates. The total cost of operating the plant for t hours is given by $C(t) = 200t + 200$. If the plant is shut down overnight, then it takes a period of time to resume production.

a. The average cost per lamp produced is given by $A(t) = C(t) \div N(t)$. Write a formula for the function $A(t)$.

b. Use this function to complete the following table.

t	2	6	10	14	18	22
$A(t)$						

c. Describe what happens to the average cost when the number of hours the factory operates is reduced to almost 1 hour.

d. Describe what happens to the average cost when the number of hours the factory operates is increased to almost 24 hours.

66. Average Cost of Lamps The number of lamps that a factory can assemble in a day is approximated by the function $N(t) = 2.5t^2 - 2.5t$, where t is the number of hours the factory operates. The total cost of operating the plant for t hours is given by $C(t) = 750t + 750$. If the plant is shut down overnight, then it takes a period of time to resume production.

a. The average cost per lamp produced is given by $A(t) = C(t) \div N(t)$. Write a formula for the function $A(t)$.

b. Use this function to complete the following table.

t	2	6	10	14	18	22
$A(t)$						

c. Describe what happens to the average cost when the number of hours the factory operates is reduced to almost 1 hour.

d. Describe what happens to the average cost when the number of hours the factory operates is increased to almost 24 hours.

Group discussion questions

67. Challenge Question Perform the indicated operations, and reduce all results to lowest terms. Assume that m is a natural number and that all values of the variables that cause division by 0 are excluded.

a. $\dfrac{x^{2m+1}}{x^m - 1} \cdot \dfrac{3x^m - 3}{x}$

b. $\dfrac{x^{2m} - 1}{x^{2m} + x^m} \div \dfrac{x^m - 1}{x^{2m}}$

c. $\dfrac{x^{2m} - 1}{x^m - 1} \cdot \dfrac{3x^m + 3}{(x^m + 1)^2}$

d. $\dfrac{x^{3m} - x^m}{x^{2m} + x^m} \div \dfrac{(x^m + 1)^3}{x^{2m} + 2x^m + 1}$

68. Challenge Question

Reduce $\dfrac{x}{x - 1} \cdot \dfrac{x - 1}{x - 2} \cdot \dfrac{x - 2}{x - 3} \cdot \ldots \cdot \dfrac{x - 8}{x - 9} \cdot \dfrac{x - 9}{x - 10}$, and list all the excluded values.

69. Error Analysis A student graphed $f(x) = \dfrac{x^2 - x}{x - 1}$ on a TI-84 Plus calculator and obtained the graph shown here.

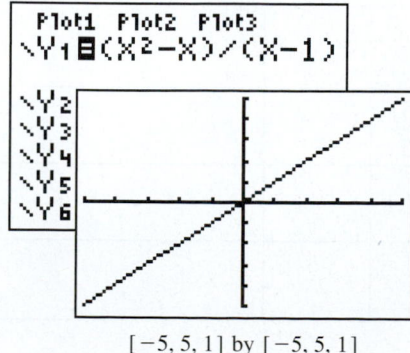

$[-5, 5, 1]$ by $[-5, 5, 1]$

Because no breaks appear in the graph of this function, the student concluded that the domain of

$$f(x) = \frac{x^2 - x}{x - 1} \text{ is } \mathbb{R}.$$

This is incorrect.

a. What is the actual domain?

b. Sketch this graph by hand to properly show the correct domain.

c. Graph this function on a TI-84 Plus calculator, using the ⟨ZOOM⟩ ④ option. Then carefully examine this graph. Using the **TRACE** feature, what do you observe at the excluded value of this function?

d. Why do you think the excluded value is not revealed by a break in the graph on the window $[-5, 5, 1]$ by $[-5, 5, 1]$?

| **9.2** | **Cumulative Review** |

Simplify each expression.

1. $2 - 5(3 + 1)^2$

2. $(2 - 5)(3 + 1)^2$

3. $2x - 5(3x + 1)^2$

4. $(2x - 5)(3x + 1)^2$

5. $((2x - 5)(3x + 1))^2$

Section 9.3 | Adding and Subtracting Rational Expressions

Objective:

1. Add and subtract rational expressions.

1. Add and Subtract Rational Expressions

The procedures for adding and subtracting rational expressions are the same as those given in Chapter 1 for adding and subtracting rational numbers. The procedures for adding and subtracting rational expressions with the same denominator are given in the following box. The result should be reduced to lowest terms.

Adding and Subtracting Rational Expressions with the Same Denominator

Algebraically	Verbally	Algebraic Example
If A, B, and C are real polynomials and $C \neq 0$, then $$\frac{A}{C} + \frac{B}{C} = \frac{A + B}{C}$$ and	To add two rational expressions with the same denominator, add the numerators and use this common denominator.	$\dfrac{7}{24x} + \dfrac{5}{24x} = \dfrac{7 + 5}{24x}$ $= \dfrac{12}{24x}$ $= \dfrac{1}{2x}$
$$\frac{A}{C} - \frac{B}{C} = \frac{A - B}{C}$$	To subtract two rational expressions with the same denominator, subtract the numerators and use this common denominator.	$\dfrac{7}{24x} - \dfrac{5}{24x} = \dfrac{7 - 5}{24x}$ $= \dfrac{2}{24x}$ $= \dfrac{1}{12x}$

It is easy to make errors in the order of operations if you are not careful when adding or subtracting rational expressions. Example 1 illustrates how parentheses can be used to avoid these errors.

Throughout this section, we will assume that the variables are restricted to values that prevent division by 0.

Example 1 Subtracting Rational Expressions with the Same Denominator

Simplify $\dfrac{5x + 7}{2x + 2} - \dfrac{3x - 1}{2x + 2}$.

Solution

$$\dfrac{5x + 7}{2x + 2} - \dfrac{3x - 1}{2x + 2} = \dfrac{(5x + 7) - (3x - 1)}{2x + 2}$$

Note: A common error is to write the numerator as $5x + 7 - 3x - 1$. To avoid this error, enclose the terms of the numerator in parentheses before combining like terms.

$$= \dfrac{5x + 7 - 3x + 1}{2x + 2}$$

$$= \dfrac{2x + 8}{2x + 2}$$

$$= \dfrac{\overset{1}{\cancel{2}}(x + 4)}{\underset{1}{\cancel{2}}(x + 1)}$$

Factor, and then reduce to lowest terms.

$$= \dfrac{x + 4}{x + 1}$$

This sum is defined for all real numbers other than -1. So $x = -1$ is the only excluded value.

Self-Check 1

Simplify $\dfrac{2x - 1}{x^2} + \dfrac{4x + 1}{x^2}$.

Example 2 Subtracting Rational Expressions with the Same Denominator

Simplify $\dfrac{x^2}{x^2 - 1} - \dfrac{2x - 1}{x^2 - 1}$.

Solution

$$\dfrac{x^2}{x^2 - 1} - \dfrac{2x - 1}{x^2 - 1} = \dfrac{x^2 - (2x - 1)}{x^2 - 1}$$

Use parentheses in the numerator to avoid an error in sign.

$$= \dfrac{x^2 - 2x + 1}{x^2 - 1}$$

Simplify the numerator.

$$= \dfrac{\overset{1}{\cancel{(x - 1)}}(x - 1)}{\underset{1}{\cancel{(x - 1)}}(x + 1)}$$

Factor, and then reduce to lowest terms. The excluded values are 1 and -1.

$$= \dfrac{x - 1}{x + 1}$$

This result is defined for all real numbers other than 1 and -1. The only excluded values are $x = -1$ and $x = 1$.

Self-Check 2

Simplify $\dfrac{9x-2}{2x+1} - \dfrac{5x-4}{2x+1}$.

Remember that $-\dfrac{a}{b} = \dfrac{-a}{b} = \dfrac{a}{-b}$. These alternative forms are often useful when you are working with two fractions whose denominators are opposites of each other.

Example 3 **Adding Rational Expressions Whose Denominators Are Opposites**

Simplify $\dfrac{5x}{3x-6} + \dfrac{-4x}{6-3x}$.

Solution

$\dfrac{5x}{3x-6} + \dfrac{-4x}{6-3x} = \dfrac{5x}{3x-6} + \dfrac{-4x}{6-3x} \cdot \dfrac{-1}{-1}$

Note that the denominators $3x-6$ and $6-3x$ are opposites. Convert $\dfrac{-4x}{6-3x}$ to the equivalent form $\left(\dfrac{-4x}{6-3x}\right) \cdot \left(\dfrac{-1}{-1}\right) = \dfrac{4x}{3x-6}$.

$= \dfrac{5x}{3x-6} + \dfrac{4x}{3x-6}$

$= \dfrac{5x+4x}{3x-6}$

Now that the denominators are the same, you can add the numerators of these like terms.

$= \dfrac{\overset{3}{9x}}{\underset{1}{3}(x-2)}$

Then factor the denominator, and reduce this result by dividing both the numerator and the denominator by 3.

$= \dfrac{3x}{x-2}$

This sum is defined for all real numbers other than 2.

Self-Check 3

Simplify $\dfrac{2y-1}{7y-3} - \dfrac{5y-2}{3-7y}$.

Fractions can be added or subtracted only if they are expressed in terms of a common denominator. Although any common denominator can be used, we can simplify our work considerably by using the least common denominator (LCD).

How Can I Determine the LCD?

Sometimes the LCD can be determined by inspection. In other cases, it may be necessary to use the procedure given in the following box.

Finding the LCD of Two or More Rational Expressions

Verbally	Algebraic Example
1. Factor each denominator completely, including constant factors. Express repeated factors in exponential form. 2. List each factor to the highest power to which it occurs in any single factorization. 3. Form the LCD by multiplying the factors listed in step 2.	Determine the LCD of $$\frac{7}{6xy^2} \text{ and } \frac{2}{15x^2y}.$$ $$6xy^2 = 2 \cdot 3 \cdot x \cdot y^2$$ $$15x^2y = 3 \cdot 5 \cdot x^2 \cdot y$$ $$\text{LCD} = 2 \cdot 3 \cdot 5 \cdot x^2 \cdot y^2$$ $$\text{LCD} = 30x^2y^2$$

The procedure for adding or subtracting rational expressions with different denominators is given in the following box. The common denominator in the example was calculated in the previous box.

Adding (Subtracting) Rational Expressions

Verbally	Algebraic Example
1. Determine the LCD of all the terms. 2. Convert each term to an equivalent rational expression whose denominator is the LCD. 3. Retain the LCD as the denominator and add (subtract) the numerators to form the sum (difference). 4. Reduce the expression to lowest terms.	$$\frac{7}{6xy^2} + \frac{2}{15x^2y} = \frac{?}{30x^2y^2} + \frac{?}{30x^2y^2}$$ $$= \frac{7}{6xy^2} \cdot \frac{5x}{5x} + \frac{2}{15x^2y} \cdot \frac{2y}{2y}$$ $$= \frac{35x}{30x^2y^2} + \frac{4y}{30x^2y^2}$$ $$= \frac{35x + 4y}{30x^2y^2}$$

The denominator of a sum or difference is usually left in factored form because this form reveals the most information about the rational expression.

Example 4 — Subtracting Two Rational Expressions with Different Denominators

Simplify $\frac{5}{x+4} - \frac{4}{x-6}$ and compare the simplified form of the difference to the original expression by using a table of values.

Solution

Algebraically

$$\frac{5}{x+4} - \frac{4}{x-6} = \frac{5}{x+4} \cdot \frac{x-6}{x-6} - \frac{4}{x-6} \cdot \frac{x+4}{x+4}$$

The LCD is $(x+4)(x-6)$. Convert each fraction so that it has this LCD as its denominator.

$$= \frac{5(x-6)}{(x+4)(x-6)} - \frac{4(x+4)}{(x+4)(x-6)}$$

$$= \frac{5(x-6) - 4(x+4)}{(x+4)(x-6)}$$

$$= \frac{5x - 30 - 4x - 16}{(x+4)(x-6)}$$

Add like terms and simplify the numerator.

$$= \frac{x - 46}{(x+4)(x-6)}$$

For many applications it is more useful to leave the denominator in factored form.

Numerical Check

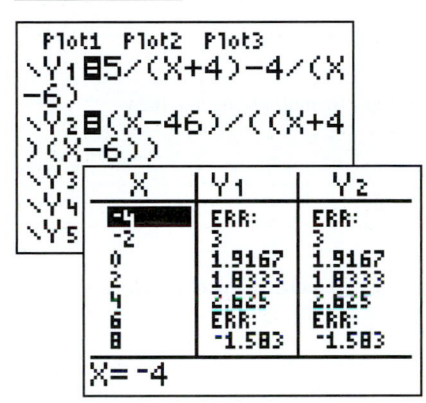

Let Y_1 represent the original expression and Y_2 represent the difference.

These tables are identical for all values of x. Both are undefined at -4 and 6.

Answer: $\dfrac{5}{x+4} - \dfrac{4}{x-6} = \dfrac{x-46}{(x+4)(x-6)}$

Self-Check 4

Simplify $\dfrac{x}{x-5} + \dfrac{3}{x+2}$ and compare the simplified sum to the original expression by using a table of values.

Some of the steps illustrated in Examples 4 and 5 can be combined to decrease the amount of writing necessary in these problems. Although most of the steps are still shown in the following examples, some have been combined. Be sure your proficiency is well developed before you try to perform too many steps mentally. The steps that are skipped are the source of many student errors—especially errors in sign.

Example 5   Combining Three Rational Expressions with Different Denominators

Simplify $\dfrac{2v + 11}{v^2 + v - 6} - \dfrac{2}{v + 3} + \dfrac{3}{2 - v}$.

Solution

$\dfrac{2v + 11}{v^2 + v - 6} - \dfrac{2}{v + 3} + \dfrac{3}{2 - v}$

$= \dfrac{2v + 11}{(v + 3)(v - 2)} - \dfrac{2}{v + 3} + \dfrac{-3}{v - 2}$

$= \dfrac{2v + 11}{(v + 3)(v - 2)} - \dfrac{2}{v + 3} \cdot \dfrac{v - 2}{v - 2} + \dfrac{-3}{v - 2} \cdot \dfrac{v + 3}{v + 3}$

$= \dfrac{2v + 11}{(v + 3)(v - 2)} - \dfrac{2(v - 2)}{(v + 3)(v - 2)} + \dfrac{-3(v + 3)}{(v + 3)(v - 2)}$

$= \dfrac{(2v + 11) - 2(v - 2) - 3(v + 3)}{(v + 3)(v - 2)}$

$= \dfrac{2v + 11 - 2v + 4 - 3v - 9}{(v + 3)(v - 2)}$

$= \dfrac{-3v + 6}{(v + 3)(v - 2)}$

$= \dfrac{-3\overset{1}{\cancel{(v - 2)}}}{(v + 3)\underset{1}{\cancel{(v - 2)}}}$

$= \dfrac{-3}{v + 3}$

$= -\dfrac{3}{v + 3}$

Factor each denominator. Noting that the factor $v - 2$ in the first denominator is the opposite of the denominator $2 - v$ in the last term, change $\dfrac{3}{2 - v}$ to the equivalent form $\left(\dfrac{3}{2 - v}\right)\left(\dfrac{-1}{-1}\right) = \dfrac{-3}{v - 2}$.
Now convert each term so that it has the LCD of $(v + 3)(v - 2)$ as its denominator.

Simplify the numerator by multiplying and then combining like terms.

Factor the numerator, and reduce by dividing out the common factor $v - 2$.

Self-Check 5

Simplify $\dfrac{3v - 1}{v^2 - 1} + \dfrac{3}{1 - v} - \dfrac{2}{v + 1}$.

Note that when we multiply or divide rational expressions, we first reduce as much as possible and then perform the multiplication or division. On the other hand, when we add or subtract rational expressions, we add or subtract first, using a common denominator, and then reduce afterward.

Self-Check Answers

1. $\dfrac{6}{x}$

2. 2

3. 1

4. $\dfrac{x^2 + 5x - 15}{(x - 5)(x + 2)}$

Check:

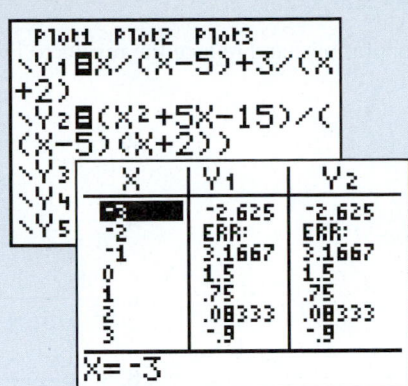

The two expressions agree for all values of x for which they are both defined.

5. $-\dfrac{2}{v - 1}$

9.3   Using the Language and Symbolism of Mathematics

1. If A, B, and C are real polynomials and $C \neq 0$, then $\dfrac{A}{C} + \dfrac{B}{C} =$ _____ and $\dfrac{A}{C} - \dfrac{B}{C} =$ _____.

2. To add two rational expressions with the same denominator, _____ the numerators and use this common denominator.

3. To subtract two rational expressions with the same denominator, _____ the numerators and use this common denominator.

4. To find the least common denominator of two or more rational expressions:
   a. Factor each _____ completely, including constant factors. Express repeated factors in _____ form.
   b. List each factor to the _____ power to which it occurs in any single factorization.
   c. Form the LCD by _____ the factors listed in step **b**.

5. To add or subtract two rational expressions:
   a. Determine the _____ of all the terms.
   b. Convert each term to an equivalent rational expression whose _____ is the LCD.
   c. Retaining the LCD as the denominator, add (subtract) the _____ to form the sum (difference).
   d. Reduce the expression to _____ _____.

9.3   Quick Review

Perform the indicated operations using only pencil and paper.

1. $\dfrac{7}{12} + \dfrac{3}{12}$

2. $\dfrac{17}{24} - \dfrac{5}{24}$

3. $\dfrac{2}{3} - \dfrac{1}{12}$

4. $\dfrac{3}{24} + \dfrac{5}{16}$

5. $\dfrac{1}{6} + \dfrac{2}{3} - \dfrac{1}{4}$

Objective 1 Add and Subtract Rational Expressions

In Exercises 1–8, perform the indicated operations, and reduce the results to lowest terms. Assume the variables are restricted to values that prevent division by 0.

1. $\dfrac{5b + 13}{3b^2} + \dfrac{b - 4}{3b^2}$

2. $\dfrac{3x^2 + 1}{8x^3} - \dfrac{1 - 3x^2}{8x^3}$

3. $\dfrac{7}{x - 7} - \dfrac{x}{x - 7}$

4. $\dfrac{a - 1}{a + 7} + \dfrac{8}{a + 7}$

5. $\dfrac{3s + 7}{s^2 - 9} + \dfrac{s + 5}{s^2 - 9}$

6. $\dfrac{5t - 22}{t^2 - 5t + 6} + \dfrac{4t - 5}{t^2 - 5t + 6}$

7. $\dfrac{x + a}{x(a + b) + y(a + b)} - \dfrac{x - b}{x(a + b) + y(a + b)}$

8. $\dfrac{2y^2 + 1}{2y^2 - 5y - 12} - \dfrac{4 - y}{2y^2 - 5y - 12}$

In Exercises 9–14, fill in the blanks to complete each addition or subtraction.

9. $\dfrac{1}{15xy^2} + \dfrac{2}{27x^2y} = \dfrac{1}{15xy^2}\left(\dfrac{?}{9x}\right) + \dfrac{2}{27x^2y}\left(\dfrac{?}{5y}\right)$

$= \dfrac{9x}{135x^2y^2} + \dfrac{?}{135x^2y^2}$

$= \dfrac{?}{135x^2y^2}$

10. $\dfrac{3}{35x^3y} + \dfrac{2}{55x^2y^3} = \dfrac{3}{35x^3y}\left(\dfrac{?}{11y^2}\right) + \dfrac{2}{55x^2y^3}\left(\dfrac{?}{7x}\right)$

$= \dfrac{?}{385x^3y^3} + \dfrac{14x}{385x^3y^3}$

$= \dfrac{?}{385x^3y^3}$

11. $\dfrac{a + b}{21(3a - b)} - \dfrac{a - b}{14(3a - b)} = \dfrac{a + b}{21(3a - b)}\left(\dfrac{?}{?}\right) - \dfrac{a - b}{14(3a - b)}\left(\dfrac{?}{?}\right)$

$= \dfrac{?}{42(3a - b)} - \dfrac{?}{42(3a - b)}$

$= \dfrac{?}{42(3a - b)}$

12. $\dfrac{2x - y}{12(x + y)} - \dfrac{x - y}{15(x + y)} = \dfrac{2x - y}{12(x + y)}\left(\dfrac{?}{?}\right) - \dfrac{x - y}{15(x + y)}\left(\dfrac{?}{?}\right)$

$= \dfrac{?}{60(x + y)} - \dfrac{?}{60(x + y)}$

$= \dfrac{?}{60(x + y)}$

13. $\dfrac{2}{(x - 3)(x - 15)} + \dfrac{1}{(x - 3)(x + 3)} = \dfrac{2}{(x - 3)(x - 15)}\left(\dfrac{?}{x + 3}\right) + \dfrac{1}{(x - 3)(x + 3)}\left(\dfrac{?}{x - 15}\right)$

$= \dfrac{?}{(x - 3)(x - 15)(x + 3)} + \dfrac{?}{(x - 3)(x - 15)(x + 3)}$

$= \dfrac{?}{(x - 3)(x - 15)(x + 3)}$

$= \dfrac{?}{(x - 15)(x + 3)}$

14. $\dfrac{3}{(2x-1)(x+4)} - \dfrac{1}{(2x-1)(x+1)} = \dfrac{3}{(2x-1)(x+4)}\left(\dfrac{?}{x+1}\right) - \dfrac{1}{(2x-1)(x+1)}\left(\dfrac{?}{x+4}\right)$

$$= \dfrac{?}{(2x-1)(x+4)(x+1)} - \dfrac{?}{(2x-1)(x+4)(x+1)}$$

$$= \dfrac{?}{(2x-1)(x+4)(x+1)}$$

$$= \dfrac{?}{(x+4)(x+1)}$$

In Exercises 15–40, perform the indicated operations, and reduce the results to lowest terms. Assume the variables are restricted to values that prevent division by 0.

15. $\dfrac{4}{9w} - \dfrac{7}{6w}$

16. $\dfrac{4}{5z} - \dfrac{1}{2z}$

17. $\dfrac{3v-1}{7v} - \dfrac{v-2}{14v}$

18. $\dfrac{12a-7}{35a} - \dfrac{a-1}{5a}$

19. $\dfrac{4}{b} + \dfrac{b}{b+4}$

20. $\dfrac{5}{c} - \dfrac{c}{c-7}$

21. $5 - \dfrac{1}{x}$

22. $7 + \dfrac{3}{y}$

23. $\dfrac{1}{x} - \dfrac{2}{x^2} + \dfrac{3}{x^3}$

24. $\dfrac{3}{y} + \dfrac{5}{y^2} - \dfrac{7}{y^3}$

25. $\dfrac{3}{x-2} - \dfrac{2}{x+3}$

26. $\dfrac{2}{x-5} - \dfrac{3}{x+6}$

27. $\dfrac{x}{x-1} - \dfrac{x}{x+1}$

28. $\dfrac{x}{x+2} - \dfrac{x}{x-2}$

29. $\dfrac{x+5}{x-4} - \dfrac{x+3}{x+2}$

30. $\dfrac{x-3}{x-2} - \dfrac{x-2}{x-3}$

31. $\dfrac{x-1}{x+2} + \dfrac{x+2}{x-1}$

32. $\dfrac{x+1}{x-3} - \dfrac{x-3}{x+1}$

33. $\dfrac{x}{77x-121y} - \dfrac{y}{49x-77y}$

34. $\dfrac{a}{6a-9b} - \dfrac{b}{4a-6b}$

35. $\dfrac{2}{(m+1)(m-2)} + \dfrac{3}{(m-2)(m+3)}$

36. $\dfrac{5}{(m-1)(m+2)} + \dfrac{2}{(m+2)(m-3)}$

37. $\dfrac{m+2}{m^2-6m+8} - \dfrac{8-3m}{m^2-5m+6}$

38. $\dfrac{3n+8}{n^2+6n+8} - \dfrac{4n+2}{n^2+n-12}$

39. $\dfrac{1}{a-3b} + \dfrac{b}{a^2-7ab+12b^2} + \dfrac{1}{a-4b}$

40. $\dfrac{5x}{5x-y} + \dfrac{2y^2}{25x^2-y^2} - \dfrac{5x}{5x+y}$

Connecting Algebra to Geometry

41. Total Area of Two Regions Find the sum of the areas of the triangle and rectangle.

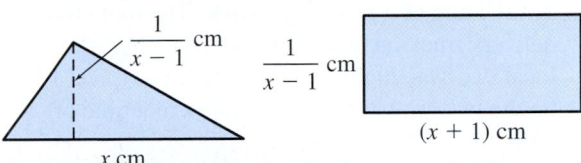

42. Total Area of Two Regions Find the sum of the areas of the parallelogram and rectangle.

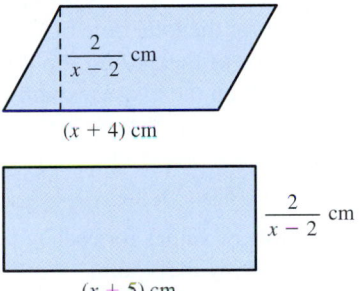

Connecting Concepts to Applications

43. Total Work by Two Painters One painter can paint a house in t hours, while a second painter would take $(t+5)$ hours to do the same house. In 8 hours the fractional portion of the job done by each is $\dfrac{8}{t}$ and $\dfrac{8}{t+5}$, respectively.

  a. Write a rational function that gives the total amount of work completed by the two painters working together for 8 hours.

  b. Use the graph of this function to describe what happens to the total amount of work done as t becomes very large. (*Hint:* See Example 3 in Section 9.1.)

44. Total Work by Two Pipes One pipe can fill a cooling tank in t hours. A second pipe would take $(t+3)$ hours to fill the same tank. If both pipes are used, the fractional portion of the tank that can be filled by each pipe in 10 hours is $\dfrac{10}{t}$ and $\dfrac{10}{t+3}$, respectively.

  a. Write a rational function that gives the total amount of the cooling tank filled by the two pipes working together for 10 hours.

b. Use the graph of this function to describe what happens to the total amount of the cooling tank filled by the two pipes as t becomes very large. (*Hint:* See Example 3 in Section 9.1.)

45. Total Time of an Automobile Trip An automobile traveled 90 mi at a rate r on a gravel road in a time of $\dfrac{90}{r}$ hours. Then the automobile traveled 165 mi on paved roads in a time of $\dfrac{165}{r + 10}$ hours.

a. Write a rational function for the total time of this trip.

b. What is the total time of the 255-mi trip if the average speed of the car on the gravel road is 45 mi/h?

46. Total Time of a Delivery Truck The route for a delivery truck includes both interstate highway and local two-way highways. The truck can travel 120 mi on the two-way highway at a rate r in a time of $\dfrac{120}{r}$ hours. It can also travel 180 mi on the interstate in a time of $\dfrac{180}{r + 15}$ hours.

a. Write a rational function for the total time of this trip.

b. Approximate the total time of the 300-mi trip if the average speed of the truck on the two-way highway is 50 mi/h.

Applying Technology

In Exercises 47–50, use the table of values for each function to match each expression with its sum.

47. $Y_1 = \dfrac{5}{x - 1} + \dfrac{3}{x + 2}$

X	Y1
-3	-4.25
-2	ERR:
-1	.5
0	-3.5
1	ERR:
2	5.75
3	3.1

X= -3

48. $Y_1 = \dfrac{5}{x - 1} - \dfrac{3}{x + 2}$

X	Y1
-3	1.75
-2	ERR:
-1	-5.5
0	-6.5
1	ERR:
2	4.25
3	1.9

X= -3

49. $Y_1 = \dfrac{2x}{x - 1} - \dfrac{2x}{x + 2}$

X	Y1
-3	-4.5
-2	ERR:
-1	3
0	0
1	ERR:
2	3
3	1.8

X= -3

50. $Y_1 = \dfrac{x}{x - 1} + \dfrac{x}{x + 2}$

X	Y1
-3	3.75
-2	ERR:
-1	-.5
0	0
1	ERR:
2	2.5
3	2.1

X= -3

A. $Y_2 = \dfrac{8x + 7}{(x - 1)(x + 2)}$

X	Y1
-3	-4.25
-2	ERR:
-1	.5
0	-3.5
1	ERR:
2	5.75
3	3.1

X= -3

B. $Y_2 = \dfrac{x(2x + 1)}{(x - 1)(x + 2)}$

X	Y1
-3	3.75
-2	ERR:
-1	-.5
0	0
1	ERR:
2	2.5
3	2.1

X= -3

C. $Y_2 = \dfrac{2x + 13}{(x - 1)(x + 2)}$

X	Y1
-3	1.75
-2	ERR:
-1	-5.5
0	-6.5
1	ERR:
2	4.25
3	1.9

X= -3

D. $Y_2 = \dfrac{6x}{(x - 1)(x + 2)}$

X	Y1
-3	-4.5
-2	ERR:
-1	3
0	0
1	ERR:
2	3
3	1.8

X= -3

Skill Development

In Exercises 51–68, perform the indicated operations, and reduce the results to lowest terms. Assume the variables are restricted to values that prevent division by 0.

51. $\dfrac{x^2 + 5}{x^2 - 3x + 4} - 2$

52. $\dfrac{2x^2 - 5x + 3}{3x^2 + 7x + 9} + 2$

53. $\dfrac{5}{2x + 2} + \dfrac{x + 5}{2x^2 - 2} - \dfrac{3}{x - 1}$

54. $\dfrac{p + 6}{p^2 - 4} - \dfrac{4}{p + 2} - \dfrac{2}{p - 2}$

55. $\dfrac{1}{x - 5} + \dfrac{1}{x + 5} - \dfrac{10}{x^2 - 25}$

56. $\dfrac{42}{x^2 - 49} + \dfrac{3}{x + 7} + \dfrac{3}{x - 7}$

57. $\dfrac{3}{s - 5t} + \dfrac{7}{s - 2t} - \dfrac{9t}{s^2 - 7st + 10t^2}$

58. $\dfrac{y + 8}{y^2 - 2y - 8} + \dfrac{1}{y + 2} - \dfrac{4}{4 - y}$

59. $\dfrac{2z + 11}{z^2 + z - 6} + \dfrac{2}{z + 3} - \dfrac{3}{z - 2}$

60. $\dfrac{3}{2w - 1} + \dfrac{4}{2w + 1} - \dfrac{10w - 3}{4w^2 - 1}$

61. $\dfrac{v + w}{vw} + \dfrac{w}{v^2 - vw} - \dfrac{1}{w}$

62. $\dfrac{5v}{v^3 - 5v^2 + 6v} - \dfrac{4}{v - 2} - \dfrac{3v}{3 - v}$

63. $\dfrac{9w + 2}{3w^2 - 2w - 8} - \dfrac{7}{4 - w - 3w^2}$

64. $\dfrac{2x + y}{(x - y)(x - 2y)} - \dfrac{x + 4y}{(x - 3y)(x - y)} - \dfrac{x - 7y}{(x - 3y)(x - 2y)}$

65. $\dfrac{2w - 7}{w^2 - 5w + 6} - \dfrac{2 - 4w}{w^2 - w - 6} + \dfrac{5w + 2}{4 - w^2}$

66. $\dfrac{6m}{m + 2} + \dfrac{2}{m - 2} - \dfrac{8m}{m^2 - 4}$

67. $\dfrac{4m}{1 - m^2} + \dfrac{2}{m + 1} - 2$

68. $\dfrac{y - 5}{(x^2 + 5x) + (xy + 5y)} + \dfrac{1}{x + y} - \dfrac{2}{x + 5}$

69. The sum of a rational expression and $\dfrac{1}{x + 5}$ is

$\dfrac{4x}{2x^2 + 5x - 25}$. Determine this rational expression.

70. The difference formed by a rational expression minus

$\dfrac{1}{x + 5}$ is $\dfrac{27}{4x^2 + 13x - 35}$. Determine this rational

expression.

Group discussion questions

71. Error Analysis Discuss the error made by the student whose work is shown here.

$$\dfrac{x}{x + 1} - \dfrac{1}{x + 1} \cdot \dfrac{2x}{x - 1} = \dfrac{x - 1}{x + 1} \cdot \dfrac{2x}{x - 1}$$

$$= \dfrac{2x}{x + 1}$$

72. Challenge Question Simplify each expression. Assume that m is a natural number and that all values of the variables that cause division by 0 are excluded.

a. $\dfrac{x^m}{x^{2m} - 4} - \dfrac{2}{x^{2m} - 4}$

b. $\dfrac{2x^m}{x^{2m} - 6x^m + 9} - \dfrac{6}{x^{2m} - 6x^m + 9}$

c. $\dfrac{3x^m}{x^{2m} - x^m - 20} - \dfrac{15}{x^{2m} - x^m - 20}$

73. Challenge Question Assume that m is an even integer.

Simplify $\dfrac{1}{m + 1} + \dfrac{2}{m + 1} + \dfrac{3}{m + 1} + \cdots + \dfrac{m}{m + 1}$.

Then test your result by evaluating both the original expression and your result for $m = 4, 6,$ and 8.

9.3   Cumulative Review

1. Give the slope of a line parallel to $y = \dfrac{2}{5}x + 1$.

2. Give the slope of a line perpendicular to $y = \dfrac{2}{5}x + 1$.

Solve each system in Exercises 3–5. If the system is a consistent system of independent equations, determine the unique ordered-pair solution. If the system is inconsistent, write "no solution." If the system is a consistent system of dependent equations, write "infinite number of solutions."

3. $\quad 2x + 3y = -10$
$\quad\ -5x + 7y = -62$

4. $2x + 3y = 5$
$\quad -6x = 9y + 15$

5. $2x + 3y = 12$
$\quad y = -\dfrac{2}{3}x + 4$

Section 9.4	**Combining Operations and Simplifying Complex Rational Expressions**

Objectives:

1. Simplify rational expressions in which the order of operations must be determined.
2. Simplify complex fractions.

1. Simplify Rational Expressions in Which the Order of Operations Must Be Determined

This section shows how to combine the operations already covered in the earlier sections of this chapter and provides further exercises involving rational expressions. The correct order of operations, from Section 1.6, is as follows:

Order of Operations

Step 1. Start with the expression within the innermost pair of grouping symbols.

Step 2. Perform all exponentiations.

Step 3. Perform all multiplications and divisions as they appear from left to right.

Step 4. Perform all additions and subtractions as they appear from left to right.

Example 1	**Using the Correct Order of Operations When Simplifying Rational Expressions**

Simplify $\dfrac{2x}{5} + \dfrac{x^3}{15} \cdot \dfrac{9}{x^2}$.

Solution

$$\frac{2x}{5} + \frac{x^3}{15} \cdot \frac{9}{x^2} = \frac{2x}{5} + \frac{\overset{3x}{9x^3}}{\underset{5}{15x^2}}$$

Multiplication has higher priority than addition, so perform the indicated multiplication first. Reduce the product to lowest terms by dividing both the numerator and the denominator by $3x^2$.

$$= \frac{2x}{5} + \frac{3x}{5}$$

$$= \frac{2x + 3x}{5}$$

Then perform the indicated addition.

$$= \frac{\overset{1}{5x}}{\underset{1}{5}}$$

Reduce this sum to lowest terms.

$$= x$$

Self-Check 1

Simplify $\dfrac{x}{3} - \dfrac{2}{x^2} \cdot \dfrac{x^3}{4}$.

The expressions in Examples 1 and 2 look very much alike. However, the pair of parentheses in Example 2 completely changes the order of operations used in Example 1.

Example 2 Using the Correct Order of Operations When Simplifying Rational Expressions

Simplify $\left(\dfrac{2x}{5} + \dfrac{x^3}{15}\right) \cdot \dfrac{9}{x^2}$.

Solution

$\left(\dfrac{2x}{5} + \dfrac{x^3}{15}\right) \cdot \dfrac{9}{x^2} = \left(\dfrac{2x}{5} \cdot \dfrac{3}{3} + \dfrac{x^3}{15}\right) \cdot \dfrac{9}{x^2}$ First add the terms inside the parentheses, using a common denominator of 15.

$= \left(\dfrac{6x}{15} + \dfrac{x^3}{15}\right) \cdot \dfrac{9}{x^2}$

$= \left(\dfrac{6x + x^3}{15}\right) \cdot \dfrac{9}{x^2}$ Then multiply these factors.

$= \dfrac{\overset{1}{\cancel{x}}(x^2 + 6)}{\underset{5}{\cancel{15}}} \cdot \dfrac{\overset{3}{\cancel{9}}}{\underset{x}{x^2}}$ Reduce the product to lowest terms.

$= \dfrac{3(x^2 + 6)}{5x}$ The answer may be written with both the numerator and the denominator in factored form or expanded as $\dfrac{3x^2 + 18}{5x}$.

Self-Check 2

Simplify $\left(\dfrac{x}{3} - \dfrac{2}{x^2}\right) \cdot \dfrac{x^3}{4}$.

Example 3, which involves division of one rational expression by another, is used as a lead-in to the topic of complex rational expressions.

Example 3 Using the Correct Order of Operations When Simplifying Rational Expressions

Simplify $\left(\dfrac{2}{x} - \dfrac{1}{3}\right) \div \left(\dfrac{2}{x} + \dfrac{1}{3}\right)$ and compare the result to the original expression by using a table of values.

Solution

Algebraically

$\left(\dfrac{2}{x} - \dfrac{1}{3}\right) \div \left(\dfrac{2}{x} + \dfrac{1}{3}\right) = \left[\dfrac{2(3) - 1(x)}{3x}\right] \div \left[\dfrac{2(3) + 1(x)}{3x}\right]$ First simplify the terms inside each set of parentheses. The LCD inside each set of parentheses is $3x$.

$= \left(\dfrac{6 - x}{3x}\right) \div \left(\dfrac{6 + x}{3x}\right)$ Simplify each numerator. Note that 0 is an excluded value to prevent division by 0.

$= \dfrac{6 - x}{\underset{1}{\cancel{3x}}} \cdot \dfrac{\overset{1}{\cancel{3x}}}{6 + x}$ To divide, invert the divisor and then multiply. Note that -6 is also an excluded value.

    Reduce by dividing out the common factor of $3x$.

$= \dfrac{6 - x}{6 + x}$

$= \dfrac{-(x - 6)}{x + 6}$ To rewrite the numerator so that x has a positive coefficient, factor -1 out of the numerator.

$= -\dfrac{x - 6}{x + 6}$

Numerical Check

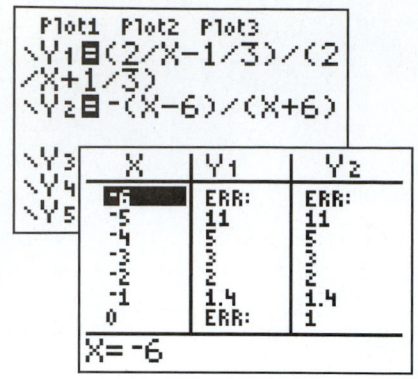

Let Y_1 represent the original expression and Y_2 represent the result.

One or both of the functions generate an error message at the excluded values of -6 and 0. The table indicates that these expressions are identical for all other real numbers. This supports the fact that these two expressions are equal.

Answer: $\left(\dfrac{2}{x} - \dfrac{1}{3}\right) \div \left(\dfrac{2}{x} + \dfrac{1}{3}\right) = -\dfrac{x-6}{x+6}$

Self-Check 3

Simplify $\left(\dfrac{2}{x} - \dfrac{1}{5}\right) \div \left(\dfrac{6}{x} - \dfrac{3}{5}\right)$ and compare the result to the original expression by using a table of values.

2. Simplify Complex Fractions

The original expression in Example 3 can be written as either

$$\left(\frac{2}{x} - \frac{1}{3}\right) \div \left(\frac{2}{x} + \frac{1}{3}\right) \qquad \text{or} \qquad \frac{\dfrac{2}{x} - \dfrac{1}{3}}{\dfrac{2}{x} + \dfrac{1}{3}}$$

The second form above is called a complex rational expression.

What Is a Complex Rational Expression?

A **complex rational expression** is a rational expression whose numerator or denominator, or both, is also a rational expression. Each of the following fractions contains more than one fraction bar and is therefore a complex fraction:

$$\frac{7 + \dfrac{1}{x}}{y} \qquad \frac{8}{19 + \dfrac{1}{2}} \qquad \frac{\dfrac{m - n}{2m + 3}}{\dfrac{m^2 - 5n}{m^2 - 2mn + n^2}}$$

Make sure the main fraction bar is longer than the other fraction bars to clearly denote the meaning of the expression.

Example 4   Simplifying Two Complex Fractions

Simplify each complex fraction.

(a) $\dfrac{\dfrac{2}{3}}{\dfrac{3}{4}}$ (b) $\dfrac{\dfrac{2}{3}}{4}$

Solution

(a) $\dfrac{2}{\dfrac{3}{4}} = 2 \div \dfrac{3}{4}$ In part (a), the main fraction bar indicates that the numerator is 2 and the denominator is $\dfrac{3}{4}$.

$\qquad = \dfrac{2}{1} \cdot \dfrac{4}{3}$ To divide by $\dfrac{3}{4}$, invert the divisor and multiply by $\dfrac{4}{3}$.

$\qquad = \dfrac{8}{3}$

(b) $\dfrac{\dfrac{2}{3}}{4} = \dfrac{2}{3} \div 4$ In part (b), the main fraction bar indicates that the numerator is $\dfrac{2}{3}$ and the denominator is 4.

$\qquad = \dfrac{\overset{1}{\cancel{2}}}{3} \cdot \dfrac{1}{\underset{2}{\cancel{4}}}$ To divide by 4, invert the divisor and multiply by $\dfrac{1}{4}$.

$\qquad = \dfrac{1}{6}$ Note that the expressions in parts (a) and (b) have different values.

Self-Check 4

Simplify each complex fraction.

a. $\dfrac{\dfrac{3}{4}}{\dfrac{5}{6}}$ **b.** $\dfrac{\dfrac{3}{4}}{5}$ **c.** $\dfrac{3}{\dfrac{4}{5}}$

How Should I Start to Simplify a Complex Rational Expression?

Because the fraction bar represents division, we can start by rewriting a complex rational expression as a division problem. We will simplify Examples 5 and 6 using this method, and then we examine a second method in Example 7.

Example 5   Simplifying a Complex Rational Expression

Simplify $\dfrac{\dfrac{a^2 - b^2}{5a + 5b}}{15a}$ by rewriting the expression as a division problem.

Solution

$\dfrac{\dfrac{a^2 - b^2}{5a + 5b}}{15a} = \left(\dfrac{a^2 - b^2}{1}\right) \div \left(\dfrac{5a + 5b}{15a}\right)$ Rewrite this complex rational expression as a division problem. A denominator of 1 is understood for $a^2 - b^2$.

$\qquad = \left(\dfrac{a^2 - b^2}{1}\right) \cdot \left(\dfrac{15a}{5a + 5b}\right)$ Then rewrite the division problem as the product of the dividend and the reciprocal of the divisor.

$\qquad = \dfrac{\overset{1}{\cancel{(a + b)}}(a - b)}{1} \cdot \dfrac{\overset{3}{\cancel{15a}}}{\underset{1}{\cancel{5}}\underset{1}{\cancel{(a + b)}}}$ Factor and reduce by dividing out the common factors.

$\qquad = \dfrac{3a(a - b)}{1}$

$\qquad = 3a^2 - 3ab$

In Example 6, we first simplify the denominator of the main fraction and then replace the main fraction bar by a division symbol.

Example 6 Simplifying a Complex Rational Expression

Simplify $\dfrac{\dfrac{x^2 + b^2}{x^2 - b^2}}{\dfrac{x - b}{x + b} + \dfrac{x + b}{x - b}}$ by rewriting the expression as a division problem.

Solution

$$\dfrac{\dfrac{x^2 + b^2}{x^2 - b^2}}{\left(\dfrac{x - b}{x + b}\right) + \left(\dfrac{x + b}{x - b}\right)} = \dfrac{\dfrac{x^2 + b^2}{x^2 - b^2}}{\dfrac{(x - b)^2 + (x + b)^2}{(x + b)(x - b)}}$$

First add the two terms in the denominator of the main fraction.

$$= \dfrac{x^2 + b^2}{x^2 - b^2} \div \dfrac{(x^2 - 2bx + b^2) + (x^2 + 2bx + b^2)}{(x + b)(x - b)}$$

Then rewrite this complex rational expression as a division problem.

$$= \dfrac{x^2 + b^2}{x^2 - b^2} \div \dfrac{2x^2 + 2b^2}{(x + b)(x - b)}$$

Simplify the numerator of the divisor.

$$= \dfrac{\overset{1}{\cancel{x^2 + b^2}}}{\cancel{(x + b)(x - b)}} \cdot \dfrac{\overset{1}{\cancel{(x + b)}}\overset{1}{\cancel{(x - b)}}}{2\underset{1}{\cancel{(x^2 + b^2)}}}$$

Rewrite the division problem by inverting the divisor and multiplying.

$$= \dfrac{1}{2}$$

Factor and reduce by dividing out the common factors.

Is There Another Method for Simplifying Complex Fractions?

Yes, we can also start by multiplying both the numerator and the denominator of the complex fraction by the LCD of all the fractions that occur in both the numerator and the denominator of the complex fraction.

Multiplying both the numerator and the denominator by the same value is equivalent to multiplying the fraction by the multiplicative identity 1. The effect of this is to remove all the individual fractions that occur in the numerator and the denominator.

Example 7   Simplifying a Complex Rational Expression

Simplify $\dfrac{1 + \dfrac{2}{x} + \dfrac{1}{x^2}}{1 - \dfrac{1}{x^2}}$ by multiplying both the numerator and the denominator by the

LCD of all terms.

Solution

$$\frac{1 + \dfrac{2}{x} + \dfrac{1}{x^2}}{1 - \dfrac{1}{x^2}} = \frac{1 + \dfrac{2}{x} + \dfrac{1}{x^2}}{1 - \dfrac{1}{x^2}} \cdot \frac{x^2}{x^2}$$

By inspection, the LCD of all the fractions in the numerator and the denominator is x^2. Multiply both the numerator and the denominator by this LCD.

$$= \frac{\left(1 + \dfrac{2}{x} + \dfrac{1}{x^2}\right)x^2}{\left(1 - \dfrac{1}{x^2}\right)x^2}$$

$$= \frac{(1)x^2 + \left(\dfrac{2}{x}\right)x^2 + \left(\dfrac{1}{x^2}\right)x^2}{(1)x^2 - \left(\dfrac{1}{x^2}\right)x^2}$$

Using the distributive property, multiply each term by x^2.

$$= \frac{x^2 + 2x + 1}{x^2 - 1}$$

Simplify the numerator and the denominator.

$$= \frac{\cancel{(x + 1)}(x + 1)}{\cancel{(x + 1)}(x - 1)}$$

Factor the numerator and denominator.

$$= \frac{x + 1}{x - 1}$$

Then reduce the fraction to lowest terms.

Self-Check 7

Simplify $\dfrac{1 - \dfrac{4}{x} + \dfrac{3}{x^2}}{1 - \dfrac{1}{x} - \dfrac{6}{x^2}}$.

Because $a^{-n} = \dfrac{1}{a^n}$, fractions that involve negative exponents can be expressed as complex fractions. One way to simplify some expressions involving negative exponents is to first rewrite these expressions in terms of positive exponents. Then you can apply the methods just described for complex fractions.

Example 8 Simplifying a Fraction with Negative Exponents

Simplify $\dfrac{x^{-1} - x^{-2}}{x^{-1} + x^{-2}}$.

Solution

$$\frac{x^{-1} - x^{-2}}{x^{-1} + x^{-2}} = \frac{\dfrac{1}{x} - \dfrac{1}{x^2}}{\dfrac{1}{x} + \dfrac{1}{x^2}}$$

Rewrite this expression as a complex fraction, using the definition of negative exponents.

$$= \frac{\dfrac{1}{x} - \dfrac{1}{x^2}}{\dfrac{1}{x} + \dfrac{1}{x^2}} \cdot \frac{x^2}{x^2}$$

Multiply both the numerator and the denominator by x^2, the LCD of all fractions that occur in the numerator and the denominator of the complex fraction.

$$= \frac{\left(\dfrac{1}{x}\right)x^2 - \left(\dfrac{1}{x^2}\right)x^2}{\left(\dfrac{1}{x}\right)x^2 + \left(\dfrac{1}{x^2}\right)x^2}$$

Use the distributive property to multiply each term by x^2.

$$= \frac{x - 1}{x + 1}$$

Simplify the numerator and the denominator.

Self-Check 8

Simplify $\dfrac{x^{-1} - x^{-3}}{x^{-1} + x^{-2}}$ by first rewriting the expression in terms of positive exponents as shown in Example 8.

In Example 8, the LCD of all the fractions within the numerator and the denominator of the complex fraction is x^2. Thus to simplify this complex fraction, we multiplied both the numerator and the denominator by x^2. This is exactly what we do in Example 9, but without going through the work of writing the expression as a complex fraction. The alternative strategy illustrated in Example 9 is particularly appropriate when negative exponents are applied only to monomials. The key is to select the appropriate expression to multiply by both the numerator and the denominator of the given fraction. The appropriate LCD to use as a factor can be determined by inspecting all the negative exponents in the original expression.

Example 9 Using an Alternative Method for Simplifying a Fraction with Negative Exponents

Simplify $\dfrac{x^{-1} - x^{-2}}{x^{-1} + x^{-2}}$.

Solution

$$\frac{x^{-1} - x^{-2}}{x^{-1} + x^{-2}} = \frac{x^{-1} - x^{-2}}{x^{-1} + x^{-2}} \cdot \frac{x^2}{x^2}$$

Multiply both the numerator and the denominator by x^2. Note that this is the lowest power of x that will eliminate all the negative exponents on x in both the numerator and the denominator.

$$= \frac{(x^{-1})x^2 - (x^{-2})x^2}{(x^{-1})x^2 + (x^{-2})x^2}$$

Use the distributive property to multiply each term by x^2.

$$= \frac{x - x^0}{x + x^0}$$

$$= \frac{x - 1}{x + 1}$$

Then simplify by replacing x^0 with 1. Note that this result is the same as the result in Example 8.

Self-Check 9

Simplify $\dfrac{x^{-1} - x^{-3}}{x^{-1} + x^{-2}}$ by multiplying the numerator and denominator by the lowest power of x that will eliminate all the negative exponents.

Self-Check Answers

1. $-\dfrac{x}{6}$

2. $\dfrac{x(x^3 - 6)}{12}$

3. $\dfrac{1}{3}$

   Check:

   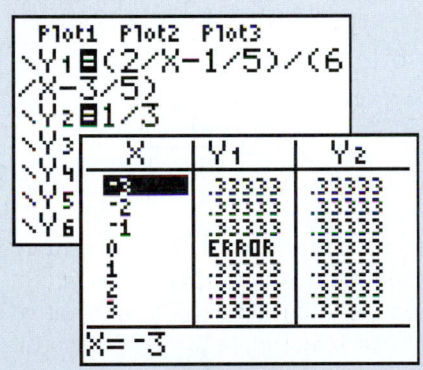

   The two expressions agree for all values of x for which they are both defined.

4. a. $\dfrac{9}{10}$ b. $\dfrac{3}{20}$ c. $\dfrac{15}{4}$

5. $\dfrac{2x(x + y)}{5y}$

6. $-\dfrac{1}{m(m + 3)}$

7. $\dfrac{x - 1}{x + 2}$

8. $\dfrac{x - 1}{x}$

9. $\dfrac{x - 1}{x}$

9.4   Using the Language and Symbolism of Mathematics

1. When one is simplifying a rational expression, the correct order of operations is essential. This order is as follows:
   a. Start with the expression within the innermost pair of _____ _____.
   b. Perform all _____.
   c. Perform all _____ and _____ as they appear from left to right.
   d. Perform all _____ and _____ as they appear from left to right.

2. A complex rational expression is a rational expression whose _____ or _____, or both, is also a rational expression.

3. Dividing by a fraction is equivalent to multiplying by the _____ of that fraction.

4. Two rational expressions must have the same denominator when combined by using the operations of _____ or _____.

9.4 Quick Review

Determine the LCD of each pair of rational expressions.

1. $\dfrac{3}{20}$ and $\dfrac{17}{45}$

2. $\dfrac{x}{x-y}$ and $\dfrac{y}{x+y}$

Simplify each expression using only pencil and paper.

3. $-5^2 + 2(11-4)$

4. $(-5)^2 + 2(11) - 4$

5. Rewrite $x^{-1} + 3x^{-2}$ by using positive exponents.

9.4 Exercises

Objective 1 Simplify Rational Expressions in Which the Order of Operations Must Be Determined

In Exercises 1–16, simplify each expression. Assume the variables are restricted to values that prevent division by 0.

1. a. $\dfrac{2x}{3} + \dfrac{x^2}{15} \cdot \dfrac{5}{x}$ **b.** $\left(\dfrac{2x}{3} + \dfrac{x^2}{15}\right) \cdot \dfrac{5}{x}$

2. a. $\dfrac{3x}{4} - \dfrac{x^2}{12} \cdot \dfrac{3}{x}$ **b.** $\left(\dfrac{3x}{4} - \dfrac{x^2}{12}\right) \cdot \dfrac{3}{x}$

3. a. $-\dfrac{2y}{3} + \dfrac{y^2}{6} \div \dfrac{y}{9}$ **b.** $\left(-\dfrac{2y}{3} + \dfrac{y^2}{6}\right) \div \dfrac{y}{9}$

4. a. $\dfrac{3y}{5} - \dfrac{2y^2}{3} \div \dfrac{y}{10}$ **b.** $\left(\dfrac{3y}{5} - \dfrac{2y^2}{3}\right) \div \dfrac{y}{10}$

5. a. $\dfrac{v}{v-6} - \dfrac{3}{v+2} \cdot \dfrac{2v+4}{v^2-6v}$

   b. $\left(\dfrac{v}{v-6} - \dfrac{3}{v+2}\right) \cdot \dfrac{2v+4}{v^2-6v}$

6. a. $\dfrac{v}{v+10} + \dfrac{5v}{v-3} \cdot \dfrac{2v-6}{v^2+10v}$

   b. $\left(\dfrac{v}{v+10} + \dfrac{5v}{v-3}\right) \cdot \dfrac{2v-6}{v^2+10v}$

7. $\dfrac{2-3x}{2x-3} + \dfrac{8-2x}{3x-6} \div \dfrac{4x-6}{3x-6}$

8. $\dfrac{6z+3}{3z-1} - \dfrac{5}{9z^2-1} \div \dfrac{1}{3z+1}$

9. $\left(\dfrac{2}{v+3} + v\right)\left(v - \dfrac{3}{v+2}\right)$

10. $\left(v - \dfrac{6v+35}{v+4}\right)\left(v - \dfrac{44}{v-7}\right)$

11. $\left(x - 2 - \dfrac{3}{x}\right) \div \left(1 + \dfrac{1}{x}\right)$

12. $\left(\dfrac{w+3}{2w+1} - 2w\right) \div \left(\dfrac{w-1}{2w+1} + 2w\right)$

13. $1 - \dfrac{4}{x} - \left(1 - \dfrac{2}{x}\right)^2$ **14.** $4 - \dfrac{12}{x} - \left(2 - \dfrac{3}{x}\right)^2$

15. $\left(1 + \dfrac{3}{5x}\right)^2 - \left(1 - \dfrac{3}{5x}\right)^2$

16. $\left(1 + \dfrac{2}{3x}\right)^2 - \left(1 - \dfrac{2}{3x}\right)^2$

Objective 2 Simplify Complex Fractions

In Exercises 17–40, simplify each expression. Assume the variables are restricted to values that prevent division by 0.

17. a. $\dfrac{\frac{4}{6}}{\frac{5}{8}}$ **b.** $\dfrac{4}{\frac{6}{5}}$

18. a. $\dfrac{\frac{3}{4}}{\frac{5}{6}}$ **b.** $\dfrac{4}{\frac{6}{5}}$

19. $\dfrac{1 + \frac{1}{5}}{1 - \frac{1}{5}}$ **20.** $\dfrac{\frac{1}{2} + \frac{2}{3}}{\frac{3}{2} + 2}$

21. $\dfrac{\frac{3}{5} - \frac{5}{3}}{\frac{1}{3} + \frac{1}{5}}$ **22.** $\dfrac{\frac{7}{10} - \frac{3}{4}}{\frac{3}{8} + \frac{1}{3}}$

23. $\dfrac{\frac{12x^2}{5y}}{\frac{16x^2}{15y^2}}$ **24.** $\dfrac{\frac{18a^3}{25x}}{\frac{24a^3}{35x^2}}$

25. $\dfrac{\frac{w-z}{x^2y^2}}{\frac{w^2-z^2}{2xy}}$ **26.** $\dfrac{\frac{a^2-b^2}{6a^2b^3}}{\frac{a+b}{9a^3b}}$

27. $\dfrac{\frac{x^2-9x+14}{34x^4}}{\frac{5x^2-20}{17x^5}}$ **28.** $\dfrac{\frac{5x^2+10xy+5y^2}{15x^2-15y^2}}{\frac{9x^2-9y^2}{25x^2-50xy+25y^2}}$

29. $\dfrac{2 - \frac{1}{x}}{4 - \frac{1}{x^2}}$ **30.** $\dfrac{\frac{1}{x^2} - 49}{7 - \frac{1}{x}}$

31. $\dfrac{vw}{\frac{1}{v} + \frac{1}{w}}$ **32.** $\dfrac{\frac{1}{v} - \frac{1}{w}}{vw}$

33. $\dfrac{\dfrac{3}{x^2} - \dfrac{6}{x} + 3}{18 - \dfrac{18}{x^2}}$

34. $\dfrac{\dfrac{1}{x} - \dfrac{8}{x^2} + \dfrac{15}{x^3}}{1 - \dfrac{5}{x}}$

35. $\dfrac{3 + \dfrac{9}{x}}{\dfrac{15}{x^3} + \dfrac{8}{x^2} + \dfrac{1}{x}}$

36. $\dfrac{x + 2 - \dfrac{6}{2x + 3}}{x + \dfrac{8x}{2x - 1}}$

37. $\dfrac{\dfrac{w - a}{w + a} - \dfrac{w + a}{w - a}}{\dfrac{w^2 + a^2}{w^2 - a^2}}$

38. $\dfrac{w - 3 + \dfrac{2}{w - 6}}{w - 2 - \dfrac{22}{w + 7}}$

39. $\dfrac{\dfrac{12}{v^2} + \dfrac{1}{v} - 1}{\dfrac{24}{v^2} - \dfrac{2}{v} - 1}$

40. $\dfrac{\dfrac{12}{v^2} - \dfrac{1}{v} - 1}{\dfrac{8}{v^2} + \dfrac{6}{v} + 1}$

49. $\dfrac{v^{-1} + w^{-1}}{v^{-1} - w^{-1}}$

50. $\dfrac{m^2 - n^2}{m^{-1} + n^{-1}}$

51. $\dfrac{m^2 - n^2}{m^{-1} - n^{-1}}$

52. $\dfrac{(x + y)^2}{x^{-1} + y^{-1}}$

53. $\dfrac{a^{-2} - b^{-2}}{a^{-1} - b^{-1}}$

54. $\dfrac{b - a^{-1}}{a - b^{-1}}$

55. $\dfrac{x^{-1}y^{-2} + x^{-2}y^{-1}}{y^{-2} - x^{-2}}$

56. $\dfrac{x^{-1}y^{-2} + x^{-2}y^{-1}}{x^{-1} + y^{-1}}$

57. $\dfrac{\dfrac{1}{a} - \dfrac{1}{b}}{ab}$

58. $\dfrac{\dfrac{1}{x} + \dfrac{1}{y}}{\dfrac{1}{x}}$

Estimate Then Calculate

In Exercises 41–44, mentally estimate the value of each expression, and then use a calculator or a spreadsheet to approximate each value to the nearest thousandth.

Problem	Mental Estimate	Approximation
41. $\dfrac{\dfrac{211}{429}}{\dfrac{107}{325}}$		
42. $\dfrac{\dfrac{211}{107}}{\dfrac{899}{298}}$		
43. $\dfrac{\dfrac{401}{99}}{\dfrac{499}{101}}$		
44. $\dfrac{\dfrac{1,022}{350}}{\dfrac{1,017}{204}}$		

59. $\dfrac{\dfrac{1 - \dfrac{a}{b}}{1 + \dfrac{a}{b}}}{\ }$ ... $\dfrac{\dfrac{1 + \dfrac{x}{y}}{1 - \dfrac{x}{y}}}{\dfrac{x + y}{x - \dfrac{y^2}{x}}}$

60. $\dfrac{\dfrac{1 + \dfrac{x}{y}}{1 - \dfrac{x}{y}}}{x - \dfrac{y^2}{x}}{x + y}$

Applying Technology

In Exercises 61–64, use the table of values for each expression to match each expression with its simplified form.

61. $\dfrac{3}{x} + \dfrac{1}{4} \div \dfrac{3}{x} - \dfrac{1}{4}$

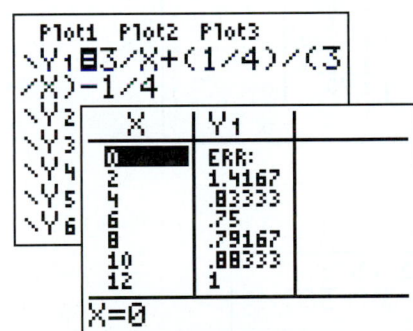

62. $\left(\dfrac{3}{x} + \dfrac{1}{4}\right) \div \left(\dfrac{3}{x} - \dfrac{1}{4}\right)$

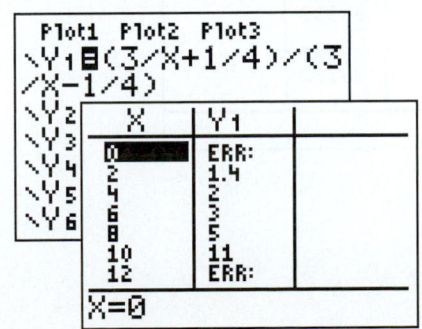

Skill Development

In Exercises 45–60, simplify each expression. Assume the variables are restricted to values that prevent division by 0.

45. $\dfrac{a - b}{ab} + \dfrac{(a + b)^2}{a^3b^3} \cdot \dfrac{a^2b^2}{4a^2 + 8ab + 4b^2}$

46. $\dfrac{a + 2}{a + 3} - \dfrac{(a - 2)^3}{a^2 - 9} \div \dfrac{a^2 - 4a + 4}{a - 3}$

47. $\dfrac{x^{-2}}{x^{-2} + y^{-2}}$

48. $\dfrac{x^{-2} - y^{-2}}{y^{-2}}$

63. $\dfrac{3}{x} + \dfrac{1}{4} \div \left(\dfrac{3}{x} - \dfrac{1}{4} \right)$

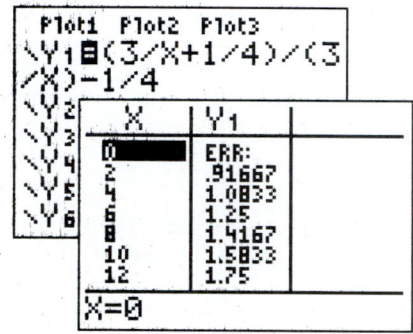

64. $\left(\dfrac{3}{x} + \dfrac{1}{4} \right) \div \dfrac{3}{x} - \dfrac{1}{4}$

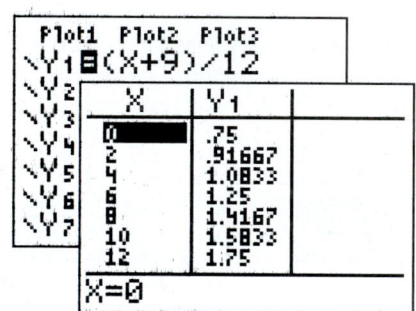

A. $\dfrac{x + 9}{12}$

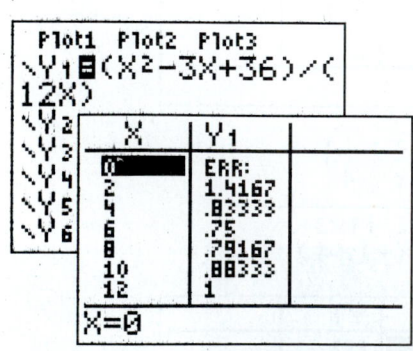

B. $\dfrac{x^2 - 3x + 36}{12x}$

C. $-\dfrac{x + 12}{x - 12}$

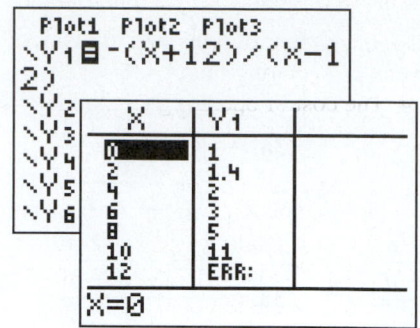

D. $-\dfrac{x^2 - 3x + 36}{x(x - 12)}$

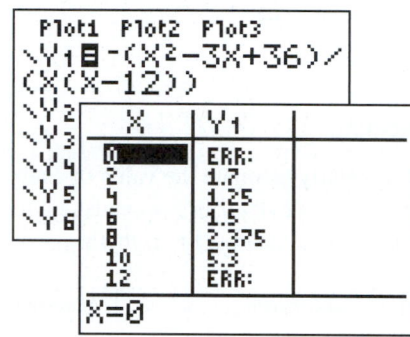

Connecting Algebra to Geometry

65. Area of a Trapezoid The formula for the area of a trapezoid is $A = \dfrac{h}{2}(a + b)$. Use this formula to write an expression that represents the area of the trapezoid.

$(x + 4)$ cm

x cm

$(x + 6)$ cm

66. Total Area of Two Regions Write an expression that represents the total area of the rectangle and triangle.

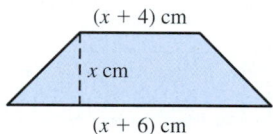

$\dfrac{x}{x^2 - 1}$ cm

x cm

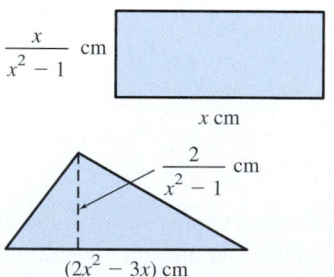

$\dfrac{2}{x^2 - 1}$ cm

$(2x^2 - 3x)$ cm

Connecting Concepts to Applications

67. Average Cost The number of units of patio chairs that a factory can make by operating t h/wk is given by $N(t) = 5t - 10$. The cost of operating the factory t h/wk is given by $C(t) = 8,400 - \dfrac{16,800}{t}$.

a. Write a rational function $A(t)$ that gives the average cost per unit when the plant operates t h/wk.

b. Determine the practical domain for the values of t.
c. Use the graph of this function to describe what happens to the average cost as t becomes larger.

68. Average Cost The number of units of patio tables a factory can make by operating t h/wk is given by $N(t) = 2t - 4$. The cost of operating the factory t h/wk is given by $C(t) = 5{,}040 - \dfrac{10{,}080}{t}$.

a. Write a rational function $A(t)$ that gives the average cost per unit when the plant operates t h/wk.
b. Determine the practical domain for the values of t.
c. Use the graph of this function to describe what happens to the average cost as t becomes larger.

69. Lens Formula The relationship between the focal length f of a lens, the distance d_o of an object from the lens, and the distance d_i of an image from the lens is $\dfrac{1}{f} = \dfrac{1}{d_o} + \dfrac{1}{d_i}$. So $f = \dfrac{1}{\dfrac{1}{d_o} + \dfrac{1}{d_i}}$.

a. Give a simplified rational expression for the focal length f.
b. Determine f when the object distance is 20 ft and the image distance is 0.5 ft. Round to the nearest thousandth.

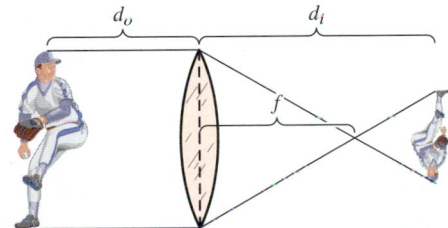

70. Electrical Resistance The total resistance R in a parallel circuit with two individual resistors r_1 and r_2 can be calculated by using the formula $\dfrac{1}{R} = \dfrac{1}{r_1} + \dfrac{1}{r_2}$. So $R = \dfrac{1}{\dfrac{1}{r_1} + \dfrac{1}{r_2}}$.

a. Give a simplified rational expression for the total resistance R.
b. Determine R when $r_1 = 8$ ohms (Ω) and $r_2 = 56\ \Omega$.

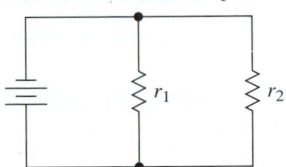

Group Discussion Questions

71. Challenge Question Simplify each expression. Assume that m and n are natural numbers and that x and y are restricted to values that prevent division by 0.

a. $\dfrac{x^{-m} + y^{-n}}{x^{-m}y^{-n}}$ **b.** $\dfrac{x^{-m} - y^{-n}}{x^m - y^n}$

c. $\dfrac{x^{-m} - x^{-2m}}{x^{-m} + x^{-2m}}$ **d.** $\dfrac{x^{-m} + 3x^{-2m}}{x^{-m} - 2x^{-2m}}$

72. Error Analysis A student simplified the expression $\dfrac{x^{-1} - y^{-1}}{x^{-2} - y^{-2}}$ as follows:

$$\frac{x^{-1} - y^{-1}}{x^{-2} - y^{-2}} = \frac{x^2 - y^2}{x - y} = \frac{(x+y)(x-y)}{x-y} = x + y$$

Correct this error and then finish simplifying the expression.

9.4   **Cumulative Review**

Use the graph of $y = -x^2 - 2x + 8$ to answer each question.

1. Determine the y-intercept.
2. Determine the x-intercepts.
3. What is the maximum y-value of this function?
4. What is the value of x when y is at a maximum?
5. Determine the interval of x-values for which the function is increasing.

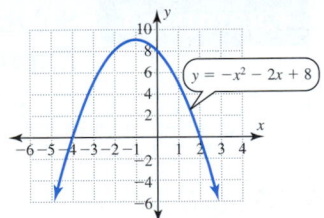

Section 9.5   Solving Equations Containing Rational Expressions

Objectives:

1. Solve equations containing rational expressions.
2. Solve a rational equation for a specified variable.

1. Solve Equations Containing Rational Expressions

In Section 2.5, we solved fractional equations by multiplying both sides of the equation by the least common denominator. This method produces an equation equivalent to the original one as long as we do not multiply by 0. Multiplying by an expression that is equal to 0 can produce an equation that is *not* equivalent to the original equation and thus can produce an extraneous value. An **extraneous value** is a value obtained in the solution procedure that is not a solution of the original equation.

How Can I Quickly Determine Whether a Possible Solution to an Equation with a Variable in the Denominator Is Extraneous?

Check to see whether this value must be excluded to prevent division by 0.

Solving an Equation Containing Rational Expressions

Verbally	Algebraic Example	
1. Multiply both sides of the equation by the LCD.	Solve: $\dfrac{21}{x} - \dfrac{15}{x} = 2$	The only excluded value is 0.
2. Solve the resulting equation.	$x\left(\dfrac{21}{x} - \dfrac{15}{x}\right) = x(2)$	Multiply both sides of the equation by the LCD, x.
3. Check the solution to determine whether it is an excluded value and therefore extraneous.	$x\left(\dfrac{21}{x}\right) - x\left(\dfrac{15}{x}\right) = 2x$	
	$21 - 15 = 2x$	
	$6 = 2x$	
	$3 = x$	Since 3 is not an excluded value, it will check.
	Check: $\dfrac{21}{3} - \dfrac{15}{3} \stackrel{?}{=} 2$	
	$7 - 5 \stackrel{?}{=} 2$	
	$2 \stackrel{?}{=} 2$ checks	
	Answer: $x = 3$	

Example 1 Solving an Equation Containing Rational Expressions

Solve $\dfrac{3}{x - 7} + 5 = \dfrac{8}{x - 7}$ algebraically and check your solution using a table of values.

Solution

Algebraically

$$\frac{3}{x - 7} + 5 = \frac{8}{x - 7}$$

Note that $x = 7$ is the only excluded value.

$$(x - 7)\left(\frac{3}{x - 7} + 5\right) = (x - 7)\left(\frac{8}{x - 7}\right)$$

Multiply both sides of the equation by the LCD, $x - 7$. The LCD is nonzero since 7 is an excluded value.

$$(x - 7)\left(\frac{3}{x - 7}\right) + (x - 7)(5) = (x - 7)\left(\frac{8}{x - 7}\right)$$

Use the distributive property to multiply each term by $x - 7$.

$$3 + 5x - 35 = 8$$

Then combine like terms.

$$5x = 40$$

$$x = 8$$

Since $x = 8$ is not an excluded value, this solution will check.

Numerical Check

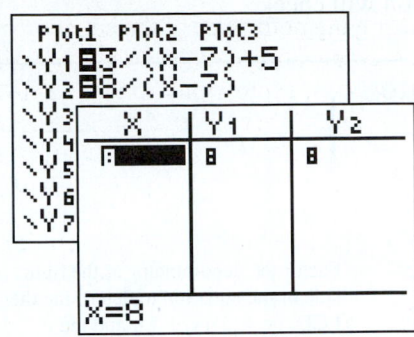

For $x = 8$, Y_1 and Y_2 are equal. This solution checks.

Answer: $x = 8$

Let Y_1 represent the left side of the equation and Y_2 represent the right side of the equation.

Use the **ASK** mode to confirm that for $x = 8$, Y_1 and Y_2 are equal. This confirms that the algebraic solution checks. *Caution:* Unless you already know how many solutions this equation has, this table will not confirm that $x = 8$ is the only solution.

Self-Check 1

Solve $\dfrac{2y}{y + 2} = \dfrac{y}{y + 2} - 3$ and check your solution using a table of values.

The solution process in Example 2 produces an excluded value for the variable. This value cannot be a solution; it is extraneous.

Example 2   Solving an Equation with an Extraneous Value

Solve $\dfrac{v}{v - 3} = 4 - \dfrac{3}{3 - v}$.

Solution

$$\frac{v}{v - 3} = 4 - \frac{3}{3 - v}$$

$$\frac{v}{v - 3} = 4 + \frac{3}{v - 3}$$

$$(v - 3)\left(\frac{v}{v - 3}\right) = (v - 3)\left(4 + \frac{3}{v - 3}\right)$$

$$\cancel{(v - 3)}\left(\frac{v}{\cancel{v - 3}}\right) = (v - 3)(4) + \cancel{(v - 3)}\left(\frac{3}{\cancel{v - 3}}\right)$$

$$v = 4v - 12 + 3$$

$$-3v = -9$$

$$v = 3$$

This value causes division by 0 in the original equation so there is no solution.

Answer: There is no solution.

Note that the only excluded value is $v = 3$. Noting that the denominators are opposites, change $-\dfrac{3}{3 - v}$ to $\dfrac{3}{v - 3}$.

Multiply both sides of the equation by the nonzero LCD, $v - 3$. The LCD is nonzero since 3 is an excluded value.

Use the distributive property to multiply each term by $v - 3$.

Solve the resulting equation for v.

Since $v = 3$ is the excluded value noted above, this value is extraneous.

Self-Check 2

Solve $\dfrac{x - 12}{x - 4} = \dfrac{2x}{4 - x} + 2$.

The solution process in Example 3 produces two values. One of these is an excluded value and cannot be a solution while the other solution will check.

Example 3 Solving an Equation with an Extraneous Value

Solve $\dfrac{y+1}{y+2} + \dfrac{y}{y-2} + 1 = \dfrac{8}{y^2-4}$.

Solution

$$\dfrac{y+1}{y+2} + \dfrac{y}{y-2} + 1 = \dfrac{8}{y^2-4}$$

Factor the denominator of the right side of the equation to determine the LCD, $(y+2)(y-2)$, and the excluded values -2 and 2.

$$\dfrac{y+1}{y+2} + \dfrac{y}{y-2} + 1 = \dfrac{8}{(y+2)(y-2)}$$

$$(y+2)(y-2)\left[\dfrac{y+1}{y+2} + \dfrac{y}{y-2} + 1\right] = (y+2)(y-2)\left[\dfrac{8}{(y+2)(y-2)}\right]$$

Multiply both sides of the equation by the LCD, $(y+2)(y-2)$. The LCD is nonzero since -2 and 2 are excluded values.

$$(y+2)(y-2)\left(\dfrac{y+1}{y+2}\right) + (y+2)(y-2)\left(\dfrac{y}{y-2}\right) + (y+2)(y-2)(1) = 8$$

Use the distributive property to multiply each term by $(y+2)(y-2)$.

$$(y-2)(y+1) + (y+2)(y) + (y+2)(y-2)(1) = 8$$

$$y^2 - y - 2 + y^2 + 2y + y^2 - 4 = 8$$

$$3y^2 + y - 14 = 0$$

Write the resulting quadratic equation in standard form, and then solve it by factoring.

$$(3y+7)(y-2) = 0$$

$$3y + 7 = 0 \quad \text{or} \quad y - 2 = 0$$

$$y = -\dfrac{7}{3} \qquad\qquad y = 2$$

Because $y = 2$ is an excluded value, it cannot be a solution. The value $y = -\dfrac{7}{3}$ is not an excluded value. Therefore it should check.

The value $y = 2$ causes division by 0 in the original equation and is therefore an extraneous value.

Answer: $y = -\dfrac{7}{3}$

Does $x = -\dfrac{7}{3}$ check?

Self-Check 3

Solve $\dfrac{x}{x-1} - \dfrac{2}{x} = \dfrac{1}{x-1}$.

How Do I Interpret Contradictions or Identities That Are Produced in the Solution Process?

If the steps in the solution of an equation produce a contradiction, then the original equation has no solution. Example 4 shows how to interpret the answer when the steps in the solution of an equation produce an identity.

Example 4   Solving an Equation That Simplifies to an Identity

Solve $\dfrac{3}{x-4} - \dfrac{2}{x-2} = \dfrac{x+2}{x^2-6x+8}$.

Solution

$$\frac{3}{x-4} - \frac{2}{x-2} = \frac{x+2}{x^2-6x+8}$$

$$\frac{3}{x-4} - \frac{2}{x-2} = \frac{x+2}{(x-4)(x-2)}$$

Factor the denominator of the right side of the equation to determine the LCD, $(x-4)(x-2)$, and the excluded values 2 and 4.

$$(x-4)(x-2)\left[\frac{3}{x-4} - \frac{2}{x-2}\right] = (x-4)(x-2)\left[\frac{x+2}{(x-4)(x-2)}\right]$$

$$(x-4)(x-2)\left(\frac{3}{x-4}\right) - (x-4)(x-2)\left(\frac{2}{x-2}\right) = x+2$$

$$3(x-2) - 2(x-4) = x+2$$

$$3x - 6 - 2x + 8 = x+2$$

$$x+2 = x+2$$

$$0 = 0 \qquad \text{(an identity)}$$

Multiply both sides of the equation by the LCD, $(x-4)(x-2)$. The LCD is nonzero since 2 and 4 are excluded values. Use the distributive property to multiply each term by $(x-4)(x-2)$. Simplify, and then solve the equation.

This simplified equation is an identity. Thus all real numbers are solutions of the original equation except for the values excluded to avoid division by 0.

Answer: $\mathbb{R} \sim \{2, 4\}$

Self-Check 4

Solve $\dfrac{2}{x} + \dfrac{1}{x-1} = \dfrac{3x-2}{x^2-x}$.

The resistance R in a parallel circuit with two individual resistors r_1 and r_2 can be calculated by using the formula $\dfrac{1}{R} = \dfrac{1}{r_1} + \dfrac{1}{r_2}$. We use this formula to determine appropriate values for r_1 and r_2.

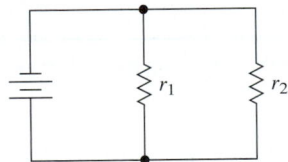

Example 5   Calculating Resistances in a Parallel Circuit

The resistance planned by an electrical technician for a parallel electric circuit with two resistors is 4 Ω. The resistance of r_2 must be 15 Ω greater than that of r_1. Determine the resistance of each resistor.

Solution

Let r = resistance of first resistor, Ω

$r + 15$ = resistance of second resistor, Ω

The resistance of the second resistor is 15 Ω greater than that of the first resistor. Note that we use one variable r rather than two variables r_1 and r_2. We use r in place of r_1 in the formula and $r + 15$ in place of r_2.

$$\frac{1}{r_1} + \frac{1}{r_2} = \frac{1}{R}$$

$$\frac{1}{r} + \frac{1}{r + 15} = \frac{1}{4}$$

Substitute the total resistance of 4 Ω and the identified variables into the resistance formula.

$$4r(r + 15)\left(\frac{1}{r} + \frac{1}{r + 15}\right) = 4r(r + 15)\left(\frac{1}{4}\right)$$

Multiply both sides of the equation by the LCD, $4r(r + 15)$.

$$4r(r + 15)\left(\frac{1}{r}\right) + 4r(r + 15)\left(\frac{1}{r + 15}\right) = r(r + 15)$$

Use the distributive property to multiply each term by $4r(r + 15)$.

$$4(r + 15) + 4r = r^2 + 15r$$

$$4r + 60 + 4r = r^2 + 15r$$

$$r^2 + 7r - 60 = 0$$

Simplify and write this quadratic equation in standard form.

$$(r - 5)(r + 12) = 0$$

Then solve this equation by factoring.

$$r - 5 = 0 \quad \text{or} \quad r + 12 = 0$$

$$r = 5 \quad \text{or} \quad r = -12$$

$$r + 15 = 20$$

The value of 5 Ω seems reasonable for r. However, the resistance of -12 Ω is not reasonable because resistance cannot be negative.

Answer: The resistance of the first resistor is 5 Ω, and the resistance of the second resistor is 20 Ω.

Self-Check 5

The resistance planned by an electrical technician for a parallel electric circuit with two resistors is 3 Ω. The resistance of r_2 must be 8 Ω greater than that of r_1. Determine the resistance of each resistor.

---

2. Solve a Rational Equation for a Specified Variable

In Section 2.6, we practiced rewriting formulas to solve for a specified variable. Example 6 involves rewriting a formula that includes a rational expression. To solve for b, we must isolate this variable in the numerator on one side of the equation.

Example 6 Solving for a Specified Variable

Solve each equation for b.

(a) $\dfrac{2}{a} = \dfrac{3}{b} - \dfrac{1}{ab}$ **(b)** $\dfrac{1}{a} + \dfrac{1}{c} = \dfrac{1}{b}$

Solution

(a) $\dfrac{2}{a} = \dfrac{3}{b} - \dfrac{1}{ab}$

$$ab\left(\frac{2}{a}\right) = ab\left(\frac{3}{b} - \frac{1}{ab}\right)$$

To get b in the numerator, multiply both sides of the equation by the LCD, ab.

$$ab\left(\frac{2}{a}\right) = ab\left(\frac{3}{b}\right) - ab\left(\frac{1}{ab}\right)$$

Use the distributive property to multiply each term by ab.

$$2b = 3a - 1$$

Simplify, and then divide both sides of the equation by 2 to isolate b on the left side with a coefficient of 1.

$$b = \frac{3a - 1}{2}$$

(b)

$$\frac{1}{a} + \frac{1}{c} = \frac{1}{b}$$

$$abc\left(\frac{1}{a} + \frac{1}{c}\right) = abc\left(\frac{1}{b}\right)$$ To get b in the numerator, multiply both sides of the equation by the LCD, abc.

$$abc\left(\frac{1}{a}\right) + abc\left(\frac{1}{c}\right) = abc\left(\frac{1}{b}\right)$$ Use the distributive property to multiply each term by abc.

$$bc + ab = ac$$ Simplify, and then use the distributive property to factor b
$$b(c + a) = ac$$ out of each term on the left side of the equation.

$$\frac{b(c + a)}{c + a} = \frac{ac}{c + a}$$ Divide both sides of the equation by $c + a$ to isolate b on the left side of the equation.

$$b = \frac{ac}{a + c}$$ Simplify the left side of the equation, and on the right side write the terms in the denominator in alphabetical order.

Self-Check 6

Solve each equation for x.

a. $\dfrac{2v}{x + 3y} = z$ **b.** $\dfrac{1}{x} + \dfrac{1}{w} = \dfrac{1}{y}$

One application of solving equations containing rational expressions is to determine the intercepts of the graph of a rational function. Recall that we can determine the y-intercept by evaluating $f(0)$. To determine the x-intercept(s), set $y = 0$ and solve for x. This is illustrated in Example 7.

Example 7   Determining the Intercepts of a Rational Function

Determine the intercepts of the rational function $f(x) = \dfrac{x + 2}{x - 3}$ whose graph is shown here.

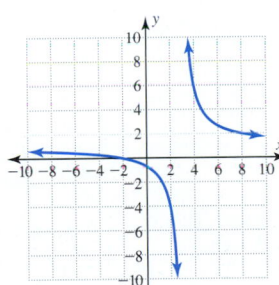

Solution

y-intercept

$$f(x) = \frac{x + 2}{x - 3}$$

$$f(0) = \frac{0 + 2}{0 - 3}$$ Substitute 0 for x and simplify.

$$f(0) = -\frac{2}{3}$$

The y-intercept of this graph is $\left(0, -\dfrac{2}{3}\right)$.   From the graph, this y-intercept seems reasonable.

x-intercept

$$f(x) = \frac{x+2}{x-3}$$

$$0 = \frac{x+2}{x-3}$$ Set $f(x) = 0$ and solve for x.

$$(x-3) \cdot 0 = \frac{x+2}{x-3} \cdot (x-3)$$

$$0 = x + 2$$

$$-2 = x$$

$$x = -2$$

The x-intercept of the graph is $(-2, 0)$.   From the graph, this x-intercept seems reasonable.

Self-Check 7

Determine the intercepts of the rational function $f(x) = \dfrac{x-4}{x+2}$ whose graph is shown.

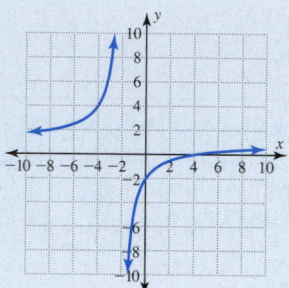

Self-Check Answers

1. $y = -\dfrac{3}{2}$

   Check:

   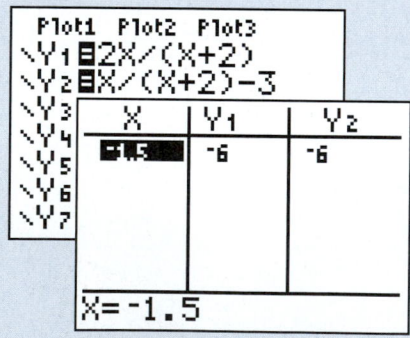

   For $x = -\dfrac{3}{2}$, Y_1 and Y_2 are equal.

   This solution checks.

2. There is no solution.
3. $x = 2$
4. $\mathbb{R} \sim \{0, 1\}$
5. The resistance of the first resistor is 4 Ω, and the resistance of the second resistor is 12 Ω.
6. **a.** $x = \dfrac{2v - 3yz}{z}$

   b. $x = \dfrac{wy}{w - y}$

7. y-intercept: $(0, -2)$;
   x-intercept: $(4, 0)$

9.5 Using the Language and Symbolism of Mathematics

1. Equivalent equations have exactly the _____ set of solutions.

2. To produce equivalent equations, we can multiply both sides of the equation by the same number as long as we do not multiply by _____.

3. To solve an equation containing rational expressions:
   a. Multiply both sides of the equation by the _____.

   b. _____ the resulting equation.

   c. Check the solution to determine whether it is an excluded value and therefore _____.

4. An excluded value for an equation containing rational expressions is one that causes division by _____.

9.5 Quick Review

Solve each linear equation.

1. $\dfrac{x}{2} - \dfrac{x}{3} = 1$

2. $x = \dfrac{36}{15} + \dfrac{x}{5}$

3. $\dfrac{x-1}{3} - \dfrac{x+3}{4} = 2$

4. $\dfrac{2x+3}{5} - 1 = \dfrac{x-2}{3}$

5. $\dfrac{x}{2} + 5 = \dfrac{x}{2}$

9.5 Exercises

Objective 1 Solve Equations Containing Rational Expressions

In Exercises 1–6, determine the values excluded from the domain of the variable because they would cause division by 0.

1. $\dfrac{3}{m-3} + 5 = \dfrac{2}{m-2}$

2. $\dfrac{2}{n+2} + 3 = \dfrac{2}{n-5}$

3. $\dfrac{3y}{(2y+3)(3y-2)} - \dfrac{7}{2y+3} = \dfrac{y-1}{3y-2}$

4. $\dfrac{y+2}{4y^2-13y+3} = \dfrac{y}{4y-1} - \dfrac{5}{y-3}$

5. $\dfrac{4y-5}{2y^2+5y-3} = \dfrac{5y-4}{6y^2-y-1}$

6. $\dfrac{7y}{6y^2-5y-6} = \dfrac{9y-1}{6y^2-11y-10}$

In Exercises 7–40, solve each equation.

7. $\dfrac{3}{z-1} + 2 = \dfrac{5}{z-1}$

8. $\dfrac{5}{z+3} - 2 = \dfrac{4}{z+3}$

9. $\dfrac{6w-1}{2w-1} - 5 = \dfrac{2w-3}{1-2w}$

10. $\dfrac{m}{m-2} - 5 = \dfrac{2}{m-2}$

11. $\dfrac{-3}{p+2} = \dfrac{-8}{p-3}$

12. $\dfrac{7}{p-4} = \dfrac{2}{p+1}$

13. $\dfrac{7}{3n-1} = \dfrac{2}{n+2}$

14. $\dfrac{10}{r-3} = \dfrac{34}{2r+1}$

15. $\dfrac{4}{k+2} = \dfrac{1}{3k+6} + \dfrac{11}{9}$

16. $\dfrac{5}{4k+1} = \dfrac{3}{8k+2} + 1$

17. $\dfrac{3y}{(y+4)(y-2)} = \dfrac{5}{y-2} + \dfrac{2}{y+4}$

18. $\dfrac{y^2+18}{(2y+3)(y-3)} = \dfrac{5}{y-3} - \dfrac{1}{2y+3}$

19. $1 - \dfrac{14}{y^2+4y+4} = \dfrac{7y}{y^2+4y+4}$

20. $1 + \dfrac{6w}{w^2-6w+9} = \dfrac{18}{w^2-6w+9}$

21. $\dfrac{4}{x-5} + \dfrac{5}{x-2} = \dfrac{x+6}{3x-6}$

22. $\dfrac{w+1}{w} + \dfrac{14}{w-7} = \dfrac{3w-7}{w^2-7w}$

23. $\dfrac{1}{(t-1)^2} - 3 = \dfrac{2}{1-t}$

24. $\dfrac{8}{(t+4)^2} + \dfrac{2}{t+4} = 3$

25. $\dfrac{z}{(z-2)(z+1)} - \dfrac{z}{(z+1)(z+3)} = \dfrac{3z}{(z+3)(z-2)}$

26. $\dfrac{4}{(z+1)(z-1)} = \dfrac{6-z}{(z+1)(z-2)} - \dfrac{8}{(z-2)(z-1)}$

27. $\dfrac{2v-5}{3v^2-v-14} + \dfrac{7}{3v-7} = \dfrac{8}{v+2}$

28. $\dfrac{v^2-2v+2}{6v^2+23v-4} + \dfrac{2}{v+4} = \dfrac{v}{6v-1}$

29. $\dfrac{m+4}{6m^2+5m-6} = \dfrac{m}{3m-2} - \dfrac{m}{2m+3}$

30. $\dfrac{2m + 17}{2m^2 + 11m + 14} + \dfrac{m - 2}{m + 2} = \dfrac{m - 3}{2m + 7}$

31. $\dfrac{x + 1}{3x^2 - 4x + 1} - \dfrac{x + 1}{2x^2 + x - 3} = \dfrac{2}{6x^2 + 7x - 3}$

32. $\dfrac{4y}{6y^2 - 7y - 3} + \dfrac{2}{3y^2 - 2y - 1} = \dfrac{y + 2}{2y^2 - 5y + 3}$

33. $\dfrac{2}{m + 2} - \dfrac{1}{m + 1} = \dfrac{1}{m}$

34. $\dfrac{m^2 - 1}{2m + 1} = \dfrac{1 - m}{3}$

35. $\dfrac{z - 2}{2z^2 - 5z + 3} + \dfrac{3}{3z^2 - 2z - 1} = \dfrac{3z}{6z^2 - 7z - 3}$

36. $\dfrac{n - 3}{n^2 + 5n + 4} + \dfrac{n - 2}{n^2 + 3n + 2} = \dfrac{n^2 - 12}{(n + 1)(n + 2)(n + 4)}$

37. $\dfrac{1}{n^2 - 5n + 6} - \dfrac{1}{n^2 - n - 2} + \dfrac{3}{n^2 - 2n - 3} = 0$

38. $\dfrac{12}{2w^2 - 13w + 6} - \dfrac{7w - 2}{2w^2 + w - 1} + \dfrac{w - 20}{w^2 - 5w - 6} = 0$

39. $\dfrac{x^2}{x^2 - x - 2} = \dfrac{2x}{x^2 + x - 6}$

40. $\dfrac{6v + 6}{2v^2 + 7v - 4} = \dfrac{3v}{v^2 + 2v - 8} - \dfrac{5v - 7}{2v^2 - 5v + 2}$

Objective 2 Solve a Rational Equation for a Specified Variable

In Exercises 41–52, solve each equation for the variable specified.

41. $\dfrac{a}{b} = \dfrac{c}{d}$ for b **42.** $a = \dfrac{bc}{d}$ for d

43. $\dfrac{a}{b - 1} = \dfrac{c}{d + 1}$ for b

44. $\dfrac{a}{b - 1} = \dfrac{c}{d + 1}$ for d

45. $I = \dfrac{k}{d}$ for d **46.** $F = \dfrac{k}{r}$ for r

47. $\dfrac{1}{R} = \dfrac{1}{r_1} + \dfrac{1}{r_2}$ for R **48.** $\dfrac{1}{R} = \dfrac{1}{r_1} + \dfrac{1}{r_2}$ for r_1

49. $h = \dfrac{2A}{B + b}$ for B **50.** $h = \dfrac{2A}{B + b}$ for b

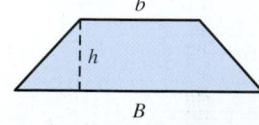

51. $\dfrac{1}{x} = \dfrac{1}{y} - \dfrac{1}{z}$ for z **52.** $I = \dfrac{E}{r_1 + r_2}$ for r_1

Connecting Concepts to Applications

53. Radius of a Window A window in an office is in the shape of a semicircle. One foot from the center of the semicircle, a 4-ft metal security bar is placed in the window. Determine the radius of this window to the nearest hundredth of a foot.

$\left(\text{Hint: } \dfrac{a}{b} = \dfrac{b}{c} \right)$

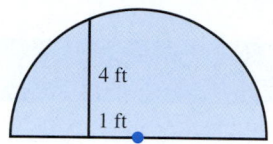
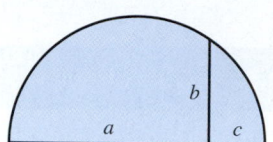

54. Length of a Security Bar A window in an office is in the shape of a semicircle with a 5-ft radius. Two feet from the center of the semicircle, a metal security bar is placed in the window. Determine the length of this security bar to the nearest hundredth of a foot. (See Exercise 53.)

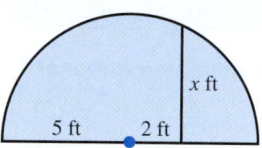

55. Resistance in a Parallel Circuit The resistance planned by an electrical technician for a parallel electric circuit with two resistors is 5 Ω. The resistance of r_2 must be 5 times the resistance of r_1. Determine the resistance of each resistor. The resistance R in a parallel circuit with two individual resistors r_1 and r_2 can be calculated by using the formula $\dfrac{1}{R} = \dfrac{1}{r_1} + \dfrac{1}{r_2}$.

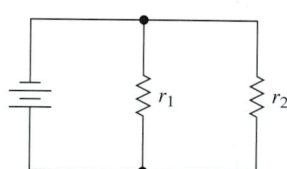

56. Resistance in a Parallel Circuit The resistance planned by an electrical technician for a parallel electric circuit with two resistors is 3 Ω. The resistance of r_2 must be 8 Ω greater than that of r_1. Determine the resistance of each resistor. (See Exercise 55.)

Review and Concept Development

In Exercises 57–62, simplify the expression in part (a) and solve the equation in part (b) for all real solutions.

Simplify	Solve

57. a. $\dfrac{1}{p-1} - \dfrac{3}{p+1}$ **b.** $\dfrac{1}{p-1} = \dfrac{3}{p+1}$

58. a. $\dfrac{m-1}{m+1} - \dfrac{m-3}{m-2}$ **b.** $\dfrac{m-1}{m+1} = \dfrac{m-3}{m-2}$

59. a. $\dfrac{x-1}{x+1} - 1 - \dfrac{x-6}{x-2}$ **b.** $\dfrac{x-1}{x+1} - 1 = \dfrac{x-6}{x-2}$

60. a. $\dfrac{x^2}{x+2} + \dfrac{x-1}{x-3}$ **b.** $\dfrac{x^2}{x+2} + \dfrac{x-1}{x-3} = 0$

61. a. $\dfrac{2x-8}{6x^2+x-2} - \dfrac{4}{3x+2} + \dfrac{2}{2x-1}$

   b. $\dfrac{2x-8}{6x^2+x-2} = \dfrac{4}{3x+2} - \dfrac{2}{2x-1}$

62. a. $\dfrac{w^2-w-3}{2w^2-9w+9} + \dfrac{1}{3-w} - \dfrac{w}{2w-3}$

   b. $\dfrac{w^2-w-3}{2w^2-9w+9} + \dfrac{1}{3-w} = \dfrac{w}{2w-3}$

In Exercises 63–66, determine the intercepts of each rational function. Use the graph to help check your work.

63.   **64.**

$f(x) = \dfrac{x+5}{x-2}$ $f(x) = \dfrac{x-3}{x+1}$

65.   **66.**

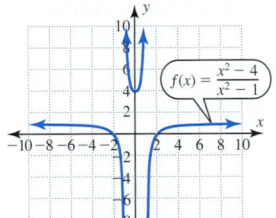

$f(x) = \dfrac{x^2-4}{x^2-1}$ $f(x) = \dfrac{x^2+x-6}{x^2-2x-3}$

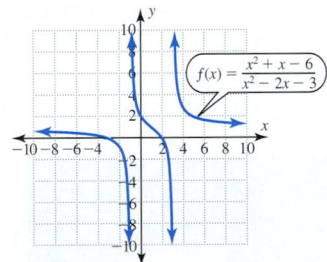

Skill and Concept Development

In Exercises 67–74, solve each equation.

67. a. $\dfrac{5v+4}{2v^2+v-15} + \dfrac{3}{5-2v} - \dfrac{1}{v+3} = 0$

   b. $\dfrac{5v+3}{2v^2+v-15} + \dfrac{3}{5-2v} - \dfrac{1}{v+3} = 0$

68. a. $\dfrac{w}{3w^2-11w+10} + \dfrac{5}{3w-5} = \dfrac{2}{w-2}$

   b. $\dfrac{w+1}{3w^2-11w+10} + \dfrac{5}{3w-5} = \dfrac{2}{w-2}$

69. $\dfrac{x^2}{2x^2+9x-5} + \dfrac{2x}{x^2+2x-15} = \dfrac{4x}{(2x-1)(x+5)(x-3)}$

70. $\dfrac{x}{x-3} + \dfrac{1}{x-2} - \dfrac{1}{x+2} = \dfrac{x-12}{x^3-3x^2-4x+12}$

71. $\dfrac{z-2}{4z^2-29z+30} - \dfrac{z+2}{5z^2-27z-18} = \dfrac{z+1}{20z^2-13z-15}$

72. $\dfrac{3n-7}{n^2-5n+6} + \dfrac{2n+8}{9-n^2} - \dfrac{n+2}{n^2+n-6} = 0$

73. $\dfrac{z-1}{z^2-2z-3} + \dfrac{z+1}{z^2-4z+3} = \dfrac{z+8}{z^2-1} + \dfrac{20}{(z-1)(z+1)(z-3)}$

74. $\dfrac{3w-8}{w^2-5w+6} + \dfrac{w+2}{w^2-6w+8} = \dfrac{5-2w}{w^2-7w+12} + \dfrac{12}{(w-4)(w-3)(w-2)}$

Group discussion questions

75. Error Analysis Determine the error in the following argument:

Let $x = 1$

Then $3x = 2x + 1$

     $3x - 3 = 2x - 2$

     $3(x-1) = 2(x-1)$

     $\dfrac{3(x-1)}{x-1} = \dfrac{2(x-1)}{x-1}$

Thus $3 = 2$

76. Challenge Question

   a. Complete this equation so that the solution is all real numbers except for -1 and 2.

       $\dfrac{5}{x-2} + \dfrac{2}{x+1} = \dfrac{?}{x^2-x-2}$

   b. Complete this equation so that the solution is all real numbers except $-\dfrac{2}{3}$ and $\dfrac{3}{2}$.

       $\dfrac{?}{6x^2-5x-6} = \dfrac{4}{3x+2} - \dfrac{1}{2x-3}$

77. Challenge Question Complete each equation so that the equation is a contradiction with no solution.

   a. $\dfrac{5}{x-2} + \dfrac{2}{x+1} = \dfrac{7x+?}{x^2-x-2}$

   b. $\dfrac{5x+?}{6x^2-5x-6} = \dfrac{4}{3x+2} - \dfrac{1}{2x-3}$

9.5 | Cumulative Review

In Exercises 1–3, construct a quadratic equation with the given solutions.

1. $x = 3$ and $x = -\dfrac{1}{5}$

2. $x = -\sqrt{6}$ and $x = \sqrt{6}$

3. $x = 1 - 2i$ and $x = 1 + 2i$

In Exercises 4 and 5, find the domain of each function.

4. $f(x) = \dfrac{x + 2}{2x - 6}$

5. $f(x) = \dfrac{2x + 16}{x^2 + 4}$

Section 9.6 | Inverse and Joint Variation and Other Applications Yielding Equations with Fractions

Objectives:

1. Distinguish between direct and inverse variation.
2. Translate statements of variation.
3. Solve problems involving variation.
4. Solve applied problems that yield equations with fractions.

The problem-solving skills developed in earlier chapters are now used to solve applications that yield equations containing rational expressions. Some of these applications illustrate the variety that you may encounter outside the classroom. Other problems are skill builders to gradually develop your expertise for these applications.

1. Distinguish Between Direct and Inverse Variation

In Section 2.7, we examined problems that involved direct variation. If y varies directly as x with a constant of variation $k \neq 0$, then $y = kx$. When two variables vary directly, an increase in magnitude of one variable will cause an increase in magnitude of the other variable. We now examine problems involving inverse variation.

What Is Meant by Inverse Variation?

A Mathematical Note

Evelyn Boyd Granville completed her Ph.D. in functional analysis in 1949 and became one of the first African-American women to earn a doctorate in mathematics. She later worked on both the Mercury and Apollo space programs.

If **y varies inversely as x** with a constant of variation $k \neq 0$, then $y = \dfrac{k}{x}$ for $x \neq 0$. When two variables vary inversely, an increase in magnitude of one variable will cause a decrease in magnitude of the other variable. Exercise 29 states that the weight of an astronaut varies inversely as the square of the distance from the center of the earth. As the distance of the astronaut from the center of the earth increases, the weight of the astronaut will decrease. Direct and inverse variation have quite contrasting behaviors, as described in the following table.

Comparison of Direct and Inverse Variation

If x and y are real variables, k is a real constant, and $k \neq 0$, then:

Verbally	Algebraically	Numerical Example	Graphical Example
y varies **directly** as x. As the magnitude of x increases, the magnitude of y increases linearly.	$y = kx$ *Example:* $\quad y = 2x$		

x	$y = 2x$
-3	-6
-1	-2
0	0
1	2
3	6

Verbally	Algebraically	Numerical Example	Graphical Example
y varies **inversely** as x. As the magnitude of x increases, the magnitude of y decreases.	$y = \dfrac{k}{x} \quad$ for $x \neq 0$ *Example:* $\quad y = \dfrac{12}{x}$		

x	$y = \dfrac{12}{x}$
1	12
2	6
3	4
4	3
5	2.4

How Can I Recognize the Contrasting Behaviors of Direct and Inverse Variation When I Analyze Real-World Relationships?

If y varies directly as x, then the magnitude of y increases linearly as the magnitude of x increases. If y varies inversely as x, then the magnitude of y decreases as the magnitude of x increases. This is illustrated in Example 1.

Example 1   Selecting Direct or Inverse Variation

Determine whether the underlying relationship illustrated is more likely to be direct variation or inverse variation.

Solution

(a)

Height of Medium-Frame Female Age 25 (in)	Recommended Weight (lb)
60	122
62	127
64	132
66	137
68	142
70	147

Direct variation

Taller people are generally larger people and also weigh more.

(b)

Number of Years over of 25 for Males	Vertical Jumping Height (cm)
10	72.0
20	36.0
30	24.0
40	18.0
50	14.4
60	12.0

Inverse variation

After maturity and past age 25, the jumping ability of people declines as they age.

Self-Check 1

Determine whether the relationship illustrated in the table is direct variation or inverse variation.

a.

x	y
1	120
2	60
3	40
4	30
5	24
6	20

b.

x	y
1	12
2	24
3	36
4	48
5	60
6	72

2. Translate Statements of Variation

The language of variation is used to describe relationships between real-world variables. In many cases, there are more than two variables involved. In these cases, we often will see some variables related directly and others inversely.

Are There Any Other Types of Variation?

Yes, you can also see examples of joint variation. To say that the variable z **varies jointly** as x and y means that z varies directly as the product of x and y.

Joint Variation

If x, y, and z are real variables, k is a real constant, and $k \neq 0$, then:

Verbally	Algebraically	Example
z varies jointly as x and y.	$z = kxy$	$z = 5xy$

In Example 2, we translate statements of variation into algebraic equations.

Example 2   Translating Statements of Variation

Translate each statement of variation into an equation.

Solution

(a) At a fixed speed, the distance D that a car travels varies directly as the time T it travels.

$D = kT$

(b) The current I in an electric circuit varies inversely as the resistance R.

$I = \dfrac{k}{R}$

(c) The pressure P of a gas varies directly as the absolute temperature T and inversely as the volume V.

$P = k\dfrac{T}{V}$

(d) The volume V of a cylinder varies jointly as the square of its radius r and its height h with constant of variation π.

$V = \pi r^2 h$

Self-Check 2

Translate each statement of variation into an equation.

a. m varies directly as n.

b. p varies inversely as q.

c. x varies directly as y and inversely as the square of w.

d. a varies jointly as b and c.

3. Solve Problems Involving Variation

Sometimes proportions (see Section 2.7) are also used to describe the relation between variables that vary directly or inversely. If x and y are **directly proportional,** then y varies directly as x. If x and y are **inversely proportional,** then y varies inversely as x.

Example 3 works through a numeric word problem involving direct and inverse variation. This allows us to compare these two types of variation before the more involved application shown in Example 4.

Example 3   Solving Problems with Direct and Inverse Variation

The variable y equals 48 when x equals 4. Find y when $x = 6$ if

(a) y varies directly as x.

(b) y varies inversely as x.

Solution

(a) Word Equation y varies directly as x

Algebraic Equation	$y = kx$	Translate the word equation into algebraic form, using k for the constant of variation.
	$48 = k(4)$	Substitute in the given values of 48 for y and 4 for x. Then solve for the constant k.
	$12 = k$	
	$y = 12x$	Use this constant to write the equation of variation. Then substitute in the value of 6 for x and solve for y.
	$y = 12(6)$	
	$y = 72$	

(b) Word Equation y varies inversely as x

Algebraic Equation	$y = \dfrac{k}{x}$	Translate the word equation into algebraic form, using k for the constant of variation.
	$48 = \dfrac{k}{4}$	Substitute in the given values of 48 for y and 4 for x.
	$192 = k$	Then solve for the constant k.
	$y = \dfrac{192}{x}$	Use this constant to write the equation of variation.
	$y = \dfrac{192}{6}$	Then substitute in the value of 6 for x and solve for y.
	$y = 32$	

Self-Check 3

a. y varies directly as x, and $y = 6$ when $x = 3$. Find y when $x = 2$.

b. y varies inversely as x, and $y = 6$ when $x = 3$. Find y when $x = 2$.

4. Solve Applied Problems That Yield Equations with Fractions

The energy sent out from a single point source often radiates in a spherical pattern. This is true whether the energy is heat, light, sound, or a radio signal. The surface area of the sphere receiving this energy at any instant is given by $S = 4\pi r^2$. As the distance r from the energy source increases, the area sharing this fixed amount of energy increases as the square of r. Thus the intensity of the energy received at any place on this sphere varies inversely as the square of the distance from the energy source. Example 4 illustrates this concept by examining the illumination given by a car's headlight.

Sphere: $S = 4\pi r^2$

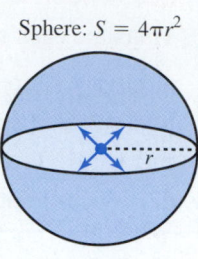

Example 4   Using Inverse Variation

The illumination provided by a car's headlight varies inversely as the square of the distance from the headlight. A headlight produces 15 foot-candles (fc) at a distance of 20 ft.

(a) Determine the constant of variation for this headlight.

(b) Write an equation relating the illumination provided by this headlight to the distance from the headlight.

(c) Use this equation to create a table of values displaying the illumination in 5-ft increments starting at 10 ft.

(d) What will the illumination be at 40 ft?

Solution

(a) Let I = illumination of headlight, fc

$\quad\quad d$ = distance of an object from the headlight, ft

$\quad\quad k$ = constant of variation for this headlight

Identify each unknown with a variable. Be sure to include units in this identification.

Word Equation

The illumination varies inversely as the square of the distance.

Precisely state an equation that can be translated into algebraic form.

Algebraic Equation

$$I = \frac{k}{d^2}$$

$$15 = \frac{k}{20^2}$$

$$15 = \frac{k}{400}$$

$$15(400) = k$$

$$k = 6,000$$

Translate the word equation into algebraic form by using the variables identified above.

Substitute in the given values for I and d, and solve for the constant k.

The constant of variation is 6,000.

(b) $I = \dfrac{6,000}{d^2}$

(c)

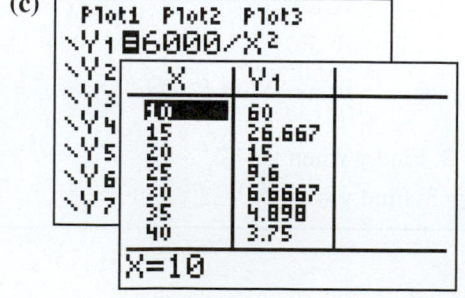

To generate a table of values, enter the equation $I = \dfrac{6,000}{d^2}$ as $Y_1 = \dfrac{6,000}{x^2}$.

Then use 10 for **TblStart** and 5 for **ΔTbl**.

(d) $I = \dfrac{6{,}000}{40^2}$

$I = 3.75$

The illumination will be 3.75 fc.

Using the constant of variation for this headlight, find I when the distance is 40 ft by substituting $d = 40$ into the equation $I = \dfrac{6{,}000}{d^2}$.

This value is also confirmed by the table.

Self-Check 4

Rework Example 4 assuming the headlight produces 18 foot-candles (fc) at a distance of 20 ft.

Many problems involving rates and times involve inverse variation. For example, the formula for distance $D = RT$ can be rewritten as $R = \dfrac{D}{T}$. This form states that the rate of travel varies directly as the distance traveled and inversely as the time traveled. As the time you have available for a trip increases, you can travel at a slower rate.

Likewise, the formula for work $W = RT$ can be rewritten as $R = \dfrac{W}{T}$. For a fixed amount of work, the rate of work varies inversely as the time required to do this work. For less time you must work faster—for more time you can work slower.

The basic strategy used to solve these word problems is summarized in the following box. It is the same strategy we employed earlier to solve word problems by using linear equations and quadratic equations.

Strategy for Solving Word Problems

Step 1. Read the problem carefully to determine what you are being asked to find.

Step 2. Select a variable to represent each unknown quantity. Specify precisely what each variable represents and note any restrictions on each variable.

Step 3. If necessary, make a sketch and translate the problem into a word equation or a system of word equations. Then translate each word equation into an algebraic equation.

Step 4. Solve the equation or the system of equations, and answer the question completely in the form of a sentence.

Step 5. Check the reasonableness of your answer.

After solving the equations that you form, always inspect the solutions to see if they are appropriate for the original problem. Check not only for extraneous values but also for answers that may not be meaningful in the application, such as negative amounts of time.

Example 5 involves the shared work of two people working together to complete a job. This problem assumes that when the painters work together, there is no gain or loss of efficiency of either painter. Both painters will continue to work at the same rate as when they were working alone. Another key point in this example is that information about the time to do a job is also information about the rate of work for this job. This is true because of the inverse relationship between the time and the rate of work.

Example 5 Determining the Time to Paint a Sign

Working alone, painter A can paint a sign in 6 hours less time than it would take painter B working alone. When painters A and B work together, the job takes only 4 hours. How many hours would it take each painter working alone to paint the sign?

Solution

Let $\quad t =$ time for painter B to paint sign working alone, hours

$\quad t - 6 =$ time for painter A to paint sign working alone, hours

$\dfrac{1}{t - 6} =$ rate of work for painter A

$\dfrac{1}{t} =$ rate of work for painter B

$\dfrac{1}{4} =$ rate of work when both painters work together

First identify the time for each painter to paint the sign while working alone.

The rate of work varies inversely as the time; that is, $R = \dfrac{W}{T}$. The work equals 1 sign painted; thus this equation becomes $R = \dfrac{1}{T}$.

Word Equation

$$\boxed{\text{A's rate of work}} + \boxed{\text{B's rate of work}} = \boxed{\text{Total rate of work when working together}}$$

The word equation is based on the mixture principle. When the two painters share the work, the total rate of work is found by adding their individual rates of work.

Algebraic Equation

$$\frac{1}{t - 6} + \frac{1}{t} = \frac{1}{4}$$

Substitute the rates of each painter into the word equation.

$$4t(t - 6)\frac{1}{t - 6} + 4t(t - 6)\frac{1}{t} = 4t(t - 6)\left(\frac{1}{4}\right)$$

Multiply both sides of the equation by the LCD, $4t(t - 6)$.

$$4t + 4(t - 6) = t(t - 6)$$
$$4t + 4t - 24 = t^2 - 6t$$
$$8t - 24 = t^2 - 6t$$
$$0 = t^2 - 14t + 24$$
$$0 = (t - 12)(t - 2)$$

Simplify this quadratic equation, and write it in standard form. Then solve this equation by factoring.

$$t - 12 = 0 \quad \text{or} \quad t - 2 = 0$$
$$t = 12 \qquad\qquad t = 2$$
$$t - 6 = 6 \qquad\quad t - 6 = -4$$

$\boxed{\text{This value is not appropriate.}}$

While 2 hours may seem meaningful for painter B, the value of -4 hours is not meaningful for painter A.

Check: The rates for each painter are $\dfrac{1}{6}$ sign/hour for A and $\dfrac{1}{12}$ sign/hour for B. In 4 hours they paint a total of

$$\frac{1}{6}(4) + \frac{1}{12}(4) = \frac{4}{6} + \frac{4}{12}$$
$$= \frac{2}{3} + \frac{1}{3}$$
$$= 1 \text{ sign.}$$

Answer: Working alone, painter A could paint the sign in 6 hours and painter B could paint the sign in 12 hours.

Self-Check 5

Working alone, a welder can weld a metal framework 5 hours faster than an apprentice can. How long will it take the apprentice to do the job alone if the welder and the apprentice can do it in 6 hours when they share the work?

Example 6 has many similarities to the $W = RT$ problem in Example 5. However, in this example, the formula is $D = RT$, and the rate involved is the rate of travel instead of the rate of work. It is wise to include the units of measurement when you identify your variables. This is especially true if there are different units within the same problem (e.g., both minutes and hours).

Example 6   Determining the Speed of a Boat in a Still Water

Two tugboats that travel at the same speed in still water travel in opposite directions in a river with a constant current of 8 mi/h. The tugboats depart at the same time from a refueling station; after a period of time, one has traveled 30 mi downstream, and the other has traveled 6 mi upstream. Determine the rate of each boat in still water.

Solution

Let $r =$ rate of each boat in still water, mi/h

$r + 8 =$ rate of boat going downstream, mi/h

$r - 8 =$ rate of boat going upstream, mi/h

$\dfrac{30}{r + 8} =$ time of boat going downstream, hours

$\dfrac{6}{r - 8} =$ time of boat going upstream, hours

First identify the quantities being sought with a variable—the rates of the two boats.

The formula $D = RT$ can also be written as $T = \dfrac{D}{R}$. Substitute the given distances and the rates $r + 8$ and $r - 8$ into this formula to also identify the time for each boat.

Word Equation

$$\boxed{\begin{array}{c} \text{Time of boat} \\ \text{going downstream} \end{array}} = \boxed{\begin{array}{c} \text{Time of boat} \\ \text{going upstream} \end{array}}$$

Although the times are unknown, we do know they are the same because the boats left at the same time.

Algebraic Equation

$$\frac{30}{r + 8} = \frac{6}{r - 8}$$

$$(r + 8)(r - 8)\frac{30}{r + 8} = (r + 8)(r - 8)\frac{6}{r - 8}$$

$$30(r - 8) = 6(r + 8)$$

$$30r - 240 = 6r + 48$$

$$24r = 288$$

$$r = 12$$

Substitute the times as identified above into the word equation.

Multiply both sides of the equation by the LCD, $(r + 8)(r - 8)$.

Then simplify this equation and solve for r.

Check: $r + 8 = 20$
             $r - 8 = 4$

Going downstream at 20 mi/h, the tugboat will take $\dfrac{30}{20} = 1.5$ hours to go 30 mi. Going upstream at 4 mi/h, the tugboat will take $\dfrac{6}{4} = 1.5$ hours to go 6 mi. These times are equal.

Answer: The rate of each tugboat in still water is 12 mi/h.

Self-Check 6

Two tugboats that travel at the same speed in still water travel in opposite directions in a river with a constant current of 5 mi/h. The tugboats depart at the same time from a dock; and after a period of time one has traveled 30 mi downstream, and the other has traveled 6 mi upstream. Determine the rate of travel of each boat in still water.

Example 7 also involves a rate problem. For this application, the rate is the interest rate charged for borrowed money. The formula $I = PRT$ can be rewritten as $P = \dfrac{I}{RT}$. For a fixed time of 1 year, this formula can be simplified to $P = \dfrac{I}{R}$, as is done in Example 7. Recall that P represents the principal, R represents the interest rate, T represents the time, and I represents the interest.

One question to ask when you are trying to form the word equation for a problem is, "What things are equal or the same?" In Example 7, the principal invested in the checking account and in the savings account would be the same.

Example 7   Determining Two Interest Rates

A local bank pays interest on both checking accounts and savings accounts. The interest rate for a savings account is 1% higher than that for a checking account. A customer calculates that a deposit in a checking account would earn yearly interest of $80, whereas this same deposit would earn yearly interest of $100 in a savings account. What is the interest rate on each account?

Solution

Let $\qquad$ r = rate of interest on checking account

$r + 0.01$ = rate of interest on savings account

$\dfrac{80}{r}$ = principal in checking account

$\dfrac{100}{r + 0.01}$ = principal in savings account

First identify the quantities being sought with a variable—the interest rates of the two accounts.

$I = PRT$; for $T = 1$ we can write $P = \dfrac{I}{R}$. Substitute the given interest for each account and the rates identified to label the principal in each account.

Word Equation

| Principal in a checking account | $=$ | Principal in a savings account |

The deposit is the same for both options.

Algebraic Equation

$$\dfrac{80}{r} = \dfrac{100}{r + 0.01}$$

Substitute the values identified above into the word equation.

$$r(r + 0.01)\left(\dfrac{80}{r}\right) = r(r + 0.01)\left(\dfrac{100}{r + 0.01}\right)$$

Multiply both sides of the equation by the LCD, $r(r + 0.01)$.

$$80(r + 0.01) = 100r$$

$$80r + 0.8 = 100r$$

Simplify, and then solve for r.

$$0.8 = 20r$$

$$0.04 = r$$

$$r = 0.04$$

$$r + 0.01 = 0.05$$

The checking account rate is 4%.
The savings account rate is 5%.

Answer: The interest rate on the checking account is 4% and on the savings account is 5%.

Do these values check?

Self-Check 7

A local bank pays interest on both checking accounts and savings accounts. The interest rate for a savings account is 0.5% higher than that for a checking account. A customer calculates that a deposit in a checking account would earn yearly interest of $80, whereas this same deposit would earn yearly interest of $100 in a savings account. What is the interest rate on each account?

In Example 8, we examine a numeric word problem involving reciprocals. These problems do not have the context of a real-world problem, but exercises like this can help to develop your problem-solving skills.

Example 8   Solving a Numeric Word Problem

The sum of the reciprocals of two consecutive even integers is $\dfrac{13}{84}$. Find these integers.

Solution

Let n = smaller integer   Identify each number and its reciprocal.

$\dfrac{1}{n}$ = reciprocal of smaller integer

$n + 2$ = larger integer   Consecutive even integers differ by 2.

$\dfrac{1}{n+2}$ = reciprocal of larger integer

Word Equation

$$\boxed{\begin{array}{c}\text{Reciprocal of}\\\text{first integer}\end{array}} + \boxed{\begin{array}{c}\text{Reciprocal of}\\\text{second integer}\end{array}} = \boxed{\dfrac{13}{84}}$$

Write the word equation.

Algebraic Equation

$$\frac{1}{n} + \frac{1}{n+2} = \frac{13}{84}$$

Substitute the values identified above into the word equation.

$$84n(n+2)\left(\frac{1}{n} + \frac{1}{n+2}\right) = 84n(n+2)\left(\frac{13}{84}\right)$$

Multiply both sides of the equation by the LCD, $84n(n+2)$.

$$84(n+2) + 84n = 13n(n+2)$$

$$84n + 168 + 84n = 13n^2 + 26n$$

$$13n^2 - 142n - 168 = 0$$

Simplify, and write the quadratic equation in standard form.

$$(n - 12)(13n + 14) = 0$$

$n - 12 = 0$ or $13n + 14 = 0$ Factor the left side of the equation.

Set each factor equal to 0.

$n = 12$ $n = -\dfrac{14}{13}$

Then solve for the smaller integer n and the larger integer $n + 2$.

$n + 2 = 14$ $\boxed{\text{This value is not an integer.}}$

Check: $\dfrac{1}{12} + \dfrac{1}{14} = \dfrac{7}{84} + \dfrac{6}{84} = \dfrac{13}{84}$.

Answer: The integers are 12 and 14.

Self-Check 8

The sum of the reciprocals of two consecutive integers is $\dfrac{15}{56}$. Find these integers.

Self-Check Answers

1. a. Inverse variation
  b. Direct variation

2. a. $m = kn$

  b. $p = \dfrac{k}{q}$

  c. $x = k\dfrac{y}{w^2}$

  d. $a = kbc$

3. a. $y = 4$
  b. $y = 9$

4. a. The constant of variation is 7,200.

  b. $I = \dfrac{7{,}200}{d^2}$

c.

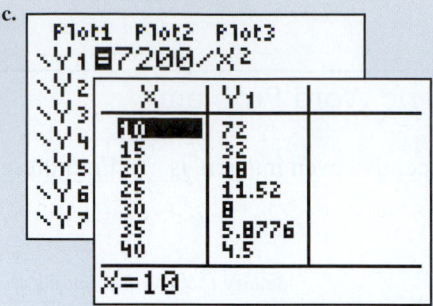

d. At a distance of 40 ft, the illumination will be 4.5 fc.

5. The apprentice can do the job alone in 15 hours.

6. The rate of each tugboat in still water is 7.5 mi/h.

7. The interest rate on a checking account is 2% and on a savings account is 2.5%.

8. The integers are 7 and 8.

9.6 Using the Language and Symbolism of Mathematics

1. If a varies directly as b, then $a =$ _____ .

2. If a varies inversely as b, then $a =$ _____ .

3. If a varies jointly as b and c, then $a =$ _____ .

4. If a varies directly as b and b increases in magnitude, then a will _____ in magnitude.

5. If a varies directly as b and b decreases in magnitude, then a will _____ in magnitude.

6. If a varies inversely as b and b increases in magnitude, then a will _____ in magnitude.

7. If a varies inversely as b and b decreases in magnitude, then a will _____ in magnitude.

8. In the formula $D = RT$, D represents distance, R represents _____ , and T represents _____ .

9. In the formula $I = PRT$, I represents interest, P represents _____ , R represents _____ , and T represents _____ .

10. In the formula $W = RT$, W represents work, R represents _____ , and T represents _____ .

9.6 Quick Review

1. Determine the time it will take to drive 240 mi while averaging 60 mi/h.

2. Determine the time it will take to drive 240 mi while averaging 50 mi/h.

3. Determine the rate of travel of a vehicle that travels 240 mi in 30 minutes.

4. Determine the interest on a $5,000 investment for 1 year at 6% annual interest.

5. A $12,000 investment earns $80 interest the first month. What is the annual rate of interest?

9.6 Exercises

Objective 1 Distinguish Between Direct and Inverse Variation

In Exercises 1–10, determine whether the relationship illustrated is direct variation or inverse variation.

1. $C = \pi d$

2. $A = \dfrac{120}{W}$

3. $P = \dfrac{250}{R}$

4. $W = 62.4V$

5.

x	y
1	3
2	6
3	9
4	12
5	15
6	18

6.

x	y
1	60
2	30
3	20
4	15
5	12
6	10

7. The relationship between the number of people passing out flyers to an entire neighborhood and the time required to pass out the flyers

N People	T Hours
1	12.0
2	6.0
3	4.0
4	3.0
5	2.4
6	2.0

8. The relationship between the number of songs on an iPod and the amount of memory occupied

N Songs	M Gigabytes
500	2
1,000	4
1,500	6
2,000	8
2,500	10
3,000	12

9. The relationship between the time T, in h, to travel 500 mi and the rate R, in mi/h

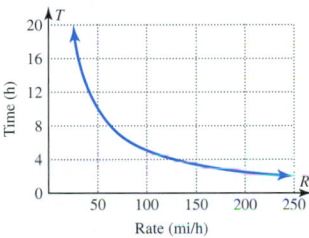

10. The relationship between the revenue R and the number of tickets n sold at a basketball game

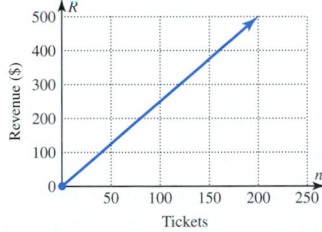

Objective 2 Translate Statements of Variation

In Exercises 11–34, write an equation for each statement of variation. Use k as the constant of variation.

11. m varies directly as n.
12. w varies directly as z.
13. m varies inversely as p.
14. w varies inversely as v.
15. v varies directly as the square root of w.
16. y varies directly as x cubed.
17. v varies inversely as the square of x.
18. y varies inversely as the cube root of z.
19. m varies directly as n and inversely as p.

20. w varies directly as z and inversely as v.
21. v varies directly as the square root of w and inversely as the square of x.
22. y varies directly as x cubed and inversely as the cube root of z.
23. m varies jointly as n and p.
24. w varies jointly as z and v.
25. v varies jointly as the square root of w and the square of x.
26. y varies jointly as x cubed and the cube root of z.
27. Ohm's Law The electric current I varies directly as the voltage V.
28. Electricity from a Windmill The number W of kilowatts of electricity that can be produced by a windmill varies directly as the cube of the speed v of the wind.
29. Weight of an Astronaut The weight w of an astronaut varies inversely as the square of the distance d from the center of the earth.

30. Travel Time The time t required to drive between two towns varies inversely as the rate r.
31. Wind Resistance The force of the wind resistance R on a moving automobile varies directly as the square of the velocity v of the automobile.
32. Period of a Pendulum The period T (time for one complete swing) of a pendulum varies directly as the square root of the length L of the pendulum.
33. Volume of a Cone The volume V of a cone varies jointly with the square of the radius r and the height h.
34. Universal Law of Gravitation Every object in the universe attracts every other object with a force F that varies jointly as their masses m_1 and m_2 and inversely as the square of the distance d between the objects.

Objective 3 Solve Problems Involving Variation

In Exercises 35–40, use the given statement of variation to solve each problem.

35. a. y varies directly as x, and $y = 24$ when $x = 8$. Find y when $x = 10$.
   b. y varies inversely as x, and $y = 24$ when $x = 8$. Find y when $x = 10$.

36. a. *a* varies directly as *b*, and $a = 12$ when $b = 3$. Find *a* when $b = 2$.
  b. *a* varies inversely as *b*, and $a = 12$ when $b = 3$. Find *a* when $b = 2$.

37. *a* varies directly as *b* and inversely as *c*, and $a = 3$ when $b = 9$ and $c = 12$. Find *a* when $b = 15$ and $c = 6$.

38. *a* varies directly as *b* and inversely as *c*, and $a = 6$ when $b = 2$ and $c = 24$. Find *a* when $b = 5$ and $c = 60$.

39. *a* varies jointly as *b* and *c*, and $a = 27$ when $b = 9$ and $c = 12$. Find *a* when $b = 15$ and $c = 6$.

40. *a* varies jointly as *b* and *c*, and $a = 8$ when $b = 2$ and $c = 24$. Find *a* when $b = 5$ and $c = 60$.

41. Use of Newsprint The use of newsprint varies directly as the number of people using the newsprint. The newsprint used to supply the needs of 1,000 people is approximately 34,800 kg. (*Source:* North American Newsprint Producers Association.)
  a. Determine the constant of variation.
  b. Write an equation relating the kilograms of newsprint to the number of people using the newsprint.
  c. Use this equation to create a table of values displaying the kilograms of newsprint required to supply the needs of various populations starting with 700,000 people and increasing in increments of 10,000 people.
  d. How many kilograms would be needed to supply Columbus, Ohio, which has a population of approximately 720,000?

42. Stopping Distance of a Car The distance required for an emergency stop for a car varies directly as the square of the car's speed. A car traveling at 50 mi/h requires 140 ft to stop.

  a. Determine the constant of variation.
  b. Write an equation relating the stopping distance to the speed of this car.
  c. Use this equation to create a table of values displaying the stopping distances required for this car traveling at various speeds starting at 45 mi/h and increasing in increments of 5 mi/h.
  d. How many feet will be required for this car to make an emergency stop if it was traveling at 70 mi/h?

43. Time to Distribute Flyers The time required to distribute a batch of political flyers in a neighborhood varies inversely as the number of people who help pass out these flyers. It will take 8 hours for 5 workers to distribute a batch of flyers.
  a. Determine the constant of variation.
  b. Write an equation relating the time required to distribute flyers to the number of people who help pass out these flyers.
  c. Use this equation to create a table of values displaying the times required for distributing flyers in this neighborhood, starting with 5 people and increasing in increments of 5 people.
  d. How long will it take to distribute these flyers in this neighborhood if 20 people are helping?

44. Force to Prevent Skidding The force required to keep a car from skidding on a curve varies inversely as the radius of the curve. A force of 7,500 lb is needed to keep a car from skidding on a curve of radius 800 ft.
  a. Determine the constant of variation.
  b. Write an equation relating the force required to keep a car from skidding to the radius of the curve.
  c. Use this equation to create a table of values displaying the force required to keep a car from skidding on various curves, starting at a radius of 500 ft and increasing in increments of 100 ft.
  d. What force will be required to keep a car from skidding on a radius of 500 ft?

45. Time to Get a Sunburn The time it takes to get a sunburn varies inversely as the UV (ultraviolet light index) rating. At a UV rating of 6, the time it takes to obtain a sunburn can be as little as 15 minutes. Use this information to determine the time it will take to get a sunburn when the UV rating is 10. (Source: CBS News online Consumer Tips.)

46. Volume of a Gas The volume of a gas varies inversely as the pressure applied to the gas. Under a pressure of 38 kg/cm², the volume of a gas is 240 cm³. Approximate the volume when the pressure is increased to 42 kg/cm² (Boyle's law).

47. Resistance of a Wire The electrical resistance of a wire varies inversely as the square of the diameter of the wire. The resistance of a wire with a diameter of 4 mm is 4.5 Ω. Find the resistance of another wire of the same length and made from similar materials with a diameter of 6 mm.

48. Illumination The illumination *I* received at an object from a light source varies inversely as the square of the distance from the source. If the illumination 5 m from a light source is 4 lumens (lm), find the illumination 10 m from this same light source.

49. Load on a Beam The load that a rectangular beam can support varies jointly as its width and the square of its height and inversely as its length. A 2 × 6 (assume actual measurements are really 2 in and 6 in) pine beam that is 120 in long can carry a load of 800 lb when installed properly with the width of 2 in. Approximate the load this beam can support if improperly installed with a width of 6 in and a height of 2 in.

50. **Volume of a Pyramid** The volume of a right pyramid varies jointly as the height and area of the base of the pyramid. The Great Pyramid of Giza has a height of 145 m and a base area of 52,500 m^2. The volume of a smaller pyramid with a height of 55 m and an area of 7,200 m^2 is 132,000 m^3. Determine the constant of variation and the volume of the Great Pyramid of Giza.

Objective 4 Solve Applied Problems That Yield Equations with Fractions

In Exercises 51–84, solve each problem.

51. The sum of the reciprocals of two consecutive integers is $\frac{11}{30}$. Find these integers.

52. The sum of the reciprocals of two consecutive even integers is $\frac{7}{24}$. Find these integers.

53. The sum of the reciprocals of two consecutive odd integers is 16 times the reciprocal of their product. Find these integers.

54. The sum of the reciprocals of two consecutive even integers is 10 times the reciprocal of their product. Find these integers.

55. The denominator of a fraction is an integer that is 4 more than the square of the numerator. If the fraction is reduced, it equals $\frac{3}{20}$. Find this numerator.

56. The denominator of a fraction is an integer that is 3 more than the square of the numerator. If the fraction is reduced, it equals $\frac{1}{4}$. Find this numerator.

57. The sum of a number and its reciprocal is $\frac{13}{6}$. Find this number.

58. The difference of a number and its reciprocal is $\frac{9}{20}$. Find this number.

59. **Ratio of Gauge Readings** The ratio of two readings from a gauge is $\frac{4}{5}$. The first reading is 3 units above normal, and the second reading is 5 units above normal. What is the normal reading?

60. **Ratio of Temperatures** The ratio of two temperature readings is $\frac{7}{8}$. The first reading is 7° below normal, whereas the second reading is 2° above normal. What is the normal reading?

61. **Wire Length** A wire 16 m long is cut into two pieces whose lengths have a ratio of 3 to 1. Find the length of each piece.

62. **Rope Length** A rope 20 m long is cut into two pieces whose lengths have a ratio of 4 to 1. Find the length of each piece.

63. **Electrical Resistance** The resistance of the parallel circuit with two resistors shown in the figure is 40 Ω. If one resistor has twice the resistance of the other, what is the resistance of each?

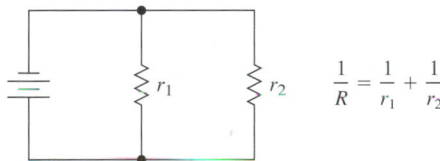

64. **Electrical Resistance** The resistance of one resistor in a parallel circuit is 3 Ω greater than that of the second resistor in the circuit. Find the resistance of each resistor if the resistance of the circuit is $5\frac{1}{7}$ Ω.

65. **Complementary Angles** The ratio of the measures of two complementary angles is 3 to 2. Find the measure of each angle.

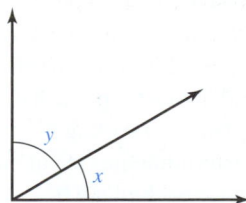

66. **Supplementary Angles** The ratio of the measures of two supplementary angles is 4 to 1. Find the measure of each angle.

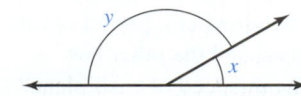

67. Investment Ratio An investor invests $12,000 in a combination of secure funds and high-risk funds. The ratio of dollars invested in secure funds to dollars invested in high-risk funds is 7 to 3. Find the amount invested in secure funds.

68. Investment Ratio By charter, a mutual fund must invest its funds in bonds and stocks in a 5 to 3 ratio, respectively. If the mutual fund has $16,000,000, how much is invested in bonds?

69. Ratio of Assets to Liabilities The assets of a small automobile dealership are approximated by $500x + 10,000$, and its liabilities are approximated by $100x + 14,000$, where x represents the number of vehicles sold. One measure of the strength of this business is the ratio of its assets to its liabilities. For the month of January the ratio was 5 to 4. How many vehicles were sold in January?

70. Ratio of Assets to Liabilities The assets of a heating and air-conditioning business are approximated by $300x + 5,000$, and its liabilities are approximated by $200x + 6,000$, where x represents the number of furnaces sold. One measure of the strength of this business is the ratio of its assets to its liabilities. For the month of February, the ratio was $\frac{7}{6}$. How many furnaces were sold in February?

71. Boats in a Flowing River Two boats that travel the same speed in still water depart simultaneously from a dock, traveling in opposite directions on a river that has a current of 6 km/h. After a period of time one boat is 51 km downstream, and the other boat is 15 km upstream. What is the speed of each boat in still water?

72. Boats in a Flowing River Two boats that travel the same speed in still water depart simultaneously from a dock, traveling in opposite directions on a river that has a current of 7 km/h. After a period of time one boat is 30 km downstream and the other boat is 9 km upstream. What is the speed of each boat in still water?

73. Airplanes Traveling in an Airstream Two planes departed from the same airport at the same time, flying in opposite directions. After a period of time, the slower plane has traveled 1,080 mi and the faster plane has traveled 1,170 mi. The faster plane is traveling 40 mi/h faster than the slower plane. Determine the rate of each plane. (*Hint:* What is the same for each plane?)

74. Airplanes Traveling in an Airstream Two planes traveling at the same airspeed depart simultaneously from an airport, flying in opposite directions. One plane flies directly into a 40-mi/h wind, and the other flies in the same direction as this wind. After a period of time, one plane has traveled 540 mi and the other has traveled 660 mi. What is the airspeed of each plane?

75. Time for Pipes to Fill a Tank One pipe can fill a tank in 15 hours, and a larger pipe can fill the same tank in

10 hours. If both pipes are used simultaneously, how many hours will it take to fill this tank?

76. Time for Assembly Line to Fill an Order Working alone, one assembly line can complete an order in 36 hours. A newer type of assembly line in this same factory can complete the same order in 18 hours. If both assembly lines are used simultaneously, how many hours will it take to complete this order?

77. Time for Pipes to Fill a Tank Working together, two pipes can fill one tank in 4 hours. Working alone, the smaller pipe would take 6 hours longer than the larger pipe to fill the tank. How many hours would it take the larger pipe to fill the tank, working alone?

78. Time for Assembly Line to Fill an Order Working together, two assembly lines can complete an order in 18 hours. Working alone, the older assembly line would take 15 hours longer than the newer assembly line to complete the order. How many hours would it take the newer assembly line to complete the order, working alone?

79. Search Time for Planes Search plane A can search an area for a crash victim in 50 hours. Planes A and B can jointly search the area in 30 hours. How many hours would it take plane B to search the area alone?

80. Generating a Mailing List Printer A takes twice as long as printer B to generate a mailing list. If a particular list can be generated in 9 hours when both printers are used, how many hours would it take each printer to generate the list, working separately?

81. Interest Rates on Two Accounts A 1-year investment will earn $120 in interest in a savings account or $200 in interest in a money market account that pays 2% more than the savings account. Determine each interest rate.

82. Interest Rates on Two Bonds Two bonds pay yearly interest. The longer-term bond pays 1.5% more than the shorter-term bond. The yearly interest on an investment will be $275 if invested in the shorter-term bond or $350 if invested in the longer-term bond. Determine each interest rate.

83. Comparison of Mortgage Rates A prospective homeowner is looking for a mortgage for a new house. For a 30-year mortgage, the interest for the first month will be $850. For a 15-year mortgage, the interest for the first month will be $700. The rate is 1.5% less for the 15-year mortgage than for the 30-year mortgage. Determine each interest rate.

84. Comparison of Mortgage Rates A prospective homeowner is looking for a mortgage for a new house. For a 30-year mortgage, the interest for the first month will be $720. For a 15-year mortgage, the interest for the first month will be $630. The rate is 1% less for the 15-year mortgage than for the 30-year mortgage. Determine each interest rate.

Group discussion questions

85. Communicating Mathematically For each of the following pairs of variables, determine whether it is more reasonable that these variables vary directly or vary inversely.

a. The volume of paint to be painted on a wall and the time to brush this paint on the wall

b. The area covered by a bucket of paint and the thickness with which the paint is applied

c. The resistance of a pipe to the flow of water and the square of the radius of the pipe

d. The volume of water flowing through a pipe and the square of the radius of the pipe

86. Challenge Question

a. Write a numerical word problem that is modeled by $\dfrac{1}{n} + \dfrac{1}{n+2} = \dfrac{12}{35}$.

b. Write a word problem involving the work done by two farm tractors that is modeled by the equation $\dfrac{4}{t} + \dfrac{4}{t+6} = 1$.

c. Write a word problem involving two investments that is modeled by the equation $\dfrac{600}{r} = \dfrac{400}{r - 0.02}$.

d. Write a word problem involving the distance traveled by two planes that is modeled by the equation $\dfrac{600}{r+50} = \dfrac{400}{r-50}$.

87. Discovery Question

a. Describe two variables that you believe should vary directly.

b. Collect 10 pairs of data involving these variables and construct a scatter diagram for these points.

c. Do these data points support your claim?

88. Discovery Question

a. Describe two variables that you believe should vary inversely.

b. Collect 10 pairs of data involving these variables and construct a scatter diagram for these points.

c. Do these data points support your claim?

9.6   Cumulative Review

In Exercises 1–3, select the choice that best describes each linear equation.

1. $x + 2 = x + 3$ **A.** A conditional equation

2. $x + 3 = x + 3$ **B.** An unconditional equation

3. $x + 3 = 2x + 3$ **C.** A contradiction

4. Amount of Acid in a Solution A chemist needs to create a 20% acid solution. She has 5 L of a 25% acid solution in a 10-L container. How much water should she mix with this solution to decrease the concentration to the desired 20%?

5. Rates of Two Trains Two trains depart simultaneously from a station, traveling in opposite directions. One averages 8 km/h less than the other. After $\dfrac{1}{2}$ hour they are 99 km apart. Determine the speed of each train.

Chapter 9   Key Concepts

1. Rational Expression A rational expression is the ratio of two polynomials. Values of the variables that would cause division by 0 must be excluded.

2. Reducing Rational Expressions If A, B, and C are real algebraic expressions and $B \neq 0$ and $C \neq 0$, then

$\dfrac{AC}{BC} = \dfrac{A}{B}$. Note $\dfrac{C}{C} = 1$, the multiplicative identity. A rational expression is in lowest terms when the numerator and the denominator have no common factor other than -1 or 1.

3. Least Common Denominator (LCD) The least common denominator of two or more fractions is the product formed by using each factor the greatest number of times it occurs in any of the denominators.

4. Operations with Rational Expressions If A, B, C, and D are real polynomials, then:

- $\dfrac{A}{C} + \dfrac{B}{C} = \dfrac{A + B}{C}$ for $C \neq 0$

- $\dfrac{A}{C} - \dfrac{B}{C} = \dfrac{A - B}{C}$ for $C \neq 0$

- $\dfrac{A}{B} \cdot \dfrac{C}{D} = \dfrac{AC}{BD}$ for $B \neq 0$ and $D \neq 0$

- $\dfrac{A}{B} \div \dfrac{C}{D} = \dfrac{A}{B} \cdot \dfrac{D}{C} = \dfrac{AD}{BC}$ for $B \neq 0$, $C \neq 0$, and $D \neq 0$

5. Signs of Rational Expressions If A and B are rational expressions, then:

- $-\dfrac{A}{B} = \dfrac{-A}{B} = \dfrac{A}{-B}$ for $B \neq 0$

- $\dfrac{1}{B - A} = -\dfrac{1}{A - B}$ for $A \neq B$

- $\dfrac{A - B}{B - A} = -1$ for $A \neq B$

6. Order of Operations

Step 1. Start with the expression within the innermost pair of grouping symbols.

Step 2. Perform all exponentiations.

Step 3. Perform all multiplications and divisions as they appear from left to right.

Step 4. Perform all additions and subtractions as they appear from left to right.

7. Complex Rational Expression A complex rational expression is a rational expression whose numerator or denominator (or both) is also a rational expression.

8. Simplifying Complex Fractions

- Complex fractions can be simplified by rewriting the complex fraction as a division problem. Divide the main numerator by the main denominator.

- Complex fractions can be simplified by multiplying both the numerator and the denominator by the LCD of all the fractions that occur in the numerator and denominator of the complex fraction.

9. Rational Function A rational function f is defined by the ratio of polynomials: $f(x) = \dfrac{P(x)}{Q(x)}$, where $P(x)$ and $Q(x)$ are polynomials and $Q(x) \neq 0$.

10. Domain of a Rational Function The domain of a rational function consists of all real numbers except those that are excluded to prevent division by 0.

11. Notation for Excluded Values If 2 is the only value excluded from the domain of a function, we can write the domain by using interval notation as $(-\infty, 2) \cup (2, \infty)$. Another notation to represent this domain is $\mathbb{R} \sim \{2\}$, which is read "the set of all real numbers except for 2."

12. Linear Asymptotes

- **Vertical asymptote** A vertical line that a graph approaches but does not touch is called a vertical asymptote.

- **Horizontal asymptote** A horizontal line that a graph approaches is called a horizontal asymptote.

13. Vertical Asymptote of a Rational Function The graph of a rational function will have a vertical asymptote of $x = c$ if $x - c$ is a factor of the denominator of the reduced form of this function.

14. Variation

- **Direct variation** If x and y are variables and $k \neq 0$ is a constant, then stating "y varies directly as x with constant of variation k" means $y = kx$.

- **Inverse variation** If x and y are variables and $k \neq 0$ is a constant, then stating "y varies inversely as x with constant of variation k" means $y = \dfrac{k}{x}$ for $x \neq 0$.

- **Joint variation** If x, y, and z are variables and $k \neq 0$ is a constant, then stating "z varies jointly as x and y with constant of variation k" means $z = kxy$.

15. Extraneous Value An extraneous value is a value produced in the solution process that does not check in the original equation.

16. Solving an Equation Containing Rational Expressions

Step 1. Multiply both sides of the equation by the LCD.

Step 2. Solve the resulting equation.

Step 3. Check the solution to determine whether it is an excluded value and therefore extraneous.

17. Strategy for Solving Word Problems

Step 1. Read the problem carefully to determine what you are being asked to find.

Step 2. Select a variable to represent each unknown quantity. Specify precisely what each variable represents and note any restrictions on each variable.

Step 3. If necessary, make a sketch and translate the problem into a word equation or a system of word equations. Then translate each word equation into an algebraic equation.

Step 4. Solve the equation or system of equations, and answer the question completely in the form of a sentence.

Step 5. Check the reasonableness of your answer.

18. Rate of Work If one job can be done in a time t, then the rate of work is $\dfrac{1}{t}$ of the job per unit of time.

Domain, Excluded Values, and Vertical Asymptotes

In Exercises 1–4, evaluate each expression for

$$f(x) = \frac{2x - 9}{x^2 - 1}.$$

1. $f(0)$ **2.** $f(1)$ **3.** $f(4)$ **4.** $f(10)$

5. Use the table below to determine the domain of

$$f(x) = \frac{3x + 1}{x^2 - x - 6}.$$

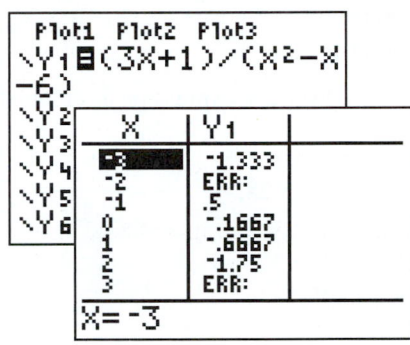

6. Use the graph below to determine the domain of

$$f(x) = \frac{x + 3}{x^2 + 2x - 15}.$$

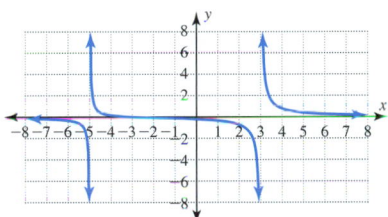

In Exercises 7 and 8, determine the domain of each function.

7. $f(x) = \dfrac{(5x + 1)(x - 3)}{(x + 5)(3x - 1)}$ **8.** $f(x) = \dfrac{x^2 - 11}{x^2 + 13}$

In Exercises 9–11, determine the vertical asymptotes of the graph of each rational function.

9. $f(x) = \dfrac{7x - 9}{4x - 2}$ **10.** $f(x) = \dfrac{5x^2 - 7x + 11}{x^2 - 36}$

11. $f(x) = \dfrac{6x^2 - 23x + 21}{x^2 - 9x}$

Reducing Rational Expressions

In Exercises 12–20, reduce each expression to lowest terms.

12. $\dfrac{36x^2 y}{12xy^3}$ **13.** $\dfrac{6a - 18b}{12b - 4a}$

14. $\dfrac{15x^2 - 15}{25x + 25}$ **15.** $\dfrac{x^2 + x - 30}{2x^2 + 11x - 6}$

16. $\dfrac{cx - cy}{ax - ay + bx - by}$ **17.** $\dfrac{3 - 7m}{14m^2 - 6m}$

18. $\dfrac{10x^2 + 29xy + 10y^2}{6x^2 + 13xy - 5y^2}$ **19.** $\dfrac{9x^2 - 24xy + 16y^2}{12x^2 - 25xy + 12y^2}$

20. $\dfrac{x^2 - 4y^2}{x^2 - 4xy + 4y^2}$

In Exercises 21 and 22, find the missing numerators.

21. $\dfrac{2x + 1}{x - 3} = \dfrac{?}{x^2 - 6x + 9}$ **22.** $\dfrac{y - 3}{y + 4} = \dfrac{?}{2y^2 + 9y + 4}$

Comparing Expressions Using Tables

In Exercises 23–26, use the table of values for each expression to match each expression with its result.

23. $\dfrac{2x - 1}{x - 2} + \dfrac{x - 2}{2x - 1}$

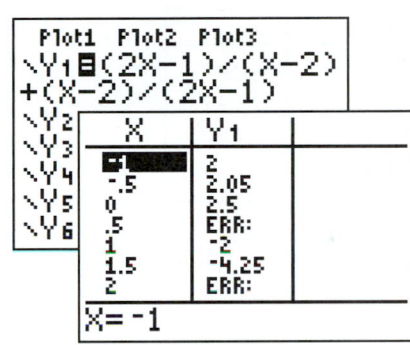

24. $\dfrac{2x - 1}{x - 2} - \dfrac{x - 2}{2x - 1}$

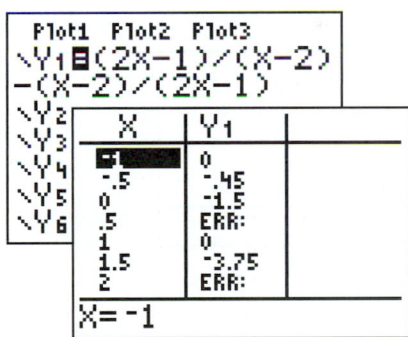

25. $\dfrac{2x - 1}{x - 2} \cdot \dfrac{x - 2}{2x - 1}$

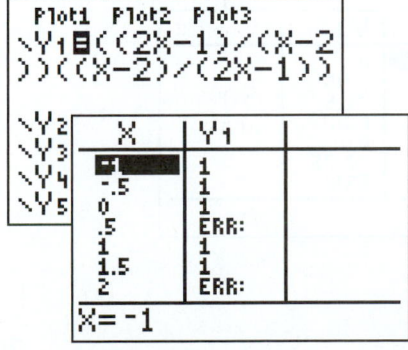

26. $\dfrac{2x - 1}{x - 2} \div \dfrac{x - 2}{2x - 1}$

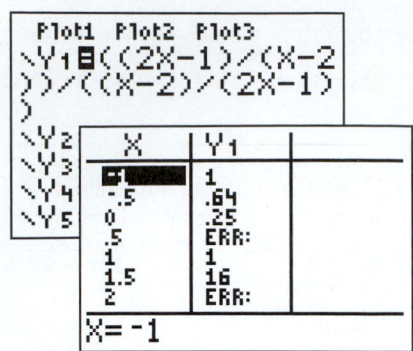

A. 1

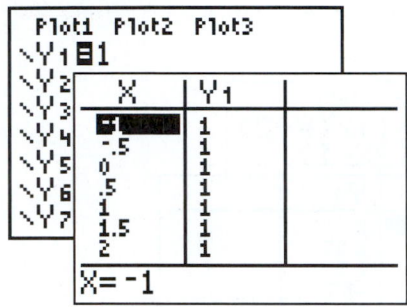

B. $\dfrac{4x^2 - 4x + 1}{(x - 2)^2}$

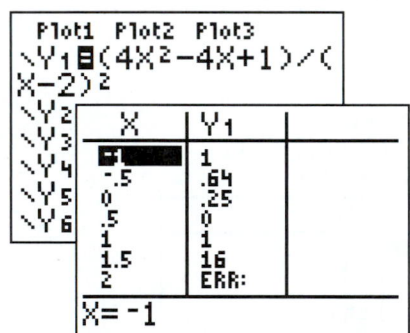

C. $\dfrac{3x^2 - 3}{(x - 2)(2x - 1)}$

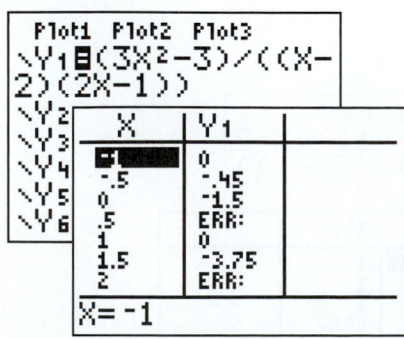

D. $\dfrac{5x^2 - 8x + 5}{(x - 2)(2x - 1)}$

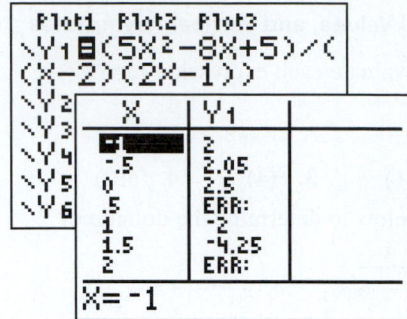

Simplifying Expressions

In Exercises 27–48, simplify each expression as completely as possible.

27. $\dfrac{36x^2 - 24x}{6x}$

28. $\dfrac{6xy}{2x - y} \cdot \dfrac{4x^2 - y^2}{3x^2}$

29. $\dfrac{9t^2 - 4}{16st} \div \dfrac{3t + 2}{16st^2 + 8st}$

30. $\dfrac{v^2 + 9vw + 8w^2}{v^2 - w^2} \div \dfrac{v^2 + 7vw - 8w^2}{v^2 + 5vw - 6w^2}$

31. $\dfrac{3x}{6x^2 + x - 1} - \dfrac{1}{6x^2 + x - 1}$

32. $\dfrac{1}{w + 1} - \dfrac{w}{w - 2} + \dfrac{w^2 + 2}{w^2 - w - 2}$

33. $\dfrac{3v}{3v^2 - 5v + 2} - \dfrac{2v}{2v^2 - v - 1}$

34. $\dfrac{6x}{2x - 3} - \dfrac{9x + 18}{4x^2 - 9} \cdot \dfrac{2x^2 - x - 6}{x^2 - 4}$

35. $\dfrac{2w + 4}{w^2 + 4w - 12} + \dfrac{w^2 - 169}{2w^2 - 13w + 21} \div \dfrac{w^2 - 15w + 26}{2w - 7}$

36. $\dfrac{6v^2 - 25v + 4}{6v^2 + 5v - 6} \div \dfrac{2v^2 - 3v - 20}{4v^2 + 16v + 15}$

37. $\left(\dfrac{y}{4} - \dfrac{4}{y}\right)\left(y - \dfrac{y^2}{y + 4}\right)$

38. $\dfrac{5z + 5}{6z^2 + 13z + 6} - \dfrac{1 - 4z}{3z^2 - 7z - 6} - \dfrac{3z}{2z^2 - 3z - 9}$

39. $\dfrac{\dfrac{1}{a} + \dfrac{1}{a + 1}}{\dfrac{1}{a} + \dfrac{1}{a^2}}$

40. $\dfrac{x + \dfrac{44}{x + 5} - 10}{x + \dfrac{33}{x + 5} - 9}$

41. $\left(\dfrac{3v^2 + 11v + 6}{3v^2 + 5v + 2}\right)^2 \div \dfrac{v^2 - 9}{v^2 + 2v + 1}$

42. $\dfrac{x^{-2} - x^{-1}}{x^{-2} + x^{-1}}$

43. $\dfrac{x + y^{-1}}{x - y^{-1}}$

44. $\dfrac{36w^{-2} + 23w^{-1} - 8}{24 - 5w^{-1} - 36w^{-2}}$

45. $\left(m - \dfrac{15}{m + 2}\right) \div \left(m - 1 - \dfrac{10}{m + 2}\right)$

46. $\dfrac{6y}{3y + 2} + \dfrac{20y + 4}{15y^2 + 7y - 2} \div \dfrac{5y^2 - 24y - 5}{5y^2 - 26y + 5}$

47. $\dfrac{3y}{25y^2 - 1} + \dfrac{2y - 1}{25y^2 - 1}$

48. $\dfrac{z^2 - 2z + 1}{z^5 - z^4} \cdot \dfrac{2z^4}{z^2 - 1} + \dfrac{2z^2 + 2z}{z^2 + 2z + 1}$

Simplify Versus Solve

In Exercises 49–52, simplify the expression in the left column, and solve the equation in the right column.

Simplify	**Solve**
49. a. $\dfrac{1}{m + 2} - \dfrac{3}{m + 4}$	**b.** $\dfrac{1}{m + 2} = \dfrac{3}{m + 4}$
50. a. $\dfrac{x - 2}{x + 1} - \dfrac{x - 4}{x - 6}$	**b.** $\dfrac{x - 2}{x + 1} = \dfrac{x - 4}{x - 6}$
51. a. $\dfrac{x}{x - 3} + \dfrac{x + 4}{x - 2}$	**b.** $\dfrac{x}{x - 3} = \dfrac{x + 4}{x - 2}$
52. a. $\dfrac{2x}{x^2 - 1} + \dfrac{2}{x^2 - 1}$	**b.** $\dfrac{2x}{x^2 - 1} = \dfrac{2}{x^2 - 1}$

Solving Rational Equations

In Exercises 53–60, solve each equation.

53. $\dfrac{7}{2x} = \dfrac{2}{x} - \dfrac{3}{2}$

54. $\dfrac{15}{w^2 + 5w} + \dfrac{w + 4}{w + 5} = \dfrac{w + 3}{w}$

55. $\dfrac{1}{y^2 + 5y + 6} - \dfrac{2}{y + 3} = \dfrac{7}{y + 2}$

56. $1 - \dfrac{14}{y^2 + 4y + 4} = \dfrac{7y}{y^2 + 4y + 4}$

57. $\dfrac{w}{w^2 - 9w + 20} + \dfrac{14}{w^2 - 3w - 4} = \dfrac{18}{w^2 - 4w - 5}$

58. $\dfrac{4}{v^2 + 7v + 10} - \dfrac{3}{v + 2} + \dfrac{8}{v + 5} = 0$

59. $\dfrac{5z + 11}{2z^2 + 7z - 4} = \dfrac{1}{z + 4} - \dfrac{3}{1 - 2z}$

60. $\dfrac{1}{2z^2 - 9z - 5} - \dfrac{1}{2z^2 - 6z - 20} =$

$\dfrac{3}{(2z + 1)(2z + 4)(z - 5)}$

61. Solve $\dfrac{a}{b + 1} = \dfrac{a + 1}{b}$ for b.

62. Solve $I = \dfrac{E}{r_1 + r_2}$ for r_2.

Estimate Then Calculate

In Exercises 63 and 64, mentally estimate the value of each expression for $f(x) = \dfrac{(x + 5)(x - 8)}{x^2 - 4}$ and then use a calculator or a spreadsheet to approximate each value to the nearest hundredth.

Problem	Mental Estimate	Approximation
63. $f(1.01)$		
64. $f(-4.99)$		

Connecting Algebra to Geometry

65. Perimeter and Area of a Trapezoid
   a. Determine the perimeter of the trapezoid shown in the figure.
   b. Determine the area of this trapezoid by using the formula $A = \dfrac{h}{2}(a + b)$.

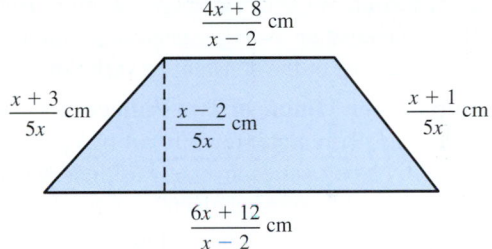

66. Area of a Region Determine the area of the region shown in the figure.

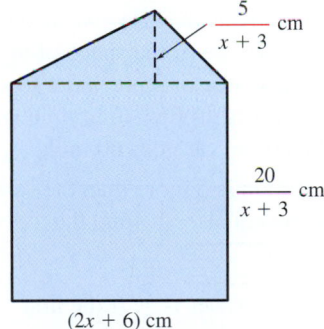

67. Volume of a Box Determine the volume of the given box.

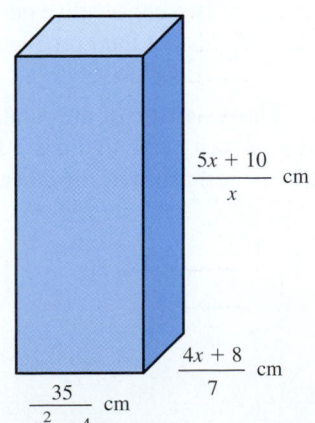

Connecting Concepts to Applications

68. Average Cost The number of units that a plant can produce by operating t h/day is given by $N(t) = 0.5t^2 + 10t$. The cost in dollars of operating the plant for t hours is given by $C(t) = 100t + 500$.

a. What is the practical domain for the values of t?

b. Determine $A(t)$, the average cost per unit when the plant is operated for t h/day.

c. Evaluate and interpret $N(10)$.

d. Evaluate and interpret $C(10)$.

e. Evaluate and interpret $A(10)$.

f. Describe what happens to the average cost when the number of hours the plant operates approaches 0 hours.

g. Describe what happens to the average cost when the plant approaches operating around the clock.

69. Travel Time The amount of time required for a bus to travel from St. Louis, Missouri, to Chicago, Illinois, varies inversely with the average speed of the bus. The trip takes 5 hours if the bus averages 60 mi/h. How long will the trip take at an average speed of 50 mi/h? What is the significance of the constant of variation k?

70. Revolutions per Minute of Two Pulleys The number of revolutions per minute (rev/min) of two pulleys connected by a belt varies inversely as their diameters. A pump is belt-driven by an electric motor that runs at 2,000 rev/min. If the pump has a pulley of 6-in diameter and must operate at 1,500 rev/min, what size pulley is required on the motor?

71. Force of the Wind The force of the wind blowing into the front of a car varies jointly as the effective area of the front of the car and the square of the speed of the wind. A car with an effective area of 25 ft^2 has a force of 80 lb exerted when the car is driving 40 mi/h in still air. Find the force exerted if the car goes 60 mi/h.

72. The denominator of a fraction is 6 more than the numerator, and the fraction equals -1. Find the numerator.

73. The ratio of one number to another is $\dfrac{5}{2}$. If the first number is 5 less than 3 times the second, find both numbers.

74. The sum of the reciprocals of two consecutive odd integers is 12 times the reciprocal of their product. Find these integers.

75. Electrical Resistance The resistance of the parallel circuit with two resistors shown in the figure is 3 Ω. If one resistor has 3 times the resistance of the other, what is the resistance of each?

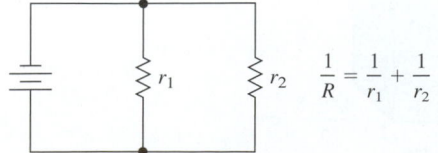

76. Filling a Bathtub A hot-water faucet takes 7 minutes longer than the cold-water faucet to fill a bathtub. If both faucets were turned on, the tub would be half full in 6 minutes. How many minutes would it take to fill an empty tub with cold water if the hot-water faucet were turned off?

77. Average Cost A furniture company can produce x units of a desk for $54 per desk, plus an initial start-up investment of $20,000. To compete in the market, it must be able to achieve an average cost of only $74 per unit. How many units must the company produce to reach an average cost of $74 per unit?

78. Dimensions of a Metal Sheet A rectangular piece of metal is 6 cm longer than it is wide. A 0.5-cm strip is cut off of each side, leaving an area thirteen-sixteenths of the original area. Find the original dimensions.

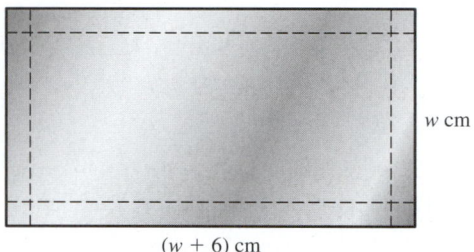

$(w + 6)$ cm

79. Speed of Boats Two boats that travel the same speed in still water depart from a dock at the same time. The boats travel in opposite directions in a river that has a current of 6 mi/h. After a period of time, one boat is 54 mi downstream, and the other boat is 30 mi upstream. What is the speed of each boat in still water?

80. Selecting the Appropriate Type of Variation For each of the following pairs of variables, determine whether it is more reasonable that these variables vary directly or inversely.

a. The distance a car must travel between two cities and the distance between these two cities on a map

b. The time interval on a train schedule and the rate of travel of the train between two stations

c. A weather balloon filled with helium is pulled under water by a small rope. The number of feet the balloon is submerged and the volume of the balloon

d. A weather balloon filled with helium is pulled under water by a small rope. The number of feet the balloon is submerged and the pressure on the helium in the balloon

Chapter 9 | Mastery Test

Objective 9.1.1 Identify Rational Functions

1. Match each function with the most appropriate description.

   a. $f(x) = -x^2 + 4x + 5$ **A.** A rational function whose graph is a line

   b. $f(x) = \dfrac{5}{x-1}$ **B.** A rational function whose graph is a parabola

   c. $f(x) = |3x + 2| - 1$ **C.** A rational function whose graph will have a break

   d. $f(x) = -\dfrac{3}{4}x - \dfrac{2}{3}$ **D.** Not a rational function; an absolute value function whose graph has a V-shape

Objective 9.1.2 Determine the Domain and Identify the Vertical Asymptotes of a Rational Function

2. Determine the domain of each rational function.

   a. $f(x) = \dfrac{x-5}{2x-12}$ **b.** $f(x) = \dfrac{x+5}{(x-1)(x+6)}$

   c. $f(x) = \dfrac{x+7}{x^2-64}$ **d.** $f(x) = \dfrac{2x+19}{x^2+1}$

Objective 9.1.3 Reduce a Rational Expression to Lowest Terms

3. Reduce each rational expression to lowest terms.

   a. $\dfrac{6x}{12x^2 - 18x}$ **b.** $\dfrac{2x-6}{x^2-9}$

   c. $\dfrac{2x^2 - 7x - 15}{x^2 - 25}$ **d.** $\dfrac{4x^2 - 12xy + 9y^2}{3ay - 6by - 2ax + 4bx}$

Objective 9.2.1 Multiply and Divide Rational Expressions

4. Perform each multiplication or division and reduce the result to lowest terms.

   a. $\dfrac{ax - bx}{x^2} \cdot \dfrac{5x}{4a - 4b}$ **b.** $\dfrac{x^2 - 1}{x + 1} \div \dfrac{x^2 - 3x + 2}{x - 2}$

   c. $\dfrac{v^2 - v - 20}{v^2 - 16} \div \dfrac{3v - 15}{2v - 8}$

   d. $\dfrac{x^2 - 4}{x^2 - 4x - 21} \cdot \dfrac{x^2 - 2x - 35}{x^2 - 7x + 10}$

Objective 9.3.1 Add and Subtract Rational Expressions

5. Perform each addition or subtraction and reduce the result to lowest terms.

   a. $\dfrac{3x - 4}{2x - 3} + \dfrac{x - 2}{2x - 3}$

   b. $\dfrac{3x + 7}{x^2 + x - 12} - \dfrac{2x + 3}{x^2 + x - 12}$

   c. $\dfrac{7}{w^2 + w - 12} + \dfrac{2}{w^2 - 8w + 15}$

   d. $\dfrac{1}{x + y} + \dfrac{3}{x - y} + \dfrac{2y}{x^2 - y^2}$

Objective 9.4.1 Simplify Rational Expressions in Which the Order of Operations Must Be Determined

6. Simplify each expression.

   a. $\dfrac{4x}{5} + \dfrac{x^2}{15} \cdot \dfrac{3}{x}$ **b.** $\left(\dfrac{4x}{5} + \dfrac{x^2}{15}\right) \cdot \dfrac{3}{x}$

   c. $\dfrac{x - y}{x + y} + \dfrac{3x - 21y}{x^2 - y^2} \div \dfrac{15x^2 - 105xy}{10x^2 y - 10xy^2}$

   d. $\left(\dfrac{w}{w - 2} + \dfrac{1}{w - 3}\right) \cdot \dfrac{w^2 - w - 2}{w^2 - 2w - 2}$

Objective 9.4.2 Simplify Complex Fractions

7. Simplify each expression.

   a. $\dfrac{\dfrac{4}{5}}{\dfrac{5}{6}}$ **b.** $\dfrac{\dfrac{x}{3} - 2 + \dfrac{3}{x}}{1 - \dfrac{3}{x}}$

   c. $\dfrac{36x^{-2} - 3x^{-1} - 18}{12x^{-2} - 25x^{-1} + 12}$ **d.** $\dfrac{\dfrac{v}{v - 4} - \dfrac{2v}{v + 3}}{\dfrac{v^2 - 4v}{2v + 6}}$

Objective 9.5.1 Solve Equations Containing Rational Expressions

8. Solve each of the following equations.

   a. $\dfrac{z - 4}{z - 2} = \dfrac{1}{z - 2}$ **b.** $\dfrac{z - 2}{z - 1} + \dfrac{z - 3}{2z - 5} = 1$

   c. $\dfrac{z + 11}{z^2 - 5z + 4} = \dfrac{5}{z - 4} + \dfrac{3}{1 - z}$

   d. $\dfrac{x}{x - 2} + \dfrac{1}{x - 4} = \dfrac{2}{x^2 - 6x + 8}$

Objective 9.5.2 Solve a Rational Equation for a Specified Variable

9. Solve each equation for the specified variable.

   a. Solve $\dfrac{2A}{h} = b$ for h.   **b.** Solve $\dfrac{1}{x} - \dfrac{1}{y} = \dfrac{1}{z}$ for z.

   c. Solve $\dfrac{x + 4}{x - 5} = y$ for x.   **d.** Solve $\dfrac{ac}{a - c} = b$ for a.

Objective 9.6.1 Distinguish Between Direct and Inverse Variation

10. Determine whether the relationship illustrated in the table is direct variation or inverse variation.

a. x	y
1	30
2	15
3	10
4	7.5
5	6
6	5

b. x	y
1	5
2	10
3	15
4	20
5	25
6	30

Objective 9.6.2 Translate Statements of Variation

11. Translate each statement of variation into an equation.
    a. *a* varies directly as *b*.
    b. *a* varies inversely as *b*.
    c. *a* varies directly as *b* and inversely as the square of *c*.
    d. *a* varies jointly as *b* and *c*.

Objective 9.6.3 Solve Problems Involving Variation

12. a. If *y* varies inversely as *x*, and *y* is 12 when *x* is 6, find *y* when *x* is 4.
    b. The volume of a gas varies inversely with the pressure when temperature is held constant. If the pressure on 4 L of a gas is 6 newtons per cm^2, what will the pressure be if the volume is 3 L?

Objective 9.6.4 Solve Applied Problems That Yield Equations with Fractions

13. Solve each of the following problems.
    a. **Ratio of Economic Indicators** The ratio of two monthly economic indicators is $\frac{2}{3}$. The first monthly indicator is 2 units below normal, and the second monthly indicator is 2 units above normal. What is the normal number of units for this indicator?

    b. **Work by Two Machines** When members of a construction crew use two end loaders at once, they can move a pile of sand in 6 hours. If they use only the larger end loader, they can do the job in 5 hours less time than it would take if they used the smaller machine. How many hours would it take to do the job using only the larger machine?

    c. **Interest Rates on Two Bonds** Two bonds pay yearly interest. The longer-term bond pays 2.5% more than the shorter-term bond. The yearly interest on an investment would be $680 if invested in the first bond or $480 if invested in the second bond. Determine each interest rate.

    d. **Airplanes Traveling in an Airstream** Two planes departed from the same airport at the same time, flying in opposite directions. After a period of time, the slower plane has traveled 950 miles and the faster plane has traveled 1,050 miles. The faster plane is traveling 50 mi/h faster than the slower plane. Determine the rate of each plane. (*Hint:* What is the same for each plane?)

Chapter 9 | Group Project

An Algebraic Model for Average Cost

A factory that manufactures custom steering wheels collected data over several weeks. For selected values, the table gives *t*, the number of hours the factory operated on a given day. The table also displays *N(t)*, the number of steering wheels produced on this day, and *C(t)*, the cost in dollars of operating the factory that day.

1. Use the **LinReg** feature on a graphing calculator to determine the equation of a linear function for *N(t)*. Round the coefficients to three significant digits.

2. Use the **LinReg** feature on a graphing calculator to determine the equation of a linear function for *C(t)*. Round the coefficients to three significant digits.

3. Use the functions from parts 1 and 2 to write an equation for the average cost per steering wheel as a function of the number of hours per day the factory operates: $A(t) = \dfrac{C(t)}{N(t)}$.

t	N(t)	C(t)
2	20	3,000
4	60	5,000
6	120	8,000
8	180	10,000
12	300	15,000
16	440	20,000
20	600	25,000
24	700	30,000

4. Assuming that the factory always operates at least two hours per day, what is the practical domain for this function?

5. Use this function to approximate and interpret *A*(10).

6. Describe what happens to the average cost when the plant approaches operating around the clock.

Square Root and Cube Root Functions and Rational Exponents

Chapter Outline

Geosynchronous Satellite

The weather information, global positioning system (GPS) mapping information, and the telecommunications that we take for granted rely on satellites orbiting the earth. A geosynchronous satellite completes one orbit around the earth each day. Thus these satellites remain above the same point on the equator of the earth. Special care must be taken when a geosynchronous satellite is placed into orbit. The speed and altitude must be just right to ensure that the satellite will remain in the proper position. Exercise 83 in Section 10.5 uses equations with rational exponents to calculate the radius needed for a geosynchronous orbit of the earth.

| Section 10.1 | Evaluating Radical Expressions and Graphing Square Root and Cube Root Functions |

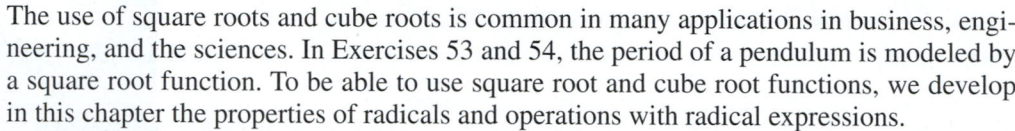

Man's unique worldwide victory of an idea is the universal adoption of the Hindu-Arabic numerals to record numbers. —HOWARD W. EVES (MATHEMATICAL HISTORIAN)

Objectives:

1. Interpret and use radical notation.
2. Graph and analyze square root and cube root functions.

1. Interpret and Use Radical Notation

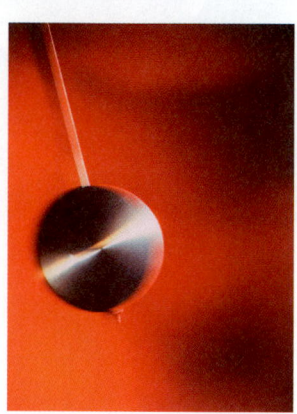

The use of square roots and cube roots is common in many applications in business, engineering, and the sciences. In Exercises 53 and 54, the period of a pendulum is modeled by a square root function. To be able to use square root and cube root functions, we develop in this chapter the properties of radicals and operations with radical expressions.

Recall from Section 1.2 that both -2 and 2 are square roots of 4. However, only the positive value of 2 is called the principal square root of 4. We defined $\sqrt{x}$ to denote the **principal square root of x** for $x \geq 0$. That is, $\sqrt{x}$ denotes the nonnegative real number r so that $r^2 = x$. For example, $\sqrt{4}$ denotes the principal square root of 4, that is, $\sqrt{4} = 2$. The other square root of 4 is denoted by $-\sqrt{4}$, that is, $-\sqrt{4} = -2$. Also recall that in Section 7.5 we defined the imaginary number $i = \sqrt{-1}$. We also simplified $\sqrt{-4}$ as $\sqrt{-4} = i\sqrt{4} = 2i$.

Example 1 reviews the evaluation of some square roots. Notice the subtle differences in notation—subtleties that result in unique values for each expression.

Example 1 Evaluating Square Roots

Evaluate each of the following expressions.

(a) $\sqrt{9}$ **(b)** $-\sqrt{9}$ **(c)** $\sqrt{-9}$
(d) $\sqrt{9 + 16}$ **(e)** $\sqrt{9} + \sqrt{16}$

Solution

(a) $\sqrt{9} = 3$ $\qquad$ $3^2 = 9$; 3 is the positive square root of 9.

(b) $-\sqrt{9} = -3$ $\qquad$ $(-3)^2 = 9$; -3 is the negative square root of 9.

(c) $\sqrt{-9} = i\sqrt{9} = 3i$ $\qquad$ This is an imaginary number, not a real number. Imaginary numbers were covered in Section 7.5.

(d) $\sqrt{9 + 16} = \sqrt{25}$ $\qquad$ Following the correct order of operations, we get different results for
$\qquad\qquad = 5$ $\qquad$ parts (d) and (e).

(e) $\sqrt{9} + \sqrt{16} = 3 + 4$ $\qquad$ In general, $\sqrt{x + y} \neq \sqrt{x} + \sqrt{y}$.
$\qquad\qquad = 7$

Self-Check 1

Evaluate each of the following expressions.

a. $\sqrt{36} + \sqrt{64}$ **b.** $\sqrt{36 + 64}$ **c.** $-\sqrt{36}$ **d.** $\sqrt{-36}$

Although square roots are the most common roots used in applications, you should also be familiar with higher-order roots. The number r is an nth root of x if $r^n = x$.

How Can I Represent the Principal nth Root of a Number?

The **principal nth root of x** is denoted by $\sqrt[n]{x}$. This notation is described in the following box.

Radical Notation

The number r is an nth root of x if $r^n = x$.
$\sqrt[n]{x}$ denotes the principal nth root of x.

- x is the **radicand.**
- $\sqrt{}$ is the **radical symbol,** or the radical.
- The natural number n is the **index** or the **order of the radical.**

A Mathematical Note

The radical symbol $\sqrt{}$ is composed of two parts: $\sqrt{}$ and $\overline{}$. The symbol $\sqrt{}$ comes from the letter r, the first letter of the Latin word *radix*, which means root. Thus the $\sqrt{}$ indicates that a root is to be taken of the quantity underneath the bar (also called the *vinculum*).

On many calculators, we can access the symbol $\sqrt{}$ but not the bar. To indicate the square root of a quantity, we must use parentheses instead of the bar. For example, $\sqrt{2x + 1}$ can be represented on calculators by $\sqrt{}(2x + 1)$.

If no index is written, as in $\sqrt{x}$, then the index is understood to be 2 and this denotes the principal square root of x. Likewise, we use $\sqrt[3]{8}$ to represent the principal cube root of 8. $\sqrt[3]{8} = 2$ because 2 is a positive real number and $2^3 = 8$. This is described further in the following table. Note the importance of distinguishing between even- and odd-order radicals when we examine principal nth roots.

Principal nth Root

The principal nth root of the real number x is denoted by $\sqrt[n]{x}$.

Verbally		Numerical Examples
For $x > 0$:	The principal nth root is positive for all natural numbers n.	$\sqrt{9} = 3$ Check: $3^2 = 9$ $\sqrt[3]{8} = 2$ Check: $2^3 = 8$
For $x = 0$:	The principal nth root of 0 is 0.	$\sqrt{0} = 0$ Check: $0^2 = 0$ $\sqrt[3]{0} = 0$ Check: $0^3 = 0$
For $x < 0$:	If n is odd, the principal nth root is negative.	$\sqrt[3]{-8} = -2$ Check: $(-2)^3 = -8$ $\sqrt[5]{-1} = -1$ Check: $(-1)^5 = -1$
	If n is even, there is no real nth root. The nth roots will be imaginary numbers, as covered in Section 7.5.	$\sqrt{-49} = 7i$ Check: $(7i)^2 = 49i^2$ $\phantom{\sqrt{-49} = 7i\ \ \text{Check: }} = 49(-1)$ $\phantom{\sqrt{-49} = 7i\ \ \text{Check: }} = -49$

When working with square roots, think of perfect squares, such as 1, 4, 9, 16, and 25. When working with cube roots, think of perfect cubes, such as 1, 8, 27, 64, and 125.

Example 2   Interpreting and Using Radical Notation

Interpret and evaluate each expression.

(a) $\sqrt{25}$ (b) $\sqrt[3]{27}$ (c) $\sqrt[3]{-27}$

(d) $\sqrt[4]{16}$ (e) $\sqrt[5]{1}$ (f) $\sqrt{-25}$

Solution

Radical Notation	Verbal Interpretation	
(a) $\sqrt{25} = 5$	The principal square root of 25 is 5.	Check: $5 \cdot 5 = 25$
(b) $\sqrt[3]{27} = 3$	The principal cube root of 27 is 3.	Check: $3 \cdot 3 \cdot 3 = 27$
(c) $\sqrt[3]{-27} = -3$	The principal cube root of -27 is -3.	Check: $(-3)(-3)(-3) = -27$
(d) $\sqrt[4]{16} = 2$	The principal fourth root of 16 is 2.	Check: $(2)(2)(2)(2) = 16$
(e) $\sqrt[5]{1} = 1$	The principal fifth root of 1 is 1.	Check: $(1)(1)(1)(1)(1) = 1$
(f) $\sqrt{-25} = i\sqrt{25}$ $= 5i$	The principal square root of -25 is $5i$.	Check: $(5i)(5i) = 25i^2$ $= 25(-1)$ $= -25$

Self-Check 2

Evaluate each of the following expressions.

a. $\sqrt{81}$ **b.** $\sqrt{-81}$ **c.** $\sqrt[4]{81}$ **d.** $\sqrt[5]{-32}$

The expressions in Example 2 were carefully selected so that they could be easily evaluated by inspection. Technology Perspective 10.1.1 illustrates how to use a calculator to approximate cube roots and fourth roots. We will illustrate how to do this with spreadsheets after we have covered rational exponents in Section 10.5. On the TI-84 Plus calculator, the **MATH** menu contains options for evaluating these radical expressions.

Technology Perspective 10.1.1   **Approximating Radical Expressions**

Evaluate $\sqrt[3]{27}$ and $\sqrt[4]{16}$ from Example 2 using a calculator.

TI-84 Plus Keystrokes

To Evaluate $\sqrt[3]{27}$:

1. Press **MATH** **4** to select the cube root option.

2. Enter 27, close with a right parenthesis, and press **ENTER**.

TI-84 Plus Calculator

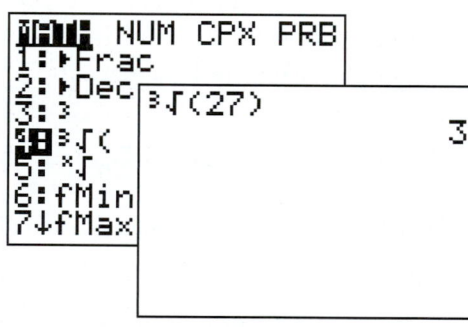

To Evaluate $\sqrt[4]{16}$:

3. First enter the index of 4.

4. Press **MATH** **5** to select the nth root option.

5. Enter a left parenthesis, then 16, close with a right parenthesis, and press **ENTER**.

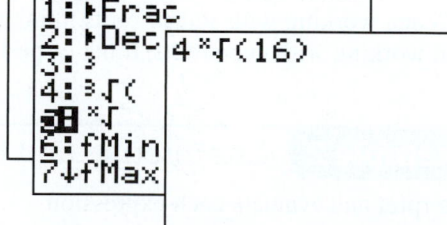

Answer: $\sqrt[3]{27} = 3$ and $\sqrt[4]{16} = 2$.

Example 3 illustrates how to evaluate a radical expression containing variables. The equation $r = \sqrt[3]{\dfrac{3V}{4\pi}}$ is another form of $V = \dfrac{4}{3}\pi r^3$, which gives the volume of a sphere with a radius r.

Sphere: $V = \dfrac{4}{3}\pi r^3$

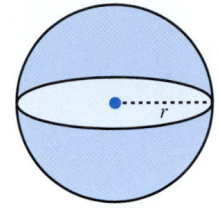

Example 3   Determining the Radius of a Spherical Container

The radius of a sphere with a given volume can be determined by using the formula $r = \sqrt[3]{\dfrac{3V}{4\pi}}$. Use a calculator to approximate to the nearest tenth of a centimeter the radius of a spherical container needed to hold 6,880 cm³ of water.

Solution

$$r = \sqrt[3]{\dfrac{3V}{4\pi}}$$

$$= \sqrt[3]{\dfrac{3(6,880)}{4\pi}}$$

$$\approx 11.8$$

Use the given equation and substitute 6,880 for V. Then use a calculator to approximate r.

Answer: The radius of the sphere should be approximately 11.8 cm.

2. Graph and Analyze Square Root and Cube Root Functions

Each family of functions has its own distinctive graph, its own numerical pattern, and its own algebraic definition. The basic shape of a square root and a cube root function is shown in the figure.

Graphs of a Square Root and a Cube Root Function

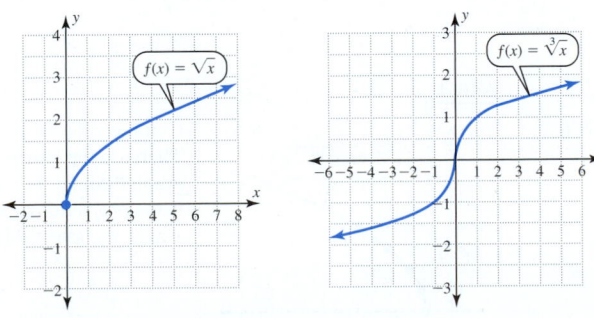

Why Are the Graphs of $f(x) = \sqrt{x}$ and $f(x) = \sqrt[3]{x}$ so Different?

The square root function is defined only for nonnegative values, while the cube root function is defined for all real numbers. We now examine why these domains are different.

In this book, we examine only functions that have real values in both the domain and the range. These functions are called **real-valued functions.** Thus we restrict the input values of square root functions to nonnegative values. This will ensure that we obtain real values as output values in the range. If we did allow negative numbers in the domain of $f(x) = \sqrt{x}$, then the output values would be imaginary numbers. By restricting the domain of $f(x) = \sqrt{x}$ to $[0, \infty)$, we can have this function be a real-valued function.

Square Root Function

Algebraically	Numerically		Graphically	Verbally
$f(x) = \sqrt{x}$	x	$\sqrt{x}$		The domain of $f(x) = \sqrt{x}$ is $[0, \infty)$.
	0	0		
	1	1		
	4	2		
	9	3		
	16	4		
	25	5		

What Is the Shape of a Square Root Function?

If this visual pattern reminds you of the parabolas that we studied in Chapter 8, there is a good reason. The equation $y = \sqrt{x}$ for $x \geq 0$ can also be written as $x = y^2$ for $y \geq 0$. This means that the graph of $y = \sqrt{x}$ is one branch of a parabola opening to the right (instead of upward or downward as in Chapter 8). All square roots share this characteristic shape. Remember when you graph $f(x) = \sqrt{2x - 6}$ in Example 4 that the domain of this function is the projection of the graph onto the x-axis.

Example 4   Graphing a Square Root Function

Use the function $f(x) = \sqrt{2x - 6}$ to complete this table and to sketch the graph of this function.

x	$f(x) = \sqrt{2x - 6}$
3.0	
3.5	
5.0	
7.5	
11.0	

Solution

Table

x	$f(x) = \sqrt{2x - 6}$
3.0	0
3.5	1
5.0	2
7.5	3
11.0	4

Graph

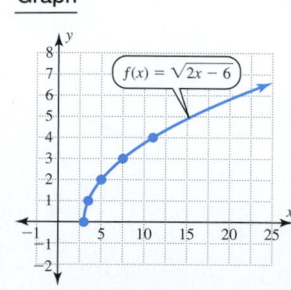

Complete the table and plot these points. Then use the known shape to sketch the graph through these points. The domain is the projection of the graph onto the x-axis, the interval $[3, \infty)$.

Self-Check 4

Sketch the graph of $f(x) = \sqrt{x + 2}$ and determine its domain.

The cube root function $f(x) = \sqrt[3]{x}$ is a real value for every real number x. Thus the domain of this function is $\mathbb{R}$, the set of all real numbers.

Cube Root Function

Algebraically	Numerically		Graphically	Verbally
$f(x) = \sqrt[3]{x}$	**x**	$f(x) = \sqrt[3]{x}$		The domain of $f(x) = \sqrt[3]{x}$ is $\mathbb{R}$.
	-27	-3		
	-8	-2		
	-1	-1		
	0	0		
	1	1		
	8	2		
	27	3		

The table of values in Example 4 contained values that could easily be computed mentally and therefore we did not show the use of a calculator to obtain the points used to sketch the graph. However, in Example 5, the table of values contains points that we would obtain by using a calculator or a spreadsheet. Thus we show a calculator screen for the points that we use to sketch the graph. This table was created using the **ASK** mode for selecting input values for a table, which was presented in Technology Perspective 8.3.1. The calculator was first set to display values rounded to the nearest tenth by using the **MODE** key.

Example 5 Graphing a Cube Root Function

Use the function $f(x) = \sqrt[3]{5x + 30}$ to complete this table and to sketch the graph of this function.

x	$f(x) = \sqrt[3]{5x + 30}$
-10	
-6	
0	
5	
10	

Solution

Table	Graph
Using the **ASK** mode on a calculator, we complete the table.	Plot these points and use the known shape to sketch the graph through these points.

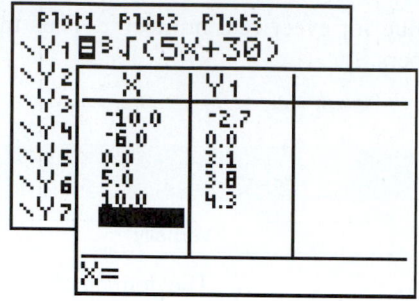
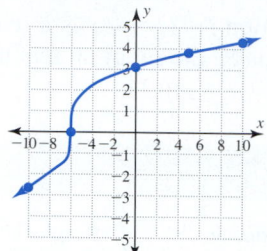

The projection of the graph onto the x-axis is the entire x-axis. The domain of this function is $\mathbb{R}$.

Self-Check 5

Sketch the graph of $f(x) = \sqrt[3]{x} - 8$ and determine its domain.

The graphs of $f(x) = \sqrt{x}$ and $f(x) = \sqrt[3]{x}$ clearly illustrate that these functions have different domains. From Chapter 9, we also know that the domain of a rational function must exclude any value that would cause division by zero. We now summarize how we can determine algebraically the domain of a real-valued function. We will examine this subject again when we examine logarithmic functions in the next chapter.

Domain of $y = f(x)$

The domain of a real-valued function $y = f(x)$ is the set of all real numbers for which $f(x)$ is also a real number. This means the domain excludes all values that

1. Cause division by 0
2. Cause a negative number under a square root symbol

Negative numbers are excluded from the radicand of any even-order radical including square roots, fourth roots, etc.

In Example 6, the domain of each function is determined algebraically. We urge you to compare the domain that is determined algebraically to the domain that can be determined from the graph of the function. Remember, the domain of a graph is its projection onto the x-axis.

Example 6   Determining the Domain of a Function

Determine the domain of each function.

(a) $f(x) = \dfrac{x + 1}{x - 2}$ **(b)** $f(x) = \sqrt{x - 2}$

Solution

Algebraically

(a) To avoid division by 0, we must determine when the denominator is 0.

$$x - 2 = 0$$
$$x = 2$$

The domain is the set of all real numbers except for 2.

$$D = \mathbb{R} \sim \{2\}$$

Graphical Check

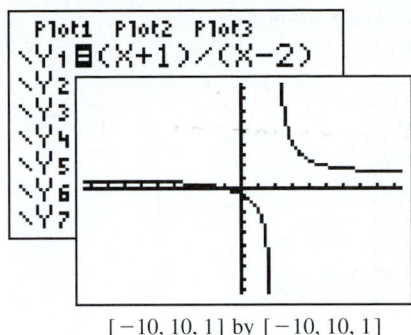

$[-10, 10, 1]$ by $[-10, 10, 1]$

The projection onto the x-axis is the set of all real numbers except for 2.

(b) The expression under a square root symbol must be nonnegative.

$$x - 2 \geq 0$$
$$x \geq 2$$
$$D = [2, \infty)$$

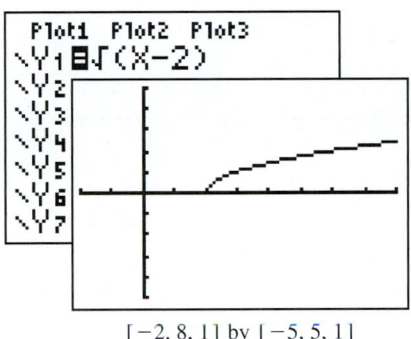

$[-2, 8, 1]$ by $[-5, 5, 1]$

The projection onto the x-axis is the interval starting at 2 and extending to the right.

Self-Check 6

Determine the domain of each of these functions.

a. $f(x) = 2x - 8$

b. $f(x) = \dfrac{1}{2x - 8}$

c. $f(x) = \sqrt{2x - 8}$

d. $f(x) = \sqrt[3]{2x - 8}$

We now summarize some of the different families of functions that we have examined so far in this book. Remember that each family of functions has its own characteristic shape.

Name of Function	Algebraic Example	Graphic Example	Domain of Example
Linear function	$f(x) = x$		$D = \mathbb{R}$
Absolute value function	$f(x) = |x|$		$D = \mathbb{R}$
Quadratic function	$f(x) = x^2$		$D = \mathbb{R}$
Rational function	$f(x) = \dfrac{1}{x}$		$D = \mathbb{R} \sim \{0\}$
Square root function	$f(x) = \sqrt{x}$		$D = [0, \infty)$
Cube root function	$f(x) = \sqrt[3]{x}$		$D = \mathbb{R}$

Self-Check Answers

1. **a.** 14   **b.** 10   **c.** −6   **d.** 6i
2. **a.** 9   **b.** 9i **c.** 3   **d.** −2
3. A pendulum of length 81 cm will have a period of 0.58 second.
4. The domain is $[-2, \infty)$.

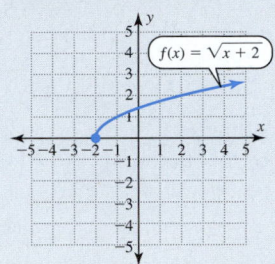

5. The domain is $\mathbb{R}$.

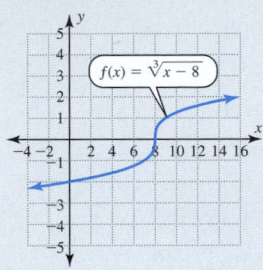

6. **a.** $\mathbb{R}$ **b.** $\mathbb{R} \sim \{4\}$ **c.** $[4, \infty)$ **d.** $\mathbb{R}$

Technology Self-Check Answer

1. **a.** 3.42   **b.** −3.42   **c.** 2.51   **d.** 2.09

10.1   Using the Language and Symbolism of Mathematics

1. In the radical notation $\sqrt[n]{x}$, x is called the _____.
2. In the radical notation $\sqrt[n]{x}$, $\sqrt{}$ is called the _____ symbol.
3. In the radical notation $\sqrt[n]{x}$, n is called the _____ or the _____ of the radical.
4. The principal nth root of x is denoted by _____.
5. $\sqrt[3]{-125} = -5$ is read "the (principal) _____ root of -125 equals -5."
6. $\sqrt[4]{16} = 2$ is read "the (principal) _____ root of 16 equals 2."

7. The principal nth root of x is not a real number if n is _____ and x is _____.
8. A real-valued function $y = f(x)$ has real values of x in the domain and is restricted to _____ numbers in the range.
9. The domain of a real-valued function excludes all values that cause division by _____.
10. The domain of a real-valued function excludes all values that cause the radicand of a square root to be _____.

10.1   Quick Review

1. Simplify $4^2 + (-4)^2$.
2. Simplify $-3^2 + (3i)^2$.
3. Determine the domain of $f(x) = \dfrac{x}{2x - 6}$.

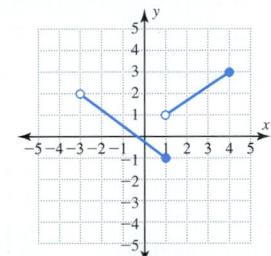

4. Determine the domain of the function with the given graph.
5. Determine the range of the function with the given graph.

10.1   Exercises

Objective 1 Interpret and Use Radical Notation

In Exercises 1–4, represent each verbal expression using radical notation. Assume that all variables are nonnegative.

1. The principal cube root of w

2. The principal fourth root of x
3. The principal seventh root of v
4. The principal sixth root of x

In Exercises 5–8, represent each verbal expression using radical notation and then evaluate this expression.

5. The principal square root of 4

6. The principal cube root of 1,000

7. The principal fifth root of 32

8. The principal square root of 100

In Exercises 9–32, mentally evaluate each radical expression. Use a calculator only to check your answer.

9. a. $\sqrt{16}$ **b.** $\sqrt[4]{16}$

10. a. $\sqrt{64}$ **b.** $\sqrt[3]{64}$

11. a. $\sqrt{36}$ **b.** $-\sqrt{36}$

12. a. $\sqrt{100}$ **b.** $-\sqrt{100}$

13. a. $\sqrt[3]{8}$ **b.** $\sqrt[3]{-8}$

14. a. $\sqrt[3]{125}$ **b.** $\sqrt[3]{-125}$

15. $\sqrt[3]{0}$ **16.** $\sqrt[4]{0}$ **17.** $\sqrt[5]{1}$

18. $\sqrt[4]{1}$ **19.** $\sqrt{10,000}$ **20.** $\sqrt[3]{1,000,000}$

21. $\sqrt[4]{10,000}$ **22.** $\sqrt[6]{1,000,000}$ **23.** $\sqrt[3]{0.125}$

24. $\sqrt[3]{0.064}$ **25.** $\sqrt{\dfrac{81}{36}}$ **26.** $\sqrt{\dfrac{16}{49}}$

27. $\sqrt[3]{-\dfrac{1}{8}}$ **28.** $\sqrt[3]{-\dfrac{27}{1,000}}$

29. $\sqrt{9} + \sqrt{16} + \sqrt{144}$ **30.** $\sqrt{25} + \sqrt{144}$

31. $\sqrt{64 + 81 + 144}$ **32.** $\sqrt{25 + 144}$

Estimate Then Calculate

In Exercises 33–36, mentally estimate the value of each radical expression, and then use a calculator to approximate each value to the nearest thousandth.

Problem	Mental Estimate	Calculator Approximation
33. $\sqrt[3]{9}$		
34. $\sqrt[3]{60}$		
35. $\sqrt[3]{80}$		
36. $\sqrt[4]{15}$		

Objective 2 Graph and Analyze Square Root and Cube Root Functions

37. Given $f(x) = \sqrt{x - 2}$:
a. Complete the table.
b. Sketch the graph of this function.
c. Determine the domain of this function.

x	f(x)
2	
3	
6	
11	

38. Given $f(x) = \sqrt{2 - x}$:
a. Complete the table.
b. Sketch the graph of this function.
c. Determine the domain of this function.

x	f(x)
-7	
-2	
1	
2	

39. Given $f(x) = \sqrt[3]{x + 3}$:
a. Complete the table.
b. Sketch the graph of this function.
c. Determine the domain of this function.

x	f(x)
-11	
-4	
-3	
-2	
5	

40. Given $f(x) = \sqrt[3]{x - 4}$:
a. Complete the table.
b. Sketch the graph of this function.
c. Determine the domain of this function.

x	f(x)
-4	
3	
4	
5	
12	

In Exercises 41–48, algebraically determine the domain of each function. (You can check your answer by graphing the function with a graphing calculator and examining the projection of the graph onto the x-axis.)

41. $f(x) = 3x - 15$ **42.** $f(x) = \sqrt[3]{3x - 15}$

43. $f(x) = \sqrt{3x - 15}$ **44.** $f(x) = \sqrt{12 - 4x}$

45. $f(x) = \dfrac{1}{3x - 15}$ **46.** $f(x) = \dfrac{-1}{4x - 12}$

47. $f(x) = |3x - 15|$ **48.** $f(x) = x^2 - 5$

Connecting Concepts to Applications

In Exercises 49 and 50, use the formula $c = \sqrt{a^2 + b^2}$ (from the Pythagorean theorem) to find the exact length of the hypotenuse c for each right triangle.

49.   **50.**

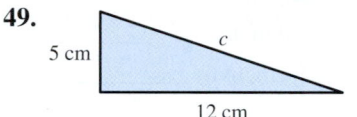

5 cm   c   20 cm   c

12 cm   21 cm

In Exercises 51 and 52, use the formula $s = \sqrt[3]{V}$ to determine to the nearest tenth of a centimeter the dimensions of a cubical container with the given volume.

51. $V = 216 \text{ cm}^3$

52. $V = 2,744 \text{ cm}^3$

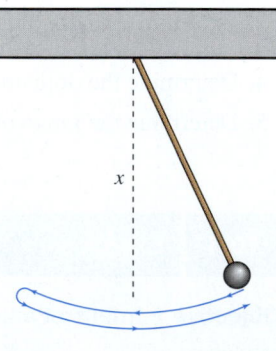

In Exercises 53 and 54, use the formula $T(x) = 0.064\sqrt{x}$ to determine to the nearest tenth of a second the time for one complete period of a pendulum with the given length x in centimeters.

53. $x = 60$ cm

54. $x = 110$ cm

In Exercises 55 and 56, use the formula $r = \sqrt[3]{\dfrac{3V}{4\pi}}$ to determine to the nearest tenth of a centimeter the radius of a spherical container with the given volume.

55. $V = 216 \text{ cm}^3$

56. $V = 2{,}744 \text{ cm}^3$

Sphere: $V = \dfrac{4}{3}\pi r^3$ $r = \sqrt[3]{\dfrac{3V}{4\pi}}$

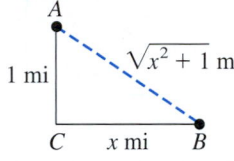

Cost to Run a Power Line The cost in dollars to run a power line from a power station on one side of a river to a factory on the other side of the river and x mi downstream is given by $C(x) = 1{,}200\sqrt{x^2 + 1}$. Evaluate and interpret each expression in Exercises 57–60.

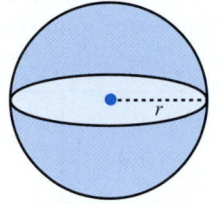

57. $C(0)$ **58.** $C(2)$

59. $C(5)$ **60.** $C(8)$

Review and Concept Development

In Exercises 61–68, match each description of a function with its graph.

61. A linear function whose graph has a positive slope

62. A linear function whose graph has a negative slope

63. An absolute value function whose graph opens upward

64. An absolute value function whose graph opens downward

65. A quadratic function whose graph opens upward

66. A quadratic function whose graph opens downward

67. A square root function

68. A cube root function

A.

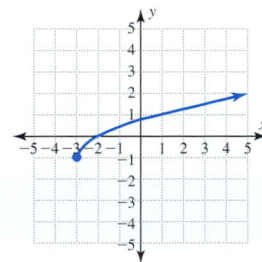

B.

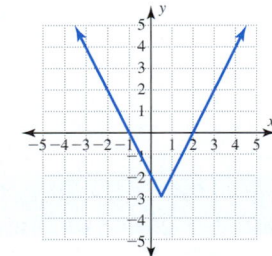

C.

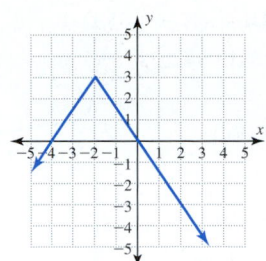

D.

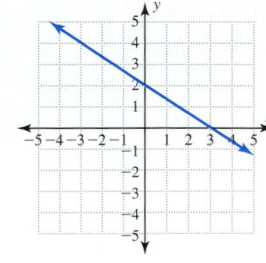

E.

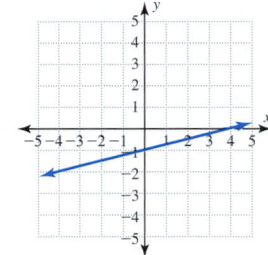

F.

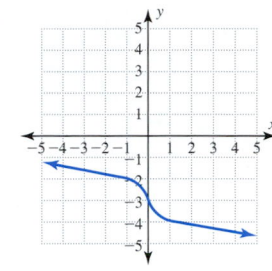

G.

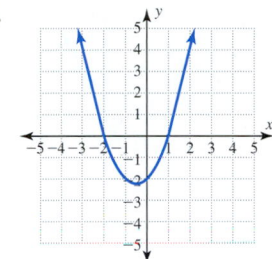

H.

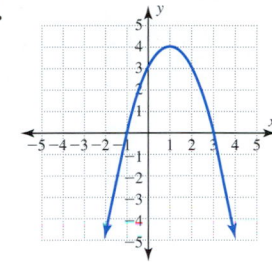

In Exercises 69–77, match each function with its graph and determine its domain. (*Hint:* Work each problem by recognizing the shape of the graph from the defining equation.) Use a graphing calculator only as a check on your work.

69. $f(x) = 2x + 1$

70. $f(x) = -2x + 1$

71. $f(x) = \sqrt{2 - x}$

72. $f(x) = \sqrt{x - 2}$

73. $f(x) = |x - 2| + 1$

74. $f(x) = -x^2 + 1$

75. $f(x) = -|x + 1| + 2$

76. $f(x) = \sqrt[3]{x} + 2$

77. $f(x) = -\sqrt[3]{x} + 2$

A.

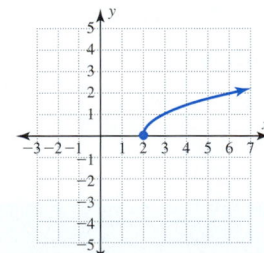

B.

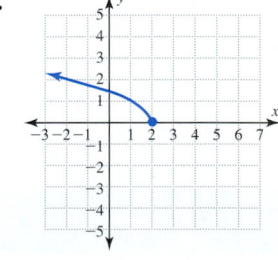

C.

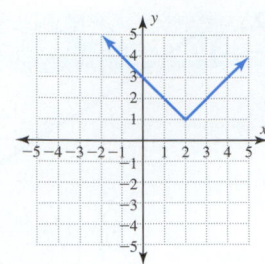

D.

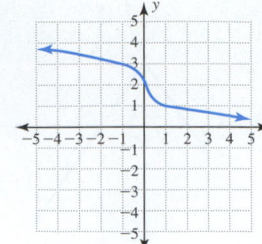

E.

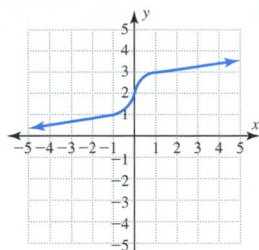

F.

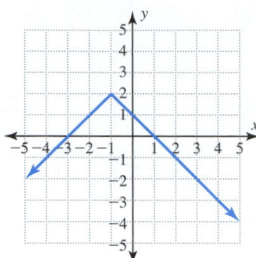

G.

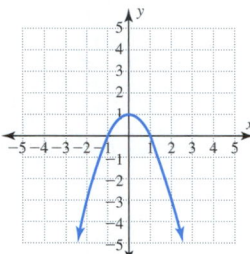

H.

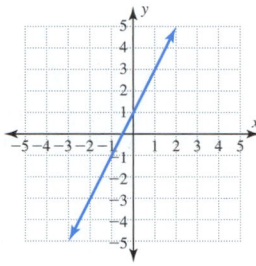

I.

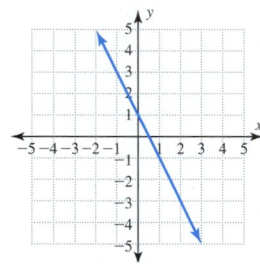

In Exercises 78–83, simplify each radical expression. Use complex numbers as necessary.

78. $\sqrt{-64}$

79. $\sqrt[3]{-64}$

80. $\sqrt[3]{-1}$

81. $\sqrt{-1}$

82. $\sqrt[3]{-1{,}000} + \sqrt{-100}$

83. $\sqrt[3]{-800} + \sqrt{-81}$

Group discussion questions

84. Calculator Discovery Use a graphing calculator to graph each of the following functions: $f(x) = \sqrt{x}$, $f(x) = \sqrt{x} + 1, f(x) = \sqrt{x} + 2, f(x) = \sqrt{x} + 3$, and $f(x) = \sqrt{x} + 4$. Describe the relationship between the graphs of $f(x) = \sqrt{x}$ and $f(x) = \sqrt{x} + c$ for $c > 0$.

85. Error Analysis A student graphing $f(x) = \sqrt[3]{x} + 2$ obtained the following display. Explain how you can tell by inspection that an error has been made.

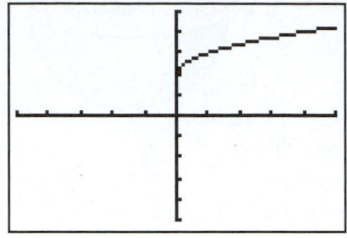

$[-5, 5, 1]$ by $[-5, 5, 1]$

86. Challenge Question

a. Give an example of x so that $\sqrt{x} = -\sqrt{x}$.

b. Give an example of x so that $\sqrt{x}$ and $\sqrt[3]{x}$ are real numbers but $\sqrt{-x}$ is not a real number.

c. Give an example of x so that $\sqrt{-x}$ and $\sqrt[3]{x}$ are real numbers but $\sqrt{x}$ is not a real number.

d. Give an example of x so that $\sqrt{x^2} \neq x$.

87. Challenge Question Match each radical expression with the appropriate simplified form. (*Hint:* If you are not sure, test some different values for x.)

a. $\sqrt{4x^2}$ for $x \geq 0$ **A.** $-4x$

b. $\sqrt{4x^2}$ for $x < 0$ **B.** $4x$

c. $\sqrt[3]{64x^3}$ for $x < 0$ **C.** $-2x$

d. $\sqrt[3]{-64x^3}$ for $x \geq 0$ **D.** $2x$

10.1 | **Cumulative Review**

Simplify each expression.

1. $5(2x - 4y) + 7(x - 3y)$

2. $-6(2x - 4y) - (x - 3y)$

3. $(2a^3b^4)(5a^6b^7)$

4. $\dfrac{12a^6b^4}{3a^3b^3}$

5. $(5a^2b^4)^3$

Section 10.2 | Adding and Subtracting Radical Expressions

Objectives:
1. Add and subtract radical expressions.
2. Simplify radical expressions.

1. Add and Subtract Radical Expressions

We add like radicals in exactly the same way that we add like terms of a polynomial.

What Are Like Radicals?

Like radicals are radical expressions that have the same index and the same radicand; only the coefficients of the like radicals can differ.

Example 1   Classifying Radicals as Like or Unlike

Classify these radicals as like or unlike.

Solution

(a) $4\sqrt[3]{2}$ and $5\sqrt[3]{2}$ — Like radicals — The radicand of each of these cube roots is 2.

(b) $7\sqrt{5}$ and $7\sqrt[3]{5}$ — Unlike radicals — These radicals are unlike because they do not have the same index. One is a square root, and the other is a cube root.

(c) $11\sqrt{7}$ and $11\sqrt{8}$ — Unlike radicals — These square roots are unlike because the radicands 7 and 8 are different.

(d) $-\sqrt{xy^2}$ and $4\sqrt{xy^2}$ — Like radicals — The radicand of each of these square roots is xy^2. The coefficients of the radicals are -1 and 4.

(e) $\sqrt[3]{xy^2}$ and $\sqrt[3]{x^2y}$ — Unlike radicals — These radicals are unlike because the radicands are not equal ($xy^2 \neq x^2y$).

Self-Check 1

Classify each pair of radicals as either like or unlike.

a. $-\dfrac{\sqrt{7}}{2}$ and $\dfrac{\sqrt{7}}{3}$ b. $\sqrt[3]{7}$ and $-\sqrt{7}$ c. $\sqrt[5]{2}$ and $\sqrt[7]{3}$

How Do I Add Like Radicals?

We add their coefficients just as we add like terms of a polynomial. This is illustrated in the box.

Comparison of the Addition of Polynomials and the Addition of Radical Terms

Addition of Polynomials	Addition of Radicals
$3x + 4x = 7x$	$3\sqrt{2} + 4\sqrt{2} = 7\sqrt{2}$
$9x + 5y - 6x = 3x + 5y$	$9\sqrt{7} + 5\sqrt{11} - 6\sqrt{7} = 3\sqrt{7} + 5\sqrt{11}$

Note the use of the distributive property in each addition. For example, $3x + 4x = (3 + 4)x = 7x$.

| **Example 2** | Adding and Subtracting Radicals |

Perform the following additions and subtractions.

(a) $\sqrt[3]{7} - 5\sqrt[3]{7} + 12\sqrt[3]{7}$

(b) $(5\sqrt{11} + 6\sqrt[3]{11}) - (2\sqrt[3]{11} - 7\sqrt{11})$

(c) $3\sqrt[3]{x^2y} - 4\sqrt[3]{x^2y}$

(d) $4\sqrt{5} + 7\sqrt[3]{5}$

Solution

(a) $\sqrt[3]{7} - 5\sqrt[3]{7} + 12\sqrt[3]{7} = (1 - 5 + 12)\sqrt[3]{7}$ Use the distributive property to combine like terms.

$\qquad\qquad\qquad\qquad\qquad\quad = 8\sqrt[3]{7}$

(b) $(5\sqrt{11} + 6\sqrt[3]{11}) - (2\sqrt[3]{11} - 7\sqrt{11})$ First remove the parentheses, and then group like terms together. Then add the coefficients of the like terms.

$\qquad = 5\sqrt{11} + 6\sqrt[3]{11} - 2\sqrt[3]{11} + 7\sqrt{11}$

$\qquad = (5\sqrt{11} + 7\sqrt{11}) + (6\sqrt[3]{11} - 2\sqrt[3]{11})$

$\qquad = (5 + 7)\sqrt{11} + (6 - 2)\sqrt[3]{11}$

$\qquad = 12\sqrt{11} + 4\sqrt[3]{11}$

(c) $3\sqrt[3]{x^2y} - 4\sqrt[3]{x^2y} = (3 - 4)\sqrt[3]{x^2y}$ Subtract the coefficients of these like terms.

$\qquad\qquad\qquad\qquad\quad = -\sqrt[3]{x^2y}$

(d) $4\sqrt{5} + 7\sqrt[3]{5}$ This expression cannot be simplified since the terms are unlike.

Self-Check 2

Perform the indicated operations.

a. $4\sqrt{6} - 9\sqrt{6}$ **b.** $8\sqrt[3]{xy} - 4\sqrt[3]{xy} + \sqrt[3]{xy}$ **c.** $11\sqrt{7} + 11\sqrt{8}$

2. Simplify Radical Expressions

One way to simplify a radical expression is to use the properties in the following table to make the radicand as small as possible. These properties are extensions of the properties of square roots that we used in Section 7.1. *Warning:* If $\sqrt[n]{x}$ is not a real number, then the properties in this table are not true.

Product and Quotient Rules for Radicals

If $\sqrt[n]{x}$ and $\sqrt[n]{y}$ are both real numbers,* then:

Radical Notation	Verbally	Radical Example
Product Rule $\sqrt[n]{xy} = \sqrt[n]{x}\,\sqrt[n]{y}$	The *n*th root of a product equals the product of the *n*th roots.	$\sqrt{50} = \sqrt{25}\sqrt{2}$
Quotient Rule $\sqrt[n]{\dfrac{x}{y}} = \dfrac{\sqrt[n]{x}}{\sqrt[n]{y}}$ for $y \neq 0$	The *n*th root of a quotient equals the quotient of the *n*th roots.	$\sqrt{\dfrac{4}{9}} = \dfrac{\sqrt{4}}{\sqrt{9}}$

*If either $\sqrt[n]{x}$ or $\sqrt[n]{y}$ is not a real number, these properties do not hold true.

   The properties in the previous box are used to simplify the expressions in Example 3. The key to using these properties is to recognize perfect nth powers. When you are working with square roots, look for perfect square factors, such as 4, 9, 16, and 25. Similarly, when you are working with cube roots, look for perfect cube factors, such as 8, 27, 64, and 125.

Example 3   Using the Product Rule for Radicals

Simplify each radical expression.

(a) $\sqrt{50}$ **(b)** $\sqrt[3]{40}$

Solution

(a) $\sqrt{50} = \sqrt{25 \cdot 2}$ 25 is a perfect square factor of 50.
$\quad\quad = \sqrt{25}\sqrt{2}$ $\sqrt{xy} = \sqrt{x}\,\sqrt{y}$
$\quad\quad\quad\quad\quad\quad\quad\quad$ $5\sqrt{2}$ is considered a simpler form than $\sqrt{50}$ because it has a smaller radicand.
$\quad\quad = 5\sqrt{2}$ You can check this result by comparing $\sqrt{50}$ and $5\sqrt{2}$ on a calculator.

(b) $\sqrt[3]{40} = \sqrt[3]{8 \cdot 5}$ 8 is a perfect cube factor of 40.
$\quad\quad = \sqrt[3]{8}\,\sqrt[3]{5}$ $\sqrt[3]{xy} = \sqrt[3]{x}\sqrt[3]{y}$
$\quad\quad\quad\quad\quad\quad\quad\quad$ *Caution:* Do not make the classic error of writing the answer incorrectly as
$\quad\quad = 2\sqrt[3]{5}$ $2\sqrt{5}$.

Self-Check 3

Simplify each radical expression.

a. $\sqrt{45}$ **b.** $\sqrt[3]{24}$

   A second way to simplify a radical expression is to use the quotient rule for radicals to remove radicals from the denominator. Sometimes this property can also be used to remove fractions from the radicand.

Example 4   Using the Quotient Rule for Radicals

Simplify each radical expression.

(a) $\sqrt[3]{\dfrac{25}{8}}$ **(b)** $\dfrac{\sqrt[3]{54}}{\sqrt[3]{2}}$

Solution

(a) $\sqrt[3]{\dfrac{25}{8}} = \dfrac{\sqrt[3]{25}}{\sqrt[3]{8}}$ $\sqrt[3]{\dfrac{x}{y}} = \dfrac{\sqrt[3]{x}}{\sqrt[3]{y}}$ The numerator $\sqrt[3]{25}$ cannot be reduced further since it has no
$\quad\quad\quad\quad\quad\quad\quad\quad\quad\quad\quad\quad$ perfect cube factor other than 1.

$\quad\quad\quad\quad = \dfrac{\sqrt[3]{25}}{2}$ $\dfrac{\sqrt[3]{25}}{2}$ is considered a simpler form than $\sqrt[3]{\dfrac{25}{8}}$ because it does not contain a radical

$\quad\quad\quad\quad\quad\quad\quad\quad\quad\quad\quad\quad$ in the denominator.

(b) $\dfrac{\sqrt[3]{54}}{\sqrt[3]{2}} = \sqrt[3]{\dfrac{54}{2}}$ Use the quotient rule for radicals and reduce the fraction.

$\quad\quad\quad\quad = \sqrt[3]{27}$ Then take the cube root of 27.

$\quad\quad\quad\quad = 3$

Self-Check 4

Simplify each radical expression.

a. $\sqrt{\dfrac{36}{169}}$ **b.** $\dfrac{\sqrt[3]{16}}{\sqrt[3]{2}}$

Why Will I Need to Simplify Radical Expressions?

One reason for always writing radicals in simplified form is to identify like terms. Radicals that appear to be unlike can sometimes be simplified so that it is clear that the terms are like terms. Remember to look for perfect square factors when you are working with square roots and for perfect cube factors when working with cube roots.

Example 5   Simplifying Radicals and Then Combining Like Radicals

Simplify each term and then combine like radicals.

(a) $7\sqrt{20} - 2\sqrt{45}$ **(b)** $11\sqrt[3]{16} - 2\sqrt[3]{54}$ **(c)** $\sqrt{18} + \sqrt{12}$

Solution

(a) $7\sqrt{20} - 2\sqrt{45} = 7\sqrt{4 \cdot 5} - 2\sqrt{9 \cdot 5}$ Note that 4 is a perfect square factor of 20 and
$\qquad\qquad\qquad = 7\sqrt{4}\sqrt{5} - 2\sqrt{9}\sqrt{5}$ that 9 is a perfect square factor of 45. Simplify by using the property that $\sqrt{xy} = \sqrt{x}\,\sqrt{y}$.
$\qquad\qquad\qquad = 7(2\sqrt{5}) - 2(3\sqrt{5})$
$\qquad\qquad\qquad = 14\sqrt{5} - 6\sqrt{5}$ Then combine like terms.
$\qquad\qquad\qquad = 8\sqrt{5}$

(b) $11\sqrt[3]{16} - 2\sqrt[3]{54} = 11\sqrt[3]{8 \cdot 2} - 2\sqrt[3]{27 \cdot 2}$ Note that 8 is a perfect cube factor of 16 and
$\qquad\qquad\qquad = 11\sqrt[3]{8}\sqrt[3]{2} - 2\sqrt[3]{27}\sqrt[3]{2}$ that 27 is a perfect cube factor of 54.
$\qquad\qquad\qquad = 11(2\sqrt[3]{2}) - 2(3\sqrt[3]{2})$ Simplify these radicals.
$\qquad\qquad\qquad = 22\sqrt[3]{2} - 6\sqrt[3]{2}$ Then combine like terms.
$\qquad\qquad\qquad = 16\sqrt[3]{2}$

(c) $\sqrt{18} + \sqrt{12} = \sqrt{9 \cdot 2} + \sqrt{4 \cdot 3}$ 9 is a perfect square factor of 18, and 4 is a
$\qquad\qquad\qquad = \sqrt{9}\,\sqrt{2} + \sqrt{4}\,\sqrt{3}$ perfect square factor of 12.
$\qquad\qquad\qquad = 3\sqrt{2} + 2\sqrt{3}$ These radicals are unlike, so the expression cannot be simplified further.

Self-Check 5

Simplify each expression.

a. $2\sqrt{50} - 3\sqrt{200}$ **b.** $\sqrt[3]{24} - \sqrt[3]{3,000}$

Expressions such as $\sqrt{-1}$, $\sqrt[4]{-1}$, and other even roots of negative values are imaginary numbers rather than real numbers. As we illustrate next, some of the properties that hold when $\sqrt[n]{x}$ is a real number will fail when $\sqrt[n]{x}$ is an imaginary number. The following situation illustrates a case where $\sqrt[n]{x^n} \neq x$:

$$\sqrt{(-5)^2} = \sqrt{25} = 5$$

Thus $\sqrt{(-5)^2} \neq -5$.

If $x < 0$ and n is an even number, then $\sqrt[n]{x^n} \neq x$. Since x^2 is positive both when x is positive and when x is negative, the principal square root of x^2 will be a positive real number. We can use absolute value notation to correctly denote the result in all cases:

$$\sqrt{x^2} = |x|$$

We now generalize this property to radicals of higher order. To determine the general nth root of both positive and negative values of x, we define $\sqrt[n]{x^n} = |x|$ when n is even. This is highlighted in the following box.

Simplifying $\sqrt[n]{x^n}$

For any real number x and natural number n:

Algebraically	Numerically	Graphically	Verbally
If n is even, $\sqrt[n]{x^n} = |x|$. *Example:* $\sqrt{x^2} = |x|$		$[-10, 10, 1]$ by $[-1, 10, 1]$	These functions are identical for all real numbers x, $\sqrt{x^2} = |x|$. The tables are the same for all values of x, and the graphs are identical.
If n is odd, $\sqrt[n]{x^n} = x$. *Example:* $\sqrt[3]{x^3} = x$		$[-10, 10, 1]$ by $[-10, 10, 1]$	These functions are identical for all real numbers x, $\sqrt[3]{x^3} = x$. The tables are the same for all values of x, and the graphs are identical.

In Example 6, compare the results of parts (c) and (d). If $x > 0$, then we do not need to use absolute value notation when we determine the sixth root. However, when x can assume negative values, then we need to use absolute value notation to represent the sixth root.

Example 6 Simplifying Radical Expressions

Simplify each radical expression. Use absolute value notation only when necessary.

(a) $\sqrt{(-7)^2}$ (b) $\sqrt[3]{(-2)^3}$ (c) $\sqrt{64x^6}$ for $x > 0$

(d) $\sqrt{64x^6}$ if x is any real number   (e) $\sqrt[3]{64x^6}$

Solution

(a) $\sqrt{(-7)^2} = |-7|$ $\sqrt[n]{x^n} = |x|$ if n is even: $\sqrt{(-7)^2} = \sqrt{49} = 7$
$\qquad\qquad\quad = 7$

(b) $\sqrt[3]{(-2)^3} = -2$ $\sqrt[n]{x^n} = x$ if n is odd: $\sqrt[3]{(-2)^3} = \sqrt[3]{-8} = -2$

(c) $\sqrt{64x^6} = \sqrt{(8x^3)^2}$ First write $64x^6$ as the perfect square of $8x^3$. Absolute value notation is
$\qquad\qquad = 8x^3$ not needed because $x > 0$. If we know that $x > 0$, then $|8x^3| = 8x^3$.

(d) $\sqrt{64x^6} = |8x^3|$ Because $\sqrt[n]{x^n} = |x|$ if n is even, we must use absolute value notation in
$\qquad\qquad = 8|x|^3$ this problem.

(e) $\sqrt[3]{64x^6} = \sqrt[3]{(4x^2)^3}$ First write $64x^6$ as the perfect cube of $4x^2$.
$\qquad\qquad = 4x^2$ $\sqrt[n]{x^n} = x$ if n is odd.

Self-Check 6

Simplify each of these radical expressions. Use absolute value notation only when necessary.

a. $\sqrt[3]{y^6}$ **b.** $\sqrt{y^6}$ **c.** $\sqrt[5]{-32y^5}$

If we restrict the variables to values that keep the radicand positive, then we do not need to use absolute value notation. This is the situation in Example 7.

Example 7 Simplifying Radical Expression

Simplify $\sqrt{x^2 + 2xy + y^2} + \sqrt{x^2 - 2xy + y^2}$, assuming that $x > y > 0$.

Solution

$$\sqrt{x^2 + 2xy + y^2} + \sqrt{x^2 - 2xy + y^2} = \sqrt{(x + y)^2} + \sqrt{(x - y)^2}$$
$$= (x + y) + (x - y)$$
$$= 2x$$

Express the radicands as perfect squares. If $x > y > 0$, then $x + y > 0$ and $x - y > 0$; thus absolute value notation is not needed for these square roots. Simplify each square root and then combine the like terms.

Self-Check 7

Simplify $\sqrt{x^2 + 4x + 4}$ assuming that $x \geq -2$.

Example 8 Simplifying a Radical Expression

Simplify $\sqrt[3]{40x^4}$.

Solution

$$\sqrt[3]{40x^4} = \sqrt[3]{(8x^3)(5x)}$$
$$= \sqrt[3]{(8x^3)}\sqrt[3]{(5x)}$$
$$= 2x\sqrt[3]{(5x)}$$

First determine the largest perfect cube factor of $40x^4$. Then use the product rule for radicals and simplify the first radical.

Self-Check 8

Simplify $\sqrt[3]{54x^5}$.

Carpenters, cooks, and many other people who make things often perform mental computations either to check their calculations or to make quick decisions about the task in front of them. Example 9 involves estimating the length of a side of a cube.

Example 9 Estimating the Dimensions of a Box

A cubical box has a volume of 67 cm³. Estimate the length of this box and then approximate this length to the nearest tenth of a cm.

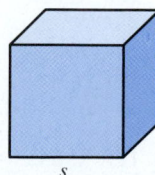

Solution

$$V = s^3 \qquad \text{Formula for the volume of a cube}$$
$$s = \sqrt[3]{V} \qquad \text{Formula for the length of a side of a cube}$$

Estimated Answer	Calculator Approximation
$s = \sqrt[3]{67}$	$s = \sqrt[3]{67}$
$\approx \sqrt[3]{64}$	≈ 4.0615481
$s \approx 4$ cm	$s \approx 4.1$ cm

Self-Check 9

Estimate the length of a cubical box with a volume of 120 cm³. Then approximate this length to the nearest tenth of a cm.

Self-Check Answers

1. **a.** Like
   b. Unlike
   c. Unlike
2. **a.** $-5\sqrt{6}$
   b. $5\sqrt[3]{xy}$
   c. $11\sqrt{7} + 11\sqrt{8}$
3. **a.** $3\sqrt{5}$ **b.** $2\sqrt[3]{3}$
4. **a.** $\dfrac{6}{13}$
   b. 2

5. **a.** $-20\sqrt{2}$
   b. $-8\sqrt[3]{3}$
6. **a.** y^2
   b. $|y^3|$ or $|y|^3$
   c. $-2y$
7. $x + 2$
8. $3x\sqrt[3]{2x^2}$
9. $s = \sqrt[3]{120} \approx \sqrt[3]{125}; s \approx 5$ cm;
   $s = \sqrt[3]{120} \approx 4.9$ cm

10.2   Using the Language and Symbolism of Mathematics

1. Like radicals must have the same _____ and the same _____ .

2. The statement that "the nth root of a product equals the product of the nth roots" states that
   $\sqrt[n]{xy}$ _____ if both $\sqrt[n]{x}$ and $\sqrt[n]{y}$ are real numbers.

3. The statement that "the nth root of a quotient equals the quotient of the nth roots" states that
   $\sqrt[n]{\dfrac{x}{y}} =$ _____ if both $\sqrt[n]{x}$ and $\sqrt[n]{y}$ are real numbers and $y \neq 0$.

4. If n is even, $\sqrt[n]{x^n} =$ _____ .

5. If n is odd, $\sqrt[n]{x^n} =$ _____ .

10.2   Quick Review

Simplify each expression.

1. $(5x + 3y) + (4x - 9y)$
2. $3(2x - y) - (4x - 9y)$
3. $|17|$
4. $|-22|$

5. Complete this table.

x	x^2	x^3
0		
1		
2		
3		
4		
5		

10.2 Exercises

Objective 1 Add and Subtract Radical Expressions

In Exercises 1 and 2, determine which radical is a like radical to the given radical.

1. $7\sqrt{5}$ **A.** $2\sqrt[3]{5}$ **B.** $3\sqrt{7}$
             C. $7\sqrt{3}$ **D.** $-2\sqrt{5}$

2. $4\sqrt[3]{7}$ **A.** $4\sqrt{7}$ **B.** $11\sqrt[3]{7}$
             C. $5\sqrt[3]{4}$ **D.** $-5\sqrt[3]{4}$

In Exercises 3–20, find the sum or difference. Assume that all variables represent positive real numbers so that absolute value notation is not necessary.

3. $\sqrt{25} + \sqrt{49}$ 4. $\sqrt{36} + \sqrt{64}$

5. $2\sqrt{9} - 3\sqrt{16}$ 6. $5\sqrt{100} - 3\sqrt{4}$

7. $19\sqrt{2} + 11\sqrt{2}$ 8. $7\sqrt{3} - 8\sqrt{3}$

9. $9\sqrt{7} - 13\sqrt{7}$ 10. $6\sqrt{5} + 8\sqrt{5}$

11. $4\sqrt[3]{6} + 2\sqrt[3]{6} - 11\sqrt[3]{6}$

12. $5\sqrt[3]{11} - 2\sqrt[3]{11} + 12\sqrt[3]{11}$

13. $5\sqrt[4]{13} + 6\sqrt[4]{13} - \sqrt[4]{13}$ 14. $\sqrt[5]{9} - 8\sqrt[5]{9} - 7\sqrt[5]{9}$

15. $7\sqrt{5} - 5\sqrt{7} - \sqrt{5} + \sqrt{7}$

16. $7\sqrt{2} - \sqrt{3} + 3\sqrt{2} + 2\sqrt{3}$

17. $6\sqrt{7x} - 9\sqrt{7x}$ 18. $4\sqrt{11y} - 7\sqrt{11y}$

19. $7\sqrt[4]{17w} - 8\sqrt[4]{17w} + \sqrt[4]{17w}$

20. $6\sqrt[7]{5w} - 11\sqrt[7]{5w} + 5\sqrt[7]{5w}$

Objective 2 Simplify Radical Expressions

In Exercises 21–28, simplify each radical expression.

21. $\sqrt{75}$ 22. $\sqrt{28}$ 23. $\sqrt{63}$ 24. $\sqrt{72}$

25. $\sqrt[3]{24}$ 26. $\sqrt[3]{40}$ 27. $\sqrt[4]{48}$ 28. $\sqrt[5]{96}$

In Exercises 29–50, find the sum or difference. Assume that all variables represent positive real numbers so that absolute value notation is not necessary.

29. $\sqrt{28} + \sqrt{63}$ 30. $\sqrt{12} - \sqrt{27}$

31. $\sqrt{75} - \sqrt{48}$ 32. $\sqrt{44} + \sqrt{99}$

33. $3\sqrt{50v} - 7\sqrt{32v}$ 34. $6\sqrt{45v} - 7\sqrt{320v}$

35. $5\sqrt{28w} - 4\sqrt{63w}$ 36. $3\sqrt{27b} - \sqrt{75b}$

37. $\sqrt[3]{24} - \sqrt[3]{375}$ 38. $\sqrt[3]{54} - \sqrt[3]{128}$

39. $20\sqrt[3]{-81t^3} - 7\sqrt[3]{24t^3}$ 40. $11\sqrt[3]{-16t^3} - 5\sqrt[3]{54t^3}$

41. $9\sqrt[3]{40z^2} - 2\sqrt[3]{5,000z^2}$

42. $7\sqrt{1,210w^3} - 5\sqrt{1,440w^3}$

43. $\sqrt{\dfrac{5}{4}} + \sqrt{\dfrac{5}{9}}$ 44. $\sqrt{\dfrac{7}{9}} - \sqrt{\dfrac{7}{25}}$

45. $2\sqrt{\dfrac{11x}{25}} - \sqrt{\dfrac{11x}{49}}$ 46. $3\sqrt{\dfrac{13x}{49}} - 5\sqrt{\dfrac{13x}{4}}$

47. $7\sqrt{0.98} + 2\sqrt{0.75} - \sqrt{0.12} + 5\sqrt{0.72}$

48. $4\sqrt{0.20} - 3\sqrt{0.90} - 7\sqrt{0.80} + \sqrt{1.60}$

49. $\sqrt{x^2 + 2xy + y^2} - \sqrt{x^2} - \sqrt{y^2}$

50. $\sqrt{x^2} + \sqrt{y^2} - \sqrt{x^2 - 2xy + y^2}$ for $x > y > 0$

In Exercises 51–58, simplify each expression. Use absolute value notation only when it is needed.

	For $x \geq 0$	For Any Real Number x
51. $\sqrt{25x^2}$	a.	b.
52. $\sqrt{36x^2}$	a.	b.
53. $\sqrt[3]{8x^3}$	a.	b.
54. $\sqrt[3]{-125x^3}$	a.	b.
55. $\sqrt{1,000,000x^6}$	a.	b.
56. $\sqrt[3]{1,000,000x^6}$	a.	b.
57. $\sqrt[5]{x^{30}}$	a.	b.
58. $\sqrt[6]{x^{30}}$	a.	b.

In Exercises 59–64, simplify each radical expression.

59. $\sqrt{50x^3}$ 60. $\sqrt{300x^5}$

61. $\sqrt[3]{16x^4}$ 62. $\sqrt[3]{40y^7}$

63. $\sqrt[3]{x^4y^8}$ 64. $\sqrt[3]{27x^3y^{10}}$

Connecting Concepts to Applications

65. **Radius of a Cylinder** The volume of a cylindrical piece of pipe is given by $V = \pi r^2 h$. A 5-ft section of pipe has a volume of 750 in³. Approximate the radius of this pipe to the nearest tenth of an inch. (*Hint:* How many inches are in 5 ft?)

66. **Radius of a Sphere** The volume of a spherical tank is given by $V = \dfrac{4}{3}\pi r^3$. If a spherical tank has a volume of 750 in³, approximate its radius to the nearest tenth of an inch.

67. **Length of a Brace** The length d of a diagonal brace from the lower corner of a rectangular storage box to the opposing upper corner is given by

$$d = \sqrt{w^2 + l^2 + h^2}$$

where w, l, and h are, respectively, the width, length, and height of the box. Determine to the nearest tenth of a meter the length of a diagonal brace of a box with width 5.2 m, length 9.4 m, and height 6.5 m.

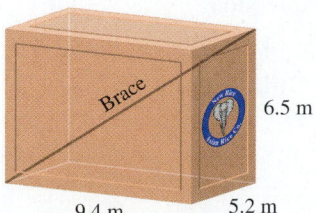

Brace   6.5 m   9.4 m   5.2 m

68. **Radius of a Cone** The volume of a cone is given by $V = \dfrac{1}{3}\pi r^2 h$. If a cone with a height of 15 in has a volume of 750 in³, approximate its radius to the nearest tenth of an inch.

Estimation Skills

69. Select the best mental estimate of the length of each side of a square of area 26 cm². (See the figure.)
  A. 3 cm   **B.** 4 cm   **C.** 5 cm
  D. 12 cm   **E.** 13 cm

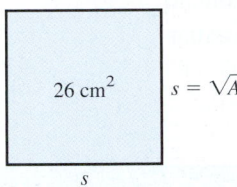

26 cm^2 $s = \sqrt{A}$

s

70. Select the best mental estimate of the length of each side of a cubic box whose volume is 26 cm³. (See the figure.)
  A. 3 cm   **B.** 4 cm   **C.** 5 cm
  D. 12 cm   **E.** 13 cm

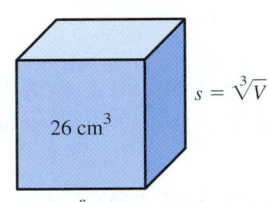

$s = \sqrt[3]{V}$

26 cm^3

s

Estimate Then Calculate

In Exercises 71–74, complete the following table.

a. Estimate each radical expression to the nearest integer.

b. Determine whether this integer estimate is less than or greater than the actual value.

c. Use a calculator to approximate each expression to the nearest thousandth.

	Integer Estimate	Inequality	Approximation
Example: $\sqrt[4]{79}$	3	$3 > \sqrt[4]{79}$	2.981
71. $\sqrt[3]{63}$			
72. $\sqrt[4]{82}$			
73. $\dfrac{1 + \sqrt{9.05}}{2}$			
74. $\dfrac{1 - \sqrt{9.05}}{2}$			

Review and Concept Development

In Exercises 75–80, match each function with its graph.

75. $f(x) = x + 3$ **76.** $f(x) = (x + 3)^2$

77. $f(x) = \sqrt{x} + 3$ **78.** $f(x) = \sqrt{x + 3}$

79. $f(x) = |x| + 3$ **80.** $f(x) = \sqrt{(x + 3)^2}$

A.

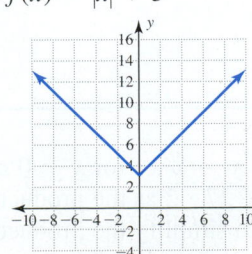

B.

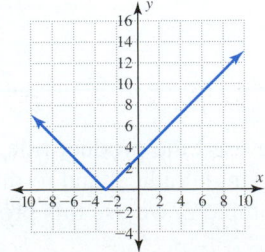

C.

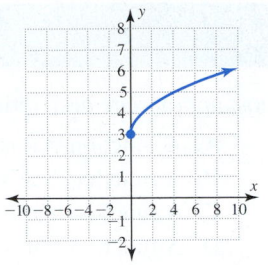

D.

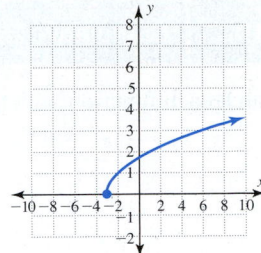

E.

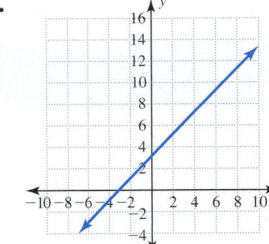

F.

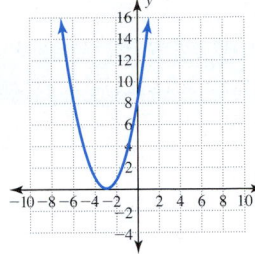

Group discussion questions

81. Discovery Question
  a. Use a graphing calculator to compare tables of values for $f(x) = \sqrt[4]{x^2}$ and $f(x) = \sqrt{x}$.
  b. Use a graphing calculator to compare tables of values for $f(x) = \sqrt[9]{x^3}$ and $f(x) = \sqrt[3]{x}$.
  c. Using your observations from parts **a** and **b,** predict how you can simplify $\sqrt[6]{x^2}$ to a radical of lower index than 6. Then use a graphing calculator to compare $f(x) = \sqrt[6]{x^2}$ to the radical function defined by the expression you predicted.

82. Discovery Question
  a. Does $\sqrt{\sqrt{256}} = \sqrt[4]{256}$? Explain how you determined your answer.
  b. Does $\sqrt{\sqrt{\sqrt{256}}}$ equal either $\sqrt[6]{256}$ or $\sqrt[8]{256}$? Explain how you determined your answer.
  c. What is another way of writing $\sqrt{\sqrt{\sqrt{\sqrt{x}}}}$? Explain your reasoning to your teacher.

83. Challenge Question
  a. Using a calculator, evaluate $\sqrt{10}$. Then take the square root of this result, then the square root of this result, and so on, until you have taken 10 square roots.
  b. Using a calculator, evaluate $\sqrt{0.1}$. Then take the square root of this result, then the square root of this result, and so on, until you have taken 10 square roots.
  c. Make a conjecture based on your observations in parts **a** and **b.**

10.2 | **Cumulative Review**

Match each function with the most appropriate description.

1. $f(x) = -3x + 35$

2. $f(x) = \sqrt{x - 4}$

3. $f(x) = -x^2 + 9$

4. $f(x) = |x + 6|$

5. $f(x) = -3$

A. A function that is increasing over its whole domain

B. A function that is decreasing over its whole domain

C. A function that is neither increasing nor decreasing

D. A function whose graph has a single highest point

E. A function whose graph has a single lowest point

Section 10.3 | Multiplying and Dividing Radical Expressions

Objectives:

1. Multiply radical expressions.

2. Divide and simplify radical expressions.

1. Multiply Radical Expressions

This section extends the multiplication and division of square roots from Section 7.1.

To multiply and divide some radical expressions, we use the product and quotient rules from the previous section.

- Product Rule: $\sqrt[n]{x}\sqrt[n]{y} = \sqrt[n]{xy}$ for $x \ge 0$ and $y \ge 0$ The restriction that $y \ne 0$ is needed

- Quotient rule: $\dfrac{\sqrt[n]{x}}{\sqrt[n]{y}} = \sqrt[n]{\dfrac{x}{y}}$ for $x \ge 0$ and $y > 0$ in the expression $\dfrac{\sqrt[n]{x}}{\sqrt[n]{y}} = \sqrt[n]{\dfrac{x}{y}}$ to prevent division by 0.

What Are Some Key Points to Remember When I Multiply or Divide Radical Expressions?

Remember two things:

1. The radicals must be of the same order before these properties can be applied.

2. Do not apply these properties if one or both of the radicands are negative and n is even.

Example 1   Multiplying Radicals

Perform each indicated multiplication, and then simplify the product. Assume $x \ge 0$.

(a) $\sqrt{9x}\sqrt{x}$ **(b)** $\sqrt[3]{9}\sqrt[3]{6}$

Solution

(a) $\sqrt{9x}\sqrt{x} = \sqrt{9x^2}$ $\sqrt{x}\sqrt{y} = \sqrt{xy}$
        Absolute value notation is not needed for the answer since $x \ge 0$.

$\qquad\qquad\ \ = 3x$

(b) $\sqrt[3]{9}\sqrt[3]{6} = \sqrt[3]{54}$

$\qquad\qquad\ = \sqrt[3]{27 \cdot 2}$ $\sqrt[3]{x}\sqrt[3]{y} = \sqrt[3]{xy}$. 27 is a perfect cube factor of 54.

$\qquad\qquad\ = \sqrt[3]{27}\sqrt[3]{2}$ $\sqrt[3]{xy} = \sqrt[3]{x}\sqrt[3]{y}$

$\qquad\qquad\ = 3\sqrt[3]{2}$

Self-Check 1

Simplify $\sqrt[3]{2}\,\sqrt[3]{12}$.

Note in Example 1(b) that we used the product rule, $\sqrt[n]{x}\sqrt[n]{y} = \sqrt[n]{xy}$ for $x \ge 0$ and $y \ge 0$, both to rewrite $\sqrt[3]{9}\sqrt[3]{6}$ as a single radical and to rewrite $\sqrt[3]{27 \cdot 2}$ as the product of two separate radical factors. In Example 2, we use this property to simplify the product of two radical expressions.

Example 2   Multiplying Radicals

Simplify each radical expression.

(a) $(4\sqrt{3})(5\sqrt{3})$ **(b)** $(\sqrt[3]{20y^2})(\sqrt[3]{50y^2})$

Solution

(a) $(4\sqrt{3})(5\sqrt{3}) = 20\sqrt{9}$ Use the product rule for radicals, $\sqrt{x}\,\sqrt{y} = \sqrt{xy}$.
Then evaluate $\sqrt{9}$.

$\qquad\qquad\qquad\quad = 20(3)$

$\qquad\qquad\qquad\quad = 60$

(b) $(\sqrt[3]{20y^2})(\sqrt[3]{50y^2}) = \sqrt[3]{1{,}000y^4}$ $\sqrt[3]{x}\,\sqrt[3]{y} = \sqrt[3]{xy}$

$\qquad\qquad\qquad\quad = \sqrt[3]{(1{,}000y^3)(y)}$ $1{,}000y^3$ is a perfect cube factor of $1{,}000y^4$.

$\qquad\qquad\qquad\quad = \sqrt[3]{1{,}000y^3}\,\sqrt[3]{y}$

$\qquad\qquad\qquad\quad = 10y\sqrt[3]{y}$

Self-Check 2

Simplify each product. Assume $x \geq 0$.

a. $\sqrt{6x}\,\sqrt{10x}$ **b.** $\sqrt[3]{4x^2}\,\sqrt[3]{2x^2}$

The multiplication of some radical expressions is similar to the multiplication of polynomials. Watch for special forms that can be multiplied by inspection, and use the distributive property when it is appropriate.

Example 3   Multiplying Radicals with More Than One Term

Perform each indicated multiplication, and then simplify the product. Assume that the variables represent positive real numbers so that absolute value notation is not necessary.

(a) $2\sqrt{3}(3\sqrt{3} - 5)$ **(b)** $(4\sqrt{11} - 2)(3\sqrt{11} + 7)$
(c) $(\sqrt{2x} + \sqrt{3y})(\sqrt{2x} - \sqrt{3y})$ **(d)** $(\sqrt[3]{x} - 1)(\sqrt[3]{x^2} + \sqrt[3]{x} + 1)$

Solution

(a) $2\sqrt{3}(3\sqrt{3} - 5) = (2\sqrt{3})(3\sqrt{3}) - (2\sqrt{3})(5)$ Distribute the multiplication of $2\sqrt{3}$,
and then simplify.

$\qquad\qquad\qquad\quad = 6(\sqrt{3})^2 - 10\sqrt{3}$

$\qquad\qquad\qquad\quad = 6(3) - 10\sqrt{3}$

$\qquad\qquad\qquad\quad = 18 - 10\sqrt{3}$

(b) $(4\sqrt{11} - 2)(3\sqrt{11} + 7)$ Use the distributive property to

$= 4\sqrt{11}(3\sqrt{11}) + 4\sqrt{11}(7) - 2(3\sqrt{11}) - 2(7)$ multiply each term in the first factor
by each term in the second factor.

$= 12(\sqrt{11})^2 + 28\sqrt{11} - 6\sqrt{11} - 14$

$= 12(11) + 22\sqrt{11} - 14$ Simplify and then add the like terms.

$= 132 + 22\sqrt{11} - 14$

$= 118 + 22\sqrt{11}$

(c) $(\sqrt{2x} + \sqrt{3y})(\sqrt{2x} - \sqrt{3y}) = (\sqrt{2x})^2 - (\sqrt{3y})^2$ This is a special product of the form
$(A + B)(A - B) = A^2 - B^2$ with

$\qquad\qquad\qquad\qquad\qquad = 2x - 3y$ $A = \sqrt{2x}$ and $B = \sqrt{3y}$.

(d) $(\sqrt[3]{x} - 1)(\sqrt[3]{x^2} + \sqrt[3]{x} + 1) = (\sqrt[3]{x})^3 - (1)^3$ This is a special product of the form
$(A - B)(A^2 + AB + B^2) = A^3 - B^3$

$\qquad\qquad\qquad\qquad\qquad = x - 1$ with $A = \sqrt[3]{x}$ and $B = 1$.

Self-Check 3

Simplify each product. Assume $x \geq 0$.

a. $(2\sqrt{3} - \sqrt{5})^2$ **b.** $(5\sqrt{x} + 4)(5\sqrt{x} - 4)$

Conjugate Radicals: The special products play an important role in simplifying many algebraic expressions. Once again we will use the product $(A + B)(A - B) = A^2 - B^2$. The radical expressions $x + \sqrt{y}$ and $x - \sqrt{y}$ are called **conjugates** of each other. Knowing that $(x + \sqrt{y})(x - \sqrt{y}) = x^2 - y$ is very useful because the product is an expression free of radicals. Note the connection between complex conjugates from Section 7.5 and conjugate radicals.

Radical Expression	Conjugate Radical Expression
$2 + \sqrt{3}$	$2 - \sqrt{3}$
$2 - \sqrt{3}$	$2 + \sqrt{3}$
$-3 + 4\sqrt{11}$	$-3 - 4\sqrt{11}$
$1 + 6\sqrt{3x}$	$1 - 6\sqrt{3x}$

Example 4   Multiplying Conjugates

Find the product of $7 + \sqrt{5}$ and its conjugate.

Solution

$$(7 + \sqrt{5})(7 - \sqrt{5}) = (7)^2 - (\sqrt{5})^2$$
$$= 49 - 5$$
$$= 44$$

The conjugate of $7 + \sqrt{5}$ is $7 - \sqrt{5}$. Note that both $7 + \sqrt{5}$ and $7 - \sqrt{5}$ are irrational numbers, but the product of these conjugates is the natural number 44.

Self-Check 4

Multiply $-5 + \sqrt{3}$ by its conjugate.

2. Divide and Simplify Radical Expressions

What Is the Preferred Form for the Answer of a Division Problem?

The preferred form is an expression whose denominator does not contain a radical expression. If there is a radical in the denominator, then the process of removing the radical from the denominator is called **rationalizing the denominator.** (Rationalizing the denominator was first covered in Section 7.1.) Because $\sqrt{x^2} = x$ for $x \geq 0$, we can multiply a square root by itself to obtain a perfect square and thus remove the radical in the denominator. This is illustrated in Example 5.

Example 5   Dividing Radicals Involving Square Roots

Perform the indicated divisions and express the quotients in rationalized form. Assume $a \geq 0$ and $b > 0$.

(a) $\sqrt{7} \div \sqrt{11}$ **(b)** $\dfrac{\sqrt{2a}}{\sqrt{3b}}$

Solution

(a) $\sqrt{7} \div \sqrt{11} = \dfrac{\sqrt{7}}{\sqrt{11}}$

Write the quotient in fractional form.
Multiply both the numerator and the denominator by $\sqrt{11}$ to make the radicand in the denominator a perfect square.

$\qquad = \dfrac{\sqrt{7}}{\sqrt{11}} \cdot \dfrac{\sqrt{11}}{\sqrt{11}}$

Multiplying by the multiplicative identity in the form $\dfrac{\sqrt{11}}{\sqrt{11}}$ does not change the value of this fraction. The denominator is now the rational number 11 instead of the irrational number $\sqrt{11}$.

$\qquad = \dfrac{\sqrt{77}}{11}$

(b) $\dfrac{\sqrt{2a}}{\sqrt{3b}} = \dfrac{\sqrt{2a}}{\sqrt{3b}} \cdot \dfrac{\sqrt{3b}}{\sqrt{3b}}$

Multiply both the numerator and the denominator by $\sqrt{3b}$ to make the radicand in the denominator a perfect square.

$\qquad = \dfrac{\sqrt{6ab}}{3b}$

Multiplying by the multiplicative identity in the form $\dfrac{\sqrt{3b}}{\sqrt{3b}}$ does not change the value of this fraction.

Self-Check 5

a. Simplify $\dfrac{\sqrt{5}}{\sqrt{3}}$.

b. Simplify $\dfrac{\sqrt{5v}}{\sqrt{3w}}$ assuming $v \geq 0$ and $w > 0$.

For a radicand to be a perfect cube, every factor in the radicand must be a perfect cube. Example 6 illustrates how to rationalize an expression involving cube roots.

Example 6   Dividing Radicals Involving Cube Roots

Perform each division and express the quotient in rationalized form. Assume $w \neq 0$ and $x \neq 0$.

(a) $\dfrac{\sqrt[3]{2}}{\sqrt[3]{25}}$

(b) $\dfrac{\sqrt[3]{v}}{\sqrt[3]{wx^2}}$

Solution

(a) $\dfrac{\sqrt[3]{2}}{\sqrt[3]{25}} = \dfrac{\sqrt[3]{2}}{\sqrt[3]{25}} \cdot \dfrac{\sqrt[3]{5}}{\sqrt[3]{5}}$

Multiply both the numerator and denominator by $\sqrt[3]{5}$ in order to produce the perfect cube 125.

$\qquad = \dfrac{\sqrt[3]{10}}{\sqrt[3]{125}}$

Then simplify the denominator.

$\qquad = \dfrac{\sqrt[3]{10}}{5}$

(b) $\dfrac{\sqrt[3]{v}}{\sqrt[3]{wx^2}} = \dfrac{\sqrt[3]{v}}{\sqrt[3]{wx^2}} \cdot \dfrac{\sqrt[3]{w^2x}}{\sqrt[3]{w^2x}}$

Multiply both the numerator and the denominator by the factors needed to produce the perfect cube w^3x^3.

$\qquad = \dfrac{\sqrt[3]{vw^2x}}{\sqrt[3]{w^3x^3}}$

Simplify the denominator.

$\qquad = \dfrac{\sqrt[3]{vw^2x}}{wx}$

This form is rationalized because the denominator is written free of radicals.

Self-Check 6

Express each quotient in rationalized form.

a. $\sqrt[3]{\dfrac{7}{4}}$

b. $\sqrt[3]{\dfrac{2v}{3w^2}}$ assuming $w \neq 0$.

We now examine division by radical expressions with two terms. The key to simplifying these expressions is to use conjugate pairs because the product of the conjugate pair $\sqrt{x} + \sqrt{y}$ and $\sqrt{x} - \sqrt{y}$ contains no radical. For $x \geq 0$ and $y \geq 0$, $(\sqrt{x} + \sqrt{y})(\sqrt{x} - \sqrt{y}) = x - y$.

Example 7   Rationalizing a Denominator with Two Terms

Simplify $\dfrac{6}{\sqrt{7} - \sqrt{5}}$ by rationalizing the denominator.

Solution

$$\dfrac{6}{\sqrt{7} - \sqrt{5}} = \dfrac{6}{\sqrt{7} - \sqrt{5}} \cdot \dfrac{\sqrt{7} + \sqrt{5}}{\sqrt{7} + \sqrt{5}}$$

Multiply the numerator and the denominator by $\sqrt{7} + \sqrt{5}$. This factor is used because $\sqrt{7} - \sqrt{5}$ and $\sqrt{7} + \sqrt{5}$ are a conjugate pair. Multiplying by the multiplicative identity 1 in the form $\dfrac{\sqrt{7} + \sqrt{5}}{\sqrt{7} + \sqrt{5}}$ does not change the value of this fraction. $(\sqrt{7} - \sqrt{5})(\sqrt{7} + \sqrt{5}) = 7 - 5$

$$= \dfrac{6(\sqrt{7} + \sqrt{5})}{7 - 5}$$

$$= \dfrac{6(\sqrt{7} + \sqrt{5})}{2}$$

$$= 3(\sqrt{7} + \sqrt{5})$$

$$= 3\sqrt{7} + 3\sqrt{5}$$

Divide the numerator and the denominator by their common factor 2.

Self-Check 7

Simplify $\dfrac{20}{\sqrt{6} - \sqrt{2}}$.

Example 8 involves a denominator with variables in the radicands. The restriction that $x \neq y$ is made to prevent division by 0.

Example 8   Rationalizing a Denominator with Variables

Simplify $\dfrac{\sqrt{x}}{\sqrt{x} + \sqrt{y}}$ by rationalizing the denominator. Assume that x and y represent positive real numbers and that $x \neq y$.

Solution

$$\dfrac{\sqrt{x}}{\sqrt{x} + \sqrt{y}} = \dfrac{\sqrt{x}}{\sqrt{x} + \sqrt{y}} \cdot \dfrac{\sqrt{x} - \sqrt{y}}{\sqrt{x} - \sqrt{y}}$$

Multiply both the numerator and the denominator by $\sqrt{x} - \sqrt{y}$. The product of the conjugate pair $\sqrt{x} + \sqrt{y}$ and $\sqrt{x} - \sqrt{y}$ yields an expression free of radicals.

$$= \dfrac{\sqrt{x}(\sqrt{x} - \sqrt{y})}{(\sqrt{x} + \sqrt{y})(\sqrt{x} - \sqrt{y})}$$

$$= \dfrac{x - \sqrt{xy}}{x - y}$$

Distribute the factor of $\sqrt{x}$ in the numerator, and multiply the conjugates in the denominator.

Self-Check 8

Simplify $\dfrac{\sqrt{2a}}{\sqrt{2a} - \sqrt{b}}$. Assume that a and b represent positive real numbers and $b \neq 2a$.

10.3 Using the Language and Symbolism of Mathematics

1. The product rule for radicals states that:
$\sqrt[n]{x}\,\sqrt[n]{y} = $ _____ for $x \geq 0$ and $y \geq 0$

2. The quotient rule for radicals states that:
$\dfrac{\sqrt[n]{x}}{\sqrt[n]{y}} = $ _____ for $x \geq 0$ and $y > 0$

3. Before the product and quotient rules can be applied, the radicals must be of the same _____.

4. The product and quotient rules cannot be applied if one or both of the radicands are negative and n is _____.

5. The radical expressions $x + \sqrt{y}$ and $x - \sqrt{y}$ are called _____ of each other.

6. The radical expressions $\sqrt{x} + \sqrt{y}$ and $\sqrt{x} - \sqrt{y}$ are a conjugate _____.

7. The process of removing the radical from the denominator of an expression is called _____ the denominator.

10.3 Quick Review

In Exercises 1–4, multiply these polynomials and simplify the results.

1. $5xy(3x - 2y)$

2. $(2x + 5y)(7x - 3y)$

3. $(4x + 5y)(4x - 5y)$

4. $(4x + 5y)^2$

5. Write the conjugate of $5 - 7i$.

10.3 Exercises

Objective 1 Multiply Radical Expressions

In Exercises 1–4, perform each indicated multiplication and simplify the product. Then use a calculator to evaluate both the original expression and your answer as a check on your answer.

1. $\sqrt{2}\sqrt{6}$

2. $\sqrt{3}\sqrt{6}$

3. $\sqrt{2}(5\sqrt{2} - 1)$

4. $\sqrt{7}(2\sqrt{7} + 1)$

In Exercises 5–38, perform the indicated multiplication and simplify the product. Assume that the variables represent nonnegative real numbers, so that absolute value notation is not necessary.

5. $(2\sqrt{3})(4\sqrt{5})$

6. $(3\sqrt{7})(4\sqrt{2})$

7. $(7\sqrt{15})(4\sqrt{21})$

8. $(4\sqrt{10})(9\sqrt{15})$

9. $\sqrt{3}(\sqrt{2} + \sqrt{5})$

10. $\sqrt{5}(\sqrt{3} + \sqrt{7})$

11. $3\sqrt{5}(2\sqrt{15} - 7\sqrt{35})$

12. $2\sqrt{7}(5\sqrt{14} - \sqrt{21})$

13. $\sqrt{2}(5\sqrt{6})(8\sqrt{3})$

14. $-\sqrt{5}(3\sqrt{15})(\sqrt{3})$

15. $\sqrt{8w}\sqrt{2w}$

16. $\sqrt{6v}\sqrt{21v}$

17. $(2\sqrt{6z})(4\sqrt{3z})$

18. $(5\sqrt{14z})(11\sqrt{7z})$

19. $3\sqrt{x}(2\sqrt{x} - 5)$

20. $4\sqrt{x}(3\sqrt{x} - 11)$

21. $(\sqrt{3x} - \sqrt{y})(\sqrt{3x} + 4\sqrt{y})$

22. $(\sqrt{2x} + \sqrt{y})(\sqrt{2x} - 5\sqrt{y})$

23. $(2\sqrt{3} - \sqrt{2})^2$

24. $(3\sqrt{2} - \sqrt{3})^2$

25. $(\sqrt{a} + 5\sqrt{3b})^2$

26. $(\sqrt{3a} + 3\sqrt{5b})^2$

27. $(\sqrt{v - 2} + 3)(\sqrt{v - 2} - 3)$

28. $(\sqrt{2v + 3} - 5)(\sqrt{2v + 3} + 5)$

29. $\sqrt[3]{7}\sqrt[3]{49}$

30. $\sqrt[3]{36}\sqrt[3]{6}$

31. $\sqrt[3]{9v}\sqrt[3]{-3v^2}$

32. $\sqrt[3]{-4v}\sqrt[3]{2v^2}$

33. $-\sqrt[3]{4}(2\sqrt[3]{2} + \sqrt[3]{5})$

34. $-\sqrt[3]{6}(3\sqrt[3]{4} - 2\sqrt[3]{9})$

35. $(\sqrt[3]{xy^2})^2$

36. $(\sqrt[3]{4ab^2})^2$

37. $-\sqrt{2}(3\sqrt{2} - 5\sqrt{6} - 7)$

38. $3(5\sqrt{2} - \sqrt{6} + 8)$

In Exercises 39–44, multiply each radical expression by its conjugate and simplify the result. Assume that the radicands represent nonnegative real numbers, so that absolute value notation is unnecessary.

39. $2 - \sqrt{2}$

40. $7 + \sqrt{6}$

41. $x + \sqrt{3y}$

42. $2x - \sqrt{y}$

43. $v + \sqrt{3v - 1}$

44. $v - \sqrt{4v - 3}$

Objective 2 Divide and Simplify Radical Expressions

In Exercises 45–58, simplify each indicated division. Rationalize the denominator only if this step is necessary. Then use a calculator to evaluate both the original expression and your answer as a check on your answer.

45. $\sqrt{\dfrac{3}{4}}$

46. $\sqrt{\dfrac{7}{36}}$

47. $\dfrac{2}{\sqrt{6}}$

48. $\dfrac{5}{\sqrt{15}}$

49. $\dfrac{\sqrt{5}}{\sqrt{8}}$

50. $\dfrac{\sqrt{11}}{\sqrt{13}}$

51. $\dfrac{\sqrt{10}}{\sqrt{2}}$

52. $\dfrac{\sqrt{33}}{\sqrt{3}}$

53. $\dfrac{12}{\sqrt[3]{4}}$

54. $\dfrac{12}{\sqrt[3]{9}}$

55. $\dfrac{\sqrt[3]{5}}{\sqrt[3]{16}}$

56. $\dfrac{\sqrt[3]{4}}{\sqrt[3]{81}}$

57. $\sqrt[3]{\dfrac{22}{99}}$

58. $\sqrt[3]{\dfrac{300}{400}}$

In Exercises 59–82, perform each indicated division by rationalizing the denominator and then simplifying. Assume that all variables represent positive real numbers.

59. $18 \div \sqrt{6}$

60. $25 \div \sqrt{5}$

61. $\dfrac{26}{\sqrt{10}}$

62. $\dfrac{9}{\sqrt{15}}$

63. $\dfrac{15}{\sqrt{3x}}$

64. $\dfrac{14}{\sqrt{7y}}$

65. $\dfrac{3}{1 + \sqrt{7}}$

66. $\dfrac{12}{\sqrt{5} - 3}$

67. $\dfrac{36}{\sqrt{13} - 5}$

68. $\dfrac{2}{1 + \sqrt{3}}$

69. $\dfrac{-15}{\sqrt{7} - \sqrt{2}}$

70. $\dfrac{-6}{\sqrt{5} - \sqrt{3}}$

71. $\dfrac{\sqrt{a}}{\sqrt{a} - \sqrt{b}}$

72. $\dfrac{\sqrt{b}}{\sqrt{a} + \sqrt{b}}$

73. $\dfrac{12}{\sqrt[3]{3}}$

74. $\dfrac{4}{\sqrt[3]{2}}$

75. $\sqrt[3]{\dfrac{3}{4}}$

76. $\sqrt[3]{\dfrac{2}{25}}$

77. $\sqrt[3]{\dfrac{v}{9w^2}}$

78. $\sqrt[3]{\dfrac{2v^2}{3w}}$

79. $\sqrt[3]{\dfrac{4v^2}{2vw^2}}$

80. $\sqrt[3]{\dfrac{6vw}{50vw^5}}$

81. $\dfrac{5}{\sqrt{3x} + \sqrt{2y}}$

82. $\dfrac{7}{\sqrt{5x} - \sqrt{3y}}$

Estimate Then Calculate

In Exercises 83 and 84, complete the following table.

a. Estimate each radical expression to the nearest integer.

b. Determine whether this integer estimate is less than or greater than the actual value.

c. Use a calculator to approximate each expression to the nearest thousandth.

	Integer Estimate	Inequality	Approximation
Example: $\dfrac{6}{\sqrt{4.1}}$	3	$3 > \dfrac{6}{\sqrt{4.1}}$	2.963
83. $\dfrac{100}{\sqrt{24}}$			
84. $\dfrac{144}{\sqrt{65}}$			

Skill and Concept Development

In Exercises 85–90, simplify each expression assuming each variable represents a positive real number.

85. $\sqrt[3]{x^7 y^5}$

86. $\sqrt[4]{81 m^5 n^7}$

87. $\dfrac{\sqrt[3]{4z}}{\sqrt[3]{z^2}}$

88. $\dfrac{\sqrt[4]{25y}}{\sqrt[4]{8z^3}}$

89. $\dfrac{10}{\sqrt[5]{8z^3}}$

90. $(\sqrt[3]{x} + 1)(\sqrt[3]{x^2} - \sqrt[3]{x} + 1)$

Group discussion questions

91. Discovery Question—Approximations of π Use a calculator to determine which of these approximations is closest to π.

A. $\dfrac{22}{7}$ **B.** $\dfrac{355}{113}$ **C.** $\sqrt{\sqrt{\dfrac{2{,}143}{22}}}$

92. Challenge Question

a. A circle and a square have the same area. Determine the ratio of the perimeter of the circle to the perimeter of the square.

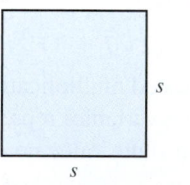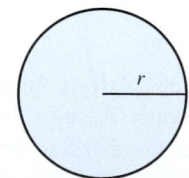

b. A circle and a square have the same perimeter. Determine the ratio of the area of the circle to the area of the square.

93. Challenge Question

a. Write each step to verify that $1 + \sqrt{2}$ is a solution of $x^2 - 2x - 1 = 0$.

b. Write each step to verify that $x = \dfrac{-b + \sqrt{b^2 - 4ac}}{2a}$ is a solution of the quadratic equation $ax^2 + bx + c = 0$.

Solve each equation.

1. $5(2v - 3) - 3(2v - 4) = 5$

2. $|2w - 3| = 5$

3. $(2x - 3)^2 = 25$

4. $(2y - 3)(y - 4) = 7$

5. $\dfrac{3}{m - 7} + 5 = \dfrac{8}{m - 7}$

Section 10.4 | Solving Equations Containing Radical Expressions

Objectives:

1. Solve equations containing radical expressions.
2. Calculate the distance between two points.

1. Solve Equations Containing Radical Expressions

Equations that contain variables in a radicand are called **radical equations.** Since these equations occur frequently in various disciplines, it is important to know how to solve them. In this section, we examine radical equations that result in either linear equations or quadratic equations. An example of such a radical equation is $\sqrt{x} = 3$. It is easy to verify that $x = 9$ is a solution of this equation because $\sqrt{9} = 3$.

What Is the Key Step in Solving a Radical Equation?

Raise both sides of the equation to the same power using the power theorem given in the following box.

> **A Mathematical Note**
>
> Niels Abel (1802–1829) was a Norwegian mathematician who was raised in extreme poverty and died of tuberculosis at the age of 26. Nonetheless, he established several new mathematical concepts. Among his accomplishments was his proof, using radicals, that the general fifth-degree polynomial equation is impossible to solve exactly.

Power Theorem

For any real numbers x and y and natural number n:

Algebraically	Verbally	Example
If $x = y$, then $x^n = y^n$.	If two expressions are equal, then their nth powers are equal.	If $\sqrt{x} = 3$, then $(\sqrt{x})^2 = (3)^2$ $x = 9$
Caution: The equations $x = y$ and $x^n = y^n$ are not always equivalent. The equation $x^n = y^n$ can have a solution that is not a solution of $x = y$.		*Caution:* $(-3)^2 = 3^2$ but $-3 \neq 3$

Because $x^n = y^n$ can be true when $x \neq y$, we must check all possible solutions in the original equation. Any value that occurs as a solution of the last equation of a solution process but is not a solution of the original equation is called an **extraneous value.**

Solving an equation with radicals is a two-stage process. Stage 1 is to solve for possible solutions. Stage 2 is to check each of these possibilities to eliminate any extraneous values.

Solving Radical Equations Containing a Single Radical

Verbally	Algebraic Example
Step 1. Isolate a radical term (of order n) on one side of the equation.	Solve $\sqrt{x-1} + 2 = 6$
	$\sqrt{x-1} = 4$
Step 2. Raise both sides to the nth power.	$(\sqrt{x-1})^2 = 4^2$
Step 3. Solve the resulting equation.	$x - 1 = 16$
Step 4. Check each possible solution in the original equation to determine whether it is a solution or an extraneous value.	$x = 17$
	Check: $\sqrt{17-1} + 2 \overset{?}{=} 6$
	$\sqrt{16} + 2 \overset{?}{=} 6$
	$4 + 2 \overset{?}{=} 6$
	$6 \overset{?}{=} 6$ is true
	Answer: $x = 17$

Example 1   Using Multiple Perspectives to Solve a Radical Equation

Solve $\sqrt{7 - x} = 4$ algebraically and graphically, and use a table to check the solution.

Solution

Algebraic Solution

$$\sqrt{7 - x} = 4$$
$$(\sqrt{7 - x})^2 = 4^2 \qquad \text{Square both sides of the equation.}$$
$$7 - x = 16 \qquad \text{Simplify, and then solve the equation for } x.$$
$$-x = 9$$
$$x = -9$$

Check: This is done in the numerical check.

Graphical Solution

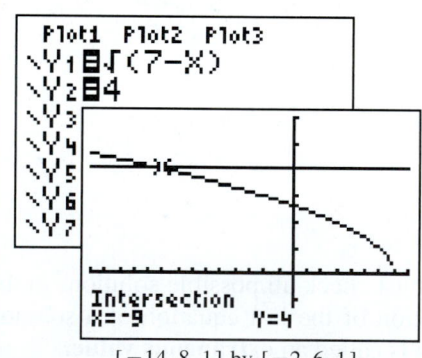

$$[-14, 8, 1] \text{ by } [-2, 6, 1]$$

Use a graphing calculator to graph $Y_1 = \sqrt{7 - x}$ and $Y_2 = 4$.
Then determine the x-value of the point of intersection of these two graphs.
These graphs intersect for an x-value of -9.

The x-coordinate of the point of intersection is $x = -9$.

Numerical Check

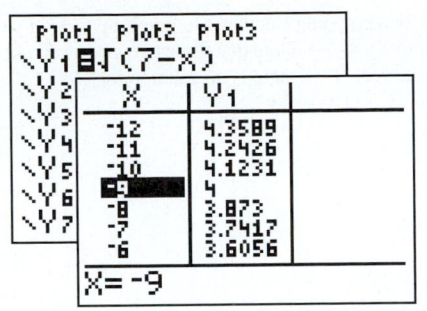

Enter $\sqrt{7-x}$ for Y_1 and examine the table to verify that an x-value of -9 results in a value of 4 for Y_1.

What this table does not answer is whether there are other solutions. The algebraic method shows that there is only one solution.

For $x = -9$, $Y_1 = 4$. $x = -9$ checks.

Answer: $x = -9$

Self-Check 1

Solve $\sqrt{t-3} = 5$.

Example 2 illustrates a problem that yields an extraneous value. By inspection, we could note that this equation cannot have a solution because the principal square root of an expression must always be nonnegative. The reason that we show the steps in the algebraic procedure is to make it clear that the step of checking a possible solution is a crucial step of this procedure.

Example 2   Solving a Radical Equation with No Solution

Solve $\sqrt{2x+3} = -5$ algebraically and graphically.

Solution

Algebraic Solution

$$\sqrt{2x+3} = -5$$

This equation cannot have a solution since a principal root is always nonnegative. However, continue the solution process to see what happens.

$$(\sqrt{2x+3})^2 = (-5)^2$$

$$2x + 3 = 25$$

Square both sides of the equation.

$$2x = 22$$

Simplify, and then solve for x.

$$x = 11$$

Check: $\sqrt{2(11)+3} \overset{?}{=} -5$

$$\sqrt{25} \overset{?}{=} -5$$

$$5 \overset{?}{=} -5 \text{ does not check}$$

$x = 11$ does not check.

Substitute 11 back into the original equation to check this value. The principal square root of 25 is $+5$, not -5. Thus 11 is an extraneous value of the original equation.

Graphical Solution

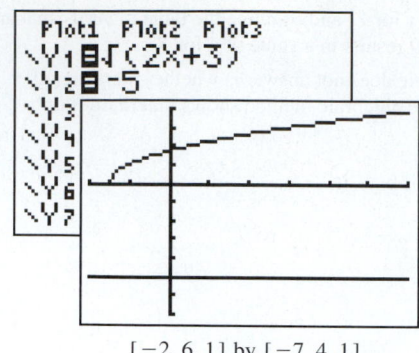

$[-2, 6, 1]$ by $[-7, 4, 1]$

Use a graphing calculator to graph $Y_1 = \sqrt{2x + 3}$ and $Y_2 = -5$. Note that these two graphs do not intersect. Thus there is no solution for $\sqrt{2x + 3} = -5$.

The graphs of the two functions do not intersect.

Answer: No solution

Self-Check 2

Solve $\sqrt{1 - p} = -1$.

Example 3 involves a cube root. The power theorem is also used to solve this equation. This time we cube both sides of the equation.

Example 3   Solving an Equation with One Radical Term

Solve $\sqrt[3]{a - 1} = 4$.

Solution

$$\sqrt[3]{a - 1} = 4$$
$$(\sqrt[3]{a - 1})^3 = 4^3 \qquad \text{Cube both sides of the equation.}$$
$$a - 1 = 64 \qquad \text{Simplify, and then solve the linear equation.}$$
$$a = 65$$

Check: $\sqrt[3]{65 - 1} \overset{?}{=} 4$
$$\sqrt[3]{64} \overset{?}{=} 4$$

Answer: $a = 65$

Self-Check 3

Solve $\sqrt[3]{x - 5} = -2$.

The solution process for solving the radical equation in Example 4 produces a quadratic equation. Although the quadratic equation has two solutions, only one of these values checks in the original equation. The other value is extraneous.

Example 4   Solving an Equation with One Radical Term

Solve $x = \sqrt{x + 6}$.

Solution

$$x = \sqrt{x + 6}$$

Square both sides of the equation.

$$x^2 = x + 6$$

Write the quadratic equation in standard form. Factor the left side of the equation.

$$x^2 - x - 6 = 0$$

$$(x - 3)(x + 2) = 0$$

Set each factor equal to 0.

$$x - 3 = 0 \quad \text{or} \quad x + 2 = 0$$

$$x = 3 \qquad\qquad x = -2$$

Solve for x.

Check:

$x = 3$: $3 \overset{?}{=} \sqrt{3 + 6}$ $x = -2$: $-2 \overset{?}{=} \sqrt{-2 + 6}$

$3 \overset{?}{=} \sqrt{9}$ $-2 \overset{?}{=} \sqrt{4}$

$3 \overset{?}{=} 3$ is true $-2 \overset{?}{=} 2$ is false -2 is an extraneous value.

Answer: $x = 3$

Self-Check 4

Solve $x = \sqrt{3x + 4}$.

In Example 4, note also that a principal square root cannot be negative. Thus we could determine by inspection that $x = -2$ is extraneous.

How Do I Start to Solve a Radical Equation?

Start by isolating a radical term on one side of the equation. Then simplify both sides by squaring if the radicals are square roots, cubing if the radicals are cube roots, etc.

Example 5   Solving an Equation with One Radical Term

Solve $\sqrt{2w - 3} + 9 = w$.

Solution

$$\sqrt{2w - 3} + 9 = w$$

First isolate the radical term on the left side of the equation by subtracting 9 from both sides.

$$\sqrt{2w - 3} = w - 9$$

Square both sides of the equation. (Be careful not to omit the middle term when you square the binomial on the right side.)

$$(\sqrt{2w - 3})^2 = (w - 9)^2$$

$$2w - 3 = w^2 - 18w + 81$$

Write the quadratic equation in standard form.

$$w^2 - 20w + 84 = 0$$

$$(w - 14)(w - 6) = 0$$

Factor, and solve for w.

$$w - 14 = 0 \quad \text{or} \quad w - 6 = 0$$

$$w = 14 \qquad\qquad w = 6$$

Check: $w = 14$: $\sqrt{2(14) - 3} + 9 \overset{?}{=} 14$

$$\sqrt{25} + 9 \overset{?}{=} 14$$

$$5 + 9 \overset{?}{=} 14$$

$$14 \overset{?}{=} 14 \text{ is true}$$

$w = 6$: $\sqrt{2(6) - 3} + 9 \overset{?}{=} 6$

$$\sqrt{9} + 9 \overset{?}{=} 6$$

$$3 + 9 \overset{?}{=} 6$$

$$12 \overset{?}{=} 6 \text{ is false}$$ 6 is an extraneous value.

Answer: $w = 14$

Self-Check 5

Solve $\sqrt{4x + 9} - 1 = x$.

Can I Determine Which Possible Solutions Are Extraneous Without Actually Checking Them?

The answer is generally no, as illustrated in Example 5.

2. Calculate the Distance Between Two Points

The Pythagorean theorem (see page 576, Section 7.4) states that, for a right triangle with sides a, b, and c, $a^2 + b^2 = c^2$. (See the figure.)

    An important formula that can be developed by using the Pythagorean theorem is the formula for the distance between two points in a plane. We develop this formula by first considering the horizontal and vertical changes between the two points. Then we compute the direct distance between them using the Pythagorean theorem.

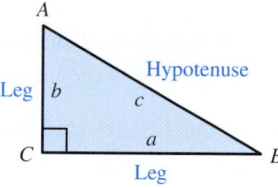

Distance Between Two Points

Special Case	General Case
Calculate the distance between $P(2, 2)$ and $Q(5, 6)$.	Calculate the distance between $P(x_1, y_1)$ and $Q(x_2, y_2)$.

Step 1 Find the horizontal change from P to Q.

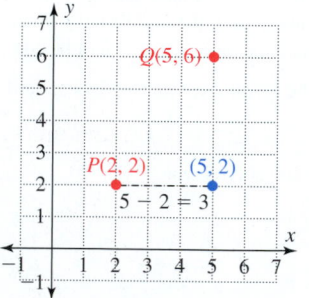

Horizontal distance $= 5 - 2 = 3$

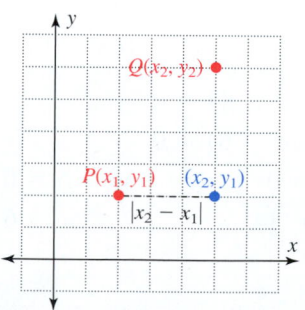

Horizontal distance $= |x_2 - x_1|$

Step 2 Find the vertical change from P to Q.

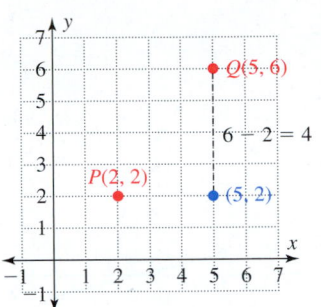
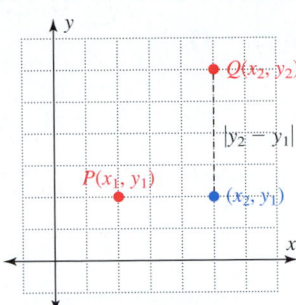

Vertical distance $= 6 - 2 = 4$ Vertical distance $= |y_2 - y_1|$

Step 3 Use the Pythagorean theorem to find the length of hypotenuse PQ.

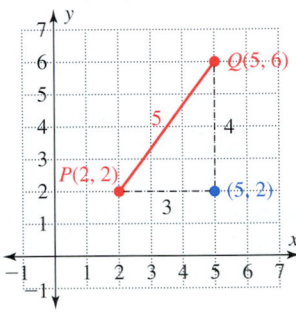
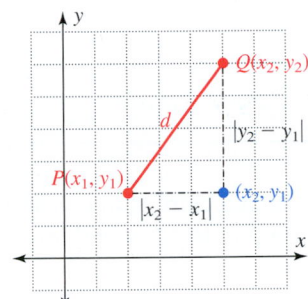

$$d^2 = 3^2 + 4^2$$
$$d = \sqrt{9 + 16}$$
$$= \sqrt{25}$$
$$= 5$$

$$d^2 = |x_2 - x_1|^2 + |y_2 - y_1|^2$$
$$d = \sqrt{|x_2 - x_1|^2 + |y_2 - y_1|^2}$$
$$d = \sqrt{(x_2 - x_1)^2 + (y_2 - y_1)^2}$$

Absolute value notation is not needed in the distance formula because the squares in this formula are nonnegative. The formula is applicable in all cases, even if P and Q are on the same horizontal or vertical line.

Distance Formula

The distance d between (x_1, y_1) and (x_2, y_2) is given by
$$d = \sqrt{(x_2 - x_1)^2 + (y_2 - y_1)^2}.$$

Why Does This Formula Refer to the Distance *Between* (x_1, y_1) and (x_2, y_2) Rather Than the Distance *From* (x_1, y_1) to (x_2, y_2)?

This terminology is used because the distance between these points is the same in either direction. In the formula, note that $(x_2 - x_1)^2 = (x_1 - x_2)^2$ and $(y_2 - y_1)^2 = (y_1 - y_2)^2$.

Example 6 Calculating the Distance Between Two Points

Calculate the distance between $(-3, 1)$ and $(5, -1)$.

Solution

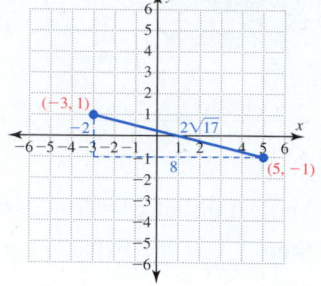

$$d = \sqrt{(x_2 - x_1)^2 + (y_2 - y_1)^2}$$

$$d = \sqrt{[5 - (-3)]^2 + (-1 - 1)^2}$$ Substitute the given values into the distance formula.

$$d = \sqrt{8^2 + (-2)^2}$$

$$d = \sqrt{64 + 4}$$

$$d = \sqrt{68}$$ This exact distance is illustrated in the figure to the right. A calculator approximation of this distance is $2\sqrt{17} \approx 8.25$.

$$d = \sqrt{4}\sqrt{17}$$

$$d = 2\sqrt{17}$$

Self-Check 6

Calculate the distance between $(2, -3)$ and $(-13, 5)$.

Engineering, architecture, and construction often involve placing new structures in a specific location relative to other building and utility lines. Example 7 is a skill builder for problems of this type.

Example 7 Applying the Distance Formula

Find all points with an x-coordinate of 6 that are 5 units from $(3, 3)$.

Solution

Identify the desired point(s) by $(6, y)$.

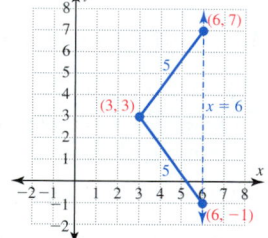

$$d = \sqrt{(x_2 - x_1)^2 + (y_2 - y_1)^2}$$ All points with an x-coordinate of 6 can be written as $(6, y)$.

$$5 = \sqrt{(6 - 3)^2 + (y - 3)^2}$$ Substitute the given values into the distance formula.

$$5 = \sqrt{9 + y^2 - 6y + 9}$$ Then simplify the radicand.

$$5 = \sqrt{y^2 - 6y + 18}$$ Solve the radical equation for y.

$$(5)^2 = (\sqrt{y^2 - 6y + 18})^2$$ Square both sides of the equation.

$$25 = y^2 - 6y + 18$$

$$0 = y^2 - 6y - 7$$ Subtract 25 from both sides of the equation.

$$(y + 1)(y - 7) = 0$$ Factor the trinomial.

$$y + 1 = 0 \qquad \text{or} \qquad y - 7 = 0$$ Set each factor equal to 0.

$$\qquad y = -1 \qquad\qquad\quad y = 7$$ Both values check.

Answer: The points $(6, -1)$ and $(6, 7)$ are both 5 units from $(3, 3)$.

Self-Check 7

Find all points with a y-coordinate of 7 that are 5 units from $(3, 3)$.

Can the Distance Formula Be Applied to a Problem If a Coordinate System Is Not Given with the Problem?

Yes, we can sometimes superimpose a coordinate system on the problem, as illustrated in Example 8.

Example 8 | Measuring an Engine Block

A machinist is measuring a metal block that is part of an automobile engine. To find the distance from the center of hole A to the center of hole B, the machinist determines the coordinates shown on the drawing with respect to a reference point at the lower left corner of the metal block. Assume the dimensions are given in cm. Approximate the distance from the center of A to the center of B to the nearest tenth of a cm.

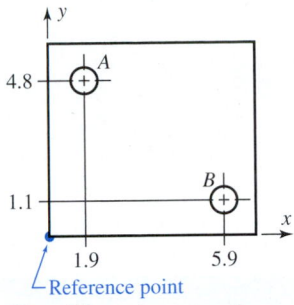

Note: Dimensions given in cm

Solution

Point A: $(1.9, 4.8)$

Point B: $(5.9, 1.1)$

$d = \sqrt{(x_2 - x_1)^2 + (y_2 - y_1)^2}$

$d = \sqrt{(5.9 - 1.9)^2 + (1.1 - 4.8)^2}$

$d = \sqrt{(4)^2 + (-3.7)^2}$

$d = \sqrt{16 + 13.69}$

$d = \sqrt{29.69}$

$d \approx 5.449$ cm

Use the figure to determine the coordinates of points A and B.

Then use the distance formula to determine the distance between point A and point B.

Answer: The distance from the center of A to the center of B is approximately 5.4 cm.

Self-Check 8

Rework Example 8 assuming hole A is at $(2.3, 4.6)$ and B is at $(5.6, 3.4)$.

Self-Check Answers

1. $t = 28$
2. No solution
3. $x = -3$

4. $x = 4$
5. $x = 4$
6. 17

7. The points $(0, 7)$ and $(6, 7)$ are both 5 units from $(3, 3)$.

8. The distance from the center of A to the center of B is approximately 3.5 cm.

10.4 | **Using the Language and Symbolism of Mathematics**

1. A radical equation is an equation that contains variables in a _____ .

2. The power theorem states that if $x = y$, then _____ , provided that x and y are real numbers and n is a natural number.

3. The reason that we must check all possible solutions of a radical equation in the original equation is that when we raise both sides to the nth power, we can introduce _____ values.

4. In a right triangle, the two shorter sides are called the legs and the longest side is called the _____ .

5. The _____ theorem states that $a^2 + b^2 = c^2$ for a right triangle with hypotenuse c and legs a and b.

6. The distance between points (x_1, y_1) and (x_2, y_2) is given by $d = $ _____ .

10.4 | Quick Review

1. Solve $5x - 1 = 64$.

2. Solve $(5x - 1)^2 = 64$.

3. Is 1 a solution of $\sqrt{2x - 1} + 2 = x$?

4. Is 5 a solution of $\sqrt{2x - 1} + 2 = x$?

5. Use this graph to solve $\dfrac{3}{4}(x + 4) = 6$.

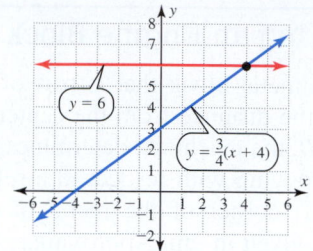

10.4 | Exercises

Objective 1 Solve Equations Containing Radical Expressions

In Exercises 1–4, use the given graphs to solve each radical equation.

1. $\sqrt{x - 4} = 1$

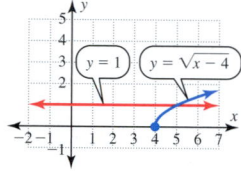

2. $\sqrt{5x - 1} = 3$

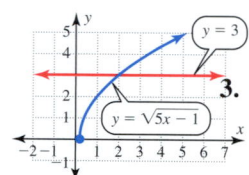

3. $\sqrt[3]{3x - 5} + 2 = 0$

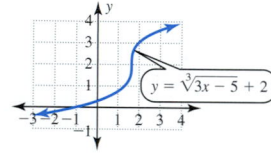

4. $\sqrt[3]{3x - 4} - 2 = 0$

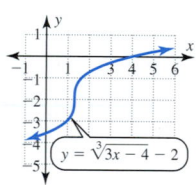

In Exercises 5–30, solve each equation.

5. $\sqrt{x} = 4$

6. $\sqrt{x} = 9$

7. $\sqrt{x} - 7 = -6$

8. $\sqrt{x} - 13 = -8$

9. $\sqrt{t - 4} = 3$

10. $\sqrt{t - 2} = 12$

11. $\sqrt{c + 7} + 23 = 43$

12. $\sqrt{c + 5} - 3 = 6$

13. $\sqrt{3x - 21} + 7 = 2$

14. $\sqrt{25 + 2x} + 11 = 5$

15. $\sqrt[3]{2w + 1} = -2$

16. $\sqrt[3]{5w + 2} = -3$

17. $\sqrt[4]{6v - 2} = 2$

18. $\sqrt[4]{7v - 3} = 3$

19. $\sqrt[5]{w^2 - 4w} = 2$

20. $\sqrt[5]{2w^2 + w - 16} = -1$

21. $\sqrt{y^2 - y + 13} - y = 1$

22. $\sqrt{y^2 + y + 11} - y = 1$

23. $\sqrt{7t + 2} = 2t$

24. $\sqrt{-9t - 2} = 2t$

25. $\sqrt{w^2 - 2w + 1} = 2w$

26. $\sqrt{2w + 1} = w + 1$

27. $\sqrt{2x + 1} + 5 = 2x$

28. $\sqrt{7x - 3} + 1 = 3x$

29. $\sqrt{6u + 7} = \sqrt{11u + 7}$

30. $\sqrt{14u - 12} = \sqrt{10u - 8}$

Objective 2 Calculate the Distance Between Two Points

In Exercises 31–38, use the distance formula to calculate the distance between each pair of points.

31. $(-2, 6)$ and $(3, 6)$

32. $(-3, 2)$ and $(1, -1)$

33. $(2, -7)$ and $(-6, 8)$

34. $(0, -20)$ and $(-9, 20)$

35. $\left(-\dfrac{1}{2}, \dfrac{2}{3}\right)$ and $\left(\dfrac{1}{2}, -\dfrac{1}{3}\right)$

36. $\left(\dfrac{4}{5}, \dfrac{3}{2}\right)$ and $\left(-\dfrac{1}{5}, \dfrac{1}{2}\right)$

37. $(0, 0)$ and $(-\sqrt{2}, \sqrt{7})$

38. $(\sqrt{11}, -\sqrt{14})$ and $(0, 0)$

Connecting Algebra to Geometry

In Exercises 39–42, use the distance formula to determine the perimeter of the triangle formed by these points. Then use the Pythagorean theorem to determine whether the triangle formed is a right triangle. (*Hint:* Does $a^2 + b^2 = c^2$?)

39. $(-3, -1)$, $(4, -4)$, and $(-1, 1)$

40. $(-6, -1)$, $(2, -1)$, and $(0, 1)$

41. $(-5, -1)$, $(2, -2)$, and $(4, 4)$

42. $(3, -3)$, $(-2, -1)$, and $(5, 2)$

In Exercises 43–48, use the Pythagorean theorem to find the length of the side that is not given.

43.
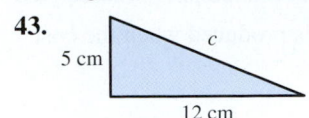
5 cm
c
12 cm

44.
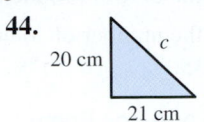
20 cm
c
21 cm

45.
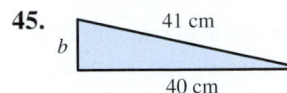
41 cm
b
40 cm

46.
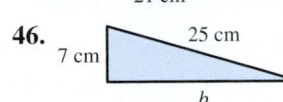
25 cm
7 cm
b

47.
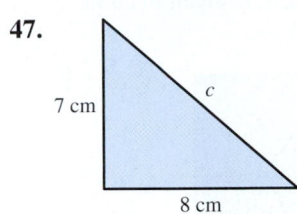
7 cm
c
8 cm

48.
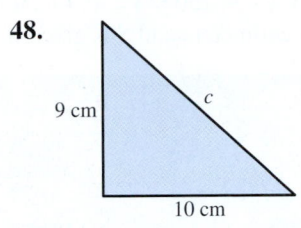
9 cm
c
10 cm

49. Find all points with an x-coordinate of 5 that are 10 units from the point $(13, 2)$.

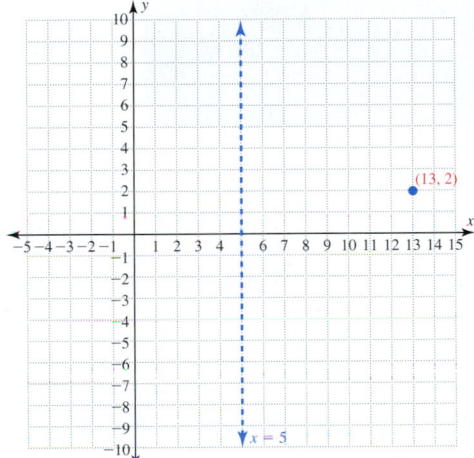

50. Find all points with a y-coordinate of 2 that are 13 units from the point $(-2, 7)$.

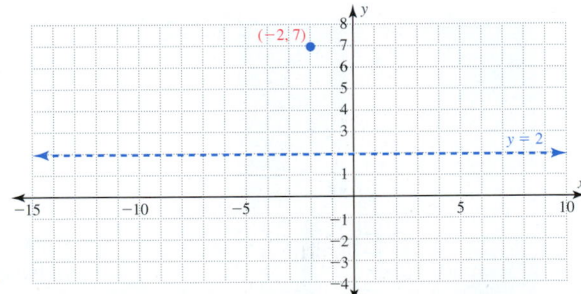

Review and Concept Development

In Exercises 51–54, solve each problem for x.

51. The square root of the quantity x plus five is equal to three.

52. The cube root of the quantity x plus two is equal to minus two.

53. Four times the square root of x equals three times the quantity x minus five.

54. If the square root of the quantity x plus four is added to two, the result equals x.

In Exercises 55–58, determine the exact x- and y-intercepts of the graph of each function.

55. $y = \sqrt{x + 9} - 2$ **56.** $y = \sqrt{x + 4} - 5$

57. $y = \sqrt[3]{x + 1} + 2$ **58.** $y = \sqrt[3]{x - 8} - 1$

In Exercises 59–66, solve each equation.

59. $\sqrt{2x + 6} = x$ **60.** $\sqrt{x + 5} = x$

61. $\sqrt{2x^2 + 3x - 1} = x$

62. $\sqrt{2x^2 + 6x + 3} - 2 = x$

63. $\sqrt[3]{x^3 - 6x^2 + 12x} = x$ **64.** $\sqrt[3]{x^3 + 9x^2 + 27x} = x$

65. $\sqrt{4w^2 + 12w + 6} = 2w$

66. $\sqrt{9w^2 - 12w + 8} = 3w$

In Exercises 67–72, there is only one solution for each equation. Use the table to determine that solution.

67. $\sqrt{2x - 11} = 3$

x	$\sqrt{2x - 11}$
6	1.00
8	2.24
10	3.00
12	3.61
14	4.12
16	4.58
18	5.00

68. $\sqrt{5x + 19} = 7$

x	$\sqrt{5x + 19}$
0	4.36
2	5.39
4	6.24
6	7.00
8	7.68
10	8.31
12	8.89

69. $\sqrt{4x + 1} = x - 1$

x	$\sqrt{4x + 1}$	$x - 1$
0	1.00	-1
2	3.00	1
4	4.12	3
6	5.00	5
8	5.74	7
10	6.40	9
12	7.00	11

70. $\sqrt{2x + 1} = x - 1$

x	$\sqrt{2x + 1}$	$x - 1$
0	1.00	-1
1	1.73	0
2	2.24	1
3	2.65	2
4	3.00	3
5	3.32	4
6	3.61	5

Connecting Concepts to Applications

71. Measuring a Buried Pipe A pipe cleaning firm contracted to clean a pipe buried in a lake. Access points to the pipe are at points A and B on the edge of the lake, as shown in the figure. The contractor placed a stake as a reference point and then measured the coordinates in meters from this reference point to A and B. The coordinates of A and B are, respectively, $(3.0, 5.2)$ and $(37.8, 29.6)$. Approximate to the nearest tenth of a meter the distance between A and B.

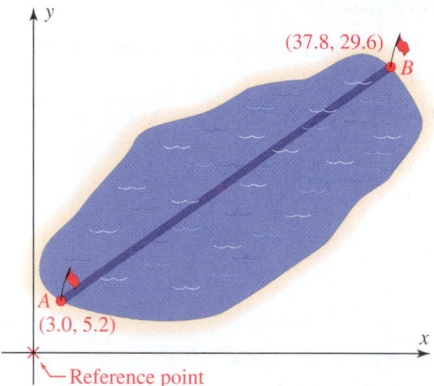

72. Measuring an Engine Block A machinist is measuring a metal block that is part of an automobile engine. To find the distance from the center of hole A to the center of hole B, the machinist determines the coordinates shown on the drawing with respect to a reference point at the lower left corner of the metal block. Approximate the distance from the center of A to the center of B. The dimensions are given in centimeters.

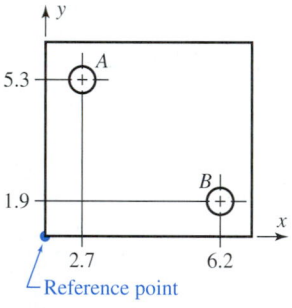

Production of Solar Cells

In Exercises 73–75, use the fact that the dollar cost C of producing n solar cells per shift is given by the formula $C = 18\sqrt[3]{n^2} + 450$.

73. Find the overhead cost for one shift.

74. Find the cost of producing 27 solar cells.

75. Find the number of solar cells produced when the cost is \$738.

Strength of a Box Beam

In Exercises 76–78, use the fact that the load strength of a square box beam is related to its volume by the function $S(V) = 750\sqrt[3]{V^2}$. In this formula, V is given in cubic centimeters and S is given in newtons.

76. Evaluate and interpret $S(0)$.

77. Evaluate and interpret $S(1,000)$.

78. If $S(V) = 750$, calculate and interpret V.

79. Height of a Wall A 17-ft ladder is leaning against a chimney.
  a. Express the distance from the base of the chimney to the top of the ladder as a function of x, the distance from the base of the ladder to the base of the chimney.
  b. If the bottom of the ladder is 8 ft from the base of the chimney, how far is it from the bottom of the chimney to the top of the ladder?

80. Length of a Guy Wire A television tower has a guy wire attached 40 ft above the base of the tower.
  a. Express the length of the guy wire as a function of x, the distance from the base of the tower to the anchor point of the guy wire.

b. If the anchor point of the guy wire is 42 ft from the base of the tower, how long is the guy wire?

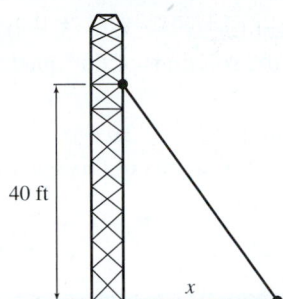

40 ft

x

81. Length of an Exercise Rope One end of a retractable exercise rope for a quarter horse is fastened to the swivel at the top of a 16-ft pole. The other end is fastened to the neck of the horse at a height of 5 ft.
a. Write a function $L(x)$ that gives the length of the rope as a function of x, where x is the distance in feet from the pole to the horse.
b. Evaluate and interpret $L(30)$.
c. Approximate the area that the horse can occupy if the rope is 35 ft.

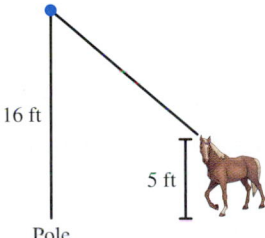

16 ft

5 ft

Pole

82. Length of an Extension Cord One end of a retractable extension cord is fastened to a swivel 12 ft above the floor on the ceiling of an automotive repair shop. The other end of the cord is an outlet that can be positioned around the floor of the shop.
a. Write a function $L(x)$ that gives the length of the cord that is extended as a function of x, where x is the distance in feet from a point directly under the swivel to the end of the cord on the shop floor.
b. Evaluate and interpret $L(15)$.
c. Solve $L(x) = 16$ for x, and interpret this result.

83. Path of a Power Cable The path of a power cable between booster stations located at A and C is shown in the figure. The maximum length of cable between booster stations is 12 mi. Determine the distance x so that the path from A to C through B will be 12 mi.

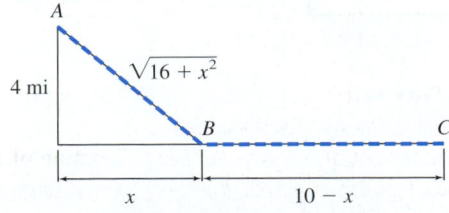

A

$\sqrt{16 + x^2}$

4 mi

B   C

x   10 − x

84. Path of a Fiber-Optic Cable The path of an underground fiber-optic cable between buildings located at A and C in a business district is shown in the figure. The maximum length of cable allowed for connections of this type is 17 km. Determine the distance x so that the path from A to C through B will be 17 km.

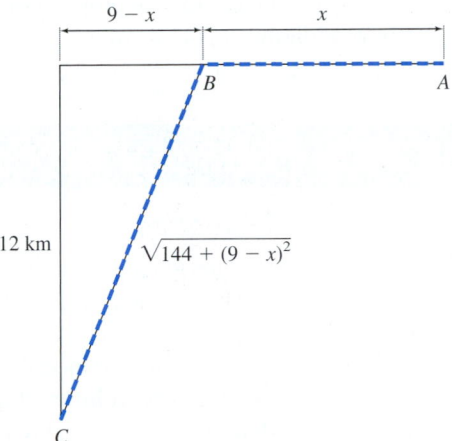

9 − x   x

B   A

12 km $\sqrt{144 + (9 - x)^2}$

C

Group discussion questions

85. Discovery Question
a. Write the equation of all points 3 units from the origin. Describe the shape formed by these points.
b. Write the equation of all points 2 units from the point $(3, 5)$. Describe the shape formed by these points.
c. Write the equation of all points r units from the point (h, k). Describe the shape formed by these points.

86. Challenge Question Solve each equation for x.
a. $y = \dfrac{\sqrt[3]{2x - z}}{3}$
b. $y = \sqrt{v^2 - w^2 + x}$
c. $5 = \sqrt{\dfrac{v + x}{v - x}}$

87. Challenge Question The square shown in the figure was etched in stone, along with the single word *BEHOLD*, by a Hindu mathematician, who meant that this figure offers visual proof of the Pythagorean theorem. Describe how this square can be used to prove the Pythagorean theorem. (*Hint:* Add the areas of the four triangles and the inner square.)

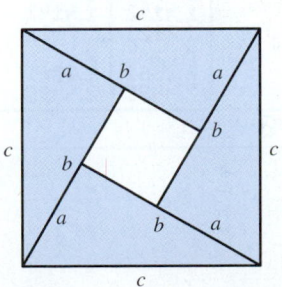

c

a   b   a

b

c   c

b

a   b   a

c

10.4 | Cumulative Review

1. Rewrite $a + b$ using the commutative property of addition.

2. Rewrite $a(b \cdot c)$ using the associative property of multiplication.

3. Rewrite $a(b + c)$ using the distributive property of multiplication over addition.

4. Calculate the discriminant for $2x^2 - 3x + 5 = 0$ and determine the nature of the solutions of this quadratic equation.

5. Calculate the discriminant for $2x^2 - 3x - 5 = 0$ and determine the nature of the solutions of this quadratic equation.

Section 10.5 | Rational Exponents and Radicals

Objectives:

1. Interpret and use rational exponents.
2. Use the properties of exponents.

One classic engineering problem that involves fractional exponents is the relationship between the volume of a box beam and the strength of the beam. Bridges built based on data from a scaled-down model may not be a good idea because the strength of the beam and weight of the beam do not grow at the same rate. We examine this problem in Example 7 at the end of this section.

1. Interpret and Use Rational Exponents

To develop a definition for fractional exponents, we start by examining $x^{1/2}$. We want $x^{1/2}$ to obey the same rules for exponents as those given in Chapter 5 for integer exponents. Thus we want to be able to use the product rule to simplify $x^{1/2} \cdot x^{1/2}$.

$$x^{1/2} \cdot x^{1/2} = x^{1/2 + 1/2}$$

Use the product rule $x^m \cdot x^n = x^{m+n}$ to add the exponents.

$$= x^1$$

A number multiplied by itself that yields x as a product is known as the square root of x. Thus it is reasonable to define $x^{1/2}$ as $\sqrt{x}$, the principal square root of x.

The following table examines the definition of $x^{1/2}$ algebraically, numerically, graphically, and verbally.

$x^{1/2}$: The Principal Square Root of x

Algebraically	Numerically	Graphically	Verbally
$x^{1/2} = \sqrt{x}$	Plot1 Plot2 Plot3 \Y1◘X^(1/2) \Y2◘√(X) \Y3 \Y4 \Y5 \Y6 \Y7 X | Y1 | Y2 0 | 0 | 0 1 | 1 | 1 2 | 1.4142 | 1.4142 3 | 1.7321 | 1.7321 4 | 2 | 2 5 | 2.2361 | 2.2361 6 | 2.4495 | 2.4495 X=0	 $[-1, 10, 1]$ by $[-1, 5, 1]$	We define $x^{1/2}$ to equal $\sqrt{x}$. In the table, the numerical values for $x^{1/2}$ and $\sqrt{x}$ are identical. The graphs of the functions $Y_1 = x^{1/2}$ and $Y_2 = \sqrt{x}$ are identical.

Can Cube Roots and Fourth Roots Also Be Defined Using Fractional Exponents?

Yes, we now generalize the meaning of fractional exponents in order to define $x^{1/n}$ for any natural number n and fraction $\dfrac{1}{n}$. We want to be able to use the power rule for exponents with $x^{1/n}$ as shown below.

$$(x^{1/n})^n = x^{n/n} \qquad \text{Use the power rule } (x^m)^n = x^{mn} \text{ to}$$
$$\phantom{(x^{1/n})^n} = x^1 \qquad \text{multiply the exponents.}$$

Because $x^{1/n}$ is used as a factor n times in the expression $(x^{1/n})^n$, it is reasonable to define $x^{1/n}$ as $\sqrt[n]{x}$, **the principal nth root of x.**

$x^{1/n}$: The Principal nth Root of x

The principal nth root of the real number x is denoted by either $x^{1/n}$ or $\sqrt[n]{x}$.

		Examples	
Verbally		**Radical Notation**	**Exponential Notation**
For $x > 0$:	The principal nth root is positive for all natural numbers n.	$\sqrt{9} = 3$ $\sqrt[3]{8} = 2$	$9^{1/2} = 3$ $8^{1/3} = 2$
For $x = 0$:	The principal nth root of 0 is 0.	$\sqrt{0} = 0$ $\sqrt[3]{0} = 0$	$0^{1/2} = 0$ $0^{1/3} = 0$
For $x < 0$:	If n is odd, the principal nth root is negative.	$\sqrt[3]{-8} = -2$ $\sqrt[5]{-1} = -1$	$(-8)^{1/3} = -2$ $(-1)^{1/5} = -1$
	If n is even, there is no real nth root. (The nth roots are imaginary.)	$\sqrt{-1}$ is not a real number.	$(-1)^{1/2}$ is not a real number.

Example 1   Representing and Evaluating Principal nth Roots

Represent each of these principal nth roots by using both radical notation and exponential notation, and evaluate each expression.

Solution

	Radical Notation	Exponential Notation
(a) The principal square root of 25	$\sqrt{25} = 5$	$25^{1/2} = 5$
(b) The principal cube root of 27	$\sqrt[3]{27} = 3$	$27^{1/3} = 3$
(c) The principal cube root of -27	$\sqrt[3]{-27} = -3$	$(-27)^{1/3} = -3$
(d) The principal fourth root of 16	$\sqrt[4]{16} = 2$	$16^{1/4} = 2$
(e) The principal fifth root of 1	$\sqrt[5]{1} = 1$	$1^{1/5} = 1$

Self-Check 1

Write each of these expressions in radical notation and evaluate each expression.

a. $64^{1/2}$ **b.** $64^{1/3}$ **c.** $64^{1/6}$ **d.** $(-64)^{1/3}$

Can Any Rational Number Be Used As an Exponent?

Yes, to extend the definition of exponents to include any rational number $\dfrac{m}{n}$, we start with $x^{1/n} = \sqrt[n]{x}$ and then use the power rule for exponents.

$$x^{1/n} = \sqrt[n]{x}$$
$$(x^{1/n})^m = (\sqrt[n]{x})^m \quad \text{Raise both sides of the equation to the } m\text{th power.}$$
$$x^{m/n} = (\sqrt[n]{x})^m \quad \text{Then use the power rule for exponents to multiply the exponents.}$$

In keeping with our earlier work with negative exponents, we also interpret $x^{-m/n}$ as $\dfrac{1}{x^{m/n}}$. These definitions are summarized in the following box.

Rational Exponents

For a real number x and natural numbers m and n:

		Examples
Algebraically	**Radical Notation**	**Exponential Notation**
If $x^{1/n}$ is a real number,* then $x^{m/n} = (x^{1/n})^m = (\sqrt[n]{x})^m$ or	$(-8)^{2/3} = (\sqrt[3]{-8})^2$ $= (-2)^2$ $= 4$	$(-8)^{2/3} = [(-8)^{1/3}]^2$ $= (-2)^2$ $= 4$
$x^{m/n} = (x^m)^{1/n} = \sqrt[n]{x^m}$	$y^{3/7} = \sqrt[7]{y^3}$	$y^{3/7} = (y^3)^{1/7}$
$x^{-m/n} = \dfrac{1}{x^{m/n}}$	$16^{-3/4} = \dfrac{1}{(\sqrt[4]{16})^3}$	$16^{-3/4} = \dfrac{1}{16^{3/4}}$
$= \dfrac{1}{(\sqrt[n]{x})^m} \quad \text{for } x \neq 0$	$= \dfrac{1}{2^3}$	$= \dfrac{1}{(16^{1/4})^3}$
	$= \dfrac{1}{8}$	$= \dfrac{1}{2^3}$
		$= \dfrac{1}{8}$

*If $x < 0$ and n is even, then $x^{1/n}$ is an imaginary number.

These definitions are illustrated in Example 2. Remember, when you use negative exponents, think "take the reciprocal."

Example 2   Evaluating Expressions with Fractional Exponents

Represent each of these expressions in radical notation and exponential notation, and evaluate each expression.

Solution

	Radical Notation	Exponential Notation
(a) $27^{2/3}$	$(\sqrt[3]{27})^2 = 3^2 = 9$	$27^{2/3} = (27^{1/3})^2 = 3^2 = 9$
(b) $32^{3/5}$	$(\sqrt[5]{32})^3 = 2^3 = 8$	$32^{3/5} = (32^{1/5})^3 = 2^3 = 8$
(c) $9^{-1/2}$	$\dfrac{1}{\sqrt{9}} = \dfrac{1}{3}$	$9^{-1/2} = \dfrac{1}{9^{1/2}} = \dfrac{1}{3}$
(d) $27^{-2/3}$	$\dfrac{1}{(\sqrt[3]{27})^2} = \dfrac{1}{3^2} = \dfrac{1}{9}$	$27^{-2/3} = \dfrac{1}{27^{2/3}} = \dfrac{1}{(27^{1/3})^2} = \dfrac{1}{3^2} = \dfrac{1}{9}$

Self-Check 2

Write each of these expressions in radical notation and evaluate each expression.

a. $64^{3/2}$ **b.** $64^{2/3}$ **c.** $(-64)^{2/3}$ **d.** $64^{-2/3}$

The expressions in Example 2 were carefully selected to illustrate the notation but also so that they could be evaluated using only pencil and paper. Technology Perspective 10.5.1 illustrates how to use exponential notation to evaluate higher-order radical expressions.

Technology Perspective 10.5.1 — Approximating an Expression Containing Rational Exponents

Approximate $27^{2/3}$.

TI-84 Plus Calculator

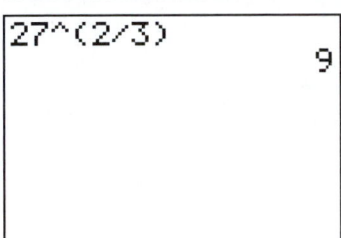

Excel Spreadsheet

A1	▼	f_x	=27^(2/3)

▲	A	B	C
1	9		
2			

1. Enter 27, press ⌃, and then enter the exponent within parentheses as shown here.

2. Press **ENTER** to evaluate this expression.

Answer: $27^{2/3} = 9$.

1. Enter "= 27^(2/3)" in cell **A1**.

2. The formula bar displays this entry.

3. The decimal approximation for $27^{2/3}$ is shown in cell **A1**.

Technology Self-Check 1

Approximate $51^{4/7}$ to the nearest hundredth.

---

2. Use the Properties of Exponents

Do the Properties of Exponents That I Studied in Chapter 5 Also Apply to Rational Exponents?

Yes, rational exponents have been defined in such a way that all the properties of integral exponents now apply to all rational exponents. We have not established it in this book, but these properties apply to all real number exponents. These properties are summarized in the following box.

Properties of Exponents

Let m and n be real numbers and x, x^m, x^n, y, y^m, and y^n be nonzero real numbers.

Product rule: $x^m \cdot x^n = x^{m+n}$

Power rule: $(x^m)^n = x^{mn}$

Product to a power: $(xy)^m = x^m y^m$

Quotient to a power: $\left(\dfrac{x}{y}\right)^m = \dfrac{x^m}{y^m}$

Quotient rule: $\dfrac{x^m}{x^n} = x^{m-n}$

Negative power: $\left(\dfrac{x}{y}\right)^{-n} = \left(\dfrac{y}{x}\right)^n$

One reason it is useful to examine radical expressions in exponential form is that we can apply the properties of exponents to simplify these expressions. This is illustrated in Example 3.

Example 3 Using the Properties of Exponents

Simplify each of the following expressions. Express all answers in terms of positive exponents.

(a) $9^{3/8} \cdot 9^{1/8}$ **(b)** $8^{1/5} \cdot 4^{1/5}$ **(c)** $\dfrac{625^{7/8}}{625^{1/8}}$

Solution

(a) $9^{3/8} \cdot 9^{1/8} = 9^{(3/8)+(1/8)}$ Use the product rule to add the exponents:

$\qquad\qquad = 9^{1/2}$ $\dfrac{3}{8} + \dfrac{1}{8} = \dfrac{4}{8} = \dfrac{1}{2}$

$\qquad\qquad = 3$ The principal square root of 9 is 3.

(b) $8^{1/5} \cdot 4^{1/5} = (8 \cdot 4)^{1/5}$ Product to a power: $x^m y^m = (xy)^m$.

$\qquad\qquad = 32^{1/5}$ The principal fifth root of 32 is 2.

$\qquad\qquad = 2$

(c) $\dfrac{625^{7/8}}{625^{1/8}} = 625^{7/8 - 1/8}$ Use the quotient rule to subtract the exponents:

$\qquad\qquad = 625^{3/4}$ $\dfrac{7}{8} - \dfrac{1}{8} = \dfrac{6}{8} = \dfrac{3}{4}$

$\qquad\qquad = (625^{1/4})^3$

$\qquad\qquad = 5^3$ The principal fourth root of 625 is 5.

$\qquad\qquad = 125$

Self-Check 3

Simplify each expression.

a. $(8^{-4/5})^{-5/2}$ **b.** $25^{5/16} \cdot 25^{3/16}$ **c.** $25^{1/3} \cdot 5^{1/3}$

To avoid the special care that must be exercised when the base is negative and we are evaluating an even-number root, we restrict the variables to positive values in Examples 4 to 7.

Example 4 Using the Properties of Exponents

Simplify each expression. Express each answer in terms of positive exponents. Assume that the variables represent only positive real numbers.

(a) $(z^{-2/3})^{-3}$ **(b)** $(a^{-8}b^{12})^{-1/4}$ **(c)** $\left(\dfrac{16x^4}{25}\right)^{-3/2}$

Solution

(a) $(z^{-2/3})^{-3} = z^{(-2/3)(-3)}$ Use the power rule to multiply the exponents:

$\qquad\qquad = z^2$ $\left(-\dfrac{2}{3}\right)(-3) = 2$.

(b) $(a^{-8}b^{12})^{-1/4} = (a^{-8})^{-1/4}(b^{12})^{-1/4}$ Use the product to a power rule for exponents

$\qquad\qquad = a^2 b^{-3}$ $(xy)^m = x^m y^m$ to take the $-\dfrac{1}{4}$ power of each factor.

$\qquad\qquad = \dfrac{a^2}{b^3}$ Use the power rule to multiply the exponents:

$\qquad\qquad\qquad$ $-8\left(-\dfrac{1}{4}\right) = 2$ and $12\left(-\dfrac{1}{4}\right) = -3$.

Rewrite the expression without negative

exponents: $b^{-3} = \dfrac{1}{b^3}$.

(c) $\left(\dfrac{16x^4}{25}\right)^{-3/2} = \left[\left(\dfrac{16x^4}{25}\right)^{1/2}\right]^{-3}$ $\qquad x^{m/n} = (x^{1/n})^m$

$\qquad\qquad = \left[\dfrac{16^{1/2}x^{4/2}}{25^{1/2}}\right]^{-3}$ Take the $\dfrac{1}{2}$ power of each factor in the numerator and the denominator, and then simplify.

$\qquad\qquad = \left[\dfrac{4x^2}{5}\right]^{-3}$ Note that $16^{1/2} = 4$ and $25^{1/2} = 5$.

$\qquad\qquad = \left[\dfrac{5}{4x^2}\right]^{3}$ Rewrite the expression without negative exponents.

$\qquad\qquad = \dfrac{5^3}{4^3(x^2)^3}$ Raise each factor to the third power. To raise a power to a power, multiply exponents.

$\qquad\qquad = \dfrac{125}{64x^6}$

Self-Check 4

Simplify each expression. Assume that x is a positive real number.

a. $\left(\dfrac{8x^6}{125}\right)^{2/3}$ **b.** $\left(\dfrac{8x^6}{125}\right)^{-2/3}$

Example 5 uses the distributive property to simplify an expression involving both fractional and negative exponents.

Example 5 Multiplying by Using the Distributive Property

Simplify $2x^{2/3}(3x^{1/3} - 5x^{-2/3})$. Express the answer in terms of positive exponents. Assume that all variables represent positive real numbers.

Solution

$2x^{2/3}(3x^{1/3} - 5x^{-2/3}) = (2x^{2/3})(3x^{1/3}) - (2x^{2/3})(5x^{-2/3})$ First use the distributive property.

$\qquad\qquad\qquad\qquad\quad = 6x^{3/3} - 10x^0$ Then simplify each term, adding the exponents of the factors with the same base. Note that $x^0 = 1$.

$\qquad\qquad\qquad\qquad\quad = 6x - 10$

Self-Check 5

Simplify $4x^{5/8}(3x^{11/8} - 5x^{3/8} - 7x^{-5/8})$. Assume that x is a positive real number.

The product of exponential expressions with two or more terms can be found by using some of the skills developed for multiplying polynomials. In particular, we will use special forms, such as the product of a sum times a difference: $(A + B)(A - B) = A^2 - B^2$.

Example 6 Multiplying a Sum by a Difference

Simplify $(a^{1/2} + b^{1/2})(a^{1/2} - b^{1/2})$. Assume that all variables represent positive real numbers.

Solution

$(a^{1/2} + b^{1/2})(a^{1/2} - b^{1/2}) = (a^{1/2})^2 - (b^{1/2})^2$ By inspection, observe that this expression fits the special form of a sum times a difference.

$\qquad\qquad\qquad\qquad\qquad = a - b$ To raise a power to a power, multiply exponents.

Self-Check 6

Simplify $(z^{1/2} - 5)(z^{1/2} + 5)$.

The history of science and engineering has many examples of "great" ideas that did not "scale up," that is, ideas that seemed to work well with a model in the lab but did not work in real-world applications. One of the problems that engineers must face is that scale models may not behave as the real product does if not all the variables involved increase at the same rate. Example 7 investigates the strength of a box beam to illustrate this problem.

Example 7   Strength of Box Beam

The strength of a box beam varies directly as the cross-sectional area. However, part of the stress on the beam comes from its own weight, which is directly related to the volume of the beam. Thus, the stress that a beam puts on itself increases at a greater rate than its strength. For the sample beam shown here, the following functions give the strength S and the volume V of the beam: $S = 180,000x^2$ and $V = 27,000x^3$. In this example, x is given in centimeters, V in cubic centimeters, and S in newtons.

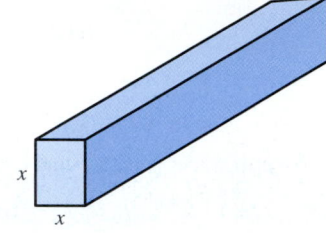

(a) Write S as a function of V.

(b) Using the function from part (a), evaluate and interpret $S(1,000)$.

Solution

(a) $V = 27,000x^3$

$\dfrac{V}{27,000} = x^3$ S is currently written as a function of x. Solve the formula $V = 27,000x^3$ for x.

$x = \left(\dfrac{V}{27,000}\right)^{1/3}$ First divide both sides of the equation by 27,000 and then raise both sides to the $\dfrac{1}{3}$ power.

$x = \dfrac{V^{1/3}}{30}$ Note that $27,000^{1/3} = 30$.

$S = 180,000x^2$

$S = 180,000\left(\dfrac{V^{1/3}}{30}\right)^2$ To write S as a function of V, substitute $\dfrac{V^{1/3}}{30}$ for x.

$S = 200V^{2/3}$

$S(V) = 200V^{2/3}$

(b) $S(V) = 200V^{2/3}$

$S(1,000) = 200(1,000)^{2/3}$ Substitute 1,000 for V and then simplify this expression.

$= 200(10)^2$

$= 200(100)$

$= 20,000$

When the volume is 1,000 cm³, the strength will be 20,000 N.

Self-Check 7

a. In Example 7, write V as a function of S.

b. Using the function from part (a), evaluate and interpret $V(20,000)$.

Self-Check Answers

1. **a.** $\sqrt{64} = 8$
   b. $\sqrt[3]{64} = 4$
   c. $\sqrt[6]{64} = 2$
   d. $\sqrt[3]{-64} = -4$

2. **a.** $64^{3/2} = (\sqrt{64})^3$
      $= 8^3 = 512$
   b. $64^{2/3} = (\sqrt[3]{64})^2$
      $= 4^2 = 16$

c. $(-64)^{2/3} = (\sqrt[3]{-64})^2$
      $= (-4)^2 = 16$

d. $64^{-2/3} = \dfrac{1}{(\sqrt[3]{64})^2}$
      $= \dfrac{1}{4^2} = \dfrac{1}{16}$

3. **a.** 64   **b.** 5   **c.** 5

4. **a.** $\dfrac{4x^4}{25}$ **b.** $\dfrac{25}{4x^4}$

5. $12x^2 - 20x - 28$

6. $z - 25$

7. **a.** $V(S) = \dfrac{S\sqrt{2S}}{4,000}$
   b. $V(20,000) = 1,000$; When the strength is 20,000 N, the volume is 1,000 cm^3.

Technology Self-Check Answer

1. 9.46

10.5 | Using the Language and Symbolism of Mathematics

1. In radical notation, the expression $x^{1/5}$ can be written as _____.

2. In exponential notation, the expression $\sqrt[7]{x}$ can be written as _____.

3. In radical notation, the expression $x^{2/5}$ can be written as _____.

4. In exponential notation, the expression $\sqrt[4]{x^3}$ can be written as _____.

5. The principal nth root of x is denoted by either $\sqrt[n]{x}$ or _____.

6. The principal nth root of x is not a real number if n is even and x is _____.

7. In the radical notation $\sqrt[n]{x}$, n is called the index or the order of the radical and x is called the _____.

10.5 | Quick Review

Simplify each expression. Assume $x \neq 0$.

1. $\left(\dfrac{4}{5}\right)^{-2}$

2. $x^7 x^{11}$

3. $(x^{11})^7$

4. $\dfrac{x^{11}}{x^7}$

5. $\left(\dfrac{x^5}{x^{-3}}\right)^2$

10.5 | Exercises

Objective 1 Interpret and Use Rational Exponents

In Exercises 1–4, represent each expression by using exponential notation. Assume that x and y are positive real numbers.

1. The principal cube root of w

2. The principal fourth root of x

3. The principal sixth root of x

4. The principal ninth root of v

In Exercises 5–8, represent each expression by using both radical notation and exponential notation, and evaluate each expression.

5. The principal cube root of 125

6. The principal fifth root of -32

7. The principal fourth root of 81

8. The principal fifth root of $-100,000$

In Exercises 9–14, represent each expression by using exponential notation, and evaluate each expression.

9. **a.** $\sqrt{16}$ **b.** $\sqrt[4]{16}$

10. **a.** $\sqrt[3]{64}$ **b.** $\sqrt[6]{64}$

11. **a.** $\sqrt[5]{0}$ **b.** $\sqrt[6]{0}$

12. **a.** $\sqrt[4]{1}$ **b.** $\sqrt[7]{1}$

13. **a.** $\sqrt[3]{-1}$ **b.** $\sqrt[5]{-1}$

14. **a.** $\sqrt{0.01}$ **b.** $\sqrt[3]{-0.001}$

In Exercises 15–32, represent each expression by using radical notation, and evaluate each expression.

15. **a.** $36^{1/2}$ **b.** $36^{-1/2}$

16. **a.** $100^{1/2}$ **b.** $100^{-1/2}$

17. **a.** $27^{1/3}$ **b.** $27^{-1/3}$

18. **a.** $16^{1/4}$ **b.** $16^{-1/4}$

19. **a.** $0.09^{1/2}$ **b.** $-0.09^{1/2}$

20. **a.** $0.49^{1/2}$ **b.** $-0.49^{1/2}$

21. **a.** $0.008^{1/3}$ **b.** $(-0.008)^{1/3}$

22. **a.** $0.027^{1/3}$ **b.** $(-0.027)^{1/3}$

23. **a.** $64^{1/3}$ **b.** $64^{-1/3}$

24. **a.** $81^{1/4}$ **b.** $81^{-1/4}$

25. **a.** $\left(\dfrac{8}{125}\right)^{2/3}$ **b.** $\left(\dfrac{8}{125}\right)^{-2/3}$

26. **a.** $\left(\dfrac{16}{81}\right)^{3/4}$ **b.** $\left(\dfrac{16}{81}\right)^{-3/4}$

27. **a.** $25^{1/2} + 144^{1/2}$ **b.** $(25 + 144)^{1/2}$

28. **a.** $9^{1/2} + 16^{1/2}$ **b.** $(9 + 16)^{1/2}$

29. **a.** $8^{1/3} + 1^{1/3}$ **b.** $(26 + 1)^{1/3}$

30. **a.** $27^{1/3} - 1^{1/3}$ **b.** $(65 - 1)^{1/3}$

31. **a.** $0.000001^{1/2}$ **b.** $0.000001^{1/3}$

32. **a.** $0.0001^{1/2}$ **b.** $0.0001^{1/4}$

Objective 2 Use the Properties of Exponents

In Exercises 33–56, simplify each expression. Express all answers in terms of positive exponents. Assume that all variables represent positive real numbers.

33. $5^{1/2} \cdot 5^{3/2}$ 34. $7^{5/3} \cdot 7^{1/3}$

35. $(8^{5/3})^{2/5}$ 36. $(32^{7/10})^{2/7}$

37. $\dfrac{11^{4/3}}{11^{1/3}}$ 38. $\dfrac{9^{5/4}}{9^{3/4}}$

39. $(27^{1/12} \cdot 27^{-5/12})^{-2}$ 40. $(4^{1/5} \cdot 4^{2/5})^{5/2}$

41. $x^{1/3} \cdot x^{1/2}$ 42. $y^{1/4} \cdot y^{1/5}$

43. $\dfrac{x^{1/2}}{x^{1/3}}$ 44. $\dfrac{y^{1/4}}{y^{1/5}}$

45. $(z^{3/4})^{2/7}$ 46. $(z^{5/12})^{4/15}$

47. $\dfrac{w^{-2/3}}{w^{-5/3}}$ 48. $\dfrac{w^{3/4}}{w^{-3/8}}$

49. $(v^{-10}w^{-15})^{-1/5}$ 50. $(v^{-26}w^{-39})^{-1/13}$

51. $(16v^{-2/5})^{3/2}$ 52. $(25v^{-4/9})^{3/2}$

53. $\left(\dfrac{16n^{2/3}}{81n^{-2/3}}\right)^{-3/4}$ 54. $\left(\dfrac{27n^{3/5}}{n^{-3/5}}\right)^{-2/3}$

55. $\dfrac{(27x^2y)^{1/2}(3xy)^{1/2}}{5x^{1/2}y^2}$ 56. $\dfrac{(25x^2y^3z)^{1/3}(5xy^2z^2)^{1/3}}{3x^2y^{-1/3}z^{-2}}$

In Exercises 57–70, perform the indicated multiplications, and express the answers in terms of positive exponents. Assume that all variables represent positive real numbers.

57. $x^{3/5}(x^{2/5} - x^{-3/5})$ 58. $x^{2/3}(x^{4/3} + x^{-2/3})$

59. $y^{-7/4}(2y^{11/4} - 3y^{7/4})$ 60. $y^{-9/7}(5y^{23/7} - 6y^{16/7})$

61. $3w^{5/11}(2w^{17/11} - 5w^{6/11} - 9w^{-5/11})$

62. $4w^{-5/3}(3w^{11/3} - 7w^{8/3} + 2w^{5/3})$

63. $(a^{1/2} + 3)(a^{1/2} - 3)$

64. $(2a^{1/2} - 3b^{1/2})(2a^{1/2} + 3b^{1/2})$

65. $(b^{3/5} - c^{5/3})(b^{3/5} + c^{5/3})$ 66. $(b^{3/5} + c^{5/3})^2$

67. $(b^{3/5} - c^{5/3})^2$ 68. $(x^{1/2} - x^{-1/2})^2$

69. $(x^{-1/2} + x^{1/2})^2$ 70. $(x^{2/3} + x)(x^{2/3} - x)$

Applying Technology

In Exercises 71–76, use a calculator or a spreadsheet to approximate each expression to the nearest thousandth.

71. $75^{1/4}$ 72. $63^{2/7}$

73. $85^{1/6}$ 74. $29^{4/9}$

75. $\sqrt[6]{361}$ 76. $\sqrt[11]{691}$

Estimation and Calculator Skills

In Exercises 77 and 78, mentally estimate the value of each expression to the nearest integer, and then use a calculator to approximate each value to the nearest thousandth.

Problem	Mental Estimate	Calculator Approximation
77. $30^{1/3}$		
78. $30^{1/5}$		

Connecting Concepts to Applications

Strength of a box beam

In Exercises 79–82, use the fact that the load strength of a square box beam is related to its volume by the function $S(V) = 750V^{2/3}$. In this formula, V is given in cubic centimeters and S is given in newtons.

79. Evaluate and interpret $S(1,000)$.

80. If $S(V) = 750$, calculate and interpret V.

81. If the volume in Exercise 79 is increased by a factor of 8 to 8,000 cm³, determine the factor by which the strength has increased.

82. If the strength in Exercise 80 is increased by a factor of 16 to 12,000 N, determine the factor by which the volume has increased.

83. Geosynchronous Satellite

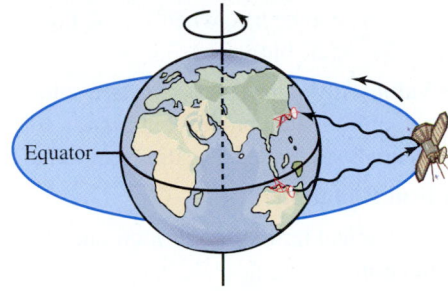

The formula $T = (3.1469 \times 10^{-7})r^{3/2}$ relates the period T of the orbit of a satellite in seconds and the radius r of the circular orbit around the center of the earth in meters. A satellite placed in a geosynchronous orbit of the earth will make 1 complete orbital revolution each day (once every 86,400 seconds). Use $T = 86,400$ to determine the radius of the geosynchronous orbit around the center of the earth. Given that the radius of the earth is 6,378,100 m, determine the distance above the earth's surface at which a geosynchronous satellite must be placed in orbit.

Group discussion questions

84. Challenge Question

  a. Give an example of a real number for which $(x^2)^{1/2} \neq x$.

  b. Give an example of a real number for which $x^{1/2}$ is defined but $x^{-1/2}$ is undefined.

85. Challenge Question Use the properties of exponents to simplify each expression, assuming m is a natural number and that x is a positive real number.

  a. $x^{m/3}x^{m/2}$ **b.** $\dfrac{16x^{m/2}}{2x^{m/3}}$ **c.** $(x^{m/3}x^{m/2})^6$

86. Challenge Question Use the distributive property to expand the first expression and to factor the second expression.

    Expand   **Factor**

  a. $x^{1/2}(x^2 + 3x - 4)$ $x^{5/2} + 3x^{3/2} - 4x^{1/2}$

$$= x^{1/2}(\underline{\quad} + \underline{\quad} - \underline{\quad})$$
$$= x^{1/2}(\underline{\quad} + \underline{\quad})(\underline{\quad} - \underline{\quad})$$

  b. $x^{-1/2}(x^2 - 25)$ $x^{3/2} - 25x^{-1/2}$

$$= x^{-1/2}(\underline{\quad} - \underline{\quad})$$
$$= x^{-1/2}(\underline{\quad} + \underline{\quad})(\underline{\quad} - \underline{\quad})$$

87. Challenge Question Use inspection to match each function with its graph.

  a. $f(x) = x^0$ **A.**

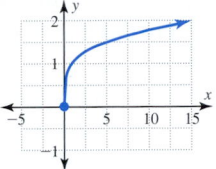

  b. $f(x) = x^{1/4}$ **B.**

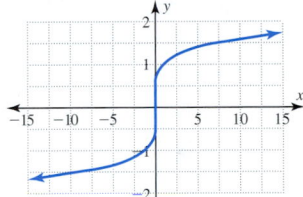

  c. $f(x) = x^{1/5}$ **C.**

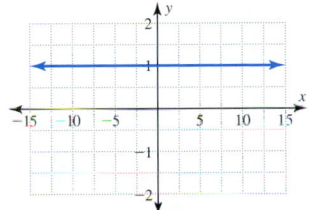

88. Challenge Question Perform the indicated multiplications by taking advantage of the factored forms for the sum and difference of two cubes.

  a. $(y^{1/3} + 2)(y^{2/3} - 2y^{1/3} + 4)$

  b. $(3y^{1/3} - 5)(9y^{2/3} + 15y^{1/3} + 25)$

10.5   **Cumulative Review**

1. Write the interval notation for the inequality $x > 5$.

2. Write an inequality for the interval $(-\infty, 4)$.

3. Write an inequality for the interval $(-2, 4)$.

4. Write an absolute value inequality for the interval $(-4, 4)$.

5. Write an absolute value inequality for the interval $(-\infty, -3) \cup (3, \infty)$.

Chapter 10 | Key Concepts

1. **Radical Notation** $\sqrt[n]{x}$ is read "the principal nth root of x." x is called the radicand, $\sqrt{}$ is called the radical symbol, and n is called the index or the order of the radical. If $\sqrt[n]{x}$ is used as a factor n times, the product is x.

2. **Principal nth Root** The principal nth root of the real number x is denoted by either $x^{1/n}$ or $\sqrt[n]{x}$.
   - For $x > 0$: The principal root is positive for all natural numbers n.
   - For $x < 0$: If n is odd, the principal root is negative. If n is even, there is no principal real nth root; it is imaginary.
   - For $x = 0$: The principal root is 0.
   - $\sqrt[n]{x^n} = |x|$ if n is even
   - $\sqrt[n]{x^n} = x$ if n is odd

3. **Rational Exponents** $x^{m/n}$ For a real number x and natural numbers m and n,
   $x^{m/n} = (x^{1/n})^m = (x^m)^{1/n}$ if $x^{1/n}$ is a real number
   $x^{-m/n} = \dfrac{1}{x^{m/n}}$ if $x \neq 0$ and $x^{1/n}$ is a real number

   Caution: If $x < 0$ and n is even, $x^{1/n}$ is not a real number and the equalities listed above are not true. For example, $\sqrt{(-5)^2} \neq (\sqrt{-5})^2$.

4. **Properties of Radicals** If x and y are real numbers and $\sqrt[n]{x}$ and $\sqrt[n]{y}$ are real numbers, then:
   - **Product rule:** $\sqrt[n]{x}\,\sqrt[n]{y} = \sqrt[n]{xy}$
   - **Quotient rule:** $\dfrac{\sqrt[n]{x}}{\sqrt[n]{y}} = \sqrt[n]{\dfrac{x}{y}}$ for $y \neq 0$

5. **Like Radicals** Like radicals have the same index and the same radicand; only the coefficients of like radicals can differ.

6. **Conjugates** The radical expressions $x + \sqrt{y}$ and $x - \sqrt{y}$ are called conjugates of each other.

7. **Operations with Radical Expressions** The addition, subtraction, and multiplication of radical expressions are similar to the corresponding operations with polynomials. Division by a radical expression is accomplished by rationalizing the denominator.

8. **Rationalizing the Denominator** If an expression contains a radical in the denominator, the process of rewriting the expression so that there is no radical in the denominator is called rationalizing the denominator.

9. **Power Theorem** For any real numbers x and y and natural number n:
   - If $x = y$, then $x^n = y^n$.
   - *Caution:* $x^n = y^n$ can be true when $x \neq y$. For example, $(-3)^2 = 3^2$ but $-3 \neq 3$.

10. **Extraneous Value** Any value that occurs as a solution of the last equation of a solution process but is not a solution of the original equation is called an extraneous value.

11. **Solving Radical Equations**

    Step 1 Isolate a radical term of order n on one side of the equation.

    Step 2 Raise both sides to the nth power.

    Step 3 Solve the resulting equation.

    Step 4 Check each possible solution in the original equation to determine whether it is a solution or an extraneous value.

12. **Formula for the Distance Between Two Points in a Plane** The distance between (x_1, y_1) and (x_2, y_2) is given by $d = \sqrt{(x_2 - x_1)^2 + (y_2 - y_1)^2}$.

13. **Real-Valued Function** A function that produces only real output values for real input values is called a real-valued function.

14. **Domain of a Real-Valued Function** The domain of a real-valued function $f(x)$ is the set of all real numbers x for which $f(x)$ is also a real number. This means the domain excludes all values that cause either of the following:
    - Division by 0
    - A negative number under a square root symbol or any other even-order radical

15. **Basic Functions Examined in This Chapter** Each type of function has an equation of a standard form and a characteristic shape.
    - Square root functions
    - Cube root functions

Chapter 10 | Review Exercises

Using Radical Notation

Evaluate each expression.

1. $\sqrt{81}$
2. $\sqrt[3]{-27}$
3. $\sqrt[4]{81}$
4. $\sqrt[5]{-32}$
5. $\sqrt{625 - 576}$
6. $\sqrt{625} - \sqrt{576}$
7. $\sqrt{\dfrac{16}{81}}$
8. $\sqrt[3]{0.125}$

9. Plot $\sqrt{27}$, $\sqrt{-27}$, $\sqrt[3]{27}$, $\sqrt[3]{-27}$ on a real number line. If the number is not located on the real number line, explain why not.

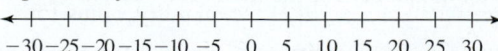

In Exercises 10–12, use a calculator or spreadsheet to approximate each expression to the nearest hundredth.

10. $\sqrt{70}$ 11. $\sqrt[3]{70}$ 12. $\sqrt[5]{-70}$

In Exercises 13–40, simplify each expression. Assume that all variables represent positive real numbers.

13. $12\sqrt{3} + 8\sqrt{3}$ **14.** $12\sqrt{3} - 8\sqrt{3}$

15. $(5\sqrt{2} - 7\sqrt{3}) - (\sqrt{2} - 4\sqrt{3})$

16. $6\sqrt[3]{5} - 2\sqrt[3]{5}$ **17.** $\dfrac{6\sqrt[3]{5}}{2\sqrt[3]{5}}$

18. $(6\sqrt[3]{5})(2\sqrt[3]{5})$ **19.** $\sqrt{20}$

20. $\sqrt[3]{24}$ **21.** $3\sqrt{72} - 2\sqrt{98}$

22. $5\sqrt{3x} + 9\sqrt{3x}$ **23.** $2\sqrt{50v} - 3\sqrt{8v}$

24. $2\sqrt[3]{8v} - \sqrt[3]{125v}$ **25.** $\dfrac{\sqrt{175}}{\sqrt{7}}$

26. $\sqrt[3]{\dfrac{27x^3}{y^6}}$ **27.** $\sqrt[3]{\dfrac{25}{27}}$

28. $3\sqrt{8} - 7\sqrt{50}$ **29.** $3\sqrt{3}(2\sqrt{12} - 9\sqrt{75})$

30. $(2\sqrt{2} - 5\sqrt{3})(2\sqrt{2} + 5\sqrt{3})$

31. $(3\sqrt{14})(15\sqrt{2})$ **32.** $\dfrac{3\sqrt{14}}{15\sqrt{2}}$

33. $\dfrac{15\sqrt{2}}{3\sqrt{14}}$ **34.** $(\sqrt{2} - \sqrt{3})^2$

35. $\dfrac{\sqrt{2} - \sqrt{3}}{\sqrt{2} + \sqrt{3}}$ **36.** $\dfrac{12}{\sqrt{7} - \sqrt{3}}$

37. $(2\sqrt{3})(3\sqrt{2})(5\sqrt{6})$ **38.** $\dfrac{(15\sqrt{2})(2\sqrt{3})}{10\sqrt{6}}$

39. $(2\sqrt{3})(3\sqrt{2}) - 5\sqrt{6}$

40. $(\sqrt[3]{5} - \sqrt[3]{2})(\sqrt[3]{25} + \sqrt[3]{10} + \sqrt[3]{4})$

In Exercises 41–44, simplify each expression. Assume that the variables represent nonnegative real numbers so that absolute value notation is not necessary.

41. $\sqrt{32x^7}$ **42.** $\sqrt[3]{32x^7}$

43. $\sqrt[4]{32x^7}$ **44.** $\sqrt{12x^3y^{11}}$

Using Rational Exponents

In Exercises 45–47, write each radical expression in exponential form.

45. $\sqrt{x}$ **46.** $\sqrt[3]{x^2}$ **47.** $\dfrac{1}{\sqrt[3]{x}}$

In Exercises 48–50, write each exponential expression in radical form.

48. $x^{1/4}$ **49.** $x^{-1/4}$ **50.** $x^{3/4}$

In Exercises 51–53, represent each expression by using both radical notation and exponential notation.

51. Twice the principal square root of w

52. The principal square root of two w

53. The principal cube root of the quantity x plus four

In Exercises 54–71, simplify each expression. Assume that all variables are positive real numbers.

54. $49^{1/2}$ **55.** $-49^{1/2}$

56. $49^{-1/2}$ **57.** $\left(\dfrac{27}{125}\right)^{2/3}$

58. $\left(\dfrac{-27}{125}\right)^{2/3}$ **59.** $\left(\dfrac{27}{125}\right)^{-2/3}$

60. $1{,}000{,}000^{1/2}$ **61.** $1{,}000{,}000^{1/3}$

62. $1{,}000{,}000^{1/6}$ **63.** $16^{5/8}16^{3/8}$

64. $\dfrac{16^{5/8}}{16^{1/8}}$ **65.** $(16^{3/8})^{2/3}$

66. $(8x^{-6}y^9)^{2/3}$ **67.** $(-32x^{5/3}y^{3/2})^{2/5}$

68. $\dfrac{(x^2y^2z^2)^{1/3}}{(xyz)^{-4/3}}$ **69.** $2x^{1/2}(3x^{1/2} - 5x^{-1/2})$

70. $(3x^{1/2} + 5)(3x^{1/2} - 5)$ **71.** $(x^2 + 2xy + y^2)^{1/2}$

Using and Interpreting Graphs and Tables

72. Match each function with its graph.

  a. $f(x) = x^2 - 4$ **b.** $f(x) = |x - 4|$
  c. $f(x) = \sqrt{x} - 4$ **d.** $f(x) = \sqrt{4 - x}$
  e. $f(x) = 4 - x$ **f.** $f(x) = x - 4$
  g. $f(x) = \sqrt[3]{x} - 4$

A.

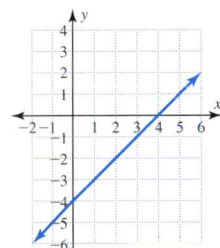

B.

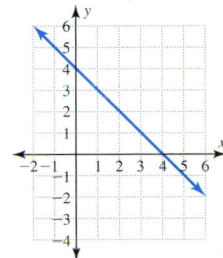

C.

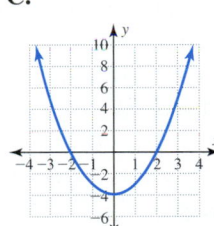

D.

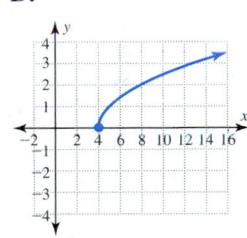

E.

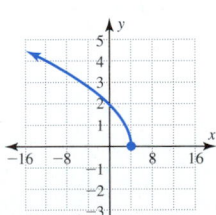

F.

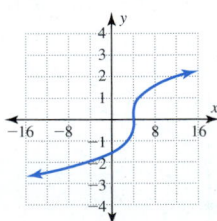

G.

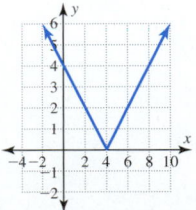

73. Use the given graphs to determine the domain of each function whose equation is given in parts (a) through (g) of Exercise 72.

74. Match each real-valued function with a table for this function.

a. $f(x) = 2x$ **b.** $f(x) = -2x$

c. $f(x) = x^2$ **d.** $f(x) = \sqrt{x}$

e. $f(x) = \sqrt[3]{x}$

A.

x	f(x)
-2	4
-1	1
0	0
1	1
2	4
3	9

B.

x	f(x)
-1	-1.0
0	0.0
1	1.0
2	1.3
3	1.4
4	1.6

C.

x	f(x)
-1	Error
0	0.0
1	1.0
2	1.4
3	1.7
4	2.0

D.

x	f(x)
-2	4
-1	2
0	0
1	-2
2	-4
3	-6

E.

x	f(x)
-2	-4
-1	-2
0	0
1	2
2	4
3	6

In Exercises 75–80, determine algebraically the domain of each function. Use a graphing calculator only to check your answer.

75. $f(x) = x^3 - 5$ **76.** $f(x) = \sqrt[3]{x - 5}$

77. $f(x) = \sqrt{x - 5}$ **78.** $f(x) = |x - 5|$

79. $f(x) = \dfrac{1}{x - 5}$ **80.** $f(x) = \dfrac{x - 5}{x^2 - 9}$

81. Use this graph to solve $\sqrt[4]{40x + 1} = 4 - \dfrac{x}{2}$.

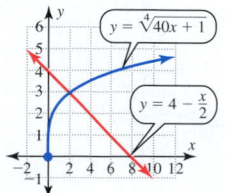

82. There is one solution of the equation $\sqrt[3]{11x + 5} = 3$. Given $f(x) = \sqrt[3]{11x + 5}$, use the table to determine this solution.

x	f(x)
-1	-1.82
0	1.71
1	2.52
2	3.00
3	3.36
4	3.66

Solving Radical Equations

In Exercises 83–86, solve each equation.

83. $\sqrt{x - 5} = 4$ **84.** $\sqrt{x + 12} - x = 0$

85. $\sqrt{2v - 1} + 2 = v$ **86.** $\sqrt[3]{4w + 5} = -3$

Connecting Concepts to Applications

87. Determine the exact x- and y-intercepts of $y = \sqrt{x + 25} - 7$.

88. Calculate the distance between $(-4, 8)$ and $(1, -4)$.

89. Perimeter of a Triangle The points $(-2, 4)$, $(4, 6)$, and $(2, 2)$ form a triangle. First calculate the perimeter of this triangle, and then determine whether it is a right triangle.

90. Dimensions of a Right Triangle The length of the hypotenuse of the right triangle shown in the figure is 2 cm more than the length of the longer leg. If the longer leg is 7 cm longer than the shorter leg, determine the length of each side.

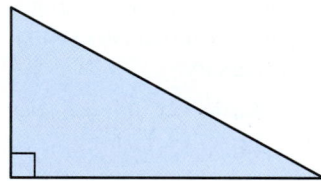

91. Location of Light Poles An electrical power source is located at the point $(4, 4)$. Two light poles are positioned along a road with an x-coordinate of 8. If these two poles are 5 units from the power source, find the points that locate these poles.

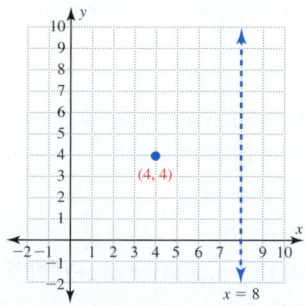

92. Radius of a Sphere An engineer is designing a spherical water tower with a capacity of 17,000 ft³. Determine the radius that the sphere will have.

93. Dimensions of a Cube What is the length of a cubical box with a volume of 3,197 cm³?

94. Period of a Pendulum The formula $T(x) = 0.064\sqrt{x}$ gives the period in seconds for a pendulum of length x cm. Determine the length of a pendulum with a period of 0.6 sec.

95. Cost of Production The dollar cost of producing n boxes for computer routers during one shift at a factory is given by $C(n) = 20\sqrt[3]{n^2} + 200$.
a. Determine the overhead cost for one shift.
b. Determine the cost of producing 512 boxes in a shift.
c. Determine the number of boxes that can be produced in a shift for a cost of $1,480.

Chapter 10 | Mastery Test

Objective 10.1.1 Interpret and Use Radical Notation

1. Represent each verbal expression by using radical notation. Assume $w > 0$.
a. The principal cube root of v
b. The principal fourth root of w

2. Mentally evaluate each radical expression.
a. $\sqrt[3]{27}$ **b.** $\sqrt[3]{-64}$
c. $\sqrt[4]{16}$ **d.** $\sqrt[5]{-100,000}$

Objective 10.1.2 Graph and Analyze Square Root and Cube Root Functions

3. Match each function with its graph.
a. $f(x) = \sqrt{x} - 3$ **b.** $f(x) = \sqrt[3]{x} - 3$
c. $f(x) = \sqrt{x} + 3$ **d.** $f(x) = \sqrt[3]{x} + 3$
A.   **B.**

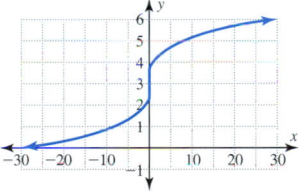   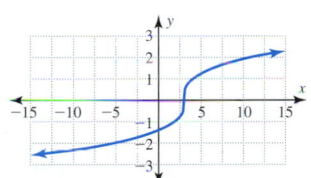

C.   **D.**

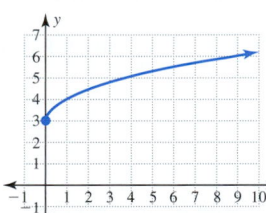   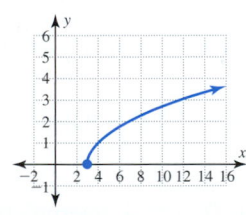

4. Determine the domain of each function.
a. $f(x) = \sqrt{x} - 3$ **b.** $f(x) = \sqrt[3]{x} - 3$
c. $f(x) = \dfrac{x + 4}{x - 3}$ **d.** $f(x) = (x - 3)^2$

Objective 10.2.1 Add and Subtract Radical Expressions

5. Determine each sum or difference. Assume $x > 0$.
a. $8\sqrt{7} - 3\sqrt{7}$
b. $(5\sqrt{2} - 3\sqrt{5}) - (2\sqrt{2} - 7\sqrt{5})$

c. $4\sqrt[3]{7} - 11\sqrt[3]{7} + 6\sqrt[3]{7}$
d. $13\sqrt{2x} - 5\sqrt{2x}$

Objective 10.2.2 Simplify Radical Expressions

6. Simplify each expression.
a. $\sqrt{40}$ **b.** $\sqrt[3]{24}$
c. $3\sqrt{28} - 5\sqrt{63}$ **d.** $\sqrt[3]{40x^5y^4}$

Objective 10.3.1 Multiply Radical Expressions

7. Perform each multiplication and then simplify the product. Assume $x > 0$.
a. $(3\sqrt{2})(5\sqrt{2})$
b. $2\sqrt{5}(3\sqrt{5} - 2\sqrt{10})$
c. $\sqrt[3]{16x^5}\sqrt[3]{-4x}$
d. $(2\sqrt{3x} - \sqrt{5})(2\sqrt{3x} + \sqrt{5})$

Objective 10.3.2 Divide and Simplify Radical Expressions

8. Perform each division and then simplify the quotient.
a. $\dfrac{\sqrt{18}}{\sqrt{2}}$ **b.** $\dfrac{18}{\sqrt{6}}$
c. $\dfrac{40}{\sqrt{7} - \sqrt{2}}$ **d.** $\dfrac{12}{2 - \sqrt{7}}$

Objective 10.4.1 Solve Equations Containing Radical Expressions

9. Solve each equation.
a. $\sqrt{x - 3} = 11$ **b.** $\sqrt[3]{2x - 17} = -3$
c. $\sqrt{x + 4} = x + 11$ **d.** $\sqrt{3x + 1} + 3 = x$

Objective 10.4.2 Calculate the Distance Between Two Points

10. Calculate the distance between each pair of points.
a. $(3, 9)$ and $(-5, 9)$ **b.** $(-4, 7)$ and $(-4, -5)$
c. $(1, -2)$ and $(7, 6)$ **d.** $(1, 1)$ and $(-4, 3)$

Objective 10.5.1 Interpret and Use Rational Exponents

11. Simplify each expression.

    a. $81^{1/2}$ **b.** $(-1{,}000)^{1/3}$

    c. $81^{3/4}$ **d.** $81^{-3/4}$

    e. $100^{1/2} - 64^{1/2}$ **f.** $(100 - 64)^{1/2}$

Objective 10.5.2 Use the Properties of Exponents

12. Simplify each expression. Assume that all variables are positive.

    a. $(8v^6w^9)^{2/3}$ **b.** $(5x^{2/3}y^{2/5})(2x^{1/3}y^{8/5})$

    c. $\dfrac{12m^{5/8}n^{-4/7}}{8m^{-3/8}n^{3/7}}$ **d.** $\left(\dfrac{320x^{-4}y^5}{40x^2y^{-4}}\right)^{-2/3}$

Chapter 10 | Group Project

Modeling the Period of a Pendulum

Supplies Needed for the Lab:

- A strong string at least 2 m long; a connector on the top to fasten to a ceiling, doorway frame, or stepladder; and a connector to attach a weight to the bottom of the string
- A weight that is small but relatively heavy, for example, a large steel nut
- A metric ruler to measure the length of the string
- Scissors to shorten the length of the string
- A stopwatch
- A graphing calculator with a Power Regression (**PwrReg**) feature

Lab: Recording the Time for the Period of a Pendulum

Length x (cm)	Time y (seconds)
200	
185	
170	
155	
140	
125	
110	
95	
80	
65	
50	

Suspend the top of the string from an elevated position such as a ceiling or the top of a doorframe. (Your physics lab may have a setup for this type of experiment.) Measure the length of string and attach a weight that is much heavier than the string. Using clock positions as a reference system, start with the top of the string in the 12:00 position and the bottom of the string in the 6:00 position. Then move the bottom of the string and the weight to approximately the 4:30 position. Simultaneously release the weight and start the stopwatch. Stop the stopwatch after this pendulum has completed 10 full periods. Divide this time by 10 to determine the time for one period. Then record the length of the string in centimeters and the time in seconds for one period. Shorten the string by about 15 cm and repeat this experiment about 11 times.

1. Record your data in a table similar to the one shown here.

2. Plot these data on a scatter diagram, using an appropriate scale for each axis.

Creating an Algebraic Model for These Data

1. Using your graphing calculator, calculate the power function of best fit. This function will be of the form $T(x) = ax^b$.

2. Let $y = T(x)$ represent the curve of best fit that you just calculated.

    a. Evaluate and interpret $T(165)$.

    b. Evaluate and interpret $T(70)$.

    c. Determine the value of x for which $T(x) = 4$. Interpret the meaning of this value.

Exponential and Logarithmic Functions

Chapter Outline

Carbon-14 Dating

Exponential and logarithmic functions are used to solve many types of growth and decay problems. We use these functions in the exercises to compute compound interest, to predict population growth, and to compute the age of ancient wood samples through carbon-14 dating. One example uses exponential growth to compute the safe shelf life of products such as milk. By calculating how long it will take bacteria to reach an unacceptable level, we can determine how many days this perishable product is safe.

This chapter examines applications of both exponential and logarithmic functions. Because exponential and logarithmic functions are inverses of each other, we also examine inverse functions in this chapter. We start by examining exponential functions and geometric sequences. We then end the chapter with a Group Project that involves collecting data on the height of a bouncing ball. An exponential function can be used to model these data. We use a graphing calculator to determine this exponential model.

Section 11.1	Geometric Sequences and Graphs of Exponential Functions

What is the most powerful force in the universe? Compound interest! —ALBERT EINSTEIN

(PHYSICIST, 1879–1955)

Objectives:

1. Identify a geometric sequence.
2. Evaluate and graph exponential functions.
3. Solve simple exponential equations.

The following tables illustrate two distinctive types of growth patterns. The first table gives the yearly salary amounts for an employee over a 10-year period, assuming a starting salary of \$40,000 and annual raises of \$2,400. The second table gives the yearly salary amounts for an employee over a 10-year period, assuming a starting salary of \$40,000 and annual raises of 6%.

Annual Salary Increase of \$2,400

Number of Prior Years x	Annual Salary y ($)
0	40,000
1	42,400
2	44,800
3	47,200
4	49,600
5	52,000
6	54,400
7	56,800
8	59,200
9	61,600

> Salary increased by **adding** a constant value

Annual Salary Increase of 6%

Number of Prior Years x	Annual Salary y ($)
0	40,000
1	42,400
2	44,944
3	47,641
4	50,499
5	53,529
6	56,741
7	60,145
8	63,754
9	67,579

> Salary increased by **multiplying** by a constant value

What Types of Mathematical Patterns Do These Examples Illustrate?

The first table shows an arithmetic sequence with a common difference $d = 2,400$. As we noted in Section 2.1, arithmetic sequences have a constant change from term to term and produce a linear pattern when graphed. The second table shows a geometric sequence. Geometric sequences have a constant factor from term to term. A geometric sequence differs significantly from an arithmetic sequence in both its numerical pattern and its graphical shape. We now examine geometric sequences.

1. Identify a Geometric Sequence

The sequence 1, 2, 4, 8, 16, 32 has a constant ratio from term to term. Each term is twice the preceding term. This is an example of a geometric sequence. A **geometric sequence** is a sequence with a constant ratio from term to term. That is, the ratio of any two consecutive terms is constant. This constant is called the **common ratio** and is usually denoted by r. The common ratio for the sequence 1, 2, 4, 8, 16, 32 is 2. The ratio of consecutive terms is

$$\frac{2}{1} = \frac{4}{2} = \frac{8}{4} = \frac{16}{8} = \frac{32}{16} = 2.$$

Example 1	Writing the Terms of Arithmetic and Geometric Sequences

Write the first five terms of each sequence.

(a) An arithmetic sequence with $a_1 = 5$ and a common difference of 4

(b) A geometric sequence with $a_1 = 5$ and a common ratio of 4.

Solution

(a) $a_1 = 5$ Each term of this arithmetic sequence is obtained by adding the common
$\quad a_2 = 5 + 4 = 9$ difference of 4 to the previous term.
$\quad a_3 = 9 + 4 = 13$
$\quad a_4 = 13 + 4 = 17$
$\quad a_5 = 17 + 4 = 21$

(b) $a_1 = 5$ Each term of this geometric sequence is obtained by multiplying the
$\quad a_2 = 5(4) = 20$ common ratio of 4 by the previous terms.
$\quad a_3 = 20(4) = 80$
$\quad a_4 = 80(4) = 320$
$\quad a_5 = 320(4) = 1,280$

Self-Check 1

a. Write an arithmetic sequence of seven terms with $a_1 = 2$ and a common difference $d = 3$.

b. Write a geometric sequence of seven terms with $a_1 = 2$ and a common ratio $r = 3$.

c. Write a geometric sequence of seven terms with $a_1 = 2$ and a common ratio $r = -3$.

In Example 2, we examine each sequence to determine whether the sequence is geometric. If a sequence is geometric, the ratio of consecutive terms will be constant. The graph of a sequence contains only discrete points. The dashed curves are shown in Example 2 to emphasize the pattern formed by these points.

Example 2	Identifying Geometric Sequences

Determine which of these sequences are geometric sequences. For those that are geometric, determine the common ratio r.

(a) 2, 3, 4.5, 6.75, 10.125, 15.1875

(b) 32, 16, 8, 4, 2, 1

(c) 2, 3.5, 5, 6.5, 8, 9.5

Solution

(a) **Numerically: Ratio of Consecutive Terms**

$$\frac{3}{2} = 1.5$$

$$\frac{4.5}{3} = 1.5$$

$$\frac{6.75}{4.5} = 1.5$$

$$\frac{10.125}{6.75} = 1.5$$

$$\frac{15.1875}{10.125} = 1.5$$

$$r = 1.5$$

Verbally

Because there is a common ratio $r = 1.5$, this is a geometric sequence.

Graphically

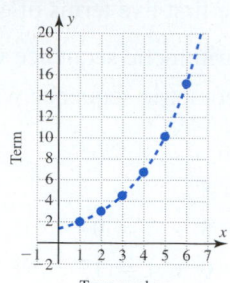

Term number

The graph of these points exhibits the shape that we will define as geometric growth.

(b) **Numerically: Ratio of Consecutive Terms**

$$\frac{16}{32} = 0.5$$

$$\frac{8}{16} = 0.5$$

$$\frac{4}{8} = 0.5$$

$$\frac{2}{4} = 0.5$$

$$\frac{1}{2} = 0.5$$

$$r = 0.5$$

Verbally

Because there is a common ratio $r = 0.5$, this is a geometric sequence.

Graphically

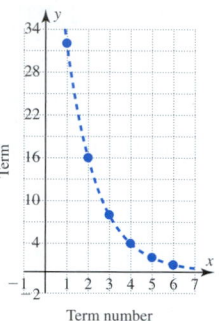

Term number

The graph of these points exhibits the shape that we will define as geometric decay.

(c) **Numerically: Ratio of Consecutive Terms**

$$\frac{3.5}{2} = 1.75$$

$$\frac{5}{3.5} \approx 1.43$$

$$1.75 \neq 1.43$$

Verbally

This is not a geometric sequence because the ratio from term to term is not constant. In fact, this is an arithmetic sequence with a common difference $d = 1.5$.

Graphically

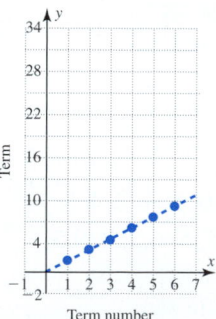

Term number

Note these points exhibit a linear pattern.

Self-Check 2

For each arithmetic sequence, determine the common difference d. For each geometric sequence, determine the common ratio r:

a. 3, 6, 12, 24, 48   **b.** 3, 6, 9, 12, 15

In Example 2, What Is Meant by Geometric Growth?

If the common ratio r of a geometric sequence is greater than 1, then we refer to the sequence as **geometric growth.** Each successive term will grow by a factor of r. If the common ratio r of a geometric sequence is positive and less than 1, then we refer to the sequence as **geometric decay.** Each successive term will decrease or decay by a factor of r.

Geometric Growth and Decay	
Growth	Decay
Algebraically	**Algebraically**
If $r > 1$, then a geometric sequence exhibits geometric growth.	If $0 < r < 1$, then a geometric sequence exhibits geometric decay.
$r = 2$	$r = 0.5$
Graphically	**Graphically**

Populations of people, bacteria, and other organisms often exhibit geometric growth. Example 3 examines an application that exhibits geometric decay.

Example 3   Determining the Height of a Bouncing Ball

A golf ball is dropped onto a concrete cart path from a height of 10 ft. Each bounce is 0.6 as high as the height of the previous bounce. Determine the height of the first four bounces.

Solution

$a_0 = 10$

$a_1 = 0.6(10) = 6$

$a_2 = 0.6(6) = 3.6$

$a_3 = 0.6(3.6) = 2.16$

$a_4 = 0.6(2.16) = 1.296$

The sequence exhibits geometric decay. The common ratio is $r = 0.6 < 1$. The height of each bounce is decreasing geometrically by a factor of 0.6, and a_0 denotes the initial position of 10 ft.

Answer: The heights of the first four bounces are 6 ft, 3.6 ft, 2.16 ft, and 1.296 ft.

Self-Check 3

One share of a particular stock on the NASDAQ is currently priced at $128.

a. Determine the price of the share at the end of each of the next 4 years if the price exhibits geometric growth with $r = 1.5$.

b. Determine the price of the share at the end of each of the next 4 years if the price exhibits geometric decay with $r = 0.5$.

2. Evaluate and Graph Exponential Functions

The terms of a geometric sequence can be obtained by repeated multiplication by the common ratio r. If a_1 is the first term, then we can write the first six terms as

$$a_1, \quad a_1r, \quad a_1r^2, \quad a_1r^3, \quad a_1r^4, \quad a_1r^5$$

Since exponentiation was originally developed to represent repeated multiplication like that just presented, we now examine exponential functions. In the work that follows, the exponent can be any real number, including an irrational number.

How Does the Graph of an Exponential Function Differ from the Graph of a Geometric Sequence?

The graph of an exponential function is connected and continuous over the whole domain of all real numbers, while the graph of a geometric sequence is a set of disconnected discrete points. Some applications such as mortgage payments occur at discrete intervals, while the growth of a tree is a continuous process.

Exponential Function: $f(x) = b^x$

Algebraically	Verbally	Algebraic Example	Graphical Example
If $b > 0$ and $b \neq 1$, then $f(x) = b^x$ is an exponential function with base b.	An exponential function has a constant for the base, and the variable is in the exponent.	$y = 2^x$ $\begin{array}{c\|c} x & y \\ \hline -2 & \frac{1}{4} \\ 0 & 1 \\ 1 & 2 \\ 4 & 16 \\ 6 & 64 \end{array}$	

Example 4   Identifying Exponential Functions

Determine which of these equations define exponential functions.

Solution

(a) $f(x) = 3^x$ An exponential function   This function has a base of 3.

(b) $f(x) = x^3$ Not an exponential function   The exponent is the constant 3, not a variable. This is called a cubic polynomial function.

(c) $f(x) = \left(\dfrac{4}{5}\right)^x$ An exponential function   This function has a base of $\dfrac{4}{5}$.

(d) $f(x) = (-5)^x$ Not an exponential function   The base of this exponential expression is -5. Since the base is negative, this equation does not represent an exponential function.

(e) $y = 1^x$ Not an exponential function $y = 1^x$ simplifies to $y = 1$, since 1 to any power equals 1. Thus it is a constant function.

Self-Check 4

Determine which of these equations defines an exponential function.

a. $f(x) = 5x$ **b.** $f(x) = 5^x$ **c.** $f(x) = x^5$

The domain of an exponential function $f(x) = b^x$ is understood to be the set of all real numbers. You may be able to evaluate some selected values such as 2^4 by hand, but you will need a calculator or a spreadsheet to approximate values such as 2^{π}.

What Is Meant by Exponential Growth?

Newspapers and magazines often use the term *exponential growth* incorrectly. They sometimes use the term simply to indicate rapid growth. If this term is to be used correctly, the data need to fit an exponential function with a base greater than 1. The exponential function $f(x) = b^x$ with $b > 1$ rises rapidly to the right and is called an **exponential growth function.** If $0 < b < 1$, then the graph of $f(x) = b^x$ declines rapidly to the right and is called an **exponential decay function.** The exponential growth function in the following box is $f(x) = 2^x$, and the exponential decay example is $f(x) = \left(\dfrac{1}{2}\right)^x$ or $f(x) = 2^{-x}$.

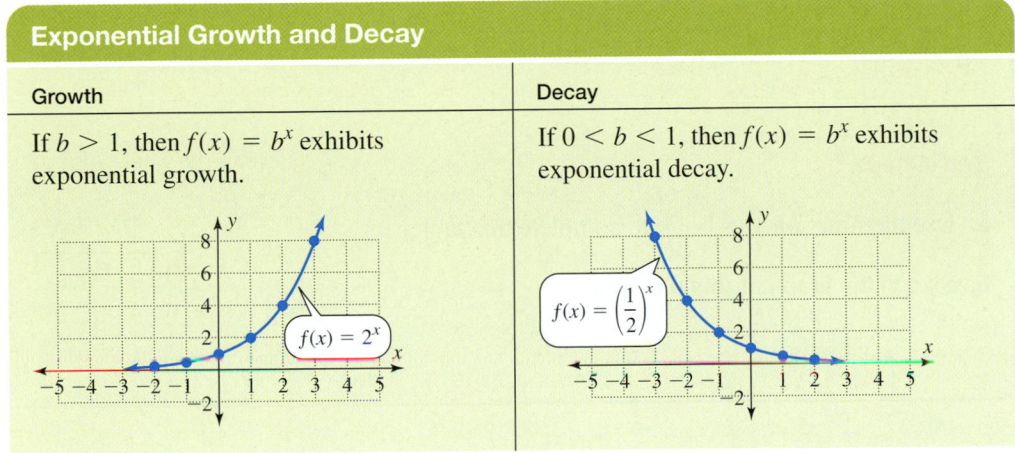

Exponential Growth and Decay

Growth	Decay
If $b > 1$, then $f(x) = b^x$ exhibits exponential growth.	If $0 < b < 1$, then $f(x) = b^x$ exhibits exponential decay.

The graphs of $f(x) = 2^x$ and $f(x) = \left(\dfrac{1}{2}\right)^x$ just given reveal several properties and features of exponential functions. We will list some of these properties below.

Recall that a function whose graph rises to the right is called an **increasing function.** Similarly, a function whose graph falls to the right is called a **decreasing function.** Also recall that the domain of a function is represented by the projection of its graph onto the x-axis, and the range is represented by the projection of its graph onto the y-axis.

Properties of the Graphs of Exponential Functions

For the exponential function $f(x) = b^x$ with $b > 0$ and $b \neq 1$:

1. The domain of f is the set of all real numbers $\mathbb{R}$.

2. The range of f is the set of all positive real numbers.

3. There is no x-intercept.

4. The y-intercept is $(0, 1)$.

5. **a.** The growth function is asymptotic to the negative portion of the x-axis (it approaches but does not touch the x-axis).
   b. The decay function is asymptotic to the positive portion of the x-axis.

6. **a.** The growth function rises (or grows) from left to right.
   b. The decay function falls (or decays) from left to right.

Example 5 │ Graphing an Exponential Function

Use the function $y = 3^x$ to complete this table and to sketch the graph of this function.

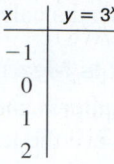

x	$y = 3^x$
-1	
0	
1	
2	

Solution

Table

x	$y = 3^x$
-1	$\dfrac{1}{3}$
0	1
1	3
2	9

Graph

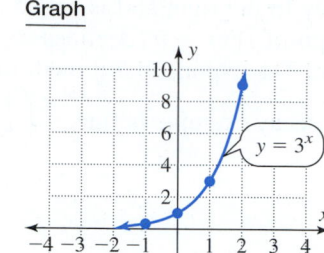

Complete the table and plot these points. Then use the known shape to sketch the graph through these points.

The domain is the projection of the graph onto the entire x-axis. The domain is $\mathbb{R}$.

Self-Check 5

a. Use the function $y = \left(\dfrac{1}{3}\right)^x$ to complete this table.

b. Sketch the graph of this function.

x	$y = \left(\dfrac{1}{3}\right)^x$
-2	
-1	
0	
1	

Some financial advisers and insurance salespeople promoting annuities and other investments refer to the "magic of compound interest." The "magic" they are referring to is exponential growth—growth in which the growth is always a factor of the previous base. Thus as the terms grow, so does the change from term to term. Example 6 involves exponential growth with a base $b = 1.0925$.

Example 6 │ Computing Compound Interest

The formula $A = P(1 + r)^t$ can be used to compute the total amount of money A that accumulates when a principal P is invested at a yearly interest rate r and left to compound annually for t years. Use this growth formula to do the following.

(a) Write a function for A for an investment of $5,000 at 9.25% for t years.

(b) Evaluate this function for $t = 8$ years.

Solution

(a) $A = P(1 + r)^t$

$A(t) = 5{,}000(1 + 0.0925)^t$

$A(t) = 5{,}000(1.0925)^t$

Substitute the given values into the compound interest formula: $P = 5{,}000$ and $r = 0.0925$. This is an exponential function of the form $f(x) = ab^x$ with a base of $b = 1.0925$.

(b) $A(8) = 5{,}000(1.0925)^8$

To determine the value of this investment after 8 years, evaluate $A(t)$ for $t = 8$.

$A(8) \approx 5{,}000(2.029418267)$

$A(8) \approx 10{,}147$

The value of this investment after 8 years is approximately $10,147.

This calculator approximation has been rounded to the nearest dollar.

Self-Check 6

Use the formula $A = P(1 + r)^t$ and a calculator to determine the value of $450 invested at 8.75% and left to compound annually for 7 years.

A Mathematical Note

Swiss-born Leonard Euler (1701–1783) was hired by Catherine the Great of Russia to write the elementary mathematics textbooks for Russian schools. He wrote prolifically on various mathematical topics. From Euler's textbook *Introducio* came many symbols, such as i for $\sqrt{-1}$, π for the ratio of the circumference of a circle to its diameter, and e for the base of natural logarithms.

Problems involving growth and decay often involve the irrational number e that is approximately equal to 2.718281828. Both π and e are fundamental constants that show up over and over in our study of the world and universe around us. To accommodate this, many calculators and spreadsheets have stored approximations for π and e. Calculator approximations are just that—approximations. Because π and e are irrational, their decimal representations never repeat or terminate.

In Technology Perspective 11.1.1, we illustrate how to use e on a calculator or in a spreadsheet. If you need to enter just e, then enter e^1. On a TI-84 Plus calculator, the e^x function is the secondary function of the LN key. On an Excel spreadsheet, this function is obtained by entering EXP(x).

Technology Perspective 11.1.1 Evaluating Expressions Involving e^x

Approximate e^2 to the nearest thousandth.

TI-84 Plus Calculator

```
e^(2)
          7.389056099
```

Excel Spreadsheet

A1 ▾	f_x	=EXP(2)

	A	B
1	7.389056	
2		

1. Press 2nd LN to obtain the exponential function base e.

2. Enter 2 and close with a right parenthesis.

2. Pressing ENTER will evaluate e^2.

1. Enter "=EXP(2)" in cell **A1**.

2. This will display the value of e^2 in cell **A1**.

Answer: $e^2 \approx 7.389$

Technology Self-Check 1

Approximate $\pi - e$ to the nearest hundredth.

Example 7 involves e and a function that models radioactive decay.

Example 7 | Determining the Amount of Radioactive Decay

The amount of a 1,000-g sample of radioactive carbon-14 remaining after t years is given by the function $A(t) = 1{,}000e^{-0.0001245t}$. How much of a 1,000-g sample will remain a century later?

Solution

$$A(t) = 1{,}000e^{-0.0001245t}$$ A century is 100 years. Substitute this value of t into the given function.

$$A(100) = 1{,}000e^{-0.0001245(100)}$$

$$A(100) = 1{,}000e^{-0.01245}$$

$$A(100) \approx 987.6$$ Approximate this value by using a calculator.

Answer: After 100 years, approximately 988 g of carbon-14 remains.

Self-Check 7

Use the formula in Example 7 to determine the amount of carbon-14 remaining after 1,000 years.

3. Solve Simple Exponential Equations

What Is an Exponential Equation?

An exponential equation is an equation that contains a variable in an exponent. The equations in Example 8 can all be solved easily without the use of a calculator because they involve familiar powers of 2, 3, 4, or other small integers. The properties we will use are given in the box.

Properties of Exponential Equations

Algebraically	Verbally
For real exponents x and y and bases $a > 0, b > 0$:	
1. For $b \neq 1$, $b^x = b^y$ if and only if $x = y$.	**1.** The exponents must be equal because the expressions are equal and have the same base.
2. For $x \neq 0$, $a^x = b^x$ if and only if $a = b$.	**2.** The bases must be equal because the expressions are equal and have the same exponents.

Example 8 | Solving Exponential Equations

Solve each equation.

(a) $3^{x-4} = 9$ **(b)** $4^x = \dfrac{1}{8}$ **(c)** $(b + 2)^3 = 125$

Solution

(a) $3^{x-4} = 9$ Substitute 3^2 for 9 in order to express both members in terms of the common base 3.

$\quad 3^{x-4} = 3^2$

$\quad x - 4 = 2$ The exponents are equal since the bases are the same.

Answer: $x = 6$

(b) $4^x = \dfrac{1}{8}$

$(2^2)^x = 2^{-3}$ Express both 4 and $\dfrac{1}{8}$ in terms of the common base 2.

$2^{2x} = 2^{-3}$ Use the power rule for exponents to simplify the left side of this equation.

$2x = -3$ The exponents are equal since the bases are the same.

Answer: $x = -\dfrac{3}{2}$

(c) $(b + 2)^3 = 125$

$(b + 2)^3 = 5^3$ Substitute 5^3 for 125.

$b + 2 = 5$ The bases are equal since the exponents are the same.

Answer: $b = 3$

Self-Check 8

Solve each equation.

a. $2^{x+7} = 16$ **b.** $27^x = 9$ **c.** $b^{-3} = \dfrac{1}{64}$

We will solve more complicated exponential equations in Section 11.6 after we develop some skills with logarithms.

Self-Check Answers

1. a. 2, 5, 8, 11, 14, 17, 20
   b. 2, 6, 18, 54, 162, 486, 1,458
   c. 2, −6, 18, −54, 162, −486, 1,458
2. a. Geometric, $r = 2$
   b. Arithmetic, $d = 3$
3. a. \$192, \$288, \$432, \$648
   b. \$64, \$32, \$16, \$8
4. a. Not an exponential function
   b. Exponential function with base 5
   c. Not an exponential function

5. a.

x	$y = \left(\dfrac{1}{3}\right)^x$
−2	9
−1	3
0	1
1	$\dfrac{1}{3}$

b.

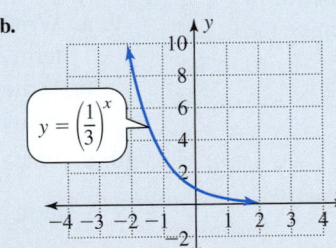

6. $A(7) \approx 809.50$; after 7 years the investment is worth \$809.50.
7. After 1,000 years, approximately 883 g of carbon-14 remains.
8. a. $x = -3$
   b. $x = \dfrac{2}{3}$
   c. $b = 4$

Technology Self-Check Answer

1. 0.42

11.1   Using the Language and Symbolism of Mathematics

1. An _____ sequence is a sequence with a constant change from term to term.

2. A geometric sequence is a sequence with a constant _____ from term to term.

3. If the common ratio r of a geometric sequence is greater than 1, then we say the sequence exhibits geometric _____.

4. If the common ratio r of a geometric sequence is between 0 and 1, then we say the sequence exhibits geometric _____.

5. An exponential function of the form $f(x) = b^x$ has a constant for the _____ and a variable in the _____.

6. An exponential function $f(x) = b^x$ with $b > 1$ is called an exponential _____ function.

7. An exponential function $f(x) = b^x$ with $0 < b < 1$ is called an exponential _____ function.

8. The graph of an exponential function $f(x) = b^x$ is asymptotic to a portion of the _____-axis.

9. The irrational number $e \approx$ _____ (to the nearest thousandth).

10. For real exponents x and y and base $b > 0$ and $b \neq 1$, $b^x = b^y$ if and if _____ = _____.

11. For a real exponent x and bases $a > 0$ and $b > 0$ and $x \neq 0$, $a^x = b^x$ if and only if _____ = _____.

11.1   Quick Review

1. The subscript notation a_n is read a _____ n.

Use the function shown to answer Exercises 2–5.

2. Determine the interval where this function is increasing.

3. Determine the interval where this function is decreasing.

4. Determine the domain of this function.

5. Determine the range of this function.

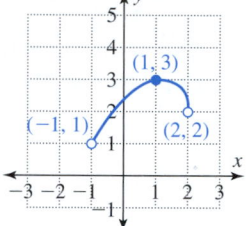

11.1   Exercises

Objective 1   Identify a Geometric Sequence

In Exercises 1–8, determine whether each sequence is geometric. If the sequence is geometric, determine the common ratio r.

1. **a.** 1, 5, 25, 125, 625, 3,125
   b. 1, −5, 25, −125, 625, −3,125

2. **a.** 4, 12, 36, 108, 324, 972
   b. −4, 12, −36, 108, −324, 972

3. **a.** 3,125, 625, 125, 25, 5, 1
   b. 3,125, −625, 125, −25, 5, −1

4. **a.** 972, 324, 108, 36, 12, 4
   b. 972, −324, 108, −36, 12, −4

5. **a.** $36, 24, 16, \dfrac{32}{3}, \dfrac{64}{9}$
   b. 36, 24, 12, 0, −12

6. **a.** 4, 8, 12, 16, 20
   b. 4, 8, 16, 32, 64

7. **a.** 2, 10, 40, 200, 1,000
   b. $2, 2\sqrt{2}, 4, 4\sqrt{2}, 8$

8. **a.** 1, 10, 100, 1,000, 10,000
   b. 100, 90, 80, 70, 60

9. **a.** Write a geometric sequence of five terms with a first term $a_1 = 10$ and a common ratio $r = 2$.
   b. Write a geometric sequence of five terms with a first term $a_1 = 10$ and a common ratio $r = -2$.

10. **a.** Write a geometric sequence of five terms with a first term of $a_1 = 12$ and a common ratio $r = \dfrac{1}{2}$.
    b. Write a geometric sequence of five terms with a first term of $a_1 = 12$ and a common ratio $r = -\dfrac{1}{2}$.

Objective 2   Evaluate and Graph Exponential Functions

11. Given $f(x) = 4^x$, mentally evaluate each expression.
    a. $f(1)$ **b.** $f(2)$ **c.** $f(0)$
    d. $f(-2)$ **e.** $f\left(\dfrac{1}{2}\right)$

12. Given $f(x) = 8^x$, mentally evaluate each expression.

   a. $f(1)$ **b.** $f(2)$ **c.** $f(0)$

   d. $f(-2)$ **e.** $f\left(\dfrac{1}{3}\right)$

In Exercises 13–16, complete each table of values and use these points to sketch a graph of the function.

x	$f(x)$
-2	
-1	
0	
1	
2	

13. $f(x) = 4^x$

14. $f(x) = \left(\dfrac{1}{4}\right)^x$

15. $f(x) = \left(\dfrac{2}{3}\right)^x$

16. $f(x) = \left(\dfrac{3}{2}\right)^x$

Objective 3 Solve Simple Exponential Equations

In Exercises 17–50, solve each equation using only pencil and paper.

17. $4^m = 4$ **18.** $2^m = 2$

19. $4^x = 16$ **20.** $2^y = 32$

21. $5^{v+1} = 125$ **22.** $7^{v+5} = 49$

23. $\left(\dfrac{2}{3}\right)^w = \dfrac{8}{27}$ **24.** $\left(\dfrac{3}{4}\right)^w = \dfrac{9}{16}$

25. $2^m = \dfrac{1}{2}$ **26.** $5^m = \dfrac{1}{5}$

27. $5^n = 1$ **28.** $2^n = 1$

29. $25^x = 5$ **30.** $27^{2y} = 3$

31. $49^x = 7$ **32.** $8^{5y} = 2$

33. $7^x = \dfrac{1}{49}$ **34.** $5^x = \dfrac{1}{25}$

35. $\left(\dfrac{2}{5}\right)^w = \dfrac{125}{8}$ **36.** $\left(\dfrac{2}{7}\right)^w = \dfrac{49}{4}$

37. $32^{3n-5} = 2$ **38.** $64^{2n+9} = 2$

39. $3^v = \sqrt{3}$ **40.** $7^v = \sqrt[3]{7}$

41. $10^x = 10{,}000$ **42.** $10^x = 100{,}000$

43. $10^y = 0.001$ **44.** $10^y = 0.00001$

45. $5^{x+7} = 25$ **46.** $7^{x-3} = 49$

47. $7^{3x} = \sqrt{7}$ **48.** $11^{2x-1} = \sqrt[3]{11}$

49. $\left(\dfrac{3}{7}\right)^x = \dfrac{49}{9}$ **50.** $\left(\dfrac{5}{4}\right)^{-x} = \dfrac{64}{125}$

Applying Technology

51. Given $f(x) = e^x$, use a calculator or a spreadsheet to approximate each expression to the nearest thousandth.

   a. $f(\sqrt{8})$ **b.** $f(-1.6)$ **c.** $f(\pi)$

52. Given $f(x) = e^{-0.2x}$, use a calculator or a spreadsheet to approximate each expression to the nearest thousandth.

   a. $f(\sqrt{7})$ **b.** $f(-2.4)$ **c.** $f(\pi)$

Estimate then Calculate

In Exercises 53–56, mentally estimate the value of each expression and then use a calculator or a spreadsheet to approximate each value to the nearest hundredth. Use the function $f(x) = (3.9)^x$.

Problem	Mental Estimate	Calculator Approximation
53. $f(2)$		
54. $f(0.99)$		
55. $f(-1)$		
56. $f(0.5)$		

Connecting Concepts to Applications

57. **Stock Prices** A share of stock on the NASDAQ is priced at $100. Determine the price per share at the end of each of the next 4 years under each of the following assumptions.

   a. Arithmetic growth of $10 per year

   b. Arithmetic decay of $10 per year

   c. Geometric growth of 10% per year

   d. Geometric decay of 10% per year

58. **Stock Prices** A share of stock on the NASDAQ is priced at $100. Determine the price per share at the end of each of the next 4 years under each of the following assumptions.

   a. Arithmetic growth of $15 per year

   b. Arithmetic decay of $15 per year

   c. Geometric growth of 15% per year

   d. Geometric decay of 15% per year

59. **Height of a Bouncing Ball** A golf ball is dropped from 36 m onto a synthetic surface that causes it to rebound to one-half its previous height on each bounce. Determine the heights of the first four bounces.

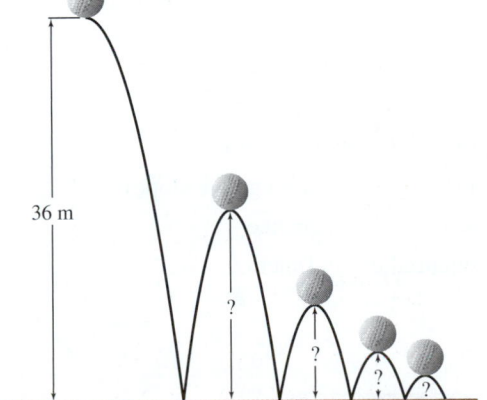

36 m

60. Radioactive Decay A nuclear chemist starts an experiment with 100 g of a radioactive material. At the end of each time period, only one-half of the amount present at the start of the period is left. Determine the amount of this material left at the end of each of the first five time periods.

In Exercises 61 and 62, use the compound interest formula $A = P(1 + r)^t$ with A representing the total amount of money that accumulates when a principal P is invested at an annual interest rate r and left to compound for a time of t years.

61. Compound Interest Write A as a function of t when $1,000 is invested at 6% for t years. Then evaluate this function for $t = 7$ years.

62. Compound Interest Write A as a function of t when $4,000 is invested at 5.5% for t years. Then evaluate this function for $t = 10$ years.

63. Radioactive Decay The strontium-90 in a nuclear reactor decays continuously. If 100 mg is present initially, the amount present after t years is given by $A(t) = 100e^{-0.0248t}$. Approximate to the nearest tenth of a milligram the amount left after 10 years.

64. Radioactive Decay The strontium-90 in a nuclear reactor decays continuously. If 100 mg is present initially, the amount present after t years is given by $A(t) = 100e^{-0.0248t}$. Approximate to the nearest tenth of a milligram the amount left after 50 years.

Review and Concept Development

65. a. Write an arithmetic sequence of five terms with a first term of 6 and a common difference of 5.
   b. Write a geometric sequence of five terms with a first term of 6 and a common ratio of 5.

66. a. Write an arithmetic sequence of five terms with a first term of 6 and a common difference of -0.5.
   b. Write a geometric sequence of five terms with a first term of 6 and a common ratio of -0.5.

In Exercises 67–72, match each description of a function with the corresponding graph.

67. A linear growth function (positive slope)

68. A geometric growth sequence

69. An exponential growth function

70. A linear decay function (negative slope)

71. A geometric decay sequence

72. An exponential decay function

A.

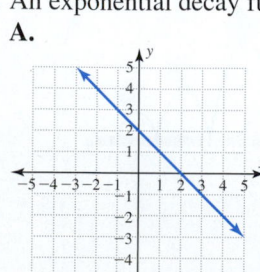

B.

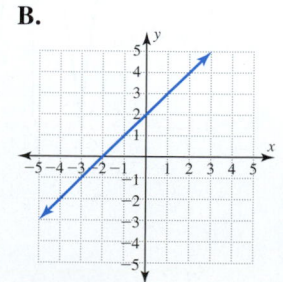

C.

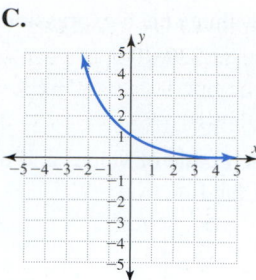

D.

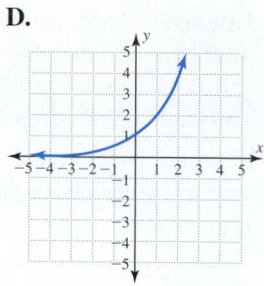

E.

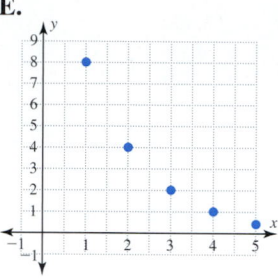

F.

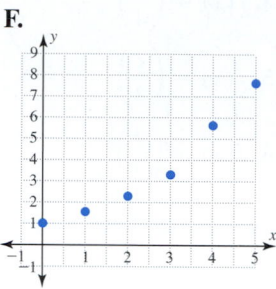

In Exercises 73–78, solve each equation.

73. $2x = 64$

74. $x^2 = 64$

75. $2^x = 64$

76. $3^v = \dfrac{1}{9}$

77. $3z = \dfrac{1}{9}$

78. $(3x - 4)(x - 1) = \dfrac{2}{3}$

Group discussion questions

79. Challenge Question $f(x) = b^x$ is an exponential function if $b > 0$ and $b \neq 1$.
   a. Discuss the nature of the function $f(x) = b^x$ if $b = 0$.
   b. Discuss the nature of the function $f(x) = b^x$ if $b = 1$.

80. Challenge Question A question showing the growth potential of geometric growth is given by the classic problem of a blacksmith shoeing a horse. A blacksmith attaches each horseshoe with eight nails. The blacksmith offers to charge by the nail for shoeing all four hooves. The cost for the first nail would be 1 cent; the second, 2 cents; the third, 4 cents; etc. At this rate, complete the following table.

Nail	1	2	4	8	10	16	20	30	32
Cost of nail, $	0.01	0.02							

81. Discovery Question An employee is offered two salary options. Option 1 starts at $20,000 per year with a $3,000 raise after each of the first 9 years. Option 2 starts at $20,000 per year with a 10% raise after each of the first 9 years. Complete the two tables.

Option 1

Year	Salary ($)	Raise ($)	Raise (%)
1	20,000		
2	23,000	3,000	15
3	26,000	3,000	13
4	29,000		
5			
6			
7			
8			
9			
10			

Option 2

Year	Salary ($)	Raise ($)	Raise (%)
1			20,000
2	22,000	2,000	10
3	24,200	2,200	10
4			26,620
5			
6			
7			
8			
9			
10			

Discuss the advantages and disadvantages of each option, using the terms *arithmetic growth* and *geometric growth* where these terms are appropriate. Which option is better for a person who plans to quit this job after 5 years? After 10 years? After 20 years?

11.1 Cumulative Review

1. Factor $2x(3x - 5) - 7(3x - 5)$.
2. Expand and simplify $2x(3x - 5) - 7(3x - 5)$.
3. The slope of a line parallel to $y = 3x - 4$ is _____ .
4. The slope of a line perpendicular to $y = 3x - 4$ is _____ .
5. Solve $-5 \le 4x - 1 < 11$ and write the answer using interval notation.

Section 11.2 Inverse Functions

Objectives:

1. Write the inverse of a function.
2. Identify a one-to-one function.
3. Graph the inverse of a function.

The purpose of this section is to provide a bridge between exponential and logarithmic functions. These functions are inverses of each other.

What Is Meant by the Inverse of a Function?

The reversing of each ordered pair of a function produces what we will call the inverse of a function. A function (see Section 8.1) is a correspondence between two sets of values; a function matches each input value with exactly one output value. Because a function clearly pairs two sets of values, the information available from this pairing can often be used both ways—from x to y and from y to x.

For example, we can write a formula for a function that for a fixed time and interest rate will pair each car loan amount with the required monthly payment. By reversing this function, we can use the monthly payment we can afford to determine the loan amount that is affordable.

Loan amount	Function $\longrightarrow$	Monthly payment
Monthly payment	Inverse $\longrightarrow$	Loan amount

Inverse of a Function

Verbally	Algebraically	Numerical Example	Graphical Example
If f is a function that matches each input value x with an output value y, then the inverse of f reverses this correspondence to match this y-value with the x-value.	For each point (x, y) of the function f, (y, x) is a point of the inverse of f.	Function: $\begin{array}{c\|c} x & y \\ \hline -2 & -5 \\ -1 & -3 \\ 0 & -1 \\ 1 & 1 \\ 2 & 3 \end{array}$ Inverse: $\begin{array}{c\|c} x & y \\ \hline -5 & -2 \\ -3 & -1 \\ -1 & 0 \\ 1 & 1 \\ 3 & 2 \end{array}$	Function: Inverse:

1. Write the Inverse of a Function

In Section 8.1, we examined several representations for functions including mapping notation, ordered pairs, tables, graphs, and function notation. We now examine the inverse of a function by using each of these representations. Note in each case that the inverse undoes what f does.

Example 1   Determining the Inverse of a Function

Determine the inverse of each function.

Solution

Function f | The Inverse of f

(a)

$\begin{array}{ccc} x & & y \\ 7 & \longrightarrow & 1 \\ 8 & \longrightarrow & 3 \\ 9 & \longrightarrow & 5 \end{array}$

$\begin{array}{ccc} x & & y \\ 1 & \longrightarrow & 7 \\ 3 & \longrightarrow & 8 \\ 5 & \longrightarrow & 9 \end{array}$

In the mapping notation, the inverse of f reverses the arrows and the order of the input-output pairs.

(b)

$\begin{array}{c\|c} x & y \\ \hline 4 & -1 \\ 7 & 2 \\ -\pi & 5 \\ 0 & 6 \end{array}$

$\begin{array}{c\|c} x & y \\ \hline -1 & 4 \\ 2 & 7 \\ 5 & -\pi \\ 6 & 0 \end{array}$

In the table, the inverse of f reverses the x-y pairing.

(c) $\{(0, 1), (2, 7), (-5, 4)$

$\{(1, 0), (7, 2), (4, -5)\}$

The inverse of f reverses the order of each (x, y) pair.

(d)

Loan Amount ($)	Monthly Payment ($)
5,000	122.06
10,000	244.13
15,000	366.19
20,000	488.26

Monthly Payment ($)	Loan Amount ($)
122.06	5,000
244.13	10,000
366.19	15,000
488.26	20,000

The given function pairs each loan amount with the payment required to pay off an 8% loan in 4 years.

The inverse pairs a payment with the amount of loan this payment will pay off.

Self-Check 1

Write the inverse of the function defined by this table.

x		y
8	$\longrightarrow$	6
4	$\longrightarrow$	11
-2	$\longrightarrow$	0

Note that all the samples in Example 1 illustrate that the inverse of f reverses the roles of the domain and range of the function f. In Example 1(c), the domain of $f = \{(0, 1), (2, 7), (-5, 4)\}$ is the set $\{0, 2, -5\}$, and the range is the set $\{1, 7, 4\}$. The domain of the inverse of f, $\{(1, 0), (7, 2), (4, -5)\}$, is the set $\{1, 7, 4\}$, and the range is the set $\{0, 2, -5\}$.

Is the Inverse of a Function Also a Function?

The answer is sometimes. Remember, a function must pair each input value with exactly one output value. The inverse of the function $f = \{(3, 4), (5, 4)\}$ is given by $\{(4, 3), (4, 5)\}$. In this case, the inverse of f does not match each input with exactly one output—4 is paired both with 3 and with 5. Examining $f = \{(3, 4), (5, 4)\}$, we note that f has two input values with an output of 4. This causes the inverse of f to have two outputs for the input of 4. Thus the inverse of f is not a function.

However, if different inputs of f have different outputs, then the inverse of f also will be a function. We call functions with this property **one-to-one functions.**

2. Identify a One-to-One Function

One-to-One Function

Verbally	Algebraically	Numerical Example	Graphical Example
A function is one-to-one if no two x-values correspond to the same y-value. That is, different input values produce different output values.	The only way that $f(x_1) = f(x_2)$ is if $x_1 = x_2$. That is, if $x_1 \neq x_2$, then $f(x_1) \neq f(x_2)$.	Function: $y = x - 1$ $\begin{array}{c\|c} x & y \\ \hline -2 & -3 \\ -1 & -2 \\ 0 & -1 \\ 1 & 0 \\ 2 & 1 \end{array}$ No two x-values correspond to the same y-value.	Function: $y = x - 1$ Since this graph is increasing, different input values will produce different output values.

If a function f is one-to-one, then the inverse of f will also be a function. In this case, we denote the inverse of f by f^{-1} and call f^{-1} an **inverse function.** Treat f^{-1} as a single symbol that undoes what the function f does. This is *not* exponential notation and does not represent $\dfrac{1}{f}$. The context of the notation should make the intent of the notation clear.

Example 2 Identifying One-to-One Functions

Determine whether the function f is one-to-one and whether the inverse of f is an inverse function.

(a) $f = \{(-5, 6), (2, 7), (8, 9), (9, 3)\}$

(b) $f = \{(-3, 9), (0, 0), (3, 9)\}$

Solution

(a) This function is one-to-one. The inverse of f is an inverse function.

$f^{-1} = \{(6, -5), (7, 2), (9, 8), (3, 9)\}$ matches each input with a unique output.

(b) This function is not one-to-one. The inverse of f is not a function.

Because the function pairs both -3 and 3 with 9, the function is not one-to-one. The inverse of f, $\{(9, -3), (0, 0), (9, 3)\}$, is not a function.

Self-Check 2

Determine whether f is one-to-one and whether the inverse of f is an inverse function.

a. $f = \{(3, \pi), (e, 4), (8, -1)\}$

b. $f = \{(5, 9), (7, 6), (-3, 9)\}$

The vertical line test covered in Section 8.1 is a quick method for determining whether a graph represents a function. This test allows us to determine visually whether each x input value is paired with exactly one y output value.

Is There Also a Quick Method for Determining Whether the Graph of a Function Represents a One-to-One Function?

Yes, we can do this with the horizontal line test.

The Horizontal Line Test

A graph of a function represents a one-to-one function if it is impossible to have any horizontal line intersect the graph at more than one point.

Example 3 Using the Horizontal Line Test

Use the horizontal line test to determine whether each graph represents a one-to-one function.

Solution

(a)

Not a one-to-one function   The horizontal line passes through more than one point of the graph of this function.

(b)

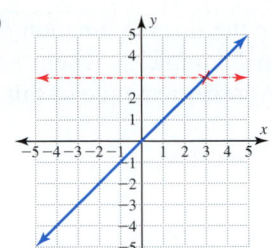

A one-to-one function   Any horizontal line will pass through exactly one point of the graph of this function.

Self-Check 3

Use the horizontal line test to determine whether each graph represents a one-to-one function.

a.

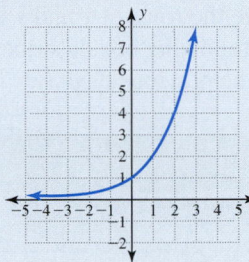

b.

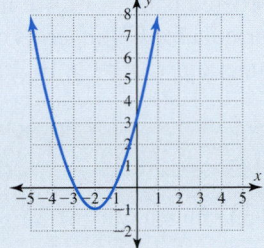

A key concept to understand about f^{-1} is that f^{-1} undoes what the function f does. This is illustrated in Example 4.

Example 4   Describing Inverse Functions

Describe each function and its inverse verbally and algebraically. Give a numerical example from the domain and range of each function.

(a) $f(x) = 2x$ **(b)** $f(x) = \sqrt[3]{x}$

Solution

Verbally	Algebraically	Numerical Example
(a) f doubles x	$f(x) = 2x$	$8 \xrightarrow{f} 16$
f^{-1} takes one-half of x	$f^{-1}(x) = \dfrac{x}{2}$	$16 \xrightarrow{f^{-1}} 8$
(b) f takes the cube root of x	$f(x) = \sqrt[3]{x}$	$8 \xrightarrow{f} 2$
f^{-1} cubes x	$f^{-1}(x) = x^3$	$2 \xrightarrow{f^{-1}} 8$

Self-Check 4

The function f increases x by 5. Describe f^{-1} verbally, and write both f and f^{-1} using function notation.

If f is a one-to-one function defined by an equation, then the inverse function f^{-1} is also indirectly given by this equation. This is illustrated in Example 4, where simple functions are used to describe how f^{-1} undoes what f does. Another example of this is as follows:

Algebraically   **Verbally**

$$f(x) = 2x + 1$$ f: $5 \longrightarrow$ | First double | $\xrightarrow{10}$ | Then add 1 | $\longrightarrow 11$

$$f^{-1}(x) = \frac{x-1}{2}$$ f^{-1}: $11 \longrightarrow$ | First subtract 1 | $\xrightarrow{10}$ | Then take one-half of this value | $\longrightarrow 5$

It is customary to use x to represent the input value for both f and f^{-1}. Thus the formulas for both f and f^{-1} are often expressed by using the variable x. Because the inverse function f^{-1} reverses the order of each ordered pair (x, y), we interchange the roles of x and y in the formula for f in order to find a formula for f^{-1} in terms of x. The following box describes this procedure and reexamines the function $f(x) = 2x + 1$ given in the previous display.

Finding an Equation for an Inverse Function

To find the inverse of a one-to-one function $y = f(x)$:

Verbally	Algebraic Example
Step 1. Replace $f(x)$ by y. Write this as a function of y in terms of x.	Function: $\qquad f(x) = 2x + 1$ $\qquad\qquad\qquad y = 2x + 1$
Step 2. To form the inverse, replace each x with y and each y with x.	Inverse function: $\qquad x = 2y + 1$
Step 3. If possible, solve the resulting equation for y.	$x - 1 = 2y$ $\dfrac{x-1}{2} = y$
Step 4. The inverse can then be written in function notation by replacing y with $f^{-1}(x)$.	$f^{-1}(x) = \dfrac{x-1}{2}$

Example 5 illustrates this procedure.

Example 5 Determining an Inverse Function

Determine the inverse of $f(x) = 3x - 5$ algebraically, and compare f and f^{-1} both verbally and with a numerical example.

Solution

Algebraically

Function:
$$f(x) = 3x - 5$$
$$y = 3x - 5 \qquad \text{First replace } f(x) \text{ by } y \text{ to express } y \text{ as a function of } x.$$

Inverse function:
$$x = 3y - 5 \qquad \text{Then interchange } x \text{ and } y \text{ in the equation to form the inverse.}$$
$$x + 5 = 3y \qquad \text{Solve for } y \text{ by adding 5 to both sides of the equation and then dividing both sides by 3.}$$
$$y = \frac{x+5}{3}$$

$$f^{-1}(x) = \frac{x+5}{3} \qquad \text{Rewrite the inverse by using functional notation.}$$

Verbally

$$f(x) = 3x - 5 \qquad\qquad f:\ \ x \rightarrow \boxed{\begin{array}{c}\text{First}\\\text{triple } x\end{array}} \xrightarrow{3x} \boxed{\begin{array}{c}\text{Then}\\\text{subtract 5}\end{array}} \rightarrow 3x - 5$$

$$f^{-1}(x) = \frac{x + 5}{3} \qquad f^{-1}:\ \ x \rightarrow \boxed{\begin{array}{c}\text{First}\\\text{add 5}\end{array}} \xrightarrow{x+5} \boxed{\begin{array}{c}\text{Then divide}\\\text{by 3}\end{array}} \rightarrow \frac{x + 5}{3}$$

Numerical Example

$$f(2) = 1:\ \ 2 \rightarrow \boxed{\begin{array}{c}\text{First}\\\text{triple 2}\end{array}} \xrightarrow{6} \boxed{\begin{array}{c}\text{Then}\\\text{subtract 5}\end{array}} \rightarrow 1$$

$$f^{-1}(1) = 2:\ \ 1 \rightarrow \boxed{\begin{array}{c}\text{First}\\\text{add 5}\end{array}} \xrightarrow{6} \boxed{\begin{array}{c}\text{Then divide}\\\text{by 3}\end{array}} \rightarrow 2$$

Self-Check 5

Determine the inverse of $f(x) = \dfrac{2x - 1}{3}$.

Is There a Relationship Between the Graphs of *f* and *f*⁻¹?

Yes, these graphs are symmetric about the line $x = y$. This is illustrated by the graphs of f and f^{-1} from Example 5.

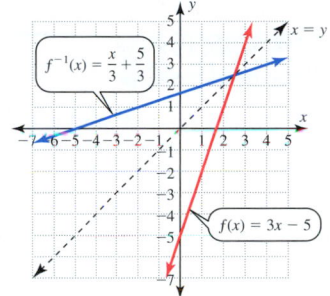

In this figure, the points (a, b) and (b, a) are mirror images about the graph of $x = y$. Thus f and f^{-1} will always be symmetric about the line $x = y$. Example 6 further illustrates this symmetry.

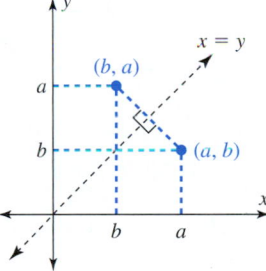

3. Graph the Inverse of a Function

Example 6   Graphing the Inverse of a Function

The graph of the function $f = \{(2, 1), (4, 2), (6, 3)\}$ is shown here. Graph the inverse of this function.

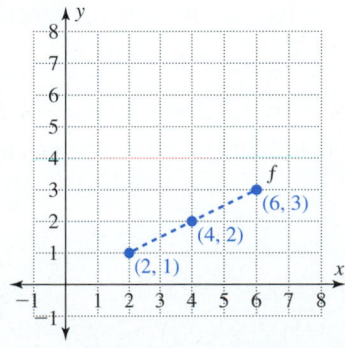

Solution

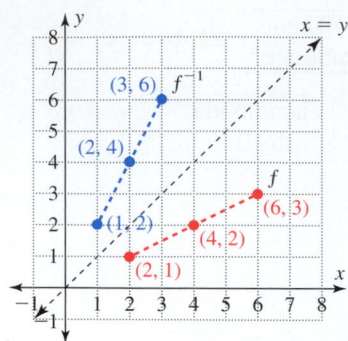

The inverse of $f = \{(2, 1), (4, 2), (6, 3)\}$ is $f^{-1} = \{(1, 2), (2, 4), (3, 6)\}$. In fact, the inverse of all the points on the dashed line segment through the points of f is the set of all the points on the dashed line segment through the points of f^{-1}. Note the symmetry of these points about the line $x = y$.

Self-Check 6

Graph both $f = \{(0, 1), (4, 3), (6, 4)\}$ and its inverse on the same coordinate system.

We can take advantage of the symmetry of f and f^{-1} about the line $x = y$. Using this symmetry, we can graph f^{-1} directly from the graph of f without knowing the equation for f^{-1}. This is illustrated in Example 7.

Example 7   Using Symmetry to Graph f^{-1}

Graph $f(x) = 2^x$ and use symmetry about the line $x = y$ to graph $y = f^{-1}(x)$.

Solution

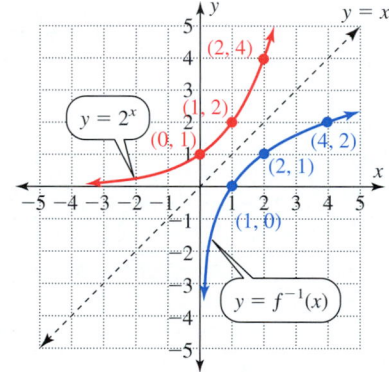

First graph the exponential function $f(x) = 2^x$ and sketch in the dashed line $x = y$. Then use symmetry to create a symmetric image for $f(x) = 2^x$ about $x = y$. This symmetric image is the graph of $y = f^{-1}(x)$.

Note that the symmetric image of $(0, 1)$ is $(1, 0)$, of $(1, 2)$ is $(2, 1)$, and of $(2, 4)$ is $(4, 2)$.

Self-Check 7

Sketch the graph of $f(x) = \sqrt{x}$ and its inverse on the same coordinate system.

In Example 7, note the shape of the inverse of $f(x) = 2^x$. This is actually the graph of $y = \log_2 x$, a logarithmic function that we will examine in the next section.

Self-Check Answers

1.

x		y
6	$\longrightarrow$	8
11	$\longrightarrow$	4
0	$\longrightarrow$	-2

2. a. f is one-to-one; the inverse of f, $\{(\pi, 3), (4, e), (-1, 8)\}$, is an inverse function.

    b. f is not one-to-one; the inverse of f, $\{(9, 5), (6, 7), (9, -3)\}$, is the inverse of a function, but it is not an inverse function.

3. a. A one-to-one function

    b. Not a one-to-one function

4. f^{-1} decreases x by 5;

    $f(x) = x + 5; f^{-1}(x) = x - 5$

5. $f^{-1}(x) = \dfrac{3x + 1}{2}$

6.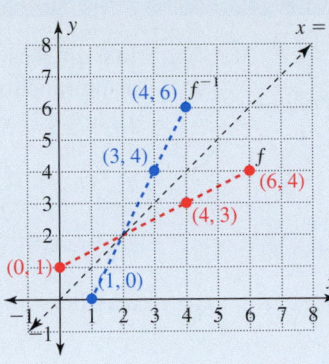

7.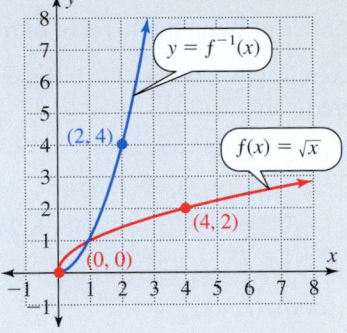

11.2 Using the Language and Symbolism of Mathematics

1. If f is a function that matches an input value 9 with an output value 4, then the inverse of f reverses this correspondence to match _____ with _____.

2. A function is one-to-one if different input values produce _____ output values.

3. A graph of a function represents a _____ -_____ -_____ function if it is impossible to

have any horizontal line intersect the graph at more than one point.

4. If f is a one-to-one function, then the inverse of f is a function denoted by _____.

5. The graphs of f and f^{-1} are symmetric about the line _____.

11.2 Quick Review

In Exercises 1–3, determine whether the graph of each relation represents a function.

In Exercises 4 and 5, use symmetry to complete the graph of each parabola.

1.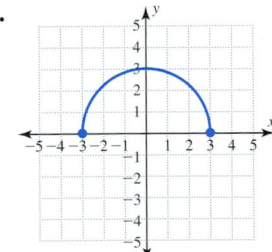

2.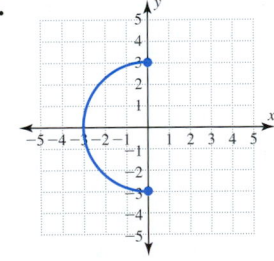

4.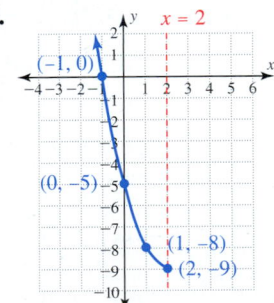

3.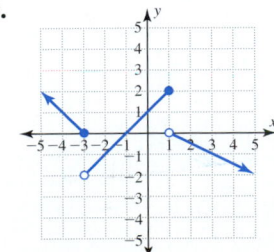

5.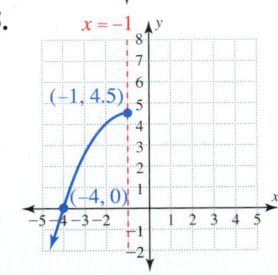

11.2 Exercises

Objective 1 Write the Inverse of a Function

In Exercises 1–10, write the inverse of each function using ordered-pair notation.

1. $\{(1, 4), (3, 11), (8, 2)\}$

2. $\{(2, -1), (0, 3), (-4, 6)\}$

3. $\{(-3, 2), (-1, 2), (0, 2), (2, 2)\}$

4. $\{(-0.7, \pi), (-0.5, \pi), (1.3, \pi)\}$

5. $\{(a, b), (c, d)\}$

6. $\{(w, x), (y, z)\}$

7.
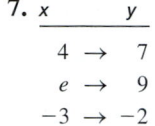

x		y
4	→	7
e	→	9
-3	→	-2

8.

x		y
-8	→	1
-9	→	3
6	→	-4

9.
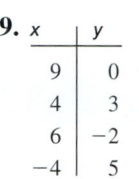

x	y
9	0
4	3
6	-2
-4	5

10.

x	y
-2	3
-1	5
0	7
1	9

In Exercises 11 and 12, first express the function f whose graph is given in ordered-pair notation. Then write f^{-1} using ordered-pair notation.

11.
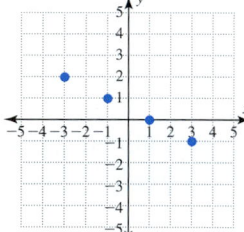

12.
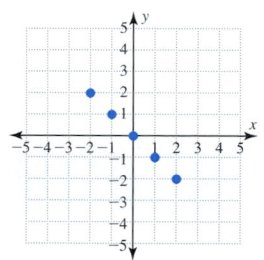

In Exercises 13–18, the function f is described verbally and algebraically. Describe f^{-1} both verbally and algebraically.

Verbally	**Algebraically**
13. f decreases x by 2	$f(x) = x - 2$
14. f increases x by 7	$f(x) = x + 7$
15. f quadruples x	$f(x) = 4x$
16. f takes one-fifth of x	$f(x) = \dfrac{x}{5}$
17. f doubles x and then subtracts 3	$f(x) = 2x - 3$
18. f takes one-half of x and then adds 5	$f(x) = \dfrac{1}{2}x + 5$

In Exercises 19–26, write the inverse of each function by using function notation.

19. $f(x) = 5x + 2$

20. $f(x) = 2x - 5$

21. $f(x) = \dfrac{1}{3}x - 7$

22. $f(x) = -\dfrac{1}{4}x + 3$

23. $f(x) = \sqrt[3]{x} + 2$

24. $f(x) = \sqrt[3]{x + 2}$

25. $f(x) = -x$

26. $f(x) = \dfrac{1}{x}$

Objective 2 Identify a One-to-One Function

In Exercises 27–30, determine whether each function is one-to-one.

27. $\{(-\pi, 4), (3, \pi), (7, -9)\}$

28. $\{(-8, 9), (7, 13), (8, 9)\}$

29. $\{(7.5, 7), (7.8, 7), (8.3, 8)\}$

30. $\{(-5, 2), (-4, -1), (3, 7), (8, 8)\}$

In Exercises 31–40, use the horizontal line test to determine whether each graph represents a one-to-one function.

31.
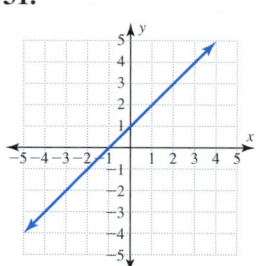

32.
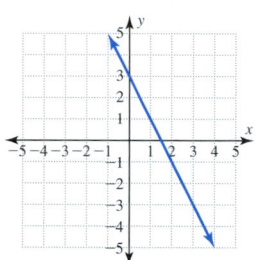

33.
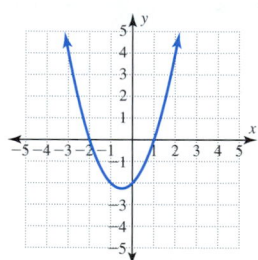

34.
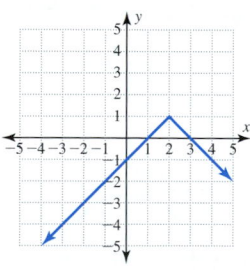

35.
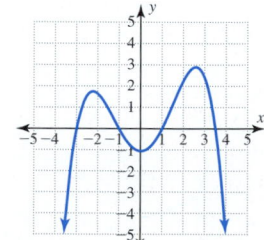

36.
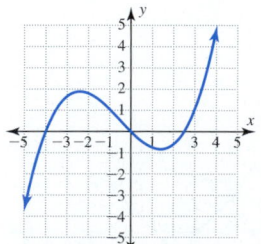

37.
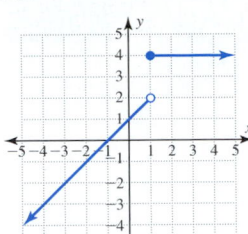

38.
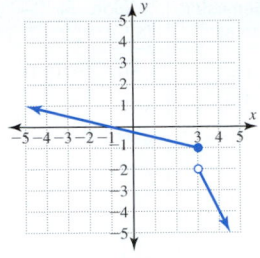

47.
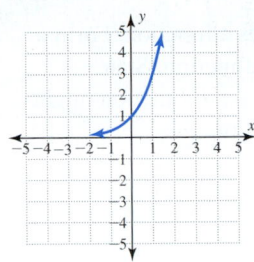

48.
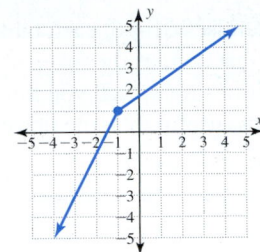

39.
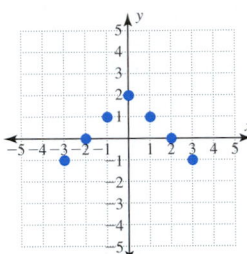

40.
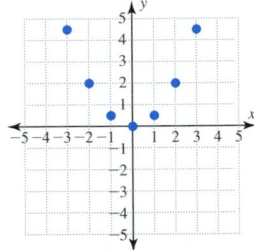

49.
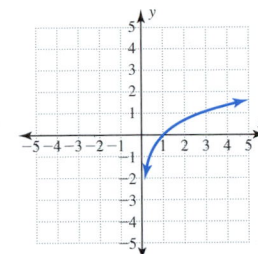

50.
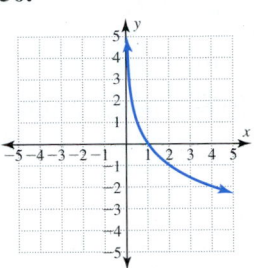

Objective 3 Graph the Inverse of a Function

In Exercises 41–50, sketch the inverse of each function by using the symmetry of f and f^{-1} about the line $x = y$. (*Hint:* For Exercises 43–50, pick a few key points on f and f^{-1} and then sketch the rest of f^{-1}.)

41.
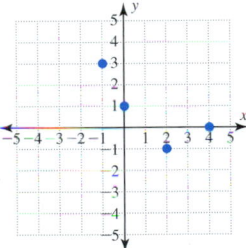

42.
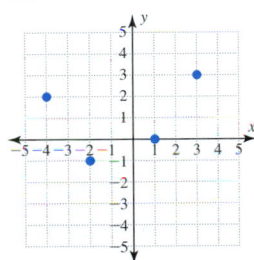

43.
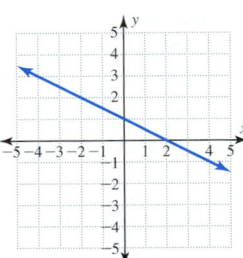

44.
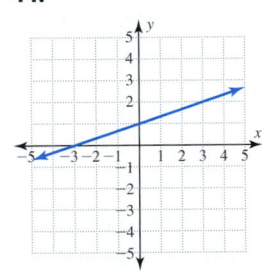

45.
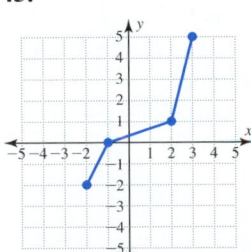

46.
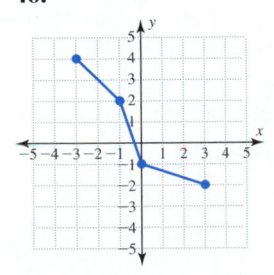

Review and Concept Development

In Exercises 51–56, use $f(x) = 4x - 3$ and $f^{-1}(x) = \dfrac{x + 3}{4}$ to evaluate each expression.

51. a. $f(5)$ **b.** $f^{-1}(17)$

52. a. $f(0)$ **b.** $f^{-1}(-3)$

53. a. $f(2)$ **b.** $f^{-1}(5)$

54. a. $f\left(\dfrac{3}{4}\right)$ **b.** $f^{-1}(0)$

55. a. $f(1)$ **b.** $f^{-1}(1)$

56. a. $f(10)$ **b.** $f^{-1}(37)$

In Exercises 57–60, use ordered-pair notation to state the domain and range of the given function f and its inverse f^{-1}.

57.

x		y
0	→	−3
1	→	−2
2	→	−1
3	→	0
4	→	1

58.

x		y
−1	→	−6
0	→	−4
1	→	−2
2	→	0
3	→	2

59.
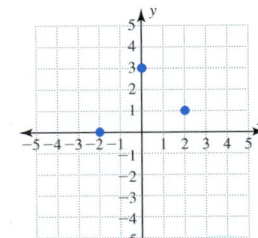

60.
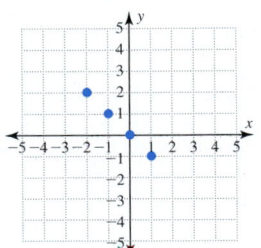

Connecting Concepts to Applications

In Exercises 61–66, each table defines a function. Form a table that defines the inverse of each function. State conditions for which you think the inverse function might be more useful than the original function.

61. Shoe Sizes

U.S. Size	European Size
7	39
8	40.5
9	42
10	43
11	44

62. Environmental Application

Year	U.S. Land Used for Recreation (Millions of Acres)
1977	66
1982	71
1987	84
1992	87
1997	96

Source: U.S. Department of Agriculture.

63. Car Loan at 7.5% for 4 Years

Loan Amount ($)	Monthly Payment ($)
5,000	120.89
10,000	241.79
15,000	362.68
20,000	483.58
25,000	604.47

64. Car Loan at 5% for 4 Years

Loan Amount ($)	Monthly Payment ($)
5,000	115.15
10,000	230.29
15,000	345.44
20,000	460.59
25,000	575.73

65. Temperature Conversions

Temperature °F	Temperature °C
0	−17.8
20	−6.7
40	4.4
60	15.6
80	26.7
100	37.8

66. Tuition at Gulf Coast Community College

Semester Hours	Tuition ($)
1	50
2	100
3	150
4	200
5	250
10	500
15	800

67. Production Cost The cost of producing x units of a product is given by the function $C(x) = 12x + 350$.
   a. What does the variable x represent in $C(x)$?
   b. What does the output of $C(x)$ represent?
   c. Determine the formula for $C^{-1}(x)$.
   d. What does the variable x represent in $C^{-1}(x)$?
   e. What does the output of $C^{-1}(x)$ represent?
   f. Determine the cost of producing 100 units.
   g. Determine the number of units that can be produced for $1,934.

68. Sales Commission A car dealership hires new salespeople and calculates the monthly salary for each person for each of the first 6 months by using the formula $S(x) = 500x + 1,000$, where x represents the number of cars this person sold for the month.
   a. Evaluate and interpret $S(0)$.
   b. Evaluate and interpret $S(8)$.
   c. Determine the formula for $S^{-1}(x)$.
   d. What does the variable x represent in $S^{-1}(x)$?
   e. Evaluate and interpret $S^{-1}(2,500)$.
   f. Evaluate and interpret $S^{-1}(6,000)$.

Group discussion questions

69. Discovery Question
   i. Make two identical copies of the graph of the function $y = 2x - 5$ on two sheets of clear plastic. Also, dash in the line $x = y$ on each sheet of plastic.
   ii. Place one sheet of plastic on an overhead projector.
   iii. Then flip the second sheet of plastic over and turn the sheet 90°, so that the upside-down sheet has its positive y-axis align with the positive x-axis on the other sheet.

   Describe to your teacher what you have observed and why you think this is so.

70. Challenge Question
   a. Is $f(x) = x^2$ a one-to-one function?
   b. If we restrict $f(x) = x^2$ to the domain $D = [0, \infty)$, is this restricted function one-to-one?
   c. Graph the restricted function in part **b** and its inverse.

d. Write the equation of the inverse of the restricted function in part **b.**

e. If we restrict $f(x) = x^2$ to the domain $D = (-\infty, 0]$, is this restricted function one-to-one?

f. Graph the restricted function in part **e** and its inverse.

g. Write the equation of the inverse of the restricted function in part **e.**

71. Discovery Question

a. Use a graphing calculator to graph $Y_1 = e^x$.

b. Use a graphing calculator to graph $Y_2 = \ln x$. (*Hint:* Use the ⬤LN key.)

c. Compare these two graphs and make a conjecture on the relationship between the functions $Y_1 = e^x$ and $Y_2 = \ln x$.

11.2 | **Cumulative Review**

1. Write the polynomial $2x^2y^2 - 3xy^3 + 4x^3y - 5x^4$ in standard form.

2. What is the degree of the polynomial $7x^3 - 2x^2y^2$?

3. A horizontal line has a slope of $m =$ _____.

4. The parabola defined by $y = ax^2 + bx + c$ opens upward if _____.

5. The parabola defined by $y = ax^2 + bx + c$ opens downward if _____.

Section 11.3 | Logarithmic Functions

Objectives:

1. Interpret and use logarithmic notation.
2. Evaluate simple logarithmic expressions.
3. Sketch the graph of a logarithmic function.

The accompanying graphs of exponential functions illustrate that an exponential function $f(x) = b^x$ is a one-to-one function for both exponential growth and exponential decay. Thus each exponential function has an inverse function.

   Also note that the projection of the graph of an exponential function onto the x-axis is the entire real axis. Thus the domain of $f(x) = b^x$ is $\mathbb{R}$, the set of all real numbers. The projection of each graph onto the y-axis is the interval $(0, \infty)$. Thus the range of $f(x) = b^x$ is $(0, \infty)$.

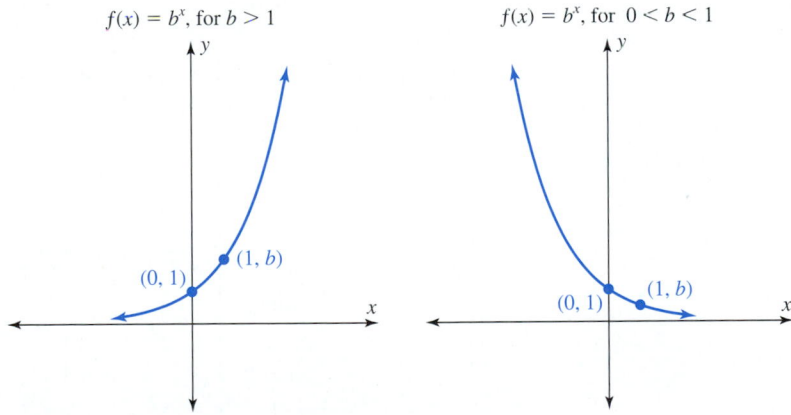

Key Facts About the Exponential Function $f(x) = b^x$ for $b > 0$ and $b \neq 1$

- The function is one-to-one and thus has an inverse function.
- The domain is $D = \mathbb{R}$.
- The range is $R = (0, \infty)$.
- The y-intercept of the graph is $(0, 1)$.
- The graph passes through $(1, b)$.

The symmetry of f and f^{-1} about the line $x = y$ is used to sketch the inverse of $f(x) = b^x$ in the following graphs. Note that the domain of each inverse function is $(0, \infty)$ and the range is $\mathbb{R}$.

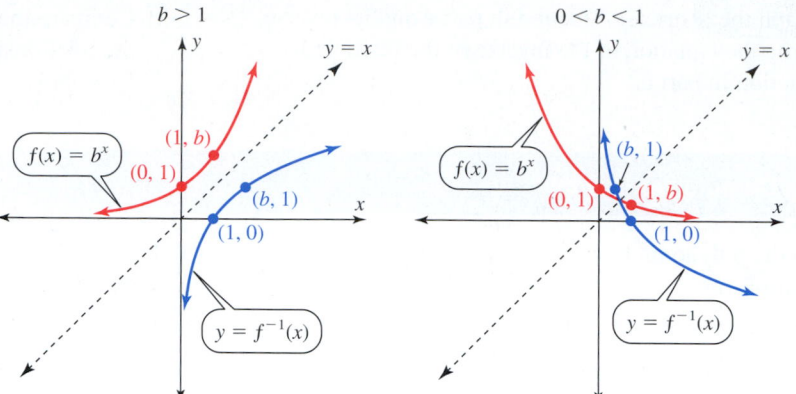

Key Facts About the Inverse of $f(x) = b^x$ for $b > 0$ and $b \neq 1$

- The inverse of this function is also a one-to-one function.
- The domain is $D = (0, \infty)$.
- The range is $R = \mathbb{R}$.
- The x-intercept of the graph is $(1, 0)$.
- The graph passes through $(b, 1)$.

How Do I Write the Inverse of an Exponential Function?

The inverse of $f(x) = b^x$ cannot be rewritten by using the basic operations to solve $x = b^y$ for y. Thus we introduce a new notation to denote the inverse of $f(x) = b^x$. This inverse is denoted by $f^{-1}(x) = \log_b x$, which is read "the log base b of x." Logarithmic functions are defined in the following box.

1. Interpret and Use Logarithmic Notation

Logarithmic Function $f(x) = \log_b x$			
Algebraically	**Verbally**	**Algebraic Example**	**Graphical Example**
For $x > 0$, $b > 0$, and $b \neq 1$, $y = \log_b x$ if and only if $b^y = x$.	$y = \log_b x$ is read "y equals the log base b of x."	$y = \log_2 x$	*(graph of $y = \log_2 x$)*

A function and its inverse both contain the same information but from different perspectives—the ordered pairs in the original function are reversed to obtain the ordered pairs in the inverse. As you work with $y = b^x$ and $y = \log_b x$, remember that these functions are inverses of each other and contain the same information.

Exponential Form   **Logarithmic Form**

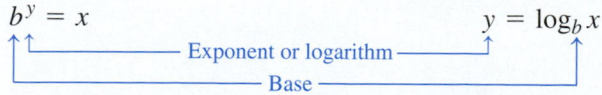

The relationship between exponential and logarithmic forms is examined further in Examples 1 and 2.

A Mathematical Note

Logarithms were invented by John Napier (1550–1617), a Scottish mathematician. His first publication on logarithms was published in 1614.

Napier's other inventions include "Napier's bones," a mechanical calculating device that is an ancestor of the slide rule, and a hydraulic screw and revolving axle for controlling the water level in coal pits. He also worked on plans to use mirrors to burn and destroy enemy ships.

| **Example 1** | Translating Logarithmic Equations to Exponential Form |

Write each logarithmic equation in verbal form, and translate this equation to the equivalent exponential form.

Solution

Logarithmic Form	Verbally	Exponential Form
(a) $\log_5 25 = 2$	The log base 5 of 25 is 2.	$5^2 = 25$
(b) $\log_2 8 = 3$	The log base 2 of 8 is 3.	$2^3 = 8$
(c) $\log_3\left(\dfrac{1}{3}\right) = -1$	The log base 3 of $\dfrac{1}{3}$ is -1.	$3^{-1} = \dfrac{1}{3}$
(d) $\log_7 \sqrt{7} = \dfrac{1}{2}$	The log base 7 of $\sqrt{7}$ is $\dfrac{1}{2}$.	$7^{1/2} = \sqrt{7}$

Self-Check 1

Write each logarithmic equation in exponential form.

a. $\log_7 49 = 2$ **b.** $\log_6 \sqrt[5]{6} = \dfrac{1}{5}$

| **Example 2** | Translating Exponential Equations to Logarithmic Form |

Translate each exponential equation to logarithmic form.

Solution

Exponential Form	Logarithmic Form
(a) $10^{-3} = 0.001$	$\log_{10} 0.001 = -3$
(b) $e^2 \approx 7.389$	$\log_e 7.389 \approx 2$
(c) $A = (1 + r)^x$	$\log_{(1+r)} A = x$

Self-Check 2

Write each exponential equation in logarithmic form.

a. $3^{-4} = \dfrac{1}{81}$ **b.** $6^0 = 1$

2. Evaluate Simple Logarithmic Expressions

To evaluate $\log_b x$, think, "What exponent on b is needed to obtain x?" Problems involving familiar powers of small integers can often be solved easily without the use of a calculator. For problems involving more complicated values, a calculator or a spreadsheet is generally used.

Example 3   Evaluating Logarithms by Inspection

Determine the value of each logarithm by inspection.

(a) $\log_5 125$ (b) $\log_2 \dfrac{1}{8}$ (c) $\log_{49} 7$ (d) $\log_5 0$ (e) $\log_5(-25)$

Solution

(a) $\log_5 125 = 3$ In exponential form, $5^3 = 125$.

(b) $\log_2 \dfrac{1}{8} = -3$ In exponential form, $2^{-3} = \dfrac{1}{8}$.

(c) $\log_{49} 7 = \dfrac{1}{2}$ In exponential form, $49^{1/2} = 7$.

(d) $\log_5 0$ is undefined.   The input of a logarithmic function cannot be 0.

(e) $\log_5(-25)$ is undefined.   The input of a logarithmic function cannot be negative.

Self-Check 3

Determine the value of each logarithm by inspection.

a. $\log_2 64$ **b.** $\log_4 64$ **c.** $\log_8 64$ **d.** $\log_7 \dfrac{1}{7}$

Four properties that result from the definition of a logarithm are given in the following box. These important properties are used often in solving the logarithmic equations that arise in the solution of some word problems.

Properties of Logarithmic Functions

For $b > 0$ and $b \neq 1$:

Logarithmic Form	Exponential Form	Numerical Example	
1. $\log_b 1 = 0$	$b^0 = 1$	$\log_5 1 = 0$	$5^0 = 1$
2. $\log_b b = 1$	$b^1 = b$	$\log_5 5 = 1$	$5^1 = 5$
3. $\log_b \dfrac{1}{b} = -1$	$b^{-1} = \dfrac{1}{b}$	$\log_5 \dfrac{1}{5} = -1$	$5^{-1} = \dfrac{1}{5}$
4. $\log_b b^x = x$	$b^x = b^x$	$\log_5 5^4 = 4$	$5^4 = 5^4$

Example 4 uses these properties to evaluate logarithms. Other properties of logarithms are examined in the following sections of this chapter.

Example 4   Using Properties to Evaluate Logarithms

Determine the value of each logarithm by inspection.

(a) $\log_{19} 1$ (b) $\log_7 7$

(c) $\log_8 \dfrac{1}{8}$ (d) $\log_8 8^{2y}$

Solution

(a) $\log_{19} 1 = 0$ In exponential form, $19^0 = 1$.

(b) $\log_7 7 = 1$ In exponential form, $7^1 = 7$.

(c) $\log_8 \dfrac{1}{8} = -1$ In exponential form, $8^{-1} = \dfrac{1}{8}$.

(d) $\log_8 8^{2y} = 2y$ In exponential form, $8^{2y} = 8^{2y}$.

Self-Check 4

Determine the value of each logarithm by inspection.

a. $\log_e 1$ **b.** $\log_e e$ **c.** $\log_e \dfrac{1}{e}$ **d.** $\log_e e^8$

The equations in Example 5 can be solved without use of a calculator because the numbers involve well-known powers. We examine equations of this type in greater detail in Section 11.6.

Example 5   Solving Equations

Solve each equation for x.

(a) $\log_3(x - 1) = 2$ **(b)** $\log_x 9 = -2$ **(c)** $\log_4 8 = x$

Solution

(a) $\log_3(x - 1) = 2$

$\quad x - 1 = 3^2$ First rewrite this equation in exponential form.

$\quad x - 1 = 9$ Substitute 9 for 3^2, and then solve for x.

$\quad\quad\quad x = 10$

(b) $\log_x 9 = -2$

$\quad x^{-2} = 9$ First rewrite this equation in exponential form.

$\quad x^{-2} = 3^2$ Now work toward expressing each side in terms of the same exponent.

$\quad x^{-2} = \left(\dfrac{1}{3}\right)^{-2}$

$\quad\quad x = \dfrac{1}{3}$ The bases are equal since the exponents are equal.

(c) $\log_4 8 = x$

$\quad 4^x = 8$ First rewrite this equation in exponential form.

$\quad (2^2)^x = 2^3$ Then express each side in terms of the common base of 2. Use the power rule for exponents to rewrite the left side of the equation.

$\quad 2^{2x} = 2^3$ Then equate the exponents since the bases are the same.

$\quad 2x = 3$

$\quad\quad x = \dfrac{3}{2}$

Self-Check 5

Solve each equation for x.

a. $\log_2 x = 3$ **b.** $\log_{11} \sqrt{11} = x$ **c.** $\log_x 125 = 3$

What Are the Characteristic Features of a Logarithmic Function?

Linear functions with a positive slope exhibit a constant rate of growth. By contrast, exponential growth functions exhibit a rate of growth that eventually becomes extremely rapid. On the other extreme, logarithmic growth functions eventually exhibit extremely slow growth. Other features of logarithmic functions are noted from the following graph.

The following graphs compare linear growth, exponential growth, and logarithmic growth. We examine these patterns again with the applications in Section 11.7.

Linear Growth	Exponential Growth	Logarithmic Growth
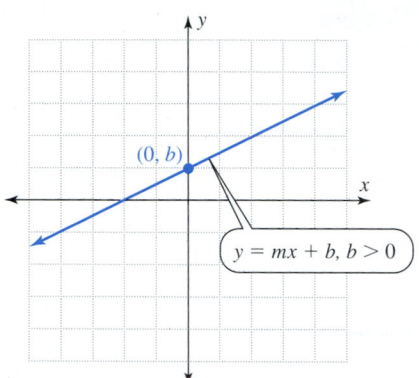	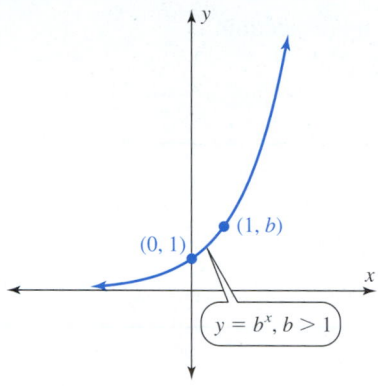	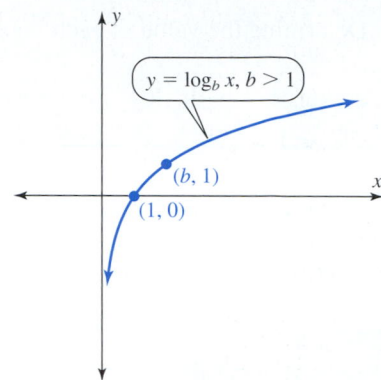

Features:

- For $m > 0$, growth
- A constant rate of growth for all values of x
- y-intercept: $(0, b)$
- x-intercept: exactly one

Features:

- For $b > 1$, growth
- For large values of x, extremely rapid growth
- x-intercept: None
- y-intercept: $(0, 1)$
- Key point: $(1, b)$
- Asymptotic to negative x-axis

Features:

- For $b > 1$, growth
- For large values of x, extremely slow growth
- x-intercept: $(1, 0)$
- y-intercept: None
- Key point: $(b, 1)$
- Asymptotic to negative y-axis

3. Sketch the Graph of a Logarithmic Function

Example 6   Graphing a Logarithmic Function

Use the function $y = \log_3 x$ to complete this table and to sketch the graph of this function.

x	$y = \log_3 x$
1/3	
1	
3	
9	

Solution

Table

x	$y = \log_3 x$
1/3	-1
1	0
3	1
9	2

Graph

Complete the table and plot these points. Then use the known shape to sketch the graph through these points.

The domain $(0, \infty)$ is the projection of the graph onto the x-axis.

Self-Check 6

a. Use the function $y = \log_2 x$ to complete this table.

b. Sketch the graph of this function.

x	$y = \log_2 x$
0.5	
1	
2	
4	

It is important to keep in mind this distinction between the rapid growth of exponential functions and the slow growth of logarithmic functions as you analyze real-world problems. This is illustrated by Example 7.

Example 7   Selecting an Exponential or Logarithmic Growth Model

Determine for each relationship whether the growth over time is more likely to be an exponential growth function or a logarithmic growth function.

Solution

(a) A frog embryo grows _____ over a period of time.

Exponentially

At the beginning, the first cell doubles to form two cells; then these cells grow and split to form four cells. This growing and splitting continues to form millions of cells and a live creature within a few weeks.

(b) The productivity gain of an employee on an assembly line grows _____.

Logarithmically

Employees normally make their greatest gains of efficiency from their novice state to becoming seasoned employees. After they have experienced all the unique features of this job, there is little new to learn and any new gains of productivity are minimal.

Self-Check 7

Determine for each relationship whether the growth over time is more likely to be an exponential growth function or a logarithmic growth function.

a. The amount of computer memory in a desktop computer since 1990

b. The yield in bushels per acre of corn on the historic Morrow plots in Champaign, Illinois

---

Self-Check Answers

1. a. $7^2 = 49$ **b.** $6^{1/5} = \sqrt[5]{6}$

2. a. $\log_3\left(\dfrac{1}{81}\right) = -4$ **b.** $\log_6 1 = 0$

3. a. 6   **b.** 3   **c.** 2   **d.** −1

4. a. 0   **b.** 1   **c.** −1   **d.** 8

5. a. $x = 8$ **b.** $x = \dfrac{1}{2}$ **c.** $x = 5$

6.

x	$y = \log_2 x$
0.5	−1
1	0
2	1
4	2

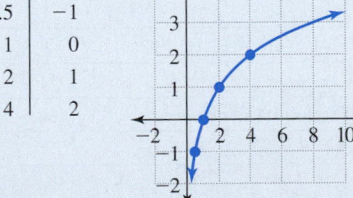

7. a. Exponential growth

b. Logarithmic growth

11.3   Using the Language and Symbolism of Mathematics

1. For $x > 0$, $b > 0$, and $b \neq 1$, the inverse of the function $f(x) = b^x$ is written as $f^{-1}(x) =$ _____.

2. $\log_b x = y$ is read "the log base _____ of _____ equals _____."

3. For $b > 0$ and $b \neq 1$, $\log_b 0$ is _____.

4. For $b > 0$ and $b \neq 1$, $\log_b 1 =$ _____.

5. For $b > 0$ and $b \neq 1$, $\log_b b =$ _____.

6. For $b > 0$ and $b \neq 1$, $\log_b \dfrac{1}{b} =$ _____.

7. For $b > 0$ and $b \neq 1$, $\log_b b^x =$ _____.

8. For $b > 1$, the graph of $f(x) = \log_b x$ is asymptotic to the negative portion of the _____-axis.

11.3 | Quick Review

In Exercises 1–3, simplify each expression.

1. x^0 for $x \neq 0$
2. x^0 for $x = 0$
3. $\left(\dfrac{2}{3}\right)^{-1}$

In Exercises 4 and 5, solve each equation.

4. $2x - 1 = 8$
5. $x^2 - 1 = 8$

11.3 | Exercises

Objective 1 Interpret and Use Logarithmic Notation

In Exercises 1–16, use the given information to complete each entry in the table, as illustrated by the example in the first row.

Logarithmic Form	Verbal Description	Exponential Form
Example: $\log_7 49 = 2$	The log base 7 of 49 is 2.	$7^2 = 49$
1. $\log_5 125 = 3$		
2. $\log_2 16 = 4$		
3. $\log_3 \sqrt{3} = \dfrac{1}{2}$		
4. $\log_5 \sqrt[3]{5} = \dfrac{1}{3}$		
5. $\log_5 \dfrac{1}{5} = -1$		
6. $\log_2 \dfrac{1}{4} = -2$		
7.	The log base 16 of 4 is $\dfrac{1}{2}$.	
8.	The log base 16 of $\dfrac{1}{16}$ is -1.	
9.		$8^{2/3} = 4$
10.		$4^{3/2} = 8$
11.		$3^{-2} = \dfrac{1}{9}$
12.		$\left(\dfrac{1}{3}\right)^{-4} = 81$
13.		$m^p = n$
14. $\log_x y = z$		
15.	The log base n of m is k.	
16. $\log_{10} 0.0001 = -4$		

Objective 2 Evaluate Simple Logarithmic Expressions

In Exercises 17–46, determine the value of each logarithm by inspection.

17. $\log_{10} 100$
18. $\log_{10} 1{,}000$
19. $\log_{10} 0.001$
20. $\log_{10} 0.01$
21. $\log_{10} \sqrt[3]{10}$
22. $\log_{10} \sqrt{10}$
23. $\log_2 32$
24. $\log_{11} 121$
25. $\log_3 \dfrac{1}{3}$
26. $\log_7 \dfrac{1}{7}$
27. $\log_{18} 1$
28. $\log_{25} 1$
29. $\log_{3/4} \dfrac{4}{3}$
30. $\log_{2/7} \dfrac{7}{2}$
31. $\log_3 81$
32. $\log_9 81$
33. $\log_2 64$
34. $\log_{64} 64$
35. $\log_3 27$
36. $\log_{27} 3$
37. $\log_{16} 2$
38. $\log_2 16$
39. $\log_{25} 125$
40. $\log_9 27$
41. $\log_{19} 19$
42. $\log_\pi \pi$
43. $\log_5 5^{2.7}$
44. $\log_3 3^{4.1}$
45. $\log_{3/4} \dfrac{16}{9}$
46. $\log_{5/7} \dfrac{49}{25}$

In Exercises 47–50, determine which expressions are defined and which are not defined. Do not evaluate these expressions.

47. a. $\log_7 10$ **b.** $\log_7 0$

48. a. $\log_0 7$ **b.** $\log_7 7$

49. a. $\log_5(-5)$ **b.** $\log_5 1$

50. a. $\log_{-3} 3$ **b.** $\log_{0.3} 3$

In Exercises 51–68, solve each equation using only pencil and paper.

51. $\log_6 36 = x$ **52.** $\log_{11} 121 = x$ **53.** $\log_6 x = 1$

54. $\log_{11} x = 1$ **55.** $\log_6 x = -1$ **56.** $\log_{11} x = -1$

57. $\log_6 x = \dfrac{1}{2}$ **58.** $\log_{11} x = \dfrac{1}{3}$ **59.** $\log_x 4 = 2$

60. $\log_x 4 = \dfrac{1}{2}$ **61.** $\log_x 125 = 3$ **62.** $\log_x 8 = 3$

63. $\log_b \sqrt[5]{b} = x$ **64.** $\log_b \sqrt[5]{b^2} = x$

65. $\log_3 1 = 2x + 1$ **66.** $\log_3 3 = 4x - 1$

67. $\log_2 (3x - 4) = 3$ **68.** $\log_5 (3x + 1) = 2$

Objective 3 Sketch the Graph of a Logarithmic Function

In Exercises 69–72, complete each table of values and use these points to sketch a graph of the function.

69. $f(x) = \log_4 x$

x	f(x)
1/4	
1	
4	
16	

70. $f(x) = \log_{2.5} x$

x	f(x)
0.40	
1.00	
2.50	
6.25	

71. $f(x) = \log_{1/4} x$

x	f(x)
1/4	
1	
4	
16	

72. $f(x) = \log_{0.4} x$

x	f(x)
0.40	
1.00	
2.50	
6.25	

In Exercises 73–76, match each function with its graph.

73. The exponential growth function $f(x) = e^x$

74. The exponential decay function $f(x) = 0.6^x$

75. The logarithmic growth function $f(x) = \log_e x$

76. The logarithmic decay function $f(x) = \log_{0.6} x$

A. 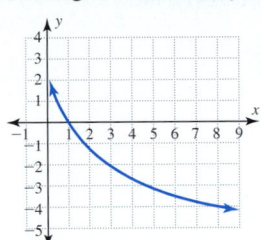   **B.**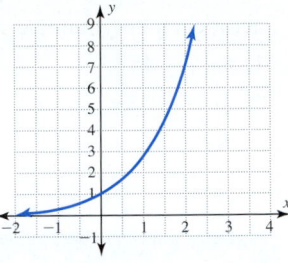

C. 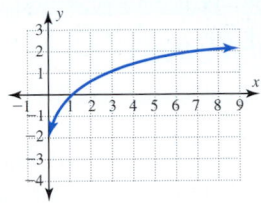   **D.**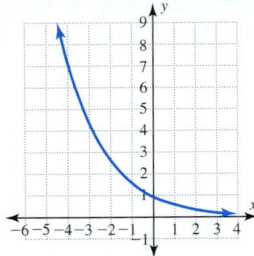

77. Use the graph of $f(x) = \log_2 x$ shown here to determine the domain and range of this function.

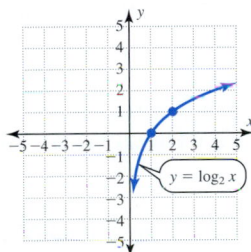

78. Use the graph of $f(x) = \log_{0.5} x$ shown here to determine the domain and range of this function.

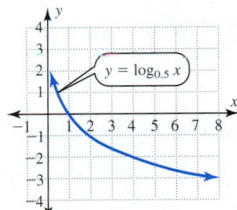

Group discussion questions

79. Challenge Question Determine for each relationship whether the growth over time is more likely to be an exponential growth function or a logarithmic growth function.

a. The value of an investment in a U.S. Treasury note over a 30-year period

b. The amount of learning by an individual over his or her 85-year lifetime

c. The population of locusts from the start of a growing season until the end of the growing season

d. The heights obtained by a pole-vaulter over a 20-year career

Factor each polynomial.

1. $5m^2 - 45$

2. $21x^2 - 14x$

3. $4v^2 - w^2$

4. $4v^2 + 4vw + w^2$

5. $a^2 - b^2 + 5a - 5b$

Section 11.4 | Evaluating Logarithms

Objective:

1. Evaluate common and natural logarithms with a calculator or a spreadsheet.

Although $\log_b x$ is defined for any base $b > 0$ and $b \neq 1$, the only two bases commonly used are base 10 and base e. Since our number system is based on powers of 10, the most convenient base for many computations is base 10. Logarithms to the base 10 are called **common logarithms.** The abbreviated form $\log x$ is used to denote $\log_{10} x$.

Why Do Mathematicians and Scientists Use Base e for Logarithms?

This base is used extensively because many formulas are stated most easily by using this base. Recall that the number e is irrational:

$$e \approx 2.718\ 281\ 828\ 459\ 045 \ldots$$

You should remember that e is approximately equal to 2.718. Logarithms to the base e are called **natural logarithms.** The abbreviated form $\ln x$ is used to denote $\log_e x$.

Common and Natural Logarithms

		Verbally	Numerical Example
Common Logarithms $\log x$ means $\log_{10} x$		$\log x$ is read "the (common) log of x" or "the log base 10 of x."	$\log 100 = 2$ because $10^2 = 100$
Natural Logarithms $\ln x$ means $\log_e x$		$\ln x$ is read "the natural log of x" or "the log base e of x."	$\ln 100 \approx 4.605$ because $e^{4.605} \approx 100$

A Mathematical Note

The earliest use of natural logarithms was in 1618. Natural logarithms are used frequently because the base of e occurs in many natural applications from astronomy to zoology.

Example 1 | Evaluating Logarithms by Inspection

Determine the value of each expression by inspection.

(a) $\log 10{,}000$ **(b)** $\log 0.1$ **(c)** $\ln e^5$ **(d)** $\ln e^{-4}$

Solution

(a) $\log 10{,}000 = 4$ In exponential form, $10^4 = 10{,}000$.

(b) $\log 0.1 = -1$ In exponential form, $10^{-1} = \dfrac{1}{10} = 0.1$.

(c) $\ln e^5 = 5$ In exponential form, $e^5 = e^5$.

(d) $\ln e^{-4} = -4$ In exponential form, $e^{-4} = e^{-4}$.

Self-Check 1

Determine the value of each expression.

a. $\log 100$ **b.** $\log 0.001$ **c.** $\ln e^3$

1. Evaluate Common and Natural Logarithms with a Calculator or a Spreadsheet

Only relatively simple logarithms, such as $\log 100 = 2$, can be determined by inspection. Thus common and natural logarithms have historically been determined through use of tables, slide rules, and other devices.

How Should I Determine Logarithmic Values?

Today calculators or spreadsheets are usually used to determine these values. The examples in this section assume calculator or spreadsheet usage.

   Technology Perspective 11.4.1 illustrates how to approximate common and natural logarithms.

Technology Perspective 11.4.1	Evaluating Common and Natural Logarithms

Approximate $\log 50$ and $\ln 50$ to the nearest thousandth.

TI-84 Plus Calculator

```
log(50)
         1.698970004
ln(50)
         3.912023005
```

Excel Spreadsheet

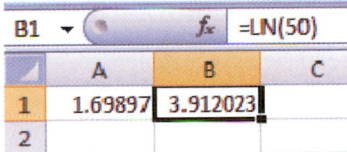

1. Press 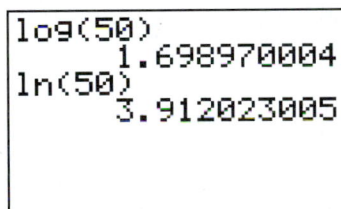 **LOG** to obtain the logarithmic function base 10. Then enter 50 and close with a right parenthesis.

    Pressing **ENTER** will evaluate $\log 50$.

2. Press **LN** to obtain the logarithmic function base e. Then enter 50 and close with a right parenthesis.

    Pressing **ENTER** will evaluate $\ln 50$.

1. Enter "=LOG(50)" into cell **A1**. This will display an approximation of $\log 50$ in cell **A1**.

2. Enter "=LN(50)" into cell **B1**. This will display an approximation of $\ln 50$ in cell **B1**.

Answer: $\log 50 \approx 1.699$; $\ln 50 \approx 3.912$

Technology Self-Check 1

Approximate both $\log 748$ and $\ln 748$ to the nearest hundredth.

Note that although the two expressions in Technology Perspective 11.4.1 look very similar, the values for common and natural logarithms are quite distinct. One way to check your answers is to use the exponential form. For example, log 50 ≈ 1.699 checks because $10^{1.699} \approx 50$. Logarithmic functions are available not only in spreadsheets but also in many computer languages. However, the names used for these functions may not mean what you expect.

How Can I Determine What Base a Logarithmic Function Is Using Without Consulting a Manual or a Help Menu?

One way is to test values such as log 10 that you can estimate mentally.

Warning for Computer Users

Some computer languages provide only the natural logarithmic function, in which case this function may be called LOG instead of ln, even though the base is *e*. If you are unsure of the meaning of LOG on a particular computer, test a known value such as LOG 10 to determine which base is being used. We know log 10 = 1 but ln 10 ≈ 2.30.

What Do I Do If I Need Several Logarithmic Values?

One way is to create a table of values that you can refer to. Example 2 illustrates a table of logarithms created by a spreadsheet. You could also create this table with a calculator. This example uses Technology Perspective 2.2.1, Creating a Table of Values.

Example 2   Completing a Table of Logarithms

Use a spreadsheet to complete this table of logarithmic values.

x	$y_1 = \log x$	$y_2 = \ln x$
10		
20		
30		
40		
50		
60		
70		

Solution

C2 f_x =LN(A2)

	A	B	C
1	x	LOG(x)	LN(x)
2	10	1.0000	2.3026
3	20	1.3010	2.9957
4	30	1.4771	3.4012
5	40	1.6021	3.6889
6	50	1.6990	3.9120
7	60	1.7782	4.0943
8	70	1.8451	4.2485

Self-Check 2

Create a table of values for ln x for x-values of 5, 10, 15, 20, 25, 30, and 35.

The input value of $y = f(x)$ is also is called the **argument** of the function. In the logarithmic function $y = \log(x + 2)$, we refer to $x + 2$ as the argument of the function. This terminology is particularly useful when we are taking a logarithm of an expression other than just x. In Example 3, we will first write the argument of the function in scientific notation and then we will take the logarithm of this number. Recall from Section 5.3 that we often use scientific notation for numbers whose magnitude is either very large or very small.

Example 3   Using Scientific Notation with Logarithmic Expressions

Use scientific notation to enter the argument of each logarithm. Use a calculator or a spreadsheet to approximate each expression to the nearest ten-thousandth.

(a) ln 894,000,000,000   **(b)** log 0.000 000 001 234

Solution

(a) ln 894,000,000,000

$= \ln(8.94 \times 10^{11})$

≈ 27.5190

   First write each argument in scientific notation. Then use a calculator or a spreadsheet to approximate the logarithm of this value.

Check: $e^{27.5190} \approx 8.94 \times 10^{11}$

(b) log 0.000 000 001 234

$= \log(1.234 \times 10^{-9})$

≈ -8.9087

Check: $10^{-8.9087} \approx 1.234 \times 10^{-9}$

Self-Check 3

Use a calculator or a spreadsheet to approximate each expression to the nearest thousandth.

a. $\log(5.81 \times 10^{4})$ **b.** $\ln(3.96 \times 10^{-5})$

To avoid order-or-operation errors with a calculator or a spreadsheet, it is wise to enclose the argument of a logarithm within parentheses. Note the similar but distinct meanings of the expressions in Example 4. It is better to use extra parentheses than not to use enough to denote the meaning that you intend.

Example 4   Evaluating Logarithmic Expressions

Rewrite each expression by using parentheses, and then use a calculator or a spreadsheet to approximate each expression to the nearest ten-thousandth.

(a) $\dfrac{\log 12}{5}$ **(b)** $\log \dfrac{12}{5}$ **(c)** $\dfrac{\log 12}{\log 5}$

Solution

(a) $\dfrac{\log 12}{5} = (\log(12)) \div 5 \approx 0.2158$

(b) $\log \dfrac{12}{5} = \log\left(\dfrac{12}{5}\right) \approx 0.3802$

(c) $\dfrac{\log 12}{\log 5} = \dfrac{\log(12)}{\log(5)} \approx 1.5440$

Self-Check 4

Approximate the value of each expression to the nearest hundredth.

a. $\dfrac{5}{\log 7}$ **b.** $\log \dfrac{5}{7}$ **c.** $\dfrac{\log 5}{\log 7}$

We will use the fact that logarithmic functions and exponential functions are inverses of each other to solve for x in Example 5.

Example 5 Solving Logarithmic Equations

Approximate the value of x to the nearest thousandth.

(a) $\log x = 0.69897$ **(b)** $\ln x = -1.045$

Solution

(a) $\log x = 0.69897$ First rewrite the equation in exponential form with a base of 10. Then

$x = 10^{0.69897}$ approximate x by using a calculator or a spreadsheet.

$x \approx 5.000$

(b) $\ln x = -1.045$ First rewrite the equation $\log_e x = -1.045$ in exponential form with a base of e.

$x = e^{-1.045}$ Then approximate x by using a calculator or a spreadsheet.

$x \approx 0.352$

Self-Check 5

Approximate x to the nearest thousandth.

a. $\log x = 1.4583$ **b.** $\ln x = 1.4583$

It is important to know that the values that we calculate are reasonable and not in error because of some careless keystroke. Example 6 illustrates two estimation problems. We examine a formula that allows us to calculate logarithms to any base in Section 11.5.

Example 6 Estimation Skills

Mentally estimate the value of each of these expressions.

(a) $\log 1{,}008$ **(b)** $\log_7 50.0017$

Solution

(a) $\log 1{,}008 \approx \log 1{,}000$ Since $1{,}008 \approx 1{,}000$, $\log 1{,}008$ is slightly larger than $\log 1{,}000 = 3$.

$\log 1{,}008 \approx 3$

(b) $\log_7 50.0017 \approx \log_7 49$ Since 50.0017 is slightly larger than 49, $\log_7 50.0017$ is slightly larger than $\log_7 49$. And $\log_7 49 = 2$ because $7^2 = 49$.

$\log_7 50.0017 \approx 2$

Self-Check 6

Mentally estimate the value of each logarithm.

a. $\log 98$ **b.** $\log_5 130$

Self-Check Answers

1. a. 2   **b.** -3 **c.** 3

2.

B2 ▾		f_x =LN(A2)	
	A	**B**	**C**
1	x	LN(x)	
2	5	1.609438	
3	10	2.302585	
4	15	2.70805	
5	20	2.995732	
6	25	3.218876	
7	30	3.401197	
8	35	3.555348	

3. a. 4.764   **b.** -10.137

4. a. 5.92   **b.** -0.15 **c.** 0.83

5. a. 28.728   **b.** 4.299

6. a. $\log 98 \approx \log 100 = 2$

   b. $\log_5 130 \approx \log_5 125 = 3$

Technology Self-Check Answer

1. 2.87, 6.62

11.4   Using the Language and Symbolism of Mathematics

1. The irrational number $e \approx$ _____ (to the nearest thousandth).

2. The common logarithm of x, denoted $\log x$, means _____.

3. The natural logarithm of x, denoted $\ln x$, means _____.

4. The argument of the logarithmic function $\ln(5x - 3)$ is _____.

11.4   Quick Review

In Exercises 1–3, simplify each expression.

1. $\dfrac{0}{10}$ **2.** $\dfrac{10}{0}$ **3.** 10^{-4}

In Exercises 4 and 5, write each number in scientific notation.

4. 75,345,247   **5.** 0.000087

11.4 Exercises

Objective 1 Evaluate Common and Natural Logarithms with a Calculator or a Spreadsheet

In Exercises 1–22, determine the value of each expression by inspection. (*Hint:* Some expressions are undefined.)

1. $\log 10$
2. $\log 10{,}000$
3. $\log 0.0001$
4. $\log 0.01$
5. $\log 10^{-8}$
6. $\log 10^{-7}$
7. $\log \sqrt[4]{10}$
8. $\log \sqrt[5]{10}$
9. $\log 1$
10. $\ln 1$
11. $\ln e^{6}$
12. $\ln e^{7}$
13. $\ln e^{-3}$
14. $\ln e^{-4}$
15. $\ln \sqrt{e}$
16. $\ln \sqrt[3]{e}$
17. $\ln 0$
18. $\log 0$
19. $\log(-10)$
20. $\ln(-e)$
21. $\ln \dfrac{1}{e}$
22. $\log \dfrac{1}{100}$

In Exercises 23–46, approximate the value of each expression to the nearest thousandth.

23. $\log 47$
24. $\ln 47$
25. $\ln 0.0034$
26. $\log 0.0034$
27. $\log 876$
28. $\ln 876$
29. $\log 4.5 \times 10^{6}$
30. $\ln 4.5 \times 10^{6}$
31. $\ln 4.5 \times 10^{-5}$
32. $\log 4.5 \times 10^{-5}$
33. $\ln \sqrt{134}$
34. $\log \sqrt{134}$
35. $\log \pi$
36. $\ln \pi$
37. $\log(\ln 932)$
38. $\ln(\log 932)$
39. $\ln \dfrac{11}{4}$
40. $\log \dfrac{11}{4}$
41. $\dfrac{\ln 11}{\ln 4}$
42. $\dfrac{\log 11}{\log 4}$
43. $\log(25 \cdot 18)$
44. $\ln(34 - 12)$
45. $(\log 25)(\log 18)$
46. $(\ln 34) - (\ln 12)$

Each of Exercises 47–52 contains one expression that is defined and one that is undefined. Determine which expression is undefined and state why.

47. **a.** $-\ln 8$ **b.** $\ln(-8)$
48. **a.** $-\log 11$ **b.** $\log(-11)$
49. **a.** $0 \cdot \log 23$ **b.** $\log(23 \cdot 0)$
50. **a.** $\ln 17 - \ln 17$ **b.** $\ln(17 - 17)$
51. **a.** $\ln 17 - \ln 19$ **b.** $\ln(17 - 19)$
52. **a.** $\dfrac{\log 41}{\log 1}$ **b.** $\dfrac{\log 41}{\log 10}$

Estimation Skills

In Exercises 53–56, mentally estimate the value of each expression to the nearest integer.

Problem	Mental Estimate
53. $\log_8 63$	
54. $\log_2 17$	
55. $2.99^{4.01}$	
56. $5.01^{1.99}$	

Review and Concept Development

In Exercises 57–60, approximate each expression to the nearest thousandth.

57. **a.** $\log 2.47$ **b.** $10^{0.393}$
58. **a.** $\ln 2.47$ **b.** $e^{0.904}$
59. **a.** $\ln 0.094$ **b.** $e^{-2.364}$
60. **a.** $\log 0.094$ **b.** $10^{-1.027}$

In Exercises 61–68, approximate the value of x to the nearest thousandth.

61. $\log x = 2.477$ 62. $\log x = 2.778$
63. $\log x = -0.301$ 64. $\log x = -0.699$
65. $\ln x = 2.079$ 66. $\ln x = 2.773$
67. $\ln x = -2.996$ 68. $\ln x = -4.711$

Group discussion questions

69. **Discovery Question**
    a. Evaluate each of these expressions:
    $10^{\log 33}$, $e^{\ln 19}$, $2^{\log_2 8}$, $3^{\log_3 81}$
    b. Based on your observations in part **a,** make a conjecture stated as an algebraic equation.
    c. Test your conjecture in part **b,** and describe it to your instructor.

70. **Discovery Question**
    a. Evaluate each of these expressions:
    $\log(5 \cdot 6)$, $\log 5 + \log 6$, $\ln(7 \cdot 11)$, $\ln 7 + \ln 11$
    b. Based on your observations in part **a,** make a conjecture stated as an algebraic equation.
    c. Test your conjecture in part **b,** and describe it to your instructor.

71. **Discovery Question**
    a. Evaluate each of these expressions:
    $\log \dfrac{15}{5}$, $\log 15 - \log 5$ $\ln \dfrac{24}{6}$, $\ln 24 - \ln 6$
    b. Based on your observations in part **a,** make a conjecture stated as an algebraic equation.
    c. Test your conjecture in part **b,** and describe it to your instructor.

72. Discovery Question
    a. Evaluate each of these expressions:
        $\log 5^7$, $7 \log 5$, $\log \pi^3$, $3 \log \pi$, $\ln 10^9$, $9 \ln 10$

b. Based on your observations in part **a**, make a conjecture stated as an algebraic equation.
c. Test your conjecture in part **b**, and describe it to your instructor.

| 11.4 | Cumulative Review |

Reduce each expression.

1. $\dfrac{12x^2y^3}{24xy}$

2. $\dfrac{5x}{10x^2 - 15x}$

3. $\dfrac{2x + 2}{7x + 7}$

4. $\dfrac{(x + 2)(5x - 1)}{(x + 2)(3x - 4)}$

5. $\dfrac{x^2 - 16}{x^2 - x - 20}$

Section 11.5 | Properties of Logarithms

Objectives:

1. Use the product rule, the quotient rule, and the power rule for logarithms.
2. Use the change-of-base formula to evaluate logarithms.

Where Do We Obtain the Properties of Logarithms?

The properties of logarithms follow directly from the definition of a logarithmic function as the inverse of an exponential function. Thus for every exponential property, there is a corresponding logarithmic property. The logarithmic properties provide a means of simplifying some algebraic expressions and make it easier to solve exponential equations.

1. Use the Product Rule, the Quotient Rule, and the Power Rule for Logarithms

The logarithmic and exponential properties are stated here side by side so that you can compare them. (The capital letters X and Y are used in the logarithmic form to show that these variables are not the same as x and y in the exponential form.)

Properties of Exponents and Logarithms

	Exponential Form	**Logarithmic Form**
Product rule:	$b^x b^y = b^{x+y}$	$\log_b XY = \log_b X + \log_b Y$
Quotient rule:	$\dfrac{b^x}{b^y} = b^{x-y}$	$\log_b \dfrac{X}{Y} = \log_b X - \log_b Y$
Power rule:	$(b^x)^p = b^{xp}$	$\log_b X^p = p \log_b X$

The proof of the product rule is presented here. The quotient rule and the power rule can be proved in a similar fashion; their proofs are left to you in Exercises 88 and 89 at the end of this section.

Proof of the Product Rule for Logarithms: Let $X = b^m$ and $Y = b^n$.

    $\log_b X = m$ and $\log_b Y = n$ Rewrite these statements in logarithmic form.

Now examine XY.

$$XY = b^m b^n \qquad \text{Substitute } b^m \text{ for } X \text{ and } b^n \text{ for } Y.$$
$$XY = b^{m+n} \qquad \text{Use the product rule for exponents.}$$
$$\log_b XY = m + n \qquad \text{Rewrite this statement in logarithmic form.}$$
$$\log_b XY = \log_b X + \log_b Y \qquad \text{Substitute } \log_b X \text{ for } m \text{ and } \log_b Y \text{ for } n.$$

Properties of Logarithms

For $x, y > 0$, $b > 0$, and $b \neq 1$:

	Algebraically	Verbally	Numerical Example
Product rule:	$\log_b xy = \log_b x + \log_b y$	The log of a product is the sum of the logs.	$\log_2 32 = \log_2 (4 \cdot 8) = \log_2 4 + \log_2 8$ $5 = 2 + 3$
Quotient rule:	$\log_b \dfrac{x}{y} = \log_b x - \log_b y$	The log of a quotient is the difference of the logs.	$\log_2 4 = \log_2 \dfrac{32}{8} = \log_2 32 - \log_2 8$ $2 = 5 - 3$
Power rule:	$\log_b x^p = p \log_b x$	The log of the pth power of x is p times the log of x.	$\log_2 64 = \log_2 4^3 = 3 \log_2 4$ $6 = 3 \cdot 2$

Example 1   Using a Calculator to Examine the Properties of Logarithms

Use the product rule for logarithms to rewrite $\ln(15 \cdot 91)$. Then use a calculator or a spreadsheet to verify that these expressions are equal.

Solution

$$\ln(15 \cdot 91) = \ln 15 + \ln 91 \qquad \text{Use the product rule: } \log_b xy = \log_b x + \log_b y$$

Check:
$$\ln(15 \cdot 91) \approx 7.219$$
$$\ln 15 + \ln 91 \approx 7.219$$

Self-Check 1

Use the quotient rule to rewrite $\ln \dfrac{86}{37}$. Then use a calculator or a spreadsheet to verify that these expressions are equal.

The properties of logarithms are used both to rewrite logarithmic expressions in a simpler form and to combine logarithms. These properties also are used to solve logarithmic equations and to simplify expressions in calculus.

Example 2 Using the Properties of Logarithms to Form Simpler Expressions

Use the properties of logarithms to write these expressions in terms of logarithms of simpler expressions. Assume that the argument of each logarithm is a positive real number.

(a) $\log xyz$ (b) $\ln x^7$ (c) $\ln x^4 y^5$ (d) $\log_5 \dfrac{x+3}{x-2}$ (e) $\ln\sqrt{\dfrac{x}{y}}$

Solution

(a) $\log xyz = \log x + \log y + \log z$ Product rule (generalized to three factors): Add the logarithms.

(b) $\ln x^7 = 7 \ln x$ Power rule: Multiply the power 7 by the logarithm.

(c) $\ln x^4 y^5 = \ln x^4 + \ln y^5$ Product rule: Add the logarithms.

$\qquad\qquad = 4 \ln x + 5 \ln y$ Power rule: Multiply each power by the logarithm.

(d) $\log_5 \dfrac{x+3}{x-2} = \log_5(x+3) - \log_5(x-2)$ Quotient rule: Subtract the logarithms.

(e) $\ln\sqrt{\dfrac{x}{y}} = \ln\left(\dfrac{x}{y}\right)^{1/2}$

$\qquad = \dfrac{1}{2} \ln \dfrac{x}{y}$ Power rule: Multiply the power $\dfrac{1}{2}$ by the logarithm.

$\qquad = \dfrac{1}{2}(\ln x - \ln y)$ Quotient rule: Subtract the logarithms.

Self-Check 2

Express each logarithm in terms of logarithms of simpler expressions. Assume that the argument of each logarithm is a positive real number.

a. $\log x^3 y^2$ **b.** $\ln \dfrac{(x+1)^5}{(y-2)^3}$

Example 3 is essentially the reverse of Example 2. The properties of logarithms are used to combine logarithmic terms.

Example 3 Using the Properties of Logarithms to Combine Expressions

Combine the logarithmic terms in each expression into a single logarithmic expression with a coefficient of 1. Assume that the argument of each logarithm is a positive real number.

(a) $2 \ln x + 3 \ln y - 4 \ln z$ (b) $5 \log x + \dfrac{1}{3} \log y$

Solution

(a) $2 \ln x + 3 \ln y - 4 \ln z = \ln x^2 + \ln y^3 - \ln z^4$ Power rule

$\qquad\qquad\qquad\qquad = \ln x^2 y^3 - \ln z^4$ Product rule

$\qquad\qquad\qquad\qquad = \ln \dfrac{x^2 y^3}{z^4}$ Quotient rule

(b) $5 \log x + \dfrac{1}{3} \log y = \log x^5 + \log y^{1/3}$ Power rule

$$= \log x^5 + \log \sqrt[3]{y}$$
$$= \log x^5 \sqrt[3]{y} \qquad \text{Product rule}$$

Self-Check 3

Combine the logarithmic terms in each expression into a single logarithmic expression with a coefficient of 1. Assume that the argument of each logarithm is a positive real number.

a. $\log(x + 3) + \log(2x - 9)$

b. $3 \ln x + 2 \ln y - \dfrac{1}{2} \ln z$

Example 4 can serve as a good source of practice in observing patterns and using the properties of logarithms. These values can also be calculated by using the change-of-base formula for logarithms. (See Example 5.) However, the purpose of Example 4 is to give practice using the product, quotient, and power rules for logarithms.

Example 4 Using Logarithmic Properties

Use $\log_5 2 \approx 0.4307$, $\log_5 3 \approx 0.6826$, and the properties of logarithms to approximate to the nearest thousandth the value of each of these logarithmic expressions.

(a) $\log_5 6$ **(b)** $\log_5 9$ **(c)** $\log_5 0.4$

Solution

(a) $\log_5 6 = \log_5(2 \cdot 3)$

$$= \log_5 2 + \log_5 3 \qquad \text{Product rule}$$
$$\approx 0.4307 + 0.6826 \qquad \text{Substitute the given values for } \log_5 2 \text{ and } \log_5 3.$$
$$\approx 1.113$$

(b) $\log_5 9 = \log_5 3^2$

$$= 2 \log_5 3 \qquad \text{Power rule}$$
$$\approx 2(0.6826) \qquad \text{Substitute the given value of } \log_5 3.$$
$$\approx 1.365$$

(c) $\log_5 0.4 = \log_5 \dfrac{2}{5}$

$$= \log_5 2 - \log_5 5 \qquad \text{Quotient rule}$$
$$\approx 0.4307 - 1 \qquad \text{The value of } \log_5 2 \text{ is given, and } \log_5 5 \text{ can be determined by inspection.}$$
$$\approx -0.569$$

Self-Check 4

Use the values given in Example 4 to approximate to the nearest thousandth the value of each expression.

a. $\log_5 10$ **b.** $\log_5 8$ **c.** $\log_5 1.5$

How Can I Calculate Logarithms to a Base Other Than 10 or *e*?

One way is to use a spreadsheet that has this capability. Another way is to use a calculator and the change-of-base formula to convert the logarithm to either base 10 or base *e*.

2. Use the Change-of-Base Formula to Evaluate Logarithms

Although any base *b* can be used in the formula $\log_a x = \dfrac{\log_b x}{\log_b a}$, we usually use either $\dfrac{\log x}{\log a}$ or $\dfrac{\ln x}{\ln a}$. One reason we use these bases is that they are available on most graphing calculators.

Change-of-Base Formula

For $a, b > 0$, $a \neq 1$, and $b \neq 1$:

For Logarithms	Examples
$\log_a x = \dfrac{\log_b x}{\log_b a} \qquad \text{for } x > 0$	$\log_2 x = \dfrac{\log x}{\log 2} \quad \text{or} \quad \log_2 x = \dfrac{\ln x}{\ln 2}$
For Exponents	
$a^x = b^{x \log_b a}$	$2^x = e^{(\ln 2)x}$

Proof of the Change-of-Base Formula for Logarithms:

Let $\log_a x = y$. Then

$x = a^y$	Rewrite this equation in exponential form.
$\log_b x = \log_b a^y$	Take the log to base *b* of both sides of this equation.
$\log_b x = y \log_b a$	Use the power rule for logarithms.
$y = \dfrac{\log_b x}{\log_b a}$	Solve for *y*.
$\log_a x = \dfrac{\log_b x}{\log_b a}$	Substitute $\log_a x$ for *y*.

The most useful forms of this identity are

$$\log_a x = \frac{\log x}{\log a} \qquad \text{and} \qquad \log_a x = \frac{\ln x}{\ln a}$$

The proof of the change-of-base formula for exponents is similar to the proof given for logarithms.

We approximated $\log_5 6$ in Example 4 using given information. In Example 5, we illustrate how to approximate this value directly using a calculator and the change-of-base formula or using a spreadsheet.

Example 5   Approximating a Logarithm Base 5

Approximate $\log_5 6$ to the nearest thousandth.

Solution

Calculator Approximation

Use the change of base formula to rewrite $\log_5 6$ as $\dfrac{\ln 6}{\ln 5}$.

```
ln(6)/ln(5)
            1.113282753
```

Spreadsheet Approximation

Use the logarithmic function LOG(argument, base) to evaluate $\log_5 6$ with an argument of 6 and a base of 5.

A1	f_x	=LOG(6,5)

	A	B	C
1	1.113283		
2			

Answer: $\log_5 6 \approx 1.113$

Self-Check 5

Approximate $\log_5 9$ to the nearest thousandth.

The change-of-base formula for exponents can be used to convert any exponential growth or decay function to base e. In scientific applications, base e is used almost exclusively.

Example 6   Using the Change-of-Base Formula for Exponents

Use the change-of-base formula for exponents to write $f(x) = 3^x$ as an exponential function with base e.

Solution

Algebraically

$f(x) = 3^x$ Use the change-of-base formula $a^x = e^{(\ln a)x}$ to convert to base e.

$f(x) = e^{(\ln 3)x}$ Then approximate $\ln 3$ with a calculator.

$f(x) \approx e^{1.099x}$

Self-Check 6

Rewrite the function $f(x) = 2^x$ as an exponential function base e.

Remember that tables of values can often be used to compare algebraic expressions. For example, in Example 6, you can compare $f(x) = 3^x$ and $f(x) = e^{1.099x}$.

Identities are often used to simplify expressions in mathematics. Some identities are used frequently and have special names. Other identities have more limited usage and are developed or presented in the context they are needed. The following display illustrates how we can prove a logarithmic identity.

Verifying an Identity Involving Logarithms: Prove that $b^{\log_b x} = x$ for any positive real number x.

Proof

Let $y = \log_b x$. Then

$$b^y = x \quad \text{and} \qquad \text{Rewrite the logarithmic expression in exponential form.}$$

$$b^{\log_b x} = x \qquad \text{Substitute } \log_b x \text{ for } y.$$

This property is summarized in the following box.

Another Property of Logarithms

For $x > 0, b > 0,$ and $b \neq 1$:

Algebraically	Numerical Example	Calculator Check
$b^{\log_b x} = x$	$10^{\log 45} = 45$	```10^(log(45))``` ``` 45```

This property of logarithms is used in Example 7 to evaluate some expressions by inspection.

Example 7   Using the Properties of Logarithms

Determine the value of each expression by inspection.

(a) $10^{\log 8}$ (b) $e^{\ln 0.179}$ (c) $a^{\log_a 231}$

Solution

(a) $10^{\log 8} = 8$

(b) $e^{\ln 0.179} = 0.179$

(c) $a^{\log_a 231} = 231$ for $a > 0, a \neq 1$

Each of these expressions can be determined by inspection, by using the identity $b^{\log_b x} = x$.

Self-Check 7

Determine the value of each expression by inspection.

a. $e^{\ln 31}$ **b.** $10^{\log 0.041}$

Self-Check Answers

1. $\ln\dfrac{86}{37} = \ln 86 - \ln 37$; $\ln\dfrac{86}{37} \approx 0.843$; $\ln 86 - \ln 37 \approx 0.843$

2. a. $3 \log x + 2 \log y$ **b.** $5 \ln(x + 1) - 3 \ln(y - 2)$

3. a. $\log(x + 3)(2x - 9)$ **b.** $\ln \dfrac{x^3 y^2}{\sqrt{z}}$

4. a. $\log_5 10 = \log_5(2 \cdot 5)$
$= \log_5 2 + \log_5 5$
≈ 1.431

b. $\log_5 8 = \log_5(2^3)$
$= 3 \log_5 2$
≈ 1.292

c. $\log_5 1.5 = \log_5\left(\dfrac{3}{2}\right)$
$= \log_5 3 - \log_5 2$
≈ 0.252

5. $\log_5 9 \approx 1.365$

6. $f(x) \approx e^{0.693x}$

7. a. 31
b. 0.041

11.5 | Using the Language and Symbolism of Mathematics

1. By the product rule for exponents, $b^x b^y = $ _____.

2. By the product rule for logarithms,
   $\log_b xy = $ _____.

3. By the quotient rule for exponents, $\dfrac{b^x}{b^y} = $ _____.

4. By the quotient rule for logarithms,
   $\log_b \dfrac{x}{y} = $ _____.

5. By the power rule for exponents, $(b^x)^p = $ _____.

6. By the power rule for logarithms,
   $\log_b x^p = $ _____.

7. The change-of-base formula for logarithms is
   $\log_a x = $ _____.

8. The change-of-base formula for exponents is
   $a^x = $ _____.

9. By the change-of-base formula for exponents,
   $e^{(\ln 7)x} = $ _____.

10. If $x > 0, b > 0$, and $b \neq 1$, then $b^{\log_b x} = $ _____.

11.5 | Quick Review

In Exercises 1–3, use the properties of exponents to simplify each expression.

1. $3x^4 y^2 (5x^3 y^7)$ **2.** $\dfrac{15x^4 y^2}{5x^3 y^7}$ **3.** $(5x^3 y^7)^2$

In Exercises 4 and 5, write each expression using exponential notation.

4. $x\sqrt{y}$ **5.** $\sqrt[3]{xy^2}$

11.5 | Exercises

Objective 1 Use the Product Rule, Quotient Rule, and the Power Rule for Logarithms

In Exercises 1–22, use the properties of logarithms to express each logarithm in terms of logarithms of simpler expressions. Assume that the argument of each logarithm is a positive real number.

1. $\log abc$ **2.** $\log xyz$

3. $\ln \dfrac{x}{11}$ **4.** $\ln \dfrac{y}{7}$

5. $\log y^5$ **6.** $\log y^4$

7. $\ln x^2 y^3$ **8.** $\ln v^6 w^8 z$

9. $\ln(x + 2)(3x - 1)$ **10.** $\ln(2x + 3)(x + 7)$

11. $\ln \dfrac{2x + 3}{x + 7}$ **12.** $\ln \dfrac{5x + 8}{3x + 4}$

13. $\log \sqrt{4x + 7}$ **14.** $\log \sqrt[3]{6x + 1}$

15. $\ln \dfrac{\sqrt{x + 4}}{(y + 5)^2}$ **16.** $\ln \dfrac{(x + 9)^3}{\sqrt{y + 1}}$

17. $\log \sqrt{\dfrac{xy}{z - 8}}$ **18.** $\log \sqrt[4]{\dfrac{x^2 y}{z^3}}$

19. $\log \dfrac{x^2 (2y + 3)^3}{z^4}$ **20.** $\log \dfrac{(x + 1)^3 (y - 2)^2}{(z + 4)^5}$

21. $\ln \left(\dfrac{xy^2}{z^3} \right)$ **22.** $\ln \left(\dfrac{x^2 y^3}{z^5} \right)^3$

In Exercises 23–40, combine the logarithmic terms into a single logarithmic expression with a coefficient of 1. Assume that the argument of each logarithm is a positive real number.

23. $\ln x + \ln y$ **24.** $\log v + \log w$

25. $\log w - \log z$ **26.** $\ln v - \ln w$

27. $3 \ln w$ **28.** $4 \ln y$

29. $2 \log x + 5 \log y$ **30.** $7 \ln x + 3 \ln y$

31. $3 \ln x + 7 \ln y - \ln z$

32. $4 \log x + 9 \log y - 5 \log z$

33. $\dfrac{1}{2} \ln x$ **34.** $\dfrac{1}{3} \log y$

35. $\dfrac{1}{2} \log(x + 1) - \log(2x + 3)$

36. $\ln(3x + 8) - \dfrac{1}{2} \ln(5x + 1)$

37. $\dfrac{1}{3} [\ln(2x + 7) + \ln(7x + 1)]$

38. $\dfrac{1}{4} [\log(x + 3) - \log(2x + 9)]$

39. $2 \log_5 x + \dfrac{2}{3} \log_5 y$

40. $\dfrac{3}{5} \log_7 x - 2 \log_7 y$

In Exercises 41–46, use the properties of logarithms to determine the value of each expression. Assume that x and y are positive.

41. $10^{\log 37}$ **42.** $10^{\log 0.0093}$

43. $e^{\ln 0.045}$ **44.** $e^{\ln 453}$

45. $e^{\ln x}$ **46.** $10^{\log y}$

Objective 2 Use the Change-of-Base Formula to Evaluate Logarithms

47. Use the change-of-base formula to rewrite $\log_3 37.1$ in terms of natural logarithms.

48. Use the change-of-base formula to rewrite $\log_7 81.8$ in terms of natural logarithms.

49. Use the change-of-base formula to rewrite $\log_{13} 7.9$ in terms of common logarithms.

50. Use the change-of-base formula to rewrite $\log_{24} 4.3$ in terms of common logarithms.

In Exercises 51–54, approximate each expression to the nearest ten-thousandth.

51. $\log_{6.8} 0.085$ **52.** $\log_{7.2} 0.0032$

53. $\log_{0.49} 3.86$ **54.** $\log_{0.61} 18.4$

Estimate Then Calculate

In Exercises 55–58, mentally estimate the value of each expression to the nearest integer, and then use a calculator or a spreadsheet to approximate each value to the nearest thousandth.

Problem	Mental Estimate	Calculator Approximation
55. $\log_4 63$		
56. $\log_2 33$		
57. $1.99^{5.02}$		
58. $10.01^{1.999}$		

Skill and Concept Development

In Exercises 59–70, use $\log_b 2 \approx 0.3562$, $\log_b 5 \approx 0.8271$, and the properties of logarithms to approximate the value of each logarithmic expression.

59. $\log_b 10$ **60.** $\log_b 4$

61. $\log_b 25$ **62.** $\log_b \sqrt{5}$

63. $\log_b \sqrt[3]{2}$ **64.** $\log_b 5b$

65. $\log_b 2b$ **66.** $\log_b 2.5$

67. $\log_b 0.4$ **68.** $\log_b \dfrac{b}{5}$

69. $\log_b \dfrac{b}{2}$ **70.** $\log_b \sqrt{10b}$

In Exercises 71–80, match each expression with an equal expression from choices A through F. Note that choices A and F can be the correct response to more than one

exercise. Assume that the argument of each logarithm is a positive real number.

71. $\log \dfrac{x}{y}$

72. $\dfrac{\log x}{\log y}$

73. $\log xy$

74. $(\log x)(\log y)$

75. $\log(x - y)$

76. $\log(x + y)$

77. $\log x^y$

78. $\log y^x$

79. $10^{\log x}$

80. $e^{\ln x}$

A. x

B. $\log x + \log y$

C. $\log x - \log y$

D. $x \log y$

E. $y \log x$

F. This expression cannot be simplified.

In Exercises 81–84, use the change-of-base formula for exponents to rewrite each exponential function as a function with base e.

81. $f(x) = 5^x$ **82.** $f(x) = 10^x$

83. $f(x) = \left(\dfrac{1}{5}\right)^x$ **84.** $f(x) = (0.1)^x$

In Exercises 85 and 86, the first column contains a student's answer to a calculus problem. The second column contains the book's answer. Use the properties of logarithms to show that these expressions are equal.

Student's Answer	Book's Answer
85. $\ln\left(\dfrac{1}{2}\right)$	$-\ln 2$
86. $\log 24^x - \log 60^x - \log 0.2^{2x}$	x

Group discussion questions

87. Challenge Question Every exponential growth or decay function can be expressed in terms of base e.
  a. Select an exponential growth function, and write it in terms of base e.
  b. Select an exponential decay function, and write it in terms of base e.
  c. Describe how one can tell by inspection whether an exponential function base e represents growth or decay.
  d. Prepare a justification for your answer in part **c** to present to your instructor.

88. Challenge Question Prove the quotient rule for logarithms. (*Hint:* See the book's proof at the beginning of this section for the product rule.)

89. Challenge Question Prove the power rule for logarithms.

90. Challenge Question Prove that $\log_b a = \dfrac{1}{\log_a b}$ for $a > 0$, $b > 0$, and $a \neq 1$ and $b \neq 1$.

Solve each equation.

1. $x^2 = 25$

2. $(x - 4)^2 = 25$

3. $(x - 4)(2x + 7) = 0$

4. $(x - 4)(2x + 7) = -25$

5. $2x^2 + 7x + 6 = 0$

Section 11.6 | Solving Exponential and Logarithmic Equations

Objective:

1. Solve exponential and logarithmic equations.

1. Solve Exponential and Logarithmic Equations

An **exponential equation** contains a variable in the exponent. A **logarithmic equation** contains a variable in the argument of a logarithm. We solve both types of equations in this section.

Relatively simple exponential equations, such as the one in Example 1, result in nice integer solutions. For other equations, we often use a calculator or a spreadsheet to approximate the solutions.

Later in this section, we also examine some identities involving exponents and logarithms.

Example 1   Solving an Exponential Equation

Solve $2^{3x+2} = 32$.

Solution

$$2^{3x+2} = 32$$

$$2^{3x+2} = 2^5 \qquad \text{Express both sides of the equation in terms of base 2.}$$

$$3x + 2 = 5 \qquad \text{The exponents are equal since the bases are the same.}$$

$$3x = 3$$

$$x = 1 \qquad \text{Does this value check?}$$

Answer: $x = 1$

Self-Check 1

Solve $3^{5x-1} = 81$.

If the Solution of an Exponential Equation Is Not a Small Integer, How Can I Solve the Equation?

This can be done either graphically or algebraically, as illustrated in Example 2. Throughout this book, we have used graphs to solve a variety of equations. In fact, this is one advantage of the graphical method—its applicability to a wide variety of equations. Another advantage is the perspective that we gain by being able to see the solution.

Nonetheless, the precision of algebraic methods is often preferred. In the algebraic method for solving an exponential equation, we can start by taking logarithms of both sides of the equation. If two expressions are equal, then their logarithms are equal.

Example 2   Solving an Exponential Equation

Algebraically determine the exact solution of $5^x = 30$. Also approximate this solution to the nearest ten-thousandth and determine this solution graphically.

Solution

Algebraically

$5^x = 30$

$\log 5^x = \log 30$ Take the common log of both sides of the equation.

$x \log 5 = \log 30$ Use the power rule for logarithms. Then divide both sides of the equation by log 5.

$x = \dfrac{\log 30}{\log 5}$ This is the exact solution.

$x \approx 2.1133$ Then approximate this value with a calculator.

Graphically

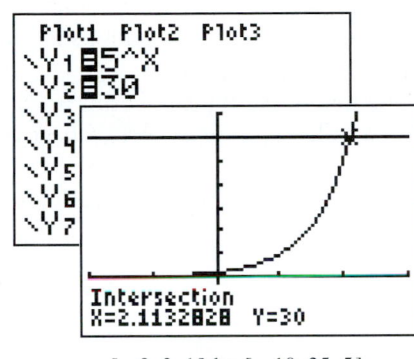

$[-2, 3, 1]$ by $[-10, 35, 5]$

Let $Y_1 = 5^x$ and $Y_2 = 30$, and use the **Intersect** feature on a calculator to determine the point of intersection.

Knowing that $5^2 = 25$ and $5^3 = 125$ can help us select a window that will display the value of x for which $5^x = 30$.

The approximate x-coordinate of the point of intersection is $x \approx 2.1133$.

Answer: $x = \dfrac{\log 30}{\log 5} \approx 2.1133$ **Check:** $5^{2.1133} \approx 30.000833 \approx 30$

Self-Check 2

Approximate the solution of $2^x = 7$ to the nearest thousandth.

The procedure used in Examples 1 and 2 is summarized in the following box.

Solving Exponential Equations Algebraically

Verbally	Algebraic Example
Step 1. a. If it is obvious that both sides of the equation are powers of the same base, express each side of the equation as a power of this base and then equate the exponents.	**a.** $2^x = 8$ $2^x = 2^3$ $x = 3$
b. Otherwise, take the logarithm of both sides of the equation and use the power rule to form an equation that does not contain variable exponents.	**b.** $2^x = 7$ $\log 2^x = \log 7$
Step 2. Solve the equation formed in step 1.	$x \log 2 = \log 7$ $x = \dfrac{\log 7}{\log 2}$

When we take the logarithm of both sides of an equation, we usually use either common logs or natural logs. We do this because these functions are available on calculators and spreadsheets. Example 3 uses natural logs.

Example 3   Solving an Exponential Equation

Solve $6^{z+3} = 8^{2z-1}$ and approximate this solution to the nearest ten-thousandth.

Solution

$$6^{z+3} = 8^{2z-1}$$

Because 6 and 8 are not powers of the same base, take the natural log of both sides of the equation. Use the power rule for logarithms.

$$\ln 6^{z+3} = \ln 8^{2z-1}$$

$$(z + 3)\ln 6 = (2z - 1)\ln 8$$

$$z \ln 6 + 3 \ln 6 = 2z \ln 8 - \ln 8$$

$$z \ln 6 - 2z \ln 8 = -\ln 8 - 3 \ln 6$$

Multiply by using the distributive property, and then rewrite the equation so that the variable terms are on the left side of the equation.

$$z(\ln 6 - 2 \ln 8) = -(\ln 8 + 3 \ln 6)$$

Factor out z, and then divide both sides of the equation by the coefficient of z.

$$z = -\frac{\ln 8 + 3 \ln 6}{\ln 6 - 2 \ln 8}$$

This is the exact solution.

$$z \approx 3.1492736$$

Use a calculator to approximate the solution.

Answer: $z \approx 3.1493$

Does this value check?

Self-Check 3

Solve $3^{2z+1} = 5^z$ to the nearest thousandth.

As Examples 2 and 3 illustrate, either common logs or natural logs can be used to solve exponential equations. If the exponential equation involves base e, then we usually use natural logarithms.

Example 4   Solving an Exponential Equation

Solve $3e^{x^2+2} = 49.287$, and then approximate this solution to the nearest ten-thousandth.

Solution

$$3e^{x^2+2} = 49.287$$

Divide both sides of the equation by 3.

$$e^{x^2+2} = 16.429$$

Rewrite this exponential equation in logarithmic form with base e.

$$x^2 + 2 = \ln 16.429$$

$$x^2 = \ln 16.429 - 2$$

Subtract 2 from both sides of the equation.

$$x = \pm\sqrt{\ln 16.429 - 2}$$

Find the exact roots of this quadratic equation by extraction of roots.

$$x \approx \pm 0.8938948853$$

Use a calculator to approximate these solutions.

Answer: $x \approx -0.8939$ or $x \approx 0.8939$

Do these values check?

Self-Check 4

Solve $2(e^{x^2-1}) = 11$, and then approximate any solutions to the nearest thousandth.

Can I Also Solve Logarithmic Equations Graphically and Algebraically?

Yes. Since exponential and logarithmic functions are inverses of each other, it is not surprising that we can use logarithms to solve some exponential equations. Likewise, we can solve some logarithmic equations by first rewriting them in exponential form.

Example 5 Solving a Logarithmic Equation with One Logarithmic Term

Solve $\log(3x + 7) = 2$ algebraically and graphically.

Solution

Algebraically

$$\log(3x + 7) = 2$$
$$3x + 7 = 10^2 \qquad \text{Rewrite this logarithmic equation in exponential form with base 10.}$$
$$3x + 7 = 100$$
$$3x = 93 \qquad \text{Then solve for } x.$$
$$x = 31$$

Graphically

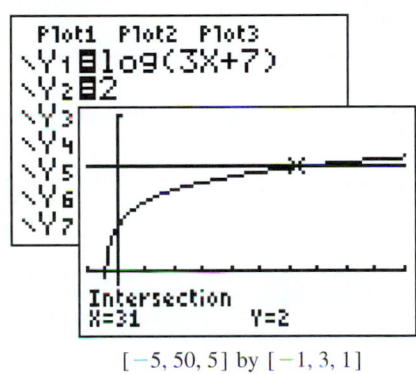

The x-coordinate of the point of intersection is $x = 31$.

$[-5, 50, 5]$ by $[-1, 3, 1]$

Answer: $x = 31$

Check: $\log[3(31) + 7] \overset{?}{=} 2$
$$\log 100 \overset{?}{=} 2$$
$$2 \overset{?}{=} 2 \text{ checks.}$$

Self-Check 5

Solve $\log(9t + 1) = 3$.

Logarithmic functions are one-to-one functions. (The horizontal line test can be applied to the graph of $y = \log_b x$ to show that this function is one-to-one.) Thus if $x, y > 0$ and $\log_b x = \log_b y$, then $x = y$. The fact that the arguments are equal when logarithms are equal allows us to replace a logarithmic equation with one that does not involve logarithms. Because real logarithms are defined only for positive arguments, each possible solution must be checked to make sure that it is not an extraneous value.

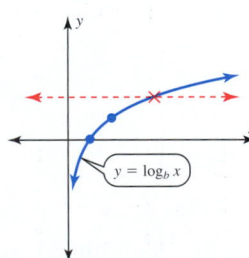

Example 6   Solving a Logarithmic Equation with Two Logarithmic Terms

Solve $\ln x = \ln(4x + 6)$.

Solution

$$\ln x = \ln(4x + 6)$$
$$x = 4x + 6 \qquad \text{Since the logarithms are equal, the arguments are equal.}$$
$$-3x = 6 \qquad \text{Now solve for } x.$$
$$x = -2$$

Check: $\ln x = \ln(4x + 6)$

$\ln(-2) \overset{?}{=} \ln[4(-2) + 6]$ Logarithms of negative argument values are not real

$\ln(-2)$ is not a real number.   numbers.

Answer: There is no solution. -2 is an extraneous value.

Self-Check 6

Solve these logarithmic equations.

a. $\ln(2y + 6) = \ln(3y + 7)$ **b.** $\log(v - 5) = \log(1 - v)$

Note that if we tried to solve the equations in Example 6 graphically, the graphs of $y = \ln x$ and $y = \ln(4x + 6)$ would not intersect.

The algebraic procedure for solving a logarithmic equation is summarized in the following box.

Solving Logarithmic Equations Algebraically

Verbally	Algebraic Example
Step 1. **a.** If possible, rewrite the logarithmic equation in exponential form.	**a.** $\log(x + 1) = 2$
b. Otherwise, use the properties of logarithms to write each side of the equation as a single logarithmic term with the same base. Then form a new equation by equating the arguments of these logarithms.	$x + 1 = 10^2$ $x = 99$ *Check:* $\log(99 + 1) \overset{?}{=} 2$ $\log(100) \overset{?}{=} 2$
Step 2. Solve the equation formed in step 1.	$2 \overset{?}{=} 2$ checks.
Step 3. Check all possible solutions for extraneous values; logarithms of zero or negative arguments are undefined.	**b.** $\ln 2x = \ln(x + 1)$ $2x = x + 1$ $x = 1$
	Check: $\ln 2(1) \overset{?}{=} \ln(1 + 1)$ $\ln 2 \overset{?}{=} \ln 2$ checks.

If a logarithmic equation has more than one logarithmic term on one side of the equation, then we may need to use the properties of logarithms to rewrite these terms as a single logarithm.

Example 7   Solving a Logarithmic Equation with Three Logarithmic Terms

Solve $\ln(3 - w) + \ln(1 - w) = \ln(11 - 6w)$.

Solution

$$\ln(3 - w) + \ln(1 - w) = \ln(11 - 6w)$$
$$\ln[(3 - w)(1 - w)] = \ln(11 - 6w)$$
$$(3 - w)(1 - w) = 11 - 6w$$
$$3 - 4w + w^2 = 11 - 6w$$
$$w^2 + 2w - 8 = 0$$
$$(w + 4)(w - 2) = 0$$
$$w + 4 = 0 \qquad \text{or} \qquad w - 2 = 0$$
$$w = -4 \qquad\qquad w = 2$$

Express the left side of the equation as a single logarithm, by using the product rule for logarithms. Since the natural logarithms are equal, the arguments are equal.

Write the quadratic equation in standard form. Then factor and solve for w.

Check: For $w = -4$,

$$\ln[3 - (-4)] + \ln[1 - (-4)] \overset{?}{=} \ln[11 - 6(-4)]$$
$$\ln 7 + \ln 5 \overset{?}{=} \ln(11 + 24)$$
$$\ln 35 \overset{?}{=} \ln 35 \text{ checks.}$$

Check: For $w = 2$,

$$\ln(3 - 2) + \ln(1 - 2) \overset{?}{=} \ln[11 - 6(2)]$$
$$\ln 1 + \ln(-1) \overset{?}{=} \ln(-1)$$
$$\ln(-1) \text{ is undefined.}$$

Thus $w = 2$ is an extraneous value.

Answer: $w = -4$

Self-Check 7

Solve $\ln(2z + 5) + \ln z = \ln 3$.

Example 8   Solving a Logarithmic Equation with Two Logarithmic Terms

Solve $\log(2v + 1) - \log(v - 4) = 1$.

Solution

$$\log(2v + 1) - \log(v - 4) = 1$$
$$\log\left(\frac{2v + 1}{v - 4}\right) = 1$$
$$\frac{2v + 1}{v - 4} = 10^1$$
$$2v + 1 = 10(v - 4)$$
$$2v + 1 = 10v - 40$$
$$-8v = -41$$
$$v = \frac{41}{8}$$

Express the left side of the equation as a single logarithm, by using the quotient rule for logarithms.

Rewrite this equation in exponential form, by using a base of 10.

Then solve for v.

Check: $\log\left[2\left(\dfrac{41}{8}\right)+1\right]-\log\left(\dfrac{41}{8}-4\right)\overset{?}{=}1$

$$\log\left(\dfrac{45}{4}\right)-\log\left(\dfrac{9}{8}\right)\overset{?}{=}1$$

$$\log\left(\dfrac{45}{4}\div\dfrac{9}{8}\right)\overset{?}{=}1$$

$$\log 10\overset{?}{=}1$$

$$1\overset{?}{=}1\text{ checks.}$$

Answer: $v=\dfrac{41}{8}$

Self-Check 8

Solve $\log 2+\log(y+4)-\log(y-94)=2$.

Remember that we can use a calculator or a spreadsheet to check our answer. In Example 8, we can store $\dfrac{41}{8}$ for the variable and then evaluate the left side of the equation.

Obtaining a value of 1 for the left side means that $v=\dfrac{41}{8}$ checks.

The exponential and logarithmic equations that we have examined in this section have been either contradictions with no solution or conditional equations with one or two solutions. We now examine some exponential and logarithmic equations that are identities.

Example 9   Verifying Logarithmic and Exponential Identities

Verify the following identities.

(a) $\log_7 14^x-\log_7 2^x=x$ **(b)** $e^{-\ln x}=\dfrac{1}{x}$ for $x>0$

Solution

(a) $\log_7 14^x-\log_7 2^x=x\log_7 14-x\log_7 2$ Rewrite the left side of the equation by using the power rule for logarithms.

$\log_7 14^x-\log_7 2^x=x(\log_7 14-\log_7 2)$ Then factor out x.

$\log_7 14^x-\log_7 2^x=x\log_7 \dfrac{14}{2}$ Simplify this expression by using the quotient rule for logarithms.

$\log_7 14^x-\log_7 2^x=x\log_7 7$

$\log_7 14^x-\log_7 2^x=x(1)$ Thus you have verified that the left side of the equation does equal the right side of the equation.

$\log_7 14^x-\log_7 2^x=x$

(b) $e^{-\ln x}=e^{\ln(x^{-1})}$ Rewrite the left side of the equation by using the power rule for logarithms: $p\ln x=\ln x^p$ with $p=-1$.

$e^{-\ln x}=x^{-1}$ Simplify by using the identity $b^{\log_b y}=y$, with $b=e$ and $y=x^{-1}$. Replace x^{-1} with $\dfrac{1}{x}$ to

$e^{-\ln x}=\dfrac{1}{x}$ obtain the right side of this identity.

Self-Check 9

Verify that $\log 25^x+\log 4^x=2x$ is an identity.

Self-Check Answers

1. $x = 1$
2. $x \approx 2.807$
3. $z \approx -1.869$
4. $x = \pm\sqrt{\ln 5.5 + 1}$
   $x \approx \pm 1.645$

5. $t = 111$
6. **a.** $y = -1$
   b. No solution; $v = 3$ is extraneous.
7. $z = \dfrac{1}{2}$

8. $y = 96$
9. $\log 25^x + \log 4^x = \log(25^x \cdot 4^x)$
   $\log 25^x + \log 4^x = \log 100^x$
   $\log 25^x + \log 4^x = x \log 100$
   $\log 25^x + \log 4^x = 2x$

11.6   Using the Language and Symbolism of Mathematics

1. An _____ equation contains a variable in the exponent.

2. A _____ equation contains a variable in the argument of a logarithm.

3. If $x > 0$, $y > 0$, and $\log_b x = \log_b y$, then _____ = _____.

4. An _____ value is a value produced in the solution process that does not check in the original equation.

11.6   Quick Review

In Exercises 1–3, use a calculator to approximate to the nearest thousandth the value of each expression.

1. $\log\left(\dfrac{48}{5}\right)$ 2. $\dfrac{\log 48}{5}$ 3. $\dfrac{48}{\ln 5}$

4. Use the tables shown to solve this system of linear equations.

$$y = 2.38x - 13.9$$
$$y = -4.71x + 21.55$$

x	$y_1 = 2.38x - 13.9$	$y_2 = -4.71x + 21.55$
0	-13.900	21.550
1	-11.520	16.840
2	-9.140	12.130
3	-6.760	7.420
4	-4.380	2.710
5	-2.000	-2.000
6	0.380	-6.710

5. Use the graph shown to solve this system of linear equations.

$$y = 2.4x + 0.1$$
$$y = -3.2x + 5.7$$

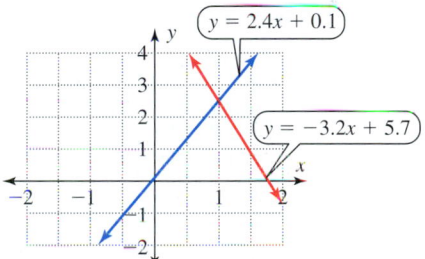

11.6   Exercises

Objective 1   Solve Exponential and Logarithmic Equations

In Exercises 1–28, solve each equation using only pencil and paper.

1. $7^x = 49$

2. $7^x = \dfrac{1}{49}$

3. $\log_5 x = -3$

4. $\log_5 x = 2$

5. $3^{w-5} = 27$

6. $7^{2w-1} = \sqrt{7}$

7. $\left(\dfrac{2}{3}\right)^{x^2} = \dfrac{16}{81}$

8. $5^{x^2 - 6} = \dfrac{1}{25}$

9. $\log(y - 5) = 1$

10. $\log(2y + 1) = 1$

11. $\log(3n - 5) = 2$

12. $\log(5n - 4) = 0$

13. $\ln(3m - 7) = \ln(2m + 9)$

14. $\log(5y + 6) = \log(2y - 9)$

15. $\log(4w + 3) = \log(8w + 5)$

16. $\ln(5y - 7) = \ln(2y + 1)$

17. $\ln(3 - x) = \ln(1 - 2x)$

18. $\log(7 - 5x) = \log(4 - 8x)$

19. $\log(t + 3) + \log(t - 1) = \log 5$

20. $\ln(7t + 3) - \ln(t + 1) = \ln(6t + 2)$

21. $\ln(v^2 - 9) - \ln(v + 3) = \ln 7$

22. $\ln(2v + 6) - \ln(v + 1) = \ln(v + 3)$

23. $\ln(5x - 7) - \ln(2x + 3) = \ln 3$

24. $\log(7x + 13) - \log(4x + 13) = \log 5$

25. $\log(1 - y) + \log(4 - y) = \log(18 - 10y)$

26. $\ln(2 - y) + \ln(1 - y) = \ln(32 - 4y)$

27. $\log(w - 3) - \log(w^2 + 9w - 32) = -1$

28. $\log(w^2 + 1) - \log(w - 2) = 1$

In Exercises 29–32, use the given tables to solve each equation.

29. $4^x = 32$

x	$y = 4^x$
0.0	1
0.5	2
1.0	4
1.5	8
2.0	16
2.5	32
3.0	64

30. $32^x = 64$

x	$y = 32^x$
0.0	1
0.2	2
0.4	4
0.6	8
0.8	16
1.0	32
1.2	64

31. $\log(5x - 9.99) = -2$

x	$y = \log(5x - 9.99)$
0	Error
1	Error
2	−2.000
3	0.700
4	1.000
5	1.176
6	1.301

32. $\log(90x + 100) = 3$

x	$y = \log(90x + 100)$
0	2.000
2	2.447
4	2.663
6	2.806
8	2.914
10	3.000
12	3.072

In Exercises 33–36, use the given graphs to solve each equation.

33. $8^{(x+2)/3} = 16$

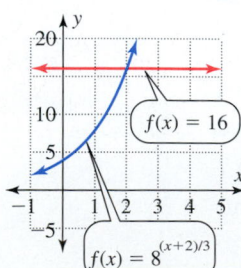

34. $27^{(2x-4)/3} = 9$

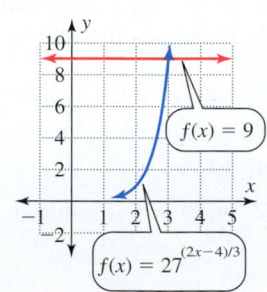

35. $\log(100x^2) = 2$

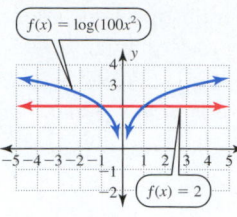

36. $\log(100x + 1,000) = 2$

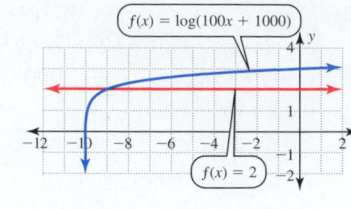

In Exercises 37–62, solve each equation and then approximate the solution to the nearest thousandth.

37. $4^v = 15$

38. $5^v = 18$

39. $3^{-w+7} = 22$

40. $6^{-w+3} = 81$

41. $9.2^{2t+1} = 11.3^t$

42. $8.7^{3t-1} = 10.8^{2t}$

43. $7.6^{-2z} = 5.3^{2z-1}$

44. $8.1^{-3z} = 6.5^{1-2z}$

45. $e^{3x} = 78.9$

46. $e^{3x+5} = 15.9$

47. $10^{2v+1} = 51.3$

48. $10^{3y-4} = 73.8$

49. $0.83^{v^2} = 0.68$

50. $0.045^{v^2} = 0.0039$

51. $3.7e^{x^2+1} = 689.7$

52. $2.5e^{x^2+4} = 193.2$

53. $\log(5x - 17) = 0.83452$

54. $\ln(11x - 3) = 1.44567$

55. $\ln(\ln y) = 1$

56. $\log(\log y) = 0.48913$

57. $\ln(v - 4) + \ln(v - 3) = \ln(5 - v)$

58. $\log(3x + 1) + \log(x + 2) = \log x$

59. $\ln(11 - 5x) - \ln(x - 2) = \ln(x - 6)$

60. $\ln(3 - x) - \ln(x - 2) = \ln(x + 1)$

61. $(\ln x)^2 = \ln x^2$

62. $(\log x)^2 = \log x^2$

Applying Technology

In Exercises 63–66, use a graphing calculator to approximate the solution(s) of each equation to the nearest thousandth.

63. $\log(398x + 5) = x + 0.5$ **64.** $\ln(18x + 3) = 3x - 5$

65. $0.1e^x = 8 - 2x$ **66.** $e^{0.5x} = 4 - x^2$

Skill and Concept Development

In Exercises 67–74, verify each identity, assuming that x is a positive real number.

67. $10^{-\log x} = \dfrac{1}{x}$

68. $100^{\log x} = x^2$

69. $e^{-x \ln 3} = \left(\dfrac{1}{3}\right)^x$

70. $e^{(\ln x)/2} = \sqrt{x}$

71. $\log 60^x - \log 6^x = x$

72. $\log 5^x + \log 2^x = x$

73. $\ln\left(\dfrac{4}{5}\right)^x + \ln\left(\dfrac{5}{3}\right)^x + \ln\left(\dfrac{3}{4}\right)^x = 0$

74. $\ln\left(\dfrac{2}{3}\right)^x + \ln\left(\dfrac{5}{2}\right)^x - \ln\left(\dfrac{5}{3}\right)^x = 0$

Connecting Concepts to Applications

75. Depreciation The value V of an industrial lathe after t years of depreciation is given by the formula $V = 35{,}000e^{-0.2t} + 1{,}000$. Approximately how many years will it take for the value to depreciate to $10,000?

76. Depreciation The value V of an irrigation system after t years of depreciation is given by the formula $V = 59{,}000e^{-0.2t} + 3{,}000$. Approximately how many years will it take for the value to depreciate to $7,000?

Group discussion questions

77. Error Analysis The steps used to solve a certain logarithmic equation produced one possible solution that was a negative number. One student claimed this number could not check in the equation, since it was negative. Discuss the logic of this student's claim.

78. Discovery Question
  a. Create an exponential equation whose only solution is 5.
  b. Create an exponential equation whose only solutions are -2 and 5.

79. Discovery Question
  a. Create a logarithmic equation with $x = -2$ as an extraneous value and that has no solution.
  b. Create a logarithmic equation with $x = -2$ as an extraneous value and with a solution of 5.
  c. Create a logarithmic equation with solutions of -2 and 5.

80. Challenge Question Which is larger, 2001^{2002} or 2002^{2001}?

11.6	**Cumulative Review**

Use the graph of $y = -|x + 2| + 4$ to answer each question.

  1. Determine the y-intercept.

  2. Determine the x-intercepts.

  3. What is the maximum y-value of this function?

  4. What is the value of x when y is at a maximum?

  5. Determine the interval of x-values for which the function is increasing.

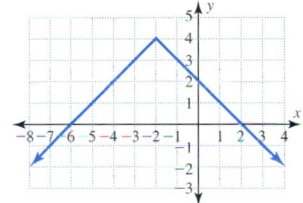

Section 11.7	Exponential Curve Fitting and Other Applications of Exponential and Logarithmic Equations

Objective:

1. Use exponential and logarithmic equations to solve applied problems.

Exponential and logarithmic equations can be used to describe an increase or a decrease in an amount of money, the growth or decline of populations, and the decay of radioactive elements. Logarithmic scales are often used to measure natural phenomena. For example, decibels measure the intensity of sounds, and the Richter scale measures the intensity of earthquakes.

1. Use Exponential and Logarithmic Equations to Solve Applied Problems

In this section we examine applications. Two important formulas that we apply are the formulas for periodic growth and continuous growth.

What Is the Difference Between Periodic Growth and Continuous Growth?

Growth that occurs at discrete intervals (such as yearly, monthly, or weekly) is called **periodic growth;** growth that occurs continually at each instant is called **continuous growth.** The formula for continuous growth is a good model for periodic growth when the number of growth periods is relatively large. For example, the population of rabbits over a period of years can be accurately estimated by using the continuous-growth formula, even though female rabbits do not have offspring continuously.

Growth and Decay Formulas

	Verbally	Algebraic Example
Periodic growth formula: $$A = P\left(1 + \frac{r}{n}\right)^{nt}$$	The amount A that is produced by an original amount P growing at an annual rate r with periodic compounding n times a year for t years.	$$A = 1{,}000\left(1 + \frac{0.07}{4}\right)^{(4)(10)}$$ This is the amount resulting from investing \$1,000 at 7% compounded quarterly for 10 years.
Continuous growth and decay formula: $$A = Pe^{rt}$$	The amount A that is produced by an original amount P growing (decaying) continuously at a rate r for a time t. If $r > 0$, there is growth. If $r < 0$, there is decay.	$$A = 1{,}000e^{(0.07)(10)}$$ This is the amount resulting from investing \$1,000 at 7% compounded continuously for 10 years.

It is customary that most statements of interest rates are given as annual rates. In Example 1, the rate of 11% means an annual rate of 11%.

Example 1   Periodic Growth of an Investment

How many years will it take an investment of \$1,000 to double in value if interest on the investment is compounded semiannually at a rate of 11%?

Solution

$$A = P\left(1 + \frac{r}{n}\right)^{nt}$$

$$2{,}000 = 1{,}000\left(1 + \frac{0.11}{2}\right)^{2t}$$

Substitute $A = 2{,}000$, $P = 1{,}000$, $r = 0.11$, and $n = 2$ into the periodic-growth formula and then simplify.

$$2 = (1.055)^{2t}$$

$$\ln 2 = \ln(1.055)^{2t}$$

Take the natural log of both sides of the equation.

$$\ln 2 = 2t \ln 1.055$$

Use the power rule for logarithms.

$$t = \frac{\ln 2}{2 \ln 1.055}$$

Solve for t by dividing both sides of the equation by $2 \ln 1.055$.

$$t \approx 6.4730785$$

Use a calculator to approximate this value.

Answer: The investment will double in value in approximately 6.5 years.

A comparison of periodic growth to continuous growth is given in Table 11.7.1. This table shows various compounding periods and the amount to which $1 accumulates in each situation after 1 year. The interest formula, with $P = 1$ and $r = 1.00$, is

$$A = 1\left(1 + \frac{1}{n}\right)^{n(1)} = \left(1 + \frac{1}{n}\right)^{n}.$$

Note that compounding monthly produces a big gain over simple interest. However, further gains are relatively small, and as n increases, the expression $\left(1 + \frac{1}{n}\right)^{n}$ approaches the irrational number e.

Table 11.7.1 Periodic Compounding and Continuous Compounding for 1 Year

Periodic Compounding	n	$A = \left(1 + \dfrac{1}{n}\right)^{n}$ (to six places)	A (to the nearest penny)
Annually	1	2.000000	$2.00
Semiannually	2	2.250000	2.25
Quarterly	4	2.441406	2.44
Monthly	12	2.613035	2.61
Weekly	52	2.692597	2.69
Daily	365	2.714567	2.71
Hourly	8,760	2.718127	2.72
Every minute	525,600	2.718279	2.72
Every second	31,536,000	2.718282	2.72
.	.	.	
.	.	.	
.	.	.	
Continuously	Infinite	e	

Continuous Compounding	$A = Pe^{rt} = e$ with $P = 1$, $r = 1$, and $t = 1$.

A key point to note is that continuous compounding provides a reasonable approximation for some periodic events; this is especially true when there are many growth periods. Because the formula $A = Pe^{rt}$ is easy to use and can provide a good approximation of periodic growth, it is used widely in growth and decay problems. Examples 2 through 4 illustrate some of the possibilities.

Example 2   Comparing Periodic Growth to Continuous Growth

Compute the value of $1,000 invested at 6% for 5 years under two conditions.

(a) The interest is compounded daily.

(b) The interest is compounded continuously.

Solution

(a) $A = P\left(1 + \dfrac{r}{n}\right)^{nt}$

Use the formula for periodic growth. Substitute in the given values. Use 365 days for each year (assume that there is no change due to leap year).

$A = 1{,}000\left(1 + \dfrac{0.06}{365}\right)^{(365)(5)}$

$A \approx 1{,}000(1.000164384)^{1{,}825}$ Use a calculator to approximate this value.

$A \approx \$1{,}349.83$

(b) $A = Pe^{rt}$

Use the formula for continuous growth. Substitute in the given values.

$A = 1{,}000e^{(0.06)(5)}$

$A = 1{,}000e^{0.30}$

$A \approx \$1{,}349.86$ Use a calculator to approximate this value.

Self-Check 2

Compute the value of $5,000 invested at 7% for 8 years under two conditions.

a. The interest is compounded monthly.

b. The interest is compounded continuously.

Note that the results in Example 2 differ by only 3 cents over a 5-year period. Thus continuous growth provided a good model for growth compounded daily.

Example 3   Continuous Growth of Bacteria

A new culture of bacteria grows continuously at the rate of 20% per day. If a culture of 10,000 bacteria isolated in a laboratory is allowed to multiply, how many days will it be before there are 1,000,000 bacteria?

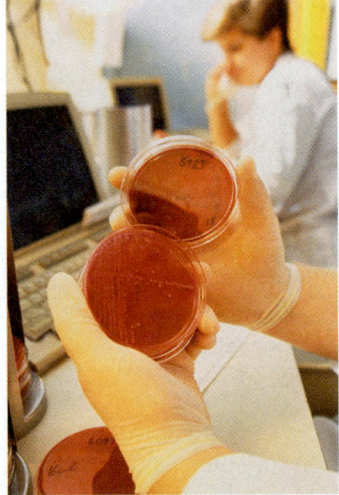

Solution

$A = Pe^{rt}$

$1{,}000{,}000 = 10{,}000e^{0.20t}$

$100 = e^{0.20t}$

$\ln 100 = 0.20\,t$

Substitute $A = 1{,}000{,}000$, $P = 10{,}000$, and $r = 0.20$ into the formula for continuous growth. Simplify and then rewrite the exponential equation in logarithmic form.

$t = \dfrac{\ln 100}{0.20}$

Solve for t and then approximate with a calculator.

$t \approx 23.0$

Answer: It will take approximately 23 days for the culture to produce 1,000,000 bacteria.

For physical phenomena such as radioactive decay, exponential decay models have been experimentally tested and provide an excellent mathematical model for predicting the future or describing the past. This is illustrated in Example 4.

Example 4   Continuous Decay of Carbon-14

Carbon-14 decays continuously at the rate of 0.01245% per year. When living tissue dies, it no longer absorbs carbon-14; any carbon-14 present decays and is not replaced. An archaeologist has determined that only 20% of the carbon-14 originally in a plant specimen remains. Estimate the age of this specimen.

Solution

$$A = Pe^{rt}$$

Substitute the given values into the continuous-decay formula. The 0.01245% decay rate means that $r = -0.0001245$, and 20% of the original amount P left means that $A = 0.20P$.

$$0.20P = Pe^{-0.0001245t}$$

$$0.20 = e^{-0.0001245t}$$

Divide both sides of the equation by P.

$$\ln 0.20 = -0.0001245t$$

Rewrite this statement in logarithmic form.

$$t = \frac{\ln 0.20}{-0.0001245}$$

Solve this linear equation for t.

$$t \approx 12,927.21215$$

Use a calculator to approximate this value.

Answer: This specimen is approximately 13,000 years old.

Logarithmic growth functions grow very slowly. For example, $\log 100{,}000 = 5$, while $\log 1{,}000{,}000 = 6$. The increase from 100,000 to 1,000,000 causes an increase in the logarithm of only 1 unit from 5 to 6. Scientists take advantage of this property to describe phenomena whose numerical measure covers a wide range of scores. By converting these measures to a logarithmic scale, the numbers become smaller and more comprehensible. Two examples of this are the Richter scale and the decibel scale. These scales are illustrated in Examples 5 and 6.

Seismologists use the Richter scale to measure the magnitude of earthquakes. The equation $R = \log \dfrac{A}{a}$ compares the amplitude A of the shock wave of an earthquake to the amplitude a of a reference shock wave of minimal intensity.

Example 5   Using the Richter Scale

In the Denver, Colorado, area in the 1960s, an earthquake with an amplitude 40,000 times the reference amplitude occurred after a liquid was injected under pressure into a well more than 2 mi deep. Calculate the magnitude of this earthquake on the Richter scale.

Solution

$$R = \log\left(\frac{A}{a}\right)$$

$$R = \log\left(\frac{40,000a}{a}\right) \quad \text{Substitute the given amplitude into the Richter scale equation.}$$

$$R = \log 40,000 \quad \text{Simplify the fraction.}$$

$$R \approx 4.6020600 \quad \text{Use a calculator to approximate this value.}$$

Answer: This earthquake measured approximately 4.6 on the Richter scale.

Self-Check 5

An earthquake has a Richter scale reading of 6.2. Compare its amplitude to that of the standard reference shock wave.

A Mathematical Note

A decibel is not a unit of loudness but rather a measure of the energy of the sound. A decibel (abbreviated dB) is one-tenth of a bel—a unit of measure named after Alexander Graham Bell, the inventor of the telephone.

The human ear can hear a vast range of sound intensities, so it is more practical to use a logarithmic scale than to use an absolute scale to measure the intensity of sound. The unit of measurement on this scale is the **decibel.** The number of decibels D of a sound is given by the formula $D = 10 \log \frac{I}{I_0}$, which compares the intensity I of the sound to the reference intensity I_0, which is at the threshold of hearing ($I_0 \approx 10^{-16}$ W/cm^2).

Example 6   Determining the Decibel Level of Music

If a store in a mall plays background music at an intensity of 10^{-14} W/cm^2, determine the number of decibels (dB) of the music you hear.

Solution

$$D = 10 \log \frac{I}{I_0}$$

$$D = 10 \log \frac{10^{-14}}{10^{-16}} \quad \text{Substitute } 10^{-14} \text{ for } I \text{ and } 10^{-16} \text{ for } I_0 \text{ into the decibel equation.}$$

$$D = 10 \log 10^2 \quad \text{Simplify and then solve for } D \text{ by inspection.}$$

$$D = 10(2)$$

$$D = 20$$

Answer: 20 dB

Self-Check 6

Compute the number of decibels of a child's cry if its intensity is 10^{-7} W/cm^2.

Chemists use pH to measure the hydrogen ion concentration in a solution. Distilled water has a pH of 7, acids have a pH of less than 7, and bases have a pH of more than 7. The formula for the pH of a solution is $pH = -\log H^+$, where H^+ measures the concentration of hydrogen ions in moles per liter.

Example 7   Determining the pH of a Beer

Determine the pH of a light beer if its H^+ is measured at 6.3×10^{-5} mol/L.

Solution

$$pH = -\log H^+$$
$$pH = -\log(6.3 \times 10^{-5}) \qquad \text{Substitute } H^+ = 6.3 \times 10^{-5} \text{ into the pH equation.}$$
$$pH \approx 4.2007 \qquad \text{Approximate this value with a calculator.}$$

Answer: $pH \approx 4.2$

Self-Check 7

Determine the pH of a dark beer if its H^+ is measured at 7.2×10^{-5} mol/L.

A Mathematical Note

In a November 2004 article in *Technology Review,* Rodney Brooks states that a stable system can become unstable when even one component experiences exponential growth. Digital memory is growing exponentially and dramatically changing the marketplace. In 2004, an iPod could store 20,000 books— more than most people will ever read in a lifetime. At this rate, a teenager in 2025 will likely be able to own a $400 device that can hold every movie ever made.

Example 8 examines data that are best modeled by exponential growth and illustrates how to use a graphing calculator to approximate the exponential curve of best fit. The exponential regression feature **ExpReg**, which is in the **CALC** submenu of the **STAT** menu on a TI-84 Plus calculator, can be used by following almost exactly the same steps given in Technology Perspective 8.6.3 on Calculating and Graphing the Line of Best Fit.

Example 8   Calculating an Exponential Curve of Best Fit

Enter these points into a graphing calculator, draw the scatter diagram, and calculate the exponential curve of best fit. Each x-value represents a year after 1970, and each y-value represents the number of transistors on an integrated circuit. Use this exponential equation to approximate the number of transistors that were on an integrated circuit in 1985.

Years	Number of Years After 1970 x	Number of Transistors y
1971	1	2,250
1972	2	2,500
1974	4	5,000
1978	8	29,000
1982	12	120,000

Solution

1   Enter the data points by using the **STAT-EDIT** feature.

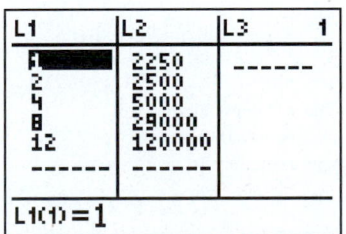

Enter all the (x, y) points.

Press 2nd Y= to access the **STAT PLOT** menu. To plot the points, use the **ZoomStat** option, which is option **9** in the ZOOM menu.

2 Turn on **Plot 1** and use **ZoomStat** to draw a scatter diagram of the data points listed in L1 and L2. Use the visual pattern to select the most appropriate family of curves. In this case, the visual pattern is exponential growth.

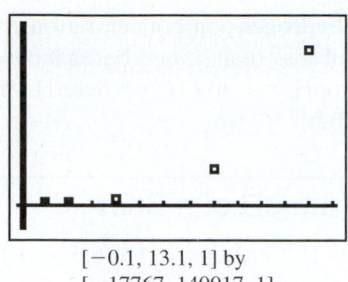

[−0.1, 13.1, 1] by
[−17767, 140017, 1]

Since the points appear to form an exponential growth curve, try to find an exponential function that best fits these data.

3 Use the **STAT-CALC** feature to calculate the exponential curve of best fit.

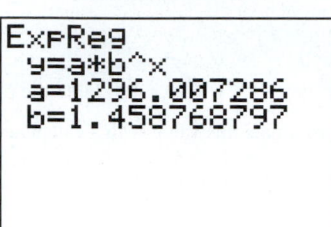

ExpReg
y=a*b^x
a=1296.007286
b=1.458768797

To calculate the exponential curve of best fit, select option **0** under the [STAT] CALC menu. This exponential curve of best fit does *not* contain each data point, but it does fit these points better than any other exponential curve. Note that the equation of the curve of best fit can be inserted into the [Y=] screen from the memory of the calculator by pressing

[VARS] [5] [›] [›] [ENTER].

4 Enter the exponential curve of best fit on the [Y=] screen, and then graph this curve. Round each coefficient to three significant digits.

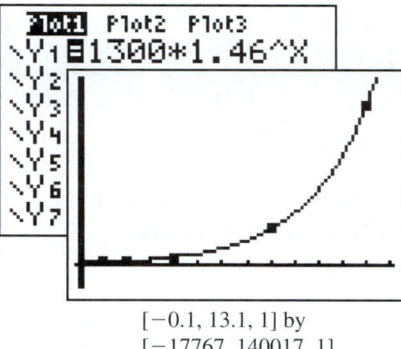

Plot1 Plot2 Plot3
\Y1◻1300*1.46^X
\Y2
\Y3
\Y4
\Y5
\Y6
\Y7

[−0.1, 13.1, 1] by
[−17767, 140017, 1]

The equation of the exponential growth function of best fit is written in $y = a \cdot b^x$ form with each coefficient rounded to three significant digits.

5 $y = 1,300(1.46^x)$

$y = 1,300(1.46^{15})$

$y \approx 380,000$

To approximate y for 1985, substitute 15 into the equation of the exponential function.

The estimate is rounded to three significant digits.

Answer: $y = 1,300(1.46^x)$ is approximately the exponential growth function of best fit. The formula predicts that in 1985 there would be 380,000 transistors on an integrated circuit. The actual number given on Intel's website is 375,000.

Self-Check 8

Determine the exponential curve of best fit for the points:
(0, 25), (1, 50), (2, 110), (3, 195), (4, 405), (5, 820).

Self-Check Answers

1. a. $2,001.60
b. Approximately 14.3 years
2. a. $8,739.13
b. $8,753.36

3. 165,000
4. The whale population will decline to 8,000 in approximately 18 years.

5. Its amplitude is approximately 1,600,000 times that of the reference shock wave.

6. 90 dB
7. pH ≈ 4.1
8. $y \approx 25(2^x)$

11.7 Using the Language and Symbolism of Mathematics

1. Growth that occurs at discrete intervals is called _____ growth.

2. Growth that occurs continually at each instant is called _____ growth.

3. The periodic-growth formula is $A =$ _____.

4. The continuous-growth formula is $A =$ _____.

5. Seismologists use the Richter scale to measure the magnitude of _____.

6. The decibel is a unit of measure for the intensity of _____.

11.7 Quick Review

1. Write $10^4 = 10,000$ in logarithmic form.

2. Write $\log 0.001 = -3$ in exponential form.

3. Write $\ln 78 \approx 4.357$ in exponential form.

4. Approximate $\log (2.58 \times 10^4)$ to the nearest hundredth.

5. Approximate $\log (2.58 \times 10^{-4})$ to the nearest hundredth.

11.7 Exercises

Objective 1 Use Exponential and Logarithmic Equations to Solve Applied Problems

Compound Interest In Exercises 1–8, use the formula for periodic growth to solve each problem. All interest rates are stated as annual rates of interest.

1. Find the value of $150 invested at 9% with interest compounded quarterly for 5 years.

2. Find the value of $210 invested at 7% with interest compounded monthly for 9 years.

3. How many years will it take an investment to double in value if interest is compounded annually at 8%?

4. How many years will it take an investment to double in value if interest is compounded semiannually at 10%?

5. How many years will it take a savings account to triple in value if interest is compounded monthly at 6%?

6. How many years will it take a savings account to triple in value if interest is compounded monthly at 7.5%?

7. How many years will it take an investment to increase in value by 50% if interest is compounded semiannually at 8%?

8. How many years will it take an investment to increase in value by 50% if interest is compounded semiannually at 6%?

In Exercises 9–22, use the formula for continuous growth and decay to solve each problem.

9. **Rate of Inflation** If prices will double in 10 years at the current rate of inflation, what is the current rate of inflation? Assume that the effect of inflation is continuous.

10. **Rate of Inflation** If prices will double in 9 years at the current rate of inflation, what is the current rate of inflation? Assume that the effect of inflation is continuous.

11. **Doubling an Investment** How many years will it take an investment to double in value if interest is compounded continuously at 7%?

12. **Doubling an Investment** How many years will it take an investment to double in value if interest is compounded continuously at 9%?

13. **Carbon-14 Dating** Carbon-14 decays continuously at the rate of 0.01245% per year. An archaeologist has determined that only 5% of the original carbon-14 from a plant specimen remains. Estimate the age of this specimen.

14. **Carbon-14 Dating** Carbon-14 decays continuously at the rate of 0.01245% per year. An archaeologist has determined that only 10% of the original carbon-14 from a plant specimen remains. Estimate the age of this specimen.

15. **Safe Storage Time for Milk** The safe level of psychrotrophic bacteria in a gallon of milk is 100 units. In a refrigerator set at 38°F, the number of units of these bacteria in a gallon of skim milk is approximated by the exponential function $B(t) = 4.0e^{0.24t}$, where t is the time in days. Evaluate and interpret each expression.
   a. $B(0)$ **b.** $B(5)$ **c.** $B(10)$ **d.** $B(t) = 100, t =$ ____

16. Safe Storage Time for Milk The safe level of psychrotrophic bacteria in a gallon of milk is 100 units. In a refrigerator set at 38°F, the number of units of these bacteria in a gallon of whole milk is approximated by the exponential function $B(t) = 4.0e^{0.26t}$, where t is the time in days. Evaluate and interpret each expression.
a. $B(0)$ **b.** $B(5)$ **c.** $B(10)$ **d.** $B(t) = 100, t = $ _____

17. Radioactive Decay The radioactive material used to power a satellite decays at a rate that decreases the available power by 0.05% per day. When the power supply reaches $\dfrac{1}{100}$ of its original level, the satellite is no longer functional. Approximately how many days should the power supply last?

18. Radioactive Decay The radioactive material used to power a satellite decays at a rate that decreases the available power by 0.03% per day. When the power supply reaches $\dfrac{1}{100}$ of its original level, the satellite is no longer functional. Approximately how many days should the power supply last?

19. Population Decline The population of a species of whales is estimated to be decreasing at a rate of 4% per year ($r = -0.04$). The current population is approximately 15,000. First estimate the population 10 years from now, and then determine how many years from now the population will have declined to 5,000.

20. Population Decline If the number of white owls in Illinois has decreased from 750 to 500 in 10 years, approximate the annual rate of decrease.

21. Population Growth If the population of a town has grown from 1,200 to 1,800 in 3 years, approximate the annual rate of increase.

22. Population Growth The human population in a remote area has doubled in the last 18 years. Approximate the annual rate of increase.

Magnitude of an Earthquake In Exercises 23–26, use the formula $R = \log \dfrac{A}{a}$ to solve each problem.

23. The amplitude of the September 19, 1985, earthquake in Mexico City was 63,100,000 times the reference amplitude. Calculate the magnitude of this earthquake on the Richter scale.

24. The amplitude of the September 20, 1965, earthquake in Mexico City was 20,000,000 times the reference amplitude. Calculate the magnitude of this earthquake on the Richter scale.

25. The damage done by an earthquake is related to its magnitude, the population of the area affected, and the quality of the buildings in this area. The August 17, 1999, earthquake that hit Izmit, Turkey, is rated by some (The Learning Channel, "Ultimate Natural Disasters") as the number 7 worst natural disaster to ever occur. At 7.4 on the Richter scale, it is the greatest earthquake to hit a major city since the 1906 San Francisco earthquake, which was estimated to be 8.25. Calculate how many times more intense the 1906 San Francisco earthquake was than the 1999 Izmit earthquake.

26. Calculate how many times more intense an earthquake with a Richter scale reading of 8.6 is than an earthquake with a Richter scale reading of 8.3.

Noise Levels In Exercises 27–30, use the formula $D = 10 \log \dfrac{I}{I_0}$ to solve each problem ($I_0 = 10^{-16}$ W/cm^2).

27. Find the number of decibels of a whisper if its intensity is 3×10^{-14} W/cm^2.

28. Find the number of decibels of city traffic if its intensity is 8.9×10^{-7} W/cm^2.

29. The noise level in a bar measures 85 dB. Calculate the intensity of this noise.

30. The decibel reading near a jet aircraft is 105 dB. Calculate the intensity of this noise.

pH Measurements In Exercises 31–34, use the formula $pH = -\log H^+$ to solve each problem.

31. Determine the pH of grape juice that has an H^+ concentration of 0.000109 mol/L.

32. Determine the pH of saccharin, a sugar substitute, which has an H^+ concentration of 4.58×10^{-7} mol/L.

33. A leading shampoo has a pH of 9.13. What is the H^+ concentration in moles per liter?

34. Blood is buffered (kept constant) at a pH of 7.35. What is the H^+ concentration in moles per liter?

The monthly payment P required to pay off a loan of amount A at an annual interest rate R in n years is given by the formula

$$P = \frac{A\left(\dfrac{R}{12}\right)}{1 - \left(1 + \dfrac{R}{12}\right)^{-12n}}$$

In Exercises 35 and 36, use this formula to calculate the monthly payment.

35. Loan Payments Determine the monthly payment necessary to pay off a $47,400 home loan that is financed for 30 years at 9.875%.

36. Loan Payments Determine the monthly payment necessary to pay off a $47,400 home loan that is financed for 30 years at 10.5%.

The number of monthly payments of amount P required to completely pay off a loan of amount A borrowed at interest rate R is given by the formula

$$n = -\frac{\log\left(1 - \dfrac{AR}{12P}\right)}{\log\left(1 + \dfrac{R}{12}\right)}$$

In Exercises 37 and 38, use this formula to calculate the number of monthly payments.

37. Car Payments Determine the number of monthly car payments of $253.59 required to pay off a $7,668.00 car loan when the interest rate is 11.7%.

38. Car Payments Determine the number of monthly car payments of $200.80 required to pay off a $7,668.00 car loan when the interest rate is 11.7%.

Newton's Cooling Law Newton's cooling law is a formula for calculating how much time it takes a hot object to cool when surrounded by a medium at some constant temperature. This formula can be applied to a hot steel ingot at a steel factory, to a heat source used to warm air, or to a murder victim at the scene of a crime. The formula is

$$t = k \ln \frac{b - r}{b_0 - r} \quad \text{where:}$$

k = a constant

t = time

r = room temperature

b = body temperature at the end of the time period

b_0 = original body temperature at the beginning of the time period

39. A coroner at a murder scene found the room temperature to be 22°C and the temperature of the victim to be 29°C. Assume that normal body temperature is 37°C and that the constant k for a body of the nature of the victim is -3.8065. Determine the number of hours the victim had been dead when these temperatures were taken.

40. Rework Exercise 39 assuming that the room temperature was 18°C.

In Exercises 41 and 42, use a graphing calculator and the points given in the table.

a. Draw a scatter diagram of these points.

b. Determine the exponential function that best fits these points. (Round the coefficients to three significant digits.)

c. Does this function represent exponential growth or exponential decay?

d. Use this equation to estimate y when x is 2.5.

41.

x	y
1	2.5
2	2.0
3	1.5
4	1.2
5	1.0
6	0.8

42.

x	y
1	6.5
2	9.5
3	14.5
4	21.5
5	32.5
6	50.0

43. Data Storage Requirements The Sweet Tooth Cookie Company recorded its data storage requirements in gigabytes (GB) for 5 years.

Years x After 2000	Storage Capacity y (GB)
0	40
1	60
2	90
3	150
4	200
5	300

a. Draw a scatter diagram for these data.

b. Determine the exponential function of best fit.

c. Use this function to estimate the storage requirements for 2006.

44. Revenue The revenue for wireless service is given for 6 years for one telecommunications company.

Years x After 2000	Revenue y ($1,000,000)
0	250
1	370
2	500
3	550
4	1,000
5	1,500

a. Draw a scatter diagram for these data.

b. Determine the exponential function of best fit.

c. Use this function to estimate the revenue for 2006.

Group discussion questions

45. Communicating Mathematically Which type of function, exponential or logarithmic, would provide a better model for the number of skills a dog acquires during the first 10 years of its life? Explain your answer.

46. Challenge Question Suppose that a bacteria culture doubles in volume every day. After 30 days, the population of these bacteria has grown to fill a laboratory jar. On what day was the jar half full of these bacteria?

11.7 | **Cumulative Review**

Simplify each complex number.

1. $5(4 - 3i)$

2. $(5 + 2i) - (4 - 3i)$

3. $(5 + 2i)(4 - 3i)$

4. $(5 + 2i)^2$

5. $\dfrac{5 + 2i}{4 - 3i}$

Chapter 11 | **Key Concepts**

1. Geometric Sequence

- A geometric sequence is a sequence with a constant ratio for any two consecutive terms. This common ratio is often denoted by r.
- If $r > 1$, the geometric sequence exhibits geometric growth.
- If $0 < r < 1$, the geometric sequence exhibits geometric decay.
- The graph of the terms of a geometric sequence consists of discrete points that lie on the graph of an exponential function.

2. Exponential Function $f(x) = b^x$

- If $b > 0$ and $b \neq 1$, then $f(x) = b^x$ is an exponential function with base b.
- If $b > 1$, then $f(x) = b^x$ is called an exponential growth function.
- If $0 < b < 1$, then $f(x) = b^x$ is called an exponential decay function.
- For $b \neq 1$, $b^x = b^y$ if and only if $x = y$.
- For $a, b > 0$, and $x \neq 0$, $a^x = b^x$ if and only if $a = b$.

3. Exponential Growth and Decay

- **Exponential Growth Function** If $b > 1$, then $f(x) = b^x$ exhibits exponential growth.

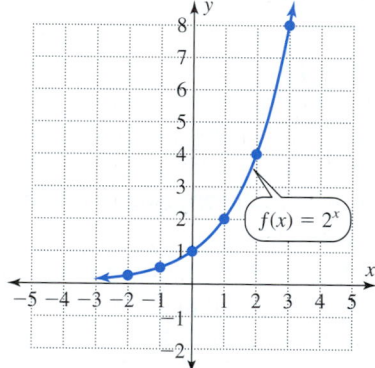

$f(x) = 2^x$

- **Exponential Decay Function** If $0 < b < 1$, then $f(x) = b^x$ exhibits exponential decay.

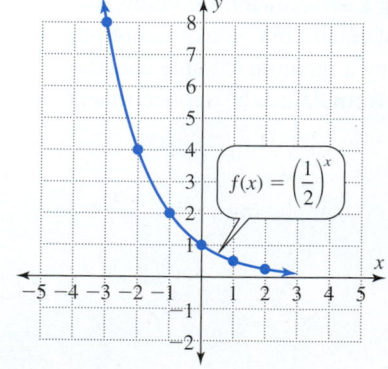

$f(x) = \left(\dfrac{1}{2}\right)^x$

4. Properties of the Graphs of Exponential Functions

For the exponential function $f(x) = b^x$ with $b > 0$ and $b \neq 1$:

- The domain of f is the set of all real numbers $\mathbb{R}$.
- The range of f is the set of all positive real numbers.
- There is no x-intercept.
- The y-intercept is $(0, 1)$.
- Key point: $(1, b)$
- The growth function is asymptotic to the negative portion of the x-axis (it approaches but does not touch the x-axis).
- The decay function is asymptotic to the positive portion of the x-axis.
- The growth function rises (or grows) from left to right.
- The decay function falls (or decays) from left to right.

5. The Irrational Constant e

$e \approx 2.718\ 281\ 828 \ldots$
Remember $e \approx 2.718$.

6. A One-to-One Function

- A function is one-to-one if different input values produce different output values.
- The vertical line test can be used to determine whether a graph represents a function.
- The horizontal line test can be used to determine whether the graph of a function represents a one-to-one function.

7. Horizontal Line Test The graph of a function represents a one-to-one function if it is impossible to have any horizontal line intersect the graph at more than one point.

8. Inverse of a Function

- If f is a function that matches each input value x with an output value y, then the inverse of f reverses this correspondence to match this y-value with the x-value.
- The inverse of a one-to-one function f is also a function. In this case, we call f^{-1} an inverse function.
- The graphs of f and f^{-1} are symmetric about the line $x = y$.

9. Finding an Equation for the Inverse of a Function

Step 1 Replace $f(x)$ by y. Write this as a function of y in terms of x.

Step 2 To form the inverse, replace each x with y and each y with x.

Step 3 If possible, solve the resulting equation for y.

Step 4 The inverse then can be written in function notation by replacing y with $f^{-1}(x)$.

10. Logarithmic and Exponential Functions

- Logarithmic and exponential functions are inverses of each other.
- For the logarithmic function $f(x) = \log_b x$, the input value x is called the argument of the function.

11. Logarithmic Function $f(x) = \log_b x$

- For $x > 0$, $b > 0$, and $b \neq 1$, $y = \log_b x$ if and only if $b^y = x$.
- Common logarithms: $\log x$ means $\log_{10} x$.
- Natural logarithms: $\ln x$ means $\log_e x$.
- $y = \log x$ is equivalent to $x = 10^y$.
- $y = \ln x$ is equivalent to $x = e^y$.

12. Properties of Logarithms For $x > 0$, $y > 0$, $b > 0$, and $b \neq 1$:

- $\log_b 1 = 0$
- $\log_b b = 1$
- $\log_b \dfrac{1}{b} = -1$
- $\log_b b^x = x$
- $b^{\log_b x} = x$
- Product rule: $\log_b xy = \log_b x + \log_b y$
- Quotient rule: $\log_b \dfrac{x}{y} = \log_b x - \log_b y$
- Power rule: $\log_b x^p = p \log_b x$

13. Logarithmic Growth Function If $x > 0$ and $b > 1$, then $f(x) = \log_b x$ exhibits logarithmic growth.

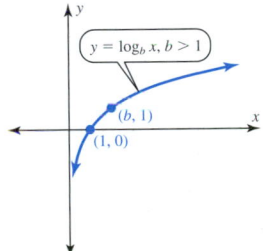

14. Properties of the Graphs of Logarithmic Functions, $f(x) = \log_b x$

For the logarithmic function $f(x) = \log_b x$ with $x > 0$, $b > 0$, and $b \neq 1$:

- The domain of f is the set of all positive real numbers $(0, \infty)$.
- The range of f is the set of all real numbers $\mathbb{R}$.
- The x-intercept is $(1, 0)$.
- There is no y-intercept.
- Key point: $(b, 1)$
- The growth function is asymptotic to the negative portion of the y-axis.
- The decay function is asymptotic to the positive portion of the y-axis.
- The growth function rises (or grows) from left to right, exhibiting extremely slow growth for large values of x.
- The decay function falls (or decays) from left to right, exhibiting extremely slow decay for large values of x.

15. Change-of-Base Formulas For $a, b > 0$ and $a, b \neq 1$:

- $\log_a x = \dfrac{\log_b x}{\log_b a}$ for $x > 0$

- $a^x = b^{x \log_b a}$

16. Extraneous Value An extraneous value is a value produced in the solution process that does not check in the original equation.

17. Solving Exponential Equations

Step 1 If it is obvious that both sides of an equation are powers of the same base, express each side as a power of this base and then equate the exponents. Otherwise, take the logarithm of both sides of the equation and use the power rule to form an equation that does not contain variable exponents.

Step 2 Then solve the equation formed.

18. Solving Logarithmic Equations

Step 1 If possible, rewrite the logarithmic equation in exponential form. Otherwise, use the properties of logarithms to write the two sides of the equation as two single logarithms with the same base. Form a new equation by equating the arguments of these logarithms.

Step 2 Then solve the equation formed.

Step 3 Check all possible solutions for extraneous values, since logarithms of 0 and negative arguments are undefined.

19. Growth and Decay Formulas
For $r > 0$ there is growth.
For $r < 0$ there is decay.

- Periodic growth decay formula: $A = P\left(1 + \dfrac{r}{n}\right)^{nt}$

- Continuous growth decay formula: $A = Pe^{rt}$

Chapter 11 | Review Exercises

In Exercises 1 and 2, determine whether each sequence is arithmetic, geometric, both, or neither. If the sequence is arithmetic, write the common difference d. If the sequence is geometric, write the common ratio r.

1. a. $2, 4, 6, 8, 10, \ldots$

  b. $2, 4, 8, 16, 32, \ldots$

  c. $2, 4, 6, 8, 10, 16, 26, \ldots$

2. a. $5, 5, 5, 5, 5, \ldots$

  b. $7, 3, -1, -5, -9, \ldots$

  c. $5, -5, 5, -5, 5, \ldots$

3. Dilution of a Mixture A tank contains 100 gal of a cleaning solvent. Then 10 gal is drained and replaced with pure water. The contents are throughly mixed. Then the process is repeated by draining 10 gal and replacing this mixture with pure water. Determine the volume of solvent in the tank after each of the first of five such drainings. 100, 90, 81, _____, _____,

_____

4. Use $f(x) = 9^x$ to evaluate each expression by inspection.
  a. $f(0)$ **b.** $f(-1)$ **c.** $f(2)$ **d.** $f\left(\dfrac{1}{2}\right)$

5. Use $f(x) = \pi^x$ and a calculator or a spreadsheet to approximate each expression to the nearest thousandth.
  a. $f(-1)$ **b.** $f(2)$ **c.** $f(\pi)$ **d.** $f(e)$

In Exercises 6–8, match each graph with the most appropriate description.

6. Not a function of x

7. A function of x but not a one-to-one function

8. A one-to-one function of x

A.   **B.**

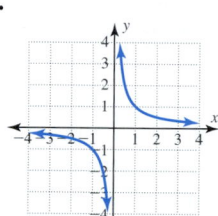   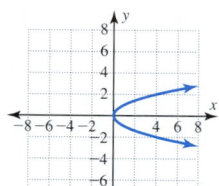

C.

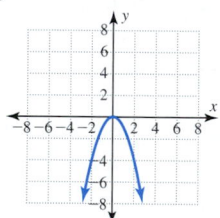

9. Graph each function and its inverse on the same coordinate system.
  a. $f(x) = \left(\dfrac{4}{3}\right)^x$ **b.** $f(x) = \ln x$

In Exercises 10 and 11, the function f is described verbally and algebraically. Describe f^{-1} both verbally and algebraically.

Verbally	**Algebraically**
10. $f(x)$ triples x and then subtracts 2.	$f(x) = 3x - 2$
11. $f(x)$ takes one-fourth of x and then adds 6.	$f(x) = \dfrac{x}{4} + 6$

In Exercises 12 and 13, write the inverse of each function by using ordered-pair notation.

12. $f = \{(-4, -5), (-3, -3), (0, 3), (2, 7), (3, 9)\}$

13.

x	-2	-1	0	1	2	3
y	0	2	4	6	8	10

In Exercises 14 and 15, write the equation of the inverse of each function.

14. $f(x) = \dfrac{1}{3}x - 4$

15. $f(x) = 3^x$

16. Product Costs The cost of producing x pizzas at a small pizzeria is given by the function
$C(x) = 2x + 250$.
   a. What does the variable x represent in $C(x)$?
   b. Determine the formula for $C^{-1}(x)$.
   c. What does the variable x represent in $C^{-1}(x)$?
   d. What does the output of $C^{-1}(x)$ represent?
   e. Determine the cost of producing 100 pizzas.
   f. Determine the number of pizzas that can be produced for $398.

In Exercises 17 and 18, graph the inverse of the given function.

17.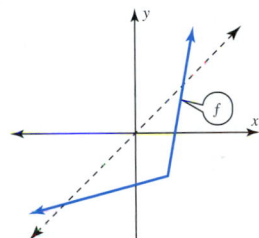

18.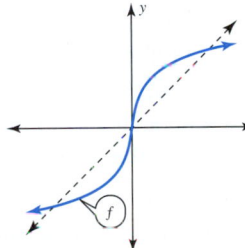

In Exercises 19 and 20, write each logarithmic equation in exponential form.

19. a. $\log_6 \sqrt{6} = \dfrac{1}{2}$ **b.** $\log_{17} 1 = 0$ **c.** $\log_8 \dfrac{1}{64} = -2$

20. a. $\log_b a = c$ **b.** $\ln a = c$ **c.** $\log c = d$

In Exercises 21 and 22, write each exponential equation in logarithmic form.

21. a. $7^3 = 343$ **b.** $19^{1/3} = \sqrt[3]{19}$ **c.** $\left(\dfrac{4}{7}\right)^{-2} = \dfrac{49}{16}$

22. a. $e^{-1} = \dfrac{1}{e}$ **b.** $10^{-4} = 0.0001$ **c.** $8^x = y$

In Exercises 23 and 24, evaluate each expression by inspection.

23. a. $\log_{12} 144$ **b.** $\log_{12} \sqrt{12}$ **c.** $\log_{12} \dfrac{1}{12}$

24. a. $\log_{12} 1$ **b.** $\log_{12} 12^{0.23}$ **c.** $\log_{12} \dfrac{1}{\sqrt[3]{12}}$

In Exercises 25–32, use a calculator or a spreadsheet to approximate each expression to the nearest thousandth.

25. a. $e^{7.83}$ **b.** $10^{7.83}$

26. a. $e^{-2.4}$ **b.** $10^{-2.4}$

27. a. $\log 113.58$ **b.** $\ln 113.58$

28. a. $\log(8.1 \times 10^{-4})$ **b.** $\ln(8.1 \times 10^{-4})$

29. a. $\dfrac{\ln 23}{\ln 19}$ **b.** $\ln \dfrac{23}{19}$

30. a. $\ln(6 + 5)$ **b.** $\ln 6 + \ln 5$

31. $\log_7 50$

32. $\log_4 63$

In Exercises 33–66, solve each equation using only pencil and paper. Note that some expressions are undefined.

33. $11^x = \dfrac{1}{121}$ **34.** $125^x = 25$

35. $2^y = \sqrt[3]{4}$ **36.** $\left(\dfrac{4}{9}\right)^y = \dfrac{3}{2}$

37. $2^{4w-1} = 8$ **38.** $9^{2v+1} = 27$

39. $2^{x^2-1} = 8$ **40.** $9^{x^2} = 3^{x+1}$

41. $\log_8 64 = z$ **42.** $\log_{16} 64 = x$

43. $\log_2 x = 2$ **44.** $\log_{13} x = \dfrac{1}{2}$

45. $\log_3 w = -2$ **46.** $\log_3(-2) = w$

47. $\log_{-2} 3 = y$ **48.** $\log_t 169 = 2$

49. $\log_t 8 = -3$ **50.** $5^{\log_5 11} = x$

51. $\log_5 5^{17} = x$ **52.** $\log_7 x = 1$

53. $\ln 0 = x$

54. $\log(3n - 4) = \log(2n - 1)$

55. $\log_3 20 + \log_3 7 = \log_3 y$

56. $\log(3x + 1) = 2$

57. $\log_3(x^2 - 19) = 4$

58. $\ln(w + 2) + \ln w = \ln 3$

59. $\ln(1 - w) + \ln(1 - 2w) = \ln(7 - 4w)$

60. $\log(5 - 2v) - \log(1 - v) = \log(3 - 2v)$

61. $\ln(5v + 3) = \ln(3v + 9)$

62. $\log_5 27 - \log_5 2 = \log_5 y$

63. $\log(2 - 6v) - \log(2 - v) = \log(1 - v)$

64. $\ln(5 - x) + \ln(x + 1) = \ln(3x - 1)$

65. $\log x + \log(x - 9) = 1$

66. $\log_2 x + \log_2(x - 2) = 3$

In Exercises 67–70, express each logarithm in terms of logarithms of simpler expressions. Assume that the argument of each logarithm is a positive real number.

67. $\log x^3 y^5$ **68.** $\ln \dfrac{7x - 9}{2x + 3}$

69. $\ln \dfrac{\sqrt{2x + 1}}{5x + 9}$ **70.** $\log \sqrt{\dfrac{x^2 y^3}{z}}$

In Exercises 71–74, combine the logarithmic terms into a single logarithmic expression with a coefficient of 1. Assume that the argument of each logarithm is a positive real number.

71. $2 \ln x + 3 \ln y$

72. $5 \ln x - 4 \ln y$

73. $\ln(x^2 - 3x - 4) - \ln(x - 4)$

74. $\frac{1}{2}(\ln x - \ln y)$

In Exercises 75 and 76, approximate the solution to each equation to the nearest thousandth.

75. $\log(5x - 2) + \log(x - 1) = \log 10$

76. $\ln(2w + 3) + \ln(w + 1) = \ln(w + 2)$

77. Convert $y = 5^x$ to an exponential function with base e.

78. Verify that $\log 50^x + \log 6^x - \log 3^x = 2x$ is an identity.

79. Verify that $1,000^{\log x} = x^3$ is an identity.

Estimate Then Calculate

In Exercises 80–83, mentally estimate the value of each expression to the nearest integer, and then use a calculator or a spreadsheet to approximate the value to the nearest thousandth.

Problem	Mental Estimate	Calculator Approximation
80. $\log 990$		
81. $\ln 3$		
82. $\log_3 30$		
83. $\log_5 120$		

84. Given $f(x) = 5x - 2$, evaluate each expression.

   a. $f(3)$ **b.** $f^{-1}(13)$ **c.** $\dfrac{1}{f(3)}$

In Exercises 85–89, match each graph with the most appropriate description.

85. A linear growth function

86. A logarithmic growth function

87. An exponential growth function

88. An exponential decay function

89. A constant function

A.

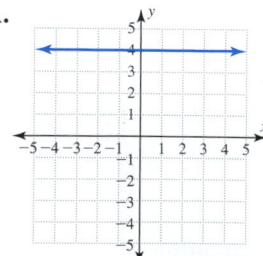

B.

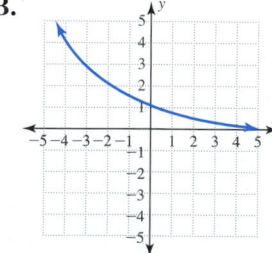

C.

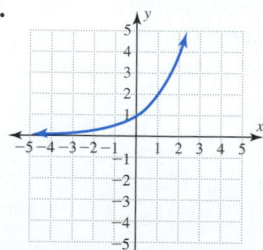

D.

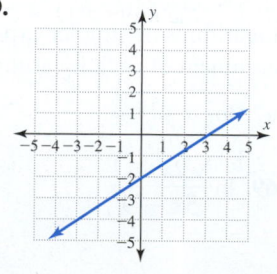

E.

90. Compound Interest A \$2,000 investment is invested at 8% for 5 years. Determine the value of the investment under each condition.
  a. The interest is compounded yearly.
  b. The interest is compounded monthly.
  c. The interest is compounded continuously.

91. Radioactive Decay The radioactive material used to power a satellite decays at a rate that decreases the available power by 0.045% per day. When the power supply reaches $\dfrac{1}{100}$ of its original level, the satellite is no longer functional. Approximately how many days should the power supply last?

92. Doubling Period for an Investment How many years will it take a savings bond to double in value if interest is compounded annually at 7.5%?

93. Rate of Interest If an investment on which interest is compounded continuously doubles in value in 8 years, what is the rate of interest?

94. Richter Scale The 2001 Kodiak Island, Alaska, earthquake had an amplitude A that was 6,310,000 times the reference amplitude a. Use the formula $R = \log \dfrac{A}{a}$ to calculate the magnitude of this earthquake on the Richter scale.

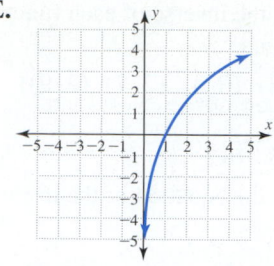

95. Modeling the Safe Storage Time for Milk The safe level of psychrotrophic bacteria in a gallon of milk is 100 units. In a refrigerator set at 40°F, the number of units of these bacteria in 1 gal of whole milk is approximated by the exponential function $B(t) = 3.0e^{0.29t}$, where t is the time in days. Evaluate and interpret each expression.
  a. $B(0)$ **b.** $B(5)$
  c. $B(10)$ **d.** $B(t) = 100, t = $ ____

Objective 11.1.1 Identify a Geometric Sequence

1. Determine whether each sequence is a geometric sequence. If the sequence is geometric, write the common ratio r.
   a. 1, 5, 25, 125, 625 **b.** 1, 5, 9, 13, 17
   c. 48, 24, 12, 6, 3 **d.** 0.9, 0.09, 0.009, 0.0009, 0.00009

Objective 11.1.2 Evaluate and Graph Exponential Functions

2. Graph each function.
   a. $f(x) = \left(\dfrac{5}{2}\right)^x$ **b.** $f(x) = \left(\dfrac{2}{5}\right)^x$

   Evaluate each expression, given $f(x) = 16^x$.

   c. $f(-1)$ **d.** $f\left(\dfrac{1}{2}\right)$

Objective 11.1.3 Solve Simple Exponential Equations

3. Solve each equation using only pencil and paper.
   a. $6^x = 36$ **b.** $6^x = 1$
   c. $6^x = \dfrac{1}{6}$ **d.** $6^x = \sqrt[5]{6}$

Objective 11.2.1 Write the Inverse of a Function

4. Write the inverse of each function. For parts **a, c,** and **d,** use ordered-pair notation.
   a. $\{(-1, 4), (8, 9), (-7, 11)\}$
   b. Write the equation for the inverse of $f(x) = 3x - 6$.
   c.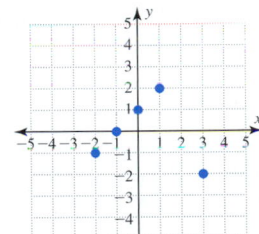
   d.

x	y
$\dfrac{1}{2}$	2
$\dfrac{1}{3}$	3
6	$\dfrac{1}{6}$
1	1

Objective 11.2.2 Identify a One-to-One Function

5. Determine whether each function is one-to-one.
   a. $\{(5, 7), (8, 3), (6, 9), (4, 2)\}$
   b. $\{(5, 7), (8, 3), (6, 7), (4, 2)\}$
   c.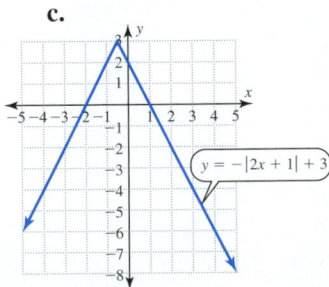
   d.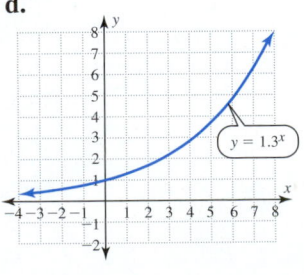

Objective 11.2.3 Graph the Inverse of a Function

6. Graph each function and its inverse on the same coordinate system.
   a. $f = \{(-3, 2), (-2, 1), (-1, 3), (2, 4)\}$
   b. $f(x) = \dfrac{3}{2}x + 2$

Objective 11.3.1 Interpret and Use Logarithmic Notation

7. Translate the following logarithmic equations to exponential form.
   a. $\log_5 \sqrt[3]{25} = \dfrac{2}{3}$
   b. $\log_5 \dfrac{1}{125} = -3$
   c. $\log_b (y + 1) = x$
   d. $\log_b x = y + 1$

Objective 11.3.2 Evaluate Simple Logarithmic Expressions

8. Determine the value of each logarithm by inspection.
   a. $\log_8 1$
   b. $\log_8 8$
   c. $\log_8 \dfrac{1}{8}$
   d. $\log_8 8^{17}$

Objective 11.3.3 Sketch the Graph of a Logarithmic Function

9. Complete this table and sketch the graph of $f(x) = \log_5 x$.

x	f(x)
0.2	
1.0	
5.0	
25.0	

Objective 11.4.1 Evaluate Common and Natural Logarithms with a Calculator or a Spreadsheet

10. Approximate each expression to the nearest ten-thousandth.
    a. $\log 19.1$
    b. $\ln 19.1$
    c. $\log(4.876 \times 10^{12})$
    d. $\ln(3.04 \times 10^{-5})$

Objective 11.5.1 Use the Product Rule, the Quotient Rule, and the Power Rule for Logarithms

11. Use the properties of logarithms to express each logarithm in terms of logarithms of simpler expressions. Assume that x and y are positive real numbers.
    a. $\log x^4 y^5$
    b. $\ln \dfrac{x^3}{\sqrt{y}}$

Use the properties of logarithms to combine these logarithmic terms into a single logarithmic expression with a coefficient of 1. Assume that the argument of each logarithm is a positive real number.

c. $2 \ln(7x + 9) - \ln x$

d. $\dfrac{1}{2} \log(x + 3) + \log x$

Objective 11.5.2 Use the Change-of-Base Formula to Evaluate Logarithms

12. Use the change-of-base formula for logarithms to approximate each expression to the nearest thousandth.

a. $\log_4 11$ **b.** $\log_\pi \sqrt{7}$

c. $\log_2 100$ **d.** $\log_\pi 5\pi$

Objective 11.6.1 Solve Exponential and Logarithmic Equations

13. Solve each equation using only pencil and paper.

a. $3^x = \dfrac{1}{81}$

b. $16^w = 64$

c. $\log_2 x = 4$

d. $\log_8 2 = y$

e. $\log(1 - 4t) - \log(5 + t) = \log 3$

f. $\ln(1 - z) + \ln(2 - z) = \ln(17 - z)$

Using a calculator or a spreadsheet, approximate the solution of each equation to the nearest thousandth.

g. $3^{4y+1} = 17.83$

h. $\ln(x + 1) + \ln(3x - 1) = \ln(6x)$

Objective 11.7.1 Use Exponential and Logarithmic Equations to Solve Applied Problems

14. a. Population Growth Assume that the population of a new space colony is growing continuously at a rate of 5% per year. Approximately how many years will it take the population to grow from 500 to 3,000?

b. Periodic Growth How many years will it take an investment to double in value if interest is compounded monthly at 8.25%?

Chapter 11 | Group Project

An Algebraic Model for Real Data

Supplies Needed for the Lab:

- A golf ball or other hard ball
- A meterstick or a metric tape measure
- A chair or a short stepladder
- Masking tape and a marking pen
- Hard, smooth floor area next to a wall
- A graphing calculator with an exponential regression feature (**ExpReg**)

Lab: Recording the Height of a Bouncing Golf Ball

Move to a location with a relatively high ceiling and a smooth concrete or other hard floor. (Tile with cracks can cause problems with the bounce of the golf ball.) Ask the tallest member of the class to get on a chair or small stepladder. Run a strip of masking tape from the base of the floor to the top of the student's reach. Stick this masking tape to the wall in a straight line. (Starting from a height of 250 to 300 cm will make the data collection easier.) Using your meterstick and the pen, label the heights on the masking tape every 10 cm, starting with 0 cm at floor level.

Placing the bottom of the golf ball level with your starting point at the top of the tape (250 to 300 cm high) and about 5 cm away from the wall, have one student drop the golf ball and have a second student note the approximate point that marks the height of the golf ball at the top of its first bounce. It would be wise to have a third student catch the golf ball at the beginning of the second bounce. Repeat this experiment now that you know approximately where to look, and have the second student mark the height of this bounce by using his or her finger. Place a pen mark on the tape at this height. Using the reference marks on the tape and the meterstick, record the height of this bounce to the nearest centimeter.

Use the height just recorded as the new release point, and repeat the experiment. You may need to repeat drops from the same height before you are comfortable that you have recorded accurately the height of the first bounce. Repeat this experiment until you have recorded the results for 8 to 10 bounces.

Bounce No. x	Height y (cm)
0	250
1	
2	
3	
4	
5	
6	
7	
8	
9	
10	

1. Record your data in a table similar to the one shown.

2. Use these data to create a scatter diagram by using an appropriate scale for each axis.

Creating an Algebraic Model for These Data

1. Using your graphing calculator, calculate the exponential function of best fit. This function will be of the form $f(x) = ab^x$.

2. Let $y = H(x)$ represent the curve of best fit that you just calculated.
   a. Evaluate and interpret $H(7)$. How does this value compare to the corresponding value in the table?
   b. Determine the value of x for which $H(x) = 100$. Interpret the meaning of this value.

3. In the function $f(x) = ab^x$ that models your data, what is the value of a? Interpret the meaning of this value in this application.

4. In the function $f(x) = ab^x$ that models your data, what is the value of b? Interpret the meaning of this value in this application.

This review is intended to help you make a realistic assessment of your areas of strength and weakness of the material covered in Chapters 6–11. A review of Chapters 1–5 is included on pages 453–456. A few questions in this review may cover topics not discussed at your school. You may wish to ask your instructor for topics that your school stresses.

Review of Chapter 6: Factoring Polynomials

1. a. Use the graph of $y = P(x)$ to factor
$P(x) = x^3 - 7x^2 + 7x + 15$.

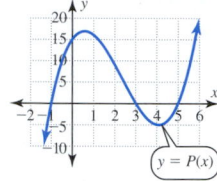

b. Use the table of values to factor
$P(x) = x^2 + 8x + 15$.

x	$P(x)$
-7	8
-6	3
-5	0
-4	-1
-3	0
-2	3
-1	8

2. Factor out the greatest common factor (GCF) from each polynomial.
a. $14x^3 - 35x^2$
b. $5x(2x - 3) - 6(2x - 3)$

3. Use grouping to factor each polynomial.
a. $2ax + 10bx + 3ay + 15by$
b. $8x^2 - 24x - 5x + 15$

4. Completely factor each trinomial.
a. $x^2 + 4x - 45$ **b.** $x^2 - 10x - 24$

5. Completely factor each trinomial.
a. $6x^2 - 17x - 10$ **b.** $6x^2 - 13x - 10$

6. Factor each perfect square trinomial.
a. $x^2 + 20x + 100$ **b.** $x^2 - 14xy + 49y^2$

7. Factor each binomial.
a. $x^2 - 64$ **b.** $x^3 - 64$

8. Use grouping to factor each polynomial.
a. $a^2 - b^2 + 5a - 5b$ **b.** $a^2 - x^2 + 10x - 25$

9. Completely factor each polynomial.
a. $5x^2 - 80$ **b.** $-3ax^2 + 21ax - 18a$

10. Solve each equation.
a. $2x^2 + 13x - 24 = 0$ **b.** $(x + 3)(x - 7) = 24$

Review of Chapter 7: Solving Quadratic Equations

11. Simplify each square root using the product rule for square roots.
a. $\sqrt{72}$ **b.** $\sqrt{500}$

12. Simplify each square root using the quotient rule for square roots.
a. $\sqrt{\dfrac{4}{25}}$ **b.** $\dfrac{\sqrt{32}}{\sqrt{8}}$

13. Simplify each expression by rationalizing the denominator.
a. $\dfrac{12}{\sqrt{3}}$ **b.** $\dfrac{\sqrt{3}}{\sqrt{5}}$

14. Solve each quadratic equation by using extraction of roots to obtain exact solutions. Approximate any irrational solutions to the nearest hundredth.
a. $(2x - 3)^2 = 7$ **b.** $(3x - 4)^2 = -9$

15. Solve each quadratic equation by completing the square to obtain exact solutions. Approximate any irrational solutions to the nearest hundredth.
a. $x^2 - 10x - 3 = 0$ **b.** $2x^2 - 5x = 6$

16. Use the discriminant to identify the nature of the solutions of each quadratic equation as distinct real solutions, a double real solution, or two imaginary solutions that are complex conjugates.
a. $4x^2 + 3x - 2 = 0$ **b.** $2x^2 - 6x - 1 = -8$
c. $4x^2 - 12x = -9$

17. Use the quadratic formula to solve each quadratic equation.
a. $4x^2 - 3x + 2 = 0$ **b.** $4x^2 - 36x + 81 = 0$
c. $3x^2 + 5x - 7 = 0$

18. Room Width Examining the blueprints for a rectangular room that is 14 feet longer than it is wide, an electrician determines that a wire run diagonally across this room will be 26 feet long. (See the figure.) What is the width of the room?

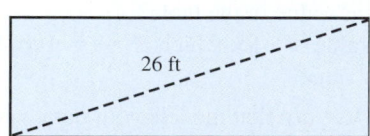

19. Write each complex number in standard $a + bi$ form.
a. $\sqrt{-36} + \sqrt{-64}$ **b.** $\sqrt{-100}$

20. Perform each operation and write the result in standard $a + bi$ form.
a. $(2 - 3i)(4 + 7i)$ **b.** $\dfrac{29 + 2i}{2 - 3i}$

Review of Chapter 8: Functions: Linear, Absolute Value, and Quadratic

21. Determine whether each relation is a function.

a.

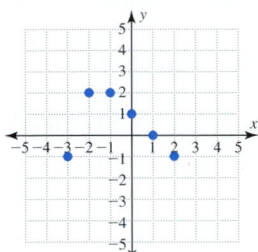

b.

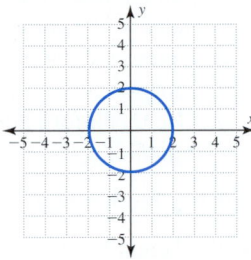

22. Determine the domain and range of each of these functions.

a. $f(x) = -3$

b.

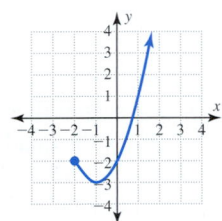

23. Evaluate each of the following expressions for $f(x) = 3x^2 + 2x - 4$.

a. $f(-2)$ **b.** $f(5)$

24. Use the given graph to determine the missing input and output values.

a. $f(0) = $ _____

b. $f(x) = 0; x = $ _____

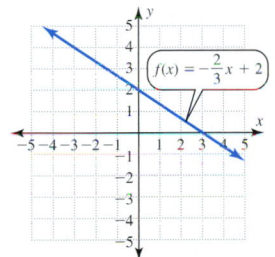

$f(x) = -\dfrac{2}{3}x + 2$

25. Use the given table of values to determine the missing input and output values.

a. $f(2) = $ _____

b. $f(x) = 2; x = $ _____

x	f(x)
-2	6
-1	2
0	-2
1	-6
2	-10
3	-14
4	-18

26. Use each equation to complete a table with input values of 1, 2, 3, 4, and 5, and then graph these points and the line through these points.

a. $y = 2x - 5$ **b.** $y = -3x + 4$

27. a. Determine the x- and y-intercepts of this line.

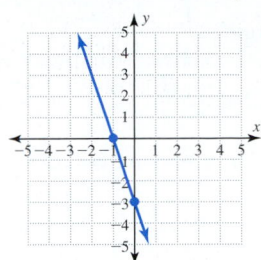

b. Determine the x- and y-intercepts of the line containing these points.

x	f(x)
-5	9
0	6
5	3
10	0
15	-3
20	-6
25	-9

c. Determine the x- and y-intercepts of the line defined by the equation $2x - 3y = 36$.

28. Ice Cream Shop Expenses The expenses required to run an ice cream shop include a fixed monthly cost of \$200 and a variable cost of \$0.25 per ice cream cone sold.

a. Write a function that gives the total cost of running this ice cream shop for a month when x ice cream cones are sold.

b. Use this function to complete this table.

x	0	50	100	150	200	250
f(x)						

c. Evaluate and interpret $f(75)$.

d. Determine the value of x for which $f(x) = 260$. Interpret this value.

29. a. Determine the slope of the line containing all the points in this table.

x	y
-3	5
0	1
3	-3
6	-7
9	-11
12	-15

b. Determine the slope of the line defined by
$$y - 5 = \frac{2}{3}(x - 3).$$

30. Write in slope-intercept form the equation of the line satisfying the given conditions.

a. y-intercept $(0, -2)$ and slope $-\dfrac{5}{3}$

b. through $(-2, 4)$ and perpendicular to $y = \dfrac{2}{3}x - 4$

31. Graph the line satisfying the given conditions.

  a. through $(0, 0)$ with slope $-\dfrac{1}{4}$

  b. y-intercept $(0, -4)$ and slope $\dfrac{7}{4}$

32. Identify each function as an increasing or a decreasing function.

  a. 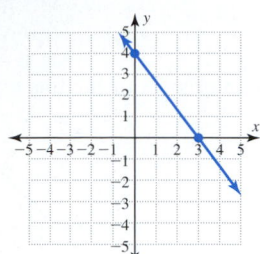 **b.** $f(x) = 4x - 5$

33. For the absolute value function $f(x) = |x - 2| - 3$:
  a. Complete the following table of values.

x	$f(x) = |x - 2| - 3$
-1	
0	
1	
2	
3	
4	
5	

  b. Graph this function.
  c. What are the x-intercepts of this graph?
  d. What is the y-intercept of this graph?
  e. What point is the vertex of this graph?
  f. What y-value is the minimum output value for this function?
  g. What is the domain of this function?
  h. What is the range of this function?
  i. For what interval of input values of x is the function decreasing?
  j. For what interval of input values of x is the function increasing?
  k. For what interval of input values of x is the function positive?
  l. For what interval of input values of x is the function negative?

34. Visually approximate the minimum value of y on this graph and give the x-value at which this minimum occurs.

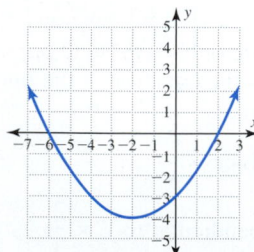

35. Use the quadratic function $f(x) = -x^2 + 2x + 8$ to complete each part of this exercise.
  a. Will the parabola open upward or open downward?
  b. Determine the coordinates of the vertex.
  c. Complete a table of values using inputs on both sides of the vertex.
  d. Determine the intercepts of the graph of this function.
  e. Use this information to sketch the graph of this function.

Applying Technology

36. The height in feet of a baseball after t seconds is given by the polynomial function $h(t) = -16t^2 + 60t + 3$. Use a calculator to determine the maximum height that this baseball reaches and the number of seconds after which this maximum height is reached. Round to the nearest tenth.

37. Draw a scatter diagram of the points in the table. Then determine the quadratic function that best fits these points. (Round the coefficients to three significant digits.) Use this function f to approximate $f(7)$.

x	-7	-3	-1	0	2	3	7	8
y	8	3	-1	-2	-3	-1	2	7

Review of Chapter 9: Rational Functions

38. Determine the domain of each rational function.

  a. $f(x) = \dfrac{x - 2}{(x - 1)(x + 3)}$ **b.** $f(x) = \dfrac{3x + 1}{x^2 + 4}$

39. Determine the vertical asymptotes of the graphs of each of these rational functions.

  a. $f(x) = \dfrac{x + 5}{(x - 3)(x + 4)}$

  b.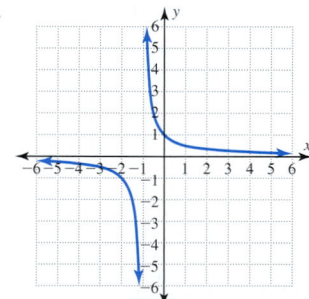

40. Reduce each rational expression to lowest terms.

  a. $\dfrac{4x}{12x^2 - 20x}$ **b.** $\dfrac{2x^2 - x - 15}{x^2 - 9}$

41. Perform each multiplication or division and reduce the result to lowest terms.

  a. $\dfrac{2x^2 - 6x}{10x^2} \cdot \dfrac{5x^2}{x - 3}$

  b. $\dfrac{x^2 - 16}{x^2 - 2x - 24} \div \dfrac{x^2 - 7x + 12}{x^2 - 3x - 18}$

42. Perform each addition or subtraction and reduce the result to lowest terms.

a. $\dfrac{5x + 4}{3x - 5} + \dfrac{x - 14}{3x - 5}$

b. $\dfrac{3x}{3x - y} - \dfrac{3x}{3x + y} - \dfrac{2y^2}{9x^2 - y^2}$

43. Simplify each expression.

a. $\dfrac{5x}{4} + \dfrac{x^3}{6} \cdot \dfrac{3}{2x^2}$

b. $\left(\dfrac{4}{5x + 25} + \dfrac{1}{5x - 25}\right) \div \dfrac{x^2 - x - 6}{x + 5}$

44. Simplify each expression.

a. $\dfrac{\dfrac{x}{4} - 2 + \dfrac{3}{x}}{1 - \dfrac{6}{x}}$

b. $\dfrac{\dfrac{3x}{x + 1} - \dfrac{2x}{x - 2}}{\dfrac{x^2 - 8x}{2x - 4}}$

45. Solve each of the following equations.

a. $\dfrac{x^2 - 13x - 10}{x^2 + 2x - 15} = \dfrac{2}{x + 5} + \dfrac{5}{3 - x}$

b. Solve $\dfrac{x - 3}{x + 2} = y$ for x.

46. a. If y varies inversely as x, and y is 2 when x is 12, find y when x is 4.

b. The volume of a gas varies inversely with the pressure when temperature is held constant. If the pressure of 6 L of a gas is 10 newtons/cm^2, what will the pressure be if the volume is 4 L?

47. Solve each of the following problems.

a. Work by Two Roofers When two members of a roofing crew work together, they can complete the roof on a shed in 2 hours. Working alone, the inexperienced worker could complete the roof in 3 more hours than it would take the more experienced worker. How many hours would it take the more experienced worker to complete the roof?

b. Airplanes Traveling in an Airstream Two planes departed from the same airport at the same time, flying in opposite directions. After a period of time, the slower plane has traveled 660 miles and the faster plane has traveled 780 miles. The faster plane is traveling 40 mi/h faster than the slower plane. Determine the rate of each plane. (*Hint:* What is the same for each plane?)

Review of Chapter 10: Square Root and Cube Root Functions and Rational Exponents

48. Match each function with its graph.

a. $f(x) = (x - 2)^2 + 1$ **b.** $f(x) = \dfrac{x}{2} + 1$

c. $f(x) = \sqrt{x - 2}$ **d.** $f(x) = \sqrt[3]{2 - x}$

e. $f(x) = |x - 2| - 1$

A.

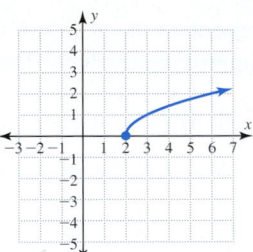

B.

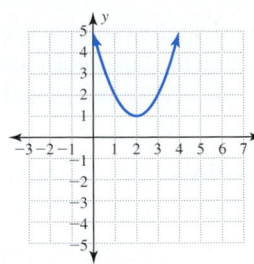

C.

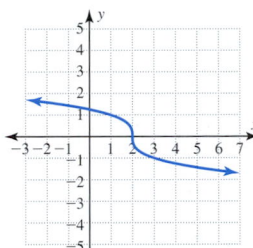

D.

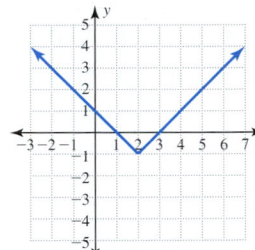

E.

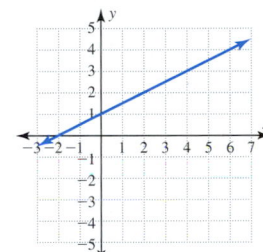

49. Algebraically determine the domain of each function.

a. $f(x) = 3x - 15$ **b.** $f(x) = \dfrac{2}{3x - 15}$

c. $f(x) = \sqrt[3]{3x - 15}$ **d.** $f(x) = \sqrt{3x - 15}$

50. Evaluate each expression without using a calculator.

a. $\sqrt{64}$ **b.** $\sqrt[3]{64}$

c. $\sqrt[3]{-0.125}$ **d.** $\sqrt[4]{16}$

51. Simplify each sum or difference using only pencil and paper. Assume $x > 0$.

a. $15\sqrt{3x} - 7\sqrt{3x}$ **b.** $5\sqrt{72} - 3\sqrt{8}$

52. Simplify each expression using only pencil and paper.

a. $\sqrt{48}$ **b.** $\sqrt[3]{48}$

53. Solve each equation.

a. $\sqrt{x + 8} = x - 4$ **b.** $\sqrt[3]{2x - 3} = 5$

54. Calculate the distance between each pair of points.

a. $(-2, 5)$ and $(3, -7)$ **b.** $(1, -3)$ and $(3, -7)$

55. Simplify each expression using only pencil and paper. Assume $x > 0$.

a. $-25^{1/2}$ **b.** $(x^{3/8}x^{1/8})^{2/3}$

c. $25^{3/2} - 9^{3/2}$ **d.** $(25 - 9)^{3/2}$

Applying Technology

56. Mentally estimate each expression to the nearest integer and then use a calculator to approximate each value to the nearest thousandth.

a. $\sqrt{26}$ b. $\sqrt[3]{26}$

Review of Chapter 11: Exponential and Logarithmic Functions

57. Determine whether each sequence is a geometric sequence. If the sequence is geometric, write the common ratio r.

a. 2, 10, 50, 250, 1250   b. 2, 5, 8, 11, 14

58. Graph each function.

a. $f(x) = 7^x$ b. $f(x) = \left(\dfrac{1}{5}\right)^x$

59. Evaluate each expression given $f(x) = 9^x$.

a. $f(-1)$ b. $f\left(\dfrac{1}{2}\right)$

60. Graph each function and its inverse on the same coordinate system.

a. $f(x) = 2x - 5$ b. $f(x) = 5^x$

61. Translate the following logarithmic equations to exponential form.

a. $\log_4 x = \dfrac{3}{2}$ b. $\log_3(y - 2) = x$

62. Determine the value of each logarithm by inspection.

a. $\log_4 1$ b. $\log_4 4$

c. $\log_4 \dfrac{1}{4}$ d. $\log_4 4^7$

63. Use the properties of logarithms to express each logarithm in terms of logarithms of simpler expressions. Assume that x and y are positive real numbers.

a. $\log x^5 y^3$ b. $\ln \dfrac{x^2}{\sqrt[3]{y}}$

64. Use the properties of logarithms to combine these logarithms into a single logarithmic expression with a coefficient of 1. Assume that the argument of each logarithm is a positive real number.

a. $2 \ln x - \ln(x - 5)$

b. $\dfrac{1}{2} \log x + \log(x + 2)$

Solve each equation using only pencil and paper.

65. a. $2^x = \dfrac{1}{32}$ b. $9^x = 27$

66. a. $\log_3 x = 4$ b. $\log_{64} 4 = y$

67. a. $\log(x - 4) + \log(x + 3) = \log 8$

b. $\ln(x - 3) + \ln(x - 7) = \ln(3 - x)$

68. Growth of an Investment How many years will it take a \$1,000 investment to increase in value to \$2,500 if interest is compounded continuously at 8.25%?

Applying Technology

69. Use a calculator to approximate each expression to the nearest thousandth.

a. $\log 25$ b. $\ln(2.89 \times 10^{-4})$

70. Use the change-of-base formula for logarithms and a calculator to approximate each expression to the nearest thousandth.

a. $\log_3 7$ b. $\log_5 26$

71. Bacteria Growth Monitoring a culture of 10,000 bacteria that has been isolated and allowed to multiply, a lab technician recorded the number of bacteria each day for seven days.

a. Draw a scatter diagram for these data points and determine the exponential function of best fit. (Round the coefficients to three significant digits.)

b. Use this function to estimate the number of bacteria in the culture after 10 days.

x (days)	y (bacteria, thousands)
1	18
2	32
3	56
4	100
5	178
6	316
7	562

Appendix A

Accuracy, Precision, Error, and Unit Conversion

Accuracy and Precision

Calculators will typically display from 8 to 12 digits, and it is tempting to copy all these digits when you are giving answers. However, it is not appropriate to copy digits that are not significant. Roughly speaking, the answer to a problem cannot be assumed to be more accurate than the measurements that produced that answer.

The **accuracy** of a number refers to the number of **significant digits.** For numbers written in scientific notation, all digits written are considered significant. For numbers obtained by measurement, the digits considered reasonably trustworthy are those that are significant.

The **precision** of a measurement refers to the smallest unit used in the measuring device and thus to the position of the last significant digit. The accuracy and precision of some sample measurements are given in the box.

Accuracy and Precision

Number of meters	Scientific Notation	Accuracy	Precision
0.045	4.5×10^{-2}	2 significant digits	thousandths of a meter
0.0450	4.50×10^{-2}	3 significant digits	ten-thousandths of a meter
45,000	4.5×10^{4}	2 significant digits	thousands of meters
123.45	1.2345×10^{2}	5 significant digits	hundredths of a meter
0.05	5×10^{-2}	1 significant digit	hundredths of a meter

There are many different rounding rules used for specific tasks and professions. All these rules are based on trying to report an answer that is reasonably trustworthy. Most of these rules are similar to the standardized rules given below.

Rounding Rules

- For calculations involving multiplication, division, or exponentiation, round the answer to the same number of significant digits as the measurement having the least number of significant digits.
- For calculations involving addition or subtraction, round the answer to the same precision as the least precise measurement.

Actual Error and Relative Error

Measurements, approximations, and estimations often introduce error into the reported results used by engineers, scientists, and statisticians. The **actual error** or **the error of the estimate** is the estimated value minus the actual value. **Relative error** is the actual error divided by the actual value. Many calculator errors are the result of keystrokes which produce incorrectly placed decimal points, missing digits, etc. These errors generate large relative errors and thus can often be detected by simply thinking about our results and asking ourselves if the results seem reasonable based on our mental estimate.

Example 1   Calculating the Relative Error

A contractor estimated the dimensions of the region shown to the right as 45 m by 60 m and calculated the approximate area. What was the relative error of this estimate?

44.1 m

58.7 m

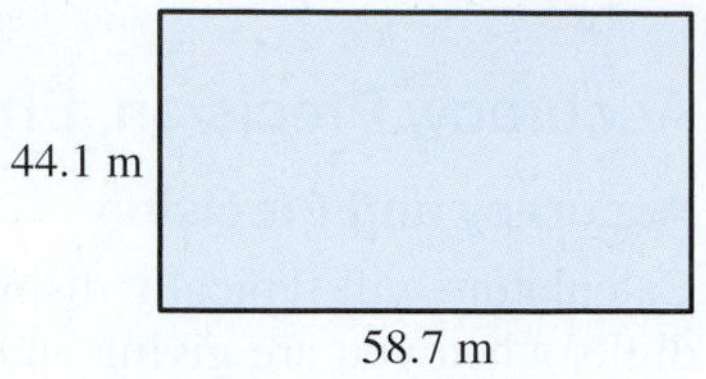

Solution

Actual area:

$A = l \cdot w$

$A = (44.1 \text{ m})(58.7 \text{ m})$

$A = 2{,}588.67 \text{ m}^2$

Use the formula for the area of a rectangle with the actual dimensions. The calculator value gives the area in square meters.

Estimated area:

$A = l \cdot w$

$A \approx (45 \text{ m})(60 \text{ m})$

$A \approx 2{,}700 \text{ m}^2$

Use the estimated values in the same formula.

$\boxed{\text{Error of estimate}} = \boxed{\text{Estimated value}} - \boxed{\text{Actual value}}$

$= 2{,}700 \text{ m}^2 \quad - \quad 2{,}588.67 \text{ m}^2$

$= +111.33 \text{ m}^2$

The estimated value is larger than the actual value.

$\boxed{\text{Relative error}} = \boxed{\text{Error of estimate}} \div \boxed{\text{Actual value}}$

$= 111.33 \text{ m}^2 \quad \div \quad 2{,}588.67 \text{ m}^2$

≈ 0.043

$\approx 4.3\%$

First, use a calculator to determine the relative error as a decimal and then convert to a percent.

Answer: The contractor's relative error was $+4.3\%$. Many contractors prefer to be a little high on their estimate so that they do not run short on materials.

Self-Check 1

Use the contractor's estimated dimensions from Example 1 and calculate the approximate perimeter of this rectangle. What is the relative error of this estimate?

Unit Conversion

To convert from one unit to another, use the method of multiplying by a fraction equal to 1. This fraction is called a conversion factor. Since 12 in is equal to 1 ft, we can write 12 in = 1 ft. If we divide both sides of this equation by 1 ft, we get $\dfrac{12 \text{ in}}{1 \text{ ft}} = 1$. This means $\dfrac{12 \text{ in}}{1 \text{ ft}}$ is a conversion factor equal to 1. Also, we could have divided both sides of the equation 12 in = 1 ft by 12 in to obtain the conversion factor $\dfrac{1 \text{ ft}}{12 \text{ in}}$. Since multiplying by 1 does not change the value of an expression, we can use conversion factors to change the units and preserve the value of a measurement.

Example 2   Using One Conversion Factor

Convert 3 ft to inches.

Solution

$$3 \text{ ft} = 3 \, \cancel{\text{ft}} \cdot \frac{12 \text{ in}}{1 \, \cancel{\text{ft}}} \qquad \text{Multiply 3 ft by the conversion factor } \frac{12 \text{ in}}{1 \text{ ft}}.$$

$$= 36 \text{ in} \qquad\qquad \text{Divide the common units of feet, leaving the desired units of inches.}$$

Self-Check 2

Convert 36 inches to centimeters using the conversion factor $\dfrac{2.54 \text{ cm}}{1 \text{ in}}$.

The table shows unit conversions for some basic units of length and time. Conversions for many other units can be found in science texts or on the Web. A search for "unit conversion" will find many useful websites for converting units. It is important to point out that the rounding of numbers can cause slight differences in these conversions.

Length	Time
1 mi = 5,280 ft	1 h = 60 min
1 yd = 3 ft	1 min = 60 s
1 ft = 12 in	
1 in = 2.54 cm	

Sometimes it is necessary to use more than one step to make a conversion. This is known as a chain of conversion factors. Example 3 shows a conversion of yards to centimeters by converting yards to feet, then feet to inches, and then inches to centimeters.

Example 3   Using a Chain of Conversion Factors

Convert 5 yd to centimeters.

Solution

$$5 \text{ yd} = 5 \, \cancel{\text{yd}} \cdot \frac{3 \, \cancel{\text{ft}}}{1 \, \cancel{\text{yd}}} \cdot \frac{12 \, \cancel{\text{in}}}{1 \, \cancel{\text{ft}}} \cdot \frac{2.54 \text{ cm}}{1 \, \cancel{\text{in}}}$$

$$= 457.2 \text{ cm}$$

Multiply 5 yd by the conversion factor $\dfrac{3 \text{ ft}}{1 \text{ yd}}$.

Divide the common units of yards leaving units of feet. Then multiply by the conversion factor $\dfrac{12 \text{ in}}{1 \text{ ft}}$ and divide the common units of feet, leaving units of inches. Finally, multiply by the conversion factor $\dfrac{2.54 \text{ cm}}{1 \text{ in}}$ and divide the common units of inches, leaving the desired units centimeters.

Self-Check 3

Convert 440 yards to miles.

Example 4 illustrates how to convert units in the numerator and in the denominator.

Example 4   Converting Units in the Numerator and Denominator

Convert 110 ft/s to miles per hour.

Solution

$$\frac{110 \text{ ft}}{\text{s}} = \frac{110 \text{ ft}}{1 \text{ s}} \cdot \frac{1 \text{ mi}}{5{,}280 \text{ ft}} \cdot \frac{60 \text{ s}}{1 \text{ min}} \cdot \frac{60 \text{ min}}{1 \text{ h}}$$

$$= \frac{396{,}000 \text{ mi}}{5{,}280 \text{ h}}$$

$$= 75 \text{ mi/h}$$

Write the original problem as $\frac{110 \text{ ft}}{1 \text{ s}}$. To convert the units in the numerator to miles, multiply by the conversion factor $\frac{1 \text{ mi}}{5{,}280 \text{ ft}}$. Divide the common units of feet, leaving the desired units of miles in the numerator. To convert the units in the denominator to hours, we use a chain of conversion factors. Multiply by the conversion factor $\frac{60 \text{ s}}{1 \text{ min}}$ and divide the common units of seconds, leaving units of minutes in the denominator. Then, multiply by the conversion factor $\frac{60 \text{ min}}{1 \text{ h}}$ and divide the common units of minutes, leaving the desired units of hours in the denominator. Multiply the values in the numerators, multiply the values in the denominators, and simplify the result.

Self-Check 4

Convert 10 miles per hour to centimeters per second.

Self-Check Answers

1. Perimeter $\approx$ 210 m. The actual perimeter is 205.6 m. The relative error of the estimate was $+2.1\%$.

2. $36'' = 91.44$ cm

3. $440 \text{ yd} = 0.25$ mi

4. $\frac{447.04 \text{ cm}}{\text{s}}$

Student's Answer Appendix

Answers to Selected Exercises

Chapter 1

Exercises 1.1
1–5. Answers can vary. **7.** 2 **9.** Answers can vary. **11.** self **13.** lecture
15. MathZone **17.** tutor **19–21.** Answers can vary.

Using the Language and Symbolism of Mathematics 1.2
1. variable **2.** opposite **3.** 0 **4.** a **5.** seven; left **6.** absolute value; distance
7. greater; equal **8.** less; left **9.** equal **10.** approximately equal
11. square root **12.** rational **13.** rational **14.** irrational **15.** pi **16.** infinity
17. is not **18.** is **19.** $\mathbb{R}$ **20.** $[-3, 2]$ **21.** (a, b)

Quick Review 1.2
1. 8 **2.** 7 **3.** 5 **4.** 4 **5.** 9,900

Exercises 1.2
1. a. -9 **b.** 13 **c.** $\dfrac{3}{5}$ **d.** -7.23 **e.** $\sqrt{13}$ **f.** 0 **3. a.** 7 **b.** -7 **c.** 7 **5. a.** 0 **b.** 0
c. 0 **d.** 0 **7. a.** 0 **b.** $6x$ **c.** 0 **d.** $-6x$ **9. a.** 29 **b.** 29 **c.** -29 **d.** -29
11. a. $>$ **b.** $<$ **13. a.** $<$ **b.** $>$ **15. a.** $>$ **b.** $>$
17. x is greater than 1;  ; $(1, \infty)$

19. $x \geq -3$; 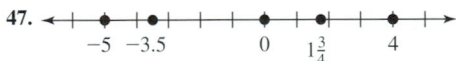 ; $[-3, \infty)$

21. $-3 < x \leq 3$; x is greater than -3 and less than or equal to 3; $(-3, 3]$
23. $x \leq 4$; x is less than or equal to 4;

25. a. 5 **b.** -5 **c.** 0.5 **d.** 50 **27.** B **29.** C **31.** 7; $7 < \sqrt{50.6}$; 7.113
33. 1; $1 > \sqrt{0.976}$; 0.988 **35. a.** 15 **b.** 0, 15 **c.** -11, $-\sqrt{9}$, 0, 15
d. -11, -4.8, $-\sqrt{9}$, 0, $1\dfrac{3}{5}$, 15 **e.** $\sqrt{5}$ **37. a.** $\sqrt{1}$, $\sqrt{16}$ **b.** $\sqrt{0}$, $\sqrt{1}$, $\sqrt{16}$
c. $-\sqrt{4}$, $\sqrt{0}$, $\sqrt{1}$, $\sqrt{16}$ **d.** $-\sqrt{4}$, $-\sqrt{\dfrac{9}{25}}$, $\sqrt{0}$, $\sqrt{1}$, $\sqrt{16}$ **e.** $-\sqrt{5}$, $\sqrt{6}$

39. integers, rational numbers **41.** natural numbers, whole numbers,
integers, rational numbers **43.** irrational numbers **45.** rational numbers

47.
![number line]

49. a. rational **b.** irrational **51. a.** rational **b.** irrational **53.** $20\% = \dfrac{1}{5}$
55. $|x| = y$ **57.** $\sqrt{x} = y$ **59.** $3.14 < \pi < 3.15$
61. a.

b.

c.
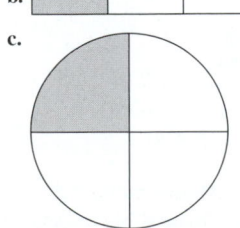

63. 7 **65.** 3 **67. a.** 0 **b.** -1, -2 **c.** $-\dfrac{1}{2}$, $-\dfrac{4}{7}$ **d.** 1.2, 1.5 (Answers will vary
for **b–d.**)

69.

x	$\sqrt{x}$
2	1.41
3	1.73
4	2.00
5	2.24

Group Discussion Questions
71. a. 16 **b.** 1 **c.** $\dfrac{1}{4}$ **d.** If $x > 1$, then $\sqrt{x} < x$.
  If $x = 0$ or $x = 1$, then $\sqrt{x} = x$.
  If $0 < x < 1$, then $\sqrt{x} > x$.
73. $+\sqrt{19}$; $-\sqrt{19}$ **75. a.** If a student answers, "Just drop the sign," you
may want to discuss what he or she would do with $-x$ for both negative
and positive values of x. **b.** $|0| = 0$, which is not positive.

Using the Language and Symbolism of Mathematics 1.3
1. a **2.** b **3.** parts; parts **4.** factor **5.** 1 **6.** prime **7.** reciprocal **8.** $\dfrac{11}{6}$
9. denominator **10.** least **11.** proper **12.** improper **13.** mixed **14.** 1 **15.** 1

Quick Review 1.3
1. 19 **2.** $5 \cdot 7$ **3.** $2 \cdot 3 \cdot 3$ **4.** $3 \cdot 3 \cdot 5$ **5.** $2 \cdot 2 \cdot 2 \cdot 11$

Exercises 1.3
1. $\dfrac{3}{7}$ **3.** $\dfrac{1}{4}$ **5.** $\dfrac{2}{9}$ **7.** $\dfrac{2}{3}$ **9.** 5 **11.** $\dfrac{1}{9}$ **13. a.** $\dfrac{11}{5}$ **b.** $\dfrac{5}{11}$ **15. a.** $\dfrac{1}{6}$ **b.** 6 **17.** $\dfrac{15}{56}$
19. $\dfrac{3}{14}$ **21.** $\dfrac{8}{45}$ **23.** 9 **25.** $\dfrac{8}{125}$ **27.** $\dfrac{2}{11}$ **29.** $\dfrac{21}{40}$ **31.** $\dfrac{1}{16}$ **33.** $\dfrac{64}{81}$ **35.** 216
37. 98 **39.** $\dfrac{50}{3}$ **41.** $\dfrac{5}{7}$ **43.** $\dfrac{1}{2}$ **45.** $\dfrac{5}{6}$ **47.** $\dfrac{32}{63}$ **49.** $\dfrac{11}{12}$ **51.** 15 **53.** 16 **55.** 45
57. $\dfrac{5}{8}$ **59.** $\dfrac{23}{30}$ **61.** $\dfrac{1}{2}$ **63.** $\dfrac{32}{63}$ **65.** $\dfrac{8}{5}$ **67.** $6\dfrac{5}{6}$ **69.** $1\dfrac{11}{12}$ **71.** 16 **73.** $12\dfrac{1}{2}$ **75.** $1\dfrac{3}{32}$
77. $\dfrac{3}{8}$ **79.** 62 lb **81.** 48 rose bushes **83.** $\dfrac{23}{24}$ **85.** $\dfrac{31}{40}$ **87.** 11 pieces of pipe

Group Discussion Questions
89. $\dfrac{1}{2}$

Cumulative Review
1. $\dfrac{5}{8}$ **2.** -4, -3, -2, -1, 0, 1, 2, 3, 4 **3.** 1, 2, 3, 4 **4.** $(-2, 3]$ **5.** 2

Using the Language and Symbolism of Mathematics 1.4
1. terms; sum **2.** commutative **3.** associative **4.** same **5.** larger
6. commutative; associative **7.** difference **8.** opposite **9.** $x + 5$ **10.** $x - 5$
11. $5 - x$ **12.** $w - 17$ **13.** $x_2 - x_1$ **14.** $y_2 - y_1$ **15.** $x + y + z$

Quick Review 1.4
1. 20 **2.** 6 **3.** -15.12 **4.** -7.35 **5.** $\dfrac{24}{35}$

Exercises 1.4
1. a. 13 **b.** -3 **c.** 3 **d.** -13 **3. a.** 0 **b.** 0 **5.** -21 **7.** -19 **9.** 1.29 **11.** $\dfrac{4}{11}$
13. $-\dfrac{13}{12}$ **15.** $\dfrac{87}{20}$ **17.** -18 **19.** -3 **21.** commutative **23.** $(11 + 12) + 13$
25. a. -6 **b.** 20 **c.** -20 **d.** 6 **27. a.** 0 **b.** 0 **29.** -39
31. 3 **33.** $-\dfrac{7}{3}$ **35.** $\dfrac{43}{30}$ **37.** 7.1 **39.** $-\dfrac{223}{20}$ **41.** Q1: \$12,000; Q2: \$2,000;
Q3: $-\$2,000$; Q4: $-\$12,000$ **43.** $+9°F$ **45.** $-7°F$ **47. a.** $+6°$ **b.** $-6°$

c. $+10°$ **d.** $-6°$ **49.** 32 million units **51.** -14 billion dollars **53.** 32 hours **55.** $+5.5$ yd **57. a.** 4 **b.** 12 **c.** 12 **d.** 24 **e.** 6 **f.** yes **59.** 70 cm

61. 76.6 m **63.** D **65.** C **67.** $+$; 5.8 **69.** $-$; $-\dfrac{13}{8}$ **71.** $b + c = a$

73. $8 + (-25) = -17$ **75. a.** $a + b = c$ **b.** $a - b = c$ **77. a.** $z - y = x$
b. $b - a = c$ **79. a.** $m = n + 11$ **b.** $m > n$ **81. a.** $(-1) + 5 = 4$; The sum of negative one and five is four. **b.** The sum of negative three and five is two.

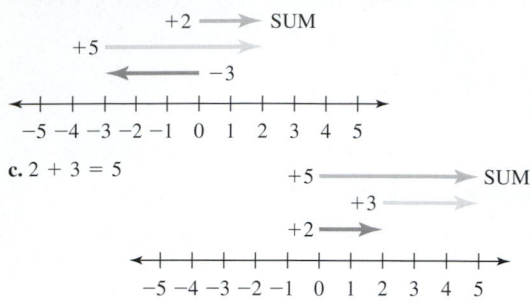

c. $2 + 3 = 5$

83. Answers can vary. **85. a.** -2 **b.** -3 **c.** -50; add 50 pairs each with a sum of -1

Cumulative Review
1. 34 **2.** -1 **3.** $-1, -2, -3; -6$ **4.** $-\dfrac{5}{7}$ **5.** $(-2, 3]$

Using the Language and Symbolism of Mathematics 1.5
1. factors; product **2.** commutative **3.** associative **4.** 0 **5.** multiplied
6. positive **7.** negative **8.** positive **9.** negative **10.** 60 **11.** inverses
12. 1 **13.** 0 **14.** 1 **15.** quotient **16.** undefined **17.** positive **18.** negative
19. ratio **20.** reciprocal **21.** sum **22.** difference **23.** product **24.** quotient

Quick Review 1.5
1. $\dfrac{2}{5}$ **2.** $\dfrac{5}{4}$ **3.** $\dfrac{4}{35}$ **4.** $\dfrac{8}{15}$ **5.** $\dfrac{9}{25}$

Exercises 1.5
1. commutative **3. a.** B **b.** A **5. a.** A **b.** B **7. a.** -77 **b.** -77 **c.** 77
d. -77 **9. a.** -60 **b.** 60 **c.** -60 **d.** 0 **11. a.** -123.4 **b.** $-123,400$
c. -1.234 **d.** $-1,234,000$ **13. a.** $\dfrac{3}{10}$ **b.** $-\dfrac{3}{10}$ **15. a.** 1 **b.** -1 **17. a.** -8
b. 8 **c.** -8 **d.** 0 **19. a.** -96 **b.** -24 **c.** 96 **d.** 24 **21. a.** 1,230 **b.** $-12,300$
c. -0.123 **d.** 12.3 **23. a.** 4 **b.** -4 **c.** $\dfrac{1}{9}$ **d.** $-\dfrac{2}{3}$ **25. a.** 0 **b.** undefined
c. undefined **d.** undefined **27.** 2:5 **29.** 2:3 **31.** 1:4 **33.** 3:1 **35. a.** 1:5
b. 1:4 **37.** 8:5 **39.** \$14.40 **41. a.** \$50 **b.** 25% **43.** \$64; \$32; \$75; \$150; total cost: \$321 **45.** 85.2 **47.** 1,260 miles **49.** 290 mi/h **51.** 300 lightbulbs/h
53. 16 cm² **55.** 3,360 in³ **57.** A **59.** C **61.** $4x$
63. $(-4) + (-4) + (-4) + (-4) + (-4)$ **65.** $ab = ba$ **67.** $3x < 9$

69. $\dfrac{a}{b} \geq c$ **71.** $2n$

73.

x	y	x + y	xy
5	8	13	40
−5	8	3	−40
5	−8	−3	−40
−5	−8	−13	40

75. The symbol * is available on all computer keyboards. The other symbols can be confused with other meanings and are not available on all keyboards. **77.** 0 **79.** The expression should be typed: (2/5)/(3/8)

Cumulative Review
1. a. $x + y = y + x$ **b.** The commutative property of addition states that the order of the terms does not change the sum. **2.** $7\dfrac{24}{35}$ **3.** 0 **4.** B **5. a.** -4 **b.** $\dfrac{1}{4}$

Using the Language and Symbolism of Mathematics 1.6
1. 4 **2.** x^7 **3.** fifth **4.** factor **5.** $-x$ **6.** x **7.** 1 **8.** multiplication
9. subtraction **10.** exponentiation **11.** x **12.** $3x$ **13.** coefficient **14.** 1
15. distributive **16.** distributive **17.** combining

Quick Review 1.6
1. 81 **2.** -81 **3.** 16 **4.** -16 **5.** 8

Exercises 1.6
1. a. 5^4 **b.** $(-4)^3$ **c.** y^5 **d.** $(3z)^6$ **3. a.** $4 \cdot 4 \cdot 4$ **b.** $(-3)(-3)(-3)(-3)$
c. $(-x)(-x)(-x)$ **d.** $(2y)(2y)(2y)$ **5. a.** 3; 4; 3 · 3 · 3 · 3 **b.** 4; 3; 4 · 4 · 4
7. a. x; 2; $-3 \cdot x \cdot x$ **b.** $-3x$; 2; $(-3x)(-3x)$ **9. a.** 9 **b.** -9 **11. a.** 0
b. 1 **13. a.** 0.001 **b.** $-1,000$ **15. a.** $\dfrac{1}{8}$ **b.** $-\dfrac{1}{8}$ **17. a.** 21 **b.** 56 **19. a.** 2 **b.** 70
21. a. 100 **b.** 52 **23. a.** 25 **b.** 1 **25. a.** -19 **b.** 0 **27. a.** -45 **b.** 9
29. a. 49 **b.** 25 **31. a.** -27 **b.** -21 **33. a.** 0 **b.** 12 **35. a.** 117 **b.** 27 **37. a.** 33
b. 23 **39. a.** $\dfrac{1}{16}$ **b.** $\dfrac{5}{12}$ **41. a.** 7 **b.** 5 **43. a.** -3 **b.** 3 **45. a.** 40 **b.** 6 **47.** 16
49. 158 **51.** $\dfrac{1}{2}$ **53. a.** $7x + 35$ **b.** $-7x - 35$ **55. a.** $-2x + 3y$ **b.** $-2x + 3y$
57. a. $7x$ **b.** $-3x$ **59. a.** $6a - 2b$ **b.** $-2a + 8b$ **61. a.** $z \cdot z \cdot z \cdot z$
b. $4z$ **63. a.** $5(x + 3y)$ **b.** $5x + 3y$ **65. a.** $(2xy)^5$ **b.** $2xy^5$ **67. a.** x squared plus y squared **b.** the square of the quantity x plus y **69.** $2(5 - (8 - 3))$
71. $(5 - 7)/(11 - 7)$ **73.** $\dfrac{x + 5y^3}{x^2 + 9y}$ **75.** $\dfrac{w + x}{y - z}$ **77.** D **79.** $-x - 19$
81. $13x - 14$ **83.** 5^4 means three multiplications are performed with four factors of 5 **85. a.** D **b.** C **c.** A **d.** B **e.** F **f.** E

Cumulative Review
1. $\dfrac{2}{5}$ **2.** $1\dfrac{31}{35}$ **3.** Division and subtraction are not commutative.
$\left(\dfrac{3}{5} \neq \dfrac{5}{3}; 5 - 3 \neq 3 - 5 \right)$ **4. a.** $<$ **b.** $>$ **5.** $>$; $\geq$; $\neq$; $\approx$

Using the Language and Symbolism of Mathematics 1.7
1. evaluate **2.** equation **3.** equation **4.** solution **5.** satisfy **6.** checking
7. solution **8.** variable **9.** subscript **10.** x_n **11.** sequence

Quick Review 1.7
1. 0.23 **2.** 0.014 **3.** 0.0004 **4.** 12.5% **5.** 140%

Exercises 1.7
1. 9 **3.** 153 **5.** 41 **7.** -16 **9.** -2 **11.** -10 **13.** 7 **15.** 4 **17.** $-\dfrac{11}{4}$ **19.** $-\dfrac{11}{40}$
21. \$280 **23.** \$595 **25.** 1,800 mi **27.** 22 m **29.** 75 mL **31.** 0.675 L **33.** 3
35. $-\dfrac{6}{7}$ **37.** $-2, 0, 2$ **39.** \$1,400, \$1,800, \$2,200 **41.** 4 is not a solution;
5 is a solution **43.** 4 is a solution; 5 is not a solution **45.** 4 is a solution; 5 is not a solution **47.** -2 is a solution; 8 is not a solution **49.** -2 is a solution;
8 is a solution **51.** -2 is not a solution; 8 is a solution **53.** $\dfrac{1}{2}$ is a solution;
$\dfrac{3}{5}$ is not a solution **55.** $\dfrac{1}{2}$ is a solution; $\dfrac{3}{5}$ is a solution **57.** $C = 13.4\pi$ cm;
$C \approx 42.1$ cm **59.** $A \approx 54$ cm²; $A = 53.68$ cm² **61.** 1,375.38975
63. 61,099.5 **65. a.** \$100 **b.** 25%

Group Discussion Questions
67. The student should write: $(-2)^2 - 5(-2) + 8 = 4 + 10 + 8 = 22$
69. The amount of alcohol in each drink is
Beer: $(12)(0.05) = 0.6$ oz
Wine: $(4)(0.15) = 0.6$ oz
Shot: $(1.5)(0.40) = 0.6$ oz
Each drink has the same amount of alcohol.

Cumulative Review
1. a. $\dfrac{5}{4}$ **b.** $1\dfrac{1}{4}$ **c.** 125% **2.** $|-6.35|$ **3.** $\dfrac{4}{15}$ **4.** -15 **5.** 81

Using the Language and Symbolism of Mathematics 1.8
1. measurement **2.** polygon **3.** triangle **4.** quadrilateral **5.** pentagon
6. hexagon **7.** octagon **8.** rhombus **9.** parallelogram **10.** rectangle
11. square **12.** regular **13.** perimeter **14.** circumference **15.** 3.14
16. area **17.** volume **18.** angles **19.** equal **20.** acute **21.** right **22.** obtuse
23. straight **24.** complementary **25.** supplementary **26.** adjacent
27. vertical **28.** transversal **29.** corresponding **30.** alternate **31.** exterior

Quick Review 1.8
1. diameter **2.** nonrepeating **3.** $A = 60$ **4.** $V = 125$ **5.** $C \approx 34.56$

Exercises 1.8
1. J **3.** D **5.** B **7.** I **9.** G **11.** 47 cm **13.** 30.2 cm **15. a.** 6.8π cm
b. 21.4 cm **17.** 8,250 m² **19.** 28.3 m² **21.** 18π m² **23.** 6,400 ft³
25. a. 14,137 ft³ **b.** 105,700 gal **27.** C **29.** I **31.** B **33.** J **35.** E
37. 144° **39.** 144° **41.** 36° **43.** 140° **45.** 159° **47.** D **49.** C

Group Discussion Questions
51. a. yes **b.** yes **c.** yes **d.** yes **e.** yes **53. a.** square **b.** hexagon **c.** As the number of sides increases, the area of the polygon will approach the area of the circle.

Cumulative Review
1. a. D **b.** C **c.** B **d.** A **2.** 25 **3.** 73 **4.** -192 **5.** 576

Review Exercises for Chapter 1
1. a. -12 **b.** -20 **c.** -20 **d.** -12 **2. a.** -64 **b.** 64 **c.** -4 **d.** 4
3. a. -7 **b.** 0 **c.** 0 **d.** undefined **4. a.** 18 **b.** 30 **c.** -144 **d.** -4
5. a. 9.01 **b.** 8.99 **c.** 0.09 **d.** 900 **6. a.** 995.5 **b.** $-1,004.5$ **c.** $-4,500$
d. -0.0045 **7. a.** $\dfrac{17}{12}$ **b.** $-\dfrac{1}{12}$ **c.** $\dfrac{1}{2}$ **d.** $\dfrac{8}{9}$ **8. a.** $-\dfrac{7}{75}$ **b.** $-\dfrac{133}{75}$ **c.** $-\dfrac{98}{125}$
d. $-\dfrac{10}{9}$ **9. a.** 0 **b.** 0 **c.** -1 **d.** 1 **10. a.** -6 **b.** -24 **11. a.** -24 **b.** 6
12. a. -18 **b.** 27 **13. a.** -27 **b.** -3 **14. a.** 100 **b.** -100 **15. a.** 49
b. 25 **16. a.** 25 **b.** 32 **17. a.** 8 **b.** -8 **18. a.** 14 **b.** 10 **19. a.** 24
b. -120 **20.** -44 **21.** 22 **22.** 79 **23.** -2 **24.** -7 **25.** $\dfrac{3}{5}$ **26.** $\dfrac{1}{3}$
27. B **28.** A **29.** D **30.** E **31.** C **32.** B **33.** C **34.** D **35.** A **36.** C
37. A **38.** D **39.** B **40. a.** $\sqrt{9}$, 15 **b.** 0, $\sqrt{9}$, 15 **c.** -7, 0, $\sqrt{9}$, 15
d. -8.1, -7, 0, $\sqrt{9}$, $\dfrac{15}{2}$, 15 **e.** π **41.** $-x = 11$ **42.** $|x| \le 7$
43. $\sqrt{26} > 5$ **44.** $x + 5 = 4$ **45.** $-3y = -1$ **46.** $\dfrac{x}{3} = 12$ **47.** $\dfrac{x}{y} = \dfrac{3}{4}$
48. $5(3x - 4) = 13$ **49.** $x_2 - x_1$ **50.** x squared minus y squared
51. the square of the quantity x minus y **52.** seven times the quantity x minus five **53.** seven times x minus five **54. a.** G **b.** F **c.** J
d. I **e.** H **f.** E **g.** D **h.** B **i.** A **j.** C **55.** $(17 + 7 \wedge 2)/(20 - 9)$
56. $13 - 4((2 - 8) - 3(7 - 2))$
57. x is less than 3;

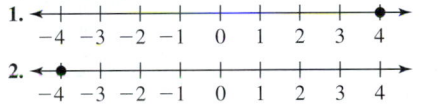 ; $(-\infty, 3)$

58. $x \ge 1$; 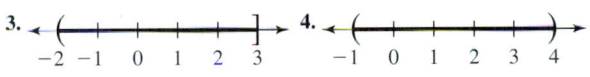 ; $[1, \infty)$

59. $4 \le x < 10$; x is greater than or equal to 4 and less than 10; $[4, 10)$
60. $-2 < x \le 3$; x is greater than -2 and less than or equal to 3;

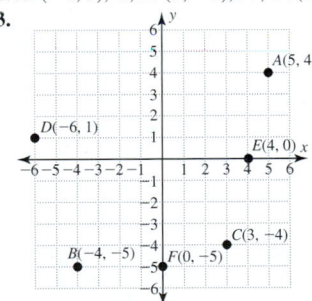

61. $-$ **62.** $+$ **63.** $+$ **64.** neither; 0 **65.** $-$ **66.** $-$ **67.** $+$ **68.** C **69.** B **70.** A
71. D **72.** C **73.** B **74.** $9x - 5y$ **75.** $2a + b$ **76.** $7a + 22$ **77.** $7x + 7$
78. 127 **79.** $-\dfrac{15}{37}$ **80.** -2 **81.** 7, 17, 27 **82.** -7 is not a solution; 4 is a
solution **83.** -7 is a solution; 4 is not a solution **84.** -7 is a solution; 4 is
a solution **85.** -7 is not a solution; 4 is a solution **86.** \$138 **87.** $+11°$
88. \$26 **89. a.** \$30 **b.** 5% **90.** 8:5 **91. a.** 40 L **b.** 30 gal
92. a. 200 golf balls/h **b.** $\dfrac{1}{4}$ Vat/h **93. a.** 54 m **b.** 162 m² **94. a.** 135 ft³
b. 159 ft² **95. a.** 190π cm **b.** $9,025\pi$ cm² **96.** 27°

Mastery Test for Chapter 1
1. Answers vary for **a–c. d.** odd **2. a.** 2 **b.** -1.5 **c.** $-\dfrac{1}{5}$ **d.** 0
3. a. 23 **b.** 23 **c.** 0 **d.** 46 **4. a.** $(-2, 3]$ **b.** $[-1, \infty)$ **c.** $(-\infty, 4)$ **d.** $[5, 9)$
5. a. 5; 5.10 **b.** 10; 9.95 **c.** 3; 2.83 **d.** 4; 4.00 **6. a.** B **b.** C **c.** A **d.** E
e. D **7. a.** $\dfrac{1}{3}$ **b.** $\dfrac{5}{8}$ **8. a.** $\dfrac{2}{3}$ **b.** $\dfrac{5}{8}$ **9. a.** $\dfrac{1}{15}$ **b.** $\dfrac{1}{3}$ **c.** $\dfrac{3}{7}$ **d.** $\dfrac{54}{49}$

10. a. $\dfrac{1}{2}$ **b.** $\dfrac{2}{3}$ **c.** $\dfrac{1}{3}$ **d.** $\dfrac{2}{7}$ **11. a.** $\dfrac{31}{40}$ **b.** $\dfrac{7}{30}$ **c.** $\dfrac{47}{42}$ **d.** $\dfrac{8}{21}$
12. a. $16\dfrac{1}{2}$ **b.** $5\dfrac{3}{4}$ **c.** 10 **d.** $1\dfrac{2}{7}$ **13. a.** 6 **b.** -28 **c.** $-\dfrac{1}{4}$ **d.** 0
14. a. commutative **b.** associative **c.** $x + (y + 5)$ **d.** $5(y + x)$
15. a. 7 **b.** -23 **c.** $\dfrac{5}{6}$ **d.** -3.6 **16. a.** commutative **b.** associative
c. $5(xy)$ **d.** $(a + b)x$ **17. a.** -30 **b.** 30 **c.** 0 **d.** $\dfrac{4}{25}$ **18. a.** -3 **b.** 3 **c.** 0
d. 4,578,910 **19. a.** 4:3 **b.** 1:13 **c.** 13:6 **d.** 1:5 **20. a.** 32 **b.** 25 **c.** -25
d. -1 **21. a.** 9 **b.** 27 **c.** 19 **d.** $-\dfrac{2}{33}$ **e.** 64 **f.** 34 **g.** -59 **h.** 2
22. a. $15x + 27$ **b.** $-32x + 20$ **c.** $2x - 3y$ **d.** $-11x + 13y$
23. a. $12x$ **b.** $9x - 33$ **c.** $-9x - 6$ **d.** $x - 7$ **24. a.** -6 **b.** 7 **c.** 8 **d.** 0
25. a. 60 m² **b.** 34 m **c.** 7.5 gal **d.** \$1,348.32 **26.** $-\dfrac{7}{2}$ **27.** 9, 7, 5
28. a. -5 is a solution **b.** -5 is not a solution **c.** -5 is a solution
d. -5 is a solution **29. a.** quadrilateral **b.** rectangle **c.** hexagon **d.** five
30. a. 21 cm **b.** $C = 18\pi$ cm; $C \approx 56.5$ cm **31. a.** 14 cm²
b. $A = 81\pi$ cm²; $A \approx 254.5$ cm² **32.** 7,500 in³ **33. a.** right **b.** 180
c. complementary **d.** adjacent **e.** alternate

Group Project for Chapter 1
a. 9, 27, 16, 43, 7; The digit check of 7 agrees with the one shown.
b. Answers can vary.

Chapter 2

Using the Language and Symbolism of Mathematics 2.1
1. Cartesian **2.** x; y **3.** origin; $(0, 0)$ **4.** y **5.** x **6.** counterclockwise
7. I **8.** III **9.** scatter **10.** arithmetic **11.** difference **12.** linear

Quick Review 2.1
1.
```
   -4 -3 -2 -1   0   1   2   3   4
```
2.
```
   -4 -3 -2 -1   0   1   2   3   4
```
3.
```
   -2 -1   0   1   2   3
```
4.
```
   -1   0   1   2   3   4
```
5. $(2, \infty)$

Exercises 2.1
1. A: $(-1, 5)$, II; B: $(5, -2)$, IV; C: $(3, 2)$, I; D: $(-4, -1)$, III
3.

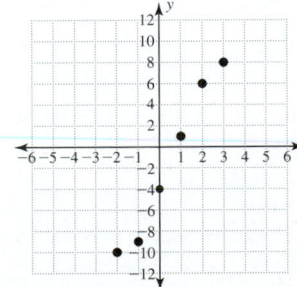

5. a. II **b.** III **c.** IV **d.** I **7. a.** y-axis **b.** x-axis **c.** x-axis **d.** y-axis
9. The points do not lie on one line, but they all lie close to a line.

11. $-4, -3, -2, -1, 0, 1, 2$

13. a. $a_1 = 5$
$a_2 = 13$
$a_3 = 19$
$a_4 = 8$
$a_5 = 6$

b.

x	y
1	5
2	13
3	19
4	8
5	6

15. a. $(1, 8), (2, 5), (3, 2), (4, -1), (5, -4)$ **b.**

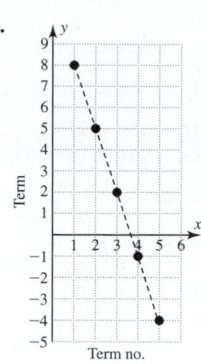

17. a. $(1, -2), (2, 4), (3, 1), (4, 1), (5, 3)$ **b.**

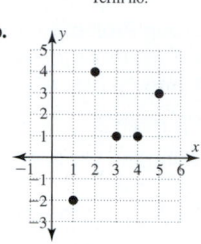

19. $4, 1, -2, -5, -8$ **21.** arithmetic; forms a linear pattern **23.** arithmetic; forms a linear pattern **25.** not arithmetic; does not form a linear pattern **27.** arithmetic; $d = 2$ **29.** not arithmetic **31. a.** 30 ft **b.** 15 ft **c.** after 2 months **d.** 10 ft **33. a.** D **b.** A **c.** B **d.** C **35.** $>; >$ **37.** $<; <$ **39.** I; III **41.** up **43.** yes; $d = 2$ assuming these integers are listed in increasing order **45. a.** $a_1 = 2,850$; the total payments after 1 month is $2,850.
b. $a_{18} = 10,500$; the total payments after 18 months is $10,500.
c. $a_{24} = 13,200$; the total payments after 24 months is $13,200.
47. area: 20 square units; perimeter: 18 units

49.

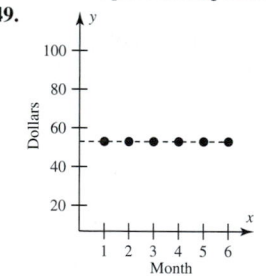

51.

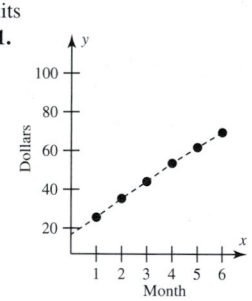

53.

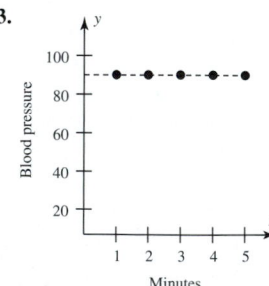

55.

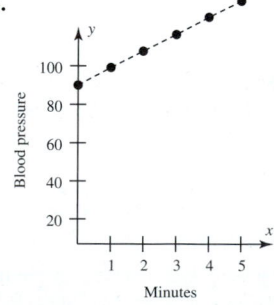

Group Discussion Questions

57. a. Option A will produces an arithmetic sequence because there is a constant difference of $1,500.

b.

Option A, $	Option B, $
30,000.00	30,000.00
31,500.00	31,500.00
33,000.00	33,075.00
34,500.00	34,728.75
36,000.00	36,465.19

c. After year 2, option B produces a greater salary. **59.** Answers can vary. You can suggest outcomes in various sports over a period of 8 events, results by a salesperson, etc.

Cumulative Review
1. $12x$ **2.** $7v$ **3.** $a + 11$ **4.** $112°$ **5.** $68°$

Using the Language and Symbolism of Mathematics 2.2
1. function **2.** $f; x$ **3.** x; output **4.** 5; 9 **5.** line **6.** linear **7.** modeling

Quick Review 2.2
1. 6 **2.** 1 **3.** -24 **4.** 13 **5.** $(x + 4)/2$

Exercises 2.2
1. 7 **3.** 4 **5.** 4 **7.** -56 **9.** 28 **11.** 3 **13.** 3 **15.** 6 **17.** 8, 6, 4, 2, 0
19. 1.5, 2, 2.5, 3   **21.**

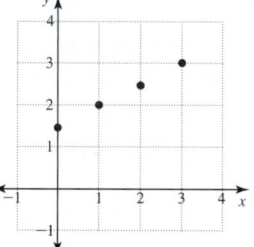

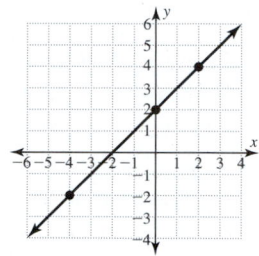

23. $f(x) = 18 - 3x$ **25.** $f(x) = 2x + 17$ **27. a.** $f(x) = 2x + 64$
b. $f(x) = 32x$ **29. a.** $f(x) = x + 5$ **b.** $f(x) = x - 5$ **c.** $f(x) = 2(x + 5)$
31. $-3, -1, 1, 3, 5$ **33.** $-2, 0, 3, 5$ **35. a.** -8 **b.** 2 **c.** 8 **d.** 0
37. a. 3 **b.** 1 **c.** 0 **d.** 4

39. a.

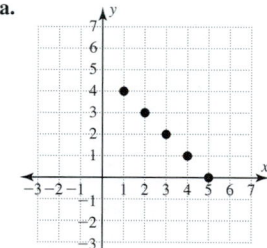

b.

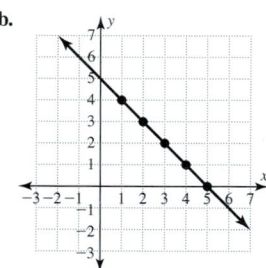

c. The graph in part **a** consists of discrete points. The graph in part **b** is a connected line.
41. a. $3, 2, 1, 0, -1, -2, -3$
b.

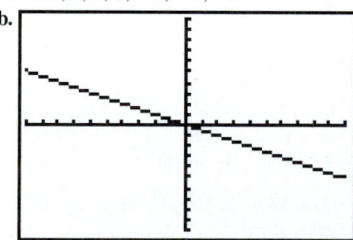

43. a. $-0.9, 1.8, 4.5, 7.2, 9.9$ **b.** 4.5, 31.5, 58.5, 85.5, 112.5
45. a. $-4, -3, -2, -1, 0$ **b.** $-2, 3, 8, 13, 18$ **47. a.** $f(x) = 0.20x$
b. $f(25) = 5$; an entree priced at $25.00 has a 20% discount of $5.00.
49. a. $f(x) = 0.85x$ **b.** $37.40, $51, $57.80, $76.50
51. a. $f(x) = 15x + 500$ **b.** 500, 2,000, 3,500, 5,000, 6,500, 8,000
c. $f(250) = 4,250$; the cost of producing 250 units is $4,250.

53. a. $f(x) = 180 - 2x$ **b.** 160, 140, 120, 100, 80, 60, 40 **c.** $f(40) = 100$; when $x = 40°$, the remaining angle is 100°. **d.** $x = 70°$; when $x = 70°$, the remaining angle is 40°. **e.** No, x must be less than 90°.

Group Discussion Questions

55. a. $-3, -2, -1, 0, 1, 2, 3$
b.

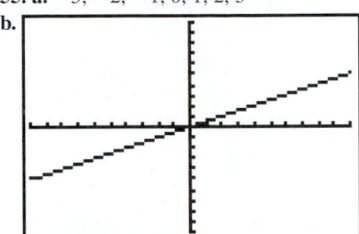

c. $+1$ **d.** 1 unit up

Cumulative Review

1. \$350 **2.** 210 mi **3.** 0 **4.** 7 **5.** 9

Using the Language and Symbolism of Mathematics 2.3

1. x-intercept **2.** y-intercept **3.** true **4.** infinite **5.** solution **6.** intersection **7.** y **8.** x **9.** y **10.** x **11.** overhead **12.** break-even

Quick Review 2.3

1. $(-4, -4)$ **2.** $(0, -2)$ **3.** $(4, 0)$ **4.** 2 **5.** 5

Exercises 2.3

1. solution **3.** not a solution **5.** solution **7.** solution **9.** not on the graph **11.** not on the graph **13.** on the graph **15.** on the graph **17.** Points A and D are solutions. **19.** Point C is a solution. **21.** x-intercept: $(3, 0)$; y-intercept: $(0, 3)$ **23.** x-intercept: $(80, 0)$; y-intercept: $(0, -500)$. The x-intercept tells us that a profit of \$0.00 occurs when 80 units are produced; this is the break-even point. The y-intercept tells us that there is a loss of \$500.00 when no items are produced; \$500 is the overhead cost. **25.** x-intercept: $(-3, 0)$; y-intercept: $(0, -3)$ **27.** $(-2, 3)$ **29.** $(-1, -2)$ **31.** $(3, 2)$

33.

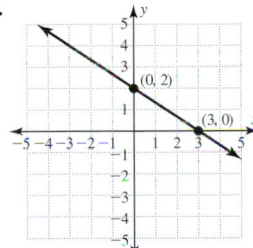

35.

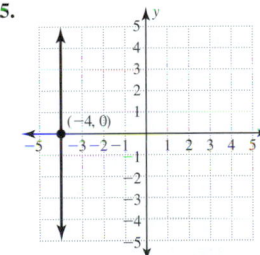

37.

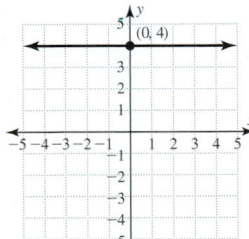

39.

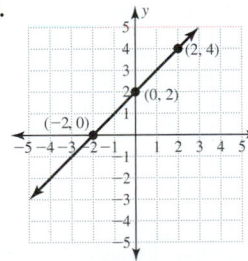

41.

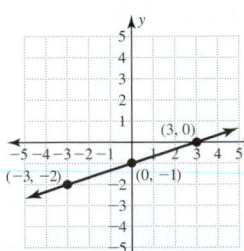

43.

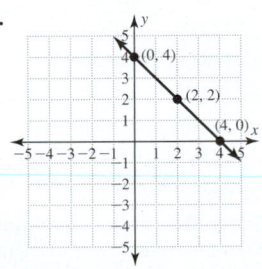

45.

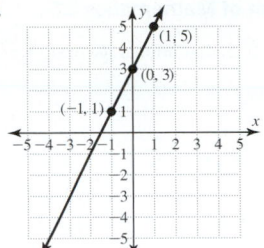

47. $(1, -3)$ **49.** $(2, 3)$

51.

x	y
-3	-18
-2	-15
-1	-12
0	-9
1	-6
2	-3
3	0

x-intercept: $(3, 0)$;
y-intercept: $(0, -9)$

53. x-intercept: $(5, 0)$; y-intercept: $(0, -6)$ **55.** $(2, 1)$

57.

x	f(x)
1	790.83
2	1,124.16
3	1,457.49
4	1,790.82
5	2,124.15
6	2,457.48

59. units: 2; cost: \$7.00 **61.** months: 2; cost: \$550.00

Group Discussion Questions

63. The lines are all parallel with y-intercept $(0, b)$. **65.** The lines are horizontal lines. If $m = 0$, the line is horizontal.

Cumulative Review

1. 1 **2.** -27 **3.** -1 **4.** -7 **5.** $-\dfrac{3}{2}$

Using the Language and Symbolism of Mathematics 2.4

1. 1 **2.** linear **3.** linear **4.** linear **5.** equivalent **6.** addition **7.** $b - c$ **8.** conditional **9.** identity **10.** contradiction

Quick Review 2.4

1. $x = 1$ is not a solution; $x = -4$ is a solution. **2.** 2 **3.** 1 **4.** 3 **5.** $3x - 5 = 13$

Exercises 2.4

1. B **3.** $x = 24$ **5.** $v = -4$ **7.** $y = -1$ **9.** $z = -11$ **11.** $y = 4$ **13.** $x = 16$ **15.** $y = 63$ **17.** $v = 2$ **19.** $y = 0$ **21.** $m = 26$ **23.** $n = 99$ **25.** $t = 0$ **27.** $v = 0$ **29.** $x = 7$ **31.** $v = 25$ **33.** $w = 22$ **35.** $m = -5$ **37.** $n = 9$ **39.** $y = -12$ **41.** $x = 1$ **43.** $x = 2$ **45.** $x = 2$ **47.** $x = 1$ **49.** $x = -1$ **51.** $x = 2$ **53.** $x = 3$ **55. a.** conditional; $x = 0$ **b.** contradiction; no solution **c.** identity; every real number is a solution **57. a.** contradiction; no solution **b.** identity; every real number is a solution **c.** conditional; $v = 4$ **59. a.** $9x - 5$ **b.** $x = -7$ **61. a.** $x - 7$ **b.** $x = 7$ **63. a.** $5.8x + 0.6$ **b.** $x = 4$ **65.** $x = 2$; $x = 1.905$ **67.** $x = 500$; $x = 500.277$ **69.** $3m + 7 = 2m - 8$; $m = -15$ **71.** $12 - 9m = 2 - 10m$; $m = -10$ **73.** $2(3m - 9) = 5(m + 13)$; $m = 83$ **75.** $a = 7$ **77.** $x = 253$

Group Discussion Questions

79. Answers can vary. **81. a.** 12 **b.** $x + k, k \neq 5$ **c.** $x + 5$

Cumulative Review

1. 150 **2.** -0.5 **3.** 2 **4.** 3 **5.** 2

Using the Language and Symbolism of Mathematics 2.5

1. equivalent **2.** multiplication **3.** $\dfrac{b}{c}$ **4.** -9 **5.** conditional **6.** zero

7. distributive **8.** least common denominator **9.** 4 **10.** -4

Quick Review 2.5

1. 60 **2.** -4 **3.** $-5x - 4$ **4.** $2x + 54$ **5.** $P = 2l + 2w$

Exercises 2.5

1. a. $x = 6$ **b.** $x = 35$ **3. a.** $v = -8$ **b.** $v = 9$ **5. a.** $t = -48$ **b.** $t = -27$

7. $t = -4$ **9.** $x = -5$ **11.** $y = -5$ **13.** $y = 3$ **15.** $z = -3$ **17.** $m = -\dfrac{1}{9}$

19. $w = 0$ **21.** $v = -15$ **23.** $x = 0$ **25.** $y = -5$ **27.** $a = 110$
29. $x = -7$ **31.** $v = -16$ **33.** $x = 3$ **35.** $w = -10$ **37.** $x = 4$

39. $x = -\dfrac{3}{2}$ **41. a.** contradiction; no solution **b.** identity; every real

number is a solution **c.** conditional; $x = 0$ **43. a.** $x - 2$ **b.** $x = 2$

45. a. $21x - 19$ **b.** $x = \dfrac{19}{21}$ **47.** 6.25 mg **49.** $s \approx 25$ cm

51. \$75.80 **53.** $x = -4; x = -4.2$ **55.** $x = 4; x = 4.1$ **57.** $x = 1.5$

59. $x = 0.7$ **61.** $5(x + 2) = 7(x - 3); x = \dfrac{31}{2}$

63. $2(3v - 2) = 4(v + 9); v = 20$
65. $2(3m - 5) = 3(m + 6) - 4; m = 8$
67. $\dfrac{1}{3}(2x + 5) = 4x + 2; x = -\dfrac{1}{10}$ **69.** $x = 6$

71. $a = \dfrac{1}{2}$ **73.** $x = 2$

Group Discussion Questions

75. $x = -2$, explanations can vary.
77. Dividing by zero produced this error.

Cumulative Review

1. 125 **2.** 175 **3.** 2005 **4.** 2007 **5.** 100

Using the Language and Symbolism of Mathematics 2.6

1. specified **2.** distributive **3.** break-even **4.** overhead **5.** x **6.** y

Quick Review 2.6

1. \$300 **2.** yes **3.** no **4.** $x = 2$ **5.** $x = -10$

Exercises 2.6

1. $y = -2x + 7$ **3.** $y = 3x - 2$ **5.** $y = 2x - 3$ **7.** $y = 2x + 4$
9. $y = -3x - 2$ **11.** $y = -4x + 4$ **13.** $y = 6x - 8$
15. $y = 2x - 3$

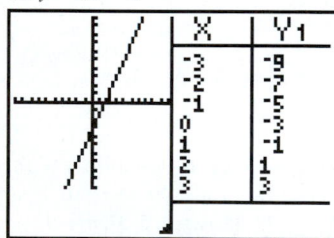

17. $y = -\dfrac{1}{2}x - 2$

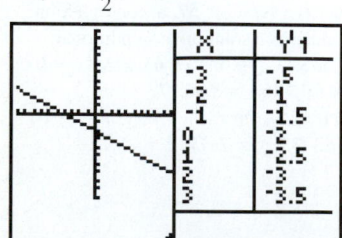

19. x-intercept: $(2, 0)$;
y-intercept: $(0, -4)$

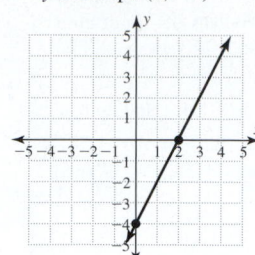

21. x-intercept: $(2, 0)$;
y-intercept: $(0, -8)$

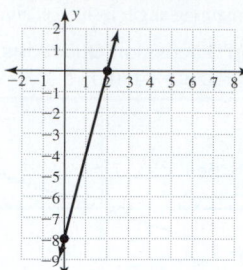

23. x-intercept: $(3, 0)$;
y-intercept: $(0, 2)$

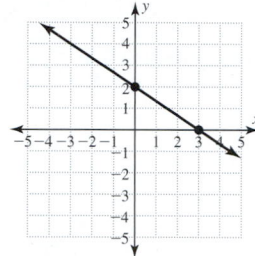

25. $l = \dfrac{A}{w}$ **27.** $r = \dfrac{C}{2\pi}$ **29.** $V_1 = \dfrac{V_2 T_1}{T_2}$ **31.** $h = \dfrac{3V}{\pi r^2}$

33. $a = P - b - c$ **35.** $C = \dfrac{5}{9}(F - 32)$ **37.** $b = \dfrac{2A}{h} - a$

39. $F = \dfrac{9}{5}C + 32$ **41.** $a = l - (n - 1)d$ **43.** $n = \dfrac{l - a}{d} + 1$

45. $m = \dfrac{y - b}{x}$ **47.** $w = \dfrac{P - 2l}{2}$

49. a. $P = \dfrac{T}{R}$ **b.** $P = \dfrac{T}{0.08}$

c.

T	P
100	1,250
200	2,500
300	3,750
400	5,000
500	6,250
600	7,500

51. a. $W = \dfrac{P - 2L}{2}$ **b.** $W = 400 - L$

c.

L	W
100	300
120	280
140	260
160	240
180	220
200	200

53. a. $y = 0.75x - 120$ **b.** x-intercept: $(160, 0)$; the ice cream shop will break even when 160 ice cream products are sold **c.** y-intercept: $(0, -120)$; the overhead for the ice cream shop is \$120. (The shop will lose \$120 if no products are sold.)

55. $R = \dfrac{I}{PT}$

P	R
10,000	0.0600
12,000	0.0500
14,000	0.0429
16,000	0.0375
18,000	0.0333
20,000	0.0300
22,000	0.0273

Group Discussion Questions

57. It is handier to have the temperature you want isolated on the left side of the equation. $212°$, $201.2°$, $37°$, $22.2°$, $32°$, $-17.8°$, $-40°$
59. a. $\approx 25{,}460$ cm^3 **b.** $\approx 15{,}920$ cm^3 **c.** No; in the formula $V = \pi r^2 h$, the radius is used as a factor twice while the height is used as a factor only once. Thus selecting the rectangular sheet so that the cylinder will have a

larger radius will result in a greater volume. **d.** 50 cm × 80 cm and 80 cm × 50 cm **e.** 4,000 cm² **f.** Yes, in the formula $A = lw$, both the length and the width are used as a factor once. Because multiplication is commutative, $lw = wl$ and both rectangles have the same area.

Cumulative Review
1. $f(x) = 50 - 6x$ **2.** 3:4 **3.** $7\frac{8}{15}$ **4.** 14 **5.** $7x - 14y = 7(x - 2y)$

Using the Language and Symbolism of Mathematics 2.7
1. proportion **2.** terms **3.** b; c **4.** a; d **5.** directly **6.** linear **7.** increases **8.** decreases **9.** arithmetic

Quick Review 2.7
1. $\frac{5}{12}$ **2.** $x = 18$ **3.** $x = 1.2$ **4.**

x	0	4	20	10
y	-10	0	40	15

5. zero

Exercises 2.7
1. $x = 12$ **3.** $x = 10$ **5.** $x = 3$ **7.** $y = 11$ **9.** $y = 16$ **11.** $a = 2$
13. $v = \frac{10}{3}$ **15.** $x = 3$ **17.** $z = -8$ **19.** $x = 9$ **21.** $v = 3$ **23.** $d = kt$
25. v varies directly as w **27.** $v = 10$ **29.** $g = 4$ **31.** $k = \frac{4}{5}$ **33.** $4\frac{1}{2}$ cups
35. 10 mm **37.** 50 bulbs **39.** 1,416 bricks **41.** 8.75 m × 7.5 m **43.** 13.5 lb
45. 9 face cards **47. a.** 0.15 **b.** $C = 0.15S$ **c.** $2,400, $2,700, $3,000, $3,300, $3,600, $3,900, $4,200
49. a. 0.78 **b.** $E = 0.78 U$

c.
USD	EUR
50	39
100	78
150	117
200	156
250	195
300	234
350	273

51. a. $W = 76.38t$ **b.** pounds per minute of fuel consumption **53.** 22.4 ft
55. 17.5 cm **57.** 17.5 ft **59.** $53\frac{1}{3}$ ft **61.** 9 **63. a.** 314.2 cm **b.** 0.00314 km
c. $d = 0.00314x$ **d.** 1.57 km **e.** 319 revolutions **65. a.** $C = 0.80x$ **b.** $400
c. $625 **67.** 0.8
69. Numerically:

x	y
1	-0.5
2	-1.0
3	-1.5
4	-2.0
5	-2.5

Graphically:

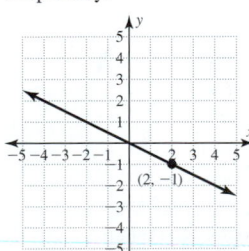

Verbally: y varies directly as x
with a constant of variation $-\frac{1}{2}$

73. $w = 3.75$ **75.** $x = 7.2$

71. Algebraically: $y = -x$
Numerically:

x	y
1	-1
2	-2
3	-3
4	-4
5	-5

Graphically:

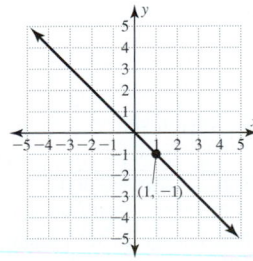

Group Discussion Questions
77. a. 10 ft **b.** 12 ft **c.** yes **d.** Urge the students to support their explanations with word equations and sketches. **79.** Form a line segment $B'C'$ parallel to BC so that B' is on AB and C' is on AC. Measure AB', AB, and $B'C'$. Then use a proportion to calculate BC.

Cumulative Review
1. $f(x) = 8x + 96$ **2.** 8 **3.** 4 **4.** 1.10P **5.** 0.90P

Using the Language and Symbolism of Mathematics 2.8
1. restrictions **2.** find **3.** variable **4.** algebraic **5.** solve **6.** reasonable **7.** mixture **8.** rate; time **9.** principal; rate; time **10.** fixed **11.** variable

Quick Review 2.8
1. 750 mi **2.** $x = 68$ **3.** conditional; $x = 13$ **4.** identity; $\mathbb{R}$
5. contradiction; no solution

Exercises 2.8
1. B **3.** D **5.** $\{0, 1, 2,..., 49, 50\}$ **7.** [0, 225] **9. a.** Let $x =$ number of months. **b.** Current thickness minus wear equals 4.0 mm. **c.** $11.2 - 0.2x = 4.0$ **d.** $x = 36$ **e.** yes **f.** The brake pads have 36 months of wear left. **11.** The bill was for 345 minutes of long distance. **13.** The manufacturer should make 50 cushions per day to have an average cost of $9.00 per cushion. **15.** The rate of return on the second investment must be 8%. **17.** The principal that is invested in the corporate bond is $57,500. **19.** The boats have been traveling for 3.2 hours (3 hours and 12 minutes). **21.** It takes 2.8 hours (2 hours and 48 minutes) for the vehicles to meet. **23.** The crop dusting service should add 150 gallons of the 1% solution. **25.** The gasoline distributor should add 800 gallons of pure ethanol. **27.** The crop dusting service should add 112.5 gallons of water. **29.** The company should add 22,500 lb of sand to the pile. **31.** The city should add 0.5 MW of windmill generation capacity. **33. a.** The manufacturer produced 7,500 tiles. **b.** The plant would need to produce 10,000 tiles per day, but it does not have this much capacity. **35. a.** The gasoline distributor must add approximately 826 gallons of ethanol. **b.** The distributor would need to add 1,267 gal, but there is not enough capacity. **37. a.** The crop dusting service would need to add 320 gal, but there is not enough capacity. **b.** The crop dusting service should add 120 gallons. **39. a.** The total cost of the mixture is $44.00. **b.** The mix should be sold for $1.10 per lb. **c.** The company should add 100 pounds of the standard mix. **d.** The barrel must hold a total of 120 pounds. **e.** The company should add 40 pounds of the standard mix.

Group Discussion Questions
41. The new weight of the watermelon is 80 lb. **43. a.** An investor has $4,000 invested in a CD that pays yearly interest at 6%. She invests an additional amount in a riskier bond that pays 10% annual interest. These two investments generate a 7% rate of return on her total investment. How much did she invest in the bond? **b.** A gasoline distributor has a 20,000-gal tank to store a gasohol mix. The tank already contains 4,000 gal of gasohol that contains 6% ethanol by volume. How many gallons of gasohol that is 10% ethanol must be added to the tank to produce a mix that is 7% ethanol?

Cumulative Review
1. regular **2.** six **3.** four **4.** complementary **5.** alternate interior

Review Exercises for Chapter 2
1. A: (4, 0); on x-axis; B: (2, 4); I; C: $(-6, 2)$, II; D: $(0, -2)$; on y-axis
2.

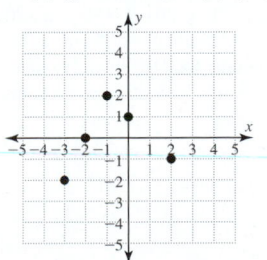

3. $(1, 1)$, $(2, -1)$, $(3, -3)$, $(4, -5)$, $(5, -7)$

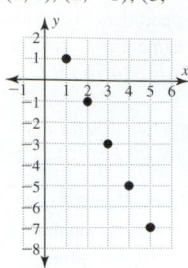

4. arithmetic; linear pattern **5.** not arithmetic; not a linear pattern **6.** not arithmetic; not a linear pattern **7.** arithmetic; linear pattern **8. a.** 3,610 ft **b.** increase **c.** 2004 **9.** Neither is a solution. **10.** $(-3, 2)$ is a solution; $(3, -2)$ is not a solution **11.** $(-3, 2)$ is a solution; $(3, -2)$ is not a solution **12.** Neither is a solution. **13.** x-intercept: $(3, 0)$; y-intercept: $(0, -2)$

14. x-intercept: $(-5, 0)$; y-intercept: $(0, 5)$

15. x-intercept: $\left(-\dfrac{11}{4}, 0\right)$; y-intercept: $(0, 11)$

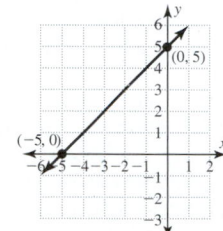

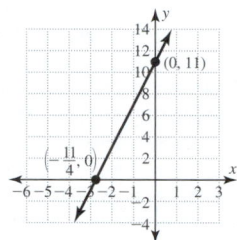

16. B **17.** C **18.** D **19.** A **20.** $(1, 2)$ **21.** $(2, -3)$ **22.** $x = 2$ is not a solution. $x = 3$ is a solution. **23.** $x = 2$ is a solution. $x = 3$ is not a solution. **24.** $x = 2.5$ **25.** $x = 24$ **26.** $x = 3$ **27.** $x = -1$ **28.** $x = 15$ **29.** $v = 2$ **30.** $m = 0$ **31.** $x = -3$ **32.** $m = 2$ **33.** $n = -36$ **34.** $y = -98$ **35.** $w = -4$ **36.** $t = 11,700$ **37.** $b = 0$ **38.** $x = 7$ **39.** $z = \dfrac{3}{16}$ **40.** $r = 0$ **41.** $x = -4$ **42.** $x = 27$ **43.** $y = 3$ **44.** $y = 0$ **45.** $w = -31.6$ **46.** $t = -24$ **47.** $m = -9$ **48.** $n = -11$ **49.** $y = -1,295$ **50.** $x = -8$ **51.** all real numbers **52.** $a = 0$ **53.** $x = 1.5$ **54.** $x = -2.4$ **55.** contradiction; no solution **56.** contradiction; no solution **57.** identity; every real number is a solution **58.** conditional; $v = 0$ **59. a.** -81 **b.** -11 **c.** 52 **d.** $7\pi - 11 \approx 10.991$ **60. a.** -74 **b.** -14 **c.** 106 **d.** 1,986 **61.** $x = v + w - y$ **62.** $x = \dfrac{vw}{y}$ **63.** $x = \dfrac{5y + 7z}{3}$ **64.** $x = -\dfrac{1}{8}y + \dfrac{19}{4}$ **65. a.** $8x + 10$ **b.** $x = -1$ **66. a.** $-2x - 2$ **b.** $x = \dfrac{1}{4}$ **67. a.** $2x - 1$ **b.** $x = \dfrac{1}{2}$ **68. a.** $-x - 7$ **b.** $x = -7$ **69.** $x = 10$; $x = 10.05$ **70.** $x = 20$; $x = 19.95$ **71.** $5m + 4 = 49$; $m = 9$ **72.** $\dfrac{1}{3}(m + 7) = 12$; $m = 29$ **73.** $2(m + 2) = -(m - 3)$; $m = -\dfrac{1}{3}$ **74.** $m + (m + 1) + (m + 2) = 93$; $m = 30$ **75.** $v = 10$ **76.** $k = \dfrac{4}{5}$ **77.** $y = 50$ **78. a.** $P = 8w + 8$ **b.** $P = 104$ cm **c.** $8w + 8 = 48$ **d.** $w = 5$ **79. a.** $C = 26x + 160$ **b.** $C = \$1,720$ **c.** $26x + 160 = 1,200$ **d.** $x = 40$ **80.** 210 mi **81.** $a = 4.5$ cm; $b = 13.5$ cm; $c = 7.5$ cm **82.** 4.4 m **83.** The speed of the plane is 290 mi/h. **84.** One should expect 20 defective panels. **85.** B **86.** C **87.** D **88.** A **89.** It would take 3.5 hours. **90.** Add 150 gal of water. There is enough room.

Mastery Test for Chapter 2

1. A: $(-3, 1)$ II; B: $(1, -3)$, IV; C: $(4, 2)$, I; D: $(-2, -2)$, III

2.

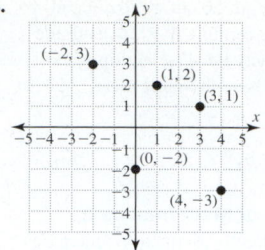

3. a. arithmetic; a linear pattern **b.** not arithmetic; not a linear pattern **c.** arithmetic; a linear pattern **4. a.** 145 **b.** 130 **c.** 3 **5. a.** -7 **b.** -18 **c.** 70 **d.** 103

6. a.

x	y
1	2
2	3
3	4
4	5
5	6

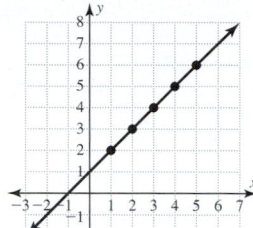

b.

x	y
1	1
2	0
3	-1
4	-2
5	-3

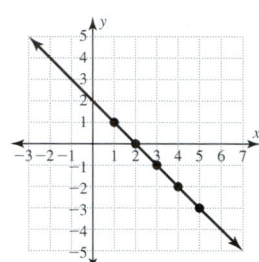

c.

x	y
1	-3
2	-1
3	1
4	3
5	5

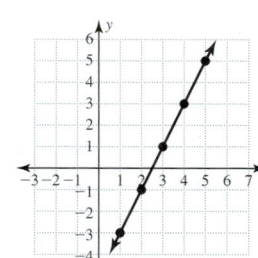

d.

x	y
1	2
2	0
3	-2
4	-4
5	-6

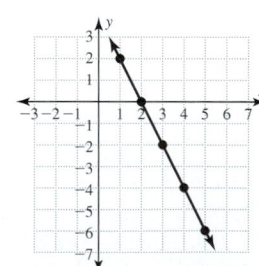

7. a. $f(x) = 375x + 800$

b.

x	0	6	12	18	24	30
$f(x)$	800	3,050	5,300	7,550	9,800	12,050

c. $f(25) = 10,175$; the total amount paid at the end of the 25th month is $10,175 **8. a.** solution **b.** not a solution **c.** solution **d.** solution **9. a.** $(-2, 0)$; $(0, 3)$ **b.** $(-3, 0)$; $(0, -1)$ **c.** no x-intercept; $(0, 2)$ **d.** $(-1, 0)$; no y-intercept **10. a.** $(2, 1)$ **b.** $(-1, 3)$ **11. a.** $x = 4$ **b.** $x = 14$ **c.** $x = 0$ **d.** $x = 6$ **12. a.** $x = 2$ **b.** $x = -2$ **13. a.** identity; every real number is a solution **b.** contradiction; no solution **c.** conditional equation; $x = 0$ **14. a.** $x = 3$ **b.** $y = -\dfrac{5}{9}$ **c.** $v = \dfrac{3}{2}$ **d.** $w = 13$

15. $y = 4x - 1$

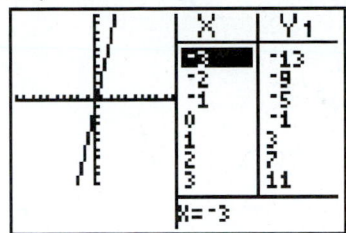

16. a. $(3, 0)$; $(0, -6)$

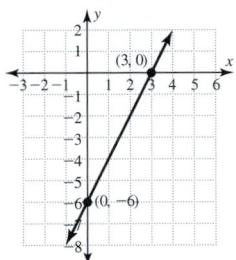

b. no x-intercept; $(0, 4)$

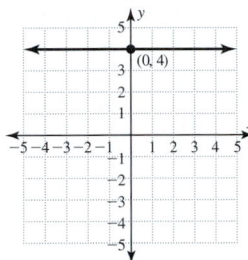

c. $(5, 0)$; $(0, -2)$

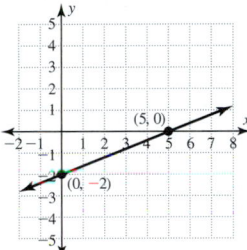

d. $(-2, 0)$; $(0, 1)$

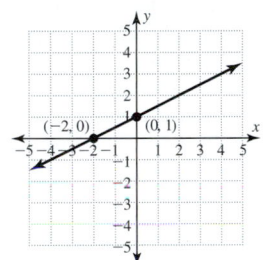

17. a. $a = \dfrac{2A - bh}{h}$ **b.** $x = \dfrac{y - b}{m}$ **18. a.** $m = -3$ **b.** $x = 11$ **c.** 187.5 km

corresponds to 5 cm. **19. a.** $y = 12$ **b.** 9.5 pesos for each U.S. dollar

20. a. $[0, 30]$ **b.** $\{0, 1, 2, 3, \ldots, 29, 30\}$ **c.** $[0, 300]$ **d.** $\{0, 1, 2, 3, \ldots, 9, 10\}$

21. a. The percent of alcohol will be 20.8%. **b.** 163 gallons of ethanol

must be added. **c.** 278 gallons of ethanol must be added.

Group Project for Chapter 2

Monthly Payment, $	Total of all Payments, $	Monthly Payment, $	Total of all Payments, $
567.79	204,404.04	580.54	69,665.09
599.55	215,838.19	449.41	80,894.54
632.07	227,544.49	387.65	93,035.87
665.30	239,508.90	353.39	106,016.88
699.21	251,717.22	332.65	119,754.45
733.76	264,155.25		

A lower interest rate contributes to a lower monthly payment as does a longer mortgage period. A lower interest rate has such a dramatic effect that one can use the same monthly payment at a lower rate and pay off the mortgage years sooner than if the mortgage were at a higher rate.

Chapter 3

Using the Language and Symbolism of Mathematics 3.1

1. y **2.** x **3.** m **4.** $m = \dfrac{y_2 - y_1}{x_2 - x_1}$ **5.** 1 **6.** 0 **7.** undefined **8.** change

9. change **10.** rate **11.** increases **12.** decreases **13.** parallel

14. perpendicular **15.** grade

Quick Review 3.1

1. $\dfrac{3}{4}$ **2.** $-\dfrac{1}{4}$ **3.** $-\dfrac{3}{4}$ **4.** 1.25 **6.** -0.875

Exercises 3.1

1. a. 2 **b.** 2 **3.** $-\dfrac{11}{9}$ **5.** 0 **7.** undefined **9.** $\dfrac{2}{5}$ **11.** $\dfrac{3}{5}$ **13. a.** $\dfrac{1}{3}$ **b.** -2

c. undefined **d.** 0 **15.** $\dfrac{3}{2}$ **17.** $-\dfrac{3}{5}$ **19.** $\dfrac{1}{7}$ **21.** $m_2 = \dfrac{3}{7}$; $m_3 = -\dfrac{7}{3}$

23. $m_2 = 0$; m_3 is undefined **25.** parallel **27.** perpendicular **29.** neither

31. perpendicular **33.** parallel **35.** 90 m **37.** The production cost is \$35 per

unit. **39.** $m = 65$; the speed of the car is 65 mi/h. **41.** $-\dfrac{8}{5}$ **43.** 2 **45.** 3 **47.** 0

49. 3.5 ft **51. a.** -3 gal/s **b.** Each second the volume of water decreases

by 3 gal. **c.** 2 min 47 s **53.** $\dfrac{3}{7}$ **55. a.** 3, 1, -1, -3 **b.** 1, -3, -7, -11

57. a. -2, 3, 8, 13, 18, 23, 28 **b.** 3 **c.** 5 **d.** $\dfrac{5}{3}$

59. Answers can vary; point $(4, 5)$ **61.** Answers can vary; point $(1, -5)$

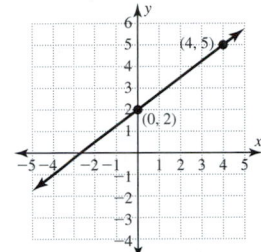

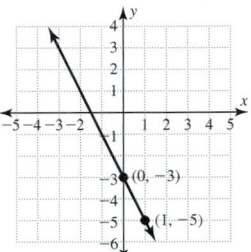

63.

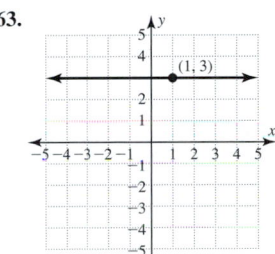

65.

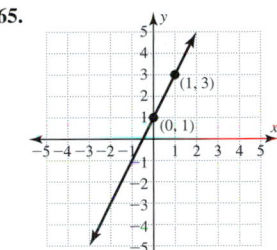

67.

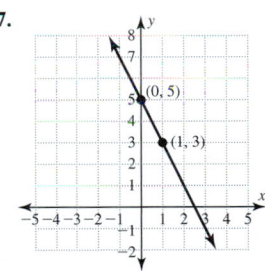

69. a. $(2, 0)$ **b.** $(0, 4)$ **c.** -2 **71.** 0 **73.** positive **75.** negative

77. positive **79.** 2

Group Discussion Questions

81. The steeper the slope, the harder it is for the bicycle rider to pedal. Therefore any road going up is more difficult than a flat road. A road going up 10 ft in 1 mi is harder to pedal than a flat road but less difficult than one going up 10 ft over 50 ft of roadway. A slope of $\dfrac{20}{200} = \dfrac{1}{10}$ is less than $\dfrac{10}{50} = \dfrac{1}{5}$; therefore a road going up 20 ft over 200 ft is less difficult to pedal than one going up 10 ft over 50 ft of roadway (however, the bicycle rider will have to do this work over a longer distance). A road going down has a negative slope, and the rider can coast (or use the brakes) on this section of the roadway.

83. No, two distinct points are needed. **85.** The slopes are not equal. Try $[-100, 100, 10]$ by $[-100, 100, 10]$ for a different perspective.

Cumulative Review

1. B **2.** A **3.** D **4.** E **5.** C

Using the Language and Symbolism of Mathematics 3.2

1. $y = mx + b$ **2.** parallel **3.** perpendicular **4.** vertical; $(5, 0)$

5. horizontal; $(0, -5)$ **6.** $Ax + By = C$ **7.** 1 **8.** point; slope; point; slope

Quick Review 3.2

1. a. C **b.** D **c.** A **d.** B   **2.** $y = -\dfrac{2}{5}x - 4$ **3.** x-intercept; $(-10, 0)$;

y-intercept: $(0, -4)$ **4.** 30   **5.** $\dfrac{15}{2}$

Exercises 3.2

1. a. $m = 2$; y-intercept: $(0, 5)$ **b.** $m = -\dfrac{3}{11}$; y-intercept: $\left(0, -\dfrac{4}{5}\right)$

c. $m = 6$; y-intercept: $(0, 0)$ **3.** $y = 4x + 7$ **5.** $y = -\dfrac{2}{11}x + 5$ **7.** $y = -6$

9. a. $m = \dfrac{2}{3}$ **b.** $(0, -4)$ **11. a.** $m = -\dfrac{5}{3}$ **b.** $(0, 2)$

c. 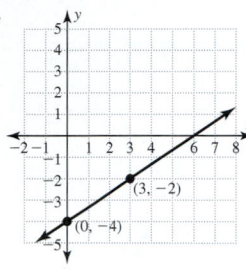   **c.**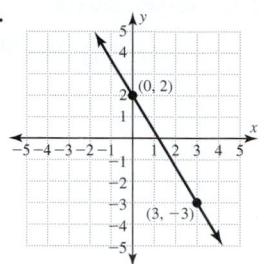

13. a. $m = 0$ **b.** $(0, 2)$

c.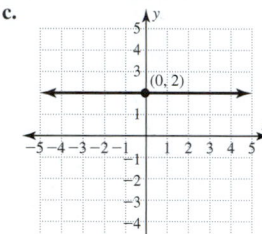

15. a. $m = \dfrac{2}{5}$ **b.** $(0, 1)$ **c.** $f(x) = \dfrac{2}{5}x + 1$ **17. a.** $m = -\dfrac{1}{4}$

b. $(0, -1)$ **c.** $f(x) = -\dfrac{1}{4}x - 1$ **19. a.** 1   **b.** 7   **c.** 7   **d.** $(0, 1)$

e. $y = 7x + 1$ **21. a.** 2   **b.** -1 **c.** $-\dfrac{1}{2}$ **d.** $(0, 8)$ **e.** $y = -\dfrac{1}{2}x + 8$

23. $y - 3 = -4(x - 2)$ **25.** $y - 4 = \dfrac{3}{5}(x + 1)$

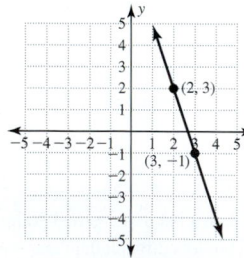   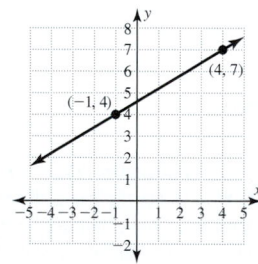

27. $m = 2$; point: $(3, 4)$ **29.** $m = -\dfrac{3}{2}$; point: $(2, -5)$

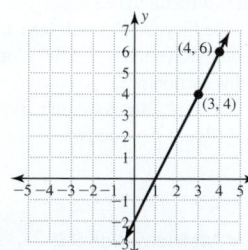   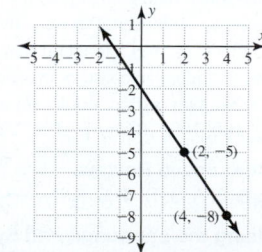

31. $y = \dfrac{2}{3}x + 6$ **33.** $y = \dfrac{1}{4}x + 3$ **35.** $y = \dfrac{1}{5}x + \dfrac{7}{5}$ **37.** $y = \dfrac{5}{3}x + \dfrac{14}{3}$

39. $y = \dfrac{1}{3}x + \dfrac{8}{3}$ **41.** $y = -\dfrac{2}{7}x + \dfrac{13}{7}$

43. $x = 2$ **45.** $y = -5$

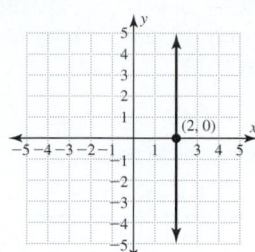   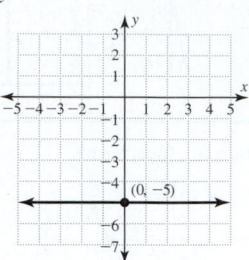

47. $y = 5$ **49.** $x = -4$ **51.** $y = -8$ **53.** neither   **55.** perpendicular

57. parallel   **59.** neither   **61.** $y = \dfrac{3}{7}x + \dfrac{15}{7}$ **63.** $y = \dfrac{3}{4}x + \dfrac{19}{4}$

65. $-2, 1, 4, 7, 10$ **67.** $y = 4x + 7$; $4x - y = -7$ **69.** $y = -\dfrac{2}{3}x - \dfrac{2}{3}$;

$2x + 3y = -2$ **71.** $f(x) = -50x + 500$; the slope gives us the speed of
the truck (50 mi/h). The y-intercept represents the initial distance from the
dispatch center (500 mi). **73. a.** -12 **b.** The jet is using 12 gal of fuel per
min from the main tank. **c.** $f(x) = 3{,}500 - 12x$ **d.** No, there is not
enough fuel in the tank for 300 min. **75.** $f(x) = 50x + 75$; the cost is $50
per hour plus a flat fee of $75 per visit **77.** $f(x) = 0.8x + 3.25$; the cost is
$0.80 per mile plus a flat fee of $3.25 per ride **79.** $y = 2x + 8$; the spring
stretches 2 cm for each kilogram attached; the length of the spring is 8 cm
when no weight is attached **81. a.** $y = \dfrac{7}{300}x + \dfrac{79}{75}$ **b.** $1.19; yes

Group Discussion Questions

83. a. $(22, 21)$ **b.** $\Delta x = 11$ in the table; Δy is 7 with the exception of the
last point **85. a.** $x - 5y = 25$ **b.** $x + 5y = 25$ **c.** $7x - y = 21$
d. $4x + y = 8$ **e.** $y = -2$ **f.** $x = -7$

Cumulative Review

1. E **2.** A **3.** D **4.** B **5.** C

Using the Language and Symbolism of Mathematics 3.3

1. solution **2.** intersection **3.** inconsistent; parallel **4.** one; one
5. infinite; coincide **6. a.** infinitely many **b.** are; zero **c.** are not; one

Quick Review 3.3

1. $m = \dfrac{2}{3}$ **2.** $(0, -3)$

3.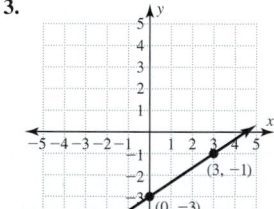

4. $-5, -3, -1, 1$ **5.** $(3, -1)$

Exercises 3.3

1. solution **3.** solution **5.** solution **7.** not a solution **9.** $(-2, 3)$ **11.** $(3, 2)$
13. $(2, 2)$ **15.** $(2, -6)$ **17.** $(5, -8)$ **19.** $(-0.5, 4)$ **21.** $(-6, 4)$
23. $(-3, 2)$ **25.** $(6, -2)$ **27.** one solution; consistent system of
independent equations **29.** no solution; inconsistent system **31.** infinite
number of solutions; consistent system of dependent equations **33.** one
solution; consistent system of independent equations **35.** consistent
system of independent equations **37.** inconsistent system **39.** $(4, 1)$

41. $\left(-\dfrac{9}{2}, -5\right)$ **43.** no solution **45.** infinite number of solutions **47.** $(1, -3)$

49. (2, 3) **51.** $\left(\dfrac{2}{7}, \dfrac{9}{7}\right)$ **53.** $\left(\dfrac{13}{15}, -\dfrac{9}{55}\right)$ **55.** (300, 700) **57.** (0.3, 0.7)

59. (1, −3) **61.** (0, 3) **63.** (30, 40) **65.** (300, 90); each company charges $90 when 300 mi are driven **67.** (2006, 8000); the European Union and the rest of the world each had a wind power capacity of 8,000 megawatts in 2006. **69.** (1, 450); at 3 P.M., each plane is 450 mi from O'Hare Airport. **71.** The smaller angle is 43.75°. The larger angle is 136.25°. **73.** The daily costs will each be $105 when 10 items are produced.

75.

x	$y_1 = 2x + 7$	$y_2 = 13 - x$
−3	1	16
−2	3	15
−1	5	14
0	7	13
1	9	12
2	11	11
3	13	10

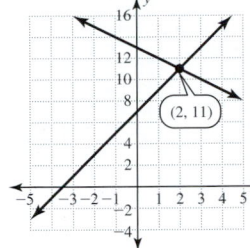

(2, 11); the numbers are 2 and 11.

77. $x = 3; y = 5$ **79.** $y = \dfrac{1}{2}x + 3; y = -4x + 3$

Group Discussion Questions
81. a. yes, graph must show two distinct parallel lines **b.** yes, graph of two lines intersecting at one point **c.** no **d.** no **e.** yes, graph of two coincident lines **83.** Although these lines do not intersect on this window, they do intersect because the slopes of 6 and 7 are different. (45, 300)

85. (3, 4); $\begin{cases} 2x - 3y = -6 \\ 4x - 3y = 0 \end{cases}$

Cumulative Review
1. −12 **2.** −3 **3.** (5, ∞) **4.** (−∞, −2] **5.** [−4, 10)

Using the Language and Symbolism of Mathematics 3.4
1. equal **2.** ordered pair **3.** false **4.** true **5.** a conditional equation **6.** an identity **7.** a contradiction **8.** consistent; independent; (3, 4) **9.** inconsistent; independent; no **10.** consistent; dependent; infinite

Quick Review 3.4
1. A **2.** C **3.** B **4.** 90° **5.** $y = \dfrac{2}{5}x + 2$

Exercises 3.4
1. (4, 5) **3.** (2, 0) **5.** (−34, 8) **7.** $\left(\dfrac{4}{7}, \dfrac{5}{7}\right)$ **9.** (0, −6) **11.** (−2, −1)

13. $\left(3, -\dfrac{3}{5}\right)$ **15.** $\left(-\dfrac{13}{3}, -7\right)$ **17.** (−1, 2) **19.** (−2, −3) **21.** (12, 9)

23. (0, −6) **25.** (3, 2) **27.** (−4, −2) **29.** (12, 15) **31.** infinite number of solutions **33.** no solution **35.** (0, 0) **37.** (3, 4) **39.** (2, −1) **41.** $\left(\dfrac{1}{2}, -\dfrac{1}{3}\right)$

43. a. $y = -2x + 2$ **b.** $y = -\dfrac{1}{2}x + \dfrac{1}{2}$ **c.** (1, 0) **45.** B; the numbers are 75 and 45. **47.** C; the numbers are 30 and 0.

49.

x	$y_1 = 8 - x$	$y_2 = x - 2$
2	6	0
3	5	1
4	4	2
5	3	3
6	2	4
7	1	5
8	0	6

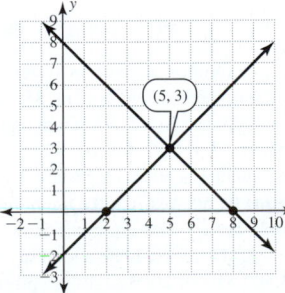

(5, 3); the two numbers are 5 and 3.

51. $x + y = 90; y = x + 18$; 36° and 54° **53.** $\dfrac{x + y}{2} = 76; x - y = 28$;

the test scores are 90 and 62. **55.** $\dfrac{3x + y}{4} = 66; x - y = 8$; Tiger scored a

68 in each of his first three rounds, and he scored a 60 in the last round. **57. a.** yes **b.** ≈1.1 years **59. a.** $y = 0.5x + 1$ **b.** $y = 0.2x + 2$

c. $\left(\dfrac{10}{3}, \dfrac{8}{3}\right)$ **d.** yes **e.** Each company charges $2.67 when 3.3 mi is driven.

61. a. $y = 180x + 1{,}500$ **b.** $y = 140x + 2{,}000$ **c.** (12.5, 3750) **d.** yes

e. After 12.5 years, each plan is worth $3,750.00. **63.** 1; 2; $\left(\dfrac{5}{3}, \dfrac{11}{3}\right)$

65. (4.65, 2.75) **67.** (−11.3, −1.1)

Group Discussion Questions
69. a. (4, 1) **b.** (4, 1) **c.** You can solve for either variable first. Select the one that will make your mental arithmetic easier. **71.** $B = -6$ **73.** $C = 6$

75. a. $\left(\dfrac{3b}{2}, \dfrac{b}{2a}\right)$ **b.** $\left(\dfrac{b}{a}, -2b\right)$

Cumulative Review
1. −2 is not a solution; 2 is a solution **2.** −2 is a solution; 2 is not a solution **3.** $x = 16$ **4.** $x = 6$ **5.** $x = 8$

Using the Language and Symbolism of Mathematics 3.5
1. equal **2.** elimination; eliminate **3.** opposites **4.** inconsistent; no **5.** consistent; dependent; infinite

Quick Review 3.5
1. 105 **2.** $-11x - 13y - 2$ **3. a.** $n + 1, n + 2, n + 3$ **b.** $n + 2, n + 4, n + 6$ **c.** $n + 1, n + 3, n + 5$ **4.** 85 **5.** positive

Exercises 3.5
1. (2, 2) **3.** (−8, 7) **5.** $\left(\dfrac{1}{6}, -1\right)$ **7.** $\left(-2, \dfrac{3}{2}\right)$ **9.** (3, −5) **11.** (4, 5)

13. (1, −1) **15.** (−9, 11) **17.** (19, 0) **19.** (0, 0) **21.** (−3, 5) **23.** (6, 6)

25. no solution **27.** infinite number of solutions **29.** $\left(\dfrac{4}{3}, -\dfrac{1}{6}\right)$ **31.** (5, 9)

33. (2, 2) **35.** $\left(\dfrac{1}{2}, -\dfrac{1}{2}\right)$ **37.** no solution **39.** (0, 3) **41.** (3, −1)

43. a. $y = -\dfrac{4}{11}x + \dfrac{13}{11}$ **b.** $y = -\dfrac{3}{17}x + \dfrac{1}{17}$ **c.** (6, −1) **45.** C; the numbers are 2 and −5 **47.** D; the numbers are −2 and 6 **49.** $x + y = 88; x - y = 28$; the numbers are 58 and 30 **51.** $x + y = 102; y = 2x$; the numbers are 34 and 68 **53.** $\dfrac{4x + y}{5} = 25; y - x = 15$; he scored 22 points in each of the first 4 games and 37 points in the fifth game. **55. a.** $y = -0.5x + 1{,}030.5$ **b.** $y = 1.5x - 2{,}959.5$ **c.** (2005, 33) **d.** In 2005, 33% attended 4-year colleges and 33% attended community colleges. **57.** 0; 1; $\left(\dfrac{5}{11}, 2\right)$

59. (10, 20) **61.** (4.4, 5.5)

Group Discussion Questions
63. a. (1.974, 14.748) **b.** In 2002, the percent of students enrolling in both systems was approximately 14.7%. **c.** the Web Method **d.** the phone method **e.** Terminate the phone system (and perhaps add more support for the Web-based system.) The shift away from the phone system indicates that the college cannot justify continuing this expense. **65. a.** The addition method would produce an identity involving only constants. **b.** The addition method would produce a contradiction involving only constants.

67. $\left(\dfrac{c}{a}, 0\right)$

Cumulative Review
1. $x \geq -3$ **2.** $x < -2$ **3.** $-2 < x \leq 6$ **4.** $-1 < x \leq 2$ **5.** $x \leq 4$

Using the Language and Symbolism of Mathematics 3.6
1. find **2.** variable **3.** algebraic **4.** solve **5.** reasonable **6.** mixture **7.** rate **8.** rate; time **9.** principal; rate; time **10.** work; time

Quick Review 3.6

1. fixed **2.** variable **3.** restrictions **4.** equivalent **5.** supplementary

Exercises 3.6

1. a. Let x = width in centimeters and y = length in centimeters.
b. The perimeter is 204 cm; the length is 6 more than the width.
c. $2x + 2y = 204$; $y = x + 6$ **d.** $(48, 54)$ **e.** yes **f.** The dimensions are 48 by 54 cm. **3.** The numbers are 28 and 72. **5.** The base is 32 cm, and the two equal sides measure 20 cm each. **7.** The numbers are 22 and 35.
9. The scores were 81 for each of the four equivalent rounds and 91 for the fifth round. **11. a.** The total grade points earned for the semester is 30.
b. The grade point average is 2.31. **c.** The student would need 9 semester hours with a grade of A to meet the goal of a GPA of 3.00.
13. The student would need 6 semester hours with a grade of A to meet the goal of a GPA of 2.75. There would be a total of 99 grade points. **15.** The fixed monthly charge is $30.00 and the charge for each pay-per-view movie is $3.50. **17.** Hire 47 union employees and 3 union stewards.
19. There were 332 adult tickets and 456 student tickets sold. **21.** The two angles are 31° and 59°. **23.** The two angles are 35° and 145°. **25.** The manager should order 60 left-handed desks and 540 right-handed desks.
27. After 20 days the total cost for using either bulb is $2.90.
29. $2,500 was invested at 9% and $9,500 was invested at 7%. **31.** $5,000 was invested at 8% and $4,000 was invested at 5%. **33.** The trains are traveling at speeds of 94 km/h and 84 km/h. **35.** The 6:30 train has a speed of 100 km/h. The 6:00 train has a speed of 85 km/h. **37.** The speed of the riverboat is 15 km/h and the speed of the current is 9 km/h. **39.** Use 20 liters of the 33% solution and 60 liters of the 5% solution. **41.** The goldsmith should use 220 grams of the 80% pure gold alloy. **43.** Use 8.4 liters of pure water. **45.** The older bricklayer can lay 280 bricks per hour while the younger bricklayer can lay 220 bricks per hour.
47. a. $(5.88, 200)$ **b.** In 1995 (5.88 years after 1990), the United States and Europe were producing 200,000 publications. **49. a.** $y = 200x + 500$
b. $y = 100x + 1,000$ **c.** $(5, 1500)$ **d.** yes **e.** In each option, $1,500 is paid after 5 months. **51. a.** fixed cost: $2,500; variable cost: $8 per pair of shoes **b.** The proposed answer does not meet the restrictions on the variable because the fixed cost cannot be negative. **53. a.** It will take the helicopter 2 h 20 min to fly to Pearl Harbor. **b.** From the point of no return, it will take 1 h to fly to Pearl Harbor. **c.** It will take the helicopter 1 h 20 min to reach the point of no return.

Group Discussion Questions

55. a. The sum of two numbers is 500. Five times the first number plus 8 times the second number gives a sum of 3,475. Find both numbers.
b. The total receipts for a basketball game are $3,475 for 500 tickets sold. Students paid $5 for admission and adults paid $8. The ticket takers counted only the number of tickets, not the type of ticket. Determine the number of each type of ticket sold. **c.** A small building contractor paid two bricklayers for laying 3,475 bricks. The younger bricklayer averaged 5 brick/min, and the older bricklayer averaged 8 bricks/min. If the total time they worked was 500 minutes, determine the number of minutes each worked. **57. a.** Yes, if you multiply both sides of the second equation in Exercise 56 by 100 you obtain the second equation in Exercise 55, Yes.
b. This is an important concept to discuss with students. By learning just a few basic principles, they can solve problems from many different sources.
c. If a student can become comfortable solving the numeric word problem in Exercise 55a, then this can become a stepping stone to all the other problems in Exercises 55 and 56.

Cumulative Review

1. $6x - 18$ **2.** $x = 3$ **3.** solution **4.** not a solution **5.** x-intercept: $(4, 0)$; y-intercept: $(0, -3)$

Review Exercises for Chapter 3

1. 2 **2.** $-\dfrac{3}{5}$ **3.** $\dfrac{4}{7}$ **4.** -1 **5.** 0 **6.** $\dfrac{2}{5}$ **7.** undefined **8.** 0 **9.** undefined **10.** 0.5

11. a. 4 **b.** 5 **c.** -4 **d.** $\dfrac{4}{5}$ **e.** $-\dfrac{4}{5}$

12. a.

x	y
0	6
5	9
10	12
15	15
20	18

b.

x	y
0	6
10	12
20	18
30	24
40	30

c.

x	y
0	6
−5	3
−10	0
−15	−3
−20	−6

13. $\dfrac{1}{6}$ **14. a.** parallel **b.** perpendicular **c.** neither **15.** $m \approx 1$; $m = 0.9$

16. $m \approx -1$; $m = -1.1$

17. 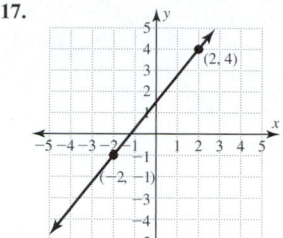 **18.**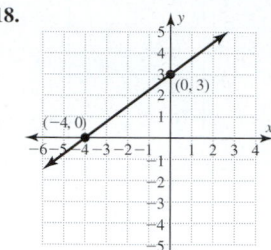

19. 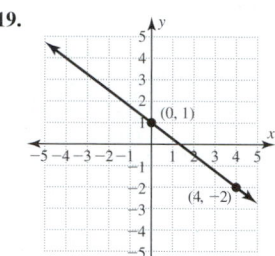 **20.**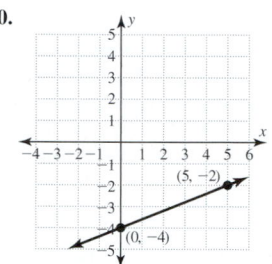

21. 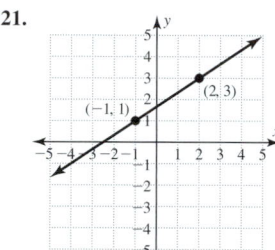 **22.**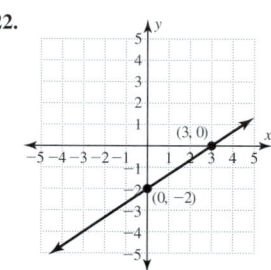

23. $y = -\dfrac{1}{2}x + 6$ **24.** $y = \dfrac{2}{3}x - 2$ **25.** $y = -3x + 11$ **26.** $y = \dfrac{3}{2}x - 4$

27. $y = 5x$ **28.** $y = \dfrac{1}{2}x + 4$ **29.** $y = 9$ **30.** $x = 4$ **31. a.** $(0, -4)$ **b.** 2 **c.** 3

d. $\dfrac{3}{2}$ **e.** $y = \dfrac{3}{2}x - 4$ **32.** A, B, D **33.** D **34.** $(2, 3)$ **35.** $(3, -1)$ **36.** $(-5, 15)$

37. consistent system of dependent equations **38.** inconsistent system
39. $(6, 5)$ **40.** infinitely many solutions **41.** no solution **42.** $(-0.5, 0.25)$

43. $(11, -7)$ **44.** no solution **45.** $\left(0, \dfrac{4}{3}\right)$ **46.** $(-1, -1)$ **47.** $(5, 11)$

48. $\left(\dfrac{18}{11}, \dfrac{1}{22}\right)$ **49.** $(-148, 225)$ **50.** $(25, 60)$ **51.** no solution **52.** $\left(-\dfrac{2}{7}, \dfrac{3}{5}\right)$

53. one solution; consistent system of independent equations **54.** no solution; inconsistent system **55.** an infinite number of solutions; consistent system of dependent equations **56.** no solution; inconsistent system **57.** B; (60, 40); the numbers are 60 and 40. **58.** D; (60, −40); the first number is 60 and the second is −40. **59.** C; (20, 0); the numbers are 20 and 0. **60.** A; (70, 50); the numbers are 70 and 50. **61.** The student would need 9 semester hours with a grade of A to meet the goal of a GPA of 3.00. There would be a total of 162 grade points. **62.** Both options cost $3,500 when 10,000 pages are printed. **63.** The angles are 46° and 134°.
64. The diameters were 24.975 cm for each of the nine equivalent wheels and 25.725 cm for the tenth wheel. **65.** The amount of the car loan is $3,000, and the amount of the educational loan is $6,000. **66.** The base is

9 centimeters. **67. a.** The trains will be 450 km apart after 2.5 hours.
b. The rates of the trains are 96 km/h and 104 km/h. **68. a.** It will take the
workers 26 hours to lay 1,170 bricks. **b.** The older bricklayer lays
26 bricks/h and the younger bricklayer lays 22 bricks/h. **69. a.** 60 liters of
petrochemicals should be used. **b.** Use 45 L worth \$6/L and 55 L worth
\$9/L. **70. a.** $y = 0.95x + 6$ **b.** The slope $m = 0.95$ represents the cost per
CD. The y-intercept $(0, 6)$ represents the initiation fee, \$6. **c.** $y = 1.95x$
d. The slope $m = 1.95$ represents the cost per CD; the y-intercept $(0, 0)$
represents the initiation fee, \$0. **e.** $(6, 11.7)$ **f.** When 6 CDs are purchased,
each plan costs \$11.70. **71. a.** $y = 30x + 20$ **b.** The slope $m = 30$
represents the charge per hour; the y-intercept $(0, 20)$ represents a \$20 flat
fee. **c.** $y = 26x + 30$ **d.** The slope $m = 26$ represents the charge per
hour; the y-intercept $(0, 30)$ represents a \$30 flat fee. **e.** $(2.5, 95)$ **f.** The
total charge for each shop is \$95 when the repairs require 2.5 hours.

Mastery Test for Chapter 3

1. a. 2 **b.** -3 **c.** $-\dfrac{3}{2}$ **d.** $\dfrac{2}{5}$ **e.** -3 **f.** 7 **2. a.** perpendicular **b.** parallel

c. perpendicular **d.** neither **3. a.** 5 **b.** 20 **c.** 4 **d.** The altitude is increasing
4 ft/s. **e.** 5 minutes

4. a. $y = \dfrac{2}{3}x + 5$ \qquad\qquad **b.** $y = -\dfrac{5}{3}x - 2$

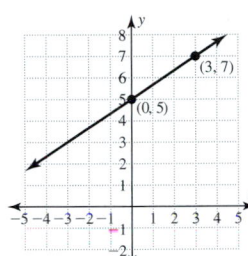 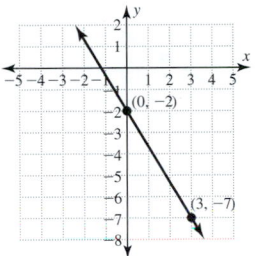

5. a. $y = 4x + 7$ \qquad\qquad **b.** $y = \dfrac{1}{3}x - \dfrac{7}{3}$

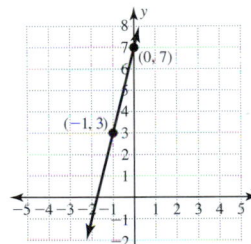 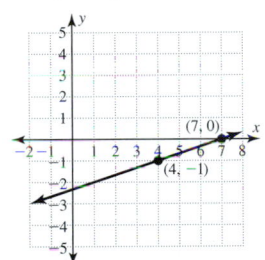

6. a. $y = 3$ **b.** $x = -4$ **c.** $x = -4$ **d.** $y = 3$ **7. a.** solution **b.** not a
solution **c.** not a solution **8. a.** $(0, 2)$ **b.** $(2, 5)$ **c.** $(-2, -1)$ **d.** $(12, 5)$
9. a. B **b.** C **c.** A **10. a.** $(-2, 7)$ **b.** $\left(\dfrac{1}{14}, -\dfrac{5}{7}\right)$ **c.** $\left(2, -\dfrac{1}{3}\right)$ **d.** $\left(\dfrac{37}{10}, \dfrac{13}{20}\right)$
11. a. $(-2, 1)$ **b.** $(-1, 1)$ **c.** infinitely many solutions **d.** no solution
12. a. The angles are 13° and 77°. **b.** \$7,500 was invested in bonds.
\$2,500 was invested in the savings account. **c.** Use 15 mL of the 30%
solution and 10 mL of the 80% solution. **d.** The jet's speed is 475 mi/h.
The tanker's speed is 225 mi/h.

Group Project for Chapter 3

1. a. yes **b.** 6 ft 8 in **c.** 7 ft $3\dfrac{3}{4}$ in (9 treads for 10 steps) **d.** 0.82

2. Answers can vary.

Chapter 4
Using the Language and Symbolism of Mathematics 4.1
1. linear **2.** 1 **3.** equivalent **4.** conditional **5.** true **6.** addition **7.** subtraction
8. a. $>$ **b.** $>$ **c.** $>$ **9.** distributive **10.** $\geq$ **11.** $\leq$ **12.** $>$ **13.** $\leq$ **14.** $\geq$

Quick Review 4.1
1. -2 **2.** parenthesis **3.** $x = 5$ **4.** $x = 4$ **5.** $x = -5$

Exercises 4.1
1. B **3. a.** $(-\infty, -4]$ **b.** $(5, \infty)$ **c.** $(-\infty, 3)$ **d.** $[-1, \infty)$ **5. a.** not a
solution **b.** solution **c.** not a solution **d.** solution **7. a.** solution **b.** not a
solution **c.** solution **d.** not a solution **9.** $(4, \infty)$ **11.** $(-\infty, 1]$ **13.** $[4, \infty)$
15. $(-\infty, 0)$ **17.** $[7, \infty)$ **19.** $(-\infty, -1)$ **21.** $(-\infty, 1)$ **23.** $[2, \infty)$
25. $(-\infty, -2)$ **27.** $[40, \infty)$ **29. a.** $x = 1$ **b.** $(-\infty, 1)$ **c.** $(1, \infty)$
31. a. $x = -3$ **b.** $(-\infty, -3]$ **c.** $[-3, \infty)$ **33. a.** $x = 500$ **b.** $(500, 800]$
c. $[0, 500)$ **d.** When 500 hours of overtime is worked, the spending limit is
met. When more than 500 hours of overtime is worked, the spending limit is
exceeded. When less than 500 hours of overtime is worked, the spending
limit is not met. **35. a.** $x = 5$ **b.** $(-\infty, 5]$ **c.** $[5, \infty)$ **37. a.** $x = -5$
b. $(-5, \infty)$ **c.** $(-\infty, -5)$ **39. a.** $x = 8$ **b.** $[0, 8)$ **c.** $(8, 24]$ **d.** Each firm
charges the same amount when 8 boxes of paper are ordered. Ace Office
Supply charges less for orders of fewer than 8 boxes of paper and more for
orders of more than 8 boxes of paper. **41.** $(1, \infty)$; $(1.112, \infty)$
43. $(-\infty, -22)$; $(-\infty, -21.93)$ **45.** $[3, \infty)$; all values of x greater than or
equal to 3 **47.** $(-\infty, 2)$; all values of x less than 2 **49.** $5 - 2x \leq 7 - 3x$;
$(-\infty, 2]$ **51.** $2w + 3 \geq w + 11$; $[8, \infty)$ **53.** $3(m + 5) > 2m + 12$;
$(-3, \infty)$ **55.** $(8, \infty)$ **57.** $(-\infty, -1]$ **59.** $[-20, \infty)$ **61.** $(1.5, \infty)$
63. $(2, \infty)$ **65.** $[2, \infty)$ **67.** He must earn at least 95 points. **69.** The length
can be from 0 to 60 ft. **71.** The company will lose money when less than 450
mice are produced; the company will have a profit if more than 450 mice are
produced.

Group Discussion Questions
73. $a + b > c$; $a + c > b$; $b + c > a$ **75.** Students answers can vary.
Example: The actual solution could be $x \geq 3$. A student could obtain an
incorrect answer of $x \geq 20$. A value of $x = 25$ would check even though
the solution is incorrect.

Cumulative Review
1. -7 **2.** -1 **3.** 3 **4.** 1 **5.** 2

Using the Language and Symbolism of Mathematics 4.2
1. multiplication **2.** multiplication **3.** $>$ **4.** $<$ **5.** least common
denominator **6.** distributive

Quick Review 4.2
1. $x = -12.5$ **2.** $x = 21$ **3.** $x = -10$ **4.** same **5.** D

Exercises 4.2
1. a. $>$ **b.** $>$ **c.** $>$ **d.** $<$ **3.** $-2x < 4$; $\dfrac{-2x}{-2} > \dfrac{4}{-2}$; $x > -2$; $(-2, \infty)$
5. $-3x + 4 \geq 7$; $-3x + 4 - 4 \geq 7 - 4$; $-3x \geq 3$;
$\dfrac{-3x}{-3} \leq \dfrac{3}{-3}$; $x \leq -1$; $(-\infty, -1]$ **7.** $[2, \infty)$ **9.** $(-\infty, -8)$ **11.** $(16, \infty)$
13. $(-\infty, 0]$ **15.** $(-\infty, 2.3)$ **17.** $(-\infty, -2)$ **19.** $(-\infty, -5]$ **21.** $(-\infty, 0]$
23. $(2, \infty)$ **25.** $(-\infty, 3]$ **27.** $(-5, \infty)$ **29.** $(-\infty, -5)$ **31.** $[-2, \infty)$
33. $(1, \infty)$ **35.** $\left[\dfrac{2}{3}, \infty\right)$ **37.** $[2, \infty)$ **39.** $\left(-\dfrac{31}{3}, \infty\right)$ **41.** $(1, \infty)$
43. $[-8, \infty)$ **45. a.** $x = -1$ **b.** $(-\infty, -1)$ **c.** $(-1, \infty)$ **47. a.** $x = -2$
b. $[-2, \infty)$ **c.** $(-\infty, -2]$ **49. a.** $x = 3$ **b.** $(-\infty, 3)$ **c.** $(3, \infty)$
51. a. $x = -5$ **b.** $[-5, \infty)$ **c.** $(-\infty, -5]$ **53.** $(-\infty, 7]$; $(-\infty, 7.1]$
55. $(-\infty, 8)$; $(-\infty, 8.9)$ **57.** $(-2, \infty)$; all values of x greater than -2
satisfy the inequality **59.** $(-\infty, 4]$; all values of x less than or equal
to 4 satisfy the inequality **61.** $2(x + 5) < 6$; $(-\infty, -2)$
63. $4(y - 7) \geq 6(y + 3)$; $(-\infty, -23]$ **65. a.** $x = 2$; each company
charges the same amount when the machines are rented for 2 hours
b. $(2, \infty)$; the Dependable Rental Company charges less when the
machines are rented for more than 2 hours **c.** $(0, 2)$; the Dependable
Rental Company charges more when the machines are rented for less than
2 hours **67. a.** $x = 30$; each store charges the same amount when 30 shirts
are ordered **b.** $(0, 30)$; Mel's Sports charges less for orders of less than
30 shirts **c.** $(30, 60)$; Mel's sports charges more for orders of more than
30 shirts **69.** The temperature must be at least 5°C. **71.** The family can
borrow at most \$60,000. **73. a.** $[0, 200)$; The company loses money when
less than 200 bricks are produced. **b.** $(200, \infty)$; The company earns a profit
when more than 200 bricks are produced.

Group Discussion Questions

75. In step 2, the inequality should be $<$. **77.** $a = 0$

Cumulative Review

1. $f(-2) = 1$ **2.** $x = 3$ **3.** $f(x) = 16 - 2x$ **4.** $(0, 8)$ **5.** $f(3) = 10$; If two 3-ft pieces are cut from the 16-ft board, a 10-ft piece will remain.

Using the Language and Symbolism of Mathematics 4.3

1. conditional **2.** unconditional **3.** contradiction **4.** compound; and **5.** intersection **6.** union

Quick Review 4.3

1. identity **2.** contradiction **3.** conditional equation **4.** $x = 12$ **5.** $x = 10$

Exercises 4.3

1. C **3.** B **5.** conditional inequality; $(-\infty, 12]$ **7.** contradiction; no solution **9.** unconditional inequality; $\mathbb{R}$ **11.** $x > 2$ and $x \le 8$
13. $5x + 2 > -13$ and $5x + 2 < 7$ **15.** $0 \le x \le 4$
17. $A \cap B = \{0, 1\}$; $A \cup B = \{-5, -4, 0, 1, 2, 3\}$ **19.** $x \le 5$ or $x > 7$
21. $-2 < x \le 5$ or $7 < x < 10$ **23.** $A \cap B = [5, 7]$; $A \cup B = (-4, 11)$
25. $A \cap B = [5, 11)$; $A \cup B = (-4, \infty)$ **27.** $A \cap B = [-4, -2]$;
$A \cup B = \mathbb{R}$ **29.** $[-1, 6)$; $\xleftarrow{\hspace{0.3cm}\underset{-1}{\bullet}\rule{1cm}{0.4pt}\underset{6}{\circ}\hspace{0.3cm}}$; x is greater than
or equal to -1 and less than 6 **31. a.** $1 \le x < 3$ **b.** $x \le 1$ or $x > 3$
33. a. $x \ge 4$ or $x \le -2$ **b.** $-2 < x < 4$ **35.** $(-6, -4)$ **37.** $[-2, 1)$
39. $[0, 4]$ **41.** $[0, 4)$ **43.** $(-1, 2)$ **45.** $[2, 4]$ **47.** $(-1, 1]$
49. $\left(-\infty, \dfrac{11}{7}\right]$ **51.** $(-1, 2)$ **53.** $[-5, 1]$ **55.** $\left(\dfrac{1}{3}, \dfrac{8}{5}\right)$
57. $(-\infty, -2) \cup (1, \infty)$ **59.** $(-\infty, -3) \cup (-2, \infty)$ **61.** $[-1, 2)$
63. $(-1, 4)$ **65.** $(1, 5)$ **67.** $(-2, 3]$ **69.** $[-5, 1)$; all real numbers greater than or equal to -5 and less than 1 **71.** $[1, 2]$; all real numbers greater than or equal to 1 and less than or equal to 2 **73.** $(2, 3)$; $(1.9, 3.1)$
75. $(-3, 2]$; $(-2.9, 2.1]$ **77.** $155 < x < 190$ **79.** $(0, 150) \cup (220, 270)$
81. a. $4,000 \le 250x \le 6,000$ **b.** $[16, 24]$; For a safety factor from 2 to 3 use 16 to 24 steel wires in the cable. **83.** $[50, 100]$; The strength requirements are met when 50 to 100 fiber strands are used. **85.** The acceptable range of temperatures is between $5°C$ and $35°C$. **87.** The third side must be between 4 and 20 m.

Group Discussion Questions

89. not possible; this interval of real numbers is infinite. **91.** The height of the can is $6r$ where r is the radius of a tennis ball. The circumference of the can is $2\pi r$. Because $2\pi > 6$, the circumference is larger. (This is a good problem to actually measure in class.)

Cumulative Review

1. 87 **2.** 10 **3.** $<$ **4.** $=$ **5.** $>$

Using the Language and Symbolism of Mathematics 4.4

1. distance **2. a.** $<$ **b.** $=$ **c.** $>$ **3.** $|x - a| = d$ **4.** tolerance **5.** isolate

Quick Review 4.4

1. 34 **2.** 34 **3.** 1 **4.** 7 **5.** -1

Exercises 4.4

1. $|x| < 5$ **3.** $|x| \ge 2$ **5.** $|x| < 4$ **7.** $|x| \le 5$ **9.** $|x - 1| = 3$
11. $|x + 7| < 3$ **13.** $|x - 4| \le 2$ **15.** $|x - 1| \ge 2$ **17.** $|x| \le 3$
19. $|x| > 3$ **21.** $|x| < 7$ **23. a.** $(-3, 3)$ **b.** $[-3, 3]$
c. $(-\infty, -3] \cup [3, \infty)$ **25.** $x = 6$ or $x = -6$ **27.** $x = 0$
29. $x = -3$ or $x = 7$ **31.** $x = -7$ or $x = 3$ **33.** $(-\infty, -2] \cup \left[-\dfrac{2}{3}, \infty\right)$
35. $[-4, 4]$ **37.** $(-4, 1)$ **39.** $\left(-\dfrac{3}{2}, -\dfrac{1}{2}\right)$ **41.** $(-\infty, -4) \cup (3, \infty)$
43. $(-\infty, -97] \cup [99, \infty)$ **45.** $\left(-4, \dfrac{10}{3}\right)$ **47.** $x = \dfrac{1}{3}$ or $x = -1$
49. $x = 3$ or $x = \dfrac{1}{3}$ **51.** $x = 4$ or $x = -\dfrac{2}{5}$ **53. a.** $x = 6$ or $x = -2$
b. $(-2, 6)$ **c.** $(-\infty, -2) \cup (6, \infty)$ **55. a.** $x = 6$ or $x = 2$ **b.** $(2, 6)$
c. $(-\infty, 2] \cup [6, \infty)$ **57. a.** $x = -4$ or $x = 0$ **b.** $[-4, 0]$
c. $(-\infty, -4] \cup [0, \infty)$ **59. a.** $x = -5$ or $x = 15$ **b.** $(-5, 15)$
c. $(-\infty, -5] \cup [15, \infty)$ **61.** $(-2, 0)$; values greater than -2 and less

than 0 **63.** $(-1, 2)$; values greater than -1 and less than 2
65. $|x + 3| < 9$ **67.** $|x - 11| \le 15$ **69.** $|x - 2| > 4$
71. $x \approx \pm 5$; $x = \pm 4.9$ **73.** $x \approx 7$ or $x \approx -23$; $x = 7.1$ or $x = -22.96$
75. $(16, 28)$ **77.** $|x - 15| \le 0.001$; $[14.999, 15.001]$
79. $|x - 16| \le 0.25$; $[15.75, 16.25]$ **81.** $|x - 25| \le 2$; $[23, 27]$
83. a. $3,000 \le 500x + 250 < 9,000$ where x is the number of synthetic strands. **b.** from 6 to 17 strands **85. a.** $100 < 250x - 150 < 250$
b. $1.0 < x < 1.6$; Use a strand diameter between 1.0 and 1.6 inches to produce a cable that can support a load with a safety factor between 2 and 5 (between 100,000 and 250,000 lb) for this bridge.

Group Discussion Questions

87. $|3x - 5| = 26$ is equivalent to $3x - 5 = -26$ or $3x - 5 = 26$. This absolute value equation has two solutions, $x = -7$ and $x = \dfrac{31}{3}$. Also note that the absolute value of a quantity cannot be applied to the expression term by term.

Cumulative Review

1.
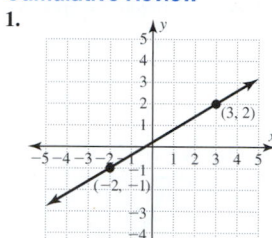

2.
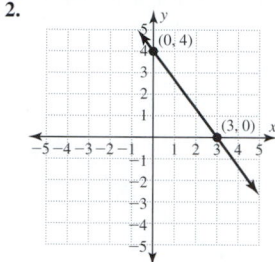

3.
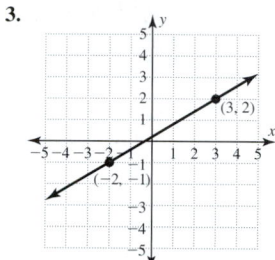

4.
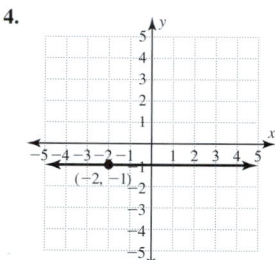

5.
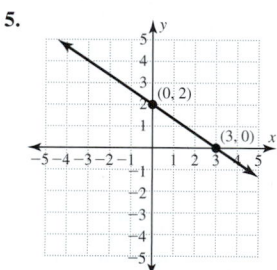

Using the Language and Symbolism of Mathematics 4.5

1. ordered pair; true **2.** half-planes **3.** solid **4.** dashed **5.** contains
6. not contain **7.** upper **8.** lower

Quick Review 4.5

1. x-intercept: $(5, 0)$; y-intercept: $(0, 2)$

2. $y = -\dfrac{2}{5}x + 2$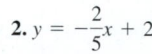

3.
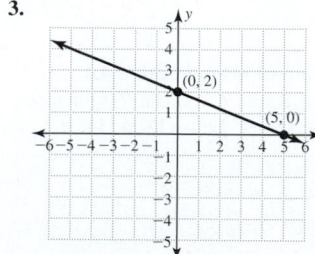

4.
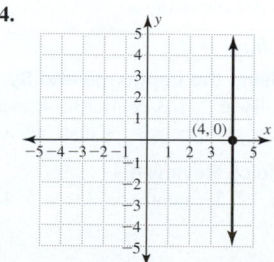

5.

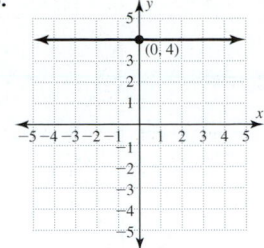

Exercises 4.5

1. a. solution **b.** solution **c.** not a solution **d.** not a solution **3. a.** not a
solution **b.** solution **c.** not a solution **d.** solution **5.** *A* and *B* are solutions
7. *B* and *C* are solutions **9.** A

11.

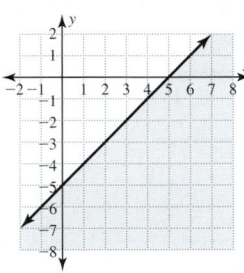

13.

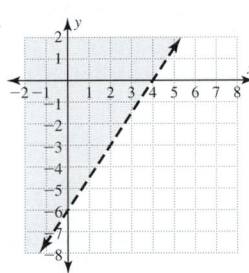

15.

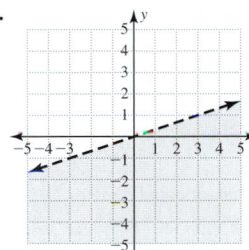

17.

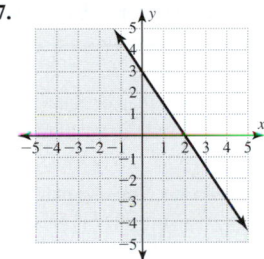

19. *A* and *C*

21. a.

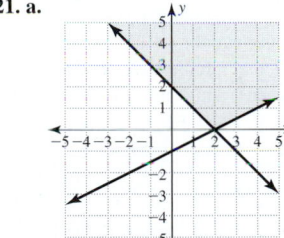

b.

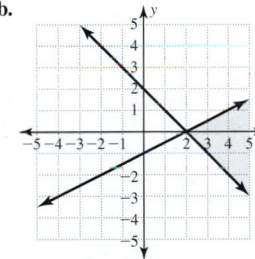

c.

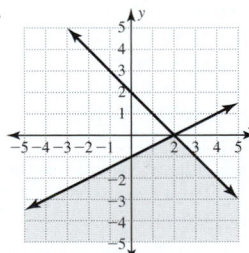

d.

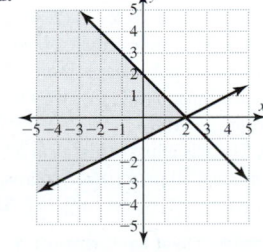

23. a.

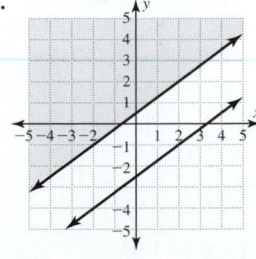

b. no solution

c.

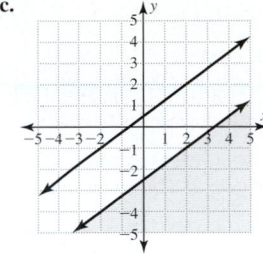

d.

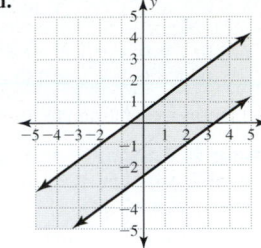

25.

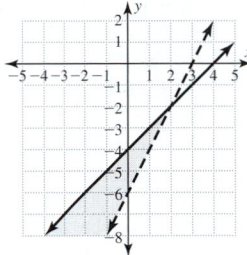

27.

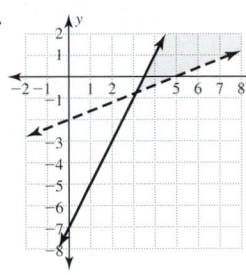

29.

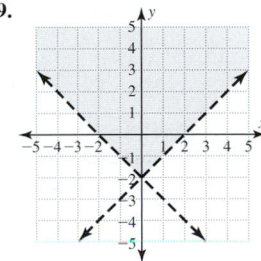

31.

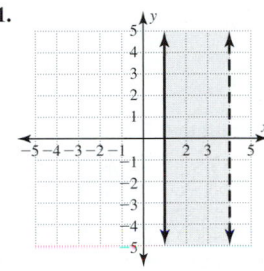

33.

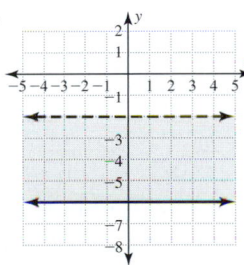

35.

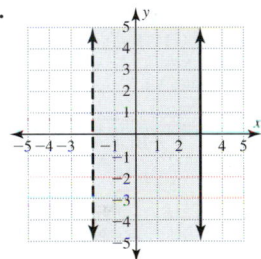

37. no solution

39.

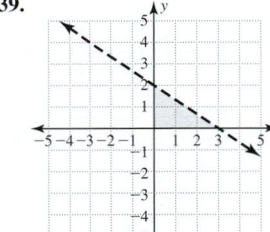

41.

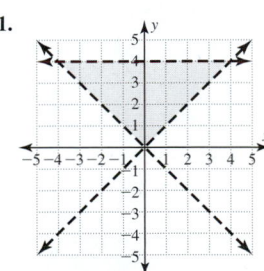

43.

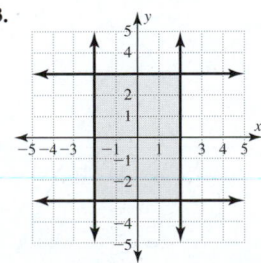

45.

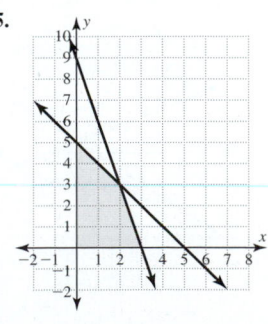

47. $y \leq x - 4; y \leq \dfrac{x}{2} - 3$ **49.** $x \geq y + 2$ **51.** $y \leq x + 4$

53. $x > 0; y > 0; x + y \leq 10$ **55.** $x \geq 0; y \geq 0; x + 2y \leq 5$

57. $x > 0; y > 0; x + y \leq 150$ **59.** $x \geq 0; y \geq 0; 40x + 50y \leq 4,400$

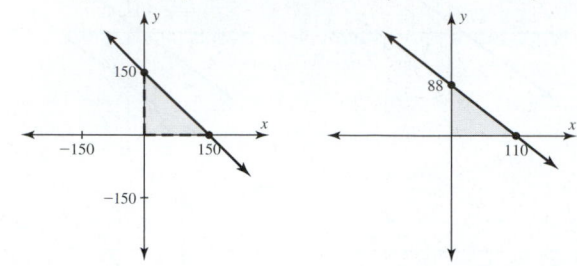

Group Discussion Questions

61. a. A hardware store sold two pieces of steel cable of lengths x and y from a spool containing 100 m. Write a system of inequalities to describe the possible lengths x and y. **b.** A lecture hall can seat at most 100 students. If x represents the number of female students and y the number of male students, write a system of inequalities to describe the possible values of x and y.

63. $\begin{cases} y \leq 4 \\ x - y \leq -1 \\ x + y \geq 2 \end{cases}$

Cumulative Review

1. 36 **2.** 14 **3.** 9 **4.** 33 **5.** $\dfrac{36}{25}$

Review Exercises for Chapter 4

1. a. not a solution **b.** solution **c.** solution **d.** not a solution **2.** C and D
3. B and D **4. a.** not a solution **b.** solution **c.** not a solution **d.** solution
5. a. $x = 3$ **b.** $(3, \infty)$ **c.** $(-\infty, 3)$ **6. a.** $x = -1$ **b.** $(-\infty, -1)$ **c.** $(-1, \infty)$
7. $(-\infty, -2)$ **8.** $(1, \infty)$ **9.** $(-\infty, -5]$ **10.** $[-4, \infty)$ **11.** $(-\infty, -3)$

12. $[0, \infty)$ **13.** $\left(-\dfrac{1}{2}, \infty\right)$ **14.** $(-\infty, 3)$ **15.** $[2, \infty)$ **16.** $(-\infty, -3]$

17. $(-3, \infty)$ **18.** $\left(-\infty, -\dfrac{4}{7}\right)$ **19.** $(-\infty, -3)$ **20.** $(-\infty, -13]$ **21.** $\mathbb{R}$

22. $\left(-\infty, \dfrac{25}{6}\right]$ **23.** $(-\infty, -22)$ **24.** $(-24, \infty)$ **25.** $(2, 6]$ **26.** $[0, 25)$

27. $(-140, -98]$ **28.** $[4, 14]$ **29.** $[-2, 2]$ **30.** $(-1, 5)$ **31.** conditional;
$(0, \infty)$ **32.** conditional; $(-\infty, 0]$ **33.** unconditional; $\mathbb{R}$ **34.** contradiction;
no solution **35.** $[-1, 5]$ **36.** $[-1, 3)$

37.
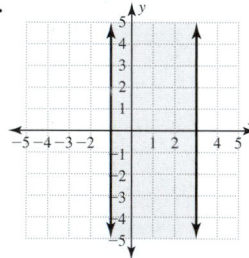
38.
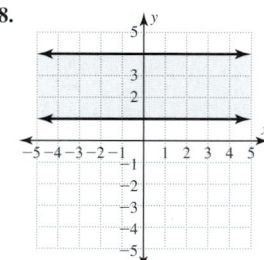

39.
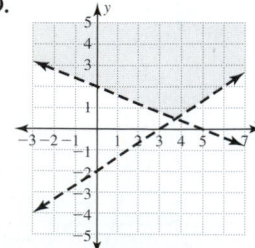
40.
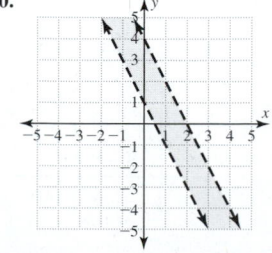

41.
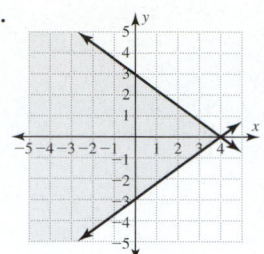
42.
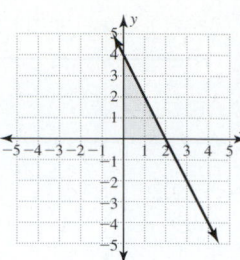

43. $x = 7$ or $x = -1$ **44.** $x = 4$ or $x = 1$ **45.** $\left(-\infty, -\dfrac{6}{5}\right) \cup (2, \infty)$

46. $(2, 8)$ **47.** $|x| < 5$ **48.** $|x| \leq 4$ **49.** $|x - 4| < 6$ **50.** $|x - 8| \leq 12$
51. (Verbal answers can vary.)
a. $x > 3; (3, \infty);$

b. x is at most 4; $(-\infty, 4];$

c. x is at least 2; $x \geq 2;$

d. x is less than 5; $x < 5; (-\infty, 5)$ **52.** (Verbal answers can vary.)
a. $2 < x < 6; (2, 6);$

b. x is greater than -3 and less than or equal to 4; $(-3, 4];$

c. x is greater than or equal to -2 and less than 0; $-2 \leq x < 0;$

d. x is greater than or equal to -1 and less than or equal to 3;
$-1 \leq x \leq 3; [-1, 3]$ **53. a.** $x + 7 \leq 11$ **b.** $2(x - 1) > 13$
c. $3x - 5 \geq 7$ **d.** $4x + 9 \leq 21$ **54.** $4(x + 3) > 8; (-1, \infty)$
55. $3(x - 17) \leq 2(x + 11); (-\infty, 73]$ **56.** $7 \leq 3x - 2 < 19; [3, 7)$
57. $-3 < x \leq 4$ **58.** $0 \leq x < \pi$ **59.** $(-2, \infty);$ all real numbers greater
than -2 **60.** $(-\infty, 4];$ all real numbers less than or equal to 4
61. $[-3, \infty);$ all real numbers greater than or equal to -3 **62.** $(-\infty, 3);$
all real numbers less than 3 **63.** $[1, \infty);$ all real numbers greater than or
equal to 1 **64.** $(0, 4);$ all real numbers between 0 and 4 **65.** $A \cap B = (2, 4);$
$A \cup B = (1, 7)$ **66.** $A \cap B = [3, 5); A \cup B = [2, 6]$
67. $A \cap B = [-3, 2]; A \cup B = \mathbb{R}$ **68.** $A \cap B = [-2, 5);$

$A \cup B = (-\infty, 6)$ **69.** $(-\infty, 1)$ **70.** $(-\infty, -3) \cup \left(-\dfrac{4}{3}, \infty\right)$

71. a. $x = 1998$ **b.** $[1990, 1998)$ **c.** $(1998, 2000]$ **d.** Before 1998, the
community used less electricity generated by wind than by coal. This
changed in 1998, and after 1998 more was generated by wind.
72. a. $x = 2$ **b.** $(0, 2)$ **c.** $(2, 4]$ **d.** If 2 hours are required, both shops
charge \$85. For less than 2 hours, shop A charges less. For more than
2 hours, shop A charges more. **73.** She must score at least 18 points.
74. She must make at least \$510. **75.** The length must be between 10 and
20 cm. **76.** $160 < x < 200$ **77.** $|x - 7.2| \leq 0.25; [6.95, 7.45]$
78. There is a loss if less than 50 posters are ordered; there is a profit if
more than 50 posters are ordered. **79.** $(-\infty, 10); (-\infty, 10.05)$
80. $(-\infty, -2]; (-\infty, -2.01]$ **81.** $(1, 3]; (1.1, 2.99]$

Mastery Test for Chapter 4

1. a. solution **b.** solution **c.** not a solution **d.** not a solution
2. a. $(-\infty, 5]$ **b.** $(-\infty, -2)$ **c.** $(-11, \infty)$ **d.** $[4, \infty)$ **3. a.** $(-\infty, 3)$
b. $[3, \infty)$ **c.** $(-\infty, 4]$ **d.** $(4, \infty)$ **4. a.** $[-15, \infty)$ **b.** $(245, \infty)$

c. $(-\infty, -9)$ **d.** $\left[-\dfrac{3}{2}, \infty\right)$ **5. a.** conditional; $(-\infty, 0)$ **b.** unconditional; $\mathbb{R}$

c. contradiction; no solution **d.** conditional; $[0, \infty)$ **6. a.** $[-3, 9]$
b. $[2, 7)$ **c.** $(-3, 5)$ **d.** $(-2, 5]$ **7. a.** $[-8, 5)$ **b.** $[-24, 54)$
c. $[-1, 4]$ **d.** $(-\infty, -4] \cup [3, \infty)$ **8. a.** $|x| = 2$ **b.** $|x| < 2$ **c.** $|x| \geq 2$
9. a. $x = 8$ or $x = -8$ **b.** $x = 22$ or $x = -27$ **c.** $(-\infty, 1) \cup (9, \infty)$

d. $\left[-1, \dfrac{11}{5}\right]$ **e.** $(0, 3)$ **f.** $(-\infty, 0) \cup (3, \infty)$ **g.** $[-4, 0]$

h. $(-\infty, -4] \cup [0, \infty)$ **10.** $(3, -5)$ is not a solution; $\left(\dfrac{2}{3}, -\dfrac{4}{5}\right)$ is a solution

11. a.

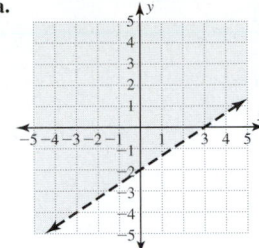

b.

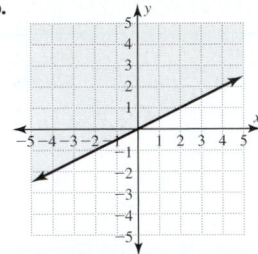

c.

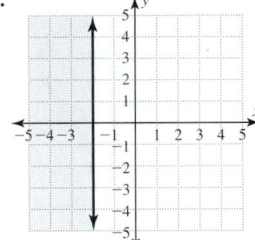

d.

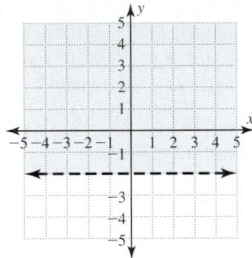

12. a.

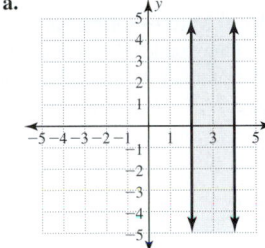

b.

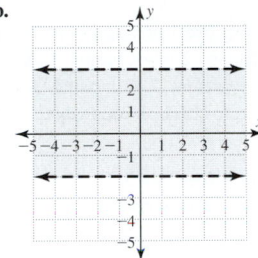

c.

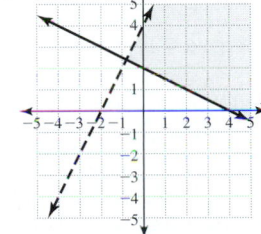

d.

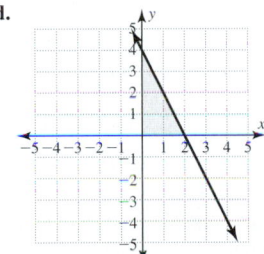

Group Project for Chapter 4

1. a. $2.485 < x < 2.515$ **b.** 2.485 cm **c.** 2.515 cm

2. a.

Roller No.	Measurement (cm)	Acceptable	Unacceptable
1	2.51	✓	
2	2.48		✓
3	2.48		✓
4	2.49	✓	
5	2.50	✓	
6	2.50	✓	
7	2.49	✓	
8	2.50	✓	
9	2.51	✓	
10	2.52		✓
11	2.50	✓	
12	2.49	✓	
13	2.48		✓
14	2.50	✓	
15	2.50	✓	
16	2.49	✓	
17	2.47		✓
18	2.50	✓	
19	2.51	✓	
20	2.49	✓	

b. 15 **c.** 5 **d.** 75% **3.** Option C is slightly cheaper than option B and guards against the production problem getting worse. It also helps fix the equipment for the next production schedule.

Chapter 5

Using the Language and Symbolism of Mathematics 5.1

1. base; exponent **2.** factors **3.** terms; addends **4.** x **5.** $-x$ **6.** xy **7.** y

8. exponential; expanded **9.** x^{m+n} **10.** $x^{m \cdot n}$ **11.** $x^m y^m$ **12.** $\dfrac{x^m}{y^m}$

Quick Review 5.1

1. -25 **2.** 25 **3.** 14 **4.** 49 **5.** 25

Exercises 5.1

1. a. x^5 **b.** xy^4 **c.** $(xy)^3$ **3. a.** $(a+b)^3$ **b.** $a^3 + b^3$ **c.** $a + b^3$

5. a. $m \cdot m \cdot m \cdot m \cdot m \cdot m$ **b.** $(-m)(-m)(-m)(-m)(-m)(-m)$

c. $-m \cdot m \cdot m \cdot m \cdot m \cdot m$ **7. a.** $(m+n)(m+n)$ **b.** $m \cdot m + n \cdot n$

c. $m + n \cdot n$ **9. a.** $\dfrac{a \cdot a}{b \cdot b \cdot b}$ **b.** $\left(\dfrac{a}{b}\right)\left(\dfrac{a}{b}\right)\left(\dfrac{a}{b}\right)\left(\dfrac{a}{b}\right)$ **c.** $\dfrac{a}{b \cdot b \cdot b \cdot b}$

11. $(x \cdot x \cdot x \cdot x \cdot x)(x \cdot x \cdot x) = x^8$ **13.** x^{20} **15.** $15m^{10}$ **17.** $-6x^6$

19. $(x \cdot x \cdot x)(x \cdot x \cdot x) = x^6$ **21.** x^{10} **23.** $9x^2 y^2$ **25.** n^{12} **27.** $5a^2 b^2 c^2$

29. $x^3 y^6$ **31.** $\dfrac{9}{16}$ **33.** $\dfrac{64}{w^3}$ **35.** $\dfrac{x^{22}}{y^{33}}$ **37. a.** -1 **b.** 1 **39. a.** -1 **b.** -1

41. $-125x^6 y^{12}$ **43.** $6x^3 y^6$ **45.** $200x^{13}$ **47.** $\dfrac{27x^3}{8y^6}$ **49.** $\dfrac{16x^8}{81y^{12}}$ **51. a.** -4 **b.** 4

53. a. 36 **b.** 18 **55. a.** 35 **b.** 125 **57. a.** 180 **b.** 900 **59. a.** $-\dfrac{4}{3}$ **b.** $\dfrac{4}{9}$

61. a. $3v$ **b.** v^3 **63. a.** $10m^3$ **b.** $24m^6$ **65. a.** $3x^2 + 2x$ **b.** $6x^3$

67. $-49; -48.7204$ **69.** $4; 4.0804$ **71.** $28; 29.0301$ **73.** $\$2.16$

75. a. $f(20) = 80$; a wind speed of 20 mi/h generates 80 kW of electricity.

b. $f(27) = 196.83$; a wind speed of 27 mi/h generates 196.83 kW of electricity. **77.** Doubling the wind speed increases the power by a factor of 8. **79. a.** x^2 **b.** y^2 **c.** $x^2 - y^2$ **81.** The volume of the larger cube is 8 times the volume of the smaller cube.

83.

x	Y_1	Y_2	Y_3
-3	729	-243	729
-2	64	-32	64
-1	1	-1	1
0	0	0	0
1	1	1	1
2	64	32	64
3	729	243	729

$$Y_1 = Y_3$$

Group Discussion Questions

85. $P(n) = 5^n$

Cumulative Review

1. $\dfrac{3}{4}$ **2. a.** 9 **b.** 9 **c.** 0 **d.** 0 **e.** 0 **f.** undefined **3.** $x = 18$ **4.** $\left(-\infty, \dfrac{17}{8}\right]$

5. $(-\infty, -4] \cup [5, \infty)$

Using the Language and Symbolism of Mathematics 5.2

1. terms; addends **2.** factors **3.** quotient **4.** x^{m-n}; 0 **5.** $\dfrac{1}{x^{n-m}}$ **6.** subtract

7. $1; 0$ **8.** undefined

Quick Review 5.2

1. -12 **2.** 36 **3.** 36 **4.** 64 **5.** $\dfrac{3}{4}$

Exercises 5.2

1. x^3 **3.** x^{10} **5.** m^{24} **7.** $-x^4$ **9.** $\dfrac{1}{t^4}$ **11.** $\dfrac{1}{v^5}$ **13.** $-\dfrac{1}{a^3}$ **15.** $\dfrac{a^3}{b^4}$ **17.** $\dfrac{2m^3}{3}$

19. $\dfrac{6}{11v^3}$ **21.** $\dfrac{3b}{4}$ **23.** $\dfrac{4n^5}{3m^2}$ **25.** $-\dfrac{3w^3}{5v^2}$ **27. a.** 1 **b.** -1 **c.** 1 **29. a.** 2 **b.** 1 **c.** 0

31. a. 1 **b.** 0 **c.** -1 **33. a.** 1 **b.** undefined **35. a.** -3 **b.** undefined **37. a.** 1

b. 4 **c.** 5 **39. a.** 1 **b.** 0 **c.** 1 **41.** 11 **43.** 1 **45.** x^6 **47.** $\dfrac{1}{m^{12}}$ **49.** $36x^{18}$

51. $36x^{10}$ **53.** $32x^{15}$ **55.** $144m^{18}$ **57.** $200x^{21}$ **59.** $6x^6$ **61.** $3a^4b^5$

63. $\dfrac{x^4}{y}$ **65. a.** 4 **b.** $\dfrac{1}{4}$ **c.** 1 **67. a.** -5 **b.** 0 **c.** 1 **d.** 0 **e.** undefined

69. a. 1,000 **b.** 1 **c.** $\dfrac{1}{1,000}$ **71. a.** $36x$ **b.** 19 **73. a.** $3x^2 - 3x$ **b.** x

75. 8; 8.04 **77.** 200; 198.59 **79.** \$14,519.23

81.

x	Y_1	Y_2	Y_3
-2	16	16	-8
-1	1	1	-1
0	ERROR	0	0
1	1	1	1
2	16	16	8

$$Y_1 = Y_2 \ (x \neq 0)$$

Group Discussion Questions

83. a. x^2 **b.** x^{2m} **c.** x^{3m+3} **d.** x^{8m+12} **e.** 1 **85. a.** 1 **b.** 0 **c.** error
d. Answers can vary.

Cumulative Review

1. $\dfrac{11}{20}$ **2.** $y = -\dfrac{4}{3}x + \dfrac{7}{3}$ **3.** $(2, 3)$

4.

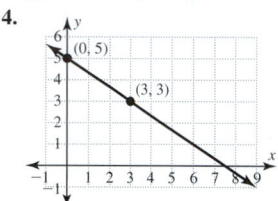

5. $(-5, 2]$

Using the Language and Symbolism of Mathematics 5.3

1. $\dfrac{1}{x^n}$ **2.** x^{m+n} **3.** x^{mn} **4.** x^{m-n} **5.** 10 **6.** 10

Quick Review 5.3

1. 10,000 **2.** $\dfrac{1}{1,000}$ **3.** 1,110 **4.** $-6x^5$ **5.** $\dfrac{4x^6}{y^{10}}$

Exercises 5.3

1. a. $\dfrac{1}{32}$ **b.** -32 **c.** 25 **3. a.** $\dfrac{5}{6}$ **b.** $\dfrac{25}{36}$ **c.** 1 **5. a.** $\dfrac{1}{100}$ **b.** $\dfrac{1}{1,000}$ **c.** $-\dfrac{1}{100}$

7. a. $\dfrac{1}{7}$ **b.** $\dfrac{7}{10}$ **c.** $-\dfrac{9}{5}$ **9. a.** $\dfrac{10}{7}$ **b.** 7 **c.** 2 **11. a.** $\dfrac{y}{x}$ **b.** $\dfrac{y^2}{x^2}$ **c.** $\dfrac{1}{x^2y}$

13. a. $\dfrac{1}{9x^2}$ **b.** $\dfrac{3}{x^2}$ **c.** $-3x^2$ **15. a.** $\dfrac{1}{m+n}$ **b.** $\dfrac{1}{m} + \dfrac{1}{n}$ **c.** $m + \dfrac{1}{n}$

17. a. v^9 **b.** v^{15} **c.** $\dfrac{1}{v^{15}}$ **19. a.** $\dfrac{1}{x^4}$ **b.** 1 **c.** $\dfrac{1}{x^{10}}$ **21. a.** $\dfrac{3}{x}$ **b.** $12x$ **c.** $\dfrac{12}{x^7}$

23. a. $\dfrac{9x^2}{25y^2}$ **b.** $\dfrac{25y^2}{9x^2}$ **c.** $225x^2y^2$ **25.** $-\dfrac{1}{20x}$ **27.** $\dfrac{72y^8}{x}$ **29.** $-\dfrac{3x^{10}}{2}$ **31.** $-\dfrac{4x^{15}}{y^{32}}$

33. $4x^{10}$ **35.** $\dfrac{m^{12}}{n^{12}}$ **37.** $\dfrac{3y^6}{2x^7}$ **39. a.** 45,800 **b.** 0.000458 **c.** 4,580,000
d. 0.00000458 **41. a.** 8,100 **b.** 0.0081 **c.** $-8,100$ **d.** -0.00000081
43. a. 9.7×10^3 **b.** 9.7×10^7 **c.** 9.7×10^{-1} **d.** 9.7×10^{-4}
45. a. 3.5×10^{10} **b.** 3.5×10^{-9} **c.** -3.5×10^3 **d.** -3.5×10^{-2}
47. 4,493,000,000 **49.** 0.000000000001 **51.** 1.4×10^7 **53.** 5.68×10^{-5}
55. a. 2,300 **b.** 220,000,000 **c.** The claim is accurate. **57.** It will take the
signal approximately 950 sec or approximately 16 min to reach Earth.
59. 0.8; 0.7953 **61.** 0.05; 0.05088 **63.** 0.000027; 0.00002628

65. $\dfrac{1}{27}$; -27 **67.** 25; 1; $-\dfrac{1}{12}$; 1 **69.** 0.0001764 **71.** 0.00351

73. a. 0.367 **b.** 51,300,000

75.

x	Y_1	Y_2	Y_3
-3	0.111	0.111	-0.667
-2	0.25	0.25	-1
-1	1	1	-2
0	ERROR	ERROR	ERROR
1	1	1	2
2	0.25	0.25	1
3	0.111	0.111	0.667

$$Y_1 = Y_2 \ (x \neq 0)$$

Group Discussion Questions

77. a. You will have lived approximately 31.7 years. **b.** You will have lived
approximately 2.5×10^9 seconds. **c.** This individual has lost
approximately 4.7×10^8 seconds of their life. **d.** Each cigarette has cost
this person approximately 1,300 seconds of their life. **79. a.** yes; if m is
even **b.** yes; if m is odd **c.** no **d.** yes; any negative value

Cumulative Review

1. a. $4y$ **b.** $y \cdot y \cdot y \cdot y$ **2.** $12x + 15$ **3.** Use 10 mL of the 4% solution and
30 mL of the 20% solution. **4.** $m = -\dfrac{3}{4}$ **5.** $A \cup B = (-\infty, 8)$;

$A \cap B = [-3, 6]$

Using the Language and Symbolism of Mathematics 5.4

1. factors **2.** plus; minus **3.** $0, 1, 2, 3, \ldots$ **4.** monomial **5.** polynomial
6. coefficient **7.** monomial **8.** binomial **9.** trinomial **10.** sum **11.** highest
12. zero **13.** 0 **14.** alphabetical; descending **15.** 1 **16.** -1 **17.** 1
18. identical

Quick Review 5.4

1. sum **2.** difference **3.** -15 **4.** -19 **5.** -7

Exercises 5.4

1. a. monomial; coefficient: -12; degree: 0 **b.** monomial; coefficient:
-12; degree: 1 **c.** monomial; coefficient: -1; degree: 12 **d.** not a
monomial **3. a.** monomial; coefficient: $\dfrac{2}{5}$; degree: 4 **b.** monomial;

coefficient: $\dfrac{5}{2}$; degree: 4 **c.** monomial; coefficient: $-\dfrac{5}{2}$; degree: 4

d. not a monomial **5. a.** monomial; coefficient: 5; degree: 4 **b.** monomial;
coefficient: -5; degree: 4 **c.** not a monomial **d.** not a monomial
7. a. second-degree binomial **b.** third-degree trinomial **c.** fourth-degree
monomial **d.** monomial of degree 0 **9. a.** second-degree monomial
b. first-degree binomial **c.** not a polynomial **d.** third-degree trinomial
11. a. $8x^2yz^3$ **b.** $x^2 - 5x - 9$ **c.** $-4x^3 - 2x^2 + 9x + 7$ **13. a.** like
b. unlike **c.** unlike **15. a.** $-7xy$ **b.** $-7x + y$ **c.** $2x^2 - 3xy + 7y^2$
17. a. $21x$ **b.** $-7x$ **19. a.** $7x^2 - 12x$ **b.** $-x^2 - 2x$ **21. a.** $9x^2 - x + 6$
b. $x^2 - 13x + 12$ **23.** $22x - 9$ **25.** $-17v + 58$ **27.** $9a^2 - a + 6$
29. $6x^2 - x + 5$ **31.** $3w^2 - 7w + 10$ **33.** $-m^4 + 3m^3 - 2m + 4$
35. $-4n^4 + 9n^3 + n^2 + n - 12$
37. $-11y^6 - 2y^5 - 5y^4 + 20y^2 + 9y - 25$ **39.** $-5x^2 - 20x - 3$
41. $-13a + 13$ **43.** $12x - 12$ **45.** $-15x + 28$ **47.** $11x^2 + 4xy - 8y^2$
49. $-20x^2 + 7xy + 11y^2$ **51.** $9x^3 + 6x^2 - 12x + 30$
53. $-8x^4 - 3x^3y + 3x^2y^2 + xy^3 + 2y^4$ **55. a.** -2 **b.** 24 **c.** 528 **d.** 342
57. $P(0) = -40$; When zero units are produced and sold, there is a loss of
\$40. **59.** $P(4) = 32$; When 4 units are produced and sold, there is a profit
of \$32. **61.** $x = 9$ or $x = 13$; When 9 or 13 units are produced and sold,
there is a profit of \$77. **63.** $2x^2 + 4x + 8$ **65.** $3x^2 + 10x + 21$
67. $2x^3 - 5x + 11$ **69.** $3w^6 + 5w^4$ **71.** $x^2 - y^2$ **73.** $2w + 7$
75. Approximately 350,000 tons of CFC was produced in 1965.

77.

x	Y_1	Y_2	Y_3
-3	18	38	18
-2	3	23	3
-1	-8	12	-8
0	-15	5	-15
1	-18	2	-18
2	-17	3	-17
3	-12	8	-12

$Y_1 = Y_3$

Group Discussion Questions

79. Answers can vary for **b** through **g. a.** $2x^5$ **b.** 3 **c.** $x - 3$ **d.** $x^2 - 3$ **e.** $5x$; $-5x$ **f.** $5x$; $5x$ **g.** $x^4 + 5x^3$; $-x^4 - 2x^3$

Cumulative Review

1. $x = -3$ **2.** $y = 4$ **3.** 1 **4.** 0 **5.** 2

Using the Language and Symbolism of Mathematics 5.5

1. 2; 5 **2.** x^{m+n} **3.** multiplied; added **4.** distributive **5.** each **6.** binomial **7.** trinomial **8.** first, outer, inner, last **9. a.** 2 **b.** 2 **c.** 3 **10. a.** 1 **b.** 1 **c.** 2 **11.** x

Quick Review 5.5

1. $-12x + 21$ **2.** $10x + 45$ **3.** $-18x + 30y - 48$ **4. a.** $12x^3$ **b.** $3x^2 + 4x$ is simplified. These are not like terms. **5. a.** $-10x^4$ **b.** $3x^2$

Exercises 5.5

1. $63v^5$ **3.** $4a^9b^5$ **5.** $24x^3 - 20x^2$ **7.** $-32v^3 - 12v^2$ **9.** $14x^3 - 18x^2 - 8x$ **11.** $-44x^4 + 55x^3 + 88x^2$ **13.** $10x^3y - 5x^2y^2 - 5xy^3$ **15.** $6a^4b + 12a^3b^2 - 3a^2b^3$ **17.** $y^2 + 5y + 6$ **19.** $6v^2 + 7v - 20$ **21.** $54x^2 + 63x + 15$ **23.** $2x^2 + xy - y^2$ **25.** $10x^2 + 6xy - 28y^2$ **27.** $x^3 - x^2 - 10x - 8$ **29.** $2m^3 + m^2 - 11m - 10$ **31.** $x^3 + y^3$ **33.** $6x^4 - 7x^3 - 5x^2 + 89x - 99$ **35.** $5x^3 + 5x^2 - 30x$ **37.** $2x^3 + 13x^2 + 17x - 12$ **39.** $m^2 + 7m + 12$ **41.** $n^2 + 3n - 40$ **43.** $20x^2 + 13x - 21$ **45.** $9y^2 - 64y + 7$ **47.** $12a^2 + 13ab - 14b^2$ **49.** $90x^2 - 43xy - 21y^2$ **51.** $(8x^2 + 10x - 3)$ cm^2 **53.** $(x^2 + x + 1)$ cm^2 **55.** $(4x^2 - 4x + 1)$ cm^2 **57.** $\left(\frac{3}{2}x^2 + \frac{5}{2}x + 1\right)$ cm^2 **59.** $(4x^2 + 4x - 3)$ cm^2 **61.** $(x^3 + 6x^2 + 5x)$ cm^3

63. $Y_1 \neq Y_2$

x	Y_1	Y_2
-3	117	96
-2	50	36
-1	7	0
0	-12	-12
1	-7	0
2	22	36
3	75	96

65. $Y_1 = Y_2$

x	Y_1	Y_2
-3	5	5
-2	0	0
-1	-3	-3
0	-4	-4
1	-3	-3
2	0	0
3	5	5

67. a. $40v^2 - 35v$ **b.** $5v(8v - 7)$ **69. a.** $2m^3n - 2m^2n^2 + 10mn^3$ **b.** $2mn(m^2 - mn + 5n^2)$ **71. a.** $x^2 - x - 6$ **b.** $(x + 2)(x - 3)$ **73. a.** $15x^2 - 26x + 8$ **b.** $(3x - 4)(5x - 2)$ **75.** $-6x^4 + 7x^2 + 20$ **77.** $-x^4 + 2x^3 - 2x^2 + x + 20$ **79.** $v^2 + 10v + 25$ **81.** $x^3 + 6x^2 + 12x + 8$ **83.** $V = 4x^3 - 72x^2 + 288x$ **85.** $A = \pi x^2 + 30\pi x$ **87. a.** $R(x) = 280x - 2x^2$

b.

x	$R(x)$
40	8,000
45	8,550
50	9,000
55	9,350
60	9,600
65	9,750
70	9,800

c. $R(50) = 9,000$; When the price is $50 per unit, the revenue will be $9,000.

Group Discussion Questions

89. a. whole: $x(x + 3)$; parts: $x^2 + 3x$ **b.** whole: $(x + 1)(x + 3)$; parts: $x^2 + 3x + x + 3$ **c.** whole: $(x + 2)(x + 4)$; parts: $x^2 + 4x + 2x + 8$; In all three cases, the two polynomials are equal. **91. a. i.** $x^2 - 1$ **ii.** $x^2 - 4$ **iii.** $x^2 - 9$ **iv.** $x^2 - 16$ **v.** $x^2 - 25$ **b.** $(x + y)(x - y) = x^2 - y^2$ **c.** $x^2 - 144$ **d.** $4v^2 - 9$

Cumulative Review

1. $\frac{43}{60}$ **2.** $\frac{1}{8}$ **3.** $f(x) = \frac{1}{2}x - 1$ **4. a.** $x = -4$, $x = 14$ **b.** $[-4, 14]$ **c.** $(-\infty, -4) \cup (14, \infty)$ **5.** $\frac{25y^{18}}{9x^{20}}$

Using the Language and Symbolism of Mathematics 5.6

1. a. C **b.** D **c.** A **d.** B **2.** 2 **3.** 3 **4.** factoring **5.** distributive **6.** $x^2 - 4$ **7.** factoring

Quick Review 5.6

1. 100 **2.** 58 **3.** 20 **4.** $x^2 - x - 20$ **5.** $4x^2 + 16x + 15$

Exercises 5.6

1. $49a^2 - 1$ **3.** $z^2 - 100$ **5.** $4w^2 - 9$ **7.** $81a^2 - 16b^2$ **9.** $16x^2 - 121y^2$ **11.** $x^4 - 4$ **13.** $16m^2 + 8m + 1$ **15.** $n^2 - 18n + 81$ **17.** $9t^2 + 12t + 4$ **19.** $25v^2 - 80v + 64$ **21.** $16x^2 + 40xy + 25y^2$ **23.** $36a^2 - 132ab + 121b^2$ **25.** $a^2 + 2abc + b^2c^2$ **27.** $x^4 - 6x^2 + 9$ **29.** $x^4 + 2x^2y + y^2$ **31. a.** $-25x - 36$ **b.** $25x^2 - 36$ **33. a.** $16x^2 - 72x + 81$ **b.** $16x^2 - 81$ **35.** $14v + 98$ **37.** $4xy$ **39.** $2vw - 2w^2$ **41.** 0 **43.** $9x^2 - 36xy + 4y^2$ **45.** $(3x + 7y)^2 = 9x^2 + 42xy + 49y^2$ **47.** $(x + 8)^2 - (x - 6)^2 = 28x + 28$ **49.** $(a^2 + 1)(a + 1)(a - 1) = a^4 - 1$

51. $Y_1 = Y_2$

x	Y_1	Y_2
-3	35	35
-2	15	15
-1	3	3
0	-1	-1
1	3	3
2	15	15
3	35	35

53. $Y_1 \neq Y_2$

x	Y_1	Y_2
-3	9	45
-2	1	25
-1	1	13
0	9	9
1	25	13
2	49	25
3	81	45

55. $81x^2 - 1$; $(9x - 1)(9x + 1)$ **57.** $100w^2 - 60wx + 9x^2$; $(10w - 3x)^2$

Group Discussion Questions

59. a. 9,991 **b.** 9,984 **c.** 9,801 **d.** 10,201 **e.** Answers can vary. **61. a.** The values that make each factor equal zero correspond to the x-intercepts. **b.** $(x - 7)(x + 7)$

Cumulative Review

1. A **2.** C **3.** B **4.** $y = -\frac{2}{5}x + \frac{11}{5}$ **5.** $y = \frac{5}{2}x + 8$

Using the Language and Symbolism of Mathematics 5.7

1. quotient **2.** $x - 3$ **3.** product **4.** $(x - 3)$; $(x + 2)$; $x^2 - x - 6$ **5.** standard **6.** zero; divisor **7.** 0 **8.** average

Quick Review 5.7

1. $\frac{7}{3}$ **2.** $\frac{5x^5}{2}$ **3.** $-12x^4y^5 + 15x^3y^6 - 21x^2y^7$ **4.** $8x^2 - 22x - 30$ **5.** $10\frac{17}{19}$

Exercises 5.7

1. $3x^2$ **3.** $-5a^2b$ **5.** $3a - 4$ **7.** $3a - 5 + \frac{2}{a}$ **9.** $8m^3 - 4m^2 + 5m + 3$ **11.** $5x^3 - 6x^2y^3 + 11y^2$ **13.** $6v^3 + 3v^2 - 9v - \frac{2}{v}$ **15.** $5v^2 - v - 4w$ **17.** $10x^{11} - 5x^3 + 3$ **19.** $x + 7$; $(x + 2)(x + 7) = x^2 + 9x + 14$ **21.** $v + 6$; $(v - 4)(v + 6) = v^2 + 2v - 24$ **23.** $4m^2 + 3m + 19 + \frac{64}{m - 3}$; $(m - 3)\left(4m^2 + 3m + 19 + \frac{64}{m - 3}\right) = 4m^3 - 9m^2 + 10m + 7$

25. $3w - 4$ **27.** $4m - 7$ **29.** $9y - 7$ **31.** $7x^3 - 9x^2 + 6x + 2$
33. $x - 7$ **35.** $3a + 2$ **37.** $3v^2 + 2v + 4$ **39.** $x^2 + 2x + 4$
41. $4y^2 + 8y + 9$ **43.** $7x^2 - 9$ **45.** $(2x - 1)$ cm **47.** $(6x + 2)$ cm
49. $(2x + 3)$ cm

51. $Y_1 = Y_2 \ (x \neq 3)$

x	Y_1	Y_2
-3	11	11
-2	9	9
-1	7	7
0	5	5
1	3	3
2	1	1
3	ERR	-1

53. $Y_1 \neq Y_2$

x	Y_1	Y_2
-3	0	-6
-2	1	-5
-1	2	-4
0	3	-3
1	4	-2
2	5	-1
3	ERR	0

55. $x^3 - 14x - 15$ **57.** $3x + 5$ **59. a.** 4 **b.** 6 **c.** 2 **61.** $2x^2 + 7x + 10$
63. $4x + 5$ **65. a.** $3x^2 - 19x + 20$ **b.** $x - 5$ **67.** $5k + 4$

69. a. $A(x) = \dfrac{125x + 960}{x}$ **b.** $A(50) = 144.2$; when 50 units are

produced, the average cost per unit is \$144.20. **71. a.** $A(x) = \dfrac{8x + 1{,}200}{x}$
b. $C(200) = 2{,}800$; the total cost of producing 200 chairs is \$2,800.
c. $A(200) = 14$; the average cost of producing 200 chairs is \$14.00 per
chair. **d.** $C(250) = 3{,}200$; the total cost of producing 250 chairs is
\$3,200. **e.** $A(250) = 12.8$; the average cost of producing 250 chairs is
\$12.80 per chair. **f.** The cost increases by \$400. **g.** The average cost
decreases by \$1.20 per chair.

Group Discussion Questions
73. Answers may vary. **a.** $(x^2 - 2x) \div (x - 2) = x$
b. $(x^2 - 4) \div (x - 2) = x + 2$ **c.** $(x^3 - 8) \div (x - 2) = x^2 + 2x + 4$
75. a. $-2, -1, 0, 1, 2,$ ERR:, 4. **b.** No y-value shown for $x = 2$. **c.** undefined
d. There is a hole in the graph. **e.** There is a hole in the graph. **f.** Same
except Y_1 has a hole in the graph.

Cumulative Review
1. A **2.** C **3.** B
4.

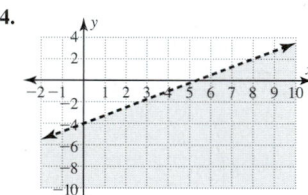

5. $4x^2 - 20x + 25$

Review Exercises for Chapter 5
1. a. xy^3 **b.** $(xy)^3$ **c.** $x^2 + y^2$ **d.** $(x + y)^2$ **2. a.** $x \cdot x + y \cdot y$
b. $(x + y)(x + y)$ **c.** $-x \cdot x \cdot x \cdot x$ **d.** $(-x)(-x)(-x)(-x)$ **3. a.** 9
b. $\dfrac{1}{9}$ **c.** -9 **d.** $\dfrac{1}{9}$ **4. a.** 1 **b.** 0 **c.** 2 **d.** 1 **5. a.** 5 **b.** $\dfrac{6}{5}$ **c.** $\dfrac{3}{2}$ **d.** $\dfrac{1}{6}$ **6. a.** -1
b. 1 **c.** $\dfrac{1}{6}$ **d.** $-\dfrac{1}{6}$ **7. a.** $2x^3$ **b.** x^6 **c.** 1 **d.** 0 **8. a.** 100 **b.** 100 **c.** $\dfrac{1}{100}$
d. $-\dfrac{1}{100}$ **9. a.** 3 **b.** 2 **c.** 1 **d.** 0 **10. a.** 9 **b.** 1 **c.** 3 **d.** 3 **11. a.** x^7 **b.** x^3
c. x^{10} **d.** x^8 **12. a.** $48x^8y^6$ **b.** $3x^2y^2$ **c.** $16x^6y^4$ **d.** $27x^6y^6$ **13. a.** -9 **b.** 9
14. a. 48 **b.** 144 **15. a.** 25 **b.** 49 **16. a.** $\dfrac{7}{12}$ **b.** $\dfrac{1}{7}$ **17. a.** 2 **b.** 1 **18. a.** 48
b. $\dfrac{16}{9}$ **19.** $30x^3y^7z^{11}$ **20.** $25x^4y^6z^8$ **21.** $\dfrac{3ac^2}{5b^3}$ **22.** $\dfrac{n^4}{9m^6}$ **23.** $60x^{16}$ **24.** $9x^3$
25. $-200x^{21}$ **26.** $56y^2$ **27.** $2w^2$ **28.** $\dfrac{m^{24}}{625}$ **29.** $\dfrac{6}{x^7}$ **30.** 1 **31.** a^3b^4 **32.** $\dfrac{y^7z^{10}}{x^8}$
33. 0.000000000001 L **34.** 22,000 droplets per second **35.** 5,230,000
36. 0.00000523 **37.** 4 h 11 min **38.** B **39.** monomial of degree 0
40. sixth-degree binomial **41.** third-degree polynomial with four terms
42. sixth-degree trinomial **43.** $7x^5 + 11x^4 + 9x^3 - 3x^2 - 8x + 4$

44. $-7x^5$ **45.** $x^2 - 3$ **46.** $11x^2 - 3x + 2$ **47.** $5x^3 - 5x^2 - 8x + 4$
48. $7x^4 - 8x^3 + 19x^2 + 9x + 14$
49. $-5x^5 + 9x^4 + 30x^3 + 3x^2 + 9x - 27$ **50.** $2x^2 - 11x - 12$
51. $35x^5 - 45x^4 + 15x^3 + 5x^2$ **52.** $35v^2 + 2v - 1$
53. $25y^2 - 70y + 49$ **54.** $81y^2 + 90y + 25$ **55.** $9a^2 - 25b^2$
56. $6m^3 - 8m^2 - 9m + 35$ **57.** $-\dfrac{4b^3}{a}$ **58.** $5m^3n - 7m^2n^2 - mn^3$
59. $v - 2$ **60.** $7w + 3$ **61.** $a + b$ **62.** $24xy + 32y^2$ **63.** $24xy$ **64.** $x - 5$
65. $(x + 4)(x^2 - 4x + 16) = x^3 + 64$ **66.** $\left(x + 6 + \dfrac{7}{x - 2}\right)(x - 2) =$
$(x + 6)(x - 2) + 7 = x^2 + 4x - 12 + 7 = x^2 + 4x - 5$
67. a. $3x^2 - 12x$ **b.** $3x(x - 4)$ **68. a.** $5x^2 + 8x - 21$ **b.** $(5x - 7)(x + 3)$
69. a. $16x^2 - 9$ **b.** $(4x + 3)(4x - 3)$ **70. a.** $36x^2 - 84x + 49$
b. $(6x - 7)^2$ **71.** $45x^2 + 17x - 6$ **72.** $20x^2 + 9xy - 18y^2$
73. $25x^2 - 36y^2$ **74.** $25x^2 - 60xy + 36y^2$ **75.** $49x^2 + 28xy + 4y^2$
76. $x^4 - 81$ **77. a.** $-25x - 7$ **b.** $12x^2 - 25x - 7$ **78. a.** $64x^2 - 25y^2$
b. $64x^2 - 80xy + 25y^2$ **79.** $98a^2 + 8b^2$ **80.** $9a^2 + 6ab + 90b^2$
81. $4x - 1$ **82.** $27x^3 + 18x^2 + 12x + 8$ **83.** $(6x^2 - 7x - 3)$ ft^2
84. $(x + 1)$ cm **85. a.** $R(x) = x(350 - 4x) = -4x^2 + 350x$
b. $R(0) = 0$; when the price is \$0 per unit, the revenue is \$0
c. $R(40) = 7{,}600$; when the price is \$40 per unit, the revenue is \$7,600
d. $R(80) = 2{,}400$; when the price is \$80 per unit, the revenue is \$2,400
86. \$17,834.78 **87. a.** $A(x) = \dfrac{15x + 1{,}000}{x}$ **b.** $C(100) = 2{,}500$; the total
cost of producing 100 range finders is \$2,500. **c.** $A(100) = 25$; the
average cost of producing 100 range finders is \$25 per range finder.
d. $C(200) = 4{,}000$; the total cost of producing 200 range finders is
\$4,000. **e.** $A(200) = 20$; the average cost of producing 200 range finders
is \$20 per range finder. **f.** The cost increases by \$1,500. **g.** The average
cost decreases by \$5 per range finder.
88. $Y_1 = Y_4$; $Y_2 = Y_3$

x	Y_1	Y_2	Y_3	Y_4
-3	-28	-26	-26	-28
-2	-9	-7	-7	-9
-1	-2	0	0	-2
0	-1	1	1	-1
1	0	2	2	0
2	7	9	9	7
3	26	28	28	26

Mastery Test for Chapter 5
1. a. x^4 **b.** x^3y^4 **c.** $-2 \cdot x \cdot x \cdot x \cdot y \cdot y$ **d.** $(-2x)(-2x)(-2x) \cdot y \cdot y$
2. a. x^{13} **b.** $40y^7$ **c.** $-v^{15}$ **d.** m^4n^3 **3. a.** w^{40} **b.** $32y^{20}$ **c.** v^8w^{12}
d. $-\dfrac{27m^3}{8n^3}$ **4. a.** z^6 **b.** $2x^6$ **c.** -1 **d.** $-2a^2b^4$ **5. a.** 1 **b.** 1 **c.** 1 **d.** 8
6. a. x^5y^{11} **b.** x^3y^6 **c.** $\dfrac{4}{3x^5}$ **d.** $36x^{12}$ **7. a.** $\dfrac{1}{3}$ **b.** $\dfrac{1}{49}$ **c.** $\dfrac{9}{4}$ **d.** 9 **8. a.** $\dfrac{x^9}{y^6}$
b. $\dfrac{16x^{10}y^{32}}{25}$ **9. a.** 357,000 **b.** 0.0000735 **c.** 5.09×10^{-4} **d.** 9.305×10^7
10. a. third-degree binomial **b.** monomial of degree 0 **c.** second-degree
trinomial **d.** fifth-degree polynomial with four terms **11. a.** $7x - y$
b. $-3x^2 + 6x + 18$ **c.** $9x^4 - 9x^3 + 19x^2 + 9x + 8$
d. $3x^3 - 12x^2 - 16x + 6$ **12. a.** $55x^6y^{10}$ **b.** $-15x^5 + 6x^4 - 21x^3 + 27x^2$
c. $x^3 + 8x^2 + 16x + 5$ **d.** $x^3 - 2x^2y - 2xy^2 - 3y^3$ **13. a.** $x^2 + 19x + 88$
b. $x^2 - 14x + 45$ **c.** $6x^2 + 7xy - 20y^2$ **d.** $10x^2 + 29xy - 21y^2$
14. a. $36x^2 - 1$ **b.** $x^2 - 81y^2$ **c.** $25x^2 - 121$ **d.** $9x^2 - 64y^2$
15. a. $36x^2 + 12x + 1$ **b.** $x^2 - 18xy + 81y^2$ **c.** $25x^2 - 110x + 121$
d. $9x^2 + 48xy + 64y^2$ **16. a.** $-6x + 18$ **b.** $-20xy$ **17. a.** $3a^2$
b. $3m^2n - 4n^3$ **18. a.** $x + 11$ **b.** $2x^2 + 6x + 8$

Group Project for Chapter 5
I. Answers may vary. **1.** 21.6 cm **2.** 28.0 cm **3.** 2 cm **7.** 850 cm^3
8. 844.8 cm^3 **9.** 0.62%; this is an acceptably small relative error.

10.

Width x (cm)	Volume y (cm³)
2.0	850
2.5	950
3.0	1,030
3.5	1,070
4.0	1,090
4.5	1,080
5.0	1,040

II. 1. $(0, 10.8)$ **2.** $W(x) = 21.6 - 2x$ **3.** $L(x) = 28 - 2x$
4. $V(x) = x(21.6 - 2x)(28 - 2x)$

5.

Width x (cm)	Volume y (cm³)
1	509.6
2	844.8
3	1,029.6
4	1,088.0
5	1,044.0
6	921.6
7	744.8
8	537.6
9	324.0
10	128.0

6.

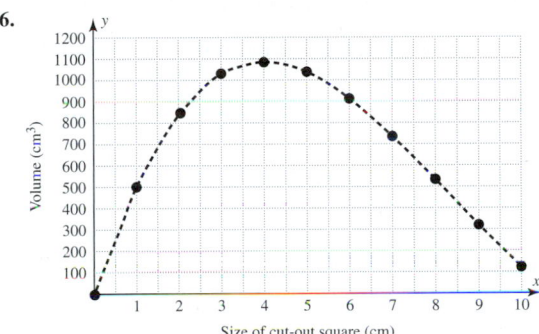

7. No. Not all values produce the same volume; $x = 4$ produces a volume larger than any other x-value.

A Review of Beginning Algebra—A Cumulative Review of Chapters 1–5

1. a. 12 **b.** 18 **c.** -45 **d.** -5 **2. a.** -0.58 **b.** -0.62 **c.** -0.012 **d.** -30
3. a. $\dfrac{11}{12}$ **b.** $-\dfrac{7}{12}$ **c.** $\dfrac{1}{8}$ **d.** $\dfrac{2}{9}$ **4. a.** 3 **b.** 3 **c.** 0 **d.** undefined **5. a.** 18 **b.** 14
6. a. $\dfrac{1}{4}$ **b.** $\dfrac{1}{13}$ **7. a.** $\left\{-5, -3.28, -\dfrac{1}{3}, 0, \sqrt{9}, 11\right\}$ **b.** $\{-\sqrt{3}\}$
c. $\{-5, 0, \sqrt{9}, 11\}$ **d.** $\{\sqrt{9}, 11\}$ **8. a.** 0.2; 20% **b.** $\dfrac{7}{20}$; 35% **c.** $\dfrac{3}{5}$; 0.6
9. a. the distributive property of multiplication over addition
b. commutative property of addition **c.** commutative property of
multiplication **d.** associative property of addition **10. a.** 0 **b.** -5
11. a. 1 **b.** $\dfrac{1}{5}$ **12. a.** $(-\infty, 5]$ **b.** $(-2, \infty)$ **c.** $[-3, 0)$ **d.** $(3, 9]$
13. a. area: 9π cm² ≈ 28.27 cm²; perimeter: 6π cm ≈ 18.85 cm
b. area: 36 in²; perimeter: 26 in **14. a.** $62°$ **b.** $152°$ **15. a.** $x = 5$ is not a
solution. **b.** $x = 5$ is a solution. **16. a.** $-2x + 14$ **b.** $x = 7$ **17. a.** $x + 7$
b. $x = -7$ **18. a.** $x = -8$ **b.** $x = 9$ **19. a.** $x = -\dfrac{20}{3}$ **b.** $x = \dfrac{5}{3}$
20. a. $y = \dfrac{2}{3}x - 4$ **b.** $y = \dfrac{5}{3}x - 5$ **21.** $x = 4$ **22.** $x = -6$
23. a. y-intercept: $(0, 4)$; x-intercept: $(3, 0)$ **b.** y-intercept: $(0, -9)$;
x-intercept: $(6, 0)$ **24. a.** $y = 25x$ **b.** The constant of variation represents a
concession average of $25 per person.

c.

Attendance	Concession Sales ($)
34,000	850,000
38,000	950,000
42,000	1,050,000

25. $x = \dfrac{9}{2}$ **26. a.** The sequence is arithmetic. The common difference is 4.
b. The sequence is arithmetic. The common difference is -3.
c. The sequence is not arithmetic. **d.** The sequence is arithmetic.
The common difference is 0. **27. a.** $f(0) = -2$ **b.** $f(-3) = -17$
c. $f(4) = 18$ **d.** $f(10) = 48$ **28. a.** 3 **b.** 1 **c.** 4 **d.** 2
29. a. $f(x) = 0.80x$ **b.**

x	$f(x)$
20	16.00
48	38.40
75	60.00
90	72.00
110	88.00

30. After 3 hours, the boats are 72 miles apart. **31.** She should add 1.25 L
of water. **32. a.** $(1, -3)$ is a solution. **b.** $(1, -3)$ is not a solution.
c. $(1, -3)$ is a solution. **d.** $(1, -3)$ is not a solution.
33. a.

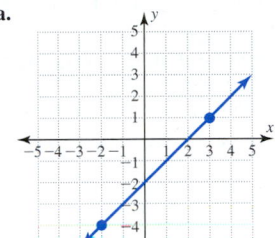

b.

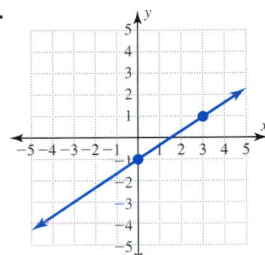

c.

d.

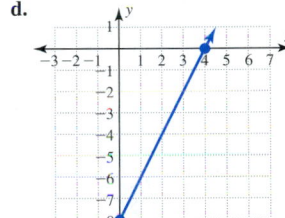

e.

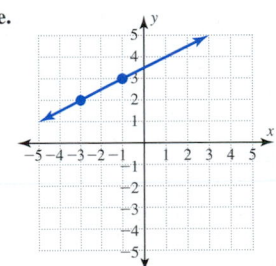

34. a. C **b.** D **c.** A **d.** B **35. a.** $m = \dfrac{6}{5}$ **b.** $m = \dfrac{3}{4}$ **c.** $m = 0$ **d.** undefined
36. $y = 3x - 2$ **37.** $y = -2x + 7$ **38.** $y = \dfrac{2}{3}x + \dfrac{5}{3}$ **39.** $y = 4$ **40.** $x = 3$
41. $y = -\dfrac{2}{5}x + 2$ **42. a.** A **b.** C **c.** B **43.** $(2, 11)$ **44.** $(3, -4)$ **45.** $(-5, 1)$
46. The trains are traveling at speeds of 103 km/h and 95 km/h. **47.** Use
2 L of the 10% solution and 4 L of the 25% solution. **48.** The job requires
10 loads of topsoil and 14 loads of fill dirt. **49. a.** $[2, \infty)$ **b.** $(-\infty, 2]$
50. a. $[-1, \infty)$ **b.** $(-\infty, -1]$ **51. a.** $[8, \infty)$ **b.** $(-\infty, 16)$
52. a. conditional inequality; $(-\infty, 0)$ **b.** contradiction; no solution
c. unconditional inequality; $\mathbb{R}$. **53. a.** $x = 4, x = -3$ **b.** $(-3, 2)$
c. $(-\infty, -2] \cup [8, \infty)$ **54.** $(2, 3]$ **55.** $(-\infty, -3] \cup [2, +\infty)$

56. 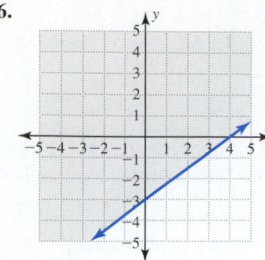 **57.**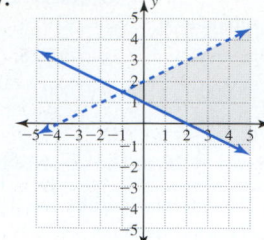

58.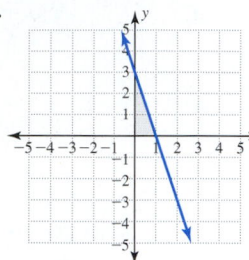

59. a. 1 **b.** 3 **c.** 1 **d.** -1 **60. a.** $\dfrac{5}{2}$ **b.** 5 **c.** $\dfrac{6}{5}$ **d.** 5 **61. a.** 1 **b.** 0 **c.** $\dfrac{1}{6}$ **d.** 1

62. a. 1 **b.** $\dfrac{9}{4}$ **c.** $\dfrac{3}{2}$ **d.** $\dfrac{4}{9}$ **63. a.** 100 **b.** 10,000 **c.** 1 **d.** $\dfrac{1}{10}$ **64. a.** 1 **b.** 2

c. 3 **d.** 10 **65. a.** $2x^3$ **b.** x^8 **c.** x^4 **d.** x^{12} **66. a.** $32x^3y^4$ **b.** $2xy^2$ **c.** $2x^3$

d. $64x^4y^6$ **67. a.** x^3 **b.** $\dfrac{1}{x^2}$ **c.** $\dfrac{x^6}{9y^8}$ **d.** $-\dfrac{y^{10}}{2x^{18}}$ **68. a.** 0.0123 **b.** 12,300

69. a. 4.35×10^{-5} **b.** 1.87×10^6 **70. a.** $5x + 3y$ **b.** $x + 11y$

c. $x + 15y$ **71. a.** $7x^2 - x - 6$ **b.** $x^2 - 9x + 12$ **c.** $3x^2 - 16x + 26$

72. a. $16x^2 - 49$ **b.** $4x^2 + 12x + 9$ **c.** $25x^2 - 60x + 36$

73. a. $x^3 + 2x^2 - 11x - 12$ **b.** $x^3 - 5x^2 - 2x + 24$

c. $6x^4 - 10x^3 - 11x^2 + 7x + 2$ **74. a.** $3x^2 + 2x$ **b.** $x - 4$ **c.** $x^3 + x + 2$

Review of Technology Perspectives

75. -9.425 **76.** 9.15 **77.** 5.916 **78.** $\dfrac{9}{10}$ **79.** $-1,296$ **80.** 4.856

81.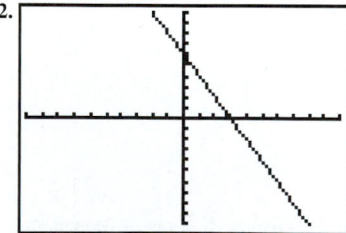

82.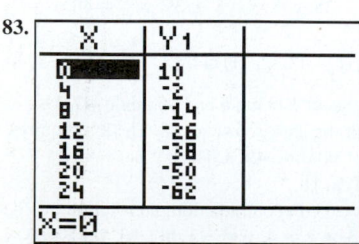

$[-10, 10, 1]$ by $[-10, 10, 1]$

83.

84. The points $(-1, 10)$ and $(4, 5)$ are on the line $y = 3x - 7$. **85.** $(4, 1)$
86. $\left(-\dfrac{8}{7}, -\dfrac{18}{7}\right)$ **87.** $(22.5, 67.5)$

88. 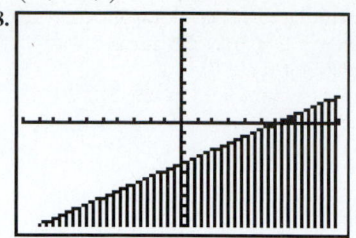 **89.** 1.8066×10^{-10}

$[-10, 10, 1]$ by $[-10, 10, 1]$

Chapter 6

Using the Language and Symbolism of Mathematics 6.1

1. greatest common **2.** coefficient; exponent **3.** distributive **4.** grouping
5. 0 **6.** zero **7.** $(-4, 0)$ **8.** -2 **9.** $x - 8$ **10.** real **11.** n

Quick Review 6.1

1. $7 \cdot 11$ **2.** $2 \cdot 3^2 \cdot 5$ **3.** $(3, 0)$ **4.** $(2, 0)$ **5.** $x^2 + x - 20$

Exercises 6.1

1. 6 **3.** $4x$ **5.** $9x^2y$ **7.** $6x^2(2x - 3)$ **9.** $2ab(5a + 3b)$
11. $5x(2x^2 - 3xy + 4y^2)$ **13.** $(2x + y)(x - 2y)$ **15.** $(7x - 9)(2x - 5)$
17. $7x(2x + 11)$ **19.** $5x^2(2x - 3)$ **21.** $4xy(2x - 5y)$
23. $2x(3x^2 - 4x + 5)$ **25.** $(5x - 2)(x + 3)$ **27.** $(x + 2y)(6x - 7y)$
29. $(a + b)(x + 2)$ **31.** $(5x - 2)(a + 3)$ **33.** $(a - b)(2x - 7)$
35. $(3x - 5)(x + 4)$ **37.** $(3x + 1)(4x - 5)$ **39. a.** 2, 5
b. $(x - 2)(x - 5)$ **41. a.** $-4, 8$ **b.** $(x + 4)(x - 8)$ **43. a.** $(2, 0), (5, 0)$
b. 2, 5 **c.** $(x - 2)(x - 5)$ **45. a.** $(-2, 0), (0, 0), (3, 0)$ **b.** $-2, 0, 3$
c. $x(x + 2)(x - 3)$ **47. a.** $-5, 9$ **b.** $(-5, 0), (9, 0)$ **49. a.** $-3, -1, 8$
b. $(-3, 0), (-1, 0), (8, 0)$ **51. a.** $-3, 17$ **b.** $(-3, 0), (17, 0)$
53. a. $300, 150, 50, 0, 0, 50, 150$ **b.** 15, 20 **c.** $(x - 15)(x - 20)$
55. a. $0, 15, 16, 9, 0, -5, 0$ **b.** 0, 4, 6 **c.** $x(x - 4)(x - 6)$
57. $(x - 8)(x + 7)$ **59.** $(x + 9)(x - 1)(x - 5)$
61. $(x - 4)(x + 4)(x - 1)(x + 1)$ **63.** 7 and -9; $(7, 0), (-9, 0)$
65. $(x + 8)(x - 10)$; -8 and 10 **67.** $(x + 2)(x + 4)$; $x^2 + 6x + 8$
69. $(2x - 7)(5x - 2)$; $10x^2 - 39x + 14$ **71. a.** $P(1 + r)$ **b.** 2,080
73. $A = 2\pi r(r + h)$

Group Discussion Questions

75. a. $A = \dfrac{1}{2}x(4x + 40) = 2x^2 + 20x$ **b.** $2x^2$ represents the sum of

the areas of the triangles. $20x$ is the area of the rectangle.

c. $A = \dfrac{1}{2}x(4x + 40) = \dfrac{1}{2}$(Height)(Sum of bases) **77.** Each tick mark on

the x-axis represents 2 units. Examining the factored form reveals that the
correct solutions are -4 and 6.

Cumulative Review

1. $y = \dfrac{2}{3}x - 2$ **2.** $(-6, 4]$ **3.** $\dfrac{1}{9x^4y^{20}}$ **4.** $2x + 5$ **5. a.** $1 \cdot 21; 3 \cdot 7$

b. $1 \cdot 20; 2 \cdot 10; 4 \cdot 5$

Using the Language and Symbolism of Mathematics 6.2

1. prime **2.** quadratic **3. a.** quadratic **b.** linear **c.** constant **4.** product;
sum **5.** same **6.** opposite **7.** inspection

Quick Review 6.2

1. $x^2 - 2x - 15$ **2.** $x^2 - 2xy - 15y^2$ **3.** $x^3 - 7x^2 + 10x$ **4.** $1 \cdot 35; 5 \cdot 7$
5. $1 \cdot 48; 2 \cdot 24; 3 \cdot 16; 4 \cdot 12; 6 \cdot 8$

Exercises 6.2

1. B **3.** A **5.** $(x - 4)(x - 2)$ **7.** $(x + 1)(x - 12)$ **9.** $(x + 2)(x + 1)$
11. $(x - 2)(x + 1)$ **13.** $(x + 5)(x - 4)$ **15.** $(x - 4)(x - 5)$
17. $(x - 9)(x + 2)$ **19.** $(x + 6)(x - 4)$ **21.** $(a - 11)(a + 1)$
23. $(y - 17)(y - 1)$ **25.** $(x - 7)(x - 1)$ **27.** $(y - 11)(y + 9)$
29. $(p - 25)(p - 4)$ **31.** $(t + 6)(t + 8)$ **33.** $(n + 9)(n - 4)$
35. a. $(x + 3)(x + 4)$ **b.** $(x + 6)(x + 2)$ **c.** prime **d.** $(x + 12)(x + 1)$

37. a. $(x + 4)(x - 7)$ **b.** $(x + 2)(x - 14)$ **c.** prime **d.** $(x + 1)(x - 28)$
39. a. $(x - 6)(x + 6)$ **b.** $(x + 12)(x - 3)$ **c.** $(x - 2)(x + 18)$ **d.** prime
41. $(x + 2y)(x + 3y)$ **43.** $(x + 3y)(x - 2y)$ **45.** $(x - 5y)(x - 5y)$
47. $(a - 18b)(a + 2b)$ **49.** $-(x - 7)(x + 5)$ **51.** $-(m - 3)(m + 6)$
53. $-(x - 15y)(x + 3y)$ **55.** $5x(x + 1)(x + 2)$ **57.** $4ax(x - 3)(x - 2)$
59. $10a(z + 4)(z + 25)$ **61.** $2ab(x - 6)(x + 8)$
63. $(a + b)(x^2 - 6x + 6)$ **65. a.** $(x - 6)(x + 3)$ **b.** prime
67. a. $(x + 12y)(x + y)$ **b.** prime **69. a.** $-9, -5$ **b.** $(x + 9)(x + 5)$
71. a. $(-10, 0), (4, 0)$ **b.** $(x + 10)(x - 4)$ **73.** $(x + 5)(x - 3); -5, 3;$
$(-5, 0), (3, 0)$ **75.** $x^2 - 5x + 6; (x - 2)(x - 3)$

Group Discussion Questions

77. a. $-\dfrac{1}{2}$ and $\dfrac{1}{3}$ **b.** $-\dfrac{1}{4}$ and $\dfrac{2}{3}$ **c.** $\dfrac{3}{4}$ and $-\dfrac{2}{5}$ **d.** $(3x - 2)(5x - 1)$
e. $(2x + 1)(5x + 3)$ **f.** $(5x - 2)(7x + 1)$ **79.** $x^2 - 35x - 36;$
$x^2 - 16x - 36; x^2 - 9x - 36; x^2 - 5x - 36; x^2 - 36; x^2 + 5x - 36;$
$x^2 + 9x - 36; x^2 + 16x - 36; x^2 + 35x - 36$

Cumulative Review

1. $x = 26$ **2.** $y = -\dfrac{4}{3}x + \dfrac{7}{3}$ **3.** $[-17, 1]$ **4.** $5x^2 - 5$
5. $3x^2y^2(4x^2y + 3xy^2 + 1)$

Using the Language and Symbolism of Mathematics 6.3
1. GCF **2.** positive **3.** negative **4.** b **5.** same **6.** opposite **7.** $1; -1$

Quick Review 6.3
1. $6x^2 + 19x - 7$ **2.** $10x^2 + 43xy + 28y^2$ **3.** $6x^2 + 7x - 20$
4. $-54x^3y + 27x^2y^2 + 15xy^3$ **5.** $(3x - 5)(2x + 7)$

Exercises 6.3
1. B **3.** A **5. a.** $(3x + 1)(x + 1)$ **b.** $(3x - 1)(x - 1)$
7. a. $(2x - 1)(x + 1)$ **b.** $(2x + 1)(x - 1)$ **9.** $(5x - 1)(x + 2)$
11. $(5x - 2)(x - 1)$ **13.** $(3v + 2)(2v - 3)$ **15.** $(6v - 1)(v - 6)$
17. $(5x + 1)(x + 1)$ **19.** $(7x - 1)(x - 1)$ **21.** $(7x + 1)(x - 1)$
23. $(11x - 1)(x + 1)$ **25.** $(2x + 3)(x + 1)$ **27.** $(3a + 2)(a + 5)$
29. $(m - 4)(3m + 2)$ **31. a.** $(m + 2)(3m + 4)$ **b.** $(3m + 8)(m + 1)$
c. $(m + 4)(3m + 2)$ **d.** $(m + 8)(3m + 1)$ **33.** $(2x + 7)(x - 1)$
35. $(w - 1)(4w - 5)$ **37.** $(3z + 2)(3z - 5)$ **39.** $(2b + 5)(5b + 2)$
41. $(3m + 5n)(4m - 7n)$ **43.** $(2y + 7)(9y - 4)$
45. $(11y + 3)(5y - 4)$ **47.** $(4x + 5y)(5x + 3y)$
49. $(7x - 2y)(2x - 5y)$ **51.** $(3m + 2n)(11m - 3n)$
53. $(2v + 3w)(3v + 2w)$ **55. a.** $(n + 10)(4n - 1)$ **b.** $(n + 4)(2n - 5)$
c. $(n + 2)(5n - 4)$ **d.** prime **57. a.** prime **b.** $(4x - 3y)(x - 3y)$
c. $(x - 2y)(2x - 9y)$ **d.** $(6x - y)(x - 6y)$ **59. a.** $(x - 5)(4x - 3)$
b. prime **61.** $-(6x - 1)(2x - 3)$ **63.** $-(7y + 2)(5y - 2)$
65. $-(3z + 8)(4z + 5)$ **67.** $-2a(19x + 3)(x - 1)$
69. $5x^3(4x - 3)(3x - 5)$ **71.** $3mn(m + n)(5m - 7n)$
73. $7av(5v + 6w)(6v - 5w)$ **75.** $(a + b)(5x - 2)(7x - 3)$
77. $\dfrac{1}{3}$ and $5; \left(\dfrac{1}{3}, 0\right), (5, 0)$ **79.** $(3x - 2)(x + 4); \left(\dfrac{2}{3}, 0\right), (-4, 0)$

Group Discussion Questions
81. a. $(x - 5)(x + 5)$ **b.** $(x - 8)(x + 8)$ **c.** $(2x - 1)(2x + 1)$
d. $(2x - 5)(2x + 5)$ **e.** $(x - 7)(x + 7)$ **f.** $(3x - 1)(3x + 1)$
g. $(3x - 4)(3x + 4)$

Cumulative Review
1. -28 **2.** 74 **3.** 7 **4.** $x = \dfrac{5}{2}; x = -\dfrac{23}{2}$ **5.** $8x^3 - 27$

Using the Language and Symbolism of Mathematics 6.4
1. D **2.** E **3.** A **4.** C **5.** B

Quick Review 6.4
1. $x^2 - 49$ **2.** $25x^2 + 20xy + 4y^2$ **3.** $36x^2 - 60x + 25$ **4.** $27x^3 - 8$
5. $1, 4, 9, 16, 25, 36, 49, 64, 81, 100; 1, 8, 27, 64, 125$

Exercises 6.4
1. $(a + 3)^2$ **3.** $(2w - 3)^2$ **5.** $(m - 1)^2$ **7.** $(6v + 1)^2$ **9.** $(5x - 2y)^2$
11. $(11m + 4n)^2$ **13.** $-(x - 8y)^2$ **15.** $(4x - 3)^2$ **17.** $2x(x + 4)^2$
19. $(w + 7)(w - 7)$ **21.** $(3v + 1)(3v - 1)$ **23.** $(9m + 5)(9m - 5)$

25. $(2a + 3b)(2a - 3b)$ **27.** $(4v + 11w)(4v - 11w)$
29. $(6 + m)(6 - m) = -(m + 6)(m - 6)$ **31.** $5(2x + 3y)(2x - 3y)$
33. $(x + 3)(x^2 - 3x + 9)$ **35.** $(m - 5)(m^2 + 5m + 25)$
37. $(4a - b)(16a^2 + 4ab + b^2)$ **39.** $(5x + 2y)(25x^2 - 10xy + 4y^2)$
41. a. $(x - 3)^2$ **b.** prime **c.** $(x - 9)(x - 1)$ **43. a.** $(x - 6y)^2$
b. $(x - 6y)(x + 6y)$ **c.** $(x - 9y)(x - 4y)$ **45. a.** prime **b.** $(2x - 3y)^2$
c. $(x + 3y)(4x + 3y)$ **47. a.** $(5x + 8y)(5x - 8y)$ **b.** $(5x - 8y)^2$
c. $(5x - 16y)(5x - 4y)$ **49. a.** $12x$ **b.** 36 **51. a.** $20x$ **b.** 100
53. $-16(t - 3)^2$ **55.** $(x^2 + 3)(x^2 - 3)$ **57.** $(5y^2 + 1)(5y^2 - 1)$
59. $(9y^2 + 1)(3y + 1)(3y - 1)$ **61.** $(x - 3y - 5)(x + 3y - 5)$
63. $(2x + y - 3)(2x - y + 3)$ **65.** $-3(a + b)(a - b)$
67. $(a - 5b)(x + y)(x - y)$ **69.** $(a + b)(x + 3)^2$
71. $(x + 10y)^2(a + 4)(a - 4)$ **73.** $(x_1 + x_2)(x_1 - x_2)$
75. $\pi h(r_1 + r_2)(r_1 - r_2)$ **77. a.** $V(x) = x(7 - 2x)(9 - 2x)$
b. The height is x, the width is $7 - 2x$, and the length is $9 - 2x$.
c. It depends on what you plan to do with the function. The factored
form displays the role of the length, width, and height.
79. a. $(-1, 0), (0, 0), (1, 0)$ **b.** $-1, 0, 1$ **c.** $x(x + 1)(x - 1)$

Group Discussion Questions
81. $a^6 + b^6$ factors as the sum of two cubes.
$a^6 + b^6 = (a^2)^3 + (b^2)^3 = (a^2 + b^2)(a^4 - a^2b^2 + b^4)$

Cumulative Review
1. $-x - 8$ **2.** $\left(2, \dfrac{13}{2}\right)$ **3.** $(-\infty, 5) \cup (8, \infty)$ **4.** $\dfrac{23}{2}$ **5.** $6x^{12}y^4$

Using the Language and Symbolism of Mathematics 6.5
1. grouping **2.** greatest common factor **3.** prime **4.** sum **5.** difference
6. cubes **7.** difference

Quick Review 6.5
1. $a(x - y - 3)$ **2.** $2b(x - 3y + 3)$ **3.** $(3a - b)(x^2 - 3x + 9)$
4. $(a + 2b + 3y)(a + 2b - 3y)$ **5.** $(5x + a - 3b)(5x - a + 3b)$

Exercises 6.5
1. $(x + 5)(x - y)$ **3.** $(x + y - 1)(x - y + 1)$ **5.** $(a + b)(c + d)$
7. $(5c + 3)(a - 2b)$ **9.** $(a + c)(b - d)$ **11.** $(v - 7)(v - w)$
13. $(2a + 4b + 3)(2a - 4b + 3)$ **15.** $(n + 5)(3m - k)$
17. $(z^3 + w^2)(a + b)$ **19.** $(3b + a - 4)(3b - a - 4)$
21. $(y + 1)(ay + a - 1)$ **23.** $(8y + 3z)(8y - 3z)$
25. prime **27.** $3(x - 1)(4x - 5)$ **29.** $(7a - 2)^2$ **31.** $(a - b)(x + y)$
33. prime **35.** prime **37.** $(2x^5 + 3y^3)^2$ **39.** $12xy(x - y)(x + y)$
41. $(c + d)(x + y)$ **43.** $(x + 1)(ax + by)$ **45.** $(x^3 + 2y)^2$
47. $8a(x - 1)(8x - 5)$ **49.** $5(x^2 - 11)$ **51.** $7ab(3a + 5)(3a - 5)$
53. $-3(2x - y)^2$ **55.** $9(16x^2 + 9)$ **57.** $(3x - 5y - 1)(3x + 5y - 1)$
59. $2a(2x - 3y)(3x + 2y)$ **61.** $7st(s^2 + t^2)(s + t)(s - t)$
63. $3a(x + y - 1)(x - y + 1)$ **65.** $2ax(x - 2)(x + 7)$
67. $(a + 1)(a + b + 1)$ **69.** $(a - 2b)(x + 4)(x - 6)$
71. $(2x - 3y + a)(2x - 3y - 25a)$ **73.** $(-3, 0), (2, 0), (4, 0);$
$-3, 2, 4; (x + 3)(x - 2)(x - 4)$ **75.** $5a(b - 1)(b^2 + b + 1)$
77. $(x + y)(x^2 + x - xy - y + y^2)$

Group Discussion Questions
79. a. $x^m(x^2 + 4)$ **b.** $x^m(x^m + 3)^2$
81. $2,025x^4 + 2,700x^2y + 900y^2 = (45x^2 + 30y)^2 =$
$[15(3x^2) + 15(2y)]^2 = 225(3x^2 + 2y)^2; 2,025x^4 + 2,700x^2y + 900y^2 =$
$225(9x^4 + 12x^2y + 4y^2) = 225(3x^2 + 2y)^2$ In most cases it is easier first
to factor out the GCF.

Cumulative Review
1. $(-5, 0), (8, 0)$ **2.** $3, -9$ **3.** $x = 0$; conditional equation **4.** all real
numbers; identity **5.** no solution; contradiction

Using the Language and Symbolism of Mathematics 6.6
1. quadratic **2.** quadratic; linear; constant **3.** $x - 3; x + 6$ **4.** LCD
5. $(c, 0)$ **6.** $x - c$ **7.** $=$ **8.** $>$ **9.** $<$

Quick Review 6.6
1. $x = 15$ **2.** $x = -7$ **3.** $x = \dfrac{7}{5}$ **4.** $x = \dfrac{2}{3}$ is a solution. **5.** $x = 3$ is not a
solution.

Exercises 6.6

1. a. D **b.** C **c.** A **d.** B **3. a.** $2x^2 - 7x + 3 = 0; a = 2, b = -7, c = 3$
b. $8x^2 - 3x = 0; a = 8, b = -3, c = 0$ **c.** $7x^2 + 5 = 0; a = 7, b = 0,$
$c = 5$ **d.** $2x^2 - x - 4 = 0; a = 2, b = -1, c = -4$ **5.** $m = 8,$
$m = -17$ **7.** $n = \dfrac{5}{2}, n = -\dfrac{1}{3}$ **9.** $z = 0, z = -\dfrac{7}{2}$ **11.** $v = -11, v = 11$

13. $x = -2, x = -1$ **15.** $y = -3, y = 6$ **17.** $v = 0, v = -\dfrac{1}{3}$

19. $w = -\dfrac{3}{2}, w = 5$ **21.** $x = -\dfrac{5}{2}, x = -\dfrac{2}{3}$ **23.** $x = 3, x = 8$

25. $z = -\dfrac{3}{7}, z = \dfrac{1}{2}$ **27.** $x = -\dfrac{5}{3}, x = \dfrac{5}{3}$ **29.** $r = -5, r = 2$

31. $x = -1, x = 2$ **33.** $x = -1, x = 3$ **35.** $m = -3, m = 6$
37. $x = -2, x = 6$ **39.** $v = 4, v = 7$ **41.** $v = 0, v = 5$

43. $w = -1, w = 4$ **45.** $x = -1, x = \dfrac{5}{2}$ **47.** $x = \dfrac{5}{2}$ **49.** $m = \dfrac{7}{3}$

51. $x^2 - 6x + 5 = 0$ **53.** $x^2 - 3x - 18 = 0$ **55.** $8x^2 - 2x - 3 = 0$
57. $15x^2 + 31x + 14 = 0$ **59. a.** $x = -3, x = 5$ **b.** $(-3, 5)$
c. $(-\infty, -3) \cup (5, \infty)$ **61. a.** $x = -5, x = 2$ **b.** $(-\infty, -5) \cup (2, \infty)$
c. $(-5, 2)$ **63. a.** $-10, 4$ **b.** $(-10, 0), (4, 0)$ **c.** $(x + 10)(x - 4)$
d. $x = -10, x = 4$ **65. a.** $(-1, 0), (0, 0), (2, 0)$ **b.** $-1, 0, 2$

c. $x(x + 1)(x - 2)$ **d.** $x = 0, x = -1, x = 2$ **67. a.** $m = \dfrac{3}{5}, m = 2$

b. $5m^2 - 13m + 6$ **69. a.** $x = 3, x = -\dfrac{2}{5}$ **b.** $20x^2 - 52x - 24$

71. $x = 2, x = -7; (-7, 2); (-\infty, -7) \cup (2, \infty)$ **73.** $x = -4, x = 1;$
$(-\infty, -4) \cup (1, \infty); (-4, 1)$ **75.** 3 cm by 11 cm **77.** 8 m **79.** 15 cm

Group Discussion Questions

81. a. $x = -\sqrt{5}, x = \sqrt{5}$ **b.** $(x + \sqrt{7})(x - \sqrt{7})$
c. $(x + \sqrt{11})(x - \sqrt{11}); x = -\sqrt{11}, x = \sqrt{11}$ **83. a.** $x = -y, x = 3y$
b. $x = -y, x = 6y$ **c.** $x = \dfrac{2y}{3}, x = -\dfrac{y}{2}$ **d.** $x = -5y, x = 3y$

Cumulative Review

1. B **2.** A **3.** D **4.** E **5.** C

Review Exercises for Chapter 6

1. $4(x - 9)$ **2.** $2x(x - 5)$ **3.** $12ax(x - 2)$ **4.** $5x(x^2 - 3)$
5. $(a + b)(x + y)$ **6.** $(a - 3)(x + 2y)$ **7.** $(2x + 7)(3x - 5)$
8. $(4x + 3)(6x - 5)$ **9.** $(x + 2)(x - 2)$ **10.** $(2x + 1)(2x - 1)$
11. $7(x + 2)(x - 2)$ **12.** $2(m + 8)(m - 8)$ **13.** $(x - 9)(x - 2)$
14. $(m + 3)(m + 7)$ **15.** $10(x + 2)(x + 7)$ **16.** $5(x - 8)(x - 1)$
17. $(x + 2)^2$ **18.** $(m + 5)^2$ **19.** $3(x - 3)^2$ **20.** $2(m - 4)^2$
21. $-(2x - 3y)^2$ **22.** $-(3x - 5y)^2$ **23.** $(a - 7)(x + 2y)$
24. $xy(x - 4)(x - 1)$ **25.** $(a + b)(x + 5)(x - 5)$
26. $11(x + y)(x - y + 3)$ **27.** $(v + w + 1)(v - w + 1)$
28. $(5a - 3b)(2x - 3y)$ **29.** Prime **30.** $(v + 3w)(v - 3w)$
31. $(10x + 7y)(10x - 7y)$ **32.** Prime **33.** $2(x + 3y)(2x - 5y)$
34. $(2x - 5y)(3x + 4y)$ **35.** $10(x + 3)(2x - 1)$ **36.** $(x + 10y)(6x + y)$
37. $(v^2 + 1)(v + 1)(v - 1)$ **38.** $(9w^2 + 4)(3w + 2)(3w - 2)$
39. $(x + 5y)(a + b)^2$ **40.** $(3x - y)(3x + y + 1)$ **41. a.** $(x - 50)(x + 2)$
b. $(x - 20)(x + 5)$ **c.** $(x + 10)(x - 10)$ **d.** Prime
42. a. $(x - 2)(x + 24)$ **b.** $(x - 3)(x + 16)$ **c.** $(x - 6)(x + 8)$
d. Prime **43. a.** $(2x + 1)(50x + 1)$ **b.** $(2x + 1)(2x + 25)$
c. $(x + 5y)(5x + 4y)$ **d.** $(5x + 2y)(2x + 5y)$ **44. a.** $(3x - 1)(16x - 1)$
b. $(x - 8)(3x - 2)$ **c.** $(x - 24y)(2x - y)$ **d.** $(4x - 3y)(4x - y)$
45. $(2x - y)(4x^2 + 2xy + y^2)$ **46.** $(x + 4)(x^2 - 4x + 16)$
47. $2x(x - 2)(x^2 + 2x + 4)$ **48.** $(v - 1)^2(v^2 + v + 1)$

49. $x = 5, x = -1$ **50.** $x = \dfrac{3}{2}, x = -\dfrac{2}{3}$ **51.** $x = 0, x = 7, x = \dfrac{2}{7}$

52. $v = 7, v = -3$ **53.** $y = \dfrac{1}{5}, y = -\dfrac{3}{2}$ **54.** $w = 3, w = -\dfrac{7}{6}$

55. $v = 5, v = -3$ **56.** $x = 3, x = -7$ **57.** $x = 0, x = 6, x = -6$

58. $x = 0, x = 5$ **59.** $z = -2, z = 6$ **60.** $x = -\dfrac{9}{2}$ **61.** $w = -4, w = -2$

62. $x = -15, x = 5$ **63. a.** $x = 3, x = -12$ **b.** $x^2 + 9x - 36$

64. a. $x = \dfrac{1}{3}, x = -\dfrac{5}{2}$ **b.** $6x^2 + 13x - 5$ **65. a.** $-6, 4$ **b.** $(-6, 0), (4, 0)$

c. $(x + 6)(x - 4)$ **d.** $x = -6, x = 4$ **66. a.** $-3, -1, 1$
b. $(-3, 0), (-1, 0), (1, 0)$ **c.** $(x + 3)(x + 1)(x - 1)$ **d.** $x = -3,$
$x = -1, x = 1$ **67. a.** $x = -5, x = 7$ **b.** $[-5, 7]$ **c.** $(-\infty, -5) \cup (7, \infty)$
68. a. $x = -5, x = 4$ **b.** $(-\infty, -5) \cup (4, \infty)$ **c.** $[-5, 4]$
69. $7x^2 - 17x - 12 = 0$ **70.** $2x^2 - x - 10 = 0$ **71.** $(-11, 0), (7, 0);$
$(x + 11)(x - 7); -11, 7; x = -11, x = 7$ **72.** $(-15, 0), (35, 0);$
$(x + 15)(x - 35); -15, 35; x = -15, x = 35$ **73.** $2\pi(r_1 - r_2)$

74. $\pi(L_1 + L_2)(L_1 - L_2)$ **75.** $\dfrac{4}{3}\pi(r_1 - r_2)(r_1^2 + r_1 r_2 + r_2^2)$

76. $h = 4$ cm **77.** either 10 cm or 20 cm

Mastery Test for Chapter 6

1. a. $9x^2(2x - 11)$ **b.** $2xy^2(2x^2 - 3xy + 5y^2)$ **c.** $(7x - 2)(3x + 4)$
d. $(2x + 3y)(x - 5y)$ **2. a.** $(a + 3)(7x + 1)$ **b.** $(3x - 2)(4a - 5b)$
c. $2(x + 2)(x - 3)$ **d.** $(5x - 6)(2x - 3)$ **3. a.** $-12, 2$
b. $(-12, 0), (2, 0)$ **c.** $(x + 12)(x - 2)$ **4. a.** $(w - 9)(w + 5)$
b. $(w + 9)(w + 5)$ **c.** $(v - 4)(v - 6)$ **d.** $(v + 9)(v - 4)$
5. a. $(x - 9)(x + 4)$ **b.** prime **c.** $(x + 2)(x + 15)$ **d.** prime
6. a. $(x - 4y)(x + 3y)$ **b.** $(x - 12y)(x - y)$ **c.** $(a + 6b)(a + 24b)$
d. $(a + 8b)(a - 6b)$ **7. a.** $(x - 7)(5x + 2)$ **b.** $(x - 12)(6x - 1)$
c. $(3x + 2)(3x + 5)$ **d.** $(3x - 4y)(4x + 3y)$ **8. a.** $(3x - 1)(3x + 8)$
b. $(3x - 2)(4x + 3)$ **c.** $(x + 3)(3x - 8)$ **d.** $(x + 4y)(18x - y)$
9. a. prime **b.** $(4x - 1)(2x - 3)$ **c.** $(3x - 1)(3x + 4)$ **d.** prime
10. a. $(x + 7y)^2$ **b.** $(x - 8y)^2$ **c.** $(3x + 10y)^2$ **d.** $(5x - 11y)^2$
11. a. $(x + 2y)(x - 2y)$ **b.** $(20a + b)(20a - b)$
c. $(4v + 7w)(4v - 7w)$ **d.** $(6x + 5y)(6x - 5y)$
12. a. $(4v - 1)(16v^2 + 4v + 1)$ **b.** $(v + 5)(v^2 - 5v + 25)$
c. $(2x + 5y)(4x^2 - 10xy + 25y^2)$ **d.** $(3a - 10b)(9a^2 + 30ab + 100b^2)$
13. a. $(2x + 3)(a + b)$ **b.** $(2x - 5y)(7a - 3b)$
c. $(a - 2b)(a + 2b + 1)$ **d.** $(x + 5y + 2)(x + 5y - 2)$
14. a. $5(x + 7)(x - 7)$ **b.** $2a(x + 5)^2$ **c.** $5a(v + x - 2y)(v - x + 2y)$
d. $19(x - 1)(x^2 + x + 1)$ **15. a.** quadratic; $x^2 - 2x + 5 = 0;$
$a = 1, b = -2, c = 5$ **b.** not quadratic **c.** not an equation **d.** quadratic;
$2x^2 - 5 = 0; a = 2, b = 0, c = -5$ **16. a.** $x = -\dfrac{1}{2}, x = 3$

b. $x = -9, x = 11$ **c.** $x = -\dfrac{11}{2}, x = 1$ **d.** $x = -3, x = -\dfrac{1}{6}$

17. a. $x^2 - 2x - 63 = 0$ **b.** $x^2 - 169 = 0$ **c.** $7x^2 + 27x - 4 = 0$
d. $21x^2 - 37x + 12 = 0$ **18. a.** $x = -5, x = 3$ **b.** $(-5, 3)$
c. $(-\infty, -5) \cup (3, \infty)$

Group Project for Chapter 6

1. $m = \dfrac{1}{4}$ **2.** The cross-sectional area is $A = 4x^2 + 42x$ square feet.

3. The volume is $V = \dfrac{1,760}{9}(4x^2 + 42x)$ cubic yards. **4.** If the cost per

cubic yard is \$12, then the total cost is $C = \dfrac{7,040}{3}(4x^2 + 42x)$ dollars.

5. If the cost per cubic yard is \$22, then the total cost is

$C = \dfrac{38,720}{9}(4x^2 + 42x)$ dollars.

6.

	A	B	C	D	E
1	Height (ft)	Area (sq ft)	Vol (cubic ft)	Cost ($) @$12	Cost ($) @$22
2	5	310	60,622	$727,467	$1,333,689
3	10	820	160,356	$1,924,267	$3,527,822
4	15	1,530	299,200	$3,590,400	$6,582,400
5	20	2,440	477,156	$5,725,867	$10,497,422

Chapter 7

Using the Language and Symbolism of Mathematics 7.1

1. quadratic **2.** $\pm$ **3.** $\pm\sqrt{k}$ **4.** square **5.** $\sqrt{x}\sqrt{y}$ **6.** $\dfrac{\sqrt{x}}{\sqrt{y}}$ **7.** rationalizing

Quick Review 7.1

1. 6 **2.** 6.245 **3.** 0.346 **4.** $\dfrac{5}{11}$ **5.** $x = 16$

Exercises 7.1

1. $m = \pm 2$ **3.** $x = \pm\sqrt{2} \approx \pm 1.41$ **5.** $y = \pm 3$ **7.** $v = \pm\sqrt{13} \approx \pm 3.61$
9. $w = \pm 6$ **11.** $z = \pm\sqrt{5} \approx \pm 2.24$ **13.** $4\sqrt{5}$ **15.** $2\sqrt{5}$ **17.** $10\sqrt{2}$
19. $x = \pm 3\sqrt{2} \approx \pm 4.24$ **21.** $y = \pm 3\sqrt{3} \approx \pm 5.20$
23. $a = \pm 2\sqrt{2} \approx \pm 2.83$ **25.** $\dfrac{2}{3}$ **27.** $\dfrac{\sqrt{6}}{7}$ **29.** 2 **31.** $\dfrac{3}{5}$ **33.** $v = \pm\dfrac{9}{2}$
35. $x = \pm\dfrac{\sqrt{7}}{6}; x \approx \pm 0.44$ **37.** $\dfrac{3\sqrt{5}}{5}$ **39.** $\dfrac{3\sqrt{10}}{5}$ **41.** $\dfrac{\sqrt{6}}{3}$ **43.** $\dfrac{\sqrt{85}}{17}$
45. $x = \pm\dfrac{\sqrt{10}}{2} \approx \pm 1.58$ **47.** $w = \pm\dfrac{\sqrt{35}}{5} \approx \pm 1.18$ **49.** $n = -5, n = 7$
51. $b = -5, b = 2$ **53.** $x = -\dfrac{4}{3}, x = 2$ **55. a.** $x = 18$ **b.** $x = \pm 6$
57. a. $m = 5$ **b.** $m = -1, m = 2$ **59. a.** $x = 0, x = 1$ **b.** $(0, 1)$
c. $(-\infty, 0) \cup (1, \infty)$ **61. a.** $x = \pm\sqrt{6}$ **b.** $\pm\sqrt{6}$ **c.** $(\pm\sqrt{6}, 0)$
63. a. $x^2 - 49 = 0$ **b.** $x^2 - 7 = 0$ **65.** $c \approx 46.3$ cm **67.** The interest rate is approximately 8.5%. **69.** It takes the hammer approximately 1.8 seconds to fall 50 ft.

Group Discussion Questions

71. a. The expression needs to be entered with grouping symbols around $4 + \sqrt{7}$ in the numerator. Try $(4 + \sqrt{7})/3$. **b.** 0.079 **c.** 0.467

73. a.
$$\left(\frac{5 - \sqrt{29}}{2}\right)^2 - 5\left(\frac{5 - \sqrt{29}}{2}\right) - 1$$
$$= \frac{25 - 10\sqrt{29} + 29}{4} - \frac{25 - 5\sqrt{29}}{2} - 1$$
$$= \frac{54 - 10\sqrt{29}}{4} - \frac{2}{2} \cdot \frac{25 - 5\sqrt{29}}{2} - 1$$
$$= \frac{54 - 10\sqrt{29}}{4} - \frac{50 - 10\sqrt{29}}{4} - 1$$
$$= \frac{54 - 10\sqrt{29} - 50 + 10\sqrt{29}}{4} - 1 = \frac{4}{4} - 1 = 1 - 1 = 0$$

b.
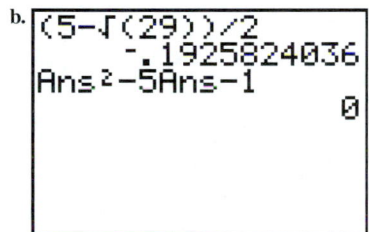

```
(5-√(29))/2
          -.1925824036
Ans²-5Ans-1
                     0
```

Cumulative Review

1. 4.7185×10^3 **2.** 0.00023 **3.** $20x^5y^7$ **4.** $25x^4y^6$ **5.** 1, 2, 3, 4

Using the Language and Symbolism of Mathematics 7.2

1. quadratic **2.** square **3.** completing **4.** equivalent

Quick Review 7.2

1. $\dfrac{49}{4}$ **2.** $x^2 - 14x + 49$ **3.** $8 - 2\sqrt{7}$ **4.** $(x + 7)^2$ **5.** $3 - 2\sqrt{5}$

Exercises 7.2

1. $(m - 1)^2 = 4$ **3.** $(y + 7)^2 = 5$ **5.** 36 **7.** 81 **9.** $4b^2$ **11.** $w = 0, w = 4$
13. $w = -2, w = 6$ **15.** $y = 0, y = -6$ **17.** $y = -5, y = -1$
19. $t = -1 + \sqrt{5}, t = -1 - \sqrt{5}; t \approx 1.24, t \approx -3.24$
21. $w = -\dfrac{11}{2}, w = -\dfrac{9}{2}$ **23.** $y = 2 + \sqrt{3}, y = 2 - \sqrt{3}; y \approx 3.73,$
$y \approx 0.27$ **25.** $x = \dfrac{-3 + \sqrt{3}}{3}, x = \dfrac{-3 - \sqrt{3}}{3}; x \approx -0.42, x \approx -1.58$
27. $m = -\dfrac{1}{2}, m = 2$ **29.** $v = \dfrac{1 + \sqrt{2}}{2}, v = \dfrac{1 - \sqrt{2}}{2}; v \approx 1.21, v \approx -0.21$
31. $v = \dfrac{1 + \sqrt{6}}{5}, v = \dfrac{1 - \sqrt{6}}{5}; v \approx 0.69, v \approx -0.29$
33. $x = \dfrac{1 + \sqrt{29}}{2}, \dfrac{1 - \sqrt{29}}{2}; x \approx 3.19, x \approx -2.19$
35. a. $x = 1$ **b.** $x = -3, x = 1$ **37. a.** $x = \pm\sqrt{3}$ **b.** $[-\sqrt{3}, \sqrt{3}]$
c. $(-\infty, -\sqrt{3}] \cup [\sqrt{3}, \infty)$ **39. a.** $x = \dfrac{1 \pm \sqrt{3}}{2}$ **b.** $\dfrac{1 \pm \sqrt{3}}{2}$

c. $\left(\dfrac{1 \pm \sqrt{3}}{2}, 0\right)$ **41. a.** $9x^2 - 4 = 0$ **b.** $9x^2 - 2 = 0$
43. The dimensions of the photo are 4 cm by 6 cm. **45.** The inner ring has a radius of 1.3 cm. **47.** $x \approx -0.73, x \approx 2.73$ **49.** $w \approx -2.33, w \approx 2.50$

Group Discussion Questions

51. $x = -5, x = \dfrac{2}{3}$; discussion can vary
53. $x = \dfrac{-19 \pm \sqrt{53}}{14}; x \approx -1.88, x \approx -0.84$; Answer cannot be determined by factoring over the integers.

Cumulative Review

1. -23 **2.** 49 **3.** $\dfrac{3}{5}$ **4.** $-\dfrac{5}{3}$ **5.** $18x^3y^8$

Using the Language and Symbolism of Mathematics 7.3

1. quadratic **2.** standard **3.** $3x^2 + 5x - 7 = 0$ **4.** $\pm$
5. discriminant **6.** 0 **7.** two **8.** no

Quick Review 7.3

1. $2\sqrt{5}$ **2.** 3 **3.** $3 + 2\sqrt{5}$ **3.** $y = -1$ **4.** 0 **5.** $x = -12, x = 2.5$

Exercises 7.3

1. $m = -1, m = 3$ **3.** $y = -\dfrac{1}{4}, y = \dfrac{1}{2}$ **5.** $x = -\dfrac{5}{2}, x = \dfrac{7}{2}$
7. $w = \dfrac{1 + \sqrt{6}}{5}, w = \dfrac{1 - \sqrt{6}}{5}; w \approx 0.69, w \approx -0.29$
9. $v = \dfrac{2 - \sqrt{3}}{3}, v = \dfrac{2 + \sqrt{3}}{3}; v \approx 0.09, v \approx 1.24$ **11.** $t = -\dfrac{5}{6}, t = 0$
13. $z = -2\sqrt{2}, z = 2\sqrt{2}; z \approx -2.83, z \approx 2.83$ **15.** $x = \dfrac{3}{2}$
17. $w = -8, w = -1$ **19.** $x = -\sqrt{3}, x = \sqrt{3}; x \approx -1.73, x \approx 1.73$
21. a. B **b.** C **c.** A **23.** $b^2 - 4ac = 0$; There is a double real solution.
25. $b^2 - 4ac = 5 > 0$; There are two distinct real solutions.
27. $b^2 - 4ac = -24 < 0$; The solutions are complex numbers with imaginary parts. **29.** $(0, -6); (\pm\sqrt{6}, 0) \approx (\pm 2.45, 0)$
31. $(0, -1); (2 \pm \sqrt{5}, 0) \approx (-0.24, 0)$ and $(4.24, 0)$
33. The numbers are either $-3 + \sqrt{17}$ and $3 + \sqrt{17}$ or $-3 - \sqrt{17}$ and $3 - \sqrt{17}$. **35.** The dimensions of the rectangle are $(-2 + 2\sqrt{3})$ cm by $(2 + 2\sqrt{3})$ cm. **37. a.** $P(0) = -61,200$, so overhead costs are $61,200. **b.** The company will break even if either 15 or 51 units are produced and sold. **39.** A car with a speed of approximately 99 km/h requires 100 m to stop. **41.** $v = 0, v = 4, v = -7$
43. $w = 0, w = \pm\sqrt{10}$ **45. a.** $(-\infty, 1 - \sqrt{7}) \cup (1 + \sqrt{7}, \infty)$
b. $(1 - \sqrt{7}, 1 + \sqrt{7})$ **47. a.** The rocket will be at ground level after approximately 0.44 second and 2.56 seconds. **b.** The rocket will be above ground level between 0.44 seconds and 2.56 seconds. **49.** $x^2 - 3x + 1 = 0$

Group Discussion Questions

51. $x = \pm 6$; discussion can vary. **53. a.** $x = \pm 2y$
b. $x = \pm\sqrt{3}y$ **c.** $x = -y, x = -3y$ **d.** $x = \dfrac{-5 \pm \sqrt{13}}{2}y$

Cumulative Review

1. $2x^3 - 5x^2y + 2xy^2 - 3y^3$ **2.** $(-4, 4)$ **3.** -1 **4.** 0 **5.** 0

Using the Language and Symbolism of Mathematics 7.4

1. find **2.** variable **3.** algebraic **4.** solve **5.** reasonable **6.** rate; time
7. legs; hypotenuse **8.** perpendicular **9.** $a^2 + b^2 = c^2$

Quick Review 7.4

1. $n + 1, n + 2$ **2.** $n + 2, n + 4$ **3.** $x = -8, x = 8$ **4.** 6.6
5. a. 0 **b.** 20 **c.** $\{0, 1, 2, 3, \ldots, 19, 20\}$

Exercises 7.4

1. The consecutive integers are either -13 and -12 or 12 and 13.
3. The consecutive even integers are either -18 and -16 or 16 and 18.
5. The consecutive integers are either -8 and -7 or 7 and 8.
7. The numbers are either $-\dfrac{25}{3}$ and -21 or 7 and 25.

9. The width of the rectangle is $\frac{7}{3}$ cm and the length is 9 cm.　**11.** The base of the triangle is 14 m.　**13.** The length of each side of the original square is approximately 7.2 cm.　**15.** The time interval in which the ball is above 50 meters is between 0.6 second and 5.4 seconds.　**17.** The company has a profit if between 10 and 60 windmills are produced.　**19.** The radius of the smaller circle is 5 cm.　**21.** They should order a 14-inch pizza to double the area of a 10-inch pizza.　**23.** The length of the shorter rope is 44 meters.　**25.** The interest rate is approximately 9.28%.　**27.** It takes approximately 2.2 seconds for the hammer to fall 75 feet.　**29.** A car with a speed of approximately 99 km/h requires 100 m to stop.　**31.** The distance to the top of the ladder is 15 feet.　**33.** The ball can be positioned a distance of approximately 18.0 feet from center court.　**35.** The beam is displaced by approximately 10.96 inches in the middle.　**37.** The height of the monitor is approximately 10.95 inches.　**39.** The dimensions of the monitor are approximately 13.6 inches by 10.2 inches. The viewable area is approximately 139 in^2.　**41.** The plane flew 7 miles south.　**43.** The speed of the plane flying south is 90 miles per hour and the speed of the plane flying east is 120 miles per hour.　**45.** The width of the room is 28 feet.　**47.** A throw from home plate must go about 37.3 feet farther to reach second base than to reach third base.　**49.** The distance to the horizon is about 180 miles.　**51.** The sides of the square are 18 cm.

Group Discussion Questions

53. Student's answers will vary. Examples: **a.** The sum of the squares of two consecutive even integers is 244. Find these integers. **b.** If each side of a square is increased by 2 cm, the total area of both the new square and the original square will be 244 cm^2. Determine the length of each side of the original square. **c.** Upon leaving an airport, an airplane flew due south and then due east. After it had flown 2 miles farther east than it had flown south, it was approximately 15.62 mi from the airport. How far south had it flown?

Cumulative Review

1. -3　**2.** $(0, 5)$　**3.** $y = 2x - 3$　**4.** $3y + 2x$　**5.** $-2 < x \le 4$

Using the Language and Symbolism of Mathematics 7.5

1. $\sqrt{-1}$　**2.** $a; bi$　**3.** sometimes　**4.** always　**5.** always　**6.** sometimes　**7.** never　**8.** $a + bi$　**9.** four　**10.** conjugate　**11.** 2　**12.** $7 - 2i$

Quick Review 7.5

1. 3.746921×10^4　**2.** 0.0000406　**3.** $10x^2 - 7x - 12$　**4.** $(5x + 4)(2x - 3)$　**5.** $25x^2 - 9y^2$

Exercises 7.5

1. a. -6　**b.** $6i$　**3. a.** $4 + 3i$　**b.** $3 + 4i$　**5.** $9 - i$　**7.** $3 + i$　**9.** $60 + 24i$　**11.** 7　**13.** $5i$　**15.** $7i$　**17.** -6　**19.** $10 + 6i$　**21.** $8 - 3i$　**23.** 53　**25.** $34 + 27i$　**27.** -10　**29.** 10　**31.** $24 + 10i$　**33.** $7 - 24i$　**35.** $56 + 33i$　**37.** $-3 + i\sqrt{6}$　**39.** $13i\sqrt{3}$　**41.** $-1 - 2i$　**43.** i　**45.** -1　**47.** 29　**49.** 169　**51.** $2 - 2i$　**53.** $\frac{12}{13} + \frac{5}{13}i$　**55.** $-2 - 3i$　**57.** $-\frac{7}{41} + \frac{22}{41}i$　**59.** $7 + 6i$　**61.** $\frac{5}{3}i$　**63.** $-\frac{5}{3}i$　**65.** $x = 2 \pm 5i$　**67.** $x = 3 \pm 2i$　**69.** $w = \frac{3}{2} \pm \frac{1}{2}i$　**71.** $x = \pm 4i$　**73.** $x = -\frac{3}{2} \pm \frac{3}{2}i$　**75.** 0; a double real solution

77. -4; complex solutions with imaginary parts　**79.** -8; complex solutions with imaginary parts　**81.** $3 - 2i$ is a solution of $x^2 - 6x + 13 = 0$.　**83.** $x^2 + 16 = 0$　**85.** $x^2 - 8x + 17 = 0$　**87.** $x = 2, x = -\frac{1}{2} \pm \frac{\sqrt{3}}{2}i$　**89. a.** $3 + i$　**b.** $-3 + i$　**c.** $\frac{3}{10} + \frac{1}{10}i$　**91.** $i^{46} = -1$

Group Discussion Questions

93. a. It is a real number. **b.** Answers can vary; $(2 + 5i) + (6 - 5i) = 8$ **c.** Answers must be conjugates; $(2 + 3i)(2 - 3i) = 13$ **d.** 0
95. a. yes **b.** yes **c.** yes **d.** 3; The cube roots of 8 are 2, $-1 + i\sqrt{3}$, and $-1 - i\sqrt{3}$. The principal cube root is 2.

Cumulative Review

1. $x = 2$　**2.** $x = \frac{11}{4}$　**3.** $x = -6, x = 8$　**4.** The perimeter is 20 cm.　**5.** The area is 25 cm^2.

Review Exercises for Chapter 7

1. $5\sqrt{7}$　**2.** $3\sqrt{11}$　**3.** $\frac{5}{7}$　**4.** $\frac{\sqrt{6}}{5}$　**5.** $\frac{3}{2}$　**6.** $\frac{2\sqrt{6}}{3}$　**7.** $\frac{\sqrt{30}}{6}$　**8.** $\frac{\sqrt{33}}{11}$　**9.** $(x - 6)^2 = 0$　**10.** 225　**11.** 100　**12.** $(v - 4)^2 = 9$　**13.** $x = -2, x = 4$　**14.** $x = -16, x = 4$　**15.** $x = 4, x = 3$　**16.** $x = \frac{5}{2}, x = -\frac{3}{4}$　**17.** $v = 5, v = 4$　**18.** $w = 2 \pm \sqrt{2}$　**19.** $m = \frac{5}{2}, m = -\frac{2}{5}$　**20.** $x = \frac{3 \pm 3\sqrt{3}}{2}$　**21.** $m = \pm 8$　**22.** $y = 2, y = 1$　**23.** $w = -12, w = 2$　**24.** $v = 5, v = 13$　**25. a.** $x = -7, x = 1$　**b.** $[-7, 1]$　**c.** $(-\infty, 7] \cup [1, \infty)$　**26.** $x^2 - 2x - 15 = 0$　**27.** $20x^2 + 3x - 2 = 0$　**28.** $x^2 - 5 = 0$　**29.** $49x^2 - 70x + 25 = 0$　**30.** $x^2 - 6x + 4 = 0$　**31.** $\left(-\frac{1}{3}, 0\right), (3, 0); -(3x + 1)(x - 3); -\frac{1}{3}, 3; x = -\frac{1}{3}, x = 3$　**32.** $(0, 0), (-2, 0), (1, 0); x(x + 2)(x - 1); 0, -2, 1; x = 0, x = -2, x = 1$　**33.** The interest rate is approximately 4.9%.　**34.** The hole should have a diameter of 4.24 cm.　**35.** A profit will be generated if x is between 5 and 40 units.　**36. a.** The ball will hit the ground after approximately 5.04 seconds. **b.** The ball's height will exceed 67 feet between 1 and 4 seconds.　**37.** The lengths of the sides are $a = 15$ cm, $b = 8$ cm, and $c = 17$ cm.　**38.** The beam has a bulge of approximately 9.8 in.　**39. a.** The distance between the speakers and the customer's ears is approximately 17.9 ft. **b.** The distance is approximately 18 ft from the reference point.　**40.** The speeds of the planes are 308 mi/h and 328 mi/h.　**41.** $10i$　**42.** $8 - 8i$　**43.** $2 - 4i$　**44.** $-2 - 36i$　**45.** $12 + 10i$　**46.** $51 - 27i$　**47.** $40 - 42i$　**48.** 29　**49.** i　**50.** $-1 - i$　**51.** i　**52.** $4 - 10i$　**53.** i　**54.** 1　**55. a.** 0; a double real solution **b.** positive; two distinct real solutions **c.** negative; complex solutions with imaginary parts　**56.** 0; a double real solution　**57.** -12; complex solutions with imaginary parts　**58.** 17; two distinct real solutions　**59.** $z = \pm 7i$　**60.** $z = -\frac{1}{2} \pm \frac{3}{2}i$　**61.** $x = 3 \pm i$　**62.** $x = 3 \pm i\sqrt{2}$　**63.** $x^2 + 121 = 0$　**64.** $x^2 - 10x + 26 = 0$　**65. a.** $2 + 5i$ **b.** $-2 + 5i$ **c.** $\frac{2}{29} + \frac{5}{29}i$　**66. a.** i **b.** This is not possible. All real numbers are complex numbers. **c.** 1 **d.** This is not possible. All imaginary numbers are complex numbers.　**67.** $i^{333} = i$　**68. a.** $x = 4$ **b.** $x = -2, x = 1$ **c.** $x = 4, x = -5$

Mastery Test for Chapter 7

1. a. $x = \pm 20$ **b.** $v = -2, v = 7$　**2. a.** $6\sqrt{2}$ **b.** $2\sqrt{7}$　**3. a.** $\frac{7}{8}$ **b.** $\frac{\sqrt{11}}{5}$ **c.** 4 **d.** $\frac{3}{7}$　**4. a.** $3\sqrt{2}$ **b.** $\frac{\sqrt{21}}{7}$　**5. a.** 100 **b.** 16　**6. a.** $y = -2, y = 8$ **b.** $w = -\frac{1}{2}, w = \frac{7}{2}$　**7. a.** $x = -3 - \sqrt{14}, x = -3 + \sqrt{14}; x \approx -6.74, x \approx 0.74$ **b.** $m = -\frac{2}{3}, m = \frac{5}{2}$ **c.** $z = \frac{7}{2}$ **d.** $x = 2 - \sqrt{5}, x = 2 + \sqrt{5}; x \approx -0.24, x \approx 4.24$　**8. a.** 5, distinct real solutions **b.** 0, a double real solution **c.** 16, distinct real solutions **d.** -3, complex solutions with imaginary parts　**9.** The company breaks even when it sells either 5 or 185 units.　**10.** The lengths of the squares are 7 cm and 12 cm.　**11.** The length of the rafter is 25 ft.　**12.** The base of the ladder is approximately 6.4 ft from the wall.　**13. a.** $9i$ **b.** $3i$ **c.** $-5 + 4i$ **d.** $-1 - i$　**14. a.** $-1 + 2i$ **b.** $-12 - 26i$ **c.** $\frac{7}{5} + \frac{3}{10}i$ **d.** $8 - 6i$　**15. a.** $y = 1 \pm 6i$ **b.** $x = 5 \pm 2i$ **c.** $v = 3 \pm 5i$

Group Project for Chapter 7

I.

1.

Width W (ft)	Length L (ft)	Area A (ft²)
5.0	140	700.0
10.0	130	1,300.0
15.0	120	1,800.0
20.0	110	2,200.0
25.0	100	2,500.0
30.0	90	2,700.0
35.0	80	2,800.0
37.5	75	2,812.5
40.0	70	2,800.0
45.0	60	2,700.0
50.0	50	2,500.0
55.0	40	2,200.0
60.0	30	1,800.0
65.0	20	1,300.0
70.0	10	700.0

2. a.

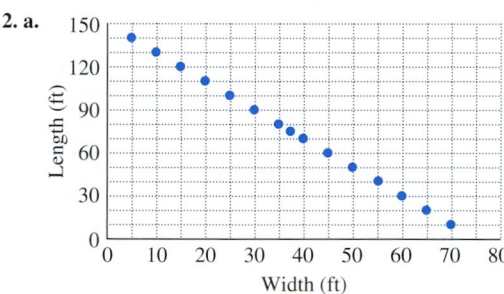

b. $L = 150 - 2W$ **c.** from 0 to 75 ft **d.** from 0 to 150 ft

3. a.

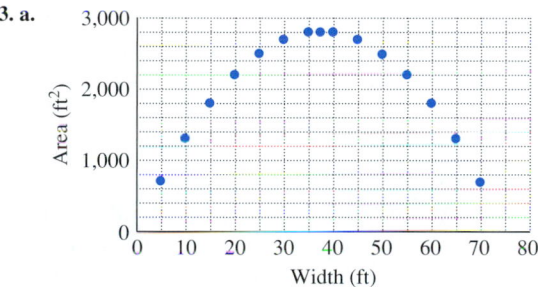

b. $A = W(150 - 2W)$ **c.** from 0 to 75 ft **d.** (0, 2,812.5)

II.

1. a. 86 ft **b.** 2,752 ft² **2. a.** 150 ft **b.** 0 ft² **3. a.** 0 ft **b.** 0 ft² **4.** 22.5 ft
5. 33 ft or 42 ft **6.** 37.5 ft by 75 ft; 2,812.5 ft² **7.** Answers can vary.

Chapter 8

Using the Language and Symbolism of Mathematics 8.1

1. One: output **2.** Domain **3.** Range **4.** x **5.** y **6.** x **7.** y **8.** x **9.** y
10. Vertical **11.** Function **12.** f; x; x; output; f

Quick Review 8.1

1. 249

2.

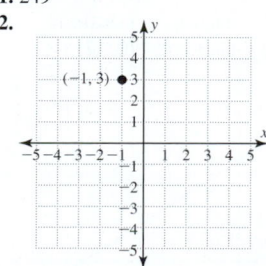

3.

4. $[-2, 2)$ **5.** $(-1, \infty)$

Exercises 8.1

1. a. Not a function **b.** Function; $D = \{-3, -1, 1, 2\}$; $R = \{1, 2, 3\}$
c. Function; $D = \{0, 1, 2, 3\}$; $R = \{-1, 0, 1, 3\}$ **3. a.** Function;
$D = \{-1, 1, 2\}$; $R = \{0, 1\}$ **b.** Not a function **c.** Function;
$D = \{-3, -2, -1, 0, 1\}$; $R = \{-1, 0, 1, 2, 3\}$ **5. a.** Function
b. Not a function **c.** Function **7. a.** Function **b.** Not a function
c. Not a function **9. a.** Function **b.** Function **c.** Not a function
11. a. $D = \{-4, -3, -2, -1, 0\}$; $R = \{8, 6, 4, 2, 0\}$
b. $D = \{1, 2, 3, 4, 5\}$; $R = \{5\}$
13. a. $D = \{2, 5, 8, 11, 14\}$; $R = \{-3, 1, 5, 9, 13\}$
b. $D = \{5, 6, 8, 12, 16\}$; $R = \{2, 4, 5, 9\}$
15. a. $D = \{-3, -2, -1, 0, 1, 2, 3\}$; $R = \{2\}$ **b.** $D = (-\pi, \pi]$;
$R = [-1, 1]$ **c.** $D = [-2, \infty)$; $R = (-\infty, 2]$
17. a. $D = [-2, 2]$; $R = [0, 2]$ **b.** $D = [-2, 4)$; $R = [-1, 4]$
c. $D = \mathbb{R}$; $R = [-2, 2]$ **19. a.** -27 **b.** -7 **c.** 3 **21. a.** 0 **b.** 3

c. $\dfrac{1}{2}$ **23. a.** 2 **b.** 7 **c.** 4 **25. a.** 0 **b.** -9 **c.** 16 **27. a.** Output value

b. Input value **c.** 3 **d.** 8 **29. a.** Input value **b.** Output value **c.** 0 **d.** 3
31. a. 8 **b.** -2 **c.** -5 **33. a.** 0 **b.** -1 **c.** -3 **35. a.** 4 **b.** 0 **c.** -3
37. a. -2 **b.** 0 **c.** 2

39. a.

D	R
$-5 \rightarrow$	4
$-3 \rightarrow$	2
$-2 \rightarrow$	0
$0 \rightarrow$	-2
$1 \rightarrow$	-3
$4 \rightarrow$	-4

c.

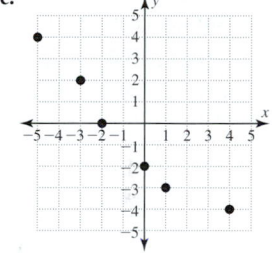

b. $\{(-5, 4), (-3, 2), (-2, 0),$
$(0, -2), (1, -3), (4, -4)\}$

41. a.

D	R
$-3 \rightarrow$	1
$-2 \rightarrow$	-1
$-1 \rightarrow$	3
$1 \rightarrow$	-2
$3 \rightarrow$	2

c.

x	y
-3	1
-2	-1
-1	3
1	-2
3	2

b. $\{(-3, 1), (-2, -1), (-1, 3),$
$(1, -2), (3, 2)\}$

43. a.

x	y
-1	1
1	3
2	-1
4	-2

c.

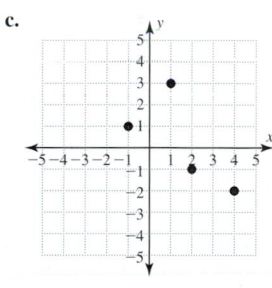

b. $\{(-1, 1), (1, 3), (2, -1), (4, -2)\}$

45. a.

x	y
-5	4
-3	4
1	4
2	4
3	4

b.

D	R
$-5 \rightarrow 4$
$-3 \rightarrow 4$
$1 \rightarrow 4$
$2 \rightarrow 4$
$3 \rightarrow 4$

c.

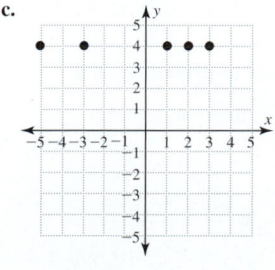

47. a. $f(x) = 2x$

b.

D	R
$25,000 →	$ 50,000
40,000 →	80,000
50,000 →	100,000
65,000 →	130,000

c.

x	f(x)
$25,000	$ 50,000
40,000	80,000
50,000	100,000
65,000	130,000

d. {(25,000, 50,000), (40,000, 80,000), (50,000, 100,000), (65,000, 130,000)}

49. a. pressure **b.** 16 in³ **c.** 15 lb/in²

51. a.

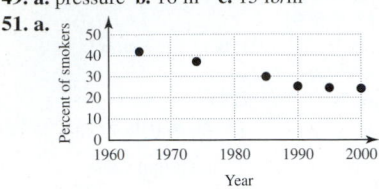

b. 29.9% **c.** 1974

53. a.

Year	Tuition
1970	1,000
1980	3,000
1990	5,000
2000	7,000

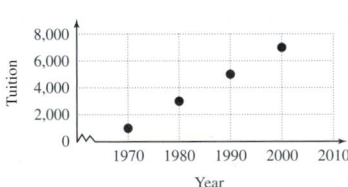

b. $9,000 **c.** 1994

55. a. $T(x) = 0.0825x$

b.

x	T(x)
0	0
100	8.25
200	16.50
300	24.75
400	33.00
500	41.25
600	49.50

57. D **59.** B

61. a.

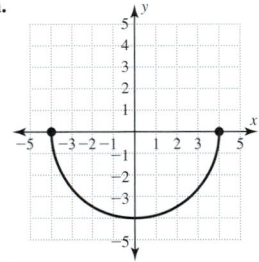

b.

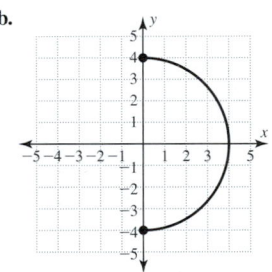

Group Discussion Questions

63. To be a function, the graph must pass the vertical line test for each x-value in the domain—not just for selected values.

65. a. $D = [0, \infty); R = [0, \infty)$ **b.** $D = [0, \infty); R = (-\infty, 0]$

c. $D = \mathbb{R}; R = (-\infty, 1]$ **d.** $D = \mathbb{R}; R = [0, \infty)$

e. $D = (-\infty, 0]; R = [0, \infty)$ **f.** $D = (-\infty, 2]; R = [0, \infty)$

67. a. $y = 4x$ **b.** $y = x^2$ **c.** $y = 4x + 1$ **d.** $y = x^2 - 1$

Cumulative Review

1. $x = \dfrac{21}{2}$ **2.** $-\dfrac{6x^6}{5y^5}$ **3.** $(5x + 6y)^2$ **4.** $(5x + 6y)(5x - 6y)$

5. $-5(x + 2)(x - 2)$

Using the Language and Symbolism of Mathematics 8.2

1. $\dfrac{y_2 - y_1}{x_2 - x_1}$ **2.** positive **3.** negative **4.** zero **5.** linear **6.** 2; (0, 5)

7. y; x **8.** x; y **9.** positive **10.** negative

Quick Review 8.2

1. $(-\infty, 6)$ **2.** $[5, \infty)$ **3.** $(-24, \infty)$

4.

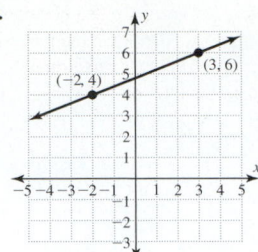

5.

x	y
−14	5
−7	8
0	11
7	14
14	17

Exercises 8.2

1. $m = 5$ **3.** $m = 0$ **5.** $m = 2$ **7.** $m = -\dfrac{4}{3}$ **9.** $m = -1$ **11.** $m = 0$

13. a.

x	f(x)
−2	4
−1	3
0	2
1	1
2	0

b. The values of y decrease by 1 unit for each 1-unit increase in x.

c.

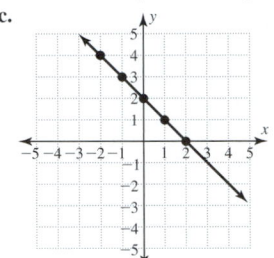

d. The line drops by 1 unit for every 1-unit move to the right.

15.

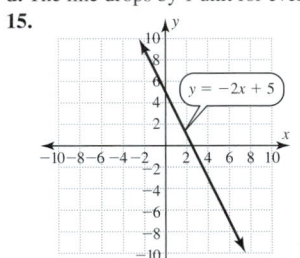

17.

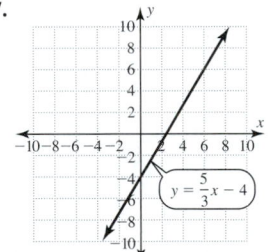

19. $f(x) = \dfrac{9}{4}x - 2$ **21.** $f(x) = \dfrac{5}{3}x - \dfrac{28}{3}$ **23. a.** $m = -2$ **b.** (0, 3)

c. $y = -2x + 3$ **25. a.** $m = \dfrac{7}{4}$ **b.** $\left(0, -\dfrac{73}{4}\right)$ **c.** $f(x) = \dfrac{7}{4}x - \dfrac{73}{4}$

27. a. $m = \dfrac{3}{2}$ **b.** (0, 3) **c.** $y = \dfrac{3}{2}x + 3$ **29. a.** $m = -\dfrac{3}{5}$ **b.** $\left(0, -\dfrac{11}{5}\right)$

c. $f(x) = -\dfrac{3}{5}x - \dfrac{11}{5}$ **31.** x-intercept: (3, 0); y-intercept: (0, 3)

33. x-intercept: (1, 0); y-intercept: (0, 4)

35. x-intercept: (2, 0); y-intercept: (0, −4)

37. x-intercept: (18, 0); y-intercept: (0, 12)

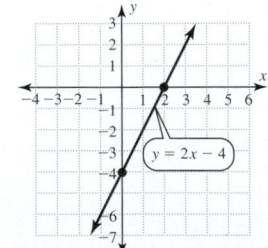

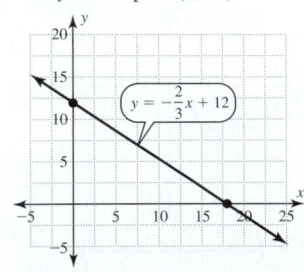

39. The linear function is negative on the interval $(-\infty, -4)$ and positive on the interval $(-4, \infty)$. **41.** The linear function is negative on the interval $(-\infty, 6)$ and positive on the interval $(6, \infty)$. **43.** The linear function is negative on the interval $(50, \infty)$ and positive on the interval $(-\infty, 50)$.

45. $y = \dfrac{3}{5}x - \dfrac{21}{5}$ **47.** $y = \dfrac{3}{2}x - 6$

49.

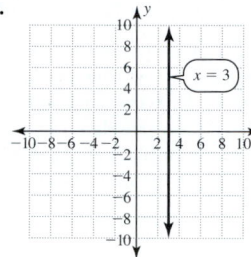

51.

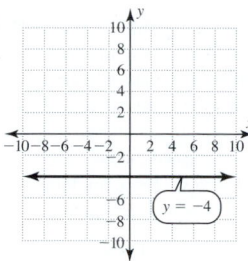

53.

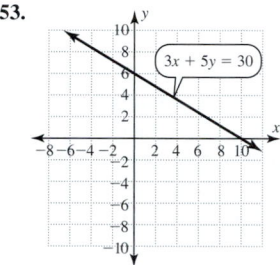

55.

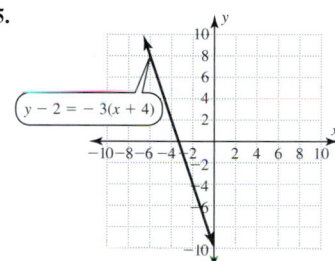

57. a.

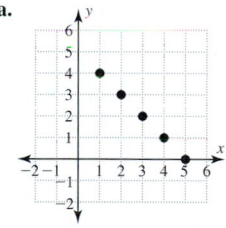

b.

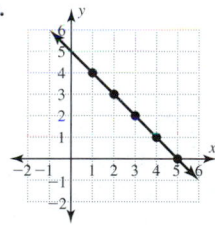

59. a. 52 **b.** 4 **61. a.** 2 **b.** 8 **63. a.** 1 **b.** 0

65. a.

x	$C(x)$
0	0
500	100
1,000	200
1,500	300
2,000	400
2,500	500
3,000	600

b. If the medical bill is $1,000, then the cost to the patient is $200.
c. If the cost to the patient is $425, then the medical bill is $2,125.

67. The profit interval is (50, 150]. The loss interval is [0, 50).
69. a. x-intercept: (10, 0); The business breaks even when 10 units are produced and sold.
y-intercept: (0, −500); The business has loss of $500 when no units are produced and sold.
b. The profit interval is (10, 30]. The loss interval is [0, 10).
71. a. $V(x) = 0.04x$ **b.** $m = 0.04$ is the annual interest rate. **c.** $b = 0$ is the amount of interest gained if the amount in the savings account is zero.

d. $V(1,100) = 44$; If there is $1,100 in the account, then the annual interest is $44. **73. a.** $V(x) = -50x + 500$ **b.** The value of the tools after 7 years is $150. **c.** The value of the tools will be $50 after 9 years.

d.

x	$V(x)$
0	500
1	450
2	400
3	350
4	300
5	250
6	200

Group Discussion Questions

75. a. x represents the amount of unpaid balance. **b.** The output value $f(x)$ represents the monthly interest charge on the unpaid balance.
c. $f(100) = 1.5$; The monthly interest charge on an unpaid balance of $100 is $1.50. **d.** $f(6,666.67) \approx 100$; The monthly interest charge on an unpaid balance of $6,666.67 is $100. **77. a.** $300, $100 **b.** The factory will lose $200 per day due to overhead costs if no tabletops are produced, and it has a revenue of $25 for each tabletop it produces.

Cumulative Review

1. $\dfrac{3}{4}$ **2.** $\dfrac{x^4}{2}$ **3.** $-3x^2 - 9x + 12$ **4.** $-2(x-6)(x+2)$ **5.** $x = -4, x = 3$

Using the Language and Symbolism of Mathematics 8.3
1. V **2.** vertex **3.** maximum **4.** minimum **5.** upward **6.** downward

Quick Review 8.3

1. −5 **2.** −5 **3.** (−3, 3) **4.** (−5, 4) **5.** $\left(-\infty, -\dfrac{8}{5}\right] \cup \left[\dfrac{12}{5}, \infty\right)$

Exercises 8.3
1. −2 **3.** 6 **5.** opens upward **7.** opens downward **9.** C **11.** D
13.

x	−3	−2	−1	0	1	2	3
y	1	0	1	2	3	4	5

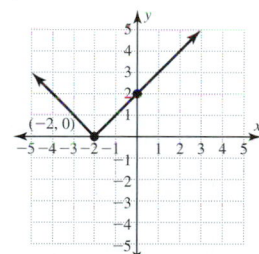

15.

x	−3	−2	−1	0	1	2	3
y	2	1	0	−1	0	1	2

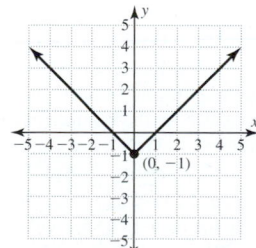

17. a. (3, 2) **b.** $y = 2$ **c.** $x = 3$ **19. a.** (−3, 4) **b.** $y = 4$ **c.** $x = -3$
21. a.

x	−3	−2	−1	0	1	2	3
y	−1	0	1	2	3	2	1

b.

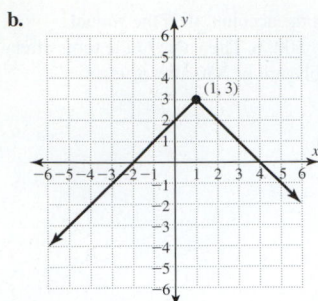

c. down **d.** (1, 3) **e.** The maximum value of y is 3. **f.** $\mathbb{R}$ **g.** $(-\infty, 3]$

23. a.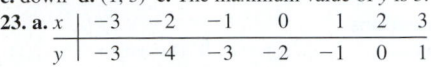

b.

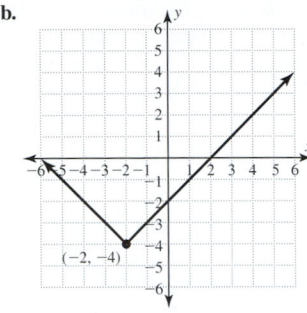

c. up **d.** $(-2, -4)$ **e.** The minimum value of y is -4. **f.** $\mathbb{R}$ **g.** $[-4, \infty)$

25.

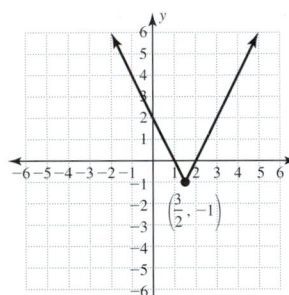

27.

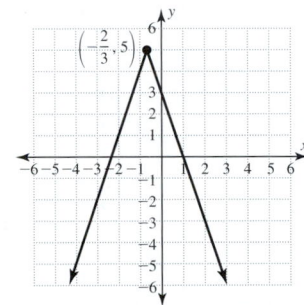

domain: $\mathbb{R}$;
range: $[-1, \infty)$

domain: $\mathbb{R}$;
range: $(-\infty, 5]$

29. x-intercepts: $(-3, 0), (1, 0)$; y-intercept: $(0, 1)$

31. x-intercepts: $(-3, 0), (1, 0)$; y-intercept: $(0, -1)$

33. x-intercepts: $(-4, 0), (10, 0)$; y-intercept: $(0, -4)$

35. x-intercepts: $(-8, 0), (6, 0)$; y-intercept: $(0, 6)$

37. The function is positive for the x-values in the interval $(-\infty, -4) \cup (2, \infty)$ and negative in the interval $(-4, 2)$.

39. The function is positive for the x-values in the interval $(-4, 2)$ and negative in the interval $(-\infty, -4) \cup (2, \infty)$. **41.** The function is positive for the x-values in the interval $(-4, -2)$ and negative in the interval $(-\infty, -4) \cup (-2, \infty)$. **43.** The function is positive for the x-values in the interval $(-3, 1)$ and negative in the interval $(-\infty, -3) \cup (1, \infty)$.

45. The function is positive for the x-values in the interval $(-\infty, -12) \cup (2, \infty)$ and negative in the interval $(-12, 2)$.

47. The function is positive for the x-values in the interval $(-5, 1)$ and negative in the interval $(-\infty, -5) \cup (1, \infty)$. **49. a.** $x = -6, x = 10$ **b.** $(-6, 10)$ **c.** $(-\infty, -6] \cup [10, \infty)$

51. a. $x = -\dfrac{4}{3}, x = 2$ **b.** $\left[-\dfrac{4}{3}, 2\right]$ **c.** $\left(-\infty, -\dfrac{4}{3}\right) \cup (2, \infty)$

53. a. $x = -16, x = 6$ **b.** $(-16, 6)$ **c.** $(-\infty, -16] \cup [6, \infty)$
55. a. -1 **b.** 0 **c.** $-4, 2$ **d.** $-2, 0$ **57. a.** 1 **b.** -3 **c.** 1, 3 **d.** $-3, 7$
59. a. 4 **b.** 1 **c.** 1, 15 **d.** $-8, 24$

61. a. $\dfrac{1}{36}$; The probability of rolling a sum of 2 with a pair of dice is $\dfrac{1}{36}$.

b. $\dfrac{1}{6}$; The probability of rolling a sum of 7 with a pair of dice is $\dfrac{1}{6}$.

c. $\dfrac{1}{36}$; The probability of rolling a sum of 12 with a pair of dice is $\dfrac{1}{36}$.

d. $\{2, 3, 4, 5, 6, 7, 8, 9, 10, 11, 12\}$ **e.** Rolling a 7 has higher probability than rolling a 2.

Group Discussion Questions
63. a. The equation has no solution since the absolute value of $x - 3$ cannot equal a negative number. **b.** The inequality has no solution since the absolute value of $x + 1$ cannot be less than -5. **c.** The solution set to the inequality is all real numbers since adding 8 to any absolute value expression will always exceed 2. **65.** This window displays only part of the V shape of this function. A window of $[-30, 10, 5]$ by $[-20, 10, 5]$ will reveal that the range is $[-10, \infty]$.

Cumulative Review

1. $\sqrt{24} = 2\sqrt{6}$; $\sqrt{24} \approx 4.90$ **2.** $x = \pm 2\sqrt{2}$ **3.** $x = \dfrac{5 \pm \sqrt{109}}{6}$

4. $x^2 - 5x - 14 = 0$ **5.** $\dfrac{3}{5} + \dfrac{14}{5}i$

Using the Language and Symbolism of Mathematics 8.4
1. parabola **2.** quadratic **3.** vertex **4.** upward **5.** axis; symmetry
6. quadratic; linear; constant **7.** linear **8.** second **9.** upward **10.** upward

11. downward **12.** $x = \dfrac{-b}{2a}$ **13.** $f\left(-\dfrac{b}{2a}\right)$

Quick Review 8.4

1. slope, intercept **2.** overhead **3.** negative **4.** III **5.** $x = \dfrac{2}{3}, x = \dfrac{3}{2}$

Exercises 8.4
1. D **3.** A **5.** line; negative slope **7.** line; positive slope **9.** parabola; opens upward **11.** parabola; opens downward **13.** $(-3, -20)$ **15.** 2

17. $(3, 2)$ **19.** $(1, 11)$ **21. a.** $(0, -2)$ **b.** $\left(-\dfrac{1}{2}, 0\right)$; $(2, 0)$ **c.** $\left(\dfrac{3}{4}, -\dfrac{25}{8}\right)$

23. a. $(0, -12)$ **b.** $\left(\dfrac{3}{2}, 0\right)$; $(-4, 0)$ **c.** $\left(-\dfrac{5}{4}, -\dfrac{121}{8}\right)$

25. a.

x	y
-3	-6
-2	-1
-1	2
0	3
1	2
2	-1
3	-6

b, c.

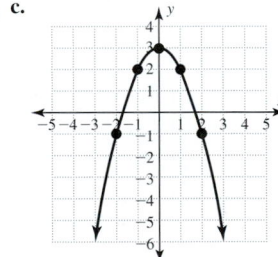

27. a.

x	y
-3	0
-2	-4
-1	-6
0	-6
1	-4
2	0
3	6

b, c.

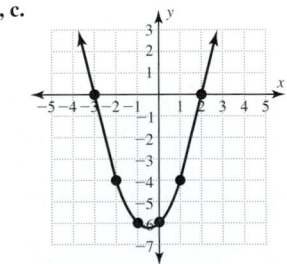

29. a. upward **b.** $(1, -4)$ **c.** $(0, -3)$ **d.** $(-1, 0), (3, 0)$ **e.** $\mathbb{R}$ **f.** $[-4, \infty)$
31. a. downward **b.** $(0, 4)$ **c.** $(0, 4)$ **d.** $(-2, 0), (2, 0)$ **e.** $\mathbb{R}$ **f.** $(-\infty, 4]$
33. a. upward **b.** $(1, 0)$ **c.** $(0, 1)$ **d.** $(1, 0)$ **e.** $\mathbb{R}$ **f.** $[0, \infty)$
35. a. upward **b.** $(-2.5, -12.25)$
　c. x-intercepts: $(-6, 0), (1, 0)$;
　　 y-intercept: $(0, -6)$

d.

x	y
−5	−6
−4	−10
−3	−12
−2.5	−12.25
−2	−12
−1	−10
0	−6

e.
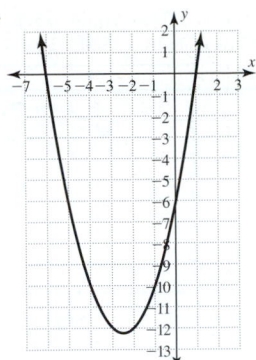

37. a. downward **b.** (0.25, 15.125)
  c. x-intercepts: (−2.5, 0), (3, 0);
  y-intercept: (0, 15)

d.

x	y
−2	5
−1	12
0	15
0.25	15.125
1	14
2	9
3	0

e.
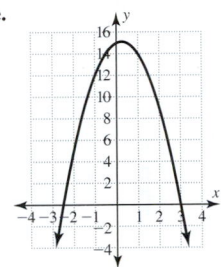

39. a. $15,000 **b.** 20 and 80 units **c.** $9,000; 50 units **41. a.** $1,000
b. 1 and 9 units **c.** $1,500; 5 units **43. a.** $240,000 **b.** 20 and 120 tons
c. $250,000; 70 tons **45. a.** $28,665 **b.** 4.5 and 45.5 tons **c.** $58,835; 25 tons
47. The ball reaches a maximum height of 147 ft after 3 seconds.
49. They should plant 75 trees to maximize the total yield.
51. a. $f(w) = -w^2 + 20w$ **b.** 100 m² **53. a.** $f(w) = -2w^2 + 60w$
b. 450 ft² **55. a.** $A(x) = -2x^2 + 100x$ **b.** 25 cm **c.** 1, 250 cm²
57. The function is positive on the interval $(-\infty, 4)$ and negative on the
interval $(4, \infty)$. **59.** The function is positive on the interval $\left(\frac{10}{3}, \infty\right)$
and negative on the interval $\left(-\infty, \frac{10}{3}\right)$. **61.** The function is positive on the
interval $(-\infty, -6) \cup (6, \infty)$ and negative on the interval $(-6, 6)$.
63. The function is positive on the interval $(-1, 5)$ and negative on the
interval $(-\infty, -1) \cup (5, \infty)$. **65. a.** $x = -3, x = 5$ **b.** $(-3, 5)$
c. $(-\infty, -3] \cup [5, \infty)$ **67. a.** 1 **b.** −3; 1 **69. a.** 6 **b.** 0; 4
71. a. 26 **b.** −2; 11

Group Discussion Questions

73. a. The graphs are identical to the graph of $y = x^2$ except they are
shifted up 1, 3, and 4 units, respectively. **b.** Shift the graph of $y = x^2$ up
6 units to obtain the graph of $y = x^2 + 6$.

Cumulative Review

1.
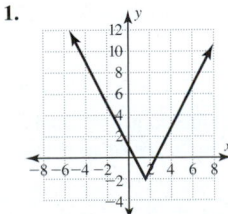

2. $(-5, 8)$ **3.** $y = \frac{1}{5}x - \frac{23}{5}$ **4.** $(x - 7)(2x + 3)$ **5.** $2x^2 - x - 15 = 0$

Using the Language and Symbolism of Mathematics 8.5

1. increasing **2.** decreasing **3.** increases **4.** decreases **5.** positive **6.** negative
7. a. i. slope **ii.** vertex; opens **b.** x **c.** domain **d.** positive **e.** decreasing

Quick Review 8.5

1.
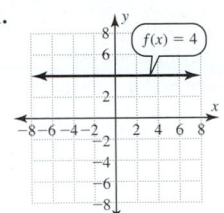

2.
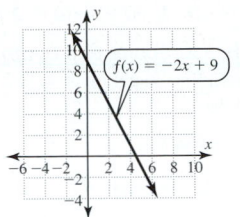

3.
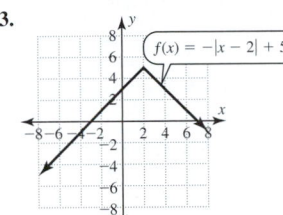

4.
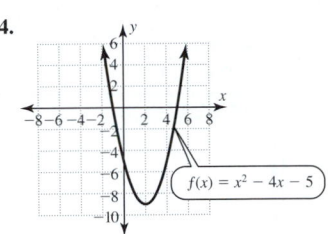

5. a. domain: $\mathbb{R}$; range: {4} **b.** domain: $\mathbb{R}$; range: $\mathbb{R}$
c. domain: $\mathbb{R}$; range: $(-\infty, 5]$ **d.** domain: $\mathbb{R}$; range: $[-9, \infty)$

Exercises 8.5

1. decreasing **3.** increasing **5.** decreasing **7.** increasing **9.** constant
11. a. during the fifth and sixth months **b.** during the first three months
c. during the fourth month **13. a.** $\frac{1}{2}$ **b.** (2, 0) **c.** (0, −1) **d.** $\mathbb{R}$ **e.** $\mathbb{R}$
f. $(2, \infty)$ **g.** $(-\infty, 2)$ **h.** The function is never decreasing. **i.** $\mathbb{R}$ **15. a.** $-\frac{4}{3}$
b. (3, 0) **c.** (0, 4) **d.** $\mathbb{R}$ **e.** $\mathbb{R}$ **f.** $(-\infty, 3)$ **g.** $(3, \infty)$ **h.** $\mathbb{R}$ **i.** The function
is never increasing. **17. a.** 5 **b.** (4, 0) **c.** (0, −20) **d.** $\mathbb{R}$ **e.** $\mathbb{R}$ **f.** $(4, \infty)$
g. $(-\infty, 4)$ **h.** The function is never decreasing. **i.** $\mathbb{R}$ **19. a.** −150
b. (2, 0) **c.** (0, 300) **d.** $\mathbb{R}$ **e.** $\mathbb{R}$ **f.** $(-\infty, 2)$ **g.** $(2, \infty)$ **h.** $\mathbb{R}$ **i.** The function
is never increasing. **21. a.** (−3, 1) **b.** downward **c.** (−4, 0), (−2, 0)
d. (0, −2) **e.** $\mathbb{R}$ **f.** $(-\infty, 1]$ **g.** (−4, −2) **h.** $(-\infty, -4) \cup (-2, \infty)$
i. $(-3, \infty)$ **j.** $(-\infty, -3)$ **23. a.** (−1, −2) **b.** upward **c.** (−3, 0), (1, 0)
d. (0, −1) **e.** $\mathbb{R}$ **f.** $[-2, \infty)$ **g.** $(-\infty, -3) \cup (1, \infty)$ **h.** (−3, 1)
i. $(-\infty, -1)$ **j.** $(-1, \infty)$ **25. a.** (−7, −12) **b.** upward **c.** (−19, 0), (5, 0)
d. (0, −5) **e.** $\mathbb{R}$ **f.** $[-12, \infty)$ **g.** $(-\infty, -19) \cup (5, \infty)$ **h.** (−19, 5)
i. $(-\infty, -7)$ **j.** $(-7, \infty)$ **27. a.** (15, 20) **b.** downward **c.** (−5, 0), (35, 0)
d. (0, 5) **e.** $\mathbb{R}$ **f.** $(-\infty, 20]$ **g.** (−5, 35) **h.** $(-\infty, -5) \cup (35, \infty)$
i. $(15, \infty)$ **j.** $(-\infty, 15)$ **29. a.** (−2, 0) **b.** downward **c.** (−2, 0) **d.** (0, −4)
e. $\mathbb{R}$ **f.** $(-\infty, 0]$ **g.** The function is never positive. **h.** $(-\infty, -2) \cup (-2, \infty)$
i. $(-2, \infty)$ **j.** $(-\infty, -2)$ **31. a.** (2, −1) **b.** upward **c.** (1, 0), (3, 0)
d. (0, 3) **e.** $\mathbb{R}$ **f.** $[-1, \infty)$ **g.** $(-\infty, 1) \cup (3, \infty)$ **h.** (1, 3) **i.** $(-\infty, 2)$
j. $(2, \infty)$ **33. a.** $\left(\frac{9}{2}, -\frac{1}{4}\right)$ **b.** upward **c.** (4, 0), (5, 0) **d.** (0, 20) **e.** $\mathbb{R}$
f. $\left[-\frac{1}{4}, \infty\right)$ **g.** $(-\infty, 4) \cup (5, \infty)$ **h.** (4, 5) **i.** $\left(-\infty, \frac{9}{2}\right)$ **j.** $\left(\frac{9}{2}, \infty\right)$
35. a. $\left(\frac{9}{4}, \frac{289}{8}\right)$ **b.** downward **c.** (−2, 0), $\left(\frac{13}{2}, 0\right)$ **d.** (0, 26) **e.** $\mathbb{R}$
f. $\left(-\infty, \frac{289}{8}\right]$ **g.** $\left(-2, \frac{13}{2}\right)$ **h.** $(-\infty, -2) \cup \left(\frac{13}{2}, \infty\right)$ **i.** $\left(\frac{9}{4}, \infty\right)$
j. $\left(-\infty, \frac{9}{4}\right)$ **37. a.** increasing **b.** The rate of increase is the profit in dollars
per snow cone. **c.** x-intercept: (1,600, 0); y-intercept: (0, −400) **d.** The
company will break even when 1,600 snow cones are sold. The company has
overhead costs of $400. **e.** $(1,600, \infty)$ **f.** The company will have a profit
when more than 1,600 snow cones are sold. **39. a.** (2.5, 30.6) **b.** The ball
reaches a maximum height of 30.6 meters 2.5 seconds after it is hit.
c. (0, 0), (5, 0) **d.** The ball is on the ground initially ($t = 0$) and after 5
seconds. **e.** (0, 2.5) **f.** The ball is rising for the first 2.5 seconds of its flight.

Group Discussion Questions

41. The graphs are all identical to $f(x) = \frac{x}{2}$ except shifted either up or down.

43. a. 2; slope **b.** −3 **d.** The change in the changes is −2.

Cumulative Review

1. 5.123×10^3 **2.** -7.8×10^{-3} **3.** $45,000$ **4.** -0.00673 **5.** 7

Using the Language and Symbolism of Mathematics 8.6

1. scatter **2.** line **3.** parabola **4.** curve **5.** regression **6.** slope; intercept; slope; $(0, b)$

Quick Review 8.6

1. $m = 15$ **2.** x-intercept is $\left(-\dfrac{4}{3}, 0\right)$; y-intercept is $(0, 20)$ **3.** $(2.5, 12.25)$

4. downward **5.** x-intercepts: $(-1, 0)$, $(6, 0)$; y-intercept: $(0, 6)$

Exercises 8.6

1. B **3.** C **5.** C **7.** B

9. a.

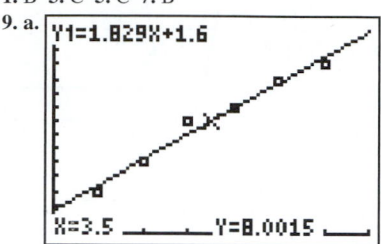

$[0, 7, 1]$ by $[0, 15, 1]$

b. $y \approx 1.829x + 1.600$ **c.** $y \approx 8.0$

11. a.

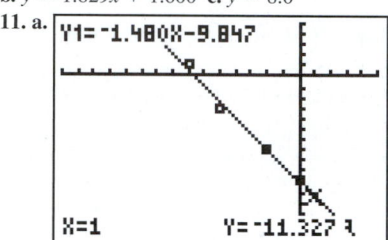

$[-15, 5, 1]$ by $[-15, 5, 1]$

b. $y \approx -1.480x - 9.847$ **c.** $y \approx -11.3$ **13. a.** $y = -0.005x + 3.201$
b. For each additional cigarette per day, the birth weight decreases by 5 g.
c. 3.11 kg **d.** 2.98 kg **15. a.** $y = -24.08x + 2,813.10$ **b.** The men's
Olympic marathon record is decreasing by approximately 96 seconds per
Olympics (that is, per 4 years). **c.** 790; 2 h 13 min 10 sec; The actual record
is 3 min 49 sec faster. **d.** 2020

17. a.

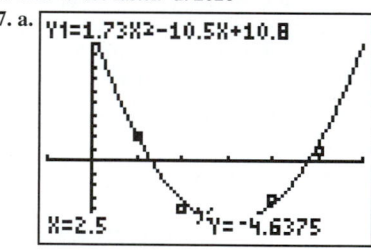

$[-1, 6, 1]$ by $[-6, 12, 1]$

b. $y \approx 1.73x^2 - 10.5x + 10.8$ **c.** $y \approx -4.6$

19. a.

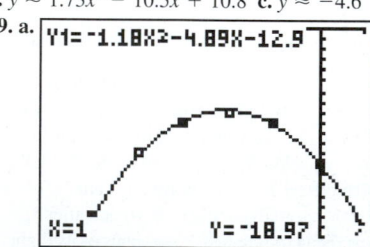

$[-6, 1, 1]$ by $[-20, 0, 1]$

b. $y \approx -1.18x^2 - 4.89x - 12.9$ **c.** $y \approx -19.0$
21. a. $y = 48.8x^2 + 25.0x - 181$ **b.** The load the beam can support is
approximately 11,000 kg. **23. a.** $y = -0.000189x^2 + 0.791x - 4.96$
b. $(2,090, 820)$ **c.** The engine has a maximum power of 820 hp when

running at 2,090 rev/min. **25. a.** $y = -0.311x^2 + 30.8x - 358$
b. 13 or 86 units **c.** \$405; 50 units

27. a.

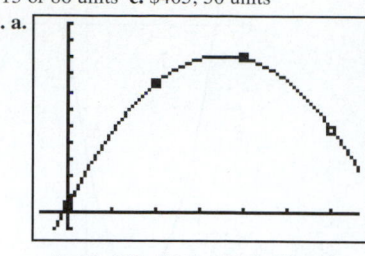

$[-30, 330, 50]$ by $[-10, 110, 10]$

b. $f(x) = -0.00280x^2 + 0.990x + 4.00$ **c.** $f(325) = 30$ **d.** The height of
the ball is 30 ft when it has traveled 325 ft horizontally.
e. $(-4, 0)$; $(357.6, 0)$ **f.** 357.6 ft is the horizontal distance the ball would
travel before reaching ground level. **g.** $(176.8, 91.5)$ **h.** The maximum
height of the ball is 91.5 ft when it is 176.8 ft from home plate.

Group Discussion Questions

29. a. $y = -2x + 4.33$ **b.** $y = -x^2 + 4$ **c.** The parabola better fits the
points. **d.** Answers may vary.

31. a.

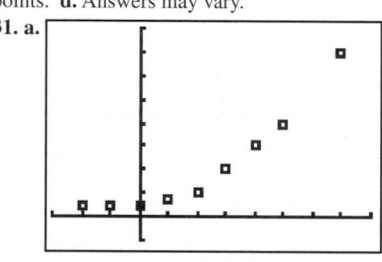

$[-2.9, 7.91, 1]$ by $[-0.722, 8.122, 1]$

b. $y = 0.695x + 0.666$

c.

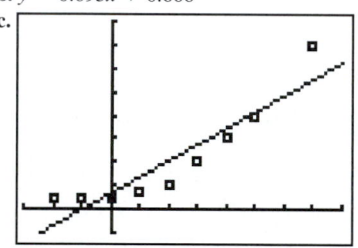

$[-2.9, 7.91, 1]$ by $[-0.722, 8.122, 1]$

d. $y = 0.109x^2 + 0.175x + 0.445$

e.

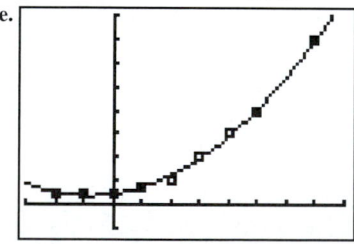

$[-2.9, 7.91, 1]$ by $[-0.722, 8.122, 1]$

f. The parabola is much closer to the points than the line is. **g.** $y = 2.404$
h. $y = 1.564$ **i.** The point on the parabola is closer to the given data points.

Cumulative Review

1. $x = 7$ **2.** $x = \dfrac{16}{7}$ **3.** $x = \dfrac{-11}{4}$, $x = \dfrac{5}{4}$ **4.** $x = \dfrac{-1 \pm 2\sqrt{3}}{3}$

5. $x = -7, x = 6$

Review Exercises for Chapter 8

1. a. function **b.** function **c.** not a function **d.** function **2. a.** not a
function **b.** function **c.** function **d.** not a function **3. a.** domain: $\{5, 6, 7\}$;
range: $\{8\}$ **b.** domain: $(-1, 3]$; range: $[0, 2]$ **c.** domain: $\mathbb{R}$; range:
$(-\infty, 4]$ **d.** domain: $\mathbb{R}$; range: $[-1, \infty)$ **4. a.** 3 **b.** 7 **c.** 12 **d.** 103

5. a. 4 **b.** 2 **c.** 2 **d.** −1 **6. a.** −2 **b.** 2 **c.** −1 **d.** −2

7. a.

D	R
−10 →	4
−5 →	3
0 →	2
5 →	1
10 →	0
15 →	−1
20 →	−2

b. $\{(-10, 4), (-5, 3), (0, 2), (5, 1), (10, 0), (15, -1), (20, -2)\}$

c.

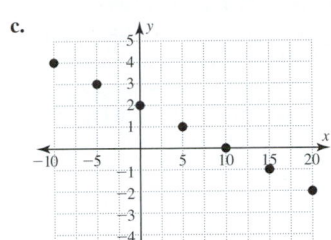

d. $y = -\dfrac{1}{5}x + 2$

8. a. $\{(-4, -1), (-2, 0), (0, 1), (2, 2), (4, 3)\}$

b.

x	y
−4	−1
−2	0
0	1
2	2
4	3

c.

D	R
−4 →	−1
−2 →	0
0 →	1
2 →	2
4 →	3

d. $y = \dfrac{1}{2}x + 1$

9. $m = -\dfrac{2}{3}$ **10.** $m = 0$

11.

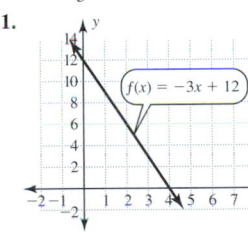

12.

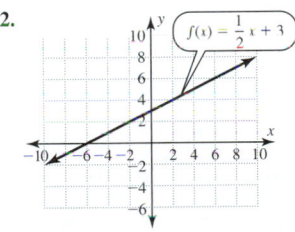

13. $y = -\dfrac{5}{4}x + 2$ **14.** $y = \dfrac{2}{3}x + \dfrac{11}{3}$ **15.** increasing **16.** decreasing

17. decreasing **18.** increasing **19. a.** $m = \dfrac{3}{2}$ **b.** $(-2, 0)$ **c.** $(0, 3)$ **d.** $\mathbb{R}$

e. $\mathbb{R}$ **f.** $(-2, \infty)$ **g.** $(-\infty, -2)$ **h.** The function is never decreasing. **i.** $\mathbb{R}$

20. a. $m = -\dfrac{2}{3}$ **b.** $(3, 0)$ **c.** $(0, 2)$ **d.** $\mathbb{R}$ **e.** $\mathbb{R}$ **f.** $(-\infty, 3)$ **g.** $(3, \infty)$ **h.** $\mathbb{R}$

i. The function is never increasing. **21. a.** $m = -12.5$ **b.** $(10, 0)$

c. $(0, 125)$ **d.** $\mathbb{R}$ **e.** $\mathbb{R}$ **f.** $(-\infty, 10)$ **g.** $(10, \infty)$ **h.** $\mathbb{R}$ **i.** The function is never increasing. **22. a.** $m = 35$ **b.** $(6, 0)$ **c.** $(0, -210)$ **d.** $\mathbb{R}$ **e.** $\mathbb{R}$

f. $(6, \infty)$ **g.** $(-\infty, 6)$ **h.** The function is never decreasing. **i.** $\mathbb{R}$

23. a.

x	−3	−2	−1	0	1	2	3
y	0	1	2	1	0	−1	−2

b.

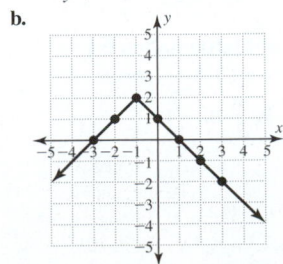

c. downward **d.** $(-1, 2)$ **e.** $y = 2$ **f.** $\mathbb{R}$ **g.** $(-\infty, 2]$

24. a.

x	−3	−2	−1	0	1	2	3
y	1	0	−1	−2	−3	−2	−1

b.

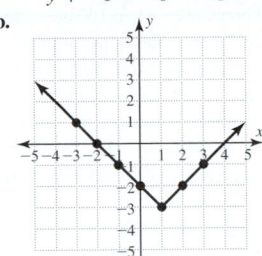

c. upward **d.** $(1, -3)$ **e.** $y = -3$ **f.** $\mathbb{R}$ **g.** $[-3, \infty)$

25.

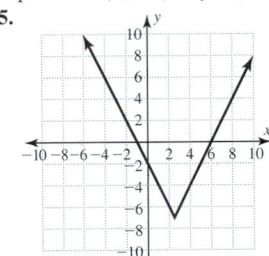

26.

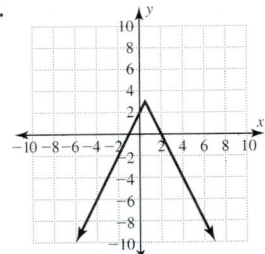

27. a. $x = -9, x = 7$ **b.** $(-9, 7)$ **c.** $(-\infty, -9] \cup [7, \infty)$
28. a. $x = -5, x = 2$ **b.** $[-5, 2]$ **c.** $(-\infty, -5) \cup (2, \infty)$ **29. a.** $(-1, 2)$
b. downward **c.** $(-3, 0), (1, 0)$ **d.** $(0, 1)$ **e.** $\mathbb{R}$ **f.** $(-\infty, 2]$ **g.** $(-3, 1)$
h. $(-\infty, -3) \cup (1, \infty)$ **i.** $(-1, \infty)$ **j.** $(-\infty, -1)$ **30. a.** $(2, -1)$ **b.** upward
c. $(1, 0), (3, 0)$ **d.** $(0, 1)$ **e.** $\mathbb{R}$ **f.** $[-1, \infty)$ **g.** $(-\infty, 1) \cup (3, \infty)$ **h.** $(1, 3)$
i. $(-\infty, 2)$ **j.** $(2, \infty)$ **31. a.** $(8, -9)$ **b.** upward **c.** $(-1, 0), (17, 0)$
d. $(0, -1)$ **e.** $\mathbb{R}$ **f.** $[-9, \infty)$ **g.** $(-\infty, -1) \cup (17, \infty)$ **h.** $(-1, 17)$
i. $(-\infty, 8)$ **j.** $(8, \infty)$ **32. a.** $(-7, 12)$ **b.** downward **c.** $(-19, 0), (5, 0)$
d. $(0, 5)$ **e.** $\mathbb{R}$ **f.** $(-\infty, 12]$ **g.** $(-19, 5)$ **h.** $(-\infty, -19) \cup (5, \infty)$
i. $(-7, \infty)$ **j.** $(-\infty, -7)$ **33.** B **34.** C **35.** A **36.** D **37.** The parabola
opens downward and the vertex is $(2, 7)$. **38.** The parabola opens upward
and the vertex is $(3, -2)$. **39.** The parabola opens upward and the vertex
is $(-5, -20)$. **40.** The parabola opens downward and the vertex is $(2, 11)$.

41.

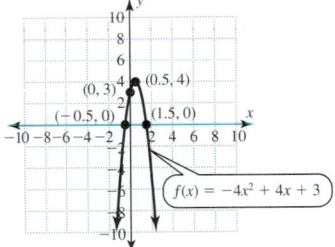

42.

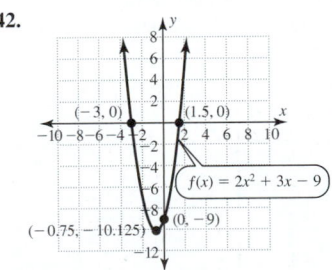

43. a. $(3, -4)$ **b.** upward **c.** $(1, 0), (5, 0)$ **d.** $(0, 5)$ **e.** $\mathbb{R}$ **f.** $[-4, \infty)$
g. $(-\infty, 1) \cup (5, \infty)$ **h.** $(1, 5)$ **i.** $(-\infty, 3)$ **j.** $(3, \infty)$ **44. a.** $(-1, 9)$
b. downward **c.** $(-4, 0), (2, 0)$ **d.** $(0, 8)$ **e.** $\mathbb{R}$ **f.** $(-\infty, 9]$ **g.** $(-4, 2)$
h. $(-\infty, -4) \cup (2, \infty)$ **i.** $(-1, \infty)$ **j.** $(-\infty, -1)$ **45. a.** $(-6, -64)$
b. upward **c.** $(-14, 0), (2, 0)$ **d.** $(0, -28)$ **e.** $\mathbb{R}$ **f.** $[-64, \infty)$
g. $(-\infty, -14) \cup (2, \infty)$ **h.** $(-14, 2)$ **i.** $(-\infty, -6)$ **j.** $(-6, \infty)$
46. a. $(3, 36)$ **b.** downward **c.** $(-3, 0), (9, 0)$ **d.** $(0, 27)$
e. $\mathbb{R}$ **f.** $(-\infty, 36]$ **g.** $(-3, 9)$ **h.** $(-\infty, -3) \cup (9, \infty)$ **i.** $(3, \infty)$

j. $(-\infty, 3)$ **47. a.** $C(x) = 0.6x + 2$ **b.** The telephone call costs $0.60 per minute. **c.** The collect call has a flat fee of $2 per call. **d.** $C(15) = 11$; The total cost of a 15-minute phone call is $11. **48. a.** x-intercept: (160, 0); The business will break even when 160 units are produced and sold. y-intercept: (0, −640); The business has a loss of $640 when zero units are produced and sold. **b.** Profit interval: (160, 300]; Loss interval: [0, 160) **49. a.** $150,000 **b.** $x = 25, x = 60$ **c.** The company has a maximum profit of $30,625 when 42.5 tons are produced and sold. **d.** (25, 60) **e.** The company has a profit when between 25 and 60 tons are produced. **50. a.** $y = 0.080x + 0.057$ **b.** For each 1 kg of mass added, the spring stretches 0.08 cm. **c.** 1.0 cm **d.** 29.3 kg **51. a.** $y = -0.224x^2 + 52.5x - 144$ **b.** $260,000 **c.** 3 units and 232 units **d.** Although the equation predicts a maximum profit of $290,000 at 117 units, the table shows a maximum profit of $300,000 at 100 units. **52.** The ball reaches a maximum height of 49 feet 1.75 seconds after it is hit. **53. a.** $A(w) = -2w^2 + 120w$ **b.** 1,800 ft²

Mastery Test for Chapter 8

1. a. function; domain: {−3, 3, −4, 4}; range: {9, 16} **b.** not a function **c.** function; domain: $\mathbb{R}$; range: $[-1, \infty)$ **d.** function; domain: {−3, −2, 0, 1, 2}; range: {−2, −1, 0, 1, 2} **e.** not a function **2. a.** 2 **b.** −2 **c.** −3 **d.** 2 **e.** −18 **f.** 10

3. a. $m = \dfrac{3}{2}$ **b.** $m = 1$

4. a.

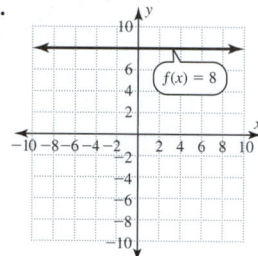

b.

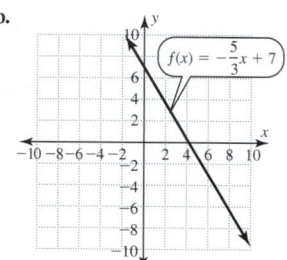

c.

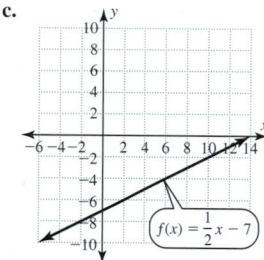

5. a. $y = -\dfrac{3}{2}x - 3$ **b.** $y = \dfrac{4}{5}x - \dfrac{21}{5}$ **6. a.** x-intercept: (8, 0); y-intercept: (0, −6) **b.** x-intercept: (48, 0); y-intercept: (0, 12) **7. a.** The function is positive on the interval $(-2, \infty)$ and negative on the interval $(-\infty, -2)$. **b.** The function is positive on the interval $(-\infty, 3)$ and negative on the interval $(3, \infty)$. **8. a.** upward **b.** (−2, −5)

c.

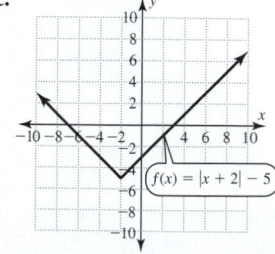

d. $y = -5$ **e.** $\mathbb{R}$ **f.** $[-5, \infty)$ **9. a.** (0, −3) **b.** (−7, 0), (3, 0) **10. a.** $(-\infty, -7) \cup (3, \infty)$ **b.** (−7, 3) **11. a.** The function is a line with a negative slope. **b.** The function is a parabola that opens upward. **c.** The function is a parabola that opens upward. **d.** The function is a line with a positive slope. **12. a.** (1, −3) **b.** (4, 1) **13. a.** upward

b. (−2.75, −55.125) **c.** x-intercepts: (−8, 0), $\left(\dfrac{5}{2}, 0\right)$; y-intercept: (0, −40)

d.

x	y
−4	−52
−3	−55
−2	−54
−1	−49

e.

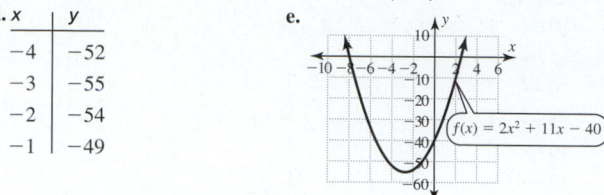

14. a. 40 ft **b.** 1.5 sec **15. a.** decreasing **b.** increasing **16. a.** (1.5, 12.25) **b.** downward **c.** (−2, 0), (5, 0) **d.** (0, 10) **e.** $\mathbb{R}$ **f.** $(-\infty, 12.25]$ **g.** (−2, 5) **h.** $(-\infty, -2) \cup (5, \infty)$ **i.** $(1.5, \infty)$ **j.** $(-\infty, 1.5)$ **17. a.** A **b.** D **c.** B **d.** C **18. a.** $y = 0.006x + 1.416$ **b.** The winning height for the Olympic women's high jump is increasing at a rate of approximately 0.006 meter per year. **c.** 1.848 m; The actual record is 0.072 m higher. **d.** 2.088 m **19.** $f(x) = 0.210x^2 - 0.497x - 3.49; f(7) \approx 3.3$

Group Project for Chapter 8

Actual data may vary.

1.

Time x (seconds)	Height y (cm)
0	20
5	18
11	16
16	14
22	12
29	10
37	8
46	6
55	4
68	2
100	0

2.

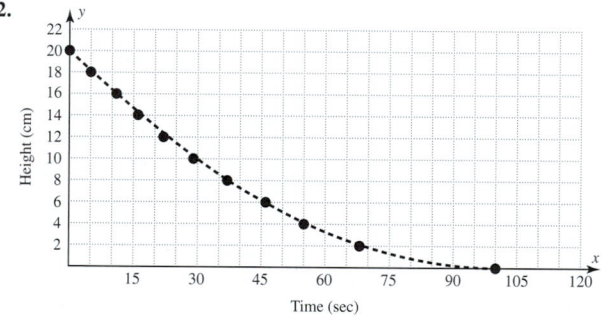

3. a. The quadratic model is the better model. **b.** The height of the water did not change at a constant rate. As time went on, the height changed more slowly. **c.** The distance the stream of water shoots out from the container reduces with time, indicating the water is not flowing at a constant rate. **4. a.** $y = 0.002x^2 - 0.4x + 20$ **b.** See graph above. **5. a.** $H(50) = 5$; after 50 seconds, the height of the water will be approximately 5 cm. **b.** $H(70) = 1.8$; after 70 seconds the height of the water will be approximately 1.8 cm. **c.** $H(x) = 11$ when $x = 25.8$; it will take approximately 25.8 seconds for the height to fall to 11 cm.

Chapter 9

Using the Language and Symbolism of Mathematics 9.1

1. ratio **2.** undefined **3.** rational **4.** denominator; 0 **5.** asymptote **6.** asymptote **7.** asymptote **8.** lowest terms **9.** $\dfrac{A}{B}$ **10.** −1

Quick Review 9.1

1. $2^3 \cdot 7$ **2.** $\dfrac{2}{5}$ **3.** $-x^2(3x - 4)$ **4.** $x(x - 10)(x + 4)$ **5.** $(2x + 5y)(3x - 7y)$

Exercises 9.1

1. a. C **b.** D **c.** A **d.** B **3. a.** -1 **b.** 3 **c.** undefined **5. a.** 4 **b.** $\mathbb{R} \sim \{4\}$

c. $x = 4$ **7.** C **9.** D **11.** D **13.** C **15. a.** $\mathbb{R} \sim \left\{\dfrac{1}{2}\right\}$ **b.** $\mathbb{R} \sim \{-9, 9\}$

17. $\mathbb{R} \sim \{1.5\}$ **19.** D **21.** B **23.** C **25.** D

27. $\dfrac{2b^2}{3a}$ **29.** $6x - 9$ **31.** $\dfrac{1}{2x - 3}$ **33.** $-\dfrac{a}{b}$ **35.** -1 **37.** $-\dfrac{a}{b}$

39. $\dfrac{x - 2y}{x + 2y}$ **41.** $\dfrac{x - y}{3}$ **43.** $-\dfrac{5x + 2}{7}$ **45.** $\dfrac{x}{x - y}$ **47.** $\dfrac{2x + 3y}{7}$

49. $\dfrac{2x + 3}{5y + 6}$ **51.** -1 **53.** $\dfrac{2a + b}{a + b}$ **55.** $\dfrac{x + y}{v + w}$ **57.** $\dfrac{x - 5}{3x - 1}$ **59.** $\dfrac{a + b}{a - b}$

61. $-\dfrac{a - b + 1}{5a}$ **63.** -1 **65.** $\dfrac{3(x + y)}{4(x - y)}$ **67.** $-\dfrac{2x - y}{y}$

69. $\dfrac{(b^2 + 1)(b - 1)}{5}$

71. a. $T(6) = 3$; when paddling at 6 mi/h, it takes 3 h to go 12 mi upstream. **b.** $T(8) = 2$; when paddling at 8 mi/h, it takes 2 h to go 12 mi upstream. **c.** $T(2)$ is undefined; paddling at 2 mi/h would only match the speed of the current and would not make progress upstream. **73. a.** The average cost increases dramatically without bound. **b.** The average cost decreases and approaches $4 per meter. **75.** 14 **77.** $14x$ **79.** $16y - 14x$ **81.** $10a - 10b$ **83.** $5x - 5$ **85.** $x^2 + 4xy + 3y^2$

Group Discussion Questions

87. a. $\dfrac{x - 3}{x + 5}$ is already reduced. **b.** $\dfrac{x + 3}{x - 1}$ is already reduced.

89. a. Answers can vary; an example is $f(x) = \dfrac{1}{(x - 2)(x - 5)}$. **b.** All functions have factors of $x - 2$ and $x - 5$ in the denominator. **c.** Answers can vary.

91.

x	$y = \dfrac{x + 3}{x - 2}$
1	-4.000
10	1.625
100	1.051
1,000	1.005

a. The values of x are approaching ∞. **b.** The values of y are approaching 1.

Cumulative Review

1.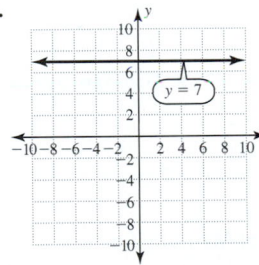

2.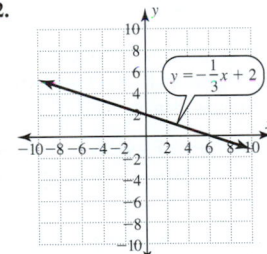

3.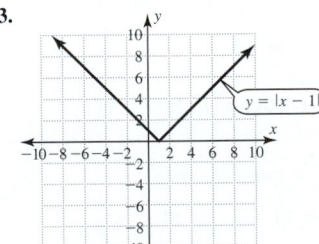

4.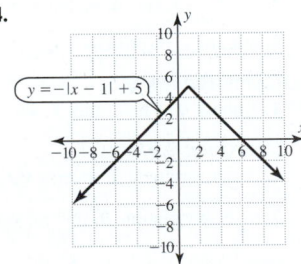

5.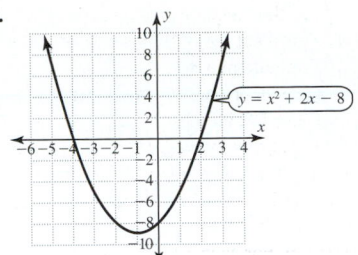

Using the Language and Symbolism of Mathematics 9.2

1. $\dfrac{AC}{BD}$ **2. a.** fraction **b.** nonzero **3.** $\dfrac{AD}{BC}$ **4.** reciprocal **5.** both

Quick Review 9.2

1. $\dfrac{16}{15}$ **2.** $\dfrac{20}{3}$ **3.** $\dfrac{125}{48}$ **4.** $\dfrac{2}{2}$ **5.** $\dfrac{7}{7}$

Exercises 9.2

1. $\dfrac{11}{9}$ **3.** $\dfrac{11}{21}$ **5.** $\dfrac{1}{3m}$ **7.** $\dfrac{9y^2}{7x^6}$ **9.** $-\dfrac{1}{2}$ **11.** $\dfrac{x - 1}{x(x + 1)}$ **13.** $\dfrac{x - 3}{2(x + 3)}$

15. $\dfrac{7(x - 3)}{3(x - y)}$ **17.** $\dfrac{3(x + 2)}{x - 1}$ **19.** $\dfrac{a - 7}{y - c}$ **21.** $-\dfrac{2(x - 1)}{3}$ **23.** $\dfrac{6(x - y)}{5}$

25. $-\dfrac{4x}{3}$ **27.** $(x + 5)(x - 5)$ **29.** $-\dfrac{1}{7x(3x - 5)}$ **31.** 2 **33.** $\dfrac{11x}{9a}$ **35.** $\dfrac{7xy}{15}$

37. $\dfrac{5}{x - 3}$ **39.** $-\dfrac{2}{ab^2}$ **41.** $\dfrac{3}{11}$ **43.** $\dfrac{7y^2}{2a}$ **45.** $\dfrac{x + y}{12y}$ **47.** 4; 3.99 **49.** 3; 2.99

51. D **53.** B **55.** 5 **57.** xy^2 **59.** $x - 8$ **61.** $\dfrac{x - 3}{x - 3}$ **63.** $\dfrac{x - 2}{x - 1}$

65. a. $A(t) = \dfrac{100(t + 1)}{t(t - 1)}$

b.

t	2	6	10	14	18	22
$A(t)$	150	23.33	12.22	8.24	6.21	4.98

c. The average cost increases without bound. **d.** The average cost decreases to approximately $4.53 per lamp.

Group Discussion Questions

67. a. $3x^{2m}$ **b.** x^m **c.** 3 **d.** $\dfrac{x^m - 1}{x^m + 1}$

69. a. $\mathbb{R} \sim \{1\}$ **b.**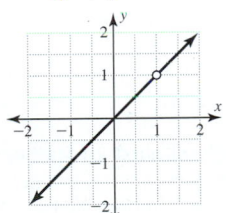

c. There is a hole at $(1, 1)$. **d.** The calculator calculates y-values for only a relatively few x-values and then connects these points. On the window $[-5, 5, 1]$ by $[-5, 5, 1]$, the calculator did not try to evaluate $f(1)$.

Cumulative Review

1. -78 **2.** -48 **3.** $-45x^2 - 28x - 5$ **4.** $18x^3 - 33x^2 - 28x - 5$ **5.** $36x^4 - 156x^3 + 109x^2 + 130x + 25$

Using the Language and Symbolism of Mathematics 9.3

1. $\dfrac{A + B}{C}$; $\dfrac{A - B}{C}$ **2.** add **3.** subtract **4. a.** denominator; exponential **b.** highest **c.** multiplying **5. a.** LCD **b.** denominator **c.** numerators **d.** lowest terms

Quick Review 9.3

1. $\dfrac{5}{6}$ **2.** $\dfrac{1}{2}$ **3.** $\dfrac{7}{12}$ **4.** $\dfrac{7}{16}$ **5.** $\dfrac{7}{12}$

Exercises 9.3

1. $\dfrac{2b+3}{b^2}$ **3.** -1 **5.** $\dfrac{4}{s-3}$ **7.** $\dfrac{1}{x+y}$

9. $\dfrac{1}{15xy^2}\left(\dfrac{9x}{9x}\right)+\dfrac{2}{27x^2y}\left(\dfrac{5y}{5y}\right)=\dfrac{9x}{135x^2y^2}+\dfrac{10y}{135x^2y^2}=\dfrac{9x+10y}{135x^2y^2}$

11. $\dfrac{a+b}{21(3a-b)}\left(\dfrac{2}{2}\right)-\dfrac{a-b}{14(3a-b)}\left(\dfrac{3}{3}\right)=\dfrac{2a+2b}{42(3a-b)}-$

$\dfrac{3a-3b}{42(3a-b)}=\dfrac{-a+5b}{42(3a-b)}$

13. $\dfrac{2}{(x-3)(x-15)}\left(\dfrac{x+3}{x+3}\right)+\dfrac{1}{(x-3)(x+3)}\left(\dfrac{x-15}{x-15}\right)=$

$\dfrac{2x+6}{(x-3)(x-15)(x+3)}+\dfrac{x-15}{(x-3)(x+3)(x-15)}=$

$\dfrac{3x-9}{(x-3)(x-15)(x+3)}=\dfrac{3}{(x-15)(x+3)}$

15. $-\dfrac{13}{18w}$ **17.** $\dfrac{5}{14}$ **19.** $\dfrac{b^2+4b+16}{b(b+4)}$

21. $\dfrac{5x-1}{x}$ **23.** $\dfrac{x^2-2x+3}{x^3}$ **25.** $\dfrac{x+13}{(x-2)(x+3)}$ **27.** $\dfrac{2x}{(x-1)(x+1)}$

29. $\dfrac{8x+22}{(x-4)(x+2)}$ **31.** $\dfrac{2x^2+2x+5}{(x+2)(x-1)}$ **33.** $\dfrac{1}{77}$

35. $\dfrac{5m+9}{(m+1)(m-2)(m+3)}$ **37.** $\dfrac{4m-13}{(m-3)(m-4)}$

39. $\dfrac{2}{a-4b}$ **41.** $\dfrac{3x+2}{2(x-1)}$ cm^2 **43. a.** $W(t)=\dfrac{8(2t+5)}{t(t+5)}$

b. If the time required to paint the house becomes very large, the fractional portion of the job completed in 8 h approaches 0.

45. a. $\dfrac{255r+900}{r(r+10)}$ **b.** 5 h **47.** A **49.** D **51.** $\dfrac{-x^2+6x-3}{x^2-3x+4}$

53. $-\dfrac{3}{(x-1)(x+1)}$ **55.** $\dfrac{2}{x+5}$ **57.** $\dfrac{10}{s-2t}$ **59.** $\dfrac{1}{z+3}$ **61.** $\dfrac{1}{v-w}$

63. $\dfrac{3w-4}{(w-1)(w-2)}$ **65.** $\dfrac{1}{w-3}$ **67.** $-\dfrac{2m}{m-1}$ **69.** $\dfrac{2x+5}{(x+5)(2x-5)}$

Group Discussion Questions

71. This is an order-of-operations error; multiplication has priority over addition. **73.** $\dfrac{m}{2}$

Cumulative Review

1. $m=\dfrac{2}{5}$ **2.** $m=-\dfrac{5}{2}$ **3.** $(4,-6)$ **4.** no solution **5.** infinite number of solutions

Using the Language and Symbolism of Mathematics 9.4

1. a. grouping symbols **b.** exponentiations **c.** multiplications; divisions **d.** additions; subtractions **2.** numerator; denominator **3.** reciprocal **4.** addition; subtraction

Quick Review 9.4

1. 180 **2.** $(x+y)(x-y)$ **3.** -11 **4.** 43 **5.** $\dfrac{1}{x}+\dfrac{3}{x^2}$

Exercises 9.4

1. a. x **b.** $\dfrac{x+10}{3}$ **3. a.** $\dfrac{5y}{6}$ **b.** $\dfrac{3(y-4)}{2}$ **5. a.** $\dfrac{v^2-6}{v(v-6)}$

b. $\dfrac{2(v^2-v+18)}{v(v-6)^2}$ **7.** -2 **9.** v^2-1 **11.** $x-3$ **13.** $-\dfrac{4}{x^2}$ **15.** $\dfrac{12}{5x}$

17. a. $\dfrac{16}{15}$ **b.** $\dfrac{10}{3}$ **19.** $\dfrac{3}{2}$ **21.** -2 **23.** $\dfrac{9y}{4}$ **25.** $\dfrac{2}{xy(w+z)}$ **27.** $\dfrac{x(x-7)}{10(x+2)}$

29. $\dfrac{x}{2x+1}$ **31.** $\dfrac{v^2w^2}{v+w}$ **33.** $\dfrac{x-1}{6(x+1)}$ **35.** $\dfrac{3x^2}{x+5}$ **37.** $-\dfrac{4aw}{w^2+a^2}$

39. $\dfrac{v+3}{v+6}$ **41.** $\dfrac{3}{2}=1.5$; 1.494 **43.** $\dfrac{4}{5}=0.8$; 0.820 **45.** $\dfrac{4a-4b+1}{4ab}$

47. $\dfrac{y^2}{x^2+y^2}$ **49.** $\dfrac{w+v}{w-v}$ **51.** $-m^2n-mn^2$ **53.** $\dfrac{a+b}{ab}$ **55.** $\dfrac{1}{x-y}$

57. $\dfrac{a+b}{a^2b^2}$ **59.** $-\dfrac{x+y}{x}$ **61.** B **63.** D **65.** $x(x+5)$ cm^2

67. a. $A(t)=\dfrac{1,680}{t}$ for $t\neq 2$ **b.** $(0,2)\cup(2,168]$ **c.** The average cost decreases to \$10 per unit when the factory operates 168 h/wk.

69. a. $f=\dfrac{d_o d_i}{d_o+d_i}$ **b.** $f\approx 0.488$ ft

Group Discussion Questions

71. a. x^m+y^n **b.** $-\dfrac{1}{x^m y^n}$ **c.** $\dfrac{x^m-1}{x^m+1}$ **d.** $\dfrac{x^m+3}{x^m-2}$

Cumulative Review

1. $(0,8)$ **2.** $(-4,0),(2,0)$ **3.** $y=9$ **4.** $x=-1$ **5.** $(-\infty,-1)$

Using the Language and Symbolism of Mathematics 9.5

1. same **2.** zero **3. a.** LCD **b.** Solve **c.** extraneous **4.** zero

Quick Review 9.5

1. $x=6$ **2.** $x=3$ **3.** $x=37$ **4.** $x=-4$ **5.** There is no solution.

Exercises 9.5

1. 2 and 3 **3.** $-\dfrac{3}{2}$ and $\dfrac{2}{3}$ **5.** -3, $-\dfrac{1}{3}$, and $\dfrac{1}{2}$ **7.** $z=2$ **9.** There is no solution.

11. $p=-5$ **13.** $n=-16$ **15.** $k=1$ **17.** There is no solution. **19.** $y=5$

21. $x=3,x=23$ **23.** $t=\dfrac{2}{3},t=2$ **25.** $z=0,z=\dfrac{2}{3}$ **27.** $v=\dfrac{13}{3}$

29. $m=2$ **31.** $x=-2,x=3$ **33.** $m=-\dfrac{2}{3}$ **35.** $z=\dfrac{11}{4}$ **37.** $n=\dfrac{2}{3}$

39. $x=-2,x=0,x=1$ **41.** $b=\dfrac{ad}{c}$ **43.** $b=\dfrac{ad+a+c}{c}$ **45.** $d=\dfrac{k}{I}$

47. $R=\dfrac{r_1 r_2}{r_1+r_2}$ **49.** $B=\dfrac{2A-bh}{h}$ **51.** $z=\dfrac{xy}{x-y}$ **53.** 4.12 ft

55. $r_1=6\,\Omega;r_2=30\,\Omega$ **57. a.** $-\dfrac{2(p-2)}{(p-1)(p+1)}$ **b.** $p=2$

59. a. $-\dfrac{(x-5)(x+2)}{(x+1)(x-2)}$ **b.** $x=-2,x=5$ **61. a.** 0 **b.** $\mathbb{R}\sim\left\{-\dfrac{2}{3},\dfrac{1}{2}\right\}$

63. x-intercept: $(-5,0)$; y-intercept: $\left(0,-\dfrac{5}{2}\right)$ **65.** x-intercepts:

$(-2,0),(2,0)$; y-intercept: $(0,4)$ **67. a.** $\mathbb{R}\sim\left\{-3,\dfrac{5}{2}\right\}$ **b.** There is no

solution. **69.** $x=-3,x=0,x=2$ **71.** $z=2$ **73.** $z=2$

Group Discussion Questions

75. When $x=1$, $x-1=0$; cannot divide by 0 **77. a.** 2 (answers may vary) **b.** 1 (answers may vary)

Cumulative Review

1. $5x^2-14x-3=0$ **2.** $x^2-6=0$ **3.** $x^2-2x+5=0$ **4.** $\mathbb{R}\sim\{3\}$ **5.** $\mathbb{R}$

Using the Language and Symbolism of Mathematics 9.6

1. kb **2.** $\dfrac{k}{b}$ **3.** kbc **4.** increase **5.** decrease **6.** decrease **7.** increase **8.** rate; time **9.** principal; rate; time **10.** rate; time

Quick Review 9.6

1. 4 hours **2.** 4 h 48 min **3.** 480 mi/h **4.** \$300 **5.** 8%

Exercises 9.6

1. direct variation **3.** inverse variation **5.** direct variation

7. inverse variation **9.** inverse variation **11.** $m=kn$ **13.** $m=\dfrac{k}{p}$

15. $v=k\sqrt{w}$ **17.** $v=\dfrac{k}{x^2}$ **19.** $m=\dfrac{kn}{p}$ **21.** $v=\dfrac{k\sqrt{w}}{x^2}$ **23.** $m=knp$

25. $v=k\sqrt{w}x^2$ **27.** $I=kV$ **29.** $w=\dfrac{k}{d^2}$ **31.** $R=kv^2$

33. $V=kr^2h;\left(k=\dfrac{1}{3}\pi\right)$ **35. a.** $y=30$ **b.** $y=19.2$ **37.** $a=10$

39. $a=22.5$

41. a. 34.8 **b.** $N = 34.8P$

c.

Population	Newsprint (kg)
700,000	24,360,000
710,000	24,708,000
720,000	25,056,000
730,000	25,404,000

d. 25,056,000 kg

43. a. 40 **b.** $t = \dfrac{40}{w}$

c.

Workers	Time (hours)
5	8.00
10	4.00
15	2.67
20	2.00

d. 2 hours **45.** When the UV rating is 10, it takes 9 minutes to get a sunburn. **47.** The resistance is 2 Ω. **49.** The beam can support a load of 270 lb. **51.** The integers are 5 and 6. **53.** The integers are 7 and 9. **55.** The numerator is 6. **57.** The number is either $\frac{2}{3}$ or $\frac{3}{2}$. **59.** The normal reading is 5. **61.** The pieces of wire are 4 meters and 12 meters long. **63.** The resistance of one resistor is 60 Ω and the other is 120 Ω. **65.** The angles are 36° and 54°. **67.** \$8,400 is invested in secure funds. **69.** 20 vehicles were sold in January. **71.** The rate of each boat in still water is 11 km/h. **73.** The rate of the slower plane is 480 mi/h and the rate of the faster plane is 520 mi/h. **75.** It would take 6 hours for both pipes to fill the tank. **77.** It would take 6 hours for the large pipe to fill the tank alone. **79.** It would take 75 hours for plane B to complete the search alone. **81.** The interest rate for the savings account is 3% and the interest for the money market account is 5%. **83.** The interest rate for the 30-year mortgage is 8.5% and the interest for the 15-year mortgage is 7%.

Group Discussion Questions

85. a. The time it takes to brush paint on the wall varies directly as the volume of paint. **b.** The thickness of paint on the wall varies inversely as the area covered. **c.** The resistance varies inversely as the square of the radius of the pipe. **d.** The volume of water varies directly as the square of the radius of the pipe. **87.** Answers will vary. Example: length of thumb and width of one's palm

Cumulative Review

1. C **2.** B **3.** A **4.** The chemist should add 1.25 L of water. **5.** The speeds of the trains are 95 km/h and 103 km/h.

Review Exercises for Chapter 9

1. 9 **2.** undefined **3.** $-\dfrac{1}{15}$ **4.** $\dfrac{1}{9}$ **5.** $\mathbb{R} \sim \{-2, 3\}$ **6.** $\mathbb{R} \sim \{-5, 3\}$

7. $\mathbb{R} \sim \left\{-5, \dfrac{1}{3}\right\}$ **8.** $\mathbb{R}$ **9.** $x = \dfrac{1}{2}$ **10.** $x = -6, x = 6$ **11.** $x = 0, x = 9$

12. $\dfrac{3x}{y^2}$ **13.** $-\dfrac{3}{2}$ **14.** $\dfrac{3(x-1)}{5}$ **15.** $\dfrac{x-5}{2x-1}$ **16.** $\dfrac{c}{a+b}$ **17.** $-\dfrac{1}{2m}$

18. $\dfrac{5x + 2y}{3x - y}$ **19.** $\dfrac{3x - 4y}{4x - 3y}$ **20.** $\dfrac{x + 2y}{x - 2y}$ **21.** $2x^2 - 5x - 3$

22. $2y^2 - 5y - 3$ **23.** D **24.** C **25.** A **26.** B **27.** $6x - 4$

28. $\dfrac{4xy + 2y^2}{x}$ **29.** $\dfrac{6t^2 - t - 2}{2}$ **30.** $\dfrac{v + 6w}{v - w}$ **31.** $\dfrac{1}{2x + 1}$ **32.** 0

33. $\dfrac{7v}{(v - 1)(2v + 1)(3v - 2)}$ **34.** 3 **35.** $\dfrac{3w^2 + 17w + 66}{(w - 3)(w - 2)(w + 6)}$

36. $\dfrac{6v - 1}{3v - 2}$ **37.** $y - 4$ **38.** $\dfrac{2}{3z + 2}$ **39.** $\dfrac{a(2a + 1)}{(a + 1)^2}$ **40.** $\dfrac{x + 1}{x + 2}$ **41.** $\dfrac{v + 3}{v - 3}$

42. $\dfrac{x - 1}{x + 1}$ **43.** $\dfrac{xy + 1}{xy - 1}$ **44.** $\dfrac{w - 4}{3w - 4}$ **45.** $\dfrac{m + 5}{m + 4}$ **46.** 2 **47.** $\dfrac{1}{5y + 1}$

48. 2 **49. a.** $\dfrac{2(m + 1)}{(m + 2)(m + 4)}$ **b.** $m = -1$ **50. a.** $\dfrac{-5x + 16}{(x + 1)(x - 6)}$

b. $x = \dfrac{16}{5}$ **51. a.** $\dfrac{2x^2 - x - 12}{(x - 3)(x - 2)}$ **b.** $x = 4$ **52. a.** $\dfrac{2}{x - 1}$

b. There is no solution. **53.** $x = -1$ **54.** There is no solution. **55.** $y = -\dfrac{8}{3}$

56. $y = 5$ **57.** $w = 1, w = 2$ **58.** $v = -1$ **59.** $\mathbb{R} \sim \left\{-4, \dfrac{1}{2}\right\}$

60. $\mathbb{R} \sim \left\{-2, -\dfrac{1}{2}, 5\right\}$ **61.** $b = -a - 1$ **62.** $r_2 = \dfrac{E - r_1 I}{I}$

63. 14; 14.10 **64.** 0; -0.01 **65. a.** $\dfrac{4(13x - 1)(x + 2)}{5x(x - 2)}$ cm **b.** $\dfrac{x + 2}{x}$ cm^2

66. 45 cm^2 **67.** $\dfrac{100(x + 2)}{x(x - 2)}$ cm^3 **68. a.** $(0, 24]$ **b.** $A(t) = \dfrac{200(t + 5)}{t(t + 20)}$

c. $N(10) = 150$; the plant can produce 150 units in 10 hours. **d.** $C(10) = 1,500$; it costs the plant \$1,500 to operate for 10 hours. **e.** $A(10) = 10$; it costs an average of \$10 per unit when the plant operates for 10 hours. **f.** The average cost increases without bound. **g.** The average cost decreases to approximately \$5.49 per unit as the time approaches 24 hours. **69.** 6 hours; the constant of variation $k = 300$ is the distance in miles between St. Louis and Chicago. **70.** The pulley should have a diameter of 4.5 in. **71.** The force is 180 lb. **72.** The numerator is -3. **73.** The numbers are 25 and 10. **74.** The integers are 5 and 7. **75.** The resistance of one resistor is 4 Ω and the other is 12 Ω. **76.** It would take 21 minutes for the cold-water faucet alone. **77.** The company should produce 1,000 desks. **78.** The original dimensions are 8 cm by 14 cm. **79.** The rate of each boat in still water is 21 miles per hour. **80. a.** directly **b.** inversely **c.** inversely **d.** directly

Mastery Test for Chapter 9

1. a. B **b.** C **c.** D **d.** A **2. a.** $\mathbb{R} \sim \{6\}$ **b.** $\mathbb{R} \sim \{1, -6\}$ **c.** $\mathbb{R} \sim \{-8, 8\}$ **d.** $\mathbb{R}$ **3. a.** $\dfrac{1}{2x - 3}$ **b.** $\dfrac{2}{x + 3}$ **c.** $\dfrac{2x + 3}{x + 5}$ **d.** $\dfrac{2x - 3y}{a - 2b}$ **4. a.** $\dfrac{5}{4}$ **b.** 1 **c.** $\dfrac{2}{3}$

d. $\dfrac{(x + 2)(x + 5)}{(x + 3)(x - 5)}$ **5. a.** 2 **b.** $\dfrac{1}{x - 3}$ **c.** $\dfrac{9}{(w - 5)(w + 4)}$ **d.** $\dfrac{4}{x - y}$

6. a. x **b.** $\dfrac{x + 12}{5}$ **c.** 1 **d.** $\dfrac{w + 1}{w - 3}$ **7. a.** $\dfrac{24}{5}$ **b.** $\dfrac{x - 3}{3}$ **c.** $\dfrac{3(2x + 3)}{4x - 3}$

d. $-\dfrac{2(v - 11)}{(v - 4)^2}$ **8. a.** $z = 5$ **b.** $z = 2, z = 4$ **c.** There is no solution.

d. $x = -1$ **9. a.** $h = \dfrac{2A}{b}$ **b.** $z = \dfrac{xy}{y - x}$ **c.** $x = \dfrac{5y + 4}{y - 1}$ **d.** $a = \dfrac{bc}{b - c}$

10. a. inverse variation **b.** direct variation **11. a.** $a = kb$ **b.** $a = \dfrac{k}{b}$

c. $a = \dfrac{kb}{c^2}$ **d.** $a = kbc$ **12. a.** $y = 18$ **b.** The pressure will be 8 N/cm^2.

13. a. The normal number of units for the indicator is 10. **b.** It would take 10 hours for the larger end loader and 15 hours for the smaller end loader to move the pile. **c.** The interest rate for the shorter-term bond is 6% and the interest for the longer-term bond is 8.5%. **d.** The rate of the slower plane is 475 miles per hour and the rate of the faster plane is 525 miles per hour.

Group Project for Chapter 9

1. $N(t) = 32.1t - 66.4$ **2.** $C(t) = 1,230t + 322$

3. $A(t) = \dfrac{1,230t + 322}{32.1t - 66.4}$ **4.** $[2, 24]$ **5.** 51; the average cost per unit is approximately \$49.60 when the factory operates 10 h/day. **6.** The average cost per unit decreases to approximately \$42.40 when the plant operates almost 24 h/day.

Chapter 10

Using the Language and Symbolism of Mathematics 10.1

1. radicand **2.** radical **3.** index; order **4.** $\sqrt[n]{x}$ **5.** cube **6.** fourth **7.** even; negative **8.** real **9.** zero **10.** negative

Quick Review 10.1

1. 32 **2.** -18 **3.** $(-\infty, 3) \cup (3, \infty)$ or $\mathbb{R} \sim \{3\}$ **4.** $(-3, 4]$ **5.** $[-1, 3]$

Exercises 10.1
1. $\sqrt[3]{w}$ **3.** $\sqrt[5]{v}$ **5.** $\sqrt{4} = 2$ **7.** $\sqrt[5]{32} = 2$ **9. a.** 4 **b.** 2 **11. a.** 6 **b.** -6
13. a. 2 **b.** -2 **15.** 0 **17.** 1 **19.** 100 **21.** 10 **23.** 0.5 **25.** $\dfrac{3}{2}$ **27.** $-\dfrac{1}{2}$
29. 19 **31.** 17 **33.** 2; 2.080 **35.** 3; 2.991
37. a.

x	$\sqrt{x-2}$
2	0
3	1
6	2
11	3

b. 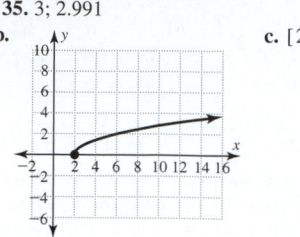 **c.** $[2, \infty)$

39. a.

x	$\sqrt[3]{x+3}$
-11	-2
-4	-1
-3	0
-2	1
5	2

b. 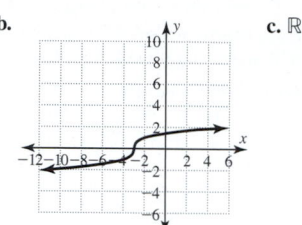 **c.** $\mathbb{R}$

41. $\mathbb{R}$ **43.** $[5, \infty)$ **45.** $\mathbb{R} \sim \{5\}$ **47.** $\mathbb{R}$ **49.** 13 cm **51.** 6 cm **53.** 0.5 sec
55. 3.7 cm **57.** $C(0) = 1,200$; The cost to run a power line directly across
the river is \$1,200. **59.** $C(5) \approx 6,120$; The cost to run a power line across
the river to a point 5 miles downstream is \$6,120. **61.** E **63.** B **65.** G
67. A **69.** H; $\mathbb{R}$ **71.** B; $(-\infty, 2]$ **73.** C; $\mathbb{R}$ **75.** F; $\mathbb{R}$ **77.** D; $\mathbb{R}$ **79.** -4
81. i **83.** $-20 + 9i$

Group Discussion Questions
85. The domain of $f(x) = \sqrt[3]{x} + 2$ is $\mathbb{R}$. This graph looks more like
$f(x) = \sqrt{x} + 2$. **87. a.** D **b.** C **c.** B **d.** A

Cumulative Review
1. $17x - 41y$ **2.** $-13x + 27y$ **3.** $10a^9 b^{11}$ **4.** $4a^3 b$ **5.** $125a^6 b^{12}$

Using the Language and Symbolism of Mathematics 10.2
1. radicand; index **2.** $\sqrt[n]{x}\,\sqrt[n]{y}$ **3.** $\dfrac{\sqrt[n]{x}}{\sqrt[n]{y}}$ **4.** $|x|$ **5.** x

Quick Review 10.2
1. $9x - 6y$ **2.** $2x + 6y$ **3.** 17 **4.** 22

5.

x	x^2	x^3
0	0	0
1	1	1
2	4	8
3	9	27
4	16	64
5	25	125

Exercises 10.2
1. D **3.** 12 **5.** -6 **7.** $30\sqrt{2}$ **9.** $-4\sqrt{7}$ **11.** $-5\sqrt[3]{6}$ **13.** $10\sqrt[4]{13}$
15. $6\sqrt{5} - 4\sqrt{7}$ **17.** $-3\sqrt{7x}$ **19.** 0 **21.** $5\sqrt{3}$ **23.** $3\sqrt{7}$ **25.** $2\sqrt[3]{3}$
27. $2\sqrt[3]{3}$ **29.** $5\sqrt{7}$ **31.** $\sqrt{3}$ **33.** $-13\sqrt{2v}$ **35.** $-2\sqrt{7w}$ **37.** $-3\sqrt[3]{3}$
39. $-74t\sqrt[3]{3}$ **41.** $-2\sqrt[3]{5z^2}$ **43.** $\dfrac{5\sqrt{5}}{6}$ **45.** $\dfrac{9\sqrt{11x}}{35}$ **47.** $7.9\sqrt{2} + 0.8\sqrt{3}$
49. 0 **51. a.** $5x$ **b.** $5|x|$ **53. a.** $2x$ **b.** $2x$ **55. a.** $1,000x^3$ **b.** $1,000|x|^3$
57. a. x^6 **b.** x^6 **59.** $5x\sqrt{2x}$ **61.** $2x\sqrt[3]{2x}$ **63.** $xy^2\sqrt[3]{xy^2}$ **65.** 2.0 in **67.** 12.6 m
69. C **71. a.** 4 **b.** $4 > \sqrt[3]{63}$ **c.** 3.979 **73. a.** 2 **b.** $2 < \dfrac{1 + \sqrt{9.05}}{2}$
c. 2.004 **75.** E **77.** C **79.** A

Group Discussion Questions
81. a. These functions are equal for $x \geq 0$. For $x < 0$, $\sqrt[4]{x^2}$ is a real
number but $\sqrt{x}$ is not a real number. **b.** These functions are equal.
c. $\sqrt[6]{x^2} = |\sqrt[3]{x}|$ **83. a.** ≈ 1.00225 **b.** ≈ 0.99775 **c.** For $x > 0$,
$\sqrt[n]{x}$ approaches 1 as n becomes large.

Cumulative Review
1. B **2.** A **3.** D **4.** E **5.** C

Using the Language and Symbolism of Mathematics 10.3
1. $\sqrt[n]{xy}$ **2.** $\sqrt[n]{\dfrac{x}{y}}$ **3.** order **4.** even **5.** conjugates **6.** pair **7.** rationalizing

Quick Review 10.3
1. $15x^2 y - 10xy^2$ **2.** $14x^2 + 29xy - 15y^2$ **3.** $16x^2 - 25y^2$
4. $16x^2 + 40xy + 25y^2$ **5.** $5 + 7i$

Exercises 10.3
1. $2\sqrt{3}$ **3.** $10 - \sqrt{2}$ **5.** $8\sqrt{15}$ **7.** $84\sqrt{35}$ **9.** $\sqrt{6} + \sqrt{15}$
11. $30\sqrt{3} - 105\sqrt{7}$ **13.** 240 **15.** $4w$ **17.** $24z\sqrt{2}$ **19.** $6x - 15\sqrt{x}$
21. $3x + 3\sqrt{3xy} - 4y$ **23.** $14 - 4\sqrt{6}$ **25.** $a + 10\sqrt{3ab} + 75b$
27. $v - 11$ **29.** 7 **31.** $-3v$ **33.** $-4 - \sqrt[3]{20}$ **35.** $y\sqrt[3]{x^2 y}$
37. $10\sqrt{3} + 7\sqrt{2} - 6$ **39.** 2 **41.** $x^2 - 3y$ **43.** $v^2 - 3v + 1$ **45.** $\dfrac{\sqrt{3}}{2}$
47. $\dfrac{\sqrt{6}}{3}$ **49.** $\dfrac{\sqrt{10}}{4}$ **51.** $\sqrt{5}$ **53.** $6\sqrt[3]{2}$ **55.** $\dfrac{\sqrt[3]{20}}{4}$ **57.** $\dfrac{\sqrt[3]{6}}{3}$ **59.** $3\sqrt{6}$
61. $\dfrac{13\sqrt{10}}{5}$ **63.** $\dfrac{5\sqrt{3x}}{x}$ **65.** $\dfrac{\sqrt{7} - 1}{2}$ **67.** $-15 - 3\sqrt{13}$
69. $-3\sqrt{7} - 3\sqrt{2}$ **71.** $\dfrac{a + \sqrt{ab}}{a - b}$ **73.** $4\sqrt[3]{9}$ **75.** $\dfrac{\sqrt[3]{6}}{2}$ **77.** $\dfrac{\sqrt[3]{3vw}}{3w}$
79. $\dfrac{\sqrt[3]{2vw}}{w}$ **81.** $\dfrac{5(\sqrt{3x} - \sqrt{2y})}{3x - 2y}$ **83.** $20; 20 < \dfrac{100}{\sqrt{24}}; 20.412$
85. $x^2 y\sqrt[3]{xy^2}$ **87.** $\dfrac{\sqrt[3]{4z^2}}{z}$ **89.** $\dfrac{5\sqrt[3]{4z^2}}{z}$

Group Discussion Questions
91. $\sqrt{\sqrt{\dfrac{2,143}{22}}}$ is closest to π.
93. a. $(1 + \sqrt{2})^2 - 2(1 + \sqrt{2}) - 1$
$= 1 + 2\sqrt{2} + 2 - 2 - 2\sqrt{2} - 1 = 0$
b. $a\left(\dfrac{-b + \sqrt{b^2 - 4ac}}{2a}\right)^2 + b\left(\dfrac{-b + \sqrt{b^2 - 4ac}}{2a}\right) + c$
$= a\left(\dfrac{b^2 - 2b\sqrt{b^2 - 4ac} + b^2 - 4ac}{4a^2}\right) + \left(\dfrac{-b^2 + b\sqrt{b^2 - 4ac}}{2a}\right) + c$
$= \dfrac{b^2 - b\sqrt{b^2 - 4ac} - 2ac - b^2 + b\sqrt{b^2 - 4ac}}{2a} + c = -c + c = 0$

Cumulative Review
1. $v = 2$ **2.** $w = -1, w = 4$ **3.** $x = -1, x = 4$ **4.** $y = \dfrac{1}{2}, y = 5$ **5.** $m = 8$

Using the Language and Symbolism of Mathematics 10.4
1. radicand **2.** $x^n = y^n$ **3.** extraneous **4.** hypotenuse **5.** Pythagorean
6. $\sqrt{(x_2 - x_1)^2 + (y_2 - y_1)^2}$

Quick Review 10.4
1. $x = 13$ **2.** $x = \dfrac{9}{5}, x = -\dfrac{7}{5}$ **3.** no **4.** yes **5.** $x = 4$

Exercises 10.4
1. $x = 5$ **3.** $x = -1$ **5.** $x = 16$ **7.** $x = 1$ **9.** $t = 13$ **11.** $c = 393$
13. no solution **15.** $w = -\dfrac{9}{2}$ **17.** $v = 3$ **19.** $w = -4, w = 8$ **21.** $y = 4$
23. $t = 2$ **25.** $w = \dfrac{1}{3}$ **27.** $x = 4$ **29.** $u = 0$ **31.** 5 **33.** 17 **35.** $\sqrt{2}$ **37.** 3
39. $7\sqrt{2} + \sqrt{58}$; right triangle **41.** $2\sqrt{10} + 5\sqrt{2} + \sqrt{106}$; not a right
triangle **43.** $c = 13$ cm **45.** $b = 9$ cm **47.** $c = \sqrt{113}$ cm **49.** $(5, -4)$ and
$(5, 8)$ **51.** $x = 4$ **53.** $x = 9$ **55.** x-intercept: $(-5, 0)$; y-intercept: $(0, 1)$
57. x-intercept: $(-9, 0)$; y-intercept: $(0, 3)$ **59.** $x = 1 + \sqrt{7}$
61. $x = \dfrac{-1 + \sqrt{5}}{2}$ **63.** $x = 0, x = 2$ **65.** no solution **67.** $x = 10$
69. $x = 6$ **71.** 42.5 m **73.** \$450 **75.** 64 **77.** $S(1,000) = 75,000$; a box
beam with a volume of 1,000 cm^3 has a strength of 75,000 N.

79. a. $f(x) = \sqrt{289 - x^2}$ **b.** 15 ft **81. a.** $L(x) = \sqrt{x^2 + 121}$
b. $L(30) \approx 32$; when the horse is 30 ft from the pole, the length of the
rope is 32 ft. **c.** The area that can be covered underneath the bridle is
approximately 3,470 ft². **83.** $x = 3$ mi

Group Discussion Questions
85. a. $x^2 + y^2 = 9$, a circle of radius 3 **b.** $(x - 3)^2 + (y - 5)^2 = 4$,
a circle of radius 2 **c.** $(x - h)^2 + (y - k)^2 = r^2$, a circle of radius r
87. Areas of four triangles + Area of inner square

$$= 4\left(\frac{1}{2}ab\right) + (b - a)^2$$
$$= 2ab + b^2 - 2ab + a^2$$
$$= a^2 + b^2$$

Area of larger square $= c^2$; $a^2 + b^2 = c^2$

Cumulative Review
1. $b + a$ **2.** $(a \cdot b)c$ **3.** $ab + ac$ **4.** -31; There are two complex solutions
with imaginary parts. **5.** 49; There are two distinct real solutions.

Using the Language and Symbolism of Mathematics 10.5
1. $\sqrt[5]{x}$ **2.** $x^{1/7}$ **3.** $\sqrt[5]{x^2}$ **4.** $x^{3/4}$ **5.** $x^{1/n}$ **6.** negative **7.** radicand

Quick Review 10.5
1. $\frac{25}{16}$ **2.** x^{18} **3.** x^{77} **4.** x^4 **5.** x^{16}

Exercises 10.5
1. $w^{1/3}$ **3.** $x^{1/6}$ **5.** $\sqrt[3]{125} = 125^{1/3} = 5$ **7.** $\sqrt[4]{81} = 81^{1/4} = 3$
9. a. $16^{1/2} = 4$ **b.** $16^{1/4} = 2$ **11. a.** $0^{1/5} = 0$ **b.** $0^{1/6} = 0$

13. a. $(-1)^{1/3} = -1$ **b.** $(-1)^{1/5} = -1$ **15. a.** $\sqrt{36} = 6$ **b.** $\frac{1}{\sqrt{36}} = \frac{1}{6}$

17. a. $\sqrt[3]{27} = 3$ **b.** $\frac{1}{\sqrt[3]{27}} = \frac{1}{3}$ **19. a.** $\sqrt{0.09} = 0.3$ **b.** $-\sqrt{0.09} = -0.3$

21. a. $\sqrt[3]{0.008} = 0.2$ **b.** $\sqrt[3]{-0.008} = -0.2$ **23. a.** $\sqrt[3]{64} = 4$

b. $\frac{1}{\sqrt[3]{64}} = \frac{1}{4}$ **25. a.** $\left(\sqrt[3]{\frac{8}{125}}\right)^2 = \frac{4}{25}$ **b.** $\left(\sqrt[3]{\frac{125}{8}}\right)^2 = \frac{25}{4}$
27. a. $\sqrt{25} + \sqrt{144} = 17$ **b.** $\sqrt{25 + 144} = 13$ **29. a.** $\sqrt[3]{8} + \sqrt[3]{1} = 3$
b. $\sqrt[3]{26 + 1} = 3$ **31. a.** $\sqrt{0.000001} = 0.001$ **b.** $\sqrt[3]{0.000001} = 0.01$
33. 25 **35.** 4 **37.** 11 **39.** 9 **41.** $x^{5/6}$ **43.** $x^{1/6}$ **45.** $z^{3/14}$ **47.** w **49.** v^2w^3

51. $\frac{64}{v^{3/5}}$ **53.** $\frac{27}{8n}$ **55.** $\frac{9x}{5y}$ **57.** $x - 1$ **59.** $2y - 3$ **61.** $6w^2 - 15w - 27$

63. $a - 9$ **65.** $b^{6/5} - c^{10/3}$ **67.** $b^{6/5} - 2b^{3/5}c^{5/3} + c^{10/3}$ **69.** $\frac{1}{x} + 2 + x$

71. 2.943 **73.** 2.097 **75.** 2.668 **77.** 3; 3.107 **79.** $S(1,000) = 75,000$; a
square box beam with a volume of 1,000 cm³ has a load strength of
75,000 N. **81.** 4 **83.** 35,864,814 m

Group Discussion Questions
85. a. $x^{5m/6}$ **b.** $8x^{m/6}$ **c.** x^{5m} **87. a.** C **b.** A **c.** B

Cumulative Review
1. $(5, \infty)$ **2.** $x < 4$ **3.** $-2 < x < 4$ **4.** $|x| < 4$ **5.** $|x| > 3$

Review Exercises for Chapter 10
1. 9 **2.** -3 **3.** 3 **4.** -2 **5.** 7 **6.** 1 **7.** $\frac{4}{9}$ **8.** 0.5

9.

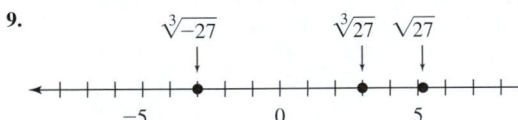

($\sqrt[3]{-27}$ is not real and cannot be located on a number line.)
10. 8.37 **11.** 4.12 **12.** -2.34 **13.** $20\sqrt{3}$ **14.** $4\sqrt{3}$ **15.** $4\sqrt{2} - 3\sqrt{3}$
16. $4\sqrt[3]{5}$ **17.** 3 **18.** $12\sqrt[3]{25}$ **19.** $2\sqrt{5}$ **20.** $2\sqrt[3]{3}$ **21.** $4\sqrt{2}$ **22.** $14\sqrt{3x}$

23. $4\sqrt{2v}$ **24.** $-\sqrt[3]{v}$ **25.** 5 **26.** $\frac{3x}{y^2}$ **27.** $\frac{\sqrt[3]{25}}{3}$ **28.** $-29\sqrt{2}$ **29.** -369

30. -67 **31.** $90\sqrt{7}$ **32.** $\frac{\sqrt{7}}{5}$ **33.** $\frac{5\sqrt{7}}{7}$ **34.** $5 - 2\sqrt{6}$ **35.** $2\sqrt{6} - 5$

36. $3\sqrt{7} + 3\sqrt{3}$ **37.** 180 **38.** 3 **39.** $\sqrt{6}$ **40.** 3 **41.** $4x^3\sqrt{2x}$ **42.** $2x^2\sqrt[3]{4x}$

43. $2x\sqrt[4]{2x^3}$ **44.** $2xy^5\sqrt{3xy}$ **45.** $x^{1/2}$ **46.** $x^{2/3}$ **47.** $x^{-1/3}$ **48.** $\sqrt[4]{x}$
49. $\frac{1}{\sqrt[4]{x}}$ **50.** $\sqrt[4]{x^3}$ **51.** $2\sqrt{w} = 2w^{1/2}$ **52.** $\sqrt{2w} = (2w)^{1/2}$

53. $\sqrt[3]{x + 4} = (x + 4)^{1/3}$ **54.** 7 **55.** -7 **56.** $\frac{1}{7}$ **57.** $\frac{9}{25}$ **58.** $\frac{9}{25}$ **59.** $\frac{25}{9}$

60. 1,000 **61.** 100 **62.** 10 **63.** 16 **64.** 4 **65.** 2 **66.** $\frac{4y^6}{x^4}$ **67.** $4x^{2/3}y^{3/5}$

68. $x^2y^2z^2$ **69.** $6x - 10$ **70.** $9x - 25$ **71.** $x + y$ **72. a.** C **b.** G **c.** D **d.** E
e. B **f.** A **g.** F **73. a.** $\mathbb{R}$ **b.** $\mathbb{R}$ **c.** $[4, \infty)$ **d.** $(-\infty, 4]$ **e.** $\mathbb{R}$ **f.** $\mathbb{R}$ **g.** $\mathbb{R}$
74. a. E **b.** D **c.** A **d.** C **e.** B **75.** $\mathbb{R}$ **76.** $\mathbb{R}$ **77.** $[5, \infty)$ **78.** $\mathbb{R}$ **79.** $\mathbb{R} \sim \{5\}$
80. $\mathbb{R} \sim \{-3, 3\}$ **81.** $x = 2$ **82.** $x = 2$ **83.** $x = 21$ **84.** $x = 4$ **85.** $v = 5$
86. $w = -8$ **87.** x-intercept: (24, 0); y-intercept: (0, -2) **88.** 13
89. $4\sqrt{5} + 2\sqrt{10}$; a right triangle **90.** 8 cm, 15 cm, 17 cm **91.** Locate the
two poles at (8, 7) and (8, 1). **92.** 16.0 ft **93.** 14.7 cm **94.** 87.9 cm
95. a. The overhead cost is $200. **b.** The cost of producing 512 boxes is
$1,480. **c.** For a cost of $1,480 the factory can produce 512 boxes.

Mastery Test for Chapter 10
1. a. $\sqrt[3]{v}$ **b.** $\sqrt[4]{w}$ **2. a.** 3 **b.** -4 **c.** 2 **d.** -10 **3. a.** D **b.** C **c.** C **d.** A
4. a. $[3, \infty)$ **b.** $\mathbb{R}$ **c.** $\mathbb{R} \sim \{3\}$ **d.** $\mathbb{R}$ **5. a.** $5\sqrt{7}$ **b.** $3\sqrt{2} + 4\sqrt{5}$
c. $-\sqrt[3]{7}$ **d.** $8\sqrt{2x}$ **6. a.** $2\sqrt{10}$ **b.** $2\sqrt[3]{3}$ **c.** $-9\sqrt{7}$ **d.** $2xy\sqrt[3]{5x^2y}$ **7. a.** 30
b. $30 - 20\sqrt{2}$ **c.** $-4x^2$ **d.** $12x - 5$ **8. a.** 3 **b.** $3\sqrt{6}$ **c.** $8\sqrt{7} + 8\sqrt{2}$
d. $-8 - 4\sqrt{7}$ **9. a.** $x = 124$ **b.** $x = -5$ **c.** no real solution **d.** $x = 8$

10. a. 8 **b.** 12 **c.** 10 **d.** $\sqrt{29}$ **11. a.** 9 **b.** -10 **c.** 27 **d.** $\frac{1}{27}$ **e.** 2 **f.** 6

12. a. $4v^4w^6$ **b.** $10xy^2$ **c.** $\frac{3m}{2n}$ **d.** $\frac{x^4}{4y^6}$

Group Project for Chapter 10
I. 1.

Length x (cm)	Time y (seconds)
200	2.8
185	2.7
170	2.6
155	2.5
140	2.4
125	2.2
110	2.1
95	2.0
80	1.8
65	1.6
50	1.4

2.

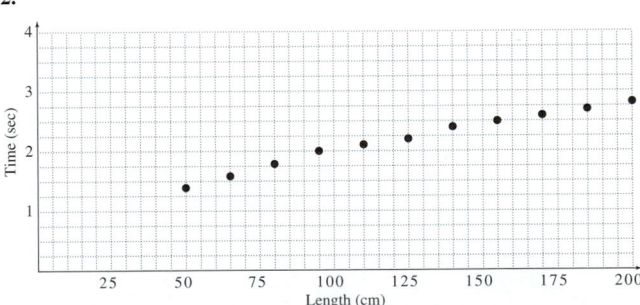

II. 1. $T(x) = 0.201x^{0.499}$
2. a. $T(165) = 2.57$; a pendulum with a length of 165 cm will have a
period of approximately 2.57 seconds. **b.** $T(70) = 1.67$; a pendulum with
a length of 70 cm will have a period of approximately 1.67 seconds.
c. $T(x) = 4$ when $x = 401$; the period of a pendulum approximately
401 cm long will be 4 seconds.

Chapter 11

Using the Language and Symbolism of Mathematics 11.1
1. arithmetic **2.** ratio **3.** growth **4.** decay **5.** base; exponent
6. growth **7.** decay **8.** x **9.** 2.718 **10.** x; y **11.** a; b

Quick Review 11.1
1. sub **2.** $(-1, 1)$ **3.** $(1, 2)$ **4.** $(-1, 2)$ **5.** $(1, 3]$

Exercises 11.1
1. a. geometric; $r = 5$ **b.** geometric; $r = -5$ **3. a.** geometric; $r = \dfrac{1}{5}$

b. geometric; $r = -\dfrac{1}{5}$ **5. a.** geometric; $r = \dfrac{2}{3}$ **b.** not geometric

7. a. not geometric **b.** geometric; $r = \sqrt{2}$ **9. a.** 10, 20, 40, 80, 160

b. 10, -20, 40, -80, 160 **11. a.** 4 **b.** 16 **c.** 1 **d.** $\dfrac{1}{16}$ **e.** 2

13.

x	$f(x)$
-2	1/16
-1	1/4
0	1
1	4
2	16

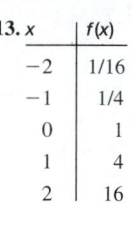

15.

x	$f(x)$
-2	9/4
-1	3/2
0	1
1	2/3
2	4/9

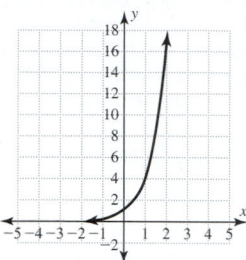

17. $m = 1$ **19.** $x = 2$ **21.** $v = 2$ **23.** $w = 3$ **25.** $m = -1$ **27.** $n = 0$
29. $x = \dfrac{1}{2}$ **31.** $x = \dfrac{1}{2}$ **33.** $x = -2$ **35.** $w = -3$ **37.** $n = \dfrac{26}{15}$ **39.** $v = \dfrac{1}{2}$

41. $x = 4$ **43.** $y = -3$ **45.** $x = -5$ **47.** $x = \dfrac{1}{6}$ **49.** $x = -2$ **51. a.** 16.919

b. 0.202 **c.** 23.141 **53.** 16; 15.21 **55.** $\dfrac{1}{4} = 0.25$; 0.26 **57. a.** \$110, \$120,

\$130, \$140 **b.** \$90, \$80, \$70, \$60 **c.** \$110, \$121, \$133.10, \$146.41
d. \$90, \$81, \$72.90, \$65.61 **59.** 18 m, 9 m, 4.5 m, 2.25 m
61. $A(t) = 1{,}000(1.06)^t$; \$1,503.63 **63.** 78.0 mg **65. a.** 6, 11, 16, 21, 26
b. 6, 30, 150, 750, 3,750 **67.** B **69.** D **71.** E **73.** $x = 32$ **75.** $x = 6$

77. $z = \dfrac{1}{27}$

Group Discussion Questions
79. a. This is the horizontal line $f(x) = 0$ with a hole at $x = 0$. **b.** This is
the horizontal line $f(x) = 1$, a constant function representing neither
growth or decay.
81. Option 1:

Year	Salary ($)	Raise ($)	Raise (%)
1	20,000		
2	23,000	3,000	15.0
3	26,000	3,000	13.0
4	29,000	3,000	11.5
5	32,000	3,000	10.3
6	35,000	3,000	9.4
7	38,000	3,000	8.6
8	41,000	3,000	7.9
9	44,000	3,000	7.3
10	47,000	3,000	6.8

Option 2:

Year	Salary ($)	Raise ($)	Raise (%)
1	20,000.00		
2	22,000.00	2,000.00	10
3	24,200.00	2,200.00	10
4	26,620.00	2,420.00	10
5	29,282.00	2,662.00	10
6	32,210.20	2,928.20	10
7	35,431.22	3,221.02	10
8	38,974.34	3,543.12	10
9	42,871.78	3,897.43	10
10	47,158.95	4,287.18	10

Option 1 is arithmetic growth at \$3,000/yr. Option 2 is geometric growth
at 10%/yr, and the raise per year will exceed \$3,000 on the seventh year.
The total dollars received for all years is higher under option 1 until the
fourteenth year.

Cumulative Review
1. $(3x - 5)(2x - 7)$ **2.** $6x^2 - 31x + 35$ **3.** $m = 3$

4. $m = -\dfrac{1}{3}$ **5.** $[-1, 3)$

Using the Language and Symbolism of Mathematics 11.2
1. an input value 4; an output value 9 **2.** different **3.** one-to-one
4. f^{-1} **5.** $x = y$

Quick Review 11.2
1. function **2.** not a function **3.** function

4.

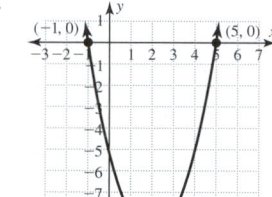

5.

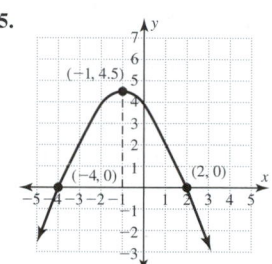

Exercises 11.2
1. $\{(4, 1), (11, 3), (2, 8)\}$ **3.** $\{(2, -3), (2, -1), (2, 0), (2, 2)\}$
5. $\{(b, a), (d, c)\}$ **7.** $\{(7, 4), (9, e), (-2, -3)\}$
9. $\{(0, 9), (3, 4), (-2, 6), (5, -4)\}$
11. $f = \{(-3, 2), (-1, 1), (1, 0), (3, -1)\}$;
$f^{-1} = \{(2, -3), (1, -1), (0, 1), (-1, 3)\}$

13. $f^{-1}(x)$ increases x by 2; $f^{-1}(x) = x + 2$ **15.** $f^{-1}(x)$ takes one-fourth
of x; $f^{-1}(x) = \dfrac{x}{4}$ **17.** $f^{-1}(x)$ increases x by 3 and then divides this

quantity by 2; $f^{-1}(x) = \dfrac{x + 3}{2}$ **19.** $f^{-1}(x) = \dfrac{x - 2}{5}$

21. $f^{-1}(x) = 3(x + 7)$ **23.** $f^{-1}(x) = (x - 2)^3$ **25.** $f^{-1}(x) = -x$
27. one-to-one **29.** not one-to-one **31.** one-to-one **33.** not one-to-one
35. not one-to-one **37.** not one-to-one **39.** not one-to-one
41.

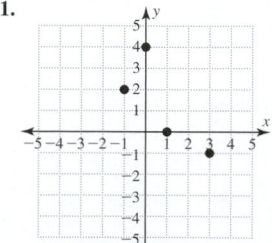

43.

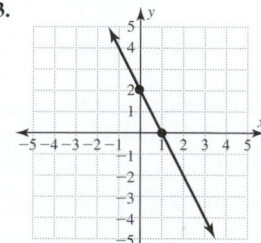

45.

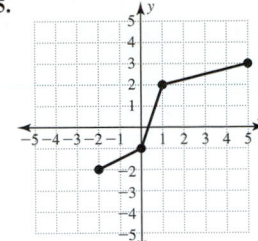

47.

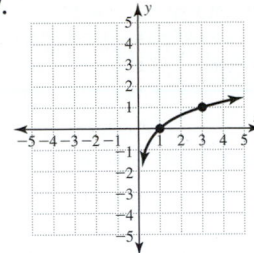

49.

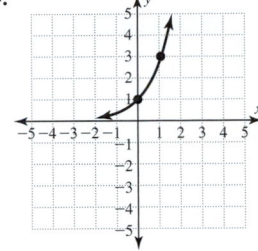

51. a. 17 **b.** 5 **53. a.** 5 **b.** 2 **55. a.** 1 **b.** 1 **57.** domain of f: $\{0, 1, 2, 3, 4\}$; range of f: $\{-3, -2, -1, 0, 1\}$; domain of f^{-1}: $\{-3, -2, -1, 0, 1\}$; range of f^{-1}: $\{0, 1, 2, 3, 4\}$ **59.** domain of f: $\{-2, 0, 2\}$; range of f: $\{0, 3, 1\}$; domain of f^{-1}: $\{0, 3, 1\}$; range of f^{-1}: $\{-2, 0, 2\}$

61.

European Size	U.S. Size
39	7
40.5	8
42	9
43	10
44	11

The inverse function is useful in converting a European shoe size to the equivalent U.S. size.

63.

Monthly Payment ($)	Loan Amount ($)
120.89	5,000
241.79	10,000
362.68	15,000
483.58	20,000
604.47	25,000

The inverse function is useful in determining the cost of the car one can afford on a given monthly budget.

65.

Celsius	Fahrenheit
$-17.8°$	$0°$
$-6.7°$	$20°$
$4.4°$	$40°$
$15.6°$	$60°$
$26.7°$	$80°$
$37.8°$	$100°$

The inverse function is useful in converting Celsius to Fahrenheit.
67. a. the number of units produced **b.** the cost of producing x units
c. $C^{-1}(x) = \dfrac{x - 350}{12}$ **d.** the cost **e.** the number of units produced
f. $C(100) = \$1,550$ **g.** $C^{-1}(1934) = 132$ units

Group Discussion Questions
69. The resulting graph is the graph of the inverse of $y = f(x)$. The two graphs are symmetric about the line $y = x$.

71. a, b.

$[-5, 13, 2]$ by $[-2, 10, 2]$
c. The graphs of these two functions indicate that these functions are inverses of each other.

Cumulative Review
1. $-5x^4 + 4x^3y + 2x^2y^2 - 3xy^3$ **2.** 4 **3.** 0 **4.** $a > 0$ **5.** $a < 0$

Using the Language and Symbolism of Mathematics 11.3
1. $\log_b x$ **2.** b; x; y **3.** undefined **4.** 0 **5.** 1 **6.** -1 **7.** x **8.** y

Quick Review 11.3
1. 1 **2.** undefined **3.** $\dfrac{3}{2}$ **4.** $x = \dfrac{9}{2}$ **5.** $x = 3, x = -3$

Exercises 11.3
1. The log base 5 of 125 is 3; $5^3 = 125$ **3.** The log base 3 of $\sqrt{3}$ is $\dfrac{1}{2}$; $3^{1/2} = \sqrt{3}$ **5.** The log base 5 of $\dfrac{1}{5}$ is -1; $5^{-1} = \dfrac{1}{5}$ **7.** $\log_{16} 4 = \dfrac{1}{2}$; $16^{1/2} = 4$ **9.** $\log_8 4 = \dfrac{2}{3}$; the log base 8 of 4 is $\dfrac{2}{3}$ **11.** $\log_3 \dfrac{1}{9} = -2$; the log base 3 of $\dfrac{1}{9}$ is -2 **13.** $\log_m n = p$; the log base m of n is p **15.** $\log_n m = k$; $n^k = m$ **17.** 2 **19.** -3 **21.** $\dfrac{1}{3}$ **23.** 5 **25.** -1 **27.** 0 **29.** -1 **31.** 4 **33.** 6 **35.** 3 **37.** $\dfrac{1}{4}$ **39.** $\dfrac{3}{2}$ **41.** 1 **43.** 2.7 **45.** -2 **47. a.** defined **b.** undefined **49. a.** undefined **b.** defined **51.** $x = 2$ **53.** $x = 6$ **55.** $x = \dfrac{1}{6}$ **57.** $x = \sqrt{6}$ **59.** $x = 2$ **61.** $x = 5$ **63.** $x = \dfrac{1}{5}$ **65.** $x = -\dfrac{1}{2}$ **67.** $x = 4$

69.

x	$f(x)$
1/4	-1
1	0
4	1
16	2

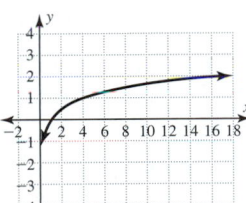

71.

x	$f(x)$
1/4	1
1	0
4	-1
16	-2

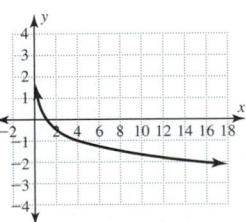

73. B **75.** C **77.** domain: $(0, \infty)$; range: $\mathbb{R}$

Group Discussion Questions
79. a. exponential growth **b.** logarithmic growth **c.** exponential growth **d.** logarithmic growth

Cumulative Review
1. $5(m + 3)(m - 3)$ **2.** $7x(3x - 2)$ **3.** $(2v + w)(2v - w)$ **4.** $(2v + w)^2$ **5.** $(a - b)(a + b + 5)$

Using the Language and Symbolism of Mathematics 11.4
1. 2.718 **2.** $\log_{10} x$ **3.** $\log_e x$ **4.** $5x - 3$

Quick Review 11.4

1. 0 **2.** undefined **3.** $\dfrac{1}{10,000}$ **4.** 7.5345247×10^7 **5.** 8.7×10^{-5}

Exercises 11.4

1. 1 **3.** -4 **5.** -8 **7.** $\dfrac{1}{4}$ **9.** 0 **11.** 6 **13.** -3 **15.** $\dfrac{1}{2}$ **17.** undefined

19. undefined **21.** -1 **23.** 1.672 **25.** -5.684 **27.** 2.943 **29.** 6.653
31. -10.009 **33.** 2.449 **35.** 0.497 **37.** 0.835 **39.** 1.012 **41.** 1.730
43. 2.653 **45.** 1.755 **47. a.** -2.079 **b.** $\ln(-8)$ is undefined. The input of a logarithmic function cannot be negative. **49. a.** 0 **b.** $\log(23 \cdot 0)$ is undefined. The input of a logarithmic function cannot be zero. **51. a.** -0.111
b. $\ln(17 - 19)$ is undefined. The input of a logarithmic function cannot be negative. **53.** 2 **55.** 81 **57. a.** 0.393 **b.** 2.472 **59. a.** -2.364 **b.** 0.094
61. $x = 299.916$ **63.** $x = 0.500$ **65.** $x = 7.996$ **67.** $x = 0.050$

Group Discussion Questions
69. a. 33, 19, 8, 81 **b.** $b^{\log_b x} = x$ for $x > 0$ **c.** If $\log_b x$ is the exponent placed on b to obtain x, then $b^{\log_b x} = x$. **71. a.** $\log\dfrac{15}{5} \approx 0.4771$;

$\log 15 - \log 5 \approx 0.4771$; $\ln\dfrac{24}{6} \approx 1.3863$; $\ln 24 - \ln 6 \approx 1.3863$

b. $\log_b \dfrac{M}{N} = \log_b M - \log_b N$ for $M, N > 0$ **c.** The log base b of a quotient is the difference of the logs base b.

Cumulative Review
1. $\dfrac{xy^2}{2}$ **2.** $\dfrac{1}{2x-3}$ **3.** $\dfrac{2}{7}$ **4.** $\dfrac{5x-1}{3x-4}$ **5.** $\dfrac{x-4}{x-5}$

Using the Language and Symbolism of Mathematics 11.5
1. b^{x+y} **2.** $\log_b x + \log_b y$ **3.** b^{x-y} **4.** $\log_b x - \log_b y$ **5.** b^{xp} **6.** $p\log_b x$
7. $\dfrac{\log_b x}{\log_b a}$ **8.** $b^{x\log_b a}$ **9.** 7^x **10.** x

Quick Review 11.5

1. $15x^7 y^9$ **2.** $\dfrac{3x}{y^5}$ **3.** $25x^6 y^{14}$ **4.** $xy^{1/2}$ **5.** $x^{1/3} y^{2/3}$

Exercises 11.5

1. $\log a + \log b + \log c$ **3.** $\ln x - \ln 11$ **5.** $5\log y$ **7.** $2\ln x + 3\ln y$
9. $\ln(x+2) + \ln(3x-1)$ **11.** $\ln(2x+3) - \ln(x+7)$
13. $\dfrac{1}{2}\log(4x+7)$ **15.** $\dfrac{1}{2}\ln(x+4) - 2\ln(y+5)$

17. $\dfrac{1}{2}[\log x + \log y - \log(z-8)]$

19. $2\log x + 3\log(2y+3) - 4\log z$ **21.** $\ln x + 2\ln y - 3\ln z$

23. $\ln xy$ **25.** $\log\dfrac{w}{z}$ **27.** $\ln w^3$ **29.** $\log x^2 y^5$ **31.** $\ln\dfrac{x^3 y^7}{z}$ **33.** $\ln\sqrt{x}$

35. $\log\left(\dfrac{\sqrt{x+1}}{2x+3}\right)$ **37.** $\ln\sqrt[3]{(2x+7)(7x+1)}$ **39.** $\log_5 x^2\sqrt[3]{y^2}$

41. 37 **43.** 0.045 **45.** x **47.** $\dfrac{\ln 37.1}{\ln 3}$ **49.** $\dfrac{\log 7.9}{\log 13}$ **51.** -1.2860
53. -1.8934 **55.** 3; 2.989 **57.** 32; 31.640 **59.** 1.1833 **61.** 1.6542
63. 0.1187 **65.** 1.3562 **67.** -0.4709 **69.** 0.6438 **71.** C **73.** B **75.** F
77. E **79.** A **81.** $f(x) = e^{x\ln 5}$ **83.** $f(x) = e^{-x\ln 5}$

85. $\ln\dfrac{1}{2} = \ln 1 - \ln 2 = 0 - \ln 2 = -\ln 2$

Group Discussion Questions
87. a. Answers can vary; $f(x) = 3^x$ or $f(x) = e^{x\ln 3} \approx e^{1.0986x}$ **b.** Answers can vary; $f(x) = \left(\dfrac{1}{3}\right)^x$ or $f(x) = e^{x\ln(1/3)} = e^{-x\ln 3} \approx e^{-1.0986x}$

c. In the form $f(x) = e^{kx}$, the function represents growth for $k > 0$ and decay for $k < 0$. **d.** For exponential growth in $f(x) = b^x$, $\ln b > 0$ for $b > 1$. For exponential decay in $f(x) = b^x$, $\ln b < 0$ for $0 < b < 1$.

89. Let $x = \log_b N$ so $b^x = N$:
$$\begin{aligned}\log_b N^p &= \log_b(b^x)^p\\ &= \log_b b^{xp}\\ &= xp\\ &= p\log_b N\end{aligned}$$

Cumulative Review
1. $x = \pm 5$ **2.** $x = -1, x = 9$ **3.** $x = 4, x = -\dfrac{7}{2}$ **4.** $x = -1, x = \dfrac{3}{2}$

5. $x = -2, x = -\dfrac{3}{2}$

Using the Language and Symbolism of Mathematics 11.6
1. exponential **2.** logarithmic **3.** x; y **4.** extraneous

Quick Review 11.6
1. 0.982 **2.** 0.336 **3.** 29.824 **4.** $(5, -2)$ **5.** $(1, 2.5)$

Exercises 11.6

1. $x = 2$ **3.** $x = \dfrac{1}{125}$ **5.** $w = 8$ **7.** $x = -2, x = 2$ **9.** $y = 15$ **11.** $n = 35$

13. $m = 16$ **15.** $w = -\dfrac{1}{2}$ **17.** $x = -2$ **19.** $t = 2$ **21.** $v = 10$

23. no solution **25.** $y = -7$ **27.** no solution **29.** $x = 2.5$ **31.** $x = 2$

33. $x = 2$ **35.** $x = -1, x = 1$ **37.** $v = \dfrac{\ln 15}{\ln 4} \approx 1.953$

39. $w = 7 - \dfrac{\ln 22}{\ln 3} \approx 4.186$ **41.** $t = \dfrac{\ln 9.2}{\ln 11.3 - 2\ln 9.2} \approx -1.102$

43. $z = \dfrac{\ln 5.3}{2\ln 5.3 + 2\ln 7.6} \approx 0.226$ **45.** $x = \dfrac{\ln 78.9}{3} \approx 1.456$

47. $y = \dfrac{\log 51.3 - 1}{2} \approx 0.355$ **49.** $v = \pm\sqrt{\dfrac{\ln 0.68}{\ln 0.83}} \approx \pm 1.439$

51. $x = \pm\sqrt{\ln\left(\dfrac{689.7}{3.7}\right) - 1} \approx \pm 2.056$ **53.** $x = \dfrac{10^{0.83452} + 17}{5} \approx 4.766$

55. $y = e^e \approx 15.154$ **57.** $v = 3 + \sqrt{2} \approx 4.414$ **59.** no solution
61. $x = 1, x = e^2 \approx 7.389$ **63.** $x = -0.005, x = 2.500$ **65.** $x = 2.998$

67. $10^{-\log x} = 10^{\log x^{-1}} = x^{-1} = \dfrac{1}{x}$

69. $e^{-x\ln 3} = e^{\ln 3^{-x}} = 3^{-x} = \dfrac{1}{3^x} = \left(\dfrac{1}{3}\right)^x$

71. $\log 60^x - \log 6^x = \log\dfrac{60^x}{6^x} = \log\left(\dfrac{60}{6}\right)^x = \log 10^x = x$

73. $\ln\left(\dfrac{4}{5}\right)^x + \ln\left(\dfrac{5}{3}\right)^x + \ln\left(\dfrac{3}{4}\right)^x = x\ln\dfrac{4}{5} + x\ln\dfrac{5}{3} + x\ln\dfrac{3}{4} =$
$x\left(\ln\dfrac{4}{5} + \ln\dfrac{5}{3} + \ln\dfrac{3}{4}\right) = x\ln\left(\dfrac{4}{5}\right)\left(\dfrac{5}{3}\right)\left(\dfrac{3}{4}\right) = x\ln 1 = x(0) = 0$

75. It will take 6.8 years for the value to depreciate to $10,000.

Group Discussion Questions
77. The argument of a logarithmic expression cannot be a negative value, but the variables within the argument can be negative.
79. a. $\ln x = \ln(3x + 4)$; answers can vary
b. $\ln(x + 3) + \ln(x - 4) = \ln(2x - 2)$; answers can vary
c. $\ln(6 - x) + \ln(7 - x) = \ln(52 - 10x)$; answers can vary

Cumulative Review
1. $(0, 2)$ **2.** $(-6, 0), (2, 0)$ **3.** $y = 4$ **4.** $x = -2$
5. $(-\infty, -2)$

Using the Language and Symbolism of Mathematics 11.7

1. periodic **2.** continuous **3.** $P\left(1 + \dfrac{r}{n}\right)^{nt}$ **4.** Pe^{rt} **5.** earthquakes

6. sound

Quick Review 11.7
1. $\log 10,000 = 4$ **2.** $10^{-3} = 0.001$ **3.** $e^{4.357} \approx 78$ **4.** 4.41
5. -3.59

Exercises 11.7

1. \$234.08 **3.** 9 years **5.** 18.4 years **7.** It will take 5.2 years for the investment to increase in value by 50%. **9.** 6.9% **11.** 9.9 years
13. 24,000 years **15. a.** $B(0) = 4$; there are 4 units of bacteria present initially. **b.** $B(5) \approx 13.3$; after 5 days there are approximately 13.3 units of bacteria present. **c.** $B(10) \approx 44$; after 10 days there are approximately 44 units of bacteria present. **d.** $B(13.4) \approx 100$; the bacteria level reaches 100 units after approximately 13 days. **17.** The power supply will last 9,200 days. **19.** There will be approximately 10,000 whales left in 10 years; it will take approximately 27.5 years for the whale population to reach 5,000. **21.** The annual rate of increase is 13.5%. **23.** The earthquake measured 7.8 on the Richter scale. **25.** The 1906 San Francisco earthquake was 7 times as intense as the 1999 Izmit earthquake. **27.** The whisper measures 24.9 dB. **29.** The intensity of this noise is 3.2×10^{-8} W/cm^2.
31. The pH of grape juice is 3.96. **33.** The H$^+$ concentration of the shampoo is 7.41×10^{-10} mol/L. **35.** A monthly payment of \$411.60 will pay off the loan. **37.** 36 payments are required to pay off the loan. **39.** The victim had been dead for approximately 2.9 hours.

41. a.
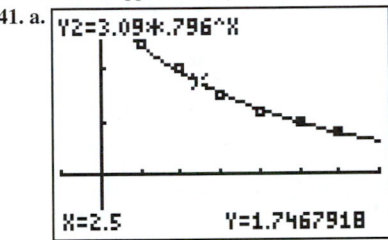
[−1, 7, 1] by [−1, 3, 1]
b. $y = 3.09(0.796^x)$ **c.** exponential decay **d.** $y \approx 1.7$

43. a.
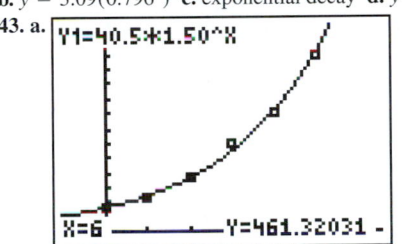
[−1, 6.5, 1] by [−5, 350, 25]
b. $f(x) = 40.5(1.50^x)$ **c.** In 2006, the storage requirements are estimated to be 460 GB.

Group Discussion Questions
45. Logarithmic growth, which has extremely slow growth in the later years, is a better model.

Cumulative Review
1. $20 - 15i$ **2.** $1 + 5i$ **3.** $26 - 7i$ **4.** $21 + 20i$
5. $\dfrac{14}{25} + \dfrac{23}{25}i$

Review Exercises for Chapter 11
1. a. arithmetic; $d = 2$ **b.** geometric; $r = 2$ **c.** neither **2. a.** geometric; $r = 1$; arithmetic; $d = 0$ **b.** arithmetic; $d = -4$ **c.** geometric; $r = -1$
3. 100, 90, 81, 72.9, 65.61, and 59.049 gal **4. a.** 1 **b.** $\dfrac{1}{9}$ **c.** 81 **d.** 3
5. a. 0.318 **b.** 9.870 **c.** 36.462 **d.** 22.459 **6.** B **7.** C **8.** A
9. a.
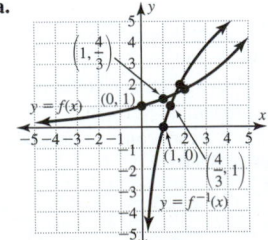
b.
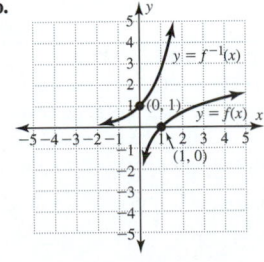

Mastery Test for Chapter 11

1. a. geometric; $r = 5$ **b.** not geometric **c.** geometric; $r = \dfrac{1}{2}$
d. geometric; $r = 0.1$

10. f^{-1} adds 2 to x and then divides the quantity by 3; $f^{-1}(x) = \dfrac{x+2}{3}$

11. f^{-1} subtracts 6 from x and then multiplies the quantity by 4; $f^{-1}(x) = 4(x - 6)$ **12.** $f^{-1} = \{(-5, -4), (-3, -3), (3, 0), (7, 2), (9, 3)\}$ **13.** $f^{-1} = \{(0, -2), (2, -1), (4, 0), (6, 1), (8, 2), (10, 3)\}$
14. $f^{-1}(x) = 3(x + 4)$ **15.** $f^{-1}(x) = \log_3 x$ **16. a.** x represents the number of pizzas. **b.** $C^{-1}(x) = \dfrac{x - 250}{2}$ **c.** the cost of producing the pizzas
d. the number of pizzas **e.** \$450 **f.** 74 pizzas
17.   **18.**
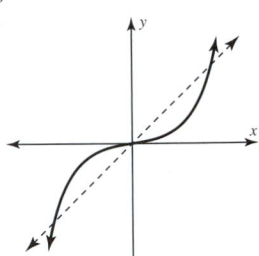

19. a. $6^{1/2} = \sqrt{6}$ **b.** $17^0 = 1$ **c.** $8^{-2} = \dfrac{1}{64}$ **20. a.** $a = b^c$ **b.** $a = e^c$
c. $c = 10^d$ **21. a.** $\log_7 343 = 3$ **b.** $\log_{19} \sqrt[3]{19} = \dfrac{1}{3}$ **c.** $\log_{4/7} \dfrac{49}{16} = -2$
22. a. $\ln \dfrac{1}{e} = -1$ **b.** $\log 0.0001 = -4$ **c.** $\log_8 y = x$ **23. a.** 2 **b.** $\dfrac{1}{2}$ **c.** −1
24. a. 0 **b.** 0.23 **c.** $-\dfrac{1}{3}$ **25. a.** 2,514.929 **b.** 67,608,297.539 **26. a.** 0.091
b. 0.004 **27. a.** 2.055 **b.** 4.733 **28. a.** −3.092 **b.** −7.118 **29. a.** 1.065
b. 0.191 **30. a.** 2.398 **b.** 3.401 **31.** 2.010 **32.** 2.989 **33.** $x = -2$
34. $x = \dfrac{2}{3}$ **35.** $y = \dfrac{2}{3}$ **36.** $y = -\dfrac{1}{2}$ **37.** $w = 1$ **38.** $v = \dfrac{1}{4}$
39. $x = 2, x = -2$ **40.** $x = -\dfrac{1}{2}, x = 1$ **41.** $z = 2$ **42.** $x = \dfrac{3}{2}$ **43.** $x = 4$
44. $x = \sqrt{13}$ **45.** $w = \dfrac{1}{9}$ **46.** no solution **47.** no solution **48.** $t = 13$
49. $t = \dfrac{1}{2}$ **50.** $x = 11$ **51.** $x = 17$ **52.** $x = 7$ **53.** no solution **54.** $n = 3$
55. $y = 140$ **56.** $x = 33$ **57.** $x = 10, x = -10$ **58.** $w = 1$ **59.** $w = -2$
60. $v = -\dfrac{1}{2}$ **61.** $v = 3$ **62.** $y = \dfrac{27}{2}$ **63.** $v = -3, v = 0$ **64.** $x = 3$
65. $x = 10$ **66.** $x = 4$ **67.** $3 \log x + 5 \log y$ **68.** $\ln(7x - 9) - \ln(2x + 3)$
69. $\dfrac{1}{2} \ln(2x + 1) - \ln(5x + 9)$ **70.** $\dfrac{1}{2}(2 \log x + 3 \log y - \log z)$
71. $\ln x^2 y^3$ **72.** $\ln \dfrac{x^5}{y^4}$ **73.** $\ln(x + 1)$ for $x > 4$ **74.** $\ln \sqrt{\dfrac{x}{y}}$
75. $x = 2.146$ **76.** $w = -0.293$ **77.** $y = e^{x \ln 5}$
78. $\log 50^x + \log 6^x - \log 3^x = x \log 50 + x \log 6 - x \log 3 =$
$x(\log 50 + \log 6 - \log 3) = x\left(\log \dfrac{50 \cdot 6}{3}\right) = x \log 100 = x(2) = 2x$
79. $1,000^{\log x} = (10^3)^{\log x} = 10^{3 \log x} = 10^{\log x^3} = x^3$ **80.** 3; 2.996
81. 1; 1.099 **82.** 3; 3.096 **83.** 3; 2.975 **84. a.** 13 **b.** 3 **c.** $\dfrac{1}{13}$ **85.** D
86. E **87.** C **88.** B **89.** A **90. a.** \$2,938.66 **b.** \$2,979.69 **c.** \$2,983.65
91. The power supply should last 10,200 days. **92.** The savings bond will double in value in 9.6 years. **93.** An interest rate of 8.7% would cause the investment to double in value in 8 years. **94.** The earthquake measured 6.8 on the Richter scale. **95. a.** $B(0) = 3$; the initial amount of bacteria is 3 units. **b.** $B(5) \approx 12.8$; after 5 days there are approximately 12.8 units of bacteria. **c.** $B(10) \approx 54.5$; after 10 days there are approximately 54.5 units of bacteria. **d.** $t \approx 12.1$; the amount of bacteria reaches 100 units after approximately 12 days.

2. a.

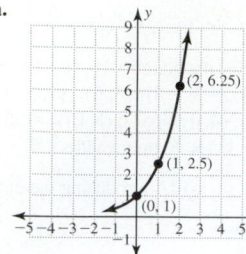

(2, 6.25)
(1, 2.5)
(0, 1)

b.

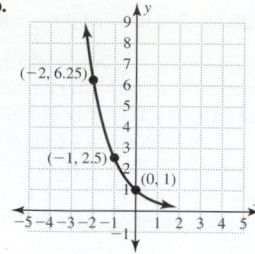

(−2, 6.25)
(−1, 2.5)
(0, 1)

c. $\dfrac{1}{16}$ **d.** 4 **3. a.** $x = 2$ **b.** $x = 0$ **c.** $x = -1$ **d.** $x = \dfrac{1}{5}$

4. a. $\{(4, -1), (9, 8), (11, -7)\}$ **b.** $f^{-1}(x) = \dfrac{x + 6}{3}$

c. $\{(-1, -2), (0, -1), (1, 0), (2, 1), (-2, 3)\}$

d. $\left\{\left(2, \dfrac{1}{2}\right), \left(3, \dfrac{1}{3}\right), \left(\dfrac{1}{6}, 6\right), (1, 1)\right\}$ **5. a.** one-to-one **b.** not one-to-one

c. not one-to-one **d.** one-to-one

6. a.

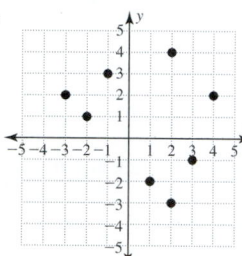

b.

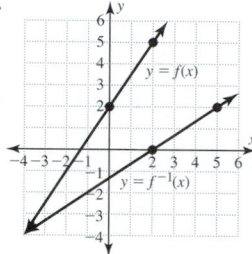

$y = f(x)$
$y = f^{-1}(x)$

7. a. $5^{2/3} = \sqrt[3]{25}$ **b.** $5^{-3} = \dfrac{1}{125}$ **c.** $b^x = y + 1$ **d.** $b^{y+1} = x$

8. a. 0 **b.** 1 **c.** −1 **d.** 17

9.

x	$\log_5 x$
0.2	−1
1.0	0
5.0	1
25.0	2

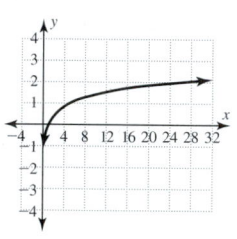

10. a. 1.2810 **b.** 2.9497 **c.** 12.6881 **d.** −10.4011 **11. a.** $4 \log x + 5 \log y$
b. $3 \ln x - \dfrac{1}{2} \ln y$ **c.** $\ln \dfrac{(7x + 9)^2}{x}$ **d.** $\log x \sqrt{x + 3}$ **12. a.** 1.730 **b.** 0.850

c. 6.644 **d.** 2.406 **13. a.** $x = -4$ **b.** $w = \dfrac{3}{2}$ **c.** $x = 16$ **d.** $y = \dfrac{1}{3}$

e. $t = -2$ **f.** $z = -3$ **g.** $y = 0.406$ **h.** $x = 1.549$ **14. a.** It will take
35.8 years for the population to reach 3000. **b.** It will take 8.4 years for
the investment to double in value.

Group Project for Chapter 11
Sample data:

Bounce x	Height y (cm)
0	250
1	200
2	165
3	130
4	100
5	80
6	60
7	50
8	40
9	30
10	25

1. $f(x) = 256(0.791^x)$ **2. a.** $H(7) \approx 49.6$ is close to the value of 50 in the
table. The height of the seventh bounce is approximately 50 cm.
b. $H(4) \approx 100$. The height of the fourth bounce is approximately 100 cm.
3. $a \approx 255$. This is approximately the original height of the golf ball.
4. $b \approx 0.791$. This is approximately the elasticity coefficient, the ratio of
the height of the bounce to the distance the ball was dropped.

A Review of Intermediate Algebra: A Cumulative Review of Chapters 6–11

1. a. $(x + 1)(x - 3)(x - 5)$ **b.** $(x + 3)(x + 5)$ **2. a.** $7x^2(2x - 5)$
b. $(2x - 3)(5x - 6)$ **3. a.** $(a + 5b)(2x + 3y)$ **b.** $(x - 3)(8x - 5)$
4. a. $(x + 9)(x - 5)$ **b.** $(x + 2)(x - 12)$ **5. a.** $(2x + 1)(3x - 10)$
b. $6x^2 - 13x - 10$ is prime. **6. a.** $(x + 10)^2$ **b.** $(x - 7y)^2$
7. a. $(x + 8)(x - 8)$ **b.** $(x - 4)(x^2 + 4x + 16)$
8. a. $(a - b)(a + b + 5)$ **b.** $(a + x - 5)(a - x + 5)$

9. a. $5(x + 4)(x - 4)$ **b.** $-3a(x - 6)(x - 1)$ **10. a.** $x = -8, x = \dfrac{3}{2}$

b. $x = -5, x = 9$ **11. a.** $6\sqrt{2}$ **b.** $10\sqrt{5}$ **12. a.** $\dfrac{2}{5}$ **b.** 2 **13. a.** $4\sqrt{3}$

b. $\dfrac{\sqrt{15}}{5}$ **14. a.** $x = \dfrac{3 \pm \sqrt{7}}{2}$; $x = 0.18, x = 2.82$ **b.** $x = \dfrac{4}{3} \pm i$

15. a. $x = 5 \pm 2\sqrt{7}$; $x = -0.29, x = 10.29$

b. $x = \dfrac{5 \pm \sqrt{73}}{4}$; $x = -0.89, x = 3.39$ **16. a.** The equation has two

distinct real solutions. **b.** The equation has two imaginary solutions that
are complex conjugates. **c.** The equation has a double real solution.

17. a. $x = \dfrac{3}{8} \pm \dfrac{i\sqrt{23}}{8}$ **b.** $x = \dfrac{9}{2}$ **c.** $x = \dfrac{-5}{6} \pm \dfrac{\sqrt{109}}{6}$

18. The width of the room is 10 ft. **19. a.** $14i$ **b.** $10i$ **20. a.** $29 + 2i$
b. $4 + 7i$ **21. a.** The graph represents a function. **b.** The graph does not
represent a function. **22. a.** domain: $\mathbb{R}$; range: $\{-3\}$ **b.** domain: $[-2, \infty)$;
range: $[-3, +\infty)$ **23. a.** $f(-2) = 4$ **b.** $f(5) = 81$ **24. a.** 2 **b.** 3
25. a. −10 **b.** −1

26. a.

x	y
1	−3
2	−1
3	1
4	3
5	5

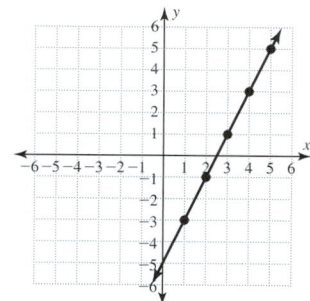

b.

x	y
1	1
2	−2
3	−5
4	−8
5	−11

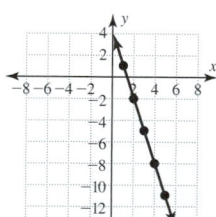

27. a. x-intercept: $(-1, 0)$; y-intercept: $(0, -3)$ **b.** x-intercept: $(10, 0)$;
y-intercept: $(0, 6)$ **c.** x-intercept: $(18, 0)$; y-intercept: $(0, -12)$
28. a. $f(x) = 0.25x + 200$

b.

x	0	50	100	150	200	250
$f(x)$	200	212.50	225	237.50	250	262.50

c. $f(75) = 218.75$; If the ice cream shop sells 75 ice cream cones in a
month, the total operating costs for the month will be $218.75.
d. $x = 240$; If the ice cream shop sells 240 ice cream cones in a month,
the total operating costs for the month will be $260.

29. a. $m = -\dfrac{4}{3}$ **b.** $m = \dfrac{2}{3}$ **30. a.** $y = -\dfrac{5}{3}x - 2$ **b.** $y = -\dfrac{3}{2}x + 1$

31. a.
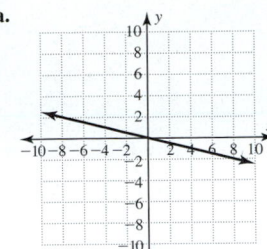
b.
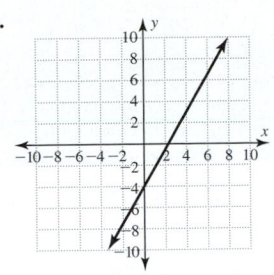

32. a. decreasing **b.** increasing

33. a.

x	$f(x) = |x - 2| - 3$
−1	0
0	−1
1	−2
2	−3
3	−2
4	−1
5	0

b.
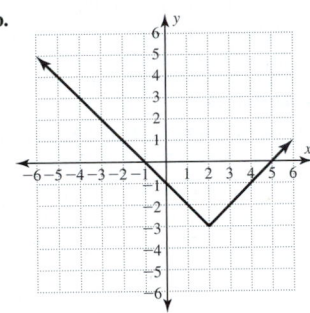

c. x-intercepts: $(-1, 0)$ and $(5, 0)$ **d.** y-intercept: $(0, -1)$ **e.** vertex: $(2, -3)$ **f.** The minimum value of y is -3. **g.** domain: $\mathbb{R}$ **h.** range: $[-3, \infty)$ **i.** interval of decrease: $(-\infty, 2)$ **j.** interval of increase: $(2, \infty)$ **k.** $f(x)$ is positive on $(-\infty, -1) \cup (5, \infty)$. **l.** $f(x)$ is negative on $(-1, 5)$.
34. The minimum value of y is -4 and occurs when $x = -2$.
35. a. downward **b.** $(1, 9)$

c.

x	y
−2	0
−1	5
0	8
1	9
2	8
3	5
4	0

d. x-intercepts: $(-2, 0)$, $(4, 0)$; y-intercept: $(0, 8)$

e.
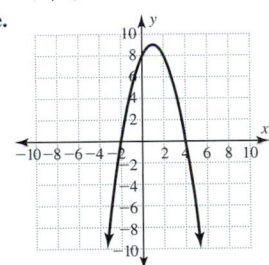

36. The maximum height of the ball is 59.3 ft and occurs 1.9 seconds after it was hit. **37.** $f(x) = 0.159x^2 - 0.383x - 1.61; f(7) = 3.5$
38. a. domain: $\mathbb{R} \sim \{-3, 1\}$ **b.** domain: $\mathbb{R}$ **39. a.** $x = 3, x = -4$
b. $x = -1$ **40. a.** $\dfrac{1}{3x - 5}$ **b.** $\dfrac{2x + 5}{x + 3}$ **41. a.** x **b.** $\dfrac{x + 3}{x - 3}$ **42. a.** 2
b. $\dfrac{2y}{3x + y}$ **43. a.** $\dfrac{3x}{2}$ **b.** $\dfrac{1}{(x - 5)(x + 2)}$ **44. a.** $\dfrac{x - 2}{4}$ **b.** $\dfrac{2}{x + 1}$

45. a. $x = 7$ **b.** $x = -\dfrac{2y + 3}{y - 1}$ **46. a.** $y = 6$ **b.** The pressure will be 15 N/cm². **47. a.** It would take 3 hours for the experienced worker to complete the roof. **b.** The rate of the slower plane is 220 miles per hour. The rate of the faster plane is 260 miles per hour. **48. a.** B **b.** E **c.** A **d.** C **e.** D **49. a.** $\mathbb{R}$ **b.** $\mathbb{R} \sim \{5\}$ **c.** $\mathbb{R}$ **d.** $[5, \infty)$ **50. a.** 8 **b.** 4 **c.** -0.5 **d.** 2
51. a. $8\sqrt{3x}$ **b.** $24\sqrt{2}$ **52. a.** $4\sqrt{3}$ **b.** $2\sqrt[3]{6}$ **53. a.** $x = 8$ **b.** $x = 64$
54. a. 13 **b.** $2\sqrt{5}$ **55. a.** -5 **b.** $x^{1/3}$ **c.** 98 **d.** 64 **56. a.** 5; 5.099 **b.** 3; 2.962
57. a. The sequence is geometric. The common ratio is $r = 5$.
b. The sequence is not geometric.

58. a.
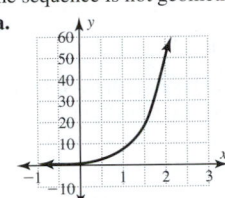
b.
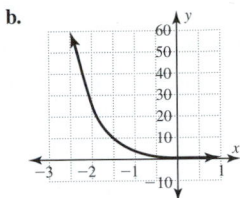

59. a. $f(-1) = \dfrac{1}{9}$ **b.** $f\left(\dfrac{1}{2}\right) = 3$

60. a.
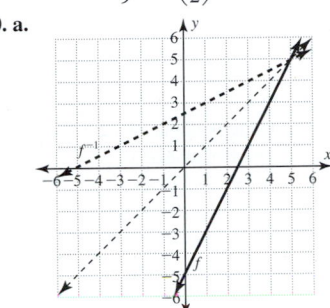

b.
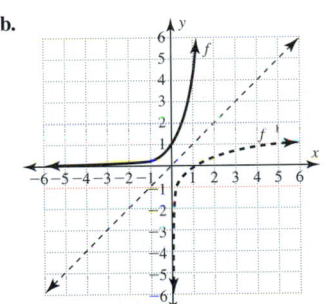

61. a. $x = 4^{3/2}$ **b.** $3^x = y - 2$ **62. a.** 0 **b.** 1 **c.** -1 **d.** 7
63. a. $5 \log x + 3 \log y$ **b.** $2 \ln x - \dfrac{1}{3} \ln y$ **64. a.** $\ln\left(\dfrac{x^2}{x - 5}\right)$
b. $\log(\sqrt{x}(x + 2))$ **65. a.** $x = -5$ **b.** $x = \dfrac{3}{2}$ **66. a.** $x = 81$ **b.** $y = \dfrac{1}{3}$
67. a. $x = 5$ **b.** no solution **68.** It will take 11.1 years for the investment to increase in value to $2,500. **69. a.** 1.398 **b.** -8.149
70. a. 1.771 **b.** 2.024 **71. a.** $f(x) = 10.1(1.77)^x$ **b.** There will be approximately 3,050,000 bacteria present after 10 days.

Photo Credits

Front Matter

© Brand X Pictures/Punchstock/RF

Chapter 1

Opener:Courtesy Brian Mercer; p. 1: Courtesy James Hall; p. 33: © Getty Images/Digital Vision/RF; p. 37/top: © PhotoLink/Photodisc/Getty Images/RF; p. 37/bottom: © BananaStock/Alamy; p. 38: © Don Farrall/Getty Images/RF; p. 49: Courtesy James Hall; p. 63: © Royalty-Free/CORBIS; p. 65/left: Courtesy G.E. Appliances; p. 65/right: © McGraw-Hill Higher Education, Inc./C.P. Hammond, photographer; p. 87: Courtesy James Hall; p. 92: © Purestock/Getty Images/RF; p. 93: © Steve Allen/Getty Images/RF; p. 96: Courtesy James Hall; p. 97: Courtesy James W. Hall; p. 102: © Steve Cole/Photodisc/Getty Images; p. 105: Courtesy James Hall.

Chapter 2

Opener: Courtesy Brian Mercer; p. 111: Courtesy James Hall; p. 119: U.S. Air Force/RF; p. 133: © Thinkstock/JupiterImages; p. 162: © Ryan McVay/Getty Images/RF; p. 163: Courtesy Resilite; p. 173: © Digital Vision/Punchstock/RF; p. 174: © Royalty-Free/CORBIS; p. 179: © Royalty-Free/CORBIS; p. 186:Courtesy James Hall; p. 197: © The McGraw-Hill Companies, Inc./Gary He, photographer; p. 198: Siede Preis/Getty Images/RF; p. 203: © Ingram Publishing/SuperStock/RF.

Chapter 3

Opener: © PhotoDisc/Vol. 1/RF; p. 208: Courtesy Manette Hall; p. 215: Courtesy James W. Hall; p. 222: © Stockbyte/RF; p. 241: © Cole Group/Getty Images/RF; p. 245: © Monty Rakusen/Getty Images/RF; p. 268: © Royalty-Free/CORBIS; p. 275: © Royalty-Free/CORBIS; p. 292: © Geostock/Getty Images/RF; p. 293: © S. Solum/PhotoLink/Getty Images/RF; p. 303: © Stair Parts, Inc. George, USA/Schilling Photography.

Chapter 4

Opener: © 1999 EyeWire, Inc./RF; p. 315: © Steve Mason/Getty Images/RF; p. 317: Courtesy Brian Mercer; p. 322: Courtesy Brian Mercer; p. 326: © Chuck Eckert/Alamy; p. 335: © Jack Hollingsworth.Getty Images/RF; p. 336: © H. Wiesenhofer/PhotoLink/Getty Images/RF; p. 337: © Royalty-Free/CORBIS; p. 346: © Royalty-Free/CORBIS; p. 347: © The McGraw-Hill Companies, Inc./Bob Coyle, photographer; p. 360: © Royalty-Free/CORBIS; p. 364: © Corbis/SuperStock/RF; p. 366: © Ryan McVay/Getty Images/RF.

Chapter 5

Opener: Photodisc/Getty Images/RF; p. 380: Courtesy James W. Hall; p. 391: © PhotoLink/Getty Images/RF; p. 401: © Digital Vision/PunchStock/RF; p. 404/left: © Comstock/PunchStock/RF; TA 5.25: p. 404/right: © Photodisc/SuperStock/RF; p. 415: Photodisc/Getty Images/RF; p. 416: Courtesy James W. Hall; p. 445: Courtesy James W. Hall.

Chapter 6

Opener: Courtesy Brian Mercer; p. 485: © Digital Vision Ltd./SuperStock/RF; p. 523: © 1998 Copyright IMS Communications Ltd./Capstone Design. All Rights Reserved/RF; p. 538: © McGraw-Hill Companies, Inc./Gary He, photographer.

Chapter 7

Opener: © PhotoLink/Photodisc/Getty Images/RF; p. 571: © Photodisc/Getty Images/RF; p. 579: Courtesy Brian Mercer; p. 582: © Manette Hall; p. 583: © The McGraw-Hill Companies, Inc./John Flournoy, photographer; p. 584/top: Courtesy Brad Leeb; p. 584/bottom: © PhotoDisc/Vol. 27.

Chapter 8

Opener: Courtesy Brad Leeb; p. 651/left © PhotoDisc/Vol. 2; p. 651/right: Courtesy James W. Hall; p. 691: © PhotoLink/Getty Images/RF; p. 693: Courtesy Brad Leeb; p. 699: Courtesy Brian Mercer.

Chapter 9

Opener: © Brand X Pictures/PunchStock; p. 706: © Royalty-Free/CORBIS; p. 728: © Ingram Publishing/Fotosearch/RF; p. 762: © Brand X Pictures/PunchStock/RF; p. 773: © Brand X Pictures/PunchStock; p. 774: © The McGraw-Hill Companies, Inc./Lars A. Niki, photographer; p. 775: © PhotoDisk/Getty/RF.

Chapter 10

Opener: © Goodshoot/RF; p. 786: © Jonnie Miles/Getty Images/RF; p. 826/left: © Royalty-Free/CORBIS; p. 826/top right: © Royalty-Free/CORBIS; p. 826/bottom right: © Manette Hall; p. 836: © Royalty-Free/CORBIS.

Chapter 11

Opener: Courtesy National Park Service; p. 903/top: © Royalty-Free/CORBIS; p. 903/bottom: © Royalty-Free/CORBIS; p. 906: © CORBIS/Vol. 154; p. 911: © PhotoDisc/Vol. 19; p. 912/left: © IT Stock/Punchstock/RF; p. 912/right: © Royalty-Free/CORBIS; p. 918: NOAA; p. 920: © Brand X Pictures/PunchStock/RF.

Index of Features

Index of Features

Index

Strategy for Solving Word Problems

Step 1. Read the problem carefully to determine what you are being asked to find.

Step 2. Select a variable to represent each unknown quantity. Specify precisely what each variable represents and note any restrictions on each variable.

Step 3. If necessary, make a sketch and translate the problem into a word equation or a system of word equations. Then translate each word equation into an algebraic equation.

Step 4. Solve the equation or the system of equations, and answer the question completely in the form of a sentence.

Step 5. Check the reasonableness of your answer.

Statements of Variation

For $k \neq 0$,

Direct variation: $y = kx$, y varies directly as x

Inverse variation: $y = \dfrac{k}{x}$, y varies inversely as x

Joint variation: $z = kxy$, z varies jointly as x and y

Absolute Value Equations and Inequalities

For $d > 0$,

$$|x| = d \text{ means } x = -d \text{ or } x = d$$

$$|x| \leq d \text{ means } -d \leq x \leq d$$

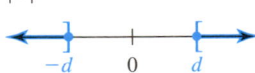

$$|x| \geq d \text{ means } x \leq -d \text{ or } x \geq d$$

Properties of Radicals

If $\sqrt[n]{x}$ and $\sqrt[n]{y}$ are both real numbers, then:

$$\sqrt[n]{xy} = \sqrt[n]{x}\sqrt[n]{y} \qquad \text{and}$$

$$\sqrt[n]{\dfrac{x}{y}} = \dfrac{\sqrt[n]{x}}{\sqrt[n]{y}} \qquad \text{for } y \neq 0$$

$$\sqrt[n]{x^m} = (\sqrt[n]{x})^m = x^{m/n}$$

For any real number x and natural number n:

$$\sqrt[n]{x^n} = |x| \qquad \text{if } n \text{ is even}$$

$$\sqrt[n]{x^n} = x \qquad \text{if } n \text{ is odd}$$

Properties of Logarithms

For $x > 0$, $y > 0$, $b > 0$, and $b \neq 1$, $\log_b x = m$ is equivalent to $b^m = x$.

Product rule: $\log_b xy = \log_b x + \log_b y$

Quotient rule: $\log_b \dfrac{x}{y} = \log_b x - \log_b y$

Power rule: $\log_b x^p = p \log_b x$

Special identities:

$\log_b 1 = 0$

$\log_b b = 1$

$\log_b \dfrac{1}{b} = -1$

$\log_b b^x = x$

$b^{\log_b x} = x$

Sequences and Series

a_n is read as "a sub n."

Arithmetic sequences: $d = a_n - a_{n-1}$

$$a_n = a_1 + (n-1)d$$

$$S_n = \dfrac{n}{2}(a_1 + a_n)$$

Geometric sequences: $r = \dfrac{a_n}{a_{n-1}}$

$$a_n = a_1 r^{n-1}$$

$$S_n = \dfrac{a_1(1 - r^n)}{1 - r}$$

Infinite geometric sequences: $S = \dfrac{a_1}{1 - r}$, $|r| < 1$.

Summation notation:

$$\sum_{i=1}^{n} a_i = a_1 + a_2 + \cdots + a_{n-1} + a_n$$

Methods of Solving Quadratic Equations

$ax^2 + bx + c = 0$ with $a \neq 0$.

Graphically

Numerically

Factoring

Extraction of roots

Completing the square

The quadratic formula: $x = \dfrac{-b \pm \sqrt{b^2 - 4ac}}{2a}$

Slope

The slope of a line through (x_1, y_1) and (x_2, y_2) with $x_1 \neq x_2$ is $m = \dfrac{\text{Change in } y}{\text{Change in } x} = \dfrac{\Delta y}{\Delta x} = \dfrac{y_2 - y_1}{x_2 - x_1} = \dfrac{\text{rise}}{\text{run}}$.

Positive slope: The line goes upward to the right.
Negative slope: The line goes downward to the right.
Zero slope: The line is horizontal.
Undefined slope: The line is vertical.

Forms of Linear Equations

Slope-intercept form: $y = mx + b$ or $f(x) = mx + b$ with slope m and y-intercept $(0, b)$

Vertical line: $x = a$ for a real constant a
Horizontal line: $y = b$ for a real constant b
General form: $Ax + By = C$
Point-slope form: $y - y_1 = m(x - x_1)$ through (x_1, y_1) with slope m

Properties of Exponents

Let $m, n, x, x^m, x^n, y, y^m,$ and y^n be real numbers.
Product rule: $x^m \cdot x^n = x^{m+n}$
Power rule: $(x^m)^n = x^{mn}$
Product to a power: $(xy)^m = x^m y^m$
Quotient to a power: $\left(\dfrac{x}{y}\right)^m = \dfrac{x^m}{y^m}$ for $y \neq 0$

Quotient rule: $\dfrac{x^m}{x^n} = x^{m-n}$ for $x \neq 0$

Negative power: $\left(\dfrac{x}{y}\right)^{-n} = \left(\dfrac{y}{x}\right)^n$ for $x \neq 0, y \neq 0$

Special identities:
$$x^0 = 1 \text{ for } x \neq 0 \quad x^1 = x \quad x^{-1} = \frac{1}{x} \text{ for } x \neq 0$$

Factoring Special Forms

Difference of two squares: $A^2 - B^2 = (A + B)(A - B)$
Perfect square trinomial: $A^2 + 2AB + B^2 = (A + B)^2$
Perfect square trinomial: $A^2 - 2AB + B^2 = (A - B)^2$
Difference of two cubes:
$$A^3 - B^3 = (A - B)(A^2 + AB + B^2)$$
Sum of two cubes:
$$A^3 + B^3 = (A + B)(A^2 - AB + B^2)$$

Systems of Two Linear Equations

One Solution

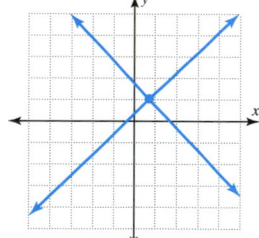

A consistent system of independent equations: The solution process will produce unique x- and y-coordinates.

No Solution

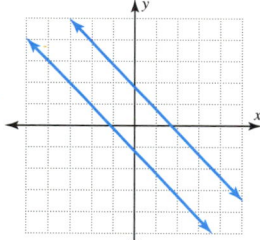

An inconsistent system: The solution process will produce a contradiction.

An Infinite Number of Solutions

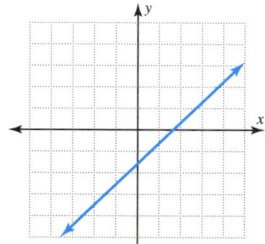

A consistent system of dependent equations: The solution process will produce an identity.

Equivalent Statements about Linear Factors of a Polynomial

For a real constant c and a real polynomial $P(x)$, the following statements are equivalent:

Algebraically

$x - c$ is a factor of $P(x)$.
$x = c$ is a solution of $P(x) = 0$.

Numerically

$P(c) = 0$; that is,
c is a zero of $P(x)$.

Graphically

$(c, 0)$ is an x-intercept
of the graph of $y = P(x)$.